Some Other Important Numbers in Su[illegible]

Errors and Error Analysis

68.3 = percent of observations that are expected within the lim[illegible] the standard deviation
0.6745 = coefficient of standard deviation for 50% error *(probable error)*
1.6449 = coefficient of standard deviation for 90% error
1.9599 = coefficient of standard deviation for 95% error *(two-sigma error)*

Electronic Distance Measurement

299,792,458 m/sec = speed of light or electromagnetic energy in a vacuum
1 Hertz (Hz) = 1 cycle per second
1 kilohertz (kHz) = 1000 Hz
1 megahertz (MHz) = 1000 kHz
1 gigahertz (GHz) = 1000 MHz
1.0003 = approximate index of atmospheric refraction (varies from 1.0001 to 1.0005)
760 mm of mercury = standard atmospheric pressure

Taping

0.00000645 = coefficient of expansion of steel tape, per 1°F
0.0000116 = coefficient of expansion of steel tape, per 1°C
29,000,000 lb/in^2 = 2,000,000 kg/cm^2 = Young's modulus of elasticity for steel
490 lb/ft^3 = density of steel for tape weight computations
15°F = change in temperature to produce a 0.01 ft length change in a 100 ft steel tape
68°F = 20°C = standard temperature for taping

Leveling

0.574 = coefficient of combined curvature and refraction (ft/miles^2)
0.0675 = coefficient of combined curvature and refraction (m/km^2)
20.6 m = 68 ft = approximate radius of a level vial having a 20″ sensitivity

Miscellaneous

6,371,000 m = 20,902,000 ft = approximate mean radius of the earth
1.15 miles = approximately 1 minute of latitude = approximately 1 nautical mile
69.1 miles = approximately 1 degree of latitude
101 ft = approximately 1 second of latitude
24 hours = 360° of longitude
15° longitude = width of one time zone, i.e., 360°/24 hr
23°26.5′ = approximate maximum declination of the sun at the solstaces
$23^h56^m04.091^s$ = length of sidereal day in mean solar time, which is $3m55.909^s$ of mean solar time short of one solar day
5,729.578 ft = radius of 1° curve, arc definition
5,729.651 ft = radius of 1° curve, chord definition
100 ft = 1 station, English system
1000 m = 1 station, metric system
6 miles = length and width of a normal township
36 = number of sections in a normal township
10,000 km = distance from equator to pole and original basis for the length of the meter

Elementary Surveying

An Introduction to Geomatics

Twelfth Edition

CHARLES D. GHILANI
The Pennsylvania State University

PAUL R. WOLF
Professor Emeritus, Civil and Environmental Engineering
University of Wisconsin–Madison

Upper Saddle River, New Jersey 07458

Library of Congress Cataloging-in-Publication Data
Ghilani, Charles D.
Elementary surveying: an introduction to geomatics/
Charles D. Ghilani, Paul R. Wolf.—Twelfth ed.
p. cm.
Includes index.
ISBN-13: 978-0-13-603100-0
1. Surveying. 2. Geomatics. I. Wolf, Paul R. II. Title.
TA545.G395 2007
526.9—dc22
2007042690

Vice President and Editorial Director, ECS: *Marcia J. Horton*
Acquisitions Editor: *Holly Stark*
Associate Editor: *Dee Bernhard*
Managing Editor: *Scott Disanno*
Marketing Manager: *Tim Galligan*
Project Manager: *Romey Bhat Wanchoo*
Media Project Manager: *Dan Sandin*
Director of Creative Services: *Paul Belfanti*
Creative Director: *Juan Lopez*
Art Director: *Kenny Beck*
Cover Designer: *Kenny Beck*
Manufacturing Manager: *Alan Fischer*
Manufacturing Buyer: *Lisa McDowell*

About the Cover: Courtesy Getty Images. Photographer: Colin Anderson. Used by permission.

Pearson Prentice Hall
Pearson Education, Inc.
Upper Saddle River, NJ 07458

Printed in the United States of America
10 9 8 7 6 5 4 3 2 1

ISBN 0-13-603100-5
978-0-13-603100-0

Pearson Education Ltd., *London*
Pearson Education Australia Pty., Limited, *Sydney*
Pearson Education Singapore, Pte. Ltd
Pearson Education North Asia Ltd., *Hong Kong*
Pearson Education Canada, Ltd., *Toronto*
Pearson Educación de Mexico, S.A. de C.V.
Pearson Education—Japan, *Tokyo*
Pearson Education Malaysia, Pte. Ltd.
Pearson Education, *Upper Saddle River, New Jersey*

Table of Contents

2 • UNITS, SIGNIFICANT FIGURES, AND FIELD NOTES 23

3 • THEORY OF ERRORS IN OBSERVATIONS 47

5 • LEVELING—FIELD PROCEDURES AND COMPUTATIONS 103

6 • DISTANCE MEASUREMENT 127

7 • ANGLES, AZIMUTHS, AND BEARINGS 165

8 • TOTAL STATION INSTRUMENTS; ANGLE MEASUREMENTS 185

10 • TRAVERSE COMPUTATIONS 239

11 • COORDINATE GEOMETRY IN SURVEYING CALCULATIONS 269

20 • STATE PLANE COORDINATES 579

21 • BOUNDARY SURVEYS 621

28 • INTRODUCTION TO GEOGRAPHIC INFORMATION SYSTEMS 831

APPENDIX A • DUMPY LEVELS, TRANSITS, AND THEODOLITES 859

Preface

This 12th edition of *Elementary Surveying: An Introduction to Geomatics* presents basic concepts and practical material in each of the areas fundamental to modern surveying (geomatics) practice. It is written primarily for an introductory course in surveying at the college level. Although the book is elementary, its depth and breadth also make it ideal for self-study. This edition includes more than 400 figures and illustrations to help clarify discussions, and numerous example problems are worked to illustrate computational procedures.

In keeping with the goal of providing an up-to-date presentation of surveying equipment and procedures, total stations are stressed as the instruments for making angle and distance observations. GPS is increasingly becoming one of the dominant tools in the surveyor's repertoire. Today, GPS is commonly used to establish control, create maps, lay out plans, and manage construction equipment. Kinematic GPS applications have become the predominant surveying method of choice. Thus, a new chapter on kinematic GPS and several new sections on machine control, localization of GPS surveys, and construction staking using GPS have been added to Chapters 15, 19, and 23. Since astronomical observations for azimuth are seldom performed today, the chapter on astronomical observations has now been moved to Appendix C. To support the additional matrix computations introduced with least-squares adjustments, a new appendix on Introduction to Matrices has been placed in Appendix E. Additionally, the *U.S. State Plane Coordinate System Defining Parameters* have been placed in Appendix F, and *Solutions to Selected Problems* can now be found in Appendix G.

As with past editions, this text continues to emphasize the theory of errors in surveying work. At the end of each chapter, common errors and mistakes related to the topic covered are listed so that students are reminded to exercise caution in all of their work. Practical suggestions resulting from the authors' many years of experience are interjected throughout the text. Many of the thousand after-chapter problems have been rewritten so that instructors can create new assignments for their students. A solutions manual is available to instructors who adopt the book by contacting your Prentice Hall representative.

Updated versions of STATS, WOLFPACK, and MATRIX are included on the CD. These programs contain options for statistical computations and many of the computations presented in this book. Mathcad® worksheets and Excel® spreadsheets are included on the compact disk (CD) that accompanies this book. These programmed computational sheets demonstrate many of the procedures discussed in this book. For those desiring additional knowledge in map projections, the Mercator, Albers Equal-Area, Oblique Stereographic, and Oblique Mercator map projections have been included with these files. Also included are hypertext markup language (html) files of the Mathcad® worksheets for use by those who do not own the software. The CD has an installation program that allows the user to select the desired installation options. It also contains a 60-day trial version of the Eagle Point® field-to-finish software package for those who do not have access to this type of software.

ACKNOWLEDGMENTS

Past editions of this book, and this current one, have benefited from the suggestions, reviews, and other input from numerous students, educators, and practitioners. For their help, the authors are extremely grateful. In this edition, those professors, practicing professionals, and graduate students who reviewed material or otherwise assisted include: Thomas Seybert, Pennsylvania State University, Lehman, PA; Bon DeWitt, University of Florida; Francis Derby, Pennsylvania State University, Lehman, PA; Brian Naberezny, Pennsylvania State University, Lehman, PA; Ron Oberlander, Topcon Positioning Systems; Charles Harpster, Pennsylvania Department of Transportation; Preston Hartzell, Eastern States Engineering; Douglas E. Smith, Montana State University; Robert G. Moynihan, University of New Hampshire; Tom Mastin, California Polytechnic, San Luis Obispo; Craig S. Moore, Virginia Tech; and Jim Bradshaw.

In addition, the authors wish to acknowledge the contributions of charts, maps, or other information from the National Geodetic Survey, the U.S. Geological Survey, and the U.S. Bureau of Land Management. Also appreciation is expressed to the many instrument manufacturers that provided pictures and other descriptive information on their equipment for use herein. Special acknowledgment is given to Eagle Point Software, Inc. for supplying a trial version of their field-to-finish software, which is included on the CD that accompanies this book.

To all of those named here, and to any others who may have been inadvertently omitted, the authors are extremely thankful.

1 Introduction

1.1 DEFINITION OF SURVEYING

Surveying, which is also interchangeably called *geomatics* (see Section 1.2), has traditionally been defined as the science, art, and technology of determining the relative positions of points above, on, or beneath the Earth's surface, or of establishing such points. In a more general sense, however, surveying (geomatics) can be regarded as that discipline which encompasses all methods for measuring and collecting information about the physical earth and our environment, processing that information, and disseminating a variety of resulting products to a wide range of clients.

Surveying has been important since the beginning of civilization. Its earliest applications were in measuring and marking boundaries of property ownership. Throughout the years its importance has steadily increased with the growing demand for a variety of maps and other spatially related types of information and the expanding need for establishing accurate line and grade to guide construction operations.

Today the importance of measuring and monitoring our environment is becoming increasingly critical as our population expands, land values appreciate, our natural resources dwindle, and human activities continue to stress the quality of our land, water, and air. Using modern ground, aerial, and satellite technologies, and computers for data processing, contemporary surveyors are now able to measure and monitor the Earth and its natural resources on literally a global basis. Never before has so much information been available for assessing current conditions, making sound planning decisions, and formulating policy in a host of land-use, resource development, and environmental preservation applications.

Recognizing the increasing breadth and importance of the practice of surveying, the *International Federation of Surveyors* (see Section 1.11) recently adopted the following definition:

"A surveyor is a professional person with the academic qualifications and technical expertise to conduct one, or more, of the following activities:

- to determine, measure and represent the land, three-dimensional objects, point-fields, and trajectories;
- to assemble and interpret land and geographically and economically related information;
- to use that information for the planning and efficient administration and management of the land, the sea and any structures thereon;
- to carryout urban and rural development and land management; and,
- to conduct research into and develop them.

Detailed Functions

The surveyor's professional tasks may involve one or more of the following activities, which may occur either on, above or below the surface of the land or the sea and may be carried out in association with other professionals.

1. The determination of the size and shape of the earth and the measurements of all data needed to define the size, position, shape and contour of any part of the earth and monitoring any change therein.
2. The positioning of objects in space and time as well as the positioning and monitoring of physical features, structures and engineering works on, above or below the surface of the earth.
3. The development, testing and calibration of sensors, instruments and systems for the above-mentioned purposes and for other surveying purposes.
4. The acquisition and use of spatial information from close range, aerial and satellite imagery and the automation of these processes.
5. The determination of the position of the boundaries of public or private land, including national and international boundaries, and the registration of those lands with the appropriate authorities.
6. The design, establishment and administration of geographic information systems (GIS) and the collection, storage, analysis, management, display and dissemination of data.
7. The analysis, interpretation and integration of spatial objects and phenomena in GIS, including the visualization and communication of such data in maps, models and mobile digital devices.
8. The study of the natural and social environment, the measurement of land and marine resources and the use of such data in the planning of development in urban, rural and regional areas.
9. The planning, development and redevelopment of property, whether urban or rural and whether land or buildings.
10. The assessment of value and the management of property, whether urban or rural and whether land or buildings.
11. The planning, measurement and management of construction works, including the estimation of costs.

In application of the foregoing activities surveyors take into account the relevant legal, economic, environmental, and social aspects affecting each project."

The breadth and diversity of the practice of surveying (geomatics), as well as its importance in modern civilization, are readily apparent from this definition.

1.2 GEOMATICS

As noted in Section 1.1, geomatics is a relatively new term that is now commonly being applied to encompass the areas of practice formerly identified as surveying. The name has gained widespread acceptance in the United States, as well as in other English-speaking countries of the world, especially in Canada, the United Kingdom, and Australia. In the United States, the *Surveying Engineering Division* of The American Society of Civil Engineers recently changed its name to the *Geomatics Division.* Many college and university programs in the United States that were formerly identified as "Surveying" or "Surveying Engineering" are now called "Geomatics" or "Geomatics Engineering."

The principal reason cited for making the name change is that the manner and scope of practice in surveying have changed dramatically in recent years. This has occurred in part because of recent technological developments that have provided surveyors with new tools for measuring and/or collecting information, for computing, and for displaying and disseminating information. It has also been driven by increasing concerns about the environment, locally, regionally, and globally, which have greatly exacerbated efforts in monitoring, managing, and regulating the use of our land, water, air, and other natural resources. These circumstances, and others, have brought about a vast increase in demands for new spatially related information.

Historically surveyors made their measurements using ground-based methods and until rather recently the transit and tape[1] were their primary instruments. Computations, analyses, and the reports, plats, and maps they delivered to their clients were prepared (in hard copy form) through tedious manual processes. Today the modern surveyor's arsenal of tools for measuring and collecting environmental information include electronic instruments for automatically measuring distances and angles, satellite surveying systems for quickly obtaining precise positions of widely spaced points, and modern aerial digital imaging and laser-scanning systems for quickly mapping and collecting other forms of data about the earth upon which we live. In addition, computer systems are available that can process the measured data and automatically produce plats, maps, and other products at speeds unheard of before the 1990s. Furthermore, these products can be prepared in electronic formats and be transmitted to remote locations via telecommunication systems.

Concurrent with the development of these new data collection and processing technologies, *geographic information systems* (GISs) have emerged and matured. These computer-based systems enable virtually any type of spatially related information about the environment to be integrated, analyzed, displayed, and disseminated.[2] The key to successfully operating geographic information systems is spatially related data of high quality, and the collection and processing of this data has placed great new demands upon the surveying community.

---

[1]These instruments are described in Appendix A and Chapter 6, respectively.

[2]Geographic information systems are briefly introduced in Section 1–9 and then described in greater detail in Chapter 28.

As a result of these new developments, and others, many feel that the name surveying no longer adequately reflects the expanded and changing role of their profession. Hence the new term geomatics has emerged. In this text, the terms surveying and geomatics are both used, although the former is used more frequently. Nevertheless students should understand that the two terms are practically synonymous, as discussed here.

■ 1.3 HISTORY OF SURVEYING

The oldest historical records in existence today that bear directly on the subject of surveying state that this science began in Egypt. Herodotus recorded that Sesostris (about 1400 B.C.) divided the land of Egypt into plots for the purpose of taxation. Annual floods of the Nile River swept away portions of these plots, and surveyors were appointed to replace the boundaries. These early surveyors were called *rope-stretchers,* since their measurements were made with ropes having markers at unit distances.

As a consequence of this work, early Greek thinkers developed the science of geometry. Their advance, however, was chiefly along the lines of pure science. Heron stands out prominently for applying science to surveying in about 120 B.C. He was the author of several important treatises of interest to surveyors, including *The Dioptra,* which related the methods of surveying a field, drawing a plan, and making related calculations. It also described one of the first pieces of surveying equipment recorded, the *diopter* [Figure 1.1(a)]. For many years Heron's work was the most authoritative among Greek and Egyptian surveyors.

Significant development in the art of surveying came from the practical-minded Romans, whose best-known writing on surveying was by Frontinus. Although the original manuscript disappeared, copied portions of his work have

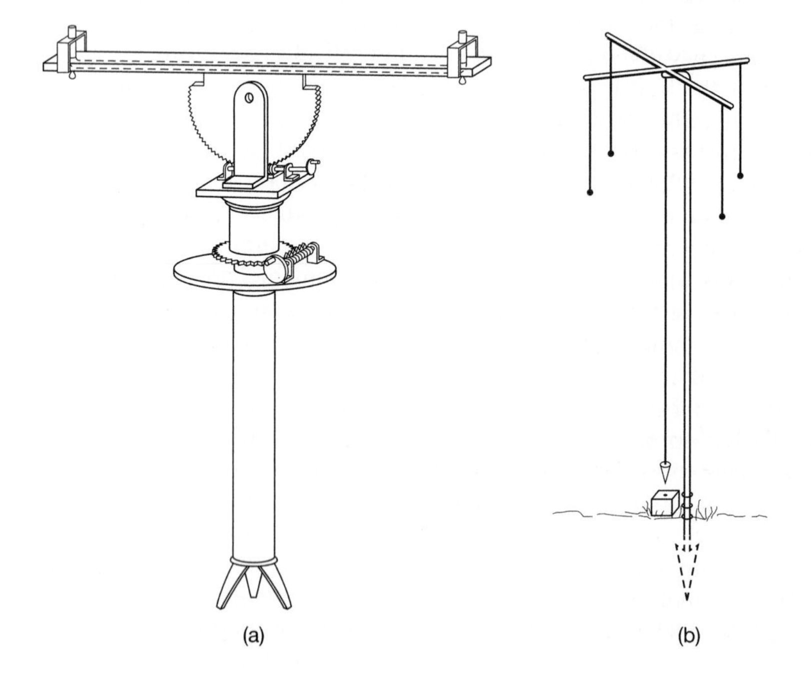

Figure 1.1 Historical surveying instruments: (a) The diopter, (b) the groma.

been preserved. This noted Roman engineer and surveyor, who lived in the first century, was a pioneer in the field, and his essay remained the standard for many years. The engineering ability of the Romans was demonstrated by their extensive construction work throughout the empire. Surveying necessary for this construction resulted in the organization of a surveyors' guild. Ingenious instruments were developed and used. Among these were the *groma* [Figure 1.1(b)], used for sighting; the *libella,* an A-frame with a plumb bob, for leveling; and the *chorobates,* a horizontal straightedge about 20 ft long with supporting legs and a groove on top for water to serve as a level.

One of the oldest Latin manuscripts in existence is the *Codex Acerianus,* written in about the sixth century. It contains an account of surveying as practiced by the Romans and includes several pages from Frontinus's treatise. The manuscript was found in the tenth century by Gerbert and served as the basis for his text on geometry, which was largely devoted to surveying.

During the Middle Ages, the Arabs kept Greek and Roman science alive. Little progress was made in the art of surveying, and the only writings pertaining to it were called "practical geometry."

In the 13th century, Von Piso wrote *Practica Geometria,* which contained instructions on surveying. He also authored *Liber Quadratorum,* dealing chiefly with the *quadrans,* a square brass frame having a 90° angle and other graduated scales. A movable pointer was used for sighting. Other instruments of the period were the *astrolabe,* a metal circle with a pointer hinged at its center and held by a ring at the top, and the *cross staff,* a wooden rod about 4 ft long with an adjustable cross-arm at right angles to it. The known lengths of the arms of the cross staff permitted distances to be measured by proportion and angles.

Early civilizations assumed the Earth to be a flat surface, but by noting the Earth's circular shadow on the moon during lunar eclipses and watching ships gradually disappear as they sailed toward the horizon, it was slowly deduced that the planet actually curved in all directions.

Determining the true size and shape of the Earth has intrigued humans for centuries. History records that a Greek named Eratosthenes was among the first to compute its dimensions. His procedure, which occurred about 200 B.C., is illustrated in Figure 1.2. Eratosthenes had concluded that the Egyptian cities of Alexandria and Syene were located approximately on the same meridian, and he had also observed that at noon on the summer solstice, the sun was directly overhead at Syene. (This was apparent because at that time of that day, the image of the sun could be seen reflecting from the bottom of a deep vertical well there.) He reasoned that at that moment, the sun, Syene, and Alexandria were in a common meridian plane, and if he could measure the arc length between the two cities, and the angle it subtended at the Earth's center, he could compute the Earth's circumference. He determined the angle by measuring the length of the shadow cast at Alexandria from a vertical staff of known length. The arc length was found from multiplying the number of caravan days between Syene and Alexandria by the average daily distance traveled. From these measurements Eratosthenes calculated the Earth's circumference to be about 25,000 mi. Subsequent precise geodetic measurements using better instruments, but techniques similar geometrically to Eratosthenes', have shown his value, though slightly too large, to be amazingly

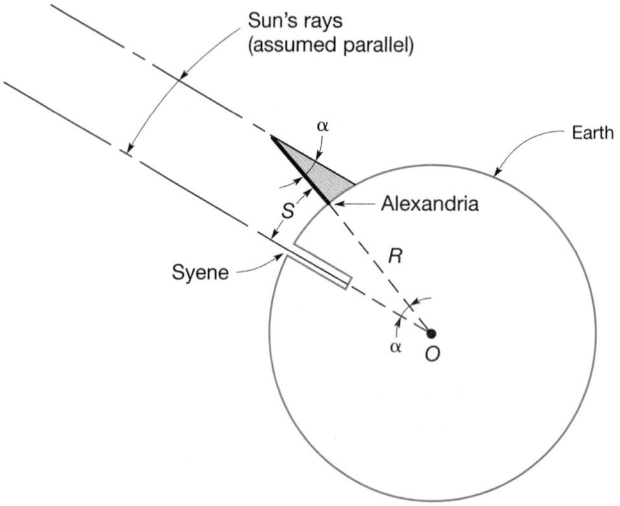

Figure 1.2 Geometry of the procedure used by Eratosthenes to determine the Earth's circumference.

close to the currently accepted one. (Actually, as explained in Chapter 19, the Earth approximates an oblate spheroid having an equatorial radius about 13.5 mi longer than the polar radius.)

In the 18th and 19th centuries, the art of surveying advanced more rapidly. The need for maps and locations of national boundaries caused England and France to make extensive surveys requiring accurate triangulation; thus geodetic surveying began. The U.S. Coast Survey (now the National Geodetic Survey of the U.S. Department of Commerce) was established by an act of Congress in 1807. Initially its charge was to perform hydrographic surveys and prepare nautical charts. Later its activities were expanded to include establishment of reference monuments of precisely known positions throughout the country.

Increased land values and the importance of precise boundaries, along with the demand for public improvements in the canal, railroad, and turnpike eras, brought surveying into a prominent position. More recently, the large volume of general construction, numerous land subdivisions with better records required, and demands posed by the fields of exploration and ecology have entailed an augmented surveying program. Surveying is still the sign of progress in the development, use, and preservation of the Earth's resources.

In addition to meeting a host of growing civilian needs, surveying has always played an important role in our nation's defense activities. World Wars I and II, the Korean and Vietnam conflicts, and the more recent conflicts in the Middle East and Europe have created staggering demands for precise measurements and accurate maps. These military operations also provided the stimulus for improving instruments and methods to meet these needs. Surveying also contributed to, and benefited from, the space program where new equipment and systems were needed to provide precise control for missile alignment and for mapping and charting portions of the moon and nearby planets.

Developments in surveying and mapping equipment have now evolved to the point where the traditional instruments that were used until about the 1960s or 1970s—the transit, theodolite, dumpy level, and steel tape—have now been

Figure 1.3 LEICA TPS 1100 total station instrument. (Courtesy Leica Geosystems, Inc.)

almost completely replaced by an array of new "high-tech" instruments. These include electronic *total station instruments,* which can be used to automatically measure and record horizontal and vertical distances and horizontal and vertical angles; and the *global positioning system* (GPS) that can provide precise location information for virtually any type of survey. Laser-scanning instruments combine automatic distance and angle measurements to compute dense grids of coordinated points. Also new aerial cameras and remote sensing instruments have been developed which provide images in digital form, and these images can be processed to obtain spatial information and maps using new *digital photogrammetric restitution instruments* (also called *softcopy plotters*). Figures 1.3, 1.4, 1.5, and 1.6, respectively, show a total station instrument, GPS receiver, laser-scanning instrument, and modern softcopy plotter.

■ 1.4 GEODETIC AND PLANE SURVEYS

Two general classifications of surveys are *geodetic* and *plane.* They differ principally in the assumptions on which the computations are based, although field measurements for geodetic surveys are usually performed to a higher order of accuracy than those for plane surveys.

In geodetic surveying, the curved surface of the Earth is considered by performing the computations on an *ellipsoid* (curved surface approximating the size and shape of the Earth—see Chapter 19). It is now becoming common to do geodetic computations in a three-dimensional, Earth-centered, Earth-fixed Cartesian coordinate system. The calculations involve solving equations derived from solid geometry and calculus. Geodetic methods are employed to determine relative positions of widely spaced monuments and to compute lengths and directions of the long lines between them. These monuments serve as the basis for referencing other subordinate surveys of lesser extents.

Figure 1.4
Global positioning system receiver. (Courtesy Spectra Precision.)

Figure 1.5
LEICA HDS 3000 laser scanner. (Courtesy of Christopher Gibbons, Leica Geosystems, Inc.)

Figure 1.6 Intergraph Image Station Z softcopy plotter. (From *Elements of Photogrammetry: With Applications in GIS,* by Wolf and Dewitt, 2000, Courtesy Intergraph, Inc., and the McGraw-Hill Companies.)

In early geodetic surveys, painstaking efforts were employed to accurately observe angles and distances. The angles were measured using precise ground-based theodolites, and the distances were measured using special tapes made from metal having a low coefficient of thermal expansion. From these basic measurements, the relative positions of the monuments were computed. Later, electronic instruments were used for observing the angles and distances. Although these latter types of instruments are still sometimes used on geodetic surveys, the global positioning system (GPS) has almost completely replaced other instruments for these types of surveys. GPS can provide the needed positions with much greater accuracy, speed, and economy. GPS receivers (see Figure 1.4) enable ground stations to be located precisely by observing distances to satellites operating in known positions along their orbits. GPS surveys are being used in all forms of surveying including geodetic, hydrographic, construction, and boundary surveying. The principles of operation of the global positioning system are given in Chapter 13, field and office procedures used in static GPS surveys are discussed in Chapter 14, and the methods used in kinematic GPS surveys are discussed in Chapter 15.

In plane surveying, except for leveling, the reference base for fieldwork and computations is assumed to be a flat horizontal surface. The direction of a plumb line (and thus gravity) is considered parallel throughout the survey region, and all observed angles are presumed to be plane angles. For areas of limited size, the surface of our vast ellipsoid is actually nearly flat. On a line 5 mi long, the ellipsoidal arc and chord lengths differ by only about 0.02 ft. A plane surface tangent to the ellipsoid departs only about 0.7 ft at 1 mi from the point of tangency. In a triangle having an area of 75 square miles, the difference between the sum of the three ellipsoidal angles and three plane angles is only about 1 second. Therefore it is evident that except in surveys covering extensive areas, the Earth's surface can be approximated as a plane, thus simplifying computations and techniques. In general, algebra, plane and analytical geometry, and plane trigonometry are used in plane surveying calculations. Even for very large areas, map projections, such as those described in Chapter 20, allow plane-surveying computations to be used without loss in accuracy. This book concentrates primarily on methods of plane surveying, an approach that satisfies the requirements of most projects.

1.5 IMPORTANCE OF SURVEYING

Surveying is one of the world's oldest and most important arts because, as noted previously, from the earliest times it has been necessary to mark boundaries and divide land. Surveying has now become indispensable to our modern way of life. The results of today's surveys are used to (1) map the Earth above and below sea level; (2) prepare navigational charts for use in the air, on land, and at sea; (3) establish property boundaries of private and public lands; (4) develop data banks of land-use and natural resource information which aid in managing our environment; (5) determine facts on the size, shape, gravity, and magnetic fields of the Earth; and (6) prepare charts of our moon and planets.

Surveying continues to play an extremely important role in many branches of engineering. For example, surveys are required to plan, construct, and maintain highways, railroads, rapid-transit systems, buildings, bridges, missile ranges, launching sites, tracking stations, tunnels, canals, irrigation ditches, dams, drainage works, urban land subdivisions, water supply and sewage systems, pipelines, and mine shafts. Surveying methods are commonly employed in laying out industrial assembly lines and jigs. These methods are also used for guiding the fabrication of large equipment, such as airplanes and ships, where separate pieces that have been assembled at different locations must ultimately be connected as a unit. Surveying is important in many related tasks in agronomy, archeology, astronomy, forestry, geography, geology, geophysics, landscape architecture, meteorology, paleontology, and seismology, but particularly in military and civil engineering.

All engineers must know the limits of accuracy possible in construction, plant design and layout, and manufacturing processes, even though someone else may do the actual surveying. In particular, surveyors and civil engineers who are called on to design and plan surveys must have a thorough understanding of the methods and instruments used, including their capabilities and limitations. This knowledge is best obtained by making observations with the kinds of equipment used in practice to get a true concept of the theory of errors and the small but recognizable differences that occur in observed quantities.

In addition to stressing the need for reasonable limits of accuracy, surveying emphasizes the value of significant figures. Surveyors and engineers must know when to work to hundredths of a foot instead of to tenths or thousandths, or perhaps the nearest foot, and what precision in field data is necessary to justify carrying out computations to the desired number of decimal places. With experience, they learn how available equipment and personnel govern procedures and results.

Neat sketches and computations are the mark of an orderly mind, which in turn is an index of sound engineering background and competence. Taking field notes under all sorts of conditions is excellent preparation for the kind of recording and sketching expected of all engineers. Performing later office computations based on the notes underscores their importance. Additional training that has a carryover value is obtained in arranging computations in an organized manner.

Engineers who design buildings, bridges, equipment, and so on are fortunate if their estimates of loads to be carried are correct within 5 percent. Then a factor of safety of 2 or more is often applied. But except for some topographic work, only exceedingly small errors can be tolerated in surveying, and there is no factor of safety. Traditionally, therefore, both manual and computational precision are stressed in surveying.

1.6 SPECIALIZED TYPES OF SURVEYS

Many types of surveys are so specialized that a person proficient in a particular discipline may have little contact with the other areas. Persons seeking careers in surveying and mapping, however, should be knowledgeable in every phase, since all are closely related in modern practice. Some important classifications are described briefly here.

Control surveys establish a network of horizontal and vertical monuments that serve as a reference framework for initiating other surveys. Many control surveys performed today are done using techniques discussed in Chapter 14 with GPS instruments.

Topographic surveys determine locations of natural and artificial features and elevations used in map making.

Land, boundary, and *cadastral surveys* establish property lines and property corner markers. The term cadastral is now generally applied to surveys of the public lands systems. There are three major categories: *original surveys* to establish new section corners in unsurveyed areas that still exist in Alaska and several western states; *retracement surveys* to recover previously established boundary lines; and *subdivision surveys* to establish monuments and delineate new parcels of ownership. *Condominium surveys,* which provide a legal record of ownership, are a type of boundary survey.

Hydrographic surveys define shorelines and depths of lakes, streams, oceans, reservoirs, and other bodies of water. *Sea surveying* is associated with port and offshore industries and the marine environment, including measurements and marine investigations made by ship borne personnel.

Alignment surveys are made to plan, design, and construct highways, railroads, pipelines, and other linear projects. They normally begin at one control point and progress to another in the most direct manner permitted by field conditions.

Construction surveys provide line, grade, control elevations, horizontal positions, dimensions, and configurations for construction operations. They also secure essential data for computing construction pay quantities.

As-built surveys document the precise final locations and layouts of engineering works and record any design changes that may have been incorporated into the construction. These are particularly important when underground facilities are constructed, so their locations are accurately known for maintenance purposes, and so that unexpected damage to them can be avoided during later installation of other underground utilities.

Mine surveys are performed above and below ground to guide tunneling and other operations associated with mining. This classification also includes geophysical surveys for mineral and energy resource exploration.

Solar surveys map property boundaries, solar access easements, position obstructions and collectors according to sun angles, and meet other requirements of zoning boards and title insurance companies.

Optical tooling (also referred to as *industrial surveying* or *optical alignment*) is a method of making extremely accurate measurements for manufacturing processes where small tolerances are required.

Except for control surveys, most other types described are usually performed using plane surveying procedures; but geodetic methods may be employed on the others if a survey covers an extensive area or requires extreme accuracy.

Ground, aerial, and *satellite surveys* are broad classifications sometimes used. Ground surveys utilize measurements made with ground-based equipment such as automatic levels and total station instruments. Aerial surveys are accomplished using either *photogrammetry* or *remote sensing.* Photogrammetry uses cameras that are carried usually in airplanes to obtain images, whereas remote sensing employs cameras and other types of sensors that can be transported either in aircraft or satellites. Procedures for analyzing and reducing the image data are described in Chapter 27. Aerial methods have been used in all the specialized types of surveys listed, except for optical tooling, and in this area *terrestrial* (ground-based) photographs are often used. Satellite surveys include the determination of ground locations from measurements made to satellites using GPS receivers, or the use of satellite images for mapping and monitoring large regions of the Earth.

1.7 SURVEYING SAFETY

Surveyors (geomatics engineers) generally are involved in both field and office work. The fieldwork consists in making observations with various types of instruments to either (a) determine the relative locations of points or (b) to set out stakes in accordance with planned locations to guide building and construction operations. The office work involves (1) research and analysis in preparing for surveys, (2) computing and processing the data obtained from field measurements, and (3) preparing maps, plats, charts, reports, and other documents according to client specifications. Sometimes the fieldwork must be performed in hostile or dangerous environments, and thus it is very important to be aware of the need to practice safety precautions.

Among the most dangerous of circumstances within which surveyors must sometimes work are job sites that are either on or near highways or railroads, or that cross such facilities. Job sites in construction zones where heavy machinery is operating are also hazardous, and the dangers are often exacerbated by poor hearing conditions from the excessive noise, and poor visibility caused by obstructions and dust, both of which are created by the construction activity. In these situations, whenever possible, the surveys should be removed from the danger areas through careful planning and/or the use of *offset* lines. If the work must be done in these hazardous areas, then certain safety precautions should be followed. Safety vests of fluorescent yellow color should always be worn in these situations, and flagging materials of the same color can be attached to the surveying equipment to make it more visible. Depending upon the circumstances, signs can be placed in advance of work areas to warn drivers of the presence of a survey party ahead, cones and/or barricades can be placed to deflect traffic around surveying activities, and flaggers can be assigned to warn drivers, or to slow or even stop them, if necessary. The *Occupational Safety and Health Administration* (OSHA), of the U.S. Department of Labor,[3] has developed safety standards and guidelines that apply to the various conditions and situations that can be encountered.

Besides the hazards described here, depending upon the location of the survey and the time of year, other dangers can also be encountered in conducting field surveys. These include problems related to weather such as overexposure to the sun's rays, which can cause skin cancers, sunburns, heat stroke, and frostbite. To help prevent these problems, surveyors should drink plenty of fluids and wear wide-brimmed hats and sunscreen, and on extremely hot days surveying should commence at dawn and terminate at midday or early afternoon. Outside work should not be done on extremely cold days, but if it is necessary, warm clothing should be worn and skin should not be exposed. Other hazards that can be encountered during field surveys include wild animals, poisonous snakes, bees, spiders, wood ticks, deer ticks (which can carry Lyme's disease), poison ivy, and poison oak. Surveyors should be knowledgeable about the types of hazards that can be expected in any local area, and always be alert and on the lookout for them. To help prevent injury from these sources, protective boots and clothing should be worn and insect sprays can be used. Certain tools can also be dangerous, such as chain saws, axes, and machetes that are sometimes necessary for clearing lines of sight. These must always be handled with care. Also, care must be exercised in handling certain surveying instruments, like long-range poles and level rods, especially when working around overhead wires, to prevent accidental electrocutions.

Many other hazards, in addition to those cited here, can be encountered when surveying in the field. Thus it is essential that surveyors always exercise caution in their work, and know and follow accepted safety standards. In addition, a first-aid kit should always accompany a survey party in the field, and it should include all of the necessary antiseptics, ointments, bandage materials, and other equipment needed to render first aid for minor accidents. The survey party should also be equipped with cell phones for more serious situations, and telephone numbers to call in emergencies should be written down and readily accessible.

■ 1.8 LAND AND GEOGRAPHIC INFORMATION SYSTEMS

Land Information Systems (LISs) and *Geographic Information Systems* (GISs) are relatively new areas of activity that have rapidly assumed positions of major prominence in surveying. These computer-based systems enable storing, integrating, manipulating, analyzing, and displaying virtually any type of spatially

---

[3]The mission of OSHA is to save lives, prevent injuries, and protect the health of America's workers. Its staff establishes protective standards, enforces those standards, and reaches out to employers and employees through technical assistance and consultation programs. For more information about OSHA and its safety standards, consult http://www.osha.gov.

related information about our environment. LISs and GISs are used at all levels of government, and by businesses, private industry, and public utilities to assist in management and decision making. Specific applications have occurred in many diverse areas and include natural resource management, facilities siting and management, land records modernization, demographic and market analysis, emergency response and fleet operations, infrastructure management, and regional, national, and global environmental monitoring. Data stored within LISs and GISs may be both natural and cultural and may be derived from new surveys or from existing sources such as maps, charts, aerial and satellite photos, tabulated data and statistics, and other documents. However in most situations, the needed information either does not exist, or it is unsatisfactory because of age, scale, or other reasons. Thus new measurements, maps, photos, or other data must be obtained.

Specific types of information (also called *themes* or *layers* of information) needed for land and geographic information systems may include political boundaries, individual property ownership, population distribution, locations of natural resources, transportation networks, utilities, zoning, hydrography, soil types, land use, vegetation types, wetlands, and many, many more. An essential ingredient of all information entered into LIS and GIS databases is that it be spatially related, that is, located in a common geographic reference framework. Only then are the different layers of information physically relatable so they can be analyzed using computers to support decision making. This geographic positional requirement will place a heavy demand upon surveyors (geomatics engineers) in the future, who will play key roles in designing, implementing, and managing these systems. Surveyors from virtually all of the specialized areas described in Section 1.6 will be involved in developing the needed databases. Their work will include establishing the required basic control framework; conducting boundary surveys and preparing legal descriptions of property ownership; performing topographic and hydrographic surveys by ground, aerial, and satellite methods; compiling and digitizing maps; and assembling a variety of other digital datafiles.

The last chapter of this book, Chapter 28, is devoted to the topic of land and geographic information systems. This subject seems appropriately covered at the end, after each of the other types of surveys needed to support these systems has been discussed.

■ 1.9 FEDERAL SURVEYING AND MAPPING AGENCIES

Several agencies of the U.S. government perform extensive surveying and mapping. Three of the major ones are:

1. The National Geodetic Survey (NGS), formerly the Coast and Geodetic Survey, was originally organized to map the coast. Its activities have included control surveys to establish a network of reference monuments throughout the United States that serve as points for originating local surveys, preparation of nautical and aeronautical charts, photogrammetric surveys, tide and current studies, collection of magnetic data, gravimetric surveys, and worldwide control survey operations. The NGS now plays a major role in coordinating and assisting in activities related to upgrading the national network of

reference control monuments, and to the development, storage, and dissemination of data used in modern LISs and GISs.

2. The U.S. Geological Survey (USGS), established in 1879, has as its mission the mapping of our nation and the survey of its resources. It provides a wide variety of maps, from topographic maps showing the geographic relief and natural and cultural features, to thematic maps that display the geology and water resources of the United States, to special maps of the moon and planets. The National Mapping Division of the USGS has the responsibility of producing topographic maps. It currently has nearly 70,000 different topographic maps available, and it distributes approximately 10 million copies annually. In recent years, the USGS has been engaged in a comprehensive program to develop a national digital cartographic database, which consists of map data in computer-readable formats.
3. The Bureau of Land Management (BLM), originally established in 1812 as the General Land Office, is responsible for managing the public lands. These lands, which total approximately 264 million acres and comprise about 1/8th of the land in the United States, exist mostly in the western states and Alaska. The BLM is responsible for surveying the land and managing its natural resources which include minerals, timber, fish and wildlife, historical sites, and other natural heritage areas. Surveys of most public lands in the conterminous United States have been completed, but much work remains in Alaska.

In addition to these three federal agencies, units of the U.S. Army Corps of Engineers have made extensive surveys for emergency and military purposes. Some of these surveys provide data for engineering projects, such as those connected with flood control. Surveys of wide extent have also been conducted for special purposes by nearly 40 other federal agencies, including the Forest Service, National Park Service, International Boundary Commission, Bureau of Reclamation, Tennessee Valley Authority, Mississippi River Commission, U.S. Lake Survey, and Department of Transportation.

All states have a surveying and mapping section for purposes of generating topographic information upon which highways are planned and designed. Likewise, many counties and cities also have surveying programs, as have various utilities.

■ 1.10 THE SURVEYING PROFESSION

The personal qualifications of surveyors are as important as their technical ability in dealing with the public. They must be patient and tactful with clients and their sometimes-hostile neighbors. Few people are aware of the painstaking research of old records required before fieldwork is started. Diligent, time-consuming effort may be needed to locate corners on nearby tracts for checking purposes as well as to find corners for the property in question.

Land or boundary surveying is classified as a learned profession because the modern practitioner needs a wide background of technical training and experience and must exercise a considerable amount of independent judgment. Registered (licensed) professional surveyors must have a thorough knowledge of mathematics (particularly geometry and trigonometry, but also calculus);

competence with computers; a solid understanding of surveying theory, instruments, and methods in the areas of geodesy, photogrammetry, remote sensing, and cartography; some competence in economics (including office management), geography, geology, astronomy, and dendrology; and a familiarity with laws pertaining to land and boundaries. They should be knowledgeable in both field operations and office computations. Above all, surveyors are governed by a professional code of ethics and are expected to charge professional-level fees for their work.

Permission to trespass on private property or to cut obstructing tree branches and shrubbery must be obtained through a proper approach. Such privileges are not conveyed by a surveying license or by employment in a state highway department or other agency (but a court order can be secured if a landowner objects to necessary surveys).

All 50 states, Guam, and Puerto Rico have registration laws for professional surveyors and engineers (as do the provinces of Canada). In general, a surveyor's license is required to make property surveys, but not for construction, topographic, or route work, unless boundary corners are set.

To qualify for registration as either a professional land surveyor (PLS) or professional engineer (PE), it is necessary to have an appropriate college degree, although some states allow relevant experience in lieu of formal education. In addition, candidates must acquire two or more years of mentored practical experience and must pass a two-day comprehensive written examination. In most states, common national examinations covering fundamentals and principles and practice of land surveying are now used. However, usually two hours of the principles and practice exam are devoted to local legal customs and aspects. As a result, transfer of registration from one state to another has become easier.

Some states also require continuing education units (CEUs) for registration renewal, and many more are considering legislation that would add this requirement. Typical state laws require that a licensed land surveyor sign all plats, assume responsibility for any liability claims, and take an *active part* in the fieldwork.

■ 1.11 PROFESSIONAL SURVEYING ORGANIZATIONS

There are many professional organizations in the United States and worldwide that serve the interests of surveying and mapping. Generally the objectives of these organizations are the advancement of knowledge in the field, encouragement of communication among surveyors, and upgrading of standards and ethics in surveying practice. The *American Congress on Surveying and Mapping* (ACSM) is the foremost professional surveying organization in the United States. Founded in 1941, ACSM regularly sponsors technical meetings at various locations throughout the country. These meetings bring together large numbers of surveyors for presentation of papers, discussion of new ideas and problems, and exhibitions of the latest in surveying equipment. ACSM publishes a quarterly journal, *Surveying and Land Information Science,* and also regularly publishes its newsletter, *The ACSM Bulletin.*

As noted in the preceding section, all states require persons who perform boundary surveys to be licensed. Most states also have professional surveyor societies or organizations with full membership open only to licensed surveyors. These state societies are generally affiliated with ACSM and offer benefits similar to those of ACSM, except that they concentrate on matters of state and local concern.

The *American Society for Photogrammetry and Remote Sensing* (ASPRS) is a sister organization of ACSM. Like ACSM, this organization is also devoted to the advancement of the fields of measurement and mapping, although its major interests are directed toward the use of aerial and satellite imagery for achieving these goals. ASPRS has been cosponsor of many technical meetings with ACSM, and its monthly journal *Photogrammetric Engineering and Remote Sensing* regularly features surveying and mapping articles.

The *Geomatics Division* of the *American Society of Civil Engineers* (ASCE) is also dedicated to professional matters related to surveying and publishes the quarterly *Journal of Surveying Engineering.*

Another organization in the United States, the *Urban and Regional Information Systems Association* (URISA), also supports the profession of surveying and mapping. This organization uses information technology to solve problems in planning, public works, the environment, emergency services, and utilities. Its *URISA Journal* is published quarterly.

The *Canadian Institute of Geomatics* (CIG) is the foremost professional organization in Canada concerned with surveying. Its objectives parallel those of ACSM. This organization, formerly the *Canadian Institute of Surveying and Mapping* (CISM), disseminates information to its members through its *CIG Journal.*

The *International Federation of Surveyors* (FIG), founded in 1878, fosters the exchange of ideas and information among surveyors worldwide. The acronym *FIG* stems from its French name, *Fédération Internationale des Géométres.* FIG membership consists of professional surveying organizations from many countries throughout the world. ACSM has been a member since 1959. FIG is organized into nine technical commissions, each concerned with a specialized area of surveying. The organization sponsors international conferences, usually at four-year intervals, and its commissions also hold periodic symposia where delegates gather for the presentation of papers on subjects of international interest.

■ 1.12 SURVEYING ON THE INTERNET

The explosion of available information on the Internet has had a significant impact upon the field of surveying (geomatics). The Internet enables the instantaneous electronic transfer of documents to any location where the necessary computer equipment is available. It brings resources directly into the office or home, where previously it was necessary to travel to obtain the information or wait for its transfer by mail. Software, educational materials, technical documents, standards, and much more useful information are available on the Internet. As an example of how surveyors can take advantage of the Internet, data from a *Continuously Operating Reference Station* (CORS) can be downloaded from the NGS website for use in a GPS survey (see Section 14.3.5).

Many agencies and institutions maintain websites that provide data free of charge on the Internet. Additionally, some educational institutions now place credit

TABLE 1.1 UNIVERSAL RESOURCE LOCATOR ADDRESSES FOR SOME SURVEYING RELATED SITES

Universal Resource Locator	Owner of Site
http://www.ngs.noaa.gov	National Geodetic Survey
http://www.usgs.gov	U.S. Geological Survey
http://www.blm.gov	Bureau of Land Management
http://www.navcen.uscg.mil	U.S. Coast Guard Navigation Center
http://www.usno.navy.mil	U.S. Naval Observatory
http://www.acsm.net	American Congress on Surveying and Mapping
http://www.asprs.org	American Society for Photogrammetry and Remote Sensing
http://www.asce.org	American Society of Civil Engineers
http://surveying.wb.psu.edu	Access to accompanying software for this book from The Pennsylvania State University Surveying Program

and noncredit courses on the Internet so that distance education can be more easily achieved. With a web browser, it is possible to research almost any topic from a convenient location, and names, addresses, and phone numbers of goods or services providers in a specific area can be identified. As an example, if it was desired to find companies offering mapping services in a certain region, a web search engine could be used to locate web pages that mention this service. Such a search may result in over a million pages if a very general term such as "mapping services" is used to search, but using more specific terms can narrow the search.

Unfortunately the addresses of particular pages and entire sites, given by their *Universal Resource Locators* (URLs), tend to change with time. However, at the risk of publishing URLs that may no longer be correct, a short list of important websites related to surveying is presented in Table 1.1.

1.13 FUTURE CHALLENGES IN SURVEYING

Surveying is currently in the midst of a revolution in the way data are measured, recorded, processed, stored, retrieved, and shared. This is in large part because of developments in computers and computer-related technologies. Concurrent with technological advancements, society continues to demand more data, with increasingly higher standards of accuracy, than ever before. Consequently in a few years the demands on surveying engineers (geomatics engineers) will likely be very different from what they are now.

In the future, the National Spatial Reference System, a network of horizontal and vertical control points, must be maintained and supplemented to meet requirements of increasingly higher-order surveys. New topographic maps with larger scales as well as digital map products are necessary for better planning. Existing maps of our rapidly expanding urban areas need revision and updating to reflect changes, and more and better map products are needed of the older parts of our cities to support urban renewal programs and infrastructure maintenance and modernization. Large quantities of data will be needed to plan and design new rapid-transit systems to connect our major cities, and surveyors will face new challenges in meeting the precise standards required in staking alignments and grades for these systems.

In the future, assessment of environmental impacts of proposed construction projects will call for more and better maps and other data. GISs and LISs that contain a variety of land-related data such as ownership, location, acreage, soil types, land uses, and natural resources must be designed, developed, and maintained. Cadastral surveys of the yet unsurveyed public lands are essential. Monuments set years ago by the original surveyors have to be recovered and remonumented for preservation of property boundaries. Appropriate surveys with very demanding accuracies will be necessary to position drilling rigs as mineral and oil explorations press further offshore. Other future challenges include making precise deformation surveys for monitoring existing structures such as dams, bridges, and skyscrapers to detect imperceptible movements that could be precursors to catastrophes caused by their failure. Timely measurements and maps of the effects of natural disasters such as earthquakes, floods, and hurricanes will be needed so that effective relief and assistance efforts can be planned and implemented. In the space program, the desire for maps of neighboring planets will continue. And we must increase our activities in measuring and monitoring natural and human-caused global changes (glacial growth and retreat, volcanic activity, large-scale deforestation, and so on) that can potentially affect our land, water, atmosphere, energy supply, and even our climate.

These and other opportunities offer professionally rewarding indoor or outdoor (or both) careers for numerous people with suitable training in the various branches of surveying.

PROBLEMS

NOTE: Answers for some of these problems, and some in later chapters, can be obtained by consulting the bibliographies, later chapters, websites, or professional surveyors.

1.1 Develop your personal definition for the practice of surveying.

1.2 Explain the difference between geodetic and plane surveys.

1.3 Describe some surveying applications in:
(a) Archeology **(b)** Mining **(c)** Agriculture

1.4 List 10 uses for surveying other than property and construction surveying.

1.5 What surveying observations does a contractor need to lay a 36-in.-diameter pipe?

1.6 Discuss the uses for topographic surveys.

1.7 What are hydrographic surveys, and why are they important?

1.8 Name and briefly describe three different surveying instruments used by early Roman engineers.

1.9 Briefly explain the procedure used by Eratosthenes in determining the Earth's circumference.

1.10 Describe the steps a land surveyor would need to do when performing a boundary survey.

1.11 Do laws in your state specify the accuracy required for surveys made to lay out a subdivision? If so, what limits are set?
1.12 What organizations in your state will furnish maps and reference data to surveyors and engineers?
1.13 List the legal requirements for registration as a land surveyor in your state.
1.14 Briefly describe the European Galileo system and discuss its similarities and differences with GPS.
1.15 List at least five nonsurveying uses for GPS.
1.16 Explain how aerial photographs and satellite images can be valuable in surveying.
1.17 Search the Internet and define a VLBI station. Discuss why these stations are important to the surveying community.
1.18 Describe how a GIS can be used in flood emergency planning.
1.19 Visit one of the surveying websites listed in Table 1.1, and write a brief summary of its contents. Briefly explain the value of the available information to surveyors.
1.20 Read one of the articles cited in the bibliography for this chapter, or another of your choosing, that describes an application where GPS was used. Write a brief summary of the article.
1.21 Same as Problem 1.20, except the article should be on safety as related to surveying.

BIBLIOGRAPHY

Bartorelli, J. 2002. "Photogrammetry 101." *Point of Beginning* 27 (No. 11): 32.
Bedini, S. A. 2003. "The History Corner: Joshua Fisher (1621–1672) Colonial Inn-keeper and Surveyor, Part 1." *Professional Surveyor Magazine* 23 (No. 9): 70.
Binge, Michael L. 2002. "Surveying GIS." *Point of Beginning* 27 (No. 6): 20.
Brock, J. F. 2001. "Superstar Surveying: Hollywood Heroes and Movietone Measurements." *Surveying and Land Information Systems* 61 (No. 3): 207.
Buhler, D. A. 2006. "Cadastral Survey Activities in the United States." *Surveying and Land Information Science.* 66 (No. 2): 115.
Dahn, R. E. and R. Lumos. 2006. "National Society of Professional Surveyors." *Surveying and Land Information Science* 66 (No. 2): 111.
Denny, Milton. 2001. "City of Philadelphia Regulators." *Point of Beginning* 27 (No. 3): 18.
DeVine, Doug. 2002. "Mapping the CSS Hunley." *Professional Surveyor Magazine* 22 (No. 3): 6.
Fields, Terry. 2003. "The Networking Nucleus." *Point of Beginning* 29 (No. 1): 20.
Finley, D., and D. Coleman. 1999. "Introducing Groupware to Distributed Geomatics Production Environments." *ASCE Journal of Surveying Engineering* 125 (No. 1): 1.
Greenfeld, J. 2006. "The Geographic and Land Information Society and GIS/LIS Activities in the United States." *Surveying and Land Information Science* 66 (No. 2): 119.
Harris, Clay. 2007. "Whole New Ball Game." *Professional Surveyor* 27 (No. 2): 26.
Hohner, Leica N. 2007. "Positioning Your Future." *Point of Beginning* 32 (No. 4): 18.
Jacobs, Geoff. 2005. "High Definition Scanning: Standard Twenty-Foot Cross Sections? No Problem." *Professional Surveyor* 25 (No. 9): 26.
Jeffress, Gary. 2001. "The Present Stage of Evolution of the Geographic Information Science Profession." *Surveying and Land Information Systems* 61 (No. 1): 37.
______. 2006. "Two Perspectives of GIS/LIS Education in the United States." *Surveying and Land Information Science* 66 (No. 2): 123.
Koon, R. 2006. "Safety Sense." *Point of Beginning* 32 (No. 2): 46.
______. 2003. "An Update on OSHA's Traffic Safety Revisions." *Point of Beginning* 29 (No. 1): 28.
______. 2002. "Constructing a Good Safety Plan." *Point of Beginning* 27 (No. 12): 30.
______. 2002. "The Weather-wise Surveyor." *Point of Beginning* 27 (No. 11): 42.

______. 2001. "Gearing Up for Safety." *Point of Beginning* 26 (No. 8): 42.

______. 2001. "Traffic Safety." *Point of Beginning* 26 (No. 12): 40.

Lathrop, W. and D. Martin. 2006. "The American Association for Geodetic Surveying: Its Continuing Role in Shaping the Profession." *Surveying and Land Information Science* 66 (No. 2): 97.

Leica, Brown. 2004. "NOAA Releases Geodesy Discovery Kit Online." *Point of Beginning* 29 (No. 10): 24.

Linklater, Andro. 2002. *Measuring America—How an Untamed Wilderness Shaped the United States and Fulfilled the Promise of Democracy*. Walker & Company, New York, NY.

Matonich, John D. 2003. "Surveyin' Da Situation—Gadget Boy." *Professional Surveyor Magazine* 23 (No. 9) 32.

National Council of Examiners for Engineering and Surveying. 2001. "Model Law." *Surveying and Land Information Systems* 61 (No. 1): 5.

Petrocchi, A. 2000. "Monitoring Oil Spills with Satellite Imagery." *Geo Info Systems* 10 (No. 5): 32.

Platz, George M. 2003. "Taming a Wilderness: Dead River Basin." *Professional Surveyor Magazine* 23 (No. 9): 15.

Robinson, Ella. 2001. "Timber! There Are Safe Ways to Cut Bushes and Trees." *Point of Beginning* 26 (No. 9): 18.

Ryerson, R., and R. Batterman. 2000. "An Approach to the Development of a Sustainable National Geomatics Infrastructure." *Photogrammetric Engineering and Remote Sensing* 66 (No. 1): 17.

Savage, Barry. 2003. "The Quintessential Surveyor." *Point of Beginning* 28 (No. 7): 42.

Schultz, R. 2005. "Education in Surveying: Bachelor Degree Surveying Programs." *Professional Surveyor* 25 (No. 9): 40.

______. 2006. "Education in Surveying: Fundamentals of Surveying Exam." *Professional Surveyor* 26 (No. 3): 38.

Slater, J. 2000. "Guide to Safe Field Surveying." *Point of Beginning* 25 (No. 7): 42.

Stenmark, John. 2003. "Save by Surveying, Satellite and Skill." *Point of Beginning* 28 (No. 5): 30.

Thurow, Glen and Steven Frank. 2001. "Coming to Terms with the Model Law: The Search for a New Definition." *Surveying and Land Information Systems* 61 (No. 1): 39.

Tolbert, Richard T. 2003. "Crossing int the Digital World." *Point of Beginning* 29 (No. 1): 24.

Vass, Emily E. 2003. "The Journey Begins." *Point of Beginning* 28 (No. 7): 36.

Wrock, Doug. 2001. "Big Dig's Precision Floatout." *Professional Surveyor Magazine* 21 (No. 8): 8.

2

Units, Significant Figures, and Field Notes

PART I • UNITS AND SIGNIFICANT FIGURES

■ 2.1 INTRODUCTION

Five types of measurements illustrated in Figure 2.1 form the basis of traditional plane surveying: (1) horizontal angles, (2) horizontal distances, (3) vertical (altitude or zenith) angles, (4) vertical distances, and (5) slope (or slant) distances. In the figure, *OAB* and *ECD* are horizontal planes, and *OACE* and *ABDC* are vertical planes. Then as illustrated, horizontal angles, such as angle *AOB,* and horizontal distances, *OA* and *OB,* are measured in horizontal planes; altitude angles, such as *AOC,* are measured in vertical planes; zenith angles, such as *EOC,* are also measured in vertical planes; vertical lines, such as *AC* and *BD,* are measured vertically (in the direction of gravity); and slope distances, such as *OC,* are determined along inclined planes. By using combinations of these basic measurements, it is possible to compute relative positions between any points. Equipment and procedures for making each of these basic kinds of measurements are described in later chapters of this book.

■ 2.2 UNITS OF MEASUREMENT

Magnitudes of measurements (or of values derived from observations) must be given in terms of specific units. In surveying, the most commonly employed units are for *length, area, volume,* and *angle.* Two different systems are in use for specifying units of observed quantities, the *English* and *metric* systems. Because of its widespread adoption, the metric system is called the *International System of Units* and abbreviated *SI.*

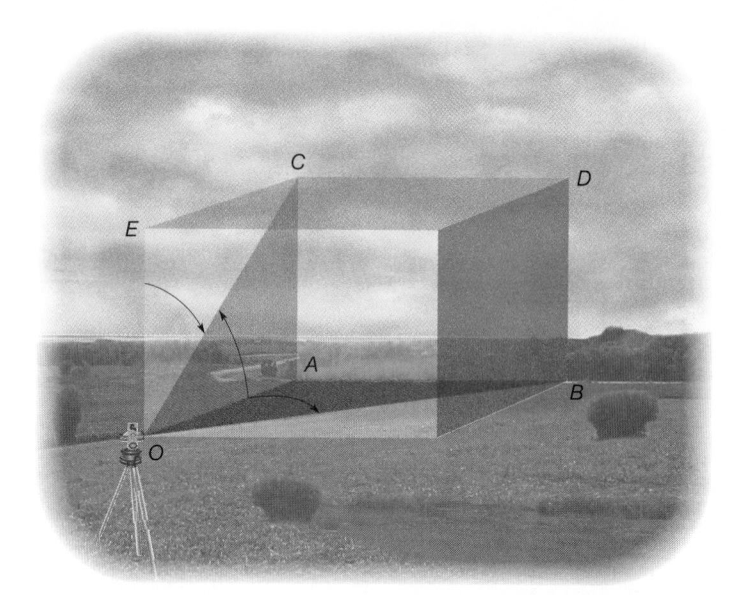

Figure 2.1 Kinds of measurements in surveying.

The basic unit employed for length measurements in the English system is the foot, whereas the meter is used in the metric system. In the past, two different definitions have been used to relate the foot and meter. Although they differ slightly, their distinction must be made clear in surveying. In 1893, the United States officially adopted a standard in which 39.37 in. was exactly equivalent to 1 m. Under this standard, the foot was approximately equal to 0.3048006 m. In 1959, a new standard was officially adopted in which the inch was equal to exactly 2.54 cm. Under this standard, 1 ft equals exactly 0.3048 m. This current unit, known as the international foot, differs from the previous one by about 1 part in 500,000, or approximately 1 foot per 100 miles. This small difference is thus important for very precise surveys conducted over long distances, and for conversions of high elevations or large coordinate values such as those used in State Plane Coordinate Systems as discussed in Chapter 20. Because of the vast number of surveys performed prior to 1959, it would have been extremely difficult and confusing to change all related documents and maps that already existed. Thus the old standard, now called the *U.S. survey foot,* is still used. Individual states have the option of officially adopting either standard. The National Geodetic Survey uses the meter in its distance measurements, thus it is unnecessary to specify the foot unit. Those making conversions from metric units, however, must know the adopted standard for their state and use the appropriate conversion factor.

Since the English system has long been the officially adopted standard for measurements in the United States, except for geodetic surveys, the linear units of feet and *decimals* of a foot are most commonly used by surveyors. In construction, feet and inches are often used. Because surveyors perform all types of surveys including geodetic and because they also provide measurements for developing construction plans and guiding building operations, they must understand all the various systems of units and be capable of making conversions between them. Caution must always be exercised to ensure that observations are recorded in their proper units and that conversions are correctly made.

A summary of the length units used in past and present surveys in the United States includes the following:

1 foot = 12 inches
1 yard = 3 feet
1 inch = 2.54 centimeters (basis of international foot)
1 meter = 39.37 inches (basis of U.S. survey foot)
1 rod = 1 pole = 1 perch = 16.5 feet
1 vara = approximately 33 inches (old Spanish unit often encountered in the southwestern United States)
1 Gunter's chain (ch) = 66 feet = 100 links (lk) = 4 rods
1 mile = 5280 feet = 80 Gunter's chains
1 nautical mile = 6076.10 feet (nominal length of a minute of latitude, or of longitude at the equator)
1 fathom = 6 feet

In the English system, areas are given in *square feet* or *square yards.* The most common unit for large areas is the *acre*. Ten square chains (Gunter's) equal 1 acre. Thus an acre contains 43,560 ft^2, which is the product of 10 and 66^2. The *arpent* (equal approximately to 0.85 acre, but varying somewhat in different states) was used in land grants of the French crown. When employed as a linear term, it refers to the length of a side of 1 square arpent.

Volumes in the English system can be given in *cubic feet* or *cubic yards.* For very large volumes, for example, the quantity of water in a reservoir, the *acre-foot* unit is used. It is equivalent to the area of an acre having a depth of 1 ft, and thus is 43,560 ft^3.

The unit of angle used in surveying is the *degree,* defined as 1/360 of a circle. One degree (1°) equals 60 min, and 1 min equals 60 sec. Divisions of seconds are given in tenths, hundredths, and thousandths. Other methods are also used to subdivide a circle, for example, 400 *grads* (with 100 *centesimal min*/grad and 100 *centesimal sec*/min. Another term, *gons,* is now used interchangeably with grads. The military services use *mils* to subdivide a circle into 6400 units.

A *radian* is the angle subtended by an arc of a circle having a length equal to the radius of the circle. Therefore, 2π rad = 360°, 1 rad $\approx$ 57°17′44.8″ $\approx$ 57.2958°, and 0.01745 rad $\approx$ 1°.

■ 2.3 INTERNATIONAL SYSTEM OF UNITS (SI)

As previously noted, the meter is the basic unit for length in the metric or SI system. Subdivisions of the meter (m) are the *millimeter* (mm), *centimeter* (cm), and *decimeter* (dm), equal to 0.001 m, 0.01 m, and 0.1 m, respectively. A kilometer (km) equals 1000 m, which is about five-eighths of a mile.

Areas in the metric system are specified using the *square meter* (m^2). Large areas, for example tracts of land, are given in *hectares* (ha), where one hectare is equivalent to a square having sides of 100 meters. Thus, there are 10,000 m^2, or about 2.471 acres per hectare. The *cubic meter* (m^3) is used for volumes in the SI system. Degrees, minutes, and seconds, or the radian, are accepted SI units for angles.

The metric system was originally developed in the 1790s in France. Although other definitions were suggested at that time, the French Academy of Sciences chose to define the meter as 1/10,000,000 of the length of the Earth's meridian through Paris from the equator to the pole. The actual length that was adopted for the meter was based on measurements that had been made up to that time to determine the Earth's size and shape. Although later measurements revealed that the initially adopted value was approximately 0.2 mm short of its intended definition related to the meridional quadrant, still the originally adopted length became the standard.

Shortly after the metric system was introduced to the world, Thomas Jefferson, who was then Secretary of State, recommended that the United States adopt it, but the proposal lost by one vote in the Congress! When the metric system was finally legalized for use (but not officially adopted) in the United States in 1866, a meter was defined as the interval under certain physical conditions between lines on an international prototype bar made of 90 percent platinum and 10 percent iridium, and accepted as equal to exactly 39.37 inches. A copy of this bar was held in Washington, D.C. and compared periodically with the international standard held in Paris. In 1960, at the General Conference on Weights and Measures (CGPM), the United States and 35 other nations agreed to redefine the meter as the length of 1,650,763.73 waves of the orange-red light produced by burning the element krypton (Kr-86). That definition permitted industries to make more accurate measurements and to check their own instruments without recourse to the standard meter-bar in Washington. The wavelength of this light is a true constant, whereas there is a risk of instability in the metal meter-bar. The CGPM met again in 1983 and established the current definition of the meter as the length of the path traveled by light in a vacuum during a time interval of 1/299,792,458 sec. Obviously, with this definition, the speed of light in a vacuum becomes exactly 299,792,458 m/sec. The advantage of this latest standard is that the meter is more accurately defined, since it is in terms of time, the most accurate of our basic measurements.

During the 1960s and 1970s, significant efforts were made towards promoting adoption of SI as the legal system for weights and measures in the United States. However, costs and frustrations associated with making the change generated substantial resistance, and the efforts were temporarily stalled. Recognizing the importance to the United States of using the metric system in order to compete in the rapidly developing global economy, in 1988 Congress enacted the *Omnibus Trade and Competitiveness Act.* It designated the metric system as the *preferred* system of weights and measures for U.S. trade and commerce. The Act, together with a subsequent *Executive Order* issued in 1991, required all federal agencies to develop definite metric conversion plans and to use SI standards in their procurements, grants, and other business-related activities to the extent economically feasible. As an example of one agency's response, the Federal Highway Administration adopted a plan calling for (1) use of metric units in all publications and correspondence after September 30, 1992, and (2) use of metric units on all plans and contracts for federal highways after September 30, 1996. Although the Act and Executive Order did not mandate states, counties, cities, or industries to convert to metric, strong incentives were provided, e.g., if SI directives were not complied with, certain federal matching funds could be withheld. In light of these developments, it appeared that the metric system would soon become the official system for use in the United States. However, again much resistance was encountered, not only from individuals but also from agencies of some state, county, and town and city governments, as well as from certain businesses. As a result, the SI still has not been officially adopted in the United States.

Besides the obvious advantage of being better able to compete in the global economy, another significant advantage that would be realized in adopting the SI standard would be the elimination of the confusion that exists in making conversions between the English system and SI. The 1999 crash of the Mars Orbiter underscores costs and frustrations associated with this confusion. This $125 million satellite was supposed to monitor the Martian atmosphere, but instead it crashed into the planet because its contractor used English units while NASA's Jet Propulsion Laboratory was giving it data in the metric system. For these reasons and others, such as the decimal simplicity of the metric system, surveyors who are presently burdened with unit conversions and awkward computations involving yard, foot, and inch units should welcome official adoption of the SI. Since this adoption has not yet occurred, however, this book uses both English and SI units in discussion and example problems.

2.4 SIGNIFICANT FIGURES

In recording observations, an indication of the accuracy attained is the number of digits (significant figures) recorded. By definition, the number of significant figures in any observed value includes the positive (certain) digits plus one (*only one*) digit that is estimated or rounded off, and therefore questionable. For example, a distance measured with a tape whose smallest graduations are 0.01 ft, and recorded as 73.52 ft, is said to have four significant figures; in this case the first three digits are certain, and the last is rounded off and is therefore questionable but still significant.

To be consistent with the theory of errors discussed in Chapter 3, it is essential that data be recorded with the correct number of significant figures. If a significant figure is dropped in recording a value, the time spent in acquiring certain precision has been wasted. On the other hand, if data are recorded with more figures than those that are significant, false precision will be implied. The number of significant figures is often confused with the number of decimal places. Decimal places may have to be used to maintain the correct number of significant figures, but in themselves they do not indicate significant figures. Some examples follow:

Two significant figures: 24, 2.4, 0.24, 0.0024, 0.020
Three significant figures: 364, 36.4, 0.000364, 0.0240
Four significant figures: 7621, 76.21, 0.0007621, 24.00

Zeros at the end of an integral value may cause difficulty because they may or may not be significant. For example, in a value expressed as 2400, it is not known how many figures are significant; there may be two, three, or four, and therefore definite rules must be followed to eliminate the ambiguity. The preferred method of eliminating this uncertainty is to express the value in terms of powers of 10. The significant figures in the measurement are then written as a number between

1 and 10, including the correct number of zeros at the end, and annexing a power of 10 places the decimal point. As an example, 2400 becomes $2.400 \times (10)^3$ if both zeros are significant, $2.40 \times (10)^3$ if one is, and $2.4 \times (10)^3$ if there are only two significant figures. Alternatively, a bar may be placed over the last significant figure, as $240\bar{0}$, $24\bar{0}0$, and $2\bar{4}00$ for 4, 3, and 2 significant figures, respectively.

When observed values are used in the mathematical processes of addition, subtraction, multiplication, and division, it is imperative that the number of significant figures given in answers be consistent with the number of significant figures in the data used and the arithmetic operations performed. The following three steps will achieve this for addition or subtraction: (1) identify the column containing the rightmost significant digit in each number being added or subtracted; (2) perform the addition or subtraction; and (3) round the answer so that its rightmost significant digit occurs in the leftmost column identified in step (1). Two examples illustrate the procedure.

(a)	**(b)**
46.7418	
+ 1.03	378.
+375.0	−2.1
422.7718	375.9
(answer 422.8)	(answer 376.)

In (**a**), the digits 8, 3, and 0 are the rightmost significant ones in the numbers 46.7418, 1.03, and 375.0, respectively. Of these, the 0 in 375.0 is leftmost with respect to the decimal. Thus the answer 422.7718 obtained in adding the numbers is rounded to 422.8, with its rightmost significant digit occurring in the same column as the 0 in 375.0. In (**b**), the digits 8 and 1 are rightmost, and of these the 8 is leftmost. Thus, the answer 375.9 is rounded to 376.

In multiplication, the number of significant figures in the answer is equal to the least number of significant figures in any of the factors. For example, $362.56 \times 2.13 = 772.2528$ when multiplied out, but the answer is correctly given as 772. Its three significant figures are governed by the three significant digits in 2.13. Likewise, in division the quotient should be rounded off to contain only as many significant figures as the least number of significant figures in either the divisor or the dividend. These rules for significant figures in computations stem from error propagation theory, which is discussed further in Section 3.17.

In surveying, four specific types of problems relating to significant figures are encountered and must be understood.

1. Field measurements are given to some specific number of significant figures, thus dictating the number of significant figures in answers derived when the measurements are used in computations. In an intermediate calculation, it is common practice to carry at least one more digit than required, and then round off the final answer to the correct number of significant figures.
2. There may be an implied number of significant figures. For instance, the length of a football field might be specified as 100 yards. But in laying out the field, such a distance would probably be measured to the nearest hundredth of a foot, not the nearest half-yard.

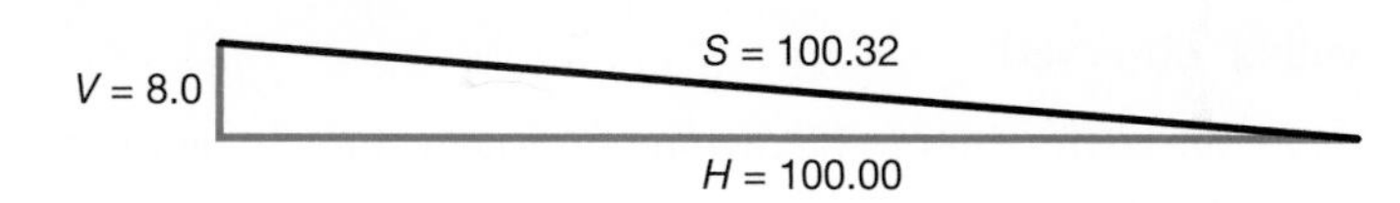

Figure 2.2
Slope correction.

3. Each factor may not cause an equal variation. For example, if a steel tape 100.00 ft long is to be corrected for a change in temperature of 15°F, one of these numbers has five significant figures while the other has only two. However, a 15° variation in temperature changes the tape length by only 0.01 ft. Therefore, an adjusted tape length to five significant figures is warranted for this type of data. Another example is the computation of a slope distance from horizontal and vertical distances, as in Figure 2.2. The vertical distance V is given to two significant figures, and the horizontal distance H is measured to five significant figures. From these data, the slope distance S can be computed to five significant figures. For small angles of slope, a considerable change in the vertical distance produces a relatively small change in the difference between slope and horizontal distances.
4. Measurements are recorded in one system of units but may have to be converted to another. A good rule to follow in making these conversions is to retain in the answer a number of significant figures equal to those in the measured value. As an example, to convert 178 ft 6-3/8 in. to meters, the number of significant figures in the measured value would first be determined by expressing it in its smallest units. In this case, 1/8th inch is the smallest unit and there are $(178 \times 12 \times 8) + (6 \times 8) + 3 = 17{,}139$ of these units in the value. Thus the measurement contains five significant figures, and the answer is $17{,}139 \div (8 \times 39.37 \text{ in./m}) = 54.416$ m, properly expressed with five significant figures. Note that 39.37 used in the conversion is an exact constant, and does not limit the number of significant figures.

One should always adopt the practice of using exact conversions in computations to avoid loss of accuracy from the conversion process. For example, converting a state plane coordinate value of 613,843.912 m to U.S. survey feet with the approximate conversion of 3.2808 ft/m will result in a loss of accuracy in the results. Using the exact conversion of 39.37 in./ft yields an equivalent coordinate value of 2,013,919.57 ft. Note that the approximate conversion value of 3.2808 yields the lower accuracy coordinate of 2,013,899.11 ft, which must be rounded to 2,013,900 ft that has only five significant figures. In this instance, the number of significant figures in the conversion is the limiting condition to the accuracy of the conversion. In fact, the approximate conversion results in an error of more than 20 ft in the resulting coordinate value! It is easy to see, from this example, why exact conversion values should always be used in surveying computations.

■ 2.5 ROUNDING OFF NUMBERS

Rounding off a number is the process of dropping one or more digits so the answer contains only those digits that are significant. In rounding off numbers to any required degree of precision in this text, the following procedures will be observed:

1. When the digit to be dropped is lower than 5, the number is written without the digit. Thus 78.374 becomes 78.37. Also 78.3749 rounded to four figures becomes 78.37.
2. When the digit to be dropped is exactly 5, the nearest even number is used for the preceding digit. Thus, 78.375 becomes 78.38 and 78.385 is also rounded to 78.38.
3. When the digit to be dropped is greater than 5, the number is written with the preceding digit increased by 1. Thus, 78.376 becomes 78.38.

Procedures 1 and 3 are standard practice. When rounding the value 78.375 in procedure 2, however, some people always take the next higher hundredth, whereas others invariably use the next lower hundredth. Using the nearest even digit, however, establishes a uniform procedure and produces better-balanced results in a series of computations. It is improper procedure to perform two-stage rounding where, for example, in rounding 78.3749 to four digits it would be first rounded to five figures, yielding 78.375, and then rounded again to 78.38. The correct answer in rounding 78.3749 to four figures is 78.37.

It is important to recognize that rounding should only occur with the final answer. Intermediate computations should be done without rounding to avoid problems that can be caused by rounding too early. Example (**a**) of Section 2.4 is repeated below to illustrate this point. The sum of 46.7418, 1.03 and 375.0 is rounded to 422.8 as shown in "correct" column. If the individual values are rounded prior to the addition as shown in the "incorrect" column, the incorrect result of 422.7 is obtained.

Correct	**Incorrect**
46.7418	46.7
+ 1.03	+ 1.0
+ 375.0	+ 375.0
422.7718	422.7
(answer 422.8)	(answer 422.7)

PART II • FIELD NOTES

■ 2.6 FIELD NOTES

Field notes are the records of work done in the field. They typically contain observations, sketches, descriptions, and many other items of miscellaneous information. In the past, field notes were prepared exclusively by hand lettering in field books or special notepads as the work progressed and data were gathered. Recently however, data collectors, also known as survey controllers, have been introduced that can interface with many different modern surveying instruments. As the work progresses, they create computer files containing a record of observed data. Data collectors are rapidly gaining popularity, but when used, manually prepared sketches and descriptions often supplement the numerical data they generate. Regardless of the manner or form in which the notes are taken, they are extremely important.

Whether prepared manually, created by a data collector, or a combination of these forms, surveying field notes are the only permanent records of work done in the field. If the data are incomplete, incorrect, lost, or destroyed, much or all of the time and money invested in making the measurements and records have been wasted. Hence, the job of data recording is frequently the most important and difficult one in a surveying party. Field books and computer files containing information gathered over a period of weeks are worth many thousands of dollars because of the costs of maintaining personnel and equipment in the field.

Recorded field data are used in the office to perform computations, make drawings, or both. The office personnel using the data are usually not the same people who took the notes in the field. Accordingly, it is essential that notes be intelligible to anyone without verbal explanations.

Property surveys are subject to court review under some conditions, so field notes become an important factor in litigation. Also, because they may be used as references in land transactions for generations, it is necessary to index and preserve them properly. The salable "goodwill" of a surveyor's business depends largely on the office library of field books. Cash receipts may be kept in an unlocked desk drawer, but field books are stored in a fireproof safe!

■ 2.7 GENERAL REQUIREMENTS OF HANDWRITTEN FIELD NOTES

The following points are considered in appraising a set of field notes:

Accuracy. This is the most important quality in all surveying operations.

Integrity. A single omitted measurement or detail can nullify use of the notes for computing or plotting. If the project was far from the office, it is time-consuming and expensive to return for a missing measurement. Notes should be checked carefully for completeness before leaving the survey site and never "fudged" to improve closures.

Legibility. Notes can be used only if they are legible. A professional-looking set of notes is likely to be professional in quality.

Arrangement. Note forms appropriate to the particular survey contribute to accuracy, integrity, and legibility.

Clarity. Advance planning and proper field procedures are necessary to ensure clarity of sketches and tabulations and to minimize the possibility of mistakes and omissions. Avoid crowding notes; paper is relatively cheap. Costly mistakes in computing and drafting are the end results of ambiguous notes.

Appendix B contains examples of handwritten field notes for a variety of surveying operations. Their plate number identifies each. Other example note forms are given at selected locations within the chapters that follow. These notes have been prepared keeping the above points in mind.

In addition to the items stressed in the foregoing, certain other guidelines must be followed to produce acceptable handwritten field notes. The notes should be lettered with a sharp pencil of at least 3-H hardness so that an indentation is made in the paper. Books so prepared will withstand damp weather in the field (or even a soaking) and still be legible, whereas graphite from a soft pencil, or ink from a pen or ballpoint, leaves an undecipherable smudge under such circumstances.

Erasures of recorded data are not permitted in field books. If a number has been entered incorrectly, a line is run through it without destroying the number's legibility, and the proper value is noted above it (see Figure 5.5). If a partial or entire page is to be deleted, diagonal lines are drawn through opposite corners, and **VOID** is lettered prominently on the page, giving the reasons.

Field notes are presumed to be "original" unless marked otherwise. Original notes are those taken at the same time the observations are being made. If the original notes are copied, they must be so marked (see Figure 5.11). Copied notes may not be accepted in court because they are open to question concerning possible mistakes, such as interchanging numbers and omissions. The value of a distance or an angle placed in the field book from memory 10 minutes after the observation is definitely unreliable. Students are tempted to scribble notes on scrap sheets of paper for later transfer in neater form to the field book. This practice may result in the loss of some or all of the original data and defeats one purpose of a surveying course—to provide experience in taking notes under actual field conditions. In a real job situation, a surveyor is not likely to spend any time at night transcribing scribbled notes. Certainly, an employer will not pay for this evidence of incompetence.

■ 2.8 TYPES OF FIELD BOOKS

Since field books contain valuable data, suffer hard wear, and must be permanent in nature, only the best should be used for practical work. Various kinds of field books as shown in Figure 2.3 are available, but bound and loose-leaf types are most common. The bound book, a standard for many years, has a sewed binding, a hard cover of leatherette, polyethylene, or covered hardboard, and contains 80 leaves. Its use ensures maximum testimony acceptability for property survey records in courtrooms. Bound duplicating books enable copies of the original notes to be made through carbon paper in the field. The alternate duplicate pages are perforated to enable their easy removal for advanced shipment to the office.

Loose-leaf books have come into wide use because of many advantages, which include (1) assurance of a flat working surface, (2) simplicity of filing

Figure 2.3
Field books.
(Courtesy Topcon Positioning Systems)

individual project notes, (3) ready transfer of partial sets of notes between field and office, (4) provision for holding pages of printed tables, diagrams, formulas, and sample forms, (5) the possibility of using different rulings in the same book, and (6) a saving in sheets and thus cost since none are wasted by filing partially filled books. A disadvantage is the possibility of losing sheets.

Stapled or spiral-bound books are not suitable for practical work. They may be satisfactory, however, for abbreviated surveying courses that have only a few field periods, because of limited service required and low cost. Special column and page rulings provide for particular needs in leveling, angle measurement, topographic surveying, cross-sectioning, and so on.

A camera is a helpful notekeeping "instrument." Moderately priced, reliable, lightweight digital cameras can be used to document monuments set or found and to provide records of other valuable information or admissible field evidence. Recorded images can become part of the final record of survey. Tape recorders, traditional or digital, can also be used in certain circumstances, particularly where lengthy written explanations would be needed to document conditions or provide detailed descriptions.

■ 2.9 KINDS OF NOTES

Four types of notes are kept in practice: (1) sketches, (2) tabulations, (3) descriptions, and (4) combinations of these. The most common type is a combination form, but an experienced recorder selects the version best fitted to the job at hand. The note forms in Appendix B illustrate some of these types and apply to field problems described in this text. Other examples are included within the text at appropriate locations. Sketches often greatly increase the efficiency with which notes can be taken. They are especially valuable to persons in the office who must interpret the notes without benefit of the notekeeper's presence. The proverb about one picture being worth a thousand words might well have been intended for notekeepers!

For a simple survey, such as measuring the distances between points on a series of lines, a sketch showing the lengths is sufficient. In measuring the length of a line forward and backward, a sketch together with tabulations properly arranged in columns is adequate, as in Plate B.1 in Appendix B. The location of a reference point may be difficult to identify without a sketch, but often a few lines of description are enough. Photos may be taken to record the location of permanent stations and the surrounding locale. The combination of a sketch with dimensions and photographic images can be invaluable in later station relocation. Benchmarks are usually briefly described, as in Figure 5.5.

In notekeeping, this axiom is always pertinent: *When in doubt about the need for any information, include it and make a sketch. It is better to have too much data than not enough.*

■ 2.10 ARRANGEMENTS OF NOTES

Note styles and arrangement depend on departmental standards and individual preference. Highway departments, mapping agencies, and other organizations engaged in surveying furnish their field personnel with sample note forms, similar to those in Appendix B and in various sections of this book, to aid in preparing uniform and complete records that can be checked quickly.

It is desirable for students to have as guides, predesigned sample sets of note forms covering their first fieldwork to set high standards and save time. The note forms shown in Appendix B are composites of several models. They stress the open style, especially helpful for beginners, in which some lines or spaces are skipped for clarity. Thus angles observed at a point *A* (see Plate B.4) are placed opposite *A* on the page, but distances observed between *A* and *B* on the ground are recorded on the line between *A* and *B* in the field book.

Left- and right-hand pages are practically always used in pairs and therefore carry the same page number. A complete title should be lettered across the top of the left page and may be extended over the right one. Titles may be abbreviated on succeeding pages for the same survey project. Location and type of work are placed beneath the title. Some surveyors prefer to confine the title on the left page and keep the top of the right one free for date, party, weather, and other items. This design is revised if the entire right page has to be reserved for sketches and benchmark descriptions. Arrangements shown in Appendix B demonstrate the flexibility of note forms. The left page is generally ruled in six columns designed for tabulation only. Column headings are placed between the first two horizontal lines at the page top and follow from left to right in the anticipated order of reading and recording. The upper part of the left or right page must contain the following items:

1. *Project name, location, date, time of day* (A.M. *or* P.M.), *and starting and finishing times.* These entries are necessary to document the notes and furnish a timetable as well as to correlate different surveys. Precision, troubles encountered, and other facts may be gleaned from the time required for a survey.
2. *Weather.* Wind velocity, temperature, and adverse weather conditions such as rain, snow, sunshine, and fog have a decided effect on accuracy in surveying operations. Surveyors are unlikely to do their best possible work at temperatures of 15°F or with rain pouring down their necks. Hence weather details are important in reviewing field notes, in applying corrections to observations due to temperature variations, and for other purposes.
3. *Party.* The names and initials of party members and their duties are required for documentation and future reference. Jobs can be described by symbols, such as $\overline{\wedge}$ for instrument operator, ϕ for rod person, and *N* for notekeeper. The party chief is frequently the notekeeper.
4. *Instrument type and number.* The type of instrument used (with its make and serial number) and its degree of adjustment affects the accuracy of a survey. Identification of the specific equipment employed may aid in isolating some errors—for example, a particular tape with an actual length that is later found to disagree with the distance recorded between its end graduations.

To permit ready location of desired data, each field book must have a table of contents that is kept current daily. In practice, surveyors cross-index their notes on days when fieldwork is impossible.

■ 2.11 SUGGESTIONS FOR RECORDING NOTES

Observing the suggestions given in preceding sections, together with those listed here, will eliminate some common mistakes in recording notes.

1. Letter the notebook owner's name and address on the cover and first inside page using permanent ink. Number all field books for record purposes.
2. Begin a new day's work on a new page. For property surveys having complicated sketches, this rule may be waived.
3. Employ any orderly, standard, familiar note form type, but, if necessary, design a special arrangement to fit the project.
4. Include explanatory statements, details, and additional measurements if they might clarify the notes for field and office personnel.
5. Record what is read without performing any mental arithmetic. Write down what you read!
6. Run notes down the page, except in route surveys, where they usually progress upward to conform to sketches made while looking in the forward direction. (See Plate B.5 in Appendix B.)
7. Use sketches instead of tabulations when in doubt. Carry a straightedge for ruling lines and a small protractor to lay off angles.
8. Make drawings to general proportions rather than to exact scale, and recognize that the usual preliminary estimate of space required is too small. To clearly show their application, letter parallel with or perpendicular to the appropriate features.
9. Exaggerate details on sketches if clarity is thereby improved, or prepare separate diagrams.
10. Line up descriptions and drawings with corresponding numerical data. For example, a benchmark description should be placed on the right-hand page opposite its elevation, as in Figure 5.5.
11. Avoid crowding. If it is helpful to do so, use several right-hand pages of descriptions and sketches for a single left-hand sheet of tabulation. Similarly, use any number of pages of tabulation for a single drawing. Paper is cheap compared with the value of time that might be wasted by office personnel in misinterpreting compressed field notes, or by requiring a party to return to the field for clarification.
12. Use explanatory notes when they are pertinent, always keeping in mind the purpose of the survey and needs of the office force. Put these notes in open spaces to avoid conflict with other parts of the sketch.
13. Employ conventional symbols and signs for compactness.
14. A meridian arrow is vital for all sketches. Have north at the top and on the left side of sketches, if possible.
15. Keep tabulated figures inside of and off column rulings, with decimal points and digits in line vertically.
16. Make a mental estimate of all measurements before receiving and recording them in order to eliminate large mistakes.
17. Repeat aloud values given for recording. For example, before writing down a distance of 124.68, call out "one, two, four, point six, eight" for verification by the person who submitted the measurement.

18. Place a zero before the decimal point for numbers smaller than 1; that is, record 0.37 instead of .37.
19. Show the precision of observations by means of significant figures. For example, record 3.80 instead of 3.8 only if the reading was actually determined to hundredths.
20. Do not superimpose one number over another or on lines of sketches, and do not try to change one figure to another, as a 3 to a 5.
21. Make all possible arithmetic checks on the notes and record them before leaving the field.
22. Compare all misclosures and error ratios while in the field. On large projects where daily assignments are made for several parties, completed work is shown by satisfactory closures.
23. Arrange essential computations made in the field so they can be checked later.
24. Title, index, and cross-reference each new job or continuation of a previous one by client's organization, property owner, and description.
25. Sign surname and initials in the lower right-hand corner of the right page on all original notes. This places responsibility just as signing a check does.

■ 2.12 INTRODUCTION TO DATA COLLECTORS

Advances in computer technology in recent years have led to the development of sophisticated data collection systems for taking field notes. These devices are about the size of a pocket calculator and are produced by a number of different manufacturers. They are available with a variety of features and capabilities. Figure 2.4 illustrates two different data collectors.

(a)

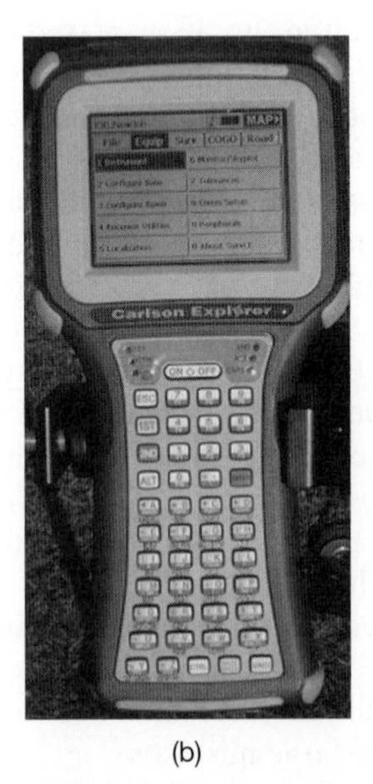

(b)

(c)

Figure 2.4 Various data collectors that are used in the field: (a) Trimble TSC2 data collector, (b) Carlson Explorer, and (c) TDS Recon. [Figure (c) Courtesy Trimble Navigation Systems]

Data collectors can be interfaced with modern surveying instruments, and when operated in that mode they can automatically receive and store data in computer compatible files as observations are taken. Control of observations and storage operations are maintained through the data collector's keyboard. For clarification of the notes, the operator inputs point identifiers and other descriptive information along with the measurements as they are being recorded. When a job is completed or at day's end, the files can be transferred directly to a computer for further processing.

In using data collectors, the usual preliminary information such as date, party, weather, time, and instrument number is entered manually into the file through the keyboard. For a given type of survey, the data collector's internal microprocessor is programmed to follow a specific sequence of steps. The operator identifies the type of survey to be performed from a menu, or by means of a code, and then follows instructions that appear on the unit's screen. Step-by-step prompts will guide the operator to either (a) input "external" data (which may include station names, descriptions, or other information), or (b) press a key to initiate the recording of observed values. Since data collectors require the users to follow specific steps when performing a survey, they are often referred to as *survey controllers*.

Data collectors store information in either binary or ASCII (American Standard Code for Information Interchange) format. Binary storage is faster and more compact, but usually the data must be translated to ASCII before they can be read or edited. Most data collectors enable an operator to scroll through stored data, displaying them on the screen for review and editing while still at the job site. The organizational structures used by different data collectors in storing information vary considerably from one manufacturer to the next. They all follow specific rules, and once they are understood, the data can be readily interpreted by both field and office personnel. The disadvantage of having varied data structures from different manufacturers is that a new system must be learned with each instrument of different make. Efforts have been made toward standardizing the data structures. The *Survey Data Management System* (SDMS), for example, has been adopted by the *American Association of State Highway and Transportation Officials* (AASHTO) and is recommended for all surveys involving highway work. The example field notes for a radial survey given in Table 17.1 of Section 17.9 are in the SDMS format.

Most manufacturers of modern surveying equipment have developed data collectors specifically to be interfaced with their own instruments, but some are flexible. The Trimble TSC2 survey controller shown in Figure 2.4(a), for example, can be interfaced with Trimble instruments but it can be used with others also. In addition to serving as a data collector, the TSC2 is able to perform a variety of time-saving calculations directly in the field. It has a Windows CE operating system and thus can run a variety of Windows software programs. Additionally it has Bluetooth technology so that it can communicate with instruments without using cables and has WiFi capabilities for connecting to the Internet.

Some data collectors can also be operated as electronic field books. In the electronic field book mode, the data collector is not interfaced with a surveying instrument. Instead of handwriting the data in a field book, the notekeeper manually enters measurements into the data collector by means of keyboard strokes after readings are taken. This has the advantage of enabling field notes to be recorded directly in a computer format ready for further processing, even though the surveying instruments being used may be older and not compatible for direct interfacing with data collectors. However, data collectors provide the utmost in efficiency when they are interfaced with surveying instruments such as total stations that have automatic readout capabilities.

The touch screen of the Carlson Explorer data collector shown in Figure 2.4(b) is a so-called third-party unit; that is, it is made by an independent company to be interfaced with instruments manufactured by others. It also utilizes a Windows® CE operating system and has both Bluetooth and WiFi capabilities. It can be either operated in the electronic field book mode or interfaced with a variety of instruments for automatic data collection.

Many instrument manufacturers incorporate data collection systems as internal components directly into their equipment. This incorporates all features of external data collectors, including the display panel, within the instrument. The Topcon GTS 800 shown in Figure 2.5 has an MS-DOS®-based operating system with the ability to run the Tripod Data System (TDS) Survey Pro Software® on board. It comes standard with 2 MB of program memory and 2 MB of internal data memory. The instrument has a PCMCIA[1] port for use with external data cards to allow for transfer of data from the field to the office without the instrument.

Data collectors currently use the Windows® CE operating system. A pen and pad arrangement enables the user to point on menus and options to run software. The data collectors in Figure 2.4 and Figure 2.6 have this type of interface. The Trimble TSC2® with Bluetooth communications shown in Figure 2.7 is being used

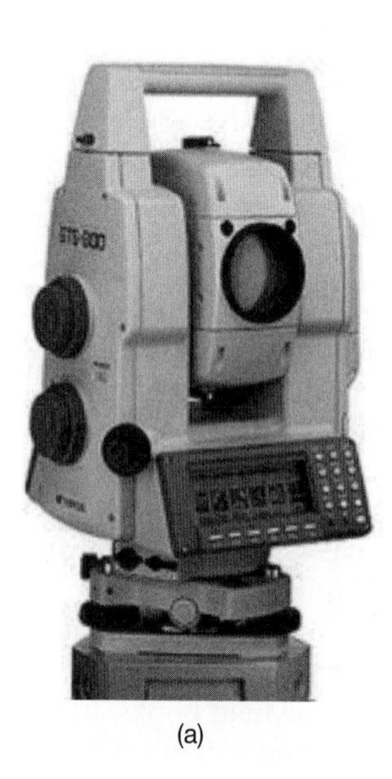

(a)

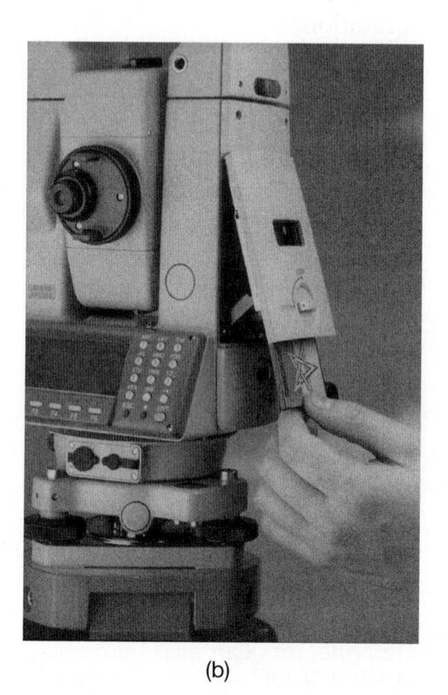

(b)

Figure 2.5 The Topcon GTS 800 total station with internal data collector. (Courtesy Topcon Positioning Systems)

[1]A PCMCIA port conforms to the *Personal Computer Memory Card International Association* standards.

Figure 2.6 Trimble TSC2 with Bluetooth technology. (Courtesy Trimble Navigation Systems)

to control a total station survey in Figure 2.6. Most modern data collectors have the capability of running advanced computer software in the field. As one example of their utility, field crews can check their data before sending it to the office.

As each new series of data collectors is developed, more sophisticated user-interfaces are being designed, and the software that accompanies the systems is being improved. These systems have resulted in increased efficiency and productivity, and have provided field personnel with new features, such as the ability to perform additional field checks. However, the increased complexity of operating surveying instruments with advanced data collectors also requires field personnel with higher levels of education and training.

■ 2.13 TRANSFER OF FILES FROM DATA COLLECTORS

At regular intervals, usually at lunchtime and at the end of a day's work, or when a survey has been completed, the information stored in files within a data collector is transferred to another device. This is a safety precaution to avoid accidentally losing substantial amounts of data. Ultimately, of course, the files are downloaded to a host computer, which will perform computations or generate maps and plots from the data. Depending on the peripheral equipment available, different procedures for data transfer can be used. In one method that is particularly convenient when surveying in remote locations, data can be returned to the home office via telephony technology using devices called *data modems*. Thus, office personnel can immediately begin using the data. In areas with cell phone coverage, this operation can be performed in the field. Another method of data transfer consists

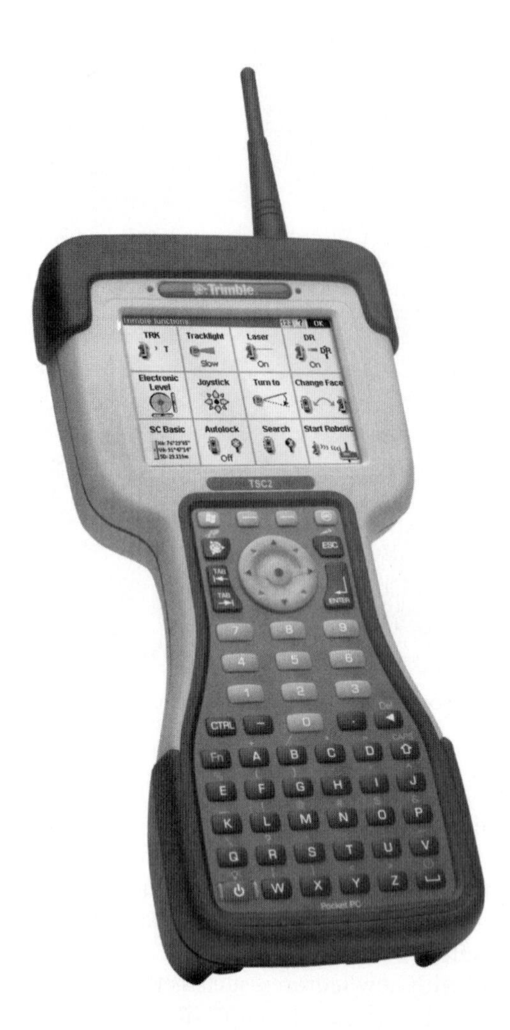

Figure 2.7 Screen of a Trimble TSC2 survey controller. (Courtesy Trimble Navigation Systems)

in downloading data straight into a computer by direct hookup via an RS-232 cable. This can be performed in the office, or it can be done in the field if a laptop computer is available. In areas with wireless Internet, data can be transferred to the office using wireless connections.

Some surveying instruments, for example, the Topcon GTS-800 Series total station shown in Figure 2.5, are capable of storing data externally on PCMCIA cards. These cards can, in turn, be taken to the office, where the files can be downloaded using a computer with a PCMCIA port. These ports are standard for most laptop computers, and thus allow field crews to download data from the PCMCIA card, external, or internal data collector to storage devices on the computer at regular intervals in the field. With the inclusion of a modem, field crews can transfer files to an office computer over phone lines. Office personnel can check field data, or compute additional points to be staked, in the office and return the results to the field crews while they are still on the site.

From the preceding discussion, and as illustrated in Figure 2.8, data collectors are central components of modern computerized surveying systems. In these

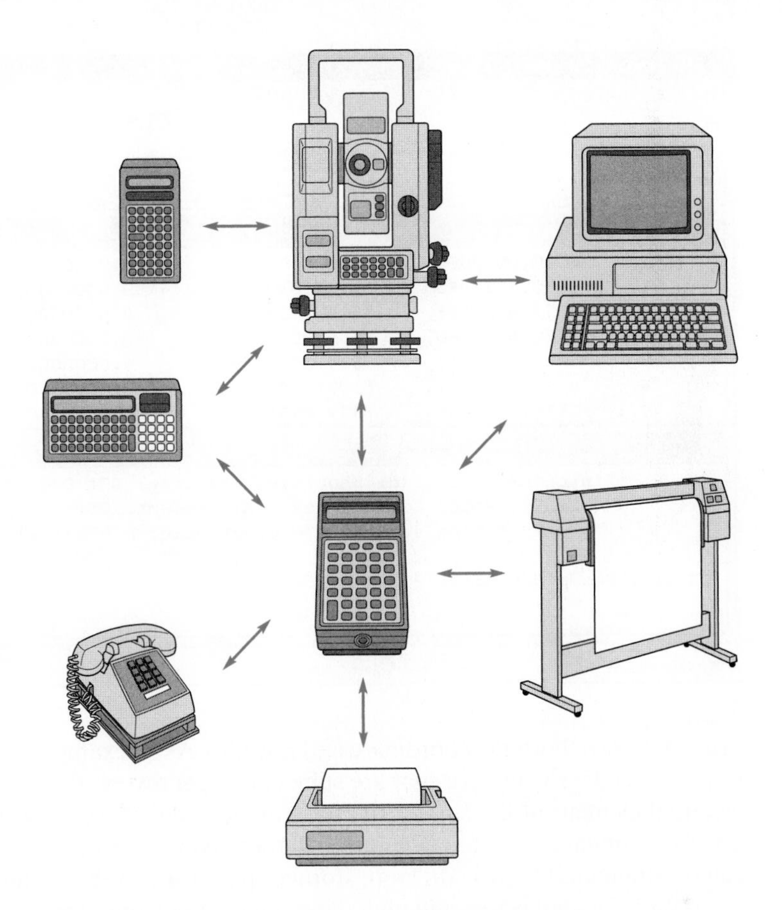

Figure 2.8 Automatic data collector—a central component in modern computerized surveying systems. (Courtesy Sokkia Corporation)

systems, data flow automatically from the field instrument through the collector to the printer, computer, plotter, and other units in the system. The term "field-to-finish systems" is often applied when this form of instrumentation and software is utilized in surveying.

■ 2.14 DIGITAL DATA FILE MANAGEMENT

Once the observing process is completed in the field, the generated data files must be transferred (downloaded) from the data collector to another secure storage device. An example of this process and the field-generated files is shown in Figure 2.9 with the SMI Transfer program. Note that the data collector generated a computed (coordinate) file (Hwy181.asc) and a raw data file (Hwy181.raw). Data collectors generally provide the option of saving these and other types of files. In this case, the coordinate file consists of computed coordinate values generated using the observations and any applied field corrections. Field corrections may include a scale factor, offsets, and Earth curvature and refraction corrections applied to distances. Field crews generally can edit and delete information from the computed file. However, the raw data file consists of the original unreduced measurements and cannot be altered in the field. The necessity for each type of data file is dependent on the intended use of the survey. In most surveys, it would be

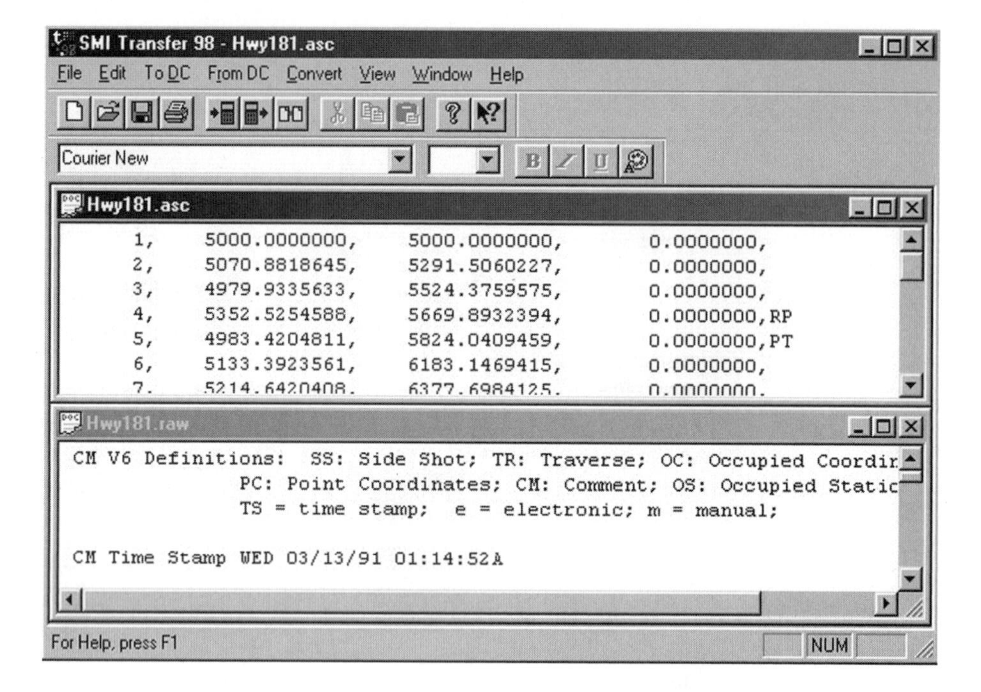

Figure 2.9 SMI Transfer program with portions of a coordinate and raw data file in its window. (Courtesy Surveyors Module Inc.)

prudent to save both the coordinate and raw files. As an example, for projects that require specific closures, or that are subject to legal review, the raw data file is an essential element of the survey. In topographic and GPS surveys, however, large quantities of data are often generated. In these types of projects, the raw data file can be eliminated to provide more storage space for coordinate files.

With data collectors and digital instruments, personnel in modern surveying offices deal with considerably more data than was customary in the past. This increased volume inevitably raises new concerns about data reliability and safe storage. Many methods can be used to provide backup of digital data. Some storage options include removable media disks and tapes. Since these tend to be magnetic, there is an inevitable danger that data could be lost due to the presence of external magnetic devices, or from the failure of the disk's surface. Because of this problem, it is wise to keep two copies of the files for all jobs. Another inexpensive solution to this problem is the use of compact disk (CD) writers. These drives will write an optical image of a project's data on a CD. Since CDs are small but have large storage capabilities, entire projects, including drawings, can be recorded in a small space that is easily archived for future reference. However, these disks can fail when scratched. Thus, care must be taken in their handling and storage.

■ 2.15 ADVANTAGES AND DISADVANTAGES OF DATA COLLECTORS

The major advantages of automatic data collection systems are that (1) mistakes in reading and manually recording observations in the field are precluded, and (2) the time to process, display, and archive the field notes in the office is reduced significantly. Systems that incorporate computers can execute some programs in the field, which adds a significant advantage. As an example, the data for a survey can be corrected for systematic errors and misclosures computed, so verification that a survey meets closure requirements is made before the crew leaves a site.

Data collectors are most useful when large quantities of information must be recorded, for example, in topographic surveys or cross-sectioning. In Section 17.9 their use in topographic surveying is described, and an example set of notes taken for that purpose is presented and discussed.

Although data collectors have many advantages, they also present some dangers and problems. There is the slight chance, for example, the files could be accidentally erased through carelessness or lost because of malfunction or damage to the unit. Some difficulties are also created by the fact that sketches cannot be entered into the computer. However, this problem can be overcome by supplementing files with sketches made simultaneously with the observations that include field codes. These field codes can instruct the drafting software to draw a map of the data complete with lines, curves, and mapping symbols. It is important to realize that not all information can be stored in digital form, and thus is it important to keep a traditional field book to enter sketches, comments, and additional notes when necessary.

Data collectors are available from numerous manufacturers. They must be capable of transferring data through various hardware in modern surveying systems such as that illustrated in Figure 2.8. Since equipment varies considerably, it is important when considering the purchase of a data collector to be certain it fits the equipment owned or perhaps needed in the future.

PROBLEMS

Asterisks (*) indicate problems that have answers given in Appendix G.

2.1 State the current definition of the meter.

2.2 What is the exact conversion of length from the metric system to
(a) survey feet? **(b)** international feet?

2.3 Why was the survey foot definition maintained in the United States?

2.4 Convert the following distances given in meters to U.S. survey feet:
***(a)** 3273.027 m **(b)** 1685.753 m **(c)** 7761.083 m

2.5 Convert the following distances given in feet to meters:
***(a)** 5468.94 ft **(b)** 10,524.85 ft **(c)** 9264.06 ft

2.6 Compute the lengths in feet corresponding to the following distances measured with a Gunter's chain:
***(a)** 22 ch 37 lk **(b)** 86 ch 23 lk **(c)** 36 ch 14 lk

2.7 Express 378,980 ft^2 in:
(a) acres ***(b)** hectares **(c)** square Gunter's chains

2.8 Convert 10.249 ha to:
(a) acres **(b)** square Gunter's chains

2.9 What are the lengths in feet and decimals for the following distances shown on a building blueprint:
(a) 25 ft 6-7/8 in. **(b)** 86 ft 8-1/16 in.

2.10 What is the area in acres of a rectangular parcel of land measured with a Gunter's chain if the recorded sides are as follows:
***(a)** 19.17 ch and 15.23 ch **(b)** 60 ch 59 lk and 44 ch 98 lk

2.11 Compute the area in acres of triangular lots shown on a plat having the following recorded right-angle sides:
(a) 536.93 ft and 241.11 ft **(b)** 9 ch 25 lk and 6 ch 16 lk

2.12 A distance is expressed as 200,365.28 U.S. survey feet. What is the length in:
*__(a)__ international feet? **(b)** meters?

2.13 What are the radian and degree-minute-second equivalents for the following angles given in grads:
*__(a)__ 77.0000 grads **(b)** 180.6942 grads **(c)** 35.9467 grads

2.14 Give answers to the following problems in the correct number of significant figures:
*__(a)__ sum of 35.16, 0.1425, 216, and 9.8
(b) sum of 11.63, 0.065, 10.1, and 273.011
(c) product of 276.75 and 33.68
(d) quotient of 5320.70 divided by 3.98

2.15 Express the value or answer in powers of 10 to the correct number of significant figures:
(a) 45,637
(b) 3260
(c) square of 6629
(d) sum of (41.523 + 0.9 + 267.89) divided by 10.8

2.16 Convert the adjusted angles of a triangle to grads and show a computational check:
*__(a)__ 39°41′54″, 91°30′16″, and 48°47′50″
(b) 82°17′43″, 29°05′54″, and 68°36′23″

2.17 What information should normally be included in a good set of field notes?

2.18 Why should ink not be used in field notekeeping?

2.19* Explain why data should always be entered directly into the field book at the time measurements are made, rather than on scrap paper for neat transfer to the field book later.

2.20 Why should a new day's work begin on a new page?

2.21 Why should the field notes show the precision of the measurements?

2.22 Explain the reason for item 7 in Section 2.11 when recording field notes.

2.23 Explain why ruled vertical and horizontal lines are necessary on field book pages.

2.24 When should sketches be made instead of just recording data?

2.25 Justify the requirement to list in a field book the makes and serial numbers of all instruments used on a survey.

2.26 Discuss the advantages of survey controllers that can communicate with both a total station and a GPS receiver.

2.27 Discuss the disadvantages of survey controllers.

2.28 Search the Internet and find at least two sites related to
(a) Manufacturers of survey controllers.
(b) Manufacturers of total stations.
(c) Manufacturers of global positioning (GPS) receivers.

2.29 What advantages are offered to field personnel if the survey controller provides a map of the survey.

2.30 Prepare a brief summary of an article from a professional journal related to the subject matter of this chapter.

2.31 Describe what is meant by the phrase "field-to-finish."

2.32 Create a computational program that solves Problems 2.4 and 2.5.

2.33 Create a computational program that solves Problem 2.16.

BIBLIOGRAPHY

Alder, Ken. 2002. *The Measure of All Things—The Seven-Year Odyssey and Hidden Error That Transformed the World.* New York: The Free Press.

Bedini, S. A. 2001. "Roger Sherman's Field Survey Book." *Professional Surveyor Magazine* 21 (No. 4): 70.

Bennett. T. D. 2002. "From Operational Efficiency to Business Process Improvement." *Professional Surveyor* 22 (No. 2): 46.

Brown, Leica. 2003. "Building a Better Handheld." *Point of Beginning* 28 (No. 7): 24.

Durgiss, Ken. 2001. "Advancing Field Data Collection with Wearable Computers." *Professional Surveyor* 21 (No. 4): 14.

Ghilani, C. D. and P. R. Wolf. 2006. *Adjustment Computations: Spatial Data Analysis* Hoboken, NJ: John Wiley & Sons, Inc.

Hanson, C. 2000. "A Few Thoughts on Metric." *Point of Beginning* 26 (No. 1): 90.

Paiva, J.V.R. 2006. "The Evolution of the Data Collector." 32 (No. 2): 22.

Pasley, Robert M. 2001. "SMI Construction Five." *Point of Beginning* 27 (No. 3): 32.

Pepling, A. 2003. "TDS Recon." *Professional Surveyor* 23 (No. 9): 34.

Yarumian II, R. 2001. "The Strong Enduring Type—A Personal Review of Rugged Penn Computer Tablets." *Point of Beginning* 26 (No. 4): 48.

3 Theory of Errors in Observations

3.1 INTRODUCTION

Making observations (measurements), and subsequent computations and analyses using them, are fundamental tasks of surveyors. Good measurements require a combination of human skill and mechanical equipment applied with the utmost judgment. However no matter how carefully made, observations are never exact and will always contain errors. Surveyors (geomatics engineers), whose work must be performed to exacting standards, should therefore thoroughly understand the different kinds of errors, their sources and expected magnitudes under varying conditions, and their manner of propagation. Only then can they select instruments and procedures necessary to reduce error sizes to within tolerable limits.

Of equal importance, surveyors must be capable of assessing the magnitudes of errors in their observations so that either their acceptability can be verified or, if necessary, new ones made. The design of measurement systems is now practiced. Computers and sophisticated software are tools now commonly used by surveyors to plan measurement projects and to investigate and distribute errors after results have been obtained.

3.2 DIRECT AND INDIRECT OBSERVATIONS

Observations may be made directly or indirectly. Examples of *direct observations* are applying a tape to a line, fitting a protractor to an angle, or turning an angle with a total station instrument.

An *indirect observation* is secured when it is not possible to apply a measuring instrument directly to the quantity to be observed. The answer is therefore determined by its relationship to some other observed value or values. As an example, observing the length of a line on one side, the angle at each end of this line to a point on the other side, and then computing the distance by one of the standard trigonometric formulas can find the distance across a river. Many indirect observations are made in surveying, and since all measurements contain errors, it is inevitable that quantities computed from them will also contain errors. The manner by which errors in measurements combine to produce erroneous computed answers is called error propagation. This topic is discussed further in Section 3.17.

■ 3.3 ERRORS IN MEASUREMENTS

By definition an error is the difference between an observed value for a quantity and its true value, or

$$E = X - \overline{X} \tag{3.1}$$

where E is the error in an observation, X the observed value, and $\overline{X}$ its true value. It can be unconditionally stated that: (1) no observation is exact, (2) every observation contains errors, (3) the true value of an observation is never known, and, therefore, (4) the exact error present is always unknown. These facts are demonstrated by the following. When a distance is observed with a scale divided into tenths of an inch, the distance can be read only to hundredths (by interpolation). If a better scale graduated in hundredths of an inch was available and read under magnification, however, the same distance might be estimated to thousandths of an inch. And with a scale graduated in thousandths of an inch, a reading to ten-thousandths might be possible. Obviously, accuracy of observations depends on the scale's division size, reliability of equipment used, and human limitations in estimating closer than about one-tenth of a scale division. As better equipment is developed, observations more closely approach their true values, but they can never be exact. Note that observations, not counts (of cars, pennies, marbles, or other objects), are under consideration here.

■ 3.4 MISTAKES

These are observer blunders and are usually caused by misunderstanding the problem, carelessness, fatigue, missed communication, or poor judgment. Examples include transposition of numbers, such as recording 73.96 instead of the correct value of 73.69; reading an angle counterclockwise, but indicating it as a clockwise angle in the field notes; sighting the wrong target; or recording a measured distance as 682.38 instead of 862.38. Large mistakes such as these are not considered in the succeeding discussion of errors. They must be detected by careful and systematic checking of all work, and eliminated by repeating some or all of the measurements. It is very difficult to detect small mistakes because they merge with errors. When not exposed, these small mistakes will therefore be incorrectly treated as errors.

■ 3.5 SOURCES OF ERRORS IN MAKING OBSERVATIONS

Errors in observations stem from three sources, and are classified accordingly.

Natural errors are caused by variations in wind, temperature, humidity, atmospheric pressure, atmospheric refraction, gravity, and magnetic declination. An example is a steel tape whose length varies with changes in temperature.

Instrumental errors result from any imperfection in the construction or adjustment of instruments and from the movement of individual parts. For example, the graduations on a scale may not be perfectly spaced or the scale may be warped. The effect of many instrumental errors can be reduced, or even eliminated, by adopting proper surveying procedures or applying computed corrections.

Personal errors arise principally from limitations of the human senses of sight and touch. As an example, a small error occurs in the observed value of a horizontal angle if the vertical crosshair in a total station instrument is not aligned perfectly on the target, or if the target is the top of a rod which is being held slightly out of plumb.

■ 3.6 TYPES OF ERRORS

Errors in observations are of two types: *systematic* and *random.*

Systematic errors, also known as *biases,* result from factors that comprise the "measuring system" and include the environment, instrument, and observer. So long as system conditions remain constant, the systematic errors will likewise remain constant. If conditions change, the magnitudes of systematic errors also change. Because systematic errors tend to accumulate, they are sometimes called *cumulative errors.*

Conditions producing systematic errors conform to physical laws that can be modeled mathematically. Thus if the conditions are known to exist and can be observed, a correction can be computed and applied to observed values. An example of a constant systematic error is the use of a 100-ft steel tape that has been calibrated and found to be 0.02 ft too long. It introduces a 0.02-ft error each time it is used, but applying a correction readily eliminates the error. An example of variable systematic error is the change in length of a steel tape resulting from temperature differentials that occur during the period of the tape's use. If the temperature changes are observed, length corrections can be computed by a simple formula, as explained in Chapter 6.

Random errors are those that remain in measured values after mistakes and systematic errors have been eliminated. They are caused by factors beyond the control of the observer, obey the laws of probability, and are sometimes called *accidental errors.* They are present in all surveying observations.

The magnitudes and algebraic signs of random errors are matters of chance. There is no absolute way to compute or eliminate them, but they can be estimated using adjustment procedures known as *least squares* (see Section 3.21 and Chapter 16). Random errors are also known as *compensating errors,* since they tend to partially cancel themselves in a series of observations. For example, a person interpolating to hundredths of a foot on a tape graduated only to tenths, or reading a level rod marked in hundredths, will presumably estimate too high on some values and too low on others. However, individual personal characteristics may nullify such partial compensation since some people are inclined to interpolate high, others interpolate low, and many favor certain digits—for example, 7 instead of 6 or 8, 3 instead of 2 or 4, and particularly 0 instead of 9 or 1.

■ 3.7 PRECISION AND ACCURACY

A *discrepancy* is the difference between two observed values of the same quantity. A small discrepancy indicates there are probably no mistakes and random errors are small. However, small discrepancies do not preclude the presence of systematic errors.

Precision refers to the degree of refinement or consistency of a group of observations and is evaluated on the basis of discrepancy size. If multiple observations are made of the same quantity and small discrepancies result, this indicates high precision. The degree of precision attainable is dependent on equipment sensitivity and observer skill.

Accuracy denotes the absolute nearness of observed quantities to their true values. The difference between precision and accuracy is perhaps best illustrated with reference to target shooting. In Figure 3.1(a), for example, all five shots exist in a small group, indicating a precise operation; that is, the shooter was able to repeat the procedure with a high degree of consistency. However, the shots are far from the bull's-eye and therefore not accurate. This probably results from misaligned rifle sights. Figure 3.1(b) shows randomly scattered shots that are neither precise nor accurate. In Figure 3.1(c) the closely spaced grouping, in the bull's-eye, represents both precision and accuracy. The shooter who obtained the results in (a) was perhaps able to produce the shots of (c) after aligning the rifle sights. In surveying, this would be equivalent to the calibration of measuring instruments.

As with the shooting example, a survey can be precise without being accurate. To illustrate, if refined methods are employed and readings taken carefully, say to 0.001 ft, but there are instrumental errors in the measuring device and corrections are not made for them, the survey will not be accurate. As a numerical example, two observations of a distance with a tape assumed to be 100.000 ft long that is actually 100.050 ft might give results of 453.270 and 453.272 ft. These values are precise, but they are not accurate, since there is a systematic error of approximately $4.53 \times 0.050 = 0.23$ ft in each. The precision obtained would be expressed as $(453.272 - 453.270)/453.271 = 1/220{,}000$, which is excellent, but

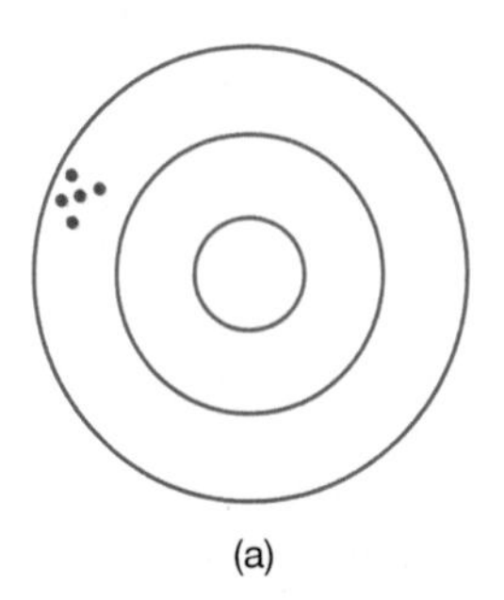

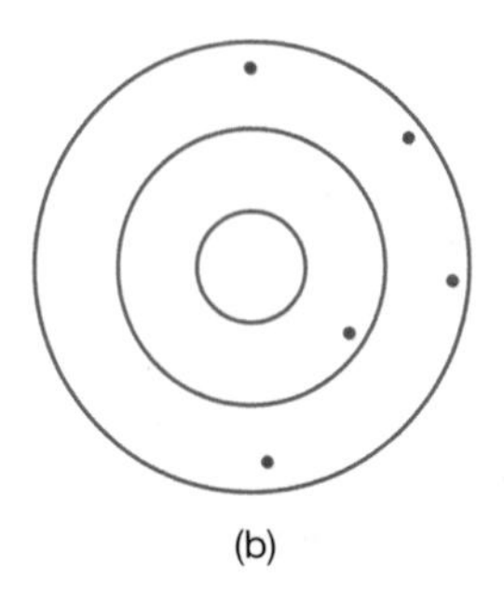

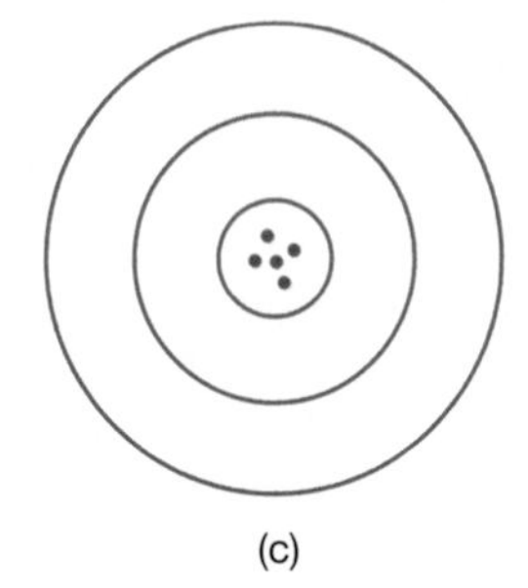

Figure 3.1 Examples of precision and accuracy: (a) Results are precise but not accurate. (b) Results are neither precise nor accurate. (c) Results are both precise and accurate.

accuracy of the distance is only $0.23/453.271 = 1$ part in 2000. Also, a survey may appear to be accurate when rough observations have been taken. For example, the angles of a triangle may be read with a compass to only the nearest 1/4 degree and yet produce a sum of exactly 180°, or a zero misclosure error. On good surveys, precision and accuracy are consistent throughout.

■ 3.8 ELIMINATING MISTAKES AND SYSTEMATIC ERRORS

All field operations and office computations are governed by a constant effort to eliminate mistakes and systematic errors. Of course it would be preferable if mistakes never occurred, but because humans are fallible, this is not possible. In the field, experienced observers who alertly perform their observations using standardized repetitive procedures can minimize mistakes. Mistakes that do occur can be corrected only if discovered. Comparing several observations of the same quantity is one of the best ways to identify mistakes. Making a commonsense estimate and analysis is another. Assume that five observations of a line are recorded as follows: 567.91, 576.95, 567.88, 567.90, and 567.93. The second value disagrees with the others, apparently because of a transposition of figures in reading or recording. An observant recorder should capture this mistake at the time the observation is recorded. However, if it is not caught, the second observation must be handled by either casting out the doubtful value, or preferably repeating the observation.

When a mistake is detected, it is usually best to repeat the observation. However if a sufficient number of other observations of the quantity are available and in agreement, as in the foregoing example, the widely divergent result may be discarded. Serious consideration must be given to the effect on an average before discarding a value. It is seldom safe to change a recorded number, even though there appears to be a simple transposition in figures. Tampering with physical data is always bad practice and will certainly cause trouble, even if done infrequently.

Systematic errors can be calculated and proper corrections applied to the observations. Procedures for making these corrections to all basic surveying observations are described in the chapters that follow. In some instances it may be possible to adopt a field procedure that automatically eliminates systematic errors. For example, as explained in Chapter 5, a leveling instrument out of adjustment causes incorrect readings, but if all backsights and foresights are made the same length, the errors cancel in differential leveling.

■ 3.9 PROBABILITY

At one time or another, everyone has had an experience with games of chance, such as coin flipping, card games, or dice, which involve probability. In basic mathematics courses, laws of combinations and permutations are introduced. It is shown that events that happen randomly or by chance are governed by mathematical principles referred to as probability.

Probability may be defined as the ratio of the number of times a result should occur to its total number of possibilities. For example, in the toss of a fair die there is a one-sixth probability that a 2 will come up. This simply means that there are six possibilities and only one of them is a 2. In general, if a result may occur in

m ways and fail to occur in n ways, then the probability of its occurrence is $m/(m + n)$. The probability that any result will occur is a fraction between 0 and 1; 0 indicating impossibility and 1 denoting absolute certainty. Since any result must either occur or fail, the sum of the probabilities of occurrence and failure is 1. Thus if 1/6 is the probability of throwing a 2 with one toss of a die, then $(1 - 1/6)$, or 5/6 is the probability that a 2 will not come up.

The theory of probability is applicable in many sociological and scientific observations. In Section 3.6, it was pointed out that random errors exist in all surveying work. This can perhaps be better appreciated by considering the measuring process, which generally involves executing several elementary tasks. Besides instrument selection and calibration, these tasks may include setup, centering, aligning, or pointing the equipment: setting, matching, or comparing index marks; and reading or estimating values from graduated scales, dials, or gauges. Because of equipment and observer imperfections, observations cannot be made exactly, so they will always contain random errors. The magnitudes of these errors, and the frequency with which errors of a given size occur, follow the laws of probability.

For convenience, the term error will be used to mean only random error for the remainder of this chapter. It will be assumed that all mistakes and systematic errors have been eliminated before random errors are considered.

■ 3.10 MOST PROBABLE VALUE

It has been stated earlier that in physical observations, the true value of any quantity is never known. However, its *most probable value* can be calculated if redundant observations have been made. *Redundant observations* are measurements in excess of the minimum needed to determine a quantity. For a single unknown, such as a line length that has been directly and independently observed a number of times using the same equipment and procedures,[1] the first observation establishes a value for the quantity and all additional observations are redundant. The most probable value in this case is simply the arithmetic mean, or

$$\overline{M} = \frac{\Sigma M}{n} \tag{3.2}$$

where $\overline{M}$ is the most probable value of the quantity, ΣM the sum of the individual measurements M, and n the total number of observations. Equation (3.2) can be derived using the principle of least squares, which is based on the theory of probability.

As discussed in Chapter 16, in more complicated problems, where the observations are not made with the same instruments and procedures, or if several interrelated quantities are being determined through indirect observations, most probable values are calculated by employing least-squares methods. The

[1]The significance of using the same equipment and procedures is that observations are of equal reliability or weight. The subject of unequal weights is discussed in Section 3.20.

treatment here relates to multiple direct observations of the same quantity using the same equipment and procedures.

■ 3.11 RESIDUALS

Having determined the most probable value of a quantity, it is possible to calculate *residuals*. A residual is simply the difference between the most probable value and any observed value of a quantity, which in equation form is

$$v = \overline{M} - M \tag{3.3}$$

where v is the residual in any observation M, and $\overline{M}$ is the most probable value for the quantity. Residuals are theoretically identical to errors, with the exception that residuals can be calculated whereas errors cannot because true values are never known. Thus, residuals rather than errors are the values actually used in the analysis and adjustment of survey data.

■ 3.12 OCCURRENCE OF RANDOM ERRORS

To analyze the manner in which random errors occur, consider the data of Table 3.1, which represents 100 repetitions of an angle observation made with a precise total station instrument (described in Chapter 8). Assume these observations are free from mistakes and systematic errors. For convenience in analyzing the data, except for the first value, only the seconds' portions of the observations are tabulated. The data have been rearranged in column (1) so that entries begin with the smallest observed value and are listed in increasing size. If a certain value was obtained more than once, the number of times it occurred, or its *frequency,* is tabulated in column (2).

From Table 3.1, it can be seen that the *dispersion* (range in observations from smallest to largest) is 30.8 − 19.5 = 11.3 sec. However, it is difficult to analyze the distribution pattern of the observations by simply scanning the tabular values, that is, beyond assessing the dispersion and noticing a general trend for observations toward the middle of the range to occur with greater frequency. To assist in studying the data, a *histogram* can be prepared. This is simply a bar graph showing the sizes of the observations (or their residuals) versus their frequency of occurrence. It gives an immediate visual impression of the distribution pattern of the observations (or their residuals).

For the data of Table 3.1, a histogram showing the frequency of occurrence of the residuals has been developed and is plotted in Figure 3.2. To plot a histogram of residuals, it is first necessary to compute the most probable value for the observed angle. This has been done with Equation (3.2). As shown at the bottom of Table 3.1, its value is 27°43′24.9″. Then using Equation (3.3), residuals for all observed values are computed. These are tabulated in column (3) of Table 3.1. The residuals vary from 5.4″ to −5.9″. (The sum of the absolute value of these two extremes is the dispersion, or 11.3″.)

To obtain a histogram with an appropriate number of bars for portraying the distribution of residuals adequately, the interval of residuals represented by

TABLE 3.1 ANGLE OBSERVATIONS FROM PRECISE THEODOLITE

Observed Value (1)	No. (2)	Residual (sec) (3)	Observed Value (1 cont.)	No. (2 cont.)	Residual (sec) (3 cont.)
27°43′19.5″	1	5.4	27°43′25.1″	3	−0.2
20.0	1	4.9	25.2	1	−0.3
20.5	1	4.4	25.4	1	−0.5
20.8	1	4.1	25.5	2	−0.6
21.2	1	3.7	25.7	3	−0.8
21.3	1	3.6	25.8	4	−0.9
21.5	1	3.4	25.9	2	−1.0
22.1	2	2.8	26.1	1	−1.2
22.3	1	2.6	26.2	2	−1.3
22.4	1	2.5	26.3	1	−1.4
22.5	2	2.4	26.5	1	−1.6
22.6	1	2.3	26.6	3	−1.7
22.8	2	2.1	26.7	1	−1.8
23.0	1	1.9	26.8	2	−1.9
23.1	2	1.8	26.9	1	−2.0
23.2	2	1.7	27.0	1	−2.1
23.3	3	1.6	27.1	3	−2.2
23.6	2	1.3	27.4	1	−2.5
23.7	2	1.2	27.5	2	−2.6
23.8	2	1.1	27.6	1	−2.7
23.9	3	1.0	27.7	2	−2.8
24.0	5	0.9	28.0	1	−3.1
24.1	3	0.8	28.6	2	−3.7
24.3	1	0.6	28.7	1	−3.8
24.5	2	0.4	29.0	1	−4.1
24.7	3	0.2	29.4	1	−4.5
24.8	3	0.1	29.7	1	−4.8
24.9	2	0.0	30.8	1	−5.9
25.0	2	−0.1	Σ = 2494.0	Σ = 100	

Mean = 2494.0/100 = 24.9″

Most Probable Value = 27°43′24.9″

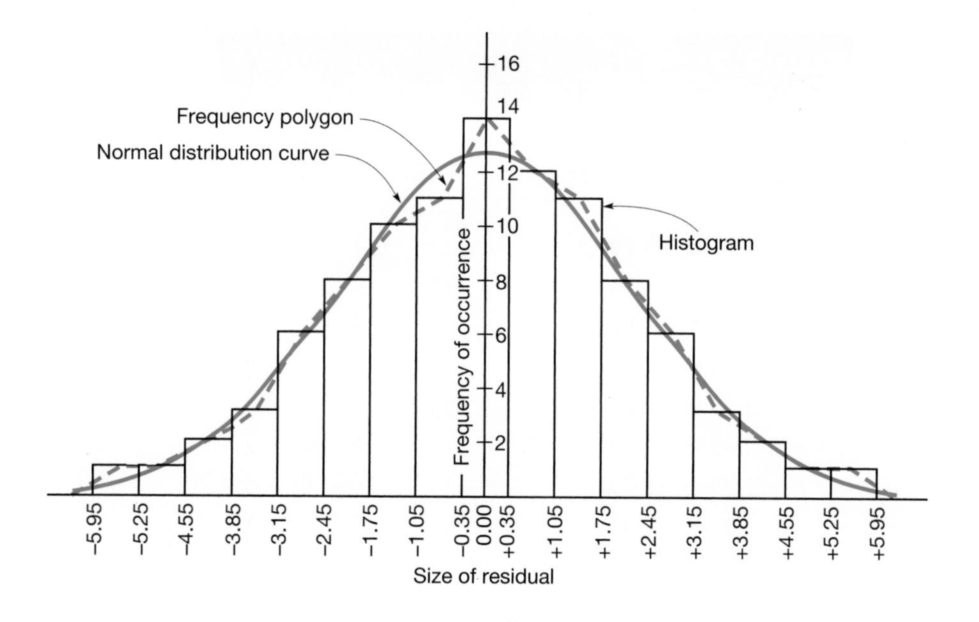

Figure 3.2 Histogram, frequency polygon, and normal distribution curve of residuals from angle measurements made with total station.

each bar, or the *class interval,* was chosen as 0.7″. This produced 17 bars on the graph. The range of residuals covered by each interval and the number of residuals that occur within each interval, are listed in Table 3.2. By plotting class intervals on the abscissa against the number (frequency of occurrence) of residuals in each interval on the ordinate, the histogram of Figure 3.2 was obtained.

If the adjacent top center points of the histogram bars are connected with straight lines, the so-called *frequency polygon* is obtained. The frequency polygon for the data of Table 3.1 is superimposed as a heavy, dashed blue line in Figure 3.2. It graphically displays essentially the same information as the histogram.

If the number of observations being considered in this analysis were increased progressively, and accordingly the histogram's class interval taken smaller and smaller, ultimately the frequency polygon would approach a smooth continuous curve, symmetrical about its center like the one shown with the heavy solid blue line in Figure 3.2. For clarity this curve is shown separately in Figure 3.3. The curve's "bell shape" is characteristic of a normally distributed group of errors, and thus it is often referred to as the *normal distribution curve.* Statisticians frequently call it the *normal density curve,* since it shows the densities of errors having various sizes. In surveying, normal or very nearly normal error distributions are expected, and henceforth in this book that condition is assumed.

In practice, histograms and frequency polygons are seldom used to represent error distributions. Instead, normal distribution curves that approximate them are preferred. (Note how closely the normal distribution curve superimposed on Figure 3.2 agrees with the histogram and the frequency polygon.)

As demonstrated with the data of Table 3.1, the histogram for a set of observations shows the probability of occurrence of an error of a given size graphically by bar areas. For example, 14 of the 100 residuals (errors) in Figure 3.2 are between −0.35″ and +0.35″. This represents 14% of the errors, and the center histogram bar, which corresponds to this interval, is 14% of the total area of all bars.

TABLE 3.2 RANGES OF CLASS INTERVALS AND NUMBER OF RESIDUALS IN EACH INTERVAL

Histogram Interval (Sec.)	Number of Residuals in Interval
−5.95 to −5.25	1
−5.25 to −4.55	1
−4.55 to −3.85	2
−3.85 to −3.15	3
−3.15 to −2.45	6
−2.45 to −1.75	8
−1.75 to −1.05	10
−1.05 to −0.35	11
−0.35 to +0.35	14
+0.35 to +1.05	12
+1.05 to +1.75	11
+1.75 to +2.45	8
+2.45 to +3.15	6
+3.15 to +3.85	3
+3.85 to +4.55	2
+4.55 to +5.25	1
+5.25 to +5.95	1
	$\Sigma = 100$

Likewise, the area between ordinates constructed at any two abscissas of a normal distribution curve represents the percent probability that an error of that size exists. Since the area sum of all bars of a histogram represents all errors, it therefore represents all probabilities, and thus its sum equals 1. Likewise, the total area beneath a normal distribution curve is also 1.

If the same observations of the preceding example had been taken using better equipment and more caution, smaller errors would be expected and the normal distribution curve would be similar to that in Figure 3.4(a). Compared to Figure 3.3, this curve is taller and narrower, showing that a greater percentage of values have smaller errors, and fewer observations contain big ones. For this comparison, the same ordinate and abscissa scales must be used for both curves. Thus, the observations of Figure 3.4(a) are more precise. For readings taken less precisely, the opposite effect is produced, as illustrated in Figure 3.4(b), which shows a shorter and wider curve. In all three cases, however, the curve maintained its characteristic symmetric bell shape.

From these examples, it is seen that relative precisions of groups of observations become readily apparent by comparing their normal distribution curves. The

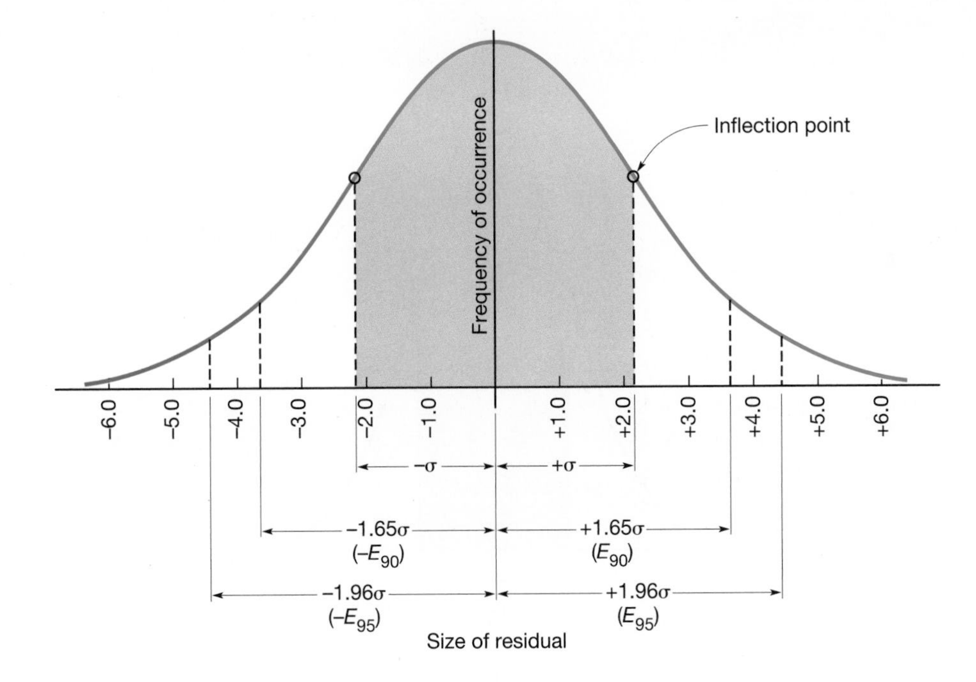

Figure 3.3 Normal distribution curve.

normal distribution curve for a set of observations can be computed using parameters derived from the residuals, but the procedure is beyond the scope of this text.

■ 3.13 GENERAL LAWS OF PROBABILITY

From an analysis of the data in the preceding section and the curves in Figures 3.2 through 3.4, some general laws of probability can be stated:

1. Small residuals (errors) occur more often than large ones; that is, they are more probable.
2. Large errors happen infrequently and are therefore less probable; for normally distributed errors, unusually large ones may be mistakes rather than random errors.
3. Positive and negative errors of the same size happen with equal frequency; that is, they are equally probable. [This enables an intuitive deduction of Equation (3.2) to be made: that is, the most probable value for a group of repeated observations, made with the same equipment and procedures, is the mean.]

■ 3.14 MEASURES OF PRECISION

As shown in Figures 3.3 and 3.4, although the curves have similar shapes, there are significant differences in their dispersions; that is, their abscissa widths differ. The magnitude of dispersion is an indication of the relative precisions of the observations. Other statistical terms more commonly used to express precisions of groups

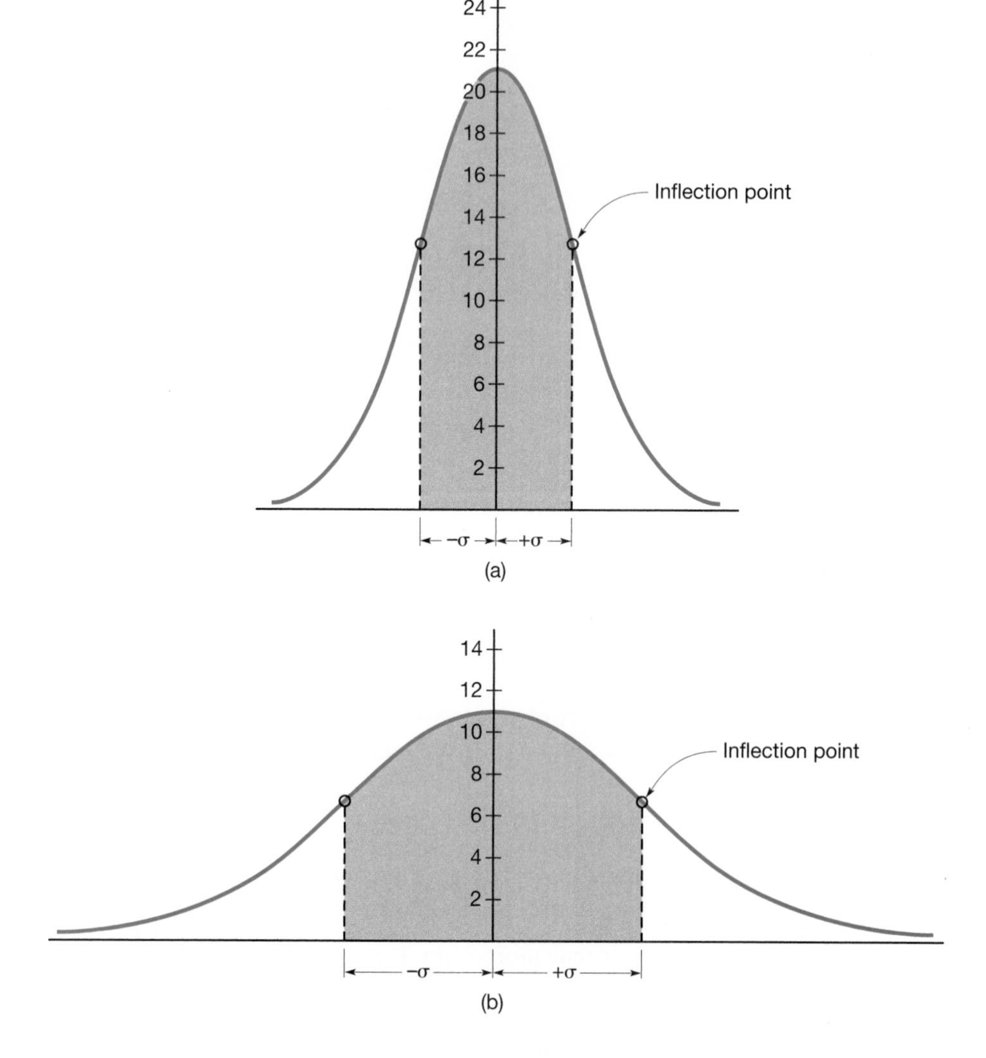

Figure 3.4 Normal distribution curves for: (a) increased precision, (b) decreased precision.

of observations are *standard deviation* and *variance.* The equation for the standard deviation is

$$\sigma = \pm\sqrt{\frac{\Sigma v^2}{n - 1}} \tag{3.4}$$

where σ is the standard deviation of a group of observations of the same quantity, v the residual of an individual observation, Σv^2 the sum of squares of the individual residuals, and n the number of observations. *Variance* is equal to σ^2, the square of the standard deviation.

Note that in Equation (3.4), the standard deviation has both plus and minus values. On the normal distribution curve, the numerical value of the standard deviation is the abscissa at the inflection points (locations where the curvature changes from concave downward to concave upward). In Figures 3.3 and 3.4 these inflection points are shown. Note the closer spacing between them for the more precise observations of Figure 3.4(a) as compared to Figure 3.4(b).

Figure 3.5 is a graph showing the percentage of the total area under a normal distribution curve that exists between ranges of residuals (errors) having equal positive and negative values. The abscissa scale is shown in multiples of the standard deviation. From this curve, the area between residuals of $+\sigma$ and $-\sigma$ equals 68.27% (round to 68.3%) of the total area under the normal distribution curve. Hence it gives the range of residuals that can be expected to occur 68.3% of the time. This relation is shown more clearly on the curves in Figures 3.3 and 3.4, where the areas between $\pm\sigma$ are shown shaded. The percentages shown in Figure 3.5 apply to all normal distributions, regardless of curve shape or the numerical value of the standard deviation.

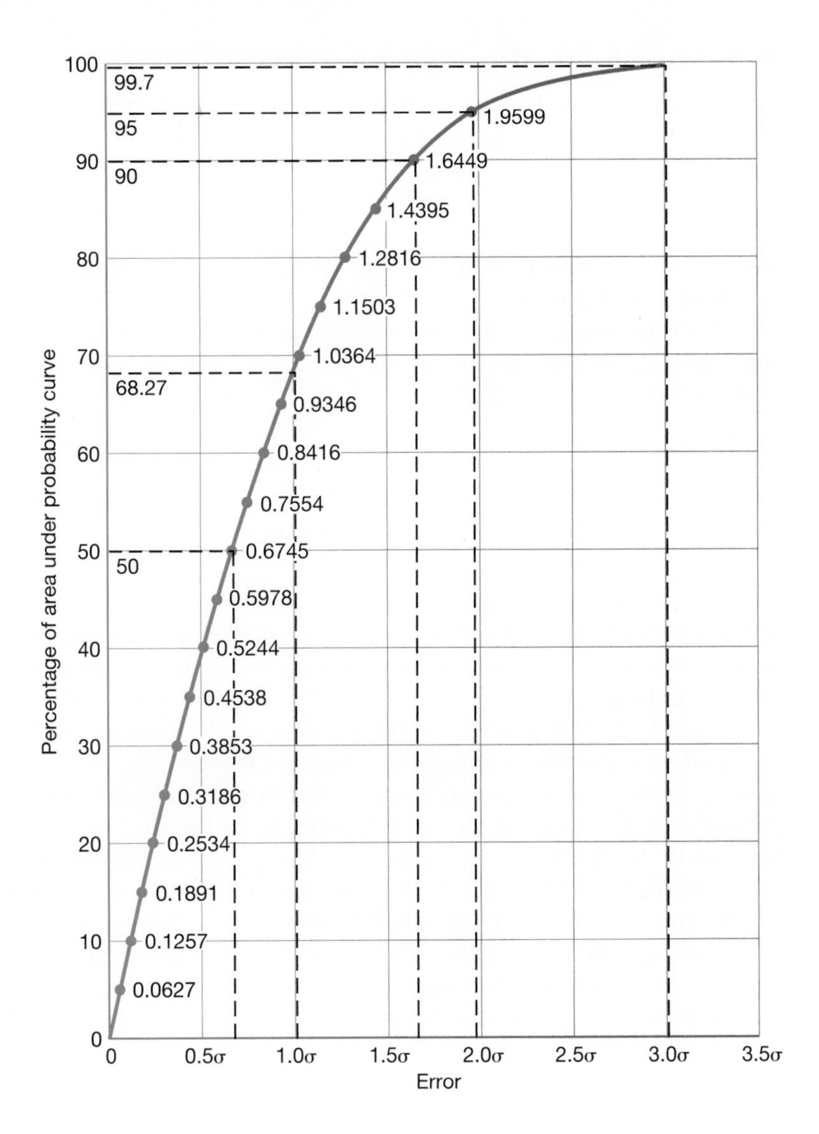

Figure 3.5 Relation between error and percentage of area under normal distribution curve.

3.15 INTERPRETATION OF STANDARD DEVIATION

It has been shown that the standard deviation establishes the limits within which observations are expected to fall 68.3 percent of the time. In other words, if an observation is repeated ten times, it will be expected that about seven of the results will fall within the limits established by the standard deviation, and conversely about three of them will fall anywhere outside these limits. Another interpretation is that one additional observation will have a 68.3 percent chance of falling within the limits set by the standard deviation.

When Equation (3.4) is applied to the data of Table 3.1, a standard deviation of $\pm 2.19''$ is obtained. In examining the residuals in the table, 70 of the 100 values, or 70%, are actually smaller than 2.19 sec. This illustrates that the theory of probability closely approximates reality.

3.16 THE 50, 90, AND 95 PERCENT ERRORS

From the data given in Figure 3.5, the probability of an error of any percentage likelihood can be determined. The general equation is

$$E_P = C_P \sigma \tag{3.5}$$

where E_P, is a certain percentage error and C_P, the corresponding numerical factor taken from Figure 3.5.

By Equation (3.5), after extracting appropriate multipliers from Figure 3.5, the following are expressions for errors that have a 50, 90, and 95 percent chance of occurring:

$$E_{50} = 0.6745\sigma \tag{3.6}$$

$$E_{90} = 1.6449\sigma \tag{3.7}$$

$$E_{95} = 1.9599\sigma \tag{3.8}$$

The 50 percent error, or E_{50}, is the so-called *probable error*. It establishes limits within which the observations should fall 50% of the time. In other words, an observation has the same chance of coming within these limits as it has of falling outside of them.

The 90 and 95 percent errors are commonly used to specify precisions required on surveying (geomatics) projects. Of these, the 95% error, also frequently called the *two-sigma* (2σ) *error,* is most often specified. As an example, a particular project may call for the 95% error to be less than or equal to a certain value for the work to be acceptable. For the data of Table 3.1, applying Equations (3.7) and (3.8), the 90 and 95 percent errors are ± 3.60 sec and ± 4.29 sec respectively. These errors are shown graphically in Figure 3.3.

The so-called *three-sigma* (3σ) *error* is also often used as a criterion for rejecting individual observations. From Figure 3.5, there is a 99.7% probability that an error will be less than this amount. Thus within a group of observations, any value whose residual exceeds 3σ is considered to be a mistake, and either a new observation must be taken or the computations based on one less value.

The x-axis is an asymptote of the normal distribution curve, so the 100 percent error cannot be evaluated. This means that no matter what size error is found, a larger one is theoretically possible.

Example 3.1

To clarify definitions and use the equations given in Sections 3.10 through 3.16, suppose that a line has been observed ten times using the same equipment and procedures. The results are shown in column (1) of the following table. It is assumed that no mistakes exist, and that the observations have already been corrected for all systematic errors. Compute the most probable value for the line length, its standard deviation, and errors having 50, 90, and 95 percent probability.

Length (ft) (1)	Residual ν (ft) (2)	ν^2 (3)
538.57	+0.12	0.0144
538.39	−0.06	0.0036
538.37	−0.08	0.0064
538.39	−0.06	0.0036
538.48	+0.03	0.0009
538.49	+0.04	0.0016
538.33	−0.12	0.0144
538.46	+0.01	0.0001
538.47	+0.02	0.0004
538.55	+0.10	0.0100
$\Sigma = 5384.50$	$\Sigma = 0.00$	$\Sigma\nu^2 = 0.0554$

Solution

By Equation (3.2), $\overline{M} = \dfrac{5384.50}{10} = 538.45$ ft

By Equation (3.3), the residuals are calculated. These are tabulated in column (2) and their squares listed in column (3). Note that in column (2) the algebraic sum of residuals is zero. (For observations of equal reliability, except for round off, this column should always total zero and thus provide a computational check.)

By Equation (3.4), $\sigma = \pm\sqrt{\dfrac{\Sigma\nu^2}{n-1}} = \sqrt{\dfrac{0.0554}{9}} = \pm 0.078 = \pm 0.08$ ft

By Equation (3.6), $E_{50} = \pm 0.6745\sigma = \pm 0.6745(0.078) = \pm 0.05$ ft.

By Equation (3.7), $E_{95} = \pm 1.6449(0.078) = \pm 0.13$ ft.

By Equation (3.8), $E_{99} = \pm 1.9599(0.078) = \pm 0.15$ ft.

The following conclusions can be drawn concerning this example.

1. The most probable line length is 538.45 ft.
2. The standard deviation of a single observation is ± 0.08 ft. Accordingly, the normal expectation is that 68 percent of the time a recorded length will lie between $538.45 - 0.08$ and $538.45 + 0.08$ or between 538.37 and 538.53 ft; that is, about seven values should lie within these limits. (Actually seven of them do.)
3. The probable error (E_{50}) is ± 0.05 ft. Therefore it can be anticipated that half, or five of the observations, will fall in the interval 538.40 to 538.50 ft. (Four values do.)
4. The 90 percent error is ± 0.13 ft, and thus nine of the observed values can be expected to be within the range of 538.32 and 538.58 ft.
5. The 95 percent error is ± 0.15 ft, so the length can be expected to lie between 538.30 and 538.60, 95 percent of the time. (Note that all observations indeed are within the limits of both the 90 and 95 percent errors.)

■ 3.17 ERROR PROPAGATION

It was stated earlier that because all observations contain errors, any quantities computed from them will likewise contain errors. The process of evaluating errors in quantities computed from observed values that contain errors is called *error propagation.* The propagation of random errors in mathematical formulas can be computed using the general law of the propagation of variances. Typically in surveying (geomatics), this formula can be simplified since the observations are usually mathematically independent. For example, let $a, b, c, \ldots, n$ be observed values containing errors $E_a, E_b, E_c, \ldots, E_n$, respectively. Also let Z be a quantity derived by computation using these observed quantities in a function f, such that

$$Z = f(a, b, c, \ldots, n) \tag{3.9}$$

Then assuming that $a, b, c, \ldots, n$ are independent observations, the error in the computed quantity Z is

$$E_Z = \pm\sqrt{\left(\frac{\partial f}{\partial a}E_a\right)^2 + \left(\frac{\partial f}{\partial b}E_b\right)^2 + \left(\frac{\partial f}{\partial c}E_c\right)^2 + \cdots + \left(\frac{\partial f}{\partial n}E_n\right)^2} \tag{3.10}$$

where the terms $\partial f/\partial a, \partial f/\partial b, \partial f/\partial c, \ldots, \partial f/\partial n$ are the partial derivatives of the function f with respect to the variables $a, b, c, \ldots, n$. In the subsections that follow, specific cases of error propagation common in surveying are discussed and examples are presented.

3.17.1 Error of a Sum

Assume the sum of independently observed observations $a, b, c, \ldots$ is Z. The formula for the computed quantity Z is

$$Z = a + b + c + \cdots$$

The partial derivatives of Z with respect to each observed quantity are $\partial Z/\partial a = \partial Z/\partial b = \partial Z/\partial c = \cdots = 1$. Substituting these partial derivatives into Equation (3.10), the following formula is obtained which gives the propagated error in the sum of quantities, each of which contains a different random error:

$$E_{Sum} = \pm\sqrt{E_a^2 + E_b^2 + E_c^2 + \cdots} \tag{3.11}$$

where E represents any specified percentage error (such as σ, E_{50}, E_{90}, or E_{95}), and a, b, and c are the separate, independent observations.

The error of a sum can be used to explain the rules for addition and subtraction using significant figures. Recall the addition of 46.7418, 1.03, and 375.0 from Example (a) from Section 2.4. Significant figures indicate that there is uncertainty in the last digit of each number. Thus assume estimated errors of ±0.0001, ±0.01, and ±0.1 respectively for each number. The error in the sum of these three numbers is $\sqrt{0.0001^2 + 0.01^2 + 0.1^2} = \pm 0.1$. The sum of the three numbers is 422.7718, which was rounded, using the rules of significant figures, to 422.8. Its precision matches the estimated accuracy produced by the error in the sum of the three numbers. Note how the least accurate number controls the accuracy in the summation of the three values.

Example 3.2

Assume that a line is observed in three sections, with the individual parts equal to (753.81, ±0.012), (1238.40, ±0.028), and (1062.95, ±0.020) ft, respectively. Determine the line's total length and its anticipated standard deviation.

Solution

Total length $= 753.81 + 1238.40 + 1062.95 = 3055.16$ ft.

By Equation (3.11), $E_{Sum} = \pm\sqrt{0.012^2 + 0.028^2 + 0.020^2} = \pm 0.036$ ft

3.17.2 Error of a Series

Sometimes a series of similar quantities, such as the angles within a closed polygon, are read with each observation being in error by about the same amount. The total error in the sum of all observed quantities of such a series is called the *error of the series, des*ignated as E_{Series}. If the same error E in each observation is assumed and Equation (3.11) applied, the series error is

$$E_{Series} = \pm\sqrt{E^2 + E^2 + E^2 + \cdots} = \pm\sqrt{nE^2} = \pm E\sqrt{n} \tag{3.12}$$

where E represents the error in each individual observation and n the number of observations.

This equation shows that when the same operation is repeated, random errors tend to balance out and the resulting error of a series is proportional to the square root of the number of observations. This equation has extensive use—for instance, to determine the allowable misclosure error for angles of a traverse, as discussed in Chapter 9.

Example 3.3

Assume that any distance of 100 ft can be taped with an error of ± 0.02 ft if certain techniques are employed. Determine the error in taping 5000 ft using these skills.

Solution

Since the number of 100-ft lengths in 5000 ft is 50, then by Equation (3.12)

$$E_{Series} = \pm E\sqrt{n} = \pm 0.02\sqrt{50} = \pm 0.14 \text{ ft}$$

Example 3.4

A distance of 1000 ft is to be taped with an error of not more than ± 0.10 ft. Determine how accurately each 100-ft length must be observed to ensure that the error will not exceed the permissible limit.

Solution

Since by Equation (3.12), $E_{Series} = \pm E\sqrt{n}$ and $n = 10$, the allowable error E in 100 ft is

$$E = \pm\frac{E_{Series}}{\sqrt{n}} \pm \frac{0.10}{\sqrt{10}} = \pm 0.03 \text{ ft}$$

Example 3.5

Suppose it is required to tape a length of 2500 ft with an error of not more than ± 0.10 ft. How accurately must each tape length be observed?

Solution

Since 100 ft is again considered the unit length, $n = 25$, and by Equation (3.12), the allowable error E in 100 ft is

$$E = \pm\frac{0.10}{\sqrt{25}} = \pm 0.02 \text{ ft}$$

Analyzing Examples 3.4 and 3.5 shows that the larger the number of possibilities, the greater the chance for the errors to cancel out.

3.17.3 Error in a Product

The equation for propagated AB, where E_a and E_b, are the respective errors in A and B, is

$$E_{prod} = \pm\sqrt{A^2E_b^2 + B^2E_a^2} \tag{3.13}$$

The physical significance of the error propagation formula for a product is illustrated in Figure 3.6, where A and B are shown to be observed sides of a rectangular parcel of land with errors E_a and E_b, respectively. The product AB is the parcel area. In Equation (3.13), $\sqrt{A^2E_b^2} = AE_b$, represents either of the longer (horizontal) crosshatched bars and is the error caused by either $-E_b$ or $+E_b$. The term $\sqrt{B^2E_a^2} = BE_a$ is represented by the shorter (vertical) crosshatched bars, which is the error resulting from either $-E_a$ or $+E_a$.

Example 3.6

For the rectangular lot illustrated in Figure 3.6, observations of sides A and B with their 95% errors are (252.46, ±0.053) and (605.08, ±0.072) ft, respectively. Calculate the parcel area and the expected 95% error in the area.

Solution

Area = 252.46 × 605.08 = 152,760 ft^2

By Equation (3.13),

$$E_{95} = \pm\sqrt{(252.46)^2(0.072)^2 + (605.08)^2(0.053)^2} = \pm 36.9 \text{ ft}^2$$

Example 3.6 can also be used to demonstrate the validity of one of the rules of significant figures in computation. The computed area is actually 152,758.4968 ft^2. However, the rule for significant figures in multiplication (see Section 2.4) states that there cannot be more significant figures in the answer than

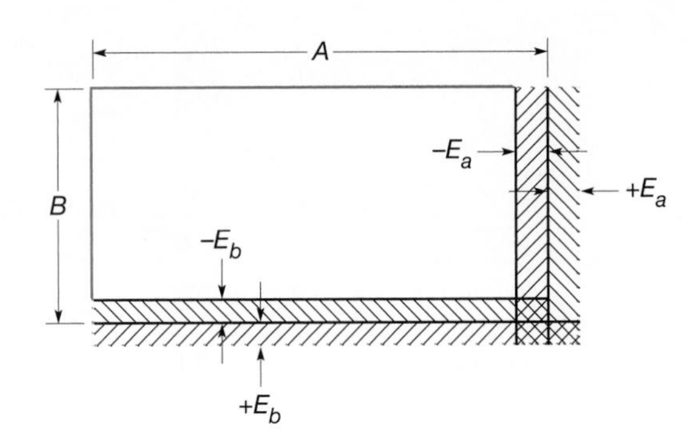

Figure 3.6 Error of area.

in any of the individual factors used. Accordingly, the area should be rounded off to 152,760 (five significant figures). From Equation (3.13), with an error of ± 36.9 ft^2 the answer could be 152,758.4968 $\pm$ 36.9, or from 152,721.6 to 152,795.4 ft^2. Thus the fifth digit in the answer is seen to be questionable, and hence the number of significant figures specified by the rule is verified.

3.17.4 Error in the Mean

Equation (3.2) stated that the most probable value of a group of repeated observations of equal weight is the arithmetic mean. Since the mean is computed from individual observed values, each of which contains an error, the mean is also subject to error. By applying Equation (3.12), it is possible to find the error for the sum of a series of observations where each one has the same error. Since the sum divided by the number of observations gives the mean, the error in the mean is found by the relation

$$E_m = \frac{E_{series}}{n}$$

Substituting Equation (3.12) for E_{series},

$$E_m = \frac{E\sqrt{n}}{n} = \frac{E}{\sqrt{n}} \tag{3.14}$$

where E is the specified percentage error of a single observation, E_m the corresponding percentage error of the mean, and n the number of observations.

The error of the mean at any percentage probability can be determined and applied to all criteria that have been developed. For example, the standard deviation of the mean, $(E_{68})_m$ or σ_m is

$$(E_{68})_m = \sigma_m = \frac{\sigma}{\sqrt{n}} = \pm\sqrt{\frac{\Sigma v^2}{n(n-1)}} \tag{3.15a}$$

and the 90 and 95 percent errors of the mean are

$$(E_{90})_m = \frac{E_{90}}{\sqrt{n}} = \pm 1.6449\sqrt{\frac{\Sigma v^2}{n(n-1)}} \tag{3.15b}$$

$$(E_{95})_m = \frac{E_{95}}{\sqrt{n}} = \pm 1.9599\sqrt{\frac{\Sigma v^2}{n(n-1)}} \tag{3.15c}$$

These equations show that *the error of the mean varies inversely as the square root of the number of repetitions*. Thus to double the accuracy—that is, to reduce the error by one-half—four times as many observations must be made.

Example 3.7

Calculate the standard deviation of the mean and the 90 percent error of the mean for the observations of Example 3.1.

Solution

By Equation (3.15a), $\sigma_m = \frac{\sigma}{\sqrt{n}} = \pm\frac{0.078}{\sqrt{10}} = \pm 0.025$ ft

Also, by Equation (3.15b), $(E_{90})_m = \pm 1.6449(0.025) = \pm 0.041$ ft.

These values show the error limits of 68 and 90 percent probability for the line's length. It can be said that the true line length has a 68 percent chance of being within ±0.025 of the mean, and a 90 percent likelihood of falling not farther than ±0.041 ft from the mean.

3.18 APPLICATIONS

The preceding example problems show that the equations of error probability are applied in two ways:

1. To analyze observations already made, for comparison with other results or with specification requirements.
2. To establish procedures and specifications in order that the required results will be obtained.

The application of the various error probability equations must be tempered with judgment and caution. Recall that they are based on the assumption that the errors conform to a smooth and continuous normal distribution curve, which in turn is based on the assumption of a large number of observations. Frequently in surveying only a few observations—often from four to eight—are taken. If these conform to a normal distribution, then the answer obtained using probability equations will be reliable; if they do not, the conclusions could be misleading. In the absence of knowledge to the contrary, however, an assumption that the errors are normally distributed is still the best available.

3.19 CONDITIONAL ADJUSTMENT OF OBSERVATIONS

In Section 3.3, it was emphasized that the true value of any observed quantity is never known. In some types of problems, however, the sum of several observations must equal a fixed value; for example, the sum of the three angles in a plane triangle has to total 180°. In practice, therefore, the observed angles are adjusted to make them add to the required amount. Correspondingly, distances—either horizontal or vertical—must often be adjusted to meet certain conditional requirements. The methods used will be explained in later chapters, where the operations are taken up in detail.

3.20 WEIGHTS OF OBSERVATIONS

It is evident that some observations are more precise than others because of better equipment, improved techniques, and superior field conditions. In making adjustments, it is consequently desirable to assign *relative weights* to individual observations. It can logically be concluded that if an observation is very precise, it will have a small standard deviation or variance, and thus should be weighted more heavily (held closer to its observed value) in an adjustment than an observation of lower precision. From this reasoning, it is deduced that weights of observations should bear an inverse relationship to precision. In fact, it can be shown that relative weights are inversely proportional to variances, or

$$W_a \propto \frac{1}{\sigma_a^2} \tag{3.16}$$

where W_a is the weight of an observation a, which has a variance of σ_a^2. Thus the higher the precision (the smaller the variance), the larger should be the relative weight of the observed value being adjusted. In some cases, variances are unknown originally and weights must be assigned to observed values based on estimates of their relative precision. If a quantity is observed repeatedly and the individual observations have varying weights, the weighted mean can be computed from the expression

$$\overline{M}_W = \frac{\Sigma WM}{\Sigma W} \tag{3.17}$$

where $\overline{M}_W$ is the weighted mean, ΣWM the sum of the individual weights times their corresponding observations, and ΣW the sum of the weights.

Example 3.8

Suppose four observations of a distance are recorded as 482.16, 482.17, 482.20, and 482.18 and given weights of 1, 2, 2, and 4, respectively, by the surveyor. Determine the weighted mean.

Solution

By Equation (3.17),

$$\overline{M}_W = \frac{482.16 + 487.17(2) + 482.20(2) + 482.14(4)}{1 + 2 + 2 + 4} = 482.18 \text{ ft}$$

In computing adjustments involving unequally weighted observations, corrections applied to observed values should be made *inversely proportional to the relative weights*.

Example 3.9

Assume the observed angles of a certain plane triangle, and their relative weights, are $A = 49°51'15''$, $W_a = 1$; $B = 60°32'08''$, $W_b = 2$; and $C = 69°36'33''$, $W_c = 3$. Compute the weighted mean of the angles.

Solution

The sum of the three angles is computed first and found to be 4″ less than the required geometrical condition of exactly 180 degrees. The angles are therefore adjusted in inverse proportion to their relative weights, as illustrated in the accompanying tabulation. Angle *C* with the greatest weight (3) gets the smallest correction, $2x$; *B* receives $3x$; and *A*, $6x$.

	Observed Angle	Wt	Correction	Numerical Corr.	Rounded Corr.	Adjusted Angle
A	49°51′15″	1	$6x$	+2.18″	+2″	49°51′17″
B	.60°32′08″	2	$3x$	+1.09″	+1″	60°32′09″
C	69°36′33″	3	$2x$	+0.73″	+1″	69°36′34″
Sum	179°59′56″	$\Sigma = 6$	$11x$	+4.00″	+4″	180°00′00″

$11x = 4''$ and $x = +0.36''$

It must be emphasized again that adjustment computations based on the theory of probability are valid only if systematic errors and employing proper procedures, equipment, and calculations eliminate mistakes.

3.21 LEAST SQUARES ADJUSTMENT

As explained in Section 3.19, most surveying observations must conform to certain geometrical conditions. The amounts by which they fail to meet these conditions are called misclosures, and they indicate the presence of random errors. In Example 3.9, for example, the misclosure was 4″. Various procedures are used to distribute these misclosure errors to produce mathematically perfect geometrical conditions. Some simply apply corrections of the same size to all observed values, where each correction equals the total misclosure (with its algebraic sign changed), divided by the number of observations. Others introduce corrections in proportion to assigned weights. Still others employ rules of thumb, for example, the "compass rule" described in Chapter 10 for adjusting closed traverses.

Because random errors in surveying conform to the mathematical laws of probability and are "normally distributed," the most appropriate adjustment procedure should be based upon these laws. Least squares is such a method. It is not a new procedure, having been applied by the German mathematician

Karl Gauss as early as the latter part of the 18th century. However until the advent of computers, it was only used sparingly because of the lengthy calculations involved.

Least squares is suitable for adjusting any of the basic types of surveying observations described in Section 2.1 and is applicable to all of the commonly used surveying procedures. The method enforces the condition that *the sum of the weights of the observations times their corresponding squared residuals is minimized.* This fundamental condition, which is developed from the equation for the normal error distribution curve, provides most probable values for the adjusted quantities. In addition, it also (a) enables the computation of precisions of the adjusted values, (b) reveals the presence of mistakes so steps can be taken to eliminate them, and (c) makes possible the optimum design of survey procedures in the office before going into the field to take observations.

The basic assumptions that underlie least squares theory are as follows: (1) Mistakes and systematic errors have been eliminated so only random errors remain; (2) the number of observations being adjusted is large; and (3) the frequency distribution of errors is normal. Although these assumptions are not always met, the least squares adjustment method still provides the most rigorous error treatment available, and hence it has become very popular and important in modern surveying. A more detailed discussion of the subject is presented in Chapter 16.

PROBLEMS

Asterisks (*) indicate problems that have answers given in Appendix G.

3.1 Explain the difference between *direct* and *indirect measurements* in surveying. Give two examples of each.

3.2 Define the term *systematic errors,* and give two surveying examples of a systematic error.

3.3 Define the term *random errors,* and give two surveying examples of a random error.

3.4 Explain the difference between accuracy and precision.

3.5 Discuss what is meant by the precision of an observation.

A distance *AB* is observed repeatedly using the same equipment and procedures, and the results, in meters, are listed in Problems 3.6 through 3.10. Calculate (a) the line's most probable length, (b) the standard deviation, and (c) the standard deviation of the mean for each set of results.

3.6* 125.474, 125.471, 125.478, 125.476, 125.472, 125.473, 125.480, 125.478, 125.478, and 125.474

3.7 Same as Problem 3.6, but discard one observation, 125.480.

3.8 Same as Problem 3.6, but discard two observations, 125.480 and 125.471.

3.9 Same as Problem 3.6, but include two additional observations, 125.477 and 125.475.

3.10 Same as Problem 3.6, but include three additional observations, 125.470, 125.479, and 125.475.

In Problems 3.11 through 3.14, determine the range within which observations should fall (a) 90 percent of the time and (b) 95 percent of the time. List the percentage of values that actually fall within these ranges.

3.11* For the data of Problem 3.6.
3.12 For the data of Problem 3.7.
3.13 For the data of Problem 3.8.
3.14 For the data of Problem 3.9.

In Problems 3.15 through 3.17, an angle is observed repeatedly using the same equipment and procedures. Calculate (a) the angle's most probable value, (b) the standard deviation, and (c) the standard deviation of the mean.

3.15* 130°32′36″, 130°32′42″, 130°32′44″, and 130°32′38″.

3.16 Same as Problem 3.15, but with three additional observations, 130°32′34″, 130°32′40″, and 130°32′36″.

3.17 Same as Problem 3.16, but with two additional observations, 130°32′40″ and 130°32′38″.

3.18* A field party is capable of making taping observations with a standard deviation of ±0.015 ft per 100-ft tape length. What standard deviation would be expected in a distance of 300 ft taped by this party?

3.19 Repeat Problem 3.18, except that the standard deviation per 30-m tape length is ±0.003 m and a distance of 90 m is taped. What is the expected 95% error in 90 m?

3.20 A distance of 200 ft must be taped in a manner to ensure a standard deviation smaller than ±0.03 ft. What must be the standard deviation per 100-ft tape length to achieve the desired precision?

3.21 Lines of levels were run requiring n instrument setups. If the rod reading for each backsight and foresight has a standard deviation σ, what is the standard deviation in each of the following level lines?
(a) $n = 40$, $\sigma = \pm 0.005$ ft **(b)** $n = 20$, $\sigma = \pm 1.00$ mm

3.22 A line AC was observed in 2 sections AB and BC, with lengths and standard deviations listed below. What is the total length AC and its standard deviation?
***(a)** $AB = 80.00 \pm 0.015$ ft; $BC = 46.13 \pm 0.008$ ft
(b) $AB = 90.000 \pm 0.009$ m; $BC = 85.413 \pm 0.008$ m

3.23 Line AD is observed in three sections, AB, BC, and CD, with lengths and standard deviations as listed below. What is the total length AD and its standard deviation?
(a) $AB = 1086.23 \pm 0.05$ ft; $BC = 569.08 \pm 0.03$ ft; $CD = 863.19 \pm 0.04$ ft
(b) $AB = 325.808 \pm 0.004$ m; $BC = 801.193 \pm 0.008$ m; $CD = 751.843 \pm 0.006$ m

3.24 A distance AB was observed four times as 193.13, 193.06, 193.18, and 193.15 ft. The observations were given weights of 3, 1, 2, and 3, respectively, by the observer. ***(a)** Calculate the weighted mean for distance AB. **(b)** What difference results if later judgment revises the weights to 1, 1, 3, and 3, respectively?

3.25 Determine the weighted mean for the following angles:
(a) 168°43′26″, wt 3; 168°43′16″, wt 1; 168°43′22″, wt 4
(b) 36°58′32″ ± 4″; 36°58′28″ ± 3″; 36°58′42″ ± 6″; 36°58′30″ ± 2″

3.26 Specifications for observing angles of an n-sided polygon limit the total angular misclosure to E. How accurately must each angle be observed for the following values of n and E?
(a) $n = 8$, $E = 6''$
(b) $n = 12$, $E = 24''$

3.27 What is the area of a rectangular field and its estimated error for the following recorded values:

*(a) 548.37 ± 0.05 ft by 345.59 ± 0.04 ft

(b) 346.98 ± 0.01 ft by 853.49 ± 0.02 ft

(c) 108.014 ± 0.003 m by 287.010 ± 0.005 m

3.28 Adjust the angles of triangle *ABC* for the following angular values and weights:

*(a) A = 49°27′36″, wt 2; B = 61°42′18″, wt 1; C = 68°50′17″, wt 3

(b) A = 80°14′00″, wt 2; B = 38°37′46″, wt 4; C = 61°07′54″, wt 2

3.29 Determine relative weights and perform a weighted adjustment (to the nearest second) for angles *A*, *B*, and *C* of a plane triangle, given the following four observations for each angle:

Angle *A*	Angle *B*	Angle *C*
38°47′58″	71°22′26″	69°50′04″
38°47′44″	71°22′22″	69°50′16″
38°48′12″	71°22′12″	69°50′30″
38°48′02″	71°22′12″	69°50′10″

3.30 A line of levels was run from benchmarks *A* to *B*, *B* to *C*, and *C* to *D*. The elevation differences obtained between benchmarks, with their standard deviations, are listed below. What is the difference in elevation from benchmark *A* to *D* and the standard deviation of that elevation difference?

(a) BM *A* to BM *B* = +37.78 ± 0.12 ft; BM *B* to BM *C* = −73.50 ± 0.16 ft; and BM *C* to BM *D* = −84.09 ± 0.08 ft

(b) BM *A* to BM *B* = −60.821 ± 0.015 m; BM *B* to BM *C* = +94.378 ± 0.020 m; and BM *C* to BM *D* = +56.805 ± 0.015 m

(c) BM *A* to BM *B* = −22.812 ± 0.013 m; BM *B* to BM *C* = +25.837 ± 0.008 m; and BM *C* to BM *D* = +35.542 ± 0.016 m

3.31 Create a computational program that solves Problem 3.9.

3.32 Create a computational program that solves Problem 3.17.

3.33 Create a computational program that solves Problem 3.29.

BIBLIOGRAPHY

Alder, K. 2002. *The Measure of All Things—The Seven-Year Odyssey and Hidden Error that Transformed the World*. New York: The Free Press.

Bell, J. 2001. "Hands On: TDS for Windows CE on the Ranger." *Professional Surveyor* 21 (No. 1): 33.

Foster, R. 2003. "Uncertainty about Positional Uncertainty." *Point of Beginning* 28 (No. 11): 40.

Geomatics Industry of America. 2004. "Tool Tips." *Point of Beginning* 29 (No. 10): 68.

Ghilani, C. D. and P. R. Wolf. 2006. *Adjustment Computations: Spatial Data Analysis*. New York: John Wiley & Sons.

Ghilani, C. D. 2003. "Statistics and Adjustments Explained. Part 1: Basic Concepts." *Surveying and Land Information Systems* 63 (No. 2): 62.

______. 2003. "Statistics and Adjustments Explained. Part 2: Sample Sets and Reliability." *Surveying and Land Information Systems* 63 (No. 3): 141.

Hintz, Raymond J. 2003. "Using Positional Tolerance in Simplifying GPS Control Survey Quality Requirements." *Surveying and Land Information Science*. 63 (No. 1).

Linklater, Andro. 2002. *Measuring America—How an Untamed Wilderness Shaped the United States and Fulfilled the Promise of Democracy*. New York: Walker and Company.

Zeiske, K. 2002. "International Standards for Surveying." *Professional Surveyor* 22 (No. 8): 53.

4

Leveling—Theory, Methods, and Equipment

PART I • LEVELING—THEORY AND METHODS

4.1 INTRODUCTION

Leveling is the general term applied to any of the various processes by which elevations of points or differences in elevation are determined. It is a vital operation in producing necessary data for mapping, engineering design, and construction. Leveling results are used to (1) design highways, railroads, canals, sewers, water supply systems, and other facilities having grade lines that best conform to existing topography; (2) lay out construction projects according to planned elevations; (3) calculate volumes of earthwork and other materials; (4) investigate drainage characteristics of an area; (5) develop maps showing general ground configurations; and (6) study earth subsidence and crustal motion.

4.2 DEFINITIONS

Basic terms in leveling are defined in this section, some of which are illustrated in Figure 4.1.

Vertical line. A line that follows the local direction of gravity as indicated by a plumb line.

Level surface. A curved surface that at every point is perpendicular to the local plumb line (the direction in which gravity acts). Level surfaces are approximately spheroidal in shape. A body of still water is the closest example of a level surface. Within local areas, level surfaces at different

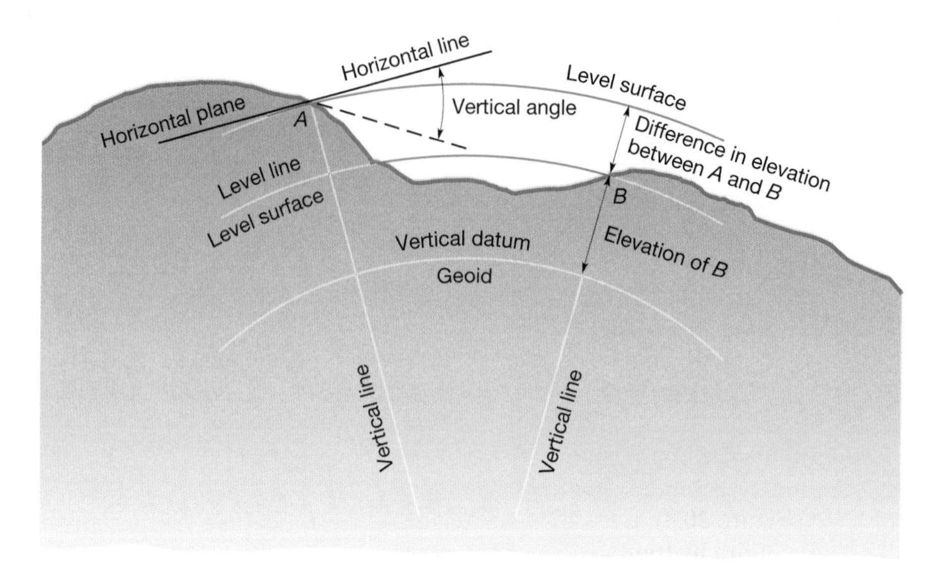

Figure 4.1
Leveling terms.

heights are considered to be concentric.[1] Level surfaces are also known as *equipotential surfaces* since, for a particular surface, the potential of gravity is equal at every point on the surface.

Level line. A line in a level surface—therefore, a curved line.

Horizontal plane. A plane perpendicular to the local direction of gravity. In plane surveying, it is a plane perpendicular to the local vertical line.

Horizontal line. A line in a horizontal plane. In plane surveying, it is a line perpendicular to the local vertical.

Vertical datum. Any level surface to which elevations are referenced. This is the surface that is arbitrarily assigned an elevation of zero (see Section 19.6). This level surface is also known as a reference datum since points using this datum have heights relative to this surface.

Elevation. The distance measured along a vertical line from a vertical datum to a point or object. If the elevation of point A is 802.46 ft, A is 802.46 ft above the reference datum. The elevation of a point is also called its height above the datum.

Geoid. A particular level surface that serves as a datum for elevations and astronomical observations.

Mean sea level (MSL). The average height of the sea's surface for all stages of the tide over a 19-year period. It was arrived at from readings, usually taken at hourly intervals, at 26 gaging stations along the Atlantic and Pacific oceans and the Gulf of Mexico. The elevation of the sea differs from station to station depending on local influences of the tide; for example, at two points 0.5 mi apart on opposite sides of an island in the Florida Keys, it varies by 0.3 ft. Mean sea level was accepted as the vertical datum for North

[1] Due to flattening of the Earth in the polar direction, level surfaces at different elevations and different latitudes are not truly concentric. This topic is discussed in more detail in Chapter 19.

America for many years. However, the current vertical datum uses a single benchmark as a reference (see Section 4.3).

Tidal datums. The vertical datums used in coastal areas for establishing property boundaries of lands bordering waters subject to tides. Tidal datums also provide the basis for locating fishing and oil drilling rights in tidal waters, and the limits of swamp and overflowed lands. Various definitions of tidal datums have been used in different areas, but the one most commonly employed is the *mean high water* (MHW) line. Others applied include *mean higher high water* (MHHW), *mean low water* (MLW), and *mean lower low water* (MLLW). Interpretations of tidal datums, and the methods by which they are determined, have been and continue to be the subject of numerous court cases.

Benchmark (BM). A relatively permanent object, natural or artificial, having a marked point whose elevation above or below a reference datum is known or assumed. Common examples are metal disks set in concrete (see Figure 20.8), reference marks chiseled on large rocks, nonmovable parts of fire hydrants, curbs, etc.

Leveling. The process of finding elevations of points or their differences in elevation.

Vertical control. A series of benchmarks or other points of known elevation established throughout an area, also termed *basic control* or *level control.* The basic vertical control for the United States was derived from first- and second-order leveling. Less precise third-order leveling has been used to fill gaps between second-order benchmarks, as well as for many other specific projects (see Section 19.10).

■ 4.3 NORTH AMERICAN VERTICAL DATUM

Precise leveling operations to establish a distributed system of reference benchmarks throughout the United States began in the 1850s. This work was initially concentrated along the eastern seaboard, but in 1887 the U.S. Coast and Geodetic Survey (USC&GS) began its first transcontinental leveling across the country's midsection. That project was completed in the early 1900s. By 1929, thousands of benchmarks had been set. In that year, the USC&GS began a general least squares adjustment of all leveling thus far completed in the United States and Canada. The adjustment involved over 100,000 km of leveling and incorporated long-term data from the 26 tidal gaging stations; hence, it was related to mean sea level. In fact, that network of benchmarks with their resulting adjusted elevations defined the mean sea level datum. It was called the *National Geodetic Vertical Datum of 1929* (*NGVD29*).

Through the years after 1929, the NGVD29 deteriorated somewhat due to various causes including changes in sea level and shifting of the Earth's crust. Also, more than 625,000 km of additional leveling was completed. To account for these changes and incorporate the additional leveling, the National Geodetic Survey (NGS) performed a new general readjustment. Work on this adjustment, which included more than 1.3 million observed elevation differences, began in 1978. Although not finished until 1991, its planned completion date was 1988, and thus it has been named the *North American Vertical Datum of 1988* (*NAVD88*). Besides the United States and Canada, Mexico was also included in this general readjustment. This adjustment shifted the position of the reference surface from the mean of the 26 tidal gage stations to a single tidal gage benchmark known as *Father Point/Rimouski,* which is in Quebec, Canada, along the St. Lawrence Seaway. Thus, elevations in NAVD88 are no longer referenced to mean sea level. Benchmark elevations that were defined by the NGVD29 datum have changed by relatively small, but nevertheless significant, amounts in the eastern half of the continental United States (see Figure 20.7). However, the changes are much greater in the western part of the country and reach 1.5 m in the Rocky Mountain region. It is therefore imperative that surveyors positively identify the datum to which their elevations are referred. Listings of the new elevations are available from the NGS.[2]

4.4 CURVATURE AND REFRACTION

From the definitions of a level surface and a horizontal line, it is evident that the horizontal plane departs from a level surface because of curvature of the Earth. In Figure 4.2, the deviation *DB* from a horizontal line through point *A* is expressed approximately by the formulas

$$C_f = 0.667\,M^2 = 0.0239F^2 \quad (4.1a)$$

or

$$C_m = 0.0785K^2 \quad (4.1b)$$

where the departure of a level surface from a horizontal line is C_f in feet or C_m, in meters, M is the distance AB in miles, F the distance in *thousands of feet,* and K the distance in kilometers.

Since points A and B are on a level line, they have the same elevation. If a graduated rod was held vertically at B and a reading was taken on it by means of

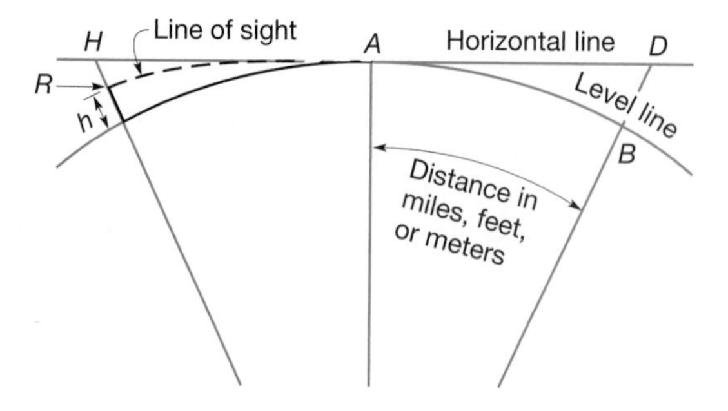

Figure 4.2 Curvature and refraction.

[2]Descriptions and NAVD88 elevations of benchmarks can be obtained from the National Geodetic Information Center at http://www.ngs.noaa.gov/datasheet.html. Information can also be obtained by email at info_center@ngs.noaa.gov, or by writing to the National Geodetic Information Center, NOAA, National Geodetic Survey, 1315 East West Highway, Silver Spring, MD 20910; telephone: (301) 713–3242.

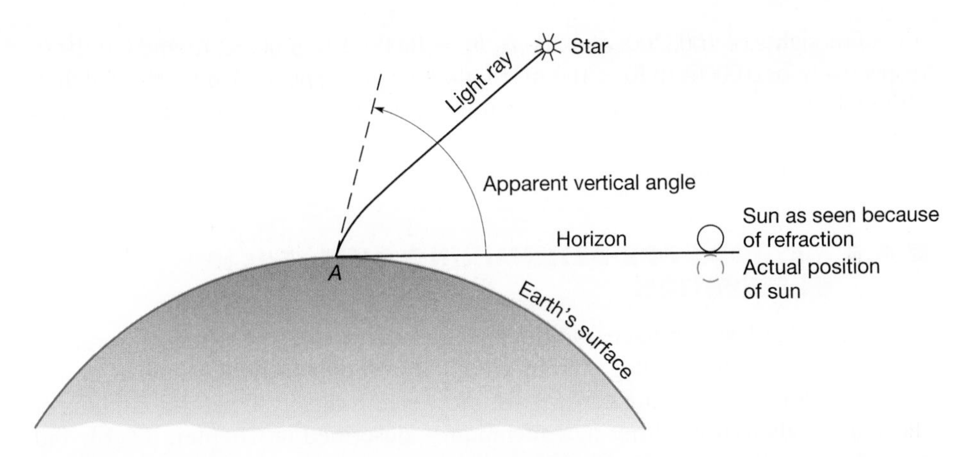

Figure 4.3 Refraction.

a telescope with its line of sight *AD* horizontal, the Earth's curvature would cause the reading to be read too high by length *BD*.

Light rays passing through the Earth's atmosphere are bent or refracted toward the Earth's surface, as shown in Figure 4.3. Thus a theoretically horizontal line of sight, like *AH* in Figure 4.2, is bent to the curved form *AR*. Hence, the reading on a rod held at *R* is diminished by length *RH*.

The effects of refraction in making objects appear higher than they really are (and therefore rod readings too small) can be remembered by noting what happens when the sun is on the horizon, as in Figure 4.3. At the moment when the sun has just passed below the horizon, it is seen just above the horizon. The sun's diameter of approximately 32 min is roughly equal to the average refraction on a horizontal sight. Since the red wavelength of light bends the greatest, it is not uncommon to see a red sun in a clear sky at dusk and dawn.

Displacement resulting from refraction is variable. It depends on atmospheric conditions, length of line, and the angle a sight line makes with the vertical. For a horizontal sight, refraction R_f, in feet or R_m, in meters is expressed approximately by the formulas

$$R_f = 0.093\ M^2 = 0.0033\ F^2 \tag{4.2a}$$

or

$$R_m = 0.011\,K^2 \tag{4.2b}$$

This is about one-seventh the effect of curvature of the Earth, but in the opposite direction.

The combined effect of curvature and refraction, h in Figure 4.2, is approximately

$$h_f = 0.574\ M^2 = 0.0206\ F^2 \tag{4.3a}$$

or

$$h_m = 0.0675\ K^2 \tag{4.3b}$$

where h_f is in feet and h_m is in meters.

For sights of 100, 200, and 300 ft, h_f = 0.00021 ft, 0.00082 ft, and 0.0019 ft, respectively, or 0.00068 m for a 100-m length. It will be explained in Section 5.4 that, although the combined effects of curvature and refraction produce rod readings that are slightly too large, proper field procedures can practically eliminate the error due to these causes.

4.5 METHODS FOR DETERMINING DIFFERENCES IN ELEVATION

Differences in elevation have traditionally been determined by taping, differential leveling, barometric leveling, and indirectly by trigonometric leveling. A newer method involves measuring vertical distances electronically. Brief descriptions of these methods follow. Other new techniques, described in Chapters 13, 14, and 15, utilize satellite systems. Elevation differences can also be determined using photogrammetry, as discussed in Chapter 27.

4.5.1 Measuring Vertical Distances by Taping or Electronic Methods

Application of a tape to a vertical line between two points is sometimes possible. This method is used to measure depths of mine shafts, to determine floor elevations in condominium surveys, and in the layout and construction of multistory buildings, pipelines, etc. When water or sewer lines are being laid, a graduated pole or rod may replace the tape (see Section 23.4). In certain situations, especially on construction projects, reflectorless electronic distance measurement (EDM) devices (see Section 6.22) are replacing the tape for measuring vertical distances on construction sites. This concept is illustrated in Figures 4.4 and 24.4.

4.5.2 Differential Leveling

In this most commonly employed method, a telescope with suitable magnification is used to read graduated rods held on fixed points. A horizontal line of sight within the telescope is established by means of a level vial or automatic compensator.

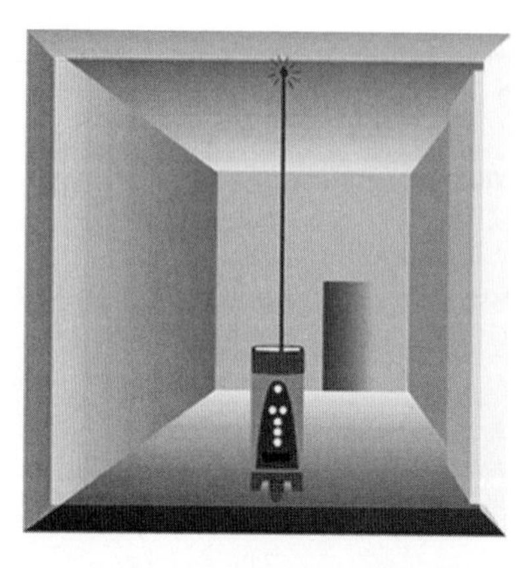

Figure 4.4 Reflectorless EDMs are being used to measure elevation differences in construction applications. (Courtesy Leica Geosystems)

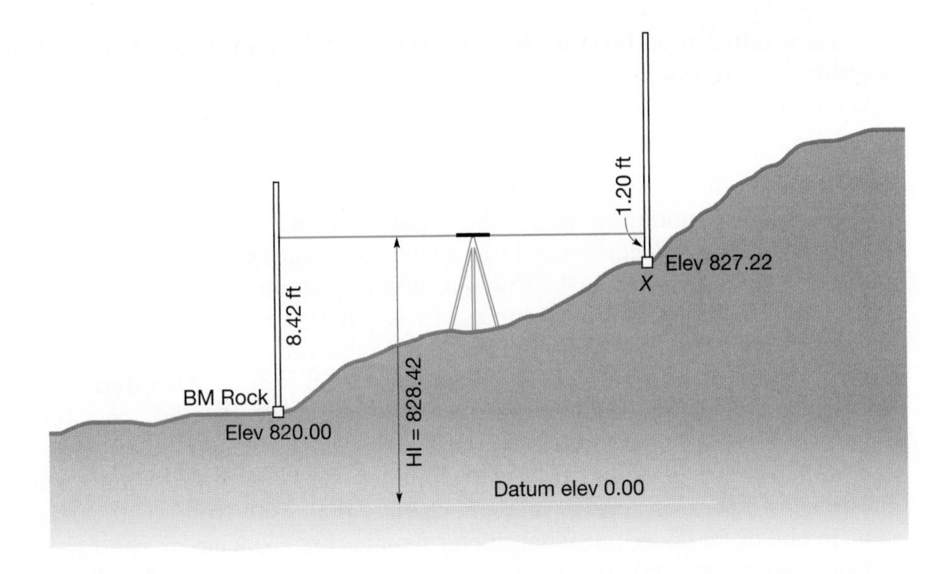

Figure 4.5
Differential leveling.

The basic procedure is illustrated in Figure 4.5. An instrument is set up approximately halfway between BM Rock and point X.[3] Assume the elevation of BM Rock is known to be 820.00 ft. After leveling the instrument, a plus sight taken on a rod held on the BM gives a reading of 8.42 ft. A *plus sight* (+S), also termed *backsight* (BS), is the reading on a rod held on a point of known or assumed elevation. This reading is used to compute the *height of instrument* (HI), defined as the vertical distance from datum to the instrument line of sight. Direction of the sight—whether forward, backward, or sideways—is not important. The term plus sight is preferable to backsight, but both are used. Adding the plus sight 8.42 ft to the elevation of BM Rock, 820.00, gives an HI of 828.42 ft.

If the telescope is then turned to bring into view a rod held on point X, a *minus sight* (−S), also called *foresight* (FS), is obtained. In this example, it is 1.20 ft. A minus sight is defined as the rod reading on a point whose elevation is desired. The term minus sight is preferable to foresight. Subtracting the minus sight, 1.20 ft, from the HI, 828.42, gives the elevation of point X as 827.22 ft.

Differential leveling theory and applications can thus be expressed by two equations, which are repeated over and over,

$$\text{HI} = \text{elev} + \text{BS} \tag{4.4}$$

and

$$\text{elev} = \text{HI} - \text{FS} \tag{4.5}$$

[3]As noted in Section 4.4, the combination of Earth curvature and atmospheric refraction causes rod readings to be too large. However for any setup, if the backsight and foresight lengths are made equal (which is accomplished with the midpoint setup) the error from these sources is eliminated, as described in Section 5.4.

Since differential leveling is by far the most commonly used method to determine differences in elevation, it will be discussed in detail in Chapter 5.

4.5.3 Barometric Leveling

The barometer, an instrument that measures air pressure, can be used to find relative elevations of points on the Earth's surface since a change of approximately 1000 ft in elevation will correspond to a change of about 1 in. of mercury (Hg) in atmospheric pressure. Figure 4.6 shows a surveying altimeter. Calibration of the scale on different models is in multiples of 1 or 2 ft, 0.5 or 1 m. Air pressures are affected by circumstances other than difference in elevation, such as sudden shifts in temperature and changing weather conditions due to storms. Also, during each day a normal variation in barometric pressure amounting to perhaps a 100-ft difference in elevation occurs. This variation is known as the *diurnal range.*

In barometric leveling, various techniques can be used to obtain correct elevation differences in spite of pressure changes that result from weather variations. In one of these, a *control* barometer remains on a benchmark (base) while a *roving* instrument is taken to points whose elevations are desired. Readings are made on the base at stated intervals of time, perhaps every 10 min, and the elevations recorded along with temperature and time. Elevation, temperature, and time readings with the roving barometer are taken at critical points and adjusted

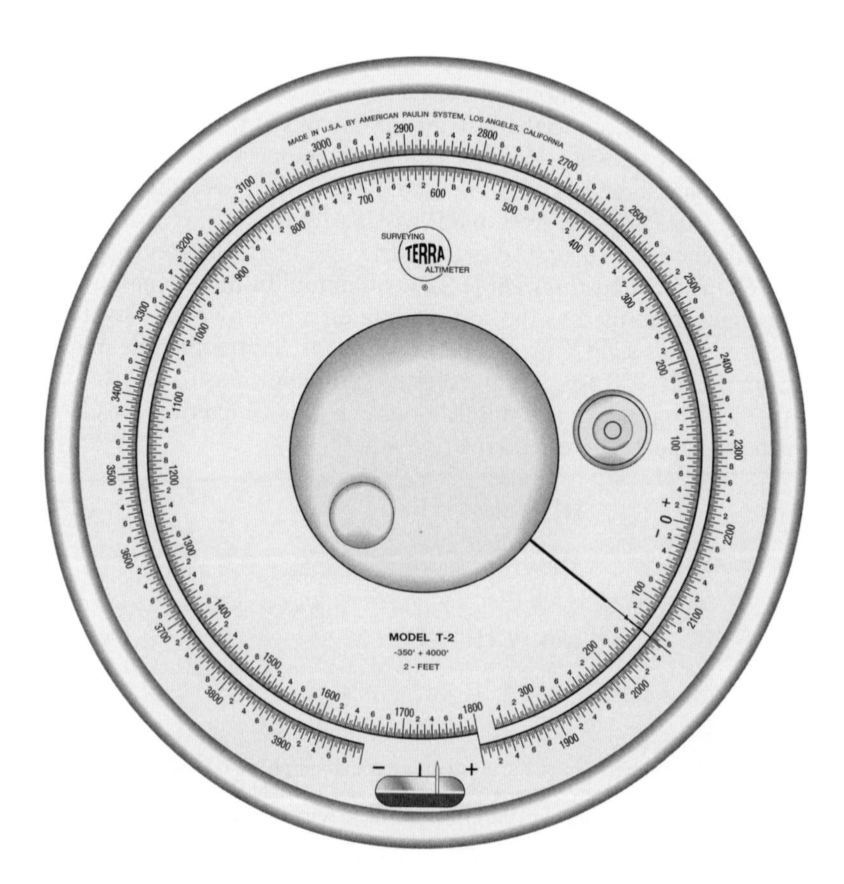

Figure 4.6
Surveying altimeter. (Courtesy American Paulin System)

later in accordance with changes observed at the control point. Methods of making field surveys using a barometer have been developed in which one, two, or three bases may be used. Other methods employ leapfrog or semi-leapfrog techniques. In stable weather conditions, and by using several barometers, elevations correct to within ±2 to 3 ft were possible.

Barometers have been used in the past for work in rough country where extensive areas had to be covered but a high order of accuracy was not required. They are seldom used today, however, having given way to other more modern and accurate equipment.

4.5.4 Trigonometric Leveling

The difference in elevation between two points can be determined by measuring (1) the inclined or horizontal distance between them and (2) the zenith angle or the altitude angle to one point from the other. (Zenith and altitude angles, described in more detail in Section 8.13, are measured in vertical planes. Zenith angles are measured downward from vertical, and altitude angles are measured up or down from horizontal.) Thus in Figure 4.7 if slope distance S and zenith angle z or altitude angle α between C and D are measured, then V, the elevation difference between C and D, is

$$V = S \cos z \tag{4.6}$$

or

$$V = S \sin \alpha \tag{4.7}$$

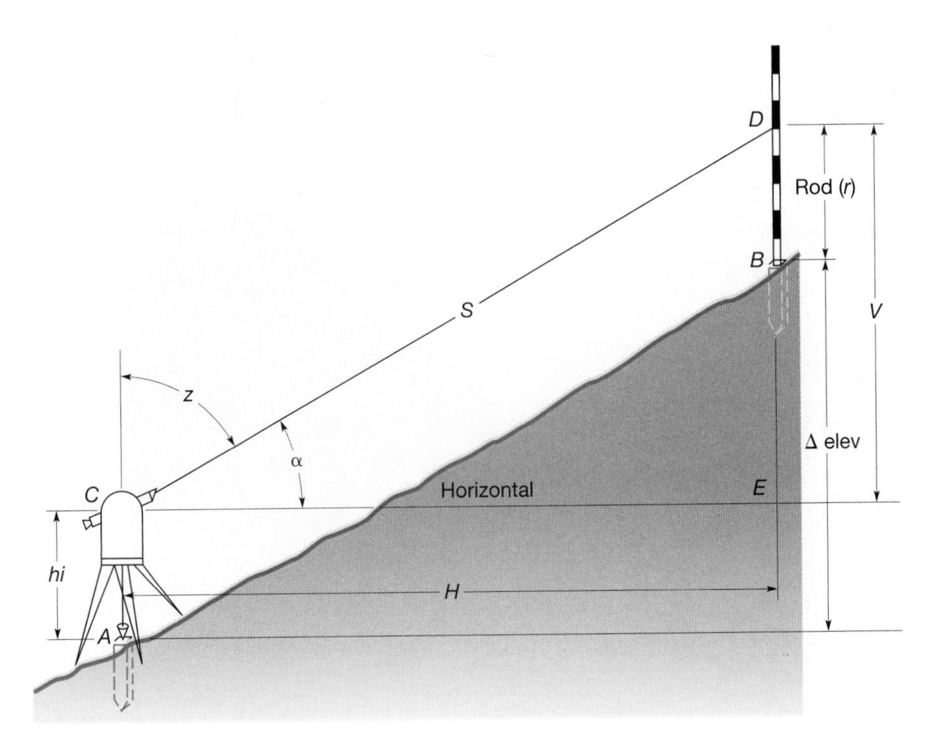

Figure 4.7 Trigonometric leveling—short lines.

Alternatively, if horizontal distance H between C and D is measured, then V is

$$V = H \cot z \tag{4.8}$$

or

$$V = H \tan \alpha \tag{4.9}$$

The difference in elevation (Δelev) between points A and B in Figure 4.7 is given by

$$\Delta\text{elev} = hi + V - r \tag{4.10}$$

where hi is the height of the instrument above point A and r is the reading on the rod held at B when zenith angle z or altitude angle α is read. If r is made equal to hi, then these two values cancel in Equation (4.10) and simplify the computations.

Note the distinction in this text between *HI* and *hi*. Although both are called height of instrument, the term *HI is the elevation of the instrument above datum,* as described in Section 4.5.2, while *hi is the height of the instrument above an occupied point,* as discussed here.

For short lines (up to about 1000 ft in length) elevation differences obtained in trigonometric leveling are appropriately depicted by Figure 4.7 and properly computed using Equations (4.6) through (4.10). However, for longer lines, Earth curvature and refraction become factors that must be considered. Figure 4.8 illustrates

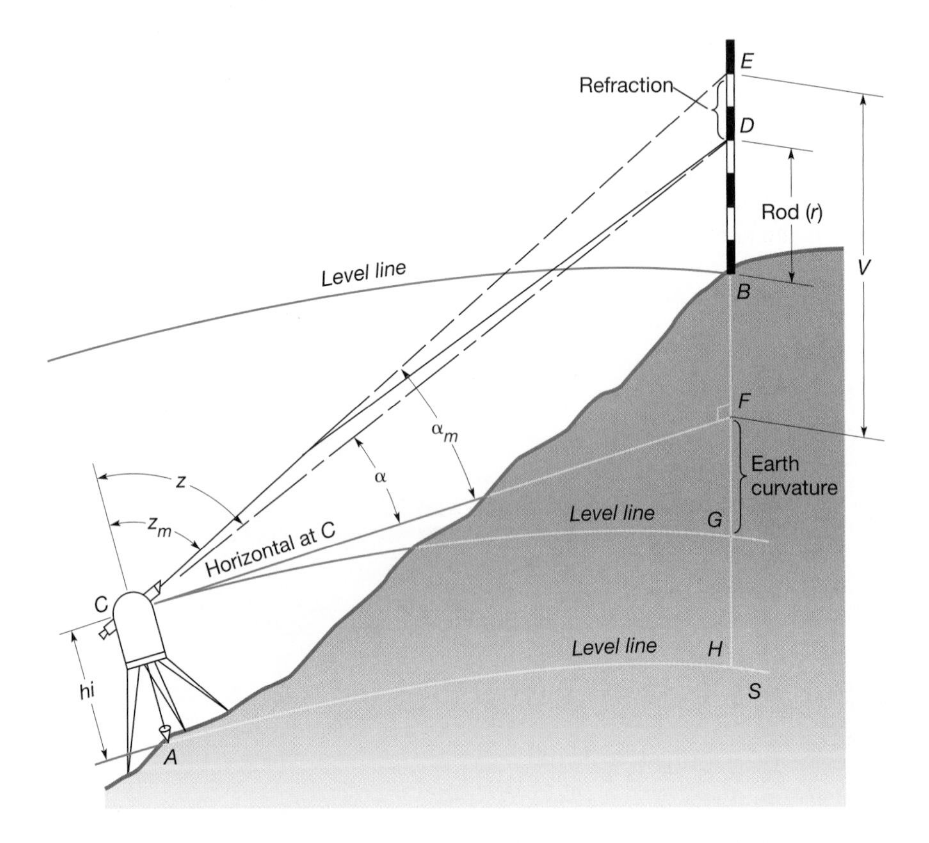

Figure 4.8 Trigonometric leveling—long lines.

the situation. Here an instrument is set up at *C* over point *A*. Sight *D* is made on a rod held at point *B*, and zenith angle z_m, or altitude angle α_m, is measured. The true difference in elevation (Δelev) between *A* and *B* is vertical distance *HB* between level lines through *A* and *B*, which is equal to $HG + GF + V - ED - r$. Since *HG* is the instrument height *hi*, *GF* is Earth curvature *C* [see Equations (4.1)], and *ED* is refraction *R* [see Equations (4.2)], the elevation difference can be written as

$$\Delta\text{elev} = hi + V + (C - R) - r \qquad \textbf{(4.11)}$$

The value of *V* in Equation (4.11) is obtained using one of Equations (4.6) through (4.9), depending on what quantities are measured. Again if *r* is made equal to *hi*, these values cancel. Also, the term $(C - R)$ is given by Equations (4.3). Thus, except for the addition of the curvature and refraction correction, long and short sights may be treated the same in trigonometric leveling computations. Note that in developing Equation (4.11), angle *F* in triangle *CFE* was assumed to be 90°. Of course as lines become extremely long, this assumption does not hold. However, for lengths within a practical range, errors caused by this assumption are negligible.

The *hi* used in Equation (4.11) can be obtained by simply measuring the vertical distance from the occupied point up to the instrument's horizontal axis (axis about which the telescope rotates) using a graduated rule or rod. An alternate method can be used to determine the elevation of a point that produces accurate results and does not require measurement of the *hi*. In this procedure, which is especially convenient if a total station instrument is used, the instrument is set up at a location where it is approximately equidistant from a point of known elevation (benchmark) and the one whose elevation is to be determined. The slope distance and zenith (or altitude) angle are measured to each point. Because the distances from the two points are approximately equal, curvature and refraction errors cancel. Also, since the same instrument setup applies to both readings, the *hi* values cancel, and if the same rod reading *r* is sighted when making both angle readings, they cancel. Thus the elevation of the unknown point is simply the benchmark elevation, minus *V* calculated for the benchmark, plus *V* computed for the unknown point, where the *V* values are obtained using either Equation (4.6) or (4.7).

Example 4.1

The slope distance and zenith angle between points *A* and *B* were measured with a total station instrument as 9585.26 ft and 81°42′20″, respectively. The *hi* and rod reading *r* were equal. If the elevation of *A* is 1238.42 ft, compute the elevation of *B*.

Solution

By Equation (4.3a), the curvature and refraction correction is

$$h_f = 0.0206\left[\frac{9585.26 \sin 81°42'20''}{1000}\right]^2 = 1.85\,\text{ft}$$

(Theoretically, the horizontal distance should be used in computing curvature and refraction. In practice, multiplying the slope distance by the sine of the zenith angle approximates it.)

By Equations (4.6) and (4.11), the elevation difference is (note that *hi* and *r* cancel)

$$V = 9585.26 \cos 81°42'20'' = 1382.77 \text{ ft}$$
$$\Delta\text{elev} = 1382.77 + 1.85 = 1384.62 \text{ ft}$$

Finally, the elevation of *B* is

$$\text{elev}_B = 1238.42 + 1384.62 = 2623.04 \text{ ft}$$

Note that if curvature and refraction had been ignored, an error of 1.85 ft would have resulted in the elevation for *B* in this calculation. Although Equation (4.11) was derived for an uphill sight, it is also applicable to downhill sights. In that case, the algebraic sign of *V* obtained in Equations (4.6) through (4.9) will be negative, however, because α will be negative or z greater than 90°.

For uphill sights curvature and refraction is added to a positive *V* to increase the elevation difference. For downhill sights, it is again added, but to a negative *V*, which decreases the elevation difference. Therefore, if "reciprocal" zenith (or altitude) angles are read (measuring the angles from both ends of a line), and *V* is computed for each and averaged, the effects of curvature and refraction cancel. Alternatively, the curvature and refraction correction can be completely ignored if one calculation of *V* is made using the average of the reciprocal angles. This assumes atmospheric conditions remain constant, so that refraction is equal for both angles. Hence they should be measured within as short a time period as possible. This method is preferred to reading the zenith (or vertical) angle from one end of the line and correcting for curvature and refraction, as in Example 4.1. The reason is that Equations (4.3) assume a standard atmosphere, which may not actually exist at the time of measurement.

Example 4.2

For Example 4.1, assume that at *B* the slope distance was measured again as 9585.26 ft and the zenith angle was read as 98°19′06″. The instrument height and *r* were equal. Compute (a) the elevation difference from this end of the line, and (b) the elevation difference using the mean of reciprocal angles.

Solution

(a) By Equation (4.3a), $h_f = 1.85$ (the same as for Example 4.1).
By Equations (4.6) and (4.11) (note that *hi* and *r* cancel),

$$\Delta\text{elev} = 9585.26 \cos 98°19'06'' + 1.85 = -1384.88 \text{ ft}$$

Note that this disagrees with the value of Example 4.1 by 0.26 ft. (The sight from B to A was downhill, hence the negative sign.) The difference of 0.26 ft is most probably due partly to measurement errors and partly to refraction changes that occurred during the time interval between vertical angle measurements. The average elevation difference for measurements made from the two ends is 1384.75 ft.

(b) The average zenith angle is $\dfrac{81°42'20'' + (180° - 98°19'06'')}{2} = 81°41'37''$

By Equation (4.10), $\Delta elev = 9585.26 \cos 81°41'37'' = 1384.75$ ft.

Note that this checks the average value obtained using the curvature and refraction correction.

With the advent of total station instruments, trigonometric leveling has become an increasingly common method for rapid and convenient measurement of elevation differences because slope distance and vertical (zenith) angle are quickly and easily measured from a single setup. Trigonometric leveling is used for topographic mapping, construction stakeout, control surveys, and other tasks. It is particularly valuable in rugged terrain. In trigonometric leveling, accurate zenith (or vertical) angle measurements are critical. For precise work, a 1″ to 3″ total station instrument is recommended and angles should be read direct and reversed from both ends of a line. Also, errors caused by uncertainties in refraction are mitigated if sight lengths are limited to about 1000 ft.

■ PART II • EQUIPMENT FOR DIFFERENTIAL LEVELING

■ 4.6 CATEGORIES OF LEVELS

Instruments used for differential leveling can be classified into four categories: *dumpy levels, tilting levels, automatic levels,* and *digital levels.* Although each differs somewhat in design, all have two common components: (1) a *telescope* to create a line of sight and enable a reading to be taken on a graduated rod, and (2) a system to orient the line of sight in a horizontal plane. Dumpy and tilting levels use level vials to orient their lines of sight, while automatic levels employ *automatic compensators.* Digital levels also employ automatic compensators, but use bar-coded rods for automated digital readings. Automatic levels are the type most commonly employed today, although tilting levels are still used especially on projects requiring very precise work. Digital levels are rapidly gaining prominence. These three types of levels are described in the sections that follow. Dumpy levels are rarely used today, having been replaced by other newer types. They are discussed in Appendix A. *Hand levels,* although not commonly used for differential leveling, have many special uses where rough elevation differences over short distances are needed. They are also discussed in this chapter. Total station instruments can also be used for differential leveling. These instruments and their uses are described in Section 8.18.

Electronic laser levels that transmit beams of either visible laser or invisible infrared light are another category of leveling instruments. They are not commonly employed in differential leveling, but are used extensively for establishing elevations on construction projects. They are described in Chapter 23.

4.7 TELESCOPES

The telescopes of leveling instruments define the line of sight and magnify the view of a graduated rod against a reference reticle, thereby enabling accurate readings to be obtained. The components of a telescope are mounted in a cylindrical tube. Its four main components are the objective lens, negative lens, reticle, and eyepiece. Two of these parts, the objective lens and eyepiece, are external to the instrument, and are shown on the automatic level illustrated in Figure 4.9.

Objective Lens. This compound lens, securely mounted in the tube's object end, has its optical axis reasonably concentric with the tube axis. Its main function is to gather incoming light rays and direct them toward the negative focusing lens.

Negative Lens. The negative lens is located between the objective lens and reticle, and mounted so its optical axis coincides with that of the objective lens. Its function is to focus rays of light that pass through the objective lens onto the reticle plane. During focusing, the negative lens slides back and forth along the axis of the tube.

Reticle. The reticle consists in a pair of perpendicular reference lines (usually called *crosshairs*) mounted at the principal focus of the objective optical system. The point of intersection of the crosshairs, together with the optical center of the objective system, forms the so-called *line of sight,* also sometimes called the *line of collimation.* The crosshairs are fine lines etched on a thin, round glass plate. The glass plate is held in place in the main cylindrical tube by two pairs of opposing screws, which are located at right angles to each other to facilitate adjusting the line of sight. Two additional lines parallel to and equidistant from the primary lines are commonly added to reticles for special purposes such as for *three-wire leveling* (see Section 5.8) and for *stadia* (see Section 17.9.2). The reticle is mounted within the main telescope tube with the lines placed in a horizontal-vertical orientation.

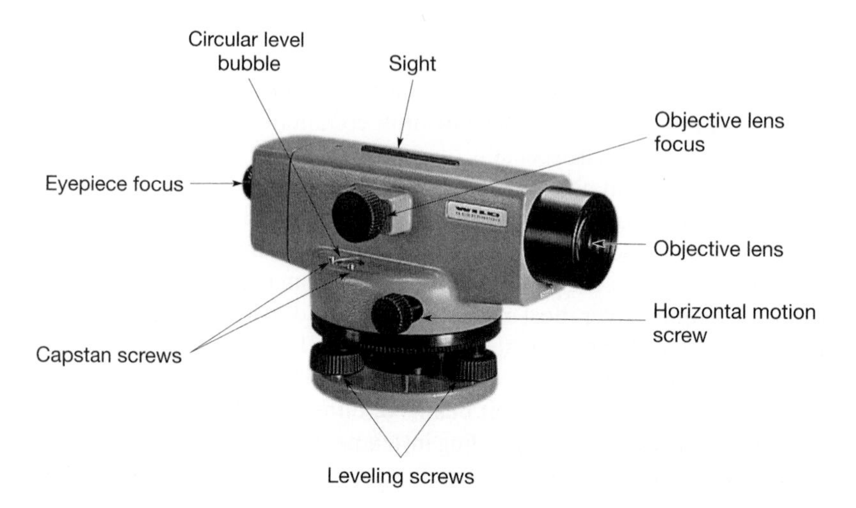

Figure 4.9 Parts of an automatic level. (Courtesy Leica Geosystems)

Eyepiece. The eyepiece is a microscope (usually with magnification from about 25 to 45×) for viewing the image.

Focusing is an important function to be performed in using a telescope. The process is governed by the fundamental principle of lenses stated in the following formula:

$$\frac{1}{f_1} + \frac{1}{f_2} = \frac{1}{f} \tag{4.12}$$

where f_1 is the distance from the lens to the image at the reticle plane, f_2 the distance from the lens to the object, and f the lens' *focal length.* The focal length of any lens is a function of the radii of the ground spherical surfaces of the lens, and of the index of refraction of the glass from which it is made. It is a constant for any particular single or compound lens. To focus for each varying f_2 distance, f_1 must be changed to maintain the equality of Equation (4.12).

Focusing the telescope of a level is a two-stage process. First the eyepiece lens must be focused. Since the position of the reticle in the telescope tube remains fixed, the distance between it and the eyepiece lens must be adjusted to suit the eye of an individual observer. This is done by bringing the crosshairs to a clear focus; that is, making them appear as black as possible when sighting at the sky or a distant, light-colored object. Once this has been accomplished, the adjustment need not be changed for the same observer, regardless of sight length, unless the eyes fatigue.

The second stage of focusing occurs after the eyepiece has been adjusted. Objects at varying distances from the telescope are brought to sharp focus at the plane of the crosshairs by turning the focusing knob. This moves the negative focusing lens to change f_1 and create the equality in Equation (4.12) for varying f_2 distances.

After focusing, if the crosshairs appear to travel over the object sighted when the eye is shifted slightly in any direction, *parallax* exists. The objective lens, the eyepiece, or both must be refocused to eliminate this effect if accurate work is to be done.

■ 4.8 LEVEL VIALS

Level vials are used to orient many different surveying instruments with respect to the direction of gravity. There are two basic types: the *tube* vial and the *circular* or so-called "bull's-eye" version. Tube vials are used on tilting levels (and also on the older dumpy levels) to precisely orient the line of sight horizontal prior to making rod readings. Bull's-eye vials are also used on tilting levels, and on automatic levels for quick, rough leveling, after which precise final leveling occurs. The principles of both types of vials are identical.

A tube level is a glass tube manufactured so that its upper inside surface precisely conforms to an arc of a given radius (see Figure 4.10). The tube is sealed at both ends, and except for a small air bubble, it is filled with a sensitive liquid. The liquid must be nonfreezing, quick acting, and maintain a bubble of relatively stable length for normal temperature variations. Purified synthetic alcohol is generally used. As the tube is tilted, the bubble moves, always to the highest point in the

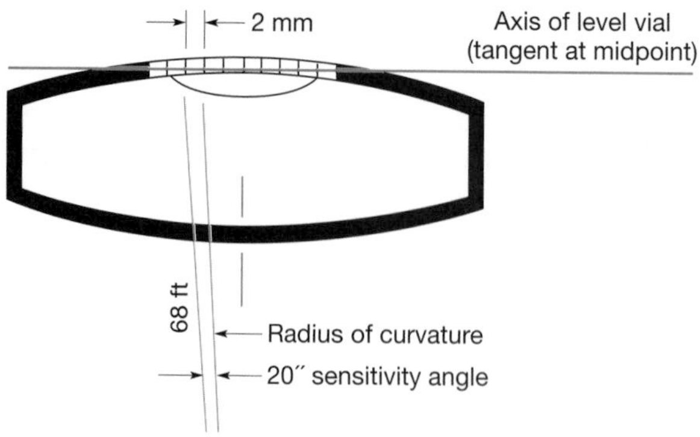

Figure 4.10
Tube-type level vial.

tube because air is lighter than the liquid. Uniformly spaced graduations etched on the tube's exterior surface, and spaced 2 mm apart, locate the bubble's relative position. The *axis of the level vial* is an imaginary longitudinal line tangent to the upper inside surface at its midpoint. When the bubble is centered in its run, the axis should be a horizontal line, as in Figure 4.10. For a leveling instrument that uses a level vial, if it is in proper adjustment, its line of sight is parallel to its level vial axis. Thus by centering the bubble, the line of sight is made horizontal.

Its radius of curvature established in manufacture determines the sensitivity of a level vial; the larger the radius, the more sensitive a bubble. A highly sensitive bubble, necessary for precise work, may be a handicap in rough surveys because more time is required to center it.

A properly designed level has a vial sensitivity correlated with the *resolving power* (resolution) of its telescope. A slight movement of the bubble should be accompanied by a small but discernible change in the observed rod reading at a distance of about 200 ft. Sensitivity of a level vial is expressed in two ways: (1) the angle, in seconds, subtended by one division on the scale, and (2) the radius of the tube's curvature. If one division subtends an angle of 20″ at the center, it is called a 20″ bubble. A 20″ bubble on a vial with 2-mm division spacings has a radius of approximately 68 ft.[4] The sensitivity of level vials on most tilting levels (and the older dumpy levels) ranges from approximately 20 to 40″.

Figure 4.11 illustrates the *coincidence-type tube level* vial used on precise equipment. A prism splits the image of the bubble and makes the two ends visible

---

[4]The relationship between sensitivity and radius is readily determined. In radian measure, an angle θ subtended by an arc whose radius and length are R and S, respectively, is given as

$$\theta = \frac{S}{R}$$

Thus for a 20″ bubble with 2-mm vial divisions, by substitution,

$$\frac{20''}{206{,}265''/rad} = \frac{2\ mm}{R}$$

Solving for R,

$$R = \frac{2\ mm \times 206{,}265''/rad}{20''} = 20{,}625\ mm = 20.6\ m = 68\ ft\ (approx.)$$

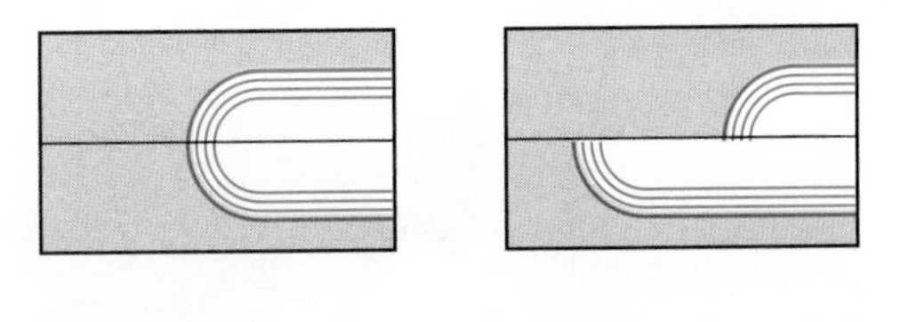

Figure 4.11 Coincidence-type level vial correctly set in left view; twice the deviation of the bubble shown in the right view.

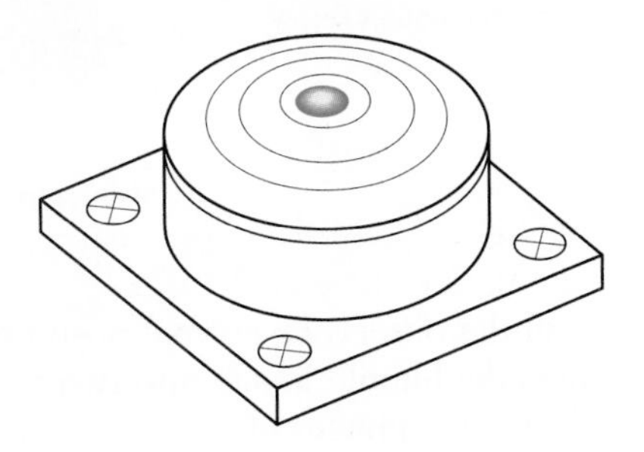

Figure 4.12 Bull's-eye level vial.

simultaneously. Bringing the two ends together to form a smooth curve centers the bubble. This arrangement enables bubble centering to be done more accurately.

Bull's-eye level vials are spherical in shape (see Figure 4.12), the inside surface of the sphere being precisely manufactured to a specific radius. Like the tube version, except for an air bubble, bull's-eye vials are filled with liquid. The vial is graduated with concentric circles having 2-mm spacings. Its axis is actually a plane tangent to the radius point of the graduated concentric circles. When the bubble is centered in the smallest circle, the axis should be horizontal. Besides their use for rough leveling of tilting and automatic levels, bull's-eye vials are also used on total station instruments, tribrachs, rod levels, prism poles, and many other surveying instruments. Their sensitivity is much lower than that of tube vials—generally in the range of from 2′ to 25′ per 2-mm division.

■ 4.9 TILTING LEVELS

Tilting levels are used for the most precise work. With these instruments, an example of which is shown in Figure 4.13, quick approximate leveling is achieved using a bull's-eye bubble and the leveling screws. On some tilting levels, a ball-and-socket arrangement (with no leveling screws) permits the head to be tilted and quickly locked nearly level. Precise level in preparation for readings is then obtained by carefully centering a telescope bubble. This is done for each sight, after aiming at the rod, by tilting or rotating the telescope slightly in a vertical plane about a fulcrum at the vertical axis of the instrument. A micrometer screw under the eyepiece controls this movement.

The tilting feature saves time and increases accuracy, since only one screw need be manipulated to keep the line of sight horizontal as the telescope is turned about a vertical axis. The telescope bubble is viewed through a system of prisms

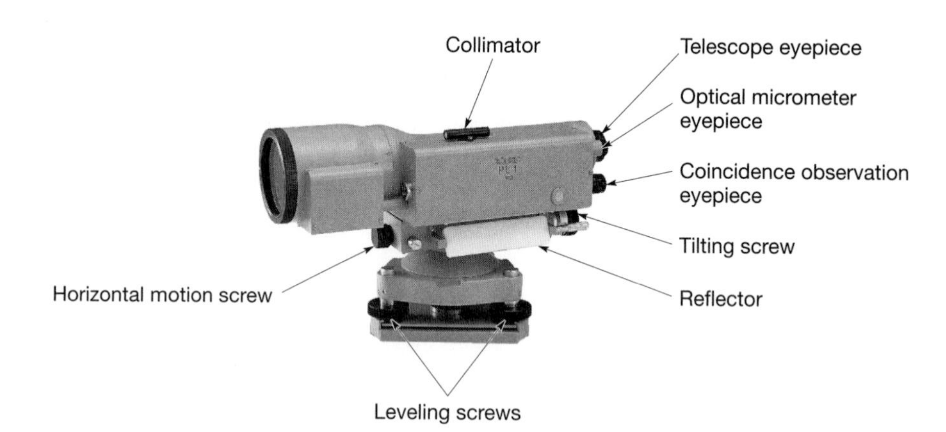

Figure 4.13 Parts of a precise tilting level. (Courtesy Sokkia Corporation)

from the observer's normal position behind the eyepiece. A prism arrangement splits the bubble image into two parts. Centering the bubble is accomplished by making the images of the two ends coincide, as in Figure 4.11.

The tilting level shown in Figure 4.13 has a three-screw leveling head, 42× magnification, and sensitivity of the level vial equal to 10″/2 mm.

■ 4.10 AUTOMATIC LEVELS

Automatic levels of the type pictured in Figure 4.14 incorporate a self-leveling feature. Most of these instruments have a three-screw leveling head, which is used to quickly center a bull's-eye bubble, although some models have a ball-and-socket arrangement for this purpose. After the bull's-eye bubble is manually centered, an *automatic compensator* takes over, levels the line of sight, and keeps it level.

The operating principle of one type of automatic compensator used in automatic levels is shown schematically in Figure 4.15. The system consists of prisms suspended from wires to create a pendulum. The wire lengths, support locations, and nature of the prisms are such that only horizontal rays reach the intersection of crosshairs. Thus a horizontal line of sight is achieved even though the telescope itself may be slightly tilted away from horizontal. Damping devices shorten the time for the pendulum to come to rest, so the operator does not have to wait.

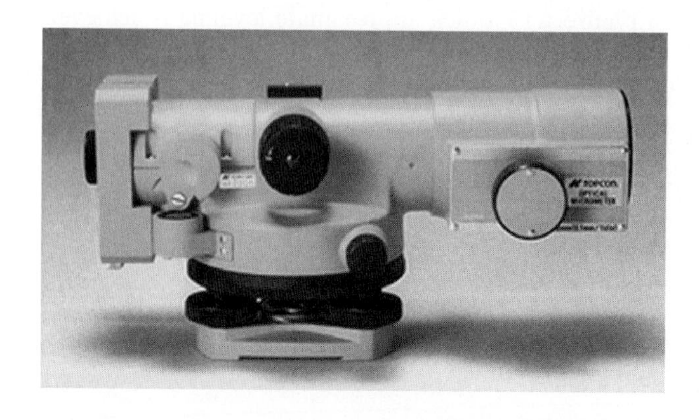

Figure 4.14 Automatic level with micrometer. (Topcon Corporation)

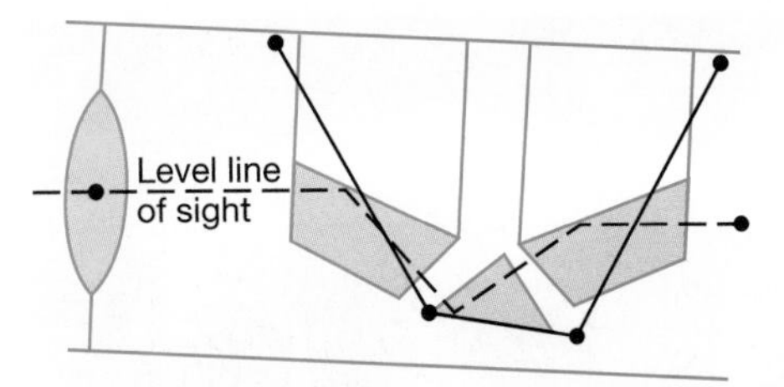

When telescope tilts up, compensator swings backward.

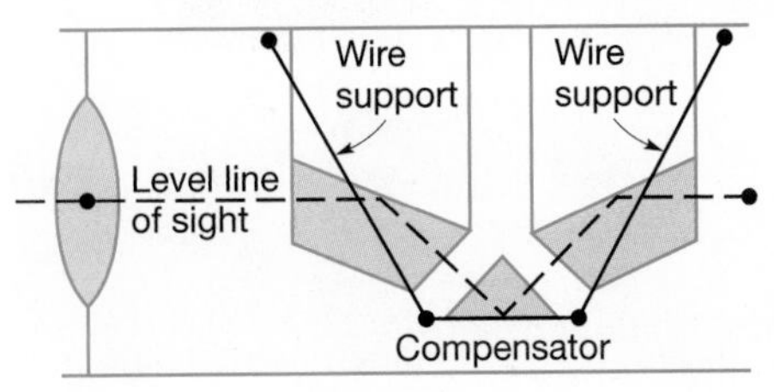

Telescope horizontal

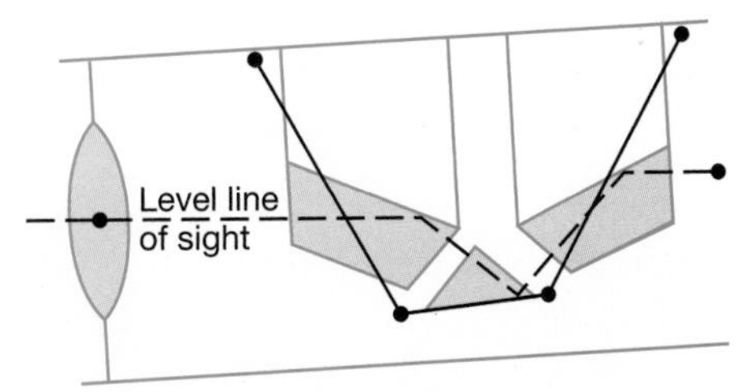

When telescope tilts down, compensator swings forward.

Figure 4.15 Compensator of self-leveling level. (Courtesy Keuffel & Esser Company)

Automatic levels have become popular for general use because of the ease and rapidity of their operation. Some are precise enough for second-order and even first-order work if a *parallel-plate micrometer* is attached to the telescope front as an accessory, as with the instrument shown in Figure 4.14. When the micrometer plate is tilted, the line of sight is displaced parallel to itself, and decimal parts of rod graduations can be measured by means of a graduated dial.

Under certain conditions, the damping devices of an automatic level compensator can stick. To check, with the instrument leveled and focused, read the rod held on a stable point, lightly tap the instrument, and after it vibrates, determine whether the same reading is obtained. Also, some unique compensator problems, such as residual stresses in the flexible links, can introduce systematic errors if not corrected by an appropriate observational routine on first-order work. Another problem is that some automatic compensators are affected by magnetic fields, which result in systematic errors in rod readings. The sizes of the errors are azimuth-dependent, maximum for lines run north and south, and can exceed 1 mm/km. Thus, it is of concern only for high-order control leveling.

■ 4.11 DIGITAL LEVELS

The newest type of automatic level, the *electronic digital level,* is pictured in Figure 4.16(a). It is classified in the automatic category because it uses a pendulum compensator to level itself, after an operator accomplishes rough leveling with

(a)

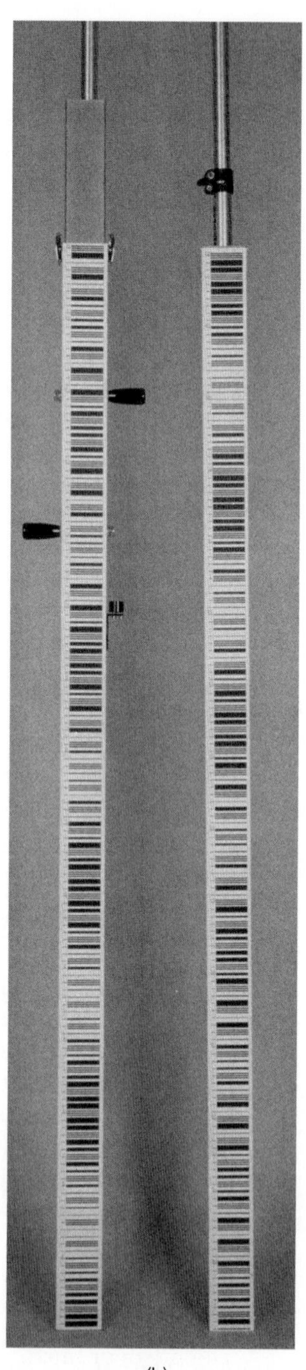

(b)

Figure 4.16 (a) Electronic digital level and (b) associated level rod. (Courtesy Topcon Corporation)

a bull's-eye bubble. With its telescope and crosshairs, the instrument could be used to obtain readings manually, just like any of the automatic levels. This instrument, however, is designed to operate by employing electronic digital image processing. After leveling the instrument, its telescope is turned toward a special bar-coded rod [Figure 4.16(b)] and focused. At the press of a button, the image of bar codes in the telescope's field of view is captured and processed. This processing consists of an on-board computer comparing the captured image to the rod's entire pattern, which is stored in memory. When a match is found, the rod reading is displayed digitally. It can be recorded manually or automatically stored in the instrument's data collector.

The length of rod appearing within the telescope's field of view is a function of the distance from the rod. Thus as a part of its image processing, the instrument is also able to automatically compute the sight length, a feature convenient for balancing backsight and foresight lengths (see Section 5.4). The instrument's maximum range is approximately 100 m, and its accuracy in rod readings is ±0.5 mm. The bar-coded rods can be obtained with English or metric graduations on the side opposite the bar code. The graduated side of the rod can be used by the operator to manually read the rod in situations that prohibit the instrument from reading the bar codes such as when the rod is in heavy brush.

■ 4.12 TRIPODS

Leveling instruments, whether tilting, automatic, or digital, are all mounted on tripods. A sturdy tripod in good condition is essential to obtain accurate results. Several types are available. The legs are made of wood or metal, may be fixed or adjustable in length, and solid or split. All models are shod with metallic conical points and hinged at the top, where they connect to a metal head. An adjustable-leg tripod is advantageous for setups in rough terrain or in a shop, but the type with a fixed-length leg may be slightly more rigid. The split-leg model is lighter than the solid type, but less rugged. (Adjustment of tripods is covered in Section 8.19.2.)

■ 4.13 HAND LEVEL

The hand level (Figure 4.17) is a hand-held instrument used on low-precision work, or to obtain quick checks on more precise work. It consists of a brass tube approximately 6 in. long, having a plain glass objective and peep-sight eyepiece. A small level vial mounted above a slot in the tube is viewed through the eyepiece by means of a prism or 45° angle mirror. A horizontal line extends across the tube's center.

The prism or mirror occupies only one-half of the tube, and the other part is open to provide a clear sight through the objective lens. Thus the rod being sighted, and the reflected image of the bubble, is visible beside each other with the horizontal cross line superimposed.

The instrument is held in one hand and leveled by raising or lowering the objective end until the cross line bisects the bubble. Resting the level against a

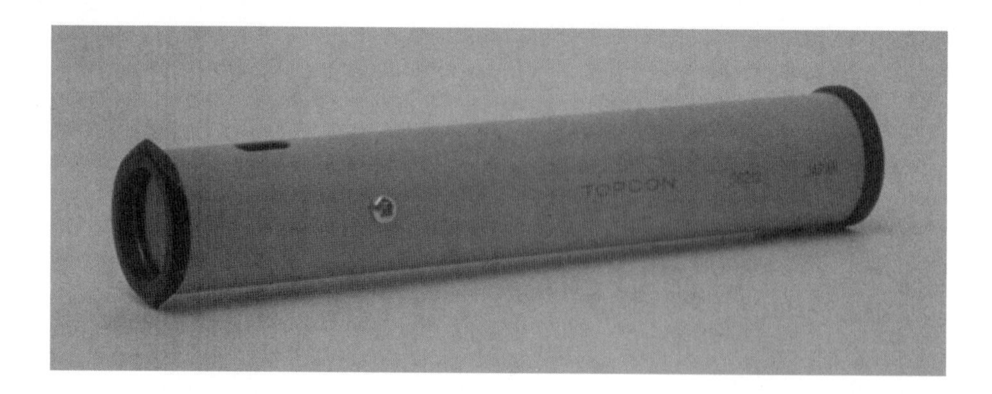

Figure 4.17 Hand level. (Courtesy Topcon Corporation)

rod or staff provides stability and increases accuracy. This instrument is especially valuable in quickly checking proposed locations for instrument setups in differential leveling.

■ 4.14 LEVEL RODS

A variety of level rods are available, some of which are shown in Figure 4.18. They are made of wood, fiberglass, or metal and have graduations in feet and decimals, or meters and decimals.

The Philadelphia rod, shown in Figure 4.18(a) and (b), is the type most commonly used in college surveying classes. It consists of two sliding sections graduated in hundredths of a foot and joined by brass sleeves *a* and *b*. The rear section can be locked in position by a clamp screw *c* to provide any length from a *short rod* for readings of 7 ft or less to a *long rod* (*high rod*) for readings up to 13 ft. *When the high rod is needed, it must be fully extended, otherwise a serious mistake will result in the reading.* Graduations on the front faces of the two sections read continuously from zero at the base to 13 ft at the top for the high-rod setting.

Rod graduations are accurately painted, alternate black and white spaces 0.01 ft wide. Spurs extending the black painting emphasize the 0.1-ft and 0.05-ft marks. Tenths are designated by black figures, and footmarks by red numbers, all straddling the proper graduation. Rodpersons should keep their hands off the painted markings, particularly in the 3- to 5-ft section, where a worn face will make the rod unfit for use. A Philadelphia rod can be read accurately with a level at distances up to about 250 ft.

A wide choice of patterns, colors, and graduations on single-piece, two-piece, three-section, and four-section leveling rods is available. The various types, usually named for cities or states, include the Philadelphia, New York, Boston, Troy, Chicago, San Francisco, and Florida rods.

Philadelphia rods can be equipped with targets [*d* in Figure 4.18(a) and (b)] for use on long sights. When employed, the rodperson sets the target at the instrument's line-of-sight height according to hand signals from the instrument operator. It is fixed using clamp *e,* then read and recorded by the rodperson. The *vernier,* at *f*, can be used to obtain readings to the nearest 0.001 ft if desired. (Verniers are described in Section A.4.2.)

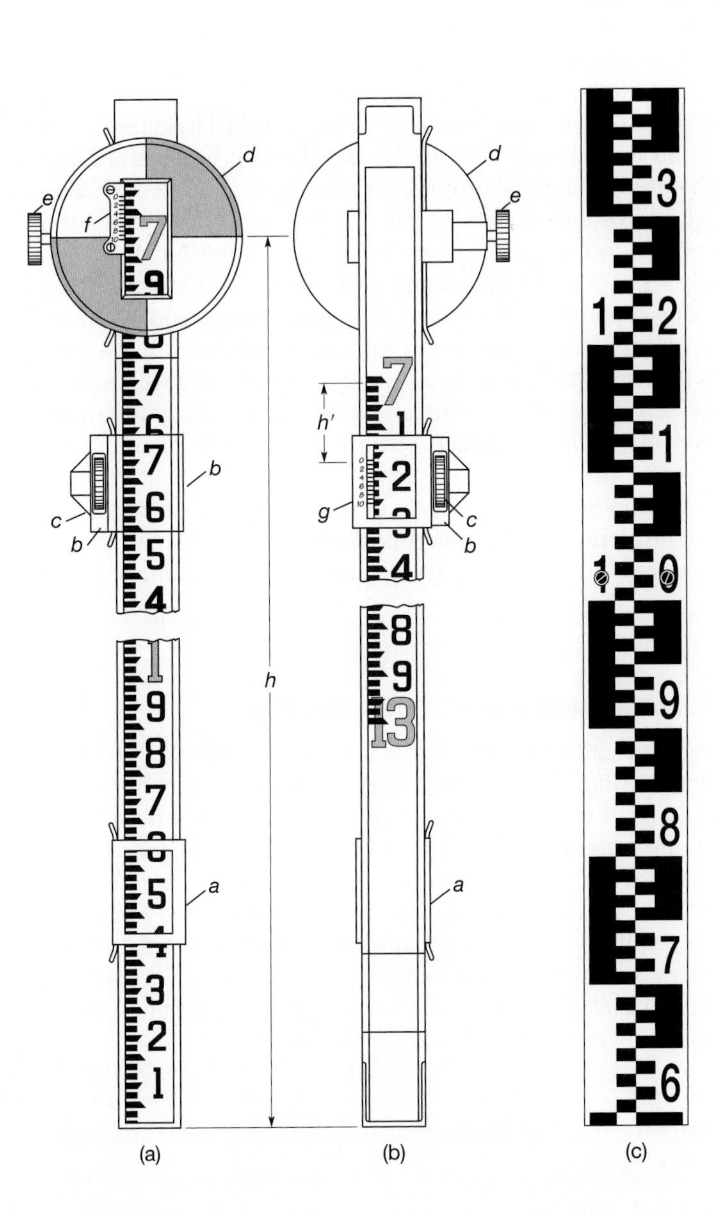

(d)

Figure 4.18
(a) Philadelphia rod (front). (b) Philadelphia rod (rear). (c) Double-faced leveling rod with metric graduations. (Courtesy Leica, Inc.) (d) Lenker direct-reading rod. (Courtesy Lenker Manufacturing Company)

The Chicago rod, consisting of independent sections (usually three) that fit together but can be disassembled, is widely used on construction surveys. The San Francisco model has separate sections that slide past each other to extend or compress its length, and is generally employed on control, land, and other surveys. Both are conveniently transported in vehicles.

The direct-reading Lenker level rod [Figure 4.18(d)] has numbers in reverse order on an endless graduated steel-band strip that can be revolved on the rod's end rollers. Figures run down the rod and can be brought to a desired reading—for example, the elevation of a benchmark. Rod readings are preset for the backsight, and then, due to the reverse order of numbers, foresight readings give elevations directly without manually adding backsights and subtracting foresights.

A rod consisting of a wooden, or fiberglass, frame and an Invar strip to eliminate the effects of humidity and temperature changes is used on precise work. The Invar strip, attached at its ends only, is free to slide in grooves on each side of the wooden frame. Rods for precise work are usually graduated in meters and often have dual scales. Readings of both scales are compared to eliminate mistakes.

As described in Section 1.8, safety in traffic and near heavy equipment is an important consideration. The Quad-pod, an adjustable stand that clamps to any leveling rod, can help to reduce traffic hazards, and in some cases also lower labor costs.

■ 4.15 TESTING AND ADJUSTING LEVELS

Through normal use and wear, all leveling instruments will likely become maladjusted from time to time. The need for some adjustments may be noticed during use, for example level vials on tilting levels. Others may not be so obvious, and therefore it is important that instruments be periodically checked to determine their state of adjustment. If the tests reveal conditions that should be adjusted, depending upon the particular instrument, and the knowledge and experience of its operator, some or all of the adjustments can be made immediately in the field. However, if the parts needing adjustment are not readily accessible, or if the operator is inexperienced in making the adjustments, it is best to send the instruments away for adjustment by qualified technicians.

4.15.1 Requirements for Testing and Adjusting Instruments

Before testing and adjusting instruments, care should be exercised to ensure that any apparent lack of adjustment is actually caused by the instrument's condition and not by test deficiencies. To properly test and adjust leveling instruments in the field, the following rules should be followed:

1. Choose terrain that permits solid setups in a nearly level area enabling sights of at least 200 ft to be made in opposite directions.
2. Perform adjustments when good atmospheric conditions prevail, preferably on cloudy days free of heat waves. No sight line should pass through alternate sun and shadow, or be directed into the sun.

3. Place the instrument in shade, or shield it from direct rays of the sun.
4. Make sure the tripod shoes are tight and the instrument is firmly screwed onto the tripod. Spread the tripod legs well apart and position them so that the tripod plate is nearly level. Press the shoes firmly into the ground.

Standard methods and a prescribed order must be followed in adjusting surveying instruments. Loosening or tightening the proper adjusting nuts and screws with special tools and pins attains correct positioning of parts. Time is wasted if each adjustment is perfected on the first trial, since some adjustments affect others. The complete series of tests may have to be repeated several times if an instrument is badly off. A final check of all adjustments should be made to ensure that all have been satisfactorily completed.

Tools and adjusting pins that fit the capstans and screws should be used, and the capstans and screws handled with care to avoid damaging the soft metal. Adjustment screws are properly set when an instrument is shipped from the factory. Tightening them too much (or not enough) nullifies otherwise correct adjustment procedures and may leave the instrument in worse condition than it was before adjusting.

4.15.2 Adjusting for Parallax

The parallax adjustment is extremely important and must be kept in mind at all times when using a leveling instrument, but especially during the testing and adjustment process. The adjustment is done by carefully focusing the objective lens and eyepiece so that the crosshairs appear clear and distinct, and so that the crosshairs do not appear to move against a background object when the eye is shifted slightly in position while viewing through the eyepiece.

4.15.3 Testing and Adjusting Level Vials

For leveling instruments that employ a level vial, the axis of the level vial should be perpendicular to the vertical axis of the instrument (axis about which the instrument rotates in azimuth). Then once the bubble is centered, the instrument can be turned about its vertical axis in any azimuth and the bubble will remain centered. Centering the bubble and revolving the telescope 180° about the vertical axis can quickly check this condition. The distance the bubble moves off the central position is twice the error. To correct any maladjustment, turn the capstan nuts at one end of the level vial to move the bubble *halfway back* to the centered position. Level the instrument using the leveling screws. Repeat the test until the bubble remains centered during a complete revolution of the telescope.

4.15.4 Preliminary Adjustment of the Horizontal CrossHair

Although it is good practice to always sight an object at the center of the crosshairs, if this is not done and the horizontal crosshair is not truly horizontal when the instrument is leveled, an error will result. To test for this condition, sight a sharply defined point with one end of the horizontal crosshair. Turn the telescope slowly on its vertical axis so that the crosshair moves across the point. If the crosshair does not remain on the point for its full length, it is out of adjustment. To correct any maladjustment, loosen the four capstan screws holding the reticle. Rotate the reticle in the telescope tube until the horizontal hair remains on the point as the telescope is turned. The screws should then be carefully tightened in their final position.

4.15.5 Testing and Adjusting the Line of Sight

For tilting levels, described in Section 4.9, when the bubble of the level vial is centered, the line of sight should be horizontal. In other words, for this type of instrument to be in perfect adjustment, the axis of the level vial and the line of sight must be parallel. If they are not, a *collimation error* exists. For the automatic levels, described in Section 4.10, after rough leveling by centering the bull's-eye bubble, the automatic compensator must define a horizontal line of sight if it is in proper adjustment. If it does not, the compensator is out of adjustment, and again a collimation error exists. *The collimation error will not cause errors in differential leveling as long as backsight and foresight distances are balanced.* However, it will cause errors when backsights and foresights are not balanced, which sometimes occurs in differential leveling, and cannot be avoided in profile leveling (see Section 5.9), and construction staking (see Chapter 23).

One method of testing a level for collimation error is to stake-out four equally spaced points, each about 100 ft apart on approximately level ground as shown in Figure 4.19. The level is then set up at point 1, leveled, and rod readings (r_A) at A, and (R_B) at B are taken. Next the instrument is moved to point 2 and releveled. Readings R_A at A, and r_B at B are then taken. As illustrated in the figure, assume that a collimation error ε exists in the rod readings of the two shorter sights. Then the error caused by this source would be 2ε in the longer sights because their length

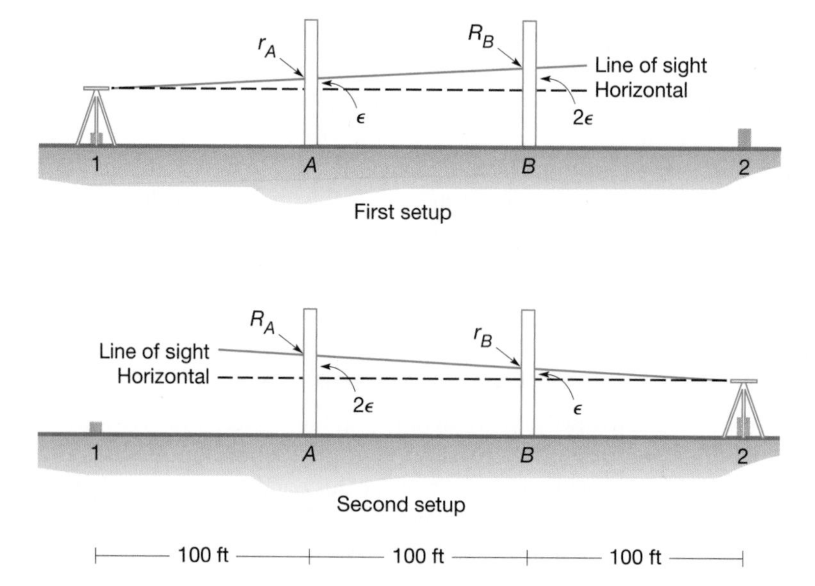

Figure 4.19 Horizontal collimation test.

is double that of the shorter ones. Whether or not there is a collimation error, the difference between the rod readings at 1 should equal the difference of the two readings at 2. Expressing this equality, with the collimation error included, gives

$$(R_B - 2\varepsilon) - (r_A - \varepsilon) = (r_B - \varepsilon) - (R_A - 2\varepsilon) \quad (4.13)$$

Solving for ε in Equation (4.13) yields

$$\varepsilon = \frac{R_B - r_A - r_B + R_A}{2} \quad (4.14)$$

The corrected reading for the level rod at point A while the instrument is still set up at point 2 should be $R_A - 2\varepsilon$. If an adjustment is necessary, it is done by loosening the top (or bottom) screw holding the reticle, and tightening the bottom (or top) screw to move the horizontal hair up or down until the required reading is obtained on the rod at A. This changes the orientation of the line of sight. Several trials may be necessary to achieve the exact setting. If the reticle is not accessible, or the operator is unqualified, then the instrument should be serviced by a qualified technician.

Example 4.3

A horizontal collimation test is performed on an automatic level following the procedures just described. With the instrument set up at point 1, the rod reading at A was 5.630 ft, and to B was 5.900 ft. After moving and leveling the instrument at point 2, the rod reading to A was determined to be 5.310 ft and to B 5.560 ft. What is the collimation error of the instrument, and the corrected reading to A from point 2?

Solution

Substituting the appropriate values into Equation (4-14), the collimation error is

$$\varepsilon = \frac{5.900 - 5.630 - 5.560 + 5.310}{2} = 0.010 \text{ ft}$$

Thus the corrected reading to A from point 2 is

$$R = 5.310 - 2 \times 0.010 = 5.290 \text{ ft}$$

As noted previously, if a collimation error exists but the instrument is not adjusted, accurate differential leveling can still be achieved when the backsight and foresight lengths are balanced. In situations where these lengths cannot be balanced, correct rod readings can still be obtained by applying collimation corrections to the rod readings. This procedure is described in Section 5.12.1.

PROBLEMS

Asterisks (*) indicate problems that have answers given in Appendix G.

4.1 Define the following leveling terms: (a) level surface, (b) elevation, and (c) geoid.

4.2* How far will a horizontal line depart from the Earth's surface in 2 km? 10 km? (Apply both curvature and refraction.)

4.3 Similar to Problem 4.2, except for 1-mi, 2-mi, and 5-mi lengths.

4.4 Visit the website of the National Geodetic Survey, and obtain a data sheet description of a benchmark in your local area.

4.5 Why is it important for a benchmark to be a stable, relatively permanent object?

4.6 Compute and tabulate the combined effect of curvature and refraction on level sights of 100, 200, 400, 800, and 1500 ft.

4.7 Similar to Problem 4.6, except for sights of 30, 100, 200, 400, 800, and 1000 m.

4.8* On a large lake without waves, how far from shore is a sailboat when the top of its 20-ft mast disappears from the view of a person lying at the water's edge?

4.9 Similar to Problem 4.8, except for a 10-m mast and a person whose eye height is 1.0 m above the water's edge.

4.10 Readings on a line of differential levels are taken to the nearest 1 mm. For what maximum distance can the Earth's curvature and refraction be neglected?

4.11 Similar to Problem 4.10, except readings are to the 0.01 ft.

Successive plus and minus sights taken on a downhill line of levels are listed in Problems 4.12 and 4.13. The values represent the horizontal distances between the instrument and either the plus or minus sights. What error results from curvature and refraction?

4.12* 50, 75; 75, 150; 25, 150; 30, 200 ft.

4.13 10, 50; 15, 40; 5, 50; 20, 75 m.

4.14 What error results if the curvature and refraction correction is neglected in trigonometric leveling for sights:
(a) 2000 ft long **(b)** 1000 m long **(c)** 4500 ft long?

4.15* The slope distance and zenith angle measured from point P to point Q were 1876.253 m and 87°33′24″, respectively. The instrument and rod target heights were equal. If the elevation of point P is 227.965 m, above datum, what is the elevation of point Q?

4.16 The slope distance and zenith angle measured from point X to point Y were 12,256.18 ft and 93°13′46″. The instrument and rod target heights were equal. If the elevation of point X is 845.08 ft above datum, what is the elevation of point Y?

4.17 Similar to Problem 4.15, except the slope distance was 1832.316 m, the zenith angle was 84°23′16″, and the elevation of point P was 487.623 m above datum.

4.18 In trigonometric leveling from point A to point B, the slope distance and zenith angle measured at A were 3852.603 m and 97°03′58″. At B these measurements were 3852.596 m and 82°55′56″ respectively. If the instrument and rod target heights were equal, calculate the difference in elevation from A to B.

4.19 Describe how parallax in the viewing system of a level can be detected and removed.

4.20 What is the sensitivity of a level vial with 2-mm divisions for: **(a)** a radius of 20 m **(b)** a radius of 41 m?

4.21* An observer fails to check the bubble, and it is off two divisions on a 150-ft sight. What error in elevation difference results with a 20-sec bubble?

4.22 An observer fails to check the bubble and it is off two divisions on a 200-m sight. What error results for a 10″ bubble?

4.23 Similar to Problem 4.22, except a 20″ bubble is off three divisions on a 150-m sight.

4.24 With the bubble centered, a 100-m-length sight gives a reading of 1.352 m. After moving the bubble four divisions off center, the reading is 1.410 m. For 2-mm vial divisions, what is: (**a**) the vial radius of curvature in meters and (**b**) the angle in seconds subtended by one division?

4.25 Similar to Problem 4.24, except the sight length was 300 ft, the initial reading was 5.132 ft, and the final reading was 5.250 ft.

4.26 Sunshine on the forward end of a 20″/2 mm level vial bubble draws it off two divisions, giving a plus sight reading of 4.362 m on a 150-m shot. Compute the correct reading.

4.27 List in tabular form, for comparison, the advantages and disadvantages of an automatic level versus a digital level.

4.28 Describe a field procedure that can be used to identify parallax.

4.29* If a plus sight of 1.097 m is taken on BM *A*, elevation 312.547 m, and a minus sight of 0.832 m is read on point *X*, calculate the HI and the elevation of point *X*.

4.30 Similar to Problem 4.29, except a plus sight of 12.03 ft is taken on BM *A* and a minus sight of 5.43 ft read on point *X*. (Assume numbers in 4.29 are in feet.)

4.31 Describe the procedure used to test if the level vial is perpendicular to the vertical axis of the instrument.

4.32 A horizontal collimation test is performed on an automatic level following the procedures described in Section 4.15.5. With the instrument set up at point 1, the rod reading at *A* was 5.261 ft, and to *B* it was 5.453 ft. After moving and leveling the instrument at point 2, the rod reading to *A* was 4.940 ft and to *B* was 5.116 ft. What is the collimation error of the instrument, and the corrected reading to *B* from point 2?

4.33 Similar to Problem 4.32 except that the rod readings are 1.677 m and 1.722 m to *A* and *B*, respectively, from point 1, and 1.543 m and 1.586 m to *A* and *B*, respectively, from point 2.

BIBLIOGRAPHY

Alsalman, A. 1999. "Evaluating the Accuracy of Differential, Trigonometric and GPS Leveling." *Surveying and Land Information Systems* 59 (No. 1): 47.

Fury, R. J. 1996. "Leveled Height Differences from Published NAVD 88 Orthometric Heights." *Surveying and Land Information Systems* 56 (No. 2): 89.

GIA. 2003. "Automatic Level Compensators." *Professional Surveyor* 23 (No. 3): 52.

______. 2003. "Tripod Performance in Geomatic Systems." *Professional Surveyor* 23 (No. 6): 40.

______. 2002. "Digital Levels." *Professional Surveyor* 22 (No. 1): 44.

Henning, W. et. al. 1998. "Baltimore County, Maryland NAVD 88 GPS-derived Orthometric Height Project." *Surveying and Land Information Systems* 58 (No. 2): 97.

Parks, W. and Dial, T. 1996. "Using GPS to Measure Leveling Section Orthometic Height Difference in a Ground Subsidence Area in Imperial Valley, California." *Surveying and Land Information Systems* 57 (No. 2): 100.

Raabe, E. A. et. al. 1996. "A Precise Vertical Network: Establishing New Orthometric Heights with Static Surveys in Florida Tidal Marshes." *Surveying and Land Information Systems* 56 (No. 4): 200.

Radcliffe, D. 1999. "How Digital Levels Work." *Professional Surveyor* 19 (No. 5): 24.

5

Leveling–Field Procedures and Computations

■ 5.1 INTRODUCTION

Chapter 4 covered the basic theory of leveling, briefly described the different procedures used in determining elevations and showed examples of most types of leveling equipment. This chapter concentrates on differential leveling, discusses handling the equipment, running and adjusting simple leveling loops, and performing some project surveys to obtain data for field and office use. Some special variations of differential leveling, useful or needed in certain situations, are presented. Profile leveling, to determine the configuration of the ground surface along some established reference line, is described in Section 5.9. Finally, errors in leveling are discussed. Leveling procedures for construction and other surveys, along with those of higher order to establish the nationwide vertical control network, will be covered in later chapters.

■ 5.2 CARRYING AND SETTING UP A LEVEL

The safest way to transport a leveling instrument in a vehicle is to leave it in the container. The case closes properly only when the instrument is set correctly in the padded supports. A level should be removed from its container by lifting from the base, not by grasping the telescope. The head must be screwed snugly on the tripod. If the head is too loose, the instrument is unstable; if it's too tight, it may "freeze." Once the instrument is removed from the container, the container should be closed to prevent dirt and moisture from entering it.

The legs of a tripod must be tightened correctly. If each leg falls slowly of its own weight after being placed in a horizontal position, it is properly adjusted. Clamping them too tightly strains the plate and screws. If the legs are loose, unstable setups result.

Except for a few instruments that employ a ball and socket arrangement, all modern levels use a three-screw leveling head for initial rough leveling. Note that each of the levels illustrated in Chapter 4 (see Figures 4.9, 4.13, 4.14, and 4.16) has this type of arrangement. In leveling a three-screw head, the telescope is rotated until it is over two screws as in the direction AB of Figure 5.1. Using the thumb and first finger of each hand to simultaneously adjust the opposite screws approximately centers the bubble. This procedure is repeated with the telescope rotated 90° so that it is over C, the remaining single screw. Time is wasted by centering the bubble exactly on the first try, since it will be thrown off during the cross leveling. Working with the same screws in succession about three times should complete the job. A simple but useful rule in centering a bubble, illustrated in Figure 5.1, is: *A bubble follows the left thumb when turning the screws.* A bull's-eye bubble is centered by alternately turning one screw and then the other two. The telescope need not be rotated during the process.

It is generally unnecessary to set up a level over any particular point. Therefore it is inexcusable to have the base plate badly out of level before using the leveling screws. On sidehill setups, placing one leg on the uphill side and two on the downhill slope eases the problem. On very steep slopes, some instrument operators prefer two legs uphill and one downhill for stability. The most convenient height of setup is one that enables the observer to sight through the telescope without stooping or stretching.

Inexperienced instrument operators running levels up or down steep hillsides are likely to find, after completing the leveling process, that the telescope is too low for sighting the upper turning point or benchmark. To avoid this, a hand level can be used to check for proper height of the setup before precisely leveling the instrument. As another alternative, the instrument can be quickly set up without attempting to level it carefully. Then the rod is sighted making sure the bubble is somewhat back of center. If it is visible for this placement, it obviously will also be seen when the instrument is leveled. Another aid in setting up the instrument is to use a hand level to determine the approximate height of the level during setup.

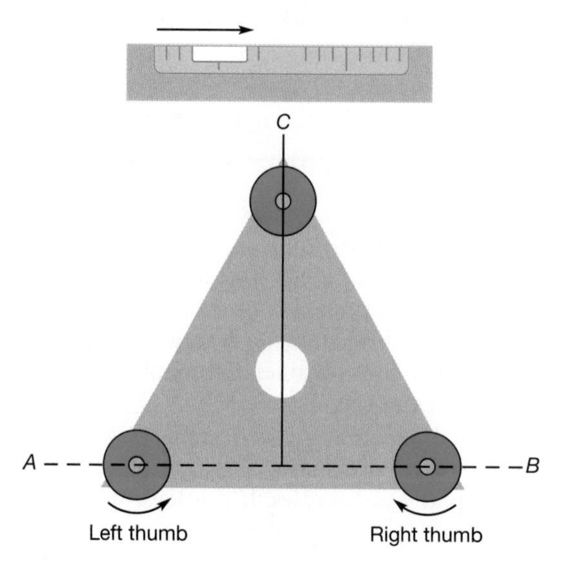

Figure 5.1 Use of leveling screws on a three-screw instrument.

■ 5.3 DUTIES OF A RODPERSON

The duties of a rodperson are relatively simple. However, a careless rodperson can nullify the best efforts of an observer by failing to follow a few basic rules.

A level rod must be held plumb on the correct monument or turning point to give the correct reading. In Figure 5.2, point A is below the line of sight by vertical distance AB. If the rod is tilted to position AD, an erroneous reading AE is obtained. It can be seen that the smallest reading possible, AB, is the correct one and is secured only when the rod is plumb.

A rod level of the type shown in Figure 5.3 ensures *fast* and *correct* rod plumbing. Its L-shape is designed to fit the rear and side faces of a rod, while the bull's-eye bubble is centered to plumb the rod in *both directions*. However if a rod level is not available, one of the following procedures can be used to plumb the rod.

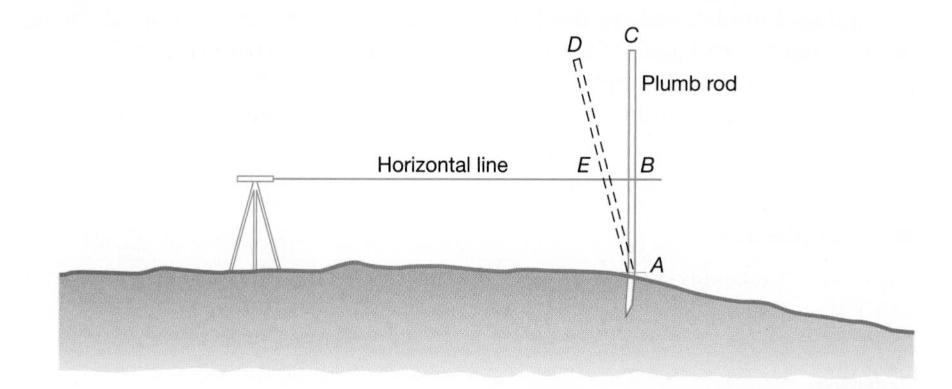

Figure 5.2
Plumbing a level rod.

Figure 5.3
Rod level.

Waving the rod is one procedure that can be used to ensure that the rod is plumb when a reading is taken. The process consists of *slowly* tilting the rod top, first perhaps a foot or two toward the instrument and then just slightly away from it. The observer watches the readings alternately increase and decrease, and then selects the minimum value, the correct one. Beginners tend to swing the rod too fast and through too long an arc. Small errors can be introduced in the process if the bottom of the rod is resting on a flat surface. A rounded-top monument, steel spike, or thin edge makes an excellent benchmark or intermediate point for leveling.

On still days the rod can be plumbed by letting it balance of its own weight while lightly supported by the fingertips. An observer makes certain the rod is plumb in the lateral direction by checking its coincidence with the vertical wire and signals for any adjustment necessary. The rodperson can save time by sighting along the side of the rod to line it up with a telephone pole, tree, or side of a building. Plumbing along the line toward the instrument is more difficult, but holding the rod against the toes, stomach, and nose will bring it close to a plumb position. A plumb bob suspended alongside the rod can also be used, and in this procedure the rod is adjusted in position until its edge is parallel with the string.

Example 5.1

In Figure 5.2, what error results if the rod is held in position AD, and if $AE = 10$ ft and $EB = 6$ in.?

Solution

Using the Pythagorean theorem, the vertical rod is

$$AB = \sqrt{10^2 - 0.5^2} = 9.987 \text{ ft}$$

Thus the error is $10.00 - 9.987 = 0.013$ ft, or 0.01 ft.

Errors of the magnitude of Example 5.1 are serious, whether the results are carried out to hundredths or thousandths. They make careful plumbing necessary, particularly for high-rod readings.

5.4 DIFFERENTIAL LEVELING

Figure 5.4 illustrates the procedure followed in differential leveling. In the figure, the elevation of new BM Oak is to be determined by originating a leveling circuit at established BM Mil. In running this circuit, the first reading, a *plus sight,* is taken on the established benchmark. From it, the HI can be computed using Equation (4.4). Then a *minus sight* is taken on the first intermediate point (called a *turning point,* and labeled TP 1 in the figure), and by Equation (4.5) its elevation

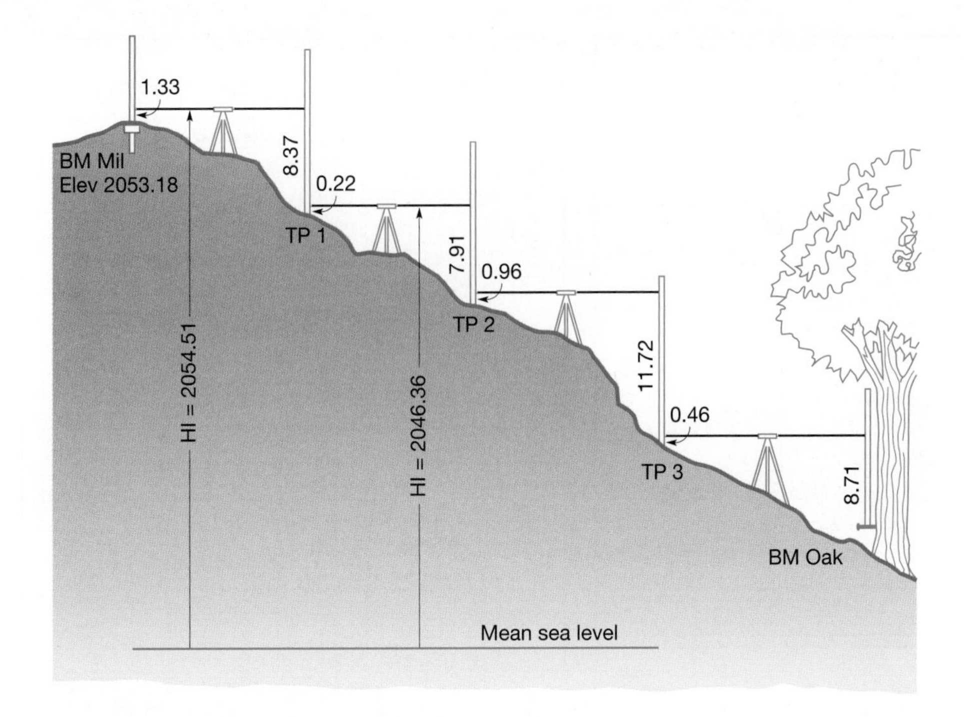

Figure 5.4
Differential leveling.

is obtained. The process of taking a plus sight, followed by a minus sight, is repeated over and over until the circuit is completed.

As shown in the example of Figure 5.4, four instrument setups were required to complete half of the circuit (the run from BM Mil to BM Oak). Field notes for the example of Figure 5.4 are given in Figure 5.5. As illustrated in this figure, a tabular form of field notes is used for differential leveling and the addition and subtraction to compute HIs and elevations is done directly in the notes. These notes also show the data for the return run from BM Oak back to BM Mil to complete the circuit. It is important in differential leveling to run closed circuits so that the accuracy of the work can be checked, as will be discussed later.

As noted, the intermediate points upon which the rod is held in running a differential leveling circuit are called *turning points* (*TPs*). Two rod readings are taken on each, a minus sight followed by a plus sight. Turning points should be solid objects with a definite high point. Careful selection of stable turning points is essential to achieve accurate results. Steel turning pins and railroad spikes driven into firm ground make excellent turning points when permanent objects are not conveniently available.

In differential leveling, horizontal lengths for the plus and minus sights should be made approximately equal. This can be done by pacing, by stadia measurements (see Section 17.9.2), by counting rail lengths or pavement joints if working along a track or roadway, or by any other convenient method. Balancing plus and minus sights will eliminate errors due to instrument maladjustment (most important) and the combined effects of the Earth's curvature and refraction, as shown in Figure 5.6. Here e_1 and e_2 are the combined curvature and refraction errors for

DIFFERENTIAL LEVELS

Sta.	+ B.S.	H.I.	− F.S.	Elev.	Adj. Elev.
BM Mil.	1.33			2053.18	2053.18
		2054.51			
TP1	0.22		8.37	2046.14	2046.14
		2046.36	7.91		
TP2	0.96		~~8.91~~	2038.45	2038.44
		2039.41			
TP3	0.46		11.72	2027.69	2027.68
		2028.15			
BM Oak	11.95		8.71	2019.44	2019.42
		2031.39			
TP4	12.55		2.61	2028.78	2028.76
		2041.33			
TP5	12.77		0.68	2040.65	2040.62
		2053.42			
BM Mil.			0.21	2053.21	2053.18
Σ =	+40.24	Σ =	−40.21		
		Page Check:			
		2053.18			
		+ 40.24			
		2093.42			
		− 40.21			
		2053.21	Check		

GRAND LAKES UNIV. CAMPUS

BM Mil. to BM Oak

BM Mil. on GLU Campus	29 Sept. 2000
SW of Engineering Bldg.	Clear, Warm 70° F
9.4 ft. north of sidewalk	T.E. Henderson N
to instrument room and	J.F. King ⌽
1.6 ft. from Bldg. Bronze	D.R. Moore ⊼
disk in concrete flush	Lietz Level #6
with ground, stamped "Mil"	

BM Oak is a temporary project bench mark located at corner of Cherry and Pine Sts., 14 ft. West of computer laboratory. Twenty penny spike in 18″ Oak tree, 1 ft. above ground.

Loop Misclosure = 2053.21 − 2053.18 = 0.03

Permissible Misclosure = $0.02\sqrt{n} = 0.02\sqrt{7}$ = 0.05 ft.

Adjustment = $\frac{0.03}{7}$ = 0.004′ per H.I.

J.E. Henderson

Figure 5.5 Differential leveling notes for Figure 5.4.

the plus and minus sights, respectively. If D_1 and D_2 are made equal, e_1 and e_2 are also equal. In calculations, e_1 is added and e_2 subtracted; thus, they cancel each other.

Figure 5.6 can also be used to illustrate the importance of balancing sight lengths if a collimation error exists in the instrument's line of sight. This condition exists if, after leveling the instrument, its line of sight is not horizontal. For example,

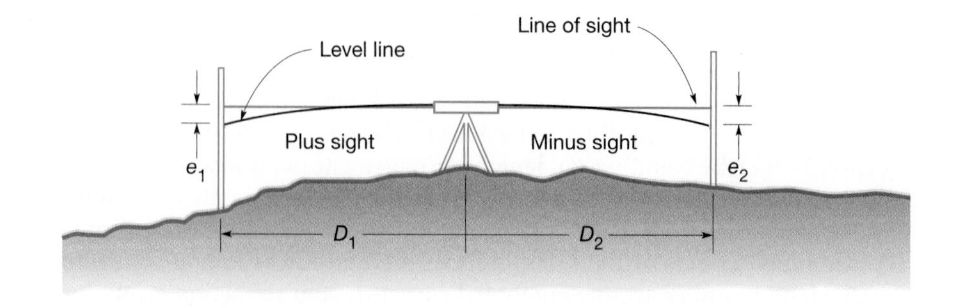

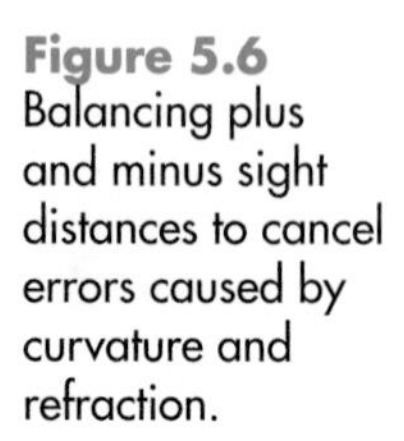

Figure 5.6 Balancing plus and minus sight distances to cancel errors caused by curvature and refraction.

suppose in Figure 5.6 that because the line of sight is systematically directed below horizontal, an error e_1 results in the plus sight. But if D_1 and D_2 are made equal, an error e_2 (equal to e_1) will result on the minus sight and the two will cancel, thus eliminating the effect of the instrumental error. On slopes it may be somewhat difficult to balance lengths of plus and minus sights, but following a zigzag path can do it usually. It should be remembered that Earth curvature, refraction, and collimation errors are systematic and will accumulate in long leveling lines if care is not taken to balance the backsight and foresight distances.

A benchmark is described in the field book the first time used, and thereafter by noting the page number on which it was recorded. Descriptions begin with the general location and must include enough details to enable a person unfamiliar with the area to find the mark readily (see the field notes of Figures 5.5 and 5.11). A benchmark is usually named for some prominent object it is on or near, to aid in describing its location; one word is preferable. Examples are BM River, BM Tower, BM Corner, and BM Bridge. On extensive surveys, benchmarks are often numbered consecutively. Although advantageous in identifying relative positions along a line, this method is more subject to mistakes in field marking or recording.

Turning points are also numbered consecutively but not described in detail, since they are merely a means to an end and usually will not have to be relocated. However, if possible, it is advisable to select turning points that can be relocated, so if reruns on long lines are necessary because of blunders, fieldwork can be reduced. Before a party leaves the field, all possible note checks must be made to detect any mistakes in arithmetic and verify achievement of an acceptable closure. *The algebraic sum of the plus and minus sights applied to the first elevation should give the last elevation.* This computation checks the addition and subtraction for all HIs and turning points unless compensating mistakes have been made. When carried out for each left-hand page of tabulations, it is termed the *page check.* In Figure 5.5, for example, note that the page check is secured by adding the sum of backsights, 40.24, to the starting elevation 2053.18, and then subtracting the sum of foresights; 40.21, to obtain 2053.21, which checks the last elevation.

As previously noted, leveling should always be checked by running closed circuits or loops. This can be done either by returning to the starting benchmark, as demonstrated with the field notes in Figure 5.5, or by ending the circuit at another benchmark of equal or higher reliability. The final elevation should agree with the starting elevation if returning to the initial benchmark. The amount by which they differ is the *loop misclosure.* Note that in Figure 5.5, a loop misclosure of 0.03 ft was obtained.

If closure is made to another benchmark, the *section misclosure* is the difference between the closing benchmark's given elevation and its elevation obtained after leveling through the section. Specifications, or purpose of the survey, fix permissible misclosures (see Section 5.5). If the allowable misclosure is exceeded, one or more additional runs must be made. When acceptable misclosure is achieved, final elevations are obtained by making an adjustment (see Section 5.6).

Note that in running a level circuit between benchmarks, *a new instrument setup has to be made before starting the return run to get a complete check.* In Figure 5.5, for example, a minus sight of 8.71 was read on BM Oak to finish the run out and a plus sight of 11.95 was recorded to start back, showing that a new setup had been made. Otherwise, an error in reading the final minus sight would be accepted for the first plus sight on the run back. An even better check is secured by tying the run to a different benchmark.

If the elevation above a particular vertical datum (i.e., NAVD88) is available for the starting benchmark, elevations then determined for all intermediate points along the circuit will also be referenced to the same datum. However, if the starting benchmark's elevation above datum is not known, an assumed value may be used and all elevations converted to the datum later by applying a constant.

A lake or pond undisturbed by wind, inflow, or outflow can serve as an extended turning point. Stakes driven flush with the water, or rocks whose high points are at this level, should be used.

Double-rodded lines of levels are sometimes used on important work. In this procedure, plus and minus sights are taken on two turning points, using two rods from each setup, and the readings carried in separate noteform columns. A check on each instrument setup is obtained if the HI agrees for both lines. This same result can be accomplished using just one set of turning points, and reading both sides of a single rod that has two-faces, i.e., one side in feet and the other in meters. These rods are often used in precise leveling.

■ 5.5 PRECISION

Precision in leveling is increased by repeating measurements, making frequent ties to established benchmarks, using high-quality equipment, keeping it in good adjustment, and performing the measurements carefully. No matter how carefully the work is executed, however, errors will exist and will be evident in the form of misclosures, as discussed in Section 5.4. To determine whether or not work is acceptable, misclosures are compared with permissible values on the basis of either number of setups or distance covered. Various organizations set precision standards based on their project requirements. For example, on a simple construction survey, an allowable misclosure of $C = 0.02\sqrt{n}$ ft might be used, where n is the number of setups. Note that this criterion was applied for the level circuit in the field notes of Figure 5.5.

The Federal Geodetic Control Subcommittee (FGCS) recommends the following formula to compute allowable misclosures:[1]

$$C = m\sqrt{K} \tag{5.1}$$

where C is the allowable loop or section[2] misclosure, in millimeters; m is a constant; and K is the total length leveled, in kilometers. For "loops" (circuits that begin and

[1]The FGCS was formerly the FGCC (Federal Geodetic Control Committee). Their complete specifications for leveling are available in a booklet entitled "Standards and Specifications for Geodetic Control Networks" (September 1984). Information on how to obtain this and other related publications can be obtained at http://www.ngs.noaa.gov. Inquiries can also be made by email at info_center@ngs.noaa.gov, or by writing to the National Geodetic Information Center, NOAA, National Geodetic Survey, 1315 East West Highway, Station 9202, Silver Spring, MD 20910; telephone: (301) 713-3242.

[2]A section consists of a line of levels that begins on one benchmark and closes on another.

end on the same benchmark), K is the total perimeter distance, and the FGCS specifies constants of 4, 5, 6, 8, and 12 mm for the five classes of leveling, designated, respectively, as (1) first-order class I, (2) first-order class II, (3) second-order class I, (4) second-order class II, and (5) third-order. For "sections" the constants are the same, except that 3 mm applies for first-order class I. The particular order of accuracy recommended for a given type of project is discussed in Section 19.7.

Example 5.2

A differential leveling loop is run from an established BM A to a point 2 mi away and back, with a misclosure of 0.056 ft. What order leveling does this represent?

Solution

$$C = \frac{0.056 \text{ ft}}{0.0028 \; ft/mm} = 17 \; mm$$

$$K = (2 \; mi + 2 \; mi) \times 1.61 \; km/mi = 6.4 \; km$$

By a rearranged form of Equation 5.1, $m = \dfrac{C}{\sqrt{K}} = \dfrac{17}{\sqrt{6.4}} = 6.7$

This leveling meets the allowable 8-mm tolerance level for second-order class II work, but does not quite meet the 6-mm level for second-order class I, and if that standard had been specified, the work would have to be repeated.

Since distance leveled is proportional to number of instrument setups, the misclosure criteria can be specified using that variable. As an example, if sights of 200 ft are taken, thereby spacing instrument setups at about 400 ft, approximately 8.2 setups/km will be made. For second-order class II leveling, the allowable misclosure will then be, again by Equation (5.1),

$$C = \frac{8}{\sqrt{8.2}}\sqrt{n} = 2.8\sqrt{n} \tag{5.2}$$

where C is the allowable misclosure, in millimeters; and n the number of times the instrument is set up.

It is important to point out that meeting FGCS misclosure criterion[3] alone does not guarantee that a certain order of accuracy has been met. Because of compensating errors, it is possible, for example, that crude instruments and low-order techniques can produce small misclosures, yet intermediate elevations along the circuit may contain large errors. To help ensure that a given level of accuracy has indeed been met, besides stating allowable misclosures, the FGCS also specifies equipment and procedures that must be used to achieve a given order of accuracy.

[3]A complete listing of the specifications for performing geodetic control leveling can be obtained at http://www.ngs.noaa.gov/fgcs/tech_pub/1984-stds-specs-geodetic-control-networks.htm.

These specifications identify calibration requirements for leveling instruments (including rods), and they also outline required field procedures that must be used. Then if the misclosure specified for a given order of accuracy has been met, while employing appropriate instruments and procedures, it can be reasonably expected that all intermediate elevations along the circuit are established to that order.

Field procedures specified by the FGCS include minimum ground clearances for the line of sight, allowable differences between the lengths of pairs of backsight and foresight distances, and maximum sight lengths. As examples, sight lengths of not more than 50 m are permitted for first-order class I, while lengths up to 90 m are allowed for third order. As noted in Section 5.8, the stadia method is convenient for measuring the lengths of backsights and foresights to verify their acceptance.

5.6 ADJUSTMENTS OF SIMPLE LEVEL CIRCUITS

Since permissible misclosures are based on the lengths of lines leveled, or number of setups, it is logical to adjust elevations in proportion to these values. Observed elevation differences d and lengths of sections L are shown for a circuit in Figure 5.7. The misclosure found by algebraic summation of the elevation differences is +0.24 ft. Adding lengths of the sections yields a total circuit length of 3.0 mi. Elevation adjustments are then (0.24 ft/3.0) multiplied by the corresponding lengths, giving corrections of −0.08, −0.06, −0.06, and −0.04 ft (shown in the figure). The adjusted elevation differences (shown in black) are used to get the final elevations of benchmarks (also shown in black in the figure). Any misclosure that fails to meet tolerances may require reruns instead of adjustment. In Figure 5.5, adjustment for misclosure was made based on the number of instrument setups. Thus after verifying that the misclosure of 0.03 ft was within tolerance, the correction per setup was $0.03/7 = 0.004$ ft. Since errors in leveling accumulate, the first point receives a correction of 1×0.004, the second 2×0.004, etc. However, the corrected elevations are rounded off to the nearest hundredth of a foot. Level circuits with different lengths and routes are sometimes run from scattered reference points to obtain the elevation of a given benchmark. The most probable value for a benchmark elevation can then be computed from a weighted mean of the observations, the weights varying inversely with line lengths.

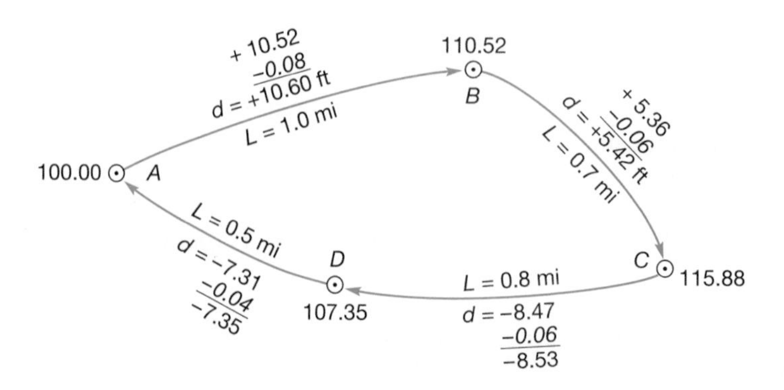

Figure 5.7 Adjustment of level circuit based on lengths of lines.

In running level circuits, especially long ones, it is recommended that some turning points or benchmarks used in the first part of the circuit be included again on the return run. This creates a multi-loop circuit, and if a blunder or large error exists, its location can be isolated to one of the smaller loops. This saves time because only the smaller loop containing the blunder or error needs to be rerun.

Although the least squares method (see Chapter 16) is the best method for adjusting circuits that contain two or more loops, an approximate procedure can also be employed. In this method each loop is adjusted separately, beginning with the one farthest from the closing benchmark.

■ 5.7 RECIPROCAL LEVELING

Sometimes in leveling across topographic features such as rivers, lakes, and canyons, it is difficult or impossible to keep plus and minus sights short and equal. Reciprocal leveling may be utilized at such locations.

As shown in Figure 5.8, a level is set up on one side of a river at X, near A, and rod readings are taken on points A and B. Since XB is very long, several readings are taken for averaging. Reading, turning the leveling screws to throw the instrument out of level, releveling, and reading again, does this. The process is repeated two, three, four, or more times. Then the instrument is moved close to Y and the same procedure followed.

The two differences in elevation between A and B, determined with an instrument first at X and then at Y, normally will not agree because of curvature, refraction, and personal and instrumental errors. In the procedure just outlined, however, the long foresight from X to B is balanced by the long backsight from Y to A. Thus the average of the two elevation differences cancels the effects of curvature, refraction, and instrumental errors, so the result is accepted as the correct

Figure 5.8
Reciprocal leveling.

value if the precision of the two differences appears satisfactory. Delays at X and Y should be minimized because refraction varies with changing atmospheric conditions.

5.8 THREE-WIRE LEVELING

As implied by its name, three-wire leveling consists in making rod readings on the upper, middle, and lower crosshairs. Formerly it was used mainly for precise work, but it can be used on projects requiring only ordinary precision. The method has the advantages of (1) providing checks against rod reading blunders, (2) producing greater accuracy because averages of three readings are available, and (3) furnishing stadia measurements of sight lengths to assist in balancing backsight and foresight distances (stadia is discussed in Section 17.9.2). In the three-wire procedure the difference between the upper and middle readings is compared with that between the middle and lower values. They must agree within one or two of the smallest units being recorded (usually 0.1 or 0.2 of the least count of the rod graduations); otherwise the readings are repeated. An average of the three readings is actually used, and as a computational check it must be very close to the middle wire figure. The difference between the upper and lower wire readings multiplied by the instrument *stadia interval factor* (see Section 17.9.2) gives the sight distances.

A sample set of field notes for the three-wire method is presented in Figure 5.9. Backsight readings on BM A of 0.718, 0.633, and 0.550 m taken on the upper, middle, and lower wires, respectively, give upper and lower differences (multiplied by 100) of 8.5 and 8.3 m, which agree within acceptable tolerance. Stadia measurement of the backsight length (the sum of the upper and lower differences) is 16.8 m. The average of the three backsight readings on BM A, 0.6337 m, agrees within 0.0007 m of the middle reading. The stadia foresight length of 15.9 m at this setup is within 0.9 m of the backsight length, and is satisfactory. The HI (104.4769 m) for the first setup is found by adding the average backsight reading to the elevation of BM A. Subtracting the average foresight reading on TP1 gives its elevation (103.4256 m). This process is repeated for each setup.

5.9 PROFILE LEVELING

Before engineers can properly design linear facilities such as highways, railroads, transmission lines, aqueducts, canals, sewers, and water mains, they need accurate information about the topography along the proposed routes. Profile leveling, which yields elevations at definite points along a reference line, provides the needed data. The subsections that follow discuss topics pertinent to profile leveling and include staking and stationing the reference line, field procedures for profile leveling, and drawing and using the profile.

5.9.1 Staking and Stationing the Reference Line

Depending on the particular project, the reference line may be a single straight segment, as in the case of a short sewer line; a series of connected straight segments

THREE-WIRE LEVELING
TAYLOR LAKE ROAD

Sta.	+ Sight	Stadia	− Sight	Stadia	Elev.
BM A					103.8432
	0.718		1.131		
	0.633	8.5	1.051	8.0	+0.6337
	0.550	8.3	0.972	7.9	104.4769
	3)1.901	16.8	3)3.154	15.9	−1.0513
	+0.6337		−1.0513		
TP 1					103.4256
	1.151		1.041		
	1.082	6.9	0.969	7.2	+1.0820
	1.013	6.9	0.897	7.2	104.5076
	3)3.246	13.8	3)2.907	14.4	−0.9690
	+1.0820		−0.9690		
TP 2					103.5386
	1.908		1.264		
	1.841	6.7	1.194	7.0	+1.8410
	1.774	6.7	1.123	7.1	105.3796
	3)5.523	13.4	3)3.581	14.1	−1.1937
	+1.8410		−1.1937		
BM B					104.1859
	Σ +3.5567		Σ −3.2140		Check
Page	Check:				
	103.8432	+3.5567	−3.2140	= 104.1	859

Figure 5.9 Sample field notes for three-wire leveling.

which change direction at angle points, as with transmission lines; or straight segments joined by curves, which occur with highways and railroads. The required alignment for any proposed facility will normally have been selected as the result of a preliminary design, which is usually based on a study of existing maps and aerial photos. The reference alignment will most often be the proposed construction centerline, although frequently offset reference lines are used.

To stake the proposed reference line, key points such as the starting and ending points and angle points will be set first. Then intermediate stakes will be placed on line, usually at 100-ft intervals if the English system of units is used, but sometimes at closer spacing. If the metric system is used, stakes are usually placed at 10-, 20-, 30-, or 40-m spacing, depending upon conditions. Distances for staking can be taped, or measured using the electronic distance measuring (EDM) component of a total station instrument operating in its tracking mode (see Sections 8.2 and 23.9).

In route surveying, a system called *stationing* is used to specify the relative horizontal position of any point along the reference line. The starting point is usually designated with some arbitrary value, for example in the English system of units, 10 + 00 or 100 + 00, although 0 + 00 can be used. If the beginning point was 10 + 00, a stake 100 ft along the line from it would be designated 11 + 00, the one 200 ft along the line 12 + 00, etc. The term *full station* is applied to each of these points set at 100-ft increments. This is the usual increment staked in rural areas. A point located between two full stations, say 84.90 ft beyond station 17 + 00, would be designated 17 + 84.90. Thus, locations of intermediate points are specified by their nearest preceding full station and their so-called *plus*. For station 17 + 84.90, the plus is 84.90. If the metric system is used, full stations are 1 km (1000 m) apart. The starting point of a reference line might be arbitrarily designated as 1 + 000 or 10 + 000, but again 0 + 000 could be used. In rural areas, intermediate points are normally set at 30- or 40-m increments along the line, and are again designated by their pluses. If the beginning point was 1 + 000, and stakes were being set at 40-m intervals, then 1 + 040, 1 + 080, 1 + 120, etc. would be set.

In rugged terrain and in urban situations, stakes are normally set closer together, for example at *half stations* (50-ft increments) or even *quarter stations* (25-ft increments) in the English system of units. In the metric system, 20-, 10-, or even 5-m increments may be staked.

Stationing not only provides a convenient unambiguous method for specifying positions of points along the reference line, but also gives the distances between points. For example, in the English system stations 24 + 18.3 and 17 + 84.9 are (2418.3 − 1784.9), or 633.4 ft, apart, and in the metric system stations 1 + 120 and 2 + 040 are 920 m apart.

5.9.2 Field Procedures for Profile Leveling

Profile leveling consists simply of differential leveling with the addition of intermediate foresights (minus sights) taken at required points along the reference line. Figure 5.10 illustrates an example of the field procedure, and the notes in Figure 5.11 relate to this example. Stationing for the example is in feet. As shown in the figure, the leveling instrument is initially set up at a convenient location and a plus sight of 10.15 ft taken on the benchmark. Adding this to the benchmark elevation yields a HI of 370.63 ft. Then intermediate minus sights are taken on points along the profile at stations as 0 + 00, 0 + 20, 1 + 00, etc. (If the reference line's beginning is far removed from the benchmark, differential levels running through several turning points may be necessary to get the instrument into position to begin taking intermediate minus sights on the profile line.) Notice that the note form for profile leveling contains all the same column headings as differential leveling, but is modified to include another column labeled "Intermediate Sight."

When distances to intermediate sights become too long, or if terrain variations or vegetation obstruct rod readings ahead, the leveling instrument must be moved. Establishing a turning point, as TP1 in Figure 5.10, does this. After reading a minus sight on the turning point, the instrument is moved ahead to a

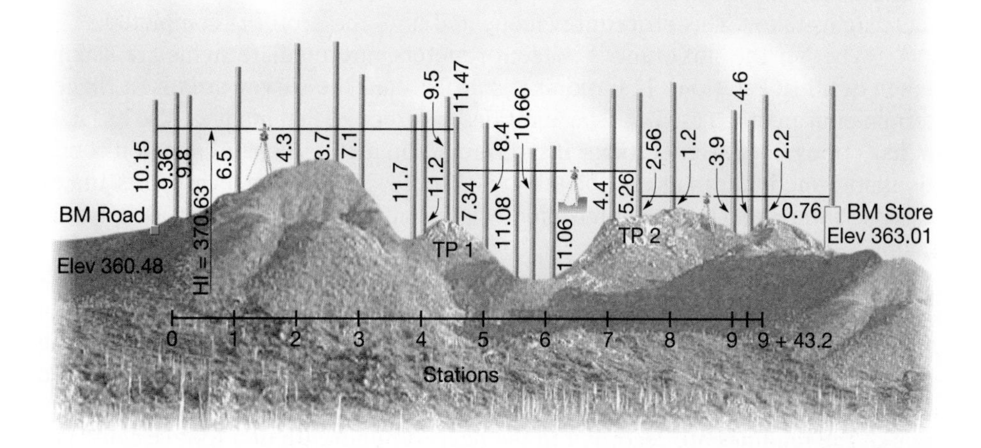

Figure 5.10 Profile leveling.

PROFILE LEVELS

Station	+ Sight	HI	− Sight	Int. Sight	Elev.
BM Road	10.15	(370.62) 370.63			360.48
0+00				9.36	361.26
0+20				9.8	360.8
1+00				6.5	364.1
2+00				4.3	366.3
2+60				3.7	366.9
3+00				7.1	363.5
3+90				11.7	358.9
4+00				11.2	359.4
4+35		(366.48)		9.5	361.1
TP 1	7.34	366.50	11.47		359.16
5+00				8.4	358.1
5+54				11.08	355.40
5+74				10.66	355.82
5+94				11.06	355.42
6+00				10.5	356.0
7+00		(362.77)		4.4	362.1
TP 2	2.56	368.80	5.26		361.24
8+00				1.2	362.6
9+00				3.9	359.9
9+25.2				3.4	360.4
9+25.3				4.6	359.2
9+43.2				2.2	361.6
BM Store			0.76		363.04
Σ	20.05		17.49		(363.01)

BM ROAD to BM STORE

BM Road 3 miles SW of Mpls. 200 yrds. N of Pine St. over pass 40ft. E of ¢Hwy. 169 Top of RW conc post No.268. — SW Minneapolis on Hwy 169

6 Oct. 2000

¢Hwy. 169, painted "X" — Cool, Sunny, 50° F

West drainage ditch — R.J. Hintz N

N.R. Olson ϕ

R.C. Perry ⊼

Summit — Wild Level #3

Sag

Summit

COPY

Page Check:

E gutter, Maple St. — +20.05

¢Maple St. — −17.49

W gutter, Maple St. — + 2.56

360.48

363.04

Summit — 363.04-363.01= Misclosure = 0.03

Top of E curb, Elm St.

Bottom of E curb, Elm St.

¢ Elm St.

BM Store. NE corner Elm St. & 4th Ave. SE corner Store foundation wall. 3″ brass disc set in grout. BM store elev. = 363.01

R.J. Hintz

Figure 5.11 Profile leveling notes for Figure 5.10.

good vantage point both for reading the backsight on the turning point and taking additional rod readings along the profile line ahead. The instrument is leveled, the plus sight taken on TP1, the new HI computed, and further intermediate sights taken. This procedure is repeated until the profile is completed.

Whether the stationing is in feet or meters, intermediate sights are usually taken at all full stations. If stationing is in feet and the survey area is in rugged terrain or in an urban area, the specifications may require that readings also be taken at half- or even quarter-stations. If stationing is in meters, depending upon conditions, intermediate sights may be taken at 40-, 30-, 20-, or 10-m increments. In any case, sights are also taken at high and low points along the alignment, as well as at changes in slope.

Intermediate sights should always be taken on "critical" points such as railroad tracks, highway centerlines, gutters, and drainage ditches. As presented in Figure 5.11, rod readings are normally only taken to the nearest 0.1 ft (English system) or nearest cm (metric system) where the rod is held on the ground, but on critical points, and for all plus and minus sights taken on turning points and benchmarks, the readings are recorded to the nearest hundredth of a foot (English) or the nearest mm (metric).

In profile leveling, lengths of intermediate minus sights vary, and in general they will not equal the plus sight length. Thus errors due to an inclined line of sight and to curvature and refraction will occur. Because errors from these sources increase with increasing sight lengths, on important work the instrument's condition of adjustment should be checked (see Section 4.15) and excessively long intermediate foresight distances should be avoided.

Instrument heights (HIs) and elevations of all turning points are computed immediately after each plus sight and minus sight. However, elevations for intermediate minus sights are not computed until after the circuit is closed on either the initial benchmark or another. Then the circuit misclosure is computed and, if acceptable, an adjustment is made and elevations of intermediate points are calculated. The procedure is described in the following subsection.

As in differential leveling, the page check should be made for each left-hand sheet. However in profile leveling, intermediate minus sights play no part in this computation. As illustrated in Figure 5.11, the page check is made by adding the algebraic sum of the column of plus sights and the column of minus sights to the beginning elevation. This should equal the last elevation tabulated on the page for either a turning point or the ending benchmark if that is the case, as it is in the example of Figure 5.11.

5.9.3 Drawing and Using the Profile

Prior to drawing the profile, it is first necessary to compute elevations along the reference line from the field notes. However, this cannot be done until an adjustment has been made to distribute any misclosure in the level circuit. In the adjustment process, HIs are adjusted, because they will affect computed profile elevations. The adjustment is made progressively in proportion to the total number of HIs in the circuit. The procedure is illustrated in Figure 5.11, where the misclosure was 0.03 ft. Since there were three HIs, the correction applied to each is $-0.03/3 = -0.01$ ft per HI.

Thus a correction of 0.01 was applied to the first HI, −0.02 ft to the second, and −0.03 ft to the third. Adjusted HIs are shown in Figure 5.11 in parentheses above their unadjusted values. It is unnecessary to correct turning point elevations since they are of no consequence. After adjusting the HIs, profile elevations are computed by subtracting intermediate minus sights from their corresponding *adjusted* HIs. The profile is then drawn by plotting elevations on the ordinate versus their corresponding stations on the abscissa. By connecting adjacent plotted points, the profile is realized.

Until recently, profiles were manually plotted, usually on special paper like the type shown in Figure 5.12. Now with computer-aided drafting and design (CADD) systems (see Section 18.14), it is only necessary to enter the stations and elevations into the computer, and this special software will plot and display the profile on the screen. Hard copies, if desired, may be obtained from plotters interfaced with a computer.

In drawing profiles, the vertical scale is generally exaggerated with respect to the horizontal scale to make differences in elevation more pronounced. A ratio of 10:1 is frequently used, but flatness or roughness of the terrain determines the desirable proportions. Thus, for a horizontal scale of 1 in. = 100 ft, the vertical scale might be 1 in. = 10 ft. The scale actually employed should be plainly marked. Plotted profiles are used for many purposes, such as (1) determining depth of cut or fill on proposed highways, railroads, and airports; (2) studying grade-crossing problems; and (3) investigating and selecting the most economical grade, location, and depth for sewers, pipelines, tunnels, irrigation ditches, and other projects.

The *rate of grade* (or *gradient* or *percent grade*) is the rise or fall in feet per 100 ft, or in meters per 100 m. Thus a grade of 2.5 percent means a 2.5-ft difference in elevation per 100 ft horizontally. Ascending grades are plus; descending grades, minus. A grade line of −0.15%, chosen to approximately equalize cuts and fills, is shown in Figure 5.12. Along this grade line, elevations drop at the rate of 0.15 ft per 100 ft. The grade begins at station 0 + 00 where it approximately meets existing

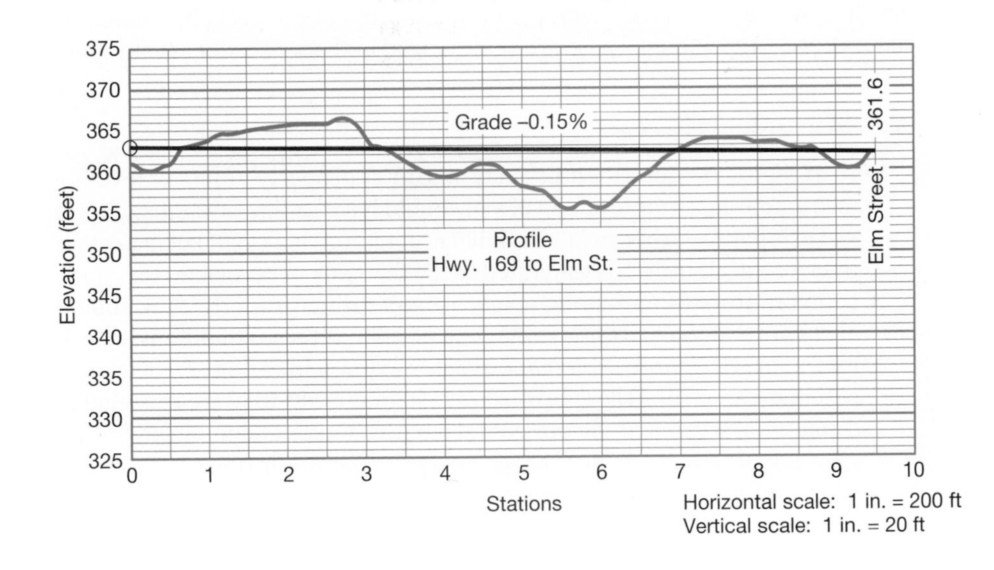

Figure 5.12
Plot of profile.

ground at elevation 363.0 ft, and ends at station 9 + 43 and elevation 361.6 ft where again it approximately meets existing ground. The process of staking grades is described in Chapter 23.

The term *grade* is also used to denote the elevation of the finished surface on an engineering project.

■ 5.10 GRID, CROSS-SECTION, OR BORROW-PIT LEVELING

Grid leveling is a method for locating contours (see Section 17.9.3). It is accomplished by staking an area in squares of 10, 20, 50, 100, or more feet (or comparable meter lengths) and determining the corner elevations by differential leveling. Rectangular blocks, say 50 by 100 ft or 20 by 30 m, that have the longer sides roughly parallel with the direction of most contour lines may be preferable on steep slopes. The grid size chosen depends on the project extent, ground roughness, and accuracy required.

The same process, termed *borrow-pit leveling,* is employed on construction jobs to ascertain quantities of earth, gravel, rock, or other material to be excavated or filled. The procedure is covered in Section 26.10 and Plate B.2.

■ 5.11 USE OF THE HAND LEVEL

A hand level can be used for some types of leveling when a low order of accuracy is sufficient. The instrument operator takes a plus and minus sight while standing in one position, and then moves ahead to repeat the process. A hand level is useful, for example, in cross-sectioning to obtain a few additional rod readings on sloping terrain where a turning point would otherwise be required.

■ 5.12 SOURCES OF ERROR IN LEVELING

All leveling measurements are subject to three sources of error: (1) instrumental, (2) natural, and (3) personal. These are summarized in the subsections that follow.

5.12.1 Instrumental Errors

Line of Sight. As described in Section 4.15, a properly adjusted leveling instrument that employs a level vial should have its line of sight and level vial axis parallel. Then, with the bubble centered, a horizontal plane, rather than a conical surface, is generated as the telescope is revolved. Also, if the compensators of automatic levels are operating properly, they should always produce a truly horizontal line of sight. If these conditions are not met, a *line of sight* (or *collimation*) error exists, and serious errors in rod readings can result. These errors are systematic, but they are canceled in differential leveling if the horizontal lengths of plus and minus sights are kept equal. The error may be serious in going up or down a steep hill where all plus sights are longer or shorter than all minus sights, unless care is taken to run a zigzag line. The size of the collimation error, ε, can be determined in a simple field procedure [see Equation (4.14) and Section 4.15.5]. If backsights and foresights cannot be balanced, a correction for this error can be made.

To apply the collimation correction, the value of ε from Equation (4.14) is divided by the length of the spaces between adjacent stakes in Figure 4.20. This yields the *collimation correction factor* in units of feet per foot, or meters per meter. Then for any backsight or foresight, the correction to be subtracted from the rod reading is obtained by multiplying the length of the sight by this correction factor. As an example, suppose that the distance between stakes in Example 4.3 was 100 ft. Then the collimation correction factor is $0.010/100 = 0.0001$ ft/ft. Suppose that a reading of 5.29 ft was obtained on a backsight of 200 ft length with this instrument. The corrected rod reading would then be $5.29 - (200 \times 0.0001) = 5.27$.

Crosshair Not Exactly Horizontal. Reading the rod near the center of the horizontal crosshair will eliminate or minimize this potential error.

Rod Not Correct Length. Inaccurate divisions on a rod cause errors in measured elevation differences similar to those resulting from incorrect markings on a measuring tape. Uniform wearing of the rod bottom makes HI values too large, but the effect is canceled when included in both plus and minus sights. Rod graduations should be checked by comparing them with those on a standardized tape.

Tripod Legs Loose. Tripod leg bolts that are too loose or too tight allow movement or strain that affects the instrument head. Loose metal tripod shoes cause unstable setups.

5.12.2 Natural Errors

Curvature of the Earth. As noted in Section 4.4, a level surface curves away from a horizontal plane at the rate of 0.667 M^2 or 0.0785 K^2, which is about 0.7 ft/mi or 8 cm/km. The effect of curvature of the Earth is to increase the rod reading. Equalizing lengths of plus and minus sights in differential leveling cancels the error due to this cause.

Refraction. Light rays coming from an object to the telescope are bent, making the line of sight a curve concave to the Earth's surface, which thereby decreases rod readings. Balancing the lengths of plus and minus sights usually eliminates errors due to refraction. However, large and sudden changes in atmospheric refraction may be important in precise work. Although errors due to refraction tend to be random over a long period of time, they could be systematic on one day's run.

Temperature Variations. Heat causes leveling rods to expand, but the effect is not important in ordinary leveling. If the level vial of a tilting level is heated, the liquid expands and the bubble shortens. This does not produce an error (although it may be inconvenient) unless one end of the tube is warmed more than the other, and the bubble therefore moves. Other parts of the instrument warp because of uneven heating, and this distortion affects the adjustment. Shading the level by means of a cover when carrying it, and by an umbrella when it is set up, will reduce or eliminate heat effects. These precautions are followed in precise leveling.

Air boiling or heat waves near the ground surface or adjacent to heated objects make the rod appear to wave and prevent accurate sighting. Raising the line of sight by high tripod setups, taking shorter sights, avoiding any that pass close to heat sources (such as buildings and stacks), and using the lower magnification of a variable-power eyepiece reduce the effect.

Wind. Strong wind causes the instrument to vibrate and makes the rod unsteady. Precise leveling should not be attempted on excessively windy days.

Settlement of the Instrument. Settlement of the instrument during the time between a plus sight reading and a minus sight makes the latter too small and therefore the recorded elevation of the next point too high. The error is cumulative in a series of setups on soft material. Therefore setups on spongy ground, blacktop, or ice should be avoided if possible, but if they are necessary, unusual care is required to reduce the resulting errors. This can include taking readings in quick order, using two rods and two observers to preclude walking around the instrument, and alternating the order of taking plus and minus sights. Additionally whenever possible, the instrument tripod's legs can be set on long hubs that are driven to refusal in the soft material.

Settlement of a Turning Point. This condition causes an error similar to that resulting from settlement of the instrument. It can be avoided by selecting firm, solid turning points or, if none are available, using a steel turning pin set firmly in the ground. A railroad spike can also be used in most situations.

5.12.3 Personal Errors

Bubble Not Centered. In working with levels that employ level vials, errors caused by the bubble not being exactly centered at the time of sighting are the most important of any, particularly on long sights. If the bubble runs between the plus and minus sights, *it must be recentered before the minus sight is taken.* Experienced observers develop the habit of checking the bubble before and after each sight, a procedure simplified with some instruments, which have a mirror-prism arrangement permitting a simultaneous view of the level vial and rod.

Parallax. Parallax caused by improper focusing of the objective or eyepiece lens results in incorrect rod readings. Careful focusing eliminates this problem.

Faulty Rod Readings. Incorrect rod readings result from parallax, poor weather conditions, long sights, improper target settings, and other causes, including mistakes such as those due to careless interpolation and transposition of figures. Short sights selected to accommodate weather and instrument conditions reduce the magnitude of reading errors. If a target is used, the rodperson should read the rod, and the observer should check it independently.

Rod Handling. Using a rod level that is in adjustment, or holding the rod parallel to a plumb bob string eliminates serious errors caused by improper plumbing of the rod. Banging the rod on a turning point for the second (plus) sight may change the elevation of a point.

Target Setting. If a target is used, it may not be clamped at the exact place signaled by the observer because of slippage. A check sight should always be taken after the target is clamped.

5.13 MISTAKES

A few common mistakes in leveling are listed here.

Improper Use of a Long Rod. If the vernier reading on the back of a damaged Philadelphia rod with English units is not exactly 6.500 ft or 7.000 ft for the short rod, the target must be set to read the same value before extending the rod.

Holding the Rod in Different Places for the Plus and Minus Sights on a Turning Point. The rodperson can avoid such mistakes by using a well-defined point or by outlining the rod base with lumber crayon, keel, or chalk.

Reading a Foot Too High. This mistake usually occurs because the incorrect footmark is in the telescope's field of view near the cross line; for example, an observer may read 5.98 instead of 4.98. Noting the footmarks both above and below the horizontal cross line will prevent this mistake.

Waving a Flat Bottom Rod While Holding It on a Flat Surface. This action produces an incorrect rod reading because rotation is about the rod edges instead of the center or front face. In precise work, plumbing with a rod level, or other means, is preferable to waving. This procedure also saves time.

Recording Notes. Mistakes in recording, such as transposing figures, entering values in the wrong column, and making arithmetic mistakes, can be minimized by having the notekeeper repeat the value called out by an observer, and by making the standard field-book checks on rod sums and elevations. Digital levels that automatically take rod readings, store the values, and compute the level notes can eliminate these mistakes.

Touching Tripod or Instrument During Reading Process. Beginners using instruments that employ level vials may center the bubble, put one hand on the tripod or instrument while reading a rod, and then remove the hand while checking the bubble, which has now returned to center but was off during the observation. Of course, the instrument should not be touched when taking readings, but detrimental effects of this bad habit are practically eliminated when using automatic levels.

5.14 REDUCING ERRORS AND ELIMINATING MISTAKES

Errors in running levels are reduced (but never eliminated) by carefully adjusting and manipulating both instrument and rod (see Section 4.15 for procedures) and establishing standard field methods and routines. The following routines prevent most large errors or quickly disclose mistakes: (1) checking the bubble before and after each reading (if an automatic level is not being used), (2) using a rod level, (3) keeping the horizontal lengths of plus and minus sights equal, (4) running lines forward and backward, and (5) making the usual field-book arithmetic checks.

PROBLEMS

Asterisks (*) indicate problems that have answers given in Appendix G.

5.1 What proper field procedures can virtually eliminate Earth curvature and refraction errors in differential leveling?

5.2 How can errors due to settlement of the instrument and rod be reduced or eliminated?

5.3 Why is it preferable to use a rod level when plumbing the rod?

5.4 Explain how errors due to lack of instrument adjustment can be practically eliminated in running a line of differential levels.

5.5 Explain why the shoes of the tripod must be snug.

5.6* What error is created by a rod leaning 5 min from plumb at a 3.513 m reading on the leaning rod?

5.7 Similar to Problem 5.6, except for a 10 ft reading.

5.8 What error results on a 100-m sight with a level if the rod reading is 3.505 m but the top of the 4 m rod is 0.2 m out of plumb?

5.9 Prepare a set of level notes for the data listed. Perform a check and adjust the misclosure. Elevation of BM 7 is 82.206 m. If the total loop length is 1 km, what order of leveling is represented? (Assume all readings are in meters.)

POINT	+S (BS)	−S (FS)
BM 7	4.388	
TP 1	6.907	4.538
BM 8	4.680	8.800
TP 2	3.730	5.978
TP 3	8.464	5.245
BM 7		3.598

5.10* Similar to Problem 5.9, except the elevation of BM 7 is 306.928 ft and the loop length 1000 ft. (Assume all readings are in feet.)

5.11 A differential leveling loop began and closed on BM Tree (elevation 323.48 ft). The plus sight and minus sight distances were kept approximately equal. Readings (in feet) listed in the order taken are 3.18 (BS) on BM Tree, 4.76 (FS) and 2.44 (BS) on TP1, 3.05 (FS) and 6.63 (BS) on BM *X*, 3.64 (FS) and 2.35 (BS) on TP2, and 3.07 (FS) on BM Tree. Prepare, check, and adjust the notes.

5.12 A differential leveling circuit began on BM Hydrant (elevation 85.35 ft) and closed on BM Rock (elevation 96.17 ft). The plus sight and minus sight distances were kept approximately equal. Readings (in feet) given in the order taken are 2.65 (BS) on BM Hydrant, 3.51 (FS) and 7.23 (BS) on TP1, 5.04 (FS) and 11.41 (BS) on BM 1, 8.58 (FS) and 7.65 (BS) on BM 2, 4.23 (FS) and 7.53 (BS), on TP2, and 4.34 (FS) on BM Rock. Prepare, check, and adjust the notes.

5.13 A differential leveling loop began and closed on BM Bridge (elevation 814.687 m). The plus sight and minus sight distances were kept approximately equal. Readings (in meters) listed in the order taken are 0.548 (BS) on BM Bridge, 1.208 (FS) and 0.843 (BS) on TP1, 1.287 (FS) and 1.482 (BS) on BM *X*, 0.743 (FS) and 0.944 (BS) on TP2, and 0.571 (FS) on BM Bridge. Prepare, check, and adjust the notes.

5.14 A differential leveling circuit began on BM Rock (elevation 543.202 m) and closed on BM Manhole (elevation 542.551 m). The plus sight and minus sight distances were kept approximately equal. Readings (in meters) listed in the order taken are 1.559 (BS) on BM Rock, 0.987 (FS) and 1.105 (BS) on TP1, 0.842 (FS) and 0.679 (BS) on BM 1, 1.846 (FS) and 0.849 (BS) on BM 2, 1.895 (FS) and 1.436 (BS) on TP2, and 0.704 (FS) on BM Manhole. Prepare, check, and adjust the notes.

5.15 A differential leveling loop started and closed on BM Juno, elevation 2485.19 ft. The plus sight and minus sight distances were kept approximately equal. Readings (in feet) listed in the order taken are 5.49 (BS), 3.46 (FS), 8.48 (BS) 5.34 (FS), 6.51 (BS), 8.27 (FS), 4.03 (BS), 9.46 (FS), 7.89 (BS), and 5.92 (FS). Prepare, check, and adjust the notes.

5.16* A level setup midway between X and Y reads 6.29 ft on X and 7.91 ft on Y. When moved within a few feet of X, readings of 5.18 ft on X and 6.76 ft on Y are recorded. What is the true elevation difference and the reading required on Y to adjust the instrument?

5.17 To test its line of sight adjustment, a level is setup near C (elev 193.436 m) and then near D. Rod readings listed in the order taken are $C = 1.315$ m, $D = 0.848$ m, $D = 1.296$ m, and $C = 1.767$ m. Compute the elevation of D and the reading required on C to adjust the instrument.

5.18* The line of sight test shows that a level's line of sight is inclined downward 0.002 m/50 m. What is the allowable difference between BS and FS distances at each setup (neglect curvature and refraction) to keep elevations correct within 0.001 m?

5.19 Reciprocal leveling gives the following readings in meters from a setup near A: on A, 4.365; on B, 7.928, 7.924, and 7.926. At the setup near B: on B, 7.253; on A, 3.687, 3.688, and 3.691. The elevation of A is 86.982 m. Determine the misclosure and elevation of B.

5.20* Reciprocal leveling across a canyon provides the data listed (in ft). The elevation of Y is 5265.879 ft. The elevation of X is required. Instrument at X: $+S = 3.182$, $-S = 9.365$, 9.370, and 9.368. Instrument at Y: $+S = 10.223$; $-S = 4.037$, 4.041, and 4.038.

5.21 Prepare a set of three-wire leveling notes for the data given and make the page check. The elevation of BM X is 185.101 m. Rod readings (in meters) are (H denotes upper cross-wire readings, M middle wire, and L lower wire): BS on BM X: $H = 0.843$, $M = 0.621$, $L = 0.397$; FS on TP 1: $H = 1.604$, $M = 1.332$, $L = 1.062$; BS on TP 1: $H = 1.459$, $M = 1.136$, $L = 0.813$; FS on BM Y: $H = 0.976$, $M = 0.646$, $L = 0.320$.

5.22 Similar to Problem 5.21, except the elevation of BM X is 638.437 ft, and rod readings (in feet) are: BS on BM X: $H = 9.467$, $M = 7.087$, $L = 4.710$; FS on TP 1: $H = 8.022$, $M = 6.597$, $L = 5.170$; BS on TP 1: $H = 6.694$, $M = 5.408$, $L = 4.122$; FS on BM Y: $H = 7.964$, $M = 5.646$, $L = 3.326$.

5.23 Prepare a set of profile leveling notes for the data listed and show the page check. All data are given in feet. The elevation of BM A is 387.90, and the elevation of BM B is 400.75. Rod readings are: +S on BM A, 10.25 intermediate foresight (IFS) on 11 + 00, 6.6; −S on TP1, 6.35; +S on TP 1, 9.91; intermediate foresight on 12 + 00, 5.3; on 12 + 50, 4.9; on 13 + 00, 5.7; on 14 + 00, 6.3; −S on TP 2, 4.44, +S on TP 2, 8.54; intermediate foresight on 14 + 73, 3.8; on 15 + 00, 2.9; on 16 + 00, 1.6; −S on BM B, 5.03.

5.24 Same as Problem 5.23, except the elevation of BM A = 1311.74 ft, the elevation of BM B = 1322.65 ft, and the +S on BM A = 8.76 ft.

5.25 Plot the profile Problem 5.23 and design a grade line between stations 11 + 00 and 15 + 00 that balances cut and fill areas.

5.26* If the elevations on a certain project at stations 8 + 50 and 13 + 00 are 4465.89 and 4479.39, respectively, what is the percent grade connecting these two points? (*Note*: Stations and elevations are in feet.)

5.27 Same as Problem 5.26, except for elevations of 837.89 and 823.02 at stations 17 + 50 and 21 + 75, respectively.

5.28 Differential leveling between BMs A, B, C, D, and A gives elevation differences (in meters) of −15.632, +32.458, +38.214, and −55.025, and distances in km of 4.0, 6.0,

5.0, and 3.0, respectively. If the elevation of *A* is 236.891, compute the adjusted elevations of BMs *B*, *C*, and *D* and the order of leveling.

5.29 Leveling from BM *X* to *W*, BM *Y* to *W*, and BM *Z* to *W* gives differences in elevation (in feet) of −30.24, +26.20, and +10.18, respectively. Distances between benchmarks are $XW = 2500$, $YW = 3000$, and $ZW = 4000$. True elevations of the benchmarks are $X = 571.93$, $Y = 515.47$, and $Z = 531.58$. What is the adjusted elevation of *W*? (*Note:* All data are given in feet.)

5.30 A 3-m level rod was calibrated and its graduated scale was found to be uniformly expanded so that the distance between its 0 and 3.000 marks was actually 2.997 m. How will this affect elevations determined with this rod for (a) circuits run on relatively flat ground, (b) circuits run downhill, and (c) circuits run uphill?

5.31* After running a line of levels between BM Sign and BM Road, examination showed that the level rod had a repaired base plate on the bottom, thus making the rod too short. Is the elevation determined for BM Road correct? Explain.

5.32* A line of levels with 22 setups (44 rod readings) was run from BM Rock to BM Pond with readings taken to the nearest 2.0 mm; hence any observed value could have an error of ±1.0 mm. For reading errors only, what total error would be expected in the elevation of BM Pond?

5.33 Same as Problem 5.32, except for 35 setups and readings to the nearest 0.01 ft with possible error of ±0.005 ft each.

5.34 Compute the permissible misclosure for the following lines of levels: (a) a 15-km loop of third-order levels; (b) a 30-km section of second-order class I levels; and (c) a 50-km loop of first-order class I levels.

5.35 Why are double-rodded lines of levels recommended for precise work?

5.36 List four considerations that govern a rodperson's selection of TPs and BMs.

5.37 What are the primary differences between "ordinary" differential leveling and "precise" three-wire leveling?

5.38 Create a computational program that solves Problem 5.9.

5.39 Create a computational program that solves Problem 5.28.

5.40 Create a computational program that solves Problem 5.29.

BIBLIOGRAPHY

Federal Geodetic Control Subcommittee. 1984. *Standards and Specifications for Geodetic Control Surveys.* Silver Spring, MD: National Geodetic Information Branch, NOAA.

Holdahl, S. R., W. E. Strange, and R. J. Harris. 1987. "Empirical Calibration of Zeiss Ni-1 Instruments to Account for Magnetic Errors." *Manuscripta Geodectica* 12: 28.

Reilly, J. P. 2004. "Tides and Their Relationship to Vertical Datums." *Point of Beginning* 29 (No. 4): 68.

Zilkoski, D., et. al. 1997. "Guidelines for Establishing GPS-derived Ellipsoid Heights (Standards 2 cm and 5 cm), Version 4.3." *NOAA Technical Memorandum NOS NGS 58.* Silver Springs, MD.

6 Distance Measurement

PART I • METHODS FOR MEASURING DISTANCES

6.1 INTRODUCTION

Distance measurement is generally regarded as the most fundamental of all surveying observations. In traditional ground surveys, even though many angles may be read, the length of at least one line must be observed to supplement the angles in locating points. In plane surveying, the distance between two points means the horizontal distance. If the points are at different elevations, the distance is the horizontal length between vertical lines at the points.

Lengths of lines may be specified in different units. In the United States, the foot, decimally divided, is usually used although meters are becoming increasingly more common. Geodetic surveys and many highway surveys employ the meter. In architectural and machine work and on some construction projects, the unit is a foot divided into inches and fractions of an inch. As discussed in Section 2.2, chains, varas, rods, and other units have been and still are utilized in some localities and for special purposes.

6.2 SUMMARY OF METHODS FOR MAKING LINEAR MEASUREMENTS

In surveying, linear measurements have been obtained by many different methods. These include (1) pacing, (2) odometer readings, (3) optical rangefinders, (4) tacheometry (stadia), (5) subtense bars, (6) taping, (7) electronic distance measurement (EDM), (8) satellite systems, and others. Of these, surveyors most commonly use taping, EDM, and satellite systems today. In particular, the satellite-supported global *navigation satellite systems* (GNSS) are rapidly replacing all other systems due to many advantages, but most notably because of their range, accuracy, and efficiency. Methods (1) through (5) are discussed briefly in the following sections. Taping is discussed in Part II of this chapter, and EDM is described in Part III of this chapter. Satellite systems are described in Chapters 13, 14, and 15.

Triangulation is a method for determining positions of points from which horizontal distances can be computed (see Section 19.12.1). In this procedure, lengths of lines are computed trigonometrically from measured base lines and angles. *Photogrammetry* can also be used to obtain horizontal distances. This topic is covered in Chapter 27. Besides these methods, distances can be estimated, a technique useful in making field note sketches and checking observations for mistakes. With practice, estimating can be done quite accurately.

6.3 PACING

Distances obtained by pacing are sufficiently accurate for many purposes in surveying, engineering, geology, agriculture, forestry, and military field sketching. Pacing is also used to detect blunders that may occur in making distance observations by more accurate methods.

Pacing consists of counting the number of steps, or paces, in a required distance. The length of an individual's pace must be determined first. This is best done by walking with natural steps back and forth over a level course at least 300 ft long and dividing the known distance by the average number of steps. For short distances, the length of each pace is needed, but the number of steps taken per 100 ft is desirable for checking long lines.

It is possible to adjust one's pace to an even 3 ft, but a person of average height finds such a step tiring if maintained for very long. The length of an individual's pace varies when going uphill or downhill and changes with age. For long distances, a pocket instrument called a *pedometer* can be carried to register the number of paces, or a *passometer* attached to the body or leg counts the steps. Some surveyors prefer to count *strides,* a stride being two paces.

Pacing is one of the most valuable things learned in surveying since it has practical applications for everybody and requires no equipment. If the terrain is open and reasonably level, experienced pacers can measure distances of 100 ft or longer with an accuracy of 1/50 to 1/100 of the distance.

6.4 ODOMETER READINGS

An odometer converts the number of revolutions of a wheel of known circumference to a distance. Lengths measured by an odometer on a vehicle are suitable for some preliminary surveys in route-location work. They also serve as rough checks on observations made by other methods. Other types of measuring wheels are available and useful for determining short distances, particularly on curved lines. Odometers give surface distances, which should be corrected to horizontal if the ground slopes severely (see Section 6.13). With odometers, an accuracy of approximately 1/200 of the distance is reasonable.

■ 6.5 OPTICAL RANGEFINDERS

These instruments operate on the same principle as rangefinders on single-lens reflex cameras. Basically, when focused, they solve for the object distance f_2 in Equation (4.12), where focal length f and image distance f_1 are known. An operator looks through the lens and adjusts the focus until a distant object viewed is focused in coincidence, whereupon the distance to that object is obtained. These instruments are capable of accuracies of 1 part in 50 at distances up to 150 ft, but accuracy diminishes as the length increases. They are suitable for reconnaissance, sketching, or checking more accurate observations for mistakes.

■ 6.6 TACHEOMETRY

Tacheometry (*stadia* is the more common term in the United States) is a surveying method used to quickly determine the horizontal distance to and elevation of a point. Stadia observations are obtained by sighting through a telescope equipped with two or more horizontal cross hairs at a known spacing. The apparent intercepted length between the top and bottom hairs is read on a graduated rod held vertically at the desired point. The distance from telescope to rod is found by proportional relationships in similar triangles. An accuracy of 1/500 of the distance is achieved with reasonable care. A brief explanation of the method is given in Section 17.9.2.

■ 6.7 SUBTENSE BAR

This indirect distance measuring procedure involves using a theodolite to read the horizontal angle subtended by two targets precisely spaced at a fixed distance apart on a subtense bar. The unknown distance is computed from the known target spacing and the measured horizontal angle. Prior to observing the angle from one end of the line, the bar is centered over the point at the other end of the line and oriented perpendicular to the line and in a horizontal plane. For sights of 500 ft (150 m) or shorter, and using a 1″ theodolite, an accuracy of 1 part in 3000 or better can be achieved. Accuracy diminishes with increased line length. Besides only being suitable for relatively short lines, this method of distance measurement is time-consuming and is seldom used today, having been replaced by electronic distance measurement.

■ PART II • DISTANCE MEASUREMENTS BY TAPING

■ 6.8 INTRODUCTION TO TAPING

Observation of horizontal distances by taping consists of applying the known length of a graduated tape directly to a line a number of times. Two types of problems arise: (1) observing an unknown distance between fixed points such as between two stakes in the ground and (2) laying out a known or required distance with only the starting mark in place.

Taping is performed in six steps: (1) lining in, (2) applying tension, (3) plumbing, (4) marking tape lengths, (5) reading the tape, and (6) recording the distance. The application of these steps in taping on level and sloping ground is detailed in Sections 6.11 and 6.12.

■ 6.9 TAPING EQUIPMENT AND ACCESSORIES

Over the years various types of tapes and other related equipment have been used for taping in the United States. Tapes in current use are described here, as are other accessories used in taping.

Surveyor's and *engineer's tapes* are made of steel 1/4 to 3/8 in. wide and weigh 2 to 3 lbs/100 ft. Those graduated in feet are most commonly 100-ft long, although they are also available in lengths of 200, 300, and 500 ft. They are marked in feet, tenths, and hundredths. Metric tapes have standard lengths of 30, 60, 100, and 150 m. All can either be wound on a reel [see Figure 6.1(a)] or done up in loops.

Invar tapes are made of a special nickel steel alloy (35 percent nickel and 65 percent steel) to reduce length variations caused by differences in temperature. The thermal coefficient of expansion and contraction of this material is only about 1/30 to 1/60 that of an ordinary steel tape. However, the metal is soft and somewhat unstable. This weakness, along with their cost of perhaps ten times that of steel tapes, made them suitable only for precise geodetic work and as a standard for comparison with working tapes. Another version, the *Lovar tape,* has properties and a cost between those of steel and Invar tapes.

Cloth (or *metallic*) *tapes* are actually made of high-grade linen, 5/8-in. wide with fine copper wires running lengthwise to give additional strength and prevent excessive elongation. Metallic tapes commonly used are 50-, 100-, and 200-ft long and come on enclosed reels [see Figure 6.1(b)]. Although not suitable for precise work, metallic tapes are convenient and practical for many purposes.

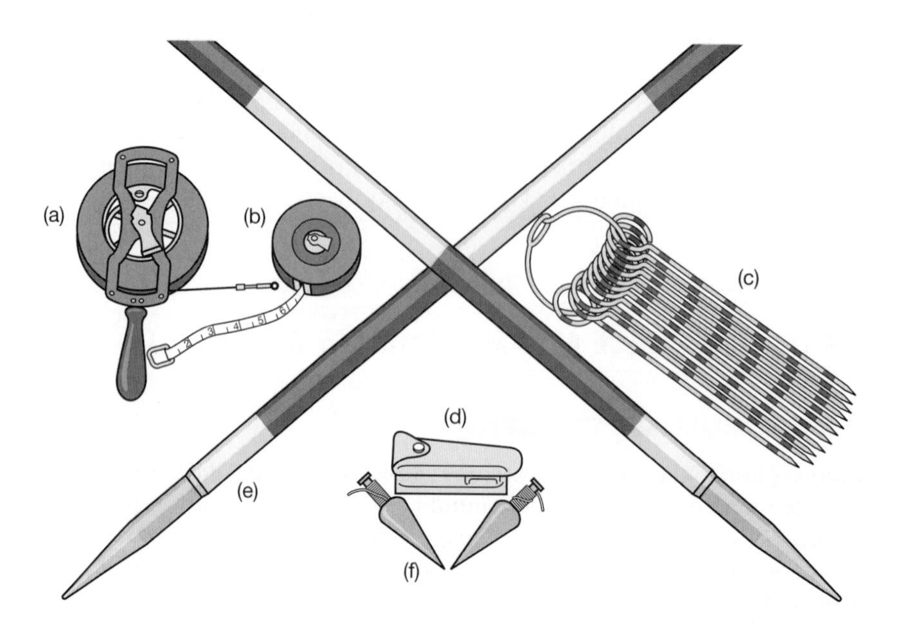

Figure 6.1 Taping equipment for field party. (Courtesy W. & L.E. Gurley)

Fiberglass tapes come in a variety of sizes and lengths and are usually wound on a reel. They can be employed for the same types of work as metallic tapes.

Chaining pins or *taping pins* are used to mark tape lengths. Most taping pins are made of number 12 steel wire, sharply pointed at one end, have a round loop at the other end, and are painted with alternate red and white bands [see Figure 6.1(c)]. Sets of 11 pins carried on a steel ring are standard.

The *hand level,* described in Section 4.13, is a simple instrument used to keep the tape ends at equal elevations when observing over rough terrain [see Figures 4.17 and 6.1(d)].

Tension handles facilitate the application of a desired standard or known tension. A complete unit consists of a wire handle, a clip to fit the ring end of the tape, and a spring balance reading up to 30 lb in 1/2-lb graduations.

Clamp handles are used to apply tension by a positive, quick grip using a scissors-type action on any part of a steel tape. They do not damage the tape and prevent injury to hands.

A *pocket thermometer* permits reading data for making temperature corrections. It is about 5-in. long, graduated from perhaps $-30°$ to $+120°F$ in 1° or 2° divisions, and kept in a protective metal case.

Range poles (lining rods) made of wood, steel, or aluminum are about 1-in. thick and 6- to 10-ft long. They are round or hexagonal in cross section and marked with alternate 1-ft long red and white bands that can be used for rough measurements [see Figure 6.1(e)]. The main utility of range poles is to mark the line being measured so that the tape's alignment can be maintained.

Plumb bobs for taping [see Figure 6.1(f)] should weigh a minimum of 8 oz and have a fine point. However, most surveyors use 24-oz plumb bobs for stability reasons. At least 6 ft of good-quality string or cord, free of knots, is necessary for convenient work with a plumb bob. The points of most plumb bobs are removable, which facilitates replacement if they become dull or broken. The string can be wound on a spring-loaded reel that is useful for rough targeting. However, in taping it is best to not use a reel.

■ 6.10 CARE OF TAPING EQUIPMENT

The following points are pertinent in the care of tapes and range poles:

1. Considering the cross-sectional area of the average surveyor's steel tape and its permissible stress, a pull of 100 lb will do no damage. If the tape is kinked, however, a pull of less than 1 lb can break it. Therefore, always check to be certain that any loops and kinks are eliminated before tension is applied.
2. If a tape gets wet, wipe it first with a dry cloth, then with an oily one.
3. Tapes should be either kept on a reel or *"thrown"* into circular loops, but not handled both ways.
4. Each tape should have an individual number or tag to identify it.
5. Broken tapes can be mended by riveting or applying a sleeve device, but a mended tape should not be used on important work.
6. Range poles are made with the metal shoe and point in line with the section above. This alignment may be lost if the pole is used improperly.

■ 6.11 TAPING ON LEVEL GROUND

The subsections that follow describe six steps in taping on level ground using a tape.

6.11.1 Lining In

Using range poles, the line to be measured should be marked at both ends, and at intermediate points where necessary, to ensure unobstructed sight lines. Taping requires a minimum of two people, a *forward tapeperson* and a *rear tapeperson*. The forward tapeperson is lined in by the rear tapeperson. Directions are given by vocal or hand signals.

6.11.2 Applying Tension

The rear tapeperson holding the 100-ft end of a tape over the first (rear) point lines in the forward tapeperson, holding the zero end of the tape. For accurate results the tape must be straight and the two ends held at the same elevation. A specified tension, generally between 10 and 25 lb, is applied. To maintain a steady pull, tapepersons wrap the leather thong at the tape's end around one hand, keep forearms against their bodies, and face at right angles to the line. In this position, they are off the line of sight. Also, the body needs only be tilted to hold, decrease, or increase the pull. Sustaining a constant tension with *outstretched* arms is difficult, if not impossible, for a pull of 15 lb or more. Good communication between forward and rear tapepersons will avoid jerking the tape, save time, and produce better results.

6.11.3 Plumbing

Weeds, brush, obstacles, and surface irregularities may make it undesirable to lay a tape on the ground. In those cases, the tape is held above ground in a horizontal position. Placing the plumb-bob string over the proper tape graduation and securing it with one thumb, mark each end point on the tape. The rear tapeperson continues to hold a plumb bob over the fixed point, while the forward tapeperson marks the length. In measuring a distance shorter than a full tape length, the forward tapeperson moves the plumb-bob string to a point on the tape over the ground mark.

6.11.4 Marking Tape Lengths

When the tape has been lined in, tension has been applied, and the rear tapeperson is over the point, "stick" is called out. The forward tapeperson then places a pin exactly opposite the zero mark of the tape and calls "stuck." The marked point is checked by repeating the measurement until certainty of its correct location is assured.

After checking the measurement, the forward tapeperson signals that the point is "OK," the rear tapeperson pulls up the rear pin, and they move ahead. The forward tapeperson drags the tape, paces roughly 100 ft, and stops. Just before the 100-ft end reaches the set pin, the rear tapeperson calls "tape" to notify the forward tapeperson that they have gone 100 ft. The process of measuring 100-ft lengths is repeated until a partial tape length is needed at the end of the line.

6.11.5 Reading the Tape

There are two common styles of graduations on 100-ft surveyor's tapes. *It is necessary to identify the type being used before starting work* to avoid making one-foot mistakes repeatedly.

The more common type of tape has a total graduated length of 101 ft. It is marked from 0 to 100 by full feet in one direction, and has an additional foot preceding the zero mark graduated from 0 to 1 ft in tenths or in tenths and hundredths in the other direction. In measuring the last partial tape length of a line with this kind of tape, a full-foot graduation is held by the rear tapeperson at the last pin set [like the 87-ft mark in Figure 6.2(a)]. The actual foot mark held is the one that causes the graduations on the extra foot between zero and the tape end to straddle the closing point. The forward tapeperson reads the additional length of 0.68 ft beyond the zero mark. In the case illustrated, to ensure correct recording, the rear tapeperson calls "87." The forward tapeperson repeats and adds the partial foot reading, calling "87.68." Since part of a foot has been added, this type of tape is known as an *add tape.*

The other kind of tape found in practice has a total graduated length of 100 ft. It is marked from 0 to 100 with full-foot increments, and the first foot at each end (from 0 to 1 and from 99 to 100) is graduated in tenths or in tenths and hundredths. With this kind of tape, the last partial tape length is measured by holding a full-foot graduation at the last chaining pin set such that the graduated section of the tape between the zero mark and the 1-ft mark straddles the closing point. This is indicated in Figure 6.2(b), where the 88-ft mark is being held on the last chaining pin and the tack marking the end of the line is opposite 0.32 ft read from the zero end of the tape. The partial tape length is then 88.00 − 0.32 = 87.68 ft. The quantity 0.32 ft is said to be *cut off;* hence this type of tape is called a *cut tape.* To ensure subtraction of a foot from the number at the full-foot graduation used, the following field procedure and calls are recommended: Rear tapeperson calls "88";

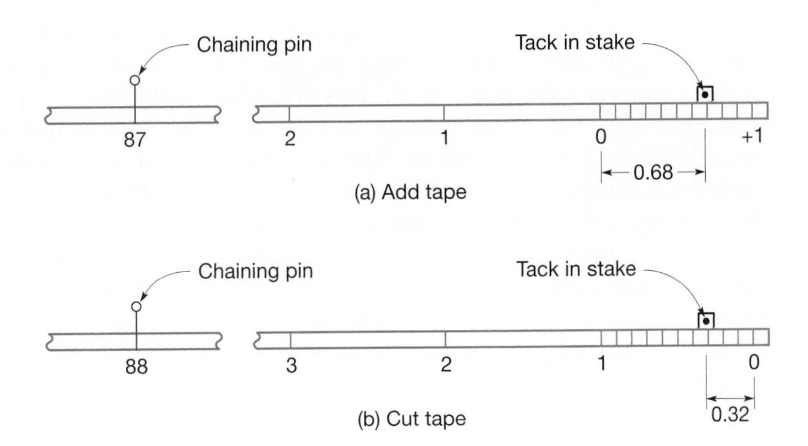

Figure 6.2 Reading partial tape lengths.

forward tapeperson says "cut point three-two"; rear tapeperson answers "eighty seven point six eight"; forward tapeperson replies "check."

An advantage of the add tape is that it is easier to use because no subtraction is needed when measuring decimal parts of a foot. Its disadvantage is that careless tapepersons will sometimes make measurements of 101.00 ft and record them as 100.00 ft. The cut tape practically eliminates this mistake. Cut tapes were most commonly used in railroad surveys where the curves were laid out within even 100-ft increments.

The same routine should be used throughout all taping by a party and the results tested in every possible way. A single mistake in subtracting the partial foot when using a cut tape will destroy the precision of a hundred other good measurements. For this reason, the add tape is more nearly foolproof. The greatest danger for mistakes in taping arises when changing from one style of tape to the other.

6.11.6 Recording the Distance

Accurate fieldwork may be canceled by careless recording. After the partial tape length is obtained at the end of a line, the rear tapeperson determines the number of full 100-ft tape lengths by counting the pins collected from the original set of 11. For distances longer than 1000 ft, a notation is made in the field book when the rear tapeperson has ten pins and one remains in the ground. This signifies a tally of 10 full tape lengths and has traditionally been called an "out." The forward tapeperson starts out again with ten pins and the process is repeated.

Although taping procedures may appear to be relatively simple, high precision is difficult to achieve, especially for beginners. Taping is a skill that can best be taught and learned by field demonstrations and practice. Except for short distances, taping has been replaced by electronic distance measurement.

■ 6.12 HORIZONTAL MEASUREMENTS ON SLOPING GROUND

In taping on uneven or sloping ground, it is standard practice to hold the tape horizontally and use a plumb bob at one or perhaps both ends. It is difficult to keep the plumb line steady for heights above the chest. Wind exaggerates this problem and may make accurate work impossible.

On steeper slopes, where a 100-ft length cannot be held horizontally without plumbing from above shoulder level, shorter distances are measured and accumulated to total a full tape length. This procedure, called *breaking tape,* is illustrated in Figure 6.3. As an example of this operation, assume that when taping downslope, the 100-ft end of the tape is held at the rear point, and the forward tapeperson can advance only 30 ft without being forced to plumb from above the chest. A pin is therefore set beneath the 70-ft mark, as in Figure 6.4. The rear tapeperson moves ahead to this pin and holds the 70-ft graduation there while another pin is set at, say, the 25-ft mark. Then, with the 25 ft graduation over the second pin, the full 100-ft distance is marked at the zero point. In this way, the partial tape lengths are added mechanically to make a full 100 ft by holding the proper graduations, and no mental arithmetic is required. The rear tapeperson returns the pins set at the intermediate points to the forward tapeperson to keep the tally

Figure 6.3
Breaking tape.

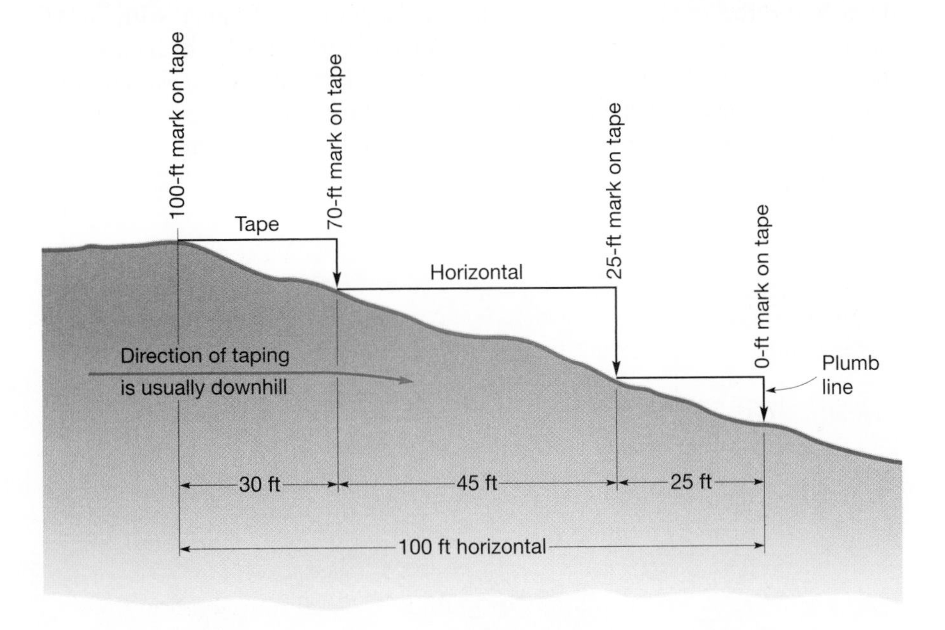

Figure 6.4
Procedure for breaking tape (when tape is not in box or on reel).

clear on the number of full tape lengths established. To avoid kinking the tape, the full 100-ft length is pulled ahead by the forward tapeperson into position for measuring the next tape length. In all cases the tape is leveled by eye or hand level, with the tapepersons remembering the natural tendency to have the downhill end of a tape too low. Practice will improve the knack of holding a tape horizontally by keeping it perpendicular to the vertical plumb-bob string.

Taping downhill is preferable to measuring uphill for two reasons. First, in taping downhill, the rear point is held steady on a fixed object while the other end is plumbed. In taping uphill, the forward point must be set while the other end is wavering somewhat. Second, if breaking tape is necessary, the head tapeperson can more conveniently use the hand level to proceed downhill a distance, which renders the tape horizontal when held comfortably at chest height.

6.13 SLOPE MEASUREMENTS

In measuring the distance between two points on a steep slope, rather than break tape every few feet, it may be desirable to tape along the slope and compute the horizontal component. This requires measurement also of either the angle of inclination α or the difference in elevation d (Figure 6.5). Breaking tape is more time consuming and generally less accurate due to the accumulation of random errors from marking tape ends and keeping the tape level and aligned for many short sections.

In Figure 6.5, if angle α is determined, the horizontal distance between points A and B can be computed from the relation

$$H = L \cos \alpha \tag{6.1}$$

where H is the horizontal distance between points, L the slope length separating them, and α the vertical angle from horizontal, usually obtained with an *Abney hand level* and *clinometer* (hand device for measuring angles of inclination). If the difference in elevation d between the ends of the tape is measured, which is done by leveling (see Chapter 5), the horizontal distance can be computed using the following expression derived from the Pythagorean Theorem:

$$H = \sqrt{L^2 - d^2} \tag{6.2a}$$

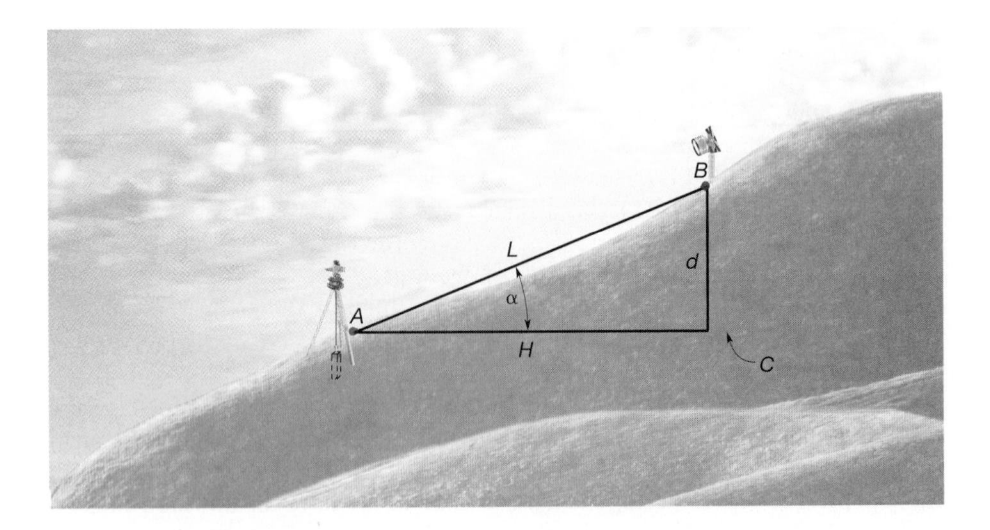

Figure 6.5
Slope measurement.

Another approximate formula, obtained from the first term of a binomial expansion of the Pythagorean Theorem, may be used in lower-order surveys to reduce slope distances to horizontal:

$$H = L - \frac{d^2}{2L} \text{(approx.)} \tag{6.2b}$$

In Equation (6.2b) the term $d^2/2L$ equals C in Figure 6.5 and is a correction to be subtracted from the measured slope length to obtain the horizontal distance. The error in using the approximate formula for a 100-ft length grows with increasing slope. Equation (6.2b) is useful for making quick estimates, without a calculator, or error sizes produced for varying slope conditions. It should not be used as an alternate method of Equation (6.2a) when reducing slope distances.

■ 6.14 SOURCES OF ERROR IN TAPING

There are three fundamental sources of error in taping:

1. *Instrumental errors.* A tape may differ in actual length from its nominal graduated length because of a defect in manufacture or repair, or as a result of kinks.
2. *Natural errors.* The horizontal distance between end graduations of a tape varies because of the effects of temperature, wind, and weight of the tape itself.
3. *Personal errors.* Tapepersons setting pins, reading the tape, or manipulating the equipment.

The most common types of taping errors are discussed in the subsections that follow. They stem from instrumental, natural, and personal sources. Some types produce systematic errors, while others produce random errors.

6.14.1 Incorrect Length of Tape

Incorrect length of a tape can be one of the most important errors. It is systematic. Tape manufacturers do not guarantee steel tapes to be exactly their graduated nominal length—for example, 100.00 ft—nor do they provide a standardization certificate unless requested and paid for as an extra. The true length is obtained by comparing it with a standard tape or distance. The National Institute of Standards and Technology (NIST)[1] of the U.S. Department of Commerce will make such a comparison and certify the exact distance between end graduations under given conditions of temperature, tension, and manner of support. A 100-ft steel tape usually is standardized for each of two sets of conditions—for example, 68°F, a 12-lb pull, with the tape lying on a flat surface (fully supported throughout); and 68°F, a 20-lb pull, with the tape supported at the ends only. Schools and surveying

---

[1]Information on tape calibration services of the National Institute of Standards and Technology can be obtained at the following website: http://www.nist.gov. Tapes can be sent for calibration to the National Institute of Standards and Technology, Building 220, Room 113, 100 Bureau Dr., Gaithersburg, Md. 20899; telephone: (301) 975–2465.

offices often have a precisely measured 100-ft line or at least one standardized tape that is used only to check other tapes subjected to wear.

An error caused by incorrect length of a tape occurs each time the tape is used. If the true length, known by standardization, is not exactly equal to its nominal value of 100.00 ft recorded for every full length, the correction can be determined as

$$C_L = \left(\frac{l - l'}{l'}\right)L \tag{6.3}$$

where C_L is the correction to be applied to the measured (recorded) length of a line to obtain the true length, l the actual tape length, l' the nominal tape length, and L the measured (recorded) length of line. Units for the terms in Equation (6.3) can be in either feet or meters.

6.14.2 Temperature Other Than Standard

Steel tapes are standardized for 68°F (20°C) in the United States. A temperature higher or lower than this value causes a change in length that must be considered. The coefficient of thermal expansion and contraction of steel used in ordinary tapes is approximately 0.00000645 per unit length per degree Fahrenheit, and 0.0000116 per unit length per degree Celsius. For any tape, the correction for temperature can be computed as

$$C_T = k(T_1 - T)L \tag{6.4}$$

where C_T is the correction in the length of a line caused by nonstandard temperature, k the coefficient of thermal expansion and contraction of the tape, T_1 the tape temperature at time of measurement, T the tape temperature when it has standard length, and L the observed (recorded) length of line. The correction C_T will have the same units as L, which can be either feet or meters. Errors caused by temperature change may be practically eliminated by either (a) measuring temperature and making corrections according to Equation (6.4), or (b) using an Invar tape.

Errors caused by temperature changes are systematic and have the same sign if the temperature is always above 68°F or always below that standard. When the temperature is above 68°F during part of the time occupied in measuring a long line, and below 68°F for the remainder of the time, the errors tend to partially balance each other, but corrections should still be computed and applied.

Temperature effects are difficult to assess in taping. The air temperature read from a thermometer may be quite different from that of the tape to which it is attached. Sunshine, shade, wind, evaporation from a wet tape, and other conditions make the tape temperature uncertain. Field experiments prove that temperatures on the ground or in the grass may be 10 to 25° higher or lower than those at shoulder height because of a 6-in. "layer of weather" (microclimate) on top of the ground. Since a temperature difference of 15°F produces a change of 0.01 ft per 100-ft tape length, the importance of such large variations is obvious.

Shop measurements made with steel scales and other devices likewise are subject to temperature effects. The precision required in fabricating a large airplane or ship can be lost by this one cause alone.

6.14.3 Inconsistent Pull

When a steel tape is pulled with a tension greater than its *standard pull* (the tension at which it was calibrated), the tape will stretch and become longer than its standard length. Conversely, if less than standard pull is used, the tape will be shorter than its standard length. The *modulus of elasticity* of the tape regulates the amount that it stretches. The correction for pull can be computed and applied using the following formula

$$C_P = (P_1 - P)\frac{L}{AE} \tag{6.5}$$

where C_P is the total elongation in tape length due to pull, in feet; P_1 the pull applied to the tape at the time of the observation, in pounds; P the standard pull for the tape in pounds; A the cross-sectional area of the tape in square inches; E the modulus of elasticity of steel in pounds per square inch; and L the observed (recorded) length of line. An average value of E is 29,000,000 lb/in.^{2} for the kind of steel typically used in tapes. In the metric system, to produce the correction C_p in meters, comparable units of P and P_1 are kilograms, L is meters, A is square centimeters, and E is kilograms per square centimeter. An average value of E for steel in these units is approximately 2,000,000 kg/cm^2.

The cross-sectional area of a steel tape can be obtained from the manufacturer, by measuring its width and thickness with calipers, or by dividing the total tape weight by the product of its length (in feet) times the unit weight of steel (490 lb/ft^2), and multiplying by 144 to convert square feet to square inches.

Errors resulting from incorrect tension can be eliminated by (a) using a spring balance to measure and maintain the standard pull, or (b) applying a pull other than standard and making corrections for the deviation from standard according to Equation (6.5).

Errors caused by incorrect pull may be either systematic or random. The pull applied by even an experienced tapeperson is sometimes greater or less than the desired value. An inexperienced person, particularly one who has not used a spring balance on a tape, is likely to apply less than the standard tension consistently.

6.14.4 Sag

A steel tape not supported along its entire length sags in the form of a *catenary curve,* a good example being the cable between two power poles. Because of sag, the horizontal distance (chord length) is less than the graduated distance between tape ends, as illustrated in Figure 6.6. Sag can be reduced by applying greater tension, but not eliminated unless the tape is supported throughout. The following formula is used to compute the sag correction:

$$C_S = -\frac{w^2 L_S^3}{24P_1^2} \tag{6.6}$$

where in the English system C_S, is the correction for sag (difference between length of curved tape and straight line from one support to the next), in feet; L_S the

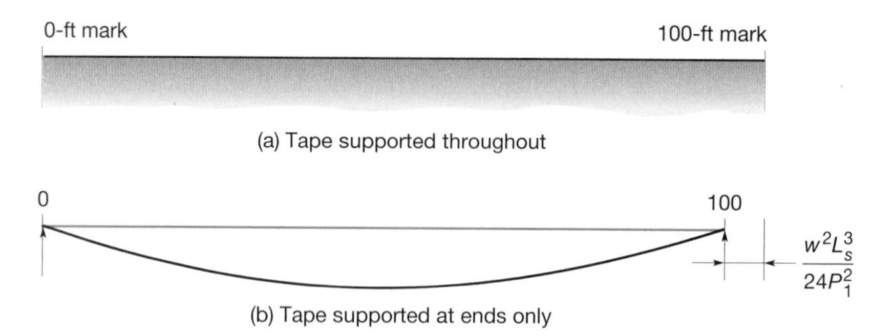

Figure 6.6 Effect of sag.

unsupported length of the tape, in feet; w the weight of the tape per foot of length, in pounds; and P_1 the pull on the tape, in pounds. Metric system units for Equation (6.6) are kg/m for w, kg for P_1, and meters for C_S and L_S.

The effects of errors caused by sag can be eliminated by (a) supporting the tape at short intervals or throughout, or (b) computing a sag correction for each unsupported segment and applying the total to the recorded length according to Equation (6.6). It is important to recognize that Equation (6.6) is nonlinear and must be applied to each unsupported section of the tape. It is incorrect to apply it to the overall length of a line unless the line was observed in one section.

As stated previously, when lines of unknown length are being measured, sag corrections are always negative, whereas positive corrections occur if the tension applied exceeds the standard pull. For any given tape, the so-called *normal tension* needed to offset these two factors can be obtained by setting Equations (6.5) and (6.6) equal to each other and solving for P_1. Although applying the normal tension does eliminate the need to make corrections for both pull and sag, it is not commonly used because the required pull is often too great for convenient application.

6.14.5 Tape Not Horizontal and Tape Off-Line

Corrections for errors caused by a tape being inclined in the vertical plane are computed in the same manner as corrections for errors resulting from it being off-line in the horizontal plane. Corrected lengths can be determined by Equation (6.2), where in the vertical plane d is the difference in elevation between the tape ends, and in the horizontal plane d is the amount that one end of the tape is off-line. In either case, L is the length of tape involved in the observation.

Errors caused by the tape not being horizontal are systematic and always make recorded lengths longer than true lengths. They are reduced by using a hand level to keep elevations of the tape ends equal, or by running differential levels (see Section 5.4) over the taping points, and applying corrections for elevation differences. Errors from the tape being off-line are also systematic and they too make recorded lengths longer than true lengths. This type of error can be eliminated by careful alignment.

6.14.6 Improper Plumbing

Practice and steady nerves are necessary to hold a plumb bob still long enough to mark a point. The plumb bob will sway, even in calm weather. On very gradual slopes and on smooth surfaces such as pavements, inexperienced tapepersons obtain better results by laying the tape on the ground instead of plumbing. Experienced tapepersons plumb most observations.

Errors caused by improper plumbing are random, since they may make distances either too long or too short. However, the errors would be systematic when taping directly against or in the direction of a strong wind. Heavier plumb bobs and touching the plumb bob on the ground, or steadying it with one foot, decreases its swing. Practice in plumbing will reduce errors.

6.14.7 Faulty Marking

Chaining pins should be set perpendicular to the taped line, but inclined 45° to the ground. This position permits plumbing to the point where the pin enters the ground without interference from the loop.

Brush, stones, and grass or weeds deflect a chaining pin and may increase the effect of incorrect marking. Errors from these sources tend to be random and are kept small by carefully locating a point, then checking it.

When taping on solid surfaces such as pavement or sidewalks, pencil marks or scratches can be used to mark taped segments. Accuracy in taping on the ground can be increased by using tacks in stakes as markers rather than chaining pins.

6.14.8 Incorrect Reading or Interpolation

The process of reading to hundredths on tapes graduated only to tenths, or to thousandths on tapes graduated to hundredths, is called interpolation. Errors from this source are random over the length of a line. They can be reduced by care in reading, employing a magnifying glass, or using a small scale to determine the last figure.

6.14.9 Summary of Effects of Taping Errors

An error of 0.01 ft is significant in many surveying measurements. Table 6.1 lists the nine types of taping errors; classifies them as instrumental (I), natural (N), or personal (P), and systematic (S) or random (R); and gives the departure from normal that produces an error of 0.01 ft in a 100-ft length.

The accepted method of reducing errors on precise work is to make separate measurements of the same line with different tapes, at different times of day, and in opposite directions. An accuracy of 1/10,000 can be obtained by careful attention to details.

■ 6.15 TAPE PROBLEMS

All tape problems develop from the fact that a tape is either longer or shorter than its graduated "nominal" length because of manufacture, temperature changes, tension applied, or some other reason. There are only two basic types of taping tasks: An unknown distance between two fixed points can be *measured,* or a required distance can be *laid off* from one fixed point. Since the tape may be too long or too short for either task, there are four possible versions of taping

TABLE 6.1 SUMMARY OF ERRORS

Error Type	Error Source*	Systematic (S) or Random (R)	Departure from Normal to Produce 0.01-ft Error for 100-ft Tape
Tape length	I	S	0.01 ft
Temperature	N	S or R	15°F
Pull	P	S or R	15 lb
Sag	N, P	S	0.6 ft at center for 100-ft tape standardized by support throughout
Alignment	P	S	1.4 ft at one end of 100-ft tape, or 0.7 ft at midpoint
Tape not level	P	S	1.4 ft elevation difference between ends of 100-ft tape
Plumbing	P	R	0.01 ft
Marking	P	R	0.01 ft
Interpolation	P	R	0.01 ft

*I—instrumental; N—natural; P—personal.

problems, which are (1) measure with a tape that is too long, (2) measure with a tape that is too short, (3) lay off with a tape that is too long, or (4) lay off with a tape that is too short. The solution of a particular problem is always simplified and verified by drawing a sketch.

Assume that the fixed distance AB in Figure 6.7 is measured with a tape that is found to be 100.03 ft as measured between the 0 and 100-ft marks. Then (the conditions in the figure are greatly exaggerated) the first tape length would extend to point 1; the next, to point 2; and the third, to point 3. Since the distance remaining from 3 to B is less than the correct distance from the true 300-ft mark to B, the *recorded* length AB is too small and must be increased by a correction. If the tape had been too short, the *recorded* distance would be too large, and the correction must be subtracted.

In laying out a required distance from one fixed point, the reverse is true. The correction must be subtracted from the desired length for tapes longer than their nominal value and added for tapes that are shorter. A simple sketch like Figure 6.7 makes clear whether the correction should be added or subtracted for any of the four cases.

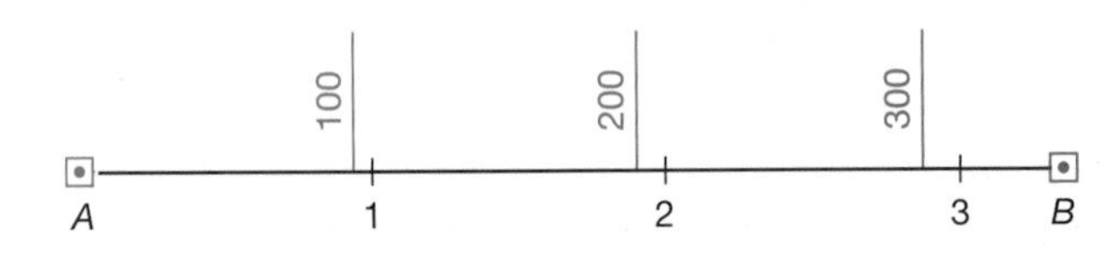

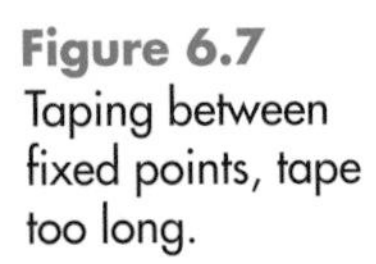

Figure 6.7 Taping between fixed points, tape too long.

6.16 COMBINED CORRECTIONS IN A TAPING PROBLEM

In taping linear distances, several types of systematic errors often occur simultaneously. The following examples illustrate procedures for computing and applying corrections for the two basic types of problems: *measurement* and *lay off.*

Example 6.1

A 30-m steel tape standardized at 20°C and supported throughout under a tension of 5.45 kg was found to be 30.012 m long. The tape had a cross-sectional area of 0.050 cm^2 and a weight of 0.03967 kg/m. This tape was held horizontal, supported at the ends only, with a constant tension of 9.09 kg, to measure a line from *A* to *B* in three segments. The data listed in the following table were recorded. Apply corrections for tape length, temperature, pull, and sag to determine the correct length of the line.

(a) The tape length correction by Equation (6.3) is:

$$C_L = \left(\frac{30.012 - 30.000}{30.000}\right) 81.151 = +0.0324 \text{ m}$$

(b) Temperature corrections by Equation (6.4) are:
(*Note:* Separate corrections are required for distances observed at different temperatures.)

Section	Measured (Recorded) Distance (m)	Temperature (°C)
A-1	30.00	14
1-2	30.00	15
2-*B*	21.151	16
	Σ81.151	

$$C_{T_1} = 0.0000116(14 - 20)30.000 = -0.0026 \text{ m}$$
$$C_{T_2} = 0.0000116(15 - 20)30.000 = -0.0017 \text{ m}$$
$$C_{T_3} = 0.0000116(16 - 20)21.151 = -0.0010 \text{ m}$$
$$\Sigma C_t = -0.0053 \text{ m}$$

(c) The pull correction by Equation (6.5) is:

$$C_P\left(\frac{9.09 - 5.45}{0.050 \times 2{,}000{,}000}\right) 81.151 = 0.0029 \text{ m}$$

(d) The sag corrections by Equation (6.6) are:
(*Note:* Separate corrections are required for the two suspended lengths.)

$$C_{S_1} = -2\left[\frac{(0.0397)^2(30.000)^3}{24 \times (9.09)^2}\right] = -0.0429 \text{ m}$$

$$C_{S_2} = -\left[\frac{(0.0397)^2(21.151)^3}{24 \times (9.09)^2}\right] = -0.0075 \text{ m}$$

$$\sum C_S = -0.0504 \text{ m}$$

(e) Finally, corrected distance AB is obtained by adding all corrections to the measured distance or

$$AB = 81.151 + 0.0324 - 0.0053 + 0.0029 - 0.0504 = 81.131 \text{ m}$$

Example 6.2

A 100-ft steel tape standardized at 68°F and supported throughout under a tension of 20 lb. was found to be 100.012 ft long. The tape had a cross-sectional area of 0.0078 in.^{2} and a weight of 0.0266 lb/ft. This tape is used to lay off a horizontal distance CD of exactly 175.00 ft. The ground is on a smooth 3% grade, thus the tape will be used fully supported. Determine the correct slope distance to lay off if a pull of 15 lb is used and the temperature is 87°F.

Solution

(a) The tape length correction, by Equation (6.3), is

$$C_L = \left(\frac{100.012 - 100.000}{100.000}\right)175.00 = +0.021 \text{ ft}$$

(b) The temperature correction, by Equation (6.4), is

$$C_T = 0.00000645(87 - 68)175.00 = +0.021 \text{ ft}$$

(c) The pull correction, by Equation (6.5), is

$$C_P = \frac{(15 - 20)}{0.078 \times 29{,}000{,}000} 175.00 = -0.0004 \text{ ft}$$

(d) Since this is a lay-off problem, all corrections are subtracted. Thus the required horizontal distance to lay off, rounded to the nearest hundredth of a foot, is

$$CD_h = 175.00 - 0.021 - 0.021 + 0.0004 = 174.96 \text{ ft}$$

(e) Finally, a rearranged form of Equation (6.2) is used to solve for the slope distance (the difference in elevation d for use in this equation, for 175 ft on a 3 percent grade, is $175(0.03) = 5.25$ ft):

$$CD_s = \sqrt{(174.96)^2 + (5.25)^2} = 175.04 \text{ ft}$$

PART III • ELECTRONIC DISTANCE MEASUREMENT

6.17 INTRODUCTION

A major advance in surveying instrumentation occurred approximately 60 years ago with the development of electronic distance measuring (EDM) instruments. These devices measure lengths by indirectly determining the number of full and partial waves of transmitted, electromagnetic energy required in traveling between the two ends of a line. In practice, the energy is transmitted from one end of the line to the other and returned to the starting point; thus it travels the double path distance. Multiplying the total number of cycles by its wavelength and dividing by 2, yields the unknown distance.

The Swedish physicist Erik Bergstrand introduced the first EDM instrument in 1948. His device, called the *geodimeter* (an acronym for geodetic distance meter), resulted from attempts to improve methods for measuring the velocity of light. The instrument transmitted visible light and was capable of accurately observing distances up to about 25 mi (40 km) at night. In 1957, a second EDM apparatus, the *tellurometer,* was introduced. Designed in South Africa by Dr. T. L. Wadley, this instrument transmitted microwaves and was capable of observing distances up to 50 mi (80 km) or more, day or night.

The potential value of these early EDM models to the surveying profession was immediately recognized. However, they were expensive and not readily portable for field operations. Furthermore, observing procedures were lengthy, and mathematical reductions to obtain distances from observed values were difficult and time consuming. Continued research and development have overcome all of these deficiencies. Prior to the introduction of EDM instruments, taping was used to make accurate distance measurements. Although seemingly a relatively simple procedure, precise taping is one of the most difficult and painstaking of all surveying tasks. Now EDM instruments have made it possible to obtain accurate distance measurements rapidly and easily. Given a line of sight, long or short lengths can be measured over bodies of water, busy freeways, or terrain that is inaccessible for taping. Accuracies of $\pm$(1 to 3) mm can be achieved by today's instruments.

In the current generation, EDM instruments are combined with *digital theodolites* and *microprocessors* to produce *total station instruments* (see Figures 1.3 and 2.5). These devices can simultaneously and automatically observe both distances and angles. The microprocessor receives the measured slope length and zenith (or vertical) angle, calculates horizontal and vertical distance components, and displays them in real time. When equipped with *data collectors* (see Section 2.12), they can record field notes electronically for transmission to computers, plotters, and other office equipment for processing. These so-called *field-to-finish* systems are gaining worldwide acceptance and changing the practice of surveying substantially.

6.18 PROPAGATION OF ELECTROMAGNETIC ENERGY

Electronic distance measurement is based on the rate and manner that electromagnetic energy propagates through the atmosphere. The rate of propagation can be expressed with the following equation

$$V = f\lambda \tag{6.7}$$

where V is the velocity of electromagnetic energy, in meters per second; f the modulated frequency of the energy, in hertz;[2] and λ the wavelength, in meters. The velocity of electromagnetic energy in a vacuum is 299,792,458 m/sec. Its speed is slowed somewhat in the atmosphere according to the following equation

$$V = c/n \tag{6.8}$$

where c is the velocity of electromagnetic energy in a vacuum, and n the atmospheric *index of refraction*. The value of n varies from about 1.0001 to 1.0005, depending on pressure and temperature, but is approximately equal to 1.0003. Thus, accurate electronic distance measurement requires that atmospheric pressure and temperature be measured so that the appropriate value of n can be determined.

Temperature, atmospheric pressure, and relative humidity all have an effect on the index of refraction. Because a light source emits light composed of many wavelengths, and since each wavelength has a different index of refraction, the group of waves has a *group index of refraction*. The value for the group refractivity N_g in *standard air*[3] for electronic distance measurement is

$$N_g = (n_g - 1)10^6 = 287.6155 + \frac{4.88660}{\lambda^2} + \frac{0.06800}{\lambda^4} \tag{6.9}$$

where λ is the wavelength of the light expressed in micrometers (μm) and n_g is the group refractive index. The wavelengths of light sources commonly used in EDMs are 0.6328 μm for red laser and 0.900 to 0.930 μm for infrared.

The actual group refractive index n_a for atmosphere at the time of observation due to variations in temperature, pressure, and humidity can be computed as

$$n_a = 1 + \left(\frac{273.15}{1013.25} \cdot \frac{N_g P}{t + 273.15} - \frac{11.27e}{t + 273.15}\right) 10^{-6} \tag{6.10}$$

where e is the partial water vapor pressure in hectopascal[4] (hPa) as defined by the temperature and relative humidity at the time of the measurement, P is the pressure in hPa, and t is the dry-bulb temperature in °C. The partial water-vapor pressure, e, can be computed with sufficient accuracy for normal operating conditions as

$$e = E \cdot h/100 \tag{6.11}$$

where $E = 10^{[(7.5t/237.3+t)+0.7858]}$ and h is the relative humidity in percent.

---

[2]The hertz (Hz) is a unit of frequency equal to 1 cycle/sec. The kilohertz (KHz), megahertz (MHz), and gigahertz (GHz) are equal to 10^3 Hz, 10^6 Hz, and 10^9 Hz, respectively.

[3]A standard air is defined with the following conditions: 0.0375 percent carbon dioxide, temperature of 0°C, pressure of 760 mm of mercury, and 0 percent humidity.

[4]1 Atmosphere = 101.325 kPa = 1013.25 hPa = 760 torr = 760 mm-Hg

Example 6.3

What is the actual wavelength and velocity of a near-infrared beam ($\lambda = 0.895\ \mu m$) of light modulated at a frequency of 320 MHz through an atmosphere with a (dry) temperature, t, of 34°C, relative humidity h of 56 percent, and an atmospheric pressure of 1041.25 hPa?

Solution

By Equation (6.9), $N_g = 287.6155 + \dfrac{4.88660}{(0.915)^2} + \dfrac{0.06800}{(0.915)^4} = 293.5491746$

By Equation (6.11),

$$a = \frac{7.5 \times 34}{(237.3 + 34)} + 0.7858 = 1.7257$$

$$E = 10^a = 53.18$$

$$e = Eh = 53.18 \times 56/100 = 29.7788$$

By Equation (6.10),

$$n_a = 1 + \left(\frac{273.15}{1013.25} \cdot \frac{293.5492 \times 1041.25}{34 + 273.15} - \frac{11.27 \times 29.7788}{34 + 273.15}\right)10^{-6}$$

$$= 1 + (268.268660 - 1.09265)10^{-6}$$

$$= 1.0002672$$

By Equation (6.8),

$$V = 299{,}792{,}458/1.0002672 = 299{,}712{,}382 \text{ m/s}$$

Rearranging Equation (6.7) yields an actual wavelength of

$$\lambda = 299{,}712{,}382/320{,}000{,}000 = 0.9366012\ \mu m$$

Note in the solution of Example 6.3 that the second parenthetical term in Equation (6.10) accounts for the effects of humidity in the atmosphere. In fact, if this term were ignored the actual index of refraction n_a would become 1.0002683, resulting in the same computed wavelength to five decimal places. This demonstrates why, in using EDM instruments that employ near-infrared light, the effects of humidity on the transmission of the wave can be ignored for all but the most precise work. The student should verify this fact.

The manner by which electromagnetic energy propagates through the atmosphere can be represented conceptually by the sinusoidal curve illustrated in Figure 6.8. This figure shows one wavelength, or *cycle*. Portions of wavelengths or the positions of points along the wavelength are given by phase angles. Thus in Figure 6.8, a 360° *phase angle* represents a full cycle, or a point at the end of a

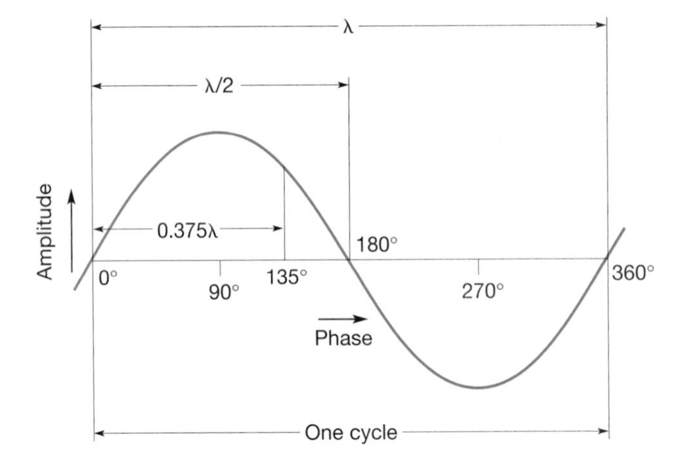

Figure 6.8 A wavelength of electromagnetic energy illustrating phase angles.

wavelength, while 180° is a half wavelength, or the midpoint. An intermediate position along a wavelength having a phase angle of, say, 135° is 135/360, or 0.375 of a wavelength.

6.19 PRINCIPLES OF ELECTRONIC DISTANCE MEASUREMENT

In Section 6.17 it was stated that distances are observed electronically by determining the number of full and partial waves of transmitted electromagnetic energy that are required in traveling the distance between the two ends of a line. In other words, this process involves determining the number of wavelengths in an unknown distance. Then, knowing the precise length of the wave, the distance can be determined. This is similar to relating an unknown distance to the calibrated length of a steel tape.

The procedure of measuring a distance electronically is depicted in Figure 6.9, where an EDM device has been centered over station A by means of a plumb bob or optical plumbing device. The instrument transmits a *carrier signal* of electromagnetic energy to station B. A *reference frequency* of a precisely regulated wavelength has been superimposed or *modulated* onto the carrier. A reflector at B returns the signal to the receiver, so its travel path is double the slope distance AB. In the figure, the modulated electromagnetic energy is represented by a series of sine waves, each having wavelength λ. The unit at A determines the number of wavelengths in the double path, multiplied by the wavelength in feet or meters, and divided by 2 to obtain distance AB.

Of course, it would be highly unusual if a measured distance was exactly an integral number of wavelengths, as illustrated in Figure 6.9. Rather, some fractional part of a wavelength would in general be expected; for example, see the partial value p shown in Figure 6.10. In that figure, distance L between the EDM instrument and reflector would be expressed as

$$L = \frac{n\lambda + p}{2} \tag{6.12}$$

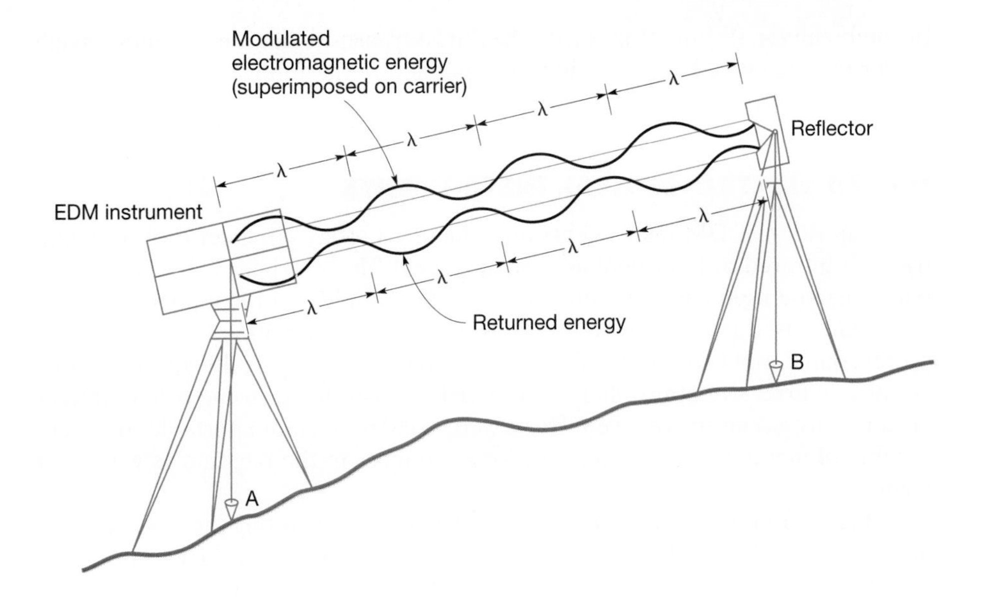

Figure 6.9
Generalized EDM procedure.

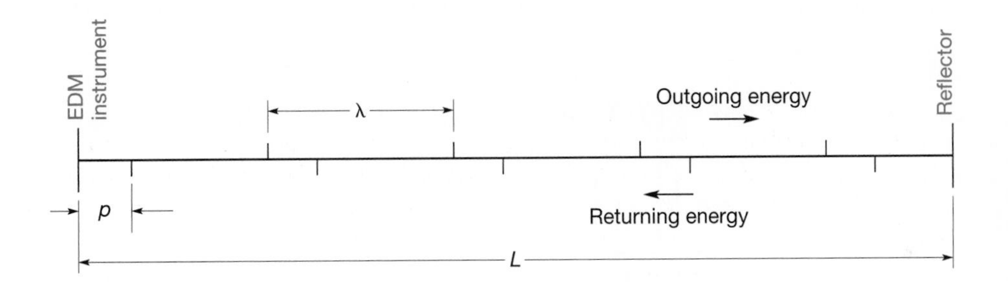

Figure 6.10
Phase difference measurement principle.

where λ is the wavelength, n the number of full wavelengths, and p the length of the fractional part. The fractional length is determined by the EDM instrument from measurement of the *phase shift* (phase angle) of the returned signal. To illustrate, assume that the wavelength for the example of Figure 6.10 was precisely 20.000 m. Assume also that the phase angle of the returned signal was 115.7°, in which case length p would be $(115.7/360) \times 20.000 = 6.428$ m. Then from the figure, since $n = 9$, by Equation (6.12), length L is

$$L = \frac{9(20.000) + 6.428}{2} = 93.214 \text{ m}$$

Considering the double path distance, the 20-m wavelength used in the example just given has an "effective wavelength" of 10 m. This is one of the fundamental wavelengths used in current EDM instruments. It is generated using a frequency of approximately 15 MHz.

EDM instruments cannot determine the number of full wavelengths in an unknown distance by transmitting only one frequency and wavelength. To resolve the ambiguity n, in Equation (6.12), they must transmit additional signals having longer wavelengths. This procedure is explained in the following section, which describes electro-optical EDM instruments.

■ 6.20 ELECTRO-OPTICAL INSTRUMENTS

The majority of EDM instruments manufactured today are electro-optical, and transmit infrared or laser light as a carrier signal. This is primarily because its intensity can be modulated directly, considerably simplifying the equipment. Earlier models used tungsten or mercury lamps. They were bulky, required a large power source, and had relatively short operating ranges, especially during the day because of excessive atmospheric scatter. EDM instruments using coherent light produced by gas lasers followed. These were smaller and more portable, and were capable of making measurements of long distances in the daytime as well as at night.

Figure 6.11 is a generalized schematic diagram illustrating the basic method of operation of one particular type of electro-optical instrument. The transmitter uses a GaAs diode that emits *amplitude-modulated* (AM) infrared light. A crystal oscillator precisely controls the frequency of modulation. The modulation process may be thought of as similar to passing light through a stovepipe in which a damper

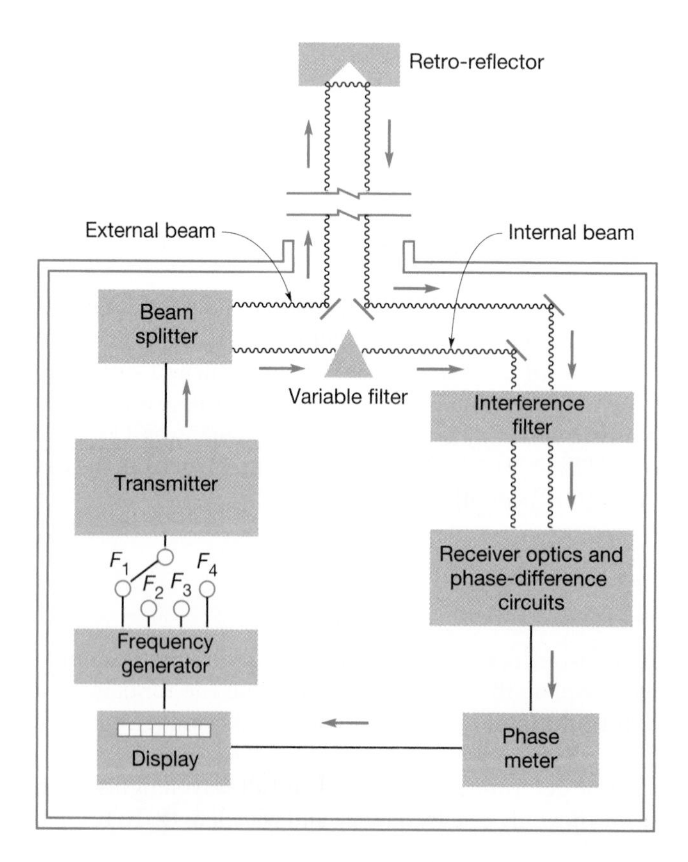

Figure 6.11 Generalized block diagram illustrating operation of electro-optical EDM instrument.

plate is spinning at a precisely controlled rate or frequency. When the damper is closed, no light passes. As it begins to open, light intensity increases to a maximum at a phase angle of 90° with the plate completely open. Intensity reduces to zero again with the damper closed at a phase angle of 180°, and so on. This intensity variation or amplitude modulation is properly represented by sine waves such as those shown in Figures 6.8 and 6.9.

As shown in Figure 6.11, a beam splitter divides the light emitted from the diode into two separate signals: an *external* measurement beam and an *internal* reference beam. By means of a telescope mounted on the EDM instrument, the external beam is carefully aimed at a retro-reflector that has been centered over the point at the line's other end. Figure 6.12 shows a triple corner cube retro-reflector of the type used to return the external beam, coaxial, to the receiver.

The internal beam passes through a variable-density filter and is reduced in intensity to a level equal to that of the returned external signal, enabling a more accurate measurement to be made. Both internal and external signals go through an interference filter, which eliminates undesirable energy such as sunlight. The internal and external beams then pass through components to convert them into electric energy while preserving the phase shift relationship resulting from their different travel path lengths. A phase meter converts this phase difference into direct current having a magnitude proportional to the differential phase. This current is connected to a *null meter* that is adjusted to null the current. The fractional wavelength is measured during the nulling process, converted to distance, and displayed.

To resolve the ambiguous number of full cycles a wave has undergone, EDM instruments transmit different modulation frequencies. The unit illustrated in the schematic of Figure 6.11 uses four frequencies: $F1$, $F2$, $F3$, and $F4$, as indicated. If

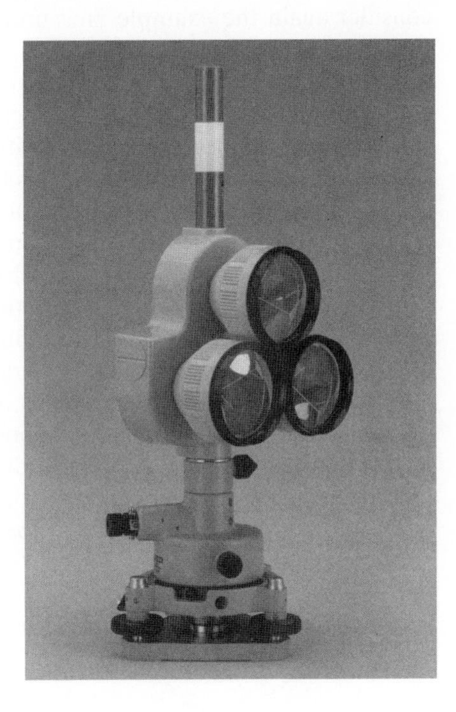

Figure 6.12
Triple retro-reflector. (Courtesy Topcon Positioning Systems)

modulation frequencies of 14.984 MHz, 1.4984 MHz, 149.84 KHz, and 14.984 KHz are used, and assuming the index of refraction is 1.0003, then their corresponding "effective" wavelengths are 10.000, 100.00, 1000.0, and 10,000 m, respectively. Assume that a distance of 3867.142 appears on the display as the result of measuring a line. The four rightmost digits, 7.142, are obtained from the phase shift measured while transmitting the 10.000-m wavelength at frequency *F*1. Frequency *F*2, having a 100.00-m wavelength, is then transmitted, yielding a fractional length of 67.14. This provides the digit 6 in the displayed distance. Frequency *F*3 gives a reading of 867.1, which provides the digit 8 in the answer, and finally, frequency *F*4 yields a reading of 3867, which supplies the digit 3, to complete the display. From this example, it should be evident that the high resolution of a measurement (nearest 0.001 m) is secured using the 10.000-m wavelength, and the others simply resolve the ambiguity of the number of these shorter wavelengths in the total distance.

With older instruments, changing of frequencies and nulling were done manually by setting dials and turning knobs. Now modern instruments incorporate microprocessors that control the entire measuring process. Once the instrument is aimed at the reflector and the measurement started, the final distance appears in the display almost instantaneously. Other changes in new instruments include improved electronics to control the amplitude modulation, and replacement of the null meter by an electronic phase detector. These changes have significantly improved the accuracy with which phase shifts can be determined, that in turn has reduced the number of different frequencies that need to be transmitted. Consequently, as few as two frequencies are now used on some instruments: one that produces a short wavelength to provide the high-resolution digits, and one with a long wavelength to provide the coarse numbers. To illustrate how this is possible, consider again the example measurement just described, which used four frequencies. Recall that a reading of 7.142 was obtained with the 10.000-m wavelength, and that 3867 was read with the 10,000-m wavelength. Note the overlap of the common digit 7 in the two readings. Assuming that both phase shift measurements are reliably made to four significant figures, the leftmost digit of the first reading should indeed be the same as the rightmost one of the second reading. If these digits are the same in the measurement, this provides a check on the operation of the instrument. Modern instruments compare these overlapping digits and will display an error message if they do not agree. If they do check, the displayed distance will take all four digits from the first (short-wavelength) reading and the first three digits from the second reading.

Manufacturers provide a full range of instruments with precisions that vary from ±(1 mm + 1 ppm) to ±(10 mm + 5 ppm).[5] Earlier versions were manufactured to stand alone on a tripod, and thus from any setup they could only measure distances. Now, as noted earlier, in most instances EDMs are combined with electronic digital theodolites to produce our modern and very versatile total station instruments. These are described in the following section.

---

[5]Accuracies in electronic distance measurements are quoted in two parts; the first part is a constant, and the second is proportional to the distance measured. The abbreviation ppm = parts per million. One ppm equals 1 mm/km. In a distance 5000 ft long, a 5-ppm error equals $5000 \times (5 \times 10^{-6}) = 0.025$ ft.

■ 6.21 TOTAL STATION INSTRUMENTS

Total station instruments (also sometimes called electronic tacheometers) combine an EDM instrument, an electronic digital theodolite, and a computer in one unit. These devices, described in more detail in Chapter 8, automatically measure horizontal and zenith (or vertical) angles, as well as distances, and transmit the results in real time to a built-in computer. The horizontal and zenith (or altitude) angle and slope distance can be displayed, and then upon keyboard commands, horizontal and vertical distance components can be instantaneously computed from these data and displayed. If the instrument is oriented in direction and the coordinates of the occupied station are input to the system, the coordinates of any point sighted can be immediately obtained. These data can all be stored within the instrument, or in a data collector, thereby eliminating manual recording.

Total station instruments are of tremendous value in all types of surveying, as will be discussed in later portions of this text. Besides automatically computing and displaying horizontal and vertical components of a slope distance, and coordinates of points sighted, total station instruments can be operated in the *tracking mode.* In this mode, sometimes also called *stakeout,* a required distance (horizontal, vertical, or slope) can be entered by means of the control panel, and the instrument's telescope aimed in the proper direction. Then as the reflector is moved forward or back in position, the difference between the desired distance and that to the reflector is rapidly updated and displayed. When the display shows the difference to be zero, the required distance has been established and a stake is set. This feature, extremely useful in construction stakeout, is described further in Section 23.9.

The total station instruments shown in Figures 2.5, 6.13, and 8.2 all have a distance range of approximately 3 km (using a single prism) with an accuracy of $\pm(2\text{ mm} + 2\text{ ppm})$ and read angles to the nearest $1''$.

Figure 6.13
The LEICA TC1101 total station. (Courtesy Leica Geosystems, Inc.)

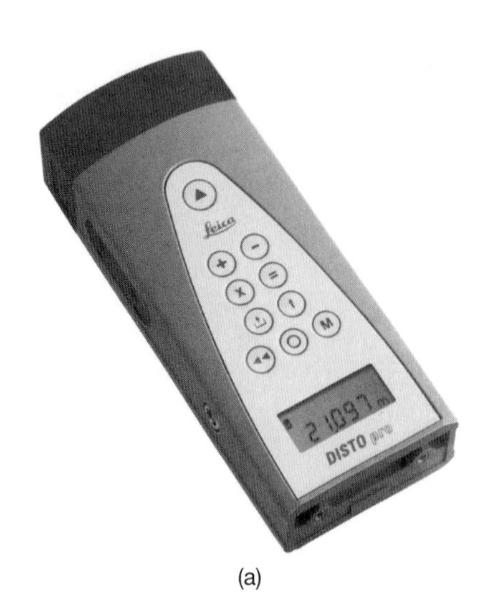

(a)

(b)

Figure 6.14 (a) The LEICA DISTO hand-held laser distance measuring instrument, (b) using the LECIA DISTO to measure to an inaccessible point. (Courtesy Leica Geosystems, Inc.)

■ 6.22 EDM INSTRUMENTS WITHOUT REFLECTORS

Recently some EDM instruments have been introduced that do not require reflectors for distance observation. These devices use time-pulsed infrared laser signals, and in their *reflectorless* mode of operation they can observe distances up to 100 m in length. Figure 6.14(a) shows a hand-held laser distance-measuring instrument.

Some total station instruments, like that shown in Figure 6.13, utilize laser signals and can also observe distances up to 100 m in the reflectorless mode. But as noted earlier, with prisms they can observe lengths up to 3 km. As described in Section 17.9.6, some reflectorless total station instruments are capable of scanning scenes automatically providing a uniform grid of observations and coordinates of all objects in the scene. These instruments are particularly useful in industrial surveying and in highway surveys where a high level of detail is required.

Using instruments in the reflectorless mode, observations can be made to inaccessible objects such as the features of a building as shown in Figure 6.14(b), faces of dams and retaining walls, structural members being assembled on bridges, and so on. These instruments can increase the speed and efficiency of surveys in any construction or fabrication project, especially when measuring to features that are inaccessible.

■ 6.23 COMPUTING HORIZONTAL LENGTHS FROM SLOPE DISTANCES

All EDM equipment measures the slope distance between two stations. As noted earlier, if the EDM unit is incorporated into a total station instrument, then it can reduce these distances to their horizontal components automatically [if the zenith (or vertical) angle is input]. With some of the earliest EDMs, this could not be

Figure 6.15 Reduction of EDM slope distance to horizontal.

done, and reductions were carried out manually. The procedures used, whether performed internally by the microprocessor or done manually, follow those outlined in this section. It is presumed, of course, that slope distances are first corrected for instrumental and atmospheric conditions.

Reduction of slope distances to horizontal can be based on elevation differences, or on zenith (or vertical) angle. Because of Earth curvature, long lines must be treated differently in reduction than short ones and will be discussed in Chapter 19.

6.23.1 Reduction of Short Lines by Elevation Differences

If difference in elevation is used to reduce slope distances to horizontal, during field operations heights h_e of the EDM instrument, and h_r of the reflector above their respective stations, are measured and recorded (see Figure 6.15). If elevations of stations A and B in the figure are known, Equation (6.2) will reduce the slope distance to horizontal, with the value of d (difference in elevation between EDM instrument and reflector) computed as follows:

$$d = (elev_A + h_e) - (elev_B + h_r) \quad \textbf{(6.13)}$$

■ Example 6.4

A slope distance of 165.360 m (corrected for meteorological conditions) was measured from A to B, whose elevations were 447.401 and 445.389 m above datum, respectively. Find the horizontal length of line AB if the heights of the

EDM instrument and reflector were 1.417 and 1.615 m above their respective stations.

Solution

By Equation (6.3), $d = (447.401 + 1.417) - (445.389 + 1.615) = 1.814$ m

By Equation (6.2), $H = \sqrt{(165.360)^2 - (1.814)^2} = 165.350$ m

6.23.2 Reduction of Short Lines by Zenith or Altitude Angle

If zenith angle z (angle measured downward from the upward direction of the plumb line) is observed to the inclined path of the transmitted energy when measuring slope distance L (see Figure 6.15), then the following equation is applicable to reduce the slope length to its horizontal component:

$$H = L\sin(z) \tag{6.14}$$

If altitude angle α (angle between horizontal and the inclined energy path) is observed (see Figure 6.15), then Equation (6.1) is applicable for the reduction. For most precise work, especially on longer lines, the zenith (or altitude) angle should be observed in both the direct and reversed modes, and averaged (see Section 8.13). Also, the mean of the vertical angles observed from both ends of the line will compensate for Earth curvature and refraction. The reduction of distance observations made on long lines is covered in Section 19.14.

■ 6.24 ERRORS IN ELECTRONIC DISTANCE MEASUREMENT

As noted earlier, accuracies of EDM instruments are quoted in two parts: a constant error and a scalar error proportional to the distance measured. Specified errors vary for different instruments, but the constant portion is usually about ±2 mm, and the proportion is generally about ±2 ppm. The constant error is most significant on short distances; for example, with an instrument having a constant error of ±2 mm, a measurement of 20 m is good to only 2/20,000 = 1/10,000, or 100 ppm. For a long distance, say 2 km, the constant error becomes negligible and the proportional part more important.

The major error components in an observed distance are instrument and target miscentering, and the specified constant and scalar errors of the EDM instrument. Using Equation (3.11), the error in an observed distance is computed as

$$E_d = \sqrt{E_i^2 + E_r^2 + E_c^2 + (ppm \times D)^2} \tag{6.15}$$

where E_i is the estimated miscentering error in the instrument; E_r is the estimated miscentering error in the reflector; E_c the specified constant error for the

EDM; *ppm* the specified scalar error for the EDM; and D the measured slope distance.

Example 6.5

A slope distance of 827.329 m was observed between two stations with an EDM instrument having specified errors of ±(2 mm + 2 ppm). The instrument was centered with an estimated error of ±3 mm. The estimated error in target miscentering was ±5 mm. What is the estimated error in the observed distance?

Solution

By Equation (6.15),

$$E_d = \sqrt{(3)^2 + (5)^2 + (2)^2 + (2 \times 10^{-6} \times 827329)^2} = \pm 6.4 \text{ mm}$$

Note in the solution that the distance of 827.329 m was converted to millimeters to obtain unit consistency. This solution results in a distance precision of 6.4/827,329, or better than 1:129,000.

From the foregoing, it is clear that except for very short distances, the order of accuracy possible with EDM instruments is very high. Errors can seriously degrade the measurements, however, and thus care should always be exercised to minimize their effects. Sources of error in EDM work may be personal, instrumental, or natural. The subsections that follow identify and describe errors from each of these sources.

6.24.1 Personal Errors

Personal errors include inaccurate setups of EDM instruments and reflectors over stations, faulty measurements of instrument and reflector heights [needed for computing horizontal lengths (see Section 6.23)], and errors in determining atmospheric pressures and temperatures. These errors are largely random. They can be minimized by exercising utmost care and by using good-quality barometers and thermometers.

Mistakes (not errors) in manually reading and recording displayed distances are common and costly. They can be eliminated with some instruments by obtaining the readings in both feet and meters and comparing them. Of course, data collectors (see Section 2.12) also circumvent this problem.

An example of a common mistake is failing to set the temperature and pressure in an EDM before obtaining an observation. Assume this occurred with the atmospheric conditions given in Example 6.3. The actual index of refraction was computed as 1.0002672. If the fundamental wavelength for a standard atmosphere was 10.000 m, then the actual wavelength produced by the EDM would be

10.000/1.0002672 = 9.9973 m. Using Equation (6.3) with an observed distance of 827.329 m, the error, *e*, in the observed distance would be

$$e = \left(\frac{9.9973 - 10.000}{10.000}\right) \times 827.329 = -0.223 \text{ m}$$

The effect of failing to account for the actual atmospheric conditions produces a precision of only $|-0.223|/827.329$, or 1:3700. This is well below the computed precision of 1:129,000 in Example 6.5.

6.24.2 Instrumental Errors

If EDM equipment is carefully adjusted and precisely calibrated, instrumental errors should be extremely small. To assure their accuracy and reliability, EDM instruments should be checked against a first-order baseline at regular time intervals. For this purpose, the National Geodetic Survey has established a number of accurate baselines in each state.[6] These are approximately a mile long and placed in relatively flat areas. Monuments are set at the ends and at intermediate points along the baseline.

Although most EDM instruments are quite stable, occasionally they become maladjusted and generate erroneous frequencies. This results in faulty wavelengths that degrade distance observations in a manner similar to using a tape of incorrect length. Periodic checking of the equipment against a calibrated baseline will detect the existence of measurement errors. It is especially important to make these checks if high-order surveys are being conducted.

The corner cube reflectors used with EDM instruments are another source of instrumental error. Since light travels at a lower velocity in glass than in air, the "effective center" of the reflector is actually behind the prism. Thus, it frequently does not coincide with the plummet, a condition that produces a systematic error in distances known as the *reflector constant.* This situation is shown in Figure 6.16. Notice that because the retroreflector is composed of mutually perpendicular faces, the light always travels a total distance of $a + b + c = 2D$ in the prism. Additionally, given a refractive index for glass, which is greater than air, the velocity of light in the prism is reduced following Equation (6.8) to create an effective distance of nD where n is the index of refraction of the glass (approximately 1.517). The dashed line in Figure 6.16 shows the effective center thus created. The reflector constant, K in the figure, can be as large as 70 mm and will vary with reflectors.

Once known, the *electrical center* of the EDM can be shifted forward to compensate for the reflector constant. However, if an EDM instrument is being regularly used with several unmatched reflectors, this shift is impractical. In this

---

[6]For locations of baselines in your area, contact the NGS National Geodetic Information Center by email at: info_center@ngs.noaa.gov; at http://www.ngs.noaa.gov/CBLINES/calibration.html; by telephone at (301)713-3242; or by writing to NOAA, National Geodetic Survey, Station 09202, 1315 East West Highway, Silver Spring, MD 20910

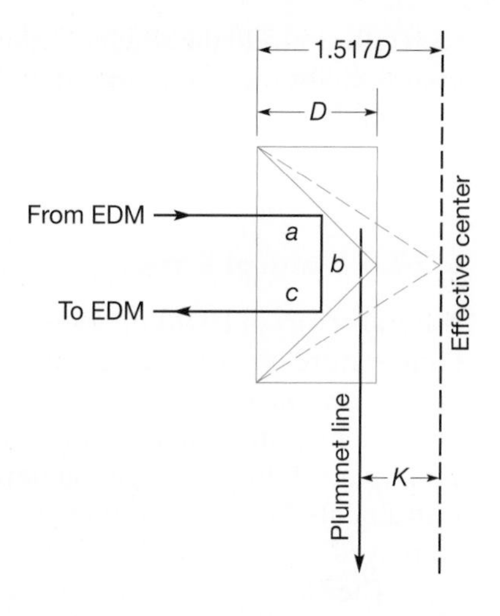

Figure 6.16 Schematic of retroreflector where D is the depth of the prism.

instance, the offset for each reflector should be subtracted from the observed distances to obtain corrected values.

With EDM instruments that are components of total stations and are controlled by microprocessors, this constant can be entered via the keyboard and included in the internally computed corrections. Equipment manufacturers also produce matching reflector sets for which the reflector constant is the same, thus allowing a single constant to be used for a set of reflectors with an instrument.

By comparing precisely known base-line lengths to observed distances, a so-called *system measurement constant* can be determined. This constant can then be applied to all subsequent measurements for proper correction. Although calibration using a base line is preferred, if one is not available, the constant can be obtained with the following procedure. Three stations, A, B, and C, should be established in a straight line on flat ground, with stations A and C a convenient distance apart, and station B approximately midway between these stations. The length of AC and the two components, AB and BC, should be observed. From these observations, the following equation can be written:

$$AC + K = (AB + K) + (BC + K)$$

from which

$$K = AC - (AB + BC) \tag{6.16}$$

where K is the system measurement constant to be added to correct observed distances.

The procedure, including centering of the EDM instrument and reflector, should be repeated several times very carefully and the average value of K adopted. Since different reflectors have varying offsets, the test should be performed with each reflector, and the results marked on each to avoid confusion later. For the most precise calibration, lengths AB and BC should be carefully laid out as even multiples of the instrument's shortest measurement wavelength. Failure to do this can cause an incorrect value of K to be obtained.

6.24.3 Natural Errors

Natural errors in EDM operations stem primarily from atmospheric variations in temperature, pressure and humidity, which affect the index of refraction, and modify the wavelength of electromagnetic energy. The values of these variables must be measured and used to correct observed distances. As demonstrated in Example 6.3, humidity can generally be neglected when using electro-optical instruments, but this variable was important when microwave instruments were employed.

The National Weather Service adjusts atmospheric pressure readings to sea level values. Since atmospheric pressure changes by approximately 1 in. per 1000 ft of elevation, under no circumstances should radio broadcast values for atmospheric pressure be used to correct distances. Instead, atmospheric pressure should be measured by an aneroid barometer that is calibrated against a locally calibrated barometer. Many high school and college physics departments have calibrated barometers.

EDM instruments within total stations have on-board microprocessors that use atmospheric variables, input through the keyboard, to compute corrected distances after making measurements, but before displaying them. For older instruments, varying the transmission frequency made corrections, or they could be computed manually after the measurement. Equipment manufacturers provided tables and charts that assisted in this process. The magnitude of error in electronic distance measurement due to errors in measuring atmospheric pressure and temperature is indicated in Figure 6.17. Note that a 10°C temperature error, or a pressure difference of 25 mm (1 in.) of mercury, each produce a distance error of about 10 ppm.

As discussed in Section 6.14.2, a microclimate can exist in the layers of atmosphere immediately above a surface such as the ground. Since this microclimate can substantially change the index of refraction, it is important to maintain a line of sight that is at least 0.5 m above the surface of the ground. On long lines of sight, the observer should be cognizant of intervening ridges or other objects, such as parked vehicles, that may exist between the instrument and reflector, which could cause problems in meeting this condition. If this condition cannot be met, the height of the reflector may be increased. Under certain conditions, it may be necessary to set an intermediate point on the encroaching surface to ensure that light from the EDM does not travel through these lower layers.

For the most precise work, on long lines, a sampling of the atmospheric conditions along the line of sight should be observed. In this case, it may be necessary to elevate the meteorological instruments. This can be difficult where the terrain becomes substantially lower than the sight line. In these cases, the atmospheric measurements for the ends of the line can be measured and averaged.

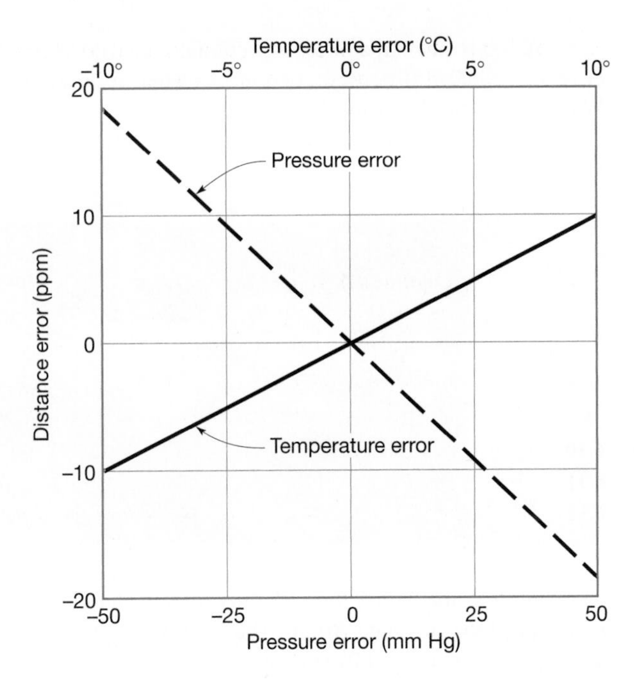

Figure 6.17 Errors in EDM produced by temperature and pressure errors (based on atmospheric temperature and pressure of 15° and 760 mm of mercury).

PROBLEMS

Asterisks (*) indicate problems that have answers given in Appendix G.

6.1 What distance in travel corresponds to 1 millisecond of time for electromagnetic energy?

6.2* A student counted 90, 89, 88, 91, 89, and 88 paces in six trials of walking along a course of 200-ft known length on level ground. Then 111, 112, 110, and 112 paces were counted in walking four repetitions of an unknown distance AB. What is (a) the pace length and (b) the length of AB?

6.3 What difference in temperature from standard, if neglected in use of a steel tape, will cause an error of 1 part in 10,000?

6.4 An add tape of 101 ft is incorrectly recorded as 100 ft for a 300-ft distance. What is the correct distance?

6.5* List five types of common errors in taping.

6.6 For the following data, compute the horizontal distance for a recorded slope distance AB,

(a) $AB = 245.387$ ft, slope angle $= -4°15'28''$

(b) $AB = 486.793$ m, difference in elevation A to $B = -12.468$ m.

A 100-ft steel tape of cross-sectional area 0.0020 in^2, weight 2.4 lb, and standardized at 68°F is 100.008 ft between end marks when supported throughout under a 15-lb pull. What is the true horizontal length of a recorded distance *AB* for the conditions given in Problems 6.7 through 6.12? (Assume horizontal taping and all full tape lengths except the last.)

	Recorded Distance *AB* (ft)	Average Temperature (°F)	Means of Support	Tension (lb)
6.7*	86.06	68	Throughout	15
6.8	249.73	45	Throughout	24
6.9	165.47	89	Throughout	29
6.10	86.35	68	Ends only	10
6.11	104.23	55	Ends only	17
6.12*	235.80	89	Ends only	22

For the tape of Problems 6.7 through 6.12, determine the true horizontal length of the recorded slope distance *BC* for the conditions shown in Problems 6.13 through 6.15. (Assume the tape was fully supported for all measurements.)

	Recorded Slope Distance *BC* (ft)	Average Temperature Per 100 ft (°F)	Tension (lb)	Elevation Difference (ft)
6.13	54.08	27	15	2.38
6.14	94.88	68	25	3.09
6.15	134.55	93	16	5.47

A 30-m steel tape measured 30.006 m when standardized fully supported under a 5.500-kg pull at a temperature of 20°C. The tape weighed 1.24 kg and had a cross-sectional area of 0.030 cm^2. What is the corrected horizontal length of a recorded distance *AB* for the conditions given in Problems 6.16 through 6.18?

	Recorded Distance *AB* (m)	Average Temperature (°C)	Tension (kg)	Means of Support
6.16	48.056	20	8.7	Throughout
6.17	26.798	30	7.2	Ends only
6.18	35.098	18	4.5	Ends only

For the conditions given in Problems 6.19 through 6.21, determine the horizontal length of *CD* that must be laid out to achieve required true horizontal distance *CD*. Assume a 100-ft steel tape will be used, with cross-sectional area 0.0020 in.^{2}, weight 2.0 lb, and standardized at 68°F to be 99.992 ft between end marks when supported throughout with a 12-lb pull. (Assume horizontal taping and all full tape lengths except the last.)

	Required Horizontal Distance *CD* (ft)	Average Temperature (°F)	Means of Support	Tension (lb)
6.19	247.16	68	Throughout	12
6.20	83.45	36	Throughout	22
6.21	123.61	93	Throughout	25

6.22* When measuring a distance *AB*, the first taping pin was placed 1.5 ft to the right of line *AB* and the second pin was set 1.5 ft left of line *AB*. The recorded distance was 236.89 ft. Calculate the corrected distance. (Assume three taped segments, the first two 100 ft each.)

6.23 Describe the advantages of making EDM distance measurements compared to taping.

6.24 Briefly describe how a distance can be measured by the method of phase comparison.

6.25 Discuss the errors that can results by sighting a reflector that is only 0.1 m off the ground.

6.26* Assume the speed of electromagnetic energy through the atmosphere is 299,784,458 m/sec for measurements with an EDM instrument. What time lag in the equipment will produce an error of 400 m in a measured distance?

6.27 What is the length of the partial wavelength for electromagnetic energy with a frequency of the 15 MHz and a phase shift of 263°?

6.28 What "actual" wavelength results from transmitting electromagnetic energy through an atmosphere having an index of refraction of 1.0003, if the frequency is:
(a)* 29.964 MHz **(b)** 2.964 MHz

6.29 How often should the atmospheric temperature and pressure be measured and used when electronically observing distances?

6.30 To calibrate an EDM instrument, distances *AC*, *AB*, and *BC* along a straight line were observed as 116.633 m, 80.320 m, and 36.281 m, respectively. What is the system measurement constant for this equipment? Compute the length of each segment corrected for the constant.

6.31 Which causes a greater error in a line measured with an EDM instrument? (a) A disregarded 20°C temperature variation from standard or (b) a neglected atmospheric pressure difference from standard of 10 mm of mercury?

6.32* In Figure 6.15, h_e, h_r, elev$_A$, elev$_B$, and the measured slope length L were 5.63, 6.02, 652.01, 789.24, and 2306.46 ft, respectively. Calculate the horizontal length between *A* and *B*.

6.33 Similar to Problem 6.32, except that the values were 1.234, 2.003, 214.098, 157.133, and 505.467 m, respectively.

6.34 In Figure 6.15, h_e, h_r, z, and the measured slope length L were 5.16 ft, 6.48 ft, 96°00′16′ and 2756.77 ft, respectively. Calculate the horizontal length between *A* and *B* if a total station measures the distance.

6.35* Similar to Problem 6.34, except that the values were 1.26 m, 1.43 m, 83°24′48″ and 708.460 m, respectively.

6.36 What is the actual wavelength and velocity of a near-infrared beam ($\lambda = 0.895\ \mu m$) of light modulated at a frequency of 330 MHz through an atmosphere with a dry bulb temperature, T, of 34°C, a relative humidity, h, of 64 percent, and an atmospheric pressure of 1033 hPa?

6.37 If the temperature and pressure at measurement time are 15°C and 760 mm Hg, what will be the error in electronic measurement of a line 5 km long if the temperature at the time of observing is recorded 10°C too low? Will the observed distance be too long or too short?

6.38 Determine the most probable length of a line AB, the standard deviation, and the 90% error of a single measurement for the following series of taped measurements made under the same conditions: 135.685, 135.680, 135.690, 135.686, 135.683, 135.684, 135.692 m.

6.39* The standard deviation of taping a 30-m distance is ±3 mm. What should it be for a 120-m distance?

6.40 If an EDM instrument has a purported accuracy capability of ±(3 mm + 3 ppm), what error can be expected in a measured distance of (a) 1978.45 ft (b) 680.963 m (c) 1975.468 m? (Assume that the instrument and target miscentering errors are equal to zero.)

6.41 The estimated instrument and target miscentering errors are ±3 mm and ±5 mm, respectively. For the EDM in Problem 6.40, what is the estimated error in the observed distances?

6.42 If a certain EDM instrument has an accuracy capability of ±(2 mm + 2 ppm), what is the precision of measurements, in terms of parts-per-million, for line lengths of: (a) 10.000 m (b) 100.000 m (c) 1000.000 m? (Assume that the instrument and target miscentering errors are equal to zero.)

6.43 The estimated instrument and target miscentering errors are ±3 mm mm and ±5 mm mm, respectively. For the EDM and distances listed in Problem 6.42, what is the estimated error in each distance? What is the precision of the measurements in terms of part-per-million?

6.44 Create a computational program that solves Problems 6.16–6.19.

6.45 Create a computational program that solves Problem 6.34.

6.46 Create a computational program that solves Problem 6.36.

6.47 Create a computational program that solves Problem 6.42.

BIBLIOGRAPHY

Cummock, M. A. 2002. "Horizontal Distance Explained, Part 1." *Professional Surveyor* 22 (No. 10): 32.

Fonczek, Charles J. 1980. *Use of Calibration Base Lines.* NOAA Technical Memorandum NOS NGS-10.

GIA. 2001. "EDM PPM Settings." *Professional Surveyor* 21 (No. 6): 26.

______. 2002. "EDM Calibration." *Professional Surveyor* 22 (No. 7): 50.

______. 2003. "Phase Resolving EDMs." *Professional Surveyor* 23 (No. 10): 34.

Hoos, D., et. al. 2003. "Growing a Business with the Help of Reflectorless Technology." *Professional Surveyor* 23 (No. 7): 6.

Scott, D. 2003. "Phase Resolving EDMs." *Professional Surveyor* 23 (No. 10): 34.

7

Angles, Azimuths, and Bearings

■ 7.1 INTRODUCTION

Determining the locations of points and orientations of lines frequently depends on measurements of angles and directions. In surveying, directions are given by *azimuths* and *bearings* (see Sections 7.5 and 7.6).

As described in Section 2.1 and illustrated in Figure 2.1, angles measured in surveying are classified as either *horizontal* or *vertical*, depending on the plane in which they are observed. Horizontal angles are the basic observations needed for determining bearings and azimuths. Vertical angles are used in trigonometric leveling, stadia (see Section 17.9.2), and for reducing slope distances to horizontal (see Section 6.23).

Angles are most often *directly* observed in the field with total station instruments, although in the past transits, theodolites, and compasses have been used. (See Appendix A for descriptions of the transit and theodolite. The surveyor's compass is described in Section 7.10.) Three basic requirements determine an angle. As shown in Figure 7.1, they are (1) *reference* or *starting line*, (2) *direction of turning*, and (3) *angular distance* (value of the angle). Methods of computing bearings and azimuths described in this chapter are based on those three elements.

■ 7.2 UNITS OF ANGLE MEASUREMENT

A purely arbitrary unit defines the value of an angle. The *sexagesimal* system used in the United States and many other countries is based on degrees, minutes, and seconds, with the last unit further divided decimally. In Europe the *grad* or *gon* is commonly used (see Section 2.2). Radians may be more suitable in computations, and in fact are employed in digital computers, but the sexagesimal system continues to be used in most U.S. surveys.

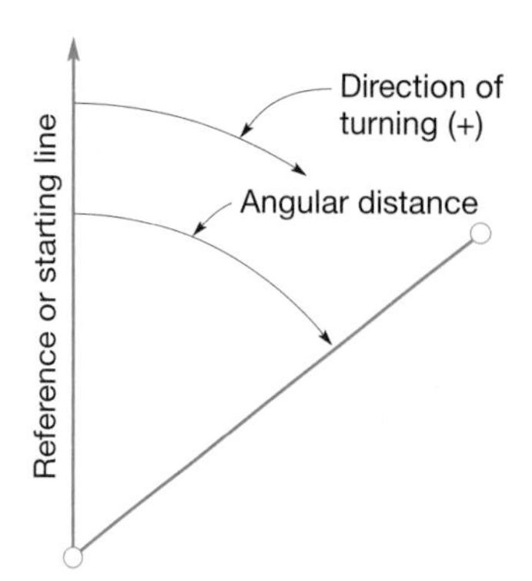

Figure 7.1 Basic requirements in determining an angle.

■ 7.3 KINDS OF HORIZONTAL ANGLES

The kinds of horizontal angles most commonly observed in surveying are (1) *interior angles*, (2) *angles to the right*, and (3) *deflection angles*. Because they differ considerably, the kind used must be clearly indicated in field notes. Interior angles, shown in Figure 7.2, are measured on the inside of a *closed polygon*. Normally the angle at each apex within the polygon is observed. Then, as discussed in Section 9.7, a check can be made on their values because the sum of all angles in any polygon must equal $(n - 2)180°$, where n is the number of angles. Polygons are commonly used for boundary surveys and many other types of work. Surveyors (geomatics engineers) normally refer to them as *closed traverses*.

Exterior angles, located outside a closed polygon, are explements of interior angles. The advantage to be gained by observing them is their use as another check, since the sum of the interior and exterior angles at any station must total 360°.

Angles to the right are measured *clockwise from the rear to the forward station*. Note: As a survey progresses, stations are commonly identified by consecutive alphabetical letters (as in Figure 7.2) or by increasing numbers. Thus, the interior angles of Figure 7.2(a) are also angles to the right. Most data collectors require that angles to the right be observed in the field. *Angles to the left*, turned counterclockwise from the rear station, are illustrated in Figure 7.2(b). Note that the polygons of Figure 7.2 are "right" and "left"—that is, similar in shape but turned over

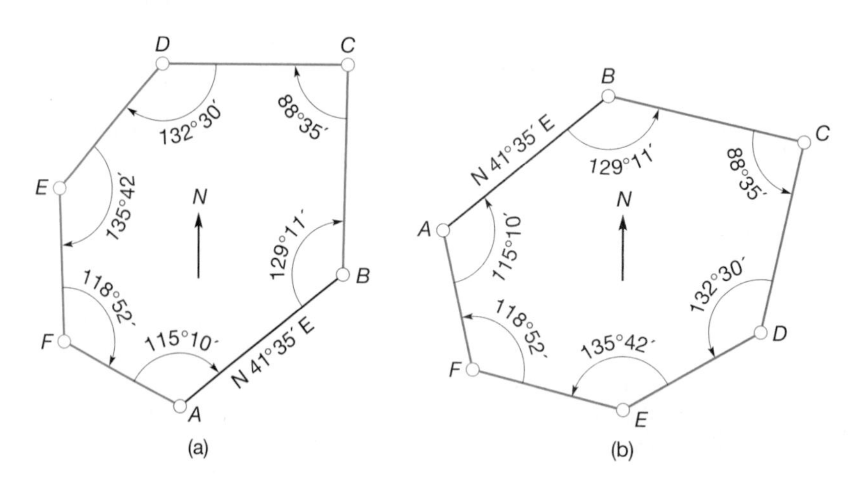

Figure 7.2 Closed polygon: (a) Clockwise interior angles (angles to the right). (b) Counterclockwise interior angles (angles to the left).

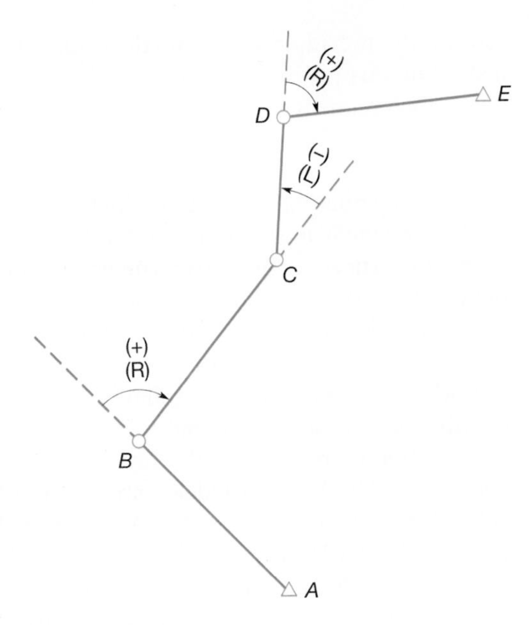

Figure 7.3
Deflection angles.

like the right and left hands. Figure 7.2(b) is shown only to emphasize a serious mistake that occurs if counterclockwise angles are observed and recorded or assumed to be clockwise. *To avoid this confusion, it is recommended that a uniform procedure of always observing angles to the right be adopted* and the direction of turning noted in the field book with a sketch.

Deflection angles (Figure 7.3) are observed from an extension of the back line to the forward station. They are used principally on the long linear alignment surveys. As illustrated in the figure, deflection angles may be observed to the right (clockwise) or to the left (counterclockwise) depending upon the direction of the alignment. Clockwise angles are considered plus, and counterclockwise ones minus, as shown in the figure. Deflection angles are always smaller than 180° and appending an R or L to the numerical value identifies the direction of turning. Thus the angle at B in Figure 7.3 is (R), and that at C is (L). Deflection angles are the only exception where counterclockwise observation of angles should be made.

■ 7.4 DIRECTION OF A LINE

The direction of a line is defined by the horizontal angle between the line and an arbitrarily chosen reference line called a *meridian*. Different meridians are used for specifying directions including (a) geodetic (also often called *true*) (b) astronomic, (c) magnetic, (d) grid, (e) record and (f) assumed.

The *geodetic* meridian is the north-south reference line that passes through a mean position of the Earth's geographic poles. The positions of the poles were defined as their mean locations between the period of 1900.0 and 1905.0 (see Section 19.3).

Wobbling of the Earth's rotational axis, also discussed in Section 19.3, causes the position of the Earth's geographic poles to change with time. At any point, the *astronomic* meridian is the north-south reference line that passes through the instantaneous position of the Earth's geographic poles. Astronomic meridians derive their name from the field operation to obtain them, which consists in making observations on the sun or stars, as described in Appendix C. Geodetic and astronomic meridians are very nearly the same, and the former can be computed from the latter by making small corrections (see Sections 19.3 and 19.5).

A *magnetic* meridian is defined by a freely suspended magnetic needle that is only influenced by the Earth's magnetic field. Magnetic meridians are discussed in Section 7.10.

Surveys based on a state or other plane coordinate system employ a *grid* meridian for reference. Grid north is the direction of geodetic north for a selected *central meridian* and held parallel to it over the entire area covered by a plane coordinate system (see Chapter 20).

In boundary surveys, the term *record* meridian refers to directional references quoted in the recorded documents from a previous survey of a particular parcel of land. Another similar term, *deed* meridian, is used in the description of a parcel of land as recorded in a property deed. Chapters 21 and 22 discuss the use of record meridians and deed meridians in boundary retracement surveys.

An *assumed* meridian can be established by merely assigning any arbitrary direction—for example, taking a certain street line to be geodetic north. The directions of all other lines are then found in relation to it. Disadvantages of using an assumed meridian are the difficulty, or perhaps impossibility, of reestablishing it if the original points are lost, and its nonconformance with other surveys and maps in an area.

From these definitions, it should be obvious that the terms north or due north, if used in a survey, must be explained, since they may not specify a unique line.

■ 7.5 AZIMUTHS

Azimuths are horizontal angles observed clockwise from any reference meridian. In plane surveying, azimuths are generally observed from north, but astronomers and the military have used south as the reference direction. The National Geodetic Survey (NGS) also used south as its reference for azimuths for NAD27, but north has been adopted for NAD83 (see Section 19.6). Examples of azimuths observed from north are shown in Figure 7.4. As illustrated, they can range from 0° to 360° in value. Thus the azimuth of *OA* is 70°; of *OB*, 145°; of *OC*, 235°; and of *OD*, 330°. Azimuths may be *geodetic, astronomic, magnetic, grid, record*, or *assumed*, depending on the reference meridian used. To avoid any confusion, it is necessary to state in the field notes, at the beginning of work, what reference meridian applies for azimuths and whether they are observed from north or south.

A line's forward direction can be given by its *forward* azimuth and its reverse direction by its *back* azimuth. In plane surveying, forward azimuths are converted to back azimuths, and vice versa, by adding or subtracting 180°. For example, if the azimuth of *OA* is 70°, the azimuth of *AO* is $70° + 180° = 250°$. If the azimuth of *OC* is 235°, the azimuth of *CO* is $235° - 180° = 55°$.

Azimuths can be read directly on the graduated circle of a total station instrument after the instrument has been oriented properly. As explained in Section 9.2.4,

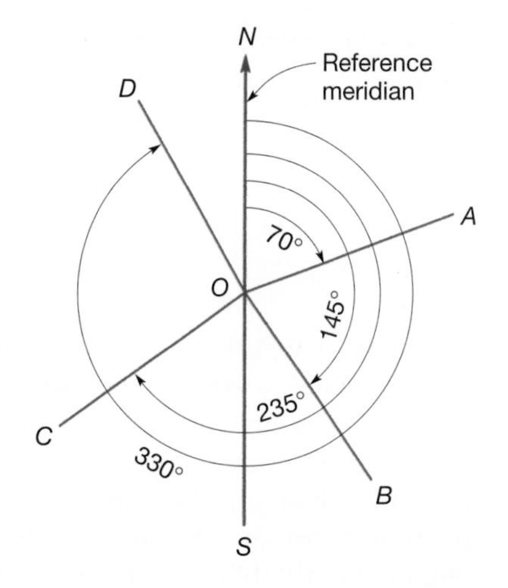

Figure 7.4
Azimuths.

this can be done by sighting along a line of known azimuth with that value indexed on the circle, and then turning to the desired course. Azimuths are used advantageously in boundary, topographic, control, and other kinds of surveys, as well as in computations.

■ 7.6 BEARINGS

Bearings are another system for designating directions of lines. *The bearing of a line is defined as the acute horizontal angle between a reference meridian and the line.* The angle is observed from either the north or south toward the east or west, to give a reading smaller than 90°. The letter N or S preceding the angle and E or W following it shows the proper quadrant. Thus, a properly expressed bearing includes quadrant letters and an angular value. An example is N80°E. In Figure 7.5, all bearings in quadrant *NOE* are measured clockwise from the meridian. Thus the bearing of line *OA* is N70°E. All bearings in quadrant *SOE* are counterclockwise from the meridian, so *OB* is S35°E. Similarly, the bearing of *OC* is S55°W and that

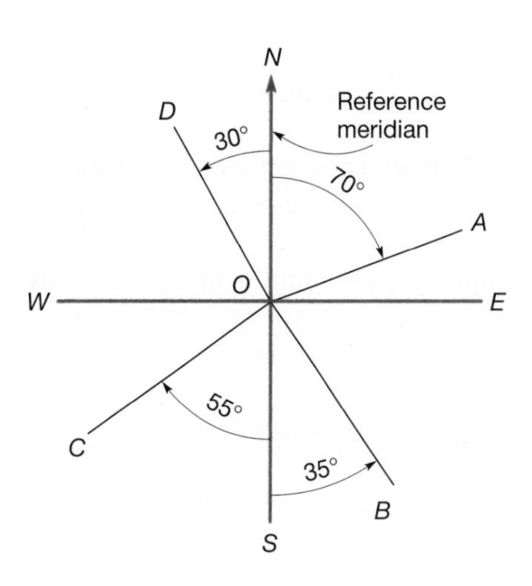

Figure 7.5
Bearing angles.

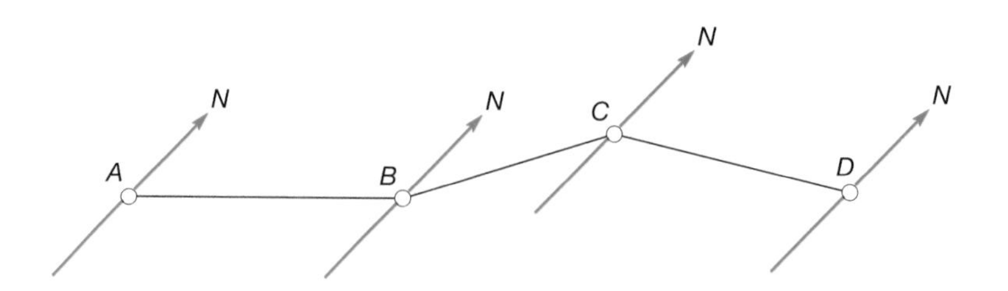

Figure 7.6 Forward and back bearings.

of OD, N30°W. When lines are in the cardinal directions, the bearings should be listed as "Due North", "Due East", "Due South", or "Due West."

Geodetic bearings are observed from the geodetic meridian, *astronomic bearings* from the local astronomic meridian, *magnetic bearings* from the local magnetic meridian, *grid bearings* from the appropriate grid meridian, and *assumed bearings* from an arbitrarily adopted meridian. The magnetic meridian can be obtained in the field by observing the needle of a compass, and used along with observed angles to get computed magnetic bearings.

In Figure 7.6 assume that a compass is set up successively at points A, B, C, and D and bearings read on lines AB, BA, BC, CB, CD, and DC. As previously noted, bearings AB, BC, and CD are *forward bearings;* those of BA, CB, and DC, *back bearings*. Back bearings should have the same numerical values as forward bearings but opposite letters. Thus if bearing AB is N44°E, bearing BA is S44°W.

7.7 COMPARISON OF AZIMUTHS AND BEARINGS

Because bearings and azimuths are encountered in so many surveying operations, the comparative summary of their properties given in Table 7.1 should be helpful. Bearings are readily computed from azimuths by noting the quadrant in which the azimuth falls, then converting as shown in the table.

Example 7.1

The azimuth of a boundary line is 128°13′46″. Convert this to a bearing.

Solution

The azimuth places the line in the southeast quadrant. Thus the bearing angle is

$$180° - 128°13'46'' = 51°46'14'',$$

and the equivalent bearing is S 51°46′14″E.

Example 7.2

The first course of a boundary survey is written as N37°13′W. What is its equivalent azimuth?

TABLE 7.1 COMPARISON OF AZIMUTHS AND BEARINGS

Azimuths	Bearings
Vary from 0 to 360°	Vary from 0 to 90°
Require only a numerical value	Require two letters and a numerical value
May be geodetic, astronomic, magnetic, grid, assumed, forward or back	Same as azimuths
Are measured clockwise only	Are measured clockwise and counterclockwise
Are measured either from north only, or from south only on a particular survey	Are measured from north and south

Quadrant	Formulas for computing bearing angles from azimuths
I (NE)	Bearing = Azimuth
II (SE)	Bearing = 180° − Azimuth
III (SW)	Bearing = Azimuth − 180°
IV (NW)	Bearing = 360° − Azimuth

Example directions for lines in the four quadrants (azimuths from north)

Azimuth	Bearing
54°	N54°E
112°	S68°E
231°	S51°W
345°	N15°W

Solution

Since the bearing is in the northwest quadrant, the azimuth is

$$360° - 37°13' = 322°47'.$$

■ 7.8 COMPUTING AZIMUTHS

Most types of surveys, but especially those which employ traverses, require computation of azimuths (or bearings). A traverse, as described in Chapter 9, is a series of connected lines whose lengths and angles at the junction points have been observed. Figures 7.2 and 7.3 illustrate examples. Traverses have many uses. For example, to survey the boundary lines of a piece of property, a "closed-polygon"

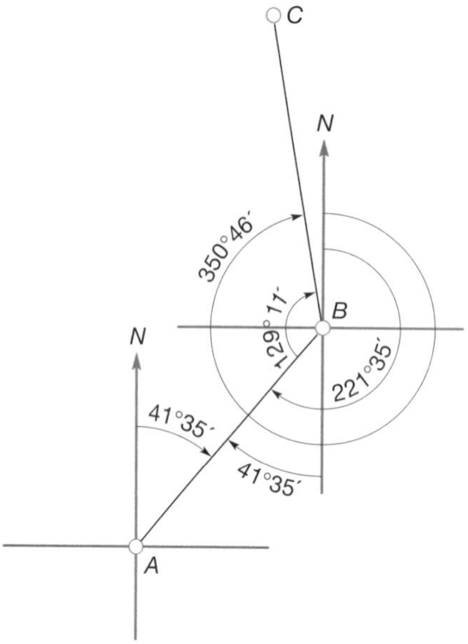

Figure 7.7 Computation of azimuth BC of Figure 7.2(a).

type traverse like that of Figure 7.2(a) would normally be used. A highway survey from one city to another would usually involve a traverse like that of Figure 7.3. Regardless of the type used, it is necessary to compute the directions of its lines.

Many surveyors prefer azimuths to bearings for directions of lines because they are easier to work with, especially when calculating traverses with computers. Also sines and cosines of azimuth angles provide correct algebraic signs for departures and latitudes as discussed in Section 10.4.

Azimuth calculations are best made with the aid of a sketch. Figure 7.7 illustrates computations for azimuth BC in Figure 7.2(a). Azimuth BA is found by adding 180° to azimuth AB: $180° + 41°35' = 221°35'$. Then the angle to the right at B, 129°11′, is added to azimuth BA to get azimuth BC: $221°35' + 129°11' = 350°46'$. This general process of adding (or subtracting) 180° to obtain the back azimuth and then adding the angle to the right is repeated for each line until the azimuth of the starting line is recomputed. If a computed azimuth exceeds 360°, then 360° is subtracted from it and the computations are continued. These calculations are conveniently handled in tabular form, as illustrated in Table 7.2. This table lists the calculations for all azimuths of Figure 7.2(a). Note that a check was secured by recalculating the beginning azimuth using the last angle. The procedures illustrated in Table 7.2 for computing azimuths are systematic and readily programmed for computer solution. The reader can view a Mathcad® worksheet on the CD accompanying this book to review these computations.

Traverse angles must be adjusted to the proper geometric total before azimuths are computed. As noted earlier, in a closed-polygon traverse, the sum of interior angles equals $(n - 2)180°$, where n is the number of angles or sides. If the traverse angles fail to close by, say 10″, and are not adjusted prior to computing azimuths, the original and computed check azimuth of AB will differ by the same 10″, assuming there are no other calculating errors. *The azimuth of any starting course should always be recomputed as a check using the last angle.* Any discrepancy shows that (a) an arithmetic error was made or (b) the angles were not

TABLE 7.2 COMPUTATION OF AZIMUTHS (FROM NORTH) FOR LINES OF FIGURE 7.2(A)

Angles to the Right [Figure 7.2(a)]	
41°35′ = AB	211°51′ = DE
+180°00′	−180°00′
221°35′ = BA	31°51′ = ED
+129°11′	+135°42′
350°46′ = BC	167°33′ = EF
−180°00′	+180°00′
170°46′ = CB	347°33′ = FE
+88°35′	+118°52′
259°21′ = CD	466°25′ − *360° = 106°25′ = FA
−180°00′	−180°00′
79°21′ = DC	286°25′ = AF
+132°30′	+115°10′
211°51′ = DE	401°35′ − *360° = 41°35′ = AB√

*When a computed azimuth exceeds 360°, the correct azimuth is obtained by merely subtracting 360°.

properly adjusted prior to computing azimuths. Common procedures for adjusting, also called "balancing," angles are covered in Section 10.2.

■ 7.9 COMPUTING BEARINGS

Drawing sketches similar to those in Figure 7.8 showing all data simplifies computations for bearings of lines. In Figure 7.8(a), the bearing of line *AB* from Figure 7.2(a) is N41°35′E and the angle at *B* turned clockwise (to the right) from known line *BA* is 129°11′. Then the bearing angle of line *BC* is 180° − (41°35′ + 129°11′) = 9°14′ and from the sketch the bearing of *BC* is N9°14′W.

In Figure 7.8(b), the clockwise angle at *C* from *B* to *D* was observed as 88°35′. The bearing of *CD* is 88°35′ − 9°14′ = S79°21′W. Continuing this technique, the bearings in Table 7.3 have been determined for all lines in Figure 7.2(a).

In Table 7.3, note that the last bearing computed is for *AB*, and it is obtained by employing the 115°10′ angle observed at *A*. It yields a bearing of N41°35′E, which agrees with the starting bearing. Students should compute each bearing of Figure 7.2(a) to verify the values given in Table 7.3.

An alternate method of computing bearings is to determine the azimuths as discussed in Section 7.8 and then convert the computed azimuths to bearings using the techniques discussed in Section 7.7. For example in Table 7.2, the azimuth of line *CD* is 259°21′. Using the procedure discussed in Section 7.7, the bearing angle is 259°21′ − 180° = 79°21′, and the bearing is S79°21′W.

Bearings, rather than azimuths, are used predominately in boundary surveying. This practice originated from the period of time when the magnetic

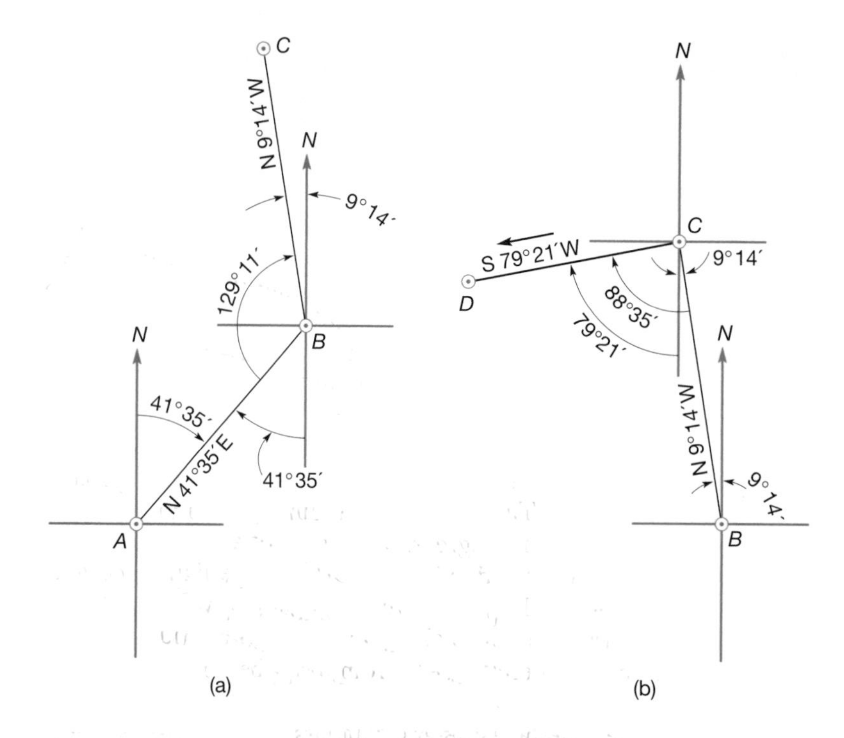

Figure 7.8 (a) Computation of bearing *BC* of Figure 7.2(a). (b) Computation of bearing *CD* of Figure 7.2(a).

TABLE 7.3 BEARINGS OF LINES IN FIGURE 7.2(A)

Course	Bearing
AB	N41°35′E
BC	N9°14′W
CD	S79°21′W
DE	S31°51′W
EF	S12°27′E
FA	S73°35′E
AB	N41°35′E√

bearings of parcel boundaries were determined directly using a surveyor's compass (see Section 7.10). Later, although other instruments (i.e., transits and theodolites) were used to observe the angles, and the astronomic meridian was more commonly used, the practice of using bearings for land surveys continued and is still in common use today. Because boundary retracement surveyors *must follow the footsteps of the original surveyor* (see Chapter 21), they need to understand magnetic directions and their nuances. The following sections discuss magnetic directions and explain how to convert directions from magnetic to other reference meridians, and vise versa.

■ 7.10 THE COMPASS AND THE EARTH'S MAGNETIC FIELD

Before transits, theodolites, and total station instruments were invented, directions of lines and angles were determined using compasses. Most of the early land-surveying work in the United States was done using these venerable instruments. Figure 7.9(a) shows the *surveyor's compass*. The instrument consists of a metal baseplate (A) with two sighting vanes (B) at the ends. The compass box (C) and two small level vials (D) are mounted on the baseplate, the level vials being perpendicular to each other. When the compass was set up and the bubbles in the vials centered, the compass box was horizontal and ready for use.

A single leg called a Jacob staff supported early compasses. A ball-and-socket joint and a clamp were used to rotate the instrument and clamp it in its horizontal position. Later versions, such as that shown in Figure 7.9(a), were mounted on a tripod. This arrangement provided greater stability.

The compass box of the surveyor's compass was covered with glass to protect the magnetized steel needle inside. The needle was mounted on a pivot at the center of a circle that was graduated in degrees. A top view of a surveyor's compass box with its graduations is illustrated in Figure 7.9(b). In the figure, the zero graduations are at the north and south points of the compass and in line with the two sight-vane slits that comprise the line of sight. Graduations are numbered in multiples of 10° clockwise and counterclockwise from 0° at the north and south, to 90° at the east and west.

In using the compass, the sighting vanes and compass box could be revolved to sight along a desired line, and then its magnetic bearing could be read directly. Note in Figure 7.9(b) for example, that the needle is pointing north and that the line of sight is directed in a northeast direction. The magnetic bearing of the line, read directly from the compass, is N 40°E. (Note that the letters *E* and *W* on the

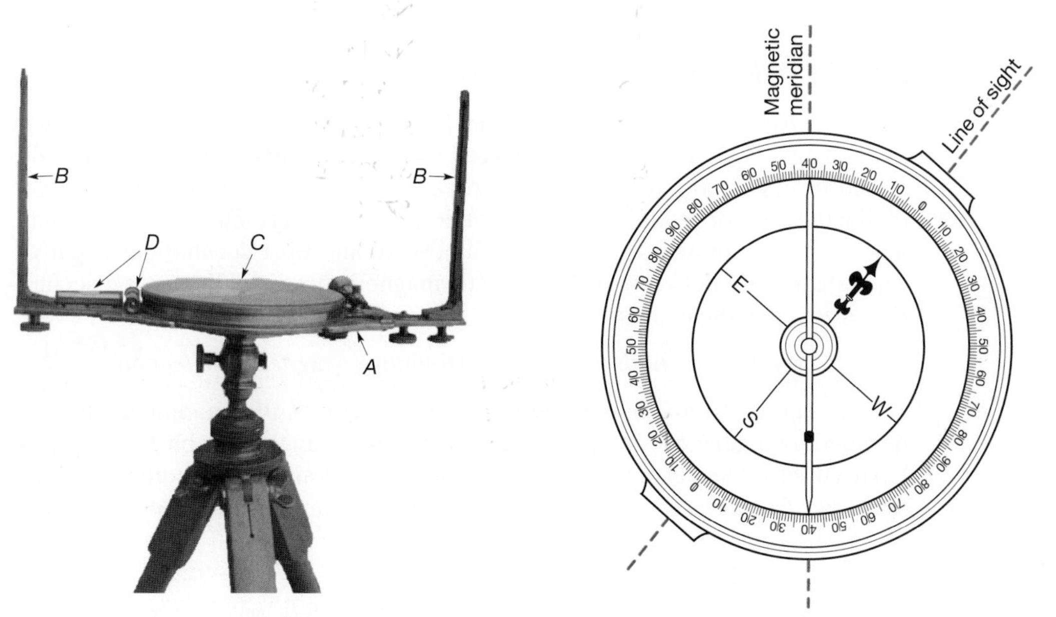

Figure 7.9 (a) Surveyor's compass. (Courtesy W. & L.E. Gurley) (b) Compass box.

face of the compass box are reversed from their normal positions to give the direct readings of bearings.)

Unless disturbed by *local attraction* (a local anomaly caused from such things as power lines, railroad tracks, metallic belt buckles, etc. that affect the direction a compass needle points at any location), a compass needle is free to spin and align itself with the Earth's magnetic field pointing in the direction of the magnetic meridian (toward the magnetic north pole in the northern hemisphere).[1]

The magnetic forces of the Earth not only align the compass needle, but they also pull or dip one end of it below the horizontal position. The *angle of dip* varies from 0° near the equator to 90° at the magnetic poles. In the northern hemisphere, the south end of the needle is weighted with a very small coil of wire to balance the dip effect and keep it horizontal. The position of the coil can be adjusted to conform to the latitude in which the compass is used. Note the coil (dark spot) on the south end of the needle of the compass of Figure 7.9(b).

The Earth's magnetic field resembles that of a huge dipole magnet located at the Earth's center, with the magnet offset from the Earth's rotational axis by about 11°. This field has been observed at about 200 magnetic observatories around the world, as well as many other temporary stations. At each observation point both the field's intensity and direction are measured. Based upon many years of this data, models of the Earth's magnetic field have been developed. These models are used to compute the *magnetic declination* and *annual change* (see Sections 7.11 and 7.12), which are elements of importance to surveyors (geomatics engineers). The accuracy of the models is affected by several items including the locations of the observations, the types of rocks at the surfaces together with the underlying geological structures in the areas, and local attractions. Today's models give magnetic declinations that are accurate to within about 30 min of arc, however, local anomalies of 3° to 4°, or more, can exist in some areas.

■ 7.11 MAGNETIC DECLINATION

Magnetic declination is the horizontal angle measured from the geodetic meridian to the magnetic meridian. Navigators call this angle *variation* of the compass; the armed forces use the term *deviation.* An east declination exists if the magnetic meridian is east of geodetic north; a west declination occurs if it is west of geodetic north. East declinations are considered positive and west declinations negative. The relationship between geodetic north, magnetic north, and magnetic declination is given by the expression

$$\textit{geodetic azimuth} = \textit{magnetic azimuth} + \textit{magnetic declination} \tag{7.1}$$

Because the magnetic pole positions are constantly changing, magnetic declinations at all locations also undergo continual changes. Establishing a meridian from astronomical or satellite (GPS) observations and then reading a compass while sighting along the observed meridian can obtain the current declination at

[1]The locations of the north and south magnetic poles are continually changing, and in 2001, they were located at approximately 81.3° north latitude and 110.8° west longitude, and 64.7° south latitude and 138.0° east longitude, respectively. It is currently moving approximately northwest at 40 km per year.

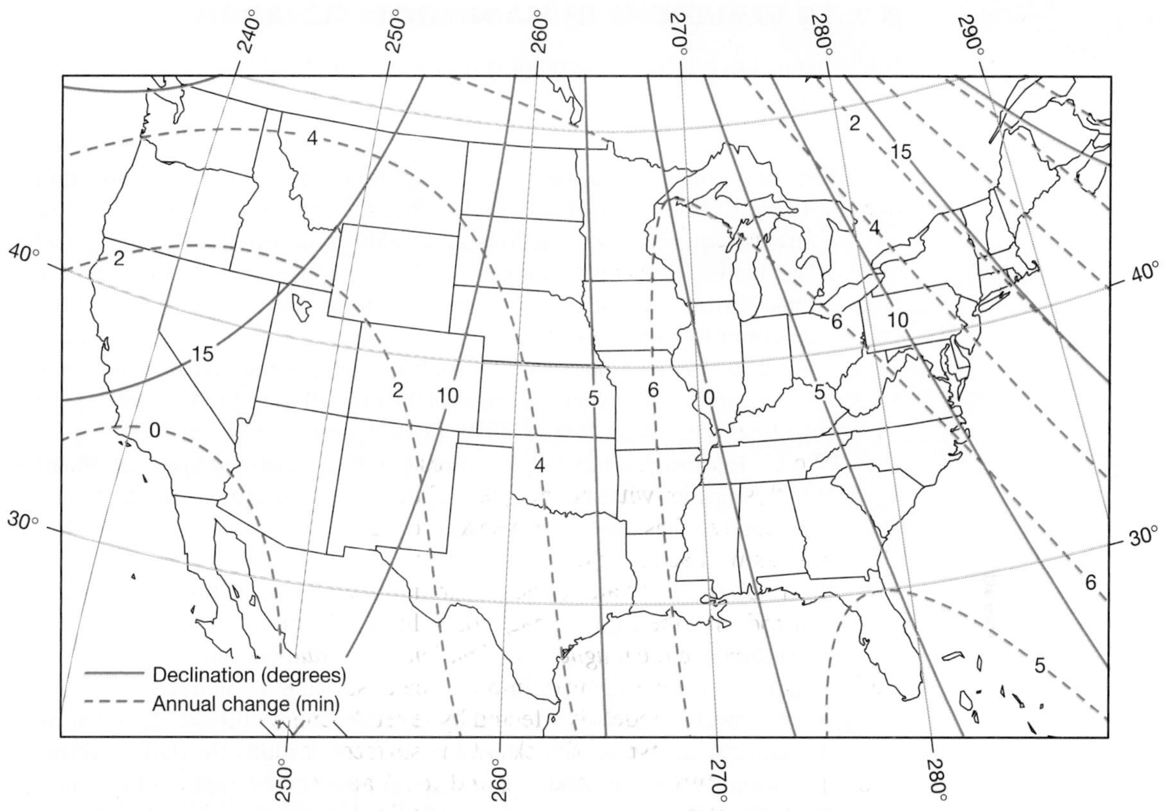

Figure 7.10 Isogonic chart showing distribution of magnetic declination for 1996.0. This image is from the NOAA National Data Centers, NGDC at http://www.ngdc.noaa.gov/seg/geomag/declination.shtml.

any location obtained barring any local attractions. Another way of determining the magnetic declination at a point is to interpolate it from an *isogonic chart.* An isogonic chart shows magnetic declinations in a certain region for a specific epoch of time. Lines on such maps connecting points that have the same declination are called *isogonic lines.* The isogonic line along which the declination is zero (where the magnetic needle defines geodetic north as well as magnetic north) is termed the *agonic line.* Figure 7.10 is an isogonic chart covering the contiguous 48 states of the United States for the year 1996. On that chart, the agonic line cuts through the central part of the United States. It is gradually moving westward. Points to the west of the agonic line have east declinations and points to the east have west declinations. As a memory aid, the needle can be thought of as pointing toward the agonic line. Note there is about a 40° difference in declination between the northeast portion of Maine and the northwest part of Washington. This is a huge change if a pilot flies by compass between the two states!

The dashed lines in Figure 7.10 show the *annual change* in declination. These lines indicate the amount of *secular change* (see Section 7.12) that is expected in magnetic declination in a period of one year. The annual change at any location can be interpolated between the lines and the value used to estimate the declination a few years before or after the chart date.

7.12 VARIATIONS IN MAGNETIC DECLINATION

It has been stated that magnetic declinations at any point vary over time. These variations can be categorized as *secular, daily, annual,* and *irregular,* and are summarized as follows.

Secular Variation. Because of its magnitude, this is the most important of the variations. Unfortunately, no physical law has been found to enable precise long-term predictions of secular variation and its past behavior can be described only by means of detailed tables and charts derived from observations. Records, which have been kept at London for four centuries, show a range in magnetic declination from 11°E in 1580, to 24°W in 1820, back to 3°W in 2000. Secular variation changed the magnetic declination at Baltimore, Maryland, from 5°11′W in 1640 to 0°35′W in 1800, 5°19′W in 1900, 7°25′W in 1950, 8°43′W in 1975, and 11°01′W in 2000.

In retracing old property lines run by compass or based on the magnetic meridian, it is necessary to allow for the difference in magnetic declination at the time of the original survey and at the present date. The difference is attributed mostly to secular variation.

Daily Variation. Daily variation of the magnetic needle's declination causes it to swing through an arc averaging approximately 8′ for the United States. The needle reaches its extreme easterly position at about 8:00 A.M. and its most westerly position at about 1:30 P.M. Mean declination occurs at around 10:30 A.M. and 8:00 P.M. These hours and the daily variation change with latitude and season of the year. Usually the daily variation is ignored since it is well within the range of error expected in compass readings.

Annual Variation. This periodic swing is less than 1 min of arc and can be neglected. It must not be confused with the annual change (the amount of secular-variation change in one year) shown on some isogonic maps.

Irregular Variations. Unpredictable magnetic disturbances and storms can cause short-term irregular variations of a degree or more.

7.13 SOFTWARE FOR DETERMINING MAGNETIC DECLINATION

As noted earlier, direct observations are only applicable for determining current magnetic declinations. In most situations, however, magnetic declinations that existed years ago, for example on the date of an old property survey, are needed in order to perform retracement surveys. Until recently these old magnetic declinations had to be interpolated from isogonic charts for the approximate time desired, and the lines of annual change used to correct to the specific year required. Now software is available that can quickly provide the needed magnetic declination values. The software uses models that were developed from historical records of magnetic declination and annual change, which have been maintained for the many observation stations throughout the United States and the world.

The program WOLFPACK on the CD that accompanies this book contains an option for computing magnetic field elements. This program uses models that

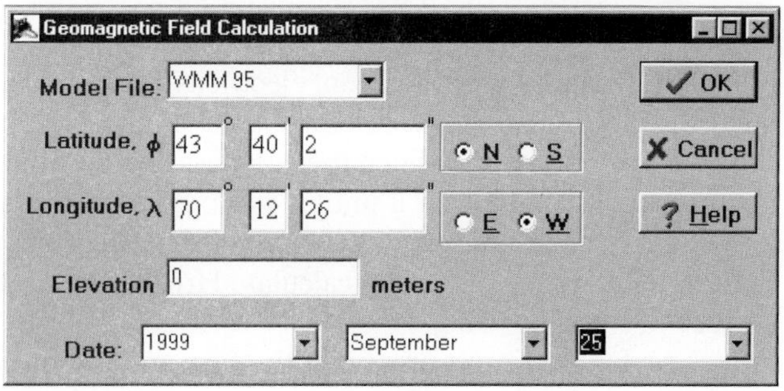

Figure 7.11 Magnetic declination data entry screen in WOLFPACK setup to compute magnetic field values for Portland, ME.

span five or more year time frames. Using the World Magnetic Model of 1995 (file: WMM-95.DAT), the declination and annual change for Portland, Maine, on September 25, 1999, were determined to be about 16°54′W[2] and 0.0′W per year, respectively (see the input data in Figure 7.11). Using this same program, the declinations for various other cities in the United States were determined for January 1, 2000, and are shown in Table 7.4. It is important when using this software to select the appropriate model file for the desired date. Select the appropriate model from a drop-down list for the "Model File." The models are given by their source, and the year. The latitude, longitude, and elevation of the station must be entered in the appropriate data boxes and the time of the desired computation is selected from the drop-down list at the bottom of the box. After computing the magnetic field elements for the particular location and time, the results are displayed for printing. Similar computations to determine magnetic declination and rates of annual change can be made by using the NOAA National Geophysical Data Centers' (NGDC) online computation page at http://www.ngdc.noaa.gov/seg/geomag/jsp/Declination.jsp. The location of any U.S. city can be found with the U.S. Gazetteer, which is linked to the software, or can be obtained at http://www.census.gov/cgibin/gazetteer on the U.S. Census Bureau website.

TABLE 7.4 MAGNETIC DECLINATION AND ANNUAL CHANGE FOR VARIOUS LOCATIONS IN THE U.S. FOR JANUARY 1, 2000

City	Magnetic Declination	Annual Change
Boston, MA	15°48′W	1.0′E
Cleveland, OH	7°45′W	4.1′W
Madison, WI	1°13′W	6.2′W
Denver, CO	10°21′E	4.9′W
San Francisco, CA	15°26′E	3.1′W
Seattle, WA	18°45′E	7.2′W

[2]The software indicates west declination as negative and east declination as positive.

7.14 LOCAL ATTRACTION

Metallic objects and direct-current electricity, both of which cause a local attraction, affect the main magnetic field. As an example, when set up beside an old-time streetcar with overhead power lines, the compass needle would swing toward the car as it approached, then follow it until it was out of effective range. If the source of an artificial disturbance is fixed, all bearings from a given station will be in error by the same amount. However, angles calculated from bearings taken at the station will be correct.

Local attraction is present if the forward and back bearings of a line differ by more than the normal observation errors. Consider the following compass bearings read on a series of lines:

AB	N24°15′W
BC	N76°40′W
CD	N60°00′E
DE	N88°35′E
BA	S24°10′E
CB	S76°40′E
DC	S61°15′W
ED	S87°25′W

Forward-bearing *AB* and back-bearing *BA* agree reasonably well, indicating that little or no local attraction exists at *A* or *B*. The same is true for point *C*. However, the bearings at *D* differ from corresponding bearings taken at *C* and *E* by roughly 1°15′ to the west of north. Local attraction therefore exists at point *D* and deflects the compass needle by approximately 1°15′ to the west of north.

It is evident that to detect local attraction, successive stations on a compass traverse have to be occupied, and forward and back bearings read, even though the directions of all lines could be determined by setting up an instrument only on alternate stations.

7.15 TYPICAL MAGNETIC DECLINATION PROBLEMS

Typical problems in boundary surveys require the conversion of geodetic bearings to magnetic bearings, magnetic bearings to geodetic bearings, and magnetic bearings to magnetic bearings for the declinations existing at different dates. The following examples illustrate two of these types of problems.

Example 7.3

Assume the magnetic bearing of a property line was recorded as S43°30′E in 1862. At that time the magnetic declination at the survey location was 3°15′W. What geodetic bearing is needed for a subdivision property plan?

Solution

A sketch similar to Figure 7.12 makes the relationship clear and should be used by beginners to avoid mistakes. Geodetic north is designated by a full-headed long

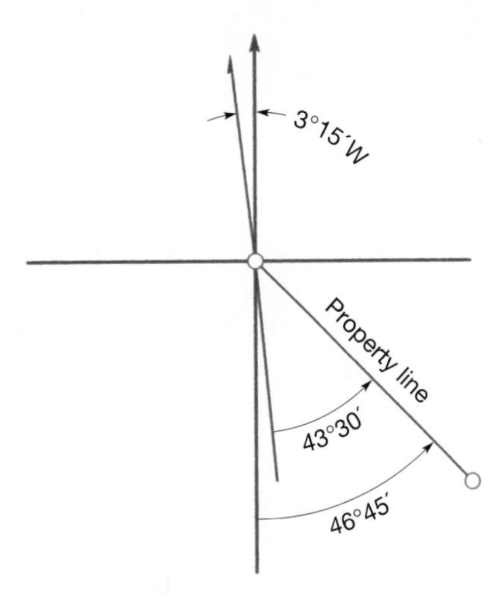

Figure 7.12 Computing geodetic bearings from magnetic bearings and declinations.

arrow and magnetic north by a half-headed shorter arrow. The geodetic bearing is seen to be S43°30′E + 3°15′ = S46°45′E. Using different colored pencils to show the direction of geodetic north, magnetic north, and lines on the ground helps clarify the sketch. Although this problem is done using bearings, Equation (7.1) could be applied by converting the bearings to azimuths. That is, the magnetic azimuth of the line is 136°30′. Applying Equation (7.1) using a negative declination angle results in a geodetic azimuth of 136°30′ − 3°15′ = 133°15′, which correctly converts to the geodetic bearing of S46°45′E.

■ Example 7.4

Assume the magnetic bearing of line *AB* read in 1878 was N26°15′E. The declination at the time and place was 7°15′W. In 2000, the declination was 4°30′E. The magnetic bearing in 2000 is needed.

Solution

The declination angles are shown in Figure 7.13. The magnetic bearing of line *AB* is equal to the earlier date bearing minus the sum of the declination angles, or

$$\text{N}26°15'\text{E} - (7°15' + 4°30') = \text{N}14°30'\text{E}$$

Again, the problem can be computed using azimuths as 26°15′ − 7°15′ − 4°30′ = 14°30′, which converts to a bearing of N14°30′E.

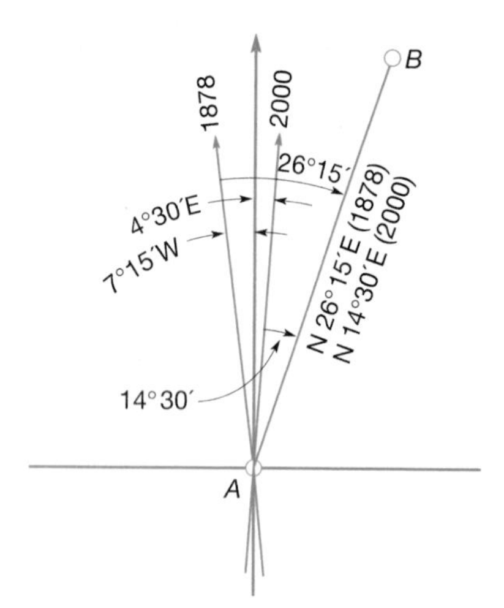

Figure 7.13 Computing magnetic bearing changes due to declination changes.

7.16 MISTAKES

Some mistakes made in using azimuths and bearings are:

1. Confusing magnetic and other reference bearings.
2. Mixing clockwise and counterclockwise angles.
3. Interchanging bearings for azimuths.
4. Listing bearings with angular values greater than 90°.
5. Failing to include both directional letters when listing a bearing.
6. Failing to change bearing letters when using the back bearing of a line.
7. Using an angle at the wrong end of a line in computing bearings—that is, using angle A instead of angle B when starting with line AB as a reference.
8. Not including the last angle to recompute the starting bearing or azimuth as a check—for example, angle A in traverse $ABCDEA$.
9. Subtracting 360°00′ as though it were 359°100′ instead of 359°60′, or using 90° instead of 180° in bearing computations.
10. Adopting an assumed reference line that is difficult to reproduce.
11. Reading degrees and decimals from a calculator as though they were degrees, minutes, and seconds.
12. Failing to adjust traverse angles before computing bearings or azimuths if there is a misclosure.

PROBLEMS

Asterisks (*) indicate problems that have answers given in Appendix G.

7.1 How would you determine the magnetic declination of a line in the field?

7.2 What are the disadvantages of using an assumed meridian for the starting course in a traverse?

7.3 Why is it important to adopt a standard angle-observing procedure, such as always measuring angles to the right?

7.4 Where on the Earth's surface are the magnetic lines of force horizontal?

7.5 Convert: *(a) 136°14′03″ to grads (b) 117.5489 grads to degrees, minutes, and seconds (c) 89°20′47″ to radians.

In Problems 7.6 through 7.7, convert the azimuths from north to bearings, and compute the angles, smaller than 180°, between successive azimuths.

7.6 38°09′23″, 144°26′58″, 219°17′37″, and 303°18′46″
7.7* 82°46′51″, 166°36′28″, 193°15′06″, and 347°11′33″

Convert the bearings in Problems 7.8 through 7.9 to azimuths from north and compute the angle, smaller than 180°, between successive bearings.

7.8 N46°24′13″E, S78°12′33″E, S52°55°05″W, and N62°08′02″W
7.9 N82°18′58″E, S88°45′48″E, S21°12′10″W, and N05°27′59″W

Compute the azimuth from north of line CD in Problems 7.10 through 7.12. (Azimuths of AB are also from north.)

7.10* Azimuth AB = 123°48′56″; angles to the right ABC = 93°47′28″, BCD = 54°47′36″.
7.11 Bearing AB = N04°26′12″W; angles to the right ABC = 68°05′22″, BCD = 101°32′28″.
7.12 Azimuth AB = 105°11′51″; angles to the right ABC = 211°36′49″, BCD = 57°27′17″.
7.13* For a bearing DE = S88°23′12″W and angles to the right, compute the bearing of FG if angle DEF = 105°17′44″ and EFG = 68°13′58″.
7.14 Similar to Problem 7.13, except the azimuth of DE is 313°07′33″ and angles to the right DEF and EFG are 63°16′56″ and 88°59′01″, respectively.

Course AB of a five-sided traverse runs due east. From the given balanced interior angles to the right, compute and tabulate the bearings and azimuths from north for each side of the traverses in Problems 7.15 through 7.17.

7.15 A = 108°46′49″, B = 115°20′10″, C = 94°30′22″, D = 99°54′37″, E = 121°28′02″
7.16* A = 126°01′28″, B = 107°54′36″, C = 86°22′33″, D = 99°02′17″, E = 120°39′06″
7.17 A = 156°23′48″, B = 41°37′02″, C = 94°30′15″, D = 154°11′50″, E = 93°17′05″

In Problems 7.18 and 7.19, compute and tabulate the bearings of the sides of a regular pentagon (polygon with five equal angles), given the starting direction of side AB.

7.18 Bearing of AB = N56°27′13″W (Station C is westerly from B.)
7.19 Azimuth of AB = 87°14′26″ (Station C is westerly from B.)
7.20 Does local attraction at a point affect the size of the angle computed from magnetic bearings read there?

Compute azimuths of all lines for a closed traverse $ABCDEFGHA$ that has the following balanced angles to the right, using the directions listed in Problems 7.21 and 7.22. HAB = 78°26′59″, ABC = 153°07′38″, BCD = 203°58′37″, CDE = 83°52′50″, DEF = 98°11′03″, EFG = 218°24′21″, FGH = 94°24′52″, GHA = 149°33′40″.

7.21 Bearing AB = N56°18′42″W.
7.22 Azimuth DE = 256°10′00″.
7.23 Similar to Problem 7.21, except that bearings are required, and fixed bearing AB = S23°36′05″E.
7.24 Similar to Problem 7.22, except that bearings are required, and fixed azimuth AB = 76°02′31″ (from north).
7.25 Geometrically show how the sum of the interior angles of a hexagon (six sides) can be computed using the formula $(n - 2)180°$.

7.26 Use WOLFPACK to determine the predicted declinations on January 1, 2008, using the World Magnetic Model for 2005 (WMM2005) at the following locations.
(a)* latitude = 42°58′28″N, longitude = 77°12′36″W, elevation = 310.0 m;
(b) latitude = 37°56′44″N, longitude = 110°50′40″W, elevation = 1500 m;
(c) latitude = 41°18′15″N, longitude = 76°00′26″W, elevation = 240 m;

7.27 Using Table 7.4, what was the total difference in magnetic declination between Cleveland, OH, and Denver, CO?

7.28 The magnetic declination at a certain place is 10°46′E. What is the magnetic bearing there: (a) of true north (b) of true south (c) of true west?

7.29 Same as Problem 7.28, except the magnetic declination at the place is 8°00′W.

For Problems 7.30 through 7.32, the observed magnetic bearing of line *AB* and its true magnetic bearing are given. Compute the amount and direction of local attraction at point *A*.

	Observed Magnetic Bearing	True Magnetic Bearing
7.30*	N32°30′E	N29°55′E
7.31	S18°02′W	S12°15′W
7.32	N3°36′W	N4°30′E

What magnetic bearing is needed to retrace a line for the conditions stated in Problems 7.33 through 7.36?

	1875 Magnetic Bearing	1875 Declination	Present Declination
7.33*	N32°45′E	8°12′W	2°30′E
7.34	S48°15′W	12°12′W	1°20′E
7.35	N88°20′W	2°30′W	8°30′W
7.36	S14°55′E	14°05′E	2°50′E

In Problems 7.37 through 7.38, calculate the magnetic declination in 1870 based on the following data from an old survey record.

	1870 Magnetic Bearing	Present Magnetic Bearing	Present Magnetic Declination
7.37	S44°20E	S41°35′E	4°15′W
7.38	S30°30′W	S44°20′W	8°30′E

7.39 An angle *APB* is measured at different times using various instruments and procedures. The results, which are assigned certain weights, are as follows: 89°43′38″, wt 3; 89°43′32″, wt 2; and 89°43′36″, wt 1. What is the most probable value of the angle?

7.40 Similar to Problem 7.39, but with an additional measurement of 89°43′30″, wt 4.

7.41 Explain why the letters E and W on a compass [see Figure 7.9(b)] are reversed from their normal positions.

7.42 Create a computational program that solves Problem 7.21.

7.43 Create a computational program that solves Problem 7.22.

BIBLIOGRAPHY

Kratz, K. E. 1990. "Compass Surveying with a Total Station." *Point of Beginning* 16 (No. 1): 30.
Sipe, F. H. 1990. "A Clinic on the Open-Sight Compass." *Surveying and Land Information Science* 50 (No. 3): 229.

8

Total Station Instruments; Angle Measurements

PART I • TOTAL STATION INSTRUMENTS

8.1 INTRODUCTION

Until recently, transits and theodolites were the most commonly used surveying instruments for making angle observations. These two devices were fundamentally equivalent and could accomplish basically the same tasks.[1] Today, total station instruments have replaced all but a few transits and theodolites. Total station instruments can accomplish all of the tasks that could be done with transits and theodolites and do them much more efficiently. In addition they can also observe distances accurately and quickly. Furthermore, they can make computations with the angle and distance observations and display the results in real time. These and many other significant advantages have made total stations the predominant instruments used in surveying practice today. They are used for all types of surveys including topographic, hydrographic, cadastral, and construction surveys. The use of total station instruments for specific types of surveys is discussed in later chapters. This chapter describes the general design and characteristics of total station instruments and also concentrates on procedures for using them in observing angles.

8.2 CHARACTERISTICS OF TOTAL STATION INSTRUMENTS

Total station instruments, as shown in Figure 8.1, combine three basic components—an electronic distance measuring (EDM) instrument, an electronic angle measuring component, and a computer or microprocessor—into one integral unit. These

[1]Discussions of the transit and theodolite are given in Appendix A.

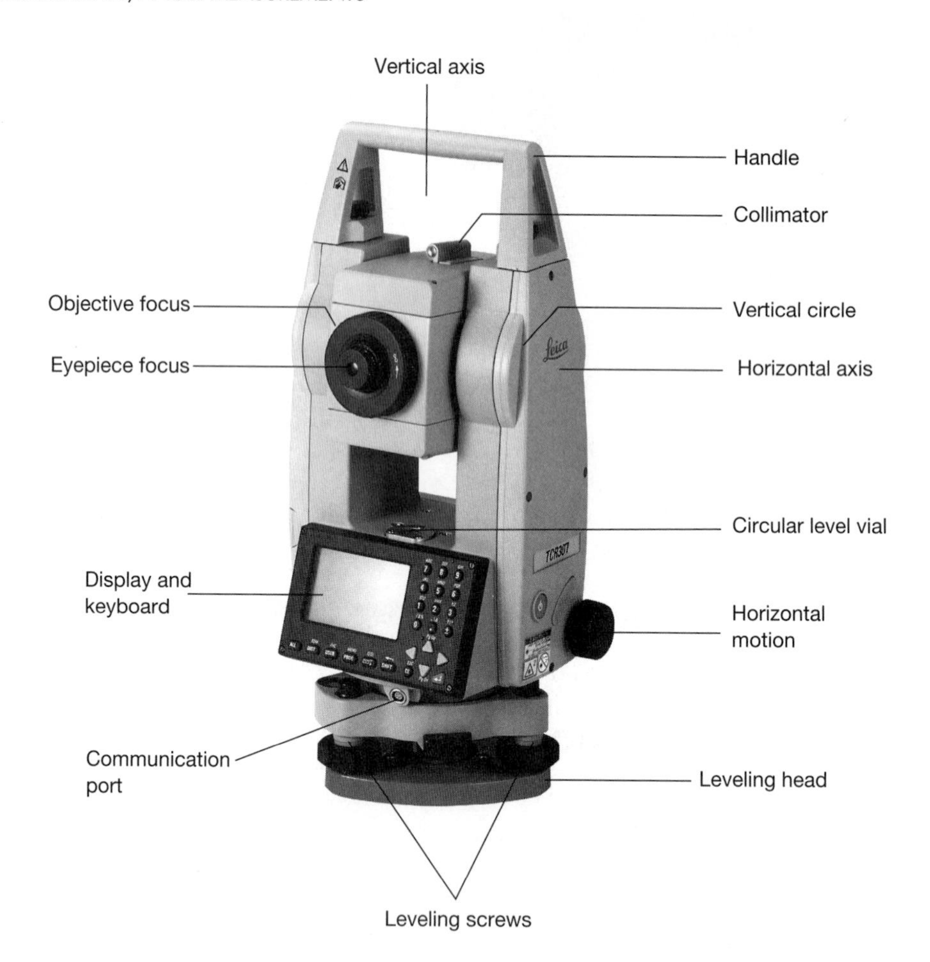

Figure 8.1 Parts of a total station instrument, with view of eyepiece end of telescope. (Courtesy Leica Geosystems, Inc.)

devices can automatically observe horizontal and vertical angles, as well as slope distances, from a single setup (see Chapter 6). From these data they can instantaneously compute horizontal and vertical distance components, elevations, and coordinates of points sighted, and display the results on a *liquid crystal display* (LCD). As discussed in Chapter 2, they can also store the data, either on board or in external data collectors connected to their communication ports.

The telescope is an important part of a total station instrument. It is mounted between the instrument's *standards* (see Figure 8.1), and after the instrument has been leveled, it can be revolved (or "plunged") so that its *axis of sight*[2] defines a vertical plane. The axis about which the telescope revolves is called the *horizontal axis*. The telescope can also be rotated in any azimuth about a vertical line called the *vertical axis*. Being able to both revolve and rotate the telescope in this manner makes it possible for an operator to aim the telescope in any azimuth, and along any slope, to sight points. This is essential in making angle observations, as

[2]The axis of sight, also often called the "line of sight," is the reference line within the telescope which an observer uses for making pointings with the instrument. As defined in Section 4.7, it is the line connecting the optical center of the objective lens and the intersection of crosshairs in the reticle.

described in Part II of this chapter. The three reference axes, the axis of sight, the horizontal axis, and the vertical axis, are illustrated in Figure 8.24.

The EDM instruments that are integrated into total station instruments (described in Section 6.21) are relatively small, and as shown in Figure 8.1, are mounted with the telescope between the standards of the instrument. Although the EDM instruments are small, they still have distance ranges adequate for most work. Lengths up to about 4 km can be observed with a single prism and even farther with a triple prism like the one shown in Figure 6.12.

Total station instruments are manufactured with two graduated circles mounted in mutually perpendicular planes. Prior to observing angles, the instrument is leveled so that its *horizontal* circle is oriented in a horizontal plane, which automatically puts the vertical circle in a vertical plane. Horizontal and zenith (or altitude) angles can then be observed directly in their respective planes of reference. To increase the precision of the final horizontal angle, repeating instruments had two vertical axes. This resulted in two horizontal motion screws. One set of motion screws allowed the instrument to be turned without changing the value on the horizontal circle. Repeating theodolites are discussed in more detail in Section A.5.2. Today's total station instruments usually have only one vertical axis and thus are considered directional instruments. However, as discussed later, angles can be repeated on a total station by following the procedures described in the instrument's manual. Most early versions of total station instruments employed level vials for orienting the circles in horizontal and vertical planes, but many newer ones now use automatic compensators, or electronic tilt-sensing mechanisms.

The angle resolution of available total stations varies from as low as a half-second for precise instruments suitable for control surveys, up to 20″ for less expensive instruments made specifically for construction work. Formats used for displaying angles also vary with different instruments. For example, the displays of some actually show the degree, minute, and second symbols, but others use only a decimal point to separate the number of degrees from the minutes and seconds. Thus, 315.1743 is read as 315°17′43″. Most instruments allow a choice of units, such as the display of angular measurements in degrees, minutes, and seconds, or in grads (gons). Distances may be shown in either English or metric units. Also, certain instruments enable the choice of displaying either zenith or altitude angles. These choices are entered through the keyboard, and the microprocessor performs the necessary conversions accordingly. The keyboard, used for instrument control and data entry, is located just above the leveling head, as shown in Figure 8.1.

Once the instrument has been set up and a sighting has been made through the telescope, the time required to make and display an angle and distance reading is approximately 2 to 4 sec when a total station instrument is being operated in the *normal mode,* and less than 0.5 sec when operated in the *tracking mode*. The normal mode, which is used in most types of surveys with the exception of construction layout, results in higher precision because multiple observations are made and averages taken. In the tracking mode, used primarily for construction layout, a prism is held on line near the anticipated final location of a stake. An observation is quickly taken to the prism and the distance that it must be moved forward or back is instantly computed and displayed. The prism is moved ahead or back according to the results of the first observation and another check of the distance is made. The process is quickly repeated as many times as necessary until the correct distance is obtained, whereupon the stake is set. This procedure is discussed in more detail in Chapter 23.

■ 8.3 FUNCTIONS PERFORMED BY TOTAL STATION INSTRUMENTS

Total station instruments, with their microprocessors, can perform a variety of functions and computations, depending on how they are programmed. Most are capable of assisting an operator, step by step, through several different types of basic surveying operations. After selecting the type of survey from a menu, prompts will automatically appear on the display to guide the operator through each step. An example illustrating a topographic survey conducted using this procedure is given in Section 17.9.1.

In addition to providing guidance to the operator, microprocessors of total stations can perform many different types of computations. The capabilities vary with different instruments, but some standard computations include (1) averaging of multiple angle and distance observations; (2) correcting electronically observed distances for prism constants, atmospheric pressure, and temperature; (3) making curvature and refraction corrections to elevations determined by trigonometric leveling; (4) reducing slope distances to their horizontal and vertical components; (5) calculating point elevations from the vertical distance components (supplemented with keyboard input of instrument and reflector heights); and (6) computing coordinates of surveyed points from horizontal angle and horizontal distance components (supplemented with keyboard input of coordinates for the occupied station, and a reference azimuth). The subject of coordinate computations is covered in Chapters 10 and 11.

Many total stations, but not all, are also capable of making corrections to observed horizontal and vertical angles for various instrumental errors. For example, by going through a simple calibration process, the *indexing error* of the vertical circle can be determined (see Section 8.13), stored in the microprocessor, and then a correction applied automatically each time a zenith angle is measured. A similar calibration and correction procedure applies to errors that exist in horizontal angles due to imperfections in the instrument (see Section 8.8). Some total stations are also able to correct for personal errors, such as imperfect leveling of the instrument. By means of tilt-sensing mechanisms, they automatically measure the amount and direction of dislevelment and then make corrections to the observed horizontal and vertical angles for this condition.

■ 8.4 PARTS OF A TOTAL STATION INSTRUMENT

The upper part of the total station instrument, called the *alidade*, includes the telescope, graduated circles, and all other elements necessary for measuring angles and distances. The basic design and appearance of these instruments (see Figures 8.1 and 8.2) are:

1. The *telescopes* are short, have reticles with crosshairs etched on glass, and are equipped with rifle sights or collimators for rough pointing. Most telescopes

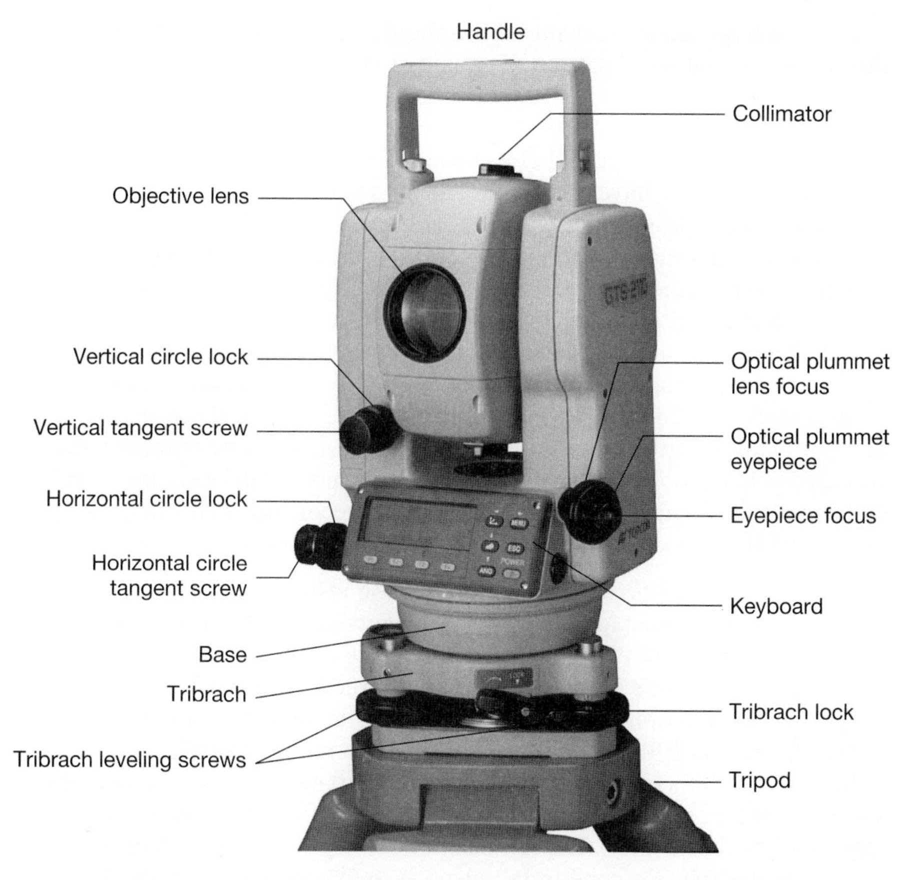

Figure 8.2 Parts of a total station instrument with view of objective-end of the telescope. (Courtesy Topcon Positioning Systems)

have two focusing controls. The objective lens control is used to focus on the object being viewed. The eyepiece control is used to focus on the reticle. If the focusing of the two lenses is not coincident, a condition known as *parallax* will exist. Parallax is the apparent motion of an object caused by a movement in the position of the observer's eye. The existence of parallax can be observed by quickly shifting one's eye position slightly and watching for movement of the object in relation to the crosshairs. Careful adjustment of the eyepiece and objective lens will result in a sharp image of both the object and the crosshairs with no visible parallax. Since the eye tends to tire through use, the presence of parallax should be checked throughout the day. A common mistake of beginners is to have a colleague "check" their pointings. This is not recommended for many reasons including the personal focusing differences that exist between different individuals.

With newer instruments, objective lens auto-focusing is available. This works in a manner similar to auto-focusing for a camera and increases the rate at which pointings can be made when objects are at variable distances from the instrument.

2. The *angle measurement system* functions by passing a beam of light through finely spaced graduations. The Topcon GTS 210 of Figure 8.2 is representative of the way total stations operate and is briefly described here. For horizontal angle measurements, two glass circles within the alidade are mounted parallel, one on top of the other, with a slight spacing between them. After the

instrument has been leveled, the circles should be in horizontal planes. The *rotor* (lower circle) contains a pattern of equally divided alternate dark lines and light spaces. The *stator* (upper circle) contains a slit-shaped pattern, which has the same pitch as that of the rotor circle. A *light-emitting diode* (LED) directs collimated light through the circles from below toward a photo detector cell above. A modern total station may have as many as 20,000 graduations!

When an angle is turned with the GTS 210, the rotor moves with respect to the stator creating alternating variations of light intensity. Photo detectors sense these variations, convert them into electrical pulses, and pass them to a microprocessor for conversion into digital values. The digits are displayed using a *liquid crystal diode* (LCD). Another separate system like that just described is also mounted within the alidade for measuring *zenith* (or *altitude*) *angles*. With the instrument leveled, this vertical circle system is aligned in a vertical plane. After making an observation, horizontal and zenith angles are both displayed, and can be manually read and recorded in field books, or alternatively, the instruments can be equipped with data collectors that eliminate manual reading and recording. (This helps eliminate mistakes!) The Topcon GTS 210 can resolve angles to an accuracy of 5″.

3. The *vertical circle* of most total station instruments is precisely indexed with respect to the direction of gravity by an *automatic compensator*. These devices are similar to those used on automatic levels (see Section 4.10) and automatically align the vertical circle so that 0° is oriented precisely upward toward the *zenith* (opposite the direction of gravity). Thus the vertical circle readings are actually *zenith angles*, that is, 0° occurs with the telescope pointing vertically upward, and either 90° or 270° is read when it is horizontal. Upon command, the microprocessor can convert zenith angles to *altitude angles* (i.e., values measured up or down from 0° at the horizontal). The *vertical motion*, which contains a *lock* and *tangent screw*, enables the telescope to be released so that it can be revolved about the horizontal axis or locked (clamped) to prevent it from revolving. To sight a point, the lock can be opened and the telescope tilted up or down about the horizontal axis as necessary to the approximate position needed to sight a point. The lock is then clamped, and fine pointing completed using the vertical tangent screw.

In servo-driven total stations (see Figure 8.7), the lock and tangent screw are replaced with a jog/shuttle mechanism. This device actuates an internal servo-drive motor that rotates the telescope about its horizontal axis. The speed at which the mechanism rotates determines the speed at which the telescope rotates.

4. Rotation of the telescope about the vertical axis occurs within a steel cylinder or on precision ball bearings, or a combination of both. The *horizontal motion*, which also contains a *lock* and *tangent screw*, controls this rotation. Clamping the lock can prevent rotation. To sight a point, the lock is released, the telescope is rotated in azimuth to the approximate direction desired, and the lock clamped again. Then the horizontal tangent screw enables a fine adjustment to be made in the direction of pointing. (Actually when sighting a point, both the vertical and horizontal locks are released so that the telescope can be simultaneously revolved and rotated. Then both are locked and fine pointing made using the two tangent screws.)

Similar to the vertical motion in servo-driven total stations, the horizontal lock and tangent screw is replaced with a jog/shuttle mechanism which actuates an internal servo-drive to rotate the instrument about its vertical axis. Again the speed

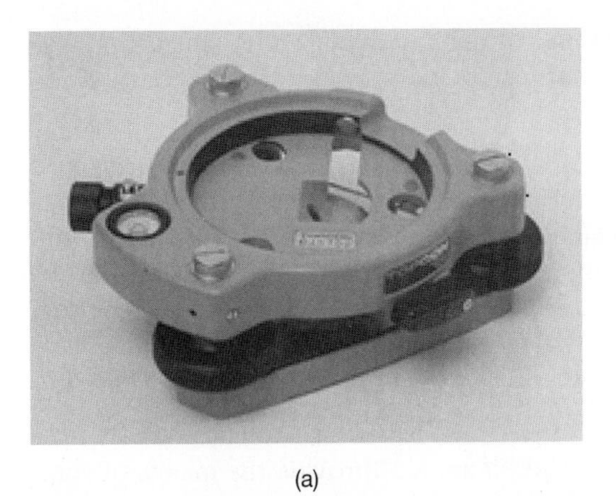

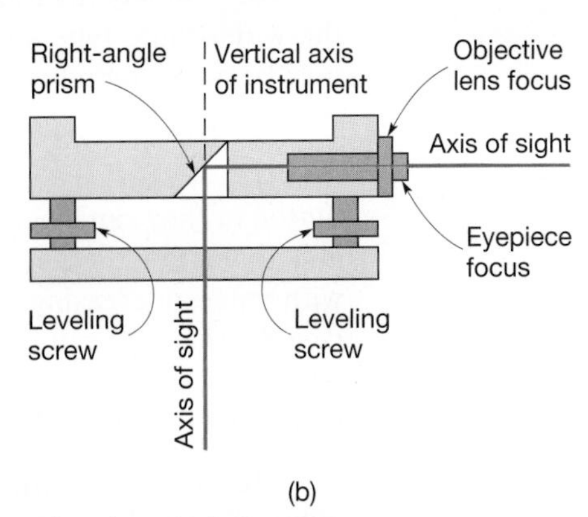

(a) (b)

Figure 8.3 (a) Tribrach with optical plummet, (b) schematic of a tribrach optical plummet. [Figure (a), Courtesy Topcon Positioning Systems]

at which the mechanism is rotated determines the speed at which the instrument rotates.

5. The *tribrach* (see Figures 8.1 and 8.2) consists of three screws or cams for leveling, a circular level (bull's-eye bubble), clamping device to secure the base of the total station or accessories (such as prisms and sighting targets), and threads to attach the tribrach to the head of a tripod. As shown in Figure 8.3, some tribrachs also have integral optical plummets (described further on) to enable centering accessories over a point without the instrument.

6. The *bases* of total stations are often designed to permit interchange of the instrument with sighting targets and prisms in tribrachs without disturbing previously established centering over survey points. This can save a considerable amount of time. Most manufacturers use a standardized "three-post" arrangement to enable interchangeability between different instruments and accessories.

7. An *optical plummet,* built into either the tribrach or alidade of total station instruments, permits accurate centering over a point. Although either type enables accurate centering, best accuracy is achieved with those that are part of the alidade of the instrument. The optical plummet provides a line of sight that is directed downward, collinear with the vertical axis of the instrument. But the total station instrument or tribrach must be leveled first for the line of sight to be *vertical.* Figures 8.3(a) and (b) show a tribrach with optical plummet and a schematic of the tribrach optical plummet, respectively. Due to the short length of the telescope in an optical plummet, it is extremely important to remove parallax before centering the instrument with this device.

In newer instruments, laser plummets have replaced the optical plummet. This device produces a beam of collimated light that coincides with the vertical axis of the instrument. Since focusing of the objective and eyepiece lens is not required with a laser plummet, this option will increase both the speed and accuracy of setups. However, the laser mark may be difficult to see in bright sunlight.

8. When being used, total station instruments sit on *tripods*. The tripods are the wide-frame type and most have adjustable legs. Their primary composition may be wood, metal, or fiberglass.

9. The *microprocessor* provides several significant advantages to surveyors. As examples, (a) the circles can be instantaneously zeroed by simply pressing a button or they can be initialized to any value by entry through the keyboard (valuable for setting the reference azimuth for a backsight); (b) angles can be observed with values increasing either left or right; and (c) angles observed by repetition (see Section 8.8) can be added, to provide the total, even though 360° may have been passed one or more times. Other advantages include reduction of mistakes in making readings and an increase in the overall speed of operation.

10. The *keyboard* and *display* (see Figure 8.2) provide the means of communicating with the microprocessor. Most total stations have a keyboard and display on both sides of the instrument, a feature which is especially convenient when operating the instrument in both the *direct* and *reversed* modes (see Section 8.8), as is usually done when observing angles. Some robotic total stations (see Section 8.6) also have a keyboard and display mounted on a remote prism pole for "one-person" operations.

11. The *communication port* (see Figure 8.1) enables external data collectors to be connected to the instrument. Some instruments have internal data collection capabilities and their communications ports permit them to be interfaced with a computer for direct downloading of data.

■ 8.5 HANDLING AND SETTING UP A TOTAL STATION INSTRUMENT

A total station instrument should be carefully lifted from its carrying case by grasping the standards or handle and the instrument securely fastened to the tripod by means of the tribrach. For most surveys, prior to observing distances and angles the instrument must first be carefully set up over a specific point. The setup process using an instrument with an optical plummet, tribrach mount with bull's-eye bubble, and adjustable-leg tripod is most easily accomplished with the following steps: (1) adjust the position of the tripod legs by lifting and moving the tripod as a whole until the point is roughly centered beneath the tripod head (beginners can drop a stone from the center of the tripod head, or use a plumb bob to check nearness to the point); (2) firmly place the legs of the tripod in the ground; (3) roughly center the tribrach leveling screws on their posts; (4) mount the tribrach approximately in the middle of the tripod head to permit maximum translation in step (9) in any direction; (5) properly focus the optical plummet on the point, making sure to check for parallax; (6) manipulate the leveling screws to aim the optical plummet telescope at the point below; (7) center the bull's-eye bubble by adjusting the lengths of the tripod extension legs; (8) and level the instrument using the plate bubble and leveling screws; and (9) if necessary, loosen the tribrach screw and translate the instrument (do not rotate it) to carefully center the plummet pointing device on the point; (10) repeat steps (8) and (9) until precise leveling and

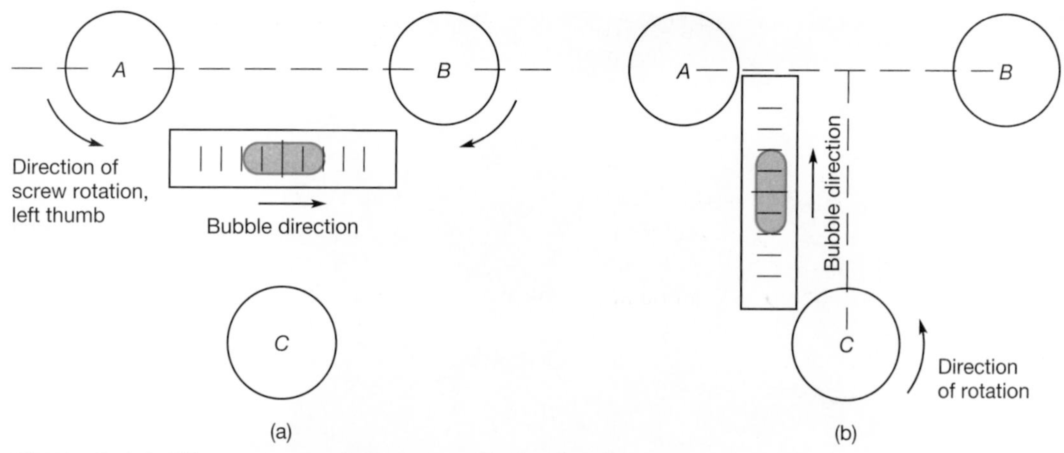

Figure 8.4 Bubble centering with three-screw leveling head.

centering are accomplished. With total stations that have their optical plummets in the tribrach, the instrument can and should be left in the case until step (8).

To level a total station instrument that has a plate level vial, the telescope is rotated to place the axis of the level vial parallel to the line through any two leveling screws, as the line through A and B in Figure 8.4(a). The bubble is centered by turning these two screws, then rotated 90°, as shown in Figure 8.4(b), and centered again using the third screw (C) only. This process is repeated and checked carefully to ensure that the bubble remains centered. As illustrated in Figure 8.4, *the bubble moves in the direction of the left thumb when the foot screws are turned.* A solid tripod setup is essential and the instrument must be shaded if set up in bright sunlight. Otherwise, the bubble will expand and run toward the warmer end as the liquid is heated.

Many instruments, such as the LEICA TPS 300 shown in Figure 8.1, do not have traditional level vials. Rather, they are equipped with an electronic, dual-axis leveling system as shown in Figure 8.5 in which four probes sense a liquid (horizontal) surface. After preliminary leveling is performed by means of the tribrach's bulls-eye bubble, signals from the probes are processed to form an image on the LCD display, which guides an operator in performing rough leveling. The three leveling screws are used, but the instrument need not be turned about its vertical axis in the leveling process. After rough leveling, the amount and direction of any residual dislevelment is automatically and continuously received by the microprocessor, which corrects observed horizontal and vertical angles accordingly in real time.

As noted earlier, total stations are controlled with entries made either through their built-in keyboards or through the keypads of hand-held data collectors. Details for operating each individual total station vary somewhat and therefore are not described here. They are covered in the manuals provided with the purchase of instruments.

When moving between setups in the field, proper care should be taken. Before the total station is removed from the tripod, the foot screws should be

Figure 8.5 The LEICA TPS 300 electronic leveling system. (Courtesy Leica Geosystems, Inc.)

(a)

(b)

Figure 8.6 (a) A proper method of transporting a total station in the field. (b) Total station in open case. (Courtesy Leica Geosystems, Inc.)

returned to the midpoints of the posts. Many instruments have a line on the screw post that indicates the halfway position. The instrument should NEVER be transported on the tripod since this causes stress to tripod head, tribrach, and instrument base. Figure 8.6(a) depicts the proper procedure for carrying equipment in the field. With adjustable-leg tripods, retracting them to their shortest positions and lightly clamping them in position can avoid stress on the legs.

When returning the total station to its case, all locking mechanisms should be released. This procedure protects the threads and reduces wear when the instrument is jostled during transport and also prevents the threads from seizing during long periods of storage. If the instrument is wet, it should be wiped down and left in an open case until it is dry as shown in Figure 8.6(b). When storing tripods, it is important to loosen or lightly clamp all legs. This is especially true with wooden tripods where the wood tends to expand and contract with humidity in the air. Failure to loosen the clamping mechanism on wooden tripods can result in crushed wood fibers, which inhibit the ability of the clamp to hold the leg during future use.

■ 8.6 SERVO-DRIVEN AND REMOTELY OPERATED TOTAL STATION INSTRUMENTS

Manufacturers also produce "robotic" total station instruments equipped with servo-drive mechanisms that enable them to aim automatically at a point to be set. A robotic total station shown on the left in Figure 8.7 is an example. With these instruments, it is only necessary to identify the point's number with a keyboard entry. The computer retrieves the direction to the point from storage or computes it and activates a servomotor to turn the telescope to that direction within a few seconds. This feature is particularly useful for construction stakeout, but it is also

Figure 8.7 The Geodimeter robotic total station. (Courtesy Spectra Precision)

convenient in control surveying when multiple observations are made in observing angles. In this instance, final precise pointing is done manually.

The *remote positioning unit* (RPU) shown on the right in Figure 8.7, which is attached to a prism pole, has a built-in telemetry link for communication with the total station. The robotic instrument is equipped with an automatic search and aim function, as well as a link for communication with the RPU. It has servomotors for automatic aiming at the prism both horizontally and vertically. With the RPU, the total station instrument can be controlled from a distance.

To operate the system, the robotic instrument must first be set up and oriented. This consists in entering the coordinates of the point where the total station is located and taking a backsight along a line of known azimuth. Once oriented, an operator carries the remote positioning unit to any convenient location and sights the robotic instrument using the telescope of the RPU. The vertical angle of sight is sent to the robotic instrument, whereupon the instrument's vertical servomotor automatically sets its telescope at the required vertical angle. Its horizontal servomotor then activates and swings around until it finds the prism. Once the total station has found the RPU, which only takes a few seconds, and locks onto it, it will automatically follow its further movements. If lock is lost, the search routine is simply repeated. The RPU not only serves as the control unit for the system, but it also operates as a data collector.

With this and similar systems, the total station instrument is completely controlled through the keyboard of the remote unit. These systems enable one person to conduct a complete survey. They are exceptionally well suited for construction surveys and topographic surveys, but can be used advantageously in other types as well. The system not only eliminates one person and speeds the work, but more importantly, it eliminates mistakes in identifying points that can occur when the prism is far from the total station and cannot be clearly seen. For safety reasons, however, another person should be present on the site to respond to emergency cases. This person may be performing some other task.

■ PART II • ANGLE MEASUREMENTS

■ 8.7 RELATIONSHIP OF ANGLES AND DISTANCES

Determining the relative positions of points often involves observing of both angles and distances. The best-quality surveys result when there is compatibility between the accuracies of these two different kinds of observations. The formula for relating distances to angles is given by the geometric relationship.

$$S = R\theta \tag{8.1}$$

In Equation (8.1), S is the arc length subtended at a distance R by an arc of θ in radians. To select instruments and survey procedures necessary for achieving consistency and to evaluate the effects of errors due to various sources, it is helpful to consider the relationships between angles and distances given here and illustrated in Figure 8.8.

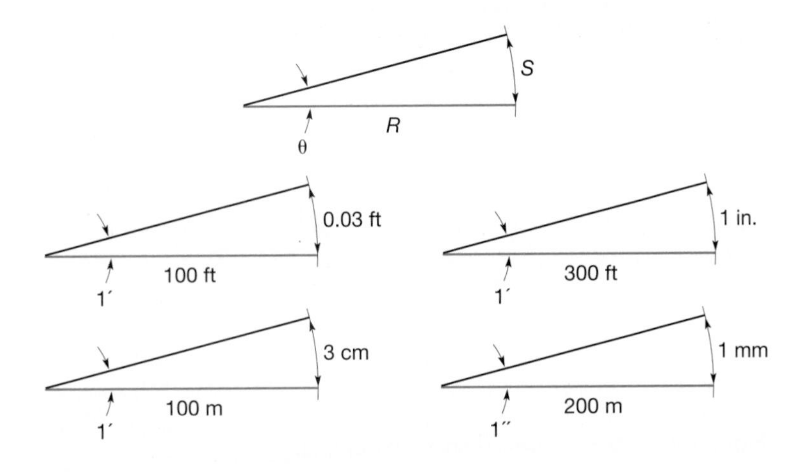

Figure 8.8 Angle and distance relationships.

$$\begin{aligned} 1'\text{of arc} &= 0.03\text{ ft at }100\text{ ft, or }3\text{ cm at }100\text{ m (approx)} \\ 1'\text{of arc} &= 1\text{ in. at }300\text{ ft (approx.; actually }340\text{ ft)} \\ 1''\text{of arc} &= 1\text{ ft at }40\text{ mi, or }0.5\text{ m at }100\text{ km, or }1\text{ mm at }200\text{ m (approx.)} \\ 1''\text{of arc} &= 0.000004848\text{ radians} \\ 1\text{ radian} &= 206{,}264.8''\text{ of arc (approx.)} \end{aligned}$$

In accordance with the relationships listed an error of approximately 1 min results in an observed angle if the line of sight is misdirected by 1 in. over a distance of 300 ft. This illustrates the importance of precisely setting the instrument and targets over their respective points, especially where short sights are involved. If an angle is expected to be accurate to within $\pm 5''$ for sights of 500 ft, then the distance must be correct to within $500 \times 5'' \times 0.000004848 = \pm 0.01$ ft for compatibility.

To appreciate the precision capabilities of a high-quality total station, an instrument reading to the nearest 0.5″ is theoretically capable of measuring the angle between two points approximately 1 cm apart and 4 km away! However, as discussed in Section 8.19 through 8.21, errors from centering the instrument, sighting the point, reading the circle, and other sources make it difficult, if not impossible, to actually accomplish this accuracy.

■ 8.8 OBSERVING HORIZONTAL ANGLES WITH TOTAL STATION INSTRUMENTS

As stated in Section 2.1, horizontal angles are observed in horizontal planes. After a total station instrument is set up and leveled, its horizontal circle is in a horizontal plane and thus in proper orientation for observing horizontal angles. To observe a horizontal angle, for example angle *JIK* of Figure 8.9(a), the instrument is first set up and centered over station *I* and leveled. Then a *backsight* is taken on station *J*. This is accomplished by releasing the horizontal and vertical locks, turning the telescope in the approximate direction of *J*, and clamping both locks. A precise pointing is then made to place the vertical crosshair on the target using the horizontal and vertical tangent screws and an initial value of 0°00′00″ is entered in the display. The horizontal motion is then unlocked and the telescope turned clockwise toward point *K* to make the *foresight*. The vertical circle lock is usually

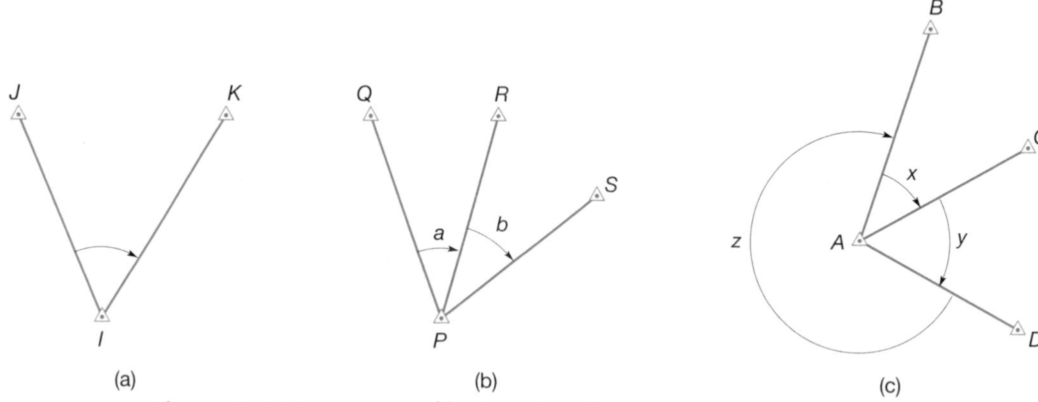

Figure 8.9 Measurement of horizontal angles.

released to rotate the telescope for sighting point K. Again the motions are clamped with the line of sight approximately on station K and precise pointing is made as before using the horizontal tangent screw. When the foresight is completed, the value of the horizontal angle will automatically appear in the display.

To eliminate instrumental errors and increase precision, angle observations should be repeated an equal number of times in each of the *direct* and *reversed* modes and the average taken. Built-in computers of total station instruments will automatically perform the averaging and display the final results. For instruments that have only a single keyboard and display, the instrument is in its direct mode when the eyepiece and keyboard are on the same side of the instrument. However, instruments do vary by manufacturer and the operator should refer to the instrument's manual to determine the proper orientation of their instrument when in the direct mode. To get from the direct mode into the reversed mode, the telescope is "plunged" (revolved 180° about the horizontal axis).

Procedures for repeating horizontal angle observations can differ with instruments of different manufacture, and operators must therefore become familiar with the features of their specific instrument by referring to its manual. The following is an example procedure that applies to some instruments. After making the first observation of angle JIK, as described previously, the angular value in the display is held by pressing a button on the keyboard of the instrument. (Assume the first observation was in the direct mode.) To repeat the angle with the instrument in the same mode, a backsight is again taken on station J using the horizontal lock and tangent screw. After the backsight is completed, with the first observed angular value still on the display, the display is released for the next angle observation by again pressing the appropriate button on the keyboard. Using the same procedures described earlier, a foresight is again taken on station K, after which the display will read the sum of the two repetitions of the angle. This procedure is repeated until the desired number of angles is observed in the direct mode, whereupon the display will show the sum of these repetitions. Then the telescope is plunged to place it in the reversed mode and the angle repeated an equal number of times using the same procedure. In the end, the sum of all angles turned, direct and reversed, will be displayed. The final angle is the mean.

HORIZONTAL ANGLE MEASUREMENT

Angle	First Reading	Fourth Reading	Mean Angle		
(1)	(2)	(3)	(4)		
	° ′ ″	° ′ ″	° ′ ″		
JIK	66 37 40	266 30 48	66 37 42		

Figure 8.10 Field notes for measuring the horizontal angle of Figure 8.9(a) by repetition.

The procedure just described for observing horizontal angles is called the *repetition method*. As noted earlier, obtaining an average value from repeated observations increases precision and by incorporating equal numbers of direct and reversed measurements, certain instrumental errors are eliminated (see Section 8.20).

An example set of field notes for observing the angle of Figure 8.9(a) by the repetition method is shown in Figure 8.10. In the example, four repetitions, two in each of the direct and reversed modes, were taken. In the notes, the identification of the angle being observed is recorded in column (1), the value of the first reading of the angle is placed in column (2) and is only recorded for checking purposes, the fourth (final) reading is tabulated in column (3), and the mean of the four readings, which produces the final angle, is given in column (4). Note that the first reading would not have to be recorded, except that it is used as a check against which the mean angle is compared. If these two values agree within tolerable limits, the mean angle is accepted, if not the work is repeated.

Special capabilities are available with many total station instruments to enhance their accuracy and expedite operation. For example, most instruments have a dual-axis automatic compensator, which senses any misorientation of the circles. This information is relayed to the built-in computer that corrects for any *indexing error* in the vertical circle (see Section 8.13) and any dislevelment of the horizontal circle, before displaying angular values. This real-time tilt sensing and correction feature makes it necessary to perform only rough leveling of the instrument, thus reducing setup time. In addition, some instruments observe angles by integration of electronic signals over the entire circle simultaneously; thus errors due to graduations and eccentricities (see Section 8.20.1) are eliminated. Furthermore, the computer also corrects horizontal angles for instrumental errors if the axis of sight is not perpendicular to the horizontal axis, or if the horizontal axis is not perpendicular to the vertical axis. (These conditions are discussed in Sections 8.15 and 8.20.1, respectively.) This feature makes it possible to obtain angle observations free from instrumental errors without averaging equal numbers of direct and reversed readings. With these advantages, and more, it is obvious why these instruments have replaced the older instruments discussed in Appendix A.

8.9 OBSERVING HORIZONTAL ANGLES BY THE DIRECTION METHOD

As an alternative to observing horizontal angles by the repetition method described in the preceding section, total station instruments can be used to determine horizontal angles by the *direction method*. This procedure consists in observing

DIRECTIONS OBSERVED FROM STATION P

Repetition No.	Station Sighted	Reading Direct	Reading Reverse	Mean	Angle
(1)	(2)	(3)	(4)	(5)	(6)
		° ′ ″	° ′ ″	° ′ ″	° ′ ″
1	Q	0 00 00	0 00 00	0 00 00	
	R	37 30 27	37 30 21	37 30 24	37 30 24
	S	74 13 42	74 13 34	74 13 38	36 43 14
2	Q	0 00 00	0 00 00	0 00 00	
	R	37 30 32	37 30 28	37 30 30	37 30 30
	S	74 13 48	74 13 42	74 13 46	36 43 16
3	Q	0 00 00	0 00 00	0 00 00	
	R	37 30 26	37 30 26	37 30 26	37 30 26
	S	74 13 36	74 13 40	74 13 38	36 43 12
4	Q	0 00 00	0 00 00	0 00 00	
	R	37 30 34	37 30 30	37 30 32	37 30 32
	S	74 13 48	74 13 44	74 13 46	36 43 14

Figure 8.11 Field notes for measuring directions for Figure 8.9(b).

directions, which are simply horizontal circle readings taken to successive stations sighted around the horizon. Then by taking the difference in directions between any two stations, the angle between them is determined. The procedure is particularly efficient when multiple angles are being measured at a station. An example of this type of situation is illustrated in Figure 8.9(b), where angles a and b must both be observed at station P. Figure 8.11 shows a set of field notes for observing these angles by direction method. The notes actually are the results of four repetitions of direction measurements in each of the direct and reversed modes. In these notes the repetition number is listed in column (1); the station sighted in column (2); direction readings taken in the direct and reversed modes in columns (3) and (4), respectively; the mean of direct and reversed readings in column (5); and the computed angles (obtained by subtracting the mean direction for station Q from that of station R, and subtracting R from S) in column (6). As a check, the four values for each angle in column (6) should be compared for agreement, and a determination made as to whether they meet acceptance criteria before leaving the occupied station, so that additional readings can be made if necessary. Final values for the two angles are taken as the averages of the four angles in column (6). These are 37°30′28″ and 37°43′14″ for angles a and b, respectively. Note that in this procedure, as was the case with the repetition method, the multiple readings increase the precisions of the angles, and by taking equal numbers of direct and reversed readings, instrumental errors are eliminated. As previously noted, this method of observing directions can significantly reduce the time at a station, especially when several angles with multiple repetitions are needed, for example in triangulation.

■ 8.10 CLOSING THE HORIZON

Closing the horizon consists in using the direction method as described in the preceding section, but including *all* angles around a point. Suppose in Figure 8.9(c) that only angles x and y are needed. However, in closing the horizon angle z is also observed thereby providing for additional checks. An example set of field notes for this operation is shown in Figure 8.12. The angles are first turned around the horizon by making a pointing and direction reading at each station with the instrument in the direct mode [see the data entries in column (3) of Figure 8.12]. A final foresight pointing is made on the initial backsight station and this provides a check because it should give the initial backsight reading (allowing for reasonable random errors). Any difference is the *horizon misclosure,* and if its value exceeds an allowable tolerance, that round of readings should be discarded and the measurements repeated. (Note that in the field notes of Figure 8.12, the maximum horizon misclosure was 4″.)

After completing the direct-mode readings, the telescope is plunged to its reversed position and all directions around the horizon observed again [see the data entries in column (4)]. A set of readings around the horizon in both the direct and

CLOSING THE HORIZON
AT STATION A

Position No.	Station Sighted	Reading Direct	Reading Reverse	Mean	Angle
(1)	(2)	(3)	(4)	(5)	(6)
		° ′ ″	° ′ ″	° ′ ″	° ′ ″
1	B	0 00 00	0 00 00	0 00 00	
	C	42 12 12	42 12 14	42 12 13	42 12 13
	D	102 08 26	102 08 28	102 08 27	59 56 14
	B	0 00 02	0 00 02	0 00 02	257 51 35
				Sum	360 00 02
2	B	0 00 00	0 00 00	0 00 00	
	C	42 12 12	42 12 14	42 12 13	42 12 13
	D	102 08 28	102 08 28	102 08 28	59 56 15
	B	0 00 04	0 00 04	0 00 04	257 51 36
				Sum	360 00 04
3	B	0 00 00	0 00 00	0 00 00	
	C	42 12 14	42 12 12	42 12 13	42 12 13
	D	102 08 28	102 08 26	102 08 27	59 56 14
	B	0 00 04	0 00 00	0 00 02	257 51 35
				Sum	360 00 02
4	B	0 00 00	0 00 00	0 00 00	
	C	42 12 14	42 12 12	42 12 13	42 12 13
	D	102 08 32	102 08 28	102 08 30	59 56 17
	B	0 00 04	0 00 04	0 00 04	257 51 34
				Sum	360 00 04

Figure 8.12 Field notes for closing the horizon at station A of Figure 8.9(c).

reversed modes constitutes a so-called *position*. The notes of Figure 8.12 contain the results of four positions.

The note reduction process consists of calculating mean values of the direct and reversed directions to each station [see column (5)], and from them, the individual angles around the horizon are computed as discussed in Section 8.9 [see column (6)]. Finally their sum is calculated and checked against (360°). Any difference reveals a mistake or mistakes in computing the individual angles. Again, repeat values for each individual angle are obtained, and as another check on the work, these should be compared for their agreement.

As an alternative to closing the horizon by observing directions, each individual angle could be independently measured using the procedures outlined in Section 8.8. After observing all angles around the horizon, their sum could also be computed and compared against 360°. However, this procedure is not as efficient as closing the horizon using directions.

8.11 OBSERVING DEFLECTION ANGLES

A deflection angle is a horizontal angle observed from the prolongation of the preceding line, right or left, to the following line. In Figure 8.13(a), the deflection angle at *F* is 12°15′10″ to the right (12°15′10″R), and the deflection angle at *G* is 16°20′27″L.

A straight line between terminal points is theoretically the most economical route to build and maintain for highways, railroads, pipelines, canals, and transmission lines. Practically, obstacles and conditions of terrain and land-use require bends in the route, but deviations from a straight line are kept as small as possible. If an instrument is in perfect adjustment (which is unlikely), the deflection angle at *F* [see Figure 8.13(a)] is observed by setting the circle to zero and backsighting on point *E* with the telescope in the direct position. The telescope is then plunged to its reversed position, which places the line of sight on *EF* extended, as shown dashed in the figure. The horizontal lock is released for the foresight, point *G* sighted, the horizontal lock clamped, the vertical crosshair carefully set on the mark by means of the horizontal tangent screw, and the angle read.

Deflection angles are subject to serious errors if the instrument is not in adjustment, particularly if the line of sight is not perpendicular to the horizontal axis (see Section 8.15). If this condition exists, deflection angles may be read larger or smaller than their correct values, depending on whether the line of sight after plunging is to the right or left of the true prolongation [see Figure 8.13(b)]. To eliminate errors from this cause, angles are usually doubled or quadrupled by the following procedure: The first backsight is taken with the circle set at zero and the telescope in the direct position. After plunging the telescope, the angle is observed

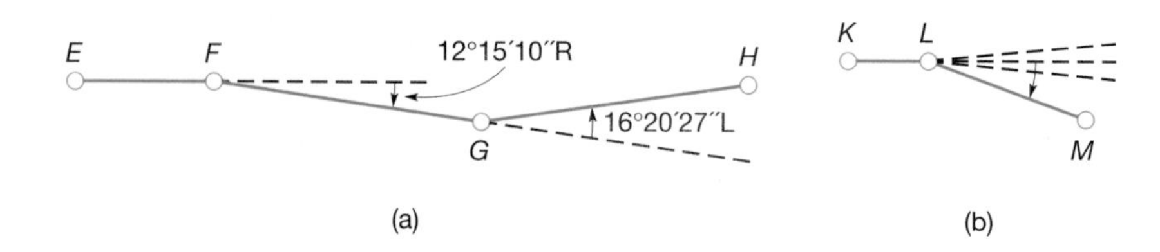

Figure 8.13 Deflection angles.

and kept in the display. Using the procedures specified by the manufacturer for holding an angle in the display, a second backsight is taken, retaining the first angle and keeping the telescope reversed. The telescope is plunged back to the normal position for the foresight, the display released, and the angle reobserved. Dividing the total angle by 2 gives an average angle from which instrumental errors have been eliminated by cancellation. In outline fashion, the method is as follows:

1. Backsight with the telescope direct. Plunge to reversed mode and observe the angle. Hold the displayed angle.
2. Backsight with the telescope still reversed. Plunge again to direct mode, release the angle display and observe the angle.
3. Read the total angle and divide by 2 for an average.

Of course, making four, six, or eight repetitions and averaging can increase the precision in direction angle observation.

Figure 8.14 shows the left-hand page of field notes for observing the deflection angles at stations *F* and *G* of Figure 8.13(a). The procedure just outlined was followed. Four repetitions of each angle were taken with the instrument alternated from direct to reverse with each repetition. Readings were recorded only after the first, second, and fourth repetitions. *The final angle is the mean obtained by dividing the last recorded value by the total number of repetitions*—four in this case. The purpose of the first two values is only to provide checks: that is, the second reading should be twice the first, and the final mean angle should be equal to the first, allowing of course for random errors.

If a mistake should occur, as in the first set of angles observed at station *G* of Figure 8.14, lines should be drawn through the incorrect data, the word "VOID" written beside them, and the observations repeated. In this voided data set, half the second recorded value (16°20′28″) agrees reasonably well with the first value (16°20′30″), but the final mean (16°20′00″) does not agree with the first recorded value. Thus that data set is discarded.

DEFLECTION ANGLES

Sta	BS/FS Sta	No. Reps.	Circle Rdg	Mean Angle	Right/Left
			° ′ ″	° ′ ″	
	E	1	12 15 12		
F		2	24 30 20		
	G	4	49 00 40	12 15 10	R
	F	1	16 20 30		
G		2	32 40 56	VOID	
	H	4	65 20 00	16 20 00	L
	F	1	16 20 24		
G		2	32 40 52		
	H	4	65 21 48	16 20 27	L

Figure 8.14 Field notes for measuring deflection angles.

8.12 OBSERVING AZIMUTHS

Azimuths are observed from a reference direction which itself must be determined from (a) a previous survey, (b) the magnetic needle, (c) a solar or star observation, (d) GPS observations, (e) a north-seeking gyro, or (f) assumption. Suppose that in Figure 8.15 the azimuth of line *AB* is known to be 137°17′ from north. The azimuth of any other line that starts at *A*, such as *AC* in the figure, can be found directly using a total station instrument. In this process, with the instrument set up and centered over station *A* and leveled, a backsight is first taken on point *B*. The azimuth of line *AB* (137°17′) is then set on the horizontal circle using the keyboard. The instrument is now "oriented," since the line of sight is in a known direction with the corresponding azimuth on the horizontal circle. If the circle were turned until it read 0°, the telescope would be pointing towards north (along the meridian). The next steps are to loosen the horizontal lock, turn the telescope clockwise to *C* and read the resultant direction, which is the azimuth of *AC*. In this case, it is 83°38′.

In Figure 8.15, if the instrument is set up at point *B* instead of *A*, the azimuth of *BA* (317°17′) or the back azimuth of *AB* is put on the circle and point *A* sighted. The horizontal lock is released, and sights taken on points whose azimuths from *B* are desired. Again, if the instrument is turned until the circle reads zero, the telescope points north (or along the reference meridian). By following this procedure at each successive station of a traverse, for example at *A*, *B*, *C*, *D*, *E*, and *F* of the traverse of Figure 7.2(a), the azimuths of all traverse lines can be determined. With a closed polygon traverse like that of Figure 7.2(a), station *A* should be occupied a second time and the azimuth of *AB* determined again to serve as a check on the work.

8.13 OBSERVING VERTICAL ANGLES

A vertical angle is the difference in direction between two intersecting lines measured in a vertical plane. Vertical angles can be measured as *altitude* or *zenith* angles. An altitude angle is the angle above or below a horizontal plane through the point of observation. Angles above the horizontal plane are called *plus angles*, or *angles of elevation*. Those below it are *minus angles*, or *angles of depression*. Zenith angles are observed with zero on the vertical circle oriented towards the zenith of the instrument and thus go from 0° to 360° in a clockwise circle about the horizontal axis of the instrument.

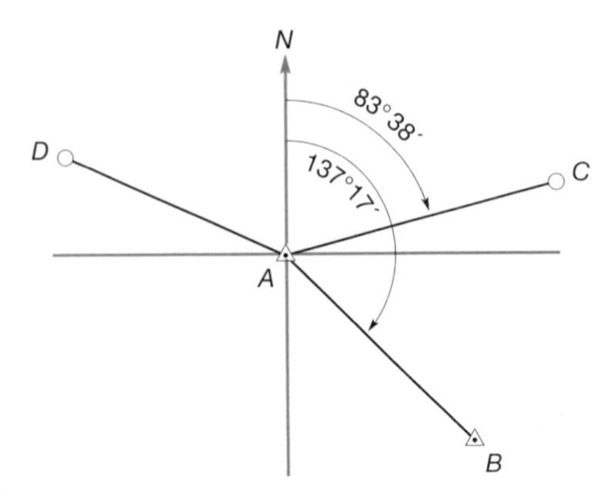

Figure 8.15 Orientation by azimuths.

Most total station instruments are designed so that *zenith angles* are displayed rather than altitude angles. In equation form, the relationship between altitude angles and zenith angles is

$$\text{Direct mode} \quad \alpha = 90^\circ - z \tag{8.2a}$$

$$\text{Reversed mode} \quad \alpha = z - 270^\circ \tag{8.2b}$$

where z and α are the zenith and altitude angles, respectively. With a total station, therefore, a reading of 0° corresponds to the telescope pointing vertically upward. In the direct mode, with the telescope horizontal, the zenith reading is 90° and if the telescope is elevated 30° above horizontal, the reading is 60°. In the reversed mode, the horizontal reading is 270° and with the telescope raised 30° above the horizon it is 300°. Altitude angles and zenith angles are observed in trigonometric leveling and in EDM work for reduction of observed slope distances to horizontal.

Observation of zenith angles with a total station instrument follows the same general procedures as those just described for horizontal angles, except that an automatic compensator orients the vertical circle. As with horizontal angles, instrumental errors in vertical angle observations are compensated for by computing the mean from an equal number of direct and reversed measurements. With zenith angles, the mean is computed from

$$\overline{z}_D = \frac{\Sigma z_D}{n} + \frac{n(360^\circ) - (\Sigma z_D + \Sigma z_R)}{2n} \tag{8.3}$$

where $\overline{z}_D$ is the mean value of the zenith angle [expressed according to its direct mode value], Σz_D the sum of direct zenith angles, Σz_R the sum of reversed angles, and n the number of z_D and z_R pairs of zenith angles read. The latter part of Equation (8.3) accounts for the *indexing error* present in the instrument.

An indexing error exists if 0° on the vertical circle is not truly at the zenith with the instrument in the direct mode. This will cause all vertical angles read in this mode to be in error by a constant amount. For any instrument, an error of the same magnitude will also exist in the reversed mode, but it will be of opposite algebraic sign. The presence of an indexing error in an instrument can be detected by observing zenith angles to a well-defined point in both modes of the instrument. If the sum of the two values does not equal 360°, an indexing error exists. To eliminate the effect of the indexing error, equal numbers of direct and reversed angle observations should be made and averaged. The averaging is normally done by the microprocessor of the total station instrument. Even though an indexing error may not exist, to be safe, *experienced surveyors always adopt field procedures that eliminate errors just in case their instruments are out of adjustment.*

With some total station instruments, indexing errors can be eliminated from zenith angles by computation, after going through a calibration procedure with the instrument. The computations are done by the microprocessor and applied to the angles before they are displayed. Procedures for performing this calibration vary with different manufacturers and are given in the manuals that accompany the equipment.

Example 8.1

A zenith angle was read twice direct giving values of 70°00′10″ and 70°00′12″, and twice reversed yielding readings of 289°59′44″ and 289°59′42″. What is the mean zenith angle?

Solution

Two pairs of zenith angles were read, thus $n = 2$. The sum of direct angles is 140°00′22″, and that of reversed values is 579°59′26″. Then by Equation (8.3):

$$\bar{z}_D = \frac{140°00'22''}{2} + \frac{2(360°) - (140°00'22'' + 579°59'26'')}{2 \times 2}$$
$$= 70°00'11'' + 0°00'03'' = 70°00'14''$$

Note that the value of 3″ from the latter part of Equation (8.3) is the indexing error.

8.14 SIGHTS AND MARKS

Objects commonly used for sights when total station instruments are being used only for angle observations include prism poles, chaining pins, pencils, plumb-bob strings, reflectors, and tripod-mounted targets. For short sights, string is preferred to a prism pole because the small diameter permits more accurate sighting. Small red and white targets of thin plastic or cardboard placed on the string extend the length of observation possible. Triangular marks placed on prisms as shown in Figure 8.16(a) provide excellent targets at both close and longer sight distances.

An error is introduced if the prism pole sighted is not plumb. The pole is kept vertical by means of a bull's-eye bubble. [The bubble should be regularly checked for adjustment and adjusted if necessary (see Section 8.19.5)]. The person holding the prism has to take special precautions in plumbing the pole, carefully watching the bull's-eye bubble on the pole. Bipods like the one shown in Figure 8.16(b) and tripods have been developed to hold the pole during multiple angle observation sessions.

The prism pole shown in Figure 8.16(b) has graduations for easy determination of the prism's height. The tripod mount shown in Figure 8.16(a) is centered over the point using the optical plummet of the tribrach. When sighting a prism pole, the vertical crosshair should bisect the pole just below the prism. Errors can result if the prism itself is sighted, especially on short lines.

In construction layout work and in topographic mapping, permanent backsights and foresights may be established. These can be marks on structures such as walls, steeples, water tanks, and bridges, or they can be fixed artificial targets. They provide definite points on which the instrument operator can check orientation without the help of a rodperson.

(a)

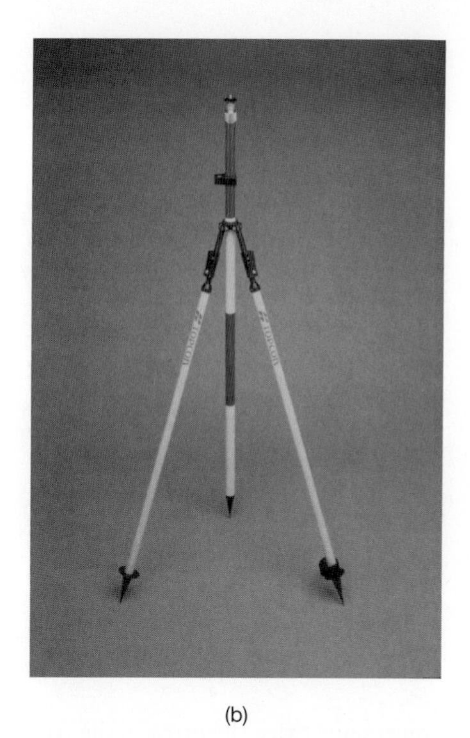

(b)

Figure 8.16 (a) Prism and sighting target with tribrach and tribrach adapter, and (b) pole and bipod, used when measuring distances and horizontal angles with total station instruments. (Courtesy Topcon Positioning Systems)

The error in a horizontal angle due to miscentering of the line of sight on a target, or too large a target, can be determined with Equation (8.1). For instance, assume a prism pole that is 20-mm wide is used as a target on a direction of only 100 m. Assuming that the pointing will be within 1/2 of the width of the pole (10 mm), then according to Equation (8.1) the error in the direction would be $(0.01/100) \times 206{,}264.8 = 21''$! For an angle where both sight distances are 100 m and assuming that the pointings are truly random, the error would propagate according to Equation (3.12), and would result in an estimated error in the angle of $21'' \times \sqrt{2}$, or approximately $30''$. From the angle-distance relationships of Section 8.7, it is easy to see why the selection of good targets in angle observations is so important.

■ 8.15 PROLONGING A STRAIGHT LINE

On route surveys, straight lines may be continued from one point through several others. To prolong a straight line from a backsight, the vertical crosshair is aligned on the back point by means of the lower motion, the telescope plunged, and a point or points, set ahead on line. In plunging the telescope, a serious error can occur if the line of sight is not perpendicular to the horizontal axis. The effects of this error can be eliminated, however, by following proper field procedures. The procedure used is known as the *principle of reversion*. The method applied, actually double reversion, is termed *double centering*. Figure 8.17 shows a simple use of the principle in drawing a right angle with a defective triangle. Lines *OX* and *OY* are drawn with the triangle in "normal" and "reversed" positions. Angle *XOY*

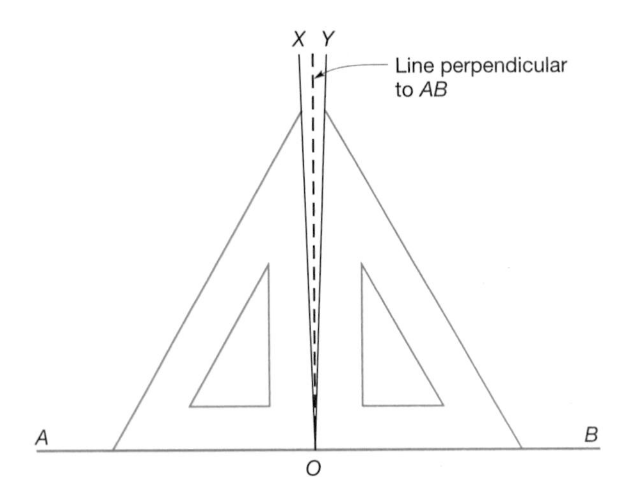

Figure 8.17 Principle of reversion.

Figure 8.18 Double centering.

represents twice the error in the triangle at the 90° corner and its bisector (shown dashed in the figure) establishes a line perpendicular to AB.

To prolong line AB of Figure 8.18 by double centering with a total station whose line of sight is not perpendicular to its horizontal axis, the instrument is set up at B. A backsight is taken on A with the telescope in the direct mode and by plunging the telescope into the reversed position the first point C' is set. The horizontal circle lock is released and the telescope turned in azimuth to take a second backsight on point A, this time with the telescope still plunged. The telescope is plunged again to its direct position and point C'' placed. Distance $C'C''$ is bisected to get point C, on line AB prolonged. In outline form, the procedure is as follows:

1. Backsight on point A with the telescope direct. Plunge to the reversed position and set point C'.
2. Backsight on point A with the telescope still reversed. Plunge to a direct position and set point C''.
3. Split the distance $C'C''$ to locate point C.

In this procedure, each time the telescope is plunged, the instrument creates twice the total error in the instrument. Thus at the end of the procedure, four times the error that exists in the instrument lies between points C' and C''. To adjust the instrument, the reticle must be shifted to bring the vertical crosshair one-fourth of the distance back from C'' towards C'. For total station instruments that have exposed capstan screws for adjusting their reticles, an adjustment can be made in the field. Generally, however, it is best to leave this adjustment to qualified experts. If the adjustment is made in the field, it must be done very carefully! Figure 8.19 depicts the condition after the adjustment is completed. Since

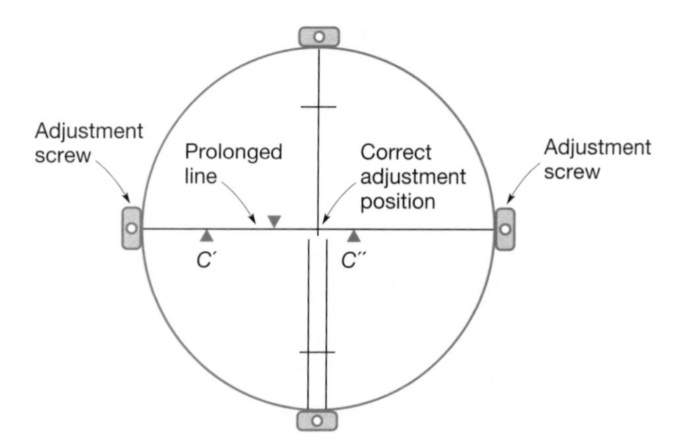

Figure 8.19 The crosshair adjustment procedure.

each crosshair has two sets of opposing capstan screws, *it is important to loosen one screw before tightening the opposing one*. After the adjustment is completed, the procedure should be repeated to check the adjustment.

■ 8.16 BALANCING-IN

Occasionally it is necessary to set up an instrument on a line between two points already established but not intervisible—for example, A and B in Figure 8.20. This can be accomplished in a process called *balancing-in* or *wiggling-in*.

Location of a trial point C' on line is estimated and the instrument set over it. A sight is taken on point A from point C' and the telescope plunged. If the line of sight does not pass through B, the instrument is moved laterally a distance CC' estimated from the proportion $CC' = BB' \times AC/AB$, and the process repeated. Several trials may be required to locate point C exactly, or close enough for the purpose at hand. The shifting head of the instrument is used to make the final small adjustment. A method for getting a close first approximation of required point C takes two persons, X able to see point A and Y having point B visible,

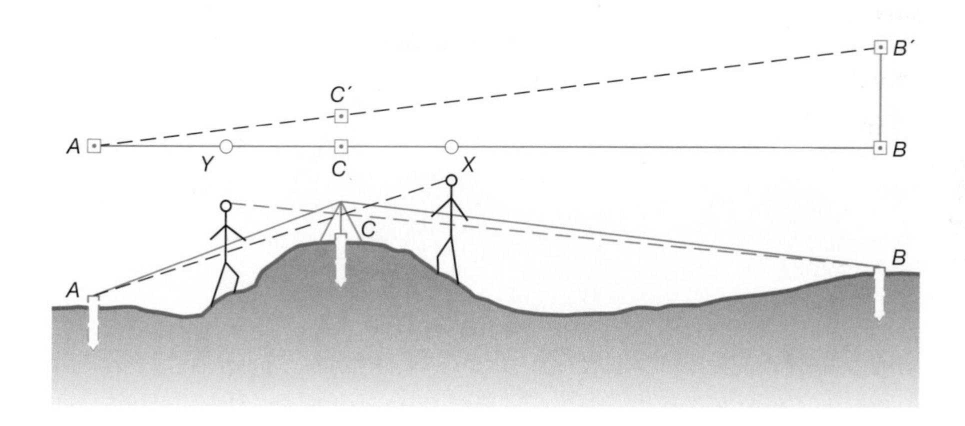

Figure 8.20 Balancing-in.

as shown in Figure 8.20. Each aligns the other in with the visible point in a series of adjustments, and two range poles are placed at least 20 ft apart on the course established. An instrument set at point *C* in line with the poles should be within a few tenths of a foot of the required location. From there, the wiggling-in process can proceed more quickly.

■ 8.17 RANDOM TRAVERSE

On many surveys it is necessary to run a line between two established points that are not intervisible because of obstructions. This situation arises repeatedly in property surveys. To solve the problem, a *random traverse* is run from one point in the approximate direction of the other. Using coordinate computation procedures presented in Chapter 10, the coordinates of the stations along the random traverse are computed. Using these same computation procedures, coordinates of the points along the "true" line are computed and observations necessary to stake out points on the line computed from the coordinates. With data collectors, the computed coordinates can be automatically determined in the field and then staked out using the functions of the data collector.

As a specific example of a random traverse, consider the case shown in Figure 8.21 where it is necessary to run line *XY*. On the basis of a compass bearing, or information from maps or other sources, the general direction to proceed is estimated and starting line *X-1* is given an assumed azimuth. Random traverse *X-1-2-3-Y* is then run and coordinates of all points determined. Based upon these computations, coordinates are also computed for points *A* and *B*, which are on line *X-Y*. The distance and direction necessary for setting *A* with an instrument set up at point 1 are then computed using procedures discussed in Chapter 10. Similarly the coordinates of *B* are determined and set from station 2. Using a data collector, these computations can be automatically performed. This procedure, known as *stake out,* is discussed in Chapter 23.

Once the angles and distances have been computed for staking points A and B, the actual stake-out procedure is aided by operating the total station instrument in its *tracking mode* (see Section 6.21 and Chapter 23). If a robotic total station instrument is available (see Section 8.6), one person can perform the lay-out procedure. *This method of establishing points on a line is only practical when direct sighting along the line is not physically possible.*

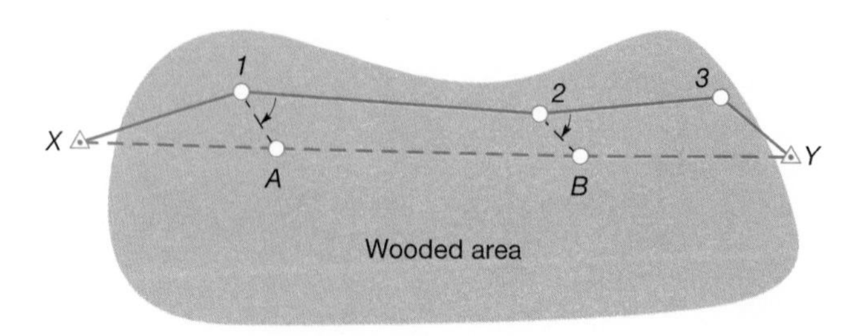

Figure 8.21 Random traverse *X-1-2-3-Y*.

8.18 TOTAL STATIONS FOR DETERMINING ELEVATION DIFFERENCES

With a total station instrument, computed vertical distances between points can be obtained in real time from observed slope distances and zenith angles. This is the basis, in fact, for *trigonometric leveling* (see Section 4.5.4). Several studies have compared the accuracies of elevation differences obtained by trigonometric leveling using modern total station instruments to those achieved by differential leveling as discussed in Chapters 4 and 5. Trigonometric leveling accuracies have always been limited by instrumental errors (discussed in Section 8.20) and the effects of refraction (see Section 4.4). Even with these problems, elevations derived from a total station survey are of sufficient accuracy for many applications such as for topographic mapping and other lower-order work.

However, studies have suggested that high-order results can be obtained in trigonometric leveling by following specific procedures. The suggested guidelines are: (1) place the instrument between two prisms so that sight distances are appropriate for the angular accuracy of the instrument, using Figure 8.22 as a guide;[3] (2) use target panels with the prisms; (3) keep rod heights equal so that their measurement is unnecessary; (4) observe the vertical distances between the prisms using two complete sets[4] of observations; (5) keep sight distances approximately equal; and (6) apply all necessary atmospheric corrections and reflector constants as discussed in Chapter 6. This type of trigonometric leveling can be done faster than differential leveling, especially in rugged terrain where sight distances are limited due to rapid changes in elevation.

A set of notes from trigonometric leveling is shown in Figure 8.23. Column (a) lists the backsight and foresight station identifiers and the positions of the telescope [direct (D) and reversed (R)] for each observation; (b) tabulates the

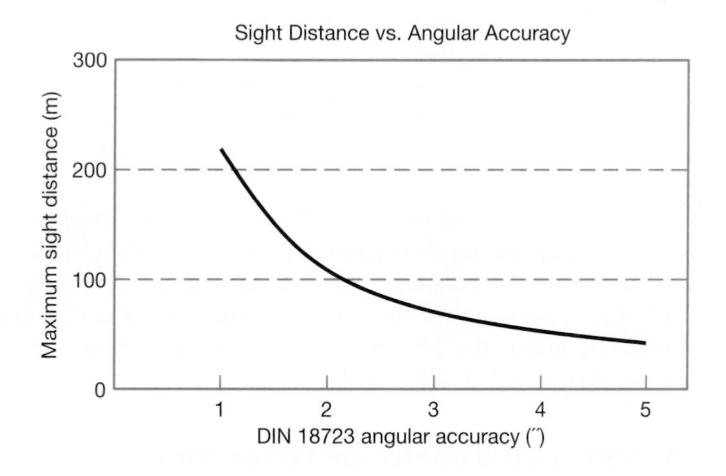

Figure 8.22 Graph of appropriate sight distance versus angular accuracy.

[3]A description of DIN18723 noted in Figure 8.22 is given in Section 8.21.

[4]One set of observations includes an elevation determination in both the direct and reversed positions.

TRIGONOMETRIC LEVELING NOTES

(a)	(b)	(c)	(d)	(e)	(f)
Sta/Pos	BS(+)	BD	FS(−)	FD	ΔElev
A					
D	1.211	98.12	1.403	86.34	
D	1.210		1.403		
R	1.211		1.404		
R	1.211		1.403		
Mean	1.2108		1.4033		−0.192
B					
D	−5.238	101.543	−9.191	93.171	
D	−5.236		−9.191		
R	−5.238		−9.193		
R	−5.237		−9.192		
Mean	−5.2373		−9.1918		3.954
C					
D	4.087	73.245	−3.849	97.392	
D	4.088		−3.851		
R	4.086		−3.849		
R	4.087		−3.849		
Mean	4.0870		−3.8495		7.936
D					
D	3.214	89.87	6.507	97.392	
D	3.214		6.507		
R	3.214		6.508		
R	3.215		6.507		
Mean	3.2143		6.5072		−3.293
E					
				Sum	8.405

Figure 8.23 Trigonometric leveling field notes.

backsight vertical distances, (BS+); (c) lists the backsight horizontal distances to the nearest decimeter; (d) gives the foresight vertical distances, (FS−); (e) lists the foresight horizontal distances to the nearest decimeter; and (f) tallies the elevation differences between the stations, computed as the difference of the BS vertical distances, minus the FS vertical distances. The observed elevation difference between station A and E is 8.405 m.

■ 8.19 ADJUSTMENT OF TOTAL STATION INSTRUMENTS AND THEIR ACCESSORIES

The accuracy achieved with total station instruments is not merely a function of their ability to resolve angles and distances. It is also related to operator procedures and the condition of the total station instrument and other peripheral equipment being used with it. Operator procedure pertains to matters such as careful centering and leveling of the instrument, accurate pointing at targets, and observing proper field procedures such as taking averages of multiple angle observations made in both direct and reversed positions.

In Section 8.2 three reference axes of a total station instrument were defined: (a) the line of sight, (b) the horizontal axis, and (c) the vertical axis. These instruments also have a fourth reference axis, (d) the *axis of the plate-level vial* (see Section 4.8). For a properly adjusted instrument, the following relationships should exist between these axes: (1) the axis of the plate-level vial should be perpendicular to the vertical axis, (2) the horizontal axis should be perpendicular to the vertical axis, and (3) the line of sight should be perpendicular to the horizontal axis. If these conditions do not exist, accurate observations can still be made by following proper procedures. However, it is more convenient if the instrument is in adjustment. Today, most total stations have calibration procedures that can electronically compensate for conditions (1) and (2) using sightings to well-defined targets with menu-defined procedures that can be performed in the field. However, if the operator is in doubt about the calibration procedures, a qualified technician should always be consulted.

The adjustment for making the line of sight perpendicular to the horizontal axis was described in Section 8.15 and the procedure for making the axis of the plate bubble perpendicular to the vertical axis is given in Section 8.19.1. The test to determine if a total station's horizontal axis is perpendicular to its vertical axis is a simple one. With the instrument in the direct mode, it is set up a convenient distance away from a high vertical surface, say the wall of a two- or three-story building. After carefully leveling the instrument, sight a well-defined point, say A, high on the wall, at an altitude angle of at least 30° and clamp the horizontal lock. Revolve (plunge) the telescope about its horizontal axis to set a point, B, on the wall below A just above ground level. Plunge the telescope to put it in reversed mode, turn the telescope 180° in azimuth, sight point A again, and clamp the horizontal lock. Plunge the telescope to set another point, C, at the same level as B. If B and C coincide, no adjustment is necessary. If the two points do not agree, then the horizontal axis is not perpendicular to the vertical axis. If an adjustment for this condition is necessary, the operator should refer to the manual that came with the instrument or send the instrument to a qualified technician.

Peripheral equipment that can affect accuracy includes tribrachs, optical plummets, prisms, and prism poles. Tribrachs must provide a snug fit without slippage. Optical plummets that are out of adjustment cause instruments to be miscentered over the point. Crooked prism poles or poles with bull's-eye bubbles that are out of adjustment also cause errors in placement of the prism over the point being observed. Prisms should be checked periodically to determine their constants (see Section 6.24.2) and their values stored for use in correcting distance measurements. Surveyors should always heed the following axiom: *In practice, instruments should always be kept in good adjustment, but used as though they might not be.*

In the following subsections procedures are described for making some relatively simple adjustments to equipment that can make observing more efficient and convenient, and also improve accuracy in the results.

8.19.1 Adjustment of Plate-Level Vials

As stated earlier, two types of leveling systems are used on total station instruments: (a) plate-level vials, and (b) electronic leveling systems. These systems control the fine level of the instrument. If an instrument is equipped with a plate-level vial, it can easily be tested for its state of adjustment. To make the test, the instrument should first be leveled following the procedures outlined in Section 8.5. Then after carefully centering the bubble, the telescope should be rotated 180° from its first position. If the level vial is in adjustment, the bubble will remain centered. If the bubble deviates from center, the axis of the plate-level vial is not perpendicular to the vertical axis. The amount of bubble run indicates twice the error that exists. Level vials usually have a capstan adjusting screw for raising or lowering one end of the tube. If the level vial is out of adjustment, it can be adjusted by bringing the bubble *halfway back* to the centered position by turning the screw. Repeat the test until the bubble remains centered during a complete revolution of the telescope. If the instrument is equipped with an electronic level, follow the procedures outlined in the operator's manual to adjust the leveling mechanism.

If a plate bubble is out of adjustment, the instrument can be used without adjusting it and accurate results can still be obtained, but specific procedures described in Section 8.20.1 must be followed.

8.19.2 Adjustment of Tripods

The nuts on the tripod legs must be snug to prevent slippage and rotation of the head. They are correctly adjusted if each tripod leg falls slowly of its own weight when placed in a horizontal position. If the nuts are overly tight, or if pressure is applied to the legs crosswise (which can break them) instead of lengthwise to fix them in the ground, the tripod is in a strained position. The result may be unnoticed movements of the instrument head after observations have begun.

Tripod legs should be well spread to furnish stability and set so that the telescope is at a convenient height for the observer. Tripod shoes must be tight. Proper field procedures can eliminate most instrument maladjustments, but there is no method that corrects a poor tripod with dried-out wooden legs, except to discard or repair it.

8.19.3 Adjustment of Tribrachs

The tribrach is an essential component of a secure and accurate setup. It consists of a minimum of three components, which are (1) a clamping mechanism, (2) leveling screws, and (3) a circular level bubble. As shown in Figure 8.3, some tribrachs also contain an optical plummet to center the tribrach over a station. The clamping mechanism consists of three slides that secure three posts that protrude from the base of the instrument or tribrach adapter. As the tribrach wears, the clamping mechanism may not sufficiently secure the instrument during observation procedures. When this happens, the instrument will move in the tribrach after it has been clamped and the tribrach should be repaired or replaced.

8.19.4 Adjustment of Optical Plummets

The line of sight in an optical plummet should coincide with the vertical axis of the instrument. Two different situations exist: (1) the optical plummet is enclosed in the alidade of the instrument and rotates with it when turned in azimuth, or (2) the optical plummet is part of the tribrach that is fastened to the tripod and does not turn in azimuth.

To adjust an optical plummet contained in the alidade, set the instrument over a fine point and aim the line of sight exactly at it by turning the leveling screws. (The instrument need not be level.) Carefully adjust for any existing parallax. Rotate the instrument 180° in azimuth. If the optical plummet reticle moves off the point, bring it *halfway* back by means of the adjusting screws provided. These screws are similar to those shown in Figure 8.19. As with any adjustment, repeat the test to check the adjustment and correct if necessary.

For the second case where the optical plummet is part of the tribrach, carefully lay the instrument, with the tribrach attached, on its side (horizontally) on a stable base such as a bench or desk, and clamp it securely. Fasten a sheet of paper on a vertical wall at least six feet away, such that it is in the field of view of the optical plummet's telescope. With the horizontal lock clamped, mark the position of the optical plummet's line of sight on the paper. Release the horizontal lock and rotate the tribrach 180°. If the reticle of the optical plummet moves off the point, bring it *halfway* back by means of the adjusting screws. Center the reticle on the point again with the leveling screws and repeat the test.

8.19.5 Adjustment of Circular Level Bubbles

If a circular-level bubble on a total station does not remain centered when the instrument is rotated in azimuth, the bubble is out of adjustment. It should be corrected, although precise adjustment is unnecessary because it does not control fine leveling of the reference axes. To adjust the bubble, carefully level the instrument using the plate bubble. Then center the bull's-eye bubble using the adjusting screws.

Bull's-eye bubbles used on prism poles and level rods must be in good adjustment for accurate work. To adjust them, carefully orient the rod or pole vertically by aligning it parallel to a long plumb line and fasten it in that position using shims and C-clamps. Then center the bubble in the vial using the adjusting screws.

For instruments such as automatic levels that do not have plate bubbles, use the following procedure. To adjust the bubble, carefully center it using the leveling screws and turn the instrument 180° in azimuth. *Half* of the bubble run is corrected by manipulating the vial-adjusting screws. Following the adjustment, the bubble should be centered using the leveling screws and the test repeated.

■ 8.20 SOURCES OF ERROR IN TOTAL STATION WORK

Errors in using total stations result from *instrumental, natural,* and *personal* sources. These are described in the subsections that follow.

8.20.1 Instrumental Errors

Figure 8.24 shows the fundamental reference axes of a total station. As discussed in Section 8.19, for a properly adjusted instrument, the four axes must bear specific relationships to each other. These are: (1) the vertical axis should be perpendicular to the axis of the plate level vial, (2) the horizontal axis should be perpendicular to the vertical axis, and (3) the axis of sight should be perpendicular to the horizontal axis. If these relationships are not true, errors will result in measured angles unless proper field procedures are observed. A discussion of errors caused by maladjustment of these axes and of other sources of instrumental errors follows.

1. *Plate bubble out of adjustment.* If the axis of the plate bubble is not perpendicular to the vertical axis, the latter will not be truly vertical when the plate bubble is centered. *This condition causes errors in observed horizontal and vertical angles that cannot be eliminated by averaging direct and reversed readings.* The plate bubble is out of adjustment if after centering, it runs when the instrument is rotated 180° in azimuth. The situation is illustrated in Figure 8.25. With the telescope initially pointing to the right and the bubble centered, the axis of the level vial is horizontal, as indicated by the solid line labeled *ALV*-1. Because the level vial is out of adjustment, it is not perpendicular to the vertical axis of the instrument, but instead makes an angle of $90° - \alpha$ with it. After turning the telescope 180°, it points left and the axis of the level vial is in the position indicated by the dashed line labeled *ALV*-2. The angle between the axis of the level vial and vertical axis is still $90° - \alpha$ but as shown in the figure, its indicated dislevelment, or bubble run, is E. From the figure's geometry, $E = 2\alpha$ is double the bubble's maladjustment. The vertical axis can be made truly vertical by bringing the bubble back *half of the bubble run,* using the foot screws. Then, even though it is not centered, the bubble should stay in the same position as the instrument is rotated in azimuth and accurate angles can be observed.

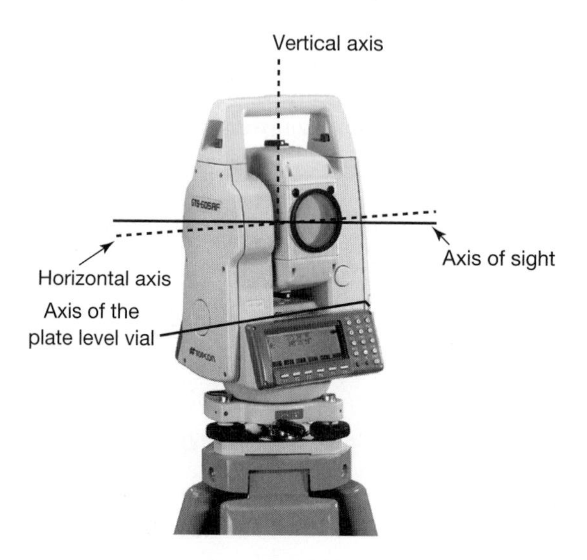

Figure 8.24 Reference axes of a total station instrument. (Courtesy Topcon Positioning Systems)

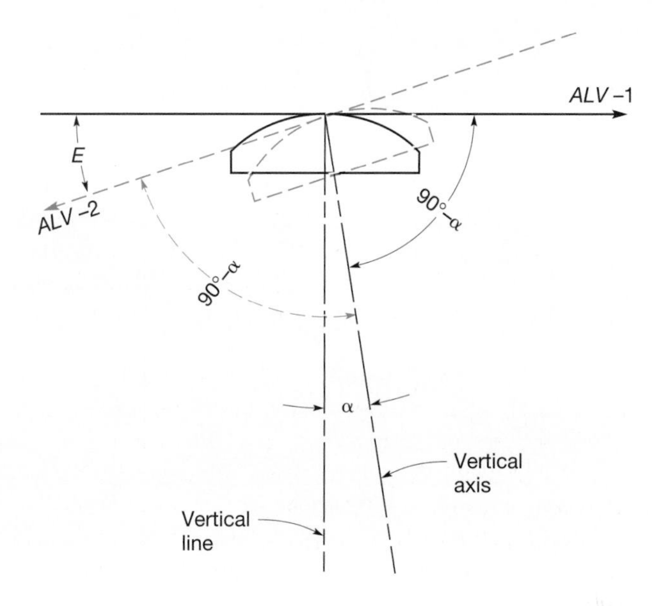

Figure 8.25
Plate bubble out of adjustment.

Although instruments can be used to obtain accurate results with their plate bubbles maladjusted, it is inconvenient and time consuming, so the required adjustment should be made as discussed in Section 8.19.1.

As noted earlier, some total stations are equipped with dual-axis compensators which are able to sense the amount and direction of vertical axis tilt automatically. They can make corrections computationally in real time to both horizontal and vertical angles for this condition. Instruments equipped with single-axis compensators can only correct vertical angles. Procedures outlined in the manuals that accompany the instruments should be followed to properly remove any error.

As was stated in Section (8.8), total station instruments with dual axis compensators can apply a mathematical correction to horizontal angles, which accounts for any dislevelment of the horizontal and vertical axes. In Figure 8.26, to sight on point S, the telescope is plunged upward. Because the instrument is misleveled, the line of sight scribes an inclined line SP' instead of the required vertical line SP. The angle between these two lines is α, the amount that the instrument is out of level. From this figure, it can be shown that the error in the horizontal direction, E_H, is

$$E_H = \alpha \tan(v) \qquad \textbf{(8.4)}$$

In Equation (8.4) v is the altitude angle to point S. For the observation of any horizontal angle if the vertical angles for both the backsight and foresight are nearly the same, the resultant error in the horizontal angle is negligible. In flat terrain, this is approximately the case and the error due to dislevelment can be small. However, in mountainous terrain where the backsight and foresight pointings can vary by large amounts, this error can become substantial. For example assume that an instrument that is 20″ out of level reads a backsight zenith angle as 93° and the

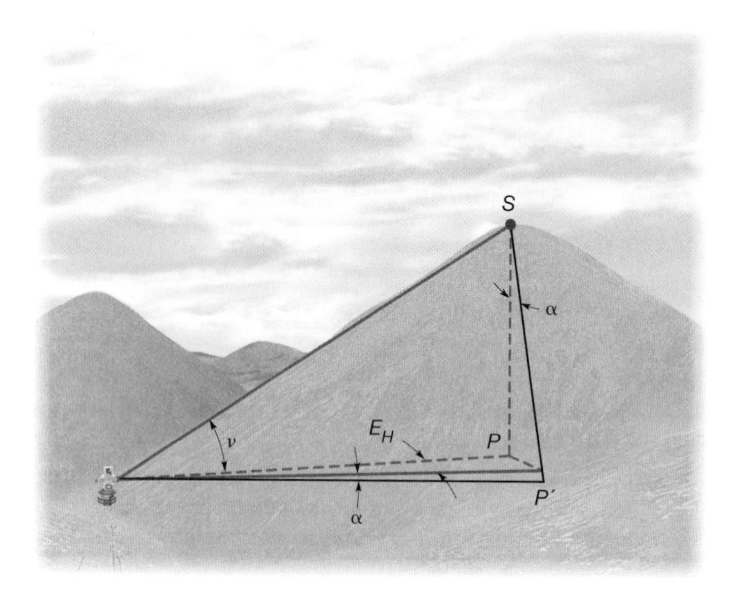

Figure 8.26 Geometry of instrument dislevelment.

foresight zenith angle as 80°. The horizontal error in the backsight direction would be $20'' \times \tan(-3°) = -1.0''$ and in the foresight is $20'' \times \tan(10°) = 3.5''$ resulting in a cumulative error in the horizontal angle of $3.5'' - (-1'') = 4.5''$. This is a systematic error that becomes more serious as larger vertical angles are observed. It is critical in astronomical observations for azimuth as discussed in Chapter 19.

Two things should be obvious from this discussion: it is important to check (1) the adjustment of the plate bubble often and (2) check the position of the bubble during the observation process.

2. *Horizontal axis not perpendicular to vertical axis.* This situation causes the axis of sight to define an inclined plane as the telescope is plunged. Therefore, if the backsight and foresight have differing angles of inclination, incorrect horizontal angles will result. Errors from this origin can be canceled by averaging an equal number of direct and reversed readings or by double centering if prolonging a straight line. With total station instruments having dual-axis compensation, this error can be determined in a calibration process that consists of carefully pointing to the same target in both direct and reversed modes. From this operation the microprocessor can compute and store a correction factor. It is then automatically applied to all horizontal angles subsequently observed.

3. *Axis of sight not perpendicular to horizontal axis.* If this condition exists, as the telescope is plunged, the axis of sight generates a cone whose axis coincides with the horizontal axis of the instrument. The greatest error from this source occurs when plunging the telescope, as in prolonging a straight line or measuring deflection angles. Also, when the angle of inclination of the backsight is not equal to that of the foresight, observed horizontal angles will be incorrect. These errors are eliminated by double centering and by averaging equal numbers of direct and reversed readings.

4. *Vertical-circle indexing error.* As noted in Section 8.13, when the axis of sight is horizontal, an altitude angle of zero, or a zenith angle of either 90° or 270°, should be read; otherwise an indexing error exists. The error can be eliminated by computing the mean from equal numbers of altitude (or zenith) angles read in the direct and reversed modes. With most newer total station instruments, the indexing error can be determined by carefully reading the same zenith angle both direct and reversed. The value of the indexing error is then computed, stored, and automatically applied to all observed zenith angles.

5. *Eccentricity of centers.* This condition exists if the geometric center of the graduated horizontal (or vertical) circle does not coincide with its center of rotation. Errors from this source are usually small. Total stations may also be equipped with systems that automatically average readings taken on opposite sides of the circles, thereby compensating for this error.

6. *Circle graduation errors.* If graduations around the circumference of a horizontal or vertical circle are nonuniform, errors in observed angles will result. These errors are generally very small. Some total stations always use readings taken from many locations around the circles for each measured horizontal and vertical angle, thus providing an elegant system for eliminating these errors.

7. *Errors caused by peripheral equipment.* Additional instrumental errors can result from worn tribrachs, optical plummets that are out of adjustment, unsteady tripods, and sighting poles with maladjusted bull's-eye bubbles. This equipment should be regularly checked and kept in good condition or adjustment. Procedures for adjusting these items are outlined in Section 8.19.

8.20.2 Natural Errors

1. *Wind.* Wind vibrates the tripod that the total station instrument rests on. On high setups, light wind can vibrate the instrument to the extent that precise pointings become impossible. Shielding the instrument, or even suspending observations on precise work, may be necessary on windy days. An optical plummet is essential for making setups in this situation.

2. *Temperature effects.* Temperature differentials cause unequal expansion of various parts of total station instruments. This causes bubbles to run, which can produce erroneous observations. Shielding instruments from sources of extreme heat or cold reduces temperature effects.

3. *Refraction.* Unequal refraction bends the line of sight and may cause an apparent shimmering of the observed object. It is desirable to keep lines of sight well above the ground and avoid sights close to buildings, smokestacks, and even large individual objects, such as vehicles, in generally open spaces. In some cases, observations may have to be postponed until atmospheric conditions have improved.

4. *Tripod settlement.* The weight of an instrument may cause the tripod to settle, particularly when set up on soft ground or asphalt highways. When a job involves crossing swampy terrain, stakes should be driven to support the tripod legs and work at a given station completed as quickly as possible. Stepping near a tripod

leg or touching one while looking through the telescope will demonstrate the effect of settlement on the position of the bubble and crosshairs. Most total station instruments have sensors that tell the operator when dislevelment has become too severe to continue the observation process.

8.20.3 Personal Errors

1. *Instrument not set up exactly over point.* Miscentering of the instrument over a point will result in an incorrect horizontal angle being observed. As shown in Figure 8.27, instrument miscentering will cause errors in both the backsight and foresight directions of an angle. The amount of error is dependent on the position of the instrument in relation to the point. For instance in Figure 8.27(a), the miscentering that is depicted will have minimal effect on the observed angle since the error on the backsight to P_1 will partially cancel the error on the foresight to P_2. However, in Figures 8.27(b) and (c), the effect of the miscentering has a maximum effect on the observed angular values. Since the position of the instrument is random in relation to the station, it is important to carefully center the instrument over the station when observing angles. The position should be checked at intervals during the time a station is occupied, to be certain it remains centered.

2. *Bubbles not centered perfectly.* The bubbles must be checked frequently, but NEVER releveled between a backsight and a foresight—only *before* starting and *after* finishing an angular position.

3. *Improper use of clamps and tangent screws.* An observer must form good operational habits and be able to identify the various clamps and tangent screws by their touch without looking at them. Final setting of tangent screws is always made with a positive motion to avoid backlash. Clamps should be tightened just once and not checked again to be certain they are secure.

4. *Poor focusing.* Correct focusing of the eyepiece on the crosshairs and of the objective lens on the target is necessary to prevent parallax. Objects sighted should be placed as near the center of the field of view as possible. Focusing affects pointing, which is an important source of error. In newer instruments like the Topcon GTS 600-AF shown in Figure 8.24, automatic focusing of the objective

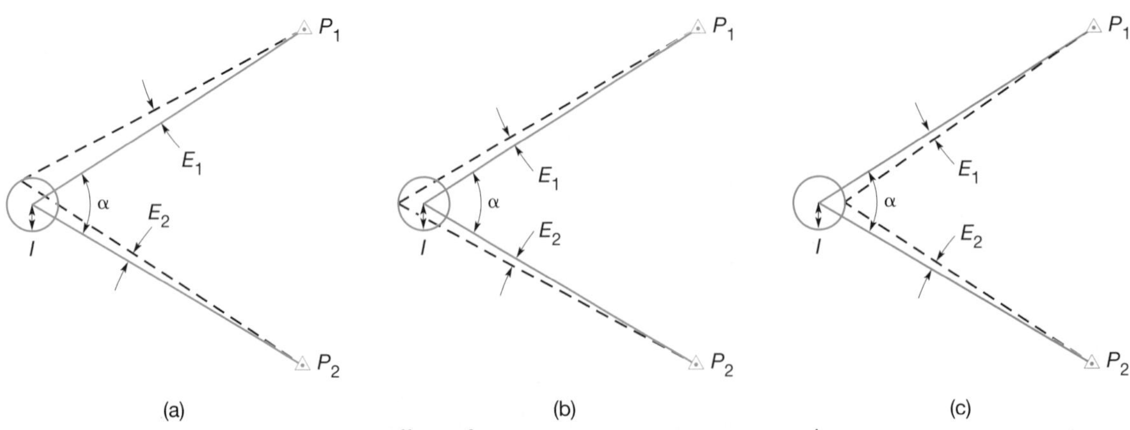

Figure 8.27 Effects of instrument miscentering on an angle.

lens is provided. These devices are similar to the modern photographic camera, and can increase the speed of the survey when sight distances to the targets vary.

5. *Overly careful sights.* Checking and double-checking the position of the crosshair setting on a target wastes time and actually produces poorer results than one fast observation. The crosshair should be aligned quickly and the next operation begun promptly.

6. *Careless plumbing and placement of rod.* One of the most common errors results from careless plumbing of a rod when the instrument operator because of brush or other obstacles in the way can only see the top. Another is caused by placing a pole off-line behind a point to be sighted.

8.21 PROPAGATION OF RANDOM ERRORS IN ANGLE OBSERVATIONS

Random errors are present in every horizontal angle observation. Whenever an instrument's circles are read, a small error is introduced into the final observed angle. Similarly, each operator will have some miscentering on the target. These error sources are random. They may be small or large, depending on the instrument, the operator, and the conditions at the time of the angle observation. Increasing the number of angle repetitions can reduce the effects of reading and pointing.

With the introduction of total station instruments, standards were developed for estimating errors in angle observations caused by reading and pointing on a well-defined target. The standards, called DIN 18723, provide values for estimated errors in the mean of two direction observations, one each in the direct and reversed modes. The Leica TPS 300 (Figure 8.1) has a DIN 18723 accuracy of ±2″, and the Topcon GTS 210A (Figure 8.2) has a DIN 18723 accuracy of ±5″.

A set of angles observed with a total station will have an estimated error of

$$E = \frac{2 \times E_{DIN}}{\sqrt{n}} \tag{8.5}$$

where E is the estimated error in the angle due to pointing and reading, n is the total number of angles read in both direct and reversed modes and E_{DIN} is the DIN 18723 error.

Example 8.2

Three sets of angles (3D and 3R) are measured with the Leica TPS 300. What is the estimated error in the angle?

Solution

By Equation (8.5), the estimated error is

$$E = \frac{2 \times 2''}{\sqrt{6}} = \pm 1.6''$$

8.22 MISTAKES

Some common mistakes in angle observation work are:

1. Sighting on or setting up over the wrong point.
2. Calling out or recording an incorrect value.
3. Improper focusing of the eyepiece and objective lenses of the instrument.
4. Leaning on the tripod or placing a hand on the instrument when pointing or taking readings.

PROBLEMS

Asterisks (*) indicate problems that have answers given in Appendix G.

8.1 What is the proper method of carrying a total station to and from each setup?

8.2 Draw a sketch of a total station identifying the *axis of sight, horizontal axis,* and *vertical axis*. What relationships should exist between these axes for an instrument in good adjustment?

8.3 What are the primary sources of random instrumental error in a total station?

8.4 Why is it important to properly focus an optical plummet when setting an instrument over a point?

8.5 What advantages does a laser-plummet have over an optical plummet?

8.6 What is the purpose of the jog/shuttle mechanism on a servo-driven total station?

8.7 Describe the parts of the telescope of a total station.

8.8 Describe the adjustment procedure for an optical plummet in a total station.

8.9 What is meant by an angular position?

8.10 What is the purpose of the horizontal tangent screw on a total station?

8.11 Why is it important to keep the bull's-eye bubble of a sighting rod in adjustment?

8.12 Determine the angles subtended for the following conditions:

(a) a 1/4-in. diameter pipe sighted by total station from 100 ft away.
(b) a 1/2-in. stake sighted by total station from 1/8 mi away.
(c) a 5-mm diameter chaining pin observed by total station from 10 m away.

8.13 What is the error in an observed direction for the situations noted?

(a) setting a total station 5 mm to the side of a tack on a 100-m sight.
(b) lining in the edge (instead of center) of an 0.02-ft diameter chaining pin at 200 ft.
(c) sighting the edge (instead of center) of a 0.15-ft diameter range pole 350 ft away.
(d) sighting the top of a 2-m range pole that is 5′ off-level on a 200-m sight.

8.14* Intervening terrain obstructs the line of sight so only the top of a 2-m long pole can be seen on a 350-m sight. If the range pole is out of plumb and leaning sideways 5 mm per vertical meter, what maximum angular error results?

8.15 Describe the procedure of centering a total station instrument, having an optical plummet with an adjustable leg tripod, over a point.

8.16 Discuss an advantage and disadvantage of a robotic total station instrument in construction stakeout.

8.17 Which instrumental errors can be compensated by averaging an equal number of observations made with the telescope direct and reversed?

8.18 How is a total station with a level bubble off by 2 graduations leveled in the field?

8.19 An interior angle x and its explement y were turned to close the horizon. Each angle was observed once direct and once reversed using the repetition method. Starting with an initial backsight setting of 0°00′00″ for each angle, the readings after the first and second turnings of angle x were 50°38′48″ and 101°17′40″, and the readings after the first and second turnings of angle y were 309°21′06″ and 258°42′10″. Calculate each angle and the horizon misclosure.

8.20* A zenith angle is measured as 94°23′06″ in the direct position. What is the equivalent zenith angle in the reversed position?

8.21 What is the average zenith angle given the following direct and reversed readings
Direct: 86°35′14″, 86°35′12″, 86°35′08″
Reversed: 273°24′50″, 273°24′49″, 273°24′52″

In Figure 8.9(c), direct and reversed directions observed with a total station instrument from A to points B, C, and D are listed in Problems 8.22 and 8.23. Determine the values of the three angles, and the horizon misclosure.

8.22 Direct: 0°00′00″, 36°19′31″, 95°57′04″, 359°59′58″
Reversed: 0°00′00″, 36°19′27″, 95°57′06″, 0°00′02″

8.23 Direct: 0°00′00″, 116°32′26″, 221°08′33″, 0°00′04″
Reversed: 0°00′00″, 116°32′24″, 221°08′31″, 0°00′01″

8.24* The angles at point X were observed with a total station instrument. Based on 4 readings, the standard deviation of the angle was ±6.8″. If the same procedure is used in observing each angle within an eight-sided polygon, what is the estimated standard deviation of closure?

8.25 The line of sight of a total station is out of adjustment by 10″.
(a) In prolonging a line by plunging the telescope between backsight and foresight, but not double centering, what angular error is introduced?
(b) What off-line linear error results on a foresight of 200 m?

8.26 A line PQ is prolonged to point R by double centering. Two foresight points R' and R'' are set. What angular error would be introduced in a single plunging based on the following lengths of QR and $R'R''$, respectively?
(a)* 650.50 ft and 0.35 ft.
(b) 212.60 m and 28 mm.

8.27 Discuss where the "principal of reversion" is used in angle measurement.

8.28 What is indexing error, and how can its value be obtained and eliminated from observed zenith angles?

8.29* A total station with a 20″/div. level bubble is one division out of level on a point with a zenith angle of 49°25′24″. What is the error in the horizontal pointing?

8.30 What is the equivalent altitude angle for a zenith angle of 266°02′56″?

8.31 What error in horizontal angles is consistent with the following linear precisions?
(a) 1/2000, 1/5000, 1/20,000, 1/50,000, and 1/100,000
(b) 1/3000, 1/8000, 1/15,000, 1/30,000, and 1/80,000

8.32 Why is it important to keep your tripod in good condition?

8.33 Discuss the procedure for testing a tripod.

8.34 List the procedures for "balancing-in" a point.

8.35 A zenith angle was read twice direct giving values of 93°32′34″ and 93°32′38″, and twice reversed yielding readings of 266°27′20″ and 266°27′02″. What is the mean zenith angle? What is the indexing error?

8.36 Write a review of an article on total station instruments written in a professional journal.

8.37 Create a computational program that takes the directions in Figure 8.12 and computes the average angles, their standard deviations, and the horizon misclosure.

8.38 Solve Problem 8.21 using computational software.

BIBLIOGRAPHY

Crawford, W. 2001. "Calibration Field Tests of Any Angle Measuring Instrument." *Point of Beginning* 26 (No. 8): 54.

Ghilani, C. D. and P. R. Wolf. 2006. *Adjustment Computations: Spatial Data Analysis.* Hoboken, NJ: John Wiley & Sons, Inc.

GIA. 2001. "Electronic Angle Measurement." *Professional Surveyor* 21 (No. 10): 47.

———. 2002. "2-axis Compensators." *Professional Surveyor* 22 (No. 9): 38.

———. 2002. "Basic Total Station Calibration." *Professional Surveyor* 22 (No. 5): 60.

———. 2005. "How Things Work: Modern Total Station and Theodolite Axes." *Professional Surveyor* 25 (No. 10): 42.

Stevens, K. 2003. "Locking in the Benefits." *Point of Beginning* 28 (No. 11): 16.

9

Traversing

9.1 INTRODUCTION

A traverse is a series of consecutive lines whose ends have been marked in the field and whose lengths and directions have been determined from observations. In traditional surveying by ground methods, *traversing,* the act of marking the lines, i.e., establishing traverse stations and making the necessary observations, is one of the most basic and widely practiced means of determining the relative locations of points.

There are two kinds of traverses: *closed* and *open.* Two categories of closed traverses exist: *polygon* and *link.* In the polygon traverse, as shown in Figure 9.1(a), the lines return to the starting point, thus forming a closed figure that is both geometrically and mathematically closed. Link traverses finish upon another station that should have a positional accuracy equal to or greater than that of the starting point. The link type (geometrically open, mathematically closed), as illustrated in Figure 9.1(b), must have a closing reference direction—for example, line $E\text{-}Az\ Mk_2$. Closed traverses provide checks on the observed angles and distances, which is an extremely important consideration. They are used extensively in control, construction, property, and topographic surveys.

If the distance between stations C and E in Figure 9.1(a) were observed, the resultant set of observations would become what is called a *network.* A network involves the interconnection of stations within the survey to create additional redundant observations. Networks offer more geometric checks than closed traverses. For instance in Figure 9.1(a), after computing coordinates on stations C and E using elementary procedures, the observed distance CE can be compared against a value obtained by inversing the coordinates (see Chapter 10 for discussion on computation of coordinates and inversing coordinates). Figure 9.7(b) shows

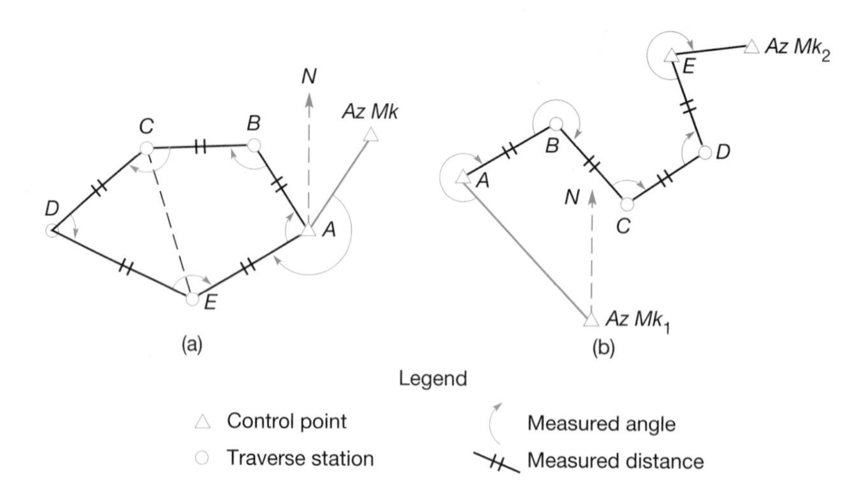

Figure 9.1 Examples of closed traverses.

another example where a network has been developed. Networks should be adjusted using the method of least squares as presented in Chapter 16.

An open traverse (geometrically and mathematically open) (Figure 9.2) consists of a series of lines that are connected but do not return to the starting point or close upon a point of equal or greater order accuracy. *Open traverses should be avoided because they offer no means of checking for observational errors and mistakes.* If they must be used, observations should be repeated carefully to guard against mistakes.

Hubs (wooden stakes with tacks to mark the points), steel stakes, or pipes are typically set at each traverse station *A*, *B*, *C*, etc., in Figures 9.1 and 9.2, where a change in direction occurs. Spikes, "P-K"[1] nails, and scratched crosses are used on blacktop pavement. Chiselled or painted marks are made on concrete. Traverse

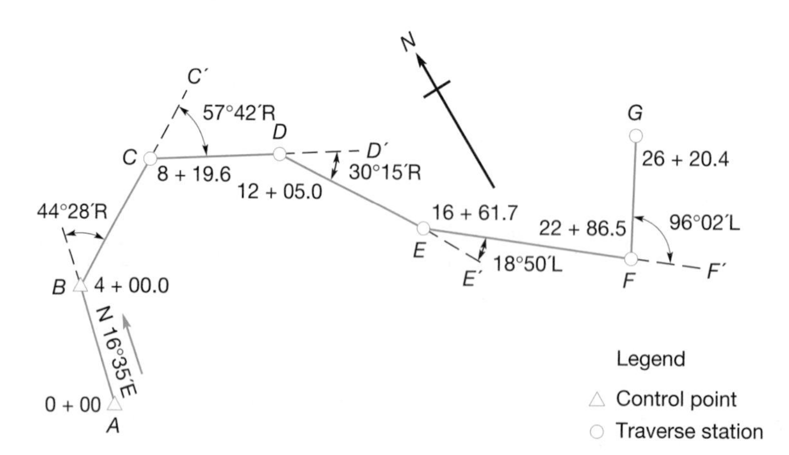

Figure 9.2 Open traverse.

[1]P-K nails are a trade name for concrete nails. The Parker-Kalon Company originally manufactured these nails. There is a small depression in the center of the nail that serves as a marker for the location of the station. Several companies now manufacture similar or better versions of this nail. Still the original name, P-K, is used to denote this type of nail.

stations are sometimes interchangeably called *angle points* because an angle is observed at each one.

■ 9.2 OBSERVATION OF TRAVERSE ANGLES OR DIRECTIONS

The methods used in observing angles or directions of traverse lines vary and include (1) interior angles, (2) angles to the right, (3) deflection angles, and (4) azimuths. These are described in the following subsections.

9.2.1 Traversing by Interior Angles

Interior-angle traverses are used for many types of work, but they are especially convenient for property surveys. Although interior angles could be observed either clockwise or counterclockwise, *to reduce mistakes in reading, recording, and computing, they should always be turned clockwise* from the backsight station to the foresight station. The procedure is illustrated in Figure 9.1(a). In this text, except for left deflection angles, clockwise turning will always be assumed. Furthermore, when angles are designated by three station letters or numbers in this text, the backsight station will be given first, the occupied station second, and the foresight station third. Thus angle *EAB* of Figure 9.1(a) was observed at station *A*, with the backsight on station *E* and the foresight at station *B*.

Interior angles may be improved by averaging equal numbers of direct and reversed readings. As a check, exterior angles may also be observed to close the horizon (see Section 8.10). In the traverse of Figure 9.1(a), a reference line *A-Az MK* of known direction exists. Thus the clockwise angle at *A* from *Az Mk* to *E* must also be observed to enable determining the directions of all other lines. This would not be necessary if, for example, the traverse contained a line of known direction, like *AB* of Figure 7.2.

9.2.2 Traversing by Angles to the Right

Angles observed clockwise from a backsight on the "rearward" traverse station to a foresight on the "forward" traverse station [see Figures 9.1(a) and (b)] are called *angles to the right.* According to this definition, to avoid ambiguity in angle-to-the-right designations, the "sense" of the forward traverse direction must be established. This is normally done by consecutive numbering or lettering of traverse stations so that they increase in the forward direction. Data collectors generally follow this convention when traversing. Thus in Figure 9.1(b), for example, the direction from *A* to *B*, *B* to *C*, *C* to *D*, etc., is forward. By averaging equal numbers of direct and reversed readings, observed angles to the right can also be checked and their accuracy improved. From the foregoing definitions of interior angles and angles to the right, it is evident that in a polygon traverse the only difference between the two types of observational procedures may be ordering of the backsight and foresight stations since both procedures observe clockwise angles.

9.2.3 Traversing by Deflection Angles

Route surveys are commonly run by deflection angles measured to the right or left from the lines extended, as indicated in Figure 9.2. A deflection angle is not

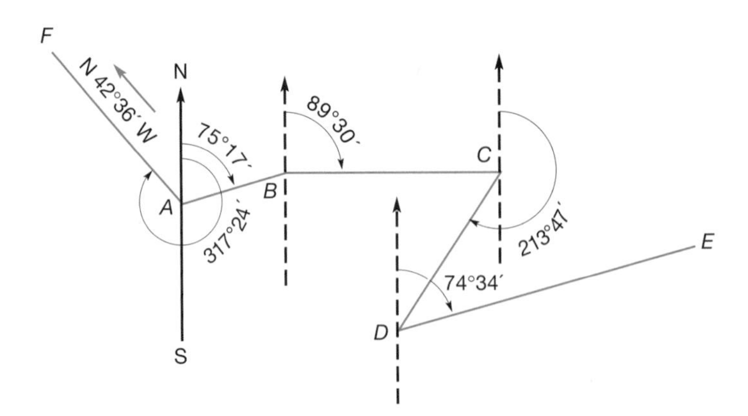

Figure 9.3
Azimuth traverse.

complete without a designation R or L, and, of course, it cannot exceed 180°. Each angle should be doubled or quadrupled and an average value determined. The angles should be observed an equal number of times in face left and face right to reduce instrumental errors. Deflection angles can be obtained by subtracting 180° from angles to the right. Positive values so obtained denote right deflection angles; negative ones are left.

9.2.4 Traversing by Azimuths

With total station instruments, traverses can be run using azimuths. This process permits reading azimuths of all lines directly and thus eliminates the need to calculate them. In Figure 9.3, azimuths are observed clockwise from the north end of the meridian through the angle points. The instrument is oriented at each setup by sighting on the previous station with either the back azimuth on the circle (if angles to the right are turned) or the azimuth (if deflection angles are turned), as described in Section 8.11. Then the forward station is sighted. The resulting reading on the horizontal circle will be the forward line's azimuth.

■ 9.3 OBSERVATION OF TRAVERSE LENGTHS

The length of each traverse line (also called a *course*) must be observed. This is usually done by the simplest and most economical method capable of satisfying the required precision of a given project. Their speed, convenience, and accuracy makes the EDM component of a total station instrument the most often used, although taping or other methods discussed in Chapter 6 could be employed. A distinct advantage of traversing with total station instruments is that both angles and distances can be observed with a single setup at each station. Averages of distances observed both forward and back will provide increased accuracy, and the repeat readings afford a check on the observations.

Sometimes state statutes govern the precision specified for a traverse to locate boundaries. On construction work, allowable limits of closure depend on the use and extent of the traverse and project type. Bridge location, for example, demands a high degree of precision.

In closed traverses, each course is observed and recorded as a separate distance. On long link traverses with highways and railroads, distances are carried along continuously from the starting point using stationing (see Section 5.9.1). In Figure 9.2, which uses stationing in feet, for example, beginning with station 0 + 00 at point *A*, 100-ft stations (1 + 00, 2 + 00, and 3 + 00) are marked until hub *B* at station 4 + 00 is reached. Then stations 5 + 00, 6 + 00, 7 + 00, 8 + 00, and 8 + 19.60 are set along course *BC* to *C*, etc. The length of a line in a stationed link traverse is the difference between stationing at its end points; thus, the length of line *BC* is 819.6 − 400.0 = 419.6 ft.

■ 9.4 SELECTION OF TRAVERSE STATIONS

Positions selected for setting traverse stations vary with the type of survey. In general, guidelines to consider in choosing them include accuracy, utility, and efficiency. Of course, intervisibility between adjacent stations, forward and back, must be maintained for angle and distance observations. The stations should also ideally be set in convenient locations that allow for easy access. Ordinarily, stations are placed to create lines that are as long as possible. This not only increases efficiency by reducing the number of instrument setups, but it also increases accuracy in angle observations. However, utility may override using very long lines because intermediate hubs, or stations at strategic locations, may be needed to complete the survey's objectives.

Often the number of stations can be reduced and the length of the sight lines increased by careful reconnaissance. It is always wise to "walk" the area being surveyed and find ideal locations for stations before the traverse stakes are set and the observation process is undertaken.

Each different type of survey will have its unique requirements concerning traverse station placement. On property surveys, for example, traverse stations are placed at each corner if the actual boundary lines are not obstructed and can be occupied. If offset lines are necessary, a stake is located near each corner to simplify the observations and computations. Long lines and rolling terrain may necessitate extra stations.

On route surveys, stations are set at each angle point and at other locations where necessary to obtain topographic data or extend the survey. Usually the centerline is run before construction begins, but it will likely be destroyed and need replacement one or more times during various phases of the project. An offset traverse can be used to avoid this problem.

A traverse run to provide control for topographic mapping serves as a control framework to which map details such as roads, buildings, streams, and hills are referenced. Station locations must be selected to permit complete coverage of the area to be mapped. *Spurs* consisting of one or more lines may branch off as open (*stub*) traverses to reach vantage points. However, their use should be discouraged since a check on their positions cannot be made.

■ 9.5 REFERENCING TRAVERSE STATIONS

Traverse stations often must be found and reoccupied months or even years after they are established. Also they may be destroyed through construction or other activity. Therefore it is important that they be referenced by creating observational *ties* to them so that they can be relocated if obscured or reestablished if destroyed.

Figure 9.4
Referencing a point.

Figure 9.4 presents a typical traverse tie. As illustrated, these ties consist of distance observations made to nearby fixed objects. Short lengths (less than 100 ft) are convenient if a steel tape is being used, but, of course, the distance to definite and unique points is a controlling factor. Two ties, preferably at about right angles to each other, are sufficient, but three should be used to allow for the possibility that one reference mark may be destroyed. Ties to trees can be observed in hundredths of a foot if nails are driven into them. However, *permission must be obtained from the landowner before driving nails into trees*.

If natural or existing features such as trees, utility poles, or corners of buildings are not available, stakes may be driven and used as ties. Figure 9.5(a) shows an arrangement of *straddle hubs* well suited to tying in a point such as H on a highway centerline or elsewhere. Reference points A and B are carefully set on the line through H, as are C and D. Lines AB and CD should be roughly perpendicular, and the four points should be placed in safe locations, outside of areas likely to be disturbed. It is recommended that a third point be placed on each line to serve as an alternate in the event one point is destroyed. The intersection of the lines of sight of two total stations set up at A and C and simultaneously aimed at B and D, respectively, will recover the point. The traverse hub H may also be found by intersecting strings stretched between diagonally opposite ties if the lengths are not too long. Hubs in the position illustrated by Figure 9.5(b) are sometimes used, but are not as desirable as straddle hubs for stringing.

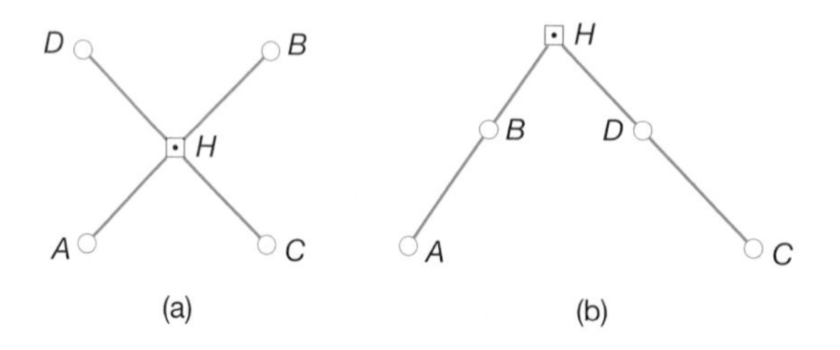

Figure 9.5
Hubs for ties.

■ 9.6 TRAVERSE FIELD NOTES

The importance of notekeeping was discussed in Chapter 2. Since a traverse is itself the end on a property survey and the basis for all other data in mapping, a single mistake or omission in recording is one too many. All possible field and office checks must therefore be made. A partial set of field notes for an interior-angle traverse run using a total station instrument is shown in Figure 9.6. Notice that details such as date, weather, instrument identifications, and party members and their duties are recorded on the right-hand page of the notes. Also a sketch with a north arrow is shown. The observed data are recorded on the left-hand page. First, each station that is occupied is identified and the heights of the total station instrument and reflector that apply at that station are recorded. Then horizontal circle readings, zenith angles, horizontal distances, and elevation differences observed at each station are recorded. Notice that each horizontal angle is measured twice in the direct mode, and twice in the reversed mode. As noted earlier, this practice eliminates instrumental errors and gives repeat angle values for checking.

TRAVERSING WITH A

Sta. Sighted	D/R	Horiz. Circle	Zenith Angle	Horiz. Dist.	Elev. Diff.
Instrument at sta 101					
h_e=5.3		h_r=5.3			
104	D	0°00′00″	86°30′01″	324.38	+19.84
102	D	82°18′19″	92°48′17″	216.02	−10.58
104	R	180°00′03″	273°30′00″		
102	R	262°18′18″	267°11′41″		
Instrument at sta 102					
h_e=5.5		h_r=5.5			
101	D	0°00′00″	87°11′19″	261.05	+10.61
103	D	95°32′10″	85°19′08″	371.65	+30.43
101	R	180°00′02″	272°48′43″		
103	R	275°32′08″	274°40′50″		
Instrument at sta 103					
h_e=5.4		h_r=5.4			
102	D	0°00′00″	94°40′48″	371.63	−30.42
104	D	49°33′46″	90°01′54″	145.03	− 0.08
102	R	180°00′00″	265°19′14″		
104	R	229°33′47″	269°58′00″		

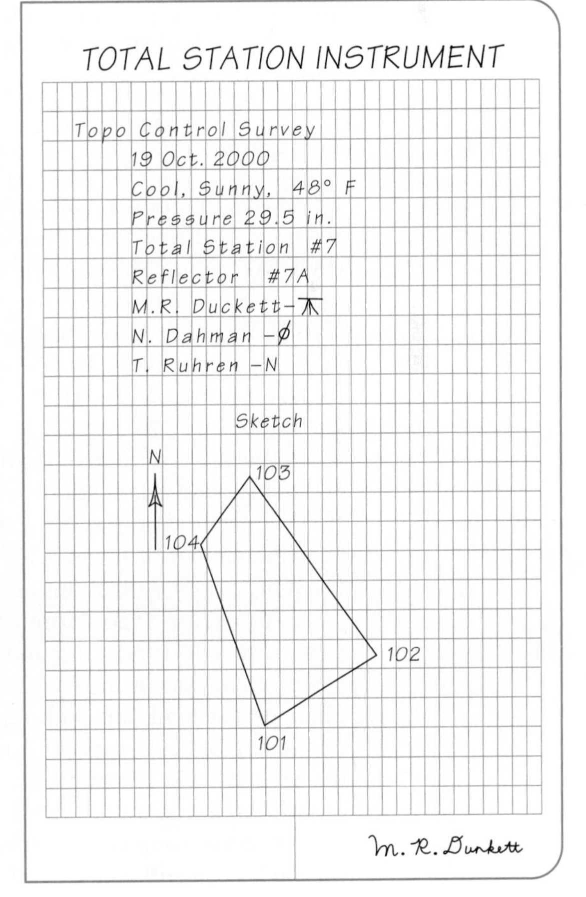

Figure 9.6 Example traverse field notes using a total station instrument.

Zenith angles were also observed twice each direct and reversed. Although not needed for traversing, they are available for checking if larger than tolerable misclosures (see Chapter 10) should exist in the traverse. Details of making traverse observations with a total station instrument are described in Section 9.8.

■ 9.7 ANGLE MISCLOSURE

The angular misclosure for an interior-angle traverse is the difference between the sum of the observed angles and the geometrically correct total for the polygon. The sum, Σ, of the interior angles of a closed polygon should be

$$\Sigma = (n - 2)180^\circ \tag{9.1}$$

where n is the number of sides, or angles, in the polygon. This formula is easily derived from known facts. The sum of the angles in a triangle is 180°; in a rectangle, 360°; and in a pentagon, 540°. Thus each side added to the three required for a triangle increases the sum of the angles by 180°.

Figure 9.1(a) shows a five-sided figure in which, if the sum of the observed interior angles equals 540°00′05″, the angular misclosure is 5″. Misclosures result from the accumulation of random errors in the angle observations. Permissible misclosure can be computed by the formula

$$c = K\sqrt{n} \tag{9.2}$$

where n is the number of angles, and K a constant that depends on the level of accuracy specified for the survey. The Federal Geodetic Control Subcommittee (FGCS) recommends constants for five different orders of traverse accuracy: *first-order, second-order class I, second-order class II, third-order class I,* and *third-order class II.* Values of K for these orders, from highest to lowest, are 1.7″, 3″, 4.5″, 10″, and 12″, respectively. Thus if the traverse of Figure 9.1(a) were being executed to second-order class II standards, its allowable misclosure error would be $4.5'' \times \sqrt{5} = \pm 10''$.

The algebraic sum of the deflection angles in a closed-polygon traverse equals 360°, clockwise (right) deflections being considered plus and counterclockwise (left) deflections, minus. This rule applies if lines do not crisscross, or if they cross an even number of times. When lines in a traverse cross an odd number of times, the sum of right deflections equals the sum of left deflections.

A closed-polygon azimuth traverse is checked by setting up on the starting point a second time, after having occupied the successive stations around the traverse, and orienting by back azimuths. The azimuth of the first side is then obtained a second time and compared with its original value. Any difference is the misclosure. If the first point is not reoccupied, the interior angles computed from the azimuths will automatically check the proper geometric total, even though one or more of the azimuths may be incorrect.

Although angular misclosures cannot be directly computed for link traverses, the angles can still be checked. The direction of the first line may be determined from two intervisible stations with a known azimuth between them, or from a sun or Polaris observation, as described in Appendix C. Observed angles are then applied to calculate the azimuths of all traverse lines. The last line's computed azimuth is compared with its known value or the result obtained from another sun or Polaris observation. On long traverses, intermediate lines can be similarly checked. In using sun or Polaris observations to check angles on traverses of long east-west extent, allowance must be made for *convergence of meridians*. This topic is discussed in Section 19.12.2.

9.8 TRAVERSING WITH TOTAL STATION INSTRUMENTS

Total station instruments, with their combined electronic angle and distance measurement components, speed the process of traversing significantly because both the angles and distances can be observed from a single setup. The observing process is further aided because angles and distances are resolved automatically and displayed. Furthermore, the microprocessors of total stations can perform traverse computations, reduce slope distances to their horizontal and vertical components, and instantaneously calculate and store station coordinates and elevations. The reduction to obtain horizontal and vertical distance components was illustrated with the traverse notes of Figure 9.6.

To illustrate a method of traversing with a total station instrument, refer to the traverse of Figure 9.1(b). With the instrument set up and leveled at station A, a backsight is carefully taken on $Az\ MK_1$. The azimuth of line A-$Az\ MK_1$ is initialized on the horizontal circle by entering it in the unit using the keyboard. The coordinates and elevation of station A are also entered in memory. Next a foresight is made on station B. The azimuth of line AB will now appear on the display, and upon keyboard command, can be stored in the microprocessor's memory. Slope distance AB is then observed and reduced to its horizontal and vertical components by the microprocessor. Then the line's departure and latitude are computed and added to the coordinates of station A to yield the coordinates of station B. (Departures, latitudes, and coordinates are described in Chapter 10.) These procedures should be performed in both the direct and reversed modes and the results averaged to account for instrumental errors.

The procedure outlined for station A is repeated at station B, except that the back azimuth BA and coordinates of station B need not be entered; rather, they are recalled from the instrument's memory. From the setup at B, azimuth BC and coordinates of C are determined and stored. This procedure is continued until a station of known coordinates is reached, as E in Figure 9.1(b). Here the known coordinates of E are entered in the unit's computer and compared to those obtained for E through the traverse observations. Their difference (or misclosure) is computed, displayed, and, if within allowable limits, distributed by the microprocessor to produce final coordinates of intermediate stations. (The process of distributing traverse misclosure errors is covered in Chapter 10.)

Mistakes in orientation can be minimized when a data collector is used in combination with a total station. In this process, the coordinates of each backsight station are checked before proceeding with the angle and distance observations to the next foresight station. For example in Figure 9.1(a), after the total station is leveled and oriented at station B, an observation is taken "back" on A. If the newly computed coordinates of A do not closely match their previously stored values, the instrument setup, leveling, and orientation should be rechecked, and the problem resolved before proceeding with any further measurements. This procedure often takes a minimal amount of time and typically identifies most field mistakes that occur during the observation process.

If desired, traverse station elevations can also be determined as a part of the procedure (usually the case for topographic surveys). Then entries *hi* (height of instrument) and *hr* (height of reflector) must be input (see Section 6.23). The microprocessor computes the vertical component of the slope distance, which includes a correction for curvature and refraction (see Section 4.5.4). The elevation difference is added to the occupied station's elevation to produce the next point's elevation. At the final station, any misclosure is determined by comparing the computed elevation with its known value, and if within tolerance, the misclosure is distributed to produce adjusted elevations of intermediate traverse stations.

All data from traversing with a total station instrument can be stored in a data collector for later printing and transfer to the office for computing and plotting (see Sections 2.12 through 2.15). Alternatively, the traverse notes can be recorded manually as illustrated with Figure 9.6.

9.9 RADIAL TRAVERSING

In certain situations, it may be most convenient to determine the relative positions of points by radial traversing. In this procedure, as illustrated in Figure 9.7(a), some point O, whose position is assumed known, is selected from which all points to be located can be seen. If a point such as O does not exist, it can be established. It is also assumed that a nearby azimuth mark, like Z in Figure 9.7(a), is available, and that reference azimuth OZ is known. With a total station instrument at point O, after backsighting on Z, horizontal angles to all stations A through F are observed. Azimuths of all radial lines from O (as OA, OB, OC, etc.) can then be calculated. The horizontal lengths of all radiating lines are also observed. By using the

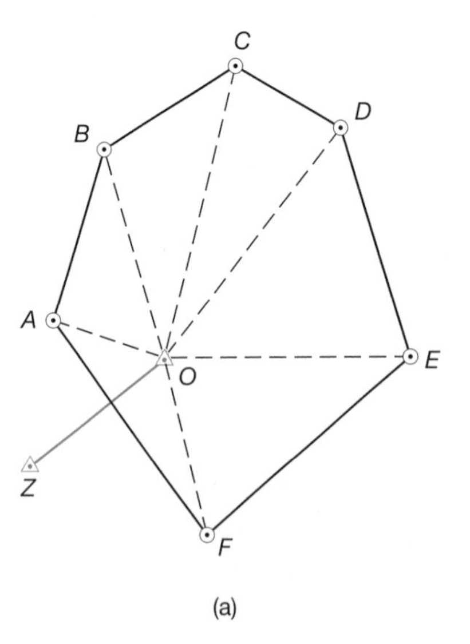

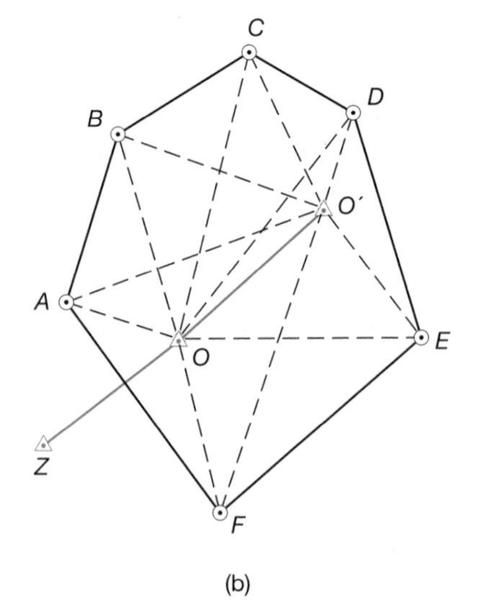

Figure 9.7 Radial traversing. (a) From one occupied station. (b) From two occupied stations.

observed lengths and azimuths, coordinates for each point can be computed. (The subject of coordinate computations is discussed in Chapter 10.)

It should be clear that in the procedure just described, each point A through F has been surveyed independently of all others and that no checks on their computed positions exist. To provide checks, lengths AB, BC, CD, etc., could be computed from the coordinates of points and then these same lengths observed. This results in many extra setups and substantially more fieldwork, thus defeating one of the major benefits of radial traversing. To solve the problem of gaining checks with a minimum of extra fieldwork, the method presented in Figure 9.7(b) is recommended. Here a second hub O' is selected from which all points can also be seen. The position of O' is determined by observations of the horizontal angle and distance from station O. This second hub O' is then occupied, and horizontal angles and distances to all stations A through F are observed as before. With the coordinates of both O and O' known, and by using the two independent sets of angles and distances, two sets of coordinates can be computed for each station, thus obtaining the checks. If the two sets for each point agree within a reasonable tolerance, the average can be taken. However, a better adjustment is obtained using the method of least squares (see Section 3.21 and Chapter 16). Although radial traversing can provide coordinates of many points in an area rapidly, the method is not as rigorous as running closed traverses.

Radial traversing is ideal for quickly establishing a large number of points in an area, especially when a total station instrument is employed. They not only enable the angle and distance observations to be made quickly, but they also perform the calculations for azimuth, horizontal distance, and station coordinates in real time. Radial methods are also very convenient for laying out planned construction projects with a total station instrument. In this application, the required coordinates of points to be staked are determined from the design, and the angles and distances that must be observed from a selected station of known position are computed. These are then laid out with a total station to set the stakes. The procedures are discussed in detail in Section 23.9.

■ 9.10 SOURCES OF ERROR IN TRAVERSING

Some sources of error in running a traverse are:

1. Poor selection of stations, resulting in bad sighting conditions caused by (a) alternate sun and shadow, (b) visibility of only the rod's top, (c) line of sight passing too close to the ground, (d) lines that are too short, and (e) sighting into the sun.
2. Errors in observations of angles and distances.
3. Failure to observe angles an equal number of times direct and reversed.

■ 9.11 MISTAKES IN TRAVERSING

Some mistakes in traversing are:

1. Occupying or sighting on the wrong station.
2. Incorrect orientation.

3. Confusing angles to the right and left.
4. Mistakes in note taking.
5. Misidentification of the sighted station.

PROBLEMS

Asterisks (*) indicate problems that have answers given in Appendix G.

9.1 Explain the differences and similarities between a polygon and link traverse.

9.2 Explain the differences between an open and closed traverse.

9.3 How can an angular check be obtained on a closed-azimuth traverse?

9.4 What similarities and differences exist between interior angles and angles to the right in a polygon traverse?

9.5 List four pertinent considerations in selecting locations for traverse stations.

9.6 Discuss the importance of reconnaissance in establishing traverse stations.

9.7 What are the two primary reasons for referencing traverse stations?

9.8 Discuss how angles in a link traverse are closed.

9.9 What should be the sum of the interior angles for a closed-polygon traverse that has: *(a) 6 sides (b) 5 sides (c) 12 sides.

9.10 Four interior angles of a five-sided polygon traverse were observed as: $A = 93°27'28''$, $B = 141°54'07''$, $C = 103°33'44''$, and $D = 98°35'15''$. The angle at E was not observed. If all observed angles are assumed to be correct, what is the value of angle E?

9.11 Similar to Problem 9.10, except the traverse had eight sides with observed angles of $A = 158°15'44''$, $B = 135°25'24''$, $C = 136°04'56''$, $D = 121°56'13''$, $E = 138°48'37''$, $F = 198°08'20''$ and $G = 81°20'26''$. Compute the angle at H, which was not observed.

9.12 What is the angular misclosure of a six-sided polygon traverse with observed angles of 98°10′19″, 133°45′48″, 98°23′20″, 152°51′04″, 134°32′01″, and 102°17′18″.

9.13 What FGCS standards would the angular misclosure in Problem 9.12 meet?

9.14* According to FGCS standards, what is the maximum acceptable angular misclosure for a second order, class I traverse having 24 angles?

9.15* What is the angular misclosure for a five-sided polygon traverse with observed angles of 102°26′37″, 99°55′13″, 127°15′53″, 116°35′02″, and 93°47′05″?

9.16 What is the angular misclosure for a seven-sided polygon traverse with observed angles of 121°36′06″, 135°16′04″, 123°21′44″, 131°09′58″, 140°30′12″, 140°00′00″, and 108°06′03″?

9.17 Discuss how a data collector can be used to check the setup of a total station in traversing.

9.18* If the standard error for each measurement of a traverse angle is ±2″, what is the expected standard error of the misclosure in the sum of the angles for a ten-sided traverse?

9.19 If the angles of a traverse are turned so that the 95 percent error of any angle is ±3″, what is the 95 percent error in a seven-sided traverse?

9.20 List three advantages of traversing with a total station.

9.21* The azimuth from station A of a link traverse to an azimuth mark is 136°45′08″. The azimuth from the last station of the traverse to an azimuth mark is 71°49′40″. Angles to the right are observed at each station: $A = 103°05'30''$, $B = 82°32'04''$, $C = 59°29'09''$, $D = 128°05'08''$, $E = 119°27'10''$, $F = 64°16'42''$, and $G = 98°08'42''$. What is the angular misclosure of this link traverse?

9.22 What FGCS order and class does the traverse in Problem 9.21 meet?

9.23* The interior angles in a five-sided closed-polygon traverse were observed as $A = 114°28'36''$, $B = 90°15'26''$, $C = 110°07'32''$, $D = 100°41'53''$, and $E = 124°26'28''$. Compute the angular misclosure. For what FGCS order and class is this survey adequate?

9.24 Similar to Problem 9.23, except for a six-sided traverse with observed angles of $A = 124°21'05''$, $B = 126°07'30''$, $C = 108°29'24''$, $D = 106°39'46''$, $E = 115°08'55''$, and $F = 139°13'10''$.

9.25 In Figure 9.6, what is the average interior angle with the instrument at station 102?

9.26 Same as Problem 9.25 except at instrument station 103.

9.27 Explain why it is advisable to use two instrument stations, as O and O' in Figure 9.7(b), when running radial traverses.

9.28 Create a computational program that computes the misclosure of angles in a closed polygon traverse. Use this program to solve Problem 9.24.

9.29 Create a computational program that computes the misclosure of angles in a closed link traverse. Use this program to solve Problem 9.21.

10

Traverse Computations

10.1 INTRODUCTION

Observed angles or directions of closed traverses are readily investigated before leaving the field. Linear measurements, even though repeated, are more likely a source of error, and must also be checked. Although the calculations are lengthier than angle checks, with today's data collectors, programmable calculators, and portable computers they can also be done in the field to determine, before leaving, whether a traverse meets the required precision. If specifications have been satisfied, the traverse is then adjusted to create perfect "closure" or geometric consistency among angles and lengths; if not, field observations must be repeated until adequate results are obtained.

Investigation of precision and acceptance or rejection of the field data are extremely important in surveying. Adjustment for geometric closure is also crucial. For example, in land surveying the law may require property descriptions to have exact geometric agreement.

Different procedures can be used for computing and adjusting traverses. These vary from elementary methods to more advanced techniques based on the method of least squares (see Chapter 16). This chapter concentrates on elementary procedures. The usual steps followed in making elementary traverse computations are (1) adjusting angles or directions to fixed geometric conditions, (2) determining preliminary azimuths (or bearings) of the traverse lines, (3) calculating departures and latitudes and adjusting them for misclosure, (4) computing rectangular coordinates of the traverse stations, and (5) calculating the lengths and azimuths (or bearings) of the traverse lines after adjustment. These procedures are all discussed in this chapter and are illustrated with several examples. Chapter 16 discusses traverse adjustment using the method of least squares.

10.2 BALANCING ANGLES

In elementary methods of traverse adjustment, the first step is to balance (adjust) the angles to the proper geometric total. For closed traverses, angle balancing is done readily since the total error is known (see Section 9.7), although its exact distribution is not. Angles of a closed traverse can be adjusted to the correct geometric total by applying one of two methods:

1. Applying an average correction to each angle where observing conditions were approximately the same at all stations. The correction for each angle is found by dividing the total angular misclosure by the number of angles.
2. Making larger corrections to angles where poor observing conditions were present.

Of these two methods, the first is almost always applied.

Example 10.1

For the traverse of Figure 10.1, the measured interior angles are given in Table 10.1. Compute the adjusted angles using methods 1 and 2.

Solution

The computations are best arranged as shown in Table 10.1. The first part of the adjustment consists of summing the interior angles and determining the misclosure according to Equation (9.1), which in this instance, as shown beneath column (2), is +11″. The remaining calculations are tabulated and the rationale for the procedures follows.

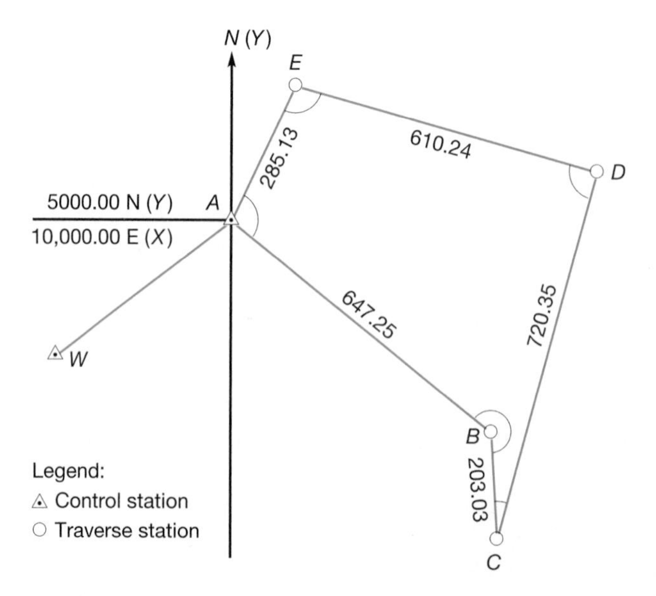

Figure 10.1
Traverse.

TABLE 10.1 ADJUSTMENT OF ANGLES

		Method 1			
Point (1)	**Measured Interior Angle (2)**	**Multiples of Average Correction (3)**	**Correction Rounded To 1″ (3)**	**Successive Differences (5)**	**Adjusted Angle (6)**
A	100°45′37″	2.2″	2″	2″	100°45′35″
B	231°23′43″	4.4″	4″	2″	231°23′41″
C	17°12′59″	6.6″	7″	3″	17°12′56″
D	89°03′28″	8.8″	9″	2″	89°03′26″
E	101°34′24″	11.0″	11″	2″	101°34′22″
	Σ = 540°00′11″			Σ = 11″	Σ = 540°00′00″

		Method 2	
Point (1)	**Measured Interior Angle (2)**	**Adjustment (7)**	**Adjusted Angle (8)**
A	100°45′37″	2″	100°45′35″
B	231°23′43″	3″	231°23′40″
C	17°12′59″	3″	17°12′56″
D	89°03′28″	1″	89°03′27″
E	101°34′24″	2″	101°34′22″
	Σ = 540°00′11″	Σ = 11″	Σ = 540°00′00″

For work of ordinary precision, it is reasonable to adopt corrections that are even multiples of the smallest recorded digit or decimal place for the angle readings. Thus in this example, corrections to the nearest 1″ will be made.

Method 1 consists of subtracting $11''/5 = 2.2''$ from each of the five angles. However, since the angles were read in multiples of 1″, applying corrections to the nearest tenth of a second would give a false impression of their precision. Therefore it is desirable to establish a pattern of corrections to the nearest 1″, as shown in Table 10.1. First multiples of the average correction of 2.2″ are tabulated in column (3). In column (4), each of these multiples has been rounded off to the nearest 1″. Then successive differences (adjustments for each angle) are found by subtracting the preceding value in column (4) from the one being considered. These are tabulated in column (5). Note that as a check, the sum of the corrections in this column must equal the angular misclosure of the traverse, which in this case is 11″. The adjusted angles obtained by applying these corrections are listed in column (6). As another check, they must total exactly the true geometric value of $(n - 2)180°$, or 540°00′00″ in this case.

In method 2, judgment is required because corrections are made to the angles expected to contain the largest errors. In this example, 3″ is subtracted from the angles at B and C, since they have the shortest sights (along line BC), and 2″ is subtracted from the angles at A and E, because they have the next shortest sights (along line AE). A 1″ correction was applied to angle D because of its long sights. The sum of the corrections must equal the total misclosure. The adjustment made in this manner is shown in columns (7) and (8) of Table 10.1.

It should be noted that, although the adjusted angles by both methods satisfy the geometric condition of a closed figure, they may be no nearer to the true values than before adjustment. Unlike corrections for linear observations (described in Section 10.7), *adjustments applied to angles are independent of the size of the angle.*

10.3 COMPUTATION OF PRELIMINARY AZIMUTHS OR BEARINGS

After balancing the angles, the next step in traverse computation is calculation of either preliminary azimuths or preliminary bearings. This requires the direction of at least one course within the traverse to be either known or assumed. For some computational purposes an assumed direction is sufficient and in that case the usual procedure is to simply assign north as the direction of one of the traverse lines. On certain traverse surveys, the magnetic bearing of one line can be determined and used as a reference for determining the other directions. However, in most instances, as in boundary surveys, true directions are needed. This requirement can be met by (1) incorporating within the traverse a line whose true direction was established through a previous survey; (2) including one end of a line of known direction as a station in the traverse [for example, station A of line A-$Az\ Mk$ of Figure 9.1(a)] and then observing an angle from that reference line to a traverse line; or (3) determining the true direction of one traverse line by astronomical observations (see Appendix C) or by GPS (see Chapters 13, 14, and 15).

If a line of known direction exists within the traverse, computation of preliminary azimuths (or bearings) proceeds as discussed in Chapter 7. Angles adjusted to the proper geometric total must be used; otherwise the azimuth or bearing of the first line, when recomputed after using all angles and progressing around the traverse, will differ from its fixed value by the angular misclosure. Azimuths or bearings at this stage are called "preliminary" because, as explained in Section 10.11, they will change after the traverse is adjusted.

Example 10.2

Compute preliminary azimuths for the traverse courses of Figure 10.1 based on a fixed azimuth of 234°17′18″ for line AW, an observed angle to the right of 151°52′24″ for WAE, and the angle adjustment by method 1 of Table 10.1.

Solution

Step 1: Compute the azimuth of course AB.

$$Az_{AB} = 234°17'18'' + 151°52'24'' + 100°45'35'' - 360° = 126°55'17''$$

TABLE 10.2 COMPUTATION OF PRELIMINARY AZIMUTH USING THE TABULAR METHOD

126°55′17″ = AB	+89°03′26″ + D
+180°	284°35′20″ = DE
306°55′17″ = BA	−180°
+231°23′41″ + B	104°35′20″ = ED
538°18′58″ − 360° = 178°18′58″ − BC	+101°34′22″ + E
−180°	206°09′42″ = EA
358°18′58″ = CD	−180°
+17°12′56″ + C	26°09′42″ = AE
375°31°54″ − 360° = 15°31′54″ = CD	+100°45′35″ + A
−180°	126°55′17″ = AB
195°31′54″ = DC	

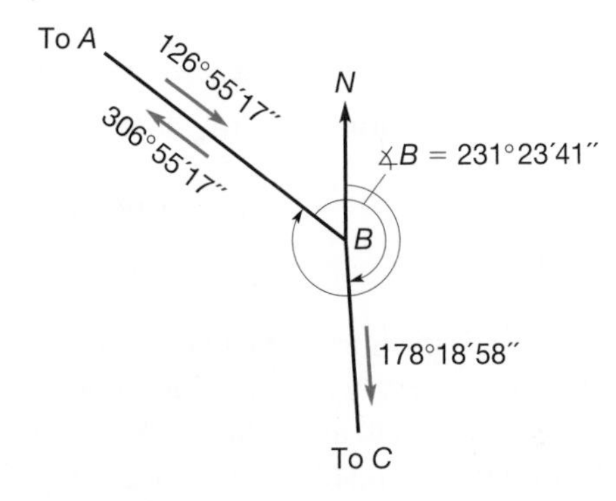

Figure 10.2 Computation of azimuth BC.

Step 2: Using the tabular method discussed in Section 7.8, compute preliminary azimuths for the remaining lines. The computations for this example are shown in Table 10.2. Figure 10.2 demonstrates the computations for line BC. Note that the azimuth of *AB* was recalculated as a check at the end of the table.

10.4 DEPARTURES AND LATITUDES

After balancing the angles and calculating preliminary azimuths (or bearings), traverse closure is checked by computing the *departure* and *latitude* of each line. As illustrated in Figure 10.3, the departure of a course is its orthographic projection on the east-west axis of the survey and is equal to the length of the course multiplied by the sine of its azimuth (or bearing) angle. Departures are sometimes called *eastings* or *westings.*

Also as shown in Figure 10.3, the latitude of a course is its orthographic projection on the north-south axis of the survey and is equal to the course length multiplied by the cosine of its azimuth (or bearing) angle. Latitude is also called *northing* or *southing.*

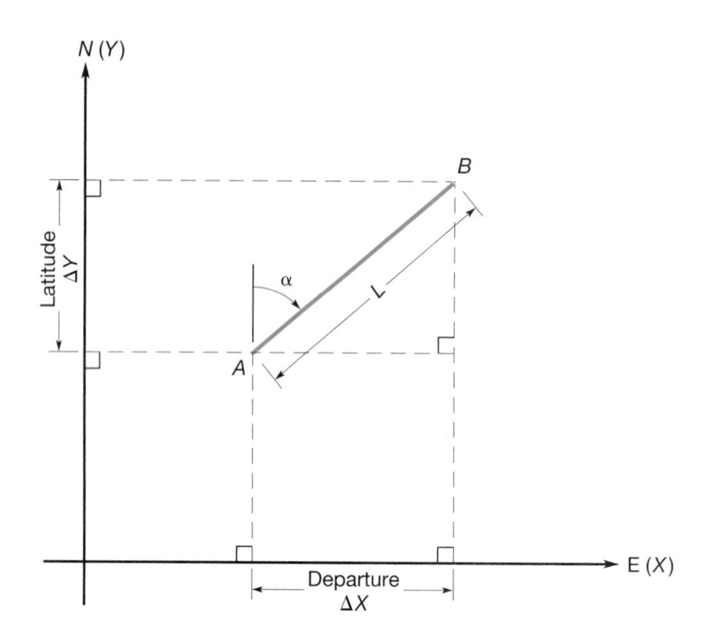

Figure 10.3 Departure and latitude of a line.

In equation form, the departure and latitude of a line are

$$\text{departure} = L \sin \alpha \tag{10.1}$$

$$\text{latitude} = L \cos \alpha \tag{10.2}$$

where L is the horizontal length and α the azimuth of the course. Departures and latitudes are merely X and Y components of a line in a rectangular grid system, sometimes referred to as ΔX and ΔY. In traverse calculations, east departures and north latitudes are considered plus; west departures and south latitudes, minus. Azimuths (from north) used in computing departures and latitudes range from 0 to 360° and the algebraic signs of sines and cosines automatically produce the proper algebraic signs of the departures and latitudes. Thus a line with an azimuth of 126°55′17″ has a positive departure and negative latitude (the sine at the azimuth is plus and the cosine minus); a course of 284°35′20″ azimuth has a negative departure and positive latitude. In using bearings for computing departures and latitudes, the angles are always between 0 and 90°; hence their sines and cosines are invariably positive. Proper algebraic signs of departures and latitudes must therefore be assigned on the basis of the bearing angle directions, so a *NE* bearing has a plus departure and latitude, a *SW* bearing gets a minus departure and latitude, and so on. Because data collectors, hand calculators, and computers automatically affix correct algebraic signs to departures and latitudes through the use of azimuth angle sines and cosines, it is more convenient to use azimuths than bearings for traverse computations.

10.5 DEPARTURE AND LATITUDE CLOSURE CONDITIONS

For a closed-polygon traverse like that of Figure 10.1, it can be reasoned that if all angles and distances were measured perfectly, the algebraic sum of the departures of all courses in the traverse should equal zero. Likewise, the algebraic sum of all latitudes should equal zero. And for closed link-type traverses like that of Figure 9.1(b), the algebraic sum of departures should equal the total difference in departure between the starting and ending control points. The same condition applies to latitudes in a link traverse. Because the observations are not perfect and errors exist in the angles and distances, the conditions just stated rarely occur. The amounts by which they fail to be met are termed *departure misclosure* and *latitude misclosure.* Their values are computed by algebraically summing the departures and latitudes and comparing the totals to the required conditions.

The magnitudes of the departure and latitude misclosures for closed-polygon-type traverses give an "indication" of the precision that exists in the observed angles and distances. Large misclosures certainly indicate that either significant errors or even mistakes exist. Small misclosures usually mean the measured data are precise and free of mistakes, but it is not a guarantee that systematic or compensating errors do not exist.

10.6 TRAVERSE LINEAR MISCLOSURE AND RELATIVE PRECISION

Because of errors in the observed traverse angles and distances, if one were to begin at point A of a closed-polygon traverse, like that of Figure 10.1, and progressively follow each course for its observed distance along its preliminary bearing or azimuth, one would finally return not to point A, but to some other nearby point A'. Point A' would be removed from A in an east-west direction by the departure misclosure and in a north-south direction by the latitude misclosure. The distance between A and A' is termed the linear misclosure of the traverse. It is calculated from the following formula:

$$\text{linear misclosure} = \sqrt{(\text{departure misclosure})^2 + (\text{latitude misclosure})^2} \quad (10.3)$$

The *relative precision* of a traverse is expressed by a fraction that has the linear misclosure as its numerator and the traverse perimeter or total length as its denominator, or

$$\text{relative precision} = \frac{\text{linear misclosure}}{\text{traverse length}} \quad (10.4)$$

The fraction that results from Equation (10.4) is then reduced to reciprocal form and the denominator rounded to the same number of significant figures as the numerator. This is illustrated in the following example.

Example 10.3

Based on the preliminary azimuths from Table 10.2 and lengths shown in Figure 10.1, calculate the departures and latitudes, linear misclosure, and relative precision of the traverse.

TABLE 10.3 COMPUTATION OF DEPARTURES AND LATITUDES

Station	Preliminary Azimuths	Length	Departure	Latitude
A	126°55′17″	647.25	517.451	−388.815
B	178°18′58″	203.03	5.966	−202.942
C	15°31′54″	720.35	192.889	694.045
D	284°35′20″	610.24	−590.565	153.708
E	206°09′42″	285.13	−125.715	−255.919
A		Σ = 2466.00	Σ = 0.026	Σ = 0.077

Solution

In computing departures and latitudes, the data and results are usually listed in a standard tabular form, such as that shown in Table 10.3. The column headings and rulings save time and simplify checking.

In Table 10.3, taking the algebraic sum of east (+) and west (−) departures gives the misclosure, 0.026 ft. Also, summing north (+) and south (−) latitudes gives the misclosure in latitude, 0.077 ft. Linear misclosure is the hypotenuse of a small triangle with sides of 0.026 ft and 0.077 ft and in this example its value is, by Equation (10.3),

$$\text{linear misclosure} = \sqrt{(0.026)^2 + (0.077)^2} = 0.081 \text{ ft}$$

The relative precision for this traverse, by Equation (10.4), is

$$\text{relative precision} = \frac{0.081}{2466.00} = \frac{1}{30{,}000}$$

10.7 TRAVERSE ADJUSTMENT

For any closed traverse, the linear misclosure must be adjusted (or distributed) throughout the traverse to "close" or "balance" the figure. This is true even though the misclosure is negligible in plotting the traverse at map scale. There are several elementary methods available for traverse adjustment, but the one most commonly used is the *compass rule* (Bowditch method). This method is known as an *arbitrary method* since the corrections to the observations are applied irrespective of their uncertainties. As noted earlier, adjustment by least squares is a more advanced technique that can also be used. The least-squares method adjusts observations based on the laws of probability. These two methods are discussed in the subsections that follow.

10.7.1 Compass (Bowditch) Rule

The compass, or Bowditch, rule adjusts the departures and latitudes of traverse courses in proportion to their lengths. Although not as rigorous as the least-squares method, it does result in a logical distribution of misclosures. Corrections by this method are made according to the following rules:

$$\text{correction in departure for } AB = -\frac{\text{(total departure misclosure)}}{\text{traverse perimeter}} \times \text{length of } AB \tag{10.5}$$

$$\text{correction in latitude for } AB = -\frac{\text{(total latitude misclosure)}}{\text{traverse perimeter}} \times \text{length of } AB \tag{10.6}$$

Note that the algebraic signs of the corrections are opposite those of the respective misclosures.

■ Example 10.4

Using the preliminary azimuths from Table 10.2 and lengths from Figure 10.1, compute departures and latitudes, linear misclosure, and relative precision. Balance the departures and latitudes using the compass rule.

Solution

A tabular solution, which is somewhat different than that used in Example 10.3, is employed for computing departures and latitudes (see Table 10.4). To compute departure and latitude corrections by the compass rule, Equation (10.5) and (10.6) are used as demonstrated. By Equation (10.5) the correction in departure for AB is

$$-\frac{0.026}{2466} \times 647.25 = -0.007 \text{ ft}$$

And by Equation (10.6) the correction for the latitude of AB is

$$-\frac{0.077}{2466} \times 647.25 = -0.020 \text{ ft}$$

The other corrections are likewise found by multiplying a constant—the ratio of misclosure in departure, and latitude, to the perimeter—by the successive course lengths.

In Table 10.4, the departure and latitude corrections are shown in parentheses above their unadjusted values. These corrections are added algebraically to their respective unadjusted values, and the corrected quantities tabulated in the "balanced" departure and latitude columns. A check is made of the computational process by algebraically summing the balanced departure and latitude

TABLE 10.4 BALANCING DEPARTURES AND LATITUDES BY THE COMPASS (BOWDITCH) RULE

Station	Preliminary Azimuths	Length (ft)	Unadjusted Departure	Unadjusted Latitude	Balanced Departure	Balanced Latitude	Coordinates* X (ft) (easting)	Coordinates* Y (ft) (northing)
A			(−0.007)	(−0.020)			10,000.00	5000.00
	126°55′17″	647.25	517.451	−388.815	517.444	−388.835		
B			(−0.002)	(−0.006)			10,517.44	4611.16
	178°18′58″	203.03	5.966	−202.942	5.964	−202.948		
C			(−0.008)	(−0.023)			10,523.41	4408.22
	15°31′54″	720.35	192.889	694.045	192.881	694.022		
D			(−0.006)	(−0.019)			10,716.29	5102.24
	284°35′20″	610.24	−590.565	153.708	−590.571	153.689		
E			(−0.003)	(−0.009)			10,125.72	5255.93
	206°09′42″	285.13	−125.715	−255.919	−125.718	−255.928		
A							10,000.00✓	5000.00✓
		Σ = 2466.00	Σ = 0.026	Σ = 0.077	Σ = 0.000	Σ = 0.000		

$$\text{Linear precision} = \sqrt{(0.026)^2 + (0.077)^2} = 0.081 \text{ ft}$$

$$\text{Relative precision} = \frac{0.081}{2466} = \frac{1}{30{,}000}$$

*Coordinates are rounded to same significance as observed lengths.

columns to verify that each is zero. In these columns, if rounding off causes a small excess or deficiency, revising one of the corrections to make the closure perfect eliminates this.

10.7.2 Least-Squares Method

As noted in Section 3.21, the method of least squares is based on the theory of probability, which models the occurrence of random errors. This results in adjusted values having the highest probability. Thus the least-squares method provides the best and most rigorous traverse adjustment, but until recently the method has not been widely used because of the lengthy computations required. The availability of computers has now made these calculations routine and consequently the least-squares method has gained popularity.

In applying the least-squares method to traverses, angle and distance observations are adjusted simultaneously. Thus no preliminary angle adjustment is made, as is done when using the compass rule. The least-squares method is valid for any type of traverse and has the advantage that observations of varying precisions can be weighted appropriately in the computations. Examples illustrating some elementary least-squares adjustments are presented in Chapter 16.

■ 10.8 RECTANGULAR COORDINATES

Rectangular X and Y coordinates of any point give its position with respect to an arbitrarily selected pair of mutually perpendicular reference axes. The X coordinate is the perpendicular distance, in feet or meters, from the point to the Y axis; the Y coordinate is the perpendicular distance to the X axis. Although the reference axes are discretionary in position, in surveying they are normally oriented so that the Y axis points north-south, with north the positive Y direction. The X axis runs east-west, with positive X being east. Given the rectangular coordinates of a number of points, their relative positions are uniquely defined.

Coordinates are useful in a variety of computations, including (1) determining lengths and directions of lines, and angles (see Section 10.11 and Chapter 11); (2) calculating areas of land parcels (see Section 12.5); (3) making certain curve calculations (see Sections 24.12 and 24.13); and (4) locating inaccessible points (see Section 11.9). Coordinates are also advantageous for plotting maps (see Section 18.8.1).

In practice, *state plane coordinate systems,* as described in Chapter 20, are most frequently used as the basis for rectangular coordinates in plane surveys. However for many calculations, any arbitrary system may be used. As an example, coordinates may be arbitrarily assigned to one traverse station. For example, to avoid negative values of X and Y an origin is assumed south and west of the traverse such that one hub has coordinates $X = 10{,}000.00$, $Y = 5{,}000.00$, or any other suitable values. In a closed traverse, assigning $Y = 0.00$ to the most southerly point and $X = 0.00$ to the most westerly station saves time in hand calculations.

Given the X and Y coordinates of any starting point A, the X coordinate of the next point B is obtained by adding the adjusted departure of line AB to X_A.

Likewise, the Y coordinate of B is the adjusted latitude of AB added to Y_A. In equation form this is

$$X_B = X_A + \text{departure } AB$$
$$Y_B = Y_A + \text{latitude } AB \tag{10.7}$$

For closed polygons, the process is continued around the traverse, successively adding departures and latitudes until the coordinates of starting point A are recalculated. If these recalculated coordinates agree exactly with the starting ones, a check on the coordinates of all intermediate points is obtained (unless compensating mistakes have been made). For link traverses, after progressively computing coordinates for each station, if the calculated coordinates of the closing control point equal that point's control coordinates, a check is obtained.

Example 10.5

Using the balanced departures and latitudes obtained in Example 10.4 (see Table 10.4) and starting coordinates $X_A = 10{,}000.00$ and $Y_A = 5{,}000.00$, calculate coordinates of the other traverse points.

Solution

The process of successively adding balanced departures and latitudes to obtain coordinates is carried out in the two rightmost columns of Table 10.4. Note that the starting coordinates $X_A = 10{,}000.00$ and $Y_A = 5{,}000.00$ are recomputed at the end to provide a check. Note also that X and Y coordinates are frequently referred to as *eastings* and *northings*, respectively, as is indicated in Table 10.4.

10.9 ALTERNATIVE METHODS FOR MAKING TRAVERSE COMPUTATIONS

Procedures for making traverse computations that vary somewhat from those described in preceding sections can be adopted. One alternative is to adjust azimuths or bearings rather than angles. Another is to apply compass rule corrections directly to coordinates. These procedures are described in the subsections that follow.

10.9.1 Balancing Angles by Adjusting Azimuths or Bearings

In this method, "unadjusted" azimuths or bearings are computed based on the observed angles. These azimuths or bearings are then adjusted to secure a geometric closure, and to obtain preliminary values for use in computing departures and latitudes. The method is equally applicable to closed-polygon traverses, like that of Figure 10.1, or to closed-link traverses, as shown in Figure 9.1(b) that begins on one control station and ends on another. The procedure of making the adjustment for angular misclosure in this manner will be explained by an example.

Example 10.6

Table 10.5 lists observed angles to the right for the traverse of Figure 9.1(b). The azimuths of lines $A\text{-}Az\ Mk_1$ and $E\text{-}Az\ Mk_2$ have known values of 139°05′45″ and 86°20′47″, respectively. Compute unadjusted azimuths and balance them to obtain geometric closure.

Solution

From the observed angles of column (2) in Table 10.5, unadjusted azimuths have been calculated and are listed in column (3). Because of angular errors, the unadjusted azimuth of the final line $E\text{-}Az\ Mk_2$ disagrees with its fixed value by 0°00′10″. This represents the angular misclosure, which is divided by 5, the number of observed angles, to yield a correction of −2″ per angle. The corrections to azimuths, which accumulate and increase by −2″ for each angle, are listed in column (4). Thus line AB, which is based on one observed angle, receives a −2″ correction; line BC which uses two observed angles, gets a −4″ correction; and so

TABLE 10.5 BALANCING TRAVERSE AZIMUTHS

Station (1)	Measured Angle* (2)	Unadjusted Azimuth (3)	Azimuth Correction (4)	Preliminary Azimuth (5)
$Az\ Mk_1$				
		319°05′45″		319°05′45″
A	283°50′10″			
		62°55′55″	−2″	62°55′53″
B	256°17′18″			
		139°13′13″	−4″	139°13′09″
C	98°12′36″			
		57°25′49″	−6″	57°25′43″
D	103°30′34″			
		340°56′23″	−8″	340°56′15″
E	285°24′34″			
		86°20′57″	−10″	86°20′47″
$Az\ Mk_2$				

86°20′57″
−86°20′47″
misclosure = 0°00′10″
correction per angle = −10″/5 = −2″

*Observed angles are angles to the right

on. The final azimuth, $E\text{-}Az\ Mk_2$, receives a $-10''$ correction because all five observed angles have been included in its calculation. The corrected preliminary azimuths are listed in column (5).

10.9.2 Balancing Departures and Latitudes by Adjusting Coordinates

In this procedure, commencing with the known coordinates of a beginning station, unadjusted departures and latitudes for each course are successively added to obtain "preliminary" coordinates for all stations. For closed-polygon traverses, after progressing around the traverse, preliminary coordinates are recomputed for the beginning station. The difference between the computed preliminary X coordinate at this station and its known X coordinate is the departure misclosure. Similarly, the disagreement between the computed preliminary Y coordinate for the beginning station and its known value is the latitude misclosure. Corrections for these misclosures can be calculated using compass-rule Equations (10.5) and (10.6) and applied directly to the preliminary coordinates to obtain adjusted coordinates. The result is exactly the same as if departures and latitudes were first adjusted and coordinates computed from them, as was done in Examples 10.4 and 10.5.

Closed traverses like the one shown in Figure 9.1(b) can be similarly adjusted. For this type of traverse, unadjusted departures and latitudes are also successively added to the beginning station's coordinates to obtain preliminary coordinates for all points, including the final closing station. Differences in preliminary X and Y coordinates, and the corresponding known values for the closing station, represent the departure and latitude misclosures, respectively. These misclosures are distributed directly to preliminary coordinates using the compass rule to obtain final adjusted coordinates. The procedure will be demonstrated by an example.

Example 10.7

Table 10.6 lists the preliminary azimuths (from Table 10.5) and measured lengths (in feet) for the traverse of Figure 9.1(b). The known coordinates of stations A and E are $X_A = 12{,}765.48$, $Y_A = 43{,}280.21$, $X_E = 14{,}797.12$, and $Y_E = 44{,}384.51$ ft. Adjust this traverse for departure and latitude misclosures by making corrections to preliminary coordinates.

Solution

From the lengths and azimuths listed in columns (2) and (3) of Table 10.6, departures and latitudes are computed and tabulated in columns (4) and (5). These unadjusted values are progressively added to the known coordinates of station A to obtain preliminary coordinates for all stations, including E, and are listed in columns (6) and (7). Comparing the preliminary X and Y coordinates of station E with its known values yields departure and latitude misclosures of $+0.179$ and

TABLE 10.6 TRAVERSE ADJUSTMENT BY COORDINATES

Station (1)	Length (ft) (2)	Preliminary Azimuth (3)	Departure (4)	Latitude (5)	Preliminary Coordinates (ft) X (6)	Preliminary Coordinates (ft) Y (7)	Corrections (ft) X (8)	Corrections (ft) Y (9)	Adjusted Coordinates* X (ft) (10)	Adjusted Coordinates* Y (ft) (11)
A					12,765.48	43,280.21			12,765.48	43,280.21
	1045.50	62°55′53″	930.978	475.762			−0.048	0.006		
B					13,696.458	43,755.972	(−0.048)	(0.006)	13,696.41	43,755.98
	1007.38	139°13′09″	657.988	−762.802			−0.046	0.006		
C					14,354.446	42,993.170	(−0.094)	(0.012)	14,354.35	42,993.18
	897.81	57°25′43″	756.604	483.336			−0.041	0.006		
D					15,111.050	43,476.506	(−0.135)	(0.018)	15,110.92	43,476.52
	960.66	340°56′15″	−313.751	907.980			−0.044	0.006		
E					14,797.299	44,384.486	(−0.179)	(0.024)	14,797.12✓	44,384.51✓
	Σ = 3911.35				−14,797.12	−44,384.51				
				misclosures	+0.179	−0.024				

$$\text{Linear precision} = \sqrt{(0.179)^2 + (-0.024)^2} = 0.181 \text{ ft}$$

$$\text{Relative precision} = \frac{0.181}{3911} = \frac{1}{21{,}000}$$

*Adjusted coordinates are rounded to same significance as observed lengths.

−0.024 ft, respectively. From these values, the linear misclosure of 0.181 ft and relative precision of 1/21,000 are computed (see Table 10.6).

Compass-rule corrections for each course are computed and listed in columns (8) and (9). Their cumulative values obtained by progressively adding the corrections are given in parentheses in columns (8) and (9). Finally, by applying the cumulative corrections to the preliminary coordinates of columns (6) and (7), final adjusted coordinates (rounded to the nearest hundredth of a foot) listed in columns (10) and (11) are obtained.

10.10 LENGTHS AND DIRECTIONS OF LINES FROM DEPARTURES AND LATITUDES, OR COORDINATES

If the departure and latitude of a line AB are known, its length and azimuth or bearing are readily obtained from the following relationships:

$$\tan \text{azimuth (or bearing) } AB = \frac{\text{departure } AB}{\text{latitude } AB} \tag{10.8}$$

$$\begin{aligned} \text{length } AB &= \frac{\text{departure } AB}{\sin \text{azimuth (or bearing) } AB} \\ &= \frac{\text{latitude } AB}{\cos \text{azimuth (or bearing) } AB} \\ &= \sqrt{(\text{departure } AB)^2 + (\text{latitude } AB)^2} \end{aligned} \tag{10.9}$$

Equations (10.7) can be written to express departures and latitudes in terms of coordinate differences ΔX and ΔY as follows:

$$\begin{aligned} \text{departure}_{AB} &= X_B - X_A = \Delta X \\ \text{latitude}_{AB} &= Y_B - Y_A = \Delta Y \end{aligned} \tag{10.10}$$

Substituting Equations (10.10) into Equations (10.8) and (10.9),

$$\tan \text{azimuth (or bearing) } AB = \frac{X_B - X_A}{Y_B - Y_A} = \frac{\Delta X}{\Delta Y} \tag{10.11}$$

$$\begin{aligned} \text{length } AB &= \frac{X_B - X_A \text{ (or } \Delta X)}{\sin \text{azimuth (or bearing) } AB} \\ &= \frac{Y_B - Y_A \text{ (or } \Delta Y)}{\cos \text{azimuth (or bearing) } AB} \\ &= \sqrt{(X_B - X_A)^2 + (Y_B - Y_A)^2} \\ &= \sqrt{(\Delta X)^2 + (\Delta Y)^2} \end{aligned} \tag{10.12}$$

Equations (10.8) through (10.12) can be applied to any line whose coordinates are known, whether or not it was actually observed in the survey. Note that X_B and Y_B

must be listed first in Equations (10.11) and (10.12), so that ΔX and ΔY will have the correct algebraic signs. Computing lengths and directions of lines from departures and latitudes, or from coordinates, is called *inversing*.

10.11 COMPUTING FINAL ADJUSTED TRAVERSE LENGTHS AND DIRECTIONS

In traverse adjustments, as illustrated in Examples 10.4 and 10.7, corrections are applied to the computed departures and latitudes to obtain adjusted values. These in turn are used to calculate X and Y coordinates of the traverse stations. By changing departures and latitudes of lines in the adjustment process, their lengths and azimuths (or bearings) also change. In many types of surveys, it is necessary to compute the changed, or "final adjusted," lengths and directions. For example, if the purpose of the traverse was to describe the boundaries of a parcel of land, the final adjusted lengths and directions would be used in the recorded deed.

The equations developed in the preceding section permit computation of final values for lengths and directions of traverse lines based either on their adjusted departures and latitudes or on their final coordinates.

Example 10.8

Calculate the final adjusted lengths and azimuths of the traverse of Example 10.4 from the adjusted departures and latitudes listed in Table 10.4.

Solution

Equations (10.8) and (10.9) are applied to calculate the adjusted length and azimuth of line AB. All others were computed in the same manner. The results are listed in Table 10.7.

By Equation (10.8),

$$\tan \text{azimuth}_{AB} = \frac{517.444}{-388.835} = -1.330755;$$

$$\text{azimuth}_{AB} = -53°04'37'' + 180° = 126°55'23''$$

TABLE 10.7 FINAL ADJUSTED LENGTHS AND DIRECTIONS FOR TRAVERSE OF EXAMPLE 10.4

	Balanced		Balanced	
Line	Departure	Latitude	Length (ft)	Azimuth
AB	517.444	−388.835	647.26	126°55′23″
BC	5.964	−202.948	203.04	178°19′00″
CD	192.881	694.022	720.33	15°31′54″
DE	−590.571	153.689	610.24	284°35′13″
EA	−125.718	−255.928	285.14	206°09′41″

By Equation (10.9),

$$\text{length}_{AB} = \sqrt{(517.444)^2 + (-388.835)^2} = 647.26 \text{ ft.}$$

Comparing the observed lengths of Table 10.4 to the final adjusted values in Table 10.7, it can be seen that, as expected, the values have undergone small changes, some increasing, others decreasing, and length *DE* remaining the same because of compensating changes. Notice that varying changes have also occurred in the azimuths.

Example 10.9

Using coordinates, calculate adjusted lengths and azimuths for the traverse of Example 10.7 (see Table 10.6).

Solution

Equations (10.11) and (10.12) are used to demonstrate calculation of the adjusted length and azimuth of line *AB*. All others were computed in the same way. The results are listed in Table 10.8. Comparing the adjusted lengths and azimuths of this table with their unadjusted values of Table 10.6 reveals that all values have undergone changes of varying amounts.

$$X_B - X_A = 13{,}696.41 - 12{,}765.48 = 930.93 = \Delta X$$

$$Y_B - Y_A = 43{,}755.98 - 43{,}280.21 = 475.77 = \Delta Y$$

By Equation (10.11), tan azimuth$_{AB}$ = 930.93/475.77 = 1.95668075; azimuth$_{AB}$ = 62°55′47″.

By Equation (10.12), length$_{AB} = \sqrt{(930.93)^2 + (475.77)^2} = 1045.46$ ft.

TABLE 10.8 FINAL ADJUSTED LENGTHS AND DIRECTIONS FOR TRAVERSE OF EXAMPLE 10.7

	Adjusted		Adjusted	
Line	ΔX	ΔY	Length (ft)	Azimuth
AB	930.93	475.77	1045.46	62°55′47″
BC	657.94	−762.80	1007.35	139°13′16″
CD	756.57	483.34	897.78	57°25′38″
DE	−313.80	907.99	960.68	340°56′06″

10.12 COORDINATE COMPUTATIONS IN BOUNDARY SURVEYS

Computation of a bearing from the known coordinates of two points on a line is commonly done in boundary surveys. If the lengths and directions of lines from traverse points to the corners of a field are known, the coordinates of the corners can be determined and the lengths and bearings of all sides calculated.

Example 10.10

In Figure 10.4, *APQDEA* is a parcel of land that must be surveyed, but because of obstructions, traverse stations cannot be set at *P* and *Q*. Therefore offset stations *B* and *C* are set nearby, and closed traverse *ABCDE* run. Lengths and azimuths of lines *BP* and *CQ* are observed as 42.50 ft, 354°50′00″, and 34.62 ft, 26°39′54″, respectively. Following procedures demonstrated in earlier examples, traverse *ABCEA* was computed and adjusted, and coordinates were determined for all stations. They are given in the following table.

Point	X (ft)	Y (ft)
A	1000.00	1000.00
B	1290.65	1407.48
C	1527.36	1322.10
D	1585.70	1017.22
E	1464.01	688.25

Compute the length and bearing of property line *PQ*.

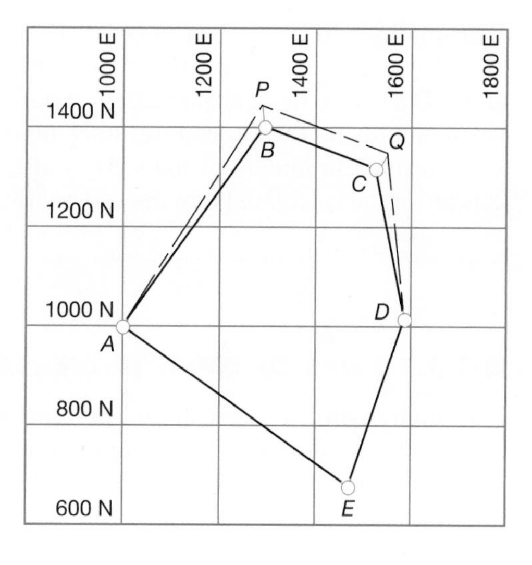

Figure 10.4 Plot of traverse for a boundary survey.

Solution

1. Using Equations (10.1) and (10.2), the departures and latitudes of lines BP and CQ are:

$$\text{Dep}_{BP} = 42.50 \sin(354°50'00'') = -3.83 \text{ ft}$$
$$\text{Dep}_{CQ} = 34.62 \sin(26°39'54'') = 15.54 \text{ ft}$$

$$\text{Lat}_{BP} = 42.50 \cos(354°50'00'') = 42.33 \text{ ft}$$
$$\text{Lat}_{CQ} = 34.62 \cos(26°39'54'') = 30.94 \text{ ft}$$

2. From the coordinates of stations B and C and the departures and latitudes just calculated, the following tabular solution yields X and Y coordinates for points P and Q:

	X	Y		X	Y
B	1290.65	1407.48	C	1527.36	1322.10
BP	−3.83	+42.33	CQ	+15.54	+30.94
P	1286.82	1449.81	Q	1542.90	1353.04

3. From the coordinates of P and Q, the length and bearing of line PQ are found in the following manner:

	X	Y
Q	1542.90	1353.04
P	−1286.88	−1449.81
PQ	ΔX = 256.02	ΔY = −96.77

By Equation (10.11), tan bearing$_{PQ}$ = 256.02/−96.77 = −2.64565; bearing$_{PQ}$ = S69°17′40″E

By Equation (10.12), length $PQ = \sqrt{(-96.77)^2 + (256.02)^2} = 273.79$ ft

By using Equations (10.11) and (10.12), lengths and bearings of lines AP and QD can also be determined. As stated earlier, extreme caution must be used when employing this procedure, since no checks are obtained on the length and azimuth measurements of lines BP and CQ, nor are there any computational checks on the calculated lengths and bearings.

10.13 USE OF OPEN TRAVERSES

Although open traverses should be used with reluctance, sometimes there are situations where it is very helpful to run one and then compute the length and direction of the "closing line." In Figure 10.5, for example, suppose that improved horizontal alignment is planned for Taylor Lake and Atkins Roads, and a new

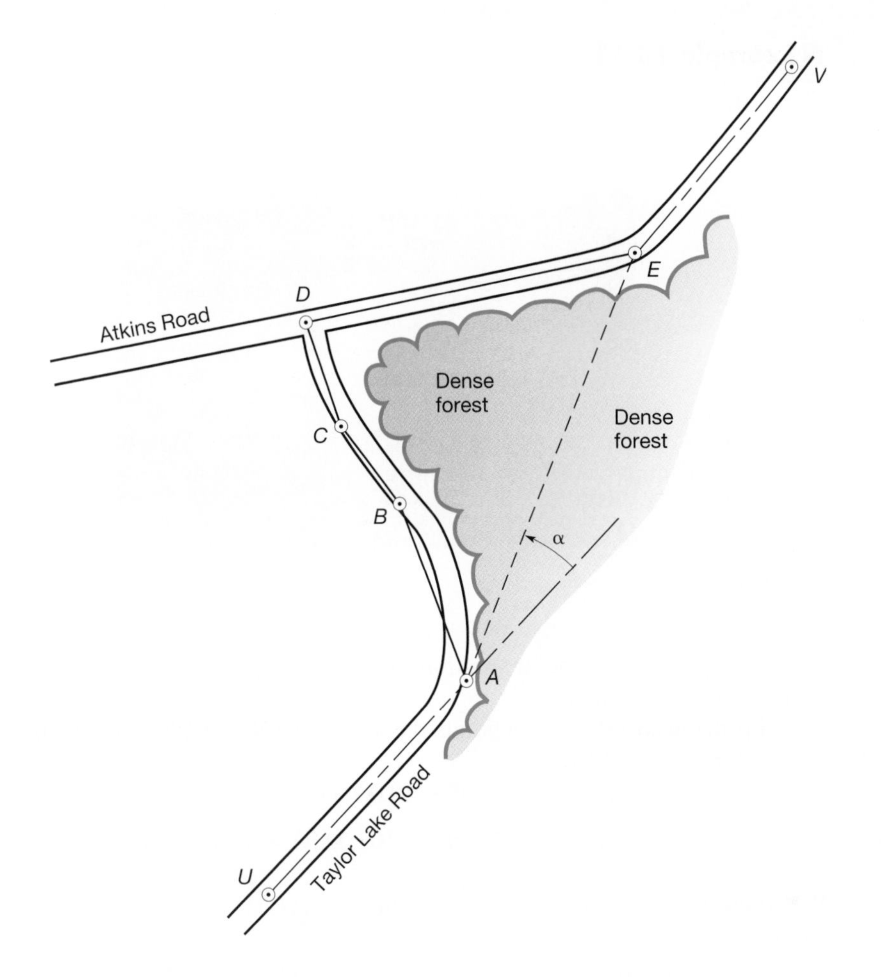

Figure 10.5
Closing line of an open traverse.

construction line AE must be laid out. Because of dense forest, visibility between points A and E is not possible. A random line (see Section 8.17) could be run from A toward E and then corrected to the desired line, but that would be very difficult and time consuming due to tree density. One solution to this problem is to run open traverse $ABCDE$, which can be done quite easily along the cleared right-of-way of existing roads.

For this problem an assumed azimuth (for example, due north) can be taken for line UA and assumed coordinates (for example, 10,000.00 and 10,000.00) can be assigned to station A. From observed lengths and angles, departures and latitudes of all lines, and coordinates of all points can be computed. From the resulting coordinates of stations A and E, the length and azimuth of closing line AE can be calculated. Finally the deflection angle α needed to reach E from A can be computed and laid off.

In running open traverses, extreme caution must be exercised in all observations, because there is no check and any errors or mistakes will result in an erroneous length and direction for the closing line. Utmost care must also be exercised in the calculations, although carefully plotting the traverse and scaling the length of the closing line and the deflection angle can secure a rough check on them.

Example 10.11

Compute the length and azimuth of closing line AE and deflection angle α of Figure 10.5, given the following observed data:

Point	Length (ft)	Angle to the Right
A		115°18′25″
	3305.78	
B		161°24′11″
	1862.40	
C		204°50′09″
	1910.22	
D		273°46′37″
	6001.83	
E		

Solution

Table 10.9 presents a tabular solution for computing azimuths, departures and latitudes, and coordinates.

From the coordinates of points A and E, the ΔX and ΔY values of line AE are

$$\Delta X = 7{,}004.05 - 10{,}000.00 = -2{,}995.95 \text{ ft}$$

$$\Delta Y = 17{,}527.05 - 10{,}000.00 = 7{,}527.05 \text{ ft}$$

By Equation (10.12), the length of closing line AE is

$$\text{length}_{AE} = \sqrt{(-2995.95)^2 + (7527.05)^2} = 8101.37 \text{ ft}$$

TABLE 10.9 COMPUTATIONS FOR CLOSING LINE

Point	Azimuth	Departure	Latitude	X (ft)	Y (ft)
U					
	North (assumed)				
A				10,000.00	10,000.00
	295°18′25″	−2988.53	1413.11		
B				7011.47	11,413.11
	276°42′36″	−1849.64	217.61		
C				5161.83	11,630.72
	301°32′45″	−1627.93	999.39		
D				3533.90	12,630.11
	35°19′22″	3470.15	4896.94		
E				7004.05	17,527.05

By Equation (10.11), the azimuth of closing line AE is

$$\tan \text{azimuth}_{AE} = \frac{-2995.95}{7527.05} = -0.39802446; \text{azimuth}_{AE} = 338°17'46''$$

(Note that with a negative ΔX and positive ΔY the bearing of AE is northwest, hence the azimuth is 338°17′46″.)

Finally, deflection angle α is the difference between the azimuths of lines AE and UA, or

$$-\alpha = 338°17'46'' - 360° = -21°42'14'' \text{ (left)}$$

With the emergence of GPS, problems like that illustrated in Example 10.11 will no longer need to be solved using open traverses. Instead, receivers could be set at points U, A, and E of Figure 10.5, and their coordinates determined. From these coordinates the azimuths of lines UA and AE can be calculated, as well as angle α.

10.14 STATE PLANE COORDINATE SYSTEMS

Under ordinary circumstances, rectangular coordinate systems for plane surveys would be limited in size due to earth curvature. However, the National Geodetic Survey (NGS) developed statewide coordinate systems for each state in the United States, which retain an accuracy of 1 part in 10,000 or better while fitting curved geodetic distances to plane grid lengths. Using methods discussed in Section 20.8, observed horizontal distances must be reduced to geodetic distances to achieve this minimal accuracy. However if reduction of observations is properly performed (see Section 20.8), no accuracy is lost due to using the State Plane Coordinate System (SPCS) in the survey.

State plane coordinates are related to latitude and longitude, so control survey stations set by the NGS, as well as those set by others, can all be tied to the systems. As additional stations are set and their coordinates determined, they too become usable reference points in the state plane systems. These monumented control stations serve as starting points for local surveys and permit accurate restoration of obliterated or destroyed marks having known coordinates. If state plane coordinates of two intervisible stations are known, like A and $Az\ Mk$ of Figure 9.1(a), the direction of line A-$Az\ Mk$ can be computed and used to orient the total station instrument at A. In this way, azimuths and bearings of traverse lines are obtained without the necessity of making astronomical observations or resorting to other means.

In the past, some cities and counties have used their own local plane coordinate systems for locating street, sewer, property, and other lines. Because of their limited extent and the resultant discontinuity at city or county lines, such local systems are less desirable than a statewide grid. Another plane coordinate system called the *Universal Transverse Mercator* (UTM) (see Section 20.12) is widely used to pinpoint the locations of objects by coordinates. The military and others use this system for a variety of purposes.

10.15 TRAVERSE COMPUTATIONS USING COMPUTERS

Computers of various types and sizes are now widely used in surveying and are particularly convenient for making traverse computations. Small programmable handheld units, data collectors, and laptop computers are commonly taken into the field and used to verify data for acceptable misclosures before returning to the office. In the office, personal computers are widely used. A variety of software is available for use by surveyors. Some manufacturers supply standard programs, which include traverse computations, with the purchase of their equipment. Various software programs are also available for purchase from a number of suppliers. Spreadsheet software can also be conveniently used with personal computers to calculate and adjust traverses. Of course, surveying and engineering firms also frequently write programs specifically for their own use. Standard programming languages employed include Fortran, Pascal, BASIC, C, and others.

A simple traverse computation program is provided in the software WOLFPACK supplied on the CD that accompanies this book. It computes departures and latitudes, linear misclosure, and relative precision, and performs adjustments by the compass (Bowditch) rule. In addition, the program calculates coordinates of the traverse points and the area within polygon traverses using the coordinate method (discussed in Section 12.5). In Figure 10.6, the input and output files from WOLFPACK are shown for Example 10.4. For the data file of Figure 10.6, the information entered to the right of the numerical data is for explanation only and need not be included in the file. The format of any data file can be found in the help screen for the desired option. A Mathcad® worksheet is included on the CD in this book to demonstrate these computations.

Besides performing routine computations such as traverse solutions, personal computers have many other valuable applications in surveying and engineering offices. Two examples are their use with *computer-aided drafting* (CAD) software for plotting maps and drawing contours (see Section 18.14), and with increasing frequency they are also being employed to operate *geographic information system* (GIS) software (see Chapter 28).

10.16 LOCATING BLUNDERS IN TRAVERSE MEASUREMENTS

A numerical or graphic analysis can often be used to determine the location of a mistake, and thereby save considerable field time in making necessary remeasurements. For example, if the sum of the interior angles of a five-sided traverse gives a large misclosure—say 10′11″— it is likely that one mistake of 10′ and several small errors accumulating to 11″ have been made. Methods of graphically locating the station or line where the mistake occurred are illustrated in Figure 10.7. The procedure is shown for a five-sided traverse, but can be used for traverses having any number of sides.

In Figure 10.7(a), a blunder in the distance BC has occurred. Notice that the mistake CC' shifts the computed coordinates of the remaining stations in such a manner that the azimuth of the linear misclosure line closely matches the azimuth of the course BC that contains the mistake. If no other errors, random or systematic,

DATA FILE

```
Figure 10.1, Example 10.4 //title line
5 1  //number of courses; 1 = angles to the right; -1 = clockwise direction
126 55 17 //azimuth of first course in traverse; degrees minutes seconds
647.25 100 45 37 //first distance and angle at control station
203.03 231 23 43 //distance and angle for second course and station, respectively
720.35 17 12 59 //and so on
610.24 89 03 28
285.13 101 34 24
10000.00 5000.00 //coordinates of first control station
```

OUTPUT FILE

```
~~~~~~~~~~~~~~~~~~~~~~~~~~~ Traverse Computation ~~~~~~~~~~~~~~~~~~~~~~~~~~~
Title: Figure 10.1, Example 10.4 //title line Type: Polygon traverse

          Angle Summary
          Station   Unadj. Angle      Adj. Angle
          --------------------------------------
              1     100°45'37.0"     100°45'34.8"
              2     231°23'43.0"     231°23'40.8"
              3      17°12'59.0"      17°12'56.8"
              4      89° 3'28.0"      89°03'25.8"
              5     101°34'24.0"     101°34'21.8"
          Angular misclosure (sec): 11"

                                                  Unbalanced
Course       Length       Azimuth          Dep          Lat
~~~~~~~~~~~~~~~~~~~~~~~~~~~~~~~~~~~~~~~~~~~~~~~~~~~~~~~~~~~~
 1-2          647.25    126°55'17.0"   517.451    -388.815
 2-3          203.03    178°18'57.8"     5.966    -202.942
 3-4          720.35     15°31'54.6"   192.891     694.044
 4-5          610.24    284°35'20.4"  -590.564     153.709
 5-1          285.13    206°09'42.2"  -125.716    -255.919
         ----------                   ---------  ---------
  Sum =   2,466.00                       0.028       0.077

           Balanced                              Coordinates
     Dep         Lat      Point          X                Y
~~~~~~~~~~~~~~~~~~~~~~~~~~~~~~~~~~~~~~~~~~~~~~~~~~~~~~~~~~~~~~
      517.443    -388.835     1       10,000.00        5,000.00
        5.964    -202.949     2       10,517.44        4,611.16
      192.883     694.022     3       10,523.41        4,408.22
     -590.571     153.690     4       10,716.29        5,102.24
     -125.719    -255.928     5       10,125.72        5,255.93

     Linear misclosure  = 0.082
     Relative Precision = 1 in 30,200

     Area: 272,600 sq. ft.
            6.258 acres {if distance units are feet}

Adjusted Observations
~~~~~~~~~~~~~~~~~~~~~
    Course    Distance    Azimuth       Point       Angle
~~~~~~~~~~~~~~~~~~~~~~~~~~~~~~~~~~~~~~~~~~~~~~~~~~~~~~~~~~~
       1-2       647.26   126°55'24"      1  100°45'42"
       2-3       203.04   178°19'00"      2  231°23'37"
       3-4       720.33    15°31'54"      3   17°12'54"
       4-5       610.24   284°35'14"      4   89°03'20"
       5-6       285.14   206°09'41"      5  101°34'28"
```

Figure 10.6 Data file and output file of traverse computations using WOLFPACK.

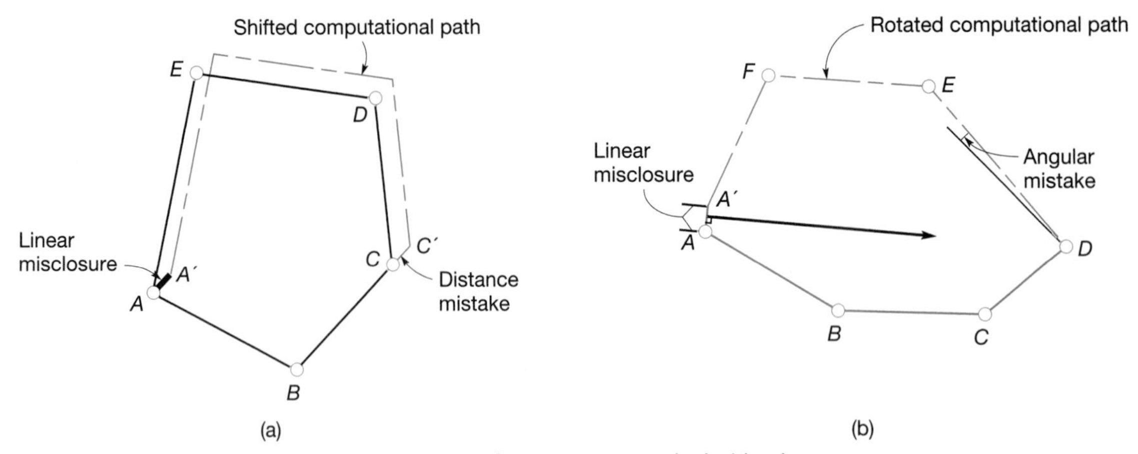

Figure 10.7 Locating a distance (a) or angle (b) blunder.

occurred in the traverse, there would be a perfect match in the directions of the two lines. However, since random errors are inevitable, the direction of the course containing the mistake and that of the linear misclosure line never matches perfectly, but will be close.

As shown in Figure 10.7(b), a mistake in an angle (such as at D) will rotate the computed coordinates of the remaining stations. When this happens, the linear misclosure line AA' is a chord of a circle with radius AD. Thus, the perpendicular bisector of the linear misclosure line will point to the center of the circle, which is the station where the angular mistake occurred. Again if no other errors occurred during the measurement process, this perpendicular bisector would point directly to the station. Since other random errors are inevitable, it will most likely point near the station.

Additional observations and careful field practice will help isolate mistakes. For instance, horizon closures often help isolate and eliminate angular mistakes in the field. A *cutoff line,* such as CE shown dashed in Figure 9.1(a), run between two stations on a traverse, produces smaller closed figures to aid in checking and isolating blunders. Additionally, the extra observations will increase the redundancy in the traverse, and hence the precision of the overall work.

■ 10.17 MISTAKES IN TRAVERSE COMPUTATIONS

Some of the more common mistakes made in traverse computations are:

1. Failing to adjust the angles before computing azimuths or bearings
2. Applying angle adjustments in the wrong direction and failing to check the angle sum for proper geometric total
3. Interchanging departures and latitudes or their signs
4. Confusing the signs of coordinates

PROBLEMS

Asterisks (*) indicate problems that have answers given in Appendix G.

10.1 What are the usual steps followed in adjusting a closed traverse?

10.2* The sum of seven interior angles of a closed-polygon traverse each read to the nearest 3″ is 899°59′39″. What is the misclosure, and what correction would be applied to each angle in balancing them by method 1 of Section 10.2?

10.3 Similar to Problem 10.2, except the angles were read to the nearest 2″, and their sum was 720°00′20″ for a six-sided polygon traverse.

10.4 Similar to Problem 10.2, except the angles were read to the nearest 1″, and their sum for a nine-sided polygon traverse was 1260°00′08″.

10.5* Compute the preliminary azimuths for each course and balance the angles in Problem 9.21.

10.6 Balance the following interior angles (angles-to-the-right) of a five-sided closed polygon traverse using method 1 of Section 10.2. If the azimuth of side AB is fixed at 48°31′43″, calculate the azimuths of the remaining sides. A = 41°09′44″; B = 200°52′14″; C = 124°57′26″; D = 64°28′16″; E = 108°32′10″. (*Note:* Line BC bears NE.)

10.7 Compute departures and latitudes, linear misclosure, and relative precision for the traverse of Problem 10.6 if the lengths of the sides (in feet) are as follows: AB = 150.50; BC = 610.39; CD = 485.14; DE = 735.35; and EA = 647.34. (*Note:* Assume units of feet for all distances.)

10.8 Using the compass (Bowditch) rule, adjust the departures and latitudes of the traverse in Problem 10.7. If the coordinates of station A are X = 20,000.00 ft and Y = 15,000.00 ft, calculate coordinates for the other stations and then the lengths and azimuths of lines AD and EB.

10.9 Balance the following interior angles-to-the-right for a polygon traverse to the nearest 1″ using method 1 of Section 10.2. Compute the azimuths assuming a fixed azimuth of 245°16′24″ for line AB. A = 151°14′41″; B = 81°50′16″; C = 128°26′46″; D = 108°12′04″; E = 70°16′28″. (*Note:* Line BC bears SE.)

10.10 Determine departures and latitudes, linear misclosure, and relative precision for the traverse of Problem 10.9 if lengths of the sides (in meters) are as follows: AB = 1352.562; BC = 1999.670; CD = 1329.127; DE = 2427.328; EA = 2163.325.

10.11 Using the compass (Bowditch) rule adjust the departures and latitudes of the traverse in Problem 10.10. If the coordinates of station A are X = 706,495.671 m and Y = 119,115.199 m, calculate coordinates for the other stations and, from them, the lengths and bearings of lines CA and BD.

10.12 Same as Problem 10.9, except assume line AB has a fixed azimuth of 208°46′08″ and line BC bears *NW*.

10.13 Using the lengths from Problem 10.10 and azimuths from Problem 10.12, calculate departures and latitudes, linear misclosure, and relative precision of the traverse.

10.14 Adjust the departures and latitudes of Problem 10.13 using the compass (Bowditch) rule, and compute coordinates of all stations if the coordinates of station A are X = 137,888.095 m and Y = 22,532.438 m. Compute the length and azimuth of line AC.

10.15 Compute and tabulate for the following closed-polygon traverse: (a) bearings (b) departures and latitudes (c) linear misclosure and (d) relative precision. (*Note:* Line *BC* bears SE.)

Line	Bearing	Length (m)	Interior Angle (Right)
AB	N82°38′16″W	1581.738	$A = 30°38'05''$
BC		442.160	$B = 31°44'44''$
CD		402.494	$C = 139°58'41''$
DE		338.148	$D = 195°33'02''$
EF		307.967	$E = 152°19'20''$
FA		213.048	$F = 169°50'05''$

10.16* In Problem 10.15, if one side and/or angle is responsible for most of the error of closure, which is it likely to be?

10.17 Adjust the traverse of Problem 10.15 using the compass rule. If the coordinates of point *A* are 10,000.000 E and 5000.000 N, determine the coordinates of all other points. Find the length and bearing of line *AE*.

For the closed-polygon traverses given in Problems 10.18 through 10.20 (lengths in feet), compute and tabulate: (a) unbalanced departures and latitudes (b) linear misclosure (c) relative precision and (d) preliminary coordinates if $X_A = 10{,}000.00$ and $Y_A = 5000.00$. Balance the traverses by coordinates using the compass rule.

	Course	AB	BC	CD	DA
10.18	Bearing	N44°59′52″E	N46°00′10″W	S42°59′58″W	S66°45′50″E
	Length	510.42	668.90	755.76	691.78
10.19	Bearing	N31°50′06″E	S58°09′52″E	S76°49′58″W	N58°09′54″W
	Length	292.80	878.40	414.14	585.60
10.20	Azimuth	111°18′10″	25°03′12″	312°43′05″	205°05′10″
	Length	385.90	1016.82	403.50	1164.47

10.21 Compute the linear misclosure, relative precision, and adjusted lengths and azimuths for the sides after the departures and latitudes are balanced by the compass rule in the following closed-polygon traverse.

Line	Length (m)	Departure (m)	Latitude (m)
AB	2119.287	−2014.123	+662.335
BC	4460.292	−1656.598	−4358.126
CA	5209.110	+3670.793	+3695.957

10.22 Balance by the compass rule the departures and latitudes listed in the following closed-polygon traverse. Calculate the linear misclosure, relative precision, and adjusted lengths and bearings.

Line	AB	BC	CD	DA
Distance (ft)	536.10	513.20	392.40	593.80
Departure (ft)	+423.95	−386.31	−356.14	+318.45
Latitude (ft)	+328.14	+337.85	−164.75	−501.19

10.23 The following data apply to a closed link traverse [like that of Figure 9.1(b)]. Compute preliminary azimuths, adjust them, and calculate departures and latitudes, misclosures in departure and latitude, and traverse relative precision. Balance the departures and latitudes using the compass rule, and calculate coordinates of points *B*, *C*, and *D*. Compute the final lengths and azimuths of lines *AB*, *BC*, *CD*, and *DE*.

Station	Measured Angle (to the right)	Adjusted Azimuth	Measured Length (ft)	X (ft)	Y (ft)
$AzMk_1$					
		310°17′20″			
A	272°40′00″			602,846.65	89,017.92
			1432.26		
B	267°27′28″				
			1380.03		
C	87°02′31″				
			1229.94		
D	109°35′39″				
			1315.62		
E	270°29′29″			64,905.59	91,251.50
		57°32′42″			
$AzMk_2$					

10.24 Similar to Problem 10.23, except use the following data:

Station	Measured Angle (to right)	Adjusted Azimuth	Measured Length (m)	X (m)	Y (m)
$AzMk_1$					
		149°09′51″			
A	282°17′18″			449,618.912	240,332.448
			436.474		
B	266°48′13″				
			420.558		
C	89°16′53″				
			374.817		
D	96°36′05″				
			400.927		
E	301°27′38″			448,812.453	240,055.372
		285°36′08″			
$AzMk_2$					

The azimuths (from north) of a polygon traverse are AB = 38°17′02″, BC = 121°26′30″, CD = 224°56′59″, and DA = 308°26′56″. If one observed distance contains a mistake, which course is most likely responsible for the closure conditions given in Problems 10.25 and 10.26? Is the course too long or too short?

10.25* Algebraic sum of departures = −4.24 ft latitudes = −4.28 ft.

10.26 Algebraic sum of departures = +2.044 m latitudes = +2.588 m.

10.27 Determine the lengths and bearings of the sides of a lot whose corners have the following X and Y coordinates (in feet): A (1320.08, 1155.40); B (1384.13, 1155.40); C (1384.13, 1281.29); D (1320.01, 1280.96).

10.28 Compute the lengths and azimuths of the sides of a closed-polygon traverse whose corners have the following X and Y coordinates (in meters): A (1198.300, 1617.109); B (1330.087, 1458.965); C (1640.903, 1718.007); D (1508.146, 1875.635).

10.29 In searching for a record of the length and true bearing of a certain boundary line which is straight between A and B, the following notes of an old random traverse were found (survey by compass and Gunter's chain, declination 4°45′ W). Compute the true bearing and length (in feet) of BA.

LINE	A-1	1-2	2-3	3-*B*
Magnetic bearing	Due north	N20°00′E	Due east	S46°30′E
Distance (ch)	8.33	25.06	16.88	8.90

10.30 Create a computational program that solves Problem 10.17.

10.31 Create a computational program that solves Problem 10.19.

10.32 Create a computational program that solves Problem 10.23.

10.33 Create a computational program that solves Problem 10.28.

BIBLIOGRAPHY

Ghilani, C. D. and P. R. Wolf. 2006. *Adjustment Computations: Spatial Data Analysis.* Hoboken, NJ: John Wiley & Sons, Inc.

11

Coordinate Geometry in Surveying Calculations

11.1 INTRODUCTION

Except for extensive geodetic control surveys, almost all other surveys are referenced to plane rectangular coordinate systems. State plane coordinates (see Chapter 20) are most frequently employed, although local arbitrary systems can be used. Advantages of referencing points in a rectangular coordinate system are (1) the relative positions of points are uniquely defined, (2) they can be conveniently plotted, (3) if lost in the field, they can readily be recovered from other available points referenced to the same system, and (4) computations are greatly facilitated.

Computations involving coordinates are performed in a variety of surveying problems. Two situations were introduced in Chapter 10, where it was shown that the length and direction (azimuth or bearing) of a line can be calculated from the coordinates of its end points. Area computation using coordinates is discussed in Chapter 12. Additional problems that are conveniently solved using coordinates are determining the point of intersection of (a) two lines, (b) a line and a circle, and (c) two circles. The solutions for these and other coordinate geometry problems are discussed in this chapter. It will be shown that the method employed to determine the intersection point of a line and a circle reduces to finding the intersection of a line of known azimuth and another line of known length. Also, the problem of finding the intersection of two circles consists of determining the intersection point of two lines having known lengths. These types of problems are regularly encountered in the horizontal alignment surveys where it is necessary to compute intersections of tangents and circular curves and in boundary and subdivision work where parcels of land are often defined by straight lines and circular arcs.

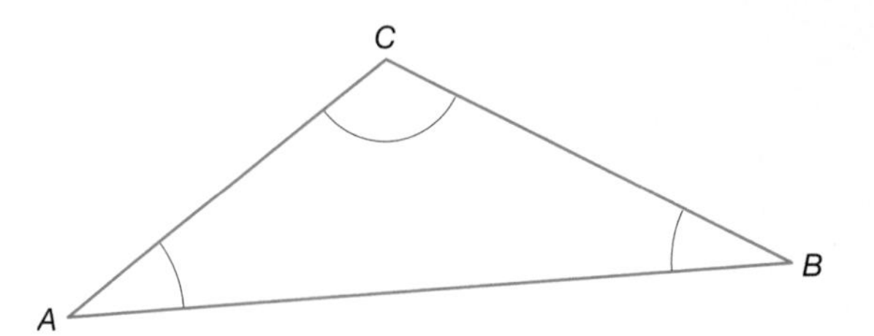

Figure 11.1
An oblique triangle.

The three types of intersection problems noted previously are conveniently solved by forming a triangle between two stations of known position from which the observations are made, and then solving for the parts of this triangle. Two important functions used in solving oblique triangles are (1) the law of sines and (2) the law of cosines. The law of sines relates the lengths of the sides of a triangle to the sines of the opposite angles. For Figure 11.1, this law is

$$\frac{BC}{\sin A} = \frac{AC}{\sin B} = \frac{AB}{\sin C} \tag{11.1}$$

where AB, BC, and AC are the lengths of the three sides of the triangle ABC, and A, B, and C are the angles. The law of cosines relates two sides and the included angle of a triangle to the length of the side opposite the angle. In Figure 11.1, the following three equations can be written that express the law of cosines:

$$\begin{aligned} BC^2 &= AC^2 + AB^2 - 2(AC)(AB)\cos A \\ AC^2 &= BA^2 + BC^2 - 2(BA)(BC)\cos B \\ AB^2 &= CB^2 + CA^2 - 2(CB)(CA)\cos C \end{aligned} \tag{11.2}$$

In some coordinate geometry solutions, the use of the quadratic formula is used. Examples where this equation is required are discussed in Sections 24.16.1 and 25.10. This formula, which gives the solution for x in any quadratic equation of form $ax^2 + bx + c = 0$, is

$$x = \frac{-b \pm \sqrt{b^2 - 4ac}}{2a} \tag{11.3}$$

In the remaining sections of this chapter, procedures using triangles and Equations (11.1) through (11.3) are presented for solving each type of standard coordinate geometry problem.

■ 11.2 COORDINATE FORMS OF EQUATIONS FOR LINES AND CIRCLES

In Figure 11.2, straight line AB is referenced in a plane rectangular coordinate system. Coordinates of end points A and B are X_A, Y_A, X_B, and Y_B. Length AB and azimuth Az_{AB} of this line in terms of these coordinates are

$$AB = \sqrt{(X_B - X_A)^2 + (Y_B - Y_A)^2} \tag{11.4}$$

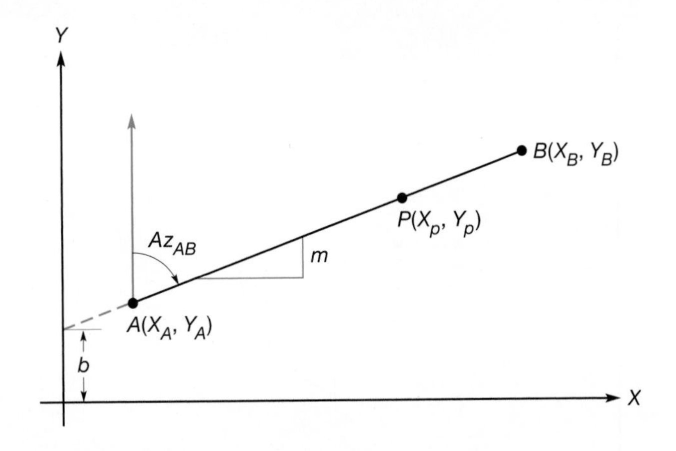

Figure 11.2 Geometry of a straight line in a plane coordinate system.

$$Az_{AB} = \tan^{-1}\left(\frac{X_B - X_A}{Y_B - Y_A}\right) + C \tag{11.5}$$

where C is 0° if both $(X_B - X_A)$ and $(Y_B - Y_A)$ are greater than zero; C is 180° if $(Y_B - Y_A)$ is less than zero, and C is 360° if $(X_B - X_A)$ is less than zero and $(Y_B - Y_A)$ is greater than zero. The general mathematical expression for a straight line is

$$Y_P = mX_P + b \tag{11.6}$$

where Y_P is the Y coordinate of any point P on the line whose X coordinate is X_P, m the slope of the line, and b the Y intercept of the line. Slope m can be expressed as

$$m = \frac{Y_B - Y_A}{X_B - X_A} = \cot(Az_{AB}) \tag{11.7}$$

From Equations (11.5) and (11.7), it can be shown that

$$Az_{AB} = \tan^{-1}\left(\frac{1}{m}\right) + C \tag{11.8}$$

The general mathematical expression for a circle in rectangular coordinates can be written as

$$R^2 = (X_P - X_O)^2 + (Y_P - Y_O)^2 \tag{11.9}$$

In Equation (11.9) and with reference to Figure 11.3, R is the radius of the circle, X_O and Y_O are the coordinates of the radius point O, and X_P and Y_P the coordinates of any point P on the circle. Another form of the circle equation is

$$X_P^2 + Y_P^2 - 2X_OX_P - 2Y_OY_P + f = 0 \tag{11.10}$$

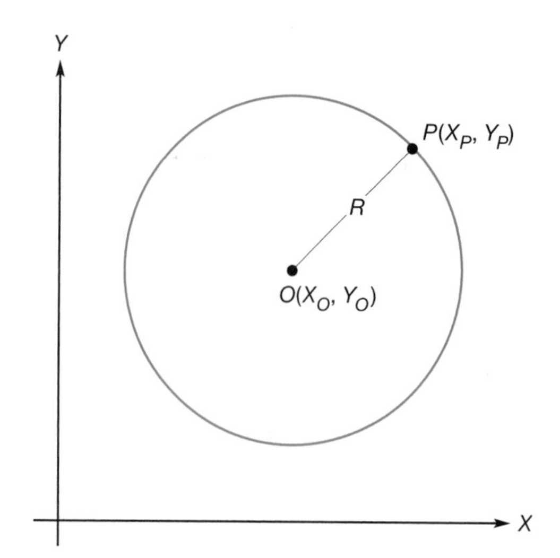

Figure 11.3 Geometry of a circle in a plane coordinate system.

where the length of the radius of the circle is given as $R = \sqrt{X_O^2 + Y_O^2 - f}$. (*Note:* Although Equations (11.9) and (11.10) are not used in solving problems in this chapter, they are applied in later chapters.)

11.3 PERPENDICULAR DISTANCE FROM A POINT TO A LINE

A common problem encountered in boundary surveying is determining the perpendicular distance of a point from a line. This procedure can be used to check the alignment of survey markers on a block and is also useful in subdivision design. Assume in Figure 11.4 that points A and B are on the line defined by two block corners whose coordinates are known. Also assume that the coordinates of point P are known. The slope, m, and y-intercept, b, of line AB are computed from the coordinates of the block corners. By assigning the X and Y coordinate axes as

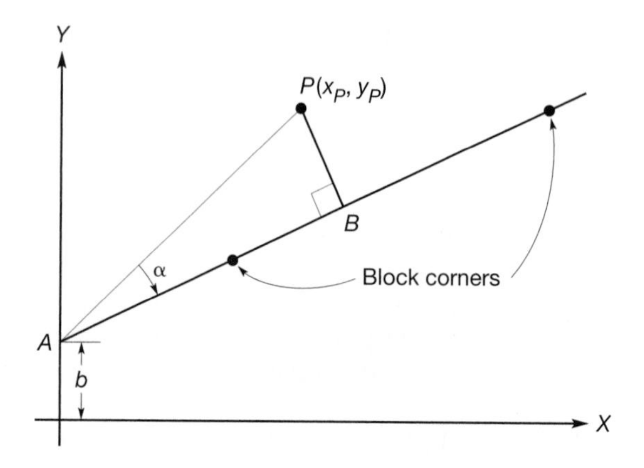

Figure 11.4 Perpendicular distance of a point from a line.

shown in the figure, the coordinates of point A are $X_A = 0$, and $Y_A = b$. Using Equations (11.4) and (11.5), the length and azimuth of line AP can be determined from its coordinates. By Equation (11.8) the azimuth of line AB can be determined from the slope of the line AB. Now angle α can be computed as the difference in the azimuth AP and AB, which for the situation depicted in Figure 11.4 is

$$\alpha = Az_{AB} - Az_{AP} \tag{11.11}$$

Recognizing that ABP is a right-triangle, length BP is

$$BP = AP \sin \alpha \tag{11.12}$$

where the length of AP is determined from the coordinates of points A and P using Equation (11.4).

Example 11.1

For Figure 11.4, assume that the coordinates (X, Y) of point P are (1123.82, 509.41) and that the coordinates of the block corners are (865.49, 416.73) and (1557.41, 669.09). What is the perpendicular distance of point P from line AB? (All units are in feet.)

Solution

By Equation (11.7), and using the block corner coordinates, the slope of line AB is

$$m = \frac{669.09 - 416.73}{1557.41 - 865.49} = 0.364724245$$

Rearranging Equation (11.6), the y-intercept of line AB is

$$b = 416.73 - 0.364724245 \times 865.49 = 101.065 \text{ ft}$$

By Equations (11.4) and (11.5), the length and azimuth of line AP is

$$AP = \sqrt{(1123.82 - 0)^2 + (509.41 - 101.065)^2} = 1195.708 \text{ ft}$$

$$Az_{AP} = \tan^{-1}\left(\frac{1123.82 - 0}{509.41 - 101.07}\right) + 0° = 70°01'52.2''$$

By Equation (11.8), the azimuth of line AB is

$$Az_{AB} = \tan^{-1}\left(\frac{1}{0.364724245}\right) + 0° = 69°57'42.7''$$

Using Equation (11.11), angle α is

$$\alpha = 70°01'52.2'' - 69°57'42.7'' = 0°04'09.5''$$

From Equation (11.12), the perpendicular distance from point P to line AB is

$$BP = 1195.708 \sin(0°04'09.5'') = 1.45 \text{ ft}$$

11.4 INTERSECTION OF TWO LINES, BOTH HAVING KNOWN DIRECTIONS

Figure 11.5 illustrates the intersection of two lines AP and BP. Each has known coordinates for one end point and each has a known direction. Determining the point of intersection for this type of situation is often called the *direction-direction* problem. A simple method of computing the intersection point P is to solve for the parts of oblique triangle ABP. Since the coordinates of A and B are known, the length and azimuth of AB (shown dashed) can be determined using Equations (11.4) and (11.5), respectively. Then, from the figure it can be seen that angle A is the difference in the azimuths of AB and AP, or

$$A = Az_{AP} - Az_{AB} \tag{11.13}$$

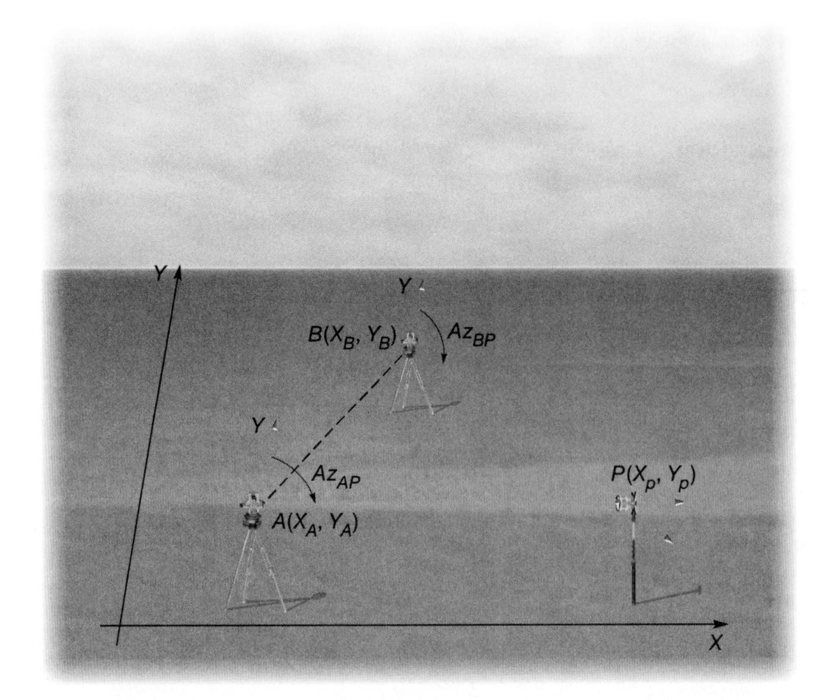

Figure 11.5 Intersection of two lines with known directions.

Similarly, angle B is the difference in the azimuths of BA and BP, or

$$B = Az_{BA} - Az_{BP} \tag{11.14}$$

With two angles of the triangle ABP computed, the remaining angle P is

$$P = 180° - A - B \tag{11.15}$$

Substituting into Equation (11.1) and rearranging, the length of side AP is

$$AP = AB\frac{\sin(B)}{\sin(P)} \tag{11.16}$$

With both the length and azimuth of AP known, the coordinates of P are

$$\begin{aligned} X_P &= X_A + AP \sin Az_{AP} \\ Y_P &= Y_A + AP \cos Az_{AP} \end{aligned} \tag{11.17}$$

A check on this solution can be obtained by solving for length BP and using it together with the azimuth of BP to compute the coordinates of P. The two solutions should agree, except for round off.

It should be noted that if the azimuths for lines AP and BP are equal, then the lines are parallel and have no intersection.

Example 11.2

In Figure 11.5, assuming the following information is known for two lines, compute coordinates X_P and Y_P of the intersection point. (Coordinates are in feet.)

$$\begin{array}{lll} X_A = 1425.07 & X_B = 7484.80 & Az_{AP} = 76°04'24'' \\ Y_A = 1971.28 & Y_B = 5209.64 & Az_{BP} = 141°30'16'' \end{array}$$

Solution

By Equations (11.4) and (11.5), the length and azimuth of side AB are

$$AB = \sqrt{(7484.80 - 1425.07)^2 + (5209.64 - 1971.28)^2} = 6870.757 \text{ ft}$$

$$Az_{AB} = \tan^{-1}\left(\frac{7484.80 - 1425.07}{5209.64 - 1971.28}\right) + 0° = 61°52'46.8''$$

By Equations (11.13) through (11.15), the three angles of triangle ABP are

$$A = 76°04'24'' - 61°52'46.8'' = 14°11'37.2''$$

$$B = (180° + 61°52'46.8'') - 141°30'16'' = 100°22'30.8''$$

$$P = 180° - 14°11'37.2'' - 100°22'30.8'' = 65°25'52.0''$$

By Equation (11.16), length AP is

$$AP = 6870.757\frac{\sin 100°22'30.8''}{\sin 65°25'52.0''} = 7431.224 \text{ ft}$$

By Equations (11.17), the coordinates of station P are

$$X_P = 1425.07 + 7431.224 \sin 76°04'24'' = 8637.85 \text{ ft}$$
$$Y_P = 1971.28 + 7431.224 \cos 76°04'24'' = 3759.83 \text{ ft}$$

Check:

$$BP = 6870.757\left[\frac{\sin 14°11'37.2''}{\sin 65°25'52''}\right] = 1852.426 \text{ ft}$$

$$X_P = 7484.80 + (1852.426) \sin 141°30'16'' = 8637.85 \, (Check!)$$
$$Y_P = 5209.64 + (1852.426) \cos 141°30'16'' = 3759.83 \, (Check!)$$

■ 11.5 INTERSECTION OF A LINE WITH A CIRCLE

Figure 11.6 illustrates the intersection of a line (AC) of known azimuth with a circle of known radius ($BP_1 = BP_2$). Finding the intersection for this situation reduces to finding the intersection of a line of known direction with another line of known length and is sometimes referred to as the *direction-distance* problem. As shown in the figure, notice that this problem has two different solutions, but as discussed later, the incorrect one can generally be detected and discarded.

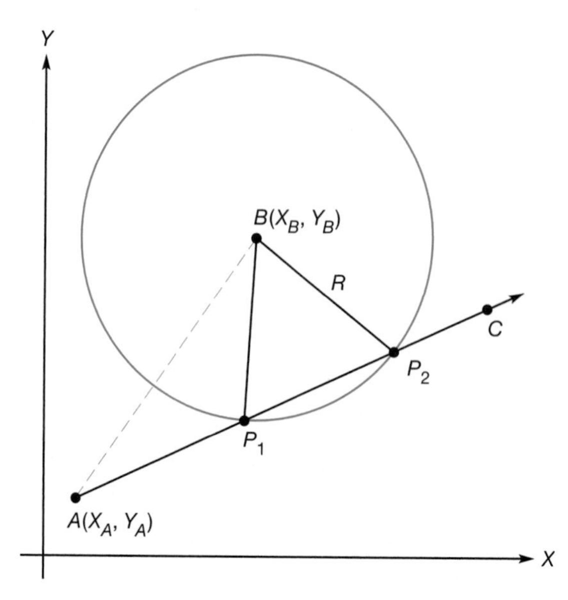

Figure 11.6
Intersection of a line and a circle.

The approach to solving this problem is similar to that employed in Section 11.4; that is, the answer is determined by solving an oblique triangle. This particular solution will demonstrate the use of the quadratic equation to obtain both solutions. In Figure 11.6, the coordinates of B (the radius point of the circle) are known. From the coordinates of points A and B, the length and azimuth of line AB (shown dashed) are determined by employing Equations (11.4) and (11.5), respectively. Then angle A is computed from the azimuths of AB and AC as follows:

$$A = Az_{AP} - Az_{AB} \tag{11.18}$$

Substituting the known values of A, AB, and BP into the law of cosines [Equation (11.2)] yields

$$BP^2 = AB^2 + AP^2 - 2(AB)(AP)\cos A \tag{11.19}$$

In Equation (11.19), AP is an unknown quantity. Rearranging this equation gives

$$AP^2 - 2(AB)(\cos A)\,AP + (AB^2 - BP^2) = 0 \tag{11.20}$$

Now Equation (11.20), which is a second degree expression, can be solved using the quadratic formula [Equation (11.3)], as follows:

$$AP = \frac{2(AB)\cos(A) \pm \sqrt{[2(AB)\cos A]^2 - 4(AB^2 - BP^2)}}{2} \tag{11.21}$$

In comparing Equation (11.21) to Equation (11.3), it can be seen that $a = 1$, $b = 2(AB)\cos A$ and $c = (AB^2 - BP^2)$. Because of the $\pm$ sign in the formula, there are two solutions for length AP. Once these two lengths are determined, the possible coordinates of station P are

$$\begin{aligned} X_{P1} &= X_A + AP_1 \sin(Az_{AP}) \\ Y_{P1} &= Y_A + AP_1 \cos(Az_{AP}) \\ X_{P2} &= X_A + AP_2 \sin(Az_{AP}) \\ Y_{P2} &= Y_A + AP_2 \cos(Az_{AP}) \end{aligned} \tag{11.22}$$

If errors exist in the given data for the problem, or if an impossible design is attempted, the circle will not intersect the line. In this case, the terms under the radical in Equation (11.21) will be negative, i.e., $[2(AB)\cos A]^2 - 4(AB^2 - BP^2) < 0$. It is therefore important when solving any of the coordinate geometry problems to be alert for these types of potential problems.

The law of sines can also be used to solve this problem. However, care must be exercised when using the sine law since the two solutions will not be readily apparent. The procedure of solving this problem using the sine law is as follows:

1. Compute the length and azimuth of line AB from the coordinates using Equations (11.4) and (11.5), respectively.

2. Compute the angle at A using Equation (11.18).
3. Using the law of sines solve for the angles at P_1 as

$$\sin P = \frac{AB \sin A}{BP} \tag{11.23}$$

4. Note that the sine function has the relationship $\sin(x) = \sin(180° - x)$. Thus the solution for the angle at B as

$$\begin{aligned} B_1 &= 180° - (A + P) \\ B_2 &= P - A \end{aligned} \tag{11.24}$$

5. Using the two solutions for angle B, determine the azimuth of line BP as

$$\begin{aligned} Az_{BP1} &= Az_{BA} - B_1 \\ Az_{BP2} &= Az_{BA} - B_2 \end{aligned} \tag{11.25}$$

6. Finally using the two azimuths and the observed length of BP determine the two possible solutions for station P as

$$\begin{aligned} X_{P1} &= X_B + BP \sin(Az_{BP1}) \\ Y_{P1} &= Y_B + BP \sin(Az_{BP1}) \\ X_{P2} &= X_B + BP \sin(Az_{BP2}) \\ Y_{P2} &= Y_B + BP \sin(Az_{BP2}) \end{aligned} \tag{11.26}$$

Example 11.3

In Figure 11.6, assume the coordinates of point A are $X = 100.00$, and $Y = 130.00$, and that the coordinates of point B are $X = 500.00$, and $Y = 600.00$. If the azimuth of AP is 70°42′36″, and the radius of the circle (length BP) is 350.00, what are the possible coordinates of point P?
(*Note:* Linear units are feet.)

Solution

By Equations (11.4) and (11.5), the length and azimuth of AB are

$$AB = \sqrt{(500 - 100)^2 + (600 - 130)^2} = 617.171 \text{ ft}$$

$$Az_{AB} = \tan^{-1}\left(\frac{500 - 100}{600 - 130}\right) + 0° = 40°23'59.7''$$

By Equation (11.18), the angle at A is

$$A = 70°42'36'' - 40°23'59.7'' = 30°18'36.3''$$

Substituting appropriate values according to Equation (11.20), the quadratic equation coefficients are

$$a = 1$$

$$b = -2 \times 617.171 \times \cos 30°18'36.3'' = -1065.616$$

$$c = 617.171^2 - 350.00^2 = 258{,}400.043$$

Substituting these values into Equation (11.21) yields

$$AP = \frac{1065.616 \pm \sqrt{1065.616^2 - 4(258{,}400.043)}}{2}$$

$$= \frac{1065.616 \pm 319.276}{2}$$

$$= 373.170 \text{ or } 692.446$$

Using the azimuth and distances for AP, the two possible solutions for the coordinates of P are

$$X_{P1} = 100.00 + 373.170 \sin 70°42'36'' = 452.22 \text{ ft}$$

$$Y_{P1} = 130.00 + 373.170 \cos 70°42'36'' = 253.28 \text{ ft}$$

or

$$X_{P2} = 100.00 + 692.446 \sin 70°42'36'' = 753.57 \text{ ft}$$

$$Y_{P2} = 130.00 + 692.446 \cos 70°42'36'' = 358.75 \text{ ft}$$

In solving a quadratic equation, the decision to add or subtract the value from the radical can be made on the basis of experience, or by using a carefully constructed scaled diagram, which also provides a check on the computations. One answer will be unreasonable and should be discarded. An arithmetic check is possible by solving for the two possible angles at B to P in triangle ABP and determining the coordinates of P from station B or by solving the problem using the second procedure. Readers should verify that the same solution can be obtained using Equations (11.23) through (11.26).

■ 11.6 INTERSECTION OF TWO CIRCLES

In Figure 11.7, the intersection of two circles is illustrated. Note that the circles are obtained by simply radiating two distances (their radius values R_A and R_B) about their radius points A and B. As shown, this geometry again results in two intersection points, P_1 and P_2. As with the two previous cases, these intersection points can again be located by solving for the parts of oblique triangle ABP. In this situation, two sides of the triangle are the known radii, and thus the problem is

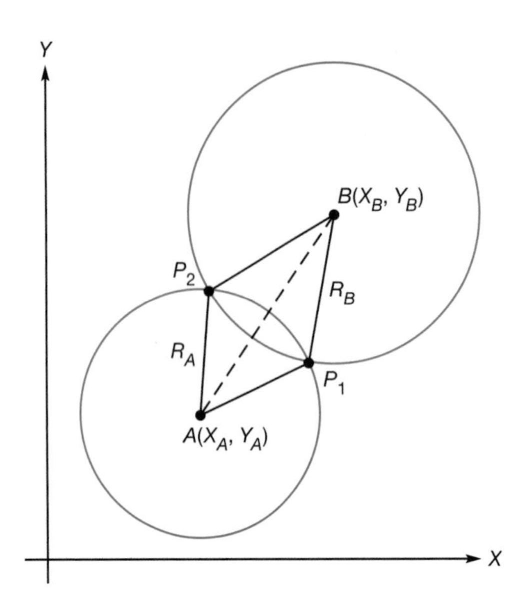

Figure 11.7
Intersection of two circles.

often called the *distance-distance* problem. The third side of the triangle, AB, can be computed from known coordinates of A and B or the distance can be observed.

The first step in solving this problem is to compute the length and azimuth of line AB using Equations (11.4) and (11.5). Then angle A can be determined using the law of cosines, (Equation 11.2). As shown in Figure 11.7, the two solutions for P at either P_1 or P_2 are derived by either adding or subtracting angle A from the azimuth of line AB to obtain the direction of AP. By rearranging Equation 11.2, angle A is

$$A = \cos^{-1}\left[\frac{(AB)^2 + (AP)^2 - (BP)^2}{2(AB)(AP)}\right] \tag{11.27}$$

Thus, the azimuth of line AP is either

$$\begin{aligned} Az_{AP1} &= Az_{AB} + A \\ Az_{AP2} &= Az_{AB} - A \end{aligned} \tag{11.28}$$

The possible coordinates of P are

$$\begin{aligned} X_{P1} &= X_A + AP_1 \sin(Az_{AP1}) \\ Y_{P1} &= Y_A + AP_1 \cos(Az_{AP1}) \end{aligned}$$

or

$$\begin{aligned} X_{P2} &= X_A + AP_2 \sin(Az_{AP2}) \\ Y_{P2} &= Y_A + AP_2 \cos(Az_{AP2}) \end{aligned} \tag{11.29}$$

The decision of whether to add or subtract angle A from the azimuth of line AB can be made on the basis of experience or through the use of a carefully constructed scaled diagram. One answer will be unreasonable and should be discarded. As can be seen from Figure 11.7, there will be no solution if length of AB is greater than the sum of R_A and R_B.

Example 11.4

In Figure 11.7, assume the following data (in meters) are available:

$$X_A = 2851.28 \quad Y_A = 299.40 \quad R_A = 2000.00$$
$$X_B = 3898.72 \quad Y_B = 2870.15 \quad R_B = 1500.00$$

Compute the X and Y coordinates of point P.

Solution

By Equations (11.4) and (11.5), the length and azimuth of AB are

$$AB = \sqrt{(3898.72 - 2851.28)^2 + (2870.15 - 299.40)^2} = 2775.948 \text{ m}$$

$$Az_{AB} = \tan^{-1}\left(\frac{3898.72 - 2851.28}{2870.15 - 299.40}\right) + 0° = 22°10'05.6''$$

By Equation (11.27), A is

$$A = \cos^{-1}\left(\frac{2775.948^2 + 2000.00^2 - 1500.00^2}{2 \times 2775.948 \times 2000.00}\right) = 31°36'53.6''$$

By combining Equations (11.28) and (11.29), the possible solutions for P are

$$X_{P_1} = 2851.28 + 2000.00 \sin(22°10'05.6'' + 31°36'53.6'') = 4464.85 \text{ m}$$
$$Y_{P_1} = 299.40 + 2000.00 \cos(22°10'05.6'' + 31°36'53.6'') = 1481.09 \text{ m}$$

or

$$X_{P_2} = 2851.28 + 2000.00 \sin(22°10'05.6'' - 31°36'53.6'') = 2523.02 \text{ m}$$
$$Y_{P_2} = 299.40 + 2000.00 \cos(22°10'05.6'' - 31°36'53.6'') = 2272.28 \text{ m}$$

An arithmetic check on this solution can be obtained by determining the angle and coordinates of P from station B.

11.7 THREE-POINT RESECTION

This procedure locates a point of unknown position by observing horizontal angles from that point to three visible stations whose positions are known. The situation is illustrated in Figure 11.8, where a total station instrument occupies station P

Figure 11.8 The resection problem.

and angles x and y are observed. A summary of the method used to compute the coordinates of station P follows (refer to Figure 11.8):

1. From the known coordinates of A, B, and C calculate lengths a and c, and angle α at station B.
2. Subtract the sum of angles x, y, and α in figure $ABCP$ from 360° to obtain the sum of angles $A + C$

$$A + C = 360° - (\alpha + x + y) \tag{11.30}$$

3. Calculate angles A and C using the following:

$$A = \tan^{-1}\left(\frac{a \sin x \sin(A + C)}{c \sin y + a \sin x \cos(A + C)}\right) \tag{11.31}$$

$$C = \tan^{-1}\left(\frac{c \sin y \sin(A + C)}{a \sin x + c \sin y \cos(A + C)}\right) \tag{11.32}$$

4. From angle A and azimuth AB, calculate azimuth AP in triangle ABP. Then solve for length AP using the law of sines, where $\alpha_1 = 180° - A - x$. Calculate the departure and latitude of AP followed by the coordinates of P.
5. In the manner outlined in step (4), use triangle BCP to calculate the coordinates of P to obtain a check.

Example 11.5

In Figure 11.8, angles x and y were observed as 48°53′12″ and 41°20′35″, respectively. Control points A, B, and C have coordinates (in feet) of $X_A = 5721.25$, $Y_A = 21{,}802.48$, $X_B = 13{,}542.99$, $Y_B = 22{,}497.95$, $X_C = 20{,}350.09$, and $Y_C = 24{,}861.22$. Calculate the coordinates of P.

Solution

1. By Equation (11.4),

$$a = \sqrt{(20{,}350.09 - 13{,}542.99)^2 + (24{,}861.22 - 22{,}497.95)^2} = 7205.67 \text{ ft}$$

$$c = \sqrt{(13{,}542.99 - 5721.25)^2 + (22{,}497.95 - 21{,}802.48)^2} = 7852.60 \text{ ft}$$

2. By Equation (11.5),

$$Az_{AB} = \tan^{-1}\left(\frac{13{,}542.99 - 5721.25}{22{,}497.95 - 21{,}802.48}\right) + 0° = 84°55'08.1''$$

$$Az_{BC} = \tan^{-1}\left(\frac{20{,}350.09 - 13{,}542.99}{24{,}861.22 - 22{,}497.95}\right) + 0° = 70°51'15.0''$$

3. Calculate angle α,

$$\alpha = 180° - (70°51'15.0'' - 88°55'08.1'') = 194°03'53.1''$$

4. By Equation (11.30),

$$A + C = 360° - 194°03'53.1'' - 48°53'12'' - 41°20'35''$$
$$= 75°42'19.9''$$

5. By Equation (11.31),

$$A = \tan^{-1}\left(\frac{7250.67 \sin 48°53'12'' \sin 75°42'19.9''}{7852.60 \sin 41°20'35'' + 7205.67 \sin 48°53'12'' \cos 75°42'19.9''}\right)$$

$$= 38°51'58.7''$$

6. By Equation (11.32),

$$C = \tan^{-1}\left(\frac{7852.60 \sin 41°20'35'' \sin 75°42'19.9''}{7205.67 \sin 48°53'12'' + 7852.60 \sin 41°20'35'' \cos 75°42'19.9''}\right)$$

$$= 36°50'21.2''$$

$$(A + C = 38°51'58.7'' + 36°50'21.2'' = 75°42'19.9'' \text{ check!})$$

7. Calculate angle α_1,

$$\alpha_1 = 180° - 38°51'58.7'' - 48°53'12'' = 92°14'49.3''$$

8. By the law of sines,

$$AP = \frac{\sin 92°14'49.3'' \, (7852.60)}{\sin 48°53'12''} = 10{,}414.72 \text{ ft}$$

$$AZ_{AP} = AZ_{AB} + A = 84°55'08.1'' + 38°51'58.7'' = 123°47'06.8''$$

9. By Equations (10.1) and (10.2),

$$\text{Dep}_{AP} = 10{,}414.72 \sin 123°47'06.8'' = 8655.97 \text{ ft}$$

$$\text{Lat}_{AP} = 10{,}414.72 \cos 123°47'06.8'' = -5791.43 \text{ ft}$$

10. By Equation (10.7),

$$X_P = 5721.25 + 8655.97 = 14{,}377.22 \text{ ft}$$

$$Y_P = 21{,}802.48 - 5791.43 = 16{,}011.05 \text{ ft}$$

11. As a check, triangle *BCP* was solved to obtain the same results.

The three-point resection problem just described provides a unique solution for the unknown coordinates of point *P*; i.e., there are no redundant observations, and thus no check can be made on the observations. This is actually a special case of the more general resection problem which provides redundancy and enables a least-squares solution. In the general resection problem, in addition to observing the angles *x* and *y*, distances from *P* to one or more control stations could also have been observed. Other possible variations in resection that provide redundancy include observing (a) one angle and two distances to two control stations; (b) two angles and one, two, or three distances to three control points; or (c) the use of more than three control stations. Then all observations can be included in a least-squares solution to obtain the most probable coordinates of point *P*. As discussed in Section 23.9, resection has become a popular method for quickly orienting total station instruments. The procedure is convenient because these instruments can readily observe both angles and distances and their on-board microprocessors can instantaneously provide the least-squares solution for the instrument's position.

It should be noted that the resection problem will not have a unique solution if points *A*, *B*, *C*, and *P* define a circle. Selecting points *B* and *P* so that they both lie on the same side of a line connecting points *A* and *C* avoids this problem. Additionally, the accuracy of the solution will decrease if the observed angles *x* and *y* become small. As a general guideline, the observed angles should be greater than 30° for best results.

11.8 TWO-DIMENSIONAL CONFORMAL COORDINATE TRANSFORMATION

It is sometimes necessary to convert coordinates of points from one survey coordinate system to another. This happens, for example, if a survey is performed in some local-assumed or arbitrary coordinate system and later it is desired to convert it to state plane coordinates. The process of making these conversions is called *coordinate transformation*. If only planimetric coordinates (i.e., *X*s and *Y*s)

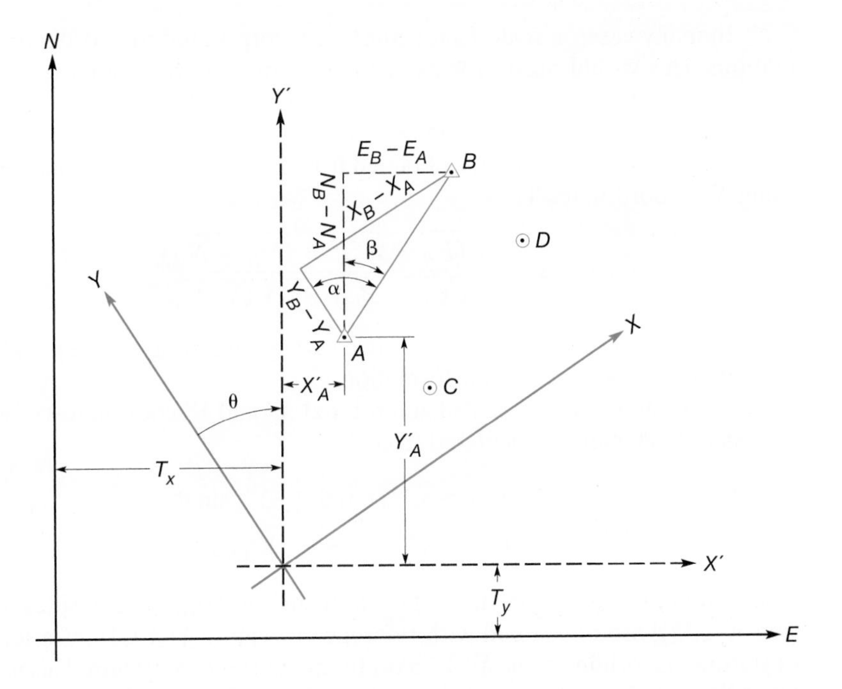

Figure 11.9 Geometry of the two-dimensional coordinate transformation.

are involved, and true shape is retained, it is called *two-dimensional conformal coordinate transformation.*

The geometry of a two-dimensional conformal coordinate transformation is illustrated in Figure 11.9. In the figure, X-Y represents a local-assumed coordinate system and E-N a state plane coordinate system. Coordinates of points A through D are known in the X-Y system and those of A and B are also known in the E-N system. Points such as A and B, whose positions are known in both systems, are termed *control points.* At least two control points are required in order to determine E-N coordinates of other points such as C and D.

In general, three steps are involved in coordinate transformation: (1) rotation, (2) scaling, and (3) translation. As shown in Figure 11.9, rotation consists in determining coordinates of points in the rotated X'-Y' axis system (shown dashed). The X'-Y' axes are parallel with E-N but the origin of this system coincides with the origin of X-Y. In the figure, the rotation angle θ, between the X-Y and X'-Y' axis systems, is

$$\theta = \alpha - \beta \tag{11.33}$$

In Equation (11.33), azimuths, α and β, are calculated from the two sets of coordinates of control points A and B using Equation (11.5) as follows:

$$\alpha = \tan^{-1}\left(\frac{X_B - X_A}{Y_B - Y_A}\right) + C$$

$$\beta = \tan^{-1}\left(\frac{E_B - E_A}{N_B - N_A}\right) + C$$

where as explained in Section 11.2, C places the azimuth in the proper quadrant.

In many cases, a scale factor must be incorporated in coordinate transformations. This would occur, for example, in transforming from a local arbitrary coordinate system into a state plane coordinate grid. The scale factor relating any two coordinate systems can be computed according to the ratio of the length of a line between two control points obtained from *E-N* coordinates to that determined using *X-Y* coordinates. Thus

$$s = \frac{\sqrt{(E_B - E_A)^2 + (N_B - N_A)^2}}{\sqrt{(X_B - X_A)^2 + (Y_B - Y_A)^2}} \tag{11.34}$$

(*Note:* If the scale factor is unity, the two surveys are of equal scale and it can be ignored in the coordinate transformation.)

With θ and s known, scaled and rotated X' and Y' coordinates of any point, for example, A, can be calculated from

$$\begin{aligned} X'_A &= sX_A \cos\theta - sY_A \sin\theta \\ Y'_A &= sX_A \sin\theta + sY_A \cos\theta \end{aligned} \tag{11.35}$$

Individual parts of the rotation formulas [right-hand sides of Equations (11.35)] are developed with reference to Figure 11.10. Translation consists of shifting the origin of the X'-Y' axes to that in the *E-N* system. This is achieved by adding translation factors T_X and T_Y (see Figure 11.9) to X' and Y' coordinates to obtain E and N coordinates. Thus for point A,

$$\begin{aligned} E_A &= X'_A + T_X \\ N_A &= Y'_A + T_Y \end{aligned} \tag{11.36}$$

Rearranging Equations (11.36) and using coordinates of one of the control points (such as A), numerical values for T_X and T_Y can be obtained as

$$\begin{aligned} T_X &= E_A - X'_A \\ T_Y &= N_A - Y'_A \end{aligned} \tag{11.37}$$

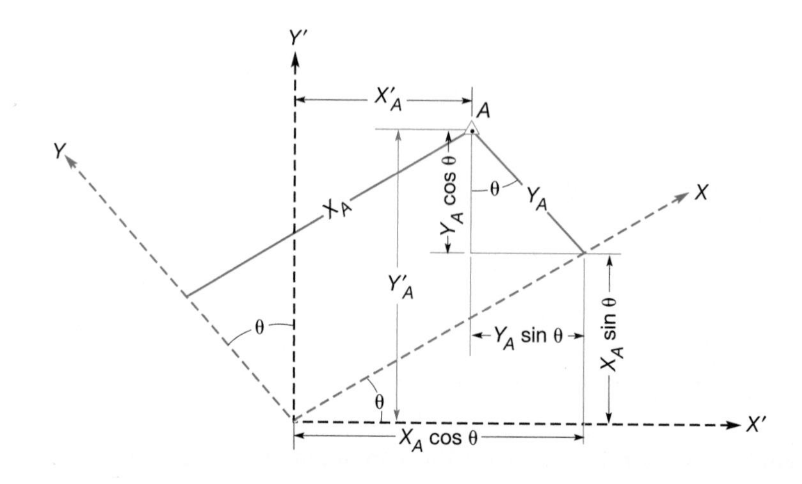

Figure 11.10 Detail of rotation formulas in two-dimensional conformal coordinate transformation.

The other control point (i.e., point B) should also be used in Equations (11.37) to calculate T_X and T_Y and thus obtain a computational check.

Substituting Equations (11.35) into Equations (11.36) and dropping subscripts, the following equations are obtained for calculating E and N coordinates of noncontrol points (such as C and D) from their X and Y values:

$$E = s(X \cos\theta - Y \sin\theta) + T_X$$
$$N = s(X \sin\theta + Y \cos\theta) + T_Y \qquad \textbf{(11.38)}$$

In summary, the procedure for performing two-dimensional conformal coordinate transformations consists of (1) calculating rotation angle θ using two control points and Equations (11.5) and (11.33); (2) solving Equations (11.34), (11.35), and (11.37) using control points to obtain scale factor s, and translation factors T_X and T_Y; and (3) applying θ, s, and T_X and T_Y in Equations (11.38) to transform all noncontrol points. If more than two control points are available, an improved solution can be obtained using least squares. Coordinate transformation calculations require a significant amount of time if done by hand, but are easily performed when programmed for computer solution.

Example 11.6

In Figure 11.9, the following *E-N* and *X-Y* coordinates are known for points *A* through *D*. Compute *E* and *N* coordinates for points *C* and *D*.

	State Plane Coordinates (ft)		Arbitrary Coordinates (ft)	
Point	**E**	**N**	**X**	**Y**
A	194,683.50	99,760.22	2848.28	2319.94
B	196,412.80	102,367.61	5720.05	3561.68
C			3541.72	897.03
D			6160.31	1941.26

Solution

1. Determine α, β, and θ from Equations (11.5) and (11.33),

$$\alpha = \tan^{-1}\left(\frac{5720.05 - 2848.28}{3561.68 - 2319.94}\right) + 0^\circ = 66^\circ 36' 59.7''$$

$$\beta = \tan^{-1}\left(\frac{196{,}412.80 - 194{,}683.50}{102{,}367.61 - 99{,}760.22}\right) + 0^\circ = 33^\circ 33' 12.7''$$

$$\theta = 66^\circ 36' 59.7'' - 33^\circ 33' 12.7'' = 33^\circ 03' 47''$$

2. Compute the scale factor from Equation (11.34),

$$s = \frac{\sqrt{(196{,}412.80 - 194{,}683.50)^2 + (102{,}367.61 - 99{,}860.22)^2}}{\sqrt{(5720.05 - 2848.28)^2 + (3561.68 - 2319.94)^2}}$$
$$= \frac{3128.73}{3128.73}$$
$$= 1.00000$$

(Since the scale factor is 1, it can be ignored.)

3. Determine T_X and T_Y from Equations (11.35) through (11.37) using point A,

$$X'_A = 2848.28 \cos 33°03'47'' - 2319.94 \sin 33°03'47'' = 1121.39 \text{ ft}$$
$$Y'_A = 2848.28 \sin 33°03'47'' + 2319.94 \cos 33°03'47'' = 3498.18 \text{ ft}$$

$$T_X = 194{,}683.50 - 1121.39 = 193{,}562.11 \text{ ft}$$
$$T_Y = 99{,}760.22 - 3498.18 = 96{,}262.04 \text{ ft}$$

4. Check T_X and T_Y using point B,

$$X'_B = 5720.05 \cos 33°03'47'' - 3561.68 \sin 33°03'47'' = 2850.69 \text{ ft}$$
$$Y'_B = 5720.05 \sin 33°03'47'' + 3561.68 \cos 33°03'47'' = 6105.58 \text{ ft}$$

$$T_X = 196{,}412.80 - 2850.69 = 193{,}562.11 \text{ ft (Check!)}$$
$$T_Y = 102{,}367.61 - 6105.58 = 96{,}262.03 \text{ ft (Check!)}$$

5. Solve Equations (11.38) for E and N coordinates of points C and D,

$$E_C = 3541.72 \cos 33°03'47'' - 897.03 \sin 33°03'47'' + 193{,}562.11$$
$$= 196{,}040.93 \text{ ft}$$

$$N_C = 3541.72 \sin 33°03'47'' + 897.03 \cos 33°03'47'' + 96{,}262.04$$
$$= 98{,}946.04 \text{ ft}$$

$$E_D = 6160.31 \cos 33°03'47'' - 1941.26 \sin 33°03'47'' + 193{,}562.11$$
$$= 197{,}665.81 \text{ ft}$$

$$N_D = 6160.31 \sin 33°03'47'' + 1941.26 \cos 33°03'47'' + 96{,}262.04$$
$$= 101{,}249.78 \text{ ft}$$

With some simple modifications, Equations (11.38) can be rewritten in matrix form as

$$sR\begin{bmatrix} X \\ Y \end{bmatrix} + \begin{bmatrix} T_X \\ T_Y \end{bmatrix} = \begin{bmatrix} E \\ N \end{bmatrix} + \begin{bmatrix} v_E \\ v_N \end{bmatrix} \qquad \textbf{(11.39)}$$

where the rotation matrix, R, is

$$R = \begin{bmatrix} \cos\theta & -\sin\theta \\ \sin\theta & \cos\theta \end{bmatrix} \tag{11.40}$$

Also v_E and v_N are residual errors which must be included if more than two control points are available. Scaling the rotation matrix by s, and substituting a for $(s\cos\theta)$, b for $(s\sin\theta)$, c for T_X and d for T_Y, Equation (11.39) can be rewritten as

$$\begin{bmatrix} a & -b \\ b & a \end{bmatrix}\begin{bmatrix} X \\ Y \end{bmatrix} + \begin{bmatrix} c \\ d \end{bmatrix} = \begin{bmatrix} E \\ N \end{bmatrix} + \begin{bmatrix} v_E \\ v_N \end{bmatrix} \tag{11.41}$$

With Equation (11.41), a least-squares adjustment (see Chapter 16) can be performed when more than two points are common in both coordinate systems. The program WOLFPACK that is supplied with this book has this software option under the Coordinate Computations submenu. It will determine the unknown parameters for the two-dimensional conformal coordinate transformation, and transform any additional points. The data file and the results of the adjustment for Example 11.6 are shown in Figure 11.11.

```
DATA FILE
Example 11.6                                  {title line}
2                                             {number of control points}
A 194683.50 99760.22 2848.28 2319.94          {Point ID, SPCS E and N, arbitrary X and Y}
B 196412.80 102367.61 5720.05 3561.68
C 3561.68 897.03                              {Point ID, arbitrary system X and Y}
D 6160.31 1941.26

RESULTS OF ADJUSTMENT

Two Dimensional Conformal Coordinate Transformation of File: Example 11.6
-------------------------------------------------------------------------

ax - by + Tx = X + VX
bx + ay + Ty = Y + VY

Transformed Control Points
   POINT        X           Y          VX         VY
--------------------------------------------------------
     A  194,683.50  99,760.22    -0.000     -0.000
     B  196,412.80  102,367.61    0.000      0.000

Transformation Parameters:
 a =  0.83807009
 b =  0.54556070
 Tx =  193562.110
 Ty =  96262.038

Rotation = 33°03'46.9"
  Scale = 1.00000

*******  Unique Solution Obtained !!  *******

   POINT       x           y           X           Y
---------------------------------------------------------
     C   3,541.72      897.03  196,040.94   98,946.04
     D   6,160.31    1,941.26  197,665.81  101,249.77
```

Figure 11.11 Data file and results of adjustment for Example 11.6 using WOLFPACK.

Note that the transformed X and Y coordinates of points C and D obtained using the computer program agree (except for round off) with those computed in Example 11.6. Note also that in this solution with two control points, there are no redundancies and thus the residuals VX and VY are zeros.

11.9 INACCESSIBLE POINT PROBLEM

It is sometimes necessary to determine the elevation of a point that is inaccessible. This task can be accomplished by establishing a baseline such that the inaccessible point is visible from both ends. As an example, assume that the elevation of the chimney shown in Figure 11.12 is desired. Baseline AB is established, its length measured and the elevations of its end points determined. Horizontal angles A and B, and altitude angles, v_1, and v_2 are observed as shown in the figure. Points I_A and I_B are vertically beneath P. Using the observed values, the law of sines is applied to compute horizontal lengths AI_A and BI_B of triangle ABI as

$$AI_A = \frac{AB\sin(B)}{\sin[180° - (A + B)]} = \frac{AB\sin(B)}{\sin(A + B)} \tag{11.42}$$

$$BI_B = \frac{AB\sin(A)}{\sin(A + B)} \tag{11.43}$$

Length IP can be derived from either triangle AI_AP or BI_BP as

$$I_AP = AI_A\tan(v_1) \tag{11.44}$$

$$I_BP = BI_B\tan(v_2) \tag{11.45}$$

The elevation of point P is computed as the average of the heights from the two triangles, which may differ because of random errors in the observation of v_1 and v_2, as

$$Elev_P = \frac{I_AP + Elev_A + hi_A + I_BP + Elev_B + hi_B}{2} \tag{11.46}$$

Figure 11.12 Geometry of the inaccessible point problem.

In Equation (11.46), hi_A and hi_B are the instrument heights at A and B, respectively.

Example 11.7

Stations A and B have elevations of 298.65 ft and 301.53 ft, respectively, and the instrument heights at A and B are $hi_A = 5.55$ ft, and $hi_B = 5.48$ ft. The other field observations are

$$AB = 136.45 \text{ ft}$$

$$A = 44°12'34'' \quad B = 39°26'56''$$

$$v_1 = 8°12'47'' \quad v_2 = 5°50'10''$$

What is the elevation of the chimney stack?

Solution

By Equations (11.42) and (11.43), the lengths of AI_A and BI_B are

$$AI_A = \frac{136.45 \sin 39°26'56''}{\sin(44°12'34'' + 39°26'56'')} = 87.233 \text{ ft}$$

$$BI_B = \frac{136.45 \sin 44°12'34''}{\sin(44°12'34'' + 39°26'56'')} = 95.730 \text{ ft}$$

From Equation (11.44), length I_AP is

$$I_AP = 87.233 \tan 8°12'47'' = 12.591 \text{ ft}$$

And from Equation (11.45), length I_BP is

$$I_BP = 95.730 \tan 5°50'10'' = 9.785 \text{ ft}$$

Finally, by Equation (11.46), the elevation of point P is

$$Elev_P = \frac{12.591 + 298.65 + 5.55 + 9.785 + 301.53 + 5.48}{2} = 316.79 \text{ ft}$$

11.10 THREE-DIMENSIONAL TWO-POINT RESECTION

The three-dimensional coordinates X_P, Y_P, and Z_P of a point such as P of Figure 11.13 can be determined based upon angle and distance observations made from that point to two other stations of known positions. This procedure is convenient for establishing coordinates of occupied stations on elevated structures, or in depressed areas such as in mines. In Figure 11.13 for example, assume

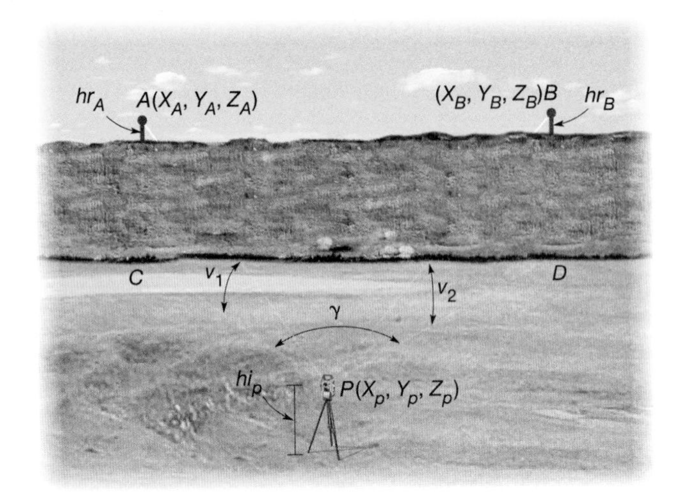

Figure 11.13 Geometry of the three-dimensional two-point resection problem.

that a total station instrument is placed at point P, whose X_P, Y_P, and Z_P coordinates are unknown, and that control points A and B are visible from P. Slope lengths PA and PB are observed along with horizontal angle γ and vertical angles v_1 and v_2. The computational process for determining X_P, Y_P, and Z_P is as follows.

1. Determine the length and azimuth of AB using Equations (11.4) and (11.5).
2. Compute horizontal distances PC and PD as

$$\begin{aligned} PC &= PA\cos(v_1) \\ PD &= PB\cos(v_2) \end{aligned} \tag{11.47}$$

   where C and D are vertically beneath A and B, respectively.
3. Using Equation (11.3), calculate horizontal angle DCP as

$$DCP = \cos^{-1}\left(\frac{AB^2 + PC^2 - PD^2}{2 \times AB \times PC}\right) \tag{11.48}$$

4. Determine the azimuth of line AP as

$$Az_{AP} = Az_{AB} + DCP \tag{11.49}$$

5. Compute the planimetric (X-Y) coordinates of point P as

$$\begin{aligned} X_P &= X_A + PC\sin Az_{AP} \\ Y_P &= Y_A + PC\cos Az_{AP} \end{aligned} \tag{11.50}$$

6. Determine elevation differences AC and BD as

$$\begin{aligned} AC &= PA\sin(v_1) \\ BD &= PB\sin(v_2) \end{aligned} \tag{11.51}$$

7. And finally calculate the elevation of P as

$$
\begin{aligned}
Elev_{P1} &= Elev_A + hr_A - AC - hi_P \\
Elev_{P2} &= Elev_B + hr_B - BD - hi_P \\
Elev_P &= \frac{Elev_{P1} + Elev_{P2}}{2}
\end{aligned}
\tag{11.52}
$$

In Equations (11.52), hi_P is the height of instrument above point P, and hr_A and hr_B are the reflector heights above stations A and B, respectively.

Example 11.8

For Figure 11.13, the X, Y, and Z coordinates (in meters) of station A are 7034.982, 5413.896, and 432.173, respectively, and those of B are 7843.745, 5807.242, and 428.795, respectively. Determine the three-dimensional position of a total station instrument at point P based upon the following observations.

$v_1 = 24°33'42''$ $PA = 667.413$ m $hr_A = 1.743$ m $\gamma = 77°48'08''$

$v_2 = 26°35'08''$ $PB = 612.354$ m $hr_B = 1.743$ m $hi_P = 1.685$ m

Solution

1. Using Equations (11.4) and (11.5), determine the length and azimuth of line AB.

$$AB = \sqrt{(7843.745 - 7034.982)^2 + (5807.242 - 5413.896)^2} = 899.3435 \text{ m}$$

$$Az_{AB} = \tan^{-1}\left(\frac{7843.745 - 7034.982}{5807.242 - 5413.896}\right) + 0° = 64°03'49.6''$$

2. By Equations (11.47), determine lengths PC and PD.

$$PC = 667.413 \cos(24°33'42'') = 607.0217 \text{ m}$$

$$PD = 612.354 \cos(26°35'08'') = 547.6080 \text{ m}$$

3. From Equation (11.48), compute angle DCP.

$$DCP = \cos^{-1}\left(\frac{899.3435^2 + 607.0217^2 - 547.6080^2}{2 \times 899.3435 \times 607.0217}\right) = 36°31'24.2''$$

Note that this computed angle can be checked by using the law of sines, Equation (11.1), as

$$DCP = \sin^{-1}\left(\frac{547.6080 \sin 77°48'08''}{899.3435}\right) = 36°31'24.2'' \text{ (Check!)}$$

4. Using Equation (11.49), find the azimuth of line AP.

$$Az_{AP} = 64°03'49.6'' + 36°31'24.2'' = 100°35'13.8''$$

5. From Equations (11.50), compute the X-Y coordinates of point P.

$$X_P = 7034.982 + 607.0217 \sin 100°35'13.8'' = 7631.670 \text{ m}$$

$$Y_P = 5413.896 + 607.0217 \cos 100°35'13.8'' = 5302.367 \text{ m}$$

6. By Equations (11.51), compute the vertical distances of AC and BD.

$$AC = 667.413 \sin 24°33'42'' = 277.425 \text{ m}$$

$$BD = 612.354 \sin 26°35'08'' = 274.049 \text{ m}$$

7. And finally, using Equations (11.52), compute and average the elevation of point P.

$$Elev_P = 432.173 + 1.743 - 277.425 - 1.685 = 154.806 \text{ m}$$

$$Elev_P = 428.795 + 1.743 - 274.049 - 1.685 = 154.804 \text{ m}$$

$$\text{Average Elevation} = 154.805 \text{ m}$$

11.11 CONCLUSIONS

Coordinate geometry provides a convenient approach to solving problems in almost all types of modern surveys. Many problems that otherwise appear difficult can be greatly simplified and readily solved by working with coordinates. Although the calculations are sometimes rather lengthy, this has become inconsequential with the advent of computers and data collectors. Many software packages are available for performing coordinate geometry calculations. However, people involved in surveying (geomatics) must understand the basis for the computations, and they must exercise all possible checks to verify the accuracy of their results. Both WOLFPACK and a Mathcad® worksheet that are included on the CD in this book demonstrate these computations.

Because of the nature of trigonometric functions, computations in some coordinate geometry problems will become numerically unstable when the angles involved approach 0° or 90°. Thus if coordinate geometry is intended to be used to determine the locations of points, it is generally prudent to design the survey so that triangles used in the solution have angles between 30° and 60°. Also, it is important to observe good surveying practices in the field, such as taking the averages of equal numbers of direct and reversed angle observations and exercising other checks and precautions.

As will be seen later, coordinate geometry plays an important role in computing highway alignments, in subdivision designs, and in the operation of geographic information systems.

PROBLEMS

Asterisks (*) indicate problems that have answers given in Appendix G.

11.1 The X and Y coordinates (in meters) of station Shore are 867.568 and 366.208, respectively, and those for station Rock are 653.283 and 142.831, respectively. What are the azimuth, bearing, and length of the line connecting station Shore to station Rock?

11.2 Same as Problem 11.1, except that the X and Y coordinates (in feet) of Shore are 4563.95 and 786.49, respectively, and those for Rock are 2831.97 and 2537.89, respectively.

11.3* What are the slope, and y-intercept for the line in Problem 11.1?

11.4 What are the slope, and the y-intercept for the line in Problem 11.2?

11.5* If the slope (XY plane) of a line is 0.63489, what is the azimuth of the line to the nearest second of arc? (XY plane)

11.6 If the slope (XY plane) of a line is −0.52849, what is the azimuth of the line to the nearest second of arc? (XY plane)

11.7* What is the perpendicular distance of a point from the line in Problem 11.1, if the X and Y coordinates (in meters) of the point are 801.507 and 298.362, respectively?

11.8 What is the perpendicular distance of a point from the line in Problem 11.2, if the X and Y coordinates (in feet) of the point are 3700.67 and 1670.31, respectively?

11.9* A line with an azimuth of 105°46′33″ from a station with X and Y coordinates of 5885.31 and 5164.15, respectively, is intersected with a line that has an azimuth of 200°31′24″ from a station with X and Y coordinates of 7337.08 and 5949.99, respectively. (All coordinates are in feet.) What are the coordinates of the intersection point?

11.10 A line with an azimuth of 306°28′17″ from a station with X and Y coordinates of 2443.94 and 3563.84, respectively, is intersected with a line that has an azimuth of 261°09′44″ from a station with X and Y coordinates of 2635.10 and 3705.15, respectively. (All coordinates are in feet.) What are the coordinates of the intersection point?

11.11 Same as Problem 11.9 except that the bearing of the first line is S 82°02′34″ E and the bearing of the second line is S 38°12′11″ W.

11.12 In the accompanying figure, the X and Y coordinates (in meters) of station A are 2084.274 and 5579.124, respectively, and those of station B are 2891.675 and 3859.513, respectively. Angle BAP was measured as 283°15′47″ and angle ABP was measured as 44°21′58″. What are the coordinates of station P?

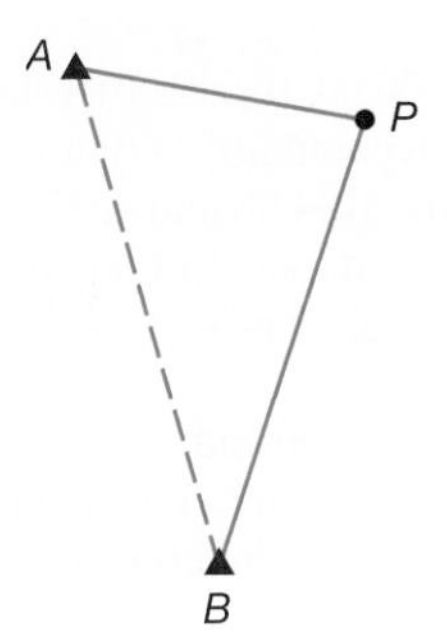

Problems 11.12 through 11.16 Field conditions for intersections.

11.13* In the figure, the X and Y coordinates (in feet) of station A are 6248.16 and 8133.35, respectively, and those of station B are 6509.15 and 6101.89, respectively. The length of BP is 2874.06 ft, and the azimuth of line AP is 95°15′40″. What are the coordinates of station P?

11.14 In the figure, the X and Y coordinates (in feet) of station A are 5393.15 and 9517.03, respectively, and those of station B are 4310.97 and 7280.60, respectively. The length of AP is 1987.54 ft, and angle ABP is 40°12′34″. What are the possible coordinates for station P?

11.15* A circle of radius 803.42 ft, centered at point A, intersects another circle of radius 1304.58 ft, centered at point B. The X and Y coordinates (in feet) of A are 3548.53 and 2836.49, respectively, and those of B are 4184.62 and 1753.52, respectively. What are the coordinates of station P in the figure?

11.16 The same as Problem 11.15, except the radii from A and B are 953.45 ft and 1098.45 ft, respectively, and the X and Y coordinates (in feet) of A are 2058.74 and 4311.23, respectively, and those of station B are 2581.52 and 2344.21, respectively.

11.17 For the subdivision in the accompanying figure, assume that lines AC, DF, GI, and JL are parallel, but that lines BK and CL are parallel to each other, but not parallel to AJ. If the X and Y coordinates (in feet) of station A are (5000.00, 5000.00), what are the coordinates of each lot corner shown?

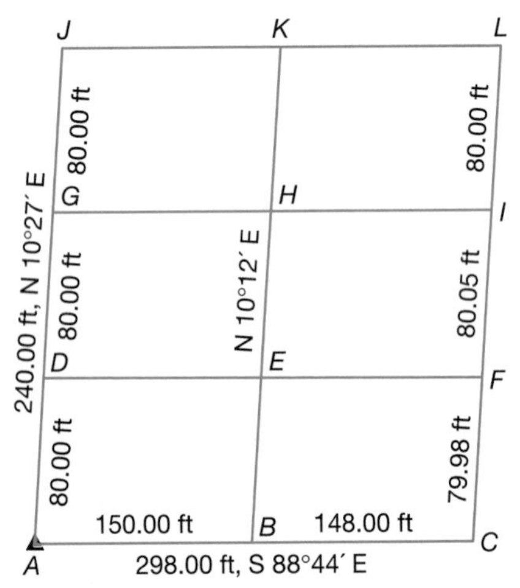

Problem 11.17 Subdivision.

11.18 If the X and Y coordinates (in feet) of station A are (10,000.00, 10,000.00), what are the coordinates of the remaining labeled corners in the accompanying figure?

11.19* In Figure 11.8, the X and Y coordinates (in feet) of A are 5134.98 and 4315.48, respectively, those of B are 5761.26 and 4140.57, respectively, and those of C are 6590.77 and 4328.31, respectively. Also angle x is 35°49′41″ and angle y is 38°45′47″. What are the coordinates of station P?

11.20 In Figure 11.8, the X and Y coordinates (in feet) of A are 5722.96 and 8643.06, those of B are 6348.05 and 7588.65, and those of C are 7335.54 and 7809.21, respectively. Also angle x is 46°51′38″ and angle y is 48°33′47″. What are the coordinates of station P?

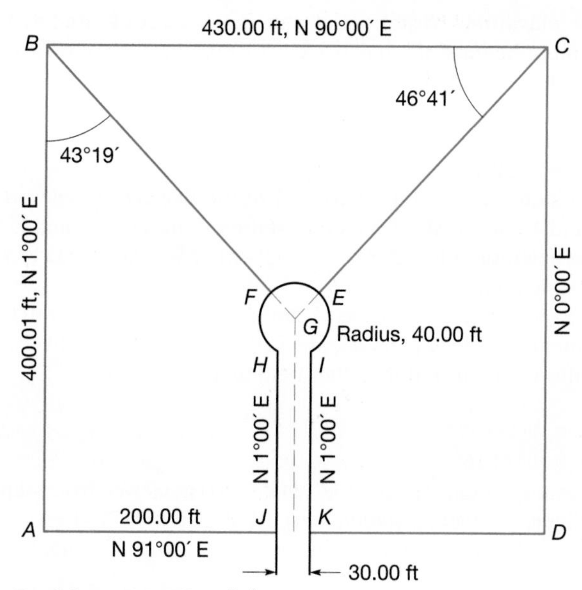

Problem 11.18 Subdivision.

11.21 In Figure 11.9, the following *EN* and *XY* coordinates for points *A* through *D* are given. In a 2-D conformal coordinate transformation, to convert the *XY* coordinates into the *EN* system, what are the

(a)* Scale factor?
(b) Rotation angle?
(c) Translations in *X* and *Y*?
(d) Coordinates of points *C* and *D* in the *EN* coordinate system?

	State Plane Coordinates (m)		Arbitrary Coordinates (ft)	
Point	***E***	***N***	***X***	***Y***
A	718,042.829	109,293.468	4873.67	6609.04
B	718,399.341	109,644.860	6402.92	7207.45
C			7041.22	6037.23
D			5405.58	5571.81

11.22 Do Problem 11.21 with the following coordinates.

	State Plane Coordinates (m)		Arbitrary Coordinates (m)	
Point	***E***	***N***	***X***	***Y***
A	677,505.266	119,651.804	6182.848	6323.893
B	678,181.136	119,752.112	5430.607	3816.422
C			3957.467	5101.501
D			7718.674	4913.441

11.23 In Figure 11.12, the elevations of stations A and B are 403.16 ft, and 410.02 ft, respectively. Instrument heights hi_A and hi_B are 5.20 ft, and 5.06 ft, respectively. What is the average elevation of point P if the other field observations are:
$AB = 282.64$ ft
$A = 56°06'34''$ $B = 45°12'07''$
$v_1 = 8°12'47''$ $v_2 = 5°50'10''$

11.24 In Problem 11.23, assume station P is to the left of the line AB, as viewed from station A. If the X and Y coordinates (in feet) of station A are 1139.92 and 1256.93, respectively, and the azimuth of line AB is 9°02′52″, what are the X and Y coordinates of the inaccessible point?

11.25 In Figure 11.12, the elevations of stations A and B are 803.96 ft, and 819.93 ft, respectively. Instrument heights hi_A and hi_B are 5.20 and 5.26 ft, respectively. What is the average elevation of point P if the other field observations are:
$AB = 368.97$ ft
$A = 43°25'17''$ $B = 48°09'57''$
$v_1 = 12°05'53''$ $v_2 = 9°37'16''$

11.26 In Problem 11.25, assume station P is to the left of line AB as viewed from station A. If the X and Y coordinates (in feet) of station A are 10438.29 and 8453.94, respectively, and the azimuth of line AB is 206°37′45″, what are the X and Y coordinates of the inaccessible point?

11.27 In Figure 11.13, the X, Y, and Z coordinates (in feet) of station A are 3108.98, 2450.05, and 1318.58, respectively, and those of B are 3624.36, 2438.96, and 1298.56, respectively. Determine the three-dimensional position of the occupied station P with the following observations:

$v_1 = 32°08'32''$ $PA = 508.96$ ft $hr_A = 6.89$ ft $\gamma = 77°43'35''$
$v_2 = 32°48'15''$ $PB = 462.79$ ft $hr_B = 6.89$ ft $hi_P = 5.35$ ft

11.28 Adapt Equations (11.43) and (11.47) so they are applicable for zenith angles.

11.29 In Figure 11.13, the X, Y, and Z coordinates (in meters) of station A are 2956.372, 3224.554, and 165.086, respectively, and those of B are 2854.615, 3451.225, and 162.129, respectively. Determine the three-dimensional position of occupied station P with the following observations:

$z_1 = 110°03'58''$ $PA = 237.894$ m $hr_A = 1.158$ m $\gamma = 70°17'13''$
$z_2 = 115°10'59''$ $PB = 198.568$ m $hr_B = 1.203$ m $hi_P = 1.520$ m

11.30 Use WOLFPACK to do Problem 11.9.
11.31 Use WOLFPACK to do Problem 11.10.
11.32 Use WOLFPACK to do Problem 11.12.
11.33 Use WOLFPACK to do Problem 11.13.
11.34 Use WOLFPACK to do Problem 11.15.
11.35 Use WOLFPACK to do Problem 11.16.
11.36 Use WOLFPACK to do Problem 11.17.
11.37 Write a computational program that solves Example 11.6 using matrices.
11.38 Write a computational program that solves Example 11.8.
11.39 Write a computational program that solves Example 11.7.

BIBLIOGRAPHY

Cederholm, Peter. 2004. "Precision of Points Computed from the Intersections of Lines or Planes." *Surveying and Land Information Science* 64 (No. 3): 163.

Ghilani, C. D., and P. R. Wolf. 2006. *Adjustment Computations: Spatial Data Analysis.* New York: John Wiley & Sons, Inc.

Tan, Willie. 2004. "The 3-point Resection Problem in Photogrammetry." *Surveying and Land Information Science* 64 (No. 3): 177.

12

Area

■ 12.1 INTRODUCTION

There are a number of important reasons for determining areas. One is to include the acreage of a parcel of land in the deed describing the property. Other purposes are to determine the acreage of fields, lakes, etc., or the number of square yards to be surfaced, paved, seeded, or sodded. Another important application is determining end areas for earthwork volume calculations (see Chapter 26).

In plane surveying, area is considered to be the orthogonal projection of the surface onto a horizontal plane. As noted in Chapter 2, in the English system the most commonly used units for specifying small areas are the ft^2 and yd^2, and for large tracts the acre is most often used, where 1 acre = 43,560 ft^2 = 10 ch^3 (Gunter's). An acre lot, if square, would thus be 208.71^+ ft on a side. In the metric system, smaller areas are usually given in m^2 and for larger tracts *hectares* are commonly used, where 1 hectare is equivalent to a square having sides of 100 m, and thus equals 10,000 m^2. In converting areas between the English and metric systems, the conversion factors given in Table 12.1 are useful.

■ 12.2 METHODS OF MEASURING AREA

Both field and map measurements are used to determine area. Field measurement methods are the more accurate and include (1) division of the tract into simple figures (triangles, rectangles, and trapezoids), (2) offsets from a straight line, (3) coordinates, and (4) double-meridian distances. Each of these methods is described in sections that follow.

Methods of determining area from map measurements include (1) counting coordinate squares, (2) dividing the area into triangles, rectangles, or other regular geometric shapes, (3) digitizing coordinates, and (4) running a planimeter over

TABLE 12.1 APPROXIMATE AREA CONVERSION FACTORS

To Convert from	To	Multiply by
ft^2	m^2	$(12/39.37)^2 \approx 0.09291$
m^2	ft^2	$(39.37/12)^2 \approx 10.76364$
yd^2	m^2	$(36/39.37)^2 \approx 0.83615$
m^2	yd^2	$(39.37/36)^2 \approx 1.19596$
acres	hectares	$[39.37/(4.356 \times 12)]^2 \approx 2.47099$
hectares	acres	$(4.356 \times 12/39.37)^2 \approx 0.40470$

the enclosing lines. These processes are described and illustrated in Section 12.9. Because maps themselves are derived from field observations, methods of area determination invariably depend on this basic source of data.

12.3 AREA BY DIVISION INTO SIMPLE FIGURES

A tract can usually be divided into simple geometric figures such as triangles, rectangles, or trapezoids. The sides and angles of these figures can be observed in the field and their individual areas calculated and totaled. An example of a parcel subdivided into triangles is shown in Figure 12.1.

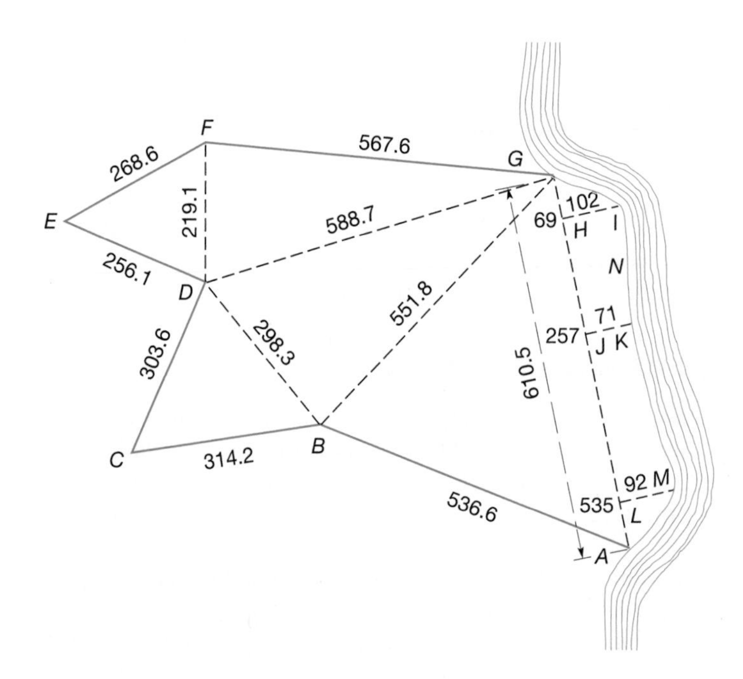

Figure 12.1 Area determination by triangles.

Formulas for computing areas of rectangles and trapezoids are well known. The area of a triangle whose lengths of sides are known can be computed by the formula

$$\text{area} = \sqrt{s(s - a)(s - b)(s - c)} \tag{12.1}$$

where a, b, and c are the lengths of sides of the triangle and $s = 1/2(a + b + c)$. Another formula for the area of a triangle is

$$\text{area} = \frac{1}{2}ab \sin C \tag{12.2}$$

where C is the angle included between sides a and b.

The choice of whether to use Equation (12.1) or (12.2) will depend on the triangle parts that are most conveniently determined, a decision ordinarily dictated by the nature of the area and the type of equipment available.

12.4 AREA BY OFFSETS FROM STRAIGHT LINES

Irregular tracts can be reduced to a series of trapezoids by observing right-angle offsets from points along a reference line. The reference line is usually marked by stationing (see Sec. 5.9.1) and positions where offsets are observed are given by their stations and pluses. The spacing between offsets may either be *regular* or *irregular* depending on conditions. These two cases are discussed in the subsections that follow.

12.4.1 Regularly Spaced Offsets

Offsets at regularly spaced intervals are shown in Figure 12.2. For this case, the area is found by the formula

$$\text{area} = b\left(\frac{h_0}{2} + h_1 + h_2 + \cdots + \frac{h_n}{2}\right) \tag{12.3}$$

where b is the length of a common interval between offsets, and $h_0, h_1, \ldots, h_n$ are the offsets. The regular interval for the example of Figure 12.2 is a half-station, or 50 ft.

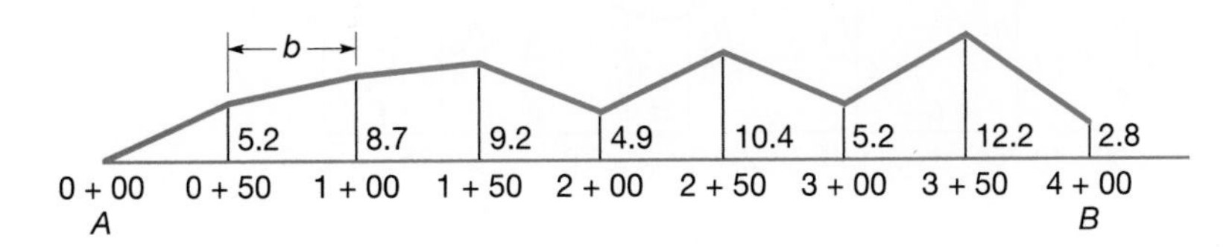

Figure 12.2 Area by offsets.

■ Example 12.1

Compute the area of the tract shown in Figure 12.2.

Solution

By Equation (12.3),

$$\begin{aligned} \text{area} &= 50\left(0 + 5.2 + 8.7 + 9.2 + 4.9 + 10.4 + 5.2 + 12.2 + \frac{2.8}{2}\right) \\ &= 2860 \text{ ft}^2 \end{aligned}$$

In this example, a summation of offsets (terms within parentheses) can be secured by the *paper-strip method,* in which the area is plotted to scale and the mid-ordinate of each trapezoid is successively added by placing tick marks on a long strip of paper. The area is then obtained by making a single measurement between the first and last tick marks, multiplying by the scale to convert it to a field distance, and then multiplying by width *b*.

12.4.2 Irregularly Spaced Offsets

For irregularly curved boundaries like that in Figure 12.3, the spacing of offsets along the reference line varies. Spacings should be selected so that the curved boundary is accurately defined when adjacent offset points on it are connected by straight lines. A formula for calculating area for this case is

$$\text{area} = \frac{1}{2}[a(h_0 + h_1) + b(h_1 + h_2) + c(h_2 + h_3) + \cdots] \tag{12.4}$$

where $a, b, c, \ldots$ are the varying offset spaces, and $h_0, h_1, h_2, \ldots$ are the observed offsets.

■ Example 12.2

Compute the area of the tract shown in Figure 12.3.

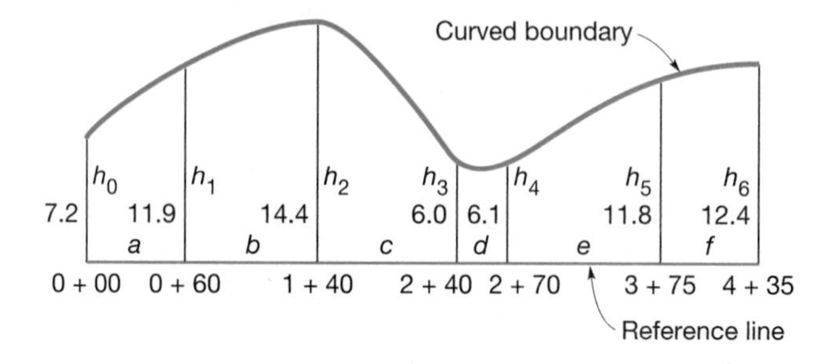

Figure 12.3
Area by offsets for a tract with a curved boundary.

Solution

By Equation (12.4),

$$\text{area} = \frac{1}{2}[60(7.2 + 11.9) + 80(11.9 + 14.4) + 100(14.4 + 6.0)$$

$$+ 30(6.0 + 6.1) + 105(6.1 + 11.8) + 60(11.8 + 12.4)]$$

$$= 4490 \text{ ft}^2$$

12.5 AREA BY COORDINATES

Computation of area within a closed polygon is most frequently done by the coordinate method. In this procedure, coordinates of each angle point in the figure must be known. They are normally obtained by traversing, although any method that yields the coordinates of these points is appropriate. If traversing is used, coordinates of the stations are computed after adjustment of the departures and latitudes, as discussed and illustrated in Chapter 10. The coordinate method is also applicable and convenient for computing areas of figures whose coordinates have been digitized using an instrument like that shown in Figure 28.8. The coordinate method is easily visualized; it reduces to one simple equation that applies to all geometric configurations of closed polygons and is readily programmed for computer solution.

The procedure for computing areas by coordinates can be developed with reference to Figure 12.4. As shown in that figure, it is convenient (but not necessary) to adopt a reference coordinate system with the X and Y axes passing through the most southerly and the most westerly traverse stations, respectively. Lines BB', CC', DD', and EE' in the figure are constructed perpendicular to the Y axis. These lines create a series of trapezoids and triangles (shown by different color shadings). The area enclosed with traverse $ABCDEA$ can be expressed in terms of the areas of these individual trapezoids and triangles as

$$\text{area}_{ABCDEA} = E'EDD'E' + D'DCC'D' - AE'EA - CC'B'BC - ABB'A \qquad (12.5)$$

The area of each trapezoid, for example $E'EDD'E'$, can be expressed in terms of lengths as

$$\text{area}_{E'EDD'E'} = \frac{E'E + DD'}{2} \times E'D'$$

In terms of coordinate values, this same area $E'EDD'E'$ is

$$\text{area}_{E'EDD'E'} = \frac{X_E + X_D}{2}(Y_E - Y_D)$$

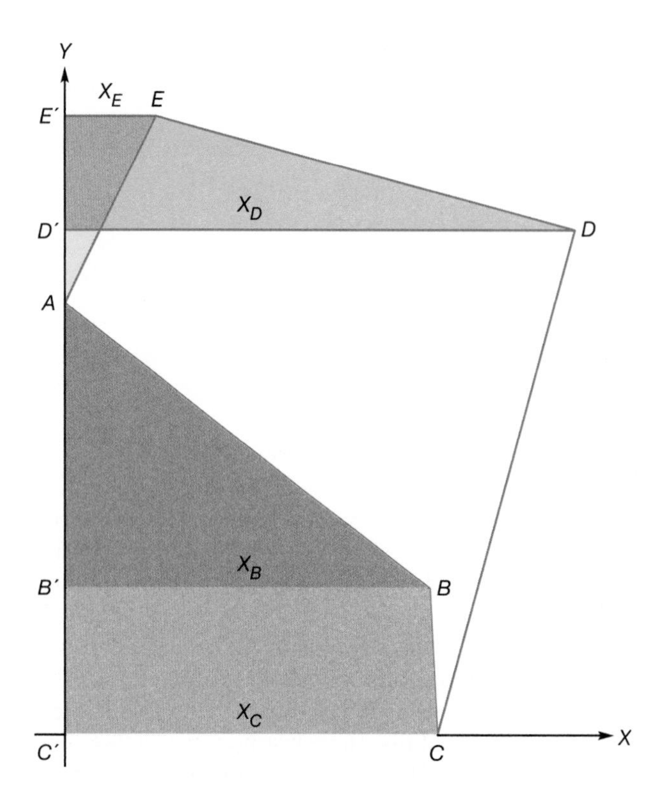

Figure 12.4
Area computation by the coordinate method.

Each of the trapezoids and triangles of Equation (12.5) can be expressed by coordinates in a similar manner. Substituting these coordinate expressions into Equation (12.5), multiplying by 2 to clear fractions, and rearranging,

$$\begin{aligned} 2(\text{area}) = &+X_AY_B + X_BY_C + X_CY_D + X_DY_E + X_EY_A \\ &-X_BY_A - X_CY_B - X_DY_C - X_EY_D - X_AY_E \end{aligned} \tag{12.6}$$

Equation (12.6) can be reduced to an easily remembered form by listing the X and Y coordinates of each point in succession in two columns, as shown in Equation (12.7), *with coordinates of the starting point repeated at the end.* The products noted by diagonal arrows are ascertained, with dashed arrows considered plus and solid ones minus. *The algebraic summation of all products is computed and its absolute value divided by 2 to get the area.*

$$\begin{matrix} X_A & & Y_A \\ X_B & \rightleftarrows & Y_B \\ X_C & \rightleftarrows & Y_C \\ X_D & \rightleftarrows & Y_D \\ X_E & \rightleftarrows & Y_E \\ X_A & & Y_A \end{matrix} \tag{12.7}$$

The procedure indicated in Equation (12.7) is applicable to calculating any size traverse. The following formula, easily derived from Equation (12.6), is a variation that can also be used,

$$\text{area} = \frac{1}{2}[X_A(Y_E - Y_B) + X_B(Y_A - Y_C) + X_C(Y_B - Y_D) + X_D(Y_C - Y_E) + X_E(Y_D - Y_A)] \tag{12.8}$$

It was noted earlier that for convenience, an axis system can be adopted in which $X = 0$ for the most westerly traverse point and $Y = 0$ for the most southerly station. Magnitudes of coordinates and products are thereby reduced and the amount of work lessened, since four products will be zero. However, selection of a special origin like that just described is of little consequence if the problem has been programmed for computer solution. Then the coordinates obtained from traverse adjustment can be used directly in the solution. However, a word of caution applies, if coordinate values are extremely large as they would normally be; for example, if state plane values are used (see Chapter 20). In those cases, to ensure sufficient precision and prevent serious round-off errors, double precision should be used. Or, as an alternative, the decimal place in each coordinate can arbitrarily be moved n places to the left, the area calculated, and then multiplied by 10^{2n}.

Either Equation (12.6) or Equation (12.8) can be readily programmed for solution by computer. The program WOLFPACK has this option under its *coordinate computations* menu. The format of the data file for this option is listed in its help screen. As was noted in Chapter 10, the "closed polygon traverse" option of WOLFPACK also computes areas using the coordinates of the adjusted traverse stations. A Mathcad® worksheet is included on the CD in this book to demonstrate the computations in Sections 12.3 through 12.5.

Example 12.3

Figure 12.5 illustrates the same traverse as Figure 12.4. The computations in Table 10.4 apply to this traverse. Coordinate values shown in Figure 12.5, however, result from shifting the axes so that $X_A = 0.00$ (A is the most westerly station) and $Y_C = 0.00$ (C is the most southerly station). This was accomplished by subtracting 10,000.00 (the value of X_A) from all X coordinates, and subtracting 4408.22 (the value of Y_C) from all Y coordinates. Compute the traverse area by the coordinate method. (Units are feet.)

Solution

These computations are best organized for tabular solution. Table 12.2 shows the procedure. Thus the area contained within the traverse is

$$\text{area} = \frac{|1{,}044{,}861 - 499{,}684|}{2} = 272{,}588 \text{ ft}^2 \text{ (say } 272{,}600 \text{ ft}^2) = 6.258 \text{ acres}$$

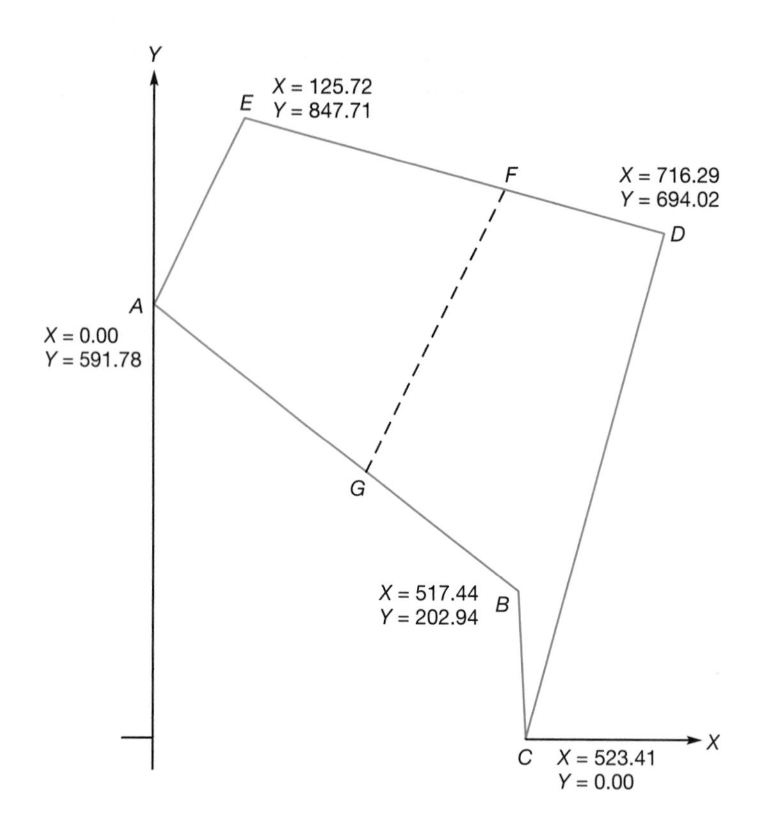

Figure 12.5 Traverse for computation of area by coordinates.

TABLE 12.2 COMPUTATION OF AREA BY COORDINATES

			Double Area (ft^2)	
Point	***X*(ft)**	***Y*(ft)**	**Plus (*XY*)**	**Minus (*YX*)**
A	0.00	591.78		
B	517.44	202.94	0	306,211
C	523.41	0.00	0	106,221
D	716.29	694.02	363,257	0
E	125.72	847.71	607,206	87,252
A	0.00	591.78	74,398	0
			Σ = 1,044,861	Σ = 499,684
			−499,684	
			545,177	

545,177 ÷ 2 = 272,588 ft^2 = 6.258 acres

Notice that the precision of the computations was limited to four digits. This is due to the propagation of errors as discussed in Section 3.17.3. As an example, consider a square that has the same area as the parcel in Table 12.2. The length of its sides would be approximately 522.1 ft. Assuming that these coordinates have uncertainties of about ±0.05 ft, the error in the product as given by Equation (3.13) would be

$$E_{\text{area}} = \sqrt{(522.1 \times 0.05)^2 + (522.1 \times 0.05)^2} = \pm 37 \text{ ft}^2$$

Thus, rounding the computed area to the nearest hundred square feet is justified. As a *rule of thumb,* the accuracy of the area should not be stated any better than

$$E_{\text{area}} = S \times \sigma_S \sqrt{2} \quad \textbf{(12.9)}$$

where S is the length of the side of a square having an area equivalent to the parcel being considered and σ_S is the uncertainty in the coordinates of the points that bound the area in question.

Because of the effects of error propagation, it is important to remember that it is better to be conservative when expressing areas, and thus a phrase such as "6.258 acres more or less" is often adopted, especially when writing property descriptions (see Chapter 21).

12.6 AREA BY DOUBLE MERIDIAN DISTANCE METHOD

The area within a closed figure can also be computed by the double-meridian distance (DMD) method. This procedure requires balanced departures and latitudes of the tract's boundary lines, which are normally obtained in traverse computations. The DMD method is not as commonly used as the coordinate method because it is not as convenient. However, given the data from an adjusted traverse, it will yield the same answer. The DMD method is useful for checking answers obtained by the coordinate method when performing hand computations.

By definition, *the meridian distance of a traverse course is the perpendicular distance from the midpoint of the course to the reference meridian.* To ease the problem of signs, a reference meridian usually is placed through the most westerly traverse station.

In Figure 12.6 the meridian distances of courses *AB, BC, CD, DE,* and *EA* are MM', PP', QQ', RR', and TT', respectively. To express PP' in terms of convenient distances, MF and BG are drawn perpendicular to PP'. Then

$$PP' = P'F + FG + GP$$

$$= \text{meridian distance of } AB + \frac{1}{2}\text{departure of } AB + \frac{1}{2}\text{departure of } BC$$

Thus, the meridian distance for any course of a traverse equals the meridian distance of the preceding course plus one-half the departure of the preceding

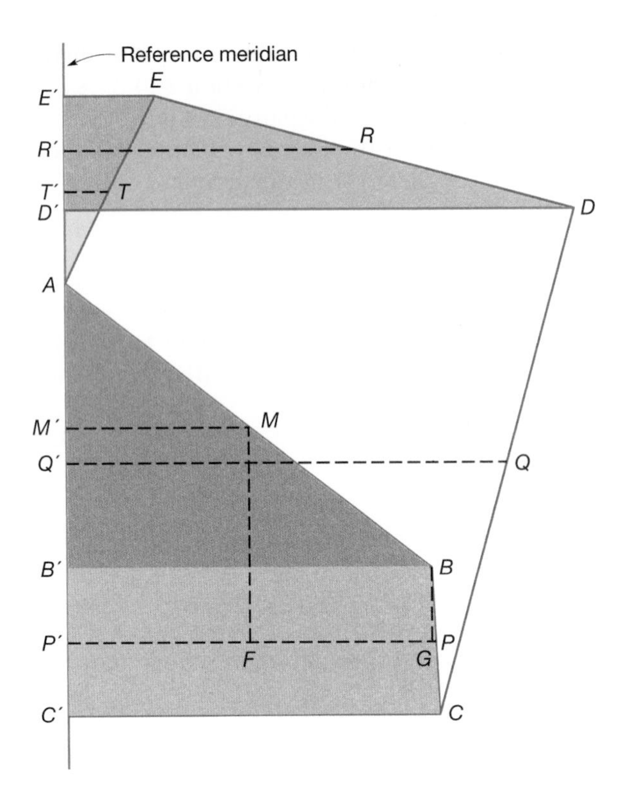

Figure 12.6 Meridian distances and traverse area computation by DMD method.

course plus half the departure of the course itself. It is simpler to employ full departures of courses. Therefore DMDs equal to twice the meridian distances are used and a single division by 2 is made at the end of the computation.

Based on the considerations described, the following general rule can be applied in calculating DMDs: *The DMD for any traverse course is equal to the DMD of the preceding course, plus the departure of the preceding course, plus the departure of the course itself.* Signs of the departures must be considered. When the reference meridian is taken through the most westerly station of a closed traverse and calculations of the DMDs are started with a course through that station, *the DMD of the first course is its departure.* Applying these rules, for the traverse in Figure 12.6,

$$\text{DMD of } AB = \text{departure of } AB$$

$$\text{DMD of } BC = \text{DMD of } AB + \text{departure of } AB + \text{departure of } BC$$

A check on all computations is obtained if the DMD of the last course, after computing around the traverse, is also equal to its departure but has the opposite sign. If there is a difference, the departures were not correctly adjusted before starting, or a mistake was made in the computations. With reference to Figure 12.6, the area enclosed by traverse $ABCDEA$ may be expressed in terms of trapezoid areas (shown by different color shadings) as

$$\text{area} = E'EDD'E' + C'CDD'C' - (AB'BA + BB'C'CB + AEE'A) \tag{12.10}$$

The area of each figure equals the meridian distance of a course times its balanced latitude. For example, the area of trapezoid $C'CDD'C' = Q'Q \times C'D'$, where $Q'Q$ and $C'D'$ are the meridian distance and latitude, respectively, of line *CD*. The DMD of a course multiplied by its latitude equals double the area. Thus the algebraic summation of all double areas gives *twice the area* inside the entire traverse. Signs of the products of DMDs and latitudes must be considered. If the reference line is passed through the most westerly station, all DMDs are positive. The products of DMDs and north latitudes are therefore plus and those of DMDs and south latitudes are minus.

Example 12.4

Using the balanced departures and latitudes listed in Table 10.4 for the traverse of Figure 12.6, compute the DMDs of all courses.

Solution

The calculations, done in tabular form following the general rule, are illustrated in Table 12.3.

Example 12.5

Using the DMDs determined in Example 12.4, calculate the area within the traverse.

TABLE 12.3 COMPUTATION OF DMDs

Departure of *AB* =	+517.444	= DMD of *AB*
Departure of *AB* =	+517.444	
Departure of *BC* =	+5.964	
	+1040.852	= DMD of *BC*
Departure of *BC* =	+5.964	
Departure of *CD* =	+192.881	
	+1239.697	= DMD of *CD*
Departure of *CD* =	+192.881	
Departure of *DE* =	−590.571	
	+842.007	= DMD of *DE*
Departure of *DE* =	−590.571	
Departure of *EA* =	−125.718	
	+125.718	= DMD of *EA* ✓

TABLE 12.4 COMPUTATION OF AREA BY DMDs

Course	Balanced Departure (ft)	Balanced Latitude (ft)	DMD (ft)	Double Areas (ft²) Plus	Minus
AB	517.44	−388.84	517.44		201,201
BC	5.96	−202.95	1040.85		211,240
CD	192.88	694.02	1239.70	860,376	
DE	−590.57	153.69	842.01	129,408	
EA	−125.72	−255.93	125.72		32,176
Total	0.00	0.00		989,784	444,617
				−444,617	
				545,167	

$545{,}167 \div 2 = 272{,}584$ ft² (say 272,600 ft²) = 6.258 acres

Solution

Computations for area by DMDs are generally arranged as in Table 12.4, although a combined form may be substituted. Sums of positive and negative double areas are obtained and the absolute value of the smaller subtracted from that of the larger. The result is divided by 2 to get the area (272,600 ft²) and by 43,560 to obtain the number of acres (6.258). Note that the answer agrees with the one obtained using the coordinate method.

If the total of minus double areas is larger than the total of plus values, it signifies only that DMDs were computed by going around the traverse in a clockwise direction.

In modern surveying and engineering offices, area calculations are seldom done by hand; rather, they are programmed for computer solution. However, if an area is computed by hand, it should be checked by using different methods or by two persons who employ the same system. As an example, an individual working alone in an office could calculate areas by coordinates and check by DMDs. Those experienced in surveying (geomatics) have learned that a half-hour spent checking computations in the field and office can eliminate lengthy frustrations at a later time.

12.7 AREA OF PARCELS WITH CIRCULAR BOUNDARIES

The area of a tract that has a circular curve for one boundary, as in Figure 12.7, can be found by dividing the figure into two parts: polygon *ABCDEGFA* and sector *EGF*. The radius $R = EG = FG$ and either central angle $\theta = EGF$ or length *EF*

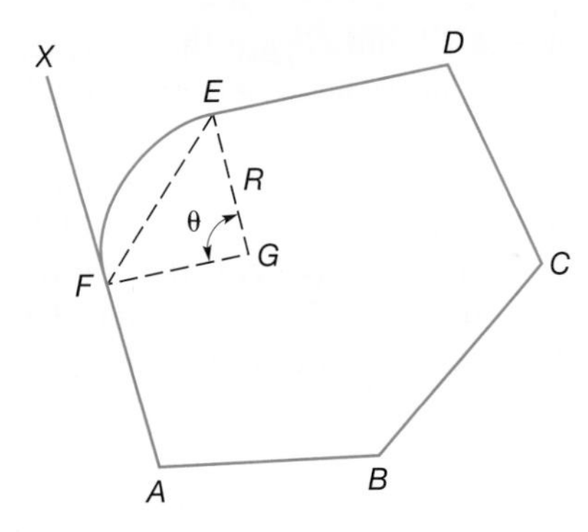

Figure 12.7
Area with circular curve as part of boundary.

must be known or computed to permit calculation of sector area *EGF*. If *R* and central angle θ are known, then the area of sector is

$$EGF = \pi R^2 \times (\theta/360^\circ) \tag{12.11}$$

If chord length *EF* is known, angle $\theta = 2 \sin^{-1}(EF/2R)$ and the preceding equation is used to calculate the sector area. To obtain the tract's total area, the sector area is added to area *ABCDEGFA* found by either the coordinate or DMD method.

Another method that can be used is to compute the area of the traverse *ABCDEFA* and then add the area of the *segment* which is the region between the arc and chord *EF*. The area of a segment is found as

$$\text{Area of segment} = (1/2)R^2(\theta - \sin\theta) \tag{12.12}$$

where θ is expressed in radian units.

■ 12.8 PARTITIONING OF LANDS

Calculations for purposes of partitioning land—that is, *cutting off* a portion of a tract for title transfer—can be aided significantly by using coordinates. For example, suppose the owner of the tract of land in Figure 12.5 wishes to subdivide the parcel with a line *GF*, parallel to *AE* and have 3.000 acres in parcel *AEFG*. This problem can be approached by three different methods. The first involves trial-and-error and works quite well given today's computing capabilities. The second consists of writing equations for simple geometric figures such as triangles, rectangles, and trapezoids that enable a unique solution to be obtained for the coordinates of points *F* and *G*. The third approach involves setting up a series of coordinate geometry equations together with an area equation, and then solving for the coordinates of *F* and *G*. The following subsections describe each of the previously mentioned procedures.

12.8.1 Trial and Error Method

In this approach, estimated coordinates for the positions of stations *F* and *G* are determined and the area of parcel *AEF′G′* is computed using Equation (12.6) where F' and G' are the estimated positions of F and G. This procedure is repeated until the area of the parcel equals 3.000 acres, or 130,680 ft^2.

Step 1: Using the final adjusted lengths and directions computed in Example 10.8 and coordinates of A and E from Example 12.3, and estimating the position of the cutoff line to be half the distance along line ED (i.e., $610.24 \div 2 = 305.12$ ft), the coordinates of stations F′ and G′ in parcel $AEF'G'$ are computed as

Station F′:

$$X = 125.72 + 305.12 \times \sin 104°35'13'' = 421.00$$

$$Y = 847.71 + 305.12 \times \cos 104°35'13'' = 770.87$$

Station G′: Is determined by direction-direction intersection using procedures discussed in Section 11.4. From WOLFPACK, the coordinates of Station G' are

$$X = 243.24 \text{ and } Y = 408.99$$

Creating a file for area computations, the area contained by these four stations is only 102,874 ft^2. Since 3.000 acres is equivalent to 130,680 ft^2, the estimated distance of 305.12 was short. It can now be increased and the process repeated.

Step 2: To estimate the amount necessary to increase the distance, an assumption that the figure $F'FGG'$ is a rectangle, with one side of length $F'G'$, or 403.18 ft, where that length is obtained by inversing the coordinates of F' and G' from Step 1. Thus the amount to move the line $F'G'$ is determined as

$$(130{,}680 - 102{,}874)/403.18 = 68.97 \text{ ft}$$

Thus for the second trial, the distance that F' is from E should be $305.12 + 68.97 = 374.09$ ft. Using the same procedure as in step 1, the area of $AEF'\,G'$ is 131,015 ft^2. The determined area is now too large and can be reduced using the same assumption that was used at the beginning of this step. Thus the distance EF' should be

$$EF' = 374.09 + (130{,}680 - 131{,}015)/$$
$$(\text{length of } F'G') = 374.09 - 0.78 = 373.31$$

This process is repeated until the final coordinates for F and G are determined. The next iteration yielded coordinates for F' of (487.00, 753.69) and for G' of (297.61, 368.14). Using these coordinates, the area of the parcel was computed to be 130,690 ft^2, or within 10 ft^2. The process is again repeated resulting in a reduction of the distance EF' of 0.02 ft, or $EF' = 373.29$ ft. The resulting area for $AEF'G'$ is 130,679 ft^2. Since this is within 1 ft^2 of the area, the coordinates are accepted as

$$F = (486.98, 753.70)$$

$$G = (297.59, 368.16)$$

The trial and error approach can be applied to solve many different types of land partitioning problems. Although the procedure may appear to involve a significant number of calculations, in many cases it provides the fastest and easiest solution when a computer program such as WOLFPACK is available for doing the coordinate geometry calculations.

12.8.2 Use of Simple Geometric Figures

As can be seen in Figure 12.8, parcel $AEFG$ is a parallelogram. Thus the formula for the area of a parallelogram $[A = 1/2(b_1 + b_2)h]$ can be employed, where b_1 is AE and b_2 is FG. In this procedure, a trigonometric relationship between the unknown length EF (denoted as d in Figure 12.8) and the missing parts h, FE', and $A'G$ must be determined. From the figure, angles α and β can be determined from azimuth differences, as

$$\alpha = AZ_{EE'} - AZ_{ED}$$

$$\beta = AZ_{AB} - AZ_{AA'}$$

Note in Table 10.7 that AZ_{EA} is 206°09′41″, and thus $AZ_{AA'}$ and $AZ_{EE'}$, which are perpendicular to line EA are 206°09′41″ − 90° = 116°09′41″. Also

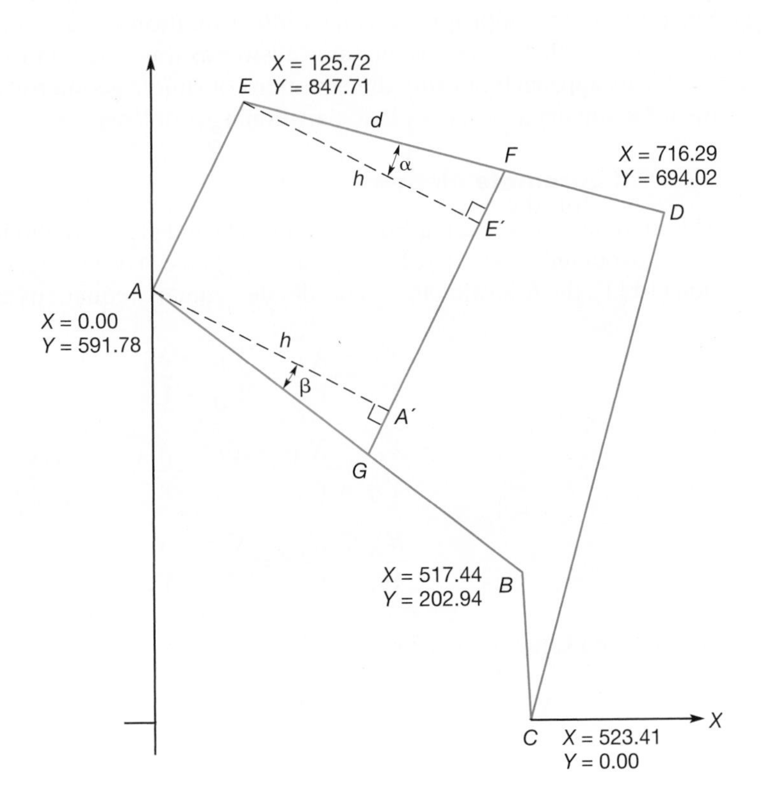

Figure 12.8 Partitioning of lands by simple geometric figures.

from Table 10.7, AZ_{ED} and AZ_{AB} are 104°35′13″ and 126°55′23″, respectively. Thus, the numerical values for α and β are:

$$\alpha = 116°09'41'' - 104°35'13'' = 11°34'28''$$

$$\beta = 126°55'23'' - 116°09'41'' = 10°45'42''$$

Now the parts h, FE', and $A'G$ can be expressed in terms of the unknown distance d as

$$\begin{aligned} h &= d \cos \alpha \\ FE' &= d \sin \alpha \\ A'G &= h \tan \beta = d \cos \alpha \tan \beta \end{aligned} \tag{12.13}$$

The formula for the area of parallelogram $AEFG$ is

$$\tfrac{1}{2}(AE + FE' + AE + A'G)h = 130{,}680 \tag{12.14}$$

Substituting Equations (12.13) into Equation (12.14) and rearranging yields

$$(\cos^2 \alpha \tan \beta + \cos \alpha \sin \alpha)d^2 + (2 \times AE \times \cos \alpha)d - 261{,}360 = 0 \tag{12.15}$$

Expression (12.15) is a quadratic equation and can be solved using Equation (11.3). Substituting the appropriate values into Equation (12.15) and solving yields $d = EF = 373.29$ ft. This is the same answer as was derived in Section 12.8.1.

This approach of using the equations of simple geometric figures is convenient for solving a variety of land partitioning problems.

12.8.3 Coordinate Method

This method involves using Equations (10.11) and (12.8) to obtain four equations with the four unknowns X_F, Y_F, X_G, and Y_G, that can be uniquely solved. By Equation (10.11), the following three coordinate geometry equations can be written:

$$\frac{X_F - X_E}{Y_F - Y_E} = \frac{X_D - X_E}{Y_D - Y_E} \tag{12.16}$$

$$\frac{X_G - X_A}{Y_G - Y_A} = \frac{X_B - X_A}{Y_B - Y_A} \tag{12.17}$$

$$\frac{X_A - X_E}{Y_A - Y_E} = \frac{X_G - X_F}{Y_G - Y_F} \tag{12.18}$$

Also by area Equation (12.8):

$$\begin{aligned} X_A(Y_G - Y_E) + X_E(Y_A - Y_F) + X_F(Y_E - Y_G) \\ + X_G(Y_F - Y_A) = 2 \times \text{area} \end{aligned} \tag{12.19}$$

Substituting the known coordinates $X_A, Y_A, X_B, Y_B, X_D, Y_D, X_E$, and Y_E into Equations (12.16) through (12.19) yields four equations that can be solved for the four unknown coordinates. The four equations can be solved simultaneously, for example by using matrix methods (see Appendix E), to determine the unknown coordinates for points F and G. (A matrix program is included on the CD that accompanies this book.)

Alternatively, the four equations can be solved by substitution. In this approach, Equations (12.16) and (12.17) are rewritten in terms of one of the unknowns, say X_F and X_G. These two new equations are then substituted into Equations (12.18) and (12.19). The resultant equations will now contain two unknowns Y_F and Y_G. The equation corresponding to Equation (12.18) is then solved in terms of unknown, say Y_F, and this can be substituted into the Equation (12.19). The resultant expression will be a quadratic equation in terms of Y_G which can be solved using Equation (11.3). This solution can then be substituted into the previous equations to derive the remaining three unknowns.

12.9 AREA BY MEASUREMENTS FROM MAPS

To determine the area of a tract of land from map measurements its boundaries must first be identified on an existing map or a plot of the parcel drawn from survey data. Then one of several available methods can be used to determine its area. Accuracy in making area determinations from map measurements is directly related to the accuracy of the maps being used. Accuracy of maps, in turn, depends on the quality of the survey data from which they were produced, map scale, and the precision of the drafting process. Therefore if existing maps are being used to determine areas, their quality should first be verified.

Even with good-quality maps, areas measured from them will not normally be as accurate as those computed directly from survey data. Map scale and the device used to extract map measurements are major factors affecting the resulting area accuracy. If a map is plotted to a scale of 1000 ft/1 in., for example, and an engineer's scale is used which produces measurements good to ± 0.02 in., distances or coordinates scaled from this map can be no better than about $(\pm 0.02 \times 1000) = \pm 20$ ft. This uncertainty can produce substantial errors in areas. Differential shrinkage or expansion of the material upon which maps are drafted is another source of error in determining areas from map measurements. Changes in dimensions of 2 to 3 percent are common for certain types of paper. (The subjects of maps and mapping are discussed in more detail in Chapters 17 and 18.)

Aerial photos can also be used as map substitutes to determine *approximate* areas if the parcel boundaries can be identified. The areas are approximate, as explained in Chapter 27, because except for flat areas the scale of an aerial photo is not uniform throughout. Aerial photos are particularly useful for determining areas of irregularly shaped tracts, such as lakes. Different procedures for determining areas from maps are described in the subsections that follow.

12.9.1 Area by Counting Coordinate Squares

A simple method for determining areas consists in overlaying the mapped parcel with a transparency having a superimposed grid. The number of grid squares included within the tract is then counted, with partial squares estimated and added to the total. Area is the product of the total number of squares times the area represented by each square. As an example, if the grids are 0.20 in. on a side, and a map at a scale of 200 ft/in. is overlaid, each square is equivalent to $(0.20 \times 200)^2 = 1600\ \text{ft}^2$.

12.9.2 Area by Scaled Lengths

If the boundaries of a parcel are identified on a map, the tract can be divided into triangles, rectangles, or other regular figures, the sides measured, and the areas computed using standard formulas and totaled.

12.9.3 Area by Digitizing Coordinates

A mapped parcel can be placed on a digitizing table which is interfaced with a computer and the coordinates of its corner points quickly and conveniently recorded. From the file of coordinates, the area can be computed using either Equation (12.6) or (12.8). It must be remembered, however, that even though coordinates may be digitized to the nearest 0.001 in., their actual accuracy can be no better than the map from which the data were extracted. Area determination by digitizing existing maps is now being practiced extensively in creating databases of geographic information systems.

12.9.4 Area by Planimeter

A planimeter measures the area contained within any closed figure that is circumscribed by its tracer. There are two types of planimeters: mechanical and electronic. The major parts of the mechanical type are a scale bar, graduated drum and disk, vernier, tracing point and guard, and anchor arm, weight, and point. The scale bar may be fixed or adjustable. For the standard fixed-arm planimeter, one revolution of the disk (dial) represents 100 in.^2 and one turn of the drum (wheel) represents 10 in.^2. The adjustable type can be set to read units of area directly for any particular map scale. The instrument touches the map at only three places: anchor point, drum, and tracing-point guard.

Because of its ease of use, the electronic planimeter (Figure 12.9) has replaced its mechanical counterpart. An electronic planimeter operates similarly to the mechanical type, except that the results are given in digital form on a display console. Areas can be measured in units of square inches or square centimeters and by setting an appropriate scale factor, they can be obtained directly in acres or hectares. Some instruments feature multipliers, which automatically compute and display volumes.

As an example of using an adjustable type of mechanical planimeter, assume the area within the traverse of Figure 12.5 will be measured. The anchor point beneath the weight is set in a position outside the traverse (if inside, a polar constant must be added) and the tracing point brought over corner A. An initial reading of 7231 is taken, the 7 coming from the disk, 23 from the drum, and 1 from the vernier. The tracing point is moved along the traverse lines from A to B, C, D, and E, and back to A. A triangle or a straightedge may guide the point, but normally it is steered

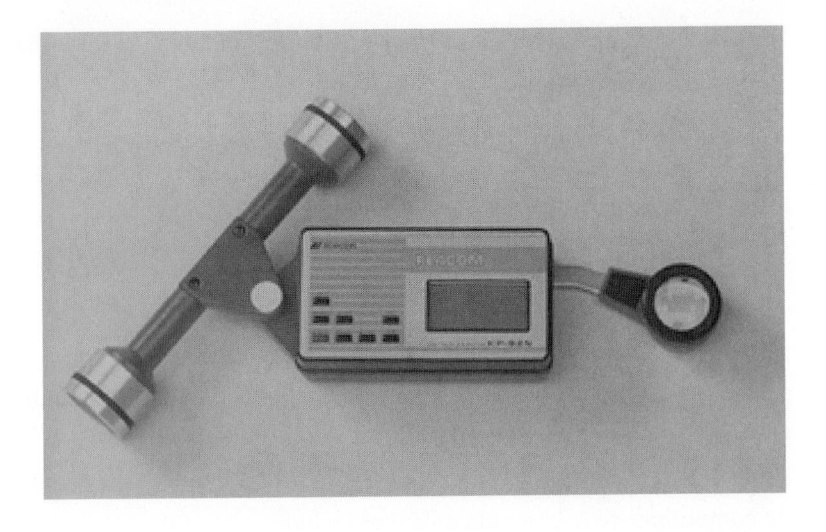

Figure 12.9 Electronic planimeter. (Courtesy Topcon Positioning Systems)

freehand. A final reading of 8596 is made. The difference between the initial and final readings, 1365, is multiplied by the planimeter constant to obtain the area. To determine the planimeter constant, a square area is carefully laid out 5 in. on a side, with diagonals of 7.07 in. and its perimeter traced with the planimeter. If the difference between initial and final readings for the 5-in. square is, for example, 1250, then

$$5 \text{ in.} \times 5 \text{ in.} = 25 \text{ in.}^2 = 1250 \text{ units}$$

Thus, the planimeter constant is

$$1 \text{ unit} = \frac{25}{1250} = 0.020 \text{ in.}^2$$

Finally the area within the traverse is

$$\text{area} = 1365 \text{ units} \times 0.020 = 27.3 \text{ in.}^2$$

If the traverse is plotted at a map scale of 1 in. = 100 ft, then 1 in.^{2} = 10,000 ft^2 and the area measured is 273,000 ft^2.

As a check on planimeter operation, the outline may be traced in the opposite direction. The initial and final readings at point *A* should agree within a limit of perhaps two to five units.

The precision obtained in using a planimeter depends on operator skill, accuracy of the plotted map, type of paper, and other factors. Results correct to within 1/2 to 1 percent can be obtained by careful work.

A planimeter is most useful for irregular areas, such as that in Figure 12.3, and has many applications in surveying and engineering. The planimeter has been widely used in highway offices for determining areas of cross sections and is also convenient for determining areas of drainage basins and lakes from measurements on aerial photos, checking computed areas in property surveys, etc.

12.10 SOURCES OF ERROR IN DETERMINING AREAS

Some sources of error in area computations are:

1. Errors in the field data from which coordinates or maps are derived.
2. Making a poor selection of intervals and offsets to fit irregular boundaries.
3. Making errors in scaling from maps.
4. Shrinkage and expansion of maps.
5. Using coordinate squares that are too large and therefore make estimation of areas of partial blocks difficult.
6. Making an incorrect setting of the planimeter scale bar.
7. Running off and on the edge of the map sheet with the planimeter drum.
8. Using different types of paper for the map and planimeter calibration sheet.

12.11 MISTAKES IN DETERMINING AREAS

In computing areas, common mistakes include:

1. Forgetting to divide by 2 in the coordinate and DMD methods.
2. Confusing signs of coordinates, departures, latitudes, and DMDs.
3. Forgetting to repeat the coordinates of the first point in the area by coordinates method.
4. Failing to check an area computation by a different method.
5. Not drawing a sketch to scale or general proportion for a visual check.
6. Not verifying the planimeter scale constant by tracing a known area.

PROBLEMS

Asterisks (*) indicate problems that have answers given in Appendix G.

12.1* Compute the area enclosed within polygon *ABDFGA* of Figure 12.1 using triangles.

12.2 Similar to Problem 12.1, except for polygon *ABCDGA* of Figure 12.1.

12.3 Compute the area enclosed between line *ABGHJLA* and the shoreline of Figure 12.1 using the offset method.

12.4 By rule of thumb, what is the estimated uncertainty in an 870,684 ft^2 if the estimated error in the coordinates was ± 0.2 ft?

12.5* Compute the area between a lake and a straight line *AG*, from which offsets are taken at irregular intervals as follows (all distances in feet):

Offset Point	*A*	*B*	*C*	*D*	*E*	*F*	*G*
Stationing	0.00	0 + 54.80	1 + 32.54	2 + 13.02	2 + 98.74	3 + 45.68	4 + 50.17
Offset	12.3	34.2	56.5	85.4	69.1	78.9	33.9

12.6 Repeat Problem 12.5 with the following offset in meters.

Offset Point	*A*	*B*	*C*	*D*	*E*	*F*	*G*
Stationing	0.00	20.000	78.940	148.963	163.651	203.687	250.447
Offset	3.15	6.32	10.04	19.57	15.43	9.42	3.65

12.7 Use the coordinate method to compute the area enclosed by the traverse of Problem 10.19.

12.8 Calculate by coordinates the area within the traverse of Problem 10.20.

12.9 Compute the area enclosed in the traverse of Problem 10.18 using DMDs.

12.10 Determine the area within the traverse of Problem 10.19 using DMDs.

12.11* By the coordinate method, find the area enclosed by the traverse of Problem 10.20.

12.12 Compute the area within the traverse of Problem 10.9 using the coordinate method. Check by DMDs.

12.13 Calculate the area inside the traverse of Problem 10.15 by coordinates and check by DMDs.

12.14 Compute the area enclosed by the traverse of Problem 10.18 using the DMD method. Check by coordinates.

12.15 Find the area within the traverse of Problem 10.21 using the DMD method. Check by coordinates.

12.16* Determine the area of the lot in Problem 10.28.

12.17 Calculate the area of Lot 16 in Figure 21.2.

12.18 Determine the area within the closed traverse of Problem 10.22 by the DMD method.

12.19 Plot the traverse of Problem 10.18 to a scale of 1 in. = 100 ft. Determine its surrounded area using a planimeter.

12.20 Similar to Problem 12.19, except for the traverse of Problem 10.19.

12.21 Plot the traverse of Problem 10.20 to a scale of 1 in. = 200 ft and find its enclosed area using a planimeter.

12.22 The (X,Y) coordinates (in feet) for a closed-polygon traverse *ABCDEFA* follow. *A* (1000.00, 1000.00), *B* (1661.73, 1002.89), *C* (1798.56, 1603.51), *D* (1289.82, 1623.69), *E* (1221.89, 1304.24) and *F* (1048.75, 1301.40). Calculate the area of the traverse by the method of coordinates.

12.23 Compute by coordinates the area in hectares within a closed-polygon traverse *ABCDEFA* by placing the *X* and *Y* axes through the most southerly and most westerly stations, respectively. Departures and latitudes (in meters) follow. *AB*: E dep. = 50, N lat. = 45; *BC*: E dep. = 60, N lat. = 55; *CD*: E dep. = 45, S lat. = 25; *DE*: W dep. = 70, S lat. = 40; *EF*: W dep. = 50, S lat. = 30; *FA*: W dep. = 35, S lat. = 5.

12.24 Calculate the area of a piece of property bounded by a traverse and circular arc with the following coordinates at angle points: *A* (1275.11,1356.11), *B* (1000.27, 1365.70), *C* (1000.00, 1000.00), *D* (1450.00, 1000.00) with a circular arc of radius *CD* starting at *D*, tangent to *CD*, and ending at *A*.

12.25 Calculate the area of a piece of property bounded by a traverse and circular arc with the following coordinates at angle points: *A* (526.68, 823.98), *B* (535.17, 745.61), *C* (745.17, 745.61), *D* (745.17, 845.61), *E* (546.62, 846.14) with a circular arc of radius 20 ft starting at *E*, tangent to *DE*, and ending at *A*.

12.26 Divide the area of the lot in Problem 12.24 into two equal parts by a line through point *B*. List in order the lengths and bearings of all sides for each parcel.

12.27 Partition the lot of Problem 12.25 into two equal areas by means of a line parallel to *BC*. Tabulate in clockwise consecutive order the lengths and bearings of all sides of each parcel.

12.28 Lot *ABCD* between two parallel street lines is 350.00 ft deep and has a 220.00 ft frontage (*AB*) on one street and a 260.00 ft frontage (*CD*) on the other. Interior angles at *A* and *B* are equal, as are those at C and D. What distances *AE* and *BF* should be laid off by a surveyor to divide the lot into two equal areas by means of a line *EF* parallel to *AB*?

12.29 Partition 1-acre parcel from the northern part of lot *ABCDEFA* in Problem 12.22 such that its southern line is parallel to the northern line.

12.30 Write a computational program for calculating areas within closed polygon traverses by the coordinate method.

12.31 Write a computational program for calculating areas within closed polygon traverses by the DMD method.

BIBLIOGRAPHY

Chrisman, N. R., and B. S. Yandell. 1988. "Effects of Point Error on Area Calculations: A Statistical Model." *Surveying and Land Information Systems* 48 (No. 4): 241.

Easa, S. M. 1988. "Area of Irregular Region with Unequal Intervals." *ASCE, Journal of Surveying Engineering* 114 (No. 2): 50.

El-Hassan, I. M. 1987. "Irregular Boundary Area Computation by Simpson's 3/8 Rule." *ASCE, Journal of the Surveying Engineering Division* 113 (No. 3): 127.

Ghilani, Charles D. 2000. "Demystifying Area Uncertainty: More or Less." *Surveying and Land Information Systems* 60 (No. 3): 177.

13

The Global Positioning System–Introduction and Principles of Operation

13.1 INTRODUCTION

In recent years, a new and unique approach to surveying, the global positioning system (GPS) has emerged. This system, which grew out of the space program, relies upon signals transmitted from satellites for its operation. It has resulted from research and development paid for by the military to produce a system for global navigation and guidance. With GPS, it is now possible to obtain precise timing and positioning information anywhere on the Earth with high reliability and low cost. The system can be operated day or night, rain or shine, and does not require cleared lines of sight between survey stations. This represents a revolutionary departure from conventional surveying procedures, which rely on observed angles and distances for determining point positions. GPS has gained worldwide acceptance and the technology is being used for virtually every type of survey. There is little doubt that GPS has affected the practice of surveying more profoundly than any other technology to date.

Development of the first generation of satellite positioning systems began in 1958. This early system, known as the *Navy Navigation Satellite System (NNSS)*, more commonly called the *TRANSIT* system, operated on the *Doppler* principle. In this system, *Doppler shifts* (changes in frequency) of signals transmitted from satellites were observed by receivers located on ground stations. The observed

Doppler shifts are a function of the distances to the satellites and their directions of movement with respect to the receivers. The transmitting frequency was known and together with accurate satellite orbital position data and precise timing of observations, the positions of the receiving stations could be determined. The constellation of satellites in the TRANSIT system, which varied between five and seven in number, operated in polar orbits at altitudes of approximately 1100 km. The objective of the TRANSIT system was to aid in the navigation of the U.S. Navy's Polaris submarine fleet. The first authorized civilian use of the system occurred in 1967 and the surveying community quickly adopted the new technology, finding it particularly useful for control surveying. Although these early instruments were bulky and expensive, the observation sessions lengthy, and the accuracy achieved moderate, the Doppler program was nevertheless an important breakthrough in satellite positioning in general, and in surveying in particular.

Because of the success of the Doppler program, the U.S. Department of Defense (DoD) began development of the ***NAV**igation **S**atellite **T**iming **a**nd **R**anging* (NAVSTAR) *Global Positioning System* (GPS). The first satellite to support the development and testing of the system was placed in orbit in 1978. Since that date many additional satellites have been launched. The global positioning system, initially developed at a cost of approximately $12 billion, became fully operational in December 1993. Like the earlier Doppler versions, the global positioning system is based on observations of signals transmitted from satellites whose positions within their orbits are precisely known. Also, the signals are picked up with *receivers* located at ground stations. However, the methods of determining distances from receivers to satellites, and of computing receiver positions, are different. These methods are described in later sections of this chapter. Current generation GPS receivers are illustrated in Figures 1.4 and 13.2. The size and cost of GPS equipment have been substantially reduced from those of the Doppler program, and field and office procedures involved in GPS surveys have been simplified so that now high accuracies can be achieved in real time.

■ 13.2 OVERVIEW OF GPS

As noted in the preceding section, precise distances from the satellites to the receivers are determined from timing and signal information, enabling receiver positions to be computed. In the global positioning system, the satellites become the reference or *control* stations, and the *ranges* (distances) to these satellites are used to compute the positions of the receiver. Conceptually, this is equivalent to resection in traditional ground surveying work, as described in Section 11.7, where distances and/or angles are observed from an unknown ground station to control points of known position.

The global positioning system can be arbitrarily broken into three parts: (a) the *space segment,* (b) the *control segment,* and (c) the *user segment.* The **space segment** consists of 24 satellites operating in six orbital planes spaced at 60° intervals around the equator. Four additional satellites are held in reserve as spares. The orbital planes are inclined to the equator at 55° [see Figure 13.1(b)]. This configuration provides 24-hour satellite coverage between the latitudes of 80°N and 80°S. The satellites travel in near-circular orbits that have a mean altitude of

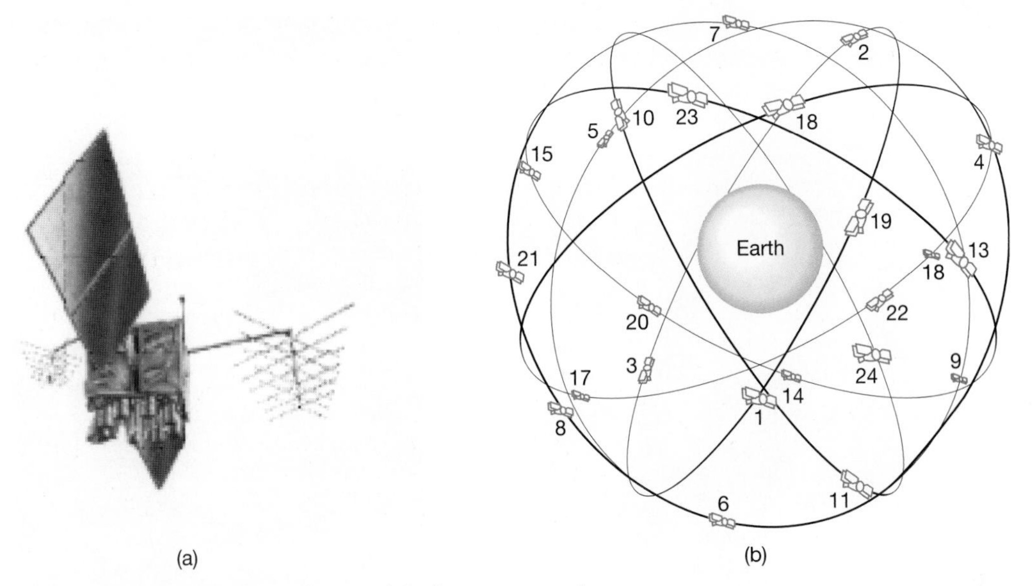

Figure 13.1 (a) A GPS satellite, and (b) the GPS constellation.

20,200 km above the Earth and an orbital period of 12 sidereal hours.[1] The individual satellites are normally identified by their *Pseudorandom Noise* (PRN) number (described later), but can also be identified by their satellite vehicle number (SVN), or orbital position.

Precise atomic clocks are used in the GPS satellites to control the timing of the signals they transmit. These are extremely accurate clocks[2] and extremely expensive as well. If the receivers used these same clocks, they would be cost prohibitive and would also require that users become trained in handling hazardous materials. Thus the clocks in the receivers are controlled by the oscillations of a quartz crystal that, although also precise, are less accurate than atomic clocks. However, these relatively low-cost timing devices produce a receiver that is also relatively inexpensive.

The **control segment** consists of twelve *monitoring stations* including those at Colorado Springs and on the islands of Hawaii, Ascension, Diego Garcia, and Kwajalein. At the monitoring stations, the signals from the satellites are monitored and their orbits tracked. The tracking information is relayed to the *master control station* in the 50th Space Wing's 2nd Space Operations Squadron located at Schriever Air Force Base in Colorado Springs. The master control station uses this data to make precise, near-future predictions of the satellite orbits and their clock correction parameters. This information is uploaded to the satellites and, in turn,

---

[1]A sidereal day is approximately 4 minutes shorter than a solar day. See Chapter 19 for more information on sidereal years and days.

[2]Atomic clocks are used, which employ either cesium or rubidium. The rubidium clocks may lose 1 second per 30,000 years, while the cesium type may lose 1 second only every 300,000 years. Hydrogen maser clocks, which may lose only 1 sec every 30,000,000 years, have been proposed for future satellites. For comparison, quartz crystal clocks used in receivers may lose a second every 30 years.

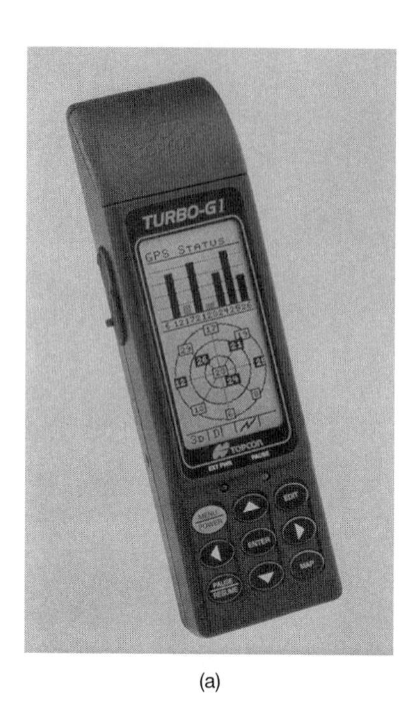

(a)

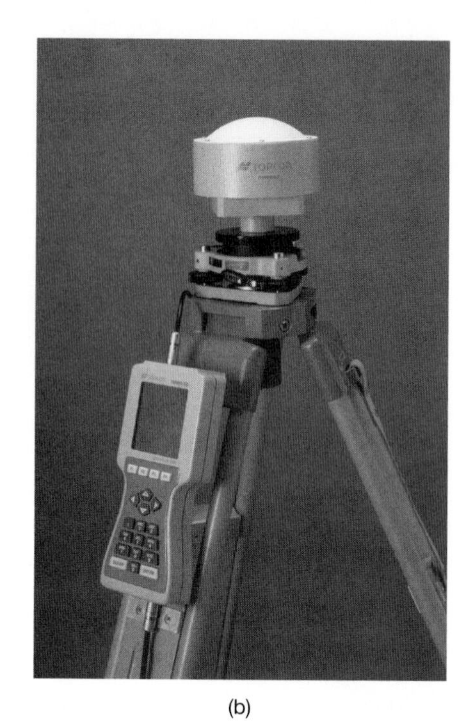

(b)

Figure 13.2
(a) The Sokkia AXIS3 and (b) GSR2700 GPS receivers. (Courtesy of Sokkia Corp.)

transmitted by them as part of their *broadcast messages* to be used by receivers to predict satellite positions and their clock *biases* (systematic errors).

The **user segment** consists of two categories of receivers that are classified by their access to two services that the system provides. These services are referred to as the *Standard Position Service* (*SPS*) and the *Precise Positioning Service* (*PPS*). The SPS is provided on the L1 broadcast frequency (see Section 13.3) at no cost to the user and was intended to provide accuracies of 100 m in horizontal positions, and 156 m in vertical positions at the 95% error level. The PPS is broadcast on both the L1 and L2 frequencies and is only available to receivers having valid cryptographic keys, which are reserved almost entirely for DoD use. This message provides a published accuracy of 18 m in the horizontal, and 28 m in the vertical at the 95% error level.

■ 13.3 THE GPS SIGNAL

As the GPS satellites are orbiting, each continually broadcasts a unique signal on two *carrier frequencies.* The carriers, which are transmitted in the L band of microwave radio frequencies, are identified as the L1 signal with a frequency of 1575.42 MHz and the L2 signal at a frequency of 1227.60 MHz. These frequencies are derived from a fundamental frequency, f_0, of 10.23 MHz. The L1 band has frequency of $154 \times f_0$, and the L2 band has a frequency of $120 \times f_0$.

Much like a radio station broadcast, several different types of information (messages) are modulated upon these carrier waves using a phase modulation technique. Some of the information included in the broadcast message is the almanac, broadcast ephemeris, satellite clock correction coefficients, ionospheric correction coefficients, and the satellite condition (also termed *satellite health*). These terms are defined later in this chapter.

In order for receivers to independently determine the ground positions of the stations they occupy in real time, it was necessary to devise a system for accurate measurement of signal travel time from satellite to receiver. This was accomplished by modulating the carriers with *pseudorandom noise* (PRN) codes. The PRN codes consist of unique sequences of binary values (zeros and ones) that appear to be random but, in fact, are generated according to a special mathematical algorithm using devices known as *tapped feedback shift registers.* Each satellite transmits two different PRN codes. The L1 signal is modulated with the *precise code,* or *P code,* and also with the *coarse/acquisition code,* or *C/A code.* The L2 signal is modulated only with the P code. Each satellite broadcasts a unique set of codes known as *GOLD codes* that allow receivers to identify the origins of received signals. This identification is important when tracking several different satellites simultaneously.

Starting in 2006, a modernized version of the satellites known as the Block IIRM is being launched. These new satellites include a second C/A (civilian) code on the L2 signal called the L2C. Additionally, the P code will be replaced by two new military codes, known as *M* codes, eventually. In 1999, the Interagency GPS Executive Board (IGEB) decided to add a third civilian signal known as the L5 to provide safety of life applications to GPS. The L5 will be broadcast at a frequency of 1176.45 MHz. Both the L2C and L5 will be added to the Block IIF satellites. The improvements in positioning due to these new codes will be discussed later in this chapter.

The C/A code has a frequency of 1.023 MHz and a wavelength of about 300 m. It is accessible to all users and is a series of 1023 binary digits (*chips*) that are unique to each satellite. This chip pattern is repeated every millisecond in the C/A code. The P code, with a frequency of 10.23 MHz and a wavelength of about 30 m, is ten times more accurate for positioning than the C/A code. The P code has a chip pattern that takes 266.4 days to repeat. Each satellite is assigned a unique single-week segment of the pattern that is reinitialized at midnight every Saturday. Table 13.1 lists the GPS frequencies and gives their factors of the fundamental frequency, f_0, of the P code.

To meet military requirements, the P code is encrypted with a W code to derive the Y code. This Y code can only be read with receivers that have the proper cryptographic keys. This encryption process is known as *anti-spoofing* (A-S). Its purpose is to deny access to the signal by potential enemies who could deliberately modify and retransmit it with the intention of "spoofing" unwary friendly users.

TABLE 13.1 FREQUENCIES TRANSMITTED BY GPS

Code Name	Frequency (MHz)	Factor of f_0
C/A	1.023	Divide by 10
P	10.23	1
L1	1575.42	Multiply by 154
L2	1227.60	Multiply by 120

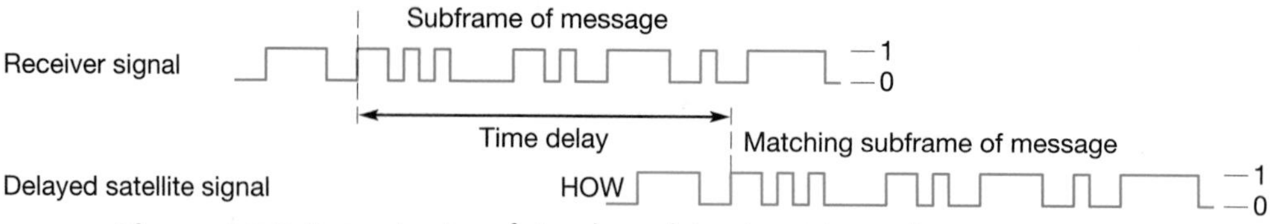

Figure 13.3 Determination of signal travel time by code matching.

Because of its need for "one-way" communication, the global positioning system depends on precise timing of the transmitted signal. This one-way system, which consists of signal transmission by satellites only, was necessary to meet military objectives; that is, the receivers could not transmit because that would give away strategic ground positions. To overcome the problem, a unique system was developed. To understand the concepts of the one-way system, consider the following. Imagine that the satellite transmits a series of audible beeps, and these beeps are broadcast in a known irregular pattern. Now imagine that this same pattern is synchronously duplicated (but not transmitted) at the receiving station. Since the signal of the satellite transmitter must travel to the receiver, its reception there will be delayed in relation to the signal being generated by the receiver. This delay can be measured and converted to a time difference.

This process is similar to that used with GPS. In GPS the chips of the PRN codes replace the beeps, and the precise time of broadcast of the satellite code is placed into the broadcast message with a starting time indicated by the front edge of one of the chips. The receiver generates a duplicate PRN code simultaneously. Matching the incoming satellite signal with the identical receiver-generated signal derives the time it takes for the signal to travel from satellite to receiver. This yields the signal delay, which is converted to travel time. From the travel time and the known signal velocity, the distance to the satellite can be computed.

To aid in matching the codes, the broadcast message from each satellite contains a *Hand-Over Word* (HOW), which consists of some identification bits, flags, and a number. This number, times four, produces the *Time of Week* (TOW), which marks the leading edge of the next section of the message. The HOW and TOW assist the receiver in matching the signal received from the satellite to that generated by the receiver, so the delay can be quickly determined. This matching process is illustrated diagrammatically in Figure 13.3.

■ 13.4 REFERENCE COORDINATE SYSTEMS FOR GPS

In determining the positions of points on Earth from satellite observations, three different reference coordinate systems are important. First of all, satellite positions at the instant they are observed are specified in the "space-related" *satellite reference coordinate systems.* These are three-dimensional rectangular systems defined by the satellite orbits. Satellite positions are then transformed into a three-dimensional rectangular *geocentric coordinate system* which is physically related to the Earth. As a result of GPS observations, the positions of new points on Earth are determined in this coordinate system. Finally, the geocentric coordinates are transformed into the more commonly used and locally oriented *geodetic coordinate system.* The following subsections describe these three coordinate systems.

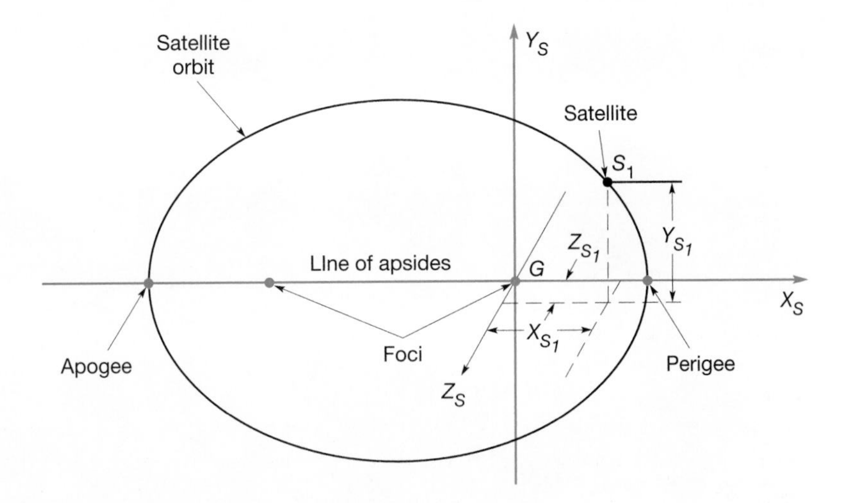

Figure 13.4 Satellite reference coordinate system.

13.4.1 The Satellite Reference Coordinate System

Once a satellite is launched into orbit, its movement thereafter within that orbit is governed primarily by the Earth's gravitational force. However, there are a number of other lesser factors involved including the gravitational forces exerted by the sun and moon, as well as forces due to solar radiation. Because of movements of the Earth, sun, and moon with respect to each other and because of variations in solar radiation, these forces are not uniform and hence satellite movements vary somewhat from their ideal paths. As shown in Figure 13.4, ignoring all forces except the Earth's gravitational pull, a satellite's idealized orbit is elliptical and has one of its two foci at G, the Earth's mass center. The figure also illustrates a satellite reference coordinate system, X_S, Y_S, Z_S. The *perigee* and *apogee* points are where the satellite is closest to and farthest away from G, respectively, in its orbit. The *line of apsides* joins these two points, passes through the two foci, and is the reference axis X_S. The origin of the X_S, Y_S, Z_S coordinate system is at G; the Y_S axis is in the mean orbital plane; and Z_S is perpendicular to this plane. Values of Z_S coordinates represent departures of the satellite from its mean orbital plane and normally are very small. A satellite at position S_1 would have coordinates X_{S1}, Y_{S1}, and Z_{S1}, as shown in Figure 13.4. For any instant of time, the satellite's position in its orbit can be calculated from its orbital parameters, which are part of the broadcast ephemeris.

13.4.2 The Geocentric Coordinate System

Because the objective of GPS surveys is to locate points on the surface of the Earth, it is necessary to have a so-called *terrestrial* frame of reference, which enables relating points physically to the Earth. The frame of reference used for this is a geocentric coordinate system. Figure 13.5 illustrates a quadrant of a *reference ellipsoid*[3] with a geocentric coordinate system (X_e, Y_e, Z_e) superimposed. This

[3]The reference ellipsoid used for most GPS work is the *World Geodetic System of 1984 (WGS 84) ellipsoid*. As explained in Section 19.1, any ellipsoid can be defined by two parameters—for example, the *semimajor axis (a)*, and the *flattening ratio (f)*. For the WGS84 ellipsoid these values are $a = 6{,}378{,}137$ m (exactly) and $f = 1/298.257223563$.

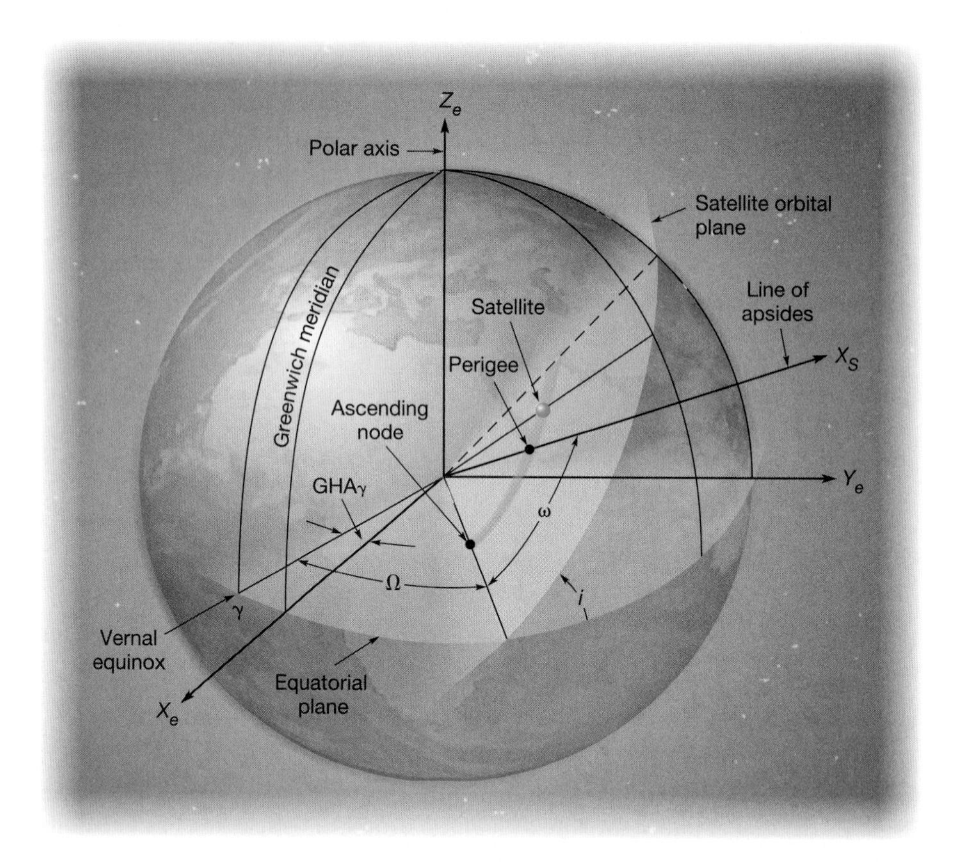

Figure 13.5 Parameters involved in transforming from the satellite reference coordinate system to the geocentric coordinate system.

three-dimensional rectangular coordinate system has its origin at the mass center of the Earth. Its X_e axis passes through the Greenwich meridian in the plane of the equator, and its Z_e axis coincides with the *Conventional Terrestrial Pole* (CTP) (see Section 20.3).

To make the conversion from the satellite reference coordinate system to the geocentric system, four angular parameters are required which define the relationship between the satellite's orbital coordinate system and key reference planes and lines on the Earth. As shown in Figure 13.5, these parameters are (1) the *inclination* angle, i, (angle between the orbital plane and the Earth's equatorial plane), (2) the *argument of perigee,* ω, (angle in the orbital plane from the equator to the line of apsides), (3) the *right ascension of the ascending node,* Ω, (angle in the plane of the Earth's equator from the vernal equinox to the line of intersection between the orbital and equatorial planes), and (4) the *Greenwich hour angle of the vernal equinox,* GHA_γ, (angle in the equatorial plane from the Greenwich meridian to the vernal equinox). These parameters are known in real-time for each satellite based upon predictive mathematical modeling of the orbits. Where higher accuracy is needed, satellite coordinates in the geocentric system for specific epochs of time are determined from observations at the tracking stations and distributed in precise ephemerides.

The equations for making conversions from satellite reference coordinate systems to the geocentric system are beyond the scope of this text. They are included in the software that accompanies GPS systems when they are purchased.

Although the equations are not presented here, through this discussion students are apprised of the nature of satellite motion, and of the fact that there are definite mathematical relationships between orbiting satellites and the positions of points located on the Earth's surface.

13.4.3 The Geodetic Coordinate System

Although the positions of points in a GPS survey are computed in the geocentric coordinate system described in the preceding subsection, in that form they are inconvenient for use by surveyors (geomatics engineers). This is the case for three reasons: (1) with their origin at the Earth's center, geocentric coordinates are typically extremely large values, (2) with the X-Y plane in the plane of the equator, the axes are unrelated to the conventional directions of north–south or east–west on the surface of the Earth, and (3) geocentric coordinates give no indication about relative elevations between points. For these reasons, the *geocentric* coordinates are converted to *geodetic* coordinates of latitude (ϕ), longitude (λ), and height (h) so that reported point positions become more meaningful and convenient for users.

Figure 13.6 also illustrates a quadrant of the reference ellipsoid and shows both the geocentric coordinate system (X, Y, Z) and the geodetic coordinate system (ϕ, λ, h). Conversions from geocentric to geodetic coordinates and vice versa are

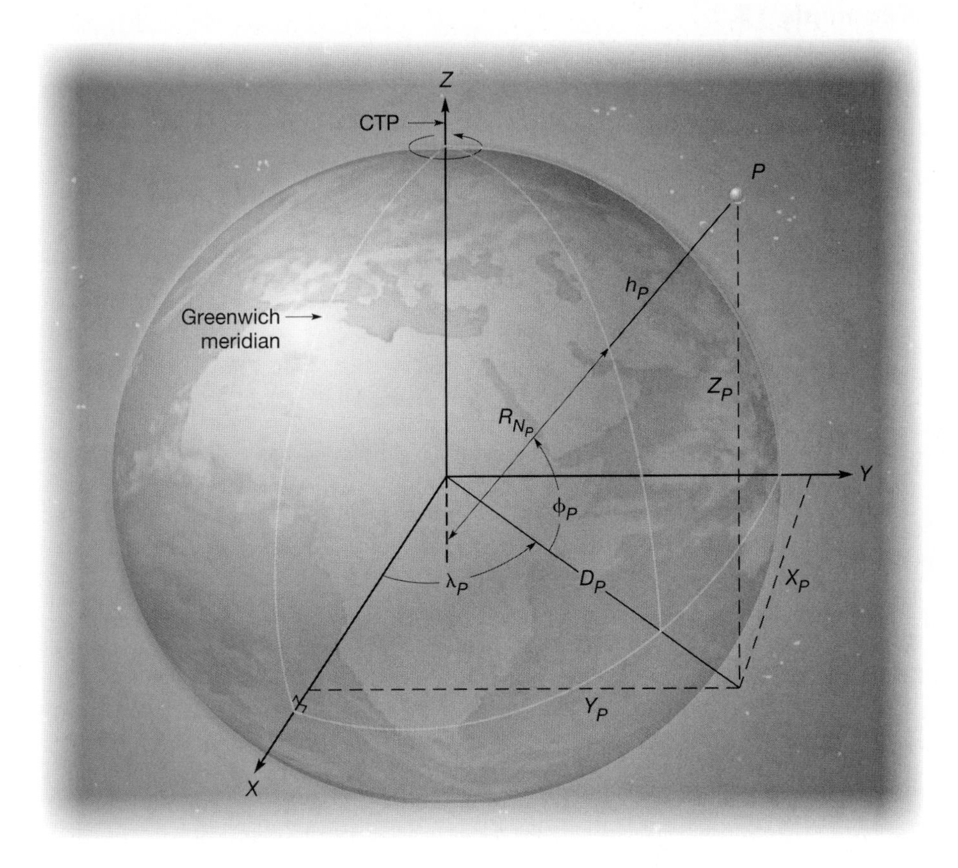

Figure 13.6 The geodetic and geocentric coordinate systems.

readily made. From the figure it can be shown that geocentric coordinates of point P can be computed from its geodetic coordinates using the following equations:

$$\begin{aligned} X_P &= (R_{N_P} + h_P)\cos\phi_P\cos\lambda_P \\ Y_P &= (R_{N_P} + h_P)\cos\phi_P\sin\lambda_P \\ Z_P &= [R_{N_P}(1 - e^2) + h_P]\sin\phi_P \end{aligned} \tag{13.1}$$

where

$$R_{N_P} = \frac{a}{\sqrt{1 - e^2\sin^2\phi_P}} \tag{13.2}$$

In Equations (13.1), X_P, Y_P, and Z_P are the geocentric coordinates of any point P, and the term e, which appears in both Equations (13.1) and (13.2), is the *eccentricity* of the WGS84 reference ellipsoid. Its value is 0.08181919084. In Equation (13.2), R_{N_P} is the *radius in the prime vertical*[4] of the ellipsoid at point P, and a, as noted earlier, is the semimajor axis of the ellipsoid. In Equations (13.1) and (13.2), north latitudes are considered positive and south latitudes negative. Similarly, east longitudes are considered positive and west longitudes negative.

Example 13.1

The geodetic latitude, longitude, and height of a point A are 41°15′18.2106″ N, 75°00′58.6127″ W, and 312.391 m, respectively. Using WGS84 values, what are the geocentric coordinates of the point?

Solution

Substituting the appropriate values into Equations (13.1) and (13.2) yields

$$R_{N_A} = \frac{6{,}378{,}137}{\sqrt{1 - 0.0066943799\sin^2(41°15'18.2106'')}} = 6{,}387{,}440.3113 \text{ m}$$

$$\begin{aligned} X_A &= (6{,}387{,}440.3113 + 312.391)\cos 41°15'18.2106''\cos(-75°00'58.6127'') \\ &= 1{,}241{,}581.343 \text{ m} \\ Y_A &= (6{,}387{,}440.3113 + 312.391)\cos 41°15'18.2106''\sin(-75°00'58.6127'') \\ &= -4{,}638{,}917.074 \text{ m} \\ Z_A &= [6{,}387{,}440.3113(1 - 0.00669437999) + 312.391)]\sin(41°15'18.2106'') \\ &= 4{,}183{,}965.568 \text{ m} \end{aligned}$$

[4]The eccentricity and radius in the prime vertical are both described in Chapter 20.

Conversion of geocentric coordinates of any point P to its geodetic values is accomplished using the following steps (refer again to Figure 13.6).

Step 1: Compute D_P as

$$D_P = \sqrt{X_P^2 + Y_P^2} \tag{13.3}$$

Step 2: Compute the longitude as[5]

$$\lambda_P = 2\tan^{-1}\left(\frac{D_P - X_P}{Y_P}\right) \tag{13.4}$$

Step 3: Calculate approximate latitude[6], ϕ_0

$$\phi_0 = \tan^{-1}\left[\frac{Z_P}{D_P(1 - e^2)}\right] \tag{13.5}$$

Step 4: Calculate the approximate radius of the prime vertical, R_N, using ϕ_0 from Step 3, and Equation (13.2).

Step 5: Calculate an improved value for the latitude from

$$\phi = \tan^{-1}\left(\frac{Z_P + e^2 R_{N_P}\sin(\phi_0)}{D_P}\right) \tag{13.6}$$

Step 6: Repeat the computations of Steps 4 and 5 until the change in ϕ between iterations becomes negligible. In this case, negligible means changes in latitude that are less than 0.00001″. This final value, ϕ_P, is the latitude of the station.

Step 7: Use the following formulas to compute the geodetic height of the station. For latitudes less than 45°, use

$$h_P = \frac{D_P}{\cos(\phi_P)} - R_{N_P} \tag{13.7a}$$

For latitudes greater than 45° use the formula

$$h_P = \left[\frac{Z_P}{\sin(\phi_P)}\right] - R_{N_P}(1 - e^2) \tag{13.7b}$$

■ Example 13.2

What are the geodetic coordinates of a point that has X, Y, Z geocentric coordinates of 1,241,581.343, −4,638,917.074, and 4,183,965.568, respectively? (*Note:* Units are meters.)

---

[5]This formula can conveniently be implemented in software with the function atan2(Y_P, X_P).

[6]A closed-form set of formulas for computing latitude of the station is demonstrated in the Mathcad® electronic book on the CD that accompanies this book.

Solution

To visualize the solution, refer to Figure 13.6. Since the X coordinate value is positive, the longitude of the point is between 0° and 90°. Also, since the Y coordinate value is negative, the point is in the western hemisphere. Similarly since the Z coordinate value is positive, the point is in the northern hemisphere.

Substituting the appropriate values into Equations (13.3) through (13.7) yields

Step 1: $D = \sqrt{(1{,}241{,}581.343)^2 + (-4{,}638{,}917.074)^2} = 4{,}802{,}194.8993$

Step 2:

$$\lambda = 2\tan^{-1}\left(\frac{4{,}802{,}194.8993 - 1{,}241{,}581.343}{-4{,}638{,}917.074}\right) = -75°00'58.6127'' \text{ (West)}$$

Step 3: $\phi_0 = \tan^{-1}\left[\dfrac{4{,}183{,}965.568}{4{,}802{,}194.8993(1 - 0.00669437999)}\right] = 41°15'18.2443''$

Step 4: $R_N = \dfrac{6{,}378{,}137}{\sqrt{1 - 0.00669437999\sin^2(41°15'18.2443'')}} = 6{,}387{,}440.3148$

Step 5: $\phi_0 = \tan^{-1}\left[\dfrac{4{,}183{,}965.568 + e^2 6{,}387{,}440.3148\sin 41°15'18.2443''}{4{,}802{,}194.8993}\right]$

$$= 41°15'18.2107''$$

Step 6: Repeat Steps 4 and 5 until the latitude converges. The values for the next iteration are

$$R_N = 6{,}387{,}440.3113$$

$$\phi_0 = 41°15'18.2106''$$

Repeating with the above values results in the same value for latitude to four decimal places, so the latitude of the station is 41°15′18.2106″ N.

Step 7: Compute the geodetic height using Equation (13.7a) as

$$h = \frac{4{,}802{,}194.8993}{\cos 41°15'18.2106''} - 6{,}387{,}440.3113 = 312.391$$

The geodetic coordinates of the station are latitude = 41°15′18.2106″ N, longitude = 75°00′58.6127″ W, and height = 312.391 m. Note that this example was the reverse computations of Example 13.1, and it reproduced the starting geodetic coordinate values for that example.

It is important to note that geodetic heights obtained with GPS are measured with respect to the ellipsoid—that is, the geodetic height of a point is the

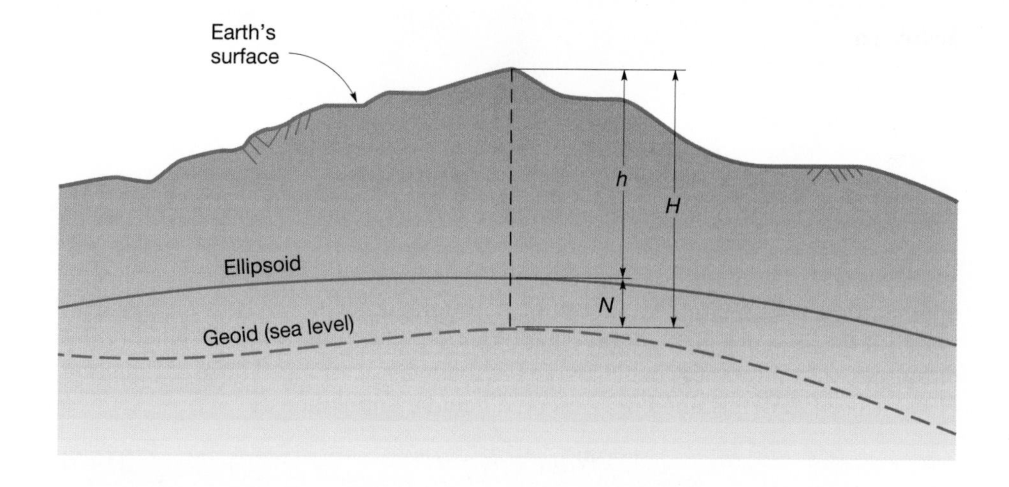

Figure 13.7 Relationships between elevation H, geodetic height h, and geoid undulation N.

vertical distance between the ellipsoid and the point as illustrated in Figure 13.7. As shown, these are not equivalent to *elevations* (also called *orthometric heights*) given with respect to the geoid. Recall from Chapter 4 that the geoid is an equipotential gravitational reference surface that is used as a datum for elevations. To convert geodetic heights to elevations, the *geoid undulation* (vertical distance between ellipsoid and geoid) must be known. Then elevations can be expressed as

$$H = h - N \tag{13.8}$$

where H is elevation above the geoid (orthometric height), h the geodetic height (determined using GPS), and N the geoidal undulation. Figure 13.7 shows the correct relationship of the geoid and the WGS 84 ellipsoid in the continental United States. Here the ellipsoid is above the geoid and geoidal undulation (*measured from the ellipsoid*) is negative. The geoidal undulation at any point can be estimated with mathematical models developed by combining gravimetric data with distributed networks of points where geoidal undulation has been observed. One such model, *GEOID03,* is a high-resolution model available from the National Geodetic Survey.[7] It uses latitude and longitude as arguments for determining geoid undulations for any locations in the conterminous United States (CONUS), Hawaii, Puerto Rico, and the Virgin Islands. *GEOID06* has been developed for Alaska.

■ Example 13.3

Compute the elevation (orthometric height) for a station whose geodetic height is 59.1 m, if the geoid undulation in the area is −21.3 m.

[7]A disk containing GEOID03 can be obtained by writing to the National Geodetic Information Center, NOAA, National Geodetic Survey, N/CG17, SSMC3 Station 09535, 1315 East West Highway, Silver Spring, Md. 20910, telephone (301) 713-3242, or it can be downloaded at http://www.ngs.noaa.gov/PC_PROD/pc_prod.shtml.

Solution

By Equation (13.8):

$$H = 59.1 - (-21.3) = 80.4 \text{ m}$$

Since the geoid undulation generally changes gradually, a value that can be applied for it over a limited area can be determined. Including NAVD 88 benchmarks in the area in a GPS survey can do this. Then with the ellipsoid heights and elevations known for these benchmarks, the following rearranged form of Equation (13.8) is used to determine GPS observed geoidal undulations:

$$N_{GPS} = h - H \tag{13.9}$$

The value for N_{GPS} obtained in this manner should be compared with that derived from the model supplied by the NGS and the difference should be computed as $\Delta N = N_{GPS} - N_{\text{model}}$. This procedure should be performed on several well-dispersed benchmarks in an area whenever possible. Then using an average ΔN for the survey area, the corrected orthometric height is:

$$H = h - (N_{\text{model}} + \Delta N_{\text{avg}}) \tag{13.10}$$

Example 13.4

The GPS observed geodetic heights of benchmark stations *Red, White,* and *Blue* are 412.345 m, 408.617 m, and 386.945 m, respectively. The model geoidal undulations for the stations are −29.894 m, −29.902 m, and −29.901 m, respectively, and their published elevations are 442.214 m, 438.490 m, and 416.822 m, respectively. What is the elevation of station *Brown,* which has a observed GPS height of 397.519 m, if the model geoidal undulation is determined to be −29.898 m?

Solution

By Equation (13.9), the observed geoidal undulations and ΔN's are

Station	N	ΔN
Red	412.345 − 442.214 = −29.869	−29.869 − (−29.894) = 0.025
White	408.617 − 438.490 = −29.873	−29.873 − (−29.902) = 0.029
Blue	386.945 − 416.822 = −29.877	−29.877 − (−29.901) = 0.024
		$\Delta N_{\text{avg}} = 0.026$

By Equation (13.10), the elevation of *Brown* is

$$Elev_{Brown} = 397.519 - (-29.898 + 0.026) = 427.391 \text{ m}$$

A word of caution should be added. Because the exact nature of the geoid is unknown, interpolated or extrapolated values of geoidal undulations from an observed network of points, or those obtained from mathematical models, are not exact. Thus orthometric heights obtained from ellipsoid heights will be close to their true values, but they may not be accurate enough to meet project requirements. Thus, for work that requires extremely accurate elevation differences, it is best to obtain them by differential leveling from nearby benchmarks as discussed in Chapters 4 and 5.

13.5 FUNDAMENTALS OF GPS POSITIONING

As discussed in Section 13.3, the precise travel time of the signal is necessary to determine the distance, or so-called *range,* to the satellite. Since the GPS satellite is in an orbit approximately 20,200 km above the Earth, the travel time of the signal will be roughly 0.07 seconds after the receiver generates the same signal. If this time-delay between the two signals is multiplied by the signal velocity (speed of light in a vacuum) c, the range to the satellite can be determined from

$$r = c \times t \tag{13.11}$$

where r is the range to the satellite and t is the elapsed time for the wave to travel from the satellite to the receiver.

GPS receivers in determining distances to satellites employ two fundamental methods; *code ranging* and *carrier phase-shift measurements.* Those that employ the former method are often called *mapping grade* receivers; those using the latter procedure are called *survey-grade* receivers. From distance observations made to multiple satellites, receiver positions can be calculated. Descriptions of the two methods and their mathematical models are presented in the subsections that follow. These mathematical models are presented to help students better understand the underlying principles of GPS operation. Computers that employ software provided by manufacturers of the equipment perform solutions of the equations.

13.5.1 Code Ranging

The code ranging (also called *code matching*) method of determining the time it takes the signals to travel from satellites to receivers was the procedure briefly described in Section 13.3. With the travel times known, the corresponding distances to the satellites can then be calculated by applying Equation (13.11). With one range known, the receiver would lie on a sphere. If the range were determined from two satellites, the results would be two intersecting spheres. As shown in Figure 13.8(a), the intersection of two spheres is a circle. Thus, two ranges from two satellites would place the receiver somewhere on this circle. Now if the range for a third satellite is added, this range would add an additional sphere which when intersected with one of the other two spheres would produce another circle of intersection. As shown in Figure 13.8(b), the intersection of two circles would leave only two possible locations for the position of the receiver. Using a "seed

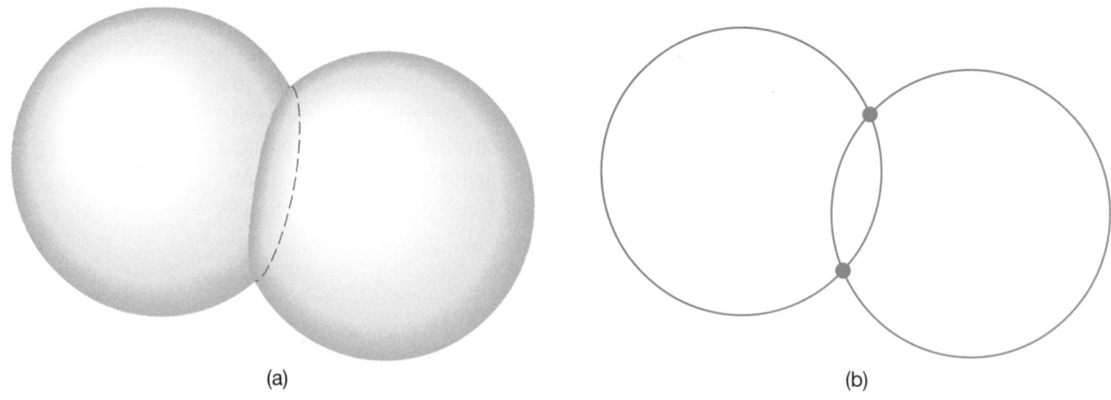

Figure 13.8 (a) The intersection of two spheres and (b) the intersection of two circles.

position" that is within a hundred kilometers of the position of the receiver will quickly eliminate one of these two intersections.

For observations taken on three satellites, the system of equations that could be used to determine the position of a receiver at station A is

$$\begin{aligned}\rho_A^1 &= \sqrt{(X^1 - X_A)^2 + (Y^1 - Y_A)^2 + (Z^1 - Z_A)^2}\\ \rho_A^2 &= \sqrt{(X^2 - X_A)^2 + (Y^2 - Y_A)^2 + (Z^2 - Z_A)^2}\\ \rho_A^3 &= \sqrt{(X^3 - X_A)^2 + (Y^3 - Y_A)^2 + (Z^3 - Z_A)^2}\end{aligned} \tag{13.12}$$

where ρ_A^n are the *geometric ranges* for the three satellites to the receiver at station A, (X^n, Y^n, Z^n) are the geocentric coordinates of the satellites at the time of the signal transmission, and (X_A, Y_A, Z_A) are the geocentric coordinates of the receiver at transmission time. Note that the variable n pertains to superscripts and takes on values of 1, 2, or 3.

However, in order to obtain a valid time observation, the systematic error (known as *bias*) in the clocks, and the refraction of the wave as it passes through the Earth's atmosphere, must also be considered. In this example, the receiver clock bias is the same for all three ranges since the same receiver is observing each range. With the introduction of a fourth satellite range, the receiver clock bias can be mathematically determined. This solution procedure allows the receiver to have a less accurate (and less expensive) clock. Algebraically, the system of equations used to solve for the position of the receiver and clock bias are:

$$\begin{aligned}R_A^1(t) &= \rho_A^1(t) + c(\delta^1(t) - \delta_A(t))\\ R_A^2(t) &= \rho_A^2(t) + c(\delta^2(t) - \delta_A(t))\\ R_A^3(t) &= \rho_A^3(t) + c(\delta^3(t) - \delta_A(t))\\ R_A^4(t) &= \rho_A^4(t) + c(\delta^4(t) - \delta_A(t))\end{aligned} \tag{13.13}$$

where is $R_A^n(t)$ is the observed *range* (also called *pseudorange*) from receiver A to satellites 1 thru 4 at epoch (time) t, $\rho_A^n(t)$ is the geometric range as defined in

Equation (13.12), c is the speed of light in a vacuum, $\delta_A(t)$ is the receiver clock bias, and $\delta^n(t)$ is the satellite clock bias, which can be modeled using the coefficients supplied in the broadcast message. These four equations can be simultaneously solved yielding the position of the receiver (X_A, Y_A, Z_A) and the receiver clock bias $\delta_A(t)$. Equations (13.13) are known as the *point positioning equations* and as noted earlier they apply to code-based GPS receivers.

As will be shown in Section 13.6, in addition to timing there are several additional sources of error that affect the satellite's signals. Because of the clock biases and other sources of error, the observed range from the satellite to receiver is not the true range and thus it is called a pseudorange. Equations (13.13) are commonly called the *code pseudorange model.*

13.5.2 Carrier Phase-Shift Measurements

Better accuracy in measuring ranges to satellites can be obtained by observing *phase-shifts* of GPS signals. In this approach, the phase-shift in the signal that occurs from the instant it is transmitted by the satellite until it is received at the ground station is observed. This procedure, which is similar to that used by EDM instruments (see Section 6.19), yields the fractional cycle of the signal from satellite to receiver.[8] However, it does not account for the number of full wavelengths or *cycles* that occurred as the signal traveled between the satellite and receiver. This number is called the *integer ambiguity* or simply *ambiguity*. Unlike EDM instruments, GPS utilizes one-way communication, but because the satellites are moving and thus their ranges are constantly changing, the ambiguity cannot be determined by simply transmitting additional frequencies. There are different techniques used to determine the ambiguity. All of these techniques require that additional observations be obtained. Some of these techniques are discussed in Section 15.2. Once the ambiguity is determined, the mathematical model for carrier phase-shift, corrected for clock biases, is

$$\Phi_i^j(t) = \frac{1}{\lambda}\rho_i^j(t) + N_i^j + f^j[\delta^j(t) - \delta_i(t)] \tag{13.14}$$

where for any particular epoch in time, t, $\Phi_i^j(t)$ is the carrier phase-shift measurement between satellite j and receiver i, f^j is the frequency of the broadcast signal generated by satellite j, $\delta^j(t)$ is the clock bias for satellite j, λ is the wavelength of the signal, $\rho_i^j(t)$ is the range as defined in Equations (13.12) between receiver i and satellite j, N_i^j is the integer ambiguity of the signal from satellite j to receiver i, and $\delta_i(t)$ is the receiver clock bias.

■ 13.6 ERRORS IN GPS OBSERVATIONS

Electromagnetic waves can be affected by several sources of error during their transmission. Some of the larger errors include (1) satellite and receiver clock biases, and (2) ionospheric and tropospheric refraction. Other errors in GPS work

---

[8]The phase-shift can be measured to approximately 1/100 of a cycle.

stem from (a) satellite ephemeris errors, (b) multipathing, (c) instrument miscentering, (d) antenna height measurements, (e) satellite geometry, and (f) before May 1, 2000, selective availability. All of these errors contribute to the total error of GPS-derived coordinates in the ground stations. These errors are discussed in the subsections that follow.

13.6.1 Clock Bias

Two errors already discussed in Section 13.5 were the satellite and receiver clock biases. The satellite clock bias can be modeled by applying coefficients that are part of the broadcast message using the polynomial

$$\delta^j(t) = a_0 + a_1(t - t_0) + a_2(t - t_0)^2 \tag{13.15}$$

where $\delta^j(t)$ is the satellite clock bias for epoch t, t_0 is the satellite clock reference epoch, and a_0, a_1, and a_2 are the satellite clock offset, drift, and frequency drift, respectively. The parameters a_0, a_1, a_2, and t_0 are part of the broadcast message. When using *relative-positioning* techniques, and specifically *single differencing* (see Section 13.8), the satellite clock bias can be mathematically removed during processing.

As was shown in Section 13.5, the receiver clock bias is treated as an unknown and computed using Equations (13.13) or (13.14). When using GPS relative positioning techniques, however, it can be eliminated through *double-differencing* during processing of the survey data. This method is discussed in Section 13.8.

13.6.2 Refraction

As discussed in Section 6.18, the velocities of electromagnetic waves change as they pass through media with different refractive indexes. The atmosphere is generally subdivided into regions. The subregions of the atmosphere that have similar composition and properties are known as *spheres*. The boundary layers between the spheres are called *pauses*. The two spheres that have the greatest effect on GPS signals are the *troposphere* and *ionosphere*. The troposphere is the lowest part of the atmosphere and is generally considered to exist up to 10–12 km in altitude. The *tropopause* separates the troposphere from the *stratosphere*. The stratosphere goes up to about 50 km. The combined refraction in the stratosphere, tropopause, and troposphere is known as *tropospheric refraction*.

There are several other layers of atmosphere above 50 km, but the one of most interest in GPS work is the *ionosphere* that extends from 50–1500 km above the Earth. As GPS signals pass through the ionosphere and troposphere, the signals are refracted. This produces range errors similar to timing errors and is one of the reasons why observed ranges are referred to as pseudoranges.

The ionosphere is primarily composed of ions—positively charged atoms and molecules and free negatively charged electrons. The free electrons affect the propagation of electromagnetic waves. The number of ions at any given time in the ionosphere is dependent on the sun's ultraviolet radiation. Solar flare activity known as *space weather* can dramatically increase the number of ions in the ionosphere, and thus can be reason for concern when working with GPS during periods of high sunspot activity, which follows a periodic peak variation of 11 years.[9] Since ionospheric refraction is the single largest error in GPS positioning, it is important to explore the space weather when performing GPS surveys. Space weather forecasts are available from the National Oceanic and Atmospheric Administration (NOAA) at http://www.sec.noaa.gov/Data/index.html. Only lower-order GPS surveys should be conducted when the k-index is 4 or greater.

A term for both the ionospheric and tropospheric refraction can be incorporated into Equations (13.13) and (13.14) to account for those errors in the signal. Letting $\Delta\delta^j$ equal the difference between the clock bias for satellite j and the receiver at A for epoch t, [i.e., $\Delta\delta^j = \delta^j(t) - \delta_A(t)$], then for any particular range listed in Equation (13.13) the incorporation of tropospheric and ionospheric refraction on the code pseudorange model yields

$$\begin{aligned} R^j_{L1}(t) &= \rho^j(t) + c\Delta\delta^j + c[\delta^{iono}_{f_{L1}} + \delta^{trop}(t)] \\ R^j_{L2}(t) &= \rho^j(t) + c\Delta\delta^j + c[\delta^{iono}_{f_{L2}} + \delta^{trop}(t)] \end{aligned} \tag{13.16}$$

where $R^j_{L1}(t)$ and $R^j_{L2}(t)$ are the observed pseudoranges as computed with frequency L1 or L2 (f_{L1} or f_{L2}) from satellite j to the receiver, $\rho^j(t)$ is the geometric range as defined in Equation (13.12) from the satellite to the receiver, c is the velocity of light in a vacuum, $\delta^{trop}(t)$ is the delay in the signal caused by the tropospheric refraction, and δ^{iono} is the ionospheric delay for the L1 and L2 frequencies, respectively.

A similar expression can be developed for the carrier phase-shift model and is

$$\begin{aligned} \Phi^j_{L1} &= \frac{1}{\lambda_{L1}}\rho^j(t) + f_{L1}\Delta\delta^j + N_{L1} - f_{L1}\delta^{iono} + f_{L1}\delta^{trop} \\ \Phi^j_{L2} &= \frac{1}{\lambda_{L2}}\rho^j(t) + f_{L2}\Delta\delta^j + N_{L2} - f_{L2}\delta^{iono} + f_{L2}\delta^{trop} \end{aligned} \tag{13.17}$$

where Φ^j_{L1} and Φ^j_{L2} are the carrier phase-shift observations from satellite j using frequencies L1 and L2, respectively, N_{L1} and N_{L2} are integer ambiguities for the two frequencies L_1 and L_2, and the other terms are as previously defined in Equations (13.14) and(13.16) for each frequency.

By taking observations on both the L1 and L2 signals and employing either Equations (13.16) or (13.17), the atmospheric refraction can be modeled and mathematically removed from the data. This is a major advantage of *dual-frequency receivers* (those which can observe both L_1 and L_2 signals) over their single-frequency counterparts, and allows them to accurately observe baselines up to

[9]1999 had a period of high solar activity. The next peak period of high solar activity should occur around 2010.

150 km accurately. The linear combination of the L1 and L2 signals for the code pseudorange model, which is free from ionospheric refraction, is

$$R_{L1,L2} = R_{L1} - \frac{(f_{L1})^2}{(f_{L2})^2} R_{L2} \quad \textbf{(13.18)}$$

where $R_{L1,L2}$ is the pseudorange observation for the combined L_1 and L_2 signals.

The carrier-phase model, which is also free of ionospheric refraction, is

$$\Phi_{L1,L2} = \Phi_{L1} - \frac{f_{L2}}{f_{L1}} \Phi_{L2} \quad \textbf{(13.19)}$$

where $\phi_{L1,L2}$ is the phase observation of the linear combination of the L1 and L2 waves. By their very nature, single frequency receivers cannot take advantage of the two GPS signals, and thus they must use ionospheric modeling data that is part of the broadcast message.

The advantage in having the satellites at approximately 20,200 km above the Earth is that signals from one satellite going to two different receivers pass through nearly the same atmosphere. Thus the atmosphere has similar effects on the signals and its effects can be practically eliminated using mathematical differencing techniques as discussed in Sections 13.7 through 13.9. For long lines, Equations (13.18) and (13.19) are typically used.

As can be seen in Figure 13.9, signals from satellites that are on the horizon of the observer must pass through considerably more atmosphere than signals

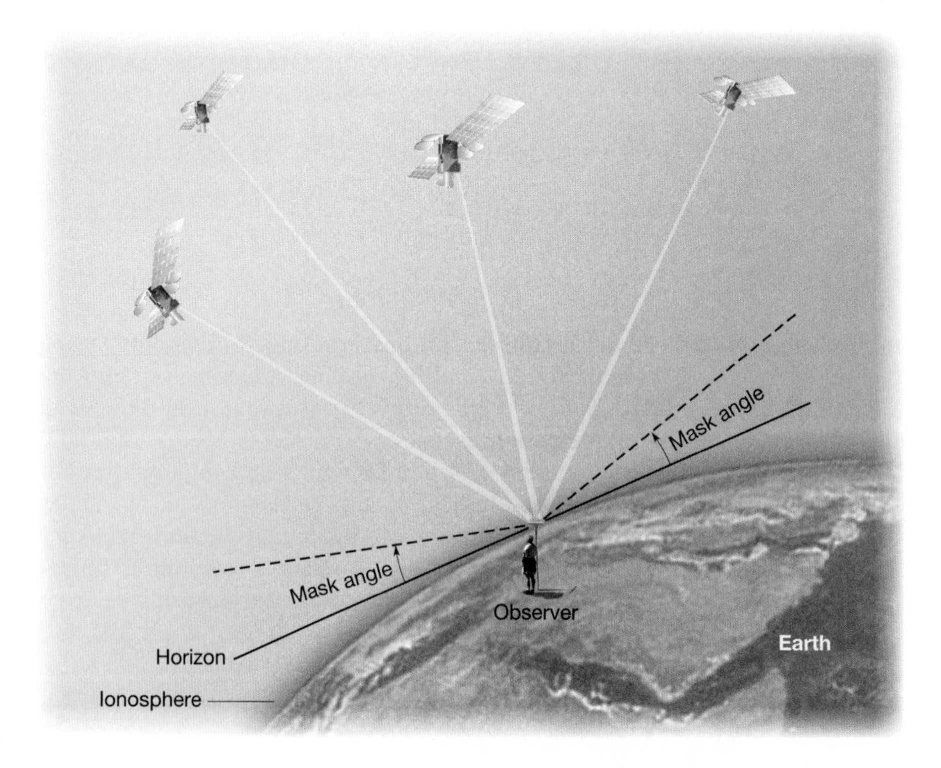

Figure 13.9 Relative positions of satellites, ionosphere, and receiver.

coming from high above the horizon. Because of the difficulty in modeling the atmosphere at low altitudes, signals from satellites below a certain threshold angle are typically omitted from the observations. The specific value for this angle (known as the *mask angle*) is somewhat arbitrary. It can vary between 10° and 20° depending on the desired accuracy of the survey. This is discussed further in Chapter 14.

13.6.3 Other Error Sources

Several other smaller error sources contribute to the positional errors of a receiver. These include (1) satellite ephemeris errors; (2) multipathing errors; (3) errors in centering the antenna over a point; (4) errors in measuring antenna height above the point; and (5) errors due to satellite geometry.

As noted earlier, the broadcast ephemeris predicts the positions of the satellites in the near future. However, because of fluctuations in gravity, solar radiation pressure, and other anomalies, these predicted orbital positions are always somewhat in error. In the GPS code matching method, these satellite position errors are translated directly into the computed positions of ground stations. This problem can be reduced by updating the orbital data using information obtained later and based on actual observations of the satellites made at tracking stations. One disadvantage of this is the delay that occurs in obtaining the updated data. One of three updated post-survey ephemerides are available: (1) ultra-rapid ephemeris, (2) the *rapid ephemeris*, and (3) the *precise ephemeris*. The ultra-rapid ephemeris is uploaded twice daily; the rapid ephemeris is available within two days after the survey; the precise ephemeris (the most accurate of the three) is not available until two weeks later. The ultra-rapid and rapid ephemerides are sufficient for most surveying applications.

As shown in Figure 13.10(a), *multipathing* occurs when a satellite signal reflects from a surface and is directed toward the receiver. This causes multiple

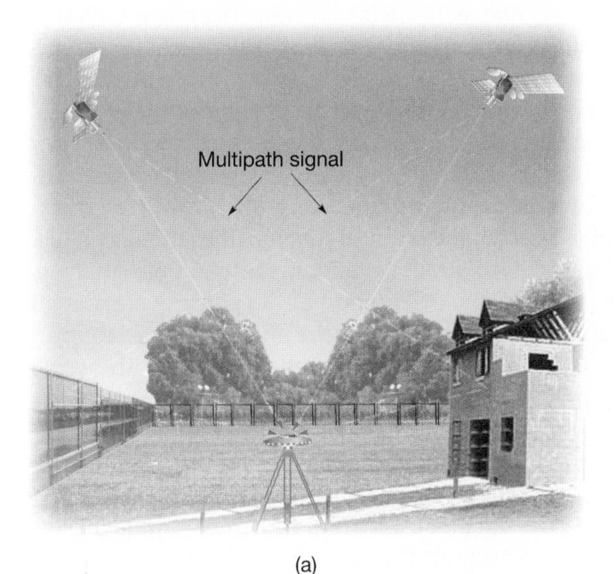

(a)

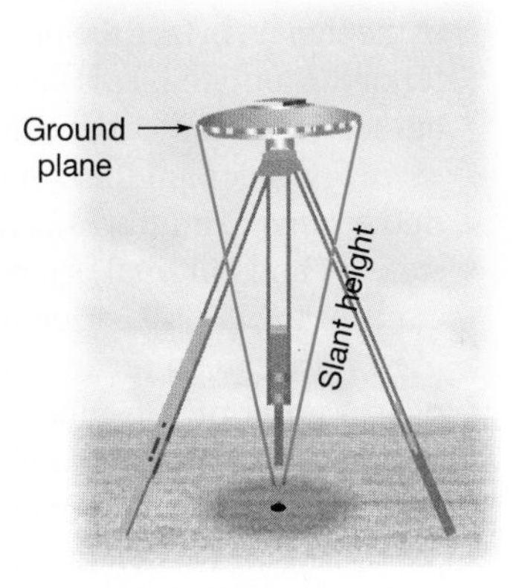

(b)

Figure 13.10 (a) Multipathing and (b) slant height measurements.

signals from a satellite to arrive at the receiver at slightly different times. Vertical structures such as buildings and chain link fences are examples of reflecting surfaces that can cause multipathing errors. Mathematical techniques have been developed to eliminate these undesirable reflections, but in extreme cases, they can cause a receiver to *lose lock* on the satellite—loss of lock is essentially a situation in which the receiver cannot use the signals from the satellite. This can be caused not only by multipathing, but also by obstructions and/or high ionospheric activity.

In GPS work, pseudoranges are observed to the receiver antennas. For precise work, the antennas are generally mounted on tripods, set up and carefully centered over a survey station, and leveled. Miscentering of the antenna over the point is another potential source of error. Setup and centering over a station should be carefully done following procedures like those described in Section 8.5. For any precise surveying work, including GPS, it is essential to have a well-adjusted tripod, tribrach, and optical plummet. Any error in miscentering of the antenna over a point will translate directly into an equal-sized error in the computed position of that point.

Observing the height of the antenna above the occupied point is another source of error in GPS work. The ellipsoid height determined from GPS observations is determined at the phase center of the antenna. Therefore, to get the ellipsoid height of the survey station, it is necessary to measure carefully, record the height of the antenna's phase center above the occupied point, and account for it in the data reduction. The distance shown in Figure 13.10(b) is known as the *slant height* and can be observed. The observations are made to the *ground plane* (a plane at the base of the antenna, which protects it from multipath signals reflecting from the ground). The slant height should be observed at several locations around the ground plane; if the observations do not agree, the instrument should be checked for level. Software within the system converts the slant height to the antenna's vertical distance above the station. Mistakes in identifying and observing heights of phase centers have caused errors as great as 10 cm in elevation. In precise GPS surveys, many surveyors use fixed-height tripods and rods that provide a constant offset from the point to the *antenna reference point* (ARP)—typically set at 1.5–2 m.

Additionally, the *phase center,* which is the electronic center of the antenna, varies with the orientation of the antenna, elevation of the satellites, and frequency of the signals. In fact, the physical center of the antenna seldom matches the phase center of the antenna. This fact is accounted for by *phase center offsets,* which are translations necessary to make the phase center and physical center of the antenna match.

For older antennas it is important to orient the antennas of multiple receivers in the same azimuth. This ensures the same orientation of the phase centers at all stations and eliminates a potential systematic error if the phase center is not precisely at the geometric center of the antenna. The same antenna should always be used with a given receiver in a precise survey, but if other antennas are used, their phase center offsets must be accounted for during post-processing. Newer antennas are directionally independent. They no longer require azimuthal alignment.

The National Geodetic Survey calibrates GPS antennas with respect to satellite elevations. When processing GPS data (see Section 14.5), users should always include the NGS calibration data to account for varying offsets due to satellite elevations when processing baselines.

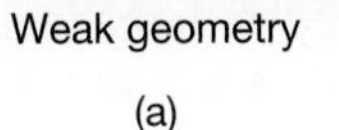

Figure 13.11 Weak and strong satellite geometry.

13.6.4 Geometry of Observed Satellites

An important additional error source in GPS surveying deals with the geometry of the visible satellite constellation at the time of observation. This is similar to the situation in traditional surveys, where the geometry of the network of observed ground stations affects the accuracies of computed positions. Figure 13.11 illustrates both weak and strong satellite geometry. As shown in Figure 13.11(a), small angles between incoming satellite signals at the receiver station produce weak geometry and generally result in larger errors in computed GPS positions. Conversely, strong geometry, as shown in Figure 13.11(b), occurs when the angles between incoming satellite signals are large, and this usually provides an improved solution. Whether conducting a GPS survey or a traditional one, by employing least-squares adjustment in the solution, the effect of the geometry upon the expected accuracy of the results is determined.

Table 13.2 lists the various categories of errors that can occur in GPS positioning. For each category, the sizes of errors that could occur in observed GPS ranges if no corrections or compensations were made are given, e.g., ±7.5 m could be expected as a result of ionospheric refraction, etc. But these error sizes assume ideal satellite geometry, i.e., no further degradation of accuracy is included for weak satellite geometry. The anticipated size of these errors with the addition of the L2C and L5 signals is shown in the third column of Table 13.2. These signals will become available to receivers as the current constellation of satellites is replaced by their modernized counterparts. Given the longevity of the current constellation, this process will take many years. At least four modernized satellites must be visible by the receiver before they can be used in the processing. Thus, the advantages of the L5 signal will not be apparent to users until a majority of the satellite constellation has been upgraded. It is anticipated that the entire satellite constellation will be upgraded with these new signals by 2030.

TABLE 13.2 ERROR SOURCES AND SIZES THAT CAN BE EXPECTED IN OBSERVED GPS RANGES

Error Source	Current Sizes of Errors (m)	Anticipated Sizes of Errors with Two or More Coded Signals (m)
Clock and ephemeris errors	±2.3	±2.3
Ionospheric Refraction	±7	±0.1
Tropospheric Refraction	±0.2	±0.2
Receiver Noise	±0.6	±0.6
Other (multipath, etc.)	±1.5	±1.5

By comparing the current errors with those anticipated with the inclusion of newer coded signals, it is obvious why the decision was made to fund the newer satellites. Using Equation (3.11), the total *User Equivalent Range Error* (UERE) is currently approximately ±7.5 m. It is anticipated that this error will drop to approximately ±2.8 m with the L2C and L5 signals.

As noted previously, by employing least squares in a GPS solution, the effect of satellite geometry can be determined. In fact, before conducting a GPS survey, the number and positions of visible satellites at a particular time and location can be evaluated in a preliminary least-squares solution to determine their estimated effect upon the resulting accuracy of the solution. This analysis produces so-called *Dilution of Precision* (DOP) factors. The DOP factors are computed through error propagation (see Section 13.7). They are simply numbers, which when multiplied by the errors of Table 13.2, give the sizes of errors that could be expected based upon the geometry of the observed constellation of satellites. For example, if the DOP factor is 2, then multiplying the sizes of errors listed in Table 13.2 by 2 would yield the estimated errors in the ranges for that time and location. Obviously, the lower the value for a DOP factor, the better the expected precision in computed positions of ground stations. If the preliminary least-squares analysis gives a higher DOP number than can be tolerated, the observations should be delayed until a more favorable satellite constellation is available.

The DOP factors that are of most concern to surveyors are PDOP (dilution of precision in position), HDOP (dilution of precision in horizontal position), and VDOP (dilution of precision in height). For the best possible constellation of satellites, the average value for HDOP is under 2 and under 5 for PDOP. Other DOP factors such as GDOP (dilution of precision in geometry) and TDOP (dilution of precision in time) can also be evaluated, but are generally of less significance in surveying. Table 13.3 lists some important categories of DOP, explains their meanings in terms of standard deviations and equations, and gives maximum values that are generally considered acceptable for most surveys.

Multiplying the DOP factor by the UERE yields the positional error in code ranging using Equations (13.13). For example, the HDOP is typically about 1.5. Recall from Equation (3.8) that the 95 percent probable error is obtained using a multiplier of about 1.96. Using the error values from Table 13.2 and a HDOP of 1.5 the

TABLE 13.3 IMPORTANT CATEGORIES OF DILUTION OF PRECISION

Category of DOP	Stand. Dev. Terms	Equation	Acceptable Value (less than)*
PDOP, Positional DOP	σ in geocentric coordinates X, Y, Z	$\sqrt{\sigma_X^2 + \sigma_Y^2 + \sigma_Z^2}$	6
HDOP, Horizontal DOP	σ in local x, y coordinates	$\sqrt{\sigma_X^2 + \sigma_Y^2}$	3
VDOP, Vertical DOP	σ in height, h	σ_h	5

*These recommended values are general guides for average types of GPS surveys, but individual project requirements may require other specific values.

current 95% probable error in horizontal positioning is ±22.5 m (1.96 × 1.5 × 7.5). When the newer coded signals are available and used by receivers, the 95% horizontal positioning error will be approximately ±8.5 m.

13.6.5 Selective Availability

Until recently, GPS signals were degraded to intentionally reduce accuracies achievable using the code-matching method. The intent was to exclude the highest accuracy attainable with GPS from nonmilitary users, especially adversaries. Two different methods were used to degrade accuracy: the *delta* process, which dithered the fundamental frequency of the satellite clock, and the *epsilon* process, which truncated the orbital parameters in the broadcast message so that the coordinates of the satellites could not be accurately computed. The errors in the coordinates of the satellites roughly translated to similar ground positional errors. The combined effect of these errors was known as *selective availability* (SA) and resulted in a positional error of approximately 100 m in the horizontal, and 156 m in the vertical, at the 95% error level. However, this error was removed by either differential or relative positioning techniques (see Sections 13.7 and 13.8).

When SA was initiated, the military and civilian communities were at odds about the need for it during times of peace. Initially, a plan was developed to turn SA off by 2006. But after agreement by the Departments of Transportation, Commerce and Defense, it was turned off at midnight on May 1, 2000, as the result of a Presidential Decision Directive (PDD). As was previously shown, with the removal of selective availability, code-based, real-time point positioning is about 20 m.

■ 13.7 DIFFERENTIAL GPS

As discussed in the two preceding sections, accuracies of observed pseudoranges are degraded by errors that stem from clock biases, atmospheric refraction, and other sources. Because of these errors, positions of points determined by point positioning techniques using a single code-based receiver can be in error by 20 m or more. While this order of accuracy is acceptable for certain uses, it is insufficient for most surveying applications. *Differential GPS* (DGPS), on the other hand, is a procedure that involves the simultaneous use of two or more code-based receivers. It can provide positional accuracies to within a few meters and thus the method is suitable for certain types of lower-order surveying work.

In DGPS, one receiver occupies a so-called *base station* (point whose coordinates are precisely known from previous surveying), and the other receiver or receivers (known as the *rovers*) are set up at stations whose positions are unknown. By placing a receiver on a station of known position, the pseudorange errors in the signal can be determined using Equation (13.16). Since this base station receiver and the rover are relatively close to each other (often less than a kilometer but seldom farther than a few hundred kilometers), the pseudorange errors at both the base station and at the rovers will have approximately the same magnitudes. After computing the corrections for each visible satellite at the base station, they can be applied to the roving receivers, thus substantially reducing or eliminating many errors listed in Table 13.2.

DGPS can be done in almost *real-time* with a radio transmitter at the base station and compatible radio receivers at the rovers. This process is known as *real-time differential GPS* (RTDGPS). The radio transmissions to the rovers contain both *pseudorange corrections* (PRCs) for particular *epochs of time* (moments in time) and *range rate corrections* (RRCs)[10] so that they can interpolate corrections to signals between each epoch. Alternatively, the errors can be eliminated from coordinates determined for rover stations during processing of the data.

To understand the mathematics in the procedure, a review of Equation (13.13) is necessary. The various error sources presented in Section 13.6 cause the observed pseudorange $R_A^j(t_0)$ to be in error by a specific amount for any epoch, t_0. Letting this error at epoch t_0 be represented by $\Delta\rho_A^j(t_0)$, the *radial orbital error*, Equation (13.13) can be rewritten as

$$R_A^j(t_0) = \rho_A^j(t_0) + \Delta\rho_A^j(t_0) + c\delta^j(t_0) - c\delta_A(t_0) \tag{13.20}$$

where the other terms are as previously defined.

Because the coordinates of the base station are known, the geometric range $\rho_A^j(t_0)$ in Equation (13.20) can be computed using Equation (13.12). Also since the pseudorange $R_A^j(t_0)$ is observed, the difference in these two values will yield the necessary correction for this particular pseudorange. The error conditions at each receiver are very similar and thus it is assumed that the error in the pseudorange observed at the base station is the same as the error at the rovers. This error at the base station is known as the code *pseudorange correction* (PRC) for satellite j at reference epoch t_0 and is represented as

$$\begin{aligned} PRC^j(t_0) &= -R_A^j(t_0) + \rho_A^j(t_0) \\ &= -\Delta\rho_A^j(t_0) - c[\delta^j(t_0) - \delta_A(t_0)] \end{aligned} \tag{13.21}$$

Because computation of the correction and transmission of the signal make it impossible to assign the PRC to the same epoch at the rovers, a range rate

[10]Pseudorange corrections (PRCs) are differences between measured ranges and ranges that are computed based upon the known coordinates of both the occupied reference station and those of the satellite. Because the satellites are moving, measured ranges to them are constantly changing. The rates of these changes per unit of time are the range rate corrections (RRCs).

correction (RRC) is approximated by numerical differentiation. This correction is used to extrapolate corrections for later epochs t. Thus, the pseudorange correction at any epoch t is given as

$$PRC^j(t) = PRC^j(t_0) + RRC^j(t_0)\,(t - t_0) \qquad \textbf{(13.22)}$$

where $RRC^j(t_0)$ is the range rate correction for satellite j determined at epoch t_0.

Now this information can be used to correct the computed ranges at the roving receiver locations. For example, at a roving station B, the corrected pseudorange, $R^j_B(t)_{\text{corrected}}$, can be computed as

$$\begin{aligned} R^j_B(t)_{\text{corrected}} &= R^j_B(t) + PRC^j(t) \\ &= \rho^j_B(t) + [\Delta\rho^j_B(t) - \Delta\rho^j_A(t)] - c[\delta_B(t) - \delta_A(t)] \qquad \textbf{(13.23)} \\ &= \rho^j_B(t) - c\Delta\delta_{AB}(t) \end{aligned}$$

where $\Delta\delta_{AB} = \delta_B(t) - \delta_A(t)$.

Notice that in the final form of Equation (13.23), it is assumed that the radial orbital errors at stations A and B, $\Delta\rho^j_A(t)$ and $\Delta\rho^j_B(t)$, respectively, are nearly the same, and are therefore eliminated mathematically. Furthermore, the satellite clock bias terms will be eliminated. Finally, assuming the signals to the base and roving receivers pass through nearly the same atmosphere (which means they should be within a few hundred kilometers of each other), the ionospheric and tropospheric refraction terms are practically eliminated.

The U.S. Coast Guard maintains a system of beacon stations along the U.S. coast and waterways. Private agencies have developed additional stations. The correction signals described above are broadcast by modulation on a frequency between 285–325 KHz using the ***R****adio* ***T****echnical* ***C****ommission for* ***M****aritime Services* ***S****pecial* ***C****ommittee* 104 (RTCM SC-104) format. Among the data contained in this broadcast are C/A code differential corrections, delta differential corrections, reference station parameters, raw carrier phase measurements, raw code range measurements, carrier phase corrections, and code range corrections.

The *Wide Area Augmentation System* (WAAS) developed by the Federal Aviation Administration has a network of ground tracking base stations that collect GPS signals and determine range errors. These errors are transmitted to geosynchronous satellites that relay the corrections to rovers. GPS software typically allows users to access the WAAS system when performing RTK-GPS surveys (see Chapter 15). This option, called *RTK with infill*, accesses the WAAS corrections when base-station radio transmissions are lost. However, these corrections will provide significantly less accuracy than relative positioning techniques typically utilized by GPS receivers using carrier phase-shift measurements.

When the WAAS is combined with a *Local Area Augmentation System* (LAAS), it is anticipated that the system will enable aircraft to key in on their destinations, after which the navigation system would develop the necessary flight paths for making landings in zero visibility. This system is expected to provide centimeter-level, real-time accuracy when implemented. Private firms have created similar systems. These systems are available as a subscription service.

Code-based receivers are used for positioning by people in all walks of life. They can be used by surveyors to gather details in situations not requiring typical survey accuracies. An example is the collection of data to update small-scale maps in a geographic information system (GIS—see Chapter 28). The use of code-based receivers in non-surveying applications includes the tracking of vehicles in transportation. The shipping industry uses code-based GPS receivers for navigation. Likewise, surveyors may use the navigation functions of a code-based GPS receiver to locate control monuments or other features where geodetic coordinates are known. Since the use of code-based receivers is so extensive and reaches far beyond the realm of the surveying community, their uses will not be covered in detail in this book.

13.8 KINEMATIC GPS METHODS

Methods similar to DGPS can also be employed with carrier phase-shift measurements to eliminate errors. The procedure, called *Kinematic* GPS surveying (see Chapter 15), again requires the simultaneous use of two or more receivers. All receivers must collect signals simultaneously from four or more of the same satellites through the entire observation process. Although single-frequency receivers can be used, kinematic GPS surveying works best with dual-frequency receivers. The method yields positions accurate to within a few centimeters, which makes it suitable for most surveying, mapping, and stakeout purposes.

As with DPGS, the fact that the base station's coordinates are known is exploited in *real-time kinematic* (RTK) surveys. Most manufacturers broadcast the observations at the base station to the rover. The roving receiver uses the relative positioning techniques discussed in Section 13.9 to determine the position of the roving receiver. However, it is possible to compute and broadcast pseudorange corrections (PRC). Once the pseudorange corrections are determined, they are used at the roving receivers to correct their pseudoranges. Multiplying Equation (13.14) by λ, and including the radial orbital error term, the carrier phase pseudorange at base station A for satellites j at epoch t_0 is

$$\lambda\Phi^j_A(t_0) = \rho^j_A(t_0) + \Delta\rho^j_A(t_0) + \lambda N^j_A + c[\delta^j(t_0) - \delta_A(t_0)] \tag{13.24}$$

where N^j_A is the initially unknown ambiguity, and all other terms were previously defined in Equation (13.20). Recalling that the base station is a point with known coordinates, the pseudorange correction at epoch t_0 is given by

$$\begin{aligned} PRC^j(t_0) &= -\lambda\Phi^j_A(t_0) + \rho^j_A(t_0) \\ &= -\Delta\rho^j_A(t_0) - \lambda N^j_A - c[\delta^j(t_0) - \delta_A(t_0)] \end{aligned} \tag{13.25}$$

and the pseudorange correction at any epoch t is

$$PRC^j(t) = PRC^j(t_0) + RRC^j(t_0)\,(t - t_0) \tag{13.26}$$

Using the same procedure as was used with code pseudoranges, the corrected phase range at the roving receiver for epoch t is

$$\lambda\,\Phi^j_B(t)_{\text{corrected}} = \rho^j_B(t) + \lambda\Delta N^j_{AB} - c\Delta\delta_{AB}(t) \tag{13.27}$$

where $\Delta N^j_{AB} = N^j_B - N^j_A$ and $\Delta\delta_{AB}(t) = \delta_B(t) - \delta_A(t)$.

These equations can be solved as long as at least four satellites are observed continuously during the survey. The pseudorange corrections and the range-rate corrections are transmitted to the receivers.

■ 13.9 RELATIVE POSITIONING

The most precise GPS positions are currently obtained using relative positioning techniques. Similar to both DGPS and kinematic GPS, this method removes most errors noted in Table 13.2 by utilizing the differences in either the code or carrier phase ranges. The objective of relative positioning is to obtain the coordinates of a point relative to another point. This can be mathematically expressed as

$$
\begin{aligned}
X_B &= X_A + \Delta X \\
Y_B &= Y_A + \Delta Y \\
Z_B &= Z_A + \Delta Z
\end{aligned}
\tag{13.28}
$$

where (X_A, Y_A, Z_A) are the geocentric coordinates at the base station A, (X_B, Y_B, Z_B) are the geocentric coordinates at the unknown station B, and $(\Delta X, \Delta Y, \Delta Z)$ are the computed *baseline vector components* (see Figure 13.12).

Relative positioning involves the use of two or more receivers observing pseudoranges simultaneously at the endpoints of lines. Simultaneity implies that the receivers are collecting observations from the same satellites at the same time. It is also important that the receivers collect data at the same epoch rate. This rate depends on the purpose of the survey and its final desired accuracy, but common intervals are 1, 2, 5, 10, or 15 sec. Assuming that simultaneous observations have been collected, different linear combinations of the equations can be produced, and in the process certain errors can be eliminated. Figure 13.13 shows three linear combinations and the required receiver-satellite combinations for each. These are described in the subsections that follow, and only carrier-phase measurements are considered.

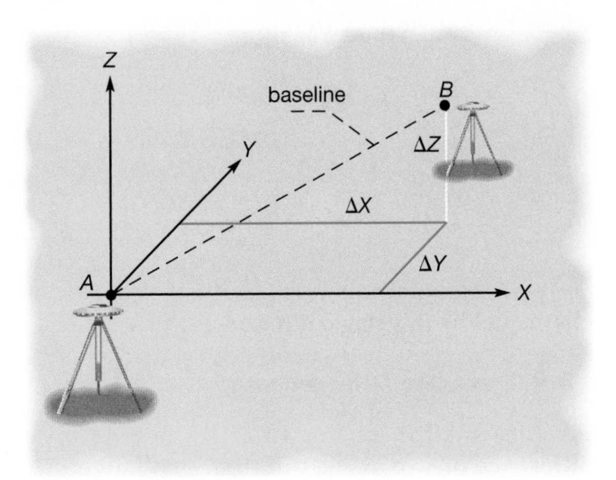

Figure 13.12 Computed baseline vector components.

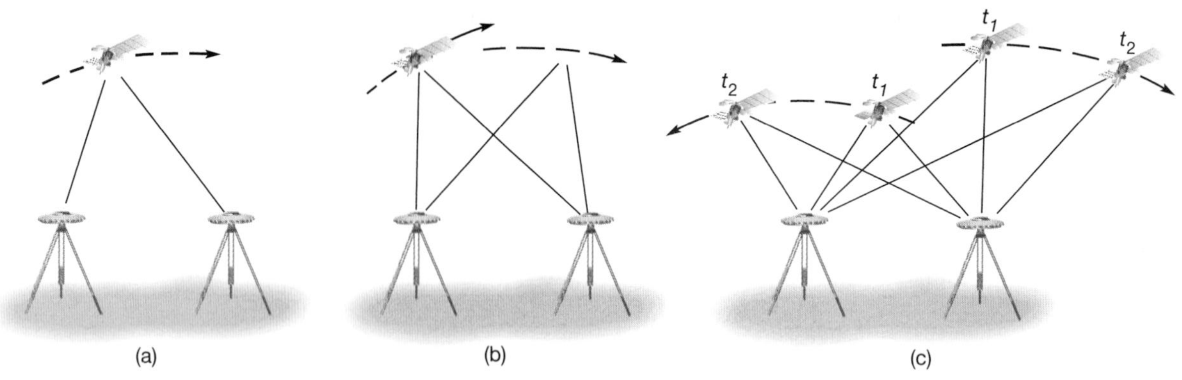

Figure 13.13 GPS differencing techniques: (a) single differencing, (b) double differencing, and (c) triple differencing.

13.9.1 Single Differencing

As illustrated in Figure 13.13(a), single differencing involves subtracting two simultaneous observations made to one satellite from two points. This difference eliminates the satellite clock bias and much of the ionospheric and tropospheric refraction from the solution. It would also eliminate the effects of *SA* if it were turned on. Following Equation (13.14), the phase equations for the two points are

$$\Phi^j_A(t) - f^j\delta^j(t) = \frac{1}{\lambda}\rho^j_A(t) + N^j_A - f^j\delta_A(t)$$

$$\Phi^j_B(t) - f^j\delta^j(t) = \frac{1}{\lambda}\rho^j_B(t) + N^j_B - f^j\delta_B(t) \tag{13.29}$$

where the terms are as noted in Equation (13.14) for stations *A* and *B*. The difference in these two equations yields

$$\Phi^j_{AB}(t) = \frac{1}{\lambda}\rho^j_{AB}(t) + N^j_{AB} - f^j\delta_{AB}(t) \tag{13.30}$$

where the individual difference terms are

$$\begin{aligned}\Phi^j_{AB}(t) &= \Phi^j_B(t) - \Phi^j_A(t),\\ \rho^j_{AB}(t) &= \rho^j_B(t) - \rho^j_A(t),\\ N^j_{AB} &= N^j_B - N^j_A, \text{ and}\\ \delta^j_{AB}(t) &= \delta^j_B(t) - \delta^j_A(t).\end{aligned}$$

Note that in Equation (13.30), the satellite clock bias error, $f^j\delta^j(t)$ has been eliminated by this single differencing procedure.

13.9.2 Double Differencing

As illustrated in Figure 13.13(b), double differencing involves taking the difference of two single differences obtained from two satellites j and k. The procedure eliminates the receiver clock bias. Assume the following two single differences:

$$\Phi^j_{AB}(t) = \frac{1}{\lambda}\rho^j_{AB}(t) + N^j_{AB} - f^j\delta^j_{AB}(t)$$
$$\Phi^k_{AB}(t) = \frac{1}{\lambda}\rho^k_{AB}(t) + N^k_{AB} - f^k\delta^k_{AB}(t) \tag{13.31}$$

Note that the receiver clock bias will be the same for observations on satellite j as it is for satellite k. Thus by taking the difference between these two single differences, the following double difference equation is obtained, in which the receiver clock bias errors, $f^j\delta^j_{AB}(t)$ and $f^k\delta^k_{AB}(t)$ are eliminated.

$$\Phi^{jk}_{AB}(t) = \frac{1}{\lambda}\rho^{jk}_{AB}(t) + N^{jk}_{AB} \tag{13.32}$$

where the difference terms are

$$\Phi^{jk}_{AB}(t) = \Phi^k_{AB}(t) - \Phi^j_{AB}(t),$$
$$\rho^{jk}_{AB}(t) = \rho^k_{AB}(t) - \rho^j_{AB}(t), \text{ and}$$
$$N^{jk}_{AB} = N^k_{AB} - N^j_{AB}.$$

13.9.3 Triple Differencing

The triple difference illustrated in Figure 13.13(c) involves taking the difference between two double differences obtained for two different epochs of time. This difference removes the integer ambiguity from Equation (13.32), leaving only the differences in the phase-shift observations and the geometric ranges. The two double-difference equations can be expressed as

$$\Phi^{jk}_{AB}(t_1) = \frac{1}{\lambda}\rho^{jk}_{AB}(t_1) + N^{jk}_{AB}$$
$$\Phi^{jk}_{AB}(t_2) = \frac{1}{\lambda}\rho^{jk}_{AB}(t_2) + N^{jk}_{AB} \tag{13.33}$$

The difference in these two double differences yields the following triple difference equation, in which the integer ambiguities have been removed. The triple difference equation is

$$\Phi^{jk}_{AB}(t_{12}) = \frac{1}{\lambda}\rho^{jk}_{AB}(t_{12}) \tag{13.34}$$

In Equation (13.34) the two difference terms are:

$$\Phi^{jk}_{AB}(t_{12}) = \Phi^{jk}_{AB}(t_2) - \Phi^{jk}_{AB}(t_1), \text{ and}$$
$$\rho^{jk}_{AB}(t_{12}) = \rho^{jk}_{AB}(t_2) - \rho^{jk}_{AB}(t_1).$$

The importance of employing the triple difference equation in the solution is that by removing the integer ambiguities, the solution becomes immune to *cycle slips*. Cycle slips are created when the receiver *loses lock* during an observation session. The three main sources of cycle slips are (1) obstructions, (2) low *signal to noise ratio* (SNR), and (3) incorrect signal processing. Signal obstructions can be minimized by careful selection of receiver stations. Low SNR can be caused by undesirable ionospheric conditions, multipathing, high receiver dynamics, or low satellite elevations. Malfunctioning satellite oscillators can also cause cycle slips, but this rarely occurs. It should be noted that due to more advanced methods of determining integer ambiguities (see Section 15.2) triple differencing is seldom, if at all, performed today.

■ 13.10 OTHER SATELLITE NAVIGATION SYSTEMS

GPS positioning is affecting all walks of life including transportation, agriculture, void and data networks, cell phone technology, sporting events, and so on. In fact, the military and economic benefits of GPS have been so great that other nations have or will be developing their own networks. This plethora of positioning satellites will greatly increase the utility and accuracy available from satellite positioning system. Because of this, the entire satellite positioning technology is now known as *Global Navigation Satellite System* (GNSS). Other implemented or planned satellite positioning systems are discussed in the following subsections.

13.10.1 The GLONASS Constellation

The ***Glo**bal **Na**vigation **S**atellite **S**ystem* (GLONASS) is the Russian equivalent of GPS. When completed, the GLONASS constellation will contain 24 satellites equally spaced in three orbital planes making a 64.8° nominal inclination angle with the equatorial plane of the Earth. The satellites orbit at a nominal altitude of 19,100 km and have a period of 11.25 hours. When the system has its full complement of satellites, at least five will always be visible to the user. The system is free from selective availability, but does not permit public access to the P code. Each satellite broadcasts two signals with frequencies that are unique. The frequencies of the satellites are determined as

$$\begin{aligned} f^j_{L1} &= 1602.0000 \text{ MHz} + j \times 0.5625 \text{ MHz} \\ f^j_{L2} &= 1246.0000 \text{ MHz} + j \times 0.4375 \text{ MHz} \end{aligned} \quad (13.35)$$

where j represents the channel number assigned to the specific satellite,[11] and varies from 1 to 24, and L1 and L2 represent the broadcast bands.

As discussed in Section 13.3, GPS satellites broadcast their positions in every repetition of the broadcast message using the WGS84 reference system as the basis for coordinates. The GLONASS satellites only broadcast their positions every 30 min and use the PZ-90 reference ellipsoid as the basis for coordinates. Thus, GNSS receivers must extrapolate positions of the satellites for real-time reductions.

[11]Some antipodal satellites use the same frequencies.

The time reference systems used in GPS and GNSS are also different. At the request of the international community, the timing of GNSS satellites has moved toward the international standard as set by the *Bureau Internationale de l'Heure* (International Bureau of Time). This standard is based on the frequency of the atom Cesium 133 in its ground state.[12] This standard differs from the orbital period of the Earth by approximately 1 sec every 6 months. To compensate, one leap second is periodically added to the atomic time (IAT) to create Universal Coordinated Time (UTC) which agrees with the solar day (see Section C.5). Currently, the GNSS system clocks differ from Universal Coordinated Time by three hours. In contrast, the GPS system clocks never account for the leap second, and differ from IAT by a constant of 19 seconds.

13.10.2 Galileo System

In 1998, the European Union decided to implement another satellite positioning system called Galileo. The Galileo system will offer five levels of service with subscriptions required for some of the services. The five levels of service are (1) *open* service (OS), (2) *commercial* service (CS), (3) *safety-of-life* service (SOL), (4) *public-regulated* service (PR), and (5) *search and rescue* (SAR) service. OS will be a free service offering positioning down to 1 m. CS is an encrypted, subscription service, which will provide positioning at the cm level. SOL will be a free service providing both guaranteed accuracy and integrity messages to warn of errors. The PR service will be available only to government agencies, which is similar to the GPS P-code. The SAR service will pick up distress beacon locations and will be able to send feedback indicating that help is on the way.

The Galileo space segment will consist of 27 satellites plus 3 spares orbiting in three planes that are inclined to the equator at 56°. The satellites will have a nominal orbital altitude of 23,222 km above the Earth. The satellites will broadcast six navigation signals denoted as L1F, L1P, E6C, E6P, E5a, and E5b. The first Galileo satellite was launched in December 2005. After a failure in the second satellite, the second launch was delayed to late 2007. Galileo should offer greater accuracy than GPS with its commercial service providing meter-level point positioning. Like the modernized GPS satellites, the strength of its signals should allow work in canopy situations. The U.S. and European Union have agreed to make their systems interoperable. Thus future receivers will be able to use satellites from either system.

13.10.3 Compass

In 2006, China confirmed that it will create a fourth GNSS. Compass[13] will contain 35 satellites. Five of these satellites will be geostationary Earth orbit (GEO) satellites with the remaining 30 satellites at about 20,000 km. Compass will offer two levels of service—an open and a commercial service with real-time positioning accuracy of 10 m. The announced completion date for the system is 2010. However,

---

[12]One second is defined as 9,192,631,770 periods of the radiation of the ground state of the cesium 133 atom.

[13]The Chinese name for their system is Beidou, which stands for North Dipper. Compass is being used in English writings on this system.

this date may be delayed since the proposed transmissions would overlay signals in both Galileo and GPS. The International Telecommunications Union is currently working on a compromise solution to the proposed system.

Even though the GLONASS and Galileo constellations are not complete at the time of this writing, manufacturers of satellite receiver technology are building receivers that will utilize these systems. The obvious advantage of using multiple systems is that many more satellites are available for observation by receivers. By combining these systems, the surveyor can expect improvements in increased speed and accuracy. Furthermore, the combination of systems will provide a viable method of bringing satellite positioning into difficult areas such as canyons, deep surface mines, and urban areas surrounded by tall buildings (urban canyons).

13.11 THE FUTURE

The overall success of GPS in the civilian sector is well documented by the number and variety of enterprises that are using this technology. This has led to increasing and improving GNSS constellations. In the near future, improvements will occur in signal acquisition and positioning. For example, signals from all three satellite-positioning systems will be able to penetrate canopy situations and may provide satellite-positioning capabilities from within buildings. The additional signals from within each system will improve both ambiguity resolution and atmospheric corrections. For example, in GPS with the addition of the L2C and L5 signals, real-time ionospheric corrections to the pseudoranges will become possible. It is anticipated that accuracies in the modernized system will be reduced to the millimeter level. The addition of GLONASS and Galileo satellites is only expected to enhance these capabilities. This will provide civilian GPS users with unprecedented real-time determination of highly accurate location anywhere on the Earth.

The use of GPS in the surveying (geomatics) community has continued to increase as the costs of the systems have decreased. This technology has and will undoubtedly continue to have considerable impact on the way data is collected and processed. In fact, as the new satellite technologies are developed, the use of conventional surveying equipment will decrease. This is due to the ease, speed, and achievable accuracies that satellite positioning technologies provide.

As is the current trend, less field time will be required for the surveyor (geomatics engineer), and more time will be used to analyze, manage, and manipulate the large volumes of data that this technology and others provide. Those engaged in surveying (geomatics) in the future will need to be knowledgeable in the areas of information management and computer science, and will provide products to clients that currently do not exist.

PROBLEMS

Asterisks (*) indicate problems that have answers given in Appendix G.

13.1 Briefly describe the orbits of GPS satellites.

13.2 Briefly describe the orbits of the Galileo satellites.

13.3 What is the difference between the standard positioning service and the precise positioning service?

13.4* Discuss the purpose of the pseudorandom noise codes.

13.5 Discuss the different frequencies used for carrier and code signals of GPS, and give their relationships to one another.

13.6 How will the modernized GPS satellites change surveying?

13.7 What information is broadcast by GPS satellites?

13.8 Describe the general procedure used in relative positioning.

13.9 What advantages does relative positioning have over point positioning?

13.10 What is meant by geodetic height? How is it different from elevations as discussed in Chapter 4?

13.11 How can a receiver's clock bias be eliminated in GPS positioning with code measurements?

13.12 How can a satellite's clock bias be eliminated in GPS positioning with code measurements?

13.13 Same as Problem 13.11, but with carrier phase measurements.

13.14 Same as Problem 13.12, but with carrier phase measurements.

13.15 What is GPS single differencing? Double differencing?

13.16 Define the term dilution of precision and its importance to GPS surveying.

13.17 List and discuss the ephemerides.

13.18 What is meant by PDOP, HDOP, and VDOP?

13.19 If the HDOP during a survey is 1.9, what is the 95 percent horizontal point-positioning error with today's GPS system?

13.20 In Problem 13.19, if the VDOP is 4.5, what is the 95 percent point-positioning error in geodetic height with today's GPS system?

13.21* What are the geocentric coordinates of a station in meters which has a latitude of 39°27′07.5894″ N, longitude of 86°16′23.4907″ W, and height of 203.245 m? (Use the WGS84 ellipsoid parameters.)

13.22 Same as Problem 13.21, except with a latitude of 55°06′57.9029″ N, longitude of 88°54′39.0646″ W, and height of 135.204 m?

13.23 Same as Problem 13.21, except with a latitude of 35°37′26.0936″ N, longitude of 109°42′15.0295″ W, and height of 202.892 m?

13.24* What are the geodetic coordinates in meters of a station with geocentric coordinates of (136,153.995, −4,859,278.535, 4,115,642.695)? (Use the WGS84 ellipsoid parameters.)

13.25 Same as Problem 13.24, except with geocentric coordinates of (1,688,140.543, −4,178,892.256, 4,498,404.776)?

13.26 Same as Problem 13.24, except with geocentric coordinates of (−1,723, 206.772, −5,302,588.623, 3,087, 329.886)?

13.27 Discuss the method of determining a local value for geoid undulation.

13.28* The GPS determined height of a station is 284.097 m. The geoid undulation at the point is −30.052 m. What is the elevation of the point?

13.29 Same as Problem 13.28, except the height is 105.684 m and the geoid undulation is −28.968.

13.30* The elevation of a point is 124.886 m. The geoid undulation of the point is 28.998 m. What is the geodetic height of the point?

13.31 Same as Problem 13.30, except the elevation is 565.904 m, and the geoid undulation is −25.649 m.

13.32 The GPS observed height of two stations is 124.687 m and 89.967 m, and their elevations are 153.106 m and 118.384 m, respectively. These stations have model-derived geoid undulations of −28.454 m and −28.457 m, respectively. What is the elevation of a station with a GPS measured height of 95.184 m and a model-derived geoid undulation of 28.360 m?

13.33 Why is the mask angle generally set to a value above 10°?

13.34 List the differences between the GLONASS, Galileo, and GPS systems.

13.35 Create a computational program that converts geocentric coordinates to geodetic coordinates.

13.36 Create a computational program that converts geodetic coordinates to geocentric coordinates.

13.37 Find at least two Internet sites that contain information about GPS. Summarize the contents of each site.

13.40 Find at least two Internet sites that contain software or data to be used with GPS. Summarize the contents of each site.

13.41 Use the Internet to identify at least two non-surveying industries not mentioned in this chapter that use GPS. Summarize the contents of each site.

13.42 Research the European Union system known as Galileo and prepare a written report on the system.

BIBLIOGRAPHY

GIA, 2001. "GPS Antennas." *Professional Surveyor* 21 (No. 4): 30.

Hofmann-Wellenhof, B., et al. 2004. *GPS Theory and Practice.* 5th ed. New York: Springer-Verlag.

Martin, D. J. 2003. "Around and Around with Orbits." *Professional Surveyor* 23 (No. 6): 50.

———. 2003. "Reaching New Heights in GPS, Part 3." *Professional Surveyor* 23 (No. 4): 42.

Reilly, J. 2002. "On Galileo, the European Satellite Navigation System." *Point of Beginning* 28 (No. 12): 46.

Reilly, J. 2003. "On Geoid Models." *Point of Beginning* 29 (No. 12): 50.

Snay, R., et al. 2002. "GPS Precision with Carrier Phase Observations: Does Distance and/or Time Matter?" *Professional Surveyor* 22 (No. 10): 20.

Vittorini, L. D. and B. Robinson. 2003. "Optimizing Indoor GPS Performance." *GPS World* 14 (No. 11): 40.

14

The Global Positioning System—Static Surveys

14.1 INTRODUCTION

Many factors have a bearing on the ultimate success of a GPS survey. Also there are many different approaches that can be taken in terms of equipment used and procedures followed. Because of these variables, GPS surveys should be carefully planned prior to going into the field. Many smaller projects of lower-order accuracy may not require a great deal of preplanning beyond selecting receiver sites and making sure they are free from overhead obstructions. On the other hand, large projects that must be executed to a high order of accuracy will require extensive preplanning to increase the probability that the survey will be successful. As an example, a survey for the purpose of establishing control for an urban rapid transit project will command the utmost care in selecting personnel, equipment, and receiver sites. It will also be necessary to make pre-survey site visits to locate existing control and identify possible overhead obstructions that could interfere with incoming satellite signals at all proposed receiver sites. Additionally, a careful preanalysis should be made to plan optimum *observation session*[1] times, the durations of the sessions, and to develop a plan for the orderly execution of the sessions. The project will probably require ground communications to coordinate survey activities, a transportation analysis to ensure reasonable itineraries for the execution of the survey, and installation of monuments to permanently mark the new points

[1]An observation session denotes the period of time during which all receivers being employed on a project have been set up on designated stations and are simultaneously engaged in receiving satellite signals. When one session is completed, all receivers except one are generally moved to different stations and another observation session is conducted. The sessions are continued until all planned project observations have been completed.

that will be located in the survey. Consideration of these factors, and others, in planning and executing GPS projects is the subject of this chapter.

This chapter concentrates on the use of carrier phase-shift receivers and the employment of relative positioning methods. This combination can provide the highest level of accuracy in determining the positions of points and thus it is the preferred approach in surveying (geomatics) applications. But as noted previously, the accuracy of a GPS survey is also dependent on several additional variables. An important one is the type of carrier phase receiver used on the survey. As noted in Chapter 13, there are two types: *dual-frequency receivers,* which can observe both the L_1 and L_2 bands, and *single-frequency receivers,* which can observe only the L_1 band. The former is preferred for several reasons: they can (a) collect the needed data faster; (b) observe longer baselines with greater accuracy; and (c) eliminate certain errors, such as ionospheric refraction, and therefore yield higher positional accuracies. Receivers also vary by the number of channels. This controls the number of satellites that they can track simultaneously. As a minimum, carrier phase-shift receivers must have at least four channels, but some are capable of tracking as many as 30 satellites from the GPS, GLONASS, and Galileo constellations simultaneously using multiple frequency bands resulting in more than 60 channels. These receivers, known as Global Navigation Satellite System (GNSS) receivers, provide higher accuracies due to the increased number of satellites and increased strength in satellite geometry.

Other important variables that bear on the accuracy of a GPS survey include the (1) accuracy of the reference station(s) to which the survey will be tied, (2) number of satellites visible during the survey, (3) geometry of the satellites during the observation sessions, (4) atmospheric conditions during the observations, (5) lengths of observation sessions, (6) number and nature of obstructions at the proposed receiver stations, (7) number of redundant observations taken in the survey, and (8) method of reduction used by the software. Some of these factors are beyond control of the surveyor (geomatics engineer), and therefore it is imperative that observational checks be made. These are made possible by the redundant observations. This chapter will discuss these checks.

The use of GPS for specific types of surveys, e.g., construction surveys, land surveys, photogrammetric surveys, etc., is covered in later chapters in this text. A GPS unit being used for a construction stakeout survey is shown in Figure 14.1. GPS is especially useful for topographic surveys and this application is covered in Section 17.9.5.

■ 14.2 FIELD PROCEDURES IN GPS SURVEYS

In practice, field procedures employed on GPS surveys depend on the capabilities of the receivers and the type of survey. Some specific field procedures currently being used in surveying include the *static, rapid static, pseudokinematic,* and *kinematic* methods. These are described in the subsections that follow. All are based on carrier phase-shift measurements and employ *relative positioning* techniques (see Section 13.9); that is, two (or more) receivers, occupying different stations and simultaneously making observations to several satellites. The vector (distance) between receivers is called a *baseline* as described in Section 13.9, and its ΔX, ΔY,

Figure 14.1 GPS receiver being used in construction stakeout. (Courtesy Ashtech, Inc.)

and ΔZ coordinate difference components (in the geocentric coordinate system described in Section 13.4.2) are computed as a result of the observations. Real-time kinematic methods are based on the computational procedures outlined in Section 13.8.

14.2.1 Static Relative Positioning

For highest accuracy, for example geodetic control surveys, static GPS procedures are used. In this procedure, two (or more) receivers are employed. The process begins with one receiver (called the *base receiver*) being located on an existing control station, while the remaining receivers (called the *roving receivers*) occupy stations with unknown coordinates. For the first observing session, simultaneous observations are made from all stations to four or more satellites for a time period of an hour or more depending on the baseline length. (Longer baselines require greater observing times.) Except for one, all the receivers can be moved upon completion of the first session. The remaining receiver now serves as the base station for the next observation session. It can be selected from any of the receivers used in the first observation session. Upon completion of the second session, the process is repeated until all stations are occupied, and the observed baselines form geometrically closed figures. As discussed in Section 14.5, for checking purposes, some repeat baseline observations should be made during the observational process.

The value for the *epoch rate*[2] in a static survey must be the same for all receivers during the survey. Typically, this rate is set to 15 sec to minimize the

[2]GPS satellites continually transmit signals, but if they were continuously collected by the receivers, the volume of data and hence storage requirements would become overwhelming. Thus the receivers are set to collect samples of the data at a certain time interval, which is called the epoch rate.

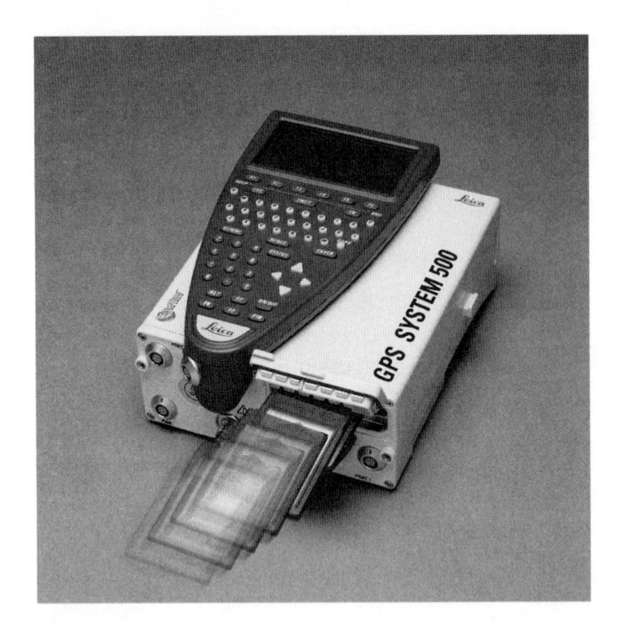

Figure 14.2 LEICA GPS System 500 with PCMCIA card. (Courtesy Leica Geosystems, Inc.)

number of observations and data storage requirements. Most receivers are connected to *controllers* that have internal memories for storing the observed data. After all observations are completed, data are transferred to a computer for post-processing. The GPS receiver shown in Figure 14.2 has a PCMCIA card for downloading field data (see Section 2.12).

Relative accuracies with static relative positioning are about ±(3 to 5 mm + 1 ppm). Typical durations for observing sessions using this technique, with both single and dual frequency receivers, are shown in Table 14.1.

14.2.2 Rapid Static Relative Positioning

This procedure is similar to static surveying, except that one receiver always remains on a control station while the other(s) are moved progressively from one unknown point to the next. An observing session is conducted for each point, but the sessions are shorter than for the static method and the epoch rates higher. Table 14.1 also shows the suggested session lengths for single and dual frequency receivers. The rapid static procedure is suitable for observing base lines up to 20 km in length under good conditions. Rapid static relative positioning can also yield accuracies on the order of about ±(3 to 5 mm + 1 ppm). However to achieve these accuracies, optimal satellite configurations (good PDOP), and favorable

TABLE 14.1 TYPICAL SESSION LENGTHS FOR VARIOUS OBSERVATION METHODS

Method of Survey	Single Frequency	Dual Frequency
Static	30 min + 3 min/km	20 min + 2 min/km
Rapid Static	20 min + 2 min/km	10 min + 1 min/km

ionospheric conditions must exist. This method is ideal for small control surveys. As with static surveys, all receivers should be set to collect data at the same epoch rate. Typically the epoch rate is set to 5 sec with this method.

14.2.3 Pseudokinematic Surveys

This procedure is also known as the *intermittent* or *reoccupation* method, and like the other static methods requires a minimum of two receivers. In pseudokinematic surveying, the base receiver always stays on a control station, while the rover goes to each point of unknown position. Two relatively short observation sessions (around 5 min each in duration) are conducted with the rover on each station. The time lapse between the first session at a station and the repeat session should be about an hour. This produces an increase in the geometric strength of the observations due to the change in satellite geometry. Reduction procedures are the same as those described in Section 13.9 and accuracies approach those of static surveying.

A disadvantage of this method, compared to other static methods, is the need to revisit the stations. This procedure requires careful pre-survey planning to ensure that sufficient time is available for site revisitation and to achieve the most efficient travel plan. Pseudokinematic surveys are most appropriately used where the points to be surveyed are along a road and rapid movement from one site to another can be readily accomplished. During the movement from one site to another, the receiver can be turned off. Some projects for which pseudokinematic surveys may be appropriate include alignment surveys (see Chapters 24 and 25), photo-control surveys (see Chapter 27), lower-order control surveys, and mining surveys.

14.2.4 Kinematic Surveys

As the name implies, during kinematic surveys one receiver, the rover, can be in continuous motion. This is the most productive of the GPS surveys, but is also the least accurate. The accuracy of a kinematic GPS survey is typically in the range of $\pm(1 \text{ to } 2 \text{ cm} + 2 \text{ ppm})$. This accuracy is sufficient for many types of surveys and thus is the most common method of surveying. Kinematic GPS is applicable for any type of survey that requires many points to be located, which makes it very appropriate for most topographic and construction surveys. It is also excellent for dynamic types of surveying, i.e., where the observation station is in motion. The range of a kinematic survey is typically limited to the broadcast range of the base radio. However, real-time networks have made kinematic surveys possible over large regions. Chapter 15 explores the procedures used in kinematic GPS surveying procedures in greater detail. The remainder of this chapter is devoted to static surveying methods.

■ 14.3 PLANNING GPS SURVEYS

As noted earlier, small GPS surveys generally do not require much in the way of project planning. However for large projects and for higher-accuracy surveys, project planning is a critical component in obtaining successful results. The subsections that follow discuss various aspects of project planning with emphasis on control surveys.

14.3.1 Preliminary Considerations

All new high-accuracy GPS projects that employ relative positioning techniques must be tied to nearby existing control points. Thus one of the first things that must be done in planning a new project is to obtain information on the availability of existing GPS control stations near the project area. For planning purposes, these should be plotted in their correct locations on an existing map or aerial photos of the area.

Another important factor that must be addressed in the preliminary stages of planning for GPS projects is the selection of the new station locations. Of course, they must be chosen so that they meet the overall project objective. But in addition, terrain, vegetation, and other factors must be considered in their selection. If possible they should be reasonably accessible by either the land vehicles or aircraft that will be used to transport the GPS hardware. The stations can be somewhat removed from vehicle access points since GPS components are relatively small and portable. Also, the receiver antenna is the only hardware component that must be accurately centered over the ground station. It is easily carried and can be separated from the other components by a length of cable, as shown in Figure 14.3. Once the preliminary station locations are selected, they should be plotted on the map or aerial photo of the area.

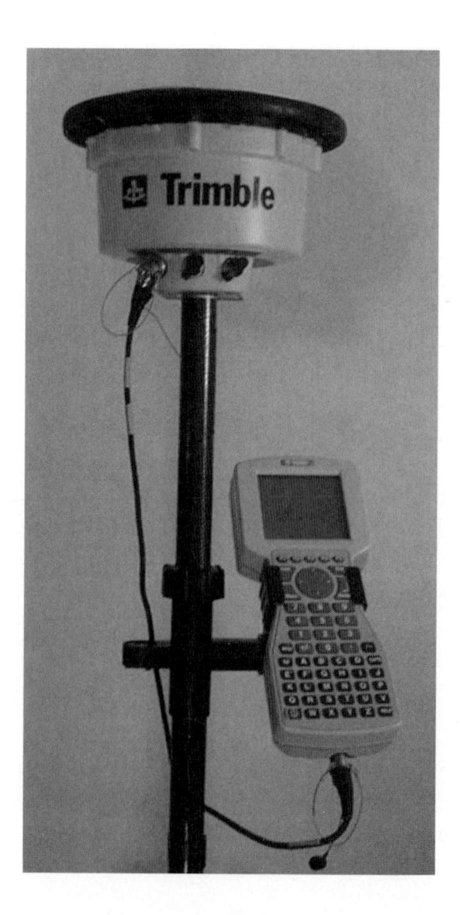

Figure 14.3 GPS antenna attached with cables to controller unit.

Another consideration in station selection is the assurance of an overhead view free of obstructions. This is known as *canopy restrictions*. As a minimum, it is recommended that visibility be clear in all directions from a mask angle (altitude angle) of 10° from the horizon. In some cases, careful station placement will enable this visibility criterion to be met without difficulty; in other situations, clearing around the stations may be necessary. Furthermore, as discussed in Section 13.6.3, potential sources that can cause interference and multipath errors should also be identified when visiting each site.

Selecting suitable *observation windows* is another important activity in planning GPS surveys. This consists of determining which satellites will be visible from a given ground station or project area during a proposed observing period. To aid in this activity, azimuth and elevation angles to each visible satellite can be predetermined by computers for times within the planned observation period. Required computer input, in addition to observing date and time, includes the station's approximate latitude and longitude, and a relatively current satellite almanac.

To aid in selecting suitable observation windows, a *satellite availability plot*, as shown in Figure 14.4, can be applied. The shaded portion of this diagram shows the number of satellites visible from station *PSU1* whose position is 41°18′00.00″ N latitude and 76°00′00.40″W longitude. The diagram is applicable for August 4, 1999, between the hours of 8:00 and 17:00 EDT. A mask angle of 15° has been used. In addition to showing the number of visible satellites, the lines running through the plot depict the predicted PDOP, HDOP, and VDOP (see Section 13.6.4) for this

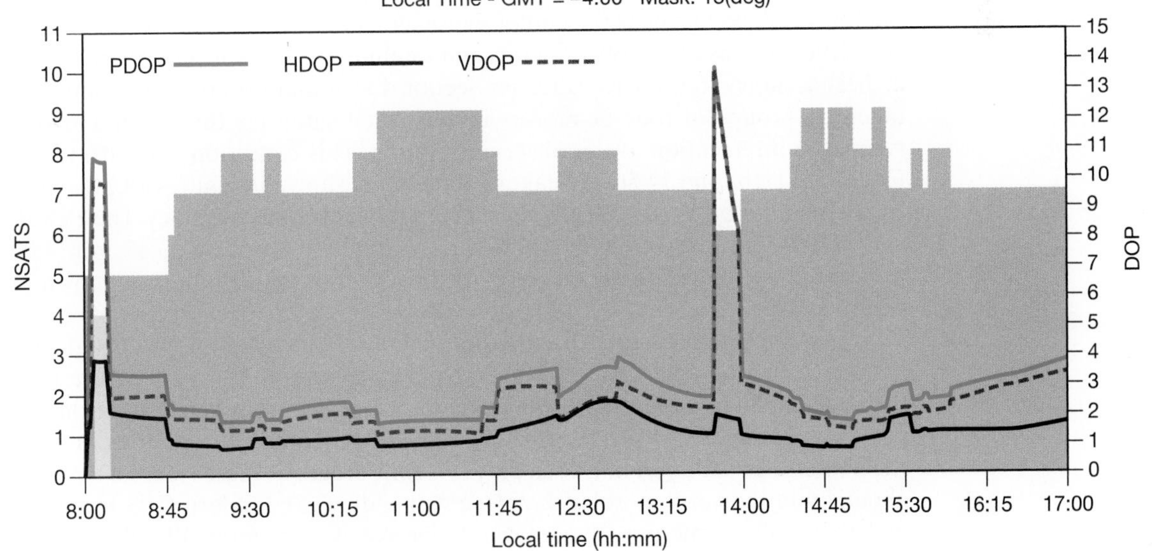

Figure 14.4 Satellite availability plot.

time period. It should be noted that, for the day shown in Figure 14.4, only two short time periods are unacceptable for data collection. DOP *spikes* occur between 8:02 and 8:12 when only four satellites are above the horizon mask angle, and between 13:45 and 14:00 when both VDOP and PDOP are unacceptable because of weak satellite geometry. However during this latter period, the HDOP is acceptable, indicating that a horizontal control survey could still be executed. Note also that one of the better times for a GPS survey is between 10:40 and 11:30 when PDOP is below 2, because 9 satellites are visible during that time. However if the receiver has less than 9 channels, then the satellite availability chart must be altered by disabling some of the usable satellites or by raising the mask angle.

Satellite visibility at any station is easily and quickly investigated using a *sky plot.* These provide a graphic representation of the azimuths and elevations to visible satellites from a given location. As illustrated in Figure 14.5(a) and (b), sky plots consist of a series of concentric circles. The circumference of the outer circle is graduated from 0 to 360° to represent satellite azimuths. Each successive concentric circle progressing toward the center represents an increment in the elevation angle with the radius point corresponding to the zenith.

For each satellite, the PRN number is plotted beside its current data point, which is its location for the selected finishing time of the survey. The arcs connect successive plotted positions for time increments after the initial time. Thus, travel paths in the sky of visible satellites are shown. Sky plots are valuable in GPS planning because they enable operators to quickly visualize not only the number of satellites available during a planned observation period, but also their geometric distribution in the sky.

In project planning, the elevations and azimuths of vertical obstructions near the station can be overlaid with the sky plot to form *obstruction diagrams.* As shown in Figure 14.5(a), a building will obscure satellite 6 briefly during the time period shown. Also note that signals from satellite 30 will experience a brief interruption caused by the presence of a nearby pole later in the session and that satellite 4 will be lost near the end of the session because of a nearby building.

The analysis of sky obstructions and satellite geometry is important for the highest accuracy in GPS. Recall from Section 13.6.4 that observations should be taken on groups of four or more widely spaced satellites that form a strong geometric intersection at the observing station. This condition is illustrated in Figures 13.11(b) and 14.5(a). Weaker geometry, as shown in Figures 13.11(a) and 14.5(b), should be avoided if possible as it will yield lower accuracy. The PDOP and VDOP spikes shown in Figure 14.4 between the times from 13:45 and 14:00 are caused by the relatively clustered distribution of the satellites as shown in Figure 14.5(b).

It is important to note that the optimal observation times will repeat four minutes earlier for each day following the planning session. That is, in Figure 14.4, the same satellite visibility chart will apply for the period from 7:56 to 16:56 on August 5, and from 7:52 to 16:52 on August 6, etc. Of course, the periods of poor PDOP will also shift by 4 min each day. This shift occurs because sidereal days are about 4 min shorter than solar days (see Appendix C.5). Modern GPS receivers, with their built-in microprocessors, can derive sky plots, compute PDOP values in the field, and display the results on their screens.

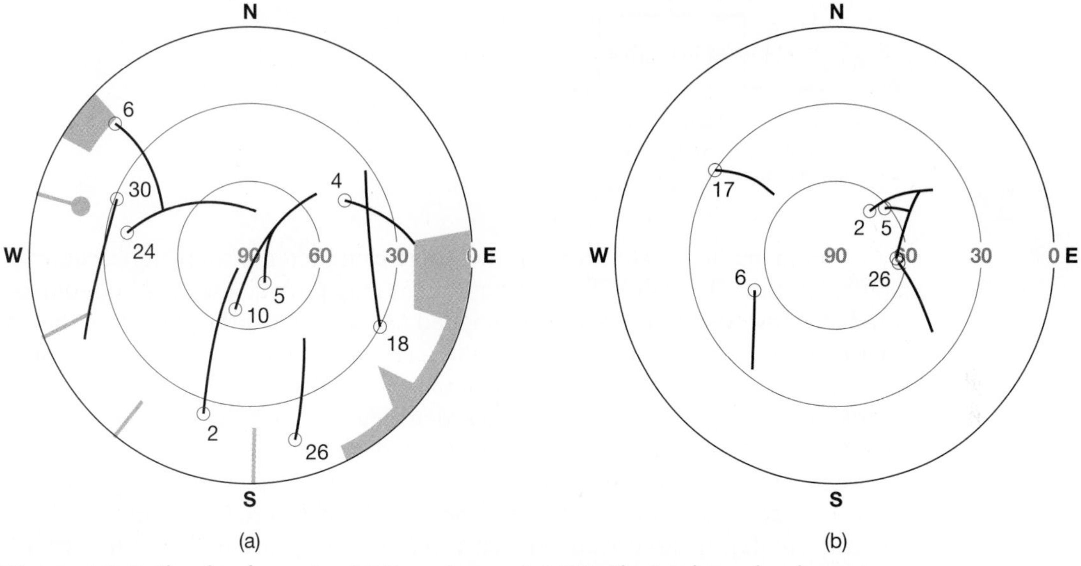

Figure 14.5 Sky plots for station PSU1 on August 4, 1999. Plot (a) shows the obstructing objects around the station from 10 to 12. Plot (b) shows the weak satellite geometry from 13:30 to 14.

14.3.2 Selecting the Appropriate Survey Method

As discussed in Section 14.2, several different methods are available with which to accomplish a GPS survey. Each method provides a unique set of procedural requirements for field personnel. In high-accuracy GPS surveys that involve long baselines, the static surveying method with dual frequency receivers is the best solution. However in typical surveys limited to small areas, a single frequency receiver using rapid static, pseudokinematic, or kinematic surveying methods (see Chapter 15) may be sufficient. Because of the variability in requirements and restrictions of GPS surveys, there is seldom only one method for accomplishing the work.

GNSS receivers will reduce the time required at each station in a static survey due to the increased number of visible satellites and the improved satellite geometry. Figure 14.6 is a schematic diagram that categorizes the various GPS survey methods. The surveying community has used those shown on the left side of the diagram traditionally.

The selection of the appropriate survey method is dependent on the (1) desired level of accuracy in the final coordinates, (2) intended use of the survey, (3) type of equipment available for the survey, (4) size of the survey, (5) canopy and other local conditions for the survey, and (6) available software for reducing the data.

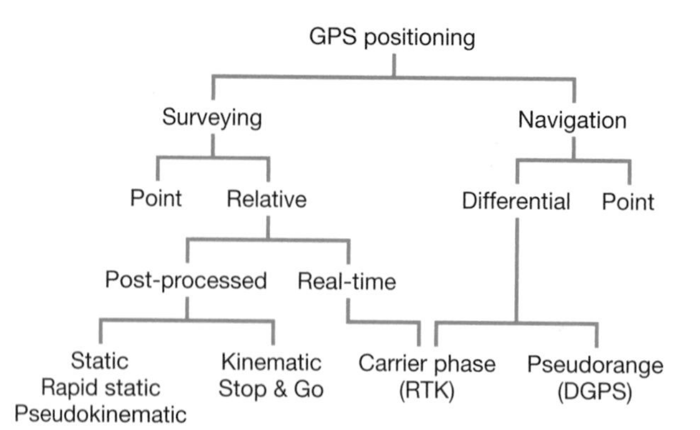

Figure 14.6 Schematic depicting GPS positioning methods.

For mapping or inventory surveys where centimeter to submeter accuracy is sufficient, an RTK or DPGS (see Chapter 15) may provide the most economical product. However if the area to be mapped has several overhead obstructions, it may only be possible to use one of the static survey procedures to bring control into the region and do limited kinematic mapping in small areas where clear overhead views to the satellites are available. Recognizing this, many manufacturers have developed equipment that allows the surveyor (geomatics engineer) to switch between a GPS receiver and a total station instrument (see Chapter 8) using the same data collector. This capability is useful in areas where GPS is not practical. A more in-depth discussion on this topic will be presented in Chapter 17. Restrictions in use of GPS surveys due to canopy restrictions will be greatly reduced when modernized GNSS is available.

The preferred approach in performing control surveys by GPS is the static method. Often a combination of the static methods will provide the most economical results for large projects. As an example, a static survey may be used to bring a sparse network of accurate control into a project area. This could be followed by a rapid static survey to densify control within the area. Finally, a pseudokinematic survey could be used to establish control along specific alignments. In smaller areas with favorable viewing conditions, the best choice for the survey may simply be the rapid static or kinematic methods. Available equipment, software, and experience often dictate the survey method of choice.

The NGS provides the *Online Positioning User Service* (OPUS). When using this service with a kinematic survey, the base station does not have to be positioned over a point with known coordinates. However, for this service to achieve its maximum accuracy, the base station must continuously collect data for a minimum of two hours. Initialization of the roving receiver is obtained using the on-the-fly method of ambiguity resolutions. After the initialization, the roving receiver can proceed with the survey. Before post-processing the data from the roving receiver, the RINEX observation file from the base station is submitted to the NGS for processing over the Internet. In a matter of minutes, OPUS returns the coordinates of the base station using the data from three nearby CORS stations. This procedure removes the need for an initial known base station. A rapid service version of OPUS (OPUS-RS) is currently being tested. This version allows sessions as short as 15 min to be processed.

14.3.3 Field Reconnaissance

Once the existing nearby control points and new stations have been located on paper, a reconnaissance trip to the field should be undertaken to check the selected observation sites for (1) overhead obstructions that rise above 10°–15° from the horizon, (2) reflecting surfaces that can cause multipathing, (3) nearby electrical installations that can interfere with the satellite's signal, and (4) other potential problems. If reconnaissance reveals that any preliminary selected point locations are unsatisfactory, adjustments in their positions should be made. For the existing control stations that will be used in the survey, ties should be made to nearby permanent objects and also photos or *rubbings*[3] of the monument caps should be created. These items will aid crews in locating the stations later during the survey, reduce the amount of time spent at each station, and minimize possible station misidentifications.

Once final sites have been selected for the new stations, permanent monuments should be set, and the positions of these stations also documented with ties to nearby objects, photos, and rubbings. At this time, an accurate horizon plot of any surrounding obstructions should be prepared and road directions and approximate driving times between stations recorded. A valuable aid in identifying locations of stations is the use of a code-based GPS receiver. These inexpensive devices will determine the geodetic coordinates of the stations with sufficient accuracy to enable plotting them on a map for later use in reaching the station.

14.3.4 Developing an Observation Scheme

For GPS projects, especially those employing relative positioning used in control surveying, once existing nearby control points have been located and the new GPS stations set, they comprise what is termed the "network." Depending upon the nature of the project and extent of the survey, the network can vary from only a few stations to very large and complicated configurations. Figure 14.7 illustrates a small network consisting of only two existing control points and four new stations.

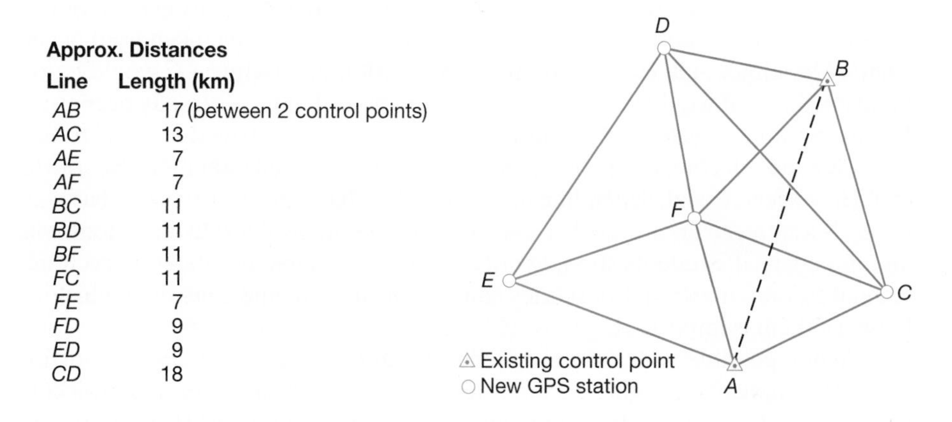

Approx. Distances

Line	Length (km)
AB	17 (between 2 control points)
AC	13
AE	7
AF	7
BC	11
BD	11
BF	11
FC	11
FE	7
FD	9
ED	9
CD	18

Figure 14.7
A GPS network.

[3]Monuments used to mark stations generally have metal caps (often brass) which give the name of the point and other information about the station. This information is stamped into the cap, and by laying a piece of paper directly over the cap and rubbing across the surface with the side of a pencil lead, an imprint of the cap is obtained. This helps to eliminate mistakes in station identification.

After the network has been established, an observation scheme is developed for performing the work. The scheme consists of a planned sequence of observing sessions that accomplishes the objectives of the survey in the most efficient manner. As a minimum, it must ensure that every station in the network is connected to at least one other station by a *nontrivial* (also called *independent*) baseline. (Nontrivial baselines are described in the following paragraph.) However, the plan should also include some redundant observations (i.e., baseline observations between existing control stations, multiple occupations of stations, and repeat observations of certain baselines) to be used for checking purposes and for improving the precision and reliability of the work. Desired accuracy is the principal factor governing the number and type of redundant observations. The *Federal Geodetic Control Subcommittee* (FGCS) has developed a set of standards and specifications for GPS relative positioning (see Section 14.5.1) that specify the number and types of redundant observations necessary for accuracy orders AA, A, B, and C. Generally on larger, high-accuracy GPS projects, these standards and specifications, or other similar ones, govern the conduct of the survey work and must be carefully followed.

In GPS relative positioning, for any observing session the number of nontrivial baselines measured is one less than the number of receivers used in the session, or

$$b = n - 1 \tag{14.1}$$

where b is the number of nontrivial baselines and n the number of GPS receivers being employed in the session. When only two receivers are used in a session, only one baseline is observed and it is nontrivial. If more than two receivers are used, both nontrivial and *trivial* (*dependent*) baselines will result. To differentiate between these two types of baselines and to understand how they occur, refer to Figure 14.8. In this figure, an observing session involving three receivers A, B, and C observing four satellites *1, 2, 3*, and *4* is shown. Pseudoranges *1A, 1B, 2A, 2B, 3A, 3B, 4A*, and *4B* are employed to compute the baseline vector AB. Also pseudoranges *1B, 1C, 2B, 2C, 3B, 3C, 4B*, and *4C* are employed in computing the baseline vector BC. Thus all possible pseudoranges in this example have been used in calculating baselines AB and BC and the computation of baseline AC would be redundant; that is, it would be based upon observations that have already been used. In this example, baselines AB and BC are considered nontrivial and AC trivial. However, the selection of the trivial baseline is arbitrary, and either AB, BC, or AC could have been trivial, depending upon which two baselines were selected as nontrivial. If four receivers are used in a session, six baselines will result: three nontrivial and three trivial. Students should verify this with a sketch. For meeting accuracy standards, only nontrivial baselines can be considered, and thus distinguishing between them is important.

When possible, at least one baseline should be observed between existing control monuments of higher accuracy for every receiver-pair used on a project to check the performance of the equipment. Also as noted previously some baselines should be observed more than once. These repeat baselines should ideally be observed at or near the beginning and at the end of the project observations to check the equipment for repeatability. The analysis of these duplicate observations will be covered in Section 14.5.

Figure 14.8
GPS observation session using three receivers. In the case shown, *AB* and *BC* are considered nontrivial baselines. Thus, *AC* is a trivial baseline.

For control surveys, the baselines should form closed, geometric figures since they are necessary to perform closure checks (see Section 14.5). The simple network of baselines, shown in Figure 14.7, will be used as an example to illustrate GPS survey planning. Assume that the project is in the area of control station *PSU1* and that the survey dates will be August 4 and 5, 1999. Thus the satellite availability plot and sky plots of Figures 14.4 and 14.5, respectively, apply. Stations *A* and *B* in Figure 14.7 are existing control monuments and a baseline observation between them will be planned to (1) verify the accuracy of the existing control and (2) confirm that the equipment is in proper working condition.

In the example of Figure 14.7 it is assumed that two dual-frequency receivers are available for the survey and that the rapid static method will be used. Following the minimum session lengths, as given in Table 14.1, of 10 min + 1 min/km, baseline *AB* would require $10 + (1 \times 17)$, or 27, minutes of observation time. The remaining baseline observation times are listed in Table 14.2 using the same computational techniques.

Two two-person crews, each working individually with separate vehicles, are assumed for conducting the survey. It is also assumed that setup and teardown times at each station are both approximately 15 min. By rounding each minimal observation session up to a nearest 5-minute interval, the following observation sessions and times were planned for a two-day data collection project.

The observation plan for this example is given in Table 14.3. The table gives the itinerary for both field crews, allowing time for setup and teardown of the equipment, travel between stations, and collecting sufficient observations. The plan includes all lines in the network as baselines. These include those for checking purposes, an observation of the control baseline *AB*, and repeat observations of *AF* and *BF*. Note that the two unfavorable times for collecting observations shown in Figure 14.4 are not scheduled as times for collecting data, but are used for other ancillary operations. In the event that operations should actually run ahead of or

TABLE 14.2 MINIMUM SESSION LENGTHS AND APPROXIMATE DRIVING TIMES BETWEEN STATIONS FOR BASELINES IN FIGURE 14.7

Line	Length (km)	Session Length (min)	Driving Time (min)
AB	17	27	15
AC	13	23	10
AE	7	17	8
AF	7	17	25
BC	11	21	15
BD	11	21	10
BF	11	21	20
FC	11	21	15
FE	7	17	15
FD	9	19	10
ED	9	19	15
CD	18	28	25

fall behind the planned schedule for some unforeseen reason, it is prudent to include a statement on the itinerary indicating that no baseline observations should be collected between the times of 8:00 to 8:15, and 13:40 to 14:00. Note that as indicated in the planned schedule, the crew with the stationary receiver should continue to collect data during the entire period of occupation of the station. This includes the periods of time during which the other crew is moving between stations. If a CORS site (see Section 14.3.5) is available, the data collected by the stationary receiver can be used to create strong baseline links with the CORS station. It is desirable to provide the field personnel with communication devices during the survey so that they can coordinate times for sessions, and handle unforeseen logistical problems that arise inevitably.

14.3.5 Availability of Reference Stations

As explained in Section 14.3.4, the availability of high-quality reference control stations is necessary to achieve the highest order of accuracy in GPS positioning. To meet this need, individual states, in cooperation with the NGS, have developed ***H****igh* ***A****ccuracy* ***R****eference* ***N****etworks* (HARNs). The HARN is a network of control points that were precisely observed using GPS under the direction of the National Geodetic Survey (NGS). These HARN points are now available to serve as reference stations for GPS surveys in their vicinity.

In recent years, the NGS with cooperation from other public and private agencies has created a national system of ***C****ontinuously* ***O****perating* ***R****eference* ***S****tations*, also called the *National CORS Network*. The location of stations in the CORS network as of 2003 is shown in Figure 14.9. As of January 2007, there

TABLE 14.3 OBSERVATION ITINERARY FOR FIGURE 14.11

DAY 1 (August 4, 1999)

Time	Crew 1	Crew 2	Baseline	Session
8:00–8:45	Drive to Station *A*	Drive to Station *C*		
9:00–9:25	Collect data	Collect data	*AC*	A1
9:40–9:55	Collect data	Drive to Station *F*		
9:55–10:15	Collect data	Collect data	*AF*	A2
10:30–10:45	Collect data	Drive to Station *E*		
11:00–11:20	Collect data	Collect data	*AE*	A3
11:35–11:50	Drive to Station *B*	Drive to Station *F*		
12:05–12:30	Collect data	Collect data	*BF*	A4
12:45–1:00	Collect data	Drive to Station *C*		
13:15–13:40	Collect data	Collect data	*BC*	A5
13:55–14:05	Collect data	Drive to Station *A*		
14:20–14:50	Collect data	Collect data	*AB*	A6
15:05–15:15	Drive to Station *D*	Drive to Station *C*		
15:30–16:00	Collect data	Collect data	*CD*	A7
16:00–17:00	Return to office	Download data		

DAY 2 (August 5, 1999)

Time	Crew 1	Crew 2	Baseline	Session
8:00–9:00	Drive to Station *A*	Drive to Station *F*		
9:15–9:35	Collect data	Collect data	*FA*	B1
9:50–10:00	Drive to Station *E*	Collect data		
10:15–10:30	Collect data	Collect data	*FE*	B2
10:45–11:00	Drive to Station *D*	Collect data		
11:00–11:20	Collect data	Collect data	*FD*	B3
11:35–11:45	Drive to Station *B*	Collect data		
12:00–12:25	Collect data	Collect data	*FB*	B4
12:40–12:55	Drive to Station *C*	Collect data		
13:10–13:35	Collect data	Collect data	*FC*	B5
13:50–14:05	Drive to Station *B*	Drive to Station *D*		
14:20–14:45	Collect data	Collect data	*BD*	B6
15:00–15:15	Drive to Station *D*	Drive to Station *E*		
15:30–15:50	Collect data	Collect data	*ED*	B7
15:50–17:00	Return to office	Download data		

Note: No baseline observations should be made between 8:00–8:15 and 13:40–14:00 on August 4, and between 7:56–8:11 and 13:36–13:56 August 5.

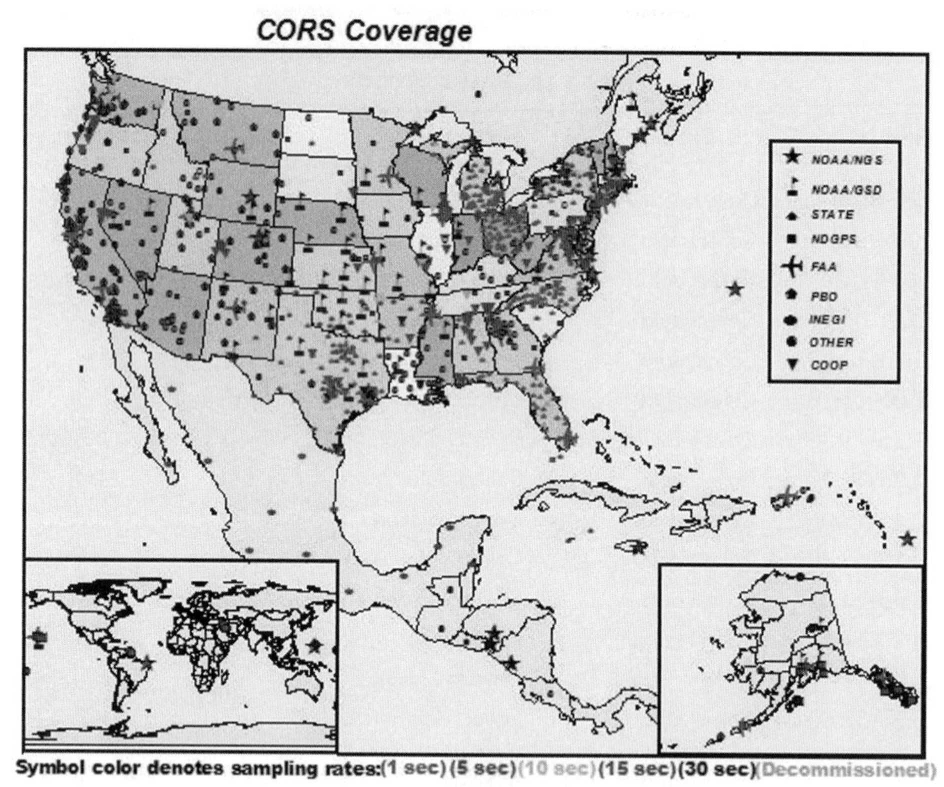

Figure 14.9 Locations that have continuously operating reference stations (CORS) and their coverages. (Courtesy National Geodetic Survey)

were 1221 CORS stations. These stations not only have their positions known to high accuracy,[4] but they also are occupied by a GPS receiver that collects GPS data continuously. The collected data is then downloaded and posted on http://www.ngs.noaa.gov/CORS/. This information can be used as base station data to support roving receivers operating in the vicinity of the CORS station. The data is stored in a format known as ***R**eceiver **IN**dependent **EX**change* (RINEX2). This format is a standard that can be read by all post-processing software. This Internet site also provides coordinates for the stations, and the ultra-rapid, rapid, and precise ephemerides (see Section 13.6.3).

Because of plate tectonics and the differences in which the reference systems handle these motions, their velocity vectors accompany the ITRF00 coordinates for the stations. While motions in the eastern United States may be less than 1 mm per year, these motions may be considerably more in the western United States. Thus coordinates derived from GPS control surveys should be accompanied

[4]At the time of this writing, the CORS and HARN sites are computed using different reference coordinate systems, and thus do not represent homogenous sets of coordinates. The NGS is currently working on readjusting both networks of points so that this problem does not exist in the future. However, current surveys should not mix HARN and CORS stations until the NGS has completed the adjustment and published the results.

by their reference frame (see Section 19.6.3) and epoch. The NGS has written software known as *horizontal time-dependent positioning* (HTDP) using 14 parameters to transform the ITRF00 positions with their velocity vectors into NAD 83 (CORS 96) equivalents. The WGS84(G1150) coordinates are essentially the same as ITRF00 coordinates. This coordinate transformation is demonstrated in the Mathcad® files that are on the CD that accompanies this book.

The CORS data files are easily downloaded using the *User-Friendly CORS* (UFCORS) option on the NGS website. This utility provides the user with an interactive form which requests the (1) starting date, (2) time, (3) duration of the survey, (4) site selection, and (5) the collection interval among other things. The request is interpreted by the server, and the appropriate data is sent, via the Internet, to the user within minutes.

Several factors can cause data not to be collected at a particular CORS site for short periods of time. These include local power outages, storm damage, and software and hardware failures. Thus, if a certain CORS station is planned for use on a particular GPS survey, it is important to check the availability of the station prior to the beginning of a survey to ensure that it is functioning properly and that the necessary data will be available following the survey.

■ 14.4 PERFORMING STATIC GPS SURVEYS

During a control survey using either the static or rapid static method, the field crew initially locates the control station and erects the antenna over the station so that it is level. To minimize setup errors, 2-m fixed height poles or tripods are often used. Cables are connected to the antenna and controller and the procedures specified by the manufacturer are followed to initiate data collection.

While the data is being logged, other ancillary information at the site can be collected and recorded. Typical ancillary data obtained during a survey includes (1) project and station names, (2) ties to the station, (3) a rubbing of the monument cap, (4) photos of the setup showing identifying background features, (5) potential obstructive or reflective surfaces, (6) date and session number, (7) start and stop times, (8) name of observer, (9) receiver and antenna serial number, (10) meteorological data, (11) PDOP value, (12) antenna height, (13) orientation of antenna, (14) rate of data collection (epoch rate), and (15) notations on any problems experienced. This ancillary data is typically entered onto a station log sheet. An example of a site log sheet is shown in Figure 14.10.

Modern GPS controllers store the GPS observations internally. At least once a day, and preferably twice a day, the data should be transferred to a computer. This can be done in the field with a laptop computer or at the end of the day in the office. File transfer is accomplished using software provided by the manufacturer. The GPS receiver shown in Figure 14.2 can transfer data from the controller to a PCMCIA card.

■ 14.5 DATA PROCESSING AND ANALYSIS

The first step in processing GPS data is to transfer the observational files from the receiver to the computer. Typically, software provided with the GPS receiver performs this function. In this process, the software will download all of the

SITE LOG

Project: ______________________________ Station: __________

Session Number: ______ Observer: ______________ Date: ___/___/___

Start Time: ____ : ____ (AM) (PM) Stop Time: ____ : ____ (AM) (PM)

Latitude: ___ ° _____ ′ _____ ″ Latitude: ___ ° _____ ′ _____ ″ Height: ______

Temperature: ______ (F) (C) Pressure: _______ Relative Humidity: _______

PDOP: ________ Epoch Rate: ________ sec

Initial Slant Height: ______ (m) (ft) Final Slant Height: ______ (m) (ft)

Receiver S/N: __________ Antenna S/N: __________

Photo #: ________ Visible Satellites: ________________

Description of Monument *(include rubbing of surface)*:

Potential Problems:

Ties to Monument:

Comments:

____
Signature of Observer

Figure 14.10
Data log sheet for a static or rapid static survey.

observational files into one subdirectory on the computer. This directory is part of the project's directory. As the observational files are downloaded, special attention should be given to checking station information that is read directly from the file with the site log sheets. Catching incorrectly entered items such as station identification, antenna heights, and antenna offsets at this point can greatly reduce later processing problems. Batch processing typically performs the reduction process. Figure 14.11 shows an individual processing GPS data with a PC.

There are three types of processing software: (1) *single baseline,* (2) *session processing,* and (3) *multipoint solutions.* The single baseline solution is most

Figure 14.11 Processing GPS data. (Courtesy Ashtech, Inc.).

commonly used and will be discussed here. The session processing software simultaneously computes all nontrivial baselines for any particular session. The most common method of using multipoint solution software is to first use the single baseline software to isolate any problems in the observations, because it provides better checks to isolate "bad" baselines. The multipoint solution software eliminates the need for the adjustment of the baselines as discussed in Section 14.5.5.

The initial processing can use the "broadcast" ephemeris, but it is recommended to use any of the "precise" ephemerides to achieve a higher level of accuracy. When CORS sites are used for a project, the most precise ephemeris available at the time of the download can be requested. Proper use of the "user-friendly" option, which is available on the NGS website, will automatically result in interpolated values for the data at the specified epoch rate to match the survey. Additionally as mentioned in Section 13.6.3, NGS antenna calibration data should be used when determining the baseline vectors to account for changing antenna offsets due to changes in satellite elevation.

As shown in Figure 13.12, the baseline is computed as changes in X, Y, and Z between two stations. Thus if station A has known coordinates, then the coordinates of station B can be computed using Equations (13.28). The single baseline processing software will (1) generate orbit files, (2) compute the best fit point positions from the code pseudoranges, (3) compute baseline components (ΔX,ΔY,ΔZ) using the double difference Equation (13.32), and (4) compute statistical information for the baseline components.

As each baseline is computed, any integer ambiguity problems should be identified and corrected. Integer ambiguity problems occur when a receiver loses lock on the satellite, which can happen because of obstructions, high solar activity, or multipathing. The only option typically provided by the software to correct this problem is to eliminate the periods of satellite data where the problem occurred. This may be done graphically or manually depending on the specific software being used.

After all baselines have been computed, their geometric closures can be analyzed. This step follows a series of procedures that check the data for internal consistency and eliminate possible blunders. No control points are needed for these analyses. Depending on the actual observations taken and the network geometry, these procedures may involve analyzing (1) differences between fixed and measured baseline components, (2) differences between repeated measurements of the same baseline components, and (3) loop closures. Procedures for making these analyses are described in the following subsections.

14.5.1 Specifications for GPS Surveys

The *Federal Geodetic Control Subcommittee* (FGCS) has published the document entitled "Geometric Geodetic Accuracy Standards and Specifications for Using GPS Relative Positioning Techniques."[5] The document specifies seven different orders of accuracy for GPS relative positioning and provides guidelines on instruments and field and office procedures to follow to achieve them. Table 14.4 lists these orders of accuracy.

The FGCS document also makes recommendations concerning categories of surveys for which the different orders of accuracy are appropriate. Some of these recommendations include: order AA for global and regional geodynamics and deformation measurements; order A for "primary" networks of the National Spatial Reference System (NSRS), and regional and local geodynamics; order B for "secondary" NSRS networks and high-precision engineering surveys; and the various classes of order C for mapping control surveys, property surveys, and

TABLE 14.4 GPS RELATIVE POSITIONING ORDERS OF ACCURACY

Order	Allowable Error Ratio	Parts-Per-Million (ppm)
AA	1:100,000,000	0.01
A	1:10,000,000	0.1
B	1:1,000,000	1.0
C-1	1:100,000	10
C-2-I	1:50,000	20
C-2-II	1:20,000	50
C-3	1:10,000	100

[5] Available from the National Geodetic Information Center, NOAA, National Geodetic Survey, N/CG17, SSMC3 Station 09535, 1315 East West Highway, Silver Spring, MD. 20910, telephone (301) 713-3242. Their email address is info_center@ngs.noaa.gov, and their website is http://www.ngs.noaa.gov.

TABLE 14.5 NGS GEODETIC HEIGHT ORDER AND CLASS

Order	Class	Maximum Height Difference Accuracy (*b*)
First	I	0.5
First	II	0.7
Second	I	1.0
Second	II	1.3
Third	I	2.0
Third	II	3.0
Fourth	I	6.0
Fourth	II	15.0
Fifth	I	30.0
Fifth	II	60.0

engineering surveys. The allowable error ratios given in these standards imply the extremely high accuracies that are now possible with GPS.

The National Geodetic Survey has created geodetic height accuracy standards that can be used for vertical GPS surveys. These standards are based on the changes in geodetic heights between control stations. Following a correctly weighted, minimally constrained, least-squares baseline adjustment (see Section 14.5.5), the geodetic height, order, and class can be determined. This standard is based on the standard deviation (s) of the geodetic height difference (in millimeters) between two points obtained from the adjustment, and the distance (d) between the two control points (in kilometers). The ellipsoid height difference accuracy (b) is computed as

$$b = \frac{s}{\sqrt{d}} \tag{14.2}$$

Table 14.5 lists classifications versus value of b computed in Equation (14.2).

14.5.2 Analysis of Fixed Baseline Measurements

As noted earlier, GPS job specifications often require that baseline observations be taken between fixed control stations. The benefit of making these observations is to verify the accuracy of both the GPS observational process and the control being held fixed. Obviously smaller discrepancies between observed and known baseline lengths mean better precisions. If the discrepancies are too large to be tolerated, the conditions causing them must be investigated before proceeding further. Table 14.6 shows the computed baseline vector data from a GPS survey of the network of Figure 14.7. Note that one fixed baseline (between control points A and B) was observed.

TABLE 14.6 OBSERVED BASELINE VECTORS FOR FIGURE 14.8

Obs. Baseline	ΔX(m)	ΔY(m)	ΔZ(m)
AC	11644.2232	3601.2165	3399.2550
AE	−5321.7164	3634.0754	3173.6652
BC	3960.5442	−6681.2467	−7279.0148
BD	−11167.6076	−394.5204	−907.9593
DC	15128.1647	−6286.7054	−6371.0583
DE	−1837.7459	−6253.8534	−6596.6697
FA	−1116.4523	−4596.1610	−4355.8962
FC	10527.7852	−994.9377	−956.6246
FE	−6438.1364	−962.0694	−1182.2305
FD	−4600.3787	5291.7785	5414.4311
FB	6567.2311	5686.2926	6322.3917
BF	−6567.2310	−5686.3033	−6322.3807
AF	1116.6883	4596.4550	4355.3008
AB	7683.6883	10282.4550	10678.3008

Assuming the geocentric coordinates of station A are (402.3509, −4, 652,995.3011, 4,349,760.7775) and those of station B are (8086.0318, −4,642, 712.8474, 4,360,439.0833), the analysis of the fixed baseline is as follows.

1. Compute the coordinate differences between stations A and B as

$$\begin{aligned}\Delta X_{AB} &= 8086.0318 - 402.3509 &&= 7683.6809\\ \Delta Y_{AB} &= -4{,}642{,}712.8474 + 4{,}652{,}995.3011 &&= 10{,}282.4537\\ \Delta Z_{AB} &= 4{,}360{,}439.0833 - 4{,}349{,}760.7775 &&= 10{,}678.3058\end{aligned}$$

2. Compute the absolute values of differences between the observed and fixed baselines as

$$\begin{aligned}dX &= |7683.6883 - 7683.6809| &&= 0.0074\\ dY &= |10{,}282.4550 - 10{,}282.4537| &&= 0.0013\\ dZ &= |10{,}678.3008 - 10{,}678.3058| &&= 0.0050\end{aligned}$$

3. Compute the length of the baseline as

$$AB = \sqrt{(7683.6809)^2 + (10{,}282.4537)^2 + (10{,}678.3058)^2} = 16{,}697.126 \text{ m}$$

4. Express the differences, as computed in Step 2, in parts per million (ppm) by dividing the difference by the length of the baseline computed in Step 3 as

$$\Delta X\text{-ppm} = 0.0074/16{,}697.126 \times 1{,}000{,}000 = 0.44$$

$$\Delta Y\text{-ppm} = 0.0013/16{,}697.126 \times 1{,}000{,}000 = 0.08$$

$$\Delta Z\text{-ppm} = 0.0050/16{,}697.126 \times 1{,}000{,}000 = 0.30$$

5. Check the computed values for the ppm against a known standard. Typically, the FGCS standard given in Section 14.5.1 is used.

14.5.3 Analysis of Repeat Baseline Measurements

Another procedure employed in evaluating the consistency of the observed data and in weeding out blunders is to make repeat observations of certain baselines. These repeat measurements are taken in different observing sessions and the results compared. For example in the data of Table 14.6, baselines *AF* and *BF* were repeated. Table 14.7 gives comparisons of these observations using the same procedure that was used in Section 14.5.2. Column (1) lists the baseline vector components to be analyzed, columns (2) and (3) show the repeat baseline vector components, column (4) lists the absolute values of the differences in these two observations, and column (5) shows the ppm values that are computed similar to the procedure given in Step 4 of Section 14.5.2.

14.5.4 Analysis of Loop Closures

Static GPS surveys consist of many interconnected closed loops typically. For example in the network of Figure 14.7, points *ACBDEA* form a closed loop. Similarly, *ACFA*, *CFBC*, *BDFB*, etc., are other closed loops. For each closed loop, the algebraic sum of the ΔX components should equal zero; the same condition exists for the ΔY and ΔZ components. An unusually large closure within any loop will indicate that either a blunder or a large error exists in one (or more) of the baselines of the loop. It is important not to include any trivial baselines (see

TABLE 14.7 ANALYSIS OF REPEAT BASELINE OBSERVATIONS

(1) Component	(2) First Observation	(3) Second Observation	(4) Difference	(5) ppm
ΔX_{AF}	1116.4577	1116.4523	0.0054	0.84
ΔY_{AF}	4596.1553	4596.1610	0.0057	0.88
ΔZ_{AF}	4355.9141	4355.9062	0.0079	1.23
ΔX_{BF}	−6567.2310	−6567.2311	0.0001	0.01
ΔY_{BF}	−5686.3033	−5686.2926	0.0107	1.00
ΔZ_{BF}	−6322.3807	−6322.3917	0.0110	1.02

Section 14.3.4) in these computations, since they can yield false accuracies for the loop.

To compute loop closures, the baseline components are added algebraically for that loop. For example, the closure in X components for loop $ACBDEA$ is

$$cx = \Delta X_{AC} + \Delta X_{CB} + \Delta X_{BD} + \Delta X_{DE} + \Delta X_{EA} \tag{14.3}$$

where cx is the loop closure in X coordinates. Similar equations apply for computing closures in Y and Z coordinates. Substituting numerical values into Equation (14.3), the closure in X coordinates for loop $ACBDEA$ is

$$\begin{aligned} cx &= 11{,}644.2232 - 3960.5442 - 11{,}167.6076 - 1837.7459 + 5321.7164 \\ &= 0.0419 \text{ m} \end{aligned}$$

Similarly, closures in Y and Z coordinates for that loop are

$$\begin{aligned} cy &= 3601.2165 + 6681.2467 - 394.5204 - 6253.8534 - 3634.0754 \\ &= 0.0140 \text{ m} \end{aligned}$$

$$\begin{aligned} cz &= 3399.2550 + 7279.0148 - 907.9593 - 6596.6697 - 3173.6652 \\ &= -0.0244 \text{ m} \end{aligned}$$

For evaluation purposes, loop closures are expressed in terms of the ratios of resultant closures to the total loop lengths. They are given in ppm. For any loop, the resultant closure is

$$lc = \sqrt{cx^2 + cy^2 + cz^2} \tag{14.4}$$

where lc is the length of the misclosure in the loop.

Using the values previously determined for loop $ACBDEA$, the length of the misclosure is 0.0505 m. The total length of a loop is computed by summing its legs, each leg being computed from the square root of the sum of the squares of its observed ΔX, ΔY, and ΔZ components. For loop $ACBDEA$, the total loop length is 50,967 m and the closure ppm ratio is therefore $(0.0505/50{,}967) \times (1{,}000{,}000) = 0.99$ ppm. Again these ppm ratios can be compared against values given in the FGCS guidelines (Table 14.4) to determine if they are acceptable for the order of accuracy of the survey. As was the case with repeat baseline observations, the FGCS guidelines also specify other criteria that must be met in loop analyses besides the ppm values.

For any network, enough loop closures should be computed so that every baseline is included within at least one loop. This should expose any large blunders that exist. If a blunder does exist, its location often can be determined through additional loop closure analyses. For example, assume that the misclosure of loop $ACDEA$ reveals the presence of a blunder. By also computing the closures of loops $AFCA$, $CFDC$, $DFED$, and $EFAE$, the exact baseline containing the blunder can be detected. In this example, if a large misclosure was found in loop $DFED$ and all other loops appeared to be blunder free, the blunder would be in line DE, because that leg was also common to loop *ACDEA*, which contained a blunder as well.

14.5.5 Baseline Network Adjustment

After the individual baselines are computed, a least-squares adjustment (see Section 16.8) of the observations is performed. This adjustment software is available from the receiver manufacturer and will provide final station coordinate values and their estimated uncertainties. If more than two receivers are used in a survey, trivial baselines will be computed during the single baseline reduction. These trivial baselines should be removed before the final network adjustment.

The observations used in the baseline network adjustment should be part of an interconnected network of baselines. Initially, a minimally constrained adjustment should be performed (see Section 16.11). The adjustment results should be analyzed both for mistakes and large errors. As an example, antenna height mistakes, which are not noticeable during a single baseline reduction, will be noticeable during the network adjustment. After the results of a minimally constrained adjustment are accepted, a fully constrained adjustment should be performed. During a fully constrained adjustment, all available control is added to the adjustment. At this time, any scaling problems between the control and the observations will become apparent by the appearance of overly large residuals. Problems that are identified should be corrected and removed before the results are accepted.

Since these computations are performed in a geocentric coordinate system, the final adjusted values can be transformed into a geodetic coordinate system using procedures as outlined in Section 13.4.3, or into a plane coordinate system (see Chapter 20). Recall that geodetic elevations are measured from the ellipsoid and thus, as discussed in Section 13.4.3, the geoid undulation must be applied to these heights to derive orthometric elevations. The GEOID03 software can be used to determine the geoid undulation. This model is included in the software typically.

Finally the horizontal and vertical accuracy of the survey can be determined based on the FGCS or NGS horizontal and vertical accuracy standards (see Section 14.5.1).

14.5.6 The Survey Report

A final survey report is helpful in documenting the project for future analysis. At a minimum, the report should contain the following items.

1. The location of the survey and a description of the project area. An area map is recommended.
2. The purpose of the survey and its intended specifications.
3. A description of the monumentation used including the tie sheets, photos, and rubbings of the monuments.
4. A thorough description of the equipment used including the serial numbers, antenna offsets, and the date the equipment was last calibrated.
5. A thorough description of the software used including name and version number.

6. The observation scheme used, including the itinerary, the names of the field crew personnel, and any problems that were experienced during the observation phase.
7. The computation scheme used to analyze the observations and the results of this analysis.
8. A list of the problems encountered in the process of performing the survey, or its analysis including unusual solar activity, potential multipathing problems, or other factors that can affect the results of the survey.
9. An appendix containing all written documentation, original observations, and analysis. Since the computer can produce volumes of printed material in a GPS survey, only the most important files should be printed. All computer files should be copied onto some backup storage. A CD-ROM provides an excellent permanent storage media that can be inserted into the back of the report.

14.6 SOURCES OF ERRORS IN GPS WORK

As is the case in any project, GPS observations are subject to instrumental, natural, and personal errors. These are summarized in the following subsections.

14.6.1 Instrumental Errors

Clock Bias As mentioned in Section 13.6.1, both the receiver and satellite clocks are subject to errors. They can be mathematically removed using differencing techniques for all forms of relative positioning.

Setup Errors As with all work involving tripods, the equipment must be in good adjustment (see Section 8.19). Careful attention should be paid to maintaining tripods that provide solid setups, and tribrachs with optical plummets that will center the antennas over the monuments. In GPS work, tribrach adapters are often used that allow the rotation of the antenna without removing it from the tribrach. If these adapters are used, they should be inspected for looseness or "play" on a regular basis. Due to the many possible errors that can occur when using a standard tripod, special fixed-height tripods and rods are often used. The fixed-height rods can be set up using either a bipod or tripod with a rod on the point. They typically are set to a height of precisely 1.5–2 m from the antenna reference point (ARP).

Non-Parallelism of the Antennas Pseudoranges are observed from the phase center of the satellite antenna to the phase center of the receiver antenna. As with EDM observations, the phase center of the antenna is not the geometric center of the antenna. Each antenna must be calibrated to determine the phase center offsets for both the $L1$ or $L2$ bands. For the most precise work in GPS, the antennas are aligned in the same direction. Generally, they are aligned according to local magnetic north using compasses.

Selective Availability On May 1, 2000, SA was turned off. However, when it was turned on it restricted the accuracy of the system to unauthorized users. As discussed in Section 13.6.5, when it was turned on, the satellite ephemeris was truncated by the so-called epsilon process (ε-process) and the frequency of the clock was dithered by the so-called delta-process (δ-process). These errors were mathematically removed by the differencing process in relative positioning. However, they did seriously restrict the accuracy of the system in point-positioning work.

Receiver Noise When working properly, the electronics of the receiver will operate within a specified tolerance. Within this tolerance, small variations occur in the generation and processing of the signals that can eventually translate into errors in the pseudorange and carrier-phase observations. Since these errors are not predictable, they are considered as part of the random errors in the system. However, periodic calibration checks and tests of receiver electronics should be made to verify that they are working within acceptable tolerances.

14.6.2 Natural Errors

Refraction Refraction due to the transit of the signal through the atmosphere plays a crucial role in delaying the signal from the satellites. The size of the error can vary from 0–10 m. Dual frequency receivers can mathematically model and remove this error using Equations (13.18) and (13.19). With single frequency receivers, this error must be modeled. For surveys involving small areas using relative positioning methods, the majority of this error will be removed by differencing. Since high solar activity affects the amount of refraction in the ionosphere, it is best to avoid these periods where the k-index is greater than 4.

Relativity GPS satellites orbit the Earth in approximately 12 hr. The speed of the satellites causes their atomic clocks to slow down, according to the theories of relativity. The master control station computes corrections for relativity and applies these to the clocks in the satellites.

Multipathing Multipathing occurs when the signal emitted by the satellite arrives at the receiver after following more than one path. It is generally caused by reflective surfaces near the receiver. As discussed in Section 13.6.3, multipathing can become so great that it will cause the receiver to lose lock on the signal. Many manufacturers use signal filters to reduce the problems of multipathing. However, these filters will not eliminate all occurrences of multipathing and are susceptible to signals that have been reflected an even number of times. Thus, the best approach to reducing this problem is to avoid setups near reflective surfaces.

14.6.3 Personal Errors

Tripod Miscentering This error will directly affect the final accuracy of the coordinates. To minimize it, check the setup carefully before data collection begins and again after it is completed.

■ 14.7 MISTAKES IN GPS WORK

A few of the more common mistakes in GPS work are listed here.

Misreading of the Antenna Height The height from the ground to the antenna ground plane or reference point should be read at several times. When measuring to a ground plane, it should be measured at several locations around the ground plane and the average recorded. To ensure that the tripod hasn't settled during the observation process and that the initial readings were correct, the slant height should be also measured at the end of the observation process. To avoid this problem, only fixed-height tripods should be used in the most precise surveys.

Incorrect Identification of the Station This mistake can cause hours of wasted time in processing of the data, and sometimes requires that the survey be repeated. To limit this possibility, each station should be located from at least three readily visible permanent objects. Also if possible, rubbings of the monument caps should be obtained and photos taken of the area showing the location of the monument. During the observation phase of the survey, a second rubbing of the monument and a photo of the setup showing the surrounding area should be taken. This data should be correlated before baseline processing.

Processing of Trivial Baselines This mistake can only occur when more than two receivers are used in a survey. While this mistake will not generate false coordinates, it will generate false accuracies for the survey. Care should be taken to remove all trivial baselines before a network adjustment is attempted.

Misidentification of Antennas Since each antenna type has different phase center offsets, misidentifying an antenna will directly result in an error in the derived pseudorange. Using antennas from only one manufacturer can reduce this mistake or by correctly identifying and entering phase center offsets into the processing software.

■ 14.8 FUTURE OUTLOOK FOR GPS

The future of GPS in the practice of surveying seems assured. The systems have already demonstrated reliability and a capability to yield extremely high accuracy. Although the most common application of GPS was originally for control work, the systems have now been used in virtually every type of survey, including property surveying, topographic mapping, and construction stakeout. GPS has been especially useful in solving some heretofore rather difficult problems, such as determining positions for hydrographic soundings, locating dredging rigs, and fixing positions of offshore oil wells. New applications are being tested on a regular basis, and research and development to improve the systems continues.

Surveying by GPS provides many advantages over other traditional methods, including speed, accuracy, and operational capability by day or night and in any weather. Also, intervisibility between stations is no longer required. For these reasons and others, GPS should be used increasingly in the future for all types of surveys.

PROBLEMS

Asterisks (*) indicate problems that have answers given in Appendix G.

14.1 Discuss the procedures used in static and rapid static GPS surveys.

14.2 When using the rapid static surveying method, what is the minimum recommended length of the session required to observe a baseline that is 15 km long for *(a) a dual frequency receiver and, (b) a single frequency receiver?

14.3 What would be the recommended epoch rate for the surveys given in Problem 14.2?

14.4 For a 27-km baseline using a dual-frequency receiver, (a) what GPS surveying method should be used, (b) for what time period should the baseline be observed, (c) and what epoch rate should be used?

14.5 Using Figure 14.7 for reference, discuss why a GPS vertical survey should not have been performed during 1:30 PM to 2:00 PM on August 4, 1999, at the specified location.

14.6 In Problem 14.5, if a vertical survey had been performed during this period, discuss why a static survey would have achieved better results than a rapid static survey for a baseline that is 12 km in length.

14.7 Explain why canopy obstructions are a problem in a GPS survey.

14.8 Why is it recommended to use a precise ephemeris when processing a static survey?

14.9 What are the recommended rates of data collection in an (a) static GPS survey, and a *(b) rapid static GPS survey?

14.10 What potential problems should a person avoid when selecting a GPS station?

14.11 How many nontrivial baselines will be observed in one session with 3 GPS receievers?

14.12 What type of GPS survey method should be used for a baseline that is 50 km long?

14.13 Same as Problem 14.12, but for a baseline that is 12 km long?

14.14 List the fundamental steps involved in planning a static GPS survey.

14.15 Explain why periods of high solar activity should be avoided when collecting GPS observations.

14.16 Explain why the survey vehicle should be parked at least 25 m from the observing station in a GPS survey?

14.17* How many nontrivial baselines will be observed during a session when using 4 GPS receivers? How many nontrivial baselines will be observed in 5 sessions?

14.18 When using 3 GPS receivers, how many sessions will it take to independently observe all the baselines of a pentagon?

14.19 Plot the following ground obstructions on a obstruction diagram.

(a) From an azimuth of 102° to 111° there is a building with an elevation of 20°.

(b) From an azimuth of 165° to 168° there is a pole with an elevation of 35°.

(c) From an azimuth of 358° to 2° there is a tree with an elevation of 20°.

14.20* In Problem 14.19, which obstruction is unlikely to interfere with satellite visibility in the northern hemisphere?

14.21 What items should be considered when deciding which method to use for a GPS survey?

14.22 How is a sky plot used in planning a GPS survey?

14.23 What is a satellite availability chart, and how is it used?

14.24 Why should GPS control surveys form closed geometric figures?

14.25 Assuming that three crews and three receivers are available for the rapid static survey of Figure 14.10, develop an itinerary for the field personnel to follow. Use the satellite visibility chart of Figure 14.7, and the session times and setup and teardown times used in the example of Section 14.3.4.

14.26 Download the coordinates for the CORS station WIL1 from the NGS website.

14.27 What are CORS and HARN stations?

14.28 Why should repeat baselines be performed in a static survey?

14.29 Use the NGS website to download the station coordinates for the nearest CORS station.

14.30 Use the NGS website to download 1-hour of data from the nearest CORS station.

14.31 What is the purpose of developing a site log sheet for each session?

14.32 A 15-km baseline has a standard deviation of ±0.8 mm. What order of survey does this line meet?

14.33 Using loop *ACFDEA* from Figure 14.7, and the data from Table 14.6, what is the

- **(a)** Misclosure in the *X* component?
- **(b)** Misclosure in the *Y* component?
- **(c)** Misclosure in the *Z* component?
- **(d)*** Length of the loop misclosure?
- **(e)** Derived ppm for the loop?

14.34 Do Problem 14.33 with loop *AFEA*.

14.35 Do Problem 14.33 with loop *DEFD*.

14.36 Why is it important to create a survey report of each survey project?

14.37 Locate a published article on an application of GPS in surveying. Give the name of the publication, the article's title and author, and write a summary of the article's content.

14.38 Download and submit two hours of data from a nearby CORS site to the OPUS website for processing.

14.39 List at least five types of surveys that could be performed using GPS.

14.40 Use your GPS mission planning software to create a sky plot for today's date during a session from 10:00 to 11:00 local time for a location with geodetic coordinates of 43°23′43.6058″ N latitude, 89°02′26.7201″ W longitude and elevation of 248.870 m. Use a mask angle of 15°. Overlay one of the sky plots developed in Problem 14.18. Discuss the effects, if any, that the ground obstructions have on this particular location and time.

14.41 Employ the user-friendly button in the NGS CORS site at http://www.ngs.noaa.gov/CORS/ to:

- **(a)** Download the navigation and observation files for station ABQ1 between the hours of 10 and 11, local time for the Monday of the current week using a 5-sec data collection rate.
- **(b)** Print the files and comment on the contents of them. (*Hint:* An explanation of the contents of the RINEX2 data file is contained at http://www.ngs.noaa.gov/CORS/ Rinex2.html.)

BIBLIOGRAPHY

Denny, Milton. 2002. "Surveying Little Egypt." *Point of Beginning* 27 (No. 8): 26.

Devine, D. 2002. "Mapping the CSS Hunley." *Professional Surveyor* 22 (No. 3): 6.

Fotopoulos, G. et. al. 2003. "How Accurately Can We Determine the Orthometric Height Differences from GPS and Geoid Data?" *Journal of Surveying Engineering* 129 (No. 1): 1.

Hartzheim, P. 2002. "No Roads Untraveled—How GPS Has Eased the Tasks of the Wisconsin Department of Transportation." *Point of Beginning* 27 (No. 12): 14.

Henstridge, F. 2001. "The National Height Modernization Program." *Professional Surveyor* 21 (No. 6): 54.
Hofmann-Wellenhof, B., et al. 1997. *GPS Theory and Practice*. 4th ed. New York: Springer-Verlag.
Katsambalos, K. and P. Savvaidis. 1996. "GPS Receiver Calibration." *Point of Beginning* 21 (No. 5): 87.
Kuang, S. et al. 2002. "GPS Control Densification Project for Illinois Department of Transportation." *Surveying and Land Information Science* 62 (No. 4): 225.
Kozlowski, J. 2000. "EDM & GPS Measurements: Why Don't They Agree? Part Three." *Professional Surveyor* 20 (No. 4): 24.
Licht, R. 2003. "A Step Ahead—Employees of a Minnesota Firm Take GPS One Step Further with the Application of Cell Phones." *Point of Beginning* 28 (No. 12): 32.
Mader, G. L., et al. 2003. "NGS Geodetic Tool Kit, Part II: The On-line Positioning User Service (OPUS)." *Professional Surveyor* 23 (No. 5): 26.
Martin, D. 2001. "GPS Basics: Where to Begin?" *Professional Surveyor* 21 (No. 9): 30.
———. 2001. "Static GPS." *Professional Surveyor* 21 (No. 11): 52.
———. 2002. "Integrating Continuously Operating Reference Stations." *Professional Surveyor* 22 (No. 10): 42.
———. 2002. "Reaching New Heights in GPS, Part 1." *Professional Surveyor* 22 (No. 12): 34.
———. 2002. "Static GPS: Baseline Reduction and Analysis." *Professional Surveyor* 22 (No. 5): 38.
———. 2002. "Static GPS: Field Procedures." *Professional Surveyor* 22 (No. 2): 38.
———. 2002. "Static GPS: Part 4." *Professional Surveyor* 22 (No. 6): 32.
———. 2002. "Static GPS: Taking the Plunge." 22 (No. 8): 42.
———. 2003. "Reaching New Heights in GPS, Part 2." *Professional Surveyor* 23 (No. 2): 33.
———. 2003. "Reaching New Heights in GPS, Part 3." *Professional Surveyor* 23 (No. 4): 42.
Meyer, T. et al. 2002. "The Effect of Broadleaf Canopies on Survey-grade Horizontal GPS/GLONASS Measurements." *Surveying and Land Information Science* 62 (No. 4): 215.
Prusky, J. 2001. "The Cooperative CORS Program." *Professional Surveyor* 21 (No. 1): 14.
Reilly, J. 2001. "Standards and Specifications for Horizontal Control." *Point of Beginning* 26 (No. 4): 24.
Speed, V. 2002. "Surveyors Dial Up for Accuracy." *Point of Beginning* 27 (No. 11): 24.
Steinberg, G. and G. Even-Tzur. 2005. "Establishment of National Grid Based on Permanent GPS Stations in Israel." *Surveying and Land Information Science* 65 (No. 1): 47.
Stevens, K. 2002. "GPS Fore the Golf Course." *Point of Beginning* 27 (No. 8): 18.
Zilkoski, David B. 1997. "Guidelines for Establishing GPS-Derived Ellipsoid Heights." *NOAA Technical Memorandum NOS NGS-58*. National Geodetic Survey.
———. 2001. "NAVD 88 GPS-Derived Orthometric Heights, Part 1." *Point of Beginning* 26 (No. 6): 67.
———. 2001. "NAVD 88 GPS-Derived Orthometric Heights, Part 2." *Point of Beginning* 26 (No. 8): 46.
———. 2001. "NAVD 88 GPS-Derived Orthometric Heights, Part 4." *Point of Beginning* 26 (No. 12): 60.

15
The Global Positioning System—Kinematic GPS

15.1 INTRODUCTION

In many areas of surveying, speed and productivity are essential elements to success. In GPS, the most productive form of surveying is kinematic GPS. It uses relative positioning techniques with carrier-phase observations as discussed in Section 13.5.2 and Section 13.8. These surveys can provide immediate values to the coordinates of points while the receiver is stationary or in motion. Its accuracy is typically less than that obtainable with static surveys, but is adequate for most forms of surveys. It has applications in many areas of surveying including mapping (Chapter 17), boundary (Chapters 21 and 22), construction (Chapters 23 and 24), and photogrammetry (Chapter 27). This chapter will look at both the post-processed and real-time kinematic GPS methods.

Kinematic GPS can provide immediate results using the real-time kinematic (RTK) mode or in the office using the post-process kinematic (PPK) mode. Kinematic GPS provides positioning while the receiver is in motion. For example, kinematic GPS has been successfully used in positioning sounding vessels during hydrographic surveys (Section 17.13) and aerial cameras during photogrammetric surveys (Section 27.16). In large construction projects, it is used in machine control to guide earthwork operations. It is also useful in nonsurveying applications such as high-precision agriculture.

It shares many commonalities with static GPS surveys. For example, a kinematic survey requires two receivers collecting observations simultaneously from a pair of stations with one receiver: the base, occupying a station of known position, and the rover collecting data on points of interest. It also uses relative positioning computational procedures similar to those used in static surveys. Thus it requires that the integer ambiguities (see Section 13.5.2) be resolved before the survey is started. The main difference between static and kinematic surveying techniques is the length of time per session. In a kinematic survey, observations from a single epoch may be all that is used to determine position of the roving receiver. Establishment of control points using static surveying methods requires much longer sessions than are typically used in kinematic surveys.

As previously stated, the accuracy of kinematic surveys does not match that of static surveys typically. Some of the limiting factors are the lack of repeated observations and the length of the session. For example, a static survey may use a 5-sec epoch rate to sample data over a one-hour session. This results in a total of 720 sets of observations per satellite where the satellite geometry will change significantly. The combination of a large set of observations with varying satellite geometry results in a better solution for the receiver coordinates. In a kinematic survey, the receiver may collect 600 observations per satellite using a 1-sec epoch rate over a 10-min interval. However, since the satellite geometry does not change significantly, the solution is often weaker than the static survey. Another accuracy degrading factor in kinematic surveys includes the motion of the rover during real-time data processing. Since, as depicted in Figure 15.1, the observations from the base receiver must be transmitted, received, and processed at the rover, any motion by the rover during this time will cause errors in its computed position. Since the motion of the rover is often small, the errors caused by this time difference, known as *data latency,* tend to be small. However, they can be significant in cases involving fast-moving rovers such as in the positioning of a camera station

Figure 15.1
A GPS receiver mounted on a fixed-height tripod.

during a photogrammetric mission. Other factors that limit the positioning accuracy of kinematic surveys are its susceptibility to errors such as DOP spikes, atmospheric and ionospheric refraction, multipathing, and obstructions to satellite signals. Often the effect of these factors can be minimized with careful planning or by advanced processing techniques.

■ 15.2 INITIALIZATION

To start a kinematic GPS survey, the receivers must be *initialized*. This process includes determining the integer ambiguity (see Section 13.5.2) for each pseudorange observation. Following any of the methods described below can yield initialization of the receivers.

One procedure for initializing the receivers uses a baseline whose ΔX, ΔY, ΔZ components are known. A very short static observing session is conducted with base and roving receivers simultaneously occupying two stations with known positions. Because the baseline coordinate differences are known, differencing of the observations will yield the unknown integer ambiguities. These differencing computations are performed in a postprocessing operation using the data from both receivers. If only one control station is available, a second one can be set using the static or rapid static surveying procedures described in Section 14.2.

An alternative initialization procedure, called *antenna swapping,* is also suitable if only one control station is available. Here receiver *A* is placed on the control point and receiver *B* on a nearby unknown point. For convenience, the unknown point can be within 30 ft (10 m) of the control station. After a few minutes of data collection with both receivers, their positions are interchanged while the receivers are still collecting data. In the interchange process, care must be exercised to make certain continuous tracking, or "lock" is maintained on at least four satellites. After a few more minutes of observations, the receivers are interchanged again, returning them to their starting positions. This procedure enables the baseline coordinate differences and the integer ambiguities to be determined, again by differencing techniques.

Finally, the most advanced technique of initialization is known as *on-the-fly* (OTF) ambiguity resolution. This method requires five usable satellites during the initialization process and dual-frequency receivers. OTF, which involves the solution of a sophisticated mathematical algorithm, has resolved ambiguities to the centimeter level in 2 min for a 20-km line. However, longer periods of time may be required since ideal conditions are seldom available. The typical period for ambiguity resolution is usually less than five minutes.

■ 15.3 EQUIPMENT USED IN KINEMATIC SURVEYS

The operator's body can be an obstruction when performing a kinematic survey. Thus as shown in Figure 15.2, the base receiver is often mounted on fixed-height tripods. Similarly as shown in Figure 15.3, the GPS antenna is often mounted on a fixed-height rod that is 2 m in length to avoid operator obstructions. In any case, the advantage of fixed-height rods and tripods in all GPS surveys is that they

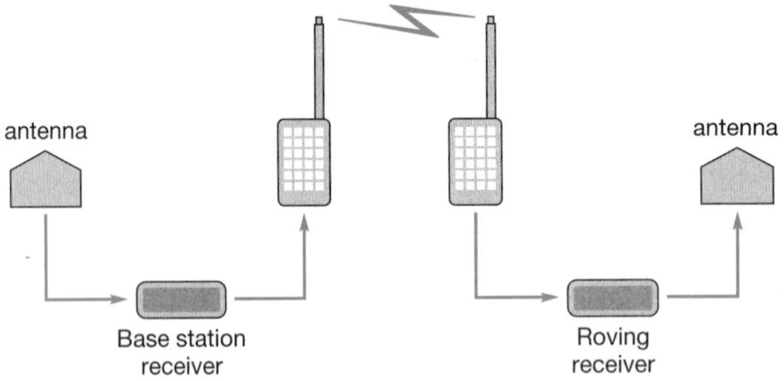

antenna
Base station receiver
antenna
Roving receiver

Figure 15.2
Note the antenna used in this stop-and-go RTK survey. (Courtesy Leica Geosystems, Inc.)

minimize measurement errors in the height of the receiver and help avoid operator-caused obstructions.

Other equipment used in kinematic surveys includes traditional adjustable-height tripods and poles. However the adjustments in these can often lead to errors in the measured heights to the receivers. Another factor to consider with traditional tripod equipment is the need to use a tribrach for mounting of the GPS receiver. As with traditional equipment, when tribrachs are used, it is extremely important to check the adjustment of the optical plummets (see Section 8.19.4). Similarly when using either fixed height or adjustable rods, it important to regularly check and adjust the level bubbles (see Section 8.19.5).

In kinematic surveys, overhead obstructions should be avoided at the base and rover stations. Additionally the base receiver location should be free of

Figure 15.3
A base receiver and rover with compatible internal radios used in RTK surveys.

reflective objects such as buildings, fences, and vehicles. Manufacturers sell 100-ft cables that enable the user to locate the vehicle away from the base receiver to avoid potential multipathing problems from the vehicle.

For the most common kinematic survey, RTK, the radio antenna at the base receiver is often mounted on a tripod. It is important to have the radio antenna vertical to match orientation of the antenna on the rover. Mounting the radio antenna high can increase the range of the base radio. However, repeater stations can also be used to extend the range of the base station radio in situations where necessary.

Several factors may determine the "best" location for the base station in a RTK survey. Since the range of the radio can be increased by increasing height of the radio antenna, it is advantageous to locate the base station on a local high point. Additionally, since the base station in a RTK survey requires the most equipment, it is also preferable to place the base station in an easily accessible location. The radios in a RTK-GPS survey are low power. Thus it is wise to avoid sources of large electromagnetic activity such as power grid substations, high-tension power lines, or buildings containing large electric motors since these items generate substantial electromagnetic fields that can overpower radio transmissions. Furthermore, the radio signals can interfere with the GPS receiver. Thus the radio antenna should be placed several meters from the GPS receiver.

Other equipment needed for RTK surveys includes a radio and its power source. Typically an external battery is used for the larger base receiver radio. When planning an RTK survey, it is important to provide some backup source of power for the inevitable event of a dead battery. A vehicle can serve as source for charging batteries and providing power to the base station radios. Finally, a survey controller (data collector) is required to control the collection of data from both the base and roving receivers.

■ 15.4 METHODS USED IN KINEMATIC SURVEYS

After initialization has been completed using one of the above techniques, the base receiver remains at the control station while the rover moves to collect data on features. Although the rover's positions can be determined at intervals as short as 0.2 sec, 1-sec epoch rates are typically used in kinematic surveys. The accuracy of intermediate points is in the range of $\pm(1 - 2\text{ cm} + 2\text{ ppm})$. In kinematic surveys, both receivers must maintain lock on at least four satellites throughout the entire session. If lock is lost, the receivers must be reinitialized. Thus, care must be taken to avoid obstructing the rover's antenna by carrying it close to buildings, beneath trees or bridges, shielding it with the operator's body, and so on. At the end of the survey, the rover should be returned to its initial control station, or another, as a check.

Kinematic surveys generally follow two forms of data collection. In *true kinematic* mode, data is collected at a specific rate. This method is useful for collecting points along an alignment, or grade elevations for topographic surveys. An alternative to the true kinematic mode is to stop for a few epochs of data at each point of interest. This method, known as *semikinematic* or the *stop-and-go* mode, is useful for mapping and construction surveys where increased accuracy is desired for a specific feature. In the semikinematic mode, the antenna is positioned over points of interest and a point identifier is entered into the survey controller for each feature. Since multiple epochs of data are usually are recorded at each point, the accuracy of this mode is greater than that obtainable in the true-kinematic mode. In both surveys, the epoch rate of data collection at the base station and rover is typically set to 1 sec.

In a kinematic survey, the rover is never at a station long enough to survive a PDOP spike (see Section 14.3.1). For kinematic surveys, PDOP values should be less than four. Additionally, since high free electron counts in the ionosphere can affect GPS signals greatly, it is important to collect data only during periods of low solar activity. During periods of high solar activity, poor positioning results can be obtained with GPS. NOAA provides daily, three-day, and weekly forecast of the space weather. In particular, GPS users should monitor the k-index values. GPS surveys should not be performed when the k-index is greater than 4. Both high PDOP values and high solar activity periods can be avoided with careful project planning.

In post-processed kinematic (PPK) GPS surveys, the collected data is stored on the survey controller or receiver until the fieldwork is completed. The data is then processed in the office using the same software and processing techniques used in static GPS surveys. Data latency is not a problem in PPK surveys since the data is post-processed. Other advantages of PPK surveys are (1) that precise ephemeris can be combined with the observational data to remove errors in the broadcast ephemeris and (2) the base station coordinates can be resolved after the fieldwork is complete. Thus the base station's coordinates do not have to be known prior to the survey. The lack of data latency and use of a precise ephemeris results in a PPK survey having higher accuracies than that obtainable from the same RTK survey.

As discussed in Section 13.8, real-time kinematic (RTK) GPS surveying, as implied by its name, enables positions of points to be determined instantaneously as the rover occupies a point. Like other kinematic methods, RTK surveying requires that two (or more) receivers be operated simultaneously, collecting data from 4 or more common satellites at the same epoch rate. The unique aspect of this procedure is that a radio is used to transmit the base receiver coordinates and its raw GPS observations to the rover. At the rover, the GPS observations from both receivers are processed in real-time by the unit's on-board computer to produce a nearly immediate determination of its location according to Equation (13.27). Like PPK surveys, the processing techniques are similar to those used in static GPS surveys. However, the epoch rate for data collection is typically set to 1 sec. Figure 15.3 shows the stop-and-go RTK survey in progress.

RTK surveys require compatible hardware at each end of the radio link. Normally, this equipment is purchased from one manufacturer. In North America and in other areas of the world, frequencies in the range of 150–174 MHz in the VHF radio spectrum, and from 450–470 MHz in the UHF radio spectrum, can be used for RTK transmissions. Typically, the messages are updated at the rover every 0.5 – 2 sec. The data link for RTK requires a minimum of 2400 baud or higher for operation. However, it is typically much higher with a baud rate of typically 38,400.

Even with higher baud rates, the application of the base receiver data to the rover data is delayed because of delays in transmitting the base receiver's observations and position to the rover and the additional time required to compute the rover's position. Typically, this delay, known as *data latency,* is between 0.05–1.0 sec. Data latency plays a role in the final accuracies of derived positions. However, these problems have been minimized since selective availability has been turned off. Selective availability was the largest random error source, and thus could only be modeled for time-delayed corrections. Even with selective availability, these errors tended to be small and did not significantly affect the final quality of most surveys.

The radio link used with RTK can limit the distance between the base receiver and rover(s) to under 10 km, or about 6 mi. As shown in Figure 15.4, this distance can be increased with more powerful transmitters, or through the use of repeater stations. A repeater station receives the signal from a transmitter such as the base radio and re-transmits it. Some transmitters require a Federal Communication Commission (FCC) license to broadcast the data. With low-power radios, line-of-sight between the transmitter and receiver is required. An advantage of repeater stations is that they can be used to survey around obstructions and increase the range of the base radio. In areas where cellular coverage is available, data modems can also be used to broadcast data from the base receiver to the rover.

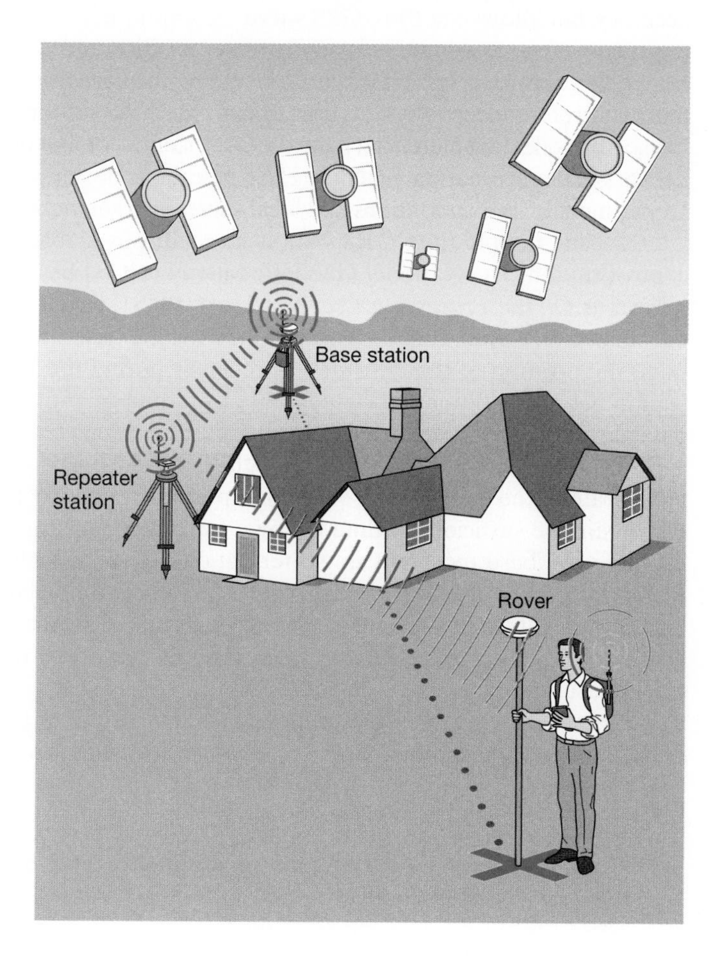

Figure 15.4
A repeater radio.

The advantages of RTK surveys over PPK surveys are the reduction in office time, and the ability to verify observations in the field. When using RTK, the data can also be downloaded immediately into a GIS (see Chapter 28) or an existing surveying project. This increases the overall productivity of the survey.

15.5 PERFORMING POST-PROCESSED KINEMATIC SURVEYS

A PPK survey requires a base receiver that is collecting data at the same epoch rate as the rover. The base receiver is usually set over a reference station established from a prior static survey. If a local CORS station is collecting data at the same rate, it can also be used as a base receiver. For example if a local CORS station is gathering data at a 5-sec epoch rate and the rover is also set to a 5-sec epoch rate, the CORS station's observation files can be downloaded and used to reduce the rover's data. Some CORS stations have an epoch rate of 1 sec for this purpose.[1]

When a local reference station is not immediately available for use as a base station, it is possible to establish the temporary coordinates of the base receiver by using its autonomous position. The autonomous position of the GPS receiver can be derived from as little as a single epoch of data.[2] This position is not high in accuracy, but allows the PPK-GPS survey to continue.

Prior to processing the roving receiver positions, accurate positioning of the base station can be achieved by post-processing the base receiver's data with an established reference network station. Ties to the base station receiver can be performed by collecting data from a local CORS station, or any other established GPS network reference station including HARN stations or stations established by state Department of Transportations and local surveying (geomatics) engineering firms.

As an example, the CORS station is a readily available established reference control station. The National Geodetic Survey (NGS) has established a website known as *Online Processing User Service* (OPUS)[3] to perform this function for the surveyor. It uses three CORS stations near the base station receiver to determine the position of the base station receiver during the PPK survey. A CORS station has the additional advantage of not requiring another receiver to gather data at the network reference station.

It is not important that data for this connection be gathered at the same epoch rate as the PPK survey. What is important is that the length of the observing session be sufficient in time to solve for the connecting baseline vector. For example, the base receiver can gather data at a 1-sec epoch rate to support the PPK survey while the GPS receiver on the network station collects data at a 5-sec epoch rate. Table 14.1 provides typical lengths of sessions for various baseline lengths. While the connection data is being collected, the rover can proceed with the PPK survey.

The NGS recommends a minimum session length of two hours at the base receiver; it has a maximum length of 24 hours. However, the NGS provides a *rapid*

---

[1]Space weather forecasts are available at http:// www.sec.noaa.gov/Data/index.html.

[2]Due to the inaccuracy of the autonomous position of a GPS receiver, it should never be used in RTK-GPS stakeout surveys.

[3]OPUS is available at http://www.ngs.noaa.gov/OPUS/.

service that can determine the position of the base station receiver with as little as 15 minutes of data under good observation conditions. In the meantime, the PPK survey can be performed. After the survey is complete, the RINEX observation file (see Section 14.3.5) from the base receiver is sent to OPUS using the Internet. OPUS solves for the position of the base receiver using nearby three CORS stations. Within minutes typically, the results of the computations are sent back to the user via e-mail. The OPUS derived position for the base station is then entered into the user's processing software to determine the locations of the rover during the simultaneous PPK survey. In order to achieve the best results, the model of the base receiver antenna and receiver antenna height in relation to the antenna reference point (see Section 13.6.3) must be known and entered along with the observation file into OPUS.

PPK is typically used to collect data for mapping surveys. They are especially useful for large surveys with minimal obstructions where the rover can be mounted on a vehicle. As mentioned in Section 15.5, features can be collected using the semikinematic or true kinematic mode. In areas where canopy obstructions are a problem, the semikinematic mode can be used to establish temporary, higher-accuracy reference points for later use by a total station. A minimum of two points is required to provide the total station with occupying and backsight points. As the new and stronger L2C and L5 signals become available, the problem caused by canopy obstructions could significantly disappear, thus removing the need for a total station to gather data in these areas.

When collecting data using the true kinematic mode, it is important to set a reasonable epoch rate based on the speed of motion of the rover. For example if the rover is being hand carried, a 1-sec epoch rate would result in data being gathered every five or six feet. This is an excessive amount of data for the typical topographic survey. Furthermore, excessive data collection in lines would result in a weak triangulated irregular network. Section 17.8 discusses the importance of properly collecting data for topographic features. Section 17.12 discusses the use of drawing designators to efficiently produce line work on a map. To avoid this, many survey controllers allow their users to set the data collection rate on a specified two- or three-dimensional distance.

As mentioned in Section 15.2, the receivers must be initialized before a kinematic survey can be started. Once this occurs, data collection can proceed as long as lock on four satellites is maintained. Thus it is important to watch the number of visible satellites and the PDOP of the solution. If the number of satellites drops below four, the rover must be reinitialized. The most common method of reinitialization is accomplished by moving the rover to a location where five satellites are visible. At this location, OTF will quickly re-establish lock on the satellites. OTF can re-establish lock on the satellites in less than 1 min in these situations. However, if this is not achievable, the user can move to a previously surveyed identifiable feature to re-establish lock on the satellites. Temporary control points can be established at the beginning of the survey to facilitate this solution. Since returning to a previously surveyed, identifiable feature can be time-consuming, most users try to maintain lock on five or more satellites at all times and avoid situations where loss-of-lock problems will occur. As previously stated, when the number of modernized L2C and L5 satellites becomes significant, loss-of-lock problems may disappear.

Since kinematic surveys use a small number of observations to establish the coordinates of points, a PDOP value that is less than four is recommended for most surveys. The user can also watch the PDOP as the survey proceeds. When weak satellite geometry is present, the PDOP value will rise. No data should be collected if the PDOP value is greater than six. A sudden change in the PDOP value is usually caused by an obstruction that has removed a key satellite from the geometric solution of the point. When this happens, the user should again proceed to a location where the PDOP is reduced and continue the survey.

After the data is collected, it is loaded into the processing software. An advantage PPK surveys have over RTK surveys is that a precise ephemeris can be used in the processing. As discussed in Section 13.6.3, this will result in a better solution for the positions of the surveyed points. The base station coordinates should be established or entered before the rover's observation file is downloaded. If the base station's coordinates are not known, the position of the base receiver should be computed in the processing software or obtained using OPUS. Having established the base station's coordinates before loading the rover's observation file ensures that the vectors to the rover will radiate from the base station. The processing of the baseline vectors is then performed. Since this is a radial survey, no checks are available on the resultant coordinate values. However, for critical features, it is possible to resurvey these points from a second base station location. This is similar to the radial traversing procedure discussed in Section 9.9.

As discussed in Section 13.4.3, the heights determined by GPS are in the geodetic coordinate system. Typically, topographic maps are produced using a map coordinate system and orthometric heights. The conversion of geodetic coordinates to map coordinates is covered in Chapter 20. As shown in Equation (13.8), the geoidal undulation at each point must be applied to the geodetic height to determine the orthometric height. If requested by the user and a geoid model is available, the software can determine the orthometric height of the points surveyed. The current geoid model for the conterminous United States is GEOID03 and GEOID06 for Alaska. These models have an accuracy of a few centimeters for most of the U.S. Thus, the GPS-derived orthometric height will be slightly worse. The software manufacturer supplies support files to upgrade both the survey controller and software to the current geoid model.

■ 15.6 COMMUNICATION IN REAL-TIME KINEMATIC SURVEYS

As depicted in Figure 15.1, roving receivers in RTK surveys require continuous communication with base receivers. These communications can be accomplished with radios, wireless Internet connections, or data modems. Using these devices, the base receiver transmits both corrections and raw data to the rover. The rover processes this data using procedures similar to those discussed in Sections 13.8 or 13.9.

The most common form of communication between the base receiver and rover is by low-powered radios. These radios are typically an integral part of the GPS receiver. The Federal Communications Commission (FCC) does not require a license for radios that broadcast in the range from 157–174 MHz. However, all other frequencies given in Section 15.4 do require a FCC license. The stronger, external GPS radio transmitters typically use the frequencies in the 450–470 MHz range and these frequencies require a FCC license. Since, by FCC regulations, voice communication takes precedence over data communication, GPS radio transmitters generally come with as many as 10 or more preset frequencies or channels. The operator must find a channel that is not in use already. Additionally using an unlicensed channel is a violation of FCC regulations, thus it is wise to license several of the channels that are available on the transmitter. The maximum power of the radios is typically 35 watts. This form of communication will work in all areas of the world although additional licensing to use the frequencies may be required. When using radios, it is important to connect the antenna to the radio before powering the transmitter to avoid equipment failure.

Another option for communication between the base receiver and rover is data modems. These require cellular coverage in the area being surveyed. When coverage is available, the data is transmitted via cellular technology to the rover. The cell provider charges a monthly service fee to use this option. Obviously, this form of communication is not available in areas that do not have cell coverage. Additionally, data latency with this form of communication will be greater than that experienced with radios.

In areas where wireless Internet connections are available, it is possible for the base receiver and rover to communicate over the Internet. This option requires that the base receiver have an Internet connection and the rover have a wireless connection. Again data latency will be greater than that experienced with radios using this form of communication.

Several problems can occur with communication equipment. Cables often develop breaks near connectors resulting in intermittent transmission problems. In severe cases, the cables fail and communication is impossible. Also the power of the radio limits its range. When using GPS receivers with internal radios, this range is often limited to small areas around the base station. As discussed in Section 15.3 and depicted in Figure 15.4, this range can be increased with repeater stations. With larger 35-watt, external radios, the achievable range of the survey is maximized, but is generally limited to areas under 6 mi in radius. Again larger ranges can be achieved with repeater stations. In one survey in Alaska under ideal conditions, the range from the base radio to the rover was 38 mi! Obviously, this was not a typical situation.

■ 15.7 REAL-TIME NETWORKS

In kinematic surveys, the accuracy of the position degrades as the rover moves away from the base station. Furthermore, a base station requires additional receivers and personnel to perform the survey. If the base receiver could be used as a rover, the work could be performed in half the time.

An option that eliminates the need for a base receiver in a RTK survey is known as a *Real-Time Network* (RTN). Both the private and public sector are implementing this technology. The RTN is a network of base stations that are connected to a central processing computer using the Internet. Using the known positions of the base receivers and their observational data, the central processor models errors in the satellite ephemerides, range errors caused by ionospheric and tropospheric refraction, and the geometric integrity of the network stations. *Virtual reference station* (VRS) and *spatial correction parameter* (FKP)[4] are examples of two methods used in modeling these errors.

Of course, these systems may not work reliably in areas that are cellularly challenged. Since the entire system involves communication from multiple base receivers to a central processor and finally to a rover, high-traffic volume on the Internet, multiple connections between network servers to the central processor, and time of transmissions in the cellular world can create greater data latency than much simpler base-to-rover radio connections. Therefore, some manufacturers wait for the corrections from the central processor before processing the data at the rover. Others extrapolate the modeled errors to process the rover's observational data at the time of reception. The application software typically stops survey operations if the data latency becomes greater than a specified time interval. However, this value may be as great as four seconds! For this reason, surveyors should use real-time networks cautiously in machine control operations (see Section 15.9).

The errors at the rover with a radio connection to a single base station are not linear with respect to distance. When the rover connects to the RTN, either the central processor or the rover interpolates these errors to a location at the survey site. When using VRS, a *virtual reference station* is created by the rover to serve as its base station. If the rover moves too far from the VRS, another virtual reference station is determined for the rover. When working with a RTN, the ppm error in GPS surveying is removed resulting in better achievable accuracies than are present with a radio and single base receiver. The accuracy of positions determined using RTN is usually within 1 cm anywhere within the bounds of the network. Another advantage of using a RTN is that the coordinates obtained from the network are referenced to a common datum,[5] and thus results from many surveys will join seamlessly.

These RTN systems are sold usually through a subscription service. Users of this service save costs since they do not need a base receiver or the additional personnel to monitor the base receiver while performing a survey. A word of caution when using a RTN network, the rover and the network base receivers must use the same tropospheric refraction models to avoid systematic errors due to modeling differences. Additionally, the system should be periodically calibrated by locating a known position in the RTN system with the rover. HARN stations can serve as good reference stations. Caution should be exercised when using a RTN outside the bounds of the network since errors increase rapidly with the extrapolation of the corrections beyond the limits of the network base stations.

■ 15.8 PERFORMING REAL-TIME KINEMATIC SURVEYS

As previously stated, the main difference between RTK surveys and PPK surveys is the fact that RTK surveys provide immediate results in the field. Thus RTK surveys are used primarily in construction stakeout. Since RTK surveys provide

[4]FKP is an acronym for Flächenkorrekturparameter, which is German for spatial correction parameter.

[5]Section 19.6 discusses the reference frames that are currently used on the North American continent.

immediate results, some form of communication as discussed in Section 15.6 must be established and maintained during the entire RTK survey. Similar to PPK surveys, the receivers must be initialized before the survey is started and initialization must be maintained during the entire survey. However, the process of surveying is similar to the methods used in a PPK survey.

Stakeout surveys using RTK have some important distinctions from conventional surveys. One important difference is the reference frame (also called the datum). As discussed in Chapter 19, conventional surveys use some form of NAD 83 as their horizontal datum and NAVD 88 for their vertical datum typically. These reference frames are considered to be regional since they were developed using observations only on the North American continent. As discussed in Section 13.4.3, the broadcast ephemeris uses a WGS 84 reference frame based on the coordinates of the tracking station. This is a worldwide reference frame. The current rendition of the WGS 84 reference frame closely approximates ITRF 2000. The difference in the origins of the NAD 83 and ITRF 2000 data is about 2.2 m. Thus when performing a stakeout survey, the coordinates for stations produced by GPS receivers can differ significantly from coordinates of the same stations in the regional reference frame that were used to create the engineering design.

As discussed in Section 19.6.6, the GPS coordinates of the points can be brought into the regional datum using a coordinate transformation. To do this, points having regional reference frame coordinate values must be established on the perimeter of the project area. A minimum of two horizontal control points and three vertical control points should exist. However, it is better to have a minimum of three horizontal control points and four vertical control points for the purpose of redundancy and checks. As discussed in Section 19.6.6, a minimal transformation considering only the translation factors between the reference frames can be performed if only a single regional control point exists. However, this will result in a significant loss in accuracy to the survey. Furthermore, as discussed in Section 19.6.6, a simplified two-dimensional coordinate transformation [see Equation (11.37)] can be used to transform GPS-derived positions into a local reference frame. Again, it is important when performing these transformations to have control on the exterior and surrounding the project area to avoid extrapolation errors.

Survey controllers have this transformation built into their software. Depending on the vendor, this transformation is known as *localization* or *site calibration*. This procedure should be performed at the beginning of each survey. The procedure involves occupying the control stations with the GPS receiver. The controller then computes the transformation parameters and allows the user to view the errors. It is wise to perform this procedure whenever questions concerning the stability of the control arise to eliminate possible errors. With this in mind, an alignment survey should be planned with control points along the entire corridor to ensure their quick availability. Following this procedure, the stakeout of the design points can proceed.

Since receivers create observation files during an RTK survey, it is possible to convert an RTK survey into a PPK survey in the office. This may be helpful when problems are experienced in the field. However, this would serve no purpose on a stakeout survey.

15.9 MACHINE CONTROL

Traditionally construction projects were executed by placing stakes at key locations in the project (see Chapter 23) to establish the levels of materials, alignments, and grades of finished work. However with RTK GPS, it is possible to load the project design, digital terrain models (see Section 18.14), and site calibration parameters into a computer that guides the vehicle during the construction process. This technology, known as *machine control,* allows the machine operator to see their position in a construction project, cut and fill levels, and finished grades of the project, all in real-time. As shown in Figure 15.5 and 15.6, this is achieved by placing RTK-GPS units on the construction equipment. One aspect of this technology is that the GPS receiver must be calibrated with respect to the construction vehicle. For instance, the distance between the GPS receiver's antenna reference point and the cutting edge of the blade on a machine must be observed and entered into the machine control system.

To accurately achieve this level of automation, the surveyor must place sufficient horizontal and vertical control about the construction project area. A range of about 10 km (6 mi) from the base station is possible with GPS. Additionally, in locations where obstructions may interfere with GPS signals, the surveyor must add sufficient control to support the use of a robotic total station. In these areas, a robotic total station can guide the construction equipment past the obstructions.

Figure 15.5 A dozer and grader using machine control to create an intersection of roads. (Courtesy Topcon Positioning Systems, Inc.)

Figure 15.6
GPS receiver mounted on grader blade. (Courtesy Topcon Positioning Systems, Inc.)

Machine control requires a digital terrain model of the existing terrain prior to the start of the project and a proposed three-dimensional surface model of the finished project. These two items are loaded into the machine control system along with the site calibration parameters. As discussed in Section 15.8, the site calibration parameters are needed to transform the GPS coordinates into the project datum. With an existing DTM, final digital surface, and localization parameters loaded into the machine control system, the construction vehicle is guided by the machine control system through the project.

The accuracy of using RTK GPS is about 1 cm in horizontal and 2 cm in vertical. This accuracy is sufficient for excavation purposes. However, finished surfaces need to have accuracies under 0.02 ft (5 mm). This accuracy is achieved by augmenting the machine control system with laser levels as shown in Figure 24.2 or robotic total stations. Some manufacturers have incorporated a rotating laser level into their machine control systems to achieve millimeter accuracies in all three-dimensions. When using this equipment, sufficient control must be placed at the perimeters of the project area to guide the construction vehicles through the project. For example, robotic total stations have a working radius of about 1000 ft from the construction vehicle and laser levels have a working radius of about 1500 ft from the construction vehicle. Thus control must be placed within the appropriate limit to provide sufficient guidance to support the machine control system. Again, real-time networks should be used with caution in machine control since data latency can be large. This can lead to significant real-time errors during the finishing process in a construction project.

As shown in Figure 15.7, another area that is utilizing RTK GPS is precision agriculture. This area does not require a surveyor's expertise, but is interesting nonetheless. In precision agriculture, crop yields are monitored with respect to positions of the harvester on the field. Additionally soil samples are located and tested for fertility, drainage, and so on to provide the farmer with a complete picture of yields versus growing conditions. In the following year, this information is fed into a guidance system that controls tillage equipment, planters, sprayers, and fertilizer spreaders so that appropriate tillage and chemicals are used as required at specific locations on the field. The end results are an economy in fuel and chemicals with

Figure 15.7 A large tractor pulling a land leveler. (Courtesy Topcon Positioning Systems, Inc.)

increased yields in crops. This is an example of surveying technology reaching into nonsurveying fields.

■ 15.10 ERRORS IN KINEMATIC SURVEYS

Kinematic surveys suffer from some same error sources that are found in conventional surveys. These include:

1. Setup errors at the base station and rover.
2. Errors in reading the height to the base station or rover antenna.

Additionally, all GPS surveys have the following errors sources.

1. Ionospheric refraction
2. Tropospheric refraction
3. Errors in ephemerides
4. Base station coordinate errors
5. Weak satellite geometry

■ 15.11 MISTAKES IN KINEMATIC SURVEYS

Some of the more common mistakes that can be made in kinematic surveys include:

1. Misidentification of stations
2. Incorrect identification of the station
3. Starting or proceeding with the survey before the integer ambiguities are resolved
4. Misidentification of the antenna
5. Surveying during high periods of solar activity
6. Incorrect settings for the radio or wireless connection
7. Surveying under overhead obstructions

PROBLEMS

Asterisks (*) indicate problems that have answers given in Appendix G.

15.1 List at least four applications of kinematic surveys.

15.2 What are the main differences between PPK and RTK GPS?

15.3* What is meant by data latency?

15.4 List three methods that can be used to resolve the integer ambiguities of the pseudorange observations.

15.5 What are the similarities between static and kinematic GPS?

15.6 Why are fixed-height tripods and rods favored in kinematic GPS surveys?

15.7 List two reasons why adjustable height tripods are not recommended in GPS surveys.

15.8 Discuss why the base receiver should not be located near a vehicle.

15.9 Explain how a CORS station can be used in a PPK survey.

15.10* List two reasons why PPK surveys usually provide better results than RTK surveys.

15.11 What is OPUS, and how can it be used in a PPK survey?

15.12 Discuss the various methods that can be used to determine the position of the base station in a PPK survey.

15.13 Discuss the differences between the stop-and-go and the true kinematic modes of surveying.
15.14 Discuss the appropriate steps used in processing PPK data.
15.15* When is an FCC license required to operate a radio transmitter?
15.16 List three possible methods of communication between a base station and rover.
15.17 What is a real-time network?
15.18 Discuss the advantages of an RTN?
15.19 When can an RTN not be used in an RTK survey?
15.20 What is meant by site calibration and why is it important in a stakeout survey?
15.21 Discuss how the coordinate differences between a regional datum and GPS coordinates can be resolved.
15.22 What three surveying elements are needed in machine control?
15.23 What surveying equipment can be used in machine control?
15.24 How are robotic total stations used in machine control?
15.25 How are finished grades determined in machine control?
15.26* What is high-precision agriculture?
15.27 What should be considered in planning a kinematic GPS survey?
15.28 How many total pseudoranges observations will be measured using a 1-sec epoch rate for a total of 8 min with 5 usable satellites?
15.29 How many pseudoranges observations will be measured using a 5-sec epoch rate for a total of 55 min with 5 usable satellites?
15.30 The baseline vector between the base and roving receivers is 8340 m long. What is the estimated uncertainty in the length of the baseline vectors if an RTK-GPS survey is performed?
15.31 In Problem 15.30, what would be the estimated uncertainty in the length of the baseline vector in a real-time network?
15.32 Discuss the importance of knowing the space weather before performing a kinematic GPS survey.
15.33 How could data latency affect machine control using RTK GPS?
15.34 How might the new L2C and L5 GPS signals help kinematic GPS?

BIBLIOGRAPHY

Barr, M. 2006. "Real-Time Connection." *Point of Beginning* 31 (No. 4): 22.

Bryant, Rod. 2005. "Assisted GPS—Using Cellular Telephone Networks for GPS Anywhere." *GPS World* 16 (No. 5): 40.

——. 2006. "Midwest Real-Time Kinematic Network Starts in St. Louis and Expands to Neighbouring States." *Professional Surveyor* 26 (No. 10): 18.

Crawford, Wesley. 2006. "What Are Your Tolerances?" *Point of Beginning* 32 (No. 3): 46.

El-Mowafy, Ahmed. 2005. "Analysis of the Design Parameters of Multi-Reference Station RTK GPS Networks." *Surveying and Land Information Science* 65 (No. 1): 17.

Glaus, R. et al. 2004. "Precise Rail Track Survey." *GPS World* 15 (No. 5): 28.

Henning, William. 2006. "The New RTK—Changing Techniques for GPS Surveying in the USA." *Surveying and Land Information Science* 66 (No. 2): 107.

Katsambalos, K. and P. Savvaidis. 1996. "GPS Receiver Calibration." *Point of Beginning* 21 (No. 5): 87.

Licht, R. 2003. "A Step Ahead–Employees of a Minnesota Firm Take GPS One Step Further with the Application of Cell Phones." *Point of Beginning* 28 (No. 12): 32.

Mosby, Mark. 2006. "Advancing with Machine Control." *Point of Beginning* 32 (No. 3): 32.

Pugh, N. 2005. "Scalable GPS Infrastructure." *Professional Surveyor* 25 (No. 1): 14.

Richert, T. and El Sheimy. 2005. "Ionosphere Modeling—The Key to GNSS Ambiguity Resolution." *GPS World* 16 (No. 6): 35.

Schrock, Gavin. 2006. "RTN-101: An Introduction to Network Corrected Real-Time GPS/GNSS (Part 1)." *The American Surveyor* 3 (No. 6): 28.
——. 2006. "RTN-101: An Introduction to Network Corrected Real-Time GPS/GNSS (Part 2)." *The American Surveyor* 3 (No. 7): 38.
——. 2006. "RTN-101: An Introduction to Network Corrected Real-Time GPS/GNSS (Part 3)." *The American Surveyor* 3 (No. 8): 38.
——. 2006. "RTN-101: On-Grid—An Initiative in Support of RTN Development (Part 4)." *The American Surveyor* 3 (No. 9): 39.
——. 2006. "The On-Grid Initiative—An Initiative to Nationally Promote the Development of Local and Regional Real-Time GNSS Infrastructure Networks to Serve as Public Amenities." American Congress on Surveying and Mapping and the National Society for Professional Surveyors. Available at http://www.aagsmo.org/textfiles/OnGrid_White-Paper42806B.pdf.
——. 2007. "RTN-101: Reference Station Communications (Part 5)." *The American Surveyor* 4 (No. 2): 38.
——. 2006. "RTN-101: Network Corrected Real-Time GPS/GNSS (Part 6)." *The American Surveyor* 4 (No. 3): 38.
Speed, V. 2002. "Surveyors Dial Up for Accuracy." *Point of Beginning* 27 (No. 11): 24.
Stevens, K. 2002. "GPS Fore the Golf Course." *Point of Beginning* 27 (No. 8): 18.
Torán, R., et al. 2004. "Positioning via Internet." *GPS World* 15 (No. 4): 28.
Wood, C. and O. Mace. 2001. "Dead Reckoning Keeps GPS in Line—Vehicle Positioning in Urban Environments." *GPS World* 12 (No. 3): 14.

16

Adjustments by Least Squares

■ 16.1 INTRODUCTION

The subject of errors in measurements was introduced in Chapter 3 where the two types of errors, systematic and random, were defined. It was noted that systematic errors follow physical laws, and if conditions producing them are observed, corrections can be computed to eliminate them. However, random errors exist in all observed values. Additionally as discussed in Chapter 3, observations can contain mistakes (blunders). Examples of mistakes are setting an instrument on the wrong point, sighting the wrong point, transcription errors in recording observed values, and so on. Mistakes should be removed when possible before the adjustment process. As further explained in Chapter 3, experience has shown that random errors in surveying follow the mathematical laws of probability and in any group of observations they are expected to conform to the laws of a *normal distribution,* as illustrated in Figure 3.3.

In surveying (geomatics), after eliminating mistakes and making corrections for systematic errors, the presence of the remaining random errors will be evident in the form of misclosures. Examples include sums of interior angles in closed polygons that do not total $(n - 2)180°$, misclosures in closed leveling circuits, and traverse misclosures in departures and latitudes. To account for these misclosures, adjustments are applied to produce mathematically perfect geometric conditions. Although various techniques are used, the most rigorous adjustments are made by the method of least squares, which is based on the laws of probability.

Although the theory of least squares was developed in the late 1700s, because of the lengthy calculations involved, the method was not commonly used prior to the availability of computers. Instead, arbitrary or "rule of thumb" methods such as the compass (Bowditch) rule were applied. Now least-squares calculations are handled routinely and making adjustments by this method is rapidly becoming indispensable in modern surveying (geomatics). The method of least squares is currently being used to adjust all kinds of observations, including differences in elevation, horizontal distances, and horizontal and vertical angles. It has become essential in the adjustment of GPS observations and is also widely used in adjusting photogrammetric data. Adjustments by the least-squares method have taken on added importance with the most recent surveying accuracy standards. These standards include the use of statistical quantities that result from a least-squares adjustment. Thus, in order to evaluate a survey for compliance with the standards, least-squares adjustments must first be performed.

Least-squares adjustments provide several advantages over other arbitrary methods. First of all, because the method is based upon the mathematical theory of probability, it is the most rigorous of adjustment procedures. It enables all observations to be simultaneously included in an adjustment, and each observation can be weighted according to its estimated precision. Furthermore, the least-squares method is applicable to any observational problem regardless of its nature or geometric configuration. In addition to these advantages, the least-squares method enables rigorous statistical analyses to be made of the results of the adjustment, i.e., the precisions of all adjusted quantities can be estimated and other factors investigated. The least-squares method even enables pre-survey planning to be done so as to ensure that required precisions of adjusted quantities are obtained in the most economical manner.

A simple example can be used to illustrate the arbitrary nature of "rule of thumb" adjustments, as compared to least squares. Consider the horizontal survey network shown in Figure 16.1. If the compass rule was used to adjust the observations in the network, several solutions would be possible. To illustrate one variation, suppose that traverse *ABCDEFGA* is adjusted first. Then holding the adjusted values of points *G* and *E*, traverse *GHKE* is adjusted, and finally, holding the adjusted values on *H* and *C*, traverse *HJC* is adjusted. This obviously would

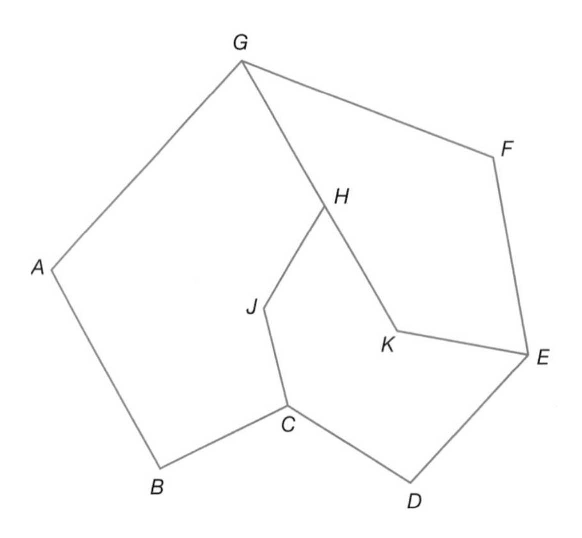

Figure 16.1 A horizontal network.

yield a solution, but there are other possible approaches. In another variation, traverse *ABCDEFGA* could be adjusted followed by *GHJC*, and then *HKE*. This sequence would result in another solution, but with different adjusted values for points *H*, *J*, and *K*. There are still other possible variations. This illustrates that the compass rule adjustment is properly referred to as an "arbitrary" method. On the other hand, the least-squares method simultaneously adjusts all observations and for a given set of weights there is only one solution—that which yields the most probable values for the given set of observations.

In the sections that follow, the fundamental condition enforced in least squares is described and elementary examples of least-squares adjustments are presented. Following this, systematic procedures for forming and solving least-squares equations using matrix methods are given and demonstrated with examples. The examples involving differential leveling, GPS baselines, and horizontal networks are performed using the software WOLFPACK, which is included within the CD that accompanies this book. For these examples, sample data files and the results of the adjustments are shown. A complete description of the data files is given in the *help* system provided with the software. Additionally, Mathcad® worksheets that demonstrate the differential leveling and plane survey adjustments are included within the CD that accompanies this book.

16.2 FUNDAMENTAL CONDITION OF LEAST SQUARES

It was shown through the discussion in Section 3.12 and the normal distribution curves illustrated in Figures 3.2 and 3.3 that small errors (residuals) have a higher probability of occurrence than large ones in a group of normally distributed observations. Also discussed was the fact that in such a set of observations there is a specific probability that an error (residual) of a certain size will exist within a group of errors. In other words, there is a direct relationship between probabilities and residual sizes in a normally distributed set of observations. The method of least-squares adjustment is derived from the equation for the normal distribution curve. *It produces that unique set of residuals for a group of observations that have the highest probability of occurrence.*

For a group of equally weighted observations, the fundamental condition enforced by the least squares method is that the *sum of the squared residuals is a minimum.* Suppose a group of m observations of equal weight were measured having residuals $v_1, v_2, v_3, \ldots, v_m$. Then, in equation form, the fundamental condition of least squares is

$$\sum_{i=1}^{m} v_i^2 = v_1^2 + v_2^2 + v_3^2 + \cdots + v_m^2 \longrightarrow \text{minimum}^1 \qquad (16.1)$$

For any group of observed values, weights may be assigned to individual observations according to *a priori* (before the adjustment) estimates of their relative worth or they may be obtained from the standard deviations of the observations,

[1]Refer to Ghilani and Wolf (2006) cited in the bibliography at the end of this chapter for a derivation of this equation.

if available. An equation expressing the relationship between standard deviations and weights, given in Section 3.20 and repeated here, is

$$w_i = \frac{1}{\sigma_i^2} \tag{16.2}$$

In Equation (16.2), w_i is the weight of the ith observed quantity and σ_i^2 is the variance of that observation. *This equation states that weights are inversely proportional to variances.*

If observed values are to be weighted in least-squares adjustment, then the fundamental condition to be enforced is that *the sum of the weights times their corresponding squared residuals is minimized* or, in equation form,

$$\sum_{i=1}^{m} w_i v_i^2 = w_1 v_1^2 + w_2 v_2^2 + w_3 v_3^2 + \cdots + w_m v_m^2 \longrightarrow \text{minimum} \tag{16.3}$$

Some basic assumptions underlying least-squares theory are that (1) mistakes and systematic errors have been eliminated, so only random errors remain in the set of observations, (2) the number of observations being adjusted is large, and (3) as stated earlier, the frequency distribution of the errors is normal. Although these basic assumptions are not always met, least-squares adjustments still provide the most rigorous error treatment available.

16.3 LEAST-SQUARES ADJUSTMENT BY THE OBSERVATION EQUATION METHOD

Two basic methods are employed in least-squares adjustments: (1) the *observation equation* method and (2) the *condition equation* method. The former is most common and is the one discussed here. In this method, "observation equations" are written relating observed values to their residual errors and the unknown parameters. One observation equation is written for each observation. For a unique solution, the number of equations must equal the number of unknowns. If redundant observations are made, the least-squares method can be applied. In that case, an expression for each residual error is obtained from every observation equation. The residuals are squared and added to obtain the function expressed in either Equation (16.1) or Equation (16.3).

To minimize the function in accordance with either Equation (16.1) or Equation (16.3), partial derivatives of the expression are taken with respect to each unknown variable and set equal to zero. This yields a set of so-called *normal equations,* which are equal in number to the number of unknowns. The normal equations are solved to obtain most probable values for the unknowns. The following elementary examples illustrate the procedures.

Example 16.1

Using least squares, compute the most probable value for the equally weighted distance observations of Example 3.1.

Solution

1. For this problem, as was done in Example 3.1, let $\overline{M}$ be the most probable value of the observed length. Then write the following observation equations that define the residual for any observed quantity as the difference between the most probable value and any individual observation:

$$
\begin{aligned}
\overline{M} &= 538.57 + v_1 \\
\overline{M} &= 538.39 + v_2 \\
\overline{M} &= 538.37 + v_3 \\
\overline{M} &= 538.39 + v_4 \\
\overline{M} &= 538.48 + v_5 \\
\overline{M} &= 538.49 + v_6 \\
\overline{M} &= 538.33 + v_7 \\
\overline{M} &= 538.46 + v_8 \\
\overline{M} &= 538.47 + v_9 \\
\overline{M} &= 538.55 + v_{10}
\end{aligned}
$$

2. Solve for the residual in each observation equation and form the function Σv^2 according to Equation (16.1),

$$
\begin{aligned}
\sum v^2 = {} & (\overline{M} - 538.57)^2 + (\overline{M} - 538.39)^2 + (\overline{M} - 538.37)^2 \\
& (\overline{M} - 538.39)^2 + (\overline{M} - 538.48)^2 + (\overline{M} - 538.49)^2 \\
& (\overline{M} - 538.33)^2 + (\overline{M} - 538.46)^2 + (\overline{M} - 538.47)^2 \\
& (\overline{M} - 538.55)^2
\end{aligned}
$$

3. Take the derivative of the function Σv^2 with respect to $\overline{M}$, set it equal to zero (this minimizes the function) and solve for $\overline{M}$,

$$
\begin{aligned}
\frac{\partial \sum v^2}{\partial \overline{M}} = 0 = {} & 2(\overline{M} - 538.57) + 2(\overline{M} - 538.39) + 2(\overline{M} - 538.37) \\
& +2(\overline{M} - 538.39) + 2(\overline{M} - 538.48) + 2(\overline{M} - 538.49) \\
& +2(\overline{M} - 538.33) + 2(\overline{M} - 538.46) + 2(\overline{M} - 538.47) \\
& +2(\overline{M} - 538.55)
\end{aligned}
$$

4. Reduce and solve for $\overline{M}$,

$$
\begin{aligned}
10\overline{M} &= 5384.50 \\
\overline{M} &= \frac{5384.50}{10} = 538.45
\end{aligned}
$$

Note that this answer agrees with the one given for Example 3.1. Note also that this procedure verifies the statement given earlier in Section 3.10 that *the most probable value for an unknown quantity, measured repeatedly using the same equipment and procedures, is simply the mean of the observations.*

Example 16.2

In Figure 8.9(c), the three horizontal angles observed around the horizon are $x = 42°12'13''$, $y = 59°56'15''$, and $z = 257°51'35''$. Adjust these angles by the least squares method so that their sum equals the required geometric total of 360°.

Solution

1. Form the observation equations

$$x = 42°12'13'' + v_1 \quad \text{(a)}$$

$$y = 59°56'15'' + v_2 \quad \text{(b)}$$

$$z = 257°51'35'' + v_3 \quad \text{(c)}$$

2. Write an expression that enforces the condition that the sum of the three adjusted angles total 360°

$$x + y + z = 360° \quad \text{(d)}$$

3. Substitute Equations (a), (b), and (c) into Equation (d), and solve for v_3,

$$(42°12'13'' + v_1) + (59°56'15'' + v_2) + (257°51'35'' + v_3) = 360°$$

$$v_3 = -3'' - v_1 - v_2 \quad \text{(e)}$$

(Because of the 360° condition, if v_1 and v_2 are fixed, v_3 is also fixed. Thus, there are only two independent residuals in the solution.)

4. Form the function Σv^2, which involves all three residuals but includes only the two independent variables v_1 and v_2,

$$\sum v^2 = v_1^2 + v_2^2 + (-3'' - v_1 - v_2)^2 \quad \text{(f)}$$

5. Take partial derivatives of Equation (f) with respect to the variables v_1 and v_2, and set them equal to zero.

$$\frac{\partial \sum v^2}{\partial v_1} = 0 = 2v_1 + 2(-3'' - v_1 - v_2)(-1); \qquad 4v_1 + 2v_2 = -6'' \quad \text{(g)}$$

$$\frac{\partial \sum v^2}{\partial v_2} = 0 = 2v_2 + 2(-3'' - v_1 - v_2)(-1); \qquad 2v_1 + 4v_2 = -6'' \quad \text{(h)}$$

6. Solve Equations (g) and (h) simultaneously,

$$v_1 = -1'' \text{ and } v_2 = -1''$$

7. Substitute v_1 and v_2 into Equation (e) to compute v_3,

$$v_3 = -3'' + 1'' + 1'' = -1''$$

8. Finally substitute the residuals into Equations (a) through (c) to get the adjusted angles,

$$x = 42°12'13'' - 1'' = 42°12'12''$$
$$y = 59°56'15'' - 1'' = 59°56'14''$$
$$z = 257°51'35'' - 1'' = \underline{257°51'34''}$$
$$\sum = 360°00'00'' \text{ (Check!)}$$

Note that this result verifies another basic procedure frequently applied in surveying (geomatics) that *for equally weighted angles observed around the horizon, corrections of equal size are applied to each angle.* The same result occurs when equally weighted interior angles in a closed polygon traverse are adjusted by least squares. That is, each receives an equal-size correction.

Examples 16.1 and 16.2 are indeed simple, hardly the type for which least squares is best suited. However, they do supply the basis for some commonly applied simple adjustments and also illustrate procedures involved in making least squares adjustments without complicating the mathematics. The following example illustrates least squares adjustment of distance observations that are functionally related.

Example 16.3

Adjust the three equally weighted distance observations taken (in feet) between points A, B, and C of Figure 16.2.

Solution

1. Let the unknown distances AB and BC be x and y, respectively. These two unknowns are related through the observations as follows

$$\begin{aligned} x + y &= 393.65 \\ x &= 190.40 \\ y &= 203.16 \end{aligned} \tag{i}$$

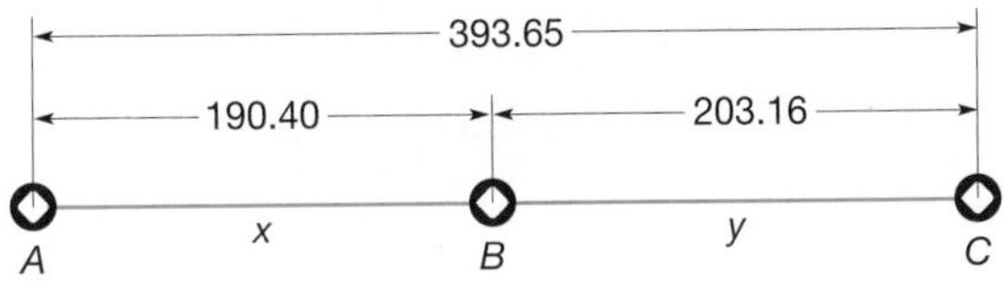

Figure 16.2 Equally weighted distance observations of Example 16.3.

2. Values for x and y could be obtained from any two of these equations so that the remaining equation is redundant. However, notice that values obtained for x and y will differ, depending on which two equations are solved. It is therefore apparent that the observations contain errors. Equations (i) may be rewritten as observation equations by including residual errors as follows

$$\begin{aligned} x + y &= 393.65 + v_1 \\ x &= 190.40 + v_2 \\ y &= 203.16 + v_3 \end{aligned} \tag{j}$$

3. To obtain the least-squares solution, the observation equations (j) are rearranged to obtain expressions for the residuals. These are squared and added to form the function given in Equation (16.1) as follows

$$\sum_{i=1}^{m} v_i^2 = (x + y - 393.65)^2 + (x - 190.40)^2 + (y - 203.16)^2 \tag{k}$$

4. Function (k) is minimized, enforcing the condition of least squares, by taking partial derivatives with respect to the unknowns x and y and setting them equal to zero. This yields the following two normal equations

$$\frac{\partial \sum v^2}{\partial x} = 0 = 2(x + y - 393.65) + 2(x - 190.40)$$

$$\frac{\partial \sum v^2}{\partial y} = 0 = 2(x + y - 393.65) + 2(y - 203.16)$$

5. Reducing the normal equations and solving yields $x = 190.43$ ft and $y = 203.19$ ft. The residuals can now be calculated by substituting x and y into the original observation equations (j),

$$\begin{aligned} v_1 &= 190.43 + 203.19 - 393.65 = -0.03 \text{ ft} \\ v_2 &= 190.43 - 190.40 = +0.03 \text{ ft} \\ v_3 &= 203.19 - 203.16 = +0.03 \text{ ft} \end{aligned}$$

16.4 MATRIX METHODS IN LEAST-SQUARES ADJUSTMENT[2]

It has been noted that least-squares computations are quite lengthy, and therefore generally performed on a computer. Their solution follows a systematic procedure that is conveniently adapted to matrix methods. In general, any group of

[2]The balance of this chapter requires a basic understanding of matrix algebra. Students who do not have this background may consult Appendix E or any mathematics book with introductory coverage on matrices. Alternatively, a good primer on matrices can be found in Appendices A and B of *Adjustment Computations: Spatial Data Analysis* by C. Ghilani and P. Wolf, John Wiley & Sons, Hoboken, NJ, 2006.

observation equations may be represented in matrix form as

$$ {}_{m}A^{n}\,{}_{n}X^{1} = {}_{m}L^{1} + {}_{m}V^{1} \tag{16.4}$$

where A is the matrix of coefficients for the unknowns, X the matrix of unknowns, L the matrix of observations, and V the matrix of residuals. The detailed structures of these matrices are

$$A = \begin{bmatrix} a_{11} & a_{12} & \cdots & a_{1n} \\ a_{21} & a_{22} & \cdots & a_{2n} \\ \vdots & \vdots & \vdots & \vdots \\ a_{m1} & a_{m2} & \cdots & a_{mn} \end{bmatrix} \quad X = \begin{bmatrix} x_1 \\ x_2 \\ \vdots \\ x_n \end{bmatrix} \quad L = \begin{bmatrix} l_1 \\ l_2 \\ \vdots \\ l_m \end{bmatrix} \quad V = \begin{bmatrix} v_1 \\ v_2 \\ \vdots \\ v_m \end{bmatrix}$$

The normal equations that result from a set of equally weighted observation equations [Equations (16.4)] are given in matrix form by

$$A^TAX = A^TL \tag{16.5}$$

In Equation (16.5) A^TA is the matrix of normal equation coefficients for the unknowns. Premultiplying both sides of Equation (16.5) by $(A^TA)^{-1}$ and reducing yields

$$X = (A^TA)^{-1}A^TL \tag{16.6}$$

Equation (16.6) is the least-squares solution for equally weighted observations. The matrix X consists of most probable values for unknowns $x_1, x_2, x_3, \ldots, x_n$.

For a system of weighted observations, the following equation provides the X matrix:

$$X = (A^TWA)^{-1}A^TWL \tag{16.7}$$

In Equation (16.7), the matrices are identical to those of the equally weighted case, except that W is a diagonal matrix of weights defined as follows[3]

$$W = \begin{bmatrix} w_1 & & & \\ & w_2 & & \text{zeros} \\ \text{zeros} & & \ddots & \\ & & & w_n \end{bmatrix}$$

If the observations in an adjustment are all of equal weight, Equation (16.7) can still be used, but the W matrix becomes an identity matrix. It therefore reduces

[3]For a group of independent and *uncorrelated observations* (a case frequently encountered in surveying), the weight matrix is diagonal, i.e., all off-diagonal elements are zeros. In certain cases, however, observations are correlated, i.e., they are related to each other. An example occurs in GPS baseline measurements, where the vector components result from least squares adjustments and thus are correlated. As will be shown in Section 16.8, this yields off-diagonal elements in the **W** matrix.

exactly to Equation (16.6). Thus, Equation (16.7) is general and can be used for both the unweighted and weighted adjustments. It is readily programmed for computer solution.

Example 16.4

Solve Example 16.3 using matrix methods.

Solution

1. The observation equations of Example 16.3 can be expressed in matrix form as follows:

$$ {}_3A^2\,{}_2X^1 = {}_3L^1 + {}_3V^1 $$

where

$$ A = \begin{bmatrix} 1 & 1 \\ 1 & 0 \\ 0 & 1 \end{bmatrix} \quad X = \begin{bmatrix} x \\ y \end{bmatrix} \quad L = \begin{bmatrix} 393.65 \\ 190.40 \\ 203.16 \end{bmatrix} \quad V = \begin{bmatrix} v_1 \\ v_2 \\ v_3 \end{bmatrix} $$

2. Solving matrix Equation (16.6),

$$ A^TA = \begin{bmatrix} 1 & 1 & 0 \\ 1 & 0 & 1 \end{bmatrix} \begin{bmatrix} 1 & 1 \\ 1 & 0 \\ 0 & 1 \end{bmatrix} = \begin{bmatrix} 2 & 1 \\ 1 & 2 \end{bmatrix} $$

$$ (A^TA)^{-1} = \frac{1}{3}\begin{bmatrix} 2 & -1 \\ -1 & 2 \end{bmatrix} \quad A^TL = \begin{bmatrix} 584.05 \\ 596.81 \end{bmatrix} $$

$$ X = (A^TA)^{-1}A^TL = \frac{1}{3}\begin{bmatrix} 2 & -1 \\ -1 & 2 \end{bmatrix} \begin{bmatrix} 584.05 \\ 596.81 \end{bmatrix} = \begin{bmatrix} 190.43 \\ 203.19 \end{bmatrix} $$

Note that this solution yields $x = 190.43$ ft and $y = 203.19$ ft, which are exactly the same values obtained through the algebraic approach of Example 16.3.

16.5 MATRIX EQUATIONS FOR PRECISIONS OF ADJUSTED QUANTITIES

The matrix equation for calculating residuals after adjustment, whether the adjustment is weighted or not, is

$$ V = AX - L \tag{16.8} $$

The standard deviation of unit weight for an unweighted adjustment is

$$ \sigma_0 = \sqrt{\frac{V^TV}{r}} \tag{16.9} $$

The standard deviation of unit weight for a weighted adjustment is

$$\sigma_0 = \sqrt{\frac{V^TWV}{r}} \tag{16.10}$$

In Equations (16.9) and (16.10), r is the *number of degrees of freedom* in an adjustment, which usually equals the number of observations minus the number of unknowns, or $r = m - n$.

Standard deviations of the individual adjusted quantities are

$$\sigma_{x_i} = \sigma_0\sqrt{q_{x_ix_i}} \tag{16.11}$$

In Equation (16.11) σ_{x_i} is the standard deviation of the ith adjusted unknown x_i, that is the value in the ith row of the X matrix; σ_0 is the standard deviation of unit weight as calculated by Equation (16.9) or (16.10); and $q_{x_ix_i}$ is the diagonal element in the ith row and ith column of matrix $(\mathbf{A^TA})^{-1}$ in the unweighted case, or matrix $(\mathbf{A^TWA})^{-1}$ in the weighted case. The $(\mathbf{A^TA})^{-1}$ and $(\mathbf{A^TWA})^{-1}$ matrices are the so-called *covariance* matrices and symbolized hereon by $\boldsymbol{Q_{xx}}$.

Example 16.5

Calculate the standard deviation of unit weight and the standard deviations of the adjusted quantities x and y for the unweighted problem of Example 16.4.

Solution

1. By Equation (16.8), the residuals are

$$V = \begin{bmatrix} 1 & 1 \\ 1 & 0 \\ 0 & 1 \end{bmatrix}\begin{bmatrix} 190.43 \\ 203.19 \end{bmatrix} - \begin{bmatrix} 393.65 \\ 190.40 \\ 203.26 \end{bmatrix} = \begin{bmatrix} -0.03 \\ 0.03 \\ 0.03 \end{bmatrix}$$

2. By Equation (16.9), the standard deviation of unit weight is

$$V^TV = [-0.03 \quad 0.03 \quad 0.03]\begin{bmatrix} -0.03 \\ 0.03 \\ 0.03 \end{bmatrix} = [0.0027]$$

$$\sigma_0 = \sqrt{\frac{0.0027}{3-2}} = \pm 0.052 \text{ ft}$$

3. Using Equation (16.11), the standard deviations of the adjusted values for x and y are

$$\sigma_x = \pm 0.052\sqrt{\frac{2}{3}} = \pm 0.042 \text{ ft}$$

$$\sigma_y = \pm 0.052\sqrt{\frac{2}{3}} = \pm 0.042 \text{ ft}$$

In Part 3, the numbers 2/3 under the radicals are the elements in row 1, column 1, and row 2, column 2, respectively, of the $(A^TA)^{-1}$ matrix of Example 16.4. The interpretation of the standard deviations computed under Step 3 of Example 16.5 is that a 68% probability exists that the adjusted values for x and y are within ±0.042 ft of their true values. Note that for this simple example, the three residuals calculated in step 1 are equal, and the standard deviations of x and y are equal in Step 3. This is caused by the symmetric nature of this particular problem (illustrated in Figure 16.2), but it is not generally the case with more complex problems.

16.6 LEAST-SQUARES ADJUSTMENT OF LEVELING CIRCUITS

When control leveling is being done to establish new benchmarks, for example, benchmarks for a construction project, it is common practice to create a network like that illustrated in Figure 16.3. This enables each new benchmark to benefit from redundant observations and least-squares adjustment. In Figure 16.3, A and B are two new project benchmarks being established near a construction site. Each could be set by running a single loop, such as from BM 1 to A and back to establish A, and from BM 3 to B and back to set B. To build redundancy into the survey, and to increase the precisions of the new benchmarks, additional lines from other nearby benchmarks can be run. Thus in Figure 16.3, five loops are run rather than the minimum of two needed to establish A and B. All observations within this leveling network can be simultaneously adjusted using the least-squares method to obtain most probable adjusted values for the two benchmarks.

In adjusting level networks, the observed difference in elevation for each course is treated as one observation containing a single random error. Observation equations are written that relate these observed elevation differences and their residual errors to the unknown elevations of the benchmarks involved. These can then be processed through the matrix equations given in Sections 16.4 and 16.5 to obtain adjusted values for the benchmarks and their standard deviations. The procedure is illustrated with the following example.

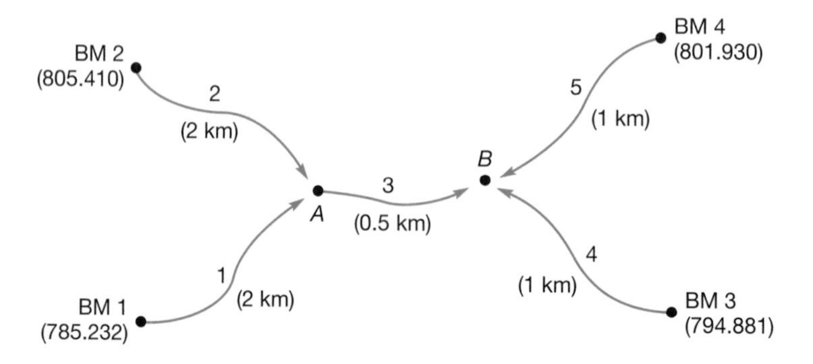

Figure 16.3 Level net for Example 16.6.

Example 16.6

Adjust the level net of Figure 16.3 by weighted least squares and compute precisions of the adjusted benchmarks. In the figure, the benchmark elevations (in meters) and course lengths (in kilometers) are shown in parentheses. Observed elevation differences for courses 1 through 5 (given in order) are +10.997 m, −9.169 m, +3.532 m, +4.858 m, and −2.202 m. Arrows on the courses in the figure indicate the direction of leveling. Thus for course 1 having a length of 2 km, leveling proceeded from BM 1 to A and the observed elevation difference was 10.970 m.

Solution

1. Observation equations are written relating each line's observed elevation difference to its residual error and the most probable values for unknown elevations A and B as follows:

$$
\begin{aligned}
A &= \text{BM }1 + 10.997 + v_1 \\
A &= \text{BM }2 - 9.169 + v_2 \\
B &= A + 3.532 + v_3 \\
B &= \text{BM }3 + 4.858 + v_4 \\
B &= \text{BM }4 - 2.202 + v_5
\end{aligned}
\qquad \text{(i)}
$$

2. Substituting the elevations of BM 1, BM 2, BM 3, and BM 4 into Equations (l) and rearranging gives

$$
\begin{aligned}
A &= 796.229 + v_1 \\
A &= 796.241 + v_2 \\
-A + B &= 3.532 + v_3 \\
B &= 799.739 + v_4 \\
B &= 799.728 + v_5
\end{aligned}
$$

3. The $\mathbf{A}$, $\mathbf{X}$, $\mathbf{L}$, and $\mathbf{V}$ matrices for this adjustment are

$$
\mathbf{A} = \begin{bmatrix} 1 & 0 \\ 1 & 0 \\ -1 & 1 \\ 0 & 1 \\ 0 & 1 \end{bmatrix} \quad \mathbf{X} = \begin{bmatrix} A \\ B \end{bmatrix} \quad \mathbf{L} = \begin{bmatrix} 796.229 \\ 796.241 \\ 3.532 \\ 799.739 \\ 799.728 \end{bmatrix} \quad \mathbf{V} = \begin{bmatrix} v_1 \\ v_2 \\ v_3 \\ v_4 \\ v_5 \end{bmatrix}
$$

4. Weights in differential leveling are inversely proportional to course lengths. Thus after inverting the lengths, the course weights are 0.5, 0.5, 2, 1, and 1,

respectively, and the weight matrix is

$$W = \begin{bmatrix} 0.5 & 0 & 0 & 0 & 0 \\ 0 & 0.5 & 0 & 0 & 0 \\ 0 & 0 & 2 & 0 & 0 \\ 0 & 0 & 0 & 1 & 0 \\ 0 & 0 & 0 & 0 & 1 \end{bmatrix}$$

5. The weighted matrix solution for the most probable values according to Equation (16.7) is

$$\mathbf{A^TW} = \begin{bmatrix} 1 & 1 & -1 & 0 & 0 \\ 0 & 0 & 1 & 1 & 1 \end{bmatrix} \begin{bmatrix} 0.5 & 0 & 0 & 0 & 0 \\ 0 & 0.5 & 0 & 0 & 0 \\ 0 & 0 & 2 & 0 & 0 \\ 0 & 0 & 0 & 1 & 0 \\ 0 & 0 & 0 & 0 & 1 \end{bmatrix} = \begin{bmatrix} 0.5 & 0.5 & -2.0 & 0.0 & 0.0 \\ 0.0 & 0.0 & 2.0 & 1.0 & 1.0 \end{bmatrix}$$

$$\mathbf{A^TWA} = \begin{bmatrix} 0.5 & 0.5 & -2.0 & 0.0 & 0.0 \\ 0.0 & 0.0 & 2.0 & 1.0 & 1.0 \end{bmatrix} \begin{bmatrix} 1 & 0 \\ 1 & 0 \\ -1 & 1 \\ 0 & 1 \\ 0 & 1 \end{bmatrix} = \begin{bmatrix} 3 & -2 \\ -2 & 4 \end{bmatrix}$$

$$Q_{xx} = \mathbf{(A^TWA)^{-1}} = \frac{1}{8}\begin{bmatrix} 4 & 2 \\ 2 & 3 \end{bmatrix}$$

$$\mathbf{A^TWL} = \begin{bmatrix} 0.5 & 0.5 & -2.0 & 0.0 & 0.0 \\ 0.0 & 0.0 & 2.0 & 1.0 & 1.0 \end{bmatrix} \begin{bmatrix} 796.229 \\ 796.241 \\ 3.532 \\ 799.739 \\ 799.728 \end{bmatrix} = \begin{bmatrix} 789.171 \\ 1606.531 \end{bmatrix}$$

$$\mathbf{X} = \mathbf{(A^TWA)^{-1}A^TWL} = \frac{1}{8}\begin{bmatrix} 4 & 2 \\ 2 & 3 \end{bmatrix}\begin{bmatrix} 789.171 \\ 1606.531 \end{bmatrix} = \begin{bmatrix} 796.218 \\ 799.742 \end{bmatrix}$$

Thus, the adjusted benchmark elevations are $A = 796.218$ m and $B = 799.742$ m.

6. The residuals by Equation (16.8) are

$$\mathbf{V} = \mathbf{AX} - \mathbf{L} = \begin{bmatrix} 1 & 0 \\ 1 & 0 \\ -1 & 0 \\ 0 & 1 \\ 0 & 1 \end{bmatrix} \begin{bmatrix} 796.218 \\ 799.742 \end{bmatrix} - \begin{bmatrix} 796.229 \\ 796.241 \\ 3.532 \\ 799.739 \\ 799.728 \end{bmatrix} = \begin{bmatrix} -0.011 \\ -0.023 \\ -0.008 \\ 0.003 \\ 0.014 \end{bmatrix}$$

7. Utilizing Equation (16.10), the estimated standard deviation of unit weight is

$$V^TWV = [-0.011 \quad -0.023 \quad -0.008 \quad 0.003 \quad 0.014] \begin{bmatrix} 0.5 & 0 & 0 & 0 & 0 \\ 0 & 0.5 & 0 & 0 & 0 \\ 0 & 0 & 2 & 0 & 0 \\ 0 & 0 & 0 & 1 & 0 \\ 0 & 0 & 0 & 0 & 1 \end{bmatrix} \begin{bmatrix} -0.011 \\ -0.023 \\ -0.008 \\ 0.003 \\ 0.014 \end{bmatrix}$$

$$= [0.00066]$$

Thus, the standard deviation of unit weight is $\sigma_0 = \sqrt{\dfrac{0.00066}{5 - 2}} = \pm 0.015$ m

8. By Equation (16.11), the estimated standard deviations of adjusted benchmark elevations of A and B are

$$\sigma_A = \sigma_0\sqrt{q_{AA}} = \pm 0.015\sqrt{\frac{4}{8}} = \pm 0.011 \text{ m}$$

$$\sigma_B = \sigma_0\sqrt{q_{BB}} = \pm 0.015\sqrt{\frac{3}{8}} = \pm 0.009 \text{ m}$$

Note in these calculations that terms in the radicals are the diagonal elements, (1, 1) and (2, 2), of the $\boldsymbol{Q}_{xx}$ matrix. Note also that B has a lower standard deviation than A, indicating that its precision is better. A check of Figure 16.3 reveals that this should be expected, because the benchmarks closest to A are both 2 km away, and those closest to B are only 1 km away. Thus, the weights of level circuits coming into B are higher than those coming into A, giving B a higher precision.

The software WOLFPACK can be used to adjust the data of Example 16.6. The format for the data file is shown in Figure 16.4. Note that the lengths of the level lines, the number of setups, or the standard deviations of the observations can

```
Differential leveling example of Section 15.6    {title line}
4 6 5                                            {Number of benchmarks; number of stations; number
                                                 of elev. diff.}
1 785.232                                        {BM identifier and elevation}
2 805.410
3 794.881
4 801.930
1 A 10.997 2                                     {from, to, elevation difference, [dist.|setups|σ}
2 A -9.169 2
A B 3.532 0.5
3 B 4.858 1
4 B -2.202 1
```

Figure 16.4 WOLFPACK data file for Example 16.6.

```
****************************
Adjusted Elevation Differences
****************************
       From         To      Elevation Difference       V        σ
==============================================================
         1           A             10.986            -0.011   0.0105
         2           A             -9.192            -0.023   0.0105
         A           B              3.524            -0.008   0.0091
         3           B              4.861             0.003   0.0091
         4           B             -2.188             0.014   0.0091


******************
Adjusted Elevations
******************
   Station       Elevation        σ
===================================
         A       796.218       0.0105
         B       799.742       0.0091


Standard Deviation of Unit Weight: 0.015
```

Figure 16.5 Results from adjustment of data in Example 16.6 using WOLFPACK.

weight the observations. In this example, course lengths were used. The results of the adjustment are shown in Figure 16.5. The programming behind this problem is contained in a Mathcad® worksheet on the CD that accompanies this book.

16.7 PROPAGATION OF ERRORS

In Section 3.17, the propagation of errors in functions using independent observations was discussed. At the completion of a least-squares adjustment, the unknowns are no longer independent as evidenced by the off-diagonal terms in the covariance matrix discussed in Section 16.5. When observations are not independent, errors propagate as

$$\boldsymbol{Q}_{\ell\ell} = \boldsymbol{A}\boldsymbol{Q}_{xx}\boldsymbol{A}^T \tag{16.12}$$

where $\boldsymbol{Q}_{xx}$ equals the matrix $(\boldsymbol{A}^T\boldsymbol{A})^{-1}$ in the unweighted case, or the matrix $(\boldsymbol{A}^T\boldsymbol{W}\boldsymbol{A})^{-1}$ in the weighted case.

The standard deviations in the computed observations are

$$\sigma_{\ell_i} = \sigma_0\sqrt{q_{\ell_i\ell_i}} \tag{16.13}$$

In Equation (16.13), σ_{ℓ_i} is the standard deviation of the ith adjusted observation; σ_0 is the standard deviation of unit weight as calculated by Equation (16.9) or (16.10); and $\boldsymbol{q}_{\ell_i\ell_i}$ is ith diagonal element of the $\boldsymbol{Q}_{\ell\ell}$ matrix of Equation (16.12).

Example 16.7

Compute the adjusted elevation differences and their standard deviations for Example 16.6.

Solution

By Equation (16.12) the $Q_{\ell\ell}$ matrix is

$$Q_{\ell\ell} = \frac{1}{8}\begin{bmatrix} 1 & 0 \\ 1 & 0 \\ -1 & 1 \\ 0 & 1 \\ 0 & 1 \end{bmatrix}\begin{bmatrix} 4 & 2 \\ 2 & 3 \end{bmatrix}\begin{bmatrix} 1 & 1 & -1 & 0 & 0 \\ 0 & 0 & 1 & 1 & 1 \end{bmatrix} = \frac{1}{8}\begin{bmatrix} 4 & 4 & -2 & 2 & 2 \\ 4 & 4 & -2 & 2 & 2 \\ -2 & -2 & 3 & 1 & 1 \\ 2 & 2 & 1 & 3 & 3 \\ 2 & 2 & 1 & 3 & 3 \end{bmatrix}$$

A tabulation of the observations, their residuals, adjusted values, and standard deviations is

From	To	Obs. ΔElev	V	Adj. ΔElev	σ
BM1	*A*	10.997	−0.011	10.986	±0.011
BM2	*A*	−9.169	−0.023	−9.192	±0.011
A	*B*	3.532	−0.008	3.524	±0.009
BM3	*B*	4.858	0.003	4.861	±0.009
BM4	*B*	−2.202	0.014	−2.188	±0.009

The contents of the third line in the above table are explained to clarify the calculations. The adjusted elevation difference for the third observation, which consisted of leveling from *A* to *B*, is obtained by adding the observed elevation difference (see the data for Example Problem 16.6) and its residual (see step 6 of Example Problem 16.6), as follows:

$$\text{Adj. }\Delta\text{Elev} = 3.532 - 0.008 = 3.524$$

Also the standard deviation, *S*, for the third observation is computed as

$$\sigma = \pm 0.015\sqrt{\frac{3}{8}} = \pm 0.009$$

where 3/8 is the third diagonal element of the $Q_{\ell\ell}$, and 0.015 is σ_0 as determined in Example 16.6. The adjusted elevations and standard deviations for the remaining observations are computed in similar fashion.

16.8 LEAST-SQUARES ADJUSTMENT OF GPS BASELINE VECTORS

It was previously noted in Section 14.5.5 that the least-squares method is essential in adjusting GPS observations. It is applied in this work in two different stages: (1) for adjusting the massive quantities of redundant data that result after several receivers have made repeated observations on multiple satellites over a period of time (this yields baseline components ΔX, ΔY, and ΔZ), and (2) it is applied in adjusting redundant observations of these baseline components to make them consistent in static differential GPS survey networks. The reduction software, which accompanies GPS equipment, is programmed to perform the first stage of these two applications and thus that development is not covered in this text. However, the second application is within the scope of this text and is illustrated with an example—the adjustment of the network of Figure 14.10.

Since the baseline vector data are the results of the first least-squares adjustment noted previously, each baseline has it own 3×3 covariance matrix. This matrix not only contains terms along the diagonal, but it also has elements in the off-diagonal locations. The covariance terms depict the amount of correlation between the adjusted ΔX, ΔY, and ΔZ values. For each baseline, the carrier-phase reduction software will list the baseline components and their covariance terms. For example for baseline *AC* in Figure 14.10, the vector and its covariance terms were listed as

$$\begin{array}{lrrrr} \Delta X & 11{,}644.2232 & 9.8E-4 & -9.6E-6 & 9.5E-6 \\ \Delta Y & 3601.2165 & & 9.4E-4 & -9.5E-6 \\ \Delta Z & 3399.2550 & & & 9.8E-4 \end{array}$$

Note that only the upper-triangular portion of the covariance matrix is shown to the right of the vector components. This is because the covariance matrix is symmetric and thus the elements of the lower-triangular portion mirror their upper-triangular values and need not be repeated. Since the baseline vectors are not independent, the weight matrix is computed as the inverse of the covariance matrix, or in matrix symbology is

$$\mathbf{W} = \Sigma^{-1} \tag{16.14}$$

where Σ is the covariance matrix for the baseline vectors, and $\mathbf{W}$ is their weight matrix. It can be shown that this equation is also valid for independent observations and thus is a general equation for weighting observations.

The baseline vector components (ΔX, ΔY, ΔZ), and their covariance terms for the GPS survey of the network of Figure 14.10 are shown in Table 16.1. In the column labeled (1) of this table, the (1,1) element of the covariance matrix is listed; column (2) lists the (1,2) element of the covariance matrix; column (3) the (1,3) element; column (4) the (2,2) element; column (5) the (2,3) element; and column (6) the (3,3) element of the covariance matrix. In the survey, two HARN stations (see Section 14.3.5) were held fixed. These stations and their coordinates are listed in Table 16.2.

From the known X, Y, and Z coordinates of stations A and B, and the observed ΔX, ΔY, and ΔZ components, coordinates of new stations C, D, E, and F can be calculated. However, an adjustment is necessary because redundant observations exist. In applying least squares to this problem, observation equations are written that relate the unknown adjusted coordinates of new stations C,

TABLE 16.1 Observed Baseline Vectors and Their Covariance Matrix Values for Figure 14.10

Baseline	ΔX	ΔY	ΔZ	(1)	(2)	(3)	(4)	(5)	(6)
AC	11644.2232	3601.2165	3399.2550	9.8E-4	−9.6E-6	9.5E-6	9.4E-4	−9.5E-6	9.8E-4
DC	15128.1647	−6286.7054	−6371.0583	1.5E-4	−1.4E-6	1.3E-6	1.6E-4	−1.4E-6	1.3E-4
AE	−5321.7164	3634.0754	3173.6652	2.2E-4	−2.1E-6	2.2E-6	1.9E-4	−2.1E-6	2.0E-4
BC	3960.5442	−6681.2467	−7279.0148	2.3E-4	−2.2E-6	2.1E-6	2.5E-4	−2.2E-6	2.2E-4
BD	−11167.6076	−394.5204	−907.9593	2.7E-4	−2.8E-6	2.8E-6	2.7E-4	−2.7E-6	2.7E-4
DE	−1837.7459	−6253.8534	−6596.6697	1.2E-4	−1.2E-6	1.2E-6	1.3E-4	−1.2E-6	1.3E-4
FA	−1116.4523	−4596.1610	−4355.9062	7.5E-5	−7.9E-7	8.8E-7	6.6E-5	−8.1E-7	7.6E-5
FC	10527.7852	−994.9370	−956.6246	2.6E-4	−2.2E-6	2.4E-6	2.2E-4	−2.3E-6	2.4E-4
FE	−6438.1364	−962.0694	−1182.2305	9.4E-5	−9.2E-7	1.0E-6	1.0E-4	−8.9E-7	8.8E-5
FD	−4600.3787	5291.7785	5414.4311	9.3E-5	−9.9E-7	9.0E-7	9.9E-5	−9.9E-7	1.2E-4
FB	6567.2311	5686.2926	6322.3917	6.6E-5	−6.5E-7	6.9E-7	7.5E-5	−6.4E-7	6.0E-5
BF	−6567.2310	−5686.3033	−6322.3807	5.5E-5	−6.3E-7	6.1E-7	7.5E-5	−6.3E-7	6.6E-5
AF	1116.4577	4596.1553	4355.9141	6.6E-5	−8.0E-7	9.0E-7	8.1E-5	−8.2E-7	9.4E-5

TABLE 16.2 HARN STATION GEOCENTRIC COORDINATES

Station	X	Y	Z
A	402.3509	−4,652,995.3011	4,349,760.7775
B	8086.0318	−4,642,712.8474	4,360,439.0833

D, E, and F to the observed ΔX, ΔY, and ΔZ values and their residual errors. As shown in Figure 14.10, excluding the check observation of fixed baseline AB, there are 11 different baselines. However, two of these, AF and BF, were repeated giving a total of 13 baseline observations. The following observation equations are written for the first two baselines:

$$
\begin{aligned}
X_C &= X_A + \Delta X_{AC} + v_1 \\
Y_C &= Y_A + \Delta Y_{AC} + v_2 \\
Z_C &= Z_A + \Delta Z_{AC} + v_3 \\
X_C - X_D &= \Delta X_{DC} + v_4 \\
Y_C - Y_D &= \Delta Y_{DC} + v_5 \\
Z_C - Z_D &= \Delta Y_{DC} + v_6
\end{aligned}
\tag{16.15}
$$

Similar equations can be written for the other 11 baseline observations, giving a total of 39 observation equations. These observation equations can be expressed in matrix form according to Equation (16.4). To illustrate the contents of the matrices, the partial matrices that result from the observation equations of Equations (16.15) are

$$
{}_{39}\mathbf{A}^{12} = \begin{bmatrix} 1 & 0 & 0 & 0 & 0 & 0 & \\ 0 & 1 & 0 & 0 & 0 & 0 & \\ 0 & 0 & 1 & 0 & 0 & 0 & \\ 1 & 0 & 0 & -1 & 0 & 0 & \cdots \\ 0 & 1 & 0 & 0 & -1 & 0 & \\ 0 & 0 & 1 & 0 & 0 & -1 & \\ & & & \vdots & & & \end{bmatrix}; \quad {}_{12}\mathbf{X}^{1} = \begin{bmatrix} X_C \\ Y_C \\ Z_C \\ X_D \\ Y_D \\ Z_D \\ \vdots \end{bmatrix};
$$

$$
{}_{39}\mathbf{L}^{1} = \begin{bmatrix} X_A + \Delta X_{AC} \\ Y_A + \Delta Y_{AC} \\ Z_A + \Delta Z_{AC} \\ \Delta X_{DC} \\ \Delta Y_{DC} \\ \Delta Z_{DC} \\ \vdots \end{bmatrix}; \quad {}_{39}\mathbf{V}^{1} = \begin{bmatrix} v_1 \\ v_2 \\ v_3 \\ v_4 \\ v_5 \\ v_6 \\ \vdots \end{bmatrix}
\tag{16.16}
$$

To complete the above $\boldsymbol{A}$ matrix, the coefficients for the remaining observation equations are entered expanding the dimensions of this matrix to 39 rows and 12 columns. Note that there are three unknowns for each of the four new ground stations, thus the $\boldsymbol{X}$ matrix has a total of 12 elements. The $\boldsymbol{L}$ and $\boldsymbol{V}$ matrices each have 39 elements, one for each observation equation.

The partial covariance matrix that results from the first two baseline observations is

$$
{}_{39}\Sigma^{39} = \begin{bmatrix}
9.8E-4 & -9.6E-6 & 9.5E-6 & 0 & 0 & 0 & \\
-9.6E-6 & 9.4E-4 & -9.5E-6 & 0 & 0 & 0 & \\
9.5E-6 & -9.5E-6 & 9.8E-4 & 0 & 0 & 0 & \\
0 & 0 & 0 & 1.5E-E & -1.4E-6 & 1.3E-6 & \cdots\cdots \\
0 & 0 & 0 & -1.4E-6 & 1.6E-4 & -1.4E-6 & \\
0 & 0 & 0 & 1.3E-6 & -1.4E-6 & 1.3E-4 & \\
 & & & \vdots & & & \ddots
\end{bmatrix}
$$

The Σ matrix, when completed, has 39 rows and 39 columns. Its formation follows the same procedure for each observed baseline as is demonstrated here. As can be seen, each observed baseline creates a 3×3 submatrix within Σ. Thus the structure of this matrix is *block diagonal,* i.e., each individual submatrix on the diagonal is 3×3, and all the remaining elements are zeros. Thus the first 3 rows and 3 columns are non-zero elements that pertain to the first baseline, the next three rows and 3 columns contain non-zero elements that pertain to the second baseline, and so on.

The weight matrix is obtained by inverting the Σ matrix according to Equation (16.14). Once all matrices are formed, the solution for the unknown station coordinates and their standard deviations can be determined using Equation (16.7). After the adjustment, the resulting geocentric coordinates can be used to determine geodetic coordinates using the procedures discussed in Section 13.4.3.

For this particular problem, the software program WOLFPACK was used to adjust the baselines using the least-squares method. The input file is shown in Figure 16.6,

```
Example of Section 15.8                       {title line}
2 13                                          {Number of control stations; Number of baselines)
A 402.3509 -4652995.3011 4349760.7775         {identifier, X, Y, Z coordinates of control station}
B 8086.0318 -4642712.8474 4360439.0833
A C 11644.2232 3601.2165 3399.2550 9.8E-4 -9.6E-6 9.5E-6 9.4E-4 -9.5E-6 9.8E-4      {baselines}
A E -5321.7164 3634.0754 3173.6652 2.2E-4 -2.1E-6 2.2E-6 1.9E-4 -2.1E-6 2.0E-4
B C 3960.5442 -6681.2467 -7279.0148 2.3E-4 -2.2E-6 2.1E-6 2.5E-4 -2.2E-6 2.2E-4
B D -11167.6076 -394.5204 -907.9593 2.7E-4 -2.8E-6 2.8E-6 2.7E-4 -2.7E-6 2.7E-4
D C 15128.1647 -6286.7054 -6371.0583 1.5E-4 -1.4E-6 1.3E-6 1.6E-4 -1.4E-6 1.3E-4
D E -1837.7459 -6253.8534 -6596.6697 1.2E-4 -1.2E-6 1.2E-6 1.3E-4 -1.2E-6 1.3E-4
F A -1116.4523 -4596.1610 -4355.9062 7.5E-5 -7.9E-7 8.8E-7 6.6E-5 -8.1E-7 7.6E-5
F C 10527.7852 -994.937 -956.6246 2.6E-4 -2.2E-6 2.4E-6 2.2E-4 -2.3E-6 2.4E-4
F E -6438.1364 -962.0694 -1182.2305 9.4E-5 -9.2E-7 1.0E-6 1.0E-4 -8.9E-7 8.8E-5
F D -4600.3787 5291.7785 5414.4311 9.3E-5 -9.9E-7 9.0E-7 9.9E-5 -9.9E-7 1.2E-4
F B 6567.2311 5686.2926 6322.3917 6.6E-5 -6.5E-7 6.9E-7 7.5E-5 -6.4E-7 6.0E-5
B F -6567.2310 -5686.3033 -6322.3807 5.5E-5 -6.3E-7 6.1E-7 7.5E-5 -6.3E-7 6.6E-5
A F 1116.4577 4596.1553 4355.9141 6.6E-5 -8.0E-7 9.00E-7 8.1E-5 -8.2E-7 9.4E-5
```

Figure 16.6 Input file for least-squares adjustment problem in Section 16.8.

and partial results of the adjustment are given in Figure 16.7. Note that the standard deviation was computed for each baseline, and that the precision of each baseline was determined as the ratio of the standard deviation, σ, over the vector length times one million.

```
              Degrees of Freedom = 27
            Reference Variance = 0.5010
   Standard Deviation of Unit Weight = ±0.71

************************
Adjusted Distance Vectors
************************
```

From	To	dX	dY	dZ	Vx	Vy	Vz
A	C	11644.2232	3601.2165	3399.2550	0.00665	0.00219	0.03197
A	E	-5321.7164	3634.0754	3173.6652	0.02665	0.00579	0.01212
B	C	3960.5442	-6681.2467	-7279.0148	0.00475	0.01169	-0.00403
B	D	-11167.6076	-394.5204	-907.9593	-0.00730	-0.00137	-0.00065
D	C	15128.1647	-6286.7054	-6371.0583	-0.00084	-0.00783	-0.00058
D	E	-1837.7459	-6253.8534	-6596.6697	-0.00985	0.00267	0.00117
F	A	-1116.4523	-4596.1610	-4355.9062	0.00197	0.00527	-0.00773
F	C	10527.7852	-994.9370	-956.6246	-0.00568	-0.00004	-0.00236
F	E	-6438.1364	-962.0694	-1182.2305	-0.00368	-0.00514	-0.00611
F	D	-4600.3787	5291.7785	5414.4311	-0.00563	-0.00230	0.00083
F	B	6567.2311	5686.2926	6322.3917	-0.00053	0.00537	0.00017
B	F	-6567.2310	-5686.3033	-6322.3807	0.00043	0.00533	-0.01117
A	F	1116.4577	4596.1553	4355.9141	-0.00737	0.00043	-0.00017

```
**********************************************
Advanced Statistical Values
**********************************************
```

From	To	±σ	Vector Length	Prec
A	C	0.0105	12,653.538	1,206,000
A	E	0.0091	7,183.255	794,000
B	C	0.0105	10,644.668	1,015,000
B	D	0.0087	11,211.408	1,282,000
D	C	0.0107	17,577.670	1,641,000
D	E	0.0097	9,273.836	960,000
F	A	0.0048	6,430.015	1,344,000
F	C	0.0104	10,617.871	1,019,000
F	E	0.0086	6,616.111	770,000
F	D	0.0083	8,859.035	1,066,000
F	B	0.0048	10,744.075	2,246,000
B	F	0.0048	10,744.075	2,246,000
A	F	0.0048	6,430.015	1,344,000

```
*******************
Adjusted Coordinates
*******************
```

Station	X	Y	Z	σx	σy	σz
A	402.3509	-4,652,995.3011	4,349,760.7775			
B	8,086.0318	-4,642,712.8474	4,360,439.0833			
C	12,046.5808	-4,649,394.0824	4,353,160.0645	0.0061	0.0061	0.0059
E	-4,919.3388	-4,649,361.2199	4,352,934.4548	0.0052	0.0053	0.0052
D	-3,081.5831	-4,643,107.3692	4,359,531.1234	0.0049	0.0051	0.0051
F	1,518.8012	-4,648,399.1454	4,354,116.6914	0.0027	0.0028	0.0028

Figure 16.7 Results of adjustment of data file in Figure 16.6 using WOLFPACK.

■ 16.9 LEAST-SQUARES ADJUSTMENT OF TRADITIONAL HORIZONTAL PLANE SURVEYS

Traversing (described in Chapters 9 and 10), and trilateration and triangulation (discussed in Chapter 19), are traditional ground surveying methods for conducting *horizontal surveys* [those surveys which establish either X and Y coordinates, usually in some grid system such as state plane coordinates (see Chapter 20), or geodetic latitudes and longitudes of points (see Section 19.4)]. The basic observations that are made in traversing, trilateration, and triangulation are azimuths, horizontal angles, and horizontal distances. As with other types of surveys, they are most appropriately adjusted by the method of least squares.

To adjust horizontal surveys by the least-squares method, it is necessary to write observation equations for the observations. These observation equations are nonlinear and to facilitate solving them they are linearized using a first-order Taylor series expansion. The procedure is described in the following subsection.

16.9.1 Linearizing Nonlinear Equations

The general form of the Taylor series linearization of a nonlinear equation is

$$F(x_1, x_2, \ldots, x_n) = F(x_{1_0}, x_{2_0}, \ldots, x_{n_0}) + \left(\frac{\partial F}{\partial x_1}\right)_0 dx_1 + \left(\frac{\partial F}{\partial x_2}\right)_0 dx_2 + \cdots + \left(\frac{\partial F}{\partial x_n}\right) dx_n + R \qquad \textbf{(16.17)}$$

where $F(x_1, x_2, \ldots, x_n)$ is a nonlinear function in terms of the unknowns $x_1, x_2, \ldots, x_n$, which represents a measured quantity; $x_{1_0}, x_{2_0}, \ldots, x_{n_0}$ are approximate values for the unknowns $x_1, x_2, \ldots, x_n$; $(\partial F/\partial x_1)_0, (\partial F/\partial x_2)_0, \ldots, (\partial F/\partial x_n)_0$ are the partial derivatives of the function F with respect to $x_1, x_2, \ldots, x_n$ evaluated using the approximate values of $x_{1_0}, x_{2_0}, \ldots, x_{n_0}$; $dx_1, dx_2, \ldots, dx_n$ are corrections to the approximate values of $x_{1_0}, x_{2_0}, \ldots, x_{n_0}$, such that $x_1 = x_{1_0} + dx_1$; $x_2 = x_{2_0} + dx_2, \ldots, x_n = x_{n_0} + d_{x_n}$; and R is a remainder. In Equation (16.17) the only unknowns are $dx_1, dx_2, \ldots, dx_n$, and R. The term R is also nonlinear, but if the values assigned for $x_{1_0}, x_{2_0}, \ldots, x_{n_0}$ are close to the true values of the unknowns, then R is small and is dropped. This process leaves a linear equation with unknowns $dx_1, dx_2, \ldots, dx_n$ However, since R was dropped, this makes the equation an approximation and thus the solution must be obtained *iteratively,* i.e., the corrections $dx_1, dx_2, \ldots, dx_n$ are computed repetitively until their sizes become negligible.

After dropping R and rearranging Equation (16.17), the following general linear form of the equation is:

$$\left(\frac{\partial F}{\partial x_1}\right)_0 dx_1 + \left(\frac{\partial F}{\partial x_2}\right)_0 dx_2 + \cdots + \left(\frac{\partial F}{\partial x_n}\right)_0 = F(x_1, x_2, \ldots, x_n) - F(x_{1_0}, x_{2_0}, \ldots, x_{n_0}) \qquad \textbf{(16.18)}$$

The subscript zeros attached to the coefficients on the left side of Equation (16.18) indicate that these coefficients are simply numbers obtained by substituting the approximate values $x_{1_0}, x_{2_0}, \ldots, x_{n_0}$ into those partial derivative functions. Also, the right side of Equation (16.18) is the observed value, $[F(x_1, x_2, \ldots, x_n)]$, minus the computed value obtained by substituting the initial approximations into the original function, $[F(x_{1_0}, x_{2_0}, \ldots, x_{n_0})]$.

The process of solving a pair of nonlinear equations using the Taylor's series will be illustrated. Suppose that the following two nonlinear functions, $F(x, y)$ and $G(x, y)$, express the relationship between observed values 115 and 75, respectively, and the unknowns x and y:

$$\begin{aligned} F(x, y) &= x^2 + 3y = 115 \\ G(x, y) &= 5x + y^2 = 75 \end{aligned} \tag{a}$$

Partial derivatives of the functions with respect to the unknowns are:

$$\frac{\partial F}{\partial x} = 2x; \quad \frac{\partial F}{\partial y} = 3; \quad \frac{\partial G}{\partial x} = 5; \quad \frac{\partial G}{\partial y} 2y \tag{b}$$

A) First Iteration

Assume that through either estimation or preliminary calculations based upon one or more observations, that values of 9 and 4 are selected as initial estimates for the unknowns, x_0 and y_0. Then using the functions of Equations (a) and substituting these initial approximations and the partial derivatives from (b) into Equation (16.18), the following two linearized equations are obtained:

$$\begin{aligned} 9^2 + 3 \times 4 + (2 \times 9)dx + 3dy &= 115 \\ 5 \times 9 + 4^2 + 5dx + (2 \times 4)dy &= 75 \end{aligned} \tag{c}$$

Equations (c) are now in linear form and contain only two unknowns, dx and dy. The solution of this pair of equations yields the following corrections: $dx = 1.04$, and $dy = 1.10$.

B) Second Iteration

Using corrections from the first iteration, new approximations x_0 and y_0 are computed as

$$\begin{aligned} x_0 &= 9.00 + 1.04 = 10.04 \\ y_0 &= 4.00 + 1.10 = 5.10 \end{aligned} \tag{d}$$

These new approximations are now used to repeat the solution. Substituting again into Equation (16.18), the following linearized equations result:

$$\begin{aligned} (10.04)^2 + 3 \times 5.10 + 2(10.04)dx + 3dy &= 115 \\ 5 \times 10.04 + (5.10)^2 + 5dx + 2(5.10)dy &= 75 \end{aligned} \tag{e}$$

Solving Equations (e) for the unknown parameters yields: $dx = -0.08$ and $dy = -0.08$.

C) Third Iteration

The corrections from the second iteration are used to get updated values for the coordinates as

$$\begin{aligned} x_0 &= 10.04 - 0.08 = 9.96 \\ y_0 &= 5.10 - 0.08 = 5.02 \end{aligned} \tag{f}$$

New linearized equations are formed by substituting these initial approximations into Equations (16.18) as follows:

$$\begin{aligned} (9.96)^2 + 3 \times 5.02 + 2(9.96)dx + 3dy &= 115 \\ 5 \times 9.96 + (5.02)^2 + 5dx + 2(5.02)dy &= 75 \end{aligned} \tag{g}$$

Solving Equations (g) for the unknown corrections gives $dx = 0.04$ and $dy = -0.02$. The updated values for x and y are therefore:

$$\begin{aligned} x_0 &= 9.96 + 0.04 = 10.00 \\ y_0 &= 5.02 - 0.02 = 5.00 \end{aligned}$$

A fourth iteration (not shown) yields zeros for both dx and dy and thus the solution has converged. The final answers are $x = 10.00$ and $y = 5.00$.

Although only two unknowns existed in this example, the first-order Taylor's series expansion is applicable to *linearizing* and solving nonlinear equations with any number of unknowns.

All that is necessary is to select an initial approximation for each unknown and take partial derivatives of the function with respect to each unknown, as indicated in Equation (16.18). As is discussed in the subsections that follow, in least-squares adjustments of horizontal surveys, up to four unknowns can appear in the distance and azimuth observation equations, and up to six unknowns can appear in angle observation equations. In more advanced types of geodesy and photogrammetry problems, many more unknowns can appear in the nonlinear equations used.

16.9.2 The Distance Observation Equation

The observation equation for an observed distance is expressed in terms of the X and Y coordinates of its end points and includes the residual error. Referring to Figure 16.8, the following distance observation equation can be written for the line whose end points are identified by I and J

$$L_{IJ} + v_{IJ} = \sqrt{(X_J - X_I)^2 + (Y_J - Y_I)^2} \tag{16.19}$$

In Equation (16.19), L_{IJ} is the observed length of the line IJ, v_{IJ} is the residual error in the observation, X_I, Y_I, X_J, and Y_J are the most probable values for

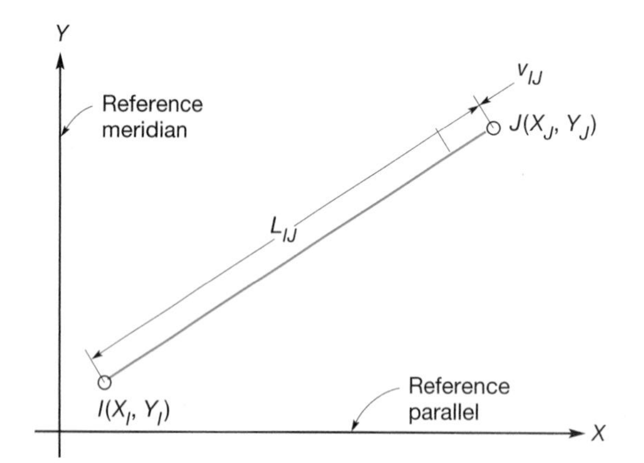

Figure 16.8 Observed distance expressed in terms of coordinates.

stations I and J, respectively. By applying Taylor's series [Equation (16.18)] to this nonlinear distance equation, the following linearized form of the equation results:[4]

$$\left(\frac{X_I - X_J}{IJ}\right)_0 dX_I + \left(\frac{Y_I - Y_J}{IJ}\right)_0 dY_J + \left(\frac{X_J - X_I}{IJ}\right)_0 dX_J + \left(\frac{Y_J - Y_I}{IJ}\right)_0 dY_J = k_{IJ} + v_{IJ} \tag{16.20}$$

In Equation (16.20), $k_{ij} = L_{ij} - (IJ)_0$, where L_{ij} is the observed length of the line, and $(IJ)_0$ is the length based upon initial approximations for the coordinates of points I and J and computed as

$$(IJ)_0 = \sqrt{(X_{J_0} - X_{I_0})^2 + (Y_{J_0} - Y_{I_0})^2} \tag{16.21}$$

For a specific distance such as AB of Figure 16.1, Equation (16.21) would be written as

$$\left(\frac{X_A - X_B}{AB}\right)_0 dX_A + \left(\frac{Y_A - Y_B}{AB}\right)_0 dY_A + \left(\frac{X_B - X_A}{AB}\right)_0 dX_B + \left(\frac{Y_B - Y_A}{AB}\right)_0 dY_B = k_{AB} + v_{AB} \tag{16.22}$$

In Equation (16.22) each coefficient of the unknown dx and dy corrections is evaluated using initial approximations selected for the unknown coordinates of stations A and B; k_{AB} is $L_{AB} - (AB)_0$, where L_{AB} is the observed distance, and

[4]The complete development of this linearized equation, and the linearized equations for azimuths and horizontal angles which are given in the following subsections, are presented in Ghilani and Wolf (2006), cited in the bibliography at the end of this chapter.

$(AB)_0$ is the distance computed using Equation (16.21) and the approximate coordinates; and v_{AB} is the residual error in the distance.

■ Example 16.8

Write the linearized observation equation for the distance AB whose observed length is 132.823 m. Assume the approximate coordinates for station A and B are (1023.151, 873.018) and (1094.310, 985.163), respectively.

Solution

Step 1: Compute the appropriate coordinate differences

$$(X_B - X_A)_0 = 1094.310 - 1023.151 = 71.159 \text{ m}$$

$$(Y_B - Y_A)_0 = 985.163 - 873.018 = 112.145 \text{ m}$$

Step 2: Compute $(AB)_0$

$$(AB)_0 = \sqrt{(71.159)^2 + (112.145)^2} = 132.816 \text{ m}$$

Step 3: Substitute the appropriate values into Equation (16.22) to develop the linearized observation equation as

$$\left(\frac{-71.159}{132.816}\right) dX_A + \left(\frac{-112.145}{132.816}\right) dY_A + \left(\frac{71.159}{132.816}\right) dX_B + \left(\frac{112.145}{132.816}\right) dY_B = (132.823 - 132.816) + v_{AB}$$

Reducing:

$$-0.53577\, dX_A - 0.84436\, dY_A + 0.53577\, dX_B + 0.84436\, dY_B = 0.007 + v_{AB}$$

16.9.3 The Azimuth Observation Equation

Equation (10.11) expressed the azimuth of a line in terms of the X and Y coordinates of the line's end points. That expression is written here in observation equation form, and to make it general, the line designation has been changed from AB to IJ, and thus subscripts I and J have replaced A and B:

$$Az_{IJ} + v_{IJ} = \tan^{-1}\left(\frac{X_J - X_I}{Y_J - Y_I}\right) + C \qquad \textbf{(16.23)}$$

In Equation (16.23), Az_{IJ} is the observed azimuth of line IJ, v_{IJ} is the residual error in the observation and the coordinates in the expression on the right side of the equation are most probable values of the line's end points. The value of the constant C depends upon the direction of the line. If the azimuth of the line is between 0° and 90°, the value of C is 0°. If the azimuth of the line is between 90° and 270°, C is 180° and if the line's azimuth is between 270° and 360°, C is 360°. Equation (16.23) is nonlinear, but again by applying Equation (16.18) the following linearized form of this equation is obtained

$$\rho\left(\frac{Y_I - Y_J}{IJ^2}\right)_0 dX_I + \rho\left(\frac{X_J - X_I}{IJ^2}\right)_0 dY_I + \rho\left(\frac{Y_J - Y_I}{IJ^2}\right)_0 dX_J + \rho\left(\frac{X_I - Y_J}{IJ^2}\right)_0 dY_J = k_{IJ} + v_{IJ} \tag{16.24}$$

As with the linearized distance observation equation, the coefficients of the unknown dx and dy terms in Equation (16.24) are obtained by using initial approximations for the coordinates of the end points of the line. The IJ^2 terms in the denominators of the coefficients are simply the squares of the line lengths as computed using initial approximations for the coordinates in Equation (16.21). The right side of Equation (16.24), which includes the constant term k_{IJ} and the residual, is given in seconds. Thus to make the units consistent on both sides of the equation, the coefficients on the left are multiplied by rho (ρ), which is 206,265 sec/rad. The constant term k_{IJ} is computed as follows

$$k_{IJ} = Az_{IJ} - \tan^{-1}\left(\frac{X_{J_0} - X_{I_0}}{Y_{J_0} - X_{I_0}}\right) + C \tag{16.25}$$

In Equation (16.25), Az_{IJ} is the observed azimuth, the arc tangent function is the computed azimuth based upon initial approximations for the coordinates and C is the constant, as previously described.

16.9.4 The Angle Observation Equation

As illustrated in Figure 16.9, an angle can be expressed as the difference between the azimuths of two lines. Thus angle BIF in the figure is simply azimuth IF minus

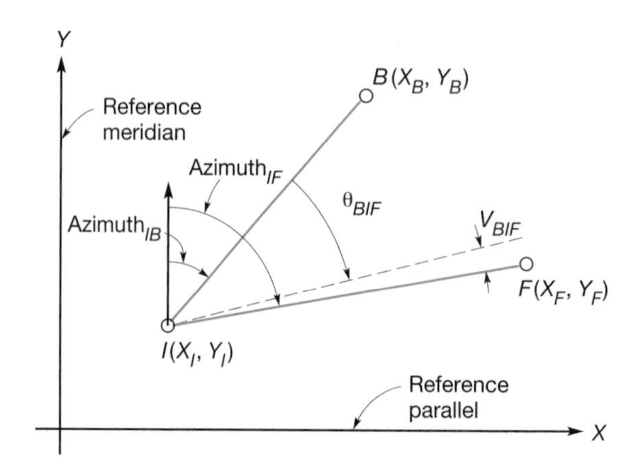

Figure 16.9 Measured angle expressed in terms of coordinates. (Note that an angle is simply the difference between the two azimuths.)

azimuth *IB*. The nonlinear observation equation for angle, *BIF*, is therefore

$$\theta_{BIF} + v_{BIF} = \tan^{-1}\left(\frac{X_F - X_I}{Y_F - Y_I}\right) - \tan^{-1}\left(\frac{X_B - X_I}{Y_B - Y_I}\right) + D \quad \textbf{(16.26)}$$

In Equation (16.26), θ_{BIF} is the observed value for angle *BIF*, and v_{BIF} is the residual error in the observation. The right side of this equation is simply the difference in the azimuths *IF* and *IB*, where these azimuths are expressed in the manner of Equation (16.23). The constant term $D = C_{IF} - C_{IB}$, where *C* was defined for Equation (16.23), and C_{IF} and C_{IB} apply to the azimuths of *IF* and *IB*, respectively. By applying Equation (16.18), the following linearized form of Equation (16.26) is obtained

$$\begin{aligned}&\rho\left(\frac{Y_I - Y_B}{IB^2}\right)_0 dX_B + \rho\left(\frac{X_B - X_I}{IB^2}\right)_0 dY_B + \rho\left(\frac{Y_B - Y_I}{IB^2} - \frac{Y_F - Y_I}{IF^2}\right)_0 dX_I \\ &+ \rho\left(\frac{X_I - X_B}{IB^2} - \frac{X_I - X_F}{IF^2}\right)_0 dY_I + \rho\left(\frac{Y_F - Y_I}{IF^2}\right)_0 dX_F \qquad \textbf{(16.27)} \\ &+ \rho\left(\frac{X_I - X_F}{IF^2}\right)_0 dY_F = k_{BIF} + v_{BIF}\end{aligned}$$

In Equation (16.27), the term k_{BIF} is

$$k_{BIF} = \theta_{BIF} - \tan^{-1}\left(\frac{Y_F - Y_I}{X_F - X_I}\right)_0 - \tan^{-1}\left(\frac{X_B - X_I}{Y_B - Y_I}\right)_0 + D$$

where θ_{BIF} is the observed value for the angle. As with the distance and azimuth equations, the coefficients of the unknown corrections *dX* and *dY* in Equation (16.27) are developed using approximate coordinates for stations *B*, *I*, and *F*. In this equation, *B* represents the backsight station, *I* the instrument station, and *F* the foresight station for clockwise angle *BIF*. In Equation (16.27), it should again be noted that the coefficients on the left side of the equation are multiplied by ρ, (206, 265″/rad), so that the units on both sides are seconds. For a specific angle, such as angle *GAB* of Figure 16.1, after substitution of corresponding subscripts into Equation (16.27), the following linearized angle observation equation results

$$\begin{aligned}&\rho\left(\frac{Y_A - Y_G}{AG^2}\right)_0 dX_G + \rho\left(\frac{X_G - X_A}{AG^2}\right)_0 dY_G + \left(\frac{Y_G - Y_A}{AG^2} - \frac{Y_B - Y_A}{AB^2}\right)_0 dX_A \\ &+ \left(\frac{X_A - X_G}{AG^2} - \frac{X_A - X_B}{AB^2}\right)_0 dY_A + \rho\left(\frac{Y_B - Y_A}{AB^2}\right)_0 dX_B \qquad \textbf{(16.28)} \\ &+ \rho\left(\frac{X_A - X_B}{AB^2}\right)_0 dY_B = k_{GAB} + v_{GAB}\end{aligned}$$

■ Example 16.9

Angle GAB was observed as 107°29′40″. The backsight station G, instrument station A, and foresight station B had the following approximate X and Y coordinates, respectively: (578.741, 1103.826); (415.273, 929.868); and (507.934, 764.652). (*Note:* All coordinate values are given in units of meters.) Write the linearized observation equation for this angle.

Solution

Step 1: Compute the appropriate coordinate differences for substitution into Equation (16.28).

$$\begin{aligned}
(X_G - X_A)_0 &= 578.741 - 415.273 = -163.468 \text{ m} \\
(Y_G - Y_A)_0 &= 1103.826 - 929.868 = 173.958 \text{ m} \\
(X_B - X_A)_0 &= 507.934 - 415.273 = 92.661 \text{ m} \\
(Y_B - Y_A)_0 &= 764.652 - 929.868 = -165.216 \text{ m}
\end{aligned}$$

Step 2: Compute the distances AG and AB, and angle GAB_0

$$\begin{aligned}
(AG)_0 &= \sqrt{(163.468)^2 + (173.958)^2} = 238.711 \text{ m} \\
(AB)_0 &= \sqrt{(92.661)^2 + (-165.216)^2} = 189.726 \text{ m}
\end{aligned}$$

$$GAB_0 = \tan^{-1}\left(\frac{92.661}{-165.216}\right) - \tan^{-1}\left(\frac{163.468}{173.958}\right) + 180° = 107°29'42''$$

Step 3: Substitute the appropriate values into Equation (16.28).

$$\begin{aligned}
&\rho\left(\frac{-173.958}{238.711^2}\right) dX_G + \rho\left(\frac{163.468}{238.711^2}\right) dY_G \\
&+ \rho\left(\frac{173.958}{238.711^2} - \frac{-165.216}{189.726^2}\right) dX_A \\
&+ \rho\left(\frac{-16.468}{238.711^2} - \frac{-92.661}{189.726^2}\right) dY_A + \rho\left(\frac{-165.216}{189.726^2}\right) dX_B \\
&+ \rho\left(\frac{-92.661}{189.726^2}\right) dY_B \\
&= (107°29'40'' - 107°29'42'') + v_{GAB}
\end{aligned}$$

Reducing:

$$\begin{aligned}
&-629.684'' \, dX_G + 591.713'' \, dY_G + 1579.405'' \, dX_A \\
&-59.065'' \, dY_A - 949.721'' \, dX_B - 532.649'' \, dY_B = -2'' + v_{GAB}
\end{aligned}$$

16.9.5 A Traverse Example Using WOLFPACK

Traverse adjustments by the least-squares method involve distance and angle observation equations, and sometimes they include azimuth observation equations as well. Because of the lengthy calculations involved in forming and solving the observation equations, and because the solution is iterative which requires repetitive computations, the least-squares adjustment of even small traverses should be done by computers. To perform a least-squares adjustment by computer, a data file must be prepared in which all observations and their identity (given by the end stations of lines for distances and azimuths, and by backsight, instrument, and foresight stations for angles) must be entered. The computer can be programmed to calculate initial approximations for the coordinates of the unknown stations by using some limited amount of the observed data. In a traverse like that of Figure 11.1(a), for example, the angles at stations *A*, *B*, *C*, and *D*, and distances *AB*, *BC*, *CD*, and *DE* could be used to compute coordinates of the four unknown stations *B*, *C*, *D*, and *E*. Once the initial approximations have been calculated, a computer can easily determine the coefficients of the unknowns in the observation equations, as well as the constant terms. By employing Equation (16.2), relative weights can be determined from the standard deviations of the observed quantities. Thus the computer is able to prepare the $\mathbf{A}$, $\mathbf{W}$, and $\mathbf{L}$ matrices, so that Equation (16.7) can be solved.

The WOLFPACK software contained within the CD that accompanies this book has been used to adjust the traverse network of Figure 16.1. For this adjustment, station *A* with coordinates of $x = 415.273$ m and $y = 929.868$ m was held fixed. Also azimuth *AB* was held fixed at 150°42′51″ by applying a heavy weight (assigning it a standard deviation of ±0.001″). The coordinates of station *A* fix the survey in position while the azimuth *AB* fixes the survey in rotation. The observed data for the traverse network, which includes distance observations and their standard deviations, and angle observations and their standard deviations, are listed in Tables 16.3 and 16.4, respectively.

Figure 16.10 shows the format and order of preparing the data file for input to the least-squares adjustment program of the WOLFPACK software. A printout giving the results of the adjustment is shown in Figure 16.11. Note that this latter table lists the adjusted coordinates of all new stations, as well as adjusted values for all distance and angle observations (obtained by adding the residuals to their corresponding observed values). The adjusted azimuth of line *AB* is also listed, which is the same as its input value. This is expected since that azimuth was held in the adjustment by assigning it a large weight.

These output results contain some of the quantities that are needed to meet newer surveying accuracy standards as was noted in Section 16.1. This example is also demonstrated with a Mathcad® worksheet on the CD accompanying this book. This worksheet demonstrates programming that is similar to that used in WOLFPACK.

■ 16.10 ERROR ELLIPSES

Error ellipses give a two-dimensional representation of the uncertainties of the adjusted coordinates of points as determined in a least-squares adjustment. They can be plotted at enlarged scales directly on scaled diagrams showing the points

TABLE 16.3 Distance Observations for Network Shown in Figure 16.1

From	To	Distance (m)	σ (m)
A	B	189.436	0.007
B	C	122.050	0.007
C	D	121.901	0.007
D	E	145.256	0.007
E	F	168.180	0.007
F	G	231.021	0.007
G	A	238.714	0.007
G	H	143.780	0.007
H	K	119.631	0.007
K	E	114.695	0.007
H	J	96.036	0.007
J	C	85.908	0.007

TABLE 16.4 Angle Observations for Network Shown in Figure 16.1

Backsight Station	Instrument Station	Foresight Station	Angle	σ
G	A	B	107°29′40″	8.9″
A	B	C	94°44′24″	11.7″
B	C	D	235°09′26″	13.7″
C	D	E	104°08′40″	12.7″
D	E	F	124°27′36″	11.2″
E	F	G	121°37′08″	9.5″
F	G	A	112°23′00″	8.3″
F	G	H	38°25′46″	9.9″
G	H	J	243°15′20″	14.6″
H	J	C	135°08′30″	18.0″
J	H	G	116°44′44″	14.6″
J	H	K	296°44′38″	15.0″
H	K	E	131°16′30″	14.3″
K	E	F	68°40′36″	12.3″

```
Example 15.9.4 data for least squares adjustment    {Title line}
12 14 1 1 10           {Number of distances, angles, azimuths, control and total stations}
A 415.273 929.868      {Control station: Identification, X, Y}
B 507.934 764.652      {Unknown station: Identification, X, Y}
C 618.952 815.353
D 723.852 753.287
E 826.128 856.438
F 794.659 1021.655
G 578.741 1103.826
H 652.221 980.245
J 600.595 899.272
K 713.362 877.418
A B 189.436 0.007      {Distance obs: Stations: From, To, Measured Distance, and Std Dev}
B C 122.050 0.007
C D 121.901 0.007
D E 145.256 0.007
E F 168.180 0.007
F G 231.021 0.007
G A 238.714 0.007
G H 143.780 0.007
H K 119.631 0.007
K E 114.695 0.007
H J 96.036 0.007
J C 85.908 0.007
G A B 107 29 40 8.9    {Angle obs: Stations: B, I, F, Measured Angle, and Std Dev}
A B C 94 44 24 11.7
B C D 235 09 26 13.4
C D E 104 08 40 12.7
D E F 124 27 36 11.2
E F G 121 37 08 9.5
F G A 112 23 00 8.3
F G H 38 25 46 9.9
G H J 243 15 20 14.6
H J C 135 08 30 18
J H G 116 44 44 14.6
J H K 296 44 38 15.0
H K E 131 16 30 14.3
K E F 68 40 36 12.3
A B 150 42 51 0.001    {Azimuth obs: Stations: I, F, Observed Azimuth, and Standard Deviation}
```

Figure 16.10 Data file for adjustment of Figure 16.1.

in the horizontal survey. When plotted in this manner, their sizes and appearances enable a quick visual analysis to be made of the overall relative precisions of all adjusted points. As discussed later in this section, this is useful in planning surveys and in analyzing the results of surveys for acceptance or rejection.

On the output listing of Figure 16.11, the adjusted coordinates for the stations in Figure 16.1 are listed, and to their right are columns titled σ_U, σ_V, and t. Respectively, these contain the semimajor axes, semiminor axes, and clockwise rotation angle from the Y-axis to the semimajor axis of the ellipse computed at each station. To compute these three terms, values from the $\boldsymbol{Q}_{xx}$ matrix (see Section 16.5), and the standard deviation of unit weight [see Equations (16.9) and (16.10)] are used with the following formulas.[5]

[5]For the derivations of these equations, see Ghilani and Wolf (2006), cited in the bibliography at the end of this chapter.

```
****************
Adjusted stations
****************
```

Station	Northing	Easting	σn	σe	σu	σv	t
B	764.645	507.938	0.0038	0.0021	0.0044	0.0000	150.06°
C	815.350	618.955	0.0049	0.0046	0.0050	0.0045	159.09°
D	753.286	723.867	0.0069	0.0064	0.0074	0.0058	36.95°
E	856.441	826.133	0.0092	0.0053	0.0093	0.0052	7.55°
F	1,021.654	794.661	0.0086	0.0058	0.0091	0.0049	156.30°
G	1,103.827	578.746	0.0045	0.0058	0.0060	0.0042	111.69°
H	980.245	652.226	0.0061	0.0049	0.0063	0.0047	157.64°
J	899.270	600.599	0.0058	0.0050	0.0058	0.0050	176.23°
K	877.418	713.370	0.0073	0.0056	0.0073	0.0056	176.91°

```
***************************
Adjusted Distance Observations
***************************
```

Station Occupied	Station Sighted	Distance	V	σ
A	B	189.434	0.0016	0.0044
B	C	122.048	0.0022	0.0045
C	D	121.895	0.0055	0.0044
D	E	145.257	-0.0005	0.0044
E	F	168.184	-0.0040	0.0042
F	G	231.024	-0.0027	0.0042

```
*************************
Adjusted Angle Observations
*************************
```

Station Backsighted	Station Occupied	Station Foresighted	Angle	V	σ
G	A	B	107°29'39"	0.8"	5.0"
A	B	C	94°44'17"	6.5"	6.4"
B	C	D	235°09'20"	5.8"	7.9"
C	D	E	104°08'39"	1.2"	7.1"
D	E	F	124°27'46"	-9.6"	5.9"
E	F	G	121°37'16"	-7.9"	5.1"
E	G	A	112°23'03"	-2.8"	4.4"

```
***************************
Adjusted Azimuth Observations
***************************
```

Station Occupied	Station Sighted	Azimuth	V	σ
A	B	150°42'51"	0.0"	0.001"

```
-----Standard Deviation of Unit Weight = 0.697667-----
```

Figure 16.11 Abbreviated results from adjustment of data file in Figure 16.10 from WOLFPACK.

1. Rotation angle, t

$$\tan(2t) = \frac{2q_{xy}}{q_{yy} - q_{xx}} \tag{16.29}$$

In Equation (16.29), the values of q_{xx} and q_{yy} are the diagonal elements from the $\boldsymbol{Q}_{xx}$ matrix, and q_{xy} is the covariance off-diagonal element in the $\boldsymbol{Q}_{xx}$

matrix for a particular station. When computing t, it is important to establish its quadrant before dividing by two.

2. *Semimajor axis:* $\sigma_U = \sigma_0\sqrt{q_{xx}\sin^2(t) + 2q_{xy}\cos(t)\sin(t) + q_{yy}\cos^2(t)}$ **(16.30)**

3. *Semiminor axis:* $\sigma_V = \sigma_0\sqrt{q_{xx}\cos^2(t) - 2q_{xy}\cos(t)\sin(t) + q_{yy}\sin^2(t)}$ **(16.31)**

To demonstrate these calculations, part of the $\boldsymbol{Q_{xx}}$ and $\boldsymbol{X}$ matrices that were generated in the least-squares adjustment of the horizontal network in Section 16.9.4 are listed below. (*Note:* These were not shown in the abbreviated output listing of Figure 16.11.) Only those parts of the matrices that pertain to stations B and C are shown. The elements in the rows of the $\boldsymbol{X}$ matrix indicate the order of the unknowns and identify the elements of $\boldsymbol{Q_{xx}}$ that apply in computing error ellipses. The upper left 2×2 submatrix of $\boldsymbol{Q_{xx}}$ (shown bold), contains the elements that apply to station B. Because X_B is in row 1 of the $\boldsymbol{X}$ matrix, q_{xx} for point B occupies the row 1 column 1, or (1, 1) position. The Y_B coordinate is in row 2 of the $\boldsymbol{X}$ matrix and thus q_{yy} is located in the (2, 2) position of X. Also, the (1, 2) element [or the (2, 1) element which is the same because of symmetry] contains q_{xy}. The following computations yield the error ellipse data for station B.

$$
{}_{18}[(\mathbf{A}^T\mathbf{W}\mathbf{A})^{-1}]^{18} = \begin{bmatrix} \mathbf{0.00009440} & \mathbf{-0.00001683} & 0.00000739 & -0.00001304 & \cdots \\ \mathbf{-0.00001683} & \mathbf{0.00003001} & -0.00001318 & 0.00002325 & \cdots \\ 0.00000739 & -0.00001318 & 0.0004332 & -0.00000294 & \cdots \\ -0.00001304 & 0.00002325 & -0.00000294 & 0.00004989 & \cdots \\ \vdots & \vdots & \vdots & \vdots & \ddots \end{bmatrix} {}_{18}X^1 = \begin{bmatrix} \boldsymbol{X_B} \\ \boldsymbol{Y_B} \\ X_C \\ Y_C \\ \vdots \end{bmatrix}
$$

Step 1: By Equation (16.29), compute rotation angle, t.

$$
\tan(2t) = \frac{2(-0.00001683)}{0.00003001 - 0.00000944} = -1.59261
$$

Since the numerator is negative and the denominator is positive, angle $2t$ is in the fourth quadrant[6] and thus angle t is

$$
t = \frac{1}{2}[\tan^{-1}(-1.59261) + 360°] = \frac{1}{2}[302.1247164°] = 151°03'44''
$$

---

[6]In Equation (16.29), $\tan 2t = \sin 2t / \cos 2t$. Thus the sines and cosines enable determining the quadrant of $2t$. If the sine and cosine are both plus, i.e., the numerator and denominator of Equation (16.29) are both plus, then $2t$ is in the first quadrant (between 0° and 90°). Similarly, if the numerator is plus and the denominator minus, $2t$ is in the second quadrant; and if the numerator and denominator are both minus, $2t$ is in the third quadrant.

Step 2: Compute the semimajor axis using Equation (16.30). In these computations, the standard deviation of unit weight, σ_0, is taken from the bottom of the printout given in Figure 16.11, and has been rounded to 0.70.

$$\sigma_U = 0.70\sqrt{0.00000944 \sin^2(t) + 2(-0.00001683)\cos(t)\sin(t) + 0.00003001\cos^2(t)}$$
$$= 0.70\sqrt{0.000039447}$$
$$= 0.004 \text{ m}$$

Step 3: Compute the semiminor axis using Equation (16.31).

$$\sigma_V = 0.70\sqrt{0.00000944 \cos^2(t) - 2(-0.00001683)\cos(t)\sin(t) + 0.00003001\sin^2(t)}$$
$$= 0.70\sqrt{0.000000002}$$
$$= 0.000 \text{ m}$$

Note again that only the upper left 2 × 2 submatrix (in bold) of the full $\boldsymbol{Q}_{xx}$ matrix was necessary for the computations. To compute the error ellipse for Station C, only the (3,3), (3,4), and (4,4) elements of $\boldsymbol{Q}_{xx}$ are needed. This pattern is continued for each station.

In the computed error ellipse for station B of this example, the semiminor axis is nearly zero. Also note that the rotation angle of the semimajor axis closely matches that of the azimuth of the line AB. This could be predicted since the azimuth AB was held fixed during the adjustment by giving it a large weight. This shows both the power, and the danger, of weights. In this example, the large weight is necessary to fix the horizontal network rotationally in azimuth. If the azimuth had not been fixed by weighting, the network would have been free to rotate about Station A and no solution could have been found. However, inappropriate application of weights can cause unwanted corrections in the adjustment. *In least-squares adjustments, it is very important to weight the observations according to their estimated uncertainties.*

As noted previously, when an adjustment is completed, it is often informative to view a graphic of the error ellipses. The error ellipses for this example, shown in Figure 16.12, have had their semi-axes magnified 200 times to make their relatively small values easily visible on the plot. If an error ellipse approximates a circle, this would indicate that point is of approximately equal precision in all directions. Long and slender error ellipses indicate low precisions in their long directions (along their semimajor axes) and higher precisions in their narrow directions (along their semiminor axes). In Figure 16.12, it can be seen that the directions of the semimajor axes for stations D, E, F, and G are aligned on circles with their centers approximately at A. This shows the rotational instability in the observational system, i.e., the held azimuth of line AB alone fixes the rotation in the network. This instability could be improved by observing an azimuth on another line such as EF, or by connecting either Stations E or F to another neighboring control point (assuming one is available). In Figure 16.12 it can also be seen that the largest uncertainties exist on the stations that are farthest from the control station. This is the usual manner that errors propagate in observational systems—i.e., the farther the unknown stations are from the control, the larger the expected errors in their coordinates. With this in mind, the sizes of the errors at all stations,

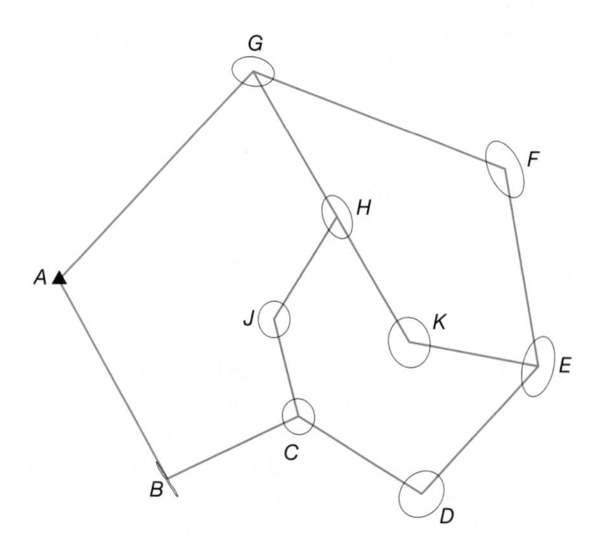

Figure 16.12 Error ellipses plotted at 200 times their actual sizes.

but especially E and F, could be reduced by connecting either of these stations to a neighboring control station (assuming one is available).

The analysis of plotted error ellipses is very useful in scrutinizing results of adjusted horizontal surveys, as illustrated by the simple analyses given in the previous paragraph. As demonstrated, both the sizes and shapes of error ellipses give an immediate visual impression about the relative accuracies of the points in a network and enable the most efficient plan to be developed for making additional observations to strengthen the observational scheme.

16.11 ADJUSTMENT PROCEDURES

Regardless of the nature of the specific adjustment problem, certain procedures should be followed. For example, before the adjustment is undertaken, all data must be carefully analyzed for blunders. Mistakes such as station misidentifications, transcription mistakes, reading blunders, and others must be identified and corrected. Failure to remove them will result in either an unsatisfactory adjustment or no adjustment at all. In several types of surveys, performing loop closures on the data can identify blunders. This is true in leveling, in GPS networks, and in horizontal surveys, including traversing. Also, in traverses the methods discussed in Section 10.16 can be employed to detect blunders.

The minimum amount of control required for making adjustments varies with the type of problem. In differential leveling, only one benchmark is needed, and in a network of GPS baseline observations, only one station with known coordinate values is necessary. For horizontal surveys such as traverses and networks, one station with known coordinates and one course with known direction must be available. If more than the minimum amount of control is present, the adjustment should be performed in two stages as a further means of detecting blunders. The first adjustment, called a *minimally constrained adjustment,* should contain only the minimum amount of control. This should then be followed by a *constrained adjustment,* in which all available control is used.

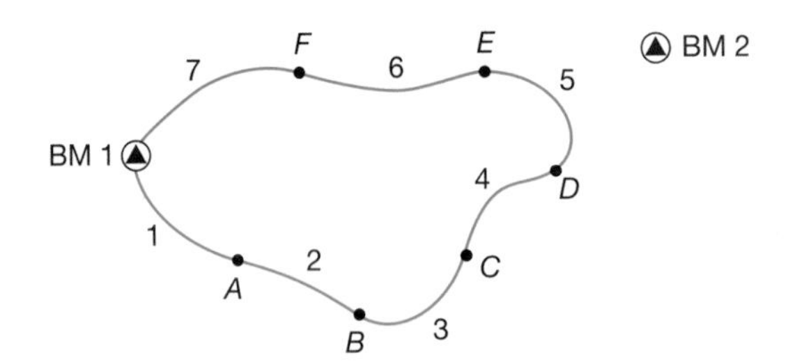

Figure 16.13 Differential leveling loop.

The minimally constrained adjustment provides checks on the geometric closures and the consistency of the observations. After the adjustment is completed, the residuals should be analyzed. Any unusually large residuals, or a preponderance of minus or plus residuals (a group of random errors should contain equal numbers of positive and negative values), will be keys to the existence of blunders in the data. However, even though a minimally constrained adjustment will check the data for *internal consistency,* it may fail to identify the presence of systematic errors and mistakes. For instance, assume during the observation of the leveling circuit shown in Figure 16.13 that a mistake of +1 m occurred during the observation of line 1 and that a −1 m error occurred during the observation of line 7. Even though every observed benchmark would have a 1-m error, the loop misclosure would not uncover this mistake and the observations would geometrically close. However, in that figure, if an additional line of differential levels is observed from BM 2 to *D*, the presence of these mistakes becomes apparent. This further demonstrates the importance of making redundant observations when in the field.

Upon acceptance of the minimally constrained adjustment, all additional available control should be added to the data and the constrained adjustment performed. This will aid in identifying compensating or systematic errors in the data and blunders in control. For instance, suppose that the coordinate values for control station *A* in Figure 16.12 are state plane coordinate values, but the distances are not properly reduced to the state plane grid (see Chapter 20). In this situation, the geometric closures of the adjustment could appear to be fine, but, because of the distance scaling error, the computed coordinates for the unknown stations would all be incorrect. In this case, the minimally constrained adjustment would fail to catch the scaling error. However, if observations had been made to connect the network to a second control station with known state plane coordinates, the scaling error would become apparent in the constrained adjustment. Similarly, a blunder in the control coordinates or benchmarks will not become apparent until the constrained adjustment is performed. Thus, it is important to perform a minimally constrained adjustment to obtain geometric checks on the data and a constrained adjustment to find possible compensating errors, systematic errors, and control blunders. Therefore it follows logically that every survey should have redundant observations to provide geometric checks and isolate mistakes caused by careless work.

Upon acceptance of the constrained adjustment, the post-adjustment statistics, as given in Sections 16.5, 16.7, and 16.10, should be computed. When possible, these statistical values should be compared against published accuracy standards. If the survey fails to meet the required standards, additional observations should be taken or the work repeated using more precise equipment.

16.12 OTHER MEASURES OF PRECISION FOR HORIZONTAL STATIONS

Since error ellipses are part of a bivariate distribution, their probability level is approximately 39%. Generally, surveyors prefer to state their results at a much higher level of confidence. For the semiminor and semimajor axes of the error ellipse, this is accomplished by using a multiplier that is based on critical values taken from the F distribution. This distribution is a function of the number of *degrees of freedom* (number of redundant observations) that existed in the adjustment. Some of the critical values from the F distribution are shown in Table 16.5. The multipliers for the semimajor and semiminor axes of an error ellipse expressed at other probabilities levels are determined from

$$c = \sqrt{2(F_{\alpha,2,\text{ degrees of freedom}})} \quad (16.32)$$

where the semimajor and semiminor axes would be computed as

$$\sigma_{U\%} = c\sigma_u \quad (16.33)$$

$$\sigma_{V\%} = c\sigma_v \quad (16.34)$$

Other measures of precision that can be used include the *circular error probable* (CEP), which is a function of the computed standard deviations for a

TABLE 16.5 $F_{\alpha,2,\text{ degrees of freedom}}$ CRITICAL STATISTIC VALUES FOR SELECTED PROBABILITIES

Degrees of Freedom	Probability 90%	95%	99%
1	49.50	199.50	4999.50
2	9.00	19.00	99.00
3	5.46	9.55	30.82
4	4.32	6.94	18.00
5	3.78	5.79	13.27
9	3.01	4.26	8.02
10	2.92	4.10	7.56
15	2.70	3.68	6.36
20	2.59	3.49	5.85
30	2.49	3.32	5.39

horizontal station, or

$$\text{CEP} = 0.5887(\sigma_X + \sigma_Y) \quad (16.35)$$

The 90% region of the CEP is called the *circular map accuracy standard* (CMAS) and is computed as

$$\text{CMAS} = 1.8227 \text{ CEP} \quad (16.36)$$

The *distance root mean square* (DRMS) error is another measure of precision. It can be computed as

$$\text{DRMS} = \sqrt{\sigma_X^2 + \sigma_Y^2} \quad (16.37)$$

Example 16.10

What are the 95% probability values for the semimajor and semiminor axes of station *B* in Section 16.10?

Solution

The adjustment of Figure 16.1 had 12 distance observations, 14 angle observations, and 1 azimuth observation for a total of 27 observations. There were 9 stations having 2 unknowns each for a total of 18 unknowns. Thus the number of degrees of freedom is 27 − 18, or 9. From Table 16.5, the appropriate *F* value for nine degrees of freedom is 4.26. By Equation (16.32), the *c*-multiplier is

$$c = \sqrt{2 \times 4.26} = 2.92$$

From Section 16.10, the values for σ_U and σ_V were 0.004 m and 0.000 m, respectively. Thus by Equations (16.33) and (16.34), the 95 percent values for the semimajor and -minor axes are

$$\sigma_{U95\%} = 2.92 \times 0.004 = 0.012 \text{ m}$$

$$\sigma_{V95\%} = 2.92 \times 0.000 = 0.000 \text{ m}$$

16.13 CONCLUSIONS

As discussed in Chapter 3, the presence of errors in observations is inevitable. However, if the method of least squares is employed, the sizes of the errors can be assessed and if they are within acceptable limits, the observations can be adjusted to determine the most probable values for the unknowns. If some of the observations contain unacceptable errors, these observations must be repeated before the final adjustment is made. The advantages of the least-squares method over other adjustment techniques are many. Some of these are that it: (1) conforms to the laws of probability, (2) provides the most probable solution for a given set of observations, (3) allows individual weighting of observations, (4) forces geometric closures on the observations, (5) simultaneously adjusts all observations, (6) provides a single unique solution for a set of data, and (7) yields estimated precisions of adjusted quantities. The least-squares method is readily programmed for computer solution and the data is easily prepared for making adjustments. Because of these advantages and the fact that data from least-squares adjustments are now necessary for assessing compliance of surveys with modern standards such as the FGCS standards and specifications for GPS Relative Positioning (see Section 14.5.1) and the ALTA-ACSM Land Title Survey Standards (see Section 21.10), all surveying offices should employ the method.

In this chapter, the basic theory of least-squares adjustment has been presented and its application to common surveying observations demonstrated. For further information on least squares, the reader is directed to the references in the bibliography.

PROBLEMS

Asterisks (*) indicate problems that have answers given in Appendix G.

16.1 What fundamental condition is enforced by the method of least squares?

16.2 List the geometric closures required for a link traverse.

16.3 Why is the compass (Bowditch) rule considered an arbitrary adjustment?

16.4* What is the most probable value for the following set of ten distance observations in meters? 532.688, 532.682, 532.682, 532.684, 532.689, 532.686, 532.690, 532.684, 532.686, 532.686

16.5 Three horizontal angles were observed around the horizon of station A. Their values are 95°17′05″, 124°16′38″, and 140°26′23″. What are the most probable values for the three angles?

16.6 In Problem 16.5, the standard deviations of the three angles are ±2.5″, ±4.0″, and ±5.9″, respectively. What are the most probable values for the three angles?

16.7* Determine the most probable values for the x and y distances of Figure 16.2, if the observed lengths of AC, AB, and BC (in meters) are 348.966, 184.307, and 164.647, respectively.

16.8 What is the standard deviation of the adjusted value in Problem 16.4?

16.9 Repeat Problem 16.8 for Problem 16.6.

16.10* Repeat Problem 16.8 for Problem 16.7.

16.11 A network of differential levels is run from existing benchmark Juniper through new stations A and B to existing benchmarks Red and Rock as shown in the accompanying figure. Develop the observation equations for adjusting this network by least squares, using the following elevation differences.

From	To	Elev. Diff. (m)	σ (m)
Juniper	A	63.508	0.023
A	B	−23.475	0.017
RED	B	−3.800	0.026
ROCK	B	18.555	0.026

Problem 16.11

16.12 For Problem 16.11, following steps outlined in Example 16.6 perform a weighted least-squares adjustment of the network. Determine weights based upon the given standard deviations. What are the:

(a) Most probable values for the elevations of A and B?

(b) Standard deviations of the adjusted elevations?

(c) Standard deviation of unit weight?

(d) Adjusted elevation differences and their residuals?

(e) Standard deviations of the adjusted elevation differences?

16.13 Repeat Problem 16.12 using distances for weighting. Assume the following course lengths for the problem.

From	To	Elev. Diff. (m)	Dist. (m)
Juniper	A	63.508	1000
A	B	−23.475	550
RED	B	−3.800	1250
ROCK	B	18.555	1345

16.14 Use WOLFPACK to do Problem 16.13.

16.15 Use the Mathcad® worksheet on the CD in the book to do Problem 16.13.

16.16 A network of differential levels is shown in the accompanying figure. The observed elevation differences and the distances between stations are shown in the following table. Using WOLFPACK, determine the

(a) Most probable values for the elevations of new benchmarks B, C, D, E, F, and H?

From	To	Elev. Diff. (ft)	σ (ft)
A	B	1.61	0.052
B	C	42.61	0.070
C	D	−25.55	0.067
D	E	35.50	0.056
E	F	−57.98	0.047
F	A	3.67	0.037
G	F	−17.59	0.042
G	B	−12.45	0.037
G	E	40.27	0.042
G	H	−0.06	0.042
H	D	4.88	0.042

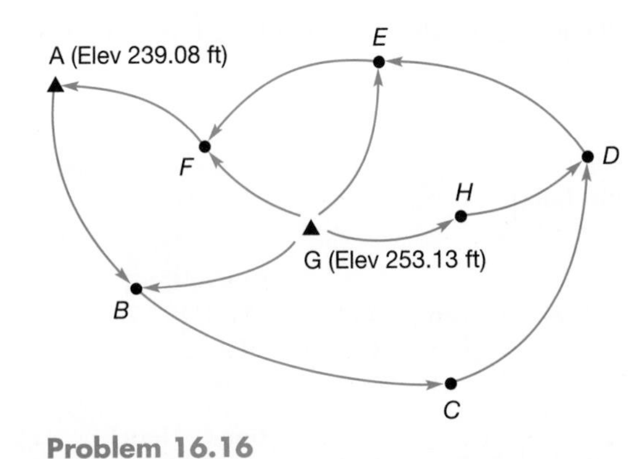

Problem 16.16

(b) Standard deviations of the adjusted elevations?
(c) Standard deviation of unit weight?
(d) Adjusted elevation differences and their residuals?
(e) Standard deviations of the adjusted elevation differences?

16.17 Develop the observation equations for Problem 16.16.

16.18 A network of GPS observations shown in the accompanying figure was made with two receivers using the static method. Known coordinates of the two control stations are in the geocentric system. Develop the observation equations for the following baseline vector components.

Jim to Troy				**Al to Troy**			
−13,024.970	5.2E-3	6.8E-6	6.7E-6	−23,335.070	5.2E-3	−6.8E-6	−6.4E-6
14,982.005		5.2E-3	6.6E-6	−12,276.803		5.3E-3	−6.2E-6
20,159.364			5.2E-3	−7219.168			5.4E-3

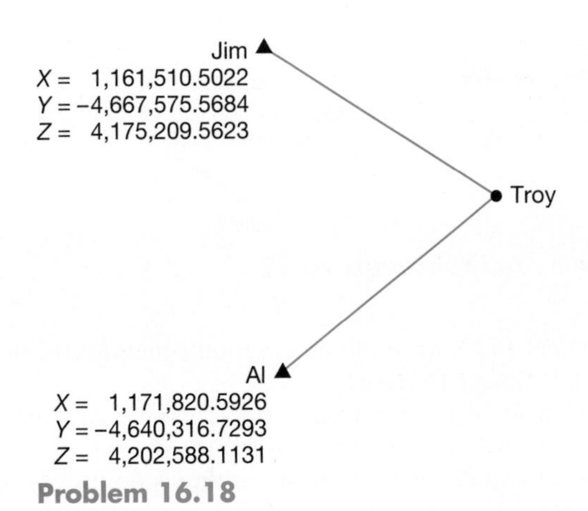

Problem 16.18

16.19 For Problem 16.18, construct the A and L matrices.

16.20 For Problem 16.18, construct the covariance matrix.

16.21 Use WOLFPACK to adjust the baselines of Problem 16.18.

16.22 Convert the geocentric coordinates obtained for station Troy in Problem 16.21 to geodetic coordinates.

16.23 A network of GPS observations shown in the accompanying figure was made with two receivers using the rapid static method. Use WOLFPACK to adjust the network, given the following data. (Geocentric coordinates are given for control stations Bonnie and Tom.)

Bonnie to Ray1				**Tom to Ray1**			
3377.79292	5.8E-3	−6.3E-6	7.8E-6	−11,899.17606	6.3E-3	6.7E-6	−9.8E-7
−4727.91291		5.9E-3	6.5E-6	−7561.27604		5.5E-3	−8.7E-6
−6172.02012			5.7E-3	−5040.14223			5.8E-3
Bonnie to Herb2				**Tom to Herb2**			
7826.23261	5.6E-3	5.9E-6	−6.3E-6	−7450.71237	5.9E-3	6.7E-6	5.3E-6
5106.72819		5.8E-3	−6.4E-6	2273.36277		5.7E-3	−6.7E-6
3521.03083			5.7E-3	4652.90063			5.8E-3
Ray1 to Herb2				**Herb2 to Ray1**			
4448.44588	5.3E-3	9.7E-6	−9.4E-6	−4448.45525	5.8E-3	6.8E-6	−6.4E-6
9834.63222		5.5E-3	9.6E-6	−9834.62002		5.9E-3	−6.4E-6
9693.05569			5.8E-3	−9693.06009			5.8E-3
Herb2 to Bonnie				**Bonnie to Tom** (Fixed line-Don't use in adjustment!)			
−7826.24915	4.5E-3	−4.3E-6	4.8E-6	15,276.97173	6.4E-3	9.8E-6	−9.4E-6
−5106.71748		4.6E-3	4.4E-6	2833.37436		9.1E-3	9.2E-6
−3521.03248			4.8E-3	−1131.86311			6.2E-3

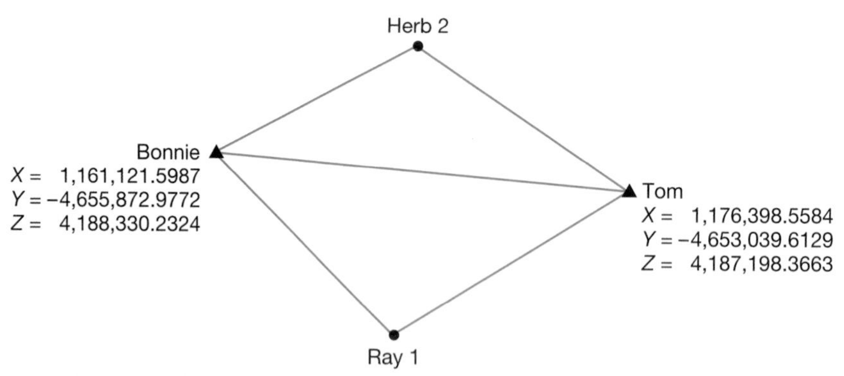

Problems 16.23 through 16.29

16.24 For Problem 16.23, write the observation equations for the baselines "Herb2 to Ray1" and "Bonnie to Herb2."

16.25 For Problem 16.24, construct the A and L matrices for the observations.

16.26 For Problem 16.24, construct the covariance matrix.

16.27* After completing Problem 16.23, convert the geocentric coordinates for station Ray1 and Herb2 to geodetic coordinates. (*Hint*: See Section 13.4.3.)

16.28* Following the procedures discussed in Section 14.5.2, analyze the fixed baseline from station Bonnie to Tom.

16.29 Following the procedures discussed in Section 14.5.3, analyze the repeat baselines Ray1 to Herb2, and Bonnie to Herb2.

16.30 For the horizontal survey of the accompanying figure, determine initial approximations for the unknown stations. The observations for the survey are

From	To	Distance (ft)	σ (ft)
Dave	Steve	2351.46	0.020
Dave	Frank	3438.80	0.020
Steve	Frank	2084.85	0.020
Steve	Wes	2612.61	0.020
Frank	Wes	1658.76	0.020
Wes	Dave	2569.06	0.020

Backsight Station	Instrument Station	Foresight Station	Angle	σ (″)
Wes	Dave	Frank	27°29′30″	3.2
Frank	Dave	Steve	36°27′32″	3.2
Dave	Steve	Wes	62°03′23″	3.2
Wes	Steve	Frank	39°22′36″	3.3
Steve	Frank	Dave	42°06′26″	3.2
Dave	Frank	Wes	45°38′14″	3.3
Frank	Wes	Steve	52°52′47″	3.4
Steve	Wes	Dave	53°59′41″	3.2

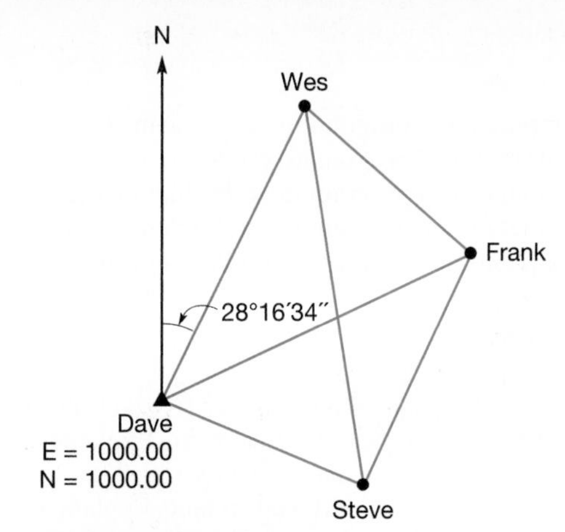

Problem 16.30

16.31* Using the data in Problem 16.30, write the linearized observation equation for the distance from Steve to Frank.

16.32 Using the data in Problem 16.30, write the linearized observation equation for the angle Steve-Frank-Dave.

16.33 Assuming a standard deviation of ±0.001″ for the azimuth line Dave-Wes, use WOLFPACK to adjust the data in Problem 16.30.

16.34* Given the following inverse matrix and a standard deviation of unit weight of 1.23, determine the parameters of the error ellipse.

$$(A^TWA)^{-1} = \begin{bmatrix} q_{xx} & q_{xy} \\ q_{xy} & q_{yy} \end{bmatrix} = \begin{bmatrix} 0.00023536 & 0.00010549 \\ 0.00010549 & 0.00033861 \end{bmatrix}$$

16.35 Compute S_x and S_y in Problem 16.34.

16.36 Given the following inverse matrix and a standard deviation of unit weight of 1.45, determine the parameters of the error ellipse.

$$(A^TWA)^{-1} = \begin{bmatrix} q_{xx} & q_{xy} \\ q_{xy} & q_{yy} \end{bmatrix} = \begin{bmatrix} 0.0004894 & 0.0000890 \\ 0.0000890 & 0.0002457 \end{bmatrix}$$

16.37 Compute S_x and S_y in Problem 16.36.

16.38 The well-known observation equation for a line is $mx + b = y + v_y$. What is the slope and y-intercept of the best fit line for a set of points with coordinates of (478.72, 3517.64), (1446.81, 2950.40), (2329.79, 2432.66), (3345.74, 1837.13), (4382.98, 1229.16)?

16.39 Create a computational program to solve Problem 16.38.

16.40 Use WOLFPACK and the following standard deviations for each observation to do a least-squares adjustment of Example 10.4, and describe any differences in the solution. What advantages are there to using the least-squares method in adjusting this traverse?

Stations	Angle ± S	Stations	Distance ± S
E-A-B	100°45′37″ ± 16.7″	AB	647.25 ± 0.027
A-B-C	231°23′43″ ± 22.1″	BC	203.03 ± 0.026
B-C-D	17°12′59″ ± 21.8″	CD	720.35 ± 0.027
C-D-E	89°03′28″ ± 10.2″	DE	610.24 ± 0.027
D-E-A	101°34′24″ ± 16.9″	EA	285.13 ± 0.026
AZIMUTH AB	126°55′17″ ± 0.001″		

16.41 Create a computational program to do Problem 16.6.

16.42 Create a computational program to do Problem 16.16.

16.43 Create a computational program to do Problem 16.22.

16.44 Create a computational program to do Problem 16.23.

16.45 Create a computational program to do Problem 16.33.

BIBLIOGRAPHY

Bell, J. 2003. "MOVE3." *Professional Surveyor* 23 (No. 11): 46.

Ghilani, Charles D. 2003. "Statistics and Adjustments Explained—Part 1: Basic Concepts." *Surveying and Land Information Science* 63 (No. 2): 73.

Ghilani, Charles D. 2003. "Statistics and Adjustments Explained—Part 2: Sample Sets and Reliability." *Surveying and Land Information Science* 63 (No. 3): 141.

Ghilani, Charles D. 2004. "Statistics and Adjustments Explained—Part 3: Error Propagation." *Surveying and Land Information Science* 64 (No. 1): 23.

Ghilani, C. D., and P. R. Wolf. 2006. *Adjustment Computations: Spatial Data Analysis.* New York. John Wiley & Sons.

Schwarz, C. R. 2005. "The Effects of Unestimated Parameters." *Surveying and Land Information Science* 65 (No. 2): 87.

Tan, Willie. 2002. "In What Sense a Free Net Adjustment?" *Surveying and Land Information Science* 62 (No. 4): 251.

17

Mapping Surveys

■ 17.1 INTRODUCTION

Mapping surveys are made to determine the locations of *natural* and *cultural* features on the Earth's surface and to define the configuration (*relief*) of that surface. Once located, these features can be represented on maps. Natural features normally shown on maps include vegetation, rivers, lakes, oceans, and so on. Cultural *(artificial)* features are the products of people and include roads, railroads, buildings, bridges, canals, boundary lines, and so on. The relief of the Earth includes its hills, valleys, plains, and other surface irregularities. Using lines and symbols depicts features shown on maps. Names and legends are added to identify the different objects shown.

Two different types of maps, *planimetric* and *topographic,* are prepared as a result of mapping surveys. The former depicts natural and cultural features in the plan (*X-Y*) views only. Objects shown are called *planimetric features.* Topographic maps also include planimetric features, but in addition they show the configuration of the Earth's surface. Both types of maps have many applications. They are used by engineers and planners to determine the most desirable and economical locations of highways, railroads, canals, pipelines, transmission lines, reservoirs, and other facilities; by geologists to investigate mineral, oil, water, and other resources; by foresters to locate access- or haul-roads, fire-control routes, and observation towers; by architects in housing and landscape design; by agriculturists in soil conservation work; and by archaeologists, geographers, and scientists in numerous fields. Maps are used extensively in geographic information system (GIS) applications (see Chapter 28). Conducting the surveys necessary for preparing maps and the production of the maps from the survey data are the mainstay of many surveying businesses.

Relief is shown on maps by using various conventions and procedures. For topographic maps, *contours* are most commonly used and are preferred by surveyors and engineers. *Digital elevation models* (DEMs) and *three-dimensional perspective models* are newer methods for depicting relief, made possible by computers. *Color, hachures, shading,* and *tinting* can also be used to show relief, but these methods are not quantitative enough and thus are generally unsuitable for surveying and engineering work. Contours, digital elevation models, and three-dimensional perspective models are discussed in later sections of this chapter and in Chapter 18.

Traditionally, maps were prepared using manual drafting methods. Now however, as described in Chapter 18, the majority of maps are produced using computers, computer-aided drafting (CAD) software, and data collectors. Currently, some data collectors include drafting software so that field personnel can display their data in the field to check for mistakes and missing elements. This chapter discusses procedures for collecting planimetric and topographic mapping data.

17.2 BASIC METHODS FOR PERFORMING MAPPING SURVEYS

Mapping surveys are conducted by one of two basic methods: *aerial* (photogrammetric) or *ground* (field) techniques, but often a combination of both is employed. Refined equipment and procedures available today have made photogrammetry very accurate and economical. Hence almost all mapping projects covering large areas now employ this method. However, ground surveys are still commonly used in preparing large-scale maps of smaller areas. Even when photogrammetry is utilized, ground surveys are necessary to establish control and to field-check mapped features for accuracy. This chapter concentrates on ground methods, and describes several field procedures for locating topographic features, both horizontally and vertically. Photogrammetry is discussed in Chapter 27.

17.3 MAP SCALE

Map scale is the ratio of the length of an object or feature on a map to the true length of the object or feature. Map scales are given in three ways: (1) by *ratio* or *representative fraction,* such as 1:2000 or 1/2000; (2) by an *equivalence,* for example, 1 in. = 200 ft; and (3) graphically using either a bar scale or labeled grid lines spaced throughout the map at uniform distances apart. Graphic scales permit accurate measurements to be made on maps, even though the paper upon which the map is printed may change dimensions.

An equivalence scale of 1 in. = 100 ft, indicates that 1 in. on the map is equivalent to 100 ft on the object. In giving scale by ratio or representative fraction, the same units are used for the map distance and the corresponding object distance, and thus 1:1200 could mean 1 in. on the map is equivalent to 1200 in. on the object, but any other units could also apply. Obviously, it is possible to convert from an equivalence scale to a ratio, and vice versa. As an example, 1 in. = 100 ft is converted to a ratio by multiplying 100 ft by 12, which converts it to inches and gives a ratio of 1:1200. Those engaged in surveying (geomatics) and engineering generally prefer an equivalence scale and grid lines on their maps, while geographers often utilize a representative fraction and bar scale.

Choice of scale depends on the purpose, size, and required precision of the finished map. Dimensions of a standard map sheet, type and number of topographic symbols used, and accuracy requirements for scaling distances from the map are some additional considerations. Maps produced using the English system of units usually have their scales selected to be compatible with one of the standard graduations on engineer's scales. These standard graduations have 10, 20, 30, 40, 50, or 60 units per inch. Thus, scales of 1 in. = 100 ft and 1 in. = 1000 ft are compatible with the 10 scale; 1 in. = 200 ft and 1 in. = 2000 ft are consistent with the 20 scale, and so on. In the metric system, ratios or representative fractions such as 1:1000, 1:2000, 1:5000, and so on are usually employed.

Map scales may be classified as *large, medium,* and *small.* Their respective scale ranges are as follows:

Large scale, 1 in. = 200 ft (1:2400) or larger
Medium scale, 1 in. = 200 ft to 1 in. = 1000 ft (1:2400 to 1:12,000)
Small scale, 1 in. = 1000 ft (1:12,000) or smaller

Large-scale maps are applied where relatively high accuracy is needed over limited areas; for example, in subdivision design and the design of engineering projects like roads, dams, airports, and water and sewage systems. Medium scales are often used for applications such as general preliminary planning where larger areas are covered, but only moderate accuracy is needed. Applications include mapping the general layout of potential construction sites, proposed transportation systems, and existing facilities. Small-scale maps are commonly used for mapping large areas where a lower order of accuracy will suffice. They are suitable for general topographic coverage, applications in site suitability analysis, preliminary layout of expansive proposed construction projects, and for special applications in forestry, geology, environmental impact and management, etc.

Maps in graphic form can have their scales enlarged or reduced photographically or by converting the maps to digital form and enlarging or reducing by computer processing. The enlargement ratios possible by either of these methods are virtually unlimited. *However, enlargements must be produced with caution since any errors in the original maps or digital data are also magnified, and the enlarged product may not meet required accuracy standards.*

The scale at which a map will be plotted directly affects the choice of instruments and procedures used in performing the mapping survey. This is because the accuracy with which the position of an object is depicted on a map is related to the map's scale, which in turn dictates the accuracy with which features must be surveyed. Consider, for example, a map plotted at a scale of 1 in. = 20 ft. If distances and locations can be scaled from the map to within say 1/50th in., this represents a scaling error of $(1/50) \times 20 = \pm 0.4$ ft. To ensure that the accuracy of the surveyed data does not limit the accuracy with which information can be scaled from a map, features must be located on the map to an accuracy better than ±0.4 ft for this map. As a safety factor, many surveying and mapping agencies apply a rule of thumb in which they require features to be located in the field to at least twice the scaling accuracy, which in this instance would require accuracy to within ±0.2 ft or better. Following this same reasoning, if map scale is 1 in. = 200 ft, then ground features should be located to an accuracy of ±2 ft so as not to limit the accuracy of the map. Another consideration regarding map scale that affects surveying accuracy is the thicknesses of lines used to plot features. Assume, for example, that line widths on a map with a scale of 1:2000 are 0.3 mm. This means that each line represents $0.3 \times 2000 = 600$ mm = 0.6 m on the object. Therefore, to accurately depict an object on a map with this line width, the survey needs to be accurate to at least half the line width, or ±0.3 m. Obviously the equipment and procedures used for the mapping work must be selected so that these accuracies are met.

While scaling factors such as those discussed above must be taken into account for each specific mapping project, it is important to also consider the possible future use of the map data being collected. Thus even though the first map produced for a particular project may be a small-scale reconnaissance map, it is possible that as the project progresses, medium-scale planning maps and large-scale design maps will be needed, and that some or all of the data collected could also be used for these maps. Thus even though relaxed accuracies may suffice for the reconnaissance map, for efficiency, the data should be collected to accuracy suitable for the other maps that may follow.

■ 17.4 CONTROL FOR MAPPING SURVEYS

Whether the mapping is done by ground or aerial methods, the first requirement for any project is good control. As discussed in Chapter 19, control is classified as either horizontal or vertical.

Horizontal control for a mapping survey is provided by two or more points on the ground, permanently or semi-permanently monumented and precisely fixed in position horizontally by distance and direction or coordinates. It is the basis for locating map features. Horizontal control can be established by the traditional ground surveying methods of *traversing, triangulation,* or *trilateration* (see Section 19.12), or by using *GPS* (see Chapters 14 and 15). For large areas, a sparse network of horizontal (and vertical) control can be densified using *photogrammetry* (see Chapter 27). For small areas, horizontal control for mapping surveys is generally established by traversing. Until recently, triangulation and trilateration were the most economical procedures available for establishing basic control for mapping projects extending over large areas such as a state or the entire United States. These techniques have now given way to GPS, which is not only highly accurate but also very efficient. Monuments whose positions have been established through higher-order control surveys and referenced in the state plane coordinate systems (see Chapter 20) are used to initiate surveys of all types, but unfortunately more are needed in most areas.

Vertical control is provided by benchmarks in or near the tract to be surveyed and becomes the foundation for correctly portraying relief on a topographic map.

Vertical control is usually established by running lines of differential levels starting from and closing on established benchmarks (see Chapters 4 and 5). Project benchmarks are established throughout the mapping area in strategic locations and their elevations determined by including them as turning points in the differential leveling lines. In rugged areas, trigonometric leveling with total station instruments is practical and frequently used to establish vertical control for mapping. GPS surveys may also be suitable for establishing vertical control for topographic mapping but the ellipsoid heights determined must be converted to orthometric heights as described in Sections 13.4.3 and 19.5 first.

Regardless of the methods used in conducting the control surveys for mapping projects, specified maximum allowable closure errors for both horizontal and vertical control should be determined in advance of the fieldwork, then used to guide it. The locations of the features, which comprise the map (often also called *map details*), are based upon the framework of control points whose positions and elevations are established. Thus, any errors in the surveyed positions or elevations of the control points will result in erroneous locations of the details on the map. Therefore, it is advisable to run, check, and adjust the horizontal and vertical control surveys before beginning to locate map details, rather than carry on both processes simultaneously. The method selected for locating map details will govern the speed, cost, and efficiency of the survey. In later sections of this chapter, the different basic field procedures and the varying equipment that can be used are described.

17.5 CONTOURS

As stated earlier, surveyors and engineers most often use contours to depict relief. The reason is that they provide an accurate quantitative representation of the terrain. Because planimetric features and contours are located simultaneously in most field topographic surveys, it is important to understand contours and their characteristics before discussing the various field procedures used to position them.

A *contour is a line connecting points of equal elevation*. Since water assumes a level surface, the shoreline of a lake is a visible contour, but in general, contours cannot be seen in nature. On maps, contours represent the planimetric locations of the traces of level surfaces for different elevations (see the plan view of Figure 17.1). Contours are drawn on maps by interpolating between points whose positions and elevations have been observed and plotted. As noted earlier, computerized mapping and contouring systems are replacing manual plotting methods, but the principles of plotting terrain points and of interpolating contours are still basically the same in either method.

The vertical distance between consecutive level surfaces forming the contours on a map (the elevation difference represented between adjacent contours) is called the *contour interval*. For the small-scale U.S. Geological Survey *quadrangle maps* (plotted at 1:24,000 scale), depending upon the nature of the terrain one of the following contour intervals is used: 5, 10, 20, 40, or 80 ft. For larger-scale maps used in engineering design, in the English system of units contour intervals of 1, 2, 5, or 10 ft are commonly used. In the metric system, a contour interval of 0.5, 1, 2, 5, or 10 m is generally selected. Figure 17.2 is a topographic map having 10-ft contours.

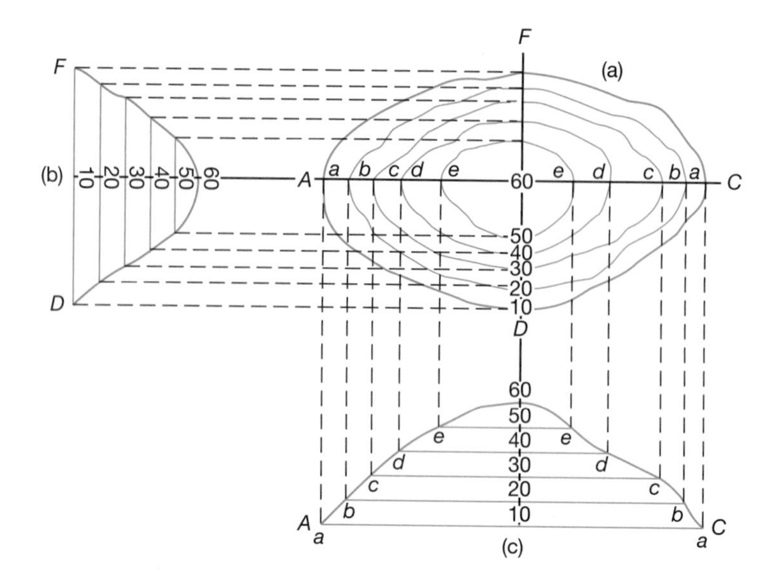

Figure 17.1
(a) Plan view of contour lines.
(b) and (c) Profile views.

The contour interval selected depends on a map's purpose and scale, and upon the diversity of relief in the area. As examples, on a map to be used for designing the streets and water and sewer systems for a subdivision, a contour interval of 1 or 2 ft would perhaps be necessary, whereas a 10- or 20-ft contour interval may be suitable for mapping a large ravine to determine the reservoir capacity that would result from constructing a dam. Also, a smaller contour interval will normally be necessary to adequately depict gently rolling terrain with only moderate elevation differences, while rugged areas with large elevation differences normally require a larger contour interval so that the contours do not become too congested on the map. In general, reducing the contour interval requires more costly and precise fieldwork. In regions where both flat coastal areas and mountainous terrain are included in a map, supplementary contours, at one-half or one-fourth the basic contour interval, are often drawn (and shown with dashed lines).

Spot elevations are used on maps to mark unique or critical points such as peaks, potholes, valleys, streams, and highway crossings. They may also be used in lieu of contours for defining elevations on relatively flat terrain that extends over a large area.

Topographic mapping convention calls for drawing only those contours that are evenly divisible by the contour interval. Thus for the 10-ft contour interval on the map in Figure 17.2, contours such as the 1100, 1110, 1120, and 1130 are shown. Elevations are shown in breaks in the contour lines, and to avoid confusion, at least every fifth contour is labeled. To aid in reading topographic maps, every fifth contour (each that is evenly divisible by five times the contour interval) is drawn using a heavier line. Thus in Figure 17.2, the 1100, 1150, 1200, and so on contours are drawn more heavily.

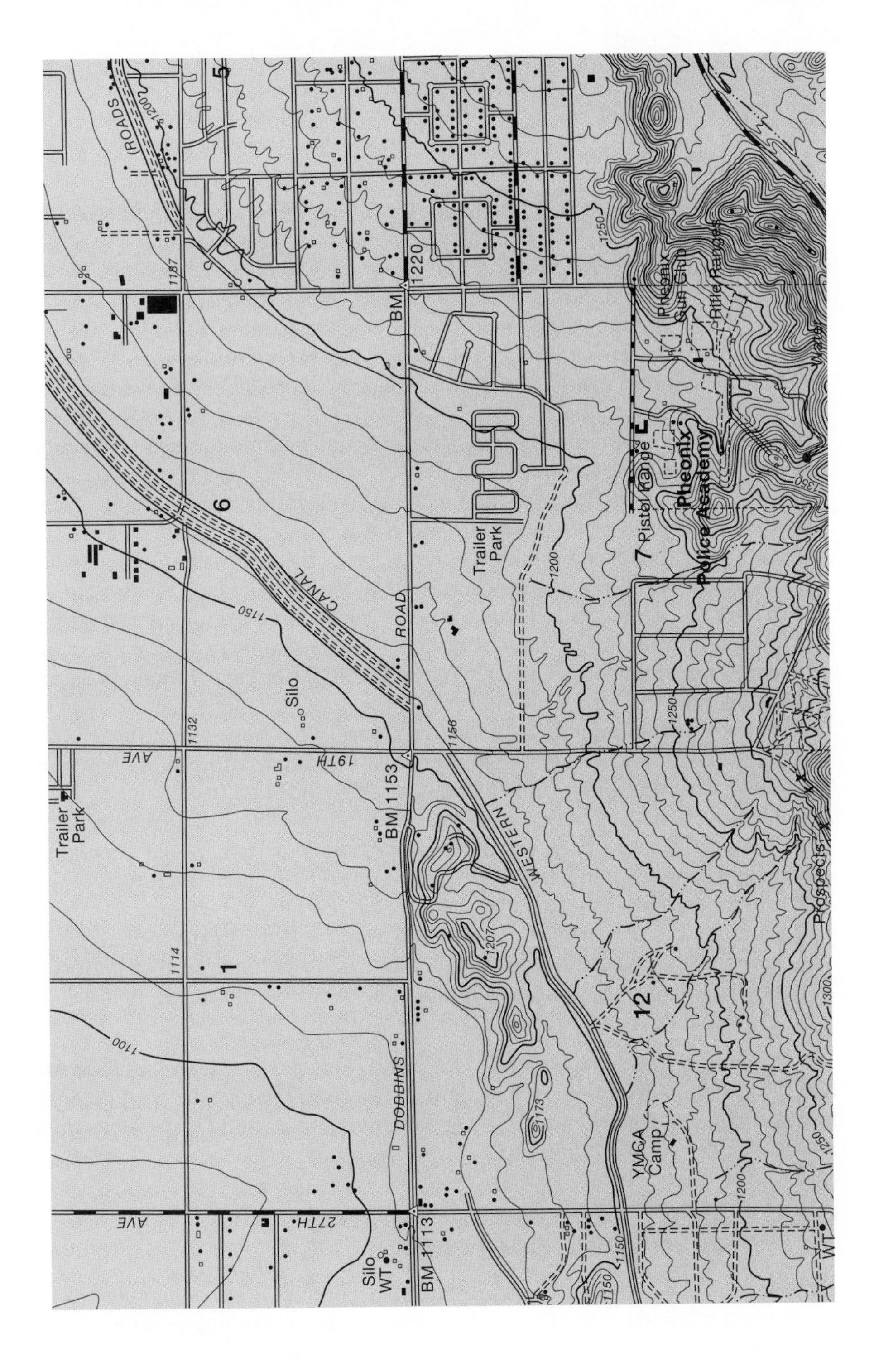

Figure 17.2 Part of U.S.G.S. Lone Butte quadrangle map. (Courtesy U.S. Geological Society)

■ 17.6 CHARACTERISTICS OF CONTOURS

Although each contour line in nature has a unique shape, all contours adhere to a set of general characteristics. Important ones, fundamental to their proper field location and correct plotting, are listed.

1. Contour lines must close on themselves, either on or off a map. They cannot dead-end.
2. Contours are perpendicular to the direction of maximum slope.
3. The slope between adjacent contour lines is assumed to be uniform. (Thus, it is necessary that breaks [changes] in grade be located in topographic surveys.)
4. The distance between contours indicates the steepness of a slope. Wide separation denotes gentle slopes; close spacing, steep slopes; even and parallel spacing, uniform slope.
5. Irregular contours signify rough, rugged country. Smooth lines imply more uniformly rolling terrain.
6. Concentric closed contours that increase in elevation represent hills. A contour forming a closed loop around lower ground is called a depression contour. (Spot elevations and hachures inside the lowest contour and pointing to the bottom of a hole or sink with no outlet make map reading easier.)
7. Contours of different elevations never meet except on a vertical surface such as a wall, cliff, or natural bridge. They cross only in the rare case of a cave or overhanging shelf. Knife-edge conditions are never found in natural formations.
8. A contour cannot branch into two contours of the same elevation.
9. Contour lines crossing a stream point upstream and form V's; they point down the ridge and form U's when crossing a ridge crest.
10. Contour lines go in pairs up valleys and along the sides of ridge tops.
11. A single contour of a given elevation cannot exist between two equal-height contours of higher or lower elevation. For example, an 820-ft contour cannot exist alone between two 810- or two 830-ft contours.
12. Cuts and fills for earth dams, levees, highways, railroads, canals, etc., produce straight or geometrically curved contour lines with uniform, or uniformly graduated, spacing. Contours cross sloping or crowned streets in typical V- or U-shaped lines.

Keeping these characteristics in mind will (1) make it easier to visualize contours when looking at an area, (2) assist in selecting the best array of points to locate in the field when conducting a topographic survey, and (3) prevent serious mistakes in drawing contours.

■ 17.7 DIRECT AND INDIRECT METHODS OF LOCATING CONTOURS

Contours can be established by either the direct method (*trace-contour method*) or the indirect method (*controlling-point method*). The controlling-point method is generally more convenient and faster, and therefore it is most often selected. It is also the most frequent choice when data will be entered into a computer for automated contouring. These two methods are described in the subsections that follow.

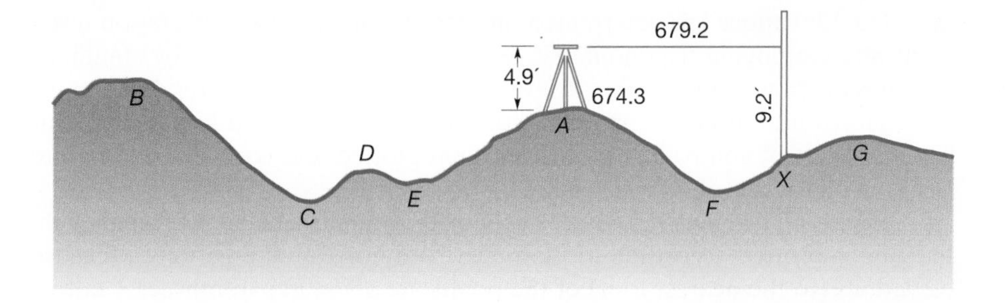

Figure 17.3 Direct method of locating contours.

17.7.1 Direct Method

This method can be performed using a total station instrument. After the instrument is set up, the *HI* is established, and the telescope oriented horizontally. Then for the existing *HI*, the rod reading (foresight) that must be subtracted to give a specific contour elevation is determined. The rodperson selects trial points expected to give this minus sight and is directed uphill or downhill by the instrument operator until the required reading is actually secured (to within perhaps 0.1 to 0.5 ft, depending upon the allowable discrepancy which is dictated by the nature of the terrain, contour interval, and specified map accuracy).

In Figure 17.3 the instrument is set up at point *A*, elevation 674.3 ft, *hi* 4.9 ft, and *HI* 679.2 ft. If 5-ft contours are being located, a reading of 4.2 or 9.2 with the telescope level will place the rod on a contour point. For example, in Figure 17.3, the 9.2-ft rod reading means that point *X* lies on the 670-ft contour. After the point, which gives the required rod reading, has been located by trial, the horizontal position of the point is determined by observing its horizontal distance and direction from the instrument. Distances are observed electronically with total station instruments (see Chapter 6, Part III), but stadia (see Section 17.9.2) or taping (see Chapter 6, Part II) can be used. The process is repeated until the entire area has been surveyed. Work is speeded by using a piece of red plastic flagging which can be moved up and down on the level rod to mark the required reading and eliminate searching for a number. The maximum distance between contour points located in this method is determined by the terrain and accuracy required. Beginners have a tendency to take more sights than necessary in ordinary terrain. Contours are sketched by connecting located points having equal elevations. This is usually done as part of the office work, but they may also be drawn in the field book to clarify unusual conditions.

The direct method is suitable in gently rolling country, but generally not practical in undulating or rough, rugged terrain. Neither is it convenient for obtaining data to be used in computer-driven automated contouring systems.

17.7.2 Indirect Method

In the indirect method, the rod is set on "controlling points" that are critical to the proper definition of the topography. They include high and low points on the terrain, and locations where changes in ground slope occur, such as *B*, *C*, *D*, *E*, *F*, and *G* in Figure 17.3. Channels of drainage features and ridge lines must be included. Elevations are determined on these points using a total station instrument and employing trigonometric leveling (see Section 4.5.4), by stadia, rod readings taken with the telescope level, or by kinematic GPS methods when possible. Horizontal distances and azimuths are also observed to locate the points. The positions of controlling points are then plotted and contours interpolated between elevations of adjacent points.

Figure 17.4(a) illustrates a set of controlling points labeled A through N that have been plotted according to their surveyed horizontal positions. Measured elevations (to the nearest foot) of the points are given in parentheses. Contours having a 10-ft interval have been sketched freehand between adjacent points by interpolation. It is improper to interpolate along lines that cross controlling features such as gullies, streams, rivers, ridge lines, roads, etc. Thus to properly draw the contours of Figure 17.4(a) with the stream located on the map, elevations were first interpolated along its thread between surveyed points E, G, I, and J. Interpolations were then made from the stream to points on either side. As an example, it would have been incorrect to interpolate across the stream between points D and F. Rather the elevation of the stream on the line between D and F (approximately 9 ft) was used to interpolate both ways from the stream to points D and F. Note in Figure 17.4(a) that the gently curved contours tend to duplicate the naturally rolling topography of nature. Note also that contours crossing the stream form V's pointing upstream.

Numerous controlling points may be needed to locate a contour in certain types of terrain. For example, in the unusual case of a nearly level field that is at or close to a contour elevation, the exact location of that contour would be time consuming or perhaps impossible to determine. In these situations, a uniform distribution of *spot elevations* can be determined in the field and plotted on the map to convey the area's relief.

■ 17.8 DIGITAL ELEVATION MODELS AND AUTOMATED CONTOURING SYSTEMS

Data for use in automated contouring systems is collected in arrays of points whose horizontal positions are given by their X and Y coordinates and whose elevations are given as Z coordinates. Such three-dimensional arrays provide a *digital* representation of the continuous variation of relief over an area and are known as *digital elevation models* (DEMs). Alternatively, the term *digital terrain model* (DTM) is sometimes used.

Two basic geometric configurations are normally used in the field for collecting DEM data: the *grid method,* and the *irregular method,* although often a combination of the two methods is employed. In the grid method, elevations are determined on points that conform to a regular square or rectangular grid. The procedure is described in Section 17.9.3 and sample field notes are given in Plate B.2 of Appendix B. From the array of grid data, the computer interpolates between points along the grid lines to locate contour points and then draws the contour lines. The major disadvantage of this method is that critical high and low points and changes in slope do not generally occur at the grid intersections, and thus they are missed in the data collection process, which results in inaccurate relief portrayal.

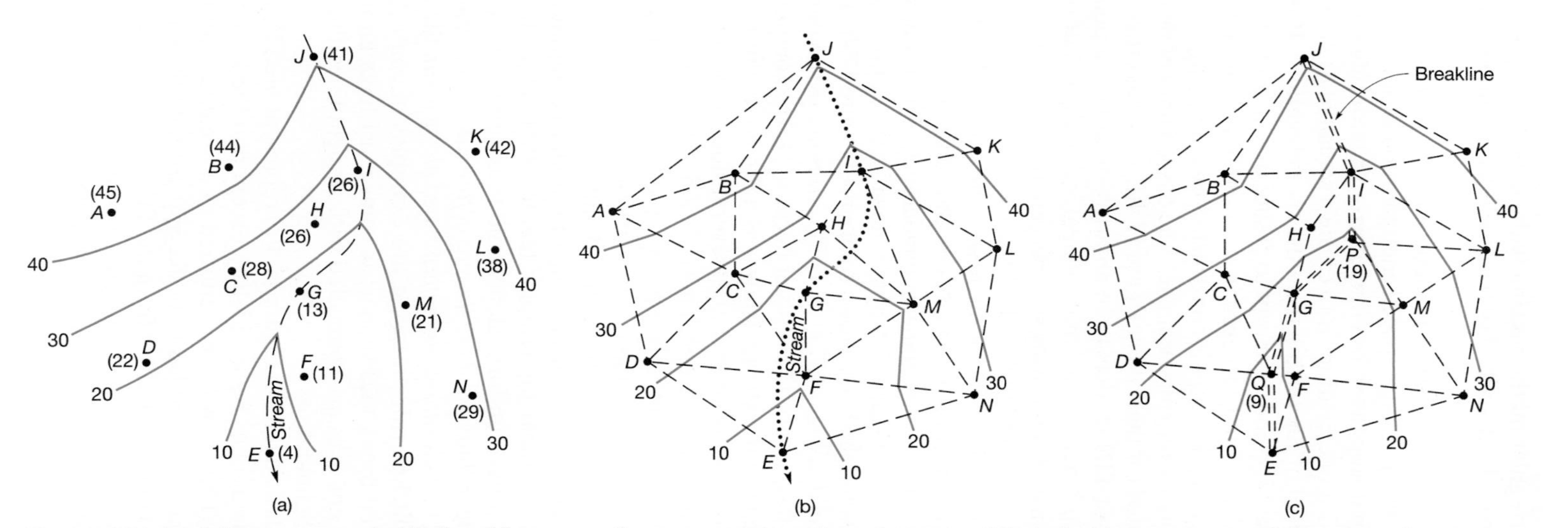

Figure 17.4 (a) Contours compiled by hand from controlling points *A* through *N*. (b) TIN model (dashed lines) constructed from data of (a), and contours (solid lines) derived from TIN model. The stream is shown with a dotted line. Note the striking differences between the 10- and 20-ft contours of (a) and (b). (c) TIN model (dashed lines) constructed from data of (a) but with the addition of two points, *P* and *Q*, and the designation of lines *EQ*, *QG*, *GP*, *PI*, and *IJ* as breaklines. Contours shown with solid lines were derived from this TIN model. Note the agreement of these contours with those of (a).

The irregular method is simply the controlling-point method, but additional information (to be described later) is included. As previously noted, the controlling-point method involves determining the elevations of all high and low points and points where slopes change (called *breaks in grade*). Of course, this produces a DEM with an irregularly spaced configuration of surveyed points.

The first step taken by computerized contouring systems that utilize irregularly spaced spot elevations is to create a so-called *triangulated irregular network,* or TIN, model of the terrain from the spot elevations. It is very important to understand the TIN model concept to ensure that an appropriate array of controlling points is selected and surveyed in the field if an automated contouring system is to be used. A TIN model is constructed by connecting points in the array to create a network of adjoining triangles. The dashed lines of Figure 17.4(b) shows a TIN model created for the data in Figure 17.4(a). Various criteria can be used in the development of TIN models from an array of surveyed points, but one commonly used standard creates the "most equilateral network" of triangles.

Automated contouring systems generally make two assumptions concerning TIN models: (1) all triangle sides have a constant slope, and (2) the surface area of any triangle is a plane. Based on these assumptions, elevations of contour crossings are interpolated along triangle edges, and contours are constructed such that they change direction only at triangle boundaries. Contours derived in this manner from the TIN model of Figure 17.4(b) are shown in the figure as solid lines. Note the disparities between the hand-drawn contours of Figure 17.4(a) and those derived from the TIN model of Figure 17.4(b). Differences are particularly obvious between the 10- and 20-ft contours. These occur because (1) the computer did not interpret the curved thread of the stream [shown as a dotted line in Figure 17.4(b)], and (2) in creating the network of triangles, several sides were constructed which cross the stream, resulting in improper interpolation across the stream.

From this example it is apparent, as noted earlier, that additional information must be provided for computer-driven systems to depict contours accurately. That important added information is the identification of controlling features, also more often called *breaklines,* or *fault lines* in modern computer mapping terminology. Breaklines are linear topographic features that delineate the intersection of two surfaces that have uniform slopes, and thus define changes in grade. *Automated mapping algorithms use these lines to define sides of the triangles that form the TIN model, and thus elevations are interpolated along them.* Streams, lake shores, roads, railroads, ditches, ridgelines, etc. are examples of controlling features or breaklines. Curved breaklines such as streams must have enough data points so that when adjacent ones are connected with straight lines, they adequately define the feature's alignment.

The dashed lines of Figure 17.4(c) represent the TIN model constructed from the same data set as Figure 17.4(b), except that the stream (shown with a double dashed line) has now been identified as a breakline, and two additional points, P and Q, have been added to better approximate the curvature of the stream. In this figure, contours derived from the TIN model are shown. Note that these now very nearly duplicate the hand-drawn contours.

The important lesson of the foregoing is that if an automated contouring system is used, field points must be selected carefully, breaklines identified, and the data properly input to meet the system's assumptions. As indicated by this example, a few more controlling points may have to be surveyed, but the benefits of automated contouring systems make it worthwhile.

In order to avoid missing significant data during topographic surveys, it is usually best to collect features in groups. That is, data should be gathered first for (1) planimetric features, followed by (2) breaklines, (3) significant controlling points of elevation, and finally (4) sufficient *grade points* (those remaining points surveyed only to enable accurate depictions to be made of slopes and grades between the other types of points). Grade points are often most efficiently collected in a grid pattern throughout the entire area to be mapped. This grid should be sufficiently dense to avoid triangles in the TIN that are geometrically weak; i.e., long and slender figures with one small angle. Varying grid sizes can be used, with larger spacing applied in areas of gradual slopes, and more dense patterns employed as the terrain becomes more undulating.

17.9 BASIC FIELD METHODS FOR LOCATING TOPOGRAPHIC DETAILS

Objects to be located in a mapping survey can range from single points or lines to meandering streams and complicated geological formations. The process of tying mapping details to the control net is called *detailing.* Regardless of their shape, all objects can be located by considering them as composed of a series of connected straight lines, with each line being determined by two points. Irregular or curved lines can be assumed straight between points sufficiently close together; thus detailing becomes a process of locating points.

Location of planimetric features and contours can be accomplished by one of the following field procedures: (1) radiation by total station instrument, (2) radiation by stadia, (3) coordinate squares or "grid" method, (4) offsets from a reference line, (5) use of portable GPS units, or (6) a combination of these methods. An explanation of each system and a discussion of their uses, advantages, and disadvantages follow.

17.9.1 Radiation by Total Station

In the radiation method, illustrated in Figure 17.5, with a total station instrument set up at a control point, the zenith angle, slope distance, and direction are observed to each desired item of mapping detail. From the zenith angle and slope distance, the elevation of the point can be determined, and by incorporating the direction, its horizontal position can be computed. As shown in the figure, the sights to details radiate from the occupied station, hence the name for the procedure. This method is especially efficient if a data collector (see Sections 2.12 through 2.15) is used to record the point identities and their associated descriptions, vertical distances, horizontal distances, and directions. The data collector permits downloading the observations directly into a computer for processing through an automated mapping system. The field procedure of radiation with a total station can be made most efficient if the instrument is placed at a good vantage point (on a hill or ridge) that overlooks a large part or all of the area to be surveyed. This permits more and longer radial lines and reduces the number of setups required.

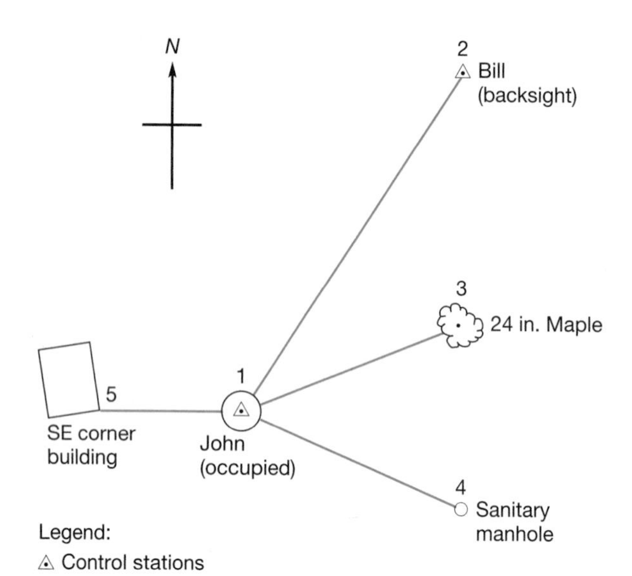

Figure 17.5 Radiation survey from control traverse.

Table 17.1 is an example illustrating the use of a total station with data collector for topographic mapping. The example relates to Figure 17.5. In Figure 17.5, a total station instrument was set up at control station 1 (John) and oriented in azimuth with a backsight on control station 2 (Bill). Observations of azimuth, zenith angle, and distance, respectively, were then taken to points 3, 4, and 5, which are *sideshots* to mapping details. From these sideshots, two- or three-dimensional coordinates can be computed that are used to locate the points on a map sheet.

When using a data collector in this process, initial data for the setup are first entered in the unit via the keyboard. These include the X and Y coordinates of stations John and Bill, the elevation of John, and the heights of both the total station instrument and the reflector. The left-hand column of Table 17.1 illustrates an actual set of field notes recorded by a data collector during the process of taking sideshots on points 3, 4, and 5. Six entries were recorded per point. On each line, the entry to the left of the colon was automatically supplied by the computer and appeared on the data collector display at the time of observation to prompt the operator. Entries to the right of the colon were supplied by the operator, either manually via the keyboard or by pressing the proper button on the total station instrument. Explanations to assist students in interpreting the data in Table 17.1 are given in parentheses.

As shown in Figure 17.6, details that have width such as trees are located with two separate observations. The first observation locates the azimuth to the object by observing an angle from a reference line to the front center of the object. The second shot measures the distance to the side center of the object. Using the azimuth of the first shot and the distance from the second shot, coordinates near the center of the object can be determined. Data collectors have various names for this collection routine such as *separate distance and angle* (SDA). This procedure should only be used when the diameter of the object is sufficiently large to cause a plotting error on the map. For smaller objects where the diameter will not

TABLE 17.1 EXCERPT OF AUTOMATIC DATA COLLECTOR FIELD NOTES OF A RADIAL SURVEY FOR TOPOGRAPHIC DETAILS

Entry	Explanation
AC:SS	(Activity: Sideshot/keyboard entry by operator)
PN:3	(Point number: 3/keyboard entry by operator)
PD:24 IN MAPLE	(Point description: 24-in. maple/keyboard entry)
HZ:16.3744	(Horizontal angle: 16°37′44″/by total station)
VT:90.2550	(Vertical "zenith" angle: 90°25′50″/by total station)
DS:565.855	(Distance: 565.855 ft/by total station)
AC:SS	(Activity: Sideshot/keyboard entry by operator)
PN:4	(Point number: 4/keyboard entry by operator)
PD:SAN MH	(Point description: Sanitary manhole/keyboard entry)
HZ:70.3524	(Horizontal angle: 70°35′24″/by total station)
VT:91.1548	(Vertical "zenith" angle: 91°15′48″/by total station)
DS:463.472	(Distance: 436.472 ft/by total station)
AC:SS	(Activity: Sideshot/keyboard entry by operator)
PN:5	(Point number: 4/keyboard entry by operator)
PD:SE COR BLDG	(Point description: SE corner building/keyboard entry)
HZ:225.1422	(Horizontal angle: 225°14′22″/by total station)
VT:88.3036	(Vertical "zenith" angle: 88°30′36″/by total station)
DS:265.934	(Distance: 265.934 ft/by total station)

Source: Courtesy ABACUS, A Division of Calculus, Inc.

noticeably displace the center of the object on the map, this procedure is unnecessary. Thus, the use of this method is dependent on the scale of the map and size of the object.

17.9.2 Radiation by Stadia

Until the advent of total station instruments and GPS, the stadia method, (referred to as *tacheometry* in Europe), was the method of choice for performing mapping surveys. This procedure enabled surveys to be completed relatively rapidly and efficiently, and it provided accuracy suitable for locating topographic details. (Stadia can also be applied in lower-order trigonometric leveling, observing lengths of backsights and foresights in differential leveling, and for making quick checks of observations made by higher-order methods.) Stadia observations can be taken with theodolites, transits and plane-table alidades (see the 8th edition of this book for a description of plane-table equipment). Now total station instruments and new portable GPS receivers have relegated stadia to the background. The method still can be used for topographic mapping, however, and therefore is discussed here.

Figure 17.6
Proper location of objects such as trees.

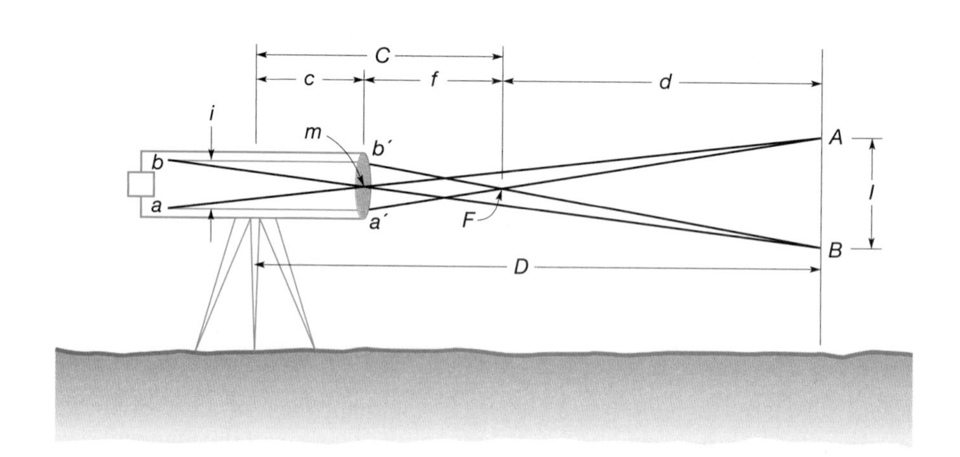

Figure 17.7
Principle of stadia.

Stadia radiation is similar to radiation by total station, except that stadia determine horizontal distances to points and their elevations. The method is based on the principle that in similar triangles, corresponding sides are proportional. In Figure 17.7, which depicts a telescope with a simple lens, light rays from points A and B pass through the lens center and form a pair of similar triangles AmB and amb. Here $AB = I$ is the rod intercept (stadia interval), and $ab = i$ is the spacing between stadia hairs.

Standard symbols used in stadia observations and their definitions are as follows (refer to Figure 17.7):

f = *focal length* of lens (a constant for any particular compound objective lens)
i = spacing between stadia hairs (ab in Figure 17.7)

f/i = *stadia interval factor*, usually 100 and denoted by K
I = rod intercept (AB in Figure 17.7), also called *stadia interval*
c = distance from instrument center (vertical axis) to objective lens center (varies slightly when focusing the objective lens for different sight lengths but is generally considered to be a constant)
C = stadia constant $= c + f$
d = distance from focal point F in front of telescope to face of rod
D = distance from instrument center to rod face $= C + d$

From similar triangles of Figure 17.7,

$$\frac{d}{f} = \frac{I}{i} \text{ or } d = \frac{f}{i}I = KI$$

Thus

$$D = KI + C \tag{17.1}$$

The geometry illustrated in Figure 17.7 pertains to a simplified type of *external focusing telescope.* It has been used because an uncomplicated drawing correctly shows the relationships and aids in deriving the stadia equation. These telescopes are now obsolete in surveying instrumentation. The objective lens of an *internal focusing telescope* (the type now used in surveying instruments) remains fixed in position, while a movable negative-focusing lens between the objective lens and the plane of the crosshairs changes directions of the light rays. As a result, the stadia constant, (C), is so small that it can be assumed equal to zero and drops out of Equation (17.1). Thus the equation for distance on a horizontal stadia sight reduces to[1]

$$D = KI \tag{17.2}$$

Fixed stadia lines in theodolites, transits, levels, and alidades are generally spaced by instrument manufacturers to make the stadia interval factor $f/i = K$ equal to 100. It should be determined the first time an instrument is used, although the manufacturer's specific value posted inside the carrying case will not change unless the crosshairs, reticle, or lenses are replaced or adjusted.

To determine the stadia interval factor K, rod intercept I for a horizontal sight of known distance D is read. Then in an alternate form of Equation 17.2, the stadia interval factor is $K = D/i$. As an example, at a measured distance of 300.0 ft, a rod interval of 3.01 was read. Then $K = 300.0/3.01 = 99.7$. Accuracy in determining K

[1]A few older transits with external focusing telescopes are still in service. They can be recognized because the objective lens is seen to move as the telescope is focused for varying lengths of sights. To determine the constant C for one of these instruments, first focus for a long sight, say 1000 ft, and measure f, the distance from the objective lens to the reticle ring. Then focus for an average sight, about 200 ft, and measure c the distance from the objective lens to the center of the instrument. The sum $f + c = C$, normally about 1 ft, is included in the stadia equations for these older external focusing instruments.

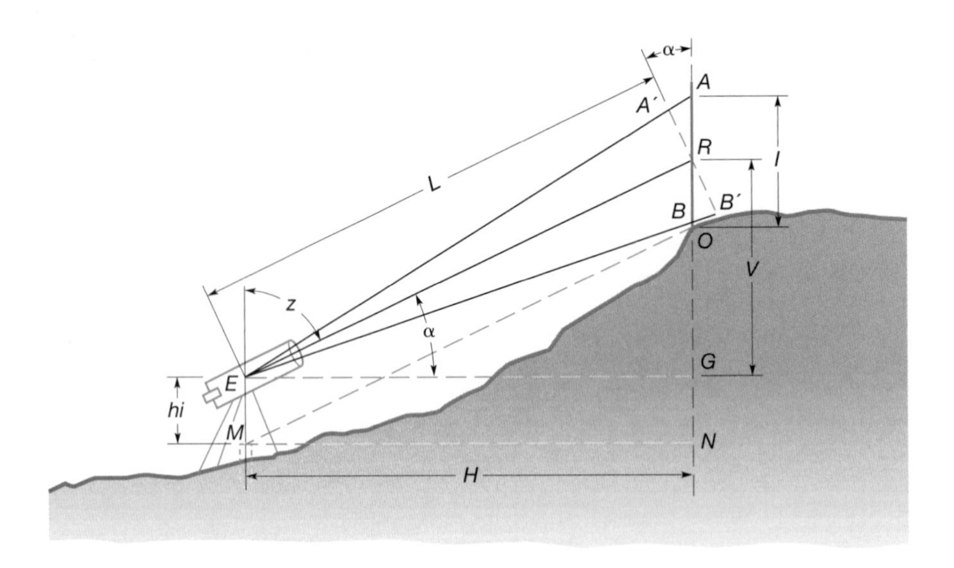

Figure 17.8 Inclined stadia measurement.

is increased by averaging values from several lines whose measured lengths vary from about 100 to 500 ft by 100 ft increments.

Most stadia shots are inclined because of varying topography, but the intercept is read on a *plumbed rod* and the slope length "reduced" to horizontal and vertical distances.

In Figure 17.8 an instrument is set over point M and the rod held at O. With the middle crosshair set on point R to make RO equal to the height of the instrument $EM = hi$, the vertical angle (angle of inclination) is α. Note that in stadia work the height of instrument *hi* is again defined as the height of the line of sight above the point occupied (*not HI,* the height above datum as in leveling).

Let L represent the slope length ER, H the horizontal distance $EG = MN$, and V the vertical distance $RG = ON$. Then

$$H = L\cos\alpha \tag{a}$$

$$V = L\sin\alpha \tag{b}$$

If the rod could be held normal to the line of sight at point O a reading $A'B'$, or I', would be obtained, making

$$L = KI' \tag{c}$$

Since it is not practical to hold the rod at an inclination angle α, it is plumbed and reading AB, or I, taken. For the small angle at R on most sights, it is sufficiently accurate to consider angle $AA'R$ as a right angle. Therefore

$$I' = I\cos\alpha \tag{d}$$

and substituting (d) into (c),

$$L = KI\cos\alpha \tag{e}$$

Finally, substituting (e) into (a), the equation for horizontal distance on an inclined stadia sight is

$$H = KI \cos^2 \alpha \tag{17.3a}$$

If zenith angles are read rather than vertical angles, then the horizontal distance is given by

$$H = KI \sin^2 z \tag{17.3b}$$

where z is the zenith angle, equal to $90° - \alpha$.

The vertical distance is found by substituting (e) into (b), which gives

$$V = KI \sin \alpha \cos \alpha \tag{17.4a}$$

or

$$V = KI \sin z \cos z \tag{17.4b}$$

In the final form generally used, K is assigned 100, and the formulas for the reduction of inclined sights to horizontal and vertical distances reduce to

$$H = 100\, I \cos^2 \alpha \tag{17.5a}$$

and

$$H = 100\, I \sin^2 z \tag{17.5b}$$

or

$$V = 100\, I \sin \alpha \cos \alpha \tag{17.6a}$$

and

$$V = 100\, I \sin z \cos z \tag{17.6b}$$

From Figure 17.8, the elevation of point O is

$$elev_O = elev_M + hi + V - R \tag{17.7}$$

From Equation 17.7, the advantage of sighting an R value that is equal to the hi when reading the vertical (or zenith) angle is evident. Since the rod reading and hi are opposite in sign, if equal in magnitude they cancel each other and can be omitted from the elevation computation. If the rod is not visible at the hi because of obstructions, any rod reading can be sighted and Equation 17.7 used.

17.9.3 Coordinate Squares or "Grid" Method

The method of coordinate squares (grid method) is better adapted to locating contours than planimetric features, but can be used for both. The area to be surveyed is staked in squares 10, 20, 50, or 100 ft (5, 10, 20, or 40 m) on a side, the size

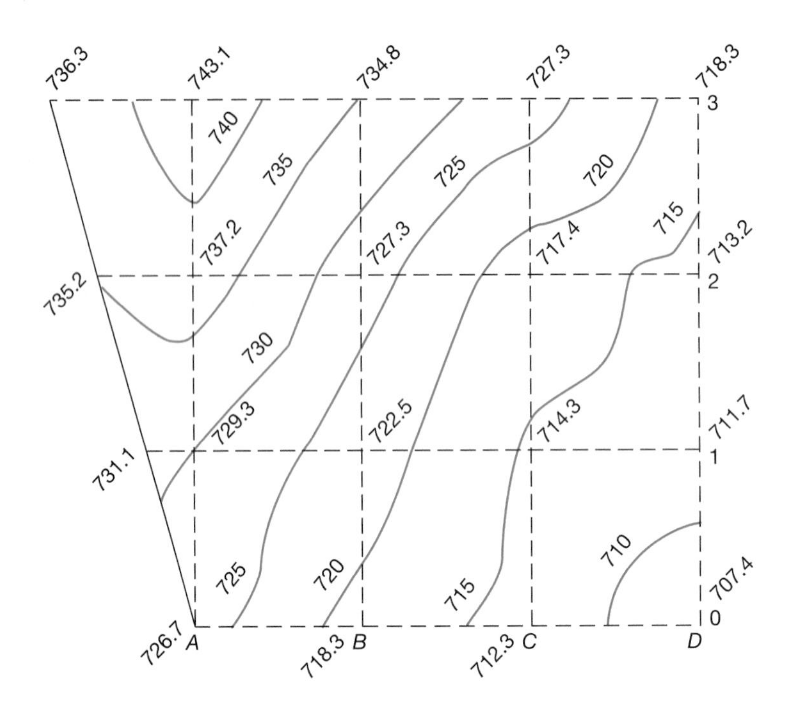

Figure 17.9
Coordinate squares.

depending on the terrain and accuracy required. A total station instrument can be used to lay out control lines at right angles to each other, such as *AD* and *D*3 in Figure 17.9. Grid lengths are marked and the other corners staked and identified by the number and letter of intersecting lines.

Elevations of the corners can be obtained by differential or trigonometric leveling. Contours are interpolated between the corner elevations (along the sides of the blocks) by estimation or by calculated proportional distances. Elevations obtained by interpolation along the diagonals will generally not agree with those from interpolation along the four sides because the ground's surface is not a plane. Except for plotting contours, this is the same procedure as that used in the borrow-pit problem in Section 26.10. In plotting contours by the grid method, a widely spaced grid can be used for gently sloping areas, but it must be made denser for areas where the relief is rolling or rugged.

A drawback of the method is that no matter how dense the grid, critical points (high and low spots and slope changes) will not generally occur at grid locations, thus degrading the accuracy of the resulting contour map.

17.9.4 Offsets from a Reference Line

This procedure is most often selected for mapping long linear features and for performing surveys necessary for route locations. After the reference line or centerline has been staked and stationed, planimetric details are located by observing perpendicular offsets from it and noting the stations from which the offsets were taken. Features such as streams, trails, fences, buildings, utilities, trees, etc., can be located in this manner. Elevations for determining contour locations

Figure 17.10 Double pentagonal prism (on side) for establishing perpendiculars to a reference line. (Courtesy Leica Geosystems, Inc.)

can also be determined by *cross-sectioning* (observing ground profiles along lines perpendicular to the reference line) as described in Section 26.3.

Shorter offset distances are generally most easily and quickly observed by taping, but longer lengths may be more efficiently obtained by electronic distance measurement using a total station instrument. Where steep slopes run transverse to the reference line, better accuracy and efficiency can often be obtained using a total station instrument. Perpendiculars to a reference line can be quickly established using a *pentagonal prism* (see Figure 17.10). This device works well when offset distances are being taped. While standing on the reference line at a location where a perpendicular is desired, the operator holds the instrument upright and looks through it sighting either ahead or back on the reference line. By means of the prisms, perpendicular views both left and right can be simultaneously seen. Alternatively, the total station instrument can be set up on the reference line at the location of a desired offset, oriented by sighting ahead or back on the line, and then used to turn a 90° angle.

Figures 17.11(a) and (b) illustrate examples of using offsets for mapping. In Figures 17.11(a), the location of Crooked Creek was determined by observing distances to the edge of the creek at intervals from the reference line. The offsets can be taken at regular intervals as illustrated in the figure, or they can be spaced at distances that permit the curved nature of the stream to be considered as a series of straight segments between successive offsets. Figure 17.11(b) illustrates an example of locating planimetric features along a road right-of-way. This type of survey would be useful to locate buildings, utilities, trees and other features along a road for highway design, or for excavation to install an underground utility. After locating at least two corners of a building by observing their offset distances, e.g., x_1 and x_2, in the figure, and their pluses on the reference line, its remaining dimensions can usually be quickly obtained by taping, e.g., a and b in the figure.

In both of these examples, it would be convenient to include a sketch in the field book, and to record the observations directly on the sketch. Since data collected by this method are based purely on distances, it is difficult to merge them with data collected by the radiation method.

17.9.5 Topographic Detailing with GPS

GPS receivers for topographic work are specially designed, small, and portable, and are interfaced with a keyboard for system control and entry of codes to identify

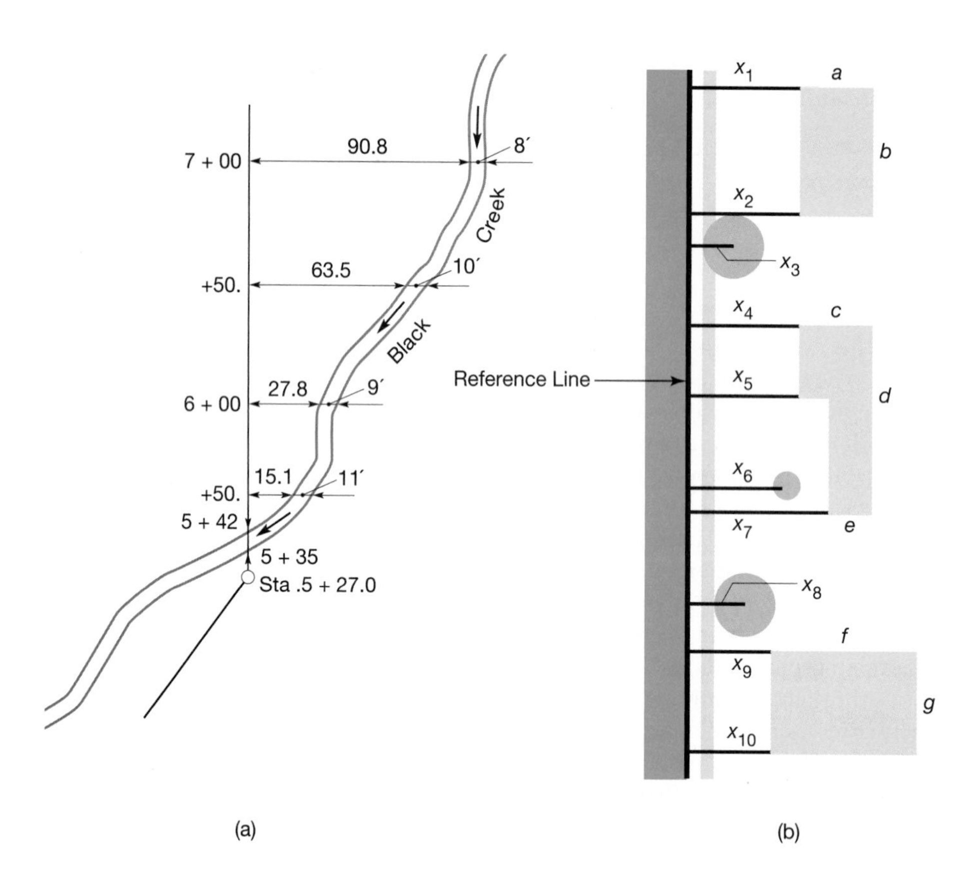

Figure 17.11 (a) Location of creek by perpendicular offsets from a reference line. (b) Locating objects by offsets from a reference line.

features surveyed. The unit shown in Figures 14. 1 is suitable for topographic work. These receivers can determine (in real-time) the coordinates of locations where the receiver antenna is placed and can store data for the points in files. The files are then directly downloaded to a computer for further processing, which could include automatic map drafting. These systems make topographic data collection a simple and very fast one-person operation. The kinematic GPS procedures discussed in Chapter 15 are most often used for mapping surveys.

The stop-and-go method (semikinematic) has the advantage over the kinematic method in that the operator can stop and collect multiple epochs of data for a point to increase positional accuracies. The stop-and-go method generally results in a smaller file size than the true kinematic methods. Additionally the stop-and-go method can be used to locate features or establish lower-order control points for occupation or sighting by conventional instruments such as total stations.

Kinematic surveys can collect data points at an operator-selected epoch (usually 1–5 sec interval), and thus can be used to quickly obtain cross-sections and locate linear features such as breaklines, roads, and streams. However since the operator has no control over the actual instance of data collection, this method does not provide a convenient means for surveying key topographical and planimetric features. However, many data collectors support collecting data at a specified 2d or 3d distance from the previous location. These options allow the user to collect data in a regularly spaced grid and are useful for collecting topographic detail.

During a mapping survey, it is possible to switch from a kinematic to semi-kinematic, or static survey (see Chapters 14 and 15). Therefore, if the conditions within a mapping project change, the operator can select the survey method that is most appropriate for the task at hand. Regardless of the method selected, it is essential that the antennas have clear visibility to the satellites. Thus GPS may not be suitable for direct location of large trees, buildings, or other objects that could obscure the view of the satellites. In these canopy situations, nearby temporary control stations can be established using a static GPS survey and traditional data collection methods with total stations used to locate these map details. However, when the modernized GPS constellation is implemented, this limitation is expected to disappear due to the stronger signals.

As discussed in Chapter 2, some data collectors work with GPS receivers as well as total station instruments. However, since orthometric heights are normally required for the vertical component of a survey, it is important to transform GPS-derived ellipsoid heights to their orthometric values using procedures discussed in Sections 13.4.3 and 19.5. Typically, GPS reduction software provides an option to perform this conversion.

Code-based GPS receivers can also be applied in topographic mapping, but their use is generally limited to lower-order work. These receivers are very affordable and currently have post-processed accuracies of fewer than 5 m. Even though this is a relatively high positional uncertainty, these units can be used to do surveys for maps with scales smaller than 1:20,000 since the plotting errors become negligibly small. As additional channels are added to the GPS satellites, it is anticipated that the accuracies of code-based GPS will be reduced to the meter-level and possibly the decimeter level.

17.9.6 Laser-Scanning

Laser scanners automate digital angle observation with reflectorless, laser EDM technologies. They can quickly produce grids of three-dimensional coordinates for user-specified scenes. A combination of rotating mirrors allows the instruments to observe distances and orthogonal angles in precise grid patterns. These instruments vary in capabilities to match the varying requirements of jobs. Characteristics of an instrument are defined by the *number of observations* they can observe per second, the observable distance from the instrument known as its *range,* the minimum spacing between observations or its *resolution, accuracy,* and *field of view*. In general, the higher the number of observations per second, the faster the data acquisition. Instruments vary from less than a hundred observations per second to 500,000 observations per second. The range of instruments can vary from several meters to several kilometers. Instrument resolutions can vary from a few millimeters to a several centimeters. Many ground-based instruments can produce finer resolutions on targets. However it must be remembered that the higher the resolution, the larger the data files. Generally, ground-based laser scanners have distance measurement accuracies of a few millimeters. The Leica HDS3000 shown in Figure 1.5 has a range accuracy of ±6 mm at 50 m and angular accuracies of

±60 microradians.[2] The field of view dictates the area that a laser scanner can observe in a single setup. Some instruments such as the Leica HDS3000 can rotate 360° in horizontal and 270° in vertical allowing it to survey the entire scene surrounding the instrument. In general, users can specify the field of view to match the area of interest. Several manufacturers have added scanning capabilities to their robotic total stations. Thus the surveyor has the option of performing a traditional radiation survey or scanning a scene. This is especially useful in rugged areas with large vertical relief.

The resulting grid of scanned, three-dimensional points can be so dense that a visual image of the scene is formed. This so-called *point cloud* differs from a photographic image in that every point has a three-dimensional coordinate assigned to it. These coordinates can be used to obtain dimensions between any two observed points in the scene. In Figure 17.12, a point cloud of piping at a refinery provided engineers with the information needed to design a new pipe addition shown in white. The detail in this image would be difficult to recreate using other surveying processes. Because of the high density of the observed points in a scene, it is sometimes referred to as *high-definition surveying*. Some instruments also capture a digital image of the scene. The digital image can then be integrated with the scanned points to create a three-dimensional image having color and texture. This process was used for the bridge shown in Figure 23.14. Laser scanning can play a significant role in as-built surveys, archaeology, and mapping of artifacts. Section 27.19 discusses the use of airborne laser mapping known as LiDAR. The Florida Department of Transportation has mapped the entire state using this technology.

Figure 17.12 Laser-scanned image of refinery showing designed pipe alignment in white. (Courtesy Christopher Gibbons, Leica Geosystems, Inc.)

---

[2]A microradian equals 0.000001 radians, which is about 0.2″.

The original point cloud has coordinates in an arbitrary three-dimensional coordinate system. If it is necessary to have coordinates in a project-based coordinate system, a traditional survey can be used to establish coordinates on targets in the scene. Control must be strategically located at the edges of each scene. A minimum of three control points per scene is required. However, additional control is often used to provide redundancy. Multiple scenes can be connected using common control targets. After determining the project coordinates of the control, a three-dimensional conformal coordinate transformation, discussed in Section 17.10, is performed to transform points from the arbitrary coordinate system to the project coordinate system.

17.10 THREE-DIMENSIONAL CONFORMAL COORDINATE TRANSFORMATION

The three-dimensional conformal coordinate transformation transfers points from one three-dimensional coordinate system (xyz) into another (XYZ). This transformation is similar to the two-dimensional coordinate transformation covered in Section 11.8. However, the three-dimensional conformal coordinate transformation involves seven unknown parameters (three rotation angles, one scale factor, and three translations). The development of the rotations is covered in Section 19.16. The mathematical model for the transformation is:

$$X = S(m_{11}x + m_{21}y + m_{31}z) + T_X \quad \textbf{(17.8a)}$$

$$Y = S(m_{12}x + m_{22}y + m_{32}z) + T_Y \quad \textbf{(17.8b)}$$

$$Z = S(m_{13}x + m_{23}y + m_{33}z) + T_Z \quad \textbf{(17.8c)}$$

where S is a scale factor; T_X, T_Y, and T_Z are the translations in x, y, and z, respectively; and m_{11} thru m_{33} are elements of the combined rotation matrix.

$$\begin{aligned}
m_{11} &= \cos(\theta_y)\cos(\theta_z)\\
m_{12} &= \sin(\theta_x)\sin(\theta_y)\cos(\theta_z) + \cos(\theta_x)\sin(\theta_z)\\
m_{13} &= -\cos(\theta_x)\sin(\theta_y)\cos(\theta_z) + \sin(\theta_x)\sin(\theta_z)\\
m_{21} &= -\cos(\theta_y)\sin(\theta_z)\\
m_{22} &= -\sin(\theta_x)\sin(\theta_y)\sin(\theta_z) + \cos(\theta_x)\cos(\theta_z)\\
m_{23} &= \cos(\theta_x)\sin(\theta_y)\sin(\theta_z) + \sin(\theta_x)\cos(\theta_z)\\
m_{31} &= \sin(\theta_x)\\
m_{32} &= -\sin(\theta_x)\cos(\theta_y)\\
m_{32} &= \cos(\theta_x)\cos(\theta_y)
\end{aligned}$$

θ_x, θ_y, and θ_z are the counterclockwise rotation angles about the X, Y, and Z axes, respectively, as viewed from their positive ends. To solve this set of nonlinear equations, methods similar to those discussed in Section 16.9.1 are employed. The linearized set of equations for this transformation given in matrix form are

$$\begin{bmatrix} \left(\frac{\partial X}{\partial S}\right)_0 & 0 & \left(\frac{\partial X}{\partial \theta_y}\right)_0 & \left(\frac{\partial X}{\partial \theta_z}\right)_0 & 1 & 0 & 0 \\ \left(\frac{\partial Y}{\partial S}\right)_0 & \left(\frac{\partial Y}{\partial \theta_x}\right)_0 & \left(\frac{\partial Y}{\partial \theta_y}\right)_0 & \left(\frac{\partial Y}{\partial \theta_z}\right)_0 & 0 & 1 & 0 \\ \left(\frac{\partial Z}{\partial S}\right)_0 & \left(\frac{\partial Z}{\partial \theta_x}\right)_0 & \left(\frac{\partial Z}{\partial \theta_y}\right)_0 & \left(\frac{\partial Z}{\partial \theta_z}\right)_0 & 0 & 0 & 1 \end{bmatrix} \begin{bmatrix} dS \\ d\theta_x \\ d\theta_y \\ d\theta_z \\ dT_X \\ dT_Y \\ dT_Z \end{bmatrix} = \begin{bmatrix} X - X_0 \\ Y - Y_0 \\ Z - Z_0 \end{bmatrix} \tag{17.9}$$

where X_0, Y_0, and Z_0 are determined using Equations (17.8a) thru (17.8c), respectively with approximations for the unknown parameters. The coefficients from the linearized equations are

$$\frac{\partial X}{\partial S} = m_{11}x + m_{21}y + m_{31}z$$

$$\frac{\partial X}{\partial \theta_y} = S(-x \sin \theta_y \cos \theta_z + y \sin \theta_y \sin \theta_z + z \cos \theta_y)$$

$$\frac{\partial X}{\partial \theta_z} = S(m_{21}x - m_{11}y)$$

$$\frac{\partial Y}{\partial S} = m_{12}x - m_{22}y + m_{33}z$$

$$\frac{\partial Y}{\partial \theta_x} = -S(m_{13}x + m_{23}y + m_{33}z)$$

$$\frac{\partial Y}{\partial \theta_y} = S(x \sin \theta_x \cos \theta_y \cos \theta_z - y \sin \theta_x \cos \theta_y \sin \theta_z + z \sin \theta_x \sin \theta_y)$$

$$\frac{\partial Y}{\partial \theta_z} = S(m_{22}x - m_{12}y)$$

$$\frac{\partial Z}{\partial S} = m_{13}x + m_{23}y + m_{33}z$$

$$\frac{\partial Z}{\partial \theta_x} = S(m_{12}x + m_{22}y + m_{32}z)$$

$$\frac{\partial Z}{\partial \theta_y} = S(-x \cos \theta_x \cos \theta_y \cos \theta_z + y \cos \theta_x \cos \theta_y \sin \theta_z - z \cos \theta_x \sin \theta_y)$$

$$\frac{\partial Z}{\partial \theta_z} = S(m_{23}x - m_{13}y)$$

As outlined in Section 16.9.1, Equation (17.9) is formed using approximate values for the unknowns and iterated until the corrections to the unknown parameters become negligibly small. This process is demonstrated in a Mathcad® worksheet on the CD that accompanies this book. The complete mathematical development of the transformation is covered in several references listed in the bibliography at the end of this chapter.

■ 17.11 SELECTION OF FIELD METHOD

Selection of the field method to be used on any topographic survey depends on many factors, including (1) purpose of the survey, (2) map use (accuracy required), (3) map scale, (4) contour interval, (5) size and type of terrain involved, (6) costs, (7) equipment and time available, and (8) experience of the survey personnel.

Items (1) to (5) are interdependent. The cost, of course, will be a minimum if the most suitable method is selected for a project. On large projects, personnel costs rather than equipment investment will usually govern. However, the equipment owned may govern the method chosen by a private surveyor making a topographic survey of 50 or 100 acres.

■ 17.12 WORKING WITH DATA COLLECTORS AND FIELD-TO-FINISH SOFTWARE

Surveying instruments equipped with data collectors can record and store field notes for electronic transmission to computers, plotters, and other office equipment for processing. Such systems, called "field-to-finish," rely on sophisticated software for their operation. Their use can tremendously increase productivity in surveying and mapping.

In using field-to-finish systems for topographic surveys, the data collector will store a point identifier and the *XYZ* coordinates for each point located. However, in addition, ancillary information must accompany each surveyed point. For instance, in Figure 17.13, points 1, 2, 14, and 15 are the corners of a building; points 5, 6, 7, and 8 mark the corners of a sidewalk; points 4, 9, 10, 11, and 12 are points

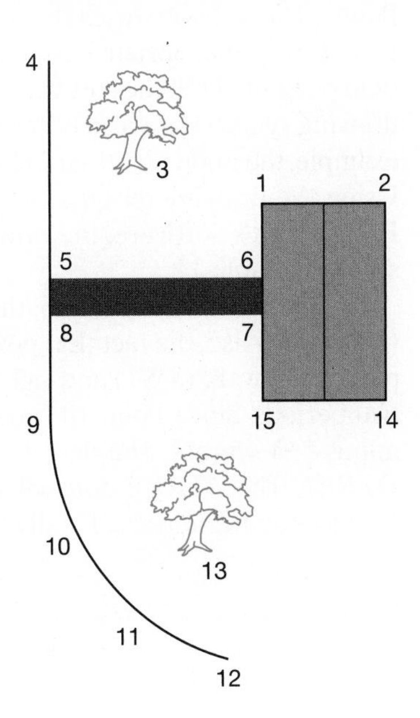

Figure 17.13 Example survey showing line work for planimetric features.

TABLE 17.2 DRAWING DESIGNATORS FOR THE CIVIL SERIES SOFTWARE FROM EAGLE POINT

Designators	System Commands
Line	. (period)
Curve	– (minus)
Close	+ (plus)
Join Last	* (asterisk)
Bearing Close	# (pound sign)
Cross Section	= (equal sign)
Stop	! (exclamation mark)
Insert	* (asterisk)

along a property line, with points 10 and 11 being on a curve; and points 3 and 13 are deciduous trees. It is possible to add and store this descriptive information in the data collector through the use of *notes*. If the notes are entered in a manner that is understandable by the field-to-finish software and the points are collected in a manner that is consistent with and supports the drafting system, the software will use appropriate symbols for plotting each feature, draw and close polygons, and create a complete and finished drawing. However, in order for the system to operate properly, the field personnel collecting the data must understand the requirements of the plotting software.

While the field-to-finish software of the various vendors use somewhat different techniques in reducing field data to a finished map, the basic concepts of these systems can be illustrated using the "Pinnacle Series" software from Eagle Point.[3] This software uses the line work *drawing designators* shown in Table 17.2. By entering appropriate designators in the notes for each point, as it is located, the field personnel tell the software how to draw the scene. Specific *node identifiers* for drawing symbols are also entered in the notes in the field as work progresses. For example, the node identifier DTREE indicates the deciduous trees of Figure 17.13. Using the drawing designators shown in Table 17.2, and the node identifiers of Eagle Point's software, the notes for the planimetric survey of Figure 17.13 are shown in Table 17.3.

As shown in Table 17.3, the note for each line element begins with a period (.) symbol. Also, the fact that points 5 and 8 are common to two lines, [i.e., they are part of sidewalk (SW1) and right-of-way (RW1)], is indicated in the table by using two periods. Since Point 10 is on the curve of the right-of-way, it is followed by a minus (−) symbol. The deciduous trees, points 3 and 13, have as their identifier *DTREE*. The DTREE note will direct the computer to use the appropriate symbol for plotting these trees. Finally note that the plus symbol (+) following point 15

[3]A demonstration version of this software and one other are included on the CD that accompanies this book.

TABLE 17.3 PROPER NOTES FOR DRAWING FEATURES IN FIGURE 16.13

Point	Notes
1	.B1
2	.B1
3	DTREE
4	.RW
5	.RW1.SW1
6	.SW1!
7	.SW1
8	.RW1.SW1!
9	.RW1
10	.RW1−
11	.RW1−
12	.RW1!
13	DTREE
14	.B1
15	.B1+

indicates that lines B1 (designating building No. 1) close back on point 1. The notes for points 6 and 8 terminate with an exclamation point (!), which indicates that SW1 stops at this location.

It is important to include the appropriate designators and notes when collecting data. It is also necessary that successive shots on any object be collected and numbered in the correct order. For example, if the field personnel had collected the building data in the order of 1, 2, 15, and 14, the line work for the building would cross, creating an hourglass shape. Similarly, it was important to collect point 4 before point 5 so that the line work for the right-of-way would be drawn in a linear fashion. From the foregoing, it should be obvious that it is essential for field personnel to understand not only what data to collect, but also the order and manner in which to collect it so that proper plotting will occur. Correctly performed, a field-to-finish survey will greatly reduce the time it takes to create a map in the office and reduce the overall time to complete the project.

With the complexities of collecting the data, selecting the locations of points, and properly noting the features, it is easy to see that some orderly plan for data collection should be developed before the instruments are taken from their cases. Also, there must be coordination between the field and drafting personnel. While each organization may develop their own procedures, some guidelines for collecting data suggest collecting planimetric feature data first, paying special attention to the sequence in which the data is collected. It is often most efficient to collect data for one feature type before beginning another; that is, locating all buildings, then the roadways, then vegetation, and so on in an orderly fashion at each instrument setup. For topographic surveys, all controlling points may be collected, followed by breaklines, and finally sufficient grade shots to allow accurate interpolation of the contours. (See Chapter 18 for a discussion on interpolation of contours.)

■ 17.13 HYDROGRAPHIC SURVEYS

Hydrographic surveys determine depths and terrain configurations of the bottoms of water bodies. Usually the survey data are used to prepare hydrographic maps. In navigation and dredging, they may be recorded in electrical formats for real-time analysis. Bodies of water surveyed include rivers, reservoirs, harbors, lakes, and oceans.

Hydrographic surveys and maps are used in a variety of ways. As examples, engineers employ them for planning and monitoring harbor and river dredging operations, and to ascertain reservoir capacities for flood control and water supply systems; petroleum engineers use them to position offshore drilling facilities and locate underwater pipelines; navigators need them to chart safe passageways and avoid reefs, bars, and other underwater hazards; biologists and conservationists find them helpful in their study and management of aquatic life; archaeologists use them to find shipwrecks; and anglers use them to locate likely fish-holding structures.

Field procedures for hydrographic surveys are similar to those for topographic work; hence, the subject is discussed in this chapter. There are some basic differences in procedures used by surveyors since the land area being mapped cannot be seen, and the depth measurements must be made in water.

Two basic tasks involved in hydrographic surveys are (1) making *soundings* (measuring depths) from the water surface to bottom, and (2) locating the positions where soundings were made. Techniques used to perform these tasks vary depending on the water body's size, accuracy required, type of equipment to be used, and number of personnel available. The subsections that follow briefly describe procedures for mapping small to moderate-sized water bodies.

17.13.1 Equipment for Making Soundings

The size of a water body and its depth control the type of equipment used to measure depths. For shallow areas of limited size, a *sounding pole* can be used. This is usually a wooden or fiberglass staff resembling a level rod. It is perhaps 15 ft long, graduated in feet or tenths of feet, with a metal shoe on the bottom. Direct depth measurements are made by lowering the pole vertically into the water until it hits bottom and then reading the graduation at the surface.

Lead lines can be used where depths are greater than can be reached with a sounding pole. These consist of a suitable length of stretch-resistant cord or other material, to which a heavy lead weight (perhaps 5–15 lb.) is attached. The cord is marked with foot graduations, and these should be checked frequently against a steel tape for their accuracy. In use, the weight is lowered into the water, being careful to keep the cord vertical. The graduation at the surface is read when the weight hits bottom.

In deep water, or for hydrographic surveys of appreciable extent, electronically operated sonic depth recorders called *echo sounders* are used to measure depths. These devices, an example of which is shown in Figure 17.14, transmit an acoustic pulse vertically downward and measure the elapsed time for the signal to travel to the bottom, be reflected, and return. The travel time is converted to depth, and displayed in either digital or graphic form. A graphic profile, such as that shown

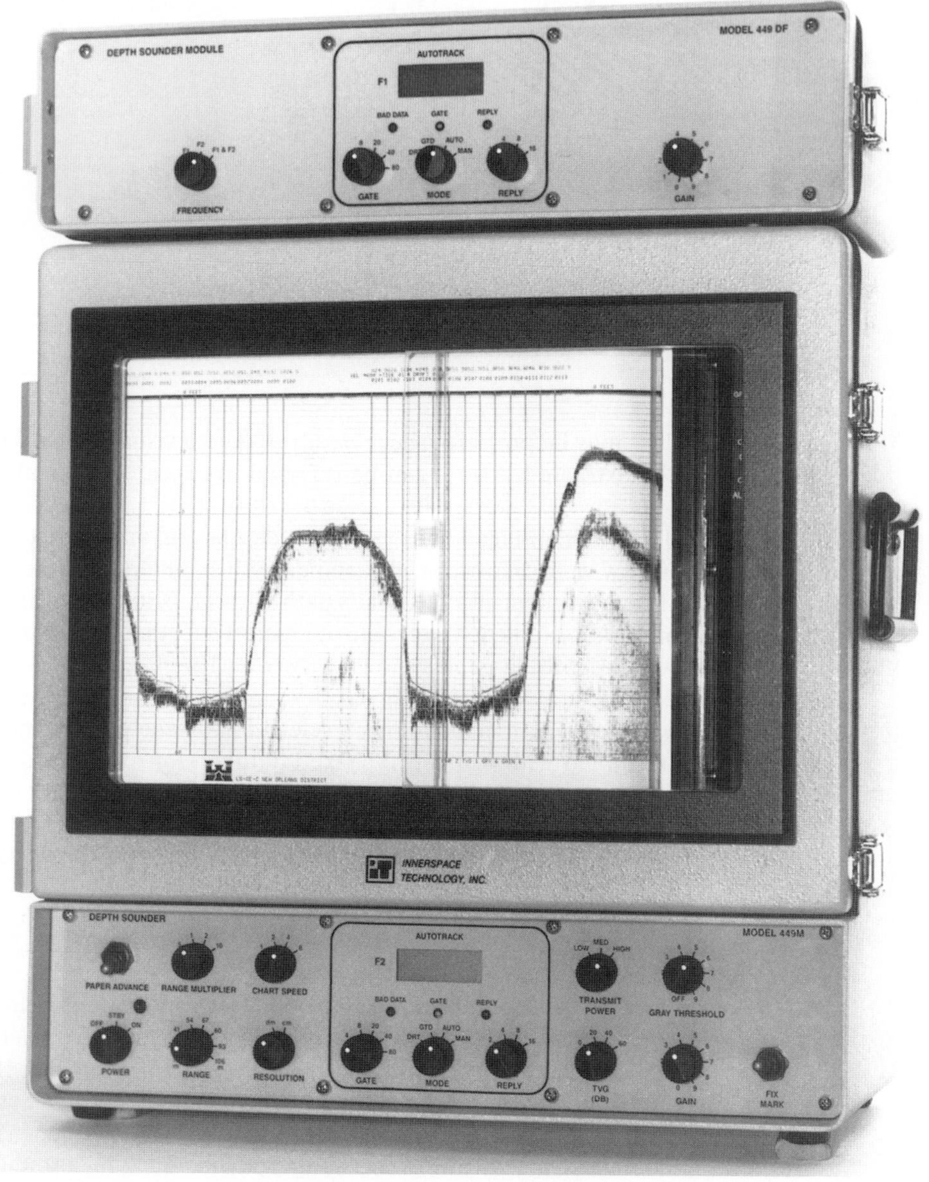

Figure 17.14 Innerspace model 455 hydrographic depth sounder. (Courtesy Innerspace Technology, Inc.)

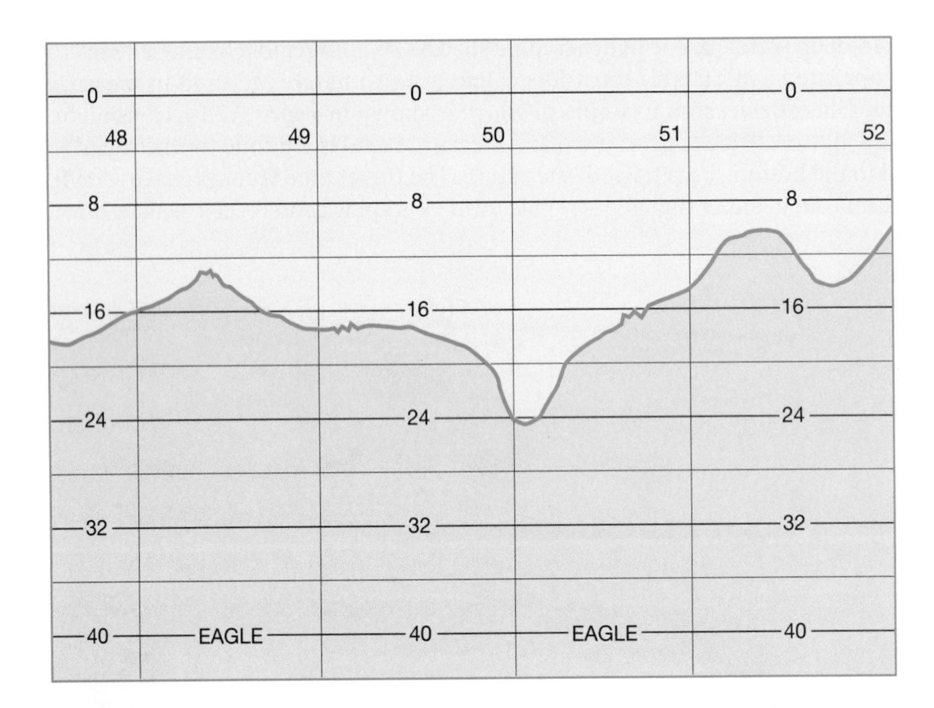

Figure 17.15 Bottom profile produced by electronic depth sounder.

in Figure 17.15, may be preferred, because it provides a visual record from which depths can be extracted. Also it can be referred to repeatedly for plotting and checking.

Sounding poles and lead lines yield spot depths and are restricted to use in relatively shallow water. However, electronic depth sounders provide continuous profiles of the surface beneath the boat's path and can be used in water of virtually any depth. For example in the profile of Figure 17.15, the chart's vertical range was set to 40 ft and profile depths shown vary from 10–24 ft.

The reference plane from which depth soundings are measured is the water surface. Because of surface fluctuations, its elevation or *stage* at the time of survey must be determined with respect to a fixed datum—for example, orthometric height. Running a level circuit to the water from a nearby benchmark can do this. In situations where soundings are repeated at regular intervals, a graduated staff can be permanently installed in the water so that its stage, in feet above the datum, can be read directly each time soundings are repeated.

17.13.2 Locating Soundings

Any of the traditional ground surveying procedures can be used to locate positions where soundings are taken. In addition to these techniques, other methods have also been applied in hydrographic surveys, for example, GPS. If ground-surveying techniques are used, some horizontal control must first be established on shore. Ideal locations for control stations are on peninsulas or in open areas that afford a wide unobstructed view of the water body for tracking a sounding boat. The coordinate positions of the control points can be established by traverse, but triangulation and trilateration are also well suited for this work.

Among the various boat-positioning methods, *radiation* and *angle intersection* are usually selected if total station instruments are used. Radiation is particularly efficient, especially if a total station instrument is used, because only one person on shore is needed to track the boat. After setting up on one control station and backsighting another, angles and distances are measured to locate each boat position. Special total station instruments and reflectors are manufactured for this work to facilitate sighting and observing distances electronically to a moving target. From angles and distances, which are automatically read, the total station's computer determines the boat's coordinates. These can either be stored in an automatic data collector for later office use in mapping, or transmitted by radio to the boat if real-time positioning is required, as in controlling ongoing dredging.

Figure 17.16 illustrates the use of angle intersections in the hydrographic survey of a lake. Here the boat travels back and forth along *range lines* while the depth sounder continuously records bottom profiles. At regular intervals, *fixes* are taken by observing angles to the boat from shore stations. Two angles establish the boat's position, but three or more provide redundancy and a check. For example in Figure 17.16, angle observations *e*, *g*, and *h* for fix number 50 (indicated by dashed lines) have been made from shore stations *E*, *G*, and *H*, respectively. Prior to observing angles, the total stations or theodolites were oriented by backsighting on another visible control station, as at station *G* from *E*.

Flag or radio signals are given from the boat to coordinate fixes and ensure that angles from all shore stations are observed simultaneously. At the precise moment of any fix, the profile is also marked and the fix number noted. For example in Figure 17.16, fixes 48 through 52 are identified and marked on the profile of Figure 17.15. This correlates bottom depths with specific locations in the water body—a necessity for mapping.

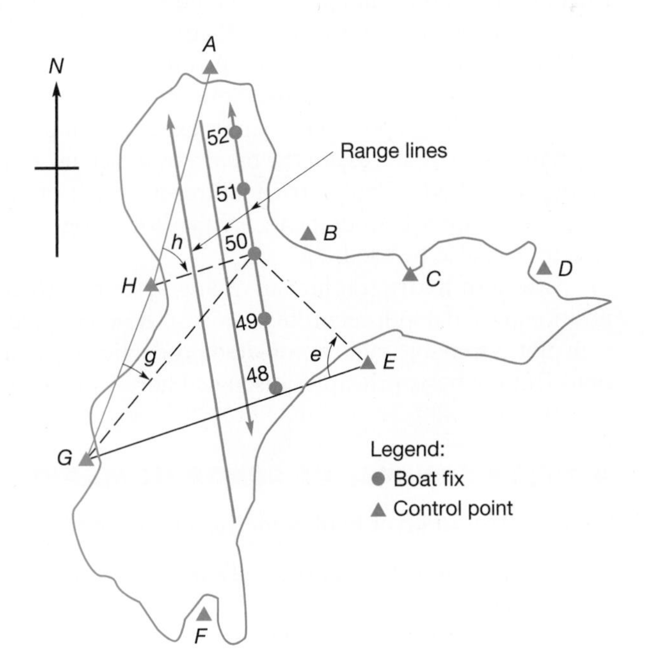

Figure 17.16 Angle intersection procedure for locating boat fixes along range lines.

If the boat is driven back and forth along parallel range lines to cover the area of interest, and then the area traversed again with perpendicular courses, a grid of profiles results from which contours can be drawn. In larger bodies of water a compass is valuable to assist in keeping the range lines parallel. Required accuracy dictates the spacing between range lines, with closer spacing yielding more accurate results. Various other boat-positioning systems can be used, depending on circumstances. One that works well for hydrographic surveys of rivers or other relatively narrow water bodies consists in laying out uniformly spaced reference lines which cross the water. Placing tall painted stakes on the bank on either side marks the lines. Then fixes can be taken as the sounding boat navigates along the reference lines. However to position each fix along the lines, either a distance must be observed from one reference point, or an azimuth to the boat from an independent control point. When the boat is moving perpendicular to the marked lines, its passage across projections between stakes locates fixes, but again a distance or angle is needed to complete the fix position.

RTK GPS surveying (see Chapter 15) is ideal for establishing sounding locations for hydrographic surveys and also for guiding the boat along planned range lines on larger water bodies. With its many advantages, it has replaced other hydrographic positioning techniques. In this case the transducer is located by measuring offset from the GPS antenna reference point. Often the GPS antenna is mounted on the sounding pole well above the top of the boat and directly over the transducer. In this case, only the distance from the antenna reference point to the transducer need by measured.

17.13.3 Hydrographic Mapping

Procedures for preparing hydrographic maps do not differ appreciably from those used in topographic mapping discussed in Chapter 18. Basically, depths are plotted in their surveyed positions and contours drawn. If an echo sounder is used, depths are interpolated from the profiles and plotted between fix locations. In addition to depth contours, the shoreline and other prominent features are also located on hydrographic maps. This is especially important for navigation and fishing maps, as the features are the means by which users align and locate themselves on the water body. Planimetric features are most often located photogrammetrically (see Chapter 27), but the techniques for topographic mapping discussed in this chapter can also be used.

Modern hydrographic surveying systems utilize sophisticated electronic positioning and depth-recording devices. These, coupled with computers interfaced with plotters, enable rapid automated production of hydrographic maps in near real time. But the basic principles discussed here still apply.

17.14 SOURCES OF ERROR IN MAPPING SURVEYS

Some sources of error in planimetric and topographic surveys are:

1. Instrumental errors, especially an index error that affects vertical and zenith angles.
2. Errors in reading instruments.

3. Control not established, checked, and adjusted before beginning to collect details.
4. Control points too far apart and poorly selected for proper coverage of an area.
5. Sights taken on detail points, which are too far away.
6. Poor selection of points for contour delineation.

17.15 MISTAKES IN MAPPING SURVEYS

Some typical mistakes in planimetric and topographic surveys are:

1. Unsatisfactory equipment or field method for particular survey and terrain conditions.
2. Mistakes in instrument reading and data recording.
3. Failure to periodically check azimuth orientation when many detail points are located from one instrument station.
4. Too few (or too many) contour points taken.
5. Failure to collect some mapping details.
6. Mistakes in entering point identifiers, drawing designators, and symbols when using field-to-finish surveying and mapping systems.

PROBLEMS

Asterisks (*) indicate problems that have answers given in Appendix G.

17.1 What are the fundamental differences between a topographic map and a planimetric map?

17.2 Name five mapping details classified as "cultural" features not mentioned in Section 17.1.

17.3 Name the different conventions or procedures that can be used to portray relief.

17.4 Give examples for which the best method of locating contours would be by the (a) grid method and the (b) indirect method.

17.5 Discuss the factors that must be considered when selecting the scale for a map.

17.6* On a map sheet having a scale of 1 in. = 360 ft, what is the smallest distance (in feet) that can be plotted with an engineer's scale? (Minimum scale graduations are 1/60th in.)

17.7 What ratio scales are suitable to replace the following equivalent scales: 1:120, 1:360, 1:2400, 1:4800, and 1:7200?

17.8 A topographic map has a contour interval of 1 ft and a scale of 1:480. If two adjacent contours are 0.5 in. apart, what is the average slope of the ground between the contours?

17.9* On a map whose scale is 1 in. = 100 ft, how far apart (in inches) would 5-ft contours be on a uniform slope (grade) of 3 percent?

17.10 On a map drawn to a scale of 1:1000 contour lines are 5 mm apart at a certain place. The contour interval is 2 m. What is the ground slope, in percent, between adjacent contours?

17.11 Similar to Problem 17.10, except for a 5-m interval, 30-mm spacing, and a map scale of 1:4000.

17.12 Sketch at a scale of 10 ft/in., the general shape of contours that cross a 20-ft wide street having a +4.00 percent grade, a 6-in. parabolic crown, and a 6-in. high curb.

17.13 Same as Problem 17.12, except a four-lane divided highway including a 30-ft wide median strip, two 12-ft wide lanes of pavement each side of the median having a +4.00 percent grade with a 1.50 percent side slope; and 10-ft wide shoulders both sides with a 2 percent side slope. The median slopes 10.0 percent toward the center.

17.14 When should points located for contours be connected by straight lines? When by smooth curves?

17.15* What conditions in the field need to exist when using kinematic GPS?

17.16 What is: (a) a digital elevation model (DEM)? (b) a triangulated irregular network (TIN)?

17.17 Discuss why it is important to locate breaks in grade with "breaklines" in the field if contours will be drawn using a computerized automated contouring system.

17.18 What are the advantages of a semikinematic method in locating planimetric features?

17.19 Using the labels given in parentheses in the legend of the accompanying figure, create a set of notes using the drawing designators listed in Table 17.2.

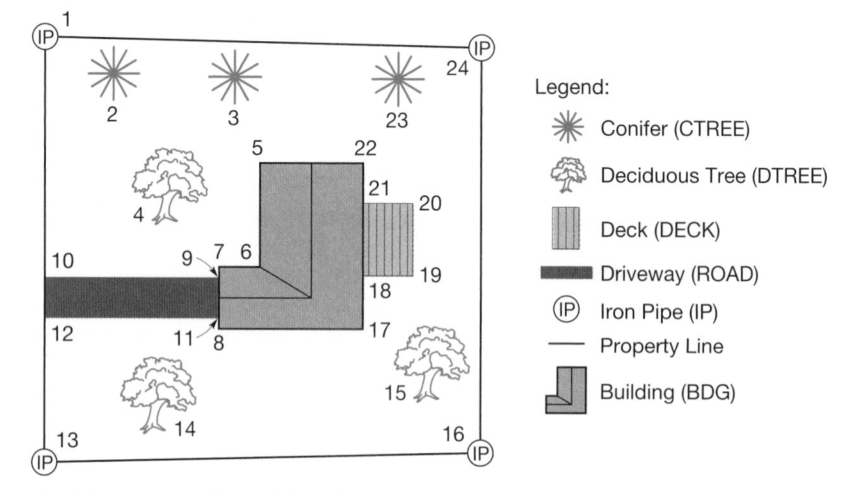

Problems 17.19 and 17.20

17.20 Why would a kinematic GPS survey of the parcel shown in the figure be difficult to perform?

17.21 Discuss the advantages and disadvantages of performing topographic surveys using kinematic GPS surveying procedures.

17.22 Discuss how data collection with a total station instrument can be combined with GPS methods to collect data for a topographic map.

17.23 Prepare a set of field notes to locate the topographic details in Figure 17.5. Scale additional distances and angles if necessary.

17.24 Why is it dangerous to run the control traverse at the same time planimetric data are being collected?

17.25 Cite the advantages of locating topographic details by radial methods using a total station instrument with a data collector.

17.26 What does the term "point cloud" describe in laser-scanning?

17.27* Distances AB, AC, AD, and AE of 121.6, 213.1, 304.2, and 406.4 ft were laid out on a level course. With an internal focusing theodolite setup on A, rod intercepts of 1.21, 2.12, 3.03, and 4.04 ft were read with the rod held on B, C, D, and E, respectively. Determine the stadia interval factor.

Compute the horizontal distances and elevation differences for the data in Problems 17.28 and 17.29. (An internal focusing instrument was used.)

17.28 Rod intercept = 1.25 ft, $z = 95°30'$ (on *hi*), $K = 99.5$
17.29 Rod intercept = 2.56 ft, $z = 93°28'$ (on *hi*), $K = 101.0$

For Problems 17.30 through 17.31, calculate horizontal distance *AB* and the elevation of *B* for the following stadia readings taken to *B* from occupied station *A*. (*Note:* The elevation of $A = 830.6$ ft, $hi = 5.5$ ft, rod intercepts *I* are in ft, and $K = 100$.)

17.30 $I = 1.88$, $z = 95°32'20''$ read to 9.7 ft on the rod
17.31 $I = 2.25$, $z = 87°22'42''$ read to 6.8 ft on the rod
17.32 List the methods for establishing control to support a topographic mapping project.

For Problems 17.33 through 17.36, calculate the *X*, *Y*, and *Z* coordinates of point *B* for radial readings taken to *B* from occupied station *A*, if the backsight azimuth at *A* is 125°32′48″, the elevation of $A = 303.058$ m, and $hi = 1.325$ m. Assume the *XY* coordinates of *A* are (10,000.000, 5000.000).

17.33* Clockwise horizontal angle = 55°37′42″, zenith angle = 92°34′18″, slope distance = 435.098 m, $hr = 2.000$ m.
17.34 Clockwise horizontal angle = 102°22′42″, zenith angle = 86°48′36″, slope distance = 48.063 m, $hr = 1.500$ m.
17.35 Clockwise horizontal angle = 205°35′46″, zenith angle = 84°25′14″, slope distance = 208.004 m, $hr = 1.500$ m.
17.36 Clockwise horizontal angle = 307°56′52″, zenith angle = 87°17′40″, slope distance = 304.902 m, $hr = 2.000$ m.
17.37 Describe how the arbitrary coordinates of a point cloud are transformed into a conventional coordinate system.
17.38 List various equipment used for making hydrographic depth soundings, and discuss the limitations, advantages, and disadvantages of each.
17.39* On a map having a scale of 200 ft/in. the distance between plotted fixes 49 and 50 of Figure 17.16 is 3.15 in. From measurements on the profile of Figure 17.15, determine how far from fix 50 the 20-ft contour (existing between fixes 49 and 50) should be plotted on the map.
17.40 Similar to Problem 17.39, except locate the 16-ft contour between fixes 50 and 51 if the corresponding map distance is 2.98 in.
17.41 Why is it important to show the shoreline and some planimetric features for navigation hydrographic maps?
17.42 Create a computational program that solves Problems 17.33 to 17.37.
17.43 Create a computational program that performs a three-dimensional conformal coordinate transformation as described in Section 17.10.

BIBLIOGRAPHY

ASPRS. 1987. *Large Scale Mapping Guidelines.* Bethesda, Md.: American Society for Photogrammetry and Remote Sensing.
Dimsdale, J. 2005. "Laser Scanners... Past, Present, and Future." *Professional Surveyor* 25 (No. 4): 18.
Francis, P. 2006. "At the Scene of the Crime." *Point of Beginning* 31 (No. 12): 28.
Gardner, Nigel. 2007. "LiDAR on a Stick." *Professional Surveyor* 27 (No. 2): 6.
Garret, Justin. 2007. "Reservoir of Lessons Learned." *Professional Surveyor* 27 (No. 2): 18.
Jacobs, Geoff. 2004. "3D Laser Scanning: An Ultra-Fast, High-Definition, Reflectorless Topographic Survey." *Professional Surveyor* 24 (No. 5): 38.

_______ 2005. "High Definition Scanning—Laser Scanning: Big Advantages, Even for Small Projects." *Professional Surveyor* 25 (No. 10): 30.

_______ 2005. "High Definition Surveying—3D Laser Scanning: Laser Scanning Capability: Ways to Get There." *Professional Surveyor* 25 (No. 3): 36.

_______ 2005. "High Definition Surveying—3D Laser Scanning: Laser Scanning Today: Another Tool in the Kit." *Professional Surveyor* 25 (No. 1): 26.

_______ 2005. "High Definition Surveying—3D Laser Scanning: Registration and Georeferencing." *Professional Surveyor* 25 (No. 7): 26.

Lyle, S. et al. 2002. "Along the Water's Edge." *Point of Beginning* 27 (No. 9): 20.

Paiva, J. V. R. 2006. "Professional Topography." *Point of Beginning* 31 (No. 10): 28.

Pesci, A. D. Conforti, and M. Bacciocchi. 2007. "Morphing Mount Vesuvius." *Professional Surveyor* 27 (No. 2): 12.

Rameriz, J. R. 2006. "A New Approach to Relief Representation." *Surveying and Land Information Science* 66 (No. 1): 19.

Rida, Ali A. and Serwan M. J. Baban. 2004. "Mapping Groundwater Level and Depth in the Azraq Basin." *Surveying and Land Information Science* 64 (No. 2): 97.

Schorr, S. M. 2006. "A Crash Course in Forensic Scanning." *Point of Beginning* 31 (No. 12): 24.

Tuck, Bobby. 2005. "Helicopter LiDAR: The Modern Response to Sudden Change." *Professional Surveyor* 25 (No. 1): 22.

Wimbush, M. "The Latest in Laser Scanners." *Point of Beginning* 32 (No. 2): 30.

18 Mapping

18.1 INTRODUCTION

Maps are visual expressions of portions of the Earth's surface. Features are depicted using various combinations of points, lines, and standard symbols. Maps have traditionally been produced in *graphic,* or "hardcopy," form—that is, printed on paper or a stable-base plastic material. However, today most mapping data are collected in digital form and are processed using *Computer-Aided Drafting and Design* (CADD) systems to develop "softcopy" maps. Softcopy maps are stored within a computer, can be analyzed, modified, enlarged, or reduced in scale, and have their contour intervals changed while being viewed on the monitors of CADD systems. Different types or "layers" of information can also be extracted from digital maps to be represented and analyzed separately and softcopy maps can be instantaneously transferred to other offices or remote locations electronically. Of course, they can also be printed in hardcopy form if desired. Softcopy maps are indispensable in the development and operation of modern *Land Information Systems* (LISs) and *Geographic Information Systems* (GISs) (see Chapter 28).

Throughout the ages, maps have had a profound impact on human activities, and today the demand for them is perhaps greater than ever. They are important in engineering, resource management, urban and regional planning, management of the environment, construction, conservation, geology, agriculture, and many other fields. Maps show various features—for example, topography, property boundaries, transportation routes, soil types, vegetation, landownership, and mineral and resource locations. Maps are especially important in engineering for planning project locations, designing facilities, and estimating contract quantities.

As noted previously, maps are essential in the development and operation of LISs and GISs. These systems for spatial data analysis and management use the computer to store, retrieve, manipulate, merge, analyze, display, and disseminate information by means of digital maps (see Chapter 28). GISs have applications in virtually every field of endeavor. Spatial databases to support the systems are generally developed either by digitizing existing graphic maps or by generating new digital maps in the computer based upon digitized ground survey or photogrammetric data. Maps of various types needed to create spatial databases for LISs and GISs include topographic maps, which show the natural and cultural features and relief in an area; cadastral maps, which give boundaries of landownership; natural resource maps, which provide the location and distribution of forest and water resources, wetlands, soil types, etc.; facilities maps, which show existing transportation networks, water and sewer mains, and distribution lines for electric power; and land-use maps, which show the various activities of humans related to the land. Applications of GIS technology have been expanding at a phenomenal rate and these activities will impose a heavy demand for high-quality maps of various types and scales in the future.

Cartography, the term applied to the overall process of map production, includes map design, preparing or compiling manuscripts, final drafting, and reproduction. These processes, which apply whether the maps are graphic or digital, are described in this chapter.

■ 18.2 AVAILABILITY OF MAPS AND RELATED INFORMATION

Maps for a variety of different purposes, prepared at scales varying from large to small, and in both graphic and digital form, are prepared by private surveying and engineering companies, industries, public utilities, cities, counties, states, and agencies of the federal government. Unfortunately, with such a wide range in organizations and agencies involved, some duplication of effort has occurred because mapping activities generally have not been coordinated. Also, the existence of available maps and related information is often unknown to potential users. However, steps have been taken to improve this situation. The U.S. Geological Survey (USGS) now coordinates all mapping activities at the federal level. Through its Earth Science Information Center (ESIC), they offer nationwide information and sales service for map products and Earth science publications. The ESIC provides information about topographic, land use, geologic and hydrologic maps, books and reports; Earth science and map data in digital format and related applications software; aerial, satellite, and radar images and related products; and geodetic data.[1] Several states have established state cartographers' offices. One of their functions is the dissemination of local maps and related products and information to surveyors, engineers, cartographers, and the general public.

---

[1]The U.S. Geological Survey ESIC can be reached by telephone at: (888) ASK-USGS, [(888) 275-8747]. Information can also be obtained and selected maps and other products can be downloaded at http://mapping.usgs.gov/esic/. Contact by mail can be made to the U.S. Dept. of the Interior, U.S. Geological Survey, 12201 Sunrise Valley Drive, Reston, VA 20192.

18.3 NATIONAL MAPPING PROGRAM

The *National Mapping Program* was established to provide maps and other cartographic products needed by the citizens of the United States. This is the responsibility of the *National Mapping Division* of the USGS. The USGS began publishing topographic maps in 1886 as an aid to scientific studies. It now produces a variety of topographic maps at differing scales; however, its standard series has a scale of 1:24,000. In this series, individual sheets cover *quadrangles* of 7 1/2-min in both latitude and longitude. Each quadrangle map is named, usually according to the most prominent feature within its bounds. Except for Alaska, the entire United States is covered at the 1:24,000 scale and over 57,000 maps are involved. (Maps covering 15′ quadrangles at a scale of 1:63,360 [1 in./mi] are standard for Alaska.) On the quadrangle maps, cultural features are shown in black, contours in brown, water features in blue, urban regions in red, and woodland areas in green. Topographic coverage of the United States is also available at scales of 1:50,000 (county maps), 1:62,500 (the older 15′ quadrangles produced until about 1950), 1:100,000, and 1:250,000. The USGS has also published a state map series. Most are at a scale of 1:500,000, but a few are at 1:1,000,000 or other scales.

As noted in Section 18.1, requirements for digital cartographic data are growing rapidly to support LISs and GISs. To meet these needs, the U.S. Geological Survey has developed two very useful types of digital data: (1) *digital line graphs* (DLGs) and (2) *digital elevation models* (DEMs). Primarily digitizing existing maps and other cartographic products is generating these products. The digital line graphs contain only linear features or planimetry in an area. Included are political boundaries, hydrography, transportation networks, and the subdivision lines of the U.S. Public Land Survey System (see Chapter 22). The digital elevation models are arrays of elevation values, produced in grids of varying dimensions, depending upon the source of the information. The horizontal positions of points in the DEMs are X and Y coordinates referenced to the Universal Transverse Mercator coordinate system (see Section 20.12). A grid of 30 m is used for DEMs generated from 7 1/2′ quadrangles, with larger spacings being used for those generated from smaller scale maps.

In addition to topographic maps, digital line graphs, and digital elevation models, a variety of other special maps and related products are published as a part of the National Mapping Program. As noted in the preceding section, information on all of their available maps and other related products is obtainable through the U.S. Geological Survey ESIC.

18.4 ACCURACY STANDARDS FOR MAPPING

To provide a set of uniform standards for guiding the production of maps and to protect consumers of maps, the United States Bureau of the Budget developed the *National Map Accuracy Standards* (NMAS). These standards, first published in 1941 and revised in 1947, provide specifications governing both the horizontal and vertical accuracy with which features are depicted on maps. Published maps meeting these accuracy requirements may have the following note in their legends: "This map complies with National Map Accuracy Standards," thereby providing assurance that the map meets these stated accuracy levels.

To meet the NMAS horizontal position specification, for maps produced at scales larger than 1:20,000, not more than 10 percent of well-defined points tested shall be in error by more than 1/30 in. (0.8 mm). Accordingly, on a map plotted to a scale of 1 in. = 100 ft, point positions would have to be correctly portrayed to within ±3.3 ft to meet this specification. On smaller scale maps, the limit of horizontal error is 1/50 in. (0.5 mm), or approximately ±40 ft on the ground at a map scale of 1:24,000. These limits of accuracy apply to positions of well-defined points only, such as monuments, benchmarks, highway intersections, and building corners.

The NMAS vertical accuracy requirements specify that not more than 10 percent of elevations tested shall be in error by more than one-half the contour interval and none can exceed the interval. To meet this requirement, contours may be shifted by distances up to the horizontal positional tolerance (discussed previously), if necessary.

The accuracy of any map can be tested by comparing the positions of points whose locations or elevations are shown on it with corresponding positions determined by surveys of a higher order of accuracy. Plotted horizontal positions of objects are checked by running an independent traverse or other survey to points selected by the person or organization for which the map was made. To check vertical accuracy, elevations obtained from field profile surveys are compared with elevations taken from profiles made from plotted contours. These procedures provide a check on both fieldwork and map drafting.

When the NMAS were developed, maps were being produced in hardcopy form. But as noted in Section 18.1, softcopy maps are now most common. To accommodate this change, the *Federal Geographic Data Committee* (FGDC) drafted a more current set of accuracy standards called the *Geospatial Positioning Accuracy Standards.*[2] The FGDC is composed of representatives from 19 federal agencies and was established to coordinate policies, standards, and procedures for producing and sharing geographic information. The new Geospatial Positioning Accuracy Standards are completed, and like NMAS, the document specifies accuracies in separate horizontal and vertical components. But unlike NMAS, accuracies are specified in terms of coordinates of points, ground distances, and elevations at the 95 percent confidence level. Thus these new standards are applicable to all types and scales of maps, including those in digital form. The test for maps intended to meet this standard involves checking a set of at least 20 well-defined points against information obtained from an independent source of higher accuracy. Root-mean-square errors are computed and converted to the 95% confidence level by using appropriate multipliers (see Section 3.16). A digital planimetric map that passes at the 1-m level, for example, would contain the statement

[2]Information on the status of the **GEOSPATIAL POSITIONING ACCURACY STANDARDS** standards can be obtained by telephone at (703) 648-5514, or at http://www.fgdc.gov. Copies of the current standards can also be downloaded at this website. Contact can also be made by mail at The Federal Geodetic Data Committee, U.S. Geological Survey, 590 National Center, Reston, VA 20192.

"Tested 1-meter horizontal accuracy at 95% confidence level" in its legend. A similar statement can be included that applies to a map's vertical accuracy.

The USGS has developed its own standards to govern the production of the maps and other products it provides through the National Mapping Program.[3] Standards have been developed not only for their hardcopy maps, but also for their digital products including digital elevation models, digital line graphs, digital orthophotos (see Section 27.15), and others.

The *American Society for Photogrammetry and Remote Sensing* (ASPRS) has also adopted its own standards to govern photogrammetric production of large-scale maps. It specifies standards for three levels of accuracy, classes 1, 2, and 3. For a map to meet its class 1 standards, the root-mean-square (rms)[4] error in both X and Y coordinates of well-defined points must not exceed ± 0.01 in. at map scale. Thus for a map scale of 500 ft/in., the allowable rms error in X and Y coordinates is ± 5.0 ft. Vertical accuracy is specified in terms of the map's contour interval (CI). For class 1, the rms error of well-defined points must not exceed $\pm(\mathrm{CI}/3)$. These horizontal and vertical standards are both relaxed by factors of 2 and 3 for class 2 and class 3 maps, respectively.

The *American Society of Civil Engineers* (ASCE) has also developed a set of standards for topographic mapping that are aimed primarily at large-scale engineering maps. In addition to suggesting accuracies for various map scales, they also provide standards for contouring, map symbols, abbreviations, lettering, and other factors important in mapping.

■ 18.5 MANUAL AND COMPUTER-AIDED DRAFTING PROCEDURES

As previously stated, maps may be drafted manually or produced with CADD systems. Manual procedures utilize standard drafting tools such as scales, protractors, compasses, triangles, and T-squares. CADD systems employ computers programmed with special software and interfaced with electronic plotting devices. With either approach, after deciding on scale and other factors that control overall map design, a manuscript is prepared. When completed, final drafting is performed.

In manual drafting, the manuscript is usually compiled in pencil. It should be prepared carefully to locate all features and contours as accurately as possible and be complete in every detail, including placement of symbols and letters. Lettering on the manuscript need not be done with extreme care, for its major purpose at this stage is to ensure good overall map design and proper placement. A well-prepared manuscript goes a long way toward achieving a good-quality final map.

[3]For information on the USGS standards, visit http://mapping.usgs.gov/standards. Contact can also be made by telephone at (888) ASK-USGS [(888) 275-8747], or by email at ask@usgs.gov.

[4]The rms error is defined as the square root of the average of squared discrepancies for points tested. Discrepancies are the differences between coordinates and elevations of points taken from the map and their values as determined by check surveys.

The completed version of the manually compiled manuscript is drafted in ink, or *scribed*. If inked, the manuscript is placed on a light table and features are traced on a stable-base transparent overlay material. Lettering is usually done first; then planimetric features and contours are made. Scribing is performed on sheets of transparent stable-base material coated with an opaque emulsion. Manuscript lines are transferred to the coating in a laboratory process. Lines representing features and contours are then made by cutting and scraping to remove the coating. Special scribing tools are used to vary line weights and make standard symbols. Scribing is generally easier and faster than inking. Reproductions are made from the finished inked or scribed product.

In drafting maps with CADD, a softcopy manuscript is compiled in the computer and displayed on its screen as work progresses. CADD software provides instructions to the computer, which basically duplicate manual drafting functions. A file containing coordinates of points, as well as specific instructions on how to plot them, must be input in preparation for mapping. An operator interactively designs and compiles the map by entering commands into the computer's keyboard or using a mouse to activate functions on a menu. Points, lines of various types, and a variety of symbols are available to the operator. Letters of differing sizes and styles can also be selected. When the manuscript is completely finished, simply activating the electronic plotter will draw the final map. Map production using CADD systems has many advantages over manual methods and therefore it has rapidly become the method of choice by most mapping firms. However, it is still important to learn the basics of manual map drafting techniques, since these are often duplicated in the CADD processes. CADD systems are described in more detail in Section 18.14.

■ 18.6 MAP DESIGN

Before beginning the design of a map, the following two basic questions should be answered: (1) What is the purpose of the map? and (2) Who is the map intended to serve? All maps have a purpose, which in turn dictates the information that the map must convey. Once the purpose of the map is fixed, emphasis should be placed on achieving the design that best meets its objectives and conveys the necessary information clearly to its users.

Maps typically depict many different types and classes of details in portraying natural and cultural features and, if properly designed, they can convey an enormous amount of information. On the other hand, maps that are carelessly designed can be confusing, difficult to read, understand, or interpret. To achieve maximum effectiveness in map design, the following elements or factors should be considered: (1) clarity, (2) order, (3) balance, (4) contrast, (5) unity, and (6) harmony. Definitions of these six elements in relation to map design and explanations of their interdependence are discussed below.

1. *Clarity* relates to the ability of a map to convey its intended information completely and unambiguously. It can only be achieved after fully examining the objectives of the map and then emphasizing the features necessary to carry out

those objectives. Maps should not be overloaded with details, as this can cause congestion and confusion. If considerable detail must be included on a map, the information could be placed in a table. Other alternatives consist of preparing larger-scale *inset maps* of areas that contain dense detail or creating an overlay to display some of the detail. The proper use of textual elements is very important in achieving clarity. (The subject of map lettering is covered in Section 18.11.)

2. *Order* refers to the logic of a map and relates to the path that a user's eye would follow when looking at one. A design should be adopted that first draws the user's attention to the subject area of the map, then the map title, and then to any notes. Never let auxiliary elements such as bar scales and directional arrows dominate the map. A common mistake made by beginners is to make bar scales and north arrows so large and bold that they attract attention away from the subject of the map.

3. All elements on a map have weight and they should be uniformly distributed around the "visual center" of the map to create good overall *balance*. The visual center is slightly above the geometrical center of the map sheet. In general, the weight of an element is affected by factors such as size, color, font, position, and line width. Map elements that appear at the center have less weight than those on the edges. Elements in the top or right half of the map will appear to have more weight than those in the bottom or left half of the map. Also map elements identified with thicker line widths will appear to have heavier weights than their slimmer counterparts. Colors such as red appear heavier than blue or yellow. Two examples illustrating balance are shown in Figure 18.1. In Figure 18.1(a), the map appears to be too heavily weighted to the left and thus has poor overall balance. A redistribution of the map features, as shown in Figure 18.1(b), produces a more visually balanced product. The use of thumbnail sketches can often help to achieve a balanced layout for a map.

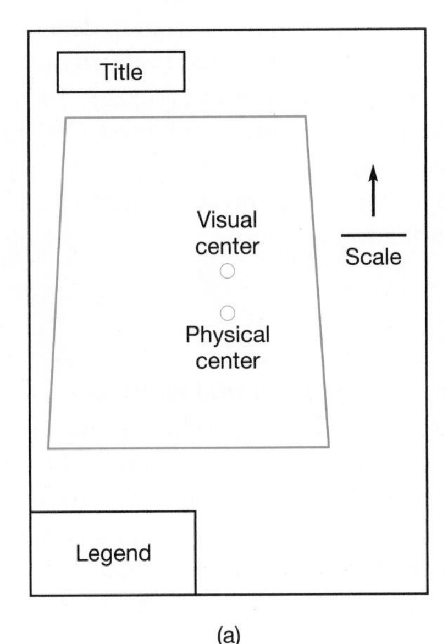

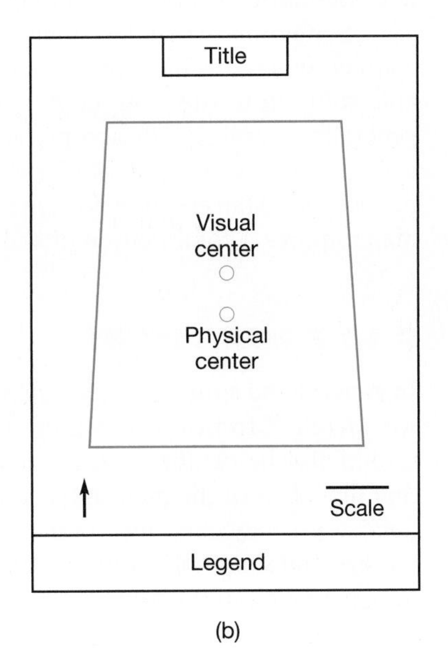

Figure 18.1 (a) The layout of a poorly balanced map sheet. (b) A better layout for the same map.

4. *Contrast* relates primarily to the use of different line weights and fonts of varying sizes. Contrast can be used to enhance balance, order, and clarity. For example, the title of the map should be displayed in a larger font than the other textual elements. This will attract the viewer's attention, thereby enhancing the order and clarity of the map. Various fonts can also be used to provide balance with other elements on the map. Another example where contrast supports the clarity of a map is in contouring. Here *index contours* (every fifth contour) should be drawn with a heavier line than the other contours. This enhances the map's clarity and facilitates the determination of elevations.

5. *Unity* refers to the interrelationships between the backgrounds, shading, and colors on a map. Again these items can enhance clarity, balance, and contrast. They can also detract from these same items. For example, yellow lettering on a white background is difficult to see and often overlooked by the reader. However this same yellow lettering on a black background will stand out and appear emphasized. A map with good unity is visualized as a unit and not as an assemblage of individual elements.

6. *Harmony* relates to the interrelationships between all elements on the map. If a map has good harmony, the elements work together. Common errors are the use of too many fonts, a north arrow that is too fancy or large, or a bar scale that is too large.

In designing maps, it is important to remember that different audiences may require different maps. For example, it would be difficult for a layperson to read and understand a map produced for an engineering project. Accordingly, maps that are developed for design professionals are not generally suitable for public hearings. In fact because laypeople often have no training in map reading, it may be best to develop specialized three-dimensional maps or models that depict relief, boundaries, proposed buildings, landscaping, and so on.

Sometimes when designing a map, certain elements of map design will be in conflict. When this happens, priorities must be established that provide a reasonable solution to the conflict. A perfect map rarely, if ever, exists, and there are generally several equally acceptable designs that could be adopted. Often there are design conflicts that cannot be resolved and a compromise solution may have to be accepted. Map creation is often subjective and the production of a well-designed map requires a combination of skill, art, and patience.

■ 18.7 MAP LAYOUT

In general, the subject area of the map should be plotted at the largest scale that will enable it to fit neatly within its borders without producing overcrowding. It should also be centered on the map sheet and, if possible, should be aligned so that the edges of the map sheet coincide with the cardinal directions. If this is not done, users may experience some confusion when viewing the map. Accordingly, the size and shape of the map sheet, the size and shape of the area to be mapped, the orientation of the subject area on the map sheet, and map scale, must be jointly considered in map layout.

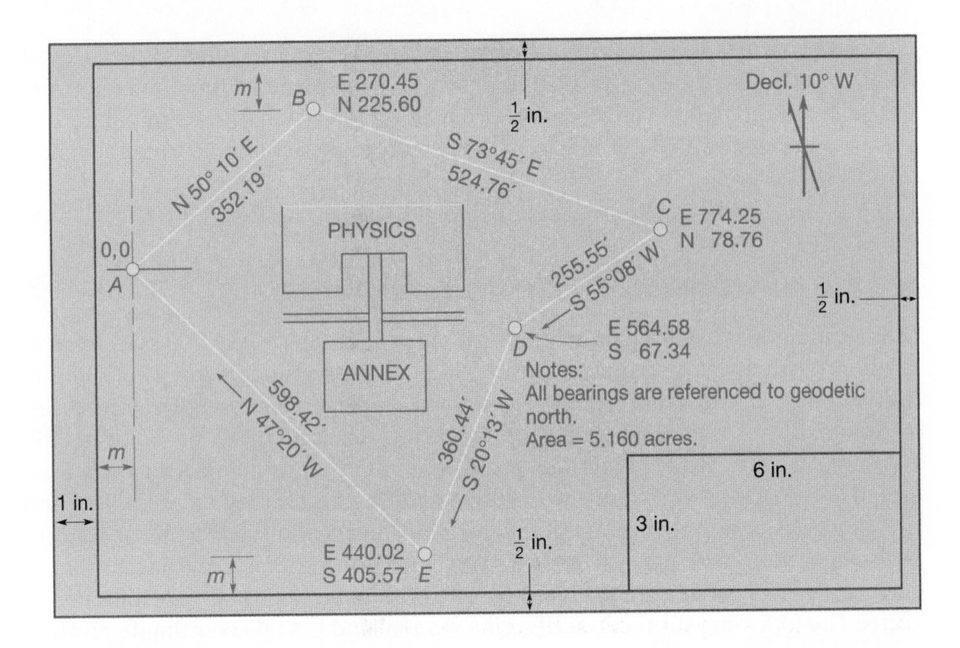

Figure 18.2
Map layout.

To illustrate, consider the example of Figure 18.2, which is a simple traverse from a planimetric survey. Before any plotting is done, the proper scale for a sheet of given size must be selected. Assume in this example that an 18- by 24-in. sheet will be used, with a 1-in. border on the left (for possible binding) and 1/2-in. borders on the other three sides. A borderline somewhat heavier than all other lines can be drawn to outline this area. If the most westerly station (*A* in the example) has been chosen as the origin of coordinates, then divide the total departure to the most easterly point *C* by the number of inches available for plotting in the east-west direction. The maximum scale possible in Figure 18.2 is 774.25 divided by 22.5, or 1 in. = 34 ft. The nearest standard scale that will fit is 1 in. = 40 ft.

This scale must be checked in the *Y* direction by dividing the total difference in *Y* coordinates, 225.60 + 405.57 = 631.17, by 40 ft, giving 15.8 in. required in the north-south direction. Since 17 in. are usable, a scale of 1 in. = 40 ft is satisfactory, although a smaller scale would yield a larger border margin. If a scale of 1 in. = 40 ft is not suitable for the map's purpose, a sheet of different size should be selected or, alternatively, more than one sheet employed to map the required area.

In Figure 18.2, the traverse is centered between the borderlines in the *Y* direction by making each distance m equal to $1/2(17 - 631.17/40)$, or 0.61 in. The same 0.61 in. can be used for the left side. Weights of the title, notes, and north arrow compensate for the traverse being to the left of the sheet center, and leave ample space for including the necessary auxiliary elements[5] of the map.

[5]Auxiliary elements include the map title block, notes, legend, bar scale, and north arrow. These elements are described in Section 18.12

If a compromise choice must be made between map scale and the sizes of auxiliary elements, it is better to maximize the map's scale and minimize the size of the auxiliary elements. Beginners should avoid using oversized auxiliary elements to use up available or leftover space since doing so detracts from the map's order and balance.

■ 18.8 BASIC MAP PLOTTING PROCEDURES

Map plotting may be done either manually or by automated CADD systems. Regardless of which method is used, the procedure consists fundamentally of plotting individual points. Lines are then drawn from point to point to portray features. Although this process may seem simple in principle, accurate work requires skill, patience, and care. While points could be laid out using angles and distances, or lines scaled and plotted directly, the most convenient method for laying out points and drafting maps involves plotting points by coordinates. Plotting by this procedure is also consistent with today's modern data collection systems, i.e., total station instruments and portable GPS units, because those devices provide coordinates directly. The following subsections describe manual and CADD coordinate plotting procedures.

18.8.1 Manually Plotting by Coordinates

To plot points by coordinates, the map sheet is first laid out precisely in a grid pattern with unit squares of appropriate size. Squares of 2, 4, or 5 in. are commonly used, and depending on map scale, they may represent 100, 200, 400, 500, or 1000 ft, or 50, 100, 200, or 500 m. The grids are constructed using a sharp, hard pencil, and are checked by carefully measuring diagonals. The grid lines are labeled with coordinate values, making sure that the range of coordinates covered on the map will accommodate the most extreme X and Y coordinates to be plotted.

Initially, coordinates for all features to be mapped must be determined. These points are then plotted by laying off their X and Y coordinates from the ruled grid lines. Mistakes in plotting can be detected by comparing scaled lengths (and directions) of lines with their measured or computed values. Since each point is plotted independently, a mistake in one will not affect the others and that point can simply be corrected.

Many mapping elements such as bar scales, legends, north arrows, etc. are prepared on separate map sheets. These items along with the map are then cut to their dimensions and manually located for optimal presentation. After the desired arrangement is achieved, the entire layout is copied onto a more stable medium as discussed in Section 18.5.

18.8.2 Plotting Using CADD

Fundamentally, CADD systems plot points and lines in a manner similar to manual drafting techniques. However, compared to manual map drafting, computer-assisted mapping offers advantages of increased accuracy, speed, flexibility, and reduced cost. Computers are capable of quickly performing many drafting chores that are tedious and time consuming if done by manual methods, e.g., drawing complicated line types and symbols and performing lettering. With CADD systems, lettering reduces to simply choosing letter sizes and styles and selecting and monitoring placement. Since these systems can often read files of coordinates, such as those from data collectors, the plotting process can become almost totally automated (see Section 17.11). For example, many common features of a map such as bar scale, north arrow, legend, and title block can be created as *blocks* and imported into any map with varied scales. This process simplifies the entire map production process and creates a standardized look for a mapping agency or company. Additionally, the digital environment of a CADD system allows for the easy arrangement of the mapping elements, which simplifies the process of map design and enables colors to be readily selected and changed.

In preparing the topographic data for computer drafting, it is advantageous to develop files that group similar categories of features in separate layers. As an example, individual layers can be created for buildings, vegetation, transportation routes, utilities, hydrology, contours, and so on. By developing the data structure in this manner, various types of maps at different scales can be produced from the same original topographic data file. This is particularly advantageous in mapping for LISs and GISs. This feature also enables specialized products to be created such as *ephemeral maps*—i.e., those produced only for current conditions and then redrawn as the conditions change. Examples of these types of maps are those used in guidance, where a map is generated showing the location of a vehicle at a particular moment in time. As the vehicle moves, the map is instantaneously redrawn on the monitor to replace its earlier version.

If digital map files developed during CADD drafting processes are suitable for importing into GISs, it is important that all closed features are actually "physically" closed in the mapping files. As shown in Figure 18.3, a frequent mistake in CADD mapping is the failure to close polygons that appear on the screen to be "visually" closed, but which in fact are not. Since GIS software packages use polygons to represent features (see Section 28.4), the visually closed but physically open features could appear as simply a series of random lines or even be viewed as errors in the drawing when imported into the GIS package.

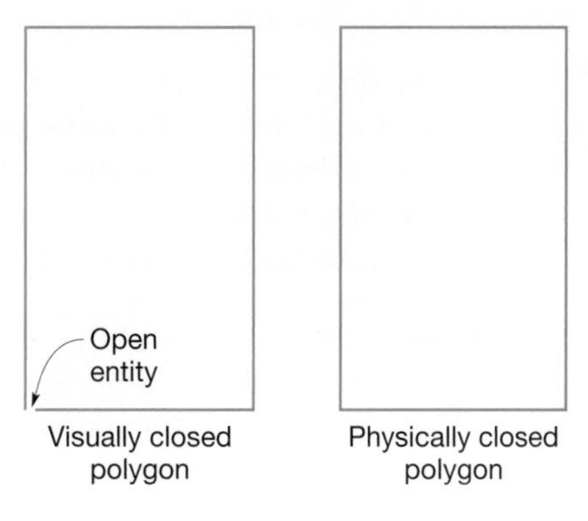

Figure 18.3 A visually closed feature versus a physically closed feature.

18.9 CONTOUR INTERVAL

As noted in Section 17.5, the choice of contour interval to be used on a topographic map depends on the map's intended use, required accuracy, type of terrain, and scale. If, according to National Map Accuracy Standards (see Section 18.4), elevations can be interpolated from a map to within one-half the contour interval, and then if elevations taken from the map must be accurate to within ±1 ft, a 2-ft maximum interval is necessary. However, if only ±10-ft accuracy is required, a 20-ft contour interval will suffice.

Terrain type and map scale combine to regulate the contour interval needed to produce a suitable density (spacing) of contours. Rugged terrain requires a larger contour interval than gently rolling country and flat ground mandates a relatively small one to portray the surface adequately. Also if the map scale is reduced, the contour interval must be increased; otherwise, lines are crowded, confuse the user, and possibly obscure other important details.

For average terrain, the following large and medium map scales and contour interval relationships generally provide suitable spacing:

English System		Metric System	
Scale (ft/in.)	**Contour Interval (ft)**	**Scale**	**Contour Interval (m)**
50	1	1:500	0.5
100	2	1:1000	1
200	5	1:2000	2
500	10	1:5000	5
1000	20	1:10,000	10

18.10 PLOTTING CONTOURS

In plotting contours, points used in locating them are first plotted on the map following techniques described in Section 18.8. Contours found by the *direct method* (see Section 17.7.1) are sketched through the plotted points. Interpolation between plotted points is necessary for the *indirect method* (see Section 17.7.2).

Interpolation to find contour locations between points of known elevation can be done in several ways:

1. Estimating.
2. Scaling the distance between points of known elevation and locating the contour points by proportion.
3. Using special devices called *variable scales,* which contain a graduated spring. The spring may be stretched to make suitable marks fall on the known elevations.
4. Using a triangle and scale, as indicated in Figure 18.4. To interpolate for the 420 ft contour between point A at elevation 415.2 and point B at elevation 423.6, first set the 152 mark on any of the engineer's scales opposite A. Then, with one side of the triangle against the scale and the 90° corner at 236, the

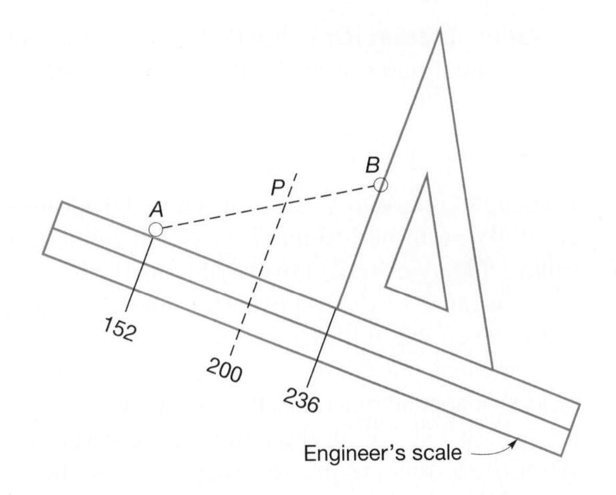

Figure 18.4 Interpolating using engineer's scale and triangle.

scale and triangle are pivoted together around A until the perpendicular edge of the triangle passes through point B. The triangle is then slid to the 200 mark and a dash drawn to intersect the line from A to B. This is the interpolated contour point P.

Contours are drawn only for elevations evenly divisible by the contour interval. Thus for a 20-ft interval, elevations of 800, 820, and 840 are shown, but 810, 830, and 850 are not. To improve legibility, every fifth line (those evenly divisible by five times the contour interval) is made heavier. So for a 20-ft interval, the 800, 900, and 1000 lines would be heavier.

■ 18.11 LETTERING

An important part of the contents of any map is its textual information. The title and all feature identifiers, coordinate values, contour elevations, and other items must be clearly identified. To produce a professional-looking drawing and one that clearly conveys the intended information, a suitable style of lettering must be selected. That style should be used consistently throughout the map, but the size varied in accordance with the importance of each particular item identified. Lettering that is too big or bold should not be used, but the letters must be large enough to be readable without difficulty.

Lettering should be carefully placed so that it is clearly associated with the item it identifies and so that letters do not interfere with other features being portrayed. Typically, the best balance results if names are centered in the objects being identified. Also both appearance and clarity are generally improved by aligning letters parallel with linear objects that run obliquely, as has been done with the traverse lengths and bearings of Figure 18.2. For ease in map reading, letters should be placed so that the map is read from either the bottom or its right side.

Text should take precedence over line work. If necessary, lines should be broken where text is placed, as this improves clarity. An example of this is in the labeling of contours, where the lines are preferably broken and the contour elevation inserted in the break. It is best to select straight, or nearly straight, sections of contours for labeling. Contours should not be labeled around tight turns since this will remove valuable topographic information expressed by the contours. When manually drafting a map from a manuscript, the text should be lettered before the line work, and then during drafting the lines can be broken where text is encountered. When using automated drafting techniques, manuscripts must be carefully examined to make sure that the text and lines do not overwrite each other and any observed overwrites corrected.

Because of the importance of lettering to a map's overall appearance and utility, even when drafting is done manually, the text is seldom hand-lettered. Rather, mechanical lettering devices that produce uniform sizes and styles of inked letters, or special machines that print letters on adhesive-backed transparent tape are usually used. With the latter device, a variety of fonts and sizes are available. After the letters are printed, they are pasted onto the map, but can be lifted and moved later if necessary.

In computer-assisted mapping, lettering is greatly simplified. A wide choice of fonts and sizes is available and letters can be easily placed, aligned, rotated, and moved. However to ensure a good-quality final product, the same rules stated previously for manual lettering should be followed when using CADD. A common mistake for beginners is to use too many fonts.

■ 18.12 CARTOGRAPHIC MAP ELEMENTS

Notes, legends, bar scales, meridian arrows, and title blocks are essential cartographic elements included on maps. *Notes* cover special features pertaining specifically to a particular map. The following are examples:

All bearings are true (or magnetic or grid).
Coordinates are based on the NAD83.
Datum for elevations is the NAVD88.
Area by calculation is X acres (or hectares).

Notes must be in a prominent place where they are certain to be seen upon even a cursory examination of the map. The best location is near the title block. This is suggested because the user of a map will find and identify a specific plot by its title and then also check any special notes beside the title before examining the drawing.

Cartographic symbols and different line types are commonly used to represent and portray different topographic features on maps and *legends* are employed to explain the meaning of those symbols and lines. Figure 18.5 gives some of the hundreds of symbols and line types employed in topographic mapping. The symbols shown in the legend should be replicas of those used on the map. In a CADD environment, it is usually expedient to copy the element to the legend from the map and reduce its scale to match the size of the font. Often legend symbols are created as blocks in CADD and recalled for later use in both the map and legend. Any symbol that is not self-explanatory should appear in the legend. Often the legend can be used to balance other map elements. Sometimes, especially if there are an

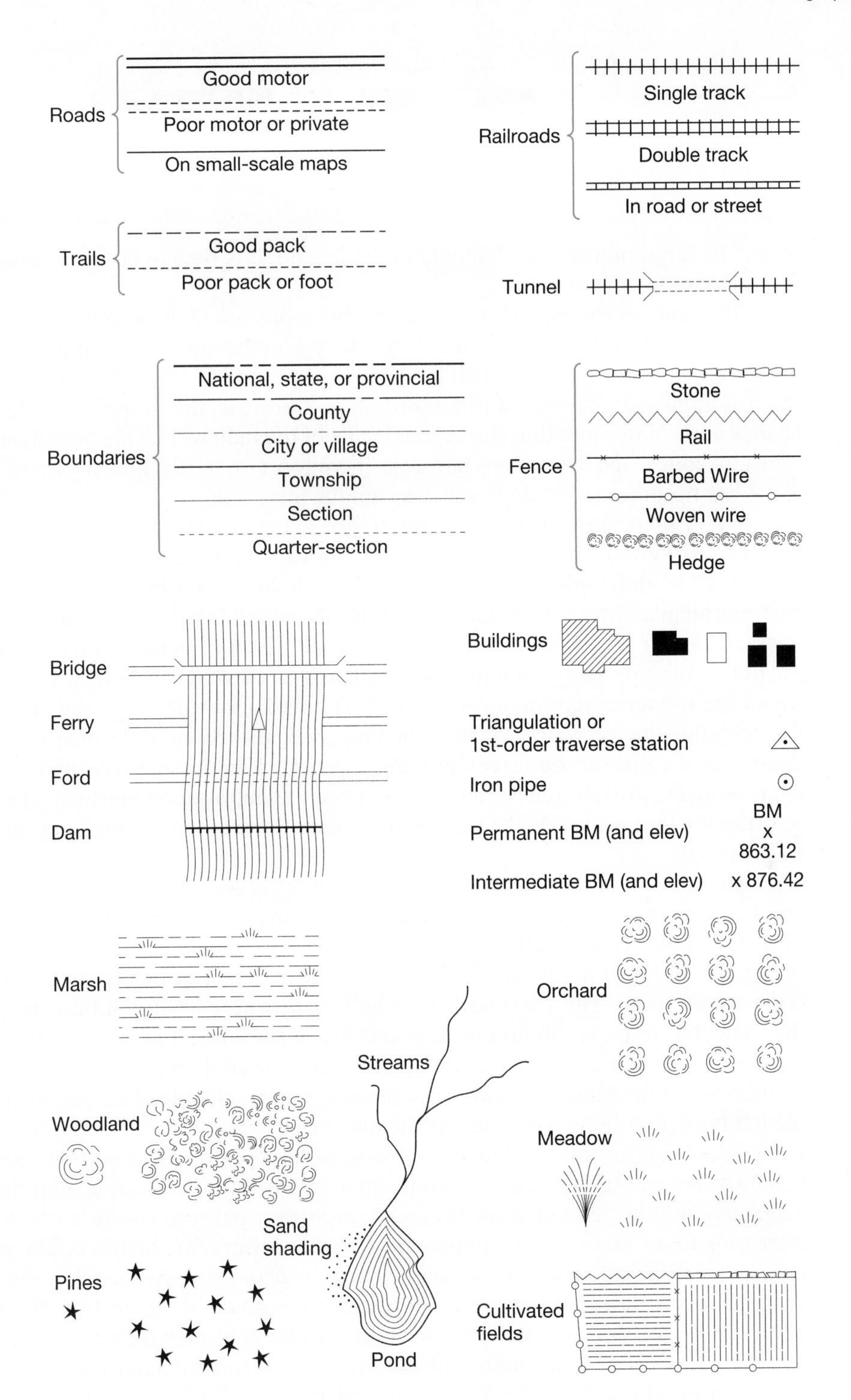

Figure 18.5 Topographic symbols.

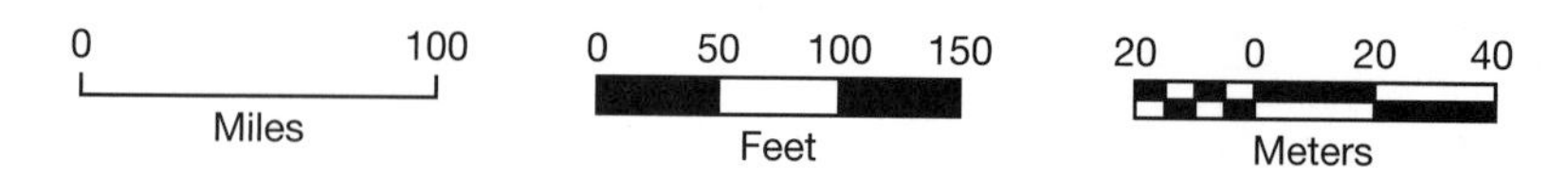

Figure 18.6 Typical graphical scales in mapping.

unusually large number of elements in the legend, it is best to create a separate legend sheet for the map.

The *scale* of the map should preferably be presented as both a representative fraction and a graphical element. A few typical examples of graphical scales are shown in Figure 18.6. Note that the units are associated with each scale. If a map sheet is enlarged or reduced in a reproduction process, the graphical scale will change accordingly and thus the original scale of the map will be preserved on the reproduction. When designing a bar scale, it is important to maintain a narrow bar. If the bar becomes too wide, it will draw undue attention.

Every map must display a *meridian arrow* for orientation purposes. However, the arrow should not be so large or elaborate that it becomes the focal point of a sheet. Geodetic, grid, or magnetic north (or all three) may be shown. Often the true-meridian arrow is identified by a full head and full feather; and a grid and/or magnetic arrow by a half-head and half-feather. The half-head and half-feather are put on the side away from the true north arrow to avoid touching it. The identity of the reference meridian used should be noted above or below the arrow in text to define the reference system. When magnetic directions are shown, the declination at the time of the survey should be indicated on the map. If a grid meridian is used, the grid system should be referenced. If an assumed meridian is used, information that will allow the reader to recreate the meridian in the field should be provided.

The *title block* should state the type of map, name of property or project and its owner or user, location or area, date completed, scale, contour interval, horizontal and vertical datums used, and for property surveys, the name of the surveyor with his or her license number. Additional data may be required on special-purpose maps. The title block may be placed wherever it will best balance the sheet, but it always should be kept outside of the subject area. Searching for a particular map in a bound set of maps, or loose pile of drawings, is facilitated if all titles are in the same location. Since sheets are normally filed flat, bound along the left border, or hung from the top, the lower right-hand corner is the most convenient location for the title block. Lettering within the title block should be simple in style rather than ornate and conform in size with the individual map sheet. Emphasis should be placed on the most important parts of the title block by increasing letter size or using uppercase (capital) letters for them. Perfect symmetry of outline about a vertical centerline is necessary since the eye tends to exaggerate any defection. An example title block is given in Figure 18.7. No part of a map better portrays the artistic ability of the compiler than a neat, well-arranged title block. Today many companies and government agencies use sheets with preprinted title forms to be filled in with individual job data or CADD systems standard title blocks are stored, retrieved, and modified as appropriate for each new project.

The Pennsylvania State University
Surveying Program

SURVEY OF CAMPUS

Scale: 1:480	Date: 16 Sept. 2000
Survey by: P. Dills and J. Tills	Map by: S. Smith

Figure 18.7
Title arrangement.

■ 18.13 DRAFTING MATERIALS

Polyester film and tracing and drawing papers are the materials commonly used for preparing maps in surveying and engineering offices. Polyesters, such as Mylar, are by far most frequently employed because they are dimensionally stable, strong, durable, and waterproof. In addition, they take pencil, ink, and stickup items, and withstand erasing, so they are ideal for manual drafting. Tracing papers are available in a variety of grades and good ones also are stable, take pencil, ink, and stickups, and endure some erasing. Both Mylar and tracing paper are transparent, so blueprints can be made from them.

Papers of different types and grades are used for printing maps made with CADD systems. When CADD is used, the paper quality can be relaxed somewhat because erasing, stickup lettering, etc., will not be necessary. *If accurate measurements are to be extracted from the maps, then material with good dimensional stability should be used.*

■ 18.14 AUTOMATED MAPPING AND COMPUTER-AIDED DRAFTING SYSTEMS

Digital computers have had a profound impact in all areas of life and surveying is certainly no exception. *Automated mapping* (AM), and *computer-aided drafting and design* (CADD) systems have now become commonplace in surveying and engineering offices throughout the world. Generic CADD systems developed for general drafting and engineering work are widely used for map drafting. In addition, special AM systems have been designed specifically for surveying, mapping, and GIS work.

The hardware necessary for AM and CADD systems varies, but as a minimum it will include a computer with a hard drive, at least one disk drive, and a high-resolution monitor; an input device such as a digitizer and/or mouse; and a plotting device. The most important component of any CADD system is its software. This enables an operator to interact with the computer and activate the system's various functions.

CADD systems enable operators to design and draw manuscript maps in real time using the computer. A visual display of the manuscript can be examined on the monitor as it is being compiled, and any additions, deletions, or changes can be made as needed. Lines can be added, deleted, or their styles altered; placement of symbols and lettering modified; and lettering sizes and styles varied. Parts of the drawing may be "picked up" and moved to other areas to resemble a "cut and paste" operation that is handy for subdivision design or placement of frequently occurring symbols. A zoom feature allows more complicated or crowded parts of the manuscript to be magnified for a better view. In the end, the map can be checked for completeness and accuracy and when the operator is satisfied that all requirements are met and the design is optimal, the final product plotted.

Required input to a computer for automated mapping includes a set of specific mapping instructions and a file of point locations and elevations. The instructions will include map scale, contour interval, line styles, lettering sizes and styles, symbols, and other items of information. Point locations are usually entered as a file of X, Y, Z coordinates, but angle and distance data can be entered and the coordinates computed. Special CADD systems for mapping with data collected by total station instruments and portable GPS units have been developed.

As explained in Section 17.8, most automated mapping systems draw contours after constructing a *triangulated irregular network* (TIN) model. These are networks of non-overlapping triangles, which represent the individual facets of the terrain. The computer interpolates contour crossings along the edges of the triangles and then draws the contours. A portion of a TIN model from an actual mapping project is illustrated in Figure 18.8(a) and the contours constructed from it are shown in Figure 18.8(b).

To understand the process of automatically constructing a contour line using a TIN, imagine contouring the TIN shown in Figure 18.9. To begin the procedure, a line on the edge of the TIN is randomly selected for contouring; in this instance line 1 has been chosen. Assume that the X, Y, Z coordinates of the end points of line 1 are (5401.08, 4369.79, 865.40) and (5434.90, 4456.90, 868.30), respectively, and that the 868-contour line is to be drawn. From the X and Y coordinate values, and applying Equations (10.11) and (10.12), the length and azimuth of line 1 are

(a)

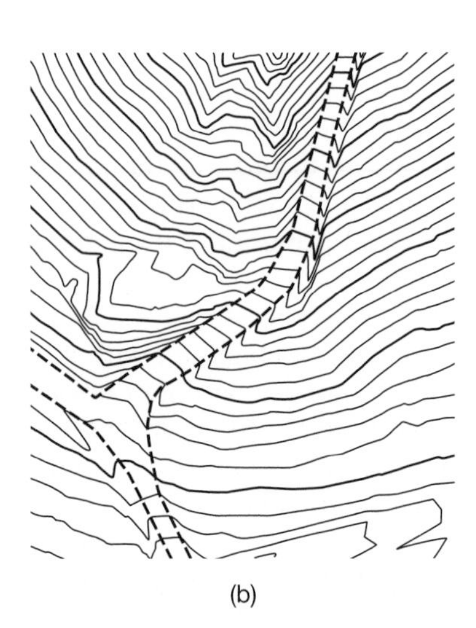

(b)

Figure 18.8 (a) Triangulated irregular network (TIN) model derived from digital-elevation model. (b) Contours derived by automated mapping system from TIN model of (a). Note the roadway edges were defined by breaklines in (a). (Courtesy Wisconsin Department of Transportation)

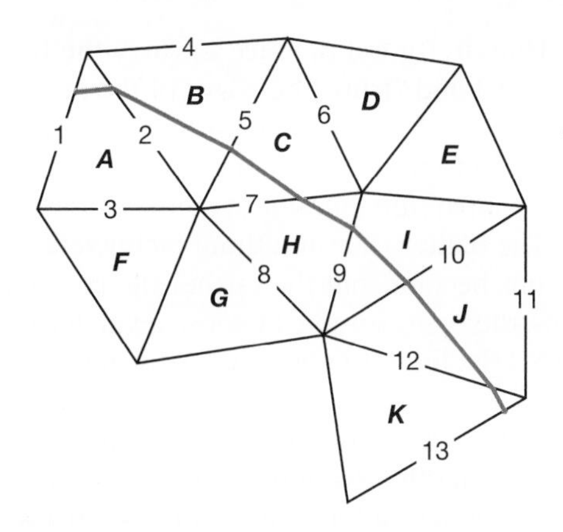

Figure 18.9 Automated contouring of TIN example.

93.45 ft and 21°13′06″, respectively. Also, from the Z coordinates, the elevation difference is $(868.30 - 865.40) = 2.9$ ft. The elevation difference from the first end point of line 1 to the 868-contour line is $(868.00 - 865.40) = 2.6$ ft. Now the following equivalent ratios can be formed:

$$\frac{2.6}{2.9} = \frac{\Delta L}{93.45}$$

$$\Delta L = 93.45 \times \frac{2.6}{2.9} = 83.78 \text{ ft}$$

This computation indicates that the distance, (ΔL) from the first end point of the line to the 868-contour line is 83.78 ft. Using the previously derived azimuth for line 1, and Equations (10.7), the X, Y, Z coordinate values for the intersection of line 1 and the 868-contour are (5431.40, 4447.89, 868.0), respectively. This is the initial point of the 868-ft contour. Now a search algorithm checks the elevations of the end points of lines 2 and 3 (the other two sides of triangle A) for the continuation of the 868-contour. Once it determines that line 2 contains the continuation of the line, it again uses the same linear interpolation procedure to determine the coordinates of the intersection of the 868-contour with line 2, and draws the contour line from line 1 to 2. The contour is now ready to enter triangle B. It proceeds to check the elevations of the end points of lines 4 and 5, and determines that the 868-contour intersects with line 5. Then it again uses linear interpolation to determine the coordinate values for the intersection and continues with the drawing of the 868-contour to line 5. It continues to draw straight-line segments for the 868-contour as it passes through each triangle until it finally exits triangle K on line 13. From the preceding, it can be seen that this algorithm is repetitive and ideally suited for computer solution.

Contours compiled automatically should be carefully edited for correctness. In certain areas, *breaklines* (see Section 17.8) may need to be added or changed to obtain the proper terrain representation. Incorrect interpolation often occurs along the outer edges of automatically contoured areas, so these areas require special processing and additional data. Thus, it is good practice to carry the field survey somewhat beyond the area of interest and "trim" the edges of the map.

Once all of the contours have been drawn as straight-line segments, the software then uses a *smoothing* algorithm to round the corners created at each intersection. The operator can usually control the amount of smoothing with a single entry called a *smoothing factor*. The higher the smoothing factor selected by the user, the smoother the intersections become, but the farther the contours depart from their computed values. Thus, the operator must choose a value carefully for the smoothing factor to ensure that the lines do not excessively depart from their original positions.

With terrain information stored in the computer in the form of TIN models, profiles and cross sections along selected lines can be derived automatically and plotted if desired. By including grade lines and design templates, earthwork computations can be made and stakeout information automatically derived for projects such as highways, railroads, and canals.

The *three-dimensional perspective grid* is an alternative form of terrain representation. It can also be produced by the computer from TIN models. The example shown in Figure 18.10 illustrates its important advantage—it gives a very vivid impression of relief.

Figure 18.11 shows a CADD workstation. An operator controls the system by making entries on the keyboard or by selecting instructions from a command board (*menu*) with a cursor.

Figure 18.12 is a portion of a topographic map for an engineering design project created with a CADD system, and Figure 18.13 is a subdivision plat, also designed and drawn using CADD. The condominium plat shown in Figure 21.5 is another example of a product designed and drafted with CADD.

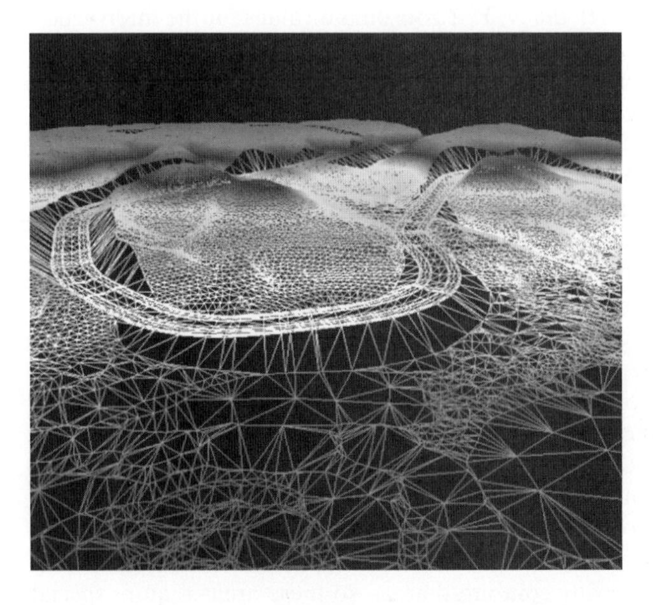

Figure 18.10 Three-dimensional perspective grid. (Courtesy Ashtech Corp.)

Figure 18.11 Computer-aided drafting and design (CADD) workstation. (Courtesy Tom Pantages)

There are numerous advantages derived from using CADD systems in map design and drafting. A major one is increased speed in completing projects. Others include reduction or elimination of errors, increased accuracy, and preparation of a consistently more uniform final product. With completed maps stored in digital form, copies can be quickly reproduced at any time and revisions easily made.

Map data compiled using CADD systems can be stored in a data bank, with different numerical codes for each of the various kinds of features. They can be retrieved later for plotting in total, or in so-called *layers,* or parts, for special-purpose maps. For example, a city engineer may only be interested in a map showing the roads and utilities, while the assessor may want only property boundaries and buildings. This concept of layered maps is fundamental to land and geographic information systems (see Chapter 28).

Another significant advantage of producing maps in digital form is that they can be transmitted electronically from one office to others at remote locations. As an example, the Wisconsin Department of Transportation produces digital maps photogrammetrically for roadway design at its central office in Madison. Using a telephone modem and/or the Internet, these maps can then be transmitted instantaneously to any of nine district offices located throughout the state, where they are immediately available to engineers there for computer-aided design, or hard copies can be printed. In spite of the many improvements made in automated mapping systems, there is still a possibility that mistakes can occur. For this reason, it is good practice to have the field party chief, who is familiar with the area, review the completed maps. The different AM and CADD systems all have varying individual capabilities. Books and brochures available from the manufacturers provide detailed descriptions.

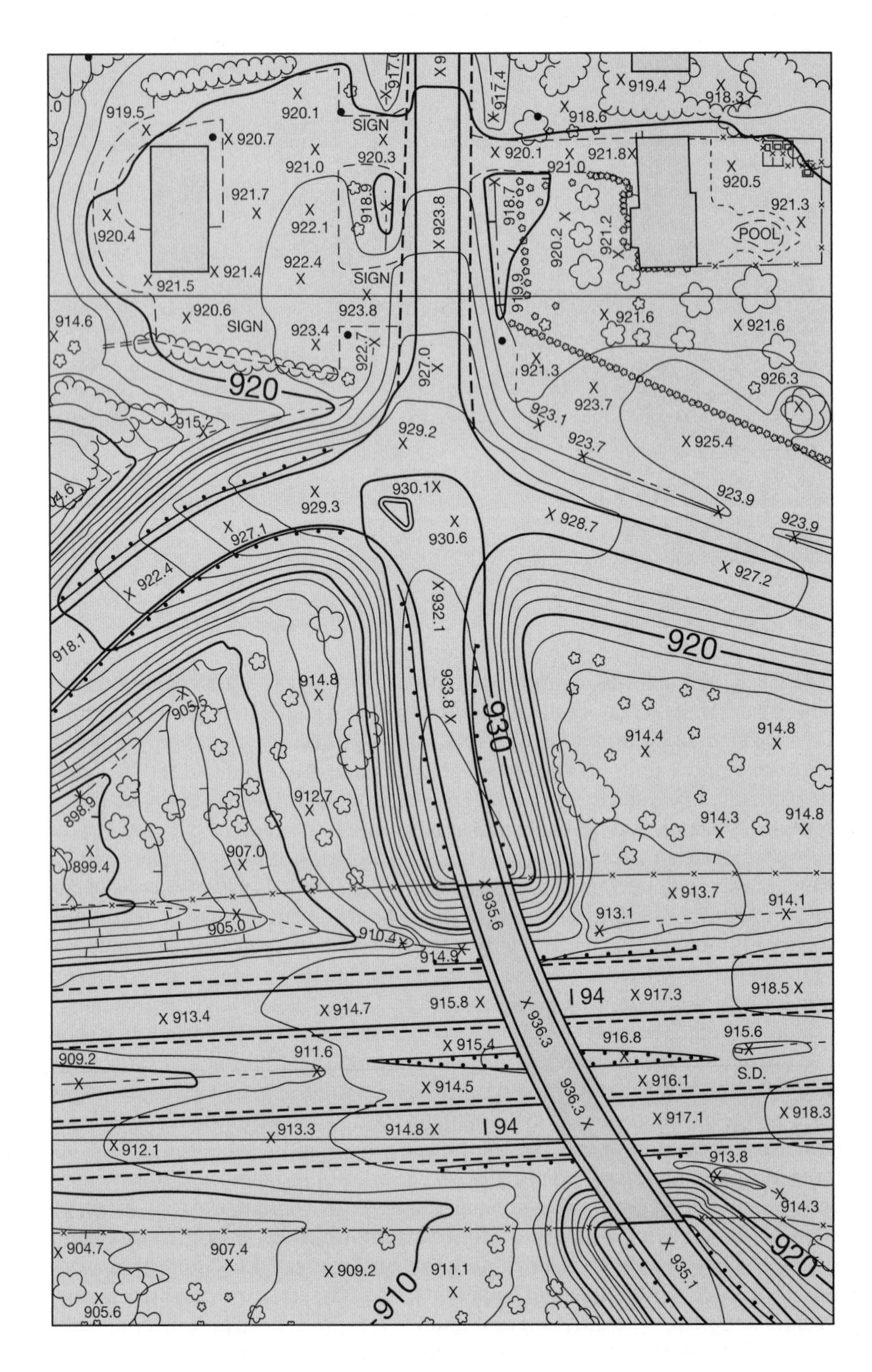

Figure 18.12 Engineering design map prepared using CADD system. (Courtesy Wisconsin Department of Transportation)

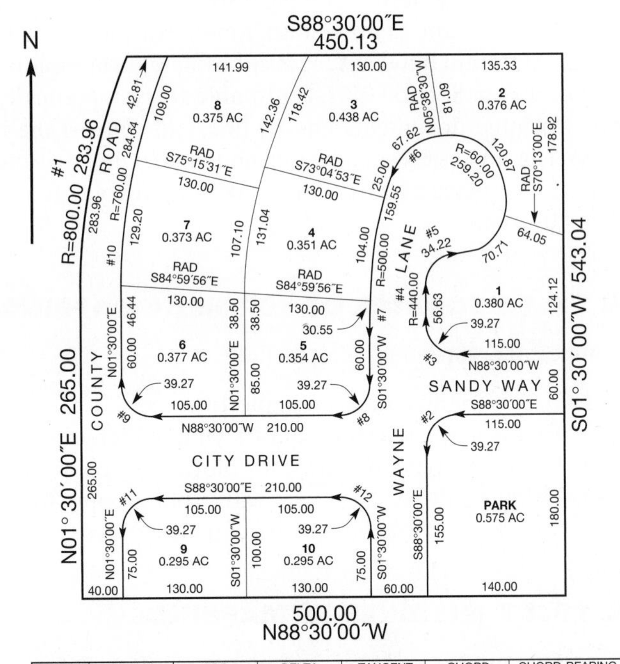

NUMBER	RADIUS	ARC	DELTA	TANGENT	CHORD	CHORD-BEARING
1	800.00	283.96	020° 20′ 14″	143.49	282.47	N11° 40′ 07″E
2	25.00	39.27	090° 00′ 00″	25.00	35.36	N46° 30′ 00″E
3	25.00	39.27	090° 00′ 00″	25.00	35.36	N43° 30′ 00″W
4	440.00	56.63	007° 22′ 26″	28.35	56.59	N05° 11′ 13″E
5	25.00	34.22	078° 25′ 48″	20.40	31.61	N48° 05′ 20″E
6	60.00	259.20	247° 31′ 14″	---	99.76	N36° 27′ 23″W
7	500.00	159.55	018° 17′ 00″	80.46	158.88	S10° 38′ 30″W
8	25.00	39.27	090° 00′ 00″	25.00	35.36	S46° 30′ 00″W
9	25.00	39.27	090° 00′ 00″	25.00	35.36	N43° 30′ 00″W
10	760.00	284.64	021° 27′ 32″	144.01	282.98	N12° 13′ 46″E
11	25.00	39.27	090° 00′ 00″	25.00	35.36	N46° 30′ 00″E
12	25.00	39.27	090° 00′ 00″	25.00	35.36	S43° 30′ 00″E
13	630.00	287.96	026° 11′ 19″	146.54	285.46	N14° 35′ 40″E

Figure 18.13 Subdivision map compiled automatically by computer-driven plotter. (Courtesy Technical Advisors, Inc.)

18.15 IMPACTS OF MODERN LAND AND GEOGRAPHIC INFORMATION SYSTEMS ON MAPPING

Land Information Systems (LISs), Geographic Information Systems (GISs), and automated mapping and facilities management (AM/FM) systems all require enormous quantities of position-related land data. From this information, maps and other special-purpose graphic displays can be made and analyzed. For example, a typical LIS or GIS may include attribute data such as political boundaries, land ownership, topography, land use, soil types, natural resources, transportation routes, utilities, and many others. From the stored information, a user can display a map of each attribute category (or *layer*) on a screen, or several layers of data can be merged to produce combination maps. This merging, or *overlaying,* concept, discussed in Chapter 28, greatly facilitates data analysis and aids significantly in management and decision-making. If printed maps of any selected layers or combinations are desired, they can be produced rapidly using automated drafting equipment.

The position-related land attribute data needed for LISs and GISs can be collected from a variety of sources and entered in the computer. These sources include field surveys (see Chapter 17), aerial photographs (see Chapter 27), and existing maps. Modern surveying instruments such as the total station instruments, GPS receivers, and digital photogrammetric plotters can produce large quantities of digital terrain data in X, Y, Z coordinate form rapidly and economically. *Raster scanners* (see Section 28.7.4) are able to systematically scan existing maps and other printed documents line-by-line, and convert the information to numerical form. The processes of collecting and digitizing data to support LISs and GISs are expected to place a heavy workload on surveyors (geomatics engineers) for many years to come.

18.16 SOURCES OF ERROR IN MAPPING

Sources of error in mapping include:

1. Errors in the data used in plotting.
2. Errors in the scales used for laying out lengths and coordinate values.
3. Errors in laying out grids for plotting by coordinates.
4. Using a soft pencil, or one with a blunt point, for plotting.
5. Variations in the dimensions of map sheets due to temperature and moisture.

18.17 MISTAKES IN MAPPING

Some common mistakes in mapping are:

1. Selecting an inappropriate scale or contour interval for the map.
2. Failing to check grids by measuring diagonals, and not checking points plotted from coordinates by measuring distances between them.
3. Using the wrong edge of an engineer's scale.
4. Making the north arrow too large or too complex.
5. Neglecting to identify the meridian of reference, i.e., geodetic, grid, magnetic, etc.
6. Omitting the scale or necessary notes.
7. Failing to balance the sheet by making a preliminary sketch.
8. Drafting the map on a poor-quality medium.
9. Failing to realize that errors are also magnified when maps are enlarged electronically or photographically.
10. Operating AM and CADD systems without sufficient prior training.

PROBLEMS

Asterisks (*) indicate problems that have answers given in Appendix G.

18.1 Give the terms to which the acronyms CADD, LIS, and GIS apply.

18.2* On a map drawn to a scale of 1:9600, a point has a plotting error of 1/30-in. What is the equivalent ground error in units of feet?

18.3 List five different types of maps.

18.4 Define the term cartography.

18.5 What is the purpose of placing bar scales on maps?

18.6 What is the purpose of placing grid lines on a map?

18.7 Discuss why a map designed for a planning board hearing may not be the same as a map designed for an engineer?

18.8 What is the content of DEMs?

18.9 List the advantages of compiling maps using field-to-finish software?

18.10* For a 20-ft contour interval, what is the greatest error in elevation expected of any definite point read from a map if it complies with National Map Accuracy Standards?

18.11 An area that varies in elevation from 463–634 ft is being mapped. What contour intervals will be drawn if a 50-ft interval is used? Which lines are emphasized?

18.12 Similar to Problem 18.11, except elevations vary from 237–383 m and a 10-m interval is used.

18.13 What are the goals in map design?

18.14 Describe how clarity can be enhanced through the use of fonts.

18.15* What is the largest acceptable error in position for 90% of the well-defined points on a map with a 1:24,000 scale that meets National Map Accuracy Standards.

18.16 Discuss the importance of balance on a map.

18.17 Discuss why insets are used on maps.

18.18* If a map is to have a 1-in. border, what is the largest nominal scale that may be used for a subject area with dimensions of 604 ft and 980 ft on a paper of dimensions 24 by 36 in?

18.19 Similar to Problem 18.18, except the dimensions of the subject area are 1210 ft and 1875 ft.

18.20 Draw 2-ft contours for the data in Plate B.2 of Appendix B.

18.21 If 90 percent of all elevations on a map must be interpolated to the nearest ±5 ft, what contour interval is necessary according to the National Map Accuracy Standards? Explain.

18.22 If an area having an average slope of 3.5 percent is mapped using a scale of 1:1000 and contour interval of 0.5 m, how far apart will contours be on the map?

18.23 Similar to Problem 18.22, except average slope is 5 percent, map scale is 60 ft/in., and contour interval is 2 ft.

18.24* Similar to Problem 18.22, except average slope is 8 percent, map scale is 1:2000, and contour interval is 1 m.

18.25* The three-dimensional (X, Y, Z) coordinates in feet of vertexes A, B, and C in Figure 18.14 are (5412.456, 4480.621, 248.14), (5463.427, 4459.660, 253.12), and (5456.081, 4514.382, 236.19), respectively. What are the coordinates of the intersection of the 250-ft contour with side AB? With side BC?

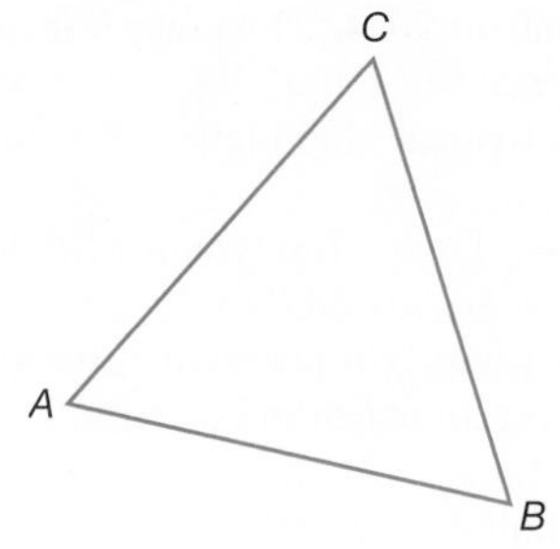

Problems 18.25 through 18.27

18.26 The three-dimensional (X, Y, Z) coordinates in meters for vertices A, B, and C in Figure 18.14 are (8649.220, 6703.670, 143.856), (8762.040, 6649.770, 165.879), and (8752.640, 6770.200, 146.844), respectively. What are the coordinates of the intersections of the 148-m contour as it passes through the sides of the triangle?

18.27 Similar to Problem 18.26, except compute the coordinates of the intersection of the 152-m contour.

18.28 Why is scale depicted graphically on a map?

The following table gives elevations at the corners of 50-ft coordinate squares, and they apply to Problems 18.29 and 18.31.

74	70	66	67	66	69
76	73	71	68	63	70
77	69	68	66	69	68

18.29 At a horizontal scale of 1 in. = 50 ft, draw 2-ft contours for the area.

18.30 Similar to Problem 18.29, except at the bottom of the table add a fourth line of elevations: 79, 78, 70, 66, 61, and 65 (from left to right).

18.31 In Problem 18.30, about which number in the table can a 2-ft closed contour be drawn?

18.32 A rectangular lot running N-S and E-W is 150 by 100 ft. To locate contours it is divided into 50-ft square blocks, and the following horizontal rod readings are taken at the corners successively along the E-W lines, proceeding from west to east, and beginning with the northernmost line: 6.7, 5.3, 3.9; 5.8, 5.9, 5.0; 5.7, 4.5, 3.6; and 3.2, 4.8, 5.8. The HI is 843.6 ft. Sketch 2-ft contours.

18.33 If a map is drawn with 10-ft contour intervals, what contours between 80 ft and 430 ft are drawn with heavier line weight?

18.34 Find the website at which USGS quadrangle sheets may be purchased.

18.35 Download and prepare a report on Part 3 of the FGDC *Geospatial Positioning Accuracy Standards.*

BIBLIOGRAPHY

American Society for Photogrammetry and Remote Sensing. 1987. "Large Scale Mapping Guidelines." Bethesda, Md.: *American Society for Photogrammetry and Remote Sensing.*

______. 1990. "ASPRS Standards for Large-Scale Maps." *Photogrammetric Engineering and Remote Sensing* 56 (No. 7): 1068.

Castagna, J. 2005. "A Picture Is Worth a Thousand Words." *Professional Surveyor* 25 (No. 11): 26.

Corbley, K. P. 2002. "Generating Online Property Maps." *Professional Surveyor* 22 (No. 1): 14.

Dronick, G. J. 2007. "Mapping Windmill Farms." *Professional Surveyor* 27 (No. 1): 14.

Felus, Y.A., A. Saalfeld, and B. Schaffrin. 2004. "Delaunay Triangulation Structured Kriging for Surface Interpolation." *Science and Land Information Science* 65 (No. 1): 27.

Keister, M. 2005. "USGS Supplies Critical Map Data in Katrina Aftermath." *Professional Surveyor* 25 (No. 11): 18.

Marin, R. M., and T. E. Martín. 2004. "Digital Bathymetric Models from Rational Profiles." *Surveying and Land Information Science* 64 (No. 4): 235.

McGray, G. L. 2005. "Maps in Our Lives." *Professional Surveyor* 25 (No. 10): 8.

Meade, Mark E. 2001. "ASPRS Accuracy Standards for Large-Scale Maps." *Point of Beginning* 26 (No. 7): 60.

Penry, J. 2007. "Denver's USGS Mapping Center." *American Surveyor* 4 (No. 1): 46.

19

Control Surveys and Geodetic Reductions

■ 19.1 INTRODUCTION

Control surveys establish precise horizontal and vertical positions of reference monuments. These serve as the basis for originating or checking subordinate surveys for projects such as topographic and hydrographic mapping; property boundary delineation; and route and construction planning, design, and layout. They are also essential as a reference framework for giving locations of data entered into Land Information Systems (LISs) and Geographic Information Systems (GISs).

Traditionally there have been two general types of control surveys: *horizontal* and *vertical.* Horizontal surveys generally establish *geodetic latitudes* and *geodetic longitudes* (see Section 19.4) of stations over large areas. From these values, plane rectangular coordinates, usually in a state plane or Universal Transverse Mercator (UTM) coordinate system (see Chapter 20) can be computed. On control surveys in smaller areas, plane rectangular coordinates may be determined directly without obtaining geodetic latitudes and longitudes.

Field procedures used in horizontal control surveys have traditionally been the ground methods of *triangulation, precise traversing, trilateration,* and combinations of these basic approaches (see Section 19.12). In addition, astronomical observations (see Appendix C) have been made to determine azimuths, latitudes, and longitudes. Rigorous photogrammetric techniques (see Chapter 27) have also been used to densify the control in an area.

The *global positioning system* (GPS) (see Chapters 13, 14, and 15) is used with increasing frequency, especially on control surveys of larger extent. GPS is rapidly replacing the other methods because of several advantages including its ease of use, speed, and extremely high-accuracy capabilities over long distances. The Federal Geodetic Control Subcommittee (FGCS) has created standards for using GPS to establish both horizontal and vertical control. The planning and execution of static GPS control surveys is covered in Chapter 14.

Vertical-control surveys establish elevations for a network of reference monuments called *benchmarks*. Depending on accuracy requirements, they have traditionally been run by either *differential leveling* or *trigonometric leveling* (see Chapters 4 and 5). As previously stated, GPS can also be used to establish vertical control, but the most accurate and widely applied method is precise differential leveling (see Section 19.13).

This chapter will define elements of the datums (geodetic reference systems) used in control surveys, describe the *National Spatial Reference System* (NSRS), discuss some of the traditional ground methods used in control surveying, and explain some basic computational methods used in making geodetic reductions.

■ 19.2 THE ELLIPSOID AND GEOID

It was noted in Section 19.1 that horizontal control surveys generally determine geodetic latitudes and geodetic longitudes of points. To explain geodetic latitude and longitude, it is necessary to first define the *geoid,* and the *ellipsoid* (sometimes called the *spheroid*). As covered in Chapter 4, the geoid is an equipotential gravitational surface located approximately at mean sea level, which is everywhere perpendicular to the direction of gravity. Because of variations in the Earth's mass distribution and the rotation of the Earth, the geoid has an irregular shape.

The ellipsoid is a mathematical surface obtained by revolving an ellipse about the Earth's polar axis. The dimensions of the ellipse are selected to give a good fit of the ellipsoid to the geoid over a large area and are based upon surveys made in the area.

A two-dimensional view, which illustrates conceptually the geoid and ellipsoid, is shown in Figure 19.1. As illustrated, the geoid contains nonuniform undulations (which are exaggerated in the figure for clarity) and is therefore not readily defined mathematically. Ellipsoids, which approximate the geoid and can be defined

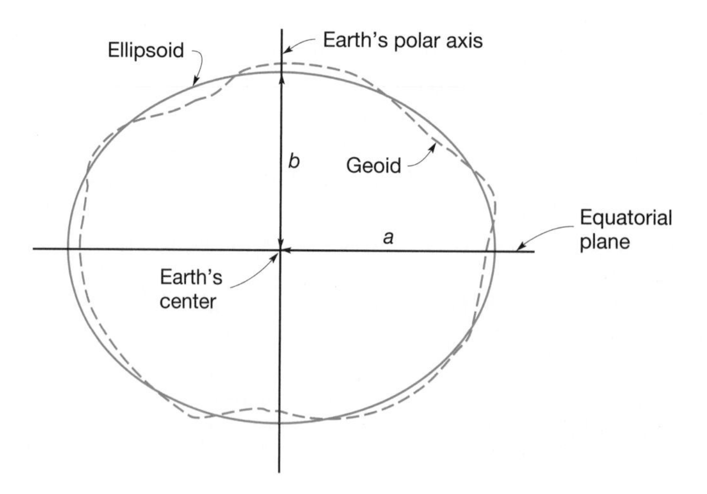

Figure 19.1
Ellipsoid and geoid.

TABLE 19.1 DEFINING ELLIPSOIDAL PARAMETERS

Ellipsoid	Semiaxis *a* (m)	Semiaxis *b* (m)	Flattening *f*
Clarke, 1866	6,378,206.4*	6,356,583.8*	1/294.978698214
GRS80	6,378,137.0*	6,356,752.3	1/298.257222101*
WGS84	6,378,137.0*	6,356,752.3	1/298.257223563*

*defining parameters for the ellipsoids

mathematically, are therefore used to compute positions of widely spaced points that are located through control surveys. The *Clarke Ellipsoid of 1866* approximates the geoid in North America very well, and from 1879 until the 1980s it was the ellipsoid used in NAD 27 as a reference surface for specifying geodetic positions of points in the United States, Canada, and Mexico. Currently, the *Geodetic Reference System of 1980* (GRS80) and *World Geodetic System of 1984* (WGS84) ellipsoids are commonly used because they provide a good worldwide fit to the geoid. This is important because of the global surveying capabilities of GPS.

Sizes and shapes of ellipsoids can be defined by two parameters. Table 19.1 lists the parameters for the three ellipsoids noted above. For the Clarke 1866 ellipsoid, the defining parameters were the semiaxes a and b. For GRS80 and WGS84, the defining parameters are the semimajor axis a and flattening f. The relationship between these three parameters is

$$f = 1 - \frac{b}{a} \tag{19.1}$$

Other quantities commonly used in ellipsoidal computations are the *first eccentricity*, e, and the second eccentricity, e', of the ellipse, where

$$e = \frac{\sqrt{a^2 - b^2}}{a} \tag{19.2a}$$

$$e = \sqrt{2f - f^2} \tag{19.2b}$$

$$e' = \frac{\sqrt{a^2 - b^2}}{b} = \frac{e}{\sqrt{1 - e^2}} \tag{19.2c}$$

Often the term eccentricity is understood to mean the first eccentricity and this book will follow that convention.

For each ellipsoid, the polar semiaxis is only about 21 km (13 mi) shorter than the equatorial semiaxis b. This means the ellipsoid is actually very nearly a sphere; hence, for some calculations involving moderate lengths (usually up to about 50 km) that assumption can be made.[1]

---

[1]In computations if the ellipsoid is assumed a sphere, its radius is usually taken such that its volume is the same as the reference ellipsoid. It is computed from $\sqrt[3]{a^2 b}$. For the GRS80 ellipsoid, its rounded value is 6,371,000 m.

Example 19.1

Using the defining parameters, what are the first eccentricities of the Clarke 1866 and GRS80 ellipsoids?

Solution

For the Clarke 1866 ellipsoid, Equation (19.2a) yields

$$e = \frac{\sqrt{6378206.4^2 - 6356583.8^2}}{6378206.4} = 0.082271854$$

For the GRS80 ellipsoid, Equation (19.2b) yields

$$e = \sqrt{\frac{2}{298.257222101} - \left(\frac{1}{298.257222101}\right)^2} = 0.081819191$$

19.3 THE CONVENTIONAL TERRESTRIAL POLE

As discussed in the preceding section, an ellipsoid is defined on the basis of the size of an ellipse that is rotated about the polar axis of the Earth. In reality, since the principal axis of inertia of the Earth does not coincide with the rotational axis of the Earth, the polar axis at any particular time is not fixed in position. Rather, as illustrated in Figure 19.2, it rotates with respect to the inertial system. This motion is generally divided into two major categories called *precession* and *nutation.* Precession is the greater of the two and is the wander of the polar axis over a long period of time. The pole makes a complete revolution once every 26,000 years. Additionally, the pole wanders in much smaller radial arcs that are superimposed upon precession. These smaller circles are known as a nutation and are completed about once every 18.6 years. By international convention, the mean rotational axis of the Earth was defined as the "mean" position of the pole between the years of 1900.0 and 1905.0. This position is known as the *Conventional Terrestrial Pole* (CTP).

The CTP defines the Z-axis of a three-dimensional global Cartesian coordinate system with the northern portion being positive. The positive X axis lies in the mean equatorial plane, begins at the mass-center of the Earth, and passes through the mean Greenwich meridian. Finally, the Y-axis also lies in the mean equatorial plane and creates a right-handed Cartesian coordinate system. This coordinate system is known as the *Conventional Terrestrial System* (CTS). The CTS is also shown in Figure 19.2.

Since 1988, the *International Earth Rotation Service* (IERS)[2] has monitored the instantaneous position of the pole with respect to the CTP using observations made by participating organizations employing advanced space methods including

[2]The instantaneous positions of the Earth's pole can be found on the IERS website at http://hpiers.obspm.ft/webiers/general/syframes/SY.htm.

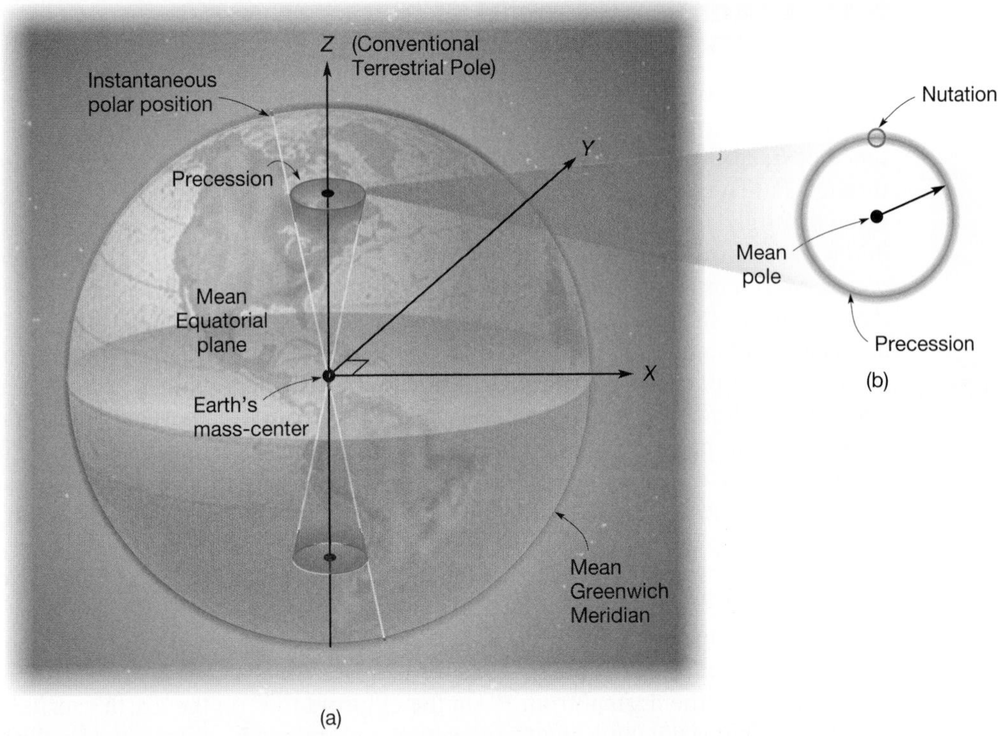

Figure 19.2 Motions of the Earth's polar axis: (a) three-dimensional and (b) plan views.

Very Long Baseline Interferometry (VLBI) and lunar and satellite laser ranging. Consequently, the CTS is now defined by a global set of stations through their instantaneous spatial coordinate positions known as the *International Terrestrial Reference Frame* (ITRF). This system is used in the computation of precise GPS orbits and is referenced through time-dependent models to other coordinate systems such as the CTS.

The instantaneous position of the pole is given in (x, y) coordinates with respect to the CTP. An application of these positions is seen in the reduction of astronomical azimuths (see Appendix C) where the observed astronomical azimuth with respect to the instantaneous position of the pole can be related to the CTP by

$$Az_A = Az_{obs} - (x \sin \lambda + y \cos \lambda) \sec \phi \tag{19.3}$$

where Az_A is the astronomic azimuth related to the position of the conventional terrestrial pole, Az_{obs} is the observed astronomical azimuth, (x, y) are the coordinates of the instantaneous pole at the time of the observation, and (ϕ, λ) are the geodetic latitude and longitude, respectively, of the observing station. Since this correction is seldom more than a few tenths of a second, and since this is below the typical observational errors associated with astronomical observations, this correction was seldom applied in practice. Today, GPS has replaced most astronomical observational procedures. Future references to the polar axis of the Earth will implicitly refer to the CTP.

19.4 GEODETIC POSITION AND ELLIPSOIDAL RADII OF CURVATURE

Figure 19.3 shows a three-dimensional view of the ellipsoid and illustrates a point P on the surface of the Earth (which in this illustration is shown to exist at a distance of h_P above the ellipsoid). Point P' is on the ellipsoid along the *normal* through P. (The normal is defined later on.) The geodetic position of point P is given by its *geodetic latitude* ϕ_P, *geodetic longitude* λ_P, and *geodetic height* h_P. To define these three terms, it is necessary to first define *meridians* and *meridian planes.* Meridians are great circles on the circumference of the ellipsoid that pass through the north and south poles. Any plane containing a meridian and the polar axis is a meridian plane. The angle in the plane of the Equator from the Greenwich meridian plane to the meridian plane passing through point P defines the geodetic longitude λ_P of the point. The plane defined by the vertical circle that passes through point P, *perpendicular to the meridian plane* on the ellipsoid, is called the plane of the *prime vertical* (also known as the *normal section*). The radius of the prime vertical at point P, R_N, is also called the *normal* since it is perpendicular to a plane that is tangent to the ellipsoid at P. The geodetic latitude ϕ_P is the angle, in the meridian plane containing P, between the equatorial plane and the normal at P.

To uniquely define the location of point P on the surface of the Earth, geodetic height h_P must be included. Geodetic height is the distance measured along the extension of the normal from P' on the ellipsoid to P on the Earth's surface. Geodetic height is not equivalent to elevation determined by differential leveling.

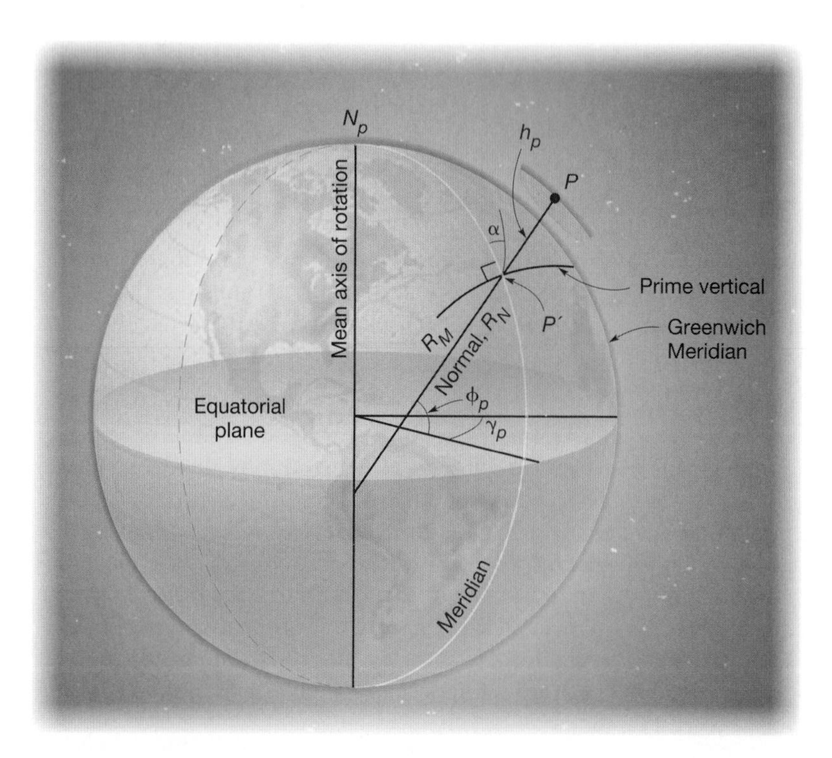

Figure 19.3 Different radii on the ellipsoid.

These differences were described in Section 13.4.3 and will be discussed further in Section 19.5.

Because the Earth is approximated by an ellipsoid and not a sphere, the great circle that defines the prime vertical at P has a radius R_N that is different than the radius in the meridian R_M at P.[3] The lengths of these two radii, which are collinear at any point, are used in many geodetic computations. Figure 19.3 shows R_N. Additionally, the radius R_α of a great circle at any azimuth α to the meridian is different from either R_N or R_M. These three radii are computed as

$$R_N = N = \frac{a}{\sqrt{1 - e^2 \sin^2 \phi}} \tag{19.4}$$

$$R_M = M = \frac{a(1 - e^2)}{(1 - e^2 \sin^2 \phi)^{\frac{3}{2}}} \tag{19.5}$$

$$R_\alpha = \frac{R_N R_M}{R_N \cos^2 \alpha + R_M \sin^2 \alpha} \tag{19.6}$$

where a and e are parameters for the ellipsoid as defined in Section 19.2 and ϕ is the geodetic latitude of the station for which the radii are computed. From an analysis of Equations (19.4) and (19.5), it can readily be shown that R_N equals R_M at the poles where ϕ is 90°. Also, since the quantity $(1 - e^2)$ is less than one, the radius of the prime vertical R_N is greater than the radius of the meridian R_M at every location other than the pole where ϕ is equal to 90°.

Example 19.2

Using the GRS80 ellipsoidal parameters, what are the radii for the meridian and prime vertical for a point of latitude 41°18′15.0132″N? What is the radius of the great circle that is at an azimuth of 142°14′36″ at this point?

Solution

From Example (19.1), $e^2 = 0.081819191^2 = 0.00669438$

By Equation (19.4), the radius of the prime vertical is

$$R_N = \frac{6378137.0}{\sqrt{1 - e^2 \sin^2(41°18'15.0132'')}} = 6{,}387{,}458.536 \text{ m}$$

By Equation (19.5), the radius of the meridian is

$$R_M = \frac{6378137(1 - e^2)}{[1 - e^2 \sin^2(41°18'15.0132'')]^{\frac{3}{2}}} = 6{,}363{,}257.346 \text{ m}$$

[3]Note that R_N is often referred to as N, and R_M is frequently designated as M.

By Equation (19.6), the radius of the great circle at an azimuth of 142°14′36″ is

$$R_\alpha = \frac{R_N R_M}{R_N \cos^2(142°14'36'') + R_M \sin^2(142°14'36'')} = 6{,}372{,}309.401 \text{ m}$$

19.5 GEOID UNDULATION AND DEFLECTION OF THE VERTICAL

As explained earlier, the geoid is an equipotential surface defined by gravity. If the Earth was a perfect ellipsoid without internal density variations, the geoid would match the ellipsoid perfectly. However, this is not the case, and thus the geoid can depart from some ellipsoids by as much as 100 m or more in certain locations. Traditional surveying instruments are oriented with respect to gravity and thus observations obtained with them are typically made with respect to the geoid. As can be seen in Figure 19.4 and discussed in Section 13.4.3, the separation between the geoid and the ellipsoid creates a difference between the height of a point above the ellipsoid (*geodetic height*) and its height above the geoid (*orthometric height* or *elevation*). This difference, known as *geoidal height*[4] (also called *geoidal separation* and *geoidal undulation*), can often be observed when comparing the geodetic height of a point derived by GPS, with its elevation as determined by differential leveling. The relationship between the orthometric height H and geodetic height h at any point is

$$h = H + N \tag{19.7}$$

where N is the geoidal height.

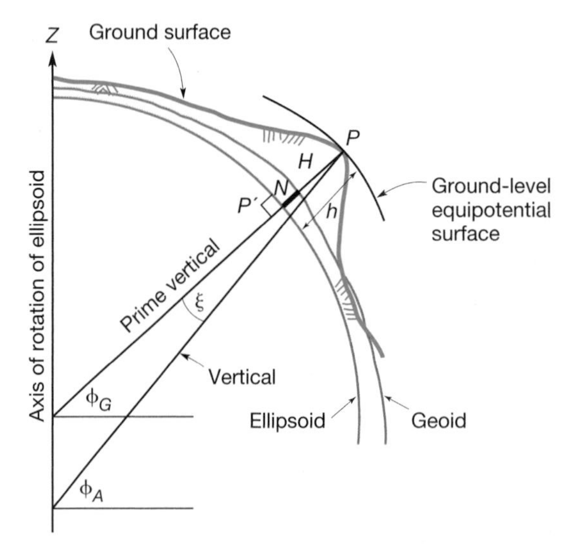

Figure 19.4 Relationships between the ellipsoid and geoid.

[4]The National Geodetic Survey regularly publishes geoid models for the U.S. Its latest version is GEOID03/. This model can be obtained from http://www.ngs.noaa.gov/PC_PROD/GEOID03/.

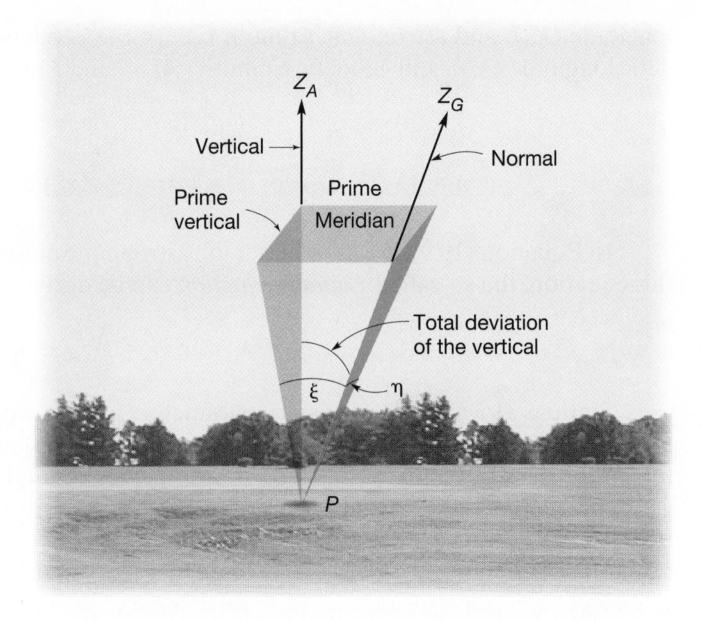

Figure 19.5
ξ and η components of deflection of the vertical.

The GRS80 and WGS84 ellipsoids were both developed with the intent of providing a good fit to the geoid worldwide. However, they yield a relatively poor fit to the geoid within the continental United States, where the average geoidal height is approximately −30 m (the negative sign means the geoid is below the ellipsoid). The Clarke 1866 ellipsoid, on the other hand, provides a very close fit to the geoid in the United States, i.e., geoidal heights are typically only a few meters.

In general, equipotential gravitational surfaces are not parallel to either the geoid or ellipsoid. This is partly due to the rotation of the Earth, which causes the surfaces to separate as they approach the equator, and partly due to density anomalies near the surface of the Earth. As a result, certain inconsistencies can occur in different types of field observations and hence corrections must be made. Some of the more significant of these corrections are discussed in Section 19.14.

As illustrated in Figure 19.5, the *deflection of the vertical* (also called *deviation of the vertical*) at any ground point P is the angle between the vertical (direction of gravity) and the normal to the ellipsoid. This angle is generally reported by giving two components: its orthogonal projections onto the meridian and normal planes. In the figure, the zenith of the ground-level equipotential surface is called the astronomical zenith Z_A since it corresponds to the direction of gravity (zenith) of a leveled instrument during astronomical observations. Also in Figure 19.5, Z_G is the normal at point P. The projected components of the total deviation of the vertical onto the meridian and normal planes are called ξ(xi) and η(eta), respectively.[5] The relationships between the astronomic latitude (ϕ_A), astronomic

[5]As with geoidal undulations, values for ξ and η can also be obtained through the National Geodetic Survey's deflection of vertical models. The latest version, DEFLEC99, can be obtained from http://www.ngs.noaa.gov/GEOID/DEFLEC99/deflec99.html.

longitude (λ_A), and astronomic azimuth (Az_A); the geodetic latitude (ϕ_G), geodetic longitude (λ_G), and geodetic azimuth (Az_G); and ξ and η are

$$\xi = \phi_A - \phi_G \tag{19.8}$$

$$\eta = (\lambda_A - \lambda_G) \cos \phi = (Az_A - Az_G) \cot \phi \tag{19.9}$$

In Equation (19.9), ϕ can be either the astronomic or geodetic latitude. From this equation, the so-called *Laplace equation* can be derived as

$$Az_G = Az_A - (\lambda_A - \lambda_G) \sin \phi = Az_A - \eta \tan \phi \tag{19.10}$$

Stations at which the necessary parameters are known, such that Equation (19.10) can be formed, are called *Laplace stations.* Note that in Equation (19.9), for points near the equator, latitude ϕ approaches 0°, and the astronomic and geodetic azimuths become essentially the same. As will be discussed in Section 19.14.3, additional corrections are needed to properly reduce an observed azimuth to the ellipsoid.

■ 19.6 U.S. REFERENCE FRAMES

Horizontal and vertical datums consist of a network of control monuments and benchmarks whose horizontal positions and/or elevations have been determined by precise geodetic control surveys. These monuments serve as reference points for originating subordinate surveys of all types and as such are known as *reference frames*. The horizontal and vertical datums used in the immediate past and at present in the United States are described in the following subsections.

19.6.1 North American Horizontal Datum of 1927 (NAD27)

In 1927, a least-squares adjustment was performed which incorporated all horizontal geodetic surveys that had been completed up to that date. This network of monumented points included in the adjustment, together with their adjusted geodetic latitudes and longitudes, was referred to as the *North American Datum of 1927* (NAD27). The adjustment utilized the Clarke ellipsoid of 1866 and held fixed the latitude and longitude of an "initial point," station *Meades Ranch* in Kansas, along with the azimuth to nearby station *Waldo.* The project yielded adjusted latitudes and longitudes for some 25,000 monuments existing at that time. Until the advent of the current datum (NAD83), positions of stations established after 1927 were adjusted in processes that held NAD27 monuments fixed.

19.6.2 North American Horizontal Datum of 1983 (NAD83)

The National Geodetic Survey (NGS) began a new program in 1974 to perform another general adjustment of the North American horizontal datum. The adjustment was deemed necessary because of the multitude of post-1927 geodetic observations that existed and because many inconsistencies had been discovered in the NAD27 network. The project was originally scheduled for completion in

1983, hence its name *North American Datum of 1983* (NAD83), but was not actually finished until 1986. The adjustment was a huge undertaking, incorporating approximately 270,000 stations and all geodetic surveying observations on record—nearly 2 million of them! About 350 person-years of effort were required to accomplish the task.

The initial point in the new adjustment is not a single station such as Meades Ranch in Kansas. Rather, the Earth's mass-center and numerous other points whose latitudes and longitudes had been precisely established from radio astronomy and satellite observations were used. The GRS80 ellipsoid was employed since, as noted earlier, it fits the Earth, in a global sense, more accurately than the Clarke ellipsoid of 1866.

Within the United States, the adjusted latitudes and longitudes of all monuments in NAD83 differ from their NAD27 values. These differences result primarily because of the different ellipsoids and origins used, but part of the change is also due to the addition of the many post-NAD27 observations in the NAD83 adjustment. The approximate magnitudes of these changes in the conterminous United States (CONUS), expressed in meters, are illustrated in Figure 19.6. Of course, these differences have a significant impact on all existing control points and map products that were based on NAD27. Various mathematical models have been developed to transform NAD27 values to their NAD83 positions (see Section 20.11).

19.6.3 Later Versions of NAD83

The internal consistency of first-order points adjusted in NAD83 was specified to be at least 1:100,000, but tests have verified that on average it is probably 1:200,000 or better. However, there are some areas where relative accuracies fall below 1:100,000. Since GPS accuracies are regularly better than 1:100,000, there was concern among GPS users about how to fit their observations to a network of reference points whose inherent accuracies were less.

State governments and the National Geodetic Survey cooperatively sought to solve this problem by adding GPS *High-Accuracy Reference Networks* (HARNs) to the National Spatial Reference System (see Section 19.8). From 1987 to 1997, HARN networks were created in every state. When each state's HARN was completely observed, an adjustment of these new stations was performed. This created an interim datum that was available to surveyors using GPS. This second version of NAD83 is known as NAD83 (HARN). For these regional reference systems, NGS retained the location of the mass-center of the Earth and orientation of the Cartesian coordinate axes,[6] but introduced a new scale that was consistent with the International Terrestrial Reference Frame of 1989 (ITRF89).

With the introduction of the CORS network in 1994 (see Section 14.3.5), the NGS was again faced with the problem of having stations of higher accuracy that were being adjusted using a different reference frame. Thus a third realization of NAD83 was obtained using the ITRF93 reference frame. This transformation

[6]Using high precision surveys, it was determined that the NAD83 (1986) Cartesian coordinate axes were misaligned by 0.03″, and its scale differed by 0.0871 ppm from the true definition of the meter.

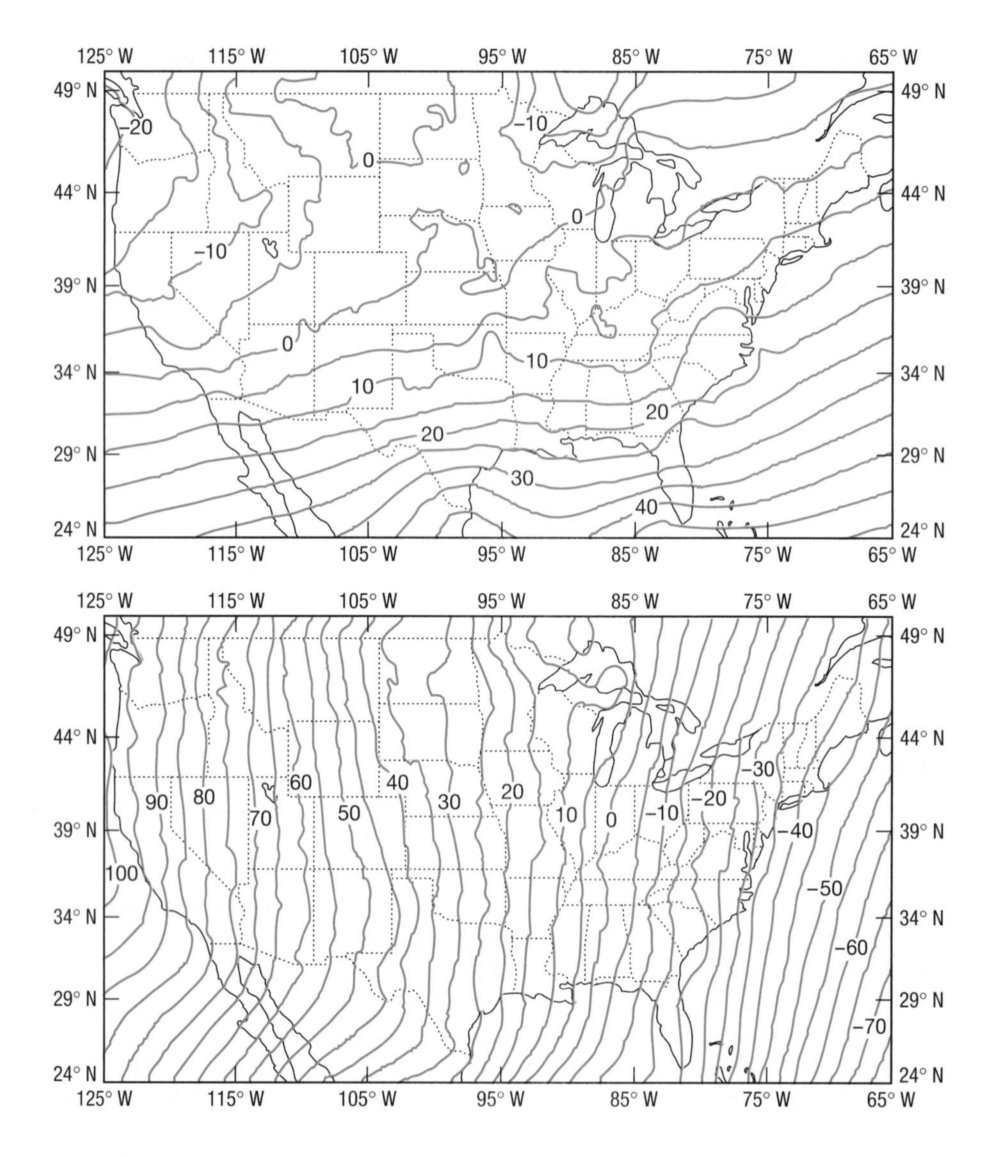

Figure 19.6 Approximate changes in latitude and longitude (in meters) in the conterminous United States from NAD27 to NAD83. (Upper figure) Latitude. (Lower figure) Longitude. (Adapted from National Geodetic Survey maps)

created another datum involving only the CORS stations and was known as NAD83 (CORS93).

In the spring of 1996, the NGS computed the positional coordinates for all existing CORS stations using ITRF94. This created the fourth realization of NAD83, and was known as NAD83 (CORS94). In 1998, the NGS computed positional coordinates for all existing CORS sites using ITRF96 as their reference frame. This version of NAD83 is known as NAD83 (CORS96). Each datum differs slightly from the previous definition of NAD83. For instance, the positions of sites in NAD83 (CORS96) differs a maximum of about 2 cm in horizontal and 4 cm in vertical from their equivalent NAD83 (CORS94) values. Furthermore, the difference between any NAD83 CORS adjustment and NAD83 (HARN) is under 10 cm in horizontal and 20 cm in vertical.

With completion of the last statewide HARN in 1997, the NGS had two spatial reference systems NAD83 (HARN) and NAD83 (CORS96). Since GPS technology and related accuracies had improved over the time of the HARN creation, the NGS decided in 1998 to reobserve all HARN stations. This process known as the *Federal Base Network* (FBN) survey was initiated in 1999. In 2007, a simultaneous readjustment of all HARN and CORS observations was completed. This created a new definition of NAD83 known as the NAD83 (NSRS2007). This system will be connected to the International Terrestrial Frame (ITRF) using the ITRF coordinates of the CORS sites. Upon completion of the adjustment in February 2007, a new and different version of NADCON (see Section 20.11) was created that allows users to convert NAD83 coordinates to the new NAD83 (NSRS2007) datum. This datum removes problems of having two different reference frames available for use in GPS surveys.

19.6.4 National Geodetic Vertical Datum of 1929 (NGVD29)

Vertical datums for referencing benchmark elevations are based on a single equipotential surface. Prior to the NAVD88 readjustment (see Section 19.6.5), the vertical datum used in the United States was the *National Geodetic Vertical Datum of 1929* (NGVD29), because that was the year that a general adjustment of benchmarks occurred. The NGVD29 was obtained from a best fit of mean sea-level observations taken at 26 tidal gauge stations in the United States and Canada and thus is often referred to as "mean sea level."

19.6.5 North American Vertical Datum of 1988 (NAVD88)

Between 1929 and 1988, more than 625,000 km of additional control leveling lines had been run. Furthermore, crustal movements and subsidence had changed the elevations of many benchmarks. To incorporate the additional leveling and correct erroneous elevations of benchmarks, a general vertical adjustment was performed. This adjustment included the new observational data, as well as an additional 81,500 km of re-leveled lines and leveling observations from both Canada and Mexico. It was originally scheduled for completion in 1988 and named the *North American Vertical Datum of 1988* (NAVD88), but it was not actually released to the public until 1991. This adjustment shifted the position of the reference equipotential surface from the mean of the 26 tidal gauge stations used in NGVD29 to a single tidal gauge benchmark known as *Father Point/Rimouski* on the Saint Lawrence Seaway in Quebec, Canada. As a result of these changes, published elevations of benchmarks in NAVD88 have shifted from their NGVD29 values. The magnitudes of these changes in the conterminous United States, expressed in millimeters, are shown in Figure 19.7. Note that the changes are largest in the western half of the country, with shifts of more than 1.5 m occurring in the Rocky Mountain region.[7]

---

[7]Those wishing to convert NGVD29 benchmark elevations to NAVD88 values can use the software VERTCON available from the NGS at http://www.ngs.noaa.gov/.

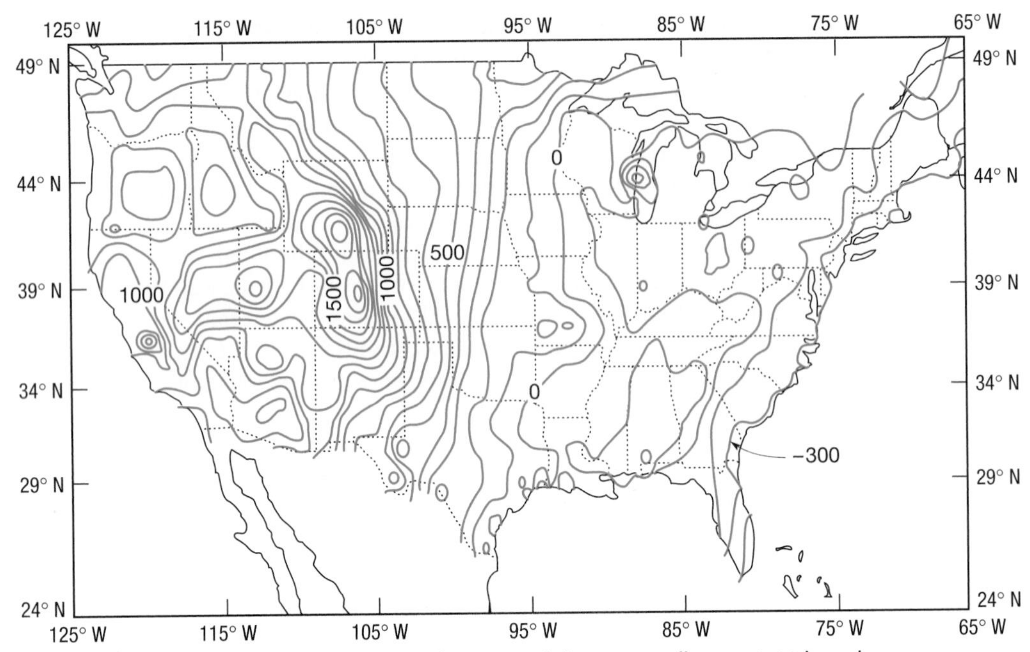

Figure 19.7 Approximate shift in vertical datum (in millimeters). Values shown are NAVD88 minus NGVD29. (Adapted from National Geodetic Survey map)

19.6.6 Transforming Coordinates Between Reference Frames

Historically, a goal of geodesy has been to obtain one common reference frame for all control coordinates. However, realistically each country or region has often developed its reference frame independently. Today, we often need to transform station coordinates from those derived using GPS and those developed in some regional reference frame such as NAD83. In Section 15.8 this process was introduced as *localization*. To do this, stations with known geodetic coordinates in both reference frames are required. If sufficient common stations are known, a three-dimensional coordinate transformation (see Section 17.10) can be used to convert the coordinates of stations from one reference frame into another. Since most reference frames have nearly aligned coordinate axes, the three-dimensional coordinate transformation can be simplified to the so-called *Helmert transformation*. The Helmert transformation is mathematically expressed as:

$$\begin{bmatrix} X \\ Y \\ Z \end{bmatrix}_2 = (1 + \Delta S)\begin{bmatrix} X \\ Y \\ Z \end{bmatrix}_1 + \begin{bmatrix} 0 & R_Z & -R_Y \\ -R_Z & 0 & R_X \\ R_Y & -R_X & 0 \end{bmatrix}\begin{bmatrix} X \\ Y \\ Z \end{bmatrix}_1 + \begin{bmatrix} T_X \\ T_Y \\ T_Z \end{bmatrix} \quad \mathbf{(19.11)}$$

where $[X\ Y\ Z]_1^T$ and $[X\ Y\ Z]_2^T$ are the geocentric coordinates of the common stations in the two reference frames derived using Equation (13.1); ΔS is the scale factor change between reference frames; R_X, R_Y, and R_Z are the rotations in radian units for the X, Y, and Z axes, respectively; T_X, T_Y, and T_Z are the translations between the two reference frames. A minimum of two stations known in horizontal position and three stations in elevation are required to perform this transformation. When insufficient common stations are not available, it is possible to perform this transformation using only the translations. However this produces considerably lower-quality results.

Besides different reference frames, the crustal plates of the Earth are constantly in motion. For example, some parts of California are moving at over 4 cm per yr. Thus the coordinates of points in any reference frame must be tied to a specific moment in time, or epoch. The National Geodetic Survey has combined the Helmert transformation with the velocity vectors of the crustal plates to produce a coordinate transformation software package known as *Horizontal Time-Dependent Positioning* (HTDP) software.[8] This software allows users to transform coordinates across time and between reference frames. Computations for the Helmert transformation as used in the HTDP software are demonstrated in a Mathcad® worksheet on the CD that accompanies this book.

It is also possible to perform the transformation in two separate transformations (horizontal and vertical). This is especially useful when the design coordinates are in a local reference frame that is arbitrarily assigned. In this case, the geodetic coordinates derived from GPS are converted to coordinates in a map projection system (see Chapter 20). Since only horizontal positioning is involved, the transformation between the reference frames is further simplified using a modified version of the two-dimensional conformal coordinate transformation shown in Equation (11.37). To perform this transformation, the centroids of the coordinates for the common stations are computed and the coordinates in both frames translated by these values. This places the origins of both reference frames at the centroid of the common points thus removing the translations from Equation (11.37). Using the translated coordinates, the remaining scale and rotation of Equation (11.37) is computed. Following these procedures, any remaining GPS coordinates can be transformed into the local reference frame. This process is demonstrated in the following example.

■ Example 19.3

A surveyor establishes a control network of points using an arbitrary coordinate system. In preparation for staking out the project using RTK-GPS (see Chapter 15), the surveyor reoccupies each station with a receiver. The resulting GPS coordinates are transformed into a two-dimensional map projection coordinate system with the common station coordinate values in both systems listed in Table 19.2. Determine the rotation and scale between the arbitrary and worldwide reference frames.

Solution

An oblique stereographic map projection (see Section 20.13.1) was used to transform the observed geodetic coordinates derived by the application software to a

[8]The HTDP software can be found at http://www.ngs.noaa.gov/PC_PROD/pc_prod.html.

TABLE 19.2 COORDINATE VALUES OF COMMON STATIONS

Station	Arbitrary Reference Frame		GPS Reference Frame	
	X (ft)	Y (ft)	E (m)	N (m)
A	5000.00	5000.00	635797.076	464685.605
B	1978.54	6075.88	625530.377	462379.464
C	6328.46	5983.64	637760.165	469740.901
D	6058.04	5000.00	638732.517	466538.417

coordinate system with the centroid of the project as its origin. The average values of the coordinates (centroid) in local reference system are

$$X_0 = \frac{5000.00 + 1978.54 + 6328.46 + 6058.04}{4} = 4841.26$$

$$Y_0 = \frac{5000.00 + 6075.88 + 5983.64 + 5000.00}{4} = 5514.88$$

These values are then used to translate the arbitrary coordinates to a common origin resulting in the following set of coordinates. The local orthometric height and GPS-derived orthometric height are also shown.

	Arbitrary			GPS		
Station	$x' = X - X_0$ **(ft)**	$y' = Y - Y_0$ **(ft)**	H **(m)**	$e' = E - E_0$ **(m)**	$n' = N - N_0$ **(m)**	H **(m)**
A	158.74	−514.88	282.486	45.212	−157.888	282.476
B	−2862.72	561.00	296.577	−869.005	−188.611	296.571
C	1487.20	468.76	313.819	456.132	133.694	313.814
D	1216.78	−514.88	304.191	367.660	−164.417	304.205

Since the two coordinate systems share a common origin, Equation (11.37) is modified as

$$\begin{bmatrix} a & -b \\ b & a \end{bmatrix} \begin{bmatrix} e' \\ n' \end{bmatrix} = \begin{bmatrix} x' \\ y' \end{bmatrix} + \begin{bmatrix} v_x \\ v_y \end{bmatrix} \tag{19.12}$$

Substituting the above coordinates into this equation results in

$$A = \begin{bmatrix} 45.212 & 157.888 \\ -157.888 & 45.212 \\ -869.005 & -188.611 \\ 188.611 & -869.005 \\ 456.132 & -133.694 \\ 133.694 & 456.132 \\ 367.660 & 164.417 \\ -164.417 & 367.660 \end{bmatrix} \quad X = \begin{bmatrix} a \\ b \end{bmatrix} \quad L = \begin{bmatrix} 158.74 \\ -514.88 \\ -2862.72 \\ 561.00 \\ 1487.20 \\ 468.76 \\ 1216.78 \\ -514.88 \end{bmatrix}$$

Using Equation (16.6), the solution of this system (AX = L + V) results in $a = 3.27987$ and $b = 0.066308$. Recognizing that the $\tan(\theta) = \frac{b}{a}$, the scale and rotation between the two systems of coordinates is s = 3.28054 and $\theta = 1°09'29.4''$, respectively. These values are now used in conjunction with Equation (19.12) to transform the GPS-derived map projection coordinates into the arbitrary local coordinate system for any additional points.

Note that the scale factor is approximately equal to the conversion factor going from meters to feet of 3.28083. Note that the computed residuals for the observations are 0.020, 0.027, −0.011, −0.003, −0.009, −0.016, −0.001, and −0.007, respectively, which are within the observational precisions of RTK-GPS for horizontal work.

The GPS-derived heights can also be transformed onto a local level plane. This process must account for the translation between the two systems and the obliquity of the two planes. The obliquity of the two level surfaces is corrected by applying two rotations in the cardinal north and east directions to bring the GPS-derived level surface parallel to the local horizontal plane. This is performed as

$$T_0 + r_e N_{GPS} + r_n E_{GPS} = H_L - H_G + v \tag{19.13}$$

where T_0 is the translation between the two level planes, N_{GPS} and E_{GPS} are the northing and easting of map projection coordinates derived from the GPS application software (see Example 19.3), respectively, r_e and r_n are the rotations about the x and y axes, respectively, H_L is the local height of the control points used in the project design, and H_G is either the geodetic height of the point as derived by GPS or the orthometric height of the point as derived by the combination of GPS and a geoid model.

As can be seen, Equation (19.13) involves three unknown parameters, T_0, r_e, and r_n. Thus a minimum of three benchmarks with local heights must be known. However it is wise to always have a fourth for the purposes of redundancy. Again, when less than three benchmarks are known, the translation can be computed from a single station. However the accuracy of this transformation will decrease significantly since it cannot account for the deflection of the vertical rotations.

Example 19.4

Using the data given in Example 19.3, determine the three transformation parameters of Equation (19.13).

Solution

Using the northing (N), easting (E), and heights from Example 19.3 in concert with Equation (19.13) yields the following observation equations.

$$A = \begin{bmatrix} 1 & -157.888 & 45.212 \\ 1 & 188.611 & -869.005 \\ 1 & 133.694 & 456.132 \\ 1 & -164.417 & 367.660 \end{bmatrix} \quad X = \begin{bmatrix} T_0 \\ r_e \\ r_n \end{bmatrix} \quad L = \begin{bmatrix} 0.025 \\ 0.021 \\ 0.018 \\ 0.001 \end{bmatrix}$$

Solving the system of equations, $AX = L + V$, using Equation (16.6) yields $T_0 = 0.016$, $r_e = 2.2''$, and $r_n = -1.3''$. With these transformation parameters and the GPS-derived map projection coordinates, a geodetic height can be transformed into a local height. The resulting residuals for the observations are -0.011, 0.003, -0.003, and 0.011, respectively. Again these values are well within the vertical accuracy of GPS-observed heights.

In order to have the distances obtained from GPS match the equivalent ground distances, the map projection plane is brought to the surface using an appropriate scaling factor. As discussed in Section 20.13.1, the oblique stereographic map projection has a defining scale factor of k_0 at its origin, which is the centroid of the projection. This makes the oblique stereographic map projection the preferred projection for this process. If this value is set to an appropriate scale, the plane surface of the map projection will be coincident with the elevation of the centroid. As an example, a scale factor of

$$k_0 = 1 + \frac{H_{centroid}}{R_e} \tag{19.14}$$

This scale factor is often used as one of the defining parameters for the oblique stereographic map projection system (see Section 20.13.1). By doing so, the distances displayed by the GPS survey controller will closely match the ground distances as observed on the project. The lengths of the distances are further refined with a scaling factor as derived from the two-dimensional conformal coordinate transformation as defined in Example 19.3. The combination of these scales applied to the observed and transformed geodetic coordinates will result in distances that closely match their equivalent ground values. Computations for this problem are demonstrated in a Mathcad® worksheet on the CD that accompanies this book.

19.7 ACCURACY STANDARDS AND SPECIFICATIONS FOR CONTROL SURVEYS

The required accuracy for a control survey depends primarily on its purpose. Some major factors that affect accuracy are type and condition of equipment used, field procedures adopted, and the experience and capabilities of personnel employed. In 1984 and again in 1998, the Federal Geodetic Control Subcommittee (FGCS) published different sets of detailed standards of accuracy and specifications for

TABLE 19.3 1998 FGCS ACCURACY STANDARDS: HORIZONTAL, ELLIPSOID HEIGHT, AND ORTHOMETRIC HEIGHT

Accuracy Classifications	95% Confidence Less than or equal to
1-millimeters	0.001 meters
2-millimeters	0.002 meters
5-millimeters	0.005 meters
1-centimeters	0.010 meters
2-centimeters	0.020 meters
5-centimeters	0.050 meters
1-decimeters	0.100 meters
2-decimeters	0.200 meters
5-decimeters	0.500 meters
1-meters	1.000 meters
2-meters	2.000 meters
5-meters	5.000 meters
10-meters	10.000 meters

geodetic surveys.[9] The rationale for both sets of standards is twofold: (1) to provide a uniform set of standards specifying minimum acceptable accuracies of control surveys for various purposes, and (2) to establish specifications for instrumentation, field procedures, and misclosure checks to ensure that the intended level of accuracy is achieved.

Table 19.3 lists the 1998 FGCS accuracy standards for control points. These standards are independent of the method of survey and based on a 95 percent confidence level (see Section 16.12). In order to meet these standards, control points in the survey must be consistent with all other points in the geodetic control network and not merely those within that particular survey. In Table 19.3, for horizontal surveys the accuracy standard specifies the radius of a circle within which the true or theoretical location of the survey point falls 95 percent of the time. The vertical accuracy standard specifies a linear value (plus or minus) within which the true or theoretical location of the point falls 95 percent of the time. Procedures leading to classification according to these standards involve four steps:

1. The survey observations, field records, sketches, and other documentation are examined to ensure their compliance with specifications for the intended accuracy of the survey.

[9]The 1998 standards are titled "Geospatial Positioning Accuracy Standards", Part 2: "Standards for Geodetic Networks". They can be downloaded at http://www.fgdc.gov. The 1984 standards, published in a booklet entitled "Standards and Specifications for Geodetic Control Networks", are available from the National Geodetic Information Center, NOAA, National Geodetic Survey, N/CG17, SSMC3 Station 09535, 1315 East-West Highway, Silver Spring, MD 20910.

2. A minimally constrained least-squares adjustment of the survey observations is analyzed to guarantee that the observations are free from blunders and have been correctly weighted.
3. The accuracy of control points in the local existing network to which the survey is tied is computed by random error propagation and weighted accordingly in the least-squares adjustment of the survey network.
4. The survey accuracy is checked at the 95 percent confidence level by comparing minimally constrained adjustment results against established control. The comparison takes into account the network accuracy of the existing control as well as systematic effects such as crustal motion or datum distortion.

Because many existing products, including control datasheets in the NAD83 datum, refer to the 1984 standards, these will also be described. This earlier set of standards established three distinct *orders of accuracy* to govern traditional control surveys, given in descending order: *first-order, second-order,* and *third-order.* For horizontal control surveys, second-order and third-order each have two separate accuracy categories, *class I* and *class II.* For vertical surveys, first-order and second-order each have class I and class II accuracy divisions. In 1985, three new orders of accuracy were defined for GPS surveys (see Section 14.5.1). These were orders AA, A, and B. Another lower order of accuracy for GPS surveys, identified as Order C, was also specified in these standards. It overlaps the three orders of accuracy applied to traditional horizontal surveys (see Tables 19.4 and 14.5).

Triangulation, trilateration, and traverse surveys are included in the 1984 horizontal control standards and specifications, and differential leveling is covered in the vertical control section.

Tables 19.4 and 19.5 give the 1984 FGCS accuracy standards required for the various orders and classes of horizontal and vertical control surveys, respectively.

TABLE 19.4 1984 AND 1985 FGCS HORIZONTAL CONTROL SURVEY ACCURACY STANDARDS

GPS Order*	Traditional Surveys** Order and Class	Relative Accuracy Required Between Points
Order AA		1 part in 100,000,000
Order A		1 part in 10,000,000
Order B		1 part in 1,000,000
Order C-1	First Order	1 part in 100,000
	Second Order	
Order C-2-I	Class I	1 part in 50,000
Order C-2-II	Class II	1 part in 20,000
	Third Order	
Order C-3	Class I	1 part in 10,000
	Class II	1 part in 5000

*Published in 1985.
**Published in 1984.

TABLE 19.5 1984 FGCS VERTICAL CONTROL SURVEY ACCURACY STANDARDS

Order and Class	Relative Accuracy Required Between Bench Marks*
First Order	
Class I	$0.5 \text{ mm} \times \sqrt{K}$
Class II	$0.7 \text{ mm} \times \sqrt{K}$
Second Order	
Class I	$1.0 \text{ mm} \times \sqrt{K}$
Class II	$1.3 \text{ mm} \times \sqrt{K}$
Third Order	$2.0 \text{ mm} \times \sqrt{K}$

*K is distance between benchmarks, in kilometers.

Values in Table 19.4 are ratios of allowable relative positional errors of a pair of horizontal control points to the horizontal distance separating them. Thus two first-order stations located 100 km (60 mi) apart are expected to be correctly located with respect to each other to within ±1 m.

Table 19.5 gives maximum relative elevation errors allowable between two benchmarks, as determined by a weighted least-squares adjustment (see Chapter 16). Thus elevations for two benchmarks 25 km apart, established by second-order class I standards, should be correct to within $\pm 1.0\sqrt{25} = \pm 5$ mm. These standards are not the same as the maximum allowable loop misclosures for the five classes of leveling given in Section 5.5. Table 19.5 values specify relative accuracies of benchmarks after adjustment, whereas loop misclosures enable assessment of results in differential leveling prior to adjustment.

The ultimate success of any engineering or mapping project depends on appropriate survey control. The higher the order of accuracy demanded, the more time and expense are required. It is therefore important to select the proper order of accuracy for a given project and carefully follow the specifications. Note that no matter how accurately a control survey is conducted, errors will exist in the computed positions of its stations, but a higher order of accuracy presumes smaller errors.

■ 19.8 THE NATIONAL SPATIAL REFERENCE SYSTEM

To meet the various local needs of surveyors, engineers, and scientists, the federal government has established a National Spatial Reference System (NSRS) consisting of more than 270,000 horizontal control monuments and approximately 600,000 benchmarks throughout the United States. The National Geodetic Survey (NGS), which began control-surveying operations as the *Survey of the Coast* in 1807, changed to *Coast Survey* in 1836, to *Coast and Geodetic Survey* in 1878, and to a division of the *National Ocean Survey* (NOS) in 1970, has primary responsibility for the NSRS. It continues to assist with and to coordinate geodetic control surveying activities with other agencies and with all states to establish new NSRS control stations and upgrade and maintain existing ones. It also disseminates a variety of publications and software related to geodetic control surveying.

The NSRS is split into horizontal and vertical divisions. All control within each part is classified in a ranking scheme based on purpose and order of accuracy. These are described in the following two sections.

19.9 HIERARCHY OF THE NATIONAL HORIZONTAL CONTROL NETWORK

The hierarchy of control stations within the NSRS Horizontal Control Network, from highest to lowest order, and their primary uses, are as follows:

Global-regional geodynamics consist of GPS surveyed points that meet the Order AA accuracy requirements. These are primarily used for international deformation studies.

Primary control consists of GPS surveyed points that meet the Order A accuracy requirements. These points are used for regional-local geodynamic and deformation studies.

Secondary control densifies the network within areas surrounded by primary control, especially in high-value land areas and for high-precision engineering surveys. Secondary-control surveys are executed to GPS Order B standards.

Terrestrial-based control is used for dependent control surveys to meet mapping, land information systems, property survey, and engineering needs. This network consists principally of stations set by traverse and triangulation to first- and second-order standards, and stations set by GPS to Order-C standards.

Local control provides reference points for local construction projects and small-scale topographic mapping. These surveys are referenced to higher-order control monuments and, depending on accuracy requirements, may be third-order class I or third-order class II.

19.10 HIERARCHY OF THE NATIONAL VERTICAL CONTROL NETWORK

The scheme of benchmarks within the National Vertical Control Network may be classified as follows:

Basic framework is a uniformly distributed nationwide network of benchmarks whose elevations are determined to the highest order of accuracy. It consists of nets A and B. In net A, adjacent level lines are ideally spaced an average of about 100–300 km apart using first-order class I standards; in net B the average separation is ideally about 50–100 km, and first-order class II standards are specified. Benchmarks are placed intermittently along the level lines at convenient locations.

Secondary network densifies the basic framework, especially in metropolitan areas and for large engineering projects. It is established to second-order class I standards.

General area control consists of vertical control for local engineering, surveying, and mapping projects. It is established to second-order class II standards.

Local control provides vertical references for minor engineering projects and small-scale topographic mapping. Benchmarks in this category satisfy third-order standards.

19.11 CONTROL POINT DESCRIPTIONS

To obtain maximum benefit from control surveys, horizontal stations and benchmarks are placed in locations favorable to their subsequent use. The points should be permanently monumented and adequately described to ensure recovery by future potential users. Reference monuments placed by the NGS are marked by bronze disks about 3.5 in. in diameter set in concrete or bedrock. Figure 19.8 shows two types of these disks.

Procedures for establishing permanent monuments vary with the type of soil or rock, climatic conditions, and intended use for the monument. In cases where soil can be excavated, monuments are commonly set in concrete that goes a foot or more below the local maximum frost depth. The bottom of the excavation is generally wider than the top to maximize monument stability during periods of freeze and thaw. Another option commonly used is to drive a stainless steel rod to refusal using powered tools. Driving depths of 10 or more feet are common when using this technique. In bedrock, holes are often drilled into the rock and the monument is simply cemented into the hole. Other variations for monumenting can be used, as long as the resulting objects will remain stable in their positions.

The NGS makes complete descriptions of all its control stations available to surveyors. As an example, a partial listing from an actual NGS horizontal control station description is given in Figure 19.9. These descriptions give each station's general placement in relation to nearby towns, instructions on how to reach the station following named or numbered roads in the area, and the monument's precise location by means of distances and directions to several nearby objects. The station's specific description, such as, "a triangulation disk set in drill hole in rock outcrop," is given, along with a record of recovery history. Data supplied with horizontal control-point descriptions include the datum(s) used and the station's geodetic latitude and longitude. Also given are the state plane coordinates,

Figure 19.8 Bronze disks used by the National Geodetic Survey to mark horizontal and vertical control stations.

```
      National Geodetic Survey,   Retrieval Date = OCTOBER 18, 1999
LZ1878 ***********************************************************************
LZ1878  DESIGNATION -  HAYFIELD NE 1974
LZ1878  PID         -  LZ1878
LZ1878  STATE/COUNTY-  PA/LUZERNE
LZ1878  USGS QUAD   -  HARVEYS LAKE (1979)
LZ1878
LZ1878                         *CURRENT SURVEY CONTROL
LZ1878  ___________________________________________________________________
LZ1878* NAD 83(1986)-  41 18 20.25410(N)    076 00 57.00239(W)     ADJUSTED
LZ1878* NAVD 88     -       398.7     (meters)    1308.      (feet)  VERTCON
LZ1878  ___________________________________________________________________
LZ1878  LAPLACE CORR-            0.27  (seconds)                    DEFLEC99
LZ1878  GEOID HEIGHT-          -31.73  (meters)                     GEOID99
LZ1878
LZ1878  HORZ ORDER  -  SECOND
LZ1878
LZ1878.The horizontal coordinates were established by classical geodetic methods
LZ1878.and adjusted by the National Geodetic Survey in July 1986.
LZ1878.No horizontal observational check was made to the station.
LZ1878
LZ1878.The NAVD 88 height was computed by applying the VERTCON shift value to
LZ1878.the NGVD 29 height (displayed under SUPERSEDED SURVEY CONTROL.)
LZ1878
LZ1878.The Laplace correction was computed from DEFLEC99 derived deflections.
LZ1878
LZ1878.The geoid height was determined by GEOID99.
LZ1878
LZ1878;                    North         East     Units   Scale      Converg.
LZ1878;SPC PA N     -   127,939.400   745,212.637   MT  0.99995873 +1 08 50.0
LZ1878;UTM  18      - 4,573,182.989   414,962.076   MT  0.99968900 -0 40 14.0
LZ1878
LZ1878                         SUPERSEDED SURVEY CONTROL
LZ1878
LZ1878  NAD 27      -  41 18 19.96597(N)    076 00 58.27835(W) AD(       ) 2
LZ1878  NGVD 29     -       398.9     (m)          1309.      (f) VERT ANG
LZ1878
LZ1878.Superseded values are not recommended for survey control.
LZ1878.NGS no longer adjusts projects to the NAD 27 or NGVD 29 datums.
LZ1878.See file dsdata.txt to determine how the superseded data were derived.
LZ1878
LZ1878_STABILITY: C = MAY HOLD, BUT OF TYPE COMMONLY SUBJECT TO
LZ1878+STABILITY: SURFACE MOTION
LZ1878
LZ1878  HISTORY     - Date      Condition         Recov. By
LZ1878  HISTORY     - 1975      MONUMENTED        NGS
LZ1878
LZ1878                          STATION DESCRIPTION
LZ1878
LZ1878'DESCRIBED BY NATIONAL GEODETIC SURVEY 1975 (CLN)
LZ1878'THE STATION IS LOCATED ABOUT 3/4 MILE SOUTHEAST OF LEHMAN AND ON THE
LZ1878'GROUNDS OF THE PENNSYLVANIA STATE UNIVERSITY (WILKES-BARRE
LZ1878'CAMPUS).
(continues...)
```

Figure 19.9 Partial listing of station data sheet in the National Spatial Reference System for horizontal control station Hayfield NE. (Courtesy National Geodetic Survey)

convergence angle and scale factor, UTM coordinates (see Chapter 20), and approximate elevation and geoidal height (in meters).

Some station descriptions give geodetic and grid azimuths (see Chapter 20) to a nearby station or stations. Geodetic and grid azimuths differ by an amount

```
NB0293 ***********************************************************************
NB0293  DESIGNATION -  F 137
NB0293  PID         -  NB0293
NB0293  STATE/COUNTY-  NY/TIOGA
NB0293  USGS QUAD   -  ENDICOTT (1978)
NB0293
NB0293                       *CURRENT SURVEY CONTROL
NB0293  ___________________________________________________________________
NB0293* NAD 83(1986)-  42 04 10.     (N)     076 07 04.     (W)     SCALED
NB0293* NAVD 88     -       252.471  (meters)      828.32   (feet)  ADJUSTED
NB0293  ___________________________________________________________________
NB0293  GEOID HEIGHT-         -32.70  (meters)                      GEOID99
NB0293  DYNAMIC HT  -         252.373 (meters)      827.99   (feet)  COMP
NB0293  MODELED GRAV-      980,231.5  (mgal)                        NAVD 88
NB0293
NB0293  VERT ORDER  -  FIRST     CLASS II
```

Figure 19.10 An excerpt from a NGS data sheet for benchmark F 137. (Courtesy National Geodetic Survey)

equal to the convergence angle, and therefore the appropriate azimuth must be selected for the particular surveying methods used.

As shown in Figure 19.10, published benchmark data include approximate station locations and adjusted elevations in both meters and feet. Again the relevant datum is identified.

Besides control within the national network set by the NGS, additional marks have been placed in various parts of the United States by other federal agencies such as the USGS, Corps of Engineers, and Tennessee Valley Authority. State, county, and municipal organizations may have also added control. This work is frequently coordinated through the NGS and descriptions of the stations involved are distributed by the NGS.

As noted earlier, complete descriptions for all points in the National Spatial Reference System can be obtained from the NGS.[10] This includes horizontal, vertical, and GPS control points. The descriptions can be obtained in hardcopy, electronic, or digital form. Only five compact disks are needed to store all NSRS data for the entire United States!

■ 19.12 FIELD PROCEDURES FOR TRADITIONAL HORIZONTAL CONTROL SURVEYS[11]

As noted earlier, in spite of the increasing prominence of GPS, horizontal-control surveys over limited areas are still being accomplished by the traditional methods of triangulation, trilateration, precise traverse, or a combination of these techniques. These methods are described briefly in the subsections that follow.

Traditional methods in horizontal control surveys require observations of horizontal distances, angles, and observations of astronomic azimuths. Basic theory,

---

[10]It is possible to obtain *data sheets* directly from the NGS at http://www.ngs.noaa.gov/datasheet.html. This website allows the user to search for data sheets of control points based on the station's name, *permanent identifier* (PID), and perform radial and rectangular searches from a location or from a clickable image map.

[11]Traditional as used here implies non-GPS ground surveying methods.

equipment, and procedures for making these observations are covered in earlier chapters and Appendix C. The following sections concentrate on procedures specific to control surveys and on matters related to obtaining the higher orders of accuracy generally required for these types of surveys.

19.12.1 Triangulation

Prior to the emergence of electronic distance-measuring equipment, triangulation was the preferred and principal method for horizontal-control surveys, especially if extensive areas were to be covered. Angles could be more easily observed compared with distances, particularly where long lines over rugged and forested terrain were involved, by erecting towers to elevate the operators and their instruments. Triangulation possesses a large number of inherent checks and closure conditions, which help detect blunders and errors in field data and increase the possibility of meeting a high standard of accuracy.

As implied by its name, triangulation utilizes geometric figures composed of triangles. Horizontal angles and a limited number of sides called *baselines* are observed. By using the angles and base-line lengths, triangles are solved trigonometrically and positions of stations (vertices) calculated.

Different geometric figures are employed for control extension by triangulation, but chains of quadrilaterals called arcs (Figure 19.11) have been most common. They are the simplest geometric figures that permit rigorous closure checks and adjustments of field observational errors and they enable point positions to be calculated by two independent routes for computational checks. More complicated figures like that illustrated in Figure 19.12 have frequently been used to establish horizontal control by triangulation in metropolitan areas.

In executing triangulation surveys, *intersection stations* can be located as part of the project. In this process, angles are observed from as many occupied points as possible to tall prominent objects in the area such as church spires, smokestacks, or water towers. The intersection stations are not occupied, but their positions are calculated; thus they become available as local reference points. An example is station *B* in Figure 19.12.

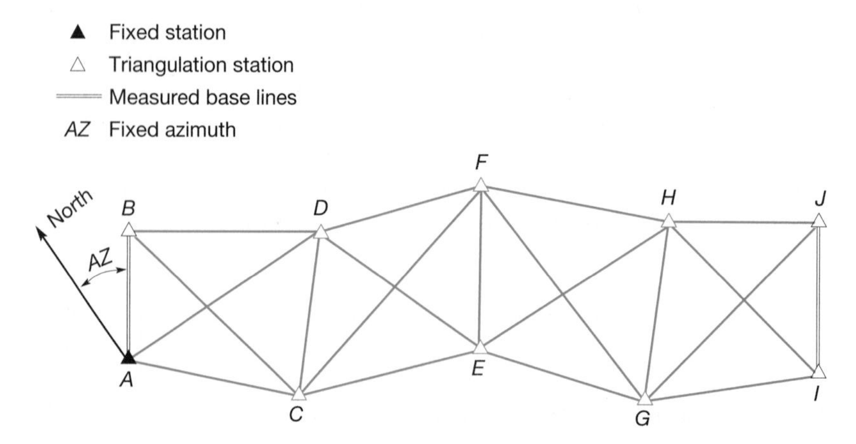

Figure 19.11 Chain of quadrilaterals.

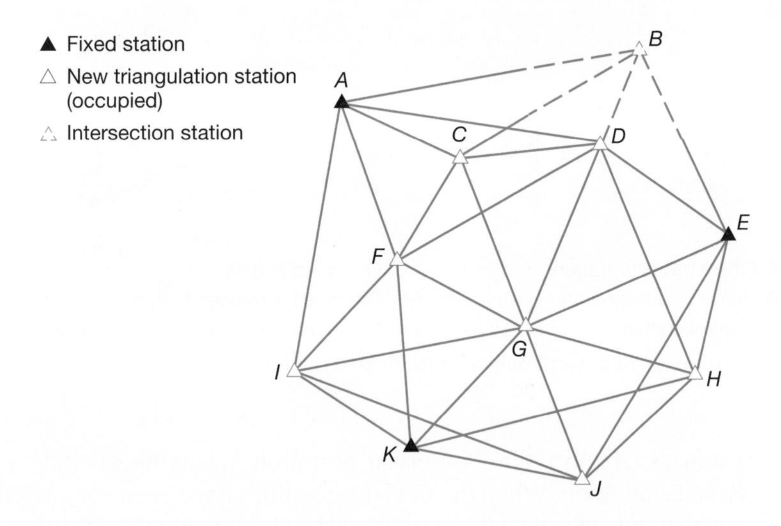

Figure 19.12 Triangulation network for a metropolitan control survey.

To compensate for the errors that occur in the observations, triangulation networks must be adjusted. The most rigorous method utilizes least squares (see Chapter 16). In that procedure all angle, distance, and azimuth observations are simultaneously included in the adjustment and given appropriate relative weights based on their precisions. The least-squares method not only yields the most probable adjusted station coordinates for a given set of data and weights, but it also gives their precisions.

19.12.2 Precise Traverse

Precise traversing is common among local surveyors for horizontal-control extension, especially for small projects. Fieldwork consists of two basic parts: observing horizontal angles at the traverse hubs and observing distances between stations. With total station instruments, these observations can be simultaneously observed. Precise traverses always begin and end on stations established by equal or higher order surveys.

Unlike triangulation, in which stations are normally widely separated and placed on the highest ridges and peaks in an area, traverse routes generally follow the cleared rights-of-way of highways and railroads, with stations located relatively close together. Besides easing fieldwork, this provides a secondary benefit in accessibility of the stations. Traverses lack the automatic checks inherent in triangulation and extreme observational caution must therefore be exercised to avoid blunders. Additionally, since traverses usually run along single lines, they are generally not as good as triangulation for establishing control over large areas.

Control traverses can be strengthened to provide additional checks in the data by establishing "offset stations" such as A', C', and E' of Figure 19.13. An offset station is set near every-other primary traverse station. In performing the field observations, instrument setups are made only at the primary traverse stations. All possible angles are observed with horizon closures at each station; thus four angles are determined at interior single primary stations and two angles are observed at primary stations with nearby offset stations. This observation scheme is shown in Figure 19.13. Additionally, all distances are observed, i.e., at station 1

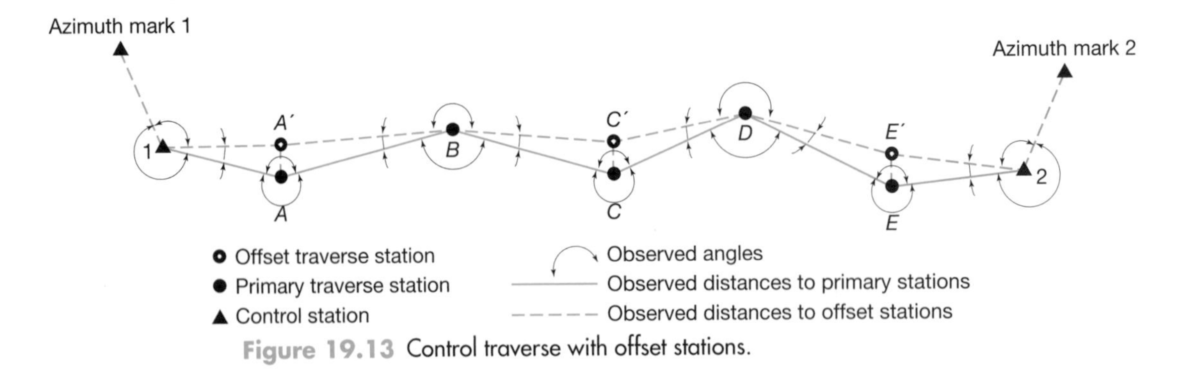

Figure 19.13 Control traverse with offset stations.

distances $1A$ and $1A'$ are observed, at station A, lengths $A1$, AA', and AB are observed, and so on. When the field observations have been completed, the network can be adjusted using all observations in a least-squares adjustment, thereby providing geometric checks for all angle and distance observations in the traverse. Additional geometric strength in the figure could be obtained by also observing angles at the offset stations.

In traversing to gain overall project efficiency and improve angle accuracy, it is always preferable to have long sight distances. Also to avoid mistakes it is advisable to avoid having nearly "flat" angles (values near 180°) whenever possible. To accomplish this, presurvey reconnaissance is recommended. An oft-made mistake is to construct the traverse while collecting the observations. This technique works in low-order surveys, but frequently results in poorly designed control traverses.

For long traverses, checks on the observed horizontal angles can be obtained by making periodic astronomical azimuth observations (see Appendix C). These should agree with the values computed from the direction of the starting line and the observed horizontal angles. However, if a traverse extends an appreciable east-west direction, as illustrated in Figure 19.14, meridian convergence will cause the two azimuths to disagree. For example, in Figure 19.14, azimuth FG obtained from direction AB and the observed horizontal angles should equal astronomic azimuth $FG + \theta$, where θ is the meridian convergence. A good approximation for meridian convergence between two points on a traverse is

$$\theta'' = \frac{\rho d \tan \phi}{R_e} \tag{19.15}$$

where θ'' is meridian convergence, in seconds; d the east-west distance between the two points in meters; R_e the mean radius of the Earth (6,371,000 m); ϕ the mean latitude of the two points; and ρ the number of seconds per radian (206,265). Because of meridian convergence, forward and back azimuths of long east-west lines do not differ by exactly 180°, but rather by $180° \pm \theta$. (If the traverse proceeds in an easterly direction the sign of θ is +; if it goes westerly θ is minus. A sketch will clarify the situation.) From Equation (19.15) an east-west traverse of 1 mi length at latitude 30° produces a convergence angle of approximately 30″. At latitude 45°, convergence is approximately 51″/mi east-west. These calculations

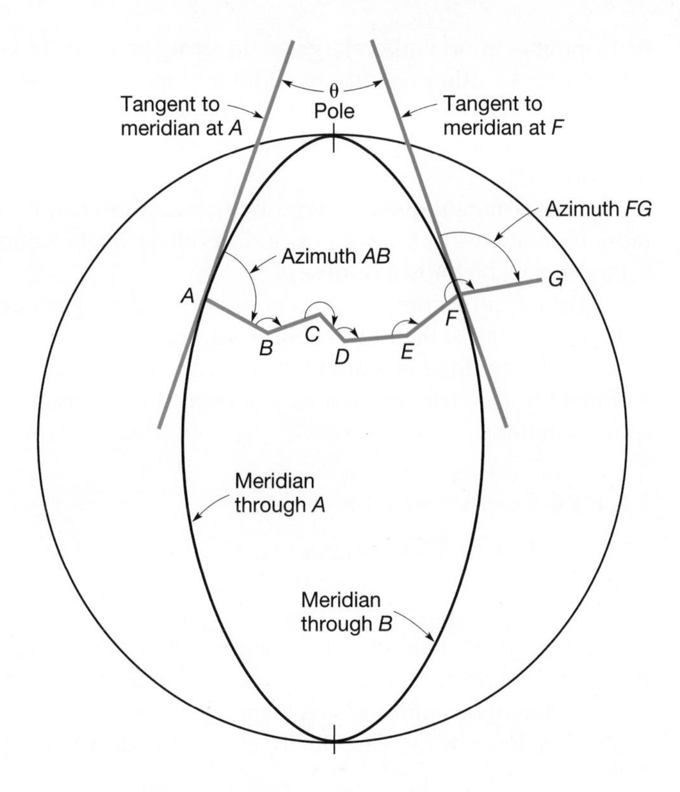

Figure 19.14 Meridian convergence on long east-west traverses.

illustrate that the magnitude of convergence can be appreciable and must be considered when astronomic observations are made in connection with plane surveys that assume the *y*-axis parallel throughout the project area.

Procedures for precise traverse computation vary depending on whether a geodetic or a plane reference coordinate system is used. In either case, it is necessary first to eliminate mistakes and compensate for systematic errors. In the adjustment, closure conditions are enforced for (1) azimuths or angles, (2) departures, and (3) latitudes. The most rigorous process, the least-squares method (see Chapter 16), should be used because it simultaneously satisfies all three conditions and yields residuals having the highest probability.

19.12.3 Trilateration

Trilateration, a method for horizontal control surveys based exclusively on observed horizontal distances, has gained acceptance because of electronic distance measuring capabilities. Both triangulation and traversing require time-consuming horizontal angle measurement. Hence, trilateration surveys often can be executed faster and produce equally acceptable accuracies.

The geometric figures used in trilateration, although not as standardized, are similar to those employed in triangulation. Stations should be intervisible and are therefore placed in elevated locations, sometimes using towers to elevate instruments and observers if necessary.

Because of intervisibility requirements and the desirability of having essentially square networks, trilateration is ideally suited to densify control in metropolitan areas and on large engineering projects. In special situations where topography or other conditions require elongated narrow figures, reading some horizontal angles can strengthen the network. Also, for long trilateration arcs, astronomic azimuth observations can help prevent the network from deforming in direction.

As in triangulation, surveys by trilateration can be extended from one or more monuments of known position. If only a single station is fixed, at least one azimuth must be known or observed.

Trilateration computations consist of reducing observed slope distances to horizontal lengths, then to the ellipsoid, and finally to grid lengths if the calculations are being done in state plane coordinate systems (see Chapter 20). Observational errors in trilateration networks must be adjusted, preferably by the least-squares method.

19.12.4 Combined Networks

With the ability to easily observe both distances and angles in the field, networks similar to that shown in Figure 19.12 are becoming increasingly popular. In a combined network, many or all angles and distances are observed. These surveys provide the greatest geometric strength and the highest coordinate accuracies for traditional survey techniques. As described in Section 19.14, all observations must be corrected to the ellipsoid or a mapping grid (see Chapter 20). The least-squares method as described in Chapter 16 is used to adjust the observations.

19.13 FIELD PROCEDURES FOR VERTICAL-CONTROL SURVEYS

Vertical-control surveys are generally run by either direct differential leveling or trigonometric leveling. The method selected will depend primarily on the accuracy required, although the type of terrain over which the leveling will be done is also a factor. Differential leveling, described in Section 5.4, produces the highest order of accuracy typically. The Global Positioning System (GPS) can be used for lower-order vertical-control surveys, but to get accurate elevations using this method, *geoidal heights* in the area must be known and applied (see Section 19.5).

Although trigonometric leveling produces a somewhat lower order of accuracy than differential leveling, the method is still suitable for many projects such as establishing vertical control for topographic mapping or for lower order construction stakeout. It is particularly convenient in hilly or mountainous terrain where large differences in elevation are encountered. Field procedures for trigonometric leveling and methods for reducing the data are discussed in Section 4.5.4.

Differential leveling can produce varying levels of accuracy, depending on the precautions taken. In this section only, *precise differential leveling,* which produces the highest quality results, is considered.

As noted in Section 19.7 and Table 19.5, the FGCS established accuracy standards and specifications for various orders of differential leveling. To achieve the higher orders, special care must be exercised to minimize errors, but the same basic principles apply.

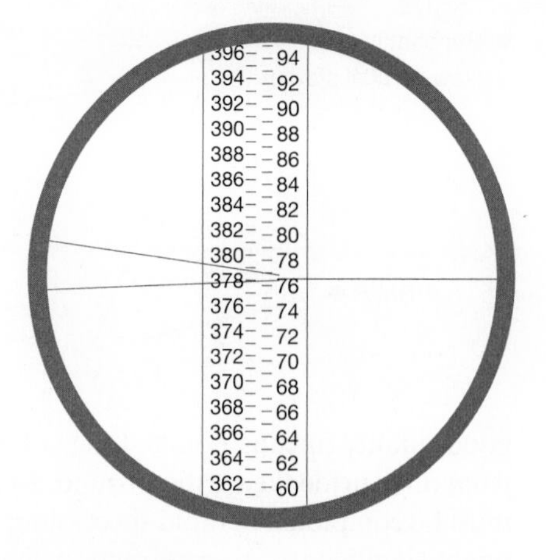

Figure 19.15 Reticle of a precise level shown with dual metric-scale precise leveling rod.

Special level rods are needed for precise work. They have scales graduated on Invar strips, which are only slightly affected by temperature variations. Precise level rods are equipped with rod levels to facilitate plumbing and special braces aid in holding the rod steady. They usually have two separate graduated scales. One type of rod is divided in centimeters on an Invar strip on the rod's front side, with a scale in feet painted on the back for checking readings and minimizing blunders. A second type of rod, shown in Figure 19.15, has two sets of centimeter graduations on the Invar strip with the right one precisely offset from the left by a constant, thereby giving checks on readings.

Cloudy weather is preferable for precise leveling, but an umbrella can be used on sunny days to shade the instrument and prevent uneven heating, which causes the bubble to run. (One design encases the vial in a Styrofoam shield.) Automatic levels are not as susceptible to errors caused by differential heating. Precise work should not be attempted on windy days. For best results, short and approximately equal backsight and foresight distances are recommended. Table 19.6 lists the maximum sight distances and allowable differences between backsight and foresight lengths for first-, second-, and third-order leveling. Rod-persons can pace or count rail lengths or highway slab joints to set sight distances, which are then checked for accuracy by three-wire stadia methods (see Section 5.8). Precise leveling demands

TABLE 19.6 RECOMMENDED FIELD CONDITIONS FOR PRECISE LEVELING

Order Class	First I	First II	Second I	Second II	Third
Maximum sight length (m)	50	60	60	70	90
Difference between foresight and backsight lengths never to exceed					
per setup (m)	2	5	5	10	10
per section (m)	4	10	10	10	10

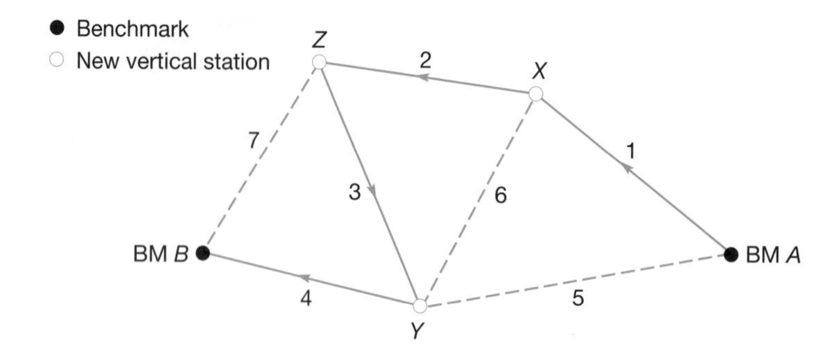

Figure 19.16 Interconnecting level network.

good-quality turning points. Lines of sight should not pass closer than about 0.5 m from any surface, e.g., the ground, to minimize refraction. Readings at any setup must be completed in rapid succession; otherwise changes in atmospheric conditions might significantly alter refraction characteristics between them.

Three-wire leveling has been employed for much of the precise work in the United States. In this procedure, as described in Section 5.8 , rod readings at the upper, middle, and lower crosshairs are taken and recorded for each backsight and foresight. The difference between the upper and middle readings is compared with that between the middle and lower values for a check, and the average of the three readings is used. A sample set of field notes for three-wire leveling is illustrated in Figure 5.9. When using a digital level, the elevation difference along with the backsight or foresight length can be digitally recorded for each sight. A second technique in precise leveling employs the parallel-plate micrometer attached to a precise leveling instrument and a pair of precise rods like those described earlier.

It is generally advisable to design large level networks so that several smaller circuits are interconnected. This enables making checks that isolate blunders or large errors. For example, in Figure 19.16, it is required to determine the elevations of points *X*, *Y*, and *Z* by commencing from BM *A* and closing on BM *B*. As a minimum, running level lines 1 through 4 could do this, but if an unacceptable misclosure were obtained at BM *B*, it would be impossible to discover in which line the blunder occurred. If additional lines 5, 6, and 7 are run, calculating differences in elevation by other routes through the network should isolate the blunder. Furthermore, by including the supplemental observations, precisions of the resulting elevations at *X*, *Y*, and *Z* are increased.

For long lines, one procedure used to help isolate mistakes and minimize field time is to run small loops with approximately five setups between temporary benchmarks. In this procedure as each loop is completed, it is checked for acceptable closure before proceeding forward to the next loop. This procedure increases the number of observations, but helps minimize the amount of time that is required to uncover mistakes. Each smaller loop is connected to subsequent loops until the entire network is observed. Figure 19.17 depicts this procedure.

In precise differential leveling, frequent calibration of the leveling instrument is necessary to determine its collimation error. A collimation error exists if, after leveling the instrument, its line of sight is inclined or depressed from horizontal. This causes errors in determining elevations when backsight and foresight distances are not equal. But they can be eliminated if the magnitude of the collimation error is known.

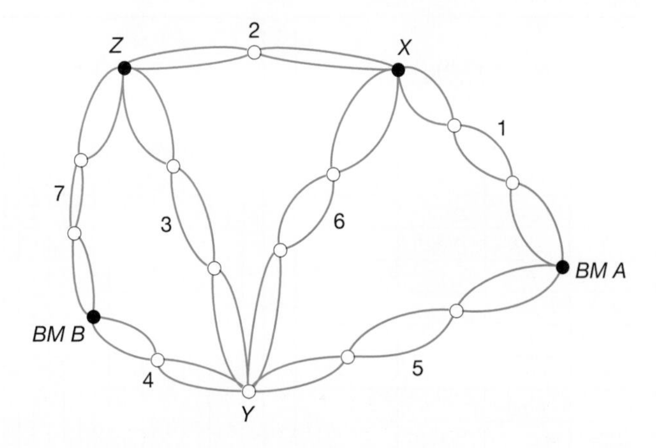

Figure 19.17 Leveling of network performed in small closing loops.

A method that originated at the National Geodetic Survey can be used to determine the collimation error. It requires a baseline approximately 300 ft (90 m) long. Stakes are set at each end of the line and at two intermediate stations located approximately 20 ft (6 m) and 40 ft (12 m) from the two ends. Figure 19.18 shows an example set of field notes for determining the collimation error and includes a sketch illustrating the base-line layout. With the instrument at station 1, middle wire r_1 and R_1 are observed on stations A and B, respectively. If there were no collimation error, the true elevation difference ΔH from these observations would be $r_1 - R_1$. However if a collimation error is present, each observation must be corrected by adding an amount proportional to the horizontal distance from the level to the rod. The horizontal distance is measured by the stadia interval (see Section 16.9.2). Introducing collimation corrections, the true elevation difference ΔH is

$$\Delta H = [r_1 + C(i_1)] - [R_1 + C(I_1)] \tag{a}$$

In Equation (a), i_1 and I_1 are the stadia intervals (differences between top and bottom crosshair values) for the rod readings on stations A and B, respectively, and C is the collimation factor (in feet per foot, or m per m, of stadia interval). A similar equation for the true elevation difference can be written for rod readings R_2 and r_2 taken on stations A and B, respectively from station 2, or

$$\Delta H = [R_2 + C(I_2)] - [r_2 + C(i_2)] \tag{b}$$

Note that in Equations (a) and (b), uppercase R and I apply to the longer sights, and lowercase r and i are for the shorter sights. Equating the right sides of Equations (a) and (b), and reducing, yields

$$C = \frac{(r_1 + r_2) - (R_1 + R_2)}{(I_1 + I_2) - (i_1 + i_2)} \tag{19.16}$$

As previously noted, the units of the collimation factor calculated by Equation (19.16) are either in feet per ft, or meters per m, of stadia interval. Computation of the factor is illustrated in Figure 19.18 for the data of the field notes (given in ft).

LEVEL CALIBRATION

STA	A	INT	B	INT
	4.516		5.567	
1	4.414	0.102	4.644	0.923
	4.312	0.102	3.721	0.923
	r_1	i_1	R_1	I_1
	4.414	0.204	4.644	1.846
	4.699		4.182	
2	3.825	0.874	4.042	0.140
	2.950	0.875	3.901	0.141
	R_2	I_2	r_2	i_2
	3.825	1.749	4.042	0.281

$$C = \frac{(4.414 + 4.042) - (4.644 + 3.825)}{(1.846 + 1.749) - (0.204 + 0.281)}$$

$$C = \frac{-0.013}{3.110} = -0.0042\ \text{m/m}$$

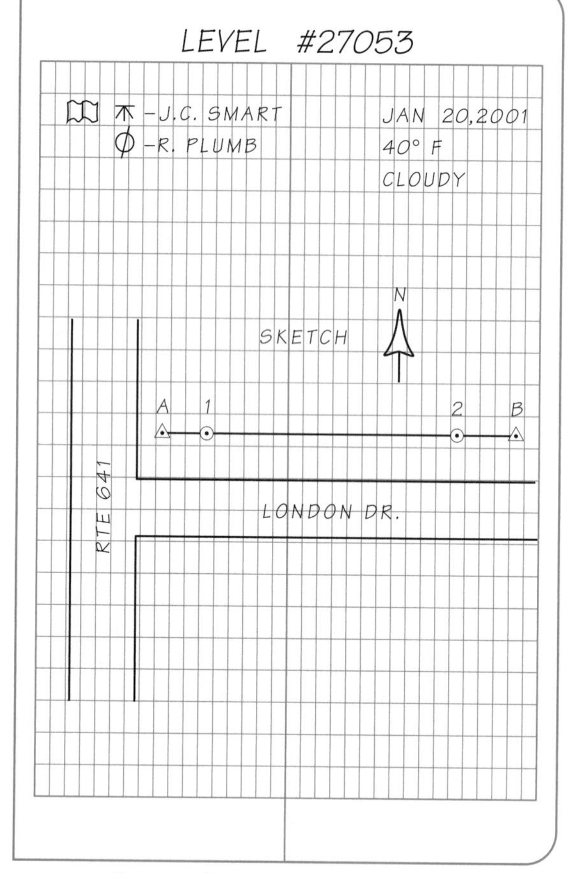

Figure 19.18 Field notes for determining collimation factor.

Because the collimation correction increases linearly with distance, it is unnecessary to apply it to each backsight and foresight. Rather the corrected elevation difference $\Delta H'$ for any loop or section leveled is computed as

$$\Delta H' = \Sigma BS - \Sigma FS + C(\Sigma I_{BS} - \Sigma I_{FS}) \qquad \textbf{(19.17)}$$

where ΣBS is the sum of the middle cross-wire readings of backsights in the loop or section; ΣFS is the middle crosshair total of foresights; and ΣI_{BS} and ΣI_{FS} are the sums of the stadia intervals for the backsights and foresights, respectively.

Example 19.5

The section from BM A to BM 3 is leveled using the instrument whose collimation factor of -0.00012 m/m of interval was determined in the field notes in Figure 19.18. The sum of the backsights is 125.590 m, and the sum of the foresights is 88.330 m. Backsight stadia intervals total 351.52 m, while the sum of foresight intervals is 548.40 m. Find the corrected elevation difference.

Solution

By Equation (19.17),

$$\Delta H' = (125.590 - 88.330) + (-0.00012)(351.52 - 548.40)$$
$$= 37.260 + 0.024 = 37.284 \text{ m}$$

Regardless of precautions taken in field observations, errors accumulate in leveling and must be adjusted to provide perfect mathematical closure. For simple level loops, adjustment procedures presented in Section 5.6 can be followed; for interconnected level networks such as those of Figures 19.16 and 19.17, the method of least squares is preferable. An example least-squares adjustment of an interconnected network is given in Section 16.6.

19.14 REDUCTION OF FIELD OBSERVATIONS TO THEIR GEODETIC VALUES

Traditional surveying instruments, such as levels and total stations, are oriented with respect to the local gravity surface. In geodetic work, since horizontal surveys are referenced to an ellipsoid and vertical surveys to the geoid, corrections must be made to field observations to obtain their equivalent geodetic values. The following subsections discuss some of these corrections.

19.14.1 Reduction of Distance Observations Using Elevations

In geodetic control survey computations, observed slope distances (sometimes called *slant* distances) must first be reduced to the surface of the ellipsoid. Observed distances in control surveys are often long and thus the short-line reduction techniques given in Section 6.13 do not provide satisfactory accuracy. This is especially true for long lines that are steeply inclined.

A procedure for reducing long slope distances to their ellipsoid lengths is discussed here. The method is based on elevation differences between the end points of the sloping line. In Figure 19.19, an EDM instrument is at A, a reflector is at B, and S is the observed slope distance from A to B. (Assume that S has been corrected for meteorological conditions.) Length D_1 is the "mark-to-mark" distance between stations A and B. Mark-to-mark distances apply for EDM calibration lines, as well as for GPS baselines. Length D_2 is the arc distance on the ellipsoid, which is also known as the *geodetic distance*. It is the length required for most geodetic computations. Distance D_3 is the ellipsoidal chord length between stations A and B.

In Figure 19.19, let $h_1' = h_1 + hi$ and $h_2' = h_2 + hr$, where h_i and h_r are instrument and reflector heights, respectively above stations A and B, and h_1 and h_2 are the geodetic heights at A and B, respectively. Expressing the relationship of the three sides of triangle ABO using the law of cosines, [see Equation (11.2)], gives

$$S^2 = (R_\alpha + h_1')^2 + (R_\alpha + h_2')^2 - 2(R_\alpha + h_1')(R_\alpha + h_2')\cos\theta \quad \textbf{(19.18)}$$

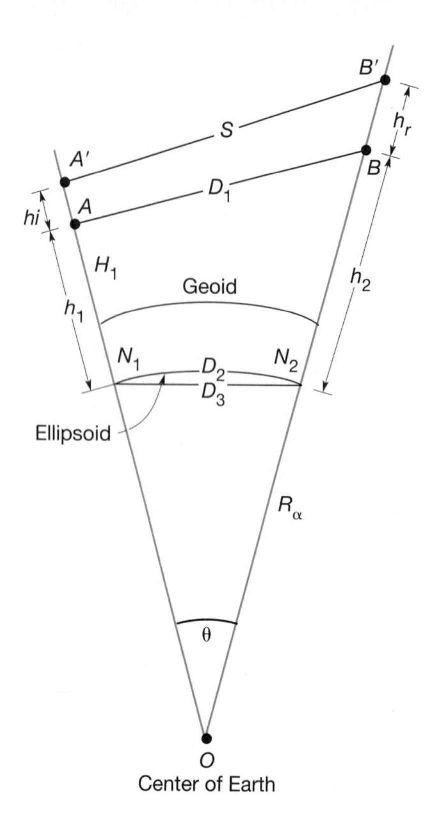

Figure 19.19 Reduction of long lengths to the ellipsoid based on elevations.

where R_α is the radius of the Earth in the azimuth of the distance from point A as defined by Equation (19.6), and θ is the angle subtended by the verticals from points A and B. Substituting the trigonometric identity of $\cos\theta = 1 - 2\sin^2(\theta/2)$ into Equation (19.18), and expanding yields

$$S^2 = (h_2' - h_1')^2 + 4R_\alpha^2\left(1 + \frac{h_1'}{R_\alpha}\right)\left(1 + \frac{h_2'}{R_\alpha}\right)\sin^2\left(\frac{\theta}{2}\right) \tag{19.19}$$

Substituting $\Delta h' = h_2' - h_1'$, and $D_3 = 2R_\alpha \sin(\theta/2)$ into Equation (19.19), the expression reduces to

$$S^2 = \Delta h'^2 + \left(1 + \frac{h_1'}{R_\alpha}\right)\left(1 + \frac{h_2'}{R_\alpha}\right)D_3^2 \tag{19.20}$$

Solving Equation (19.20) for D_3 yields the following expression for the ellipsoidal chord length:

$$D_3 = \sqrt{\frac{S^2 - \Delta h'^2}{\left(1 + \frac{h_1'}{R_\alpha}\right)\left(1 + \frac{h_2'}{R_\alpha}\right)}} \tag{19.21}$$

The arc length on the ellipsoid (*geodetic distance* or *geodetic length*) can be computed from this chord distance as

$$D_2 = 2R_\alpha \sin^{-1}\left(\frac{D_3}{2R_\alpha}\right) \quad \textbf{(19.22)}$$

Equations (19.21) and (19.22) can be used to compute the distance on any level surface by simply modifying the heights of the endpoints as appropriate. It is important to realize that the unit of the arcsine in Equation (19.22) is radians. To compute the chord distance between two points at different elevations, for example D_1 in Figure 19.19, the following equation is used:

$$D_1 = \sqrt{D_3^2\left(1 + \frac{h_1}{R_\alpha}\right)\left(1 + \frac{h_2}{R_\alpha}\right) + (h_2 - h_1)^2} \quad \textbf{(19.23)}$$

Example 19.6

A slope distance of 5000.000 m is observed between two points A and B whose orthometric heights are 451.200 m and 221.750 m, respectively. The geoidal undulation at point A is -29.7 m and is -29.5 m at point B. The height of the instrument at the time of the observation was 1.500 m, and the height of the reflector was 1.250 m. What are the geodetic and mark-to-mark distances for this observation? (Use a value of 6,386,152.318 m for R_α in the direction AB.)

Solution

By Equation (19.7), the geodetic heights at points A and B are

$$h_A = 451.200 - 29.7 = 421.500 \text{ m}$$
$$h_B = 221.750 - 29.5 = 192.250 \text{ m}$$

Thus, $h'_A = 421.500 + 1.500 = 423.000$ m, $h'_B = 192.250 + 1.250 = 193.500$, $\Delta h' = 193.500 - 423.000 = -229.500$ m, and by Equation (19.20), the ellipsoidal chord distance D_3 is

$$D_3 = \sqrt{\frac{5000^2 - 229.5^2}{\left(1 + \frac{423.0}{R_\alpha}\right)\left(1 + \frac{193.5}{R_\alpha}\right)}} = 4994.489 \text{ m}$$

By Equation (19.22), the reduced ellipsoidal arc, or geodetic length, for the line AB is

$$D_2 = 2R_\alpha \sin^{-1}\left(\frac{4994.489}{2R_\alpha}\right) = 4994.489 \text{ m}$$

Finally by Equation (19.23), the mark-to-mark distance is

$$D_1 = \sqrt{4994.489^2\left(1 + \frac{421.500}{R_\alpha}\right)\left(1 + \frac{192.250}{R_\alpha}\right) + (192.25 - 421.5)^2}$$

$$= 4999.987 \text{ m}$$

Note that the ellipsoid arc and chord lengths are the same to the nearest millimeter. As lines become longer, however, this will not necessarily be the case. Nevertheless for most geodetic observations, these arc and chord values will generally be nearly the same. Note also that the observed slope distance differs from the mark-to-mark distance by 13 mm. Finally, if the short line reduction procedure of Section (6.13) had been used, an error of more than 0.2 m would have resulted. An EDM with an accuracy of 1 mm + 1 ppm would have an estimated error for this line of approximately ±5 mm. This error is sometimes called the *noise level* of the observation. Thus the error created by a short-line reduction procedure should be considered significant.

19.14.2 Reduction of Distance Observations Using Vertical Angles

Figure 19.20 illustrates a slope distance S observed from A to B. Points A and B represent an EDM instrument and a reflector, respectively, O is the Earth's center, and R_α its radius in the direction of the azimuth as defined by Equation (19.6). Vertical angles α and β were observed at A and B, respectively. Arc AB_2, which is closely approximated by its chord, is the required horizontal distance. If the short-line reduction procedures given in Section 6.13 were used, horizontal distance AB_1 would result, which would be in error by B_1B_2. Arc $A'B'$ is the required ellipsoid distance.

From Figure 19.20 the following equations can be written to compute required horizontal (chord) distance AB_2:

$$\delta = \frac{\alpha - \beta}{2} \tag{19.24}$$

$$AB_1 = S\cos\delta \tag{19.25}$$

$$BB_1 = S\sin\delta \tag{19.26}$$

$$\psi = \frac{AB_1}{R_\alpha + h_A} \times \frac{180°}{\pi} \text{ (approx.)} \tag{19.27}$$

$$B_1B_2 = BB_1\tan\left(\frac{\psi}{2}\right) \tag{19.28}$$

$$AB_2 = AB_1 - B_1B_2 \tag{19.29}$$

Finally, ellipsoidal (chord) length $A'B'$ can be computed from

$$A'B' = AB_2\left(\frac{R_\alpha}{R_\alpha + h_A}\right) \tag{19.30}$$

where h_A is the ellipsoidal height.

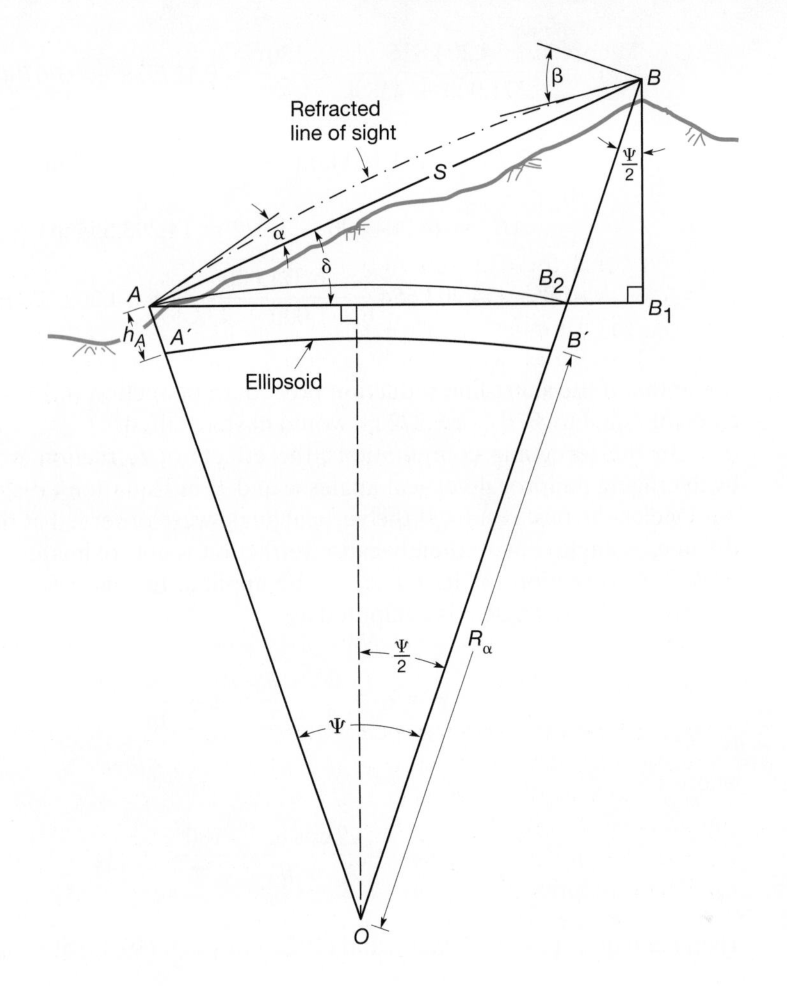

Figure 19.20 Reduction of long lengths to the ellipsoid based on vertical angles.

■ Example 19.7

In Figure 19.20, slope distance L and vertical angles α and β were observed as 14,250.590 m, 4°32′18″, and −4°38′52″, respectively. If the geodetic height at A is 438.4 m, what is distance $A'B'$ reduced to the ellipsoid? (Use the mean radius of 6,371,000 m for R_α.)

Solution

Solving Equations (19.24) through (19.30) in sequence,

$$\delta = \frac{4°32'18'' - (-4°38'52'')}{2} = 4°35'35''$$

$$AB_1 = 14{,}250.590 \cos 4°35'35'' = 14{,}204.826 \text{ m}$$

$$BB_1 = 14{,}250.590 \sin 4°35'35'' = 1141.160 \text{ m}$$

$$\psi = \frac{14{,}204.826}{6{,}371{,}000 + 438.4} \times \frac{180°}{\pi} = 0.127738° = 0°07'40''$$

$$B_1B_2 = 1141.160 \tan\left(\frac{0°07'40''}{2}\right) = 1.272 \text{ m}$$

$$AB_2 = 14{,}204.826 - 1.272 = 14{,}203.554 \text{ m}$$

$$A'B' = 14{,}203.554 \frac{6{,}371{,}000}{6{,}371{,}000 + 438.4} = 14{,}202.577 \text{ m}$$

(Note that if the short-line reduction procedure of Section (6.13) had been used, an error equal to B_1B_2, or 1.272 m, would have resulted.)

In the foregoing computations, the effects of refraction were eliminated by averaging reciprocal vertical angles α and β in Equation (19.24). This procedure yields the best results. If the vertical angle were observed at only one end of the line, as angle α at A, then because refraction is approximately 1/7 curvature, or $\psi/7$, a correction for its effect can be applied. In that case instead of using Equation (19.24), angle δ is computed as

$$\delta = \alpha + \frac{\psi}{2} - \frac{\psi}{7} = \alpha + \frac{5\psi}{14} \tag{19.31}$$

where

$$\psi = \frac{S \cos \alpha}{R_e + h_A} \times \frac{180°}{\pi}$$

Then Equations (19.25), (19.26), and (19.28) through (19.30) are solved as before.

Example 19.8

Compute the ellipsoid length of the line in Example 19.7 using only the observed vertical angle at A.

Solution

$$\psi = \frac{14{,}250.590 \cos 4°32'18''}{6{,}371{,}000 + 438.4} \times \frac{180°}{\pi} = 0.127748° = 0°07'40''$$

By Equation (19.31),

$$\delta = 4°32'18'' + \frac{5}{14}(0°07'40'') = 4°35'02''$$

Then Equations (19.25), (19.26), and (19.28) through (19.30) are solved in order,

$$AB_1 = 14{,}250.590 \cos 4°35'02'' = 14{,}205.008 \text{ m}$$

$$BB_1 = 14{,}250.590 \sin 4°35'02'' = 1138.888 \text{ m}$$

$$B_1B_2 = 1138.888 \tan = 1.270 \text{ m}$$

$$AB_2 = 14{,}205.008 - 1.270 = 14{,}203.738 \text{ m}$$

$$A'B' = 14{,}203.738 \frac{6{,}371{,}000}{6{,}371{,}000 + 438.4} = 14{,}202.761 \text{ m}$$

Note that this answer differs by 0.184 m from the one obtained in Example 19.7. This can be expected, because refraction varies with atmospheric conditions, and the correction $\psi/7$ only approximates its effects. Thus, as stated earlier, it is best to measure the vertical angles at both ends of the line if possible.

19.14.3 Reduction of Directions and Angles

As was discussed in Section 19.5, deflection of the vertical varies at different locations on the surface of the Earth. Because of this, during the angle observing process both the horizontal and vertical circles of a total station instrument are, in general, misaligned with the horizontal and vertical. Thus for geodetic calculations, the observed directions must be corrected according to Equation (19.11). Additionally, because of the sphericity of the Earth, the normals at the observing and target stations are skewed with respect to each other, and hence two additional corrections may be necessary. First, if the height of the target above the ellipsoid is substantial enough, this may necessitate a correction. Second, if the latitudes of the occupied and sighted stations differ significantly, this may also require a correction. However, for the target heights and relatively small latitude differences that apply in most surveys, corrections to azimuths for these conditions are often smaller than the observational errors. Thus except for very precise geodetic work they are not made. Finally, a correction that stems from deflection of the vertical is often significant and the procedures for making it are given here.

Correction for deflection of the vertical Since the instrument is set up with respect to the local vertical, angular measurements in the vertical and horizontal will both be affected by deflection of the vertical. The correction C''_{defl} in an observed horizontal direction in units of arc seconds for deflection of the vertical is

$$C''_{defl} = Az' - Az = \eta \tan \phi + (\xi \sin Az - \eta \cos Az) \cot z \quad \textbf{(19.32)}$$

where Az is the astronomic azimuth of the observed direction, Az' is the corrected azimuth, and z is the zenith angle.

Adding the correction determined in Equation (19.32) yields the corrected geodetic azimuth of an observed direction as

$$Az_G = Az_A + C''_{defl} \quad \textbf{(19.33)}$$

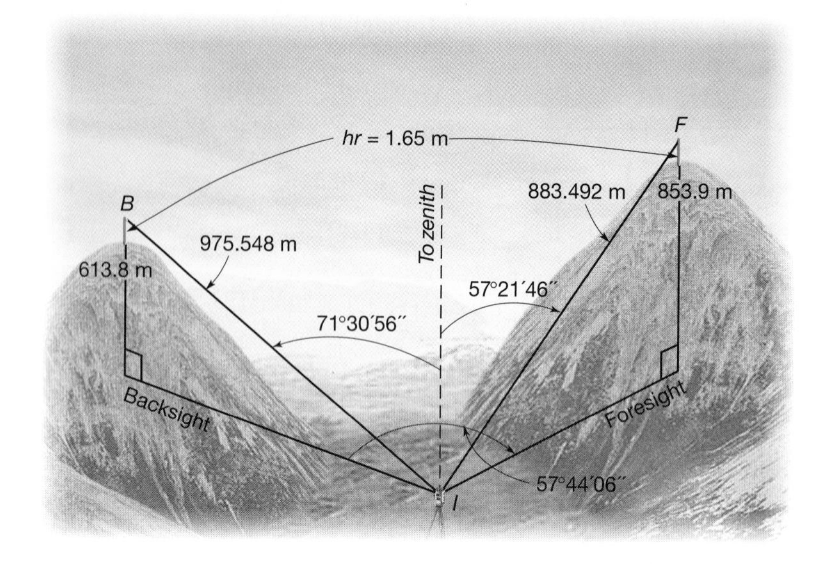

Figure 19.21 Figure for Example 19.9

For angles, the backsight and foresight directions that compose the angle are each corrected according to Equation (19.33) and subtracted according to Equation (11.11). This is demonstrated in Example 19.9.

Example 19.9

As shown in Figure 19.21, a horizontal angle *BIF* of 57°44′06″ is observed at instrument station *I*, whose latitude is 41°13′24.67″ as scaled from a topographic map. Orthometric heights were also scaled from a topographic map, and estimated as 613.8 m at backsight station *B* and 853.9 m at foresight station *F*. The geoidal separation is estimated using the NGS software GEOID03 as −29.45 m at station *B* and −29.84 m at station *F*. The reflector height *hr* at both target stations is set to 1.650 m. The azimuth from the observing station to station *B* is 23°16′24″. The components of the deflection of the vertical at the observing station are estimated using the NGS software DEFLEC99 as $\eta = 4.82''$ and $\xi = 0.29''$. The geodetic distance *IB*, as defined by Equation (19.21) is 975.548 m, and for *IF* it is 883.49 m. The zenith angle to station *B* is 71°30′56″, and to station *F* is 57°21′46″. What is the corrected geodetic angle *BIF?*

Solution

In this solution, the individual correction components are independently computed for each sight of the angle and subtracted to determine the corrected angle. *Geodetic heights:* Using Equation (19.7), the geodetic heights of h_B at the backsight station, and h_F are

$$h_B = 613.8 - 29.45 + 1.65 = 586.00 \text{ m}$$
$$h_F = 853.9 - 29.84 + 1.65 = 825.71 \text{ m}$$

Correction for deflection of the vertical: By Equation (19.32),
Backsight:

$$C''_{defl} = 4.82\tan(\phi) + [0.29\sin(Az_B) - 4.82\cos(Az_B)]\cot(71°30'56'') = 2.78''$$

where ϕ and Az_B are 41°13′24.67″ and 23°16′24″, respectively.
Foresight:

$$Az_F = 23°16'24'' + 57°44'06'' = 81°00'30''$$

$$C''_{defl} = 4.82\tan(\phi) + [0.29\sin(Az_F) - 4.82\cos(Az_F)]\cot(57°21'46'') = 3.92''$$

where ϕ again is 41°13′24.67″ and Az_F is 81°00′30″.

Corrected azimuths:

$$\text{Backsight: } 23°16'24'' + 2.78'' = 23°16'26.81''$$

$$\text{Foresight: } 81°00'30'' + 3.92'' = 81°00'33.94''$$

By Equation (11.11), the corrected angle is: 81°00′33.94″ − 23°16′26.81″ = 57°44′07.1″

Note that the correction to the angle over these short distances is 1.1″. Also note in Equation (19.32), that $\eta \tan\varphi$ is the same for both the backsight and foresight directions of an angle. Thus it did not have to be included in correction of directions for angles. The correction for deflection of vertical in angles could be rewritten as

$$C''_{\angle} = (\xi \sin Az_{FS} - \eta \cos Az_{FS})\cot z_{FS} - (\xi \sin Az_{BS} - \eta \cos Az_{BS})\cot z_{BS}$$

where *FS* represents the foresight values and *BS* represents the backsight values.

For precise geodetic control surveys, a correction due to deflection of the vertical must also be made to observed vertical angles. For zenith angles, the following equation applies:

$$z_C = z_{obs} + \xi \cos Az + \eta \sin Az \tag{19.34}$$

where z_C is the corrected zenith angle, z_{obs} is the observed zenith angle, and Az is the azimuth of the line of sight when the zenith angle is observed.

19.14.4 Leveling and Orthometric Heights

Distances observed along plumb lines (elevation differences) between equipotential gravitational surfaces provide the basis for specifying elevations, or *orthometric heights.* One of these equipotential surfaces, the geoid, is defined as the datum for observing these heights. For example, in Figure 19.22, the orthometric heights of points *A* and *B* are H_A and H_B, respectively.

The Earth spins about its rotational axis in approximately a 24-h period. From physics it is known that this spin results in a centrifugal acceleration that acts on all

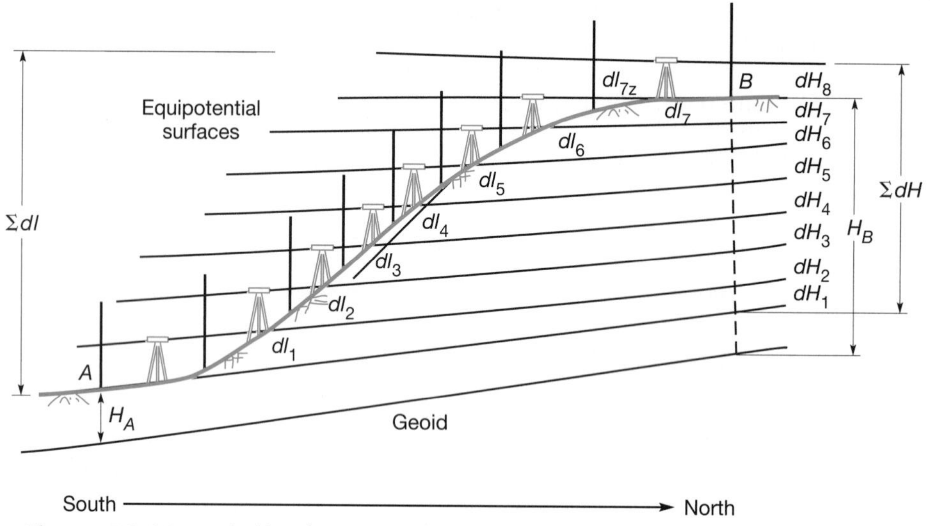

Figure 19.22 Leveled height versus orthometric height.

bodies that take part in the spin. The *gravitational force*[12] is a combination of the attractive force of the Earth's mass and the centrifugal force that acts on a body on the surface of the Earth. Since these two forces are in different directions, their combined effect results in a gravitational force that is less than the attractive force and is directed radially towards the mass center of the Earth. Thus points on the equator, which experience the greatest rotational velocity, have the weakest gravitational force. Conversely, points at the poles that are not subject to rotational velocity, do not experience any centrifugal acceleration component and thus experience the maximum gravitational force. It should be pointed out here that the difference between gravitational force at the pole and at the equator is only about 5 gals.[13]

As stated earlier, an equipotential surface has the same gravitational potential throughout its extent. From physics, potential can be defined as

$$W = mar \tag{19.35}$$

where m is the mass of the body, a is the acceleration of the body, and r is the radial distance from the mass center of the Earth. In the case of orthometric elevation differences, the acceleration a in Equation (19.35) can be replaced by the gravitational acceleration $-g$ acting on the point,[14] or

$$W = -mgr \tag{19.36}$$

From Equation (19.36) and the previous discussion concerning changes in the values of gravitational force on the Earth, it can be stated that as gravitational force increases, the radial distance must decrease in order that the potential of points on a given equipotential surface remain equal. Thus, equipotential surfaces are not concentric and instead they converge as they approach the poles, as shown in Figure 19.22.

---

[12]Force is defined from physics as mass times acceleration.

[13]A gal is a unit of acceleration where 1 gal = 1 cm/sec^2; 1 kgal = 1000 gal; 1 mgal = 0.001 gal.

[14]A negative sign is introduced in Equation (19.35) to account for the fact that gravitational attraction points radially inward while increases in elevation occur radially outward.

The difference in potential is defined as

$$\delta W = -g\,\delta l \tag{19.37}$$

where δW is the change in potential, and δl is the separation between the two equipotential surfaces. A proper expression for the potential of a point on an equipotential surface is the *geopotential number* C of a point, where C is defined as a negative difference in the potential of a point P and the potential of the geoid, and is mathematically defined as

$$C = -(W - W_0) = \int_{P_0}^{P} g\,dl \tag{19.38}$$

where W_0 is the potential of a point P_0 on the geoid. The units of the geopotential number are kgal-meters (kgals-m) where one kgal-m is called a *geopotential unit* (GPU). Since neither l nor g is known as a continuous function, in practice the difference in geopotential number is approximated as

$$\Delta C_{ij} = \sum_{k=i}^{j} \overline{g}_k\,\delta l_k \tag{19.39}$$

where $\overline{g}_k$ is the average gravitational attraction between two adjacent benchmarks i and j and δl_k is the observed level difference between the two adjacent benchmarks. Because it is impractical to observe g at every benchmark and since the units of geopotential numbers are unfamiliar to surveyors, *dynamic heights* were introduced. Dynamic heights are the geopotential number divided by a *reference gravity value,* or

$$H_i^D = C_i/g_R \tag{19.40}$$

where H_i^D is the dynamic height of a point, C_i the geopotential number of the point, and g_R the appropriate gravitational constant. In the United States, the NGS has adopted a reference gravity value of 980.6199 gal. The reader should note that the units of dynamic height are traditional distance units. In Figure 19.10, the dynamic height of point F 137 is 252.373 m where the reference gravity is 0.9806199 kgal. Thus, by rearranging Equation (19.40), the geopotential number of point F 137 is $0.9806199 \times 252.373 = 247.482$ kgal-m. It should be noted that two points that have the same dynamic height will be on the same equipotential surface. However, these same two points may not have the same orthometric heights. Conversely, as will be shown later, two points with the same orthometric heights may not lie on the same equipotential surfaces. Thus dynamic heights should be used for hydrological studies.

Height differences determined by leveling do not produce true orthometric height differences and thus a correction must be applied. As can be seen in Figure 19.22, as the leveling process proceeds from station A to B, the equipotential surfaces converge. Letting dl_i represent the leveled-height differences and dH_i represent the orthometric-height differences between incremental equipotential surfaces, it is apparent that the sum of dl_1, dl_2, dl_3, dl_4, dl_5, dl_6, dl_7, and dl_8 (Σdl_i) will not equal the sum of dH_1, dH_2, dH_3, dH_4, dH_5, dH_6, dH_7, and dH_8 (ΣdH_i) because of convergence. The difference between the leveled-height difference and the orthometric-height difference is called the *orthometric correction.* The orthometric correction O_c is added to the leveled height to obtain the orthometric height, or

$$dH = dl + O_c \tag{19.41}$$

where dH represents the orthometric-height difference between two points and dl represents the leveled-height difference between the two points.

Because gravitational surfaces converge near the Earth's poles and orthometric corrections are a function of gravity values observed along the leveling lines, the largest corrections occur along lines that are run in the north-south direction. When running a line of levels between two NSRS benchmarks, it is possible to estimate the orthometric correction from data that is published on the data sheets. Refer again to Figure 19.10, which shows an excerpt from the data sheet for benchmark F 137. Note that the NAVD88 (orthometric) height is 252.471 m, the modeled gravity value is 980,231.5 mgals, and the modeled geoidal undulation is −32.70 m. To compute the leveled-height difference, the potential heights for two control benchmark stations must be computed as

$$H_C(A) = H_A\left(g_A + \frac{0.0424}{1{,}000{,}000}H_A\right) \tag{19.42}$$

where $H_C(A)$ is the potential height of station A in units of kgal-meters, g_A the modeled gravity value at station A in units of kgals, and H_A the orthometric height of the station. Following this, the difference in the potential heights is computed and divided by the average gravity value for the two benchmarks, or,

$$dl = \frac{2[H_C(B) - H_C(A)]}{g_A + g_B} \tag{19.43}$$

Example 19.10

Given the following information from the control data sheets for Stations F 137 and J 231, what should be the leveled-height difference between stations?

Station	Orthometric Height (m)	Gravity (mgal)
F 137	252.471	980,231.5
J 231	294.548	980,143.5

Solution

By Equation (19.42)

$$H_C(F137) = 252.471\left[0.9802315 + \frac{0.0424}{1{,}000{,}000}252.471\right] = 247.4827 \text{ GPU}$$

$$H_C(J237) = 294.548\left[0.9801435 + \frac{0.0424}{1{,}000{,}000}294.548\right] = 288.7030 \text{ GPU}$$

By Equation (19.43)

$$dl(F, J) = \frac{2(288.7030 - 247.4827)}{0.9802315 + 0.9801435} = 42.053 \text{ m}$$

Note that in Example 19.8, the difference in orthometric heights is 294.548 − 252.471 = 42.077 m, but the leveled-height difference is 42.053 m yielding a difference of 2.4 cm. This difference represents the orthometric correction for the leveled line and would be seen as part of the misclosure of the line if this computation is not considered. In this example, Stations F 137 and J 231 are approximately 120 km apart in the north-south direction. As can be seen by this example, the convergence of the equipotential surfaces is very modest over long distances. Thus it is only considered in high-precision surveys involving long north-south extents.

After applying the orthometric correction, the resultant misclosure in the leveling circuit can be adjusted using least squares. However, for the most precise surveys, the gravity values at the intermediate benchmarks must also be considered. Readers who wish to learn more on this topic should consult the references at the end of this chapter.

19.15 GEODETIC POSITION COMPUTATIONS

Geodetic position computations involve two basic types of calculations, the *direct* and the *inverse* problems. In the direct problem, given the latitude and longitude of station A and the geodetic length and azimuth of line AB, the latitude and longitude of station B are computed. In the inverse problem, the geodetic length of AB and its forward and back azimuths are calculated, given the latitudes and longitudes of stations A and B.

For long lines it is necessary to account for the ellipsoidal shape of the Earth in these calculations to maintain suitable accuracy. Many formulas are available for making direct and inverse calculations, some of which are simplified approximations that only apply for shorter lines. This book will present those developed by Vincenty (1975). The procedures presented in the following subsections have a stated accuracy of a few centimeters for lines up to 20,000 km in length.

19.15.1 Direct Geodetic Problem

In the direct problem, ϕ_1 and λ_1 represent the latitude and longitude, respectively, s the geodetic length from station 1 to station 2, and α_1 the forward azimuth from station 1 to station 2. The variables a, b, and f are the defining parameters of the ellipsoid as presented in Section 19.2. The unknowns in the problem are ϕ_2 and λ_2, the geodetic latitude and longitude of the sighted station, and, α_2, the azimuth of the line from station 2 to station 1. Note that the observations used in this computation must be corrected to the ellipsoid using procedures outlined in Section 19.14.

The computational steps are as follows:[15]

1. $\tan U_1 = (1 - f) \tan \phi_1$
2. $\tan \sigma_1 = \tan U_1 / \cos \alpha_1$
3. $u = e' \cos \alpha$
4. $\sin \alpha = \cos U_1 \sin \alpha_1$
5. $A = 1 + \dfrac{u^2}{16{,}384}\{4096 + u^2[-768 + u^2(320 - 175u^2)]\}$
6. $B = \dfrac{u^2}{1024}\{256 + u^2[-128 + u^2(74 - 47u^2)]\}$
7. $2\sigma_m = 2\sigma_1 + \sigma$; where the first iteration uses $\sigma = \dfrac{s}{bA}$
8. $\Delta\sigma = B \sin \sigma \left\{ \cos(2\sigma_m) + \dfrac{1}{4} B \left[\cos \sigma(-1 + 2\cos^2 \sigma_m) - \dfrac{1}{6} B \cos(2\sigma_m)(-3 + 4\sin^2 \sigma)(-3 + 4\cos^2 \sigma_m) \right] \right\}$
9. $\sigma = \dfrac{s}{bA} + \Delta\sigma$
10. Repeat steps 7, 8, and 9 until $\Delta\sigma$ becomes negligible.
11. $\tan \phi_2 = \dfrac{\sin U_1 \cos \sigma + \cos U_1 \sin \sigma \cos \alpha_1}{(1 - f)\sqrt{\sin^2\alpha + (\sin U_1 \sin \sigma - \cos U_1 \cos \sigma \cos \alpha_1)^2}}$
12. $\tan \lambda = \dfrac{\sin \sigma \sin \alpha_1}{\cos U_1 \cos \sigma - \sin U_1 \sin \sigma \cos \alpha_1}$
13. $C = \dfrac{f}{16} \cos^2\alpha[4 + f(4 - 3\cos^2 \alpha)]$
14. $L = \lambda - (1 - C) f \sin \alpha \{\sigma + C \sin \sigma[\cos 2\sigma_m + C \cos \sigma(-1 + 2\cos^2 2\sigma_m)]\}$
15. $\lambda_2 = \lambda_1 + L$
16. $\tan \alpha_2 = \dfrac{\sin \alpha}{-\sin U_1 \sin \sigma + \cos U_1 \cos \sigma \cos \alpha_1}$

19.15.2 Inverse Geodetic Problem

In the inverse geodetic problem, ϕ_1, λ_1, ϕ_2, and λ_2 represent the latitude and longitude of the first and second stations, respectively. In the solution, the geodetic length, s, between the two points, and the forward and back azimuths of the line, α_1 and α_2 respectively, are determined. The variables a, b, and f again are the defining parameters of the ellipsoid as presented in Section 19.2.

---

[15]For the derivation of this formulation, and that of the inverse problem which follows, consult the publication by T. Vincenty cited in this chapter's bibliography.

Steps:

1. $L = \lambda = \lambda_2 - \lambda_1$
2. $\tan U_1 = (1 - f) \tan \phi_1$
3. $\tan U_2 = (1 - f) \tan \phi_2$
4. $\sin^2 \sigma = (\cos U_2 \sin \lambda)^2 + (\cos U_1 \sin U_2 - \sin U_1 \cos U_2 \cos \lambda)^2$
5. $\cos \sigma = \sin U_1 \sin U_2 + \cos U_1 \cos U_2 \cos \lambda$
6. $\sin \alpha = \cos U_1 \cos U_2 \sin \lambda / \sin \sigma$
7. $\cos 2\sigma_m = \cos \sigma - 2 \sin U_1 \sin U_2 / \cos^2 \alpha$
8. $C = \dfrac{f}{16} \cos^2 \alpha [4 + f(4 - 3 \cos^2 \alpha)]$
9. $\lambda = L - (1 - C) f \sin \alpha \{\sigma + C \sin \sigma [\cos 2\sigma_m + C \cos \sigma (-1 + 2 \cos^2 2\sigma_m)]\}$
10. Repeat steps 8 and 9 until changes in λ become negligible.
11. $s = bA(\sigma - \Delta\sigma)$ where $\Delta\sigma$ comes from the steps 12–15.
12. $u = e' \cos \alpha$
13. $A = 1 + \dfrac{u^2}{16{,}384} \{4096 + u^2[-768 + u^2(320 - 175u^2)]\}$
14. $B = \dfrac{u^2}{1024} \{256 + u^2[-128 + u^2(74 - 47u^2)]\}$
15. $\Delta\sigma = B \sin \sigma \left\{ \cos(2\sigma_m) + \dfrac{1}{4} B \left[\cos \sigma (-1 + 2 \cos^2 \sigma_m) - \dfrac{1}{6} B \cos(2\sigma_m)(-3 + 4 \sin^2 \sigma)(-3 + 4 \cos^2 \sigma_m) \right] \right\}$
16. $\tan \alpha_1 = \dfrac{\cos U_2 \sin \lambda}{\cos U_1 \sin U_2 - \sin U_1 \cos U_2 \cos \lambda}$
17. $\tan \sigma_2 = \dfrac{\cos U_1 \sin \lambda}{-\sin U_1 \cos U_2 + \cos U_1 \sin U_2 \cos \lambda}$

The software WOLFPACK, which accompanies this book, can be used to do both of these computations. Figure 19.23 shows the data entry screen for the direct geodetic problem. A similar data screen is available in WOLFPACK to compute the inverse geodetic problem. An Excel® spreadsheet and Mathcad® worksheet that demonstrate these computations are also provided on the CD that accompanies this book.

To simplify position calculations for long lines, and yet maintain geodetic accuracy, state plane coordinate systems have been developed. These are described in Chapter 20.

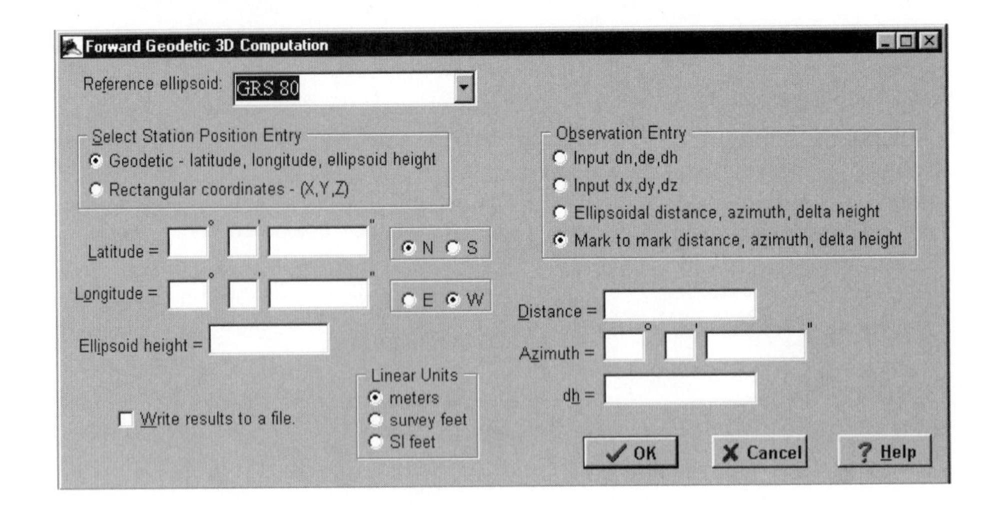

Figure 19.23 Data entry screen for forward computation from WOLFPACK.

■ 19.16 THE LOCAL GEODETIC COORDINATE SYSTEM

Surveys performed by GPS relative positioning (see Chapters 13, 14, and 15) yield three-dimensional baseline components (ΔX, ΔY, ΔZ). These vector components are in the geocentric coordinate system (see Section 13.4.3). It is common practice to transform these geocentric coordinate vector components into a *local geodetic system of easting,* (Δe), *northing* (Δn), and *local up* (Δu). The two coordinate systems are illustrated in Figure 19.24, where XYZ represents the geocentric system and enu is the local geodetic system. The local geodetic system is user oriented in that the e and n axes are in a local horizontal plane (u is coincident with the normal at the origin of the local coordinate system) and n is in the direction of local north.

To perform a transformation from the geocentric coordinate system to local geodetic, a set of three-dimensional rotation matrices must be employed. These rotation matrices are similar to their two-dimensional counterparts as shown in

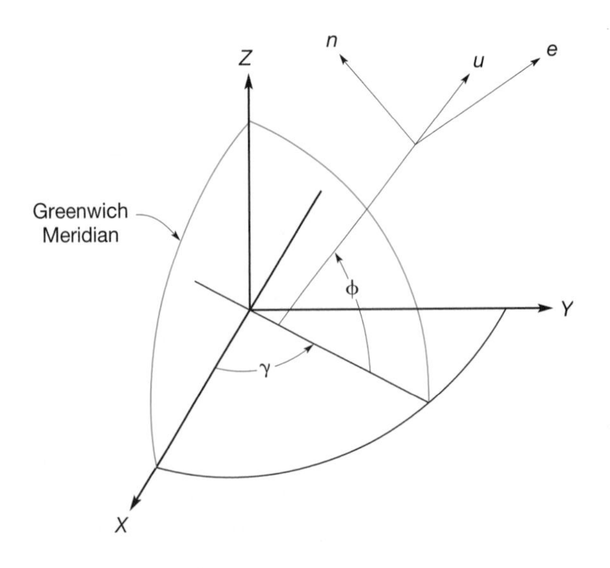

Figure 19.24 The relationship between the geocentric coordinate system and the local geodetic coordinate system.

Equation (11.36). In the transformation process, rotations occur about each of the three coordinate axes. Letting the rotation angle around the X-axis be θ_X, the rotation angle around the Y-axis be θ_Y and the rotation angle around the Z-axis be θ_Z, the three-dimensional rotation matrices are

$$R_X(\theta_X) = \begin{bmatrix} 1 & 0 & 0 \\ 0 & \cos\theta_X & \sin\theta_X \\ 0 & -\sin\theta_X & \cos\theta_X \end{bmatrix} \tag{19.44}$$

$$R_Y(\theta_Y) = \begin{bmatrix} \cos\theta_Y & 0 & -\sin\theta_Y \\ 0 & 1 & 0 \\ \sin\theta_Y & 0 & \cos\theta_Y \end{bmatrix} \tag{19.45}$$

$$R_Z(\theta_Z) = \begin{bmatrix} \cos\theta_Z & \sin\theta_Z & 0 \\ -\sin\theta_Z & \cos\theta_Z & 0 \\ 0 & 0 & 1 \end{bmatrix} \tag{19.46}$$

As shown in Figure 19.24, a rotation about the Z-axis by an amount of $\lambda - 270°$ must occur to align the X-axis with the local e-axis. Following this rotation, the once-rotated Z-axis is brought into coincidence with the u axis by a rotation of $90° - \phi$ about the once-rotated X-axis. The resultant expression is

$$\begin{bmatrix} \Delta e \\ \Delta n \\ \Delta u \end{bmatrix} = R_X(90° - \phi)R_Z(\lambda - 270°)\begin{bmatrix} \Delta X \\ \Delta Y \\ \Delta Z \end{bmatrix} \tag{19.47}$$

Performing the proper trigonometric substitutions and rearranging the equations to place them in the standard order of (n, e, u), the final transformation equations are

$$\begin{bmatrix} \Delta n \\ \Delta e \\ \Delta u \end{bmatrix} = \begin{bmatrix} -\sin\phi & 0 & \cos\phi \\ 0 & 1 & 0 \\ \cos\phi & 0 & \sin\phi \end{bmatrix}\begin{bmatrix} \cos\lambda & \sin\lambda & 0 \\ -\sin\lambda & \cos\lambda & 0 \\ 0 & 0 & 1 \end{bmatrix}\begin{bmatrix} \Delta X \\ \Delta Y \\ \Delta Z \end{bmatrix}$$

$$= \begin{bmatrix} -\sin\phi\cos\lambda & -\sin\phi\sin\lambda & \cos\phi \\ -\sin\lambda & \cos\lambda & 0 \\ \cos\phi\cos\lambda & \cos\phi\sin\lambda & \sin\phi \end{bmatrix}\begin{bmatrix} \Delta X \\ \Delta Y \\ \Delta Z \end{bmatrix}$$

$$= R(\phi, \lambda)\begin{bmatrix} \Delta X \\ \Delta Y \\ \Delta Z \end{bmatrix} \tag{19.48}$$

19.17 THREE-DIMENSIONAL COORDINATE COMPUTATIONS

Sometimes it is advantageous to compute three-dimensional changes in a local geodetic coordinate system from reduced field observations. In Figure 19.25, the

Figure 19.25 Reduction of observations in a local geodetic coordinate system.

reduced observations of azimuth Az, ellipsoid distance S, and altitude angle α can be used to derive the changes in the local geodetic coordinate system as

$$\begin{aligned}\Delta n &= S\cos(\alpha)\cos(Az)\\ \Delta e &= S\cos(\alpha)\sin(Az)\\ \Delta u &= S\sin(\alpha)\end{aligned} \tag{19.49}$$

Equations (19.49) can be modified by Equations (8.2) to incorporate a zenith angle z as

$$\begin{aligned}\Delta n &= S\sin(z)\cos(Az)\\ \Delta e &= S\sin(z)\sin(Az)\\ \Delta u &= S\cos(z)\end{aligned} \tag{19.50}$$

From Figure 19.25, the following inverse relationships can be developed:

$$\begin{aligned}S &= \sqrt{\Delta n^2 + \Delta e^2 + \Delta u^2}\\ Az &= \tan^{-1}\left(\frac{\Delta e}{\Delta n}\right)\\ \alpha &= \sin^{-1}\left(\frac{\Delta u}{S}\right) \text{ or } z = \cos^{-1}\left(\frac{\Delta u}{S}\right)\end{aligned} \tag{19.51}$$

By combining Equation (19.48) with Equations (19.51), the reduced observations can be obtained directly from changes in geocentric coordinates as

$$\begin{aligned}S &= \sqrt{\Delta X^2 + \Delta Y^2 + \Delta Z^2}\\ Az_1 &= \tan^{-1}\left(\frac{-\Delta X\sin\lambda_1 + \Delta Y\cos\lambda_1}{-\Delta X\sin\phi_1\cos\lambda_1 - \Delta Y\sin\phi_1\sin\lambda_1 + \Delta Z\cos\phi_1}\right)\\ z_1 &= \cos^{-1}\left(\frac{\Delta X\cos\phi_1\cos\lambda_1 + \Delta Y\cos\phi_1\sin\lambda_1 + \Delta Z\sin\phi_1}{S}\right)\end{aligned} \tag{19.52}$$

It is important to note that Equation (19.52) uses the latitude and longitude of the observation station P_1 in Figure 19.25. These values can be computed from the geocentric coordinate values of P_1 based on Equations (13.3) thru (13.7).

Example 19.11

The mark-to-mark distance from station Bill to station Red is 568.138 m, and the zenith angle and azimuth of this course are 92°14′25″ and 40°36′23″, respectively. If the geodetic coordinates of station Bill are 61°10′42.1058″ N latitude and 149°11′12.1033″ W longitude, what are the changes in the geocentric coordinate system?

Solution

Using Equation (19.50), the changes in the local coordinate system are

$$\begin{aligned}
\Delta n &= 568.138 \sin(92°14'25'') \cos(40°36'23'') = 431.000 \text{ m} \\
\Delta e &= 568.138 \sin(92°14'25'') \sin(40°36'23'') = 369.495 \text{ m} \\
\Delta u &= 568.138 \cos(92°14'25'') = -22.209 \text{ m}
\end{aligned}$$

To solve for the changes in geocentric coordinates in Equation (19.48), the inverse of $R(\phi, \lambda)$ must be determined. *Since $R(\phi, \lambda)$ is an orthogonal rotation matrix, its inverse is $R^T(\phi, \lambda)$*. Thus, the changes in geocentric coordinates for these observations are

$$\begin{bmatrix} \Delta X \\ \Delta Y \\ \Delta Z \end{bmatrix} = \begin{bmatrix} -\sin\phi\cos\lambda & -\sin\lambda & \cos\phi\cos\lambda \\ -\sin\phi\sin\lambda & \cos\lambda & \cos\phi\sin\lambda \\ \cos\phi & 0 & \sin\phi \end{bmatrix} \begin{bmatrix} 431.0000 \\ 369.4950 \\ -22.2087 \end{bmatrix}$$

$$= \begin{bmatrix} 522.773 \\ -118.423 \\ 188.321 \end{bmatrix}$$

Note that the changes in geodetic coordinates computed in Example 19.11 should closely agree with GPS baseline vector observations that would be obtained if the baseline between Bill and Red were observed. Equations (19.52) can be used to determine the mark-to-mark distance, geodetic azimuth, and zenith angle of the observations that would be obtained if a survey were conducted at station Bill. An advantage of local coordinate systems is that unlike plane coordinate systems as presented in Chapter 20, all three dimensions are simultaneously computed using standard observations collected with a total station. Additionally these local systems can be integrated with GPS observations for simultaneous GPS and terrestrial least-squares adjustments. However, it should always be remembered that the computed values would differ from the field-observed values if the appropriate corrections discussed in Section 19.14 were not applied.

19.18 CONCLUSIONS

In this chapter, an introductory discussion of geodetic computations was presented, reference frames (datums) were briefly discussed, and transformations between surveyed, geocentric, and geodetic observations presented. Readers who wish to explore these topics in more detail should refer to the bibliography at the end of this chapter.

PROBLEMS

Asterisks (*) indicate problems that have answers given in Appendix G.

19.1 Define the *geoid* and *ellipsoid*.

19.2 Draw a sketch that shows the *geodetic latitude, longitude*, and *height* for a point.

19.3 What is *precession?*

19.4* What is the difference between the equatorial circumference of the Clarke 1866 ellipsoid and that of the WGS84 ellipsoid?

19.5 Determine the first and second eccentricities for the WGS84 ellipsoid.

19.6 Discuss the importance of the NAD83 (NSRS 2007) datum? How will it affect GPS surveys?

19.7 Why shouldn't points defined in different versions of the NAD83 datum be mixed in a survey?

19.8 Discuss the motions of the Earth's instantaneous pole with respect to the Conventional Terrestrial Pole.

19.9* What are the radii in the meridian and prime vertical for a station with latitude 42°37′26.34584″ using the GRS80 ellipsoid?

19.10 For the station listed in Problem 19.9, what is the radius of the great circle at the station that is at an azimuth of 153°29′32″ using the GRS80 ellipsoid?

19.11* The orthometric height at Station Y 927 is 304.517 m and the geoidal undulation at that station −31.893 m. What is geodetic height?

19.12 The geodetic height at station *Roger* is 88.046 m. Its geoidal undulation is −30.113 m. What is its orthometric height?

19.13 The orthometric height of a particular benchmark is 64.24 ft. The geoidal height at the station is –29.87 m. Is the station above or below the ellipsoid? Draw a sketch depicting the geoid, ellipsoid, and benchmark.

19.14 The instantaneous position of the pole at the time of an azimuth observation is $x = -1.2''$ and $y = -1.5''$. The position of the station is (46°37′31.6436″ N, 108°25′44.3109″ W) and the observed azimuth of a line is 219°23′34″. What is the astronomic azimuth of the line corrected for polar motion?

19.15* The deflection of the vertical components ξ and η are $-2.85''$ and $5.94''$, respectively. What is the geodetic azimuth of the azimuth in Problem 19.14? (Assume a horizontal line of sight.)

19.16 To within what tolerance should the elevations of two benchmarks 15-km apart be established if first-order class II standards were used to set them? What should it be if second-order class I standards were used?

19.17 Name the orders and classes of accuracy of both horizontal and vertical control surveys, and give their relative accuracy requirements.

19.18 Obtain a USGS quadrangle map of your area. On the map, lay out a quadrilateral having sides from 5 to 8 km long. Check all lines for intervisibility by plotting their profiles, and revise station positions if necessary to provide unobstructed lines of sight.

19.19 Given the following information for stations *Fred* and *Jon,* what should be the leveled height difference between them?

Station	Height (m)	Gravity (mgal)
Fred	432.409	980,642.8
Jon	388.884	980,474.5

19.20 Similar to Problem 19.19 except that the station data is

Station	Height (m)	Gravity (mgal)
Fred	66.332	979,998.2
Jon	52.605	979,864.7

19.21* A slope distance of 2458.663 m is observed between two points *Gregg* and *Brian* whose orthometric heights are 458.966 m and 566.302 m, respectively. The geoidal undulations are −25.66 m and −25.06 m at *Gregg* and *Brian,* respectively. The height of the instrument at station *Gregg* at the time of the observation was 1.525 m and the height of the reflector at station *Brian* was 1.603 m. What are the geodetic and mark-to-mark distances for this observation? (Use an average radius for the Earth of 6,371,000 m for R_α.)

19.22 If the latitude of station *Gregg* in Problem 19.21 was 35°22′58.6430″ and the azimuth of the line was 303°33′03.8″, what are the geodetic and mark-to-mark distances for this observation? (Use the GRS80 ellipsoid.)

19.23 A slope distance of 5502.896 m is observed between two stations *A* and *B* whose geodetic heights are 184.694 m and 47.534 m, respectively. The height of the instrument at the time of the observation was 1.454 m and the height of the reflector was 1.500 m. The latitude of Station *A* is 41°18′47.0367″, and the azimuth of *AB* is 38°58′21.9″. (Use the GRS80 ellipsoid.) What are the geodetic and mark-to-mark distances for this observation?

19.24 Discuss the advantages and disadvantages of traverse control surveys as compared to triangulation for horizontal networks.

19.25* Compute the back azimuth of a line 5863 m long in the east-west direction at a mean latitude of 45°01′32.0654″, whose forward azimuth is 88°16′33.2″ from north. (Use an average radius for the Earth of 6,371,000 m.)

19.26 In Figure 19.14 azimuth of *AB* is 115°27′03″ and the angles to the right observed at *B, C, D, E,* and *F* are 132°01′04″, 241°45′12″, 141°15′01″, 162°09′24″, and 202°33′19″, respectively. An astronomic observation yielded an azimuth of 95°10′33″ for line *FG*. The mean latitude of the traverse is 39°58′20″, and the total departure between points *A* and *F* was 2022.874 m. Compute the angular misclosure and the adjusted angles. (Assume the angles and distances have already been corrected to the ellipsoid and use an average radius for the Earth of 6,371,000 m.)

19.27 In Figure 19.20 slope distance S and vertical angles α and β were observed as 8320.96 ft, $+5°26'37''$, and $-5°34'14''$, respectively. Ellipsoid height of point A is 1402.11 ft. What is length $A'B'$ on the ellipsoid? (Use an average radius for the Earth.)

19.28 In Figure 19.19 slope distance S was observed as 6844.504 m. The orthometric elevations of points A and B were 343.460 m and 632.180 m, respectively, and the geoidal undulation at both stations was -28.6 m. The instrument and reflector heights were both set at 1.500 m. Calculate geodetic distance $A'B'$. (Use an average radius for the Earth.)

19.29 In Figure 19.20, slope distance S and zenith angle β at station B were observed as 2072.33 m and $82°17'18''$, respectively to station A. If the elevation of station B is 435.967 m and the geoidal undulations at stations B and A are both -28.04 m, what is ellipsoid length $A'B'$? (Use an average radius for the Earth.)

19.30* Components of deviation of the vertical at an observing station of latitude $46°43'55.8042''$ are $\xi = -4.00''$ and $\eta = +2.54''$. If the observed zenith angle on a course with an astronomic azimuth of $204°32'44''$ is $85°56'07''$, what are the azimuth and zenith angles corrected for deviation of the vertical?

19.31 At the same observation station as Problem 19.30, the observed zenith angle on a course with an azimuth of $19°25'33''$ is $98°02'31''$, what are the azimuth and zenith angles corrected for deviation of the vertical?

19.32 Using the reduced azimuths of Problems 19.30 and 19.31, what is the reduced geodetic angle that is less than 180°?

19.33 Create a computational program that solves Problem 19.32.

19.34 Why do level rods used in precise leveling typically have two faces?

19.35 Discuss the advantages of breaking long leveling networks into shorter circuits.

19.36 Compute the collimation correction factor C for the following field data, taken in accordance with the example and sketch in the field notes of Figure 19.18. With the instrument at station 1, high, middle, and low crosshair readings were 5.612, 5.501, and 5.390 ft on station A and 4.978, 3.728, and 2.476 ft on station B. With the instrument at station 2, high, middle, and low readings were 7.211, 6.053, and 4.894 ft on A and 4.561, 4.358, and 4.155 ft on B.

19.37 A leveling instrument having a collimation factor of $+0.0009$ m/m of interval was used to run a section of three-wire differential levels from BM A to BM B. Sums of backsights and foresights for the section were 550.893 and 523.660 m, respectively. Backsight stadia intervals totaled 857.48 m, while the sum of foresight intervals was 632.08 m. What is the corrected elevation difference from BM A to BM B?

19.38 The relative error of the difference in elevation between two benchmarks directly connected in a level circuit and located 50 km apart is ± 0.008 m. What order and class of leveling does this represent?

19.39 Similar to Problem 19.38, except the relative error is ± 0.019 ft for benchmarks located 45 km apart.

19.40 The baseline components of a GPS baseline vector observed at a station A in meters are (908.652, -580.401, -245.664). The geodetic coordinates of the first base station are $39°17'57.5604''$ N latitude and $102°52'35.9669''$ W longitude. What are the changes in the local geodetic coordinate system of $(\Delta n, \Delta e, \Delta u)$? (Use the WGS 84 ellipsoid.)

19.41 In Problem 19.40, what are the geodetic distance, zenith angle, and azimuth for the baseline vector?

19.42 If the geodetic distance between two stations is 832.668 m, the zenith angle between them is $86°27'42''$ and the azimuth of the line is $321°27'06''$, what are the changes in the local geodetic coordinates?

19.43 Create a computational program to solve Problem 19.41.

19.44 Create a computational program to solve Problem 19.42.

BIBLIOGRAPHY

Czepiga, F. 2005. "Setting the Standard in New Jersey." *Professional Surveyor* 25 (No. 6): 8.

Dunn, M. 2005. "The Michigan Spatial Reference Network, Part 1." *Professional Surveyor* 25 (No. 3): 26.

______. 2005. "The Michigan Spatial Reference Network, Part 2." *Professional Surveyor* 25 (No. 4): 24.

Frakes, S. 2003. "NGS Geodetic Tool Kit, Part 4: General Geodetic Computations." *Professional Surveyor* 23 (No. 9): 30.

Heiskanen, W.A., and Moritz, H. 1967. *Physical Geodesy*. San Francisco: W. H. Freeman and Co.

Kuang, S. et. al. 2002. "GPS Control Densification Project for Illinois Department of Transportation." *Surveying and Land Information Science* 62 (No. 4): 225.

Lee, Brandon D. 2002. "Increasing the National Vertical Geodetic Network." *Surveying and Land Information Science* 62 (No. 2): 131.

McCray, J. 2001. "Geodetic Surveying Made Plain—Now Everybody's Doing It!" *Point of Beginning* 26 (No. 4): 62.

______. 2002. "Geodetic Surveying Made Plain—Where's North?" *Point of Beginning* 27 (No. 5): 52.

Meyer, T., et. al. 2004. "What Does Height Really Mean" Part I: Introduction." *Surveying and Land Information Science* 64 (No. 4): 223.

______. 2005. "What Does Height Really Mean? Part II: Physics and Gravity." *Surveying and Land Information Science* 65 (No. 1): 5.

______. 2006. "What Does Height Really Mean? Part III: Height Systems." *Surveying and Land Information Science* 66 (No. 2): 149.

______. 2006. "What Does Height Really Mean? Part IV: GPS Heighting." *Surveying and Land Information Science* 66 (No. 3): 165.

Pearson, C. 2005. "The National Spatial Reference System Readjustment of NAD 83." *Surveying and Land Information Science* 65 (No. 2): 69.

Reno, D. 2005. "Saginaw River: High Precision Vertical River Crossing." *Professional Surveyor* 25 (No. 1): 8.

Roman. D., et. al. 2004. "Assessment of the New National Geoid Height Model—GEOID03." *Surveying and Land Information Science* 64 (No. 3): 153.

Shields, Renee. 2003. "NGS Geodetic Tool Kit, Part 3: LVL_DH, and Tidal and Orthometric Elevations." *Professional Surveyor* 23 (No. 7): 31.

Snay, R. 2003. "NGS Geodetic Tool Kit, Part 5: Horizontal Time-dependent Positioning." *Professional Surveyor* 23 (No. 11): 30.

Snay, R. and T. Soler. 2000. "Modern Terrestrial Reference Systems. Part 2: The Evolution of NAD83." *Professional Surveyor* 20 (No. 2): 16.

______. 2000. "Modern Terrestrial Reference Systems. Part 3: WGS84 and ITRS." *Professional Surveyor* 20 (No. 3): 24.

______. 2000. "Modern Terrestrial Reference Systems. Part 4: Practical Considerations for Accurate Positioning." *Professional Surveyor* 20 (No. 4): 32.

Strange, W. 2000. "Do You Really Have WGS 84 Coordinates?" *Professional Surveyor* 20 (No. 9): 34.

Young, Gary M. 2003. "NGS Geodetic Tool Kit, Part 1: Using Web-based forms, users can now upload data and run programs on NGS computers." *Professional Surveyor* 23 (No. 4): 40.

20

State Plane Coordinates

■ 20.1 INTRODUCTION

Most surveys of small areas are based on the assumption that the Earth's surface is a plane. However, as explained in Chapter 19, for large-area surveys it is necessary to consider Earth curvature. Unfortunately the calculations necessary to determine geodetic positions from survey observations and get distances and azimuths from them are lengthy and practicing surveyors often are not familiar with this procedure. Clearly a system for specifying positions of geodetic stations using plane rectangular coordinates is desirable, since it allows computations to be made using simple coordinate geometry formulas, such as those presented in Chapter 11. The National Geodetic Survey met this need by developing a state plane coordinate system for each state.

A state plane coordinate system provides a common datum of reference for horizontal control of all surveys in a large area, just as the geoid furnishes a single datum for vertical control. It eliminates having individual surveys based on different assumed coordinates, unrelated to those used in other adjacent work. State plane coordinates are available for all control points in the National Spatial Reference System and for many other control points as well. They are widely used as the reference points for initiating surveys of all types, including those for highway construction projects, property boundary delineation, and photogrammetric mapping.

There are many examples illustrating the value of state plane coordinates. They make it possible for extensive surveys on highway projects to begin on one control station and close on another that is tied to the same coordinate system. On boundary surveys, if a parcel's corners are referenced to the state plane coordinate system, their locations are basically indestructible. The iron pipes, posts, or other monuments marking their positions may disappear, but their original locations can be restored from surveys initiated at other nearby monuments referenced to the state plane coordinate system. For this reason, some states require that state plane coordinates be included on all new subdivisions. State plane coordinates are highly recommended as the reference frame for entering maps and other data into Land and Geographic Information Systems. This allows all data to be referenced to a common system, and thus can be accurately registered and overlaid for analysis purposes.

20.2 PROJECTIONS USED IN STATE PLANE COORDINATE SYSTEMS

To convert geodetic positions of a portion of the Earth's surface to plane rectangular coordinates, points are projected mathematically from the ellipsoid to some imaginary *developable surface*—a surface that can conceptually be developed or "unrolled and laid out flat" without distortion of shape or size. A rectangular grid can be superimposed on the *developed* plane surface and the positions of points in the plane specified with respect to X and Y grid axes. A plane grid developed using this mathematical process is called a *map projection*.

There are several types of map projections, with the oldest known projections dating back to the ancient Greeks. Today, two of the most commonly used mapping projections are the *Lambert conformal conic* and the *Transverse Mercator* projections. Johann Heinrich Lambert initially developed both of these projections. The Transverse Mercator projection was further developed and redefined by Carl Friedrich Gauss and L. Krüger, and thus is also known as the *Gauss-Krüger* projection. These two projections are used in state plane coordinate systems. The Lambert conformal conic projection utilizes an imaginary cone as its developable surface and the Transverse Mercator employs a fictitious cylinder. These are shown in Figure 20.1(a) and (b). The cone and cylinder are secant to the ellipsoid in the state plane coordinate systems; that is, they intersect the ellipsoid along two arcs AB and CD as shown. With this placement, the conical and cylindrical surfaces conform better to the ellipsoid over larger areas than they would if placed tangent.

Figure 20.1(c) and (d) illustrate plane surfaces "developed" from the cone and cylinder. Here, points are projected mathematically from the ellipsoid to the surface of the imaginary cone or cylinder based on their geodetic latitudes and longitudes. For discussion purposes, this may be considered a radial projection from the Earth's center. Figure 20.2 illustrates this process diagrammatically and displays the relationship between the length of a line on the ellipsoid and its extent when projected onto the surface of either a cone or a cylinder. Note that distance $a'b'$ on the projection surface is greater than ab on the ellipsoid, and similarly $g'h'$ is longer than gh. From this observation it is clear that map projection scale is larger than true ellipsoid scale where the cone or cylinder is outside the ellipsoid. Conversely distance $d'e'$ on the projection is shorter than de on the ellipsoid, and thus map scale is smaller than true ellipsoid scale when the projection surface is inside the ellipsoid. Points c and f occur at the intersection of the projection and

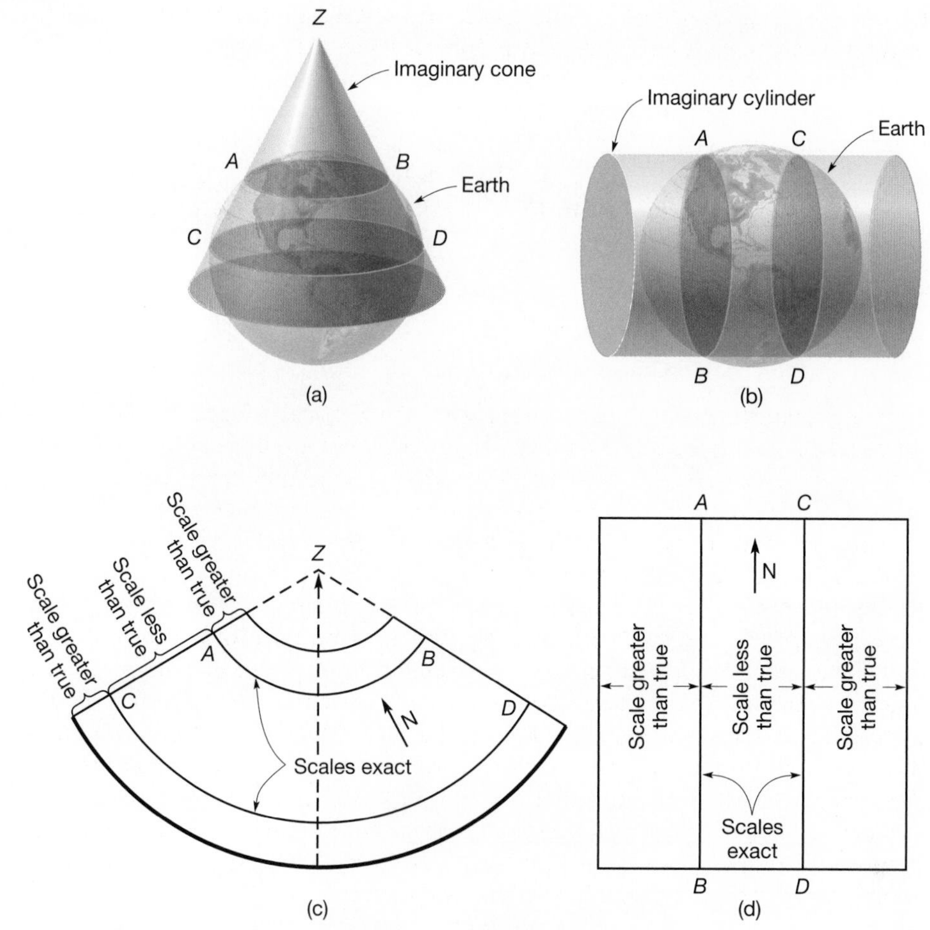

Figure 20.1 Surfaces use in state plane coordinate systems.

ellipsoid surfaces, and therefore map scale equals true ellipsoid scale along the lines of intersection. These relationships of map scale to true ellipsoid scale for various positions on the two projections are indicated in Figure 20.1(c) and (d). As will be discussed later, these length differences are accounted for by means of a *scale factor*.

From the foregoing discussion it should be clear that points couldn't be projected from the ellipsoid to developable surfaces without introducing distortions in the lengths of lines or the shapes of areas. However, these distortions are held to a minimum by selected placement of the cone or cylinder secant to the ellipsoid, by choosing a *conformal projection* (one that preserves true angular relationships around points in a small region), and also by limiting the zone size or extent of coverage on the Earth's surface for any particular map projection. If the width of zones is held to a maximum of 158 mi (254 km), and if two-thirds of this zone width is between the secant lines, distortions (differences in line lengths on the two surfaces) are kept to 1 part in 10,000 or less. The NGS intended this accuracy

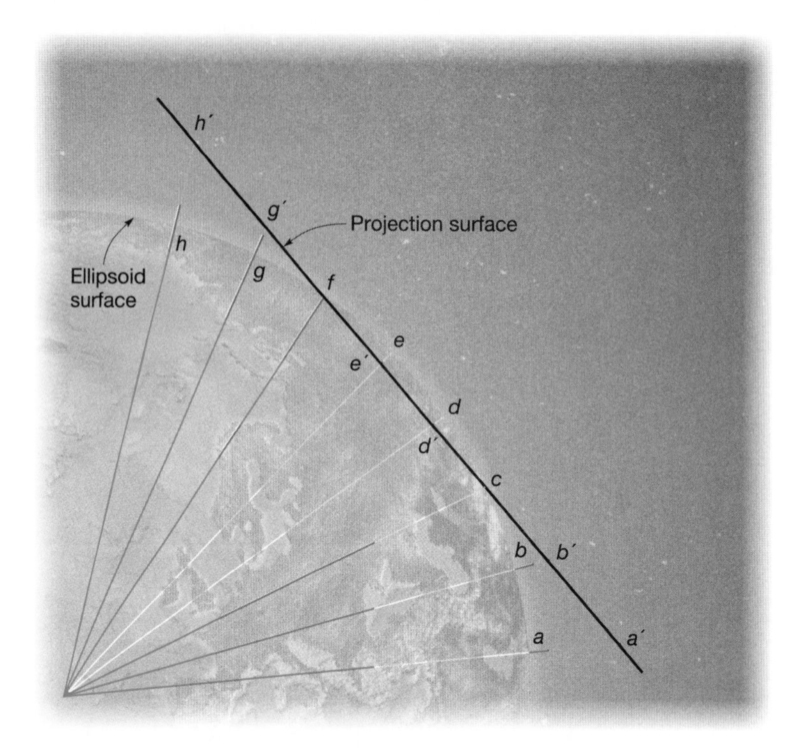

Figure 20.2 Method of projection.

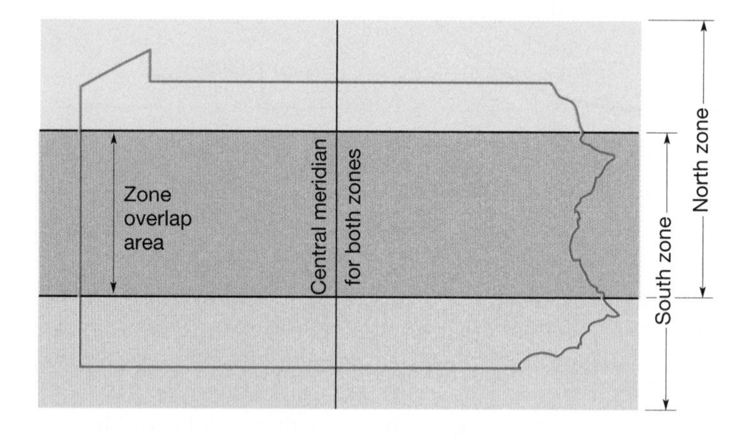

Figure 20.3 Coverages and overlap of zones in Pennsylvania's Lambert conformal conic state plane coordinate system.

in its development of the state plane coordinate systems. For small states such as Connecticut and Delaware, one state plane coordinate zone is sufficient to cover the entire state. Larger states require several zones to encompass them; for example, Alaska has 10, California has 6, and Texas has 5. Where multiple zones are needed to cover a state, adjacent zones overlap each other. As explained in Section 20.10, this is important when lengthy surveys extend from one zone to another. Figures 20.3 and 20.4 show the coverages of zones in Pennsylvania and Indiana, respectively. Both states have two zones, with Pennsylvania using the Lambert conformal conic projection and Indiana the Transverse Mercator projection.

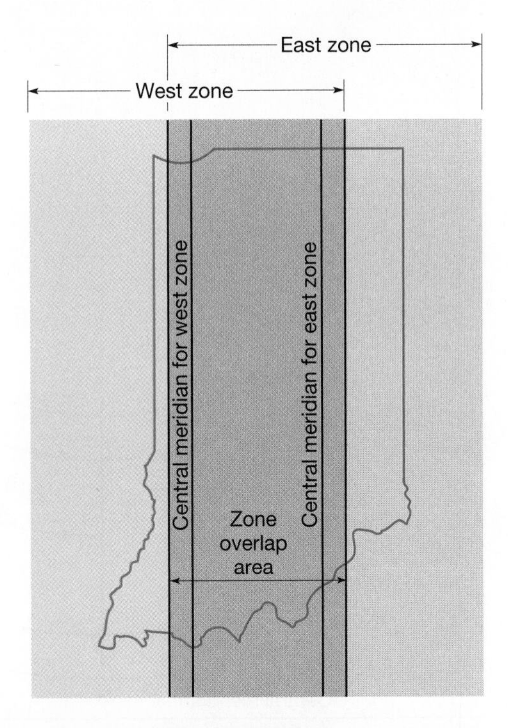

Figure 20.4 Coverages and overlap of zones in Indiana's state plane coordinate system.

■ 20.3 LAMBERT CONFORMAL CONIC PROJECTION

The Lambert conformal conic projection, as its name implies, is a projection onto the surface of an imaginary cone. The term *conformal,* as noted earlier, means that true angular relationships are retained around all points in small regions. Scale on a Lambert projection varies from north to south but not from east to west, as shown on Figure 20.1(c). Zone widths in the projection are therefore limited north-south, but not east-west. The Lambert projection is thus ideal for mapping states that are narrow north-south, but which extend long distances in an east-west direction—for example, the Carolinas, Kentucky, Massachusetts, Montana, Oregon, Pennsylvania, and Tennessee.

Figure 20.5 shows the portion of the developed cone of a Lambert projection covering an area of interest. In the Lambert projection, the cone intersects the ellipsoid along two parallels of latitude, called *standard parallels,* at one-sixth of the zone width from the north and south zone limits. All meridians are straight lines converging at Z, the apex of the cone. An example is ZM, which is the *central meridian*. All parallels of latitude are the arcs of concentric circles having centers at the apex. The projection is located in a zone in an east-west direction by assigning the central meridian a longitudinal value that is near the middle of the area to be covered. The direction of the central meridian on the projection establishes grid north. All lines parallel with the central meridian point in the direction of grid north. Therefore, except at the central meridian, directions of "true" and "grid" north do not coincide because true meridians converge. As shown in Figure 20.5, the E and N coordinates of points are measured perpendicular and parallel to the central meridian, respectively, from a reference E-N axis system that is offset to the west and south.

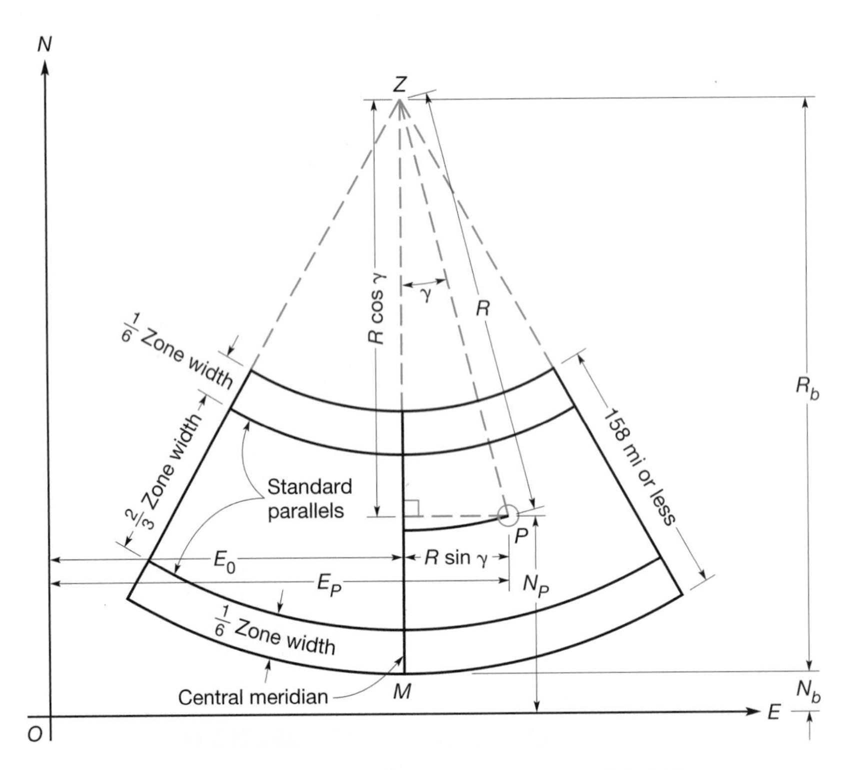

Figure 20.5 The Lambert conformal conic projection (SPCS83 symbology).

■ 20.4 TRANSVERSE MERCATOR PROJECTION

The Transverse Mercator projection is also a conformal projection, but is based on an imaginary secant cylinder as its developable surface. As illustrated in Figure 20.1(d), scale in the Transverse Mercator projection varies from east to west, but not from north to south. Thus this projection is used for states like Idaho, Illinois, Indiana, Louisiana, New Hampshire, New Jersey, and Vermont, which are narrow east-west and longer north-south.

In developing the Transverse Mercator projection, the axis of the imaginary cylinder is placed in the plane of the Earth's equator. The cylinder cuts the spheroid along two small circles equidistant from the central meridian. On the developed plane surface (Figure 20.6) all parallels of latitude, and all meridians except for the central meridian are curves (shown in light broken lines). As with the Lambert conformal conic projection, the central meridian establishes the direction of grid north and the zone is centered over an area of interest by assigning the central meridian a longitude value that applies approximately at the center of the region to be mapped. Also, the *E* and *N* coordinates of points are measured perpendicular to and parallel with the central meridian, respectively, from an *E-N* axis system offset to the west and south.

■ 20.5 STATE PLANE COORDINATES IN NAD27 AND NAD83

The first state plane coordinate system was developed by the NGS in 1933 for North Carolina. Systems for all other states followed shortly thereafter. As noted earlier, state plane coordinates of points are computed from geodetic latitudes

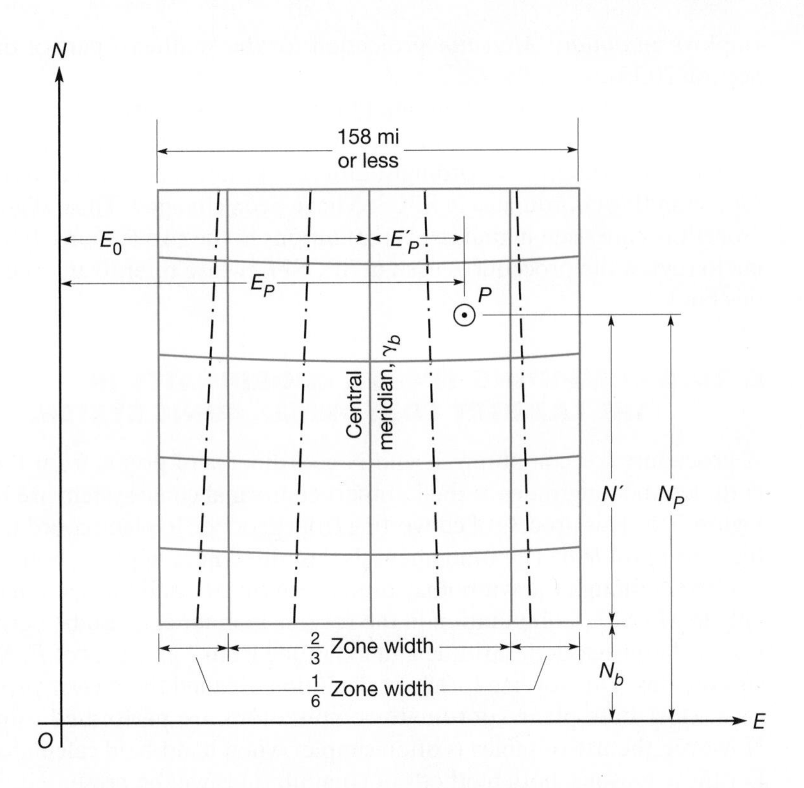

Figure 20.6
The projection.

and geodetic longitudes. Section 19.4 and Figure 19.3 define and illustrate these two terms and explain that geodetic latitudes and longitudes are defined with respect to a reference ellipsoid and associated datum. From 1927 until the inception of NAD83, the reference datum used in the United States was NAD27, based on the Clarke 1866 ellipsoid. All original state plane coordinates were therefore developed by the NGS in accordance with that ellipsoid and datum. This system is referred to as the *State Plane Coordinate System of 1927* (SPCS27).

As noted in Chapter 19, NAD83 employs a different set of defining parameters than NAD27 and it uses a reference surface of different dimensions, the GRS80 ellipsoid. Thus, the latitudes and longitudes of points in NAD83 are somewhat different from their values in NAD27. Because of these changes, the constants and variables that define the state plane coordinate systems also changed. Thus, following completion of NAD83, it was necessary to develop a new state plane coordinate system. This has been completed by the NGS and is called the *State Plane Coordinate System of 1983* (SPCS83).

In the SPCS83 most states retained the same projections they had used in SPCS27, with basically the same central meridian positioning. There were changes, however, some major ones being: (1) Montana reduced its number of zones from three to one; and (2) Nebraska and South Carolina reduced their number of zones from two to one. In SPCS83, 29 states employ the Lambert conformal conic projection, 18 states use the Transverse Mercator, and Alaska, Florida, and New York utilize both. In addition to using eight zones for its mainland and a Lambert conformal conic for the Aleutian Islands, Alaska also employs an *oblique Mercator* projection for the southeast part of the state (see Section 20.13).

Although the same fundamental approach was used to develop both the SPCS27 and the SPCS83, as noted previously, the parameters that define the two systems are different. Accordingly, different symbols are used and the equations for computing coordinates in SPCS83 have been changed. Thus, slightly different procedures are used in making computations in the two systems. For those wishing to review the procedures used in SPCS27, please refer to an earlier edition of this book.

20.6 COMPUTING SPCS83 COORDINATES IN THE LAMBERT CONFORMAL CONIC SYSTEM

A procedure for computing E and N coordinates of points from their geodetic latitudes and longitudes in the Lambert conformal conic system are illustrated in Figure 20.5. This process of converting from geodetic to plane coordinates is called the *direct problem*. The fundamentals described here apply to both SPCS27 and SPCS83, although the symbology used in Figure 20.5 and this section is applicable only to SPCS83. Computation in the reverse manner can also be performed; that is, calculating geodetic latitude and longitude from a given set of E, N state plane coordinates. This reverse form of calculation is called the *inverse problem*.

Most state plane coordinate computations are performed using software. However, the use of tables is often simpler when hand-held calculators are used. For these reasons, both methods of computations will be presented.

20.6.1 Zone Constants

A zone is defined in the Lambert conformal conic map projection by the selection of four sets of parameters. They are the defining ellipsoidal parameters a and f, grid origin (ϕ_0, λ_0), latitudes of the northern and southern standard parallels ϕ_N and ϕ_S, and false easting and northing (E_0, N_b). From these defining parameters a set of zone constants are mathematically defined that are used in both the direct and inverse problems. Common functions used in the definition of the map projection are

$$W(\phi) = \sqrt{1 - e^2 \sin^2\phi} \tag{20.1}$$

$$M(\phi) = \frac{\cos\phi}{W(\phi)} \tag{20.2}$$

$$T(\phi) = \sqrt{\left(\frac{1 - \sin\phi}{1 + \sin\phi}\right)\left(\frac{1 + e\sin\phi}{1 - e\sin\phi}\right)^e} \tag{20.3}$$

where e is the eccentricity of the ellipse as defined by Equation (19.2c). The defining zone constants for a Lambert conformal conic map projection are:

$$w_1 = W(\phi_S) \tag{20.4}$$

$$w_2 = W(\phi_N) \tag{20.5}$$

$$m_1 = M(\phi_S) \tag{20.6}$$

$$m_2 = M(\phi_N) \tag{20.7}$$

$$t_0 = T(\phi_0) \tag{20.8}$$

$$t_1 = T(\phi_S) \tag{20.9}$$

$$t_2 = T(\phi_N) \tag{20.10}$$

$$n = \sin \phi_0 = \frac{\ln(m_1) - \ln(m_2)}{\ln(t_1) - \ln(t_2)} \tag{20.11}$$

$$F = \frac{m_1}{nt_1^n} \tag{20.12}$$

$$R_b = aFt_0^n \tag{20.13}$$

20.6.2 The Direct Problem

In Figure 20.5, line ZM is the central meridian of the projection, the N axis (line ON) is parallel to ZM, and point O is the origin of the rectangular coordinate system. A constant E_0 is adopted to offset the N grid axis from the central meridian and make E coordinates of all points positive. Similarly, a constant N_b can be adopted to offset the E grid axis from the southern edge of the projection. The N coordinate of the cone's apex is the constant $(R_b + N_b)$, the numerical values of these terms being such that all N coordinates are positive.

The coordinates E and N of point P, whose geodetic latitude ϕ_P, and geodetic longitude λ_P, are known, are to be determined. Line ZP represents a portion of the meridian through point P with its length designated as R. Angle γ between the central meridian and meridian ZP represents the amount of convergence between these two meridians. In SPCS83 it is termed the *convergence angle*. In *SPCS27* it was called the *mapping angle*.

From Figure 20.5, the following equations of the direct problem can be solved for the *easting* (E) and *northing* (N) coordinates of point P:

$$\begin{aligned} E_P &= R \sin \gamma + E_0 \\ N_P &= R_b - R \cos \gamma + N_b \end{aligned} \tag{20.14}$$

To solve Equations (20.14), values for E_0, N_b, R_b, R, and γ must be known. The quantities E_0, and N_b are constants for any zone. In many states, the SPCS83 value for E_0 has been arbitrarily assigned a value of 600,000 m, and N_b assigned a value of 0.000 m; R_b is a computed zone constant defined by Equation (20.13). Values for R and γ also depend on the ellipsoid used and vary with changing locations of points in the zone; R changes with latitude, γ with longitude. The convergence angle γ and R can be computed as

$$\gamma = (\lambda_0 - \lambda)n \tag{20.15}$$

$$t = T(\phi_P) \tag{20.16}$$

$$R = aFt^n \tag{20.17}$$

TABLE 20.1 EXCERPT FROM THE PENNSYLVANIA NORTH ZONE TABLES

Zone constants: $N_b = 0.000$ m $E_0 = 600{,}000.000$ m $l_{CM} = 778459$
$R_b = 7{,}379{,}348.3668$ m $\sin \phi_0 = 0.661539733812$

Lat (ϕ)	R (meters)	Tab. Diff.	k
41°10′	7268294.836	30.84819	0.99996637
41°11′	7266443.945	30.84824	0.99996514
41°12′	7264593.050	30.84830	0.99996400
41°13′	7262742.152	30.84836	0.99996295
41°14′	7260891.251	30.84842	0.99996198
41°15′	7259040.346	30.84848	0.99996109
41°16′	7257189.437	30.84855	0.99996029
41°17′	7255338.524	30.84862	0.99995957
41°18′	**7253487.607**	**30.84869**	**0.99995893**
41°19′	7251636.685	30.84876	0.99995838

where Equation (20.15) is adjusted for western longitudes. An Excel® spreadsheet and Mathcad® worksheet that demonstrate these computations are available on the CD that accompanies this book.

To aid in hand solutions of Equations (20.14), the NGS has computed and published individual SPCS83 booklets of projection tables for each state. These give the constants for each zone and tabulate R and scale factor k values versus latitude. Thus given the geodetic latitude of any point, R for that point can be interpolated from the tables for use in Equations (20.14). Table 20.1 shows an excerpt from the north zone of the Pennsylvania tables. The column labeled *Tab. Diff.* provides the change in radius R per second of latitude. The use of this table is demonstrated in Example 20.1. The equations are solved using a calculator and the computational procedure is called the *tabular method*.

Example 20.1

Using values in Table 20.1, what are the coordinates E (easting) and N (northing) for station "Hayfield NE" that lies in Pennsylvania's north Lambert conformal conic zone? The station's geodetic latitude is 41°18′20.25410″ N, and its geodetic longitude is 76°00′57.00239″ W.

Solution

Step 1: Determine the radius to Hayfield NE.

The tabular difference (Tab. Diff. column) listed in Table 20.1 for latitude of 41°18′ is 30.84869. Thus, the change in the radius ΔR from 41°18′ to 41°18′20.25410″ is

$$\Delta R = 20.25410'' \times 30.84869 = 624.8125 \text{ m}$$

As latitude increases, the radius decreases, and thus ΔR must be subtracted from the tabulated R of 7,253,487.607 m. Thus the radius to Hayfield NE is

$$R = 7{,}253{,}487.607 - 624.8125 = 7{,}252{,}862.794 \text{ m}$$

The equivalent linear interpolation formula for the radius is

$$\begin{aligned} R &= 7{,}253{,}487.607 + (7{,}251{,}636.685 - 7{,}253{,}487.607) \times 20.25410''/60'' \\ &= 7{,}252{,}862.794 \text{ m} \end{aligned}$$

Step 2: Compute the convergence angle, γ, using Equation (20.15) as

$$\gamma = (77°45' - 76°00'57.00239'') \times 0.661539733812 = 1°08'49.991''$$

Step 3: Solve Equations (20.14) as

$$\begin{aligned} E &= 7{,}252{,}862.795 \sin 1°08'49.991'' + 600{,}000.000 \\ &= 745{,}212.637 \text{ m} \\ N &= 7{,}379{,}348.3668 - 7{,}252{,}862.795 \cos 1°08'49.991'' + 0.000 \\ &= 127{,}939.399 \text{ m} \end{aligned}$$

20.6.3 The Inverse Problem

The inverse problem in state plane coordinate computations is the determination of the geodetic latitude and geodetic longitude of a station based on its state plane coordinates and zone. The inverse equations also utilize the zone constants computed in Section 20.6.1 and given in Table 20.1. The remaining equations for accomplishing this can be derived from Equations (20.14) as

$$\begin{aligned} N' &= R_b - (N - N_b) \\ E' &= E - E_0 \\ \gamma &= \tan^{-1}\left[\frac{E'}{N'}\right] \\ R &= \frac{N'}{\cos \gamma} = \sqrt{E'^2 + N'^2} \\ \lambda &= \lambda_0 - \frac{\gamma}{n} \end{aligned} \tag{20.18}$$

The solution for the latitude of the station is iterative. This process is

$$t = \left(\frac{R}{aF}\right)^{\frac{1}{n}} \tag{20.19}$$

$$\chi = \frac{\pi}{2} - 2\tan^{-1}(t) \tag{20.20}$$

$$\phi_P = \frac{\pi}{2} - 2\tan^{-1}\left[t\left(\frac{1 - e\sin\phi_P}{1 + e\sin\phi_P}\right)^{\frac{e}{2}}\right] \tag{20.21}$$

Iterate Equation (20.21) starting with ϕ_P equal to χ on the first iteration and continuing until changes in ϕ_P are negligible. This procedure is demonstrated in the Excel® spreadsheet and Mathcad® worksheet on the CD that accompanies this book.

When using the tables, the latitude ϕ_P of the station can be interpolated, versus R, from the tables. The procedure is demonstrated in Example 20.2.

Example 20.2

What are the geodetic latitude and geodetic longitude for station Hayfield NE given the following SPCS83 coordinates? (The station lies in the north zone of Pennsylvania's Lambert conformal conic map projection.)

$$\text{Easting} = 745{,}212.637\text{ m}$$
$$\text{Northing} = 127{,}939.400\text{ m}$$

Solution

Using Table 20.1 and Equations (20.18),

$$E' = 745{,}212.637 - 600{,}000.000 = 145{,}212.637\ m$$
$$N' = 7{,}379{,}348.3668 - (127{,}939.400 - 0.000) = 7{,}251{,}408.9668\ m$$

$$\gamma = \tan^{-1}\left(\frac{145{,}212.637}{7{,}251{,}408.9668}\right) = 1°08'49.99''$$

$$R = \sqrt{145212.637^2 + 7251408.9668^2} = 7{,}252{,}862.7943$$

$$\lambda = 77°45' - \frac{1°08'49.99''}{0.661539733812} = 76°00'57.00239''$$

Now the latitude of the station can be interpolated from the values in Table 20.1. As can be seen in the table, the computed radius R is between 41°18′ and 41°19′. To determine the number of arc seconds to be added to 41°18′, the difference between the tabulated radius for 41°18′ of 7,253,487.607 and the computed radius R for the station is evaluated and divided by the tabulated difference of 30.84869. That is,

$$\Delta\phi'' = \frac{7{,}253{,}487.607 - 7{,}252{,}862.7943}{30.84869} = 20.25411''$$

Thus, the latitude and longitude of the station are computed as 41°18′ 20.25411″ and 76°00′57.00239″, respectively. Rounding errors in both the forward and inverse problems caused the slight difference of 0.00001″ in the computed and given latitude from Examples 20.1 and 20.2.

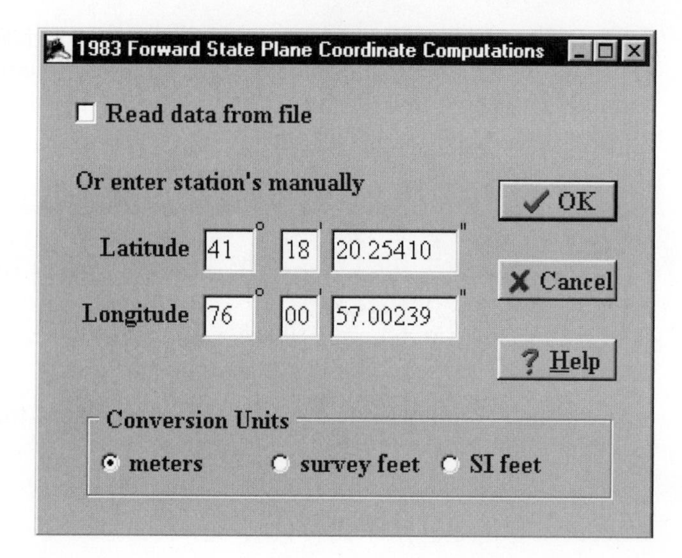

Figure 20.7 Data entry screen for forward SPCS83 computations of Example 20.1.

To aid in the solution of state plane coordinate computations, the NGS has published a booklet[1] entitled "State Plane Coordinate System of 1983" (Stem, 1989) that gives the zone constants and formulas for every zone in the United States and its territories. These constants are repeated in Appendix *F* and in an Excel® spreadsheet on the CD that accompanies this book. The computer program WOLFPACK, which accompanies this book, contains a state plane coordinate option under its coordinate computation menu. The forward data entry screen from WOLFPACK for Example 20.1 is shown in Figure 20.7. Of course the state, and zone within that state, must also be specified.

It should be emphasized that computation of coordinates in the SPCS83 should only be done using points whose geodetic positions are given in NAD83.

■ 20.7 COMPUTING SPCS83 COORDINATES IN THE TRANSVERSE MERCATOR SYSTEM

The NGS has also published SPCS83 Transverse Mercator state plane coordinate zones for those states that use this projection. All necessary constants and variables are computed and tabulated, and instructions together with sample problems are given to illustrate the computational procedures.

20.7.1 Zone Constants

A zone is defined in the Transverse Mercator map projection by the selection of four sets of parameters. They are the defining ellipsoidal parameters *a* and *f*, grid

[1]Individual booklets for each state for making SPCS83 tabular solutions, and NOAA Manual NOS NGS 5 can be obtained from the National Geodetic Information Center, NOAA, National Geodetic Survey, N/CG174, SSMC3 Station 09535, 1315 East West Highway, Silver Spring, MD 20910, telephone (301) 713-3242. Similar products are also available for SPCS27.

origin (ϕ_0, λ_0), scale factor at the central meridian λ_0, and false easting and northing (E_0, N_b). From these defining parameters a set of zone constants is mathematically defined that is used in both the direct and inverse problems. Common functions used in the definition of the map projection are

$$C(\phi) = e'^2\cos^2\phi \qquad T(\phi) = \tan^2\phi$$

$$M(\phi) = a\begin{bmatrix} \left(1 - \frac{e^2}{4} - \frac{3e^4}{64} - \frac{5e^6}{256}\right)\phi - \left(\frac{3e^2}{8} + \frac{3e^4}{32} + \frac{45e^6}{1024}\right)\sin 2\phi \\ +\left(\frac{15e^4}{256} + \frac{45e^6}{1024}\right)\sin 4\phi - \left(\frac{35e^6}{3072}\right)\sin 6\phi \end{bmatrix} \tag{20.22}$$

where e and e' are the first and second eccentricities of the ellipse as defined by Equations (19.2b) and (19.2c), respectively.

20.7.2 The Direct Problem

The computations for the direct problem using the map projection formulas are

$$m_0 = \mathrm{M}(\phi_0) \tag{20.23}$$
$$m = \mathrm{M}(\phi) \tag{20.24}$$
$$t = T(\phi) \tag{20.25}$$
$$\mathrm{c} = \mathrm{C}(\phi) \tag{20.26}$$
$$\mathrm{A} = (\lambda_0 - \lambda)\cos\phi \tag{20.27}$$

where Equation (20.27) has been adjusted for western longitudes.

$$E = k_0 R_N\left[A + (1 - t + c)\frac{A^3}{6} + (5 - 18t + t^2 + 72c - 58e'^2)\frac{A^5}{120}\right] + E_0 \tag{20.28}$$

$$N = k_0\left\{m - M_0 + R_N\tan\phi\begin{bmatrix} \frac{A^2}{2} + (5 - t + 9c + 4c^2)\frac{A^4}{24} \\ +(61 - 58t + t^2 + 600c - 330e'^2)\frac{A^6}{720} \end{bmatrix}\right\} + N_b \tag{20.29}$$

where R_N is the radius in the prime vertical for the latitude ϕ as defined by Equation (19.4). The convergence angle γ is computed as

$$c_2 = \frac{1 + 3c + 2c^2}{3} \qquad c_3 = \frac{2 - \tan^2\phi}{15}$$
$$\gamma = A\tan\phi[1 + A^2(c_2 + c_3A^2)] \tag{20.30}$$

Computations utilizing the Equations (20.23) to (20.30) are demonstrated in an Excel® spreadsheet and Mathcad® worksheet that are included on the CD that accompanies this book.

With reference to Figure 20.6 and appropriate Transverse Mercator projection tables such as those shown in Table 20.2, the following SPCS83 equations for hand computations yield the solution of the *direct problem,* that is, obtaining the E_P and N_P coordinates of any point P from its geodetic coordinates:

$$\begin{aligned} \Delta\lambda &= (\lambda_0 - \lambda)3600''/^\circ \\ p &= 10^{-4}\Delta\lambda'' \\ N_P &= (I) + (II)p^2 + (III)p^4 + N_b \\ E_P &= (IV)p + (V)p^3 + (VI)p^5 + E_0 \end{aligned} \qquad \textbf{(20.31)}$$

In Equations (20.32), $\Delta\lambda''$ is the difference in longitude between the central meridian and the point in seconds, and N and E are the state plane coordinates of the point in meters. The values for λ_0 and E_0 are zone constants, and are supplied with the table as shown in Table 20.2. The values for Roman numerals (I), (II), (III), (IV), (V), and (VI) are interpolated from the Transverse Mercator tables using the appropriate value given in the "*Difference*" column. For the first (I) and fourth (IV) column values, a small second-difference correction must also be interpolated using the numbers given at the bottom of the Table. Example 20.3 demonstrates the use of Equations (20.31) and Table 20.2.

TABLE 20.2 EXCERPT FROM THE TRANSVERSE MERCATOR PROJECTION TABLES FOR NEW JERSEY FOR THE DIRECT PROBLEM

Zone constants: E_0 = 150,000 m ϕ_b = 38°50′ λ_0 = 74°30′ N_b = 0.000 m k_0 = 0.9999

Latitude	(I) (IV)	Difference 1″	(II) (V)	Difference 1″	(III) (VI)
39°00′	18500.4650	30.834594	3670.4645	0.007613	1.902
	240604.8369	−0.910942	19.8284	−0.000975	−0.026
39°01′	20350.5407	30.834682	3670.9213	0.007592	1.901
	240548.3803	−0.941283	19.7699	−0.000975	−0.026
39°02′	**22200.6216**	**30.834771**	**3671.3768**	**0.007572**	**1.900**
	240491.9034	**−0.941623**	**19.7114**	**−0.000974**	−0.026
39°03′	24050.7079	30.834859	3671.8311	0.007551	1.899
	240435.4060	−0.941964	19.6529	−0.000974	−0.026
39°04′	25900.7994	30.834947	3672.2842	0.007530	1.898
	240378.8882	−0.942304	19.5945	−0.000974	−0.026

Second difference corrections:

	00″	10″	20″	30″		00″	10″	20″	30″
	60″	50″	40″	30″		60″	50″	40″	30″
(I)	0.0000	−0.0004	−0.0006	−0.0007	(IV)	0.0000	0.0014	0.0023	0.0025

Example 20.3

The geodetic latitude and geodetic longitude of station Stone Harbor in the state of New Jersey, which uses the Transverse Mercator projection, are 39°02′21.63632″ and 74°46′08.80133″, respectively. What are the station's SPCS83 coordinates?

Solution

The determination of the second differences and (III) and (VI) column values involves a linear interpolation. For example, the second difference for column (I) is determined as

$$-0.0006 + [-0.0007 - (-0.0006)]1.63632''/10'' = -0.00062$$

where 1.63632″ comes from the second's portion of the station's latitude. The three values (III), (IV), and (VI) are interpolated in similar fashion and shown in the computations that follow. Using Table 20.2, the appropriate column values are

$$\begin{aligned}
\text{I} &= 22{,}200.6216 + 30.834771 \times 21.63632 + (-0.00062) &&= 22{,}867.77195\\
\text{II} &= 3671.3768 + 0.007572 \times 21.63632 &&= 3671.54063\\
\text{III} &= 1.900 + (1.899 - 1.900) \times 21.63632/60 &&= 1.89964\\
\text{IV} &= 240{,}491.9034 + (-0.941623) \times 21.63632 + 0.00233 &&= 240{,}471.53247\\
\text{V} &= 19.7114 + (-0.000974) \times 21.63632 &&= 19.690326\\
\text{VI} &= -0.026 + (-0.026 + 0.026) \times 21.63632/60 &&= -0.026
\end{aligned}$$

Substituting these values into Equations (20.31) yields

$$\begin{aligned}
\Delta\lambda'' &= (74°30' - 74°46'08.80133'')3600''/° &&= -968.80133''\\
p &= -968.80133 \times 10^{-4} &&= -0.096880133\\
N &= 22{,}867.77197 + 3671.54063p^2 + 1.89964p^4 + 0 &&= 22{,}902.2323 \text{ m}\\
E &= 240{,}471.5324p + 19.690326p^3 - 0.026p^5 + E_0 &&= 126{,}703.0680 \text{ m}
\end{aligned}$$

20.7.3 The Inverse Problem

The Transverse Mercator inverse problem is computationally solved as

$$E' = E - E_0 \tag{20.32}$$

$$N' = N - N_b \tag{20.33}$$

$$e_1 = \frac{1 - \sqrt{1 - e^2}}{1 + \sqrt{1 - e^2}} \tag{20.34}$$

$$m = m_0 + \frac{N}{k_0} \tag{20.35}$$

$$\chi = \frac{m}{a\left(1 - \frac{e^2}{4} - \frac{3e^4}{64} - \frac{5e^6}{256}\right)} \tag{20.36}$$

The footpoint latitude ϕ_f is

$$\phi_f = \chi + \left(\frac{3e_1}{2} - \frac{27e_1^3}{32}\right)\sin 2\chi + \left(\frac{21e_1^2}{16} - \frac{55e_1^4}{32}\right)\sin 4\chi$$

$$+ \left(\frac{151e_1^3}{96}\right)\sin 6\chi + \left(\frac{1097e_1^4}{512}\right)\sin 8\chi \qquad \textbf{(20.37)}$$

$$c_1 = C(\phi_f) \qquad \textbf{(20.38)}$$

$$t_1 = T(\phi_f) \qquad \textbf{(20.39)}$$

$$N_1 = R_N \text{ evaluated using Equation (19.4) with } \phi_f \qquad \textbf{(20.40)}$$

$$M_1 = R_M \text{ evaluated using Equation (19.5) with } \phi_f \qquad \textbf{(20.41)}$$

$$D = \frac{E'}{N_1 k_0} \qquad \textbf{(20.42)}$$

$$B = \frac{D^2}{2} - (5 + 3t_1 + 10c_1 - 4c_1^2 - 9e')\frac{D^4}{24}$$

$$+ (61 + 90t_1 + 298c_1 + 454t_1^2 - 252e' - 3c_1^2)\frac{D^6}{720} \qquad \textbf{(20.43)}$$

$$\phi = \phi_f - \left(\frac{N_1 \tan \phi_f}{R_1}\right) B \qquad \textbf{(20.44a)}$$

$$\lambda = \lambda_0 + \frac{D - \dfrac{\left(1 + 2t_1 + c_1\right)D^3}{6} + \dfrac{\left(5 - 2c_1 + 28t_1 - 3c_1^2 + 8e'^2 + 24t_1^2\right)D^5}{120}}{\cos \varphi_f}$$

(20.44b)

Computations using Equations (20.37) to (20.44) are demonstrated in an Excel® spreadsheet and Mathcad® worksheet on the CD that accompanies this book. Similar to Table 20.2, Table 20.3 contains the necessary parameters and second differences to compute the inverse problem. The formulas necessary to perform this calculation are

$$\begin{aligned}
E' &= E - E_0 \\
N' &= N - N_b \\
q &= E' \times 10^{-6} \\
\phi_f &= \phi_{\text{table}} + \Delta\phi''(\text{interpolated from table}) \\
\Delta\phi'' &= -(\text{VII})q^2 + (\text{VIII})q^4 \\
\Delta\lambda'' &= -(\text{IX})q + (\text{X})q^3 + (\text{XI})q^5 \\
\phi &= \phi_f + \Delta\phi'' \\
\lambda &= \lambda_b + \Delta\lambda''
\end{aligned} \qquad \textbf{(20.45)}$$

In Equations (20.45) E' and N' are in meters, $\Delta\phi_f''$ and $\Delta\lambda''$ are in seconds, and ϕ_f, ϕ, and λ are in degrees, minutes and seconds. The value of ϕ_f is interpolated

TABLE 20.3 EXCERPT FROM THE TRANSVERSE MERCATOR PROJECTION TABLES FOR NEW JERSEY FOR THE INVERSE PROBLEM

Zone constants: E_0 = 150,000 m ϕ_b = 38°50′ λ_0 = 74°30′ N_b = 0.000 m k_0 = 0.9999

Latitude	(I) (IX)	Difference 1″	(VII) (X)	Difference 1″	(VIII) (XI)
39°00′	18500.4650	30.834594	2056.2443	0.020257	29.191
	41561.9242	0.162576	393.3224	0.005941	7.024
39°01′	20350.5407	30.834682	2057.4597	0.020267	29.218
	41571.6788	0.162711	393.6788	0.005949	7.035
39°02′	**22200.6216**	**30.834771**	**2058.6757**	**0.020276**	**29.245**
	41581.4415	**0.163846**	**394.0357**	**0.005957**	**7.047**
39°03′	24050.7079	30.834859	2059.8923	0.020286	29.272
	41591.2122	0.162982	394.3931	0.005965	7.058
39°04′	25900.7994	30.834947	2061.1094	0.020295	29.299
	41600.9911	0.163117	394.7510	0.005973	7.069

Second difference corrections:

	00″	10″	20″	30″		00″	10″	20″	30″
	60″	50″	40″	30″		60″	50″	40″	30″
(I)	0.0000	−0.0004	−0.0006	−0.0007	(IX)	−0.0000	−0.0006	−0.0009	−0.0010

from column (I) in Table 20.3, and is known as the *footpoint latitude*. Notice that in this table, the first (I) and the ninth (IX) values have second difference corrections. Example 20.4 demonstrates the use of these tables in the inverse problem.

Example 20.4

What are the geodetic coordinates of station Stone Harbor if its SPCS83 easting and northing coordinates are 126,703.0681 m and 22,902.2323 m, respectively? (The station lies in New Jersey's Transverse Mercator zone.)

Solution

Step 1: Calculate E', N', and q as

$$E' = 126{,}703.0680 - 150{,}000.000 = -23{,}296.9320 \text{ m}$$
$$N' = 22{,}902.2323 - 0.000 = 22{,}902.2323 \text{ m}$$
$$q = -23{,}296.9320 \times 10^{-6} = -0.023296932$$

Step 2: Looking at Table 20.3, it can be seen that the northing coordinate lies between the (I)-values of 39°02′ and 39°03′. Thus $\Delta\phi_f''$ can be interpolated from column (I) as

$$\Delta\phi_f'' = (22{,}902.2323 - 22{,}200.6216)/30.834771 = 22.75388068''$$

Hence, the footpoint latitude is

$$\phi_f = 39°02' + 22.75388068'' = 39°02'22.75388068''$$

Step 3: Using $\Delta\phi_f''$, evaluate the tabular values for VII, VIII, IX, X, and XI. Note in this procedure, the values for VIII and XI must be linearly interpolated using the tabular values for 39°02′ and 39°03′.

$$\begin{aligned}
\text{VII} &= 2058.6757 + 0.020276 \times 22.75388068 &&= 2059.13706 \\
\text{VIII} &= 29.245 + (29.272 - 29.245)/60 \times 22.75388068 &&= 29.25524 \\
\text{IX} &= 41{,}581.4415 + 0.162846 \times 22.75388068 - 0.0009 &&= 41{,}585.14688 \\
\text{X} &= 394.0357 + 0.005957 \times 22.75388068 &&= 394.17124 \\
\text{XI} &= 7.047 + (7.058 - 7.047)/60 \times 22.75388068 &&= 7.05117
\end{aligned}$$

Step 4: Applying the computed tabular values of step 3, compute the latitude and longitude of station Stone Harbor using Equations (20.45) as

$$\begin{aligned}
\Delta\phi'' &= -2059.13706q^2 + 29.25524q^4 &&= -1.11758'' \\
\Delta\lambda'' &= -41{,}585.14688q + 394.17124q^3 + 7.05117q^5 &&= 968.80135'' \\
\phi &= 39°02'22.75388068'' - 1.11758'' &&= 39°02'21.6363'' \\
\lambda &= 74°30' + 968.80135'' &&= 74°46'08.8013''
\end{aligned}$$

Again, except for small rounding errors from both the forward and inverse computations, the solution produces the geodetic latitude and longitude of Stone Harbor that were given for Example 20.3.

As with the Lambert conformal conic map conversions, the solutions of the direct and inverse problems are typically performed with computers. Several programs have been written that allow for the easy conversion from geodetic to state plane coordinates and vice versa. WOLFPACK contains this option under its coordinate computation menu. The CD accompanying this book contains both an Excel® spreadsheet and Mathcad® worksheet that demonstrate these computations.

■ 20.8 REDUCTION OF DISTANCES AND ANGLES TO STATE PLANE COORDINATE GRIDS

Ground-surveyed distances and angles must undergo corrections prior to using them when performing computations in state plane coordinate systems. As shown in Figure 20.8, distances must first be reduced from their ground-surveyed lengths to their ellipsoidal equivalents. Furthermore as shown in Figure 20.2, these ellipsoidal distances must then be reduced to the developable surface of the state

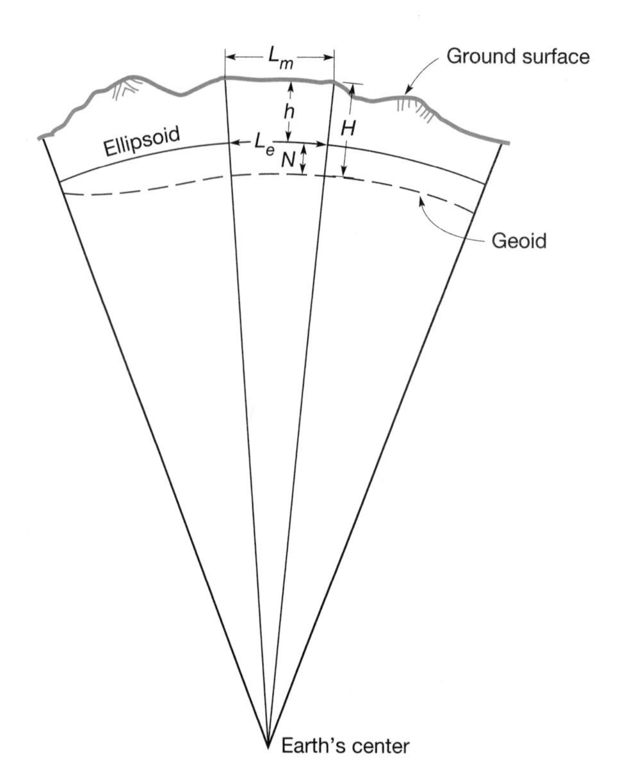

Figure 20.8 Reduction of lengths from surface observations to the ellipsoid.

plane system being used. Additionally whenever angles or azimuths are used in these computations, they can be reduced to their grid equivalents. Once these reductions have been completed, traverse computations (see Chapter 10), and adjustment procedures given in both Chapters 10 and 16 may be performed. Furthermore, accuracy of the computations will match those produced using geodetic computational methods discussed in Chapter 19. This section describes the processes of reducing these observations to state plane grids.

20.8.1 Grid Reduction of Distances

The reduction of distances is normally done in two steps: (1) reduce the observations from their ground lengths to ellipsoid lengths and (2) reduce the ellipsoid lengths to their grid equivalents. The most precise formulas for reduction of slope distances to the ellipsoid were given in Section 19.14. However in most local surveys, the lengths of the distances are short and simpler methods of reduction can be used since the horizontal distance L_m closely matches the arc distance at the surface of the Earth. With this simplification, the relationship between the ground-surveyed length and the ellipsoid length L_e is

$$L_e = L_m\left(\frac{R_\alpha}{R_\alpha + H + N}\right) \tag{20.46}$$

where R_α is the radius of the Earth in the azimuth of the line as given by Equation (19.6), H is the average orthometric height of the observed line above the geoid, and N is the geoidal separation. The ratio $R_\alpha/(R_\alpha + H + N)$ is commonly called the *elevation factor*. For all but the most rigorous surveys, acceptable results can be obtained from Equation (20.46) by substituting the mean radius of the Earth (20,902,000 ft or 6,371,000 m) for R_α.

After a distance has been reduced to its ellipsoidal equivalent, it must then be scaled to its grid equivalent. This is accomplished by multiplying the ellipsoidal length of the line by an appropriate *scale factor*. For Lambert conformal conic mapping projections, the scale factor for any latitude ϕ can be computed as

$$m = M(\phi) \tag{20.47}$$

$$k = \frac{Rn}{am} \tag{20.48}$$

where M, R, and n are previously defined in this chapter. The scale factor at a point can be also interpolated from tables. This procedure is demonstrated in Example 20.5 using Table 20.1.

Example 20.5

What is the scale factor for Hayfield NE of Example 20.1? (This station is in the north zone of Pennsylvania's Lambert conformal conic projection.)

Solution

From Example 20.1, the geodetic latitude of station Hayfield NE is 41°18′20.25410″. Using the tabulated scale factors for latitudes of 41°18′ and 41°19′ from Table 20.1, the interpolated scale factor is

$$k = 0.99995893 + (0.99995838 - 0.99995893) \times 20.25410''/60'' = 0.999958744$$

For a Transverse Mercator map projection, the scale factor k for any point is computed as

$$k = k_0\left[1 + (1 + c)\frac{A^2}{2} + (5 - 4t + 42c + 13c^2 - 23e'^2)\frac{A^4}{24} + (61 - 148t + 16t^2)\frac{A^6}{720}\right] \tag{20.49}$$

where k_0 is a zone constant and c, A, and t are previously defined. To compute the scale factor k using tables, the following formula is used:

$$k = k_0[1 + (\text{XVI})q^2 + 0.00003q^4] \tag{20.50}$$

where q is defined in Equations (20.45), and (XVI) comes from Table 20.4.

TABLE 20.4 EXCERPT FROM THE TRANSVERSE MERCATOR PROJECTION TABLES FOR NEW JERSEY FOR COMPUTATION OF THE SCALE FACTOR. (NOTE: COLUMNS ARE REARRANGED FOR PUBLICATION PURPOSES ONLY.)

Zone constants: E_0 = 150,000 m ϕ_b = 38°50′ λ_b = 74°30′ N_b = 0.000 m k_0 = 0.9999

Latitude	(I) (XIV)	Difference 1″	(XII) (XV)	Difference 1″	(XIII) (XVI)
39°00′	18500.4650	30.834594	6293.2039	0.037673	3.014
	26155.7664	0.258924	353.110		0.012311
39°01′	20350.5407	30.834682	6295.4643	0.037664	3.014
	26171.3018	0.259046	353.486		0.012311
39°02′	**22200.6216**	**30.834771**	**6297.7241**	**0.037655**	**3.014**
	26186.8446	**0.259167**	**353.863**		**0.012311**
39°03′	24050.7079	30.834859	6299.9834	0.037646	3.013
	26202.3946	0.259289	354.240		0.012310
39°04′	25900.7994	30.834947	6302.2422	0.037637	3.013
	26217.9520	0.259412	354.618		0.012310

Second difference corrections:

	00″	10″	20″	30″		00″	10″	20″	30″
	60″	50″	40″	30″		60″	50″	40″	30″
(I)	0.0000	−0.0004	−0.0006	−0.0007	(XIV)	−0.0000	−0.0005	−0.0008	−0.0009

■ Example 20.6

Compute the scale factor for station Stone Harbor of Example 20.3. (This station lies in New Jersey's Transverse Mercator projection.)

Solution

From Example 20.3, the geodetic latitude of station Stone Harbor is 39°02′21.63632″. Using Equation (20.50), the scale factor for this station is

$$k = 0.9999[1 + 0.0123106q^2 + 0.00003q^4] = 0.99990668$$

where q is −0.023296932 as determined in Example 20.4, and the value for (XVI) is interpolated from the values at the bounding latitudes of 39°02′ and 39°03′.

A line consists of many points and thus there are several approaches to computing the scale factor k of a line. The easiest and least precise involves determining an average scale factor for an entire survey project and applying this single value to all reduced distances. This approach is adequate for low-accuracy surveys and for surveys that cover small areas. To achieve a higher level of accuracy, an "average scale factor" can be applied to each individual line. In this method the values are obtained by averaging the scale factors of the end points of the lines. This method works well for moderately long distances. However, for the most precise surveys, an additional scale factor at the midpoint of the line should be computed. Then an improved scale factor k_{12} is determined as

$$k_{12} = \frac{k_1 + 4k_m + k_2}{6} \tag{20.51}$$

where k_1 and k_2 are scale factors for the endpoints of the line, and k_m is the scale factor for the midpoint of the line. Obviously, this method requires that coordinates and scale factor k_m for the midpoint of the line be computed.

The product of the elevation factor and the scale factor is the so-called *combined factor* and mathematically expressed as

$$\text{combined factor} = \text{elevation factor} \times \text{scale factor} \tag{20.52}$$

In NAD27, the combined factor was known as the *grid factor*.

It is common in lower-order surveys and surveys that cover small areas to use a single combined factor for the entire survey. Often data collectors allow for the entry of this value, so that SPCS coordinates are computed directly in the field. However, for surveys that cover larger areas or require more rigorous procedures, the scale factor and elevation factors for each line should be computed. The reduced grid distance is

$$\text{grid distance} = \text{ground distance} \times \text{combined factor} \tag{20.53}$$

Example 20.7

What are the grid lengths for the observed distances in Figure 20.9? The scale factor of station Hayfield NE in the figure was determined in Example 20.5.

Solution

For this example, elevation factors were computed using Equation (20.6) and employing the given average geoidal separation of −29.8 m and the mean radius of the Earth. Using computational procedures as given in Chapter 12, approximate coordinate values for each station were computed and then used to determine the scale factor at each station (see Example 20.5.) For this survey, it was appropriate to use scale factors obtained by averaging the end point values. Elevation factors were obtained by using average elevations for each line in Equation (20.46). Grid

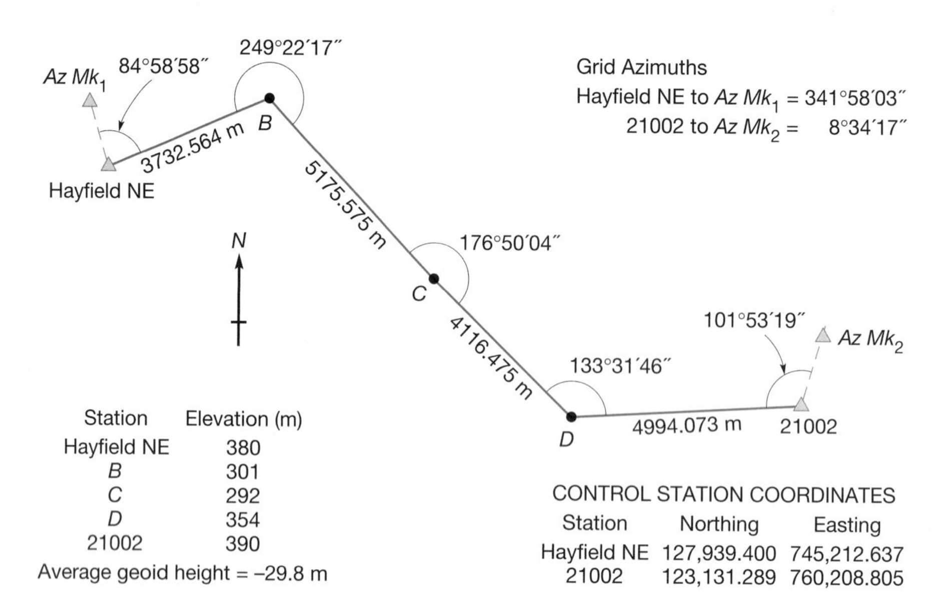

Figure 20.9 Field observed traverse.

factors for each line were then determined by multiplying the elevation factor by the average scale factor. Finally grid distances were computed by multiplying each observed distance by its corresponding combined factor. The results of these calculations are listed in Table 20.5.

20.8.2 Grid Reduction of Azimuths and Angles

As shown in Figure 20.10, all grid meridians are parallel while all geodetic meridians converge to a single point. The primary difference between these directions

TABLE 20.5 REDUCED GRID DISTANCES FOR FIGURE 20.9

Station	Elev (m)	Distance (m)	Elev. Factor	k	k_{avg}	Combined Factor	Grid Dist. (m)
Hayfield NE	380			0.99995874			
		3732.564	0.99995123		0.99995854	0.99990978	3732.2272
B	301			0.99995834			
		5175.575	0.99995814		0.99995895	0.99991709	5175.1459
C	292			0.99995956			
		4116.475	0.99995398		0.99996014	0.99991412	4116.1215
D	354			0.99996072			
		4994.073	0.99994629		0.99996068	0.99990698	4993.6084
21002	390			0.99996065			

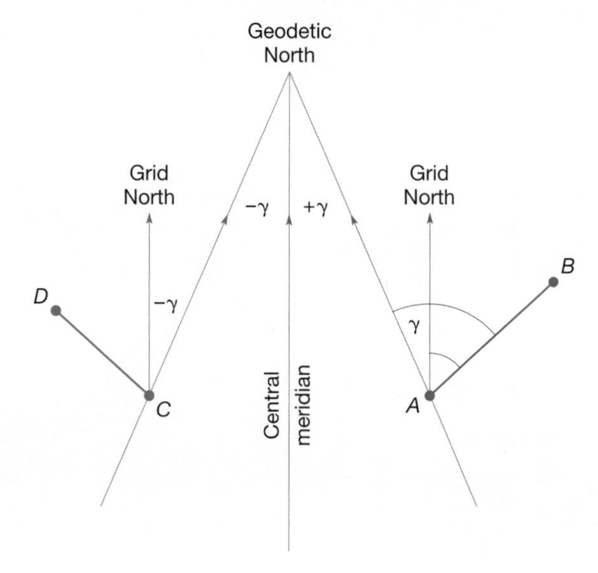

Figure 20.10 Relationship between geodetic azimuth, grid azimuth, and convergence angle γ.

is the convergence angle γ. Computation of the convergence angle for the Lambert conformal conic was demonstrated in both Example 20.1 and 20.2. The convergence angle (in seconds) for the Transverse Mercator projection is computed as

$$
\begin{aligned}
c_2 &= \frac{1 + 3c + 2c^2}{3} \\
c_3 &= \frac{2 - \tan^2\phi}{15} \\
\gamma &= A \tan\phi[1 + A^2(c_2 + c_3A^2)]
\end{aligned}
\tag{20.54}
$$

where A and c are previously defined in this chapter. It can be computed using tables as

$$\gamma'' = (\mathrm{XII})p + (\mathrm{XIII})p^3 = (\mathrm{XIV})q - (\mathrm{XV})q^3 \tag{20.55}$$

Using the value for q determined in Example 20.4 and Equation (20.55), the convergence angle at station Stone Harbor in New Jersey is

$$\gamma'' = (26{,}192.4512)q - (353.9989482)q^3 = -610.2'' = -0°10'10.2''$$

where (XIV) and (XV) were interpolated from Table 20.4.

Another factor that affects the reduction of azimuths is the projection of the geodetic azimuth onto a developable mapping surface. As can be seen in Figure 20.11, the projection of geodetic azimuths, onto a flat surface, results in an arc between the occupied and sighted stations. Letting the grid azimuth be t and the geodetic azimuth be T, the difference in these values is known as the *arc-to-chord* correction, also called the *second-term correction,* and is designated as δ.

The sign of the arc-to-chord correction is given by the location of the line and the type of map projection. For the Lambert conformal conic, the projected

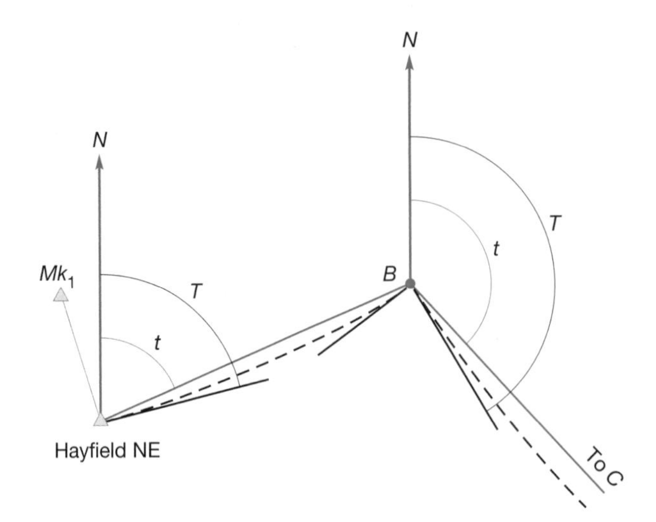

Figure 20.11 Part of Figure 20.9 showing arc-to-chord $(t - T)$ correction at stations *Hayfield NE* and *B*.

geodetic arc is always concave toward the central parallel of the zone. For the Transverse Mercator, the projected geodetic arc is always concave toward the central meridian for the zone. In Lambert conformal conic projections, the value for N_0 (the north-south center of the zone) can be determined from the derived zone constant sin ϕ_0, as

$$N_0 = R_b + N_b - R_0 \tag{20.56}$$

where R_b and N_b are zone constants, and R_0 can be interpolated from the table using the latitude of ϕ_0. Generally, the precise numerical value for N_0 is not necessary for computations because the line's location in relation to the zone's north-south center and the latitude ϕ_0 is all that is needed to determine the concavity of the projected geodetic arc. For example in the north zone of Pennsylvania, the sin ϕ_0 is given as 0.661539733812 (see Table 20.1) which yields a central parallel of approximately 41°25′. This value is sufficient to determine the concavity of the projected geodetic azimuth. Table 20.6 shows the sign of δ'' based on these criteria.

TABLE 20.6 THE SIGN OF THE δ CORRECTION

Map Projection		Azimuth of the Line from North	
Lambert:	Sign of $N - N_0$	0° to 180°	180° to 360°
	Positive	+	−
	Negative	−	+
Transverse Mercator:	Sign of $E - E_0$, or E_3	270° to 90°	90° to 270°
	Positive	−	+
	Negative	+	−

The arc-to-chord correction is a function of the positions of the endpoints of the line and the type of map projection. For a Lambert conformal conic map projection δ is given as

$$\delta_{12} = 0.5(\sin\phi_3 - \sin\phi_0)(\lambda_1 - \lambda_2) \quad \textbf{(20.57)}$$

where (ϕ_1, λ_1) and (ϕ_2, λ_2) are the geodetic positions of the endpoints of the line using positive values for western longitudes, $\sin\phi_0$ is a zone constant, and $\phi_3 = (2\phi_1 + \phi_2)/3$. For the Transverse Mercator projection, δ can be computed from the northing and easting coordinates for the endpoints of the line as

$$E' = E - E_0 \quad \textbf{(20.58a)}$$

$$\Delta N = N_2 - N_1 \quad \textbf{(20.58b)}$$

$$\eta_f^2 = e'^2\cos^2\phi_f \quad \textbf{(20.58c)}$$

$$F = (1 - e^2\sin^2\phi_f)(1 + \eta_f^2)/(k_0 a)^2 \quad \textbf{(20.58d)}$$

$$E_3 = 2E_1' + E_2' \quad \textbf{(20.58e)}$$

$$\delta_{12} = -\frac{1}{6}\Delta N E_3 F\left(1 - \frac{1}{27}E_3^2 F\right) \quad \textbf{(20.58f)}$$

An approximate value for δ_{12} can be computed as $-25.4\,\Delta N(E_3/3)\,10^{-10}$ seconds where the coordinate values are in meters.

Combining the two aforementioned corrections to the grid azimuth t, the geodetic azimuth T can be computed as

$$T = t + \gamma - \delta \quad \textbf{(20.59)}$$

In Example 20.8, the arc-to-chord correction is computed for the lines in Figure 20.9 to demonstrate the procedure. The arc-to-chord correction is generally ignored for short lines since it is typically below the error of the angle or azimuth observation. The National Geodetic Survey has recommended that this correction only be applied for lines longer than 8 km. However, each surveyor should determine the maximum arc-to-chord correction acceptable for his or her survey.

■ Example 20.8

Evaluate the arc-to-chord correction for the lines in Figure 20.9.

Solution

Using the observed angles and computational procedures as demonstrated in Chapter 10, approximate coordinates were computed for each station. From these approximate coordinate values, the magnitudes of the second term corrections were calculated using Equation (20.58). Note as can be shown from Table 20.6, the sign of all the corrections is negative since $N - N_0$ is negative for all lines in this example. The results of these computations are shown in Table 20.7.

The values in Table 20.7 are used to correct geodetic directions between the stations. Since Figure 20.9 has two grid azimuths, these azimuths do not need corrections. However when the lines are over 8 km or 5 mi in length, or a rigorous reduction is desired, the observed angles should be corrected. The arc-to-chord

TABLE 20.7 COMPUTATION OF THE ARC-TO-CHORD CORRECTION FOR FIGURE 20.9

Station	Approx. Northing	Approx. Easting	Approx. ϕ	Approx. λ	ϕ_3	δ
Hayfield NE	127,939.400	745,212.637	41°18′20.2541″	76°00′57.0024″		
					41°18′35.2908″	−0.10″
B	129,400.865	748,646.890	41°19′05.3642″	75°58′28.1096″		
					41°18′24.1269″	−0.11″
C	125,657.841	752,220.832	41°17′01.6524″	75°55′57.8495″		
					41°16′30.5490″	−0.12″
D	122,842.331	755,223.511	41°15′28.3421″	75°53′51.4269″		
					41°15′30.2932″	−0.22″
21002	123,131.289	760,208.805	41°15′34.1953″	75°50′17.0477″		

correction to observed angles can be found by taking the difference in the forward and back azimuths. The corrections for the backsight azimuths have the same magnitude but opposite signs of the computed foresight corrections in Table 20.7. For example, the correction for the azimuth from station *C* to *B* is 1.42″. Since the sight distances to both the azimuth marks was short, the corrections were assumed to be zero. The reduction of the observed angles is shown in Table 20.8.

TABLE 20.8 ARC-TO-CHORD CORRECTION FOR ANGLES IN FIGURE 20.9

Station	Obs. Angle	Backsight δ	Foresight δ	Total δ	Corr. Angle
Hayfield NE	84°58′58″	0.00″	−0.10″	−0.10″	84°58′57.9″
B	249°22′17″	0.10″	−0.11″	−0.21″	249°22′16.8″
C	176°50′04″	0.11″	−0.12″	−0.23″	176°50′03.8″
D	133°31′46″	0.12″	−0.22″	−0.34″	133°31′45.7″
21002	101°53′19″	0.22″	0.00″	−0.22″	101°53′18.8″

As can be seen in Table 20.8, the corrections are small. Often the arc-to-chord correction is ignored for traverses involving lines under 8 km and for lower-order surveys. However, the reduction of observed distances to the mapping grid is generally significant for most traverse surveys. Failure to account for these corrections will result in incorrect misclosures and subsequent incorrect adjustments. When these corrections are properly performed, the resulting adjustments will yield results similar to those achieved with geodetic computations. If adjusted ground distances are needed after an adjustment, a rearranged form of Equation (20.10) can be used to determine their values.

■ 20.9 COMPUTING STATE PLANE COORDINATES OF TRAVERSE STATIONS

Determining state plane coordinates of new traverse stations is a problem routinely solved by local surveyors. Normally it requires only that traverses (or triangulation or trilateration surveys) start and end on existing stations having known state plane coordinates and from which known grid azimuths have been established. Generally these data are available for immediate use, but if not, they can be calculated as indicated in Sections (20.6) and (20.7) when geodetic latitude and geodetic longitude are known. State plane coordinates and the grid azimuths to a nearby azimuth mark are published by the NGS for most stations in the national horizontal network. In most areas, many other stations set by local surveyors exist that also have state plane coordinates and reference grid azimuths.

It is important to note that if a survey begins with a given grid azimuth and ties into another, directions of all intermediate lines are automatically grid azimuths. Thus, corrections for convergence of meridians are not necessary when the state plane coordinate system is used throughout the survey. However, as demonstrated in Section 20.8.2, the arc-to-chord correction should be considered and applied to observed angles when appropriate. Assuming that starting stations meeting the previously described conditions are available, then there isn't any difference between making traverse computations in state plane coordinates and the procedures given for plane surveys in Chapter 10.

To illustrate the procedure of computing a traverse in the SPCS83, the following example is solved step-by-step.

■ Example 20.9

The traverse illustrated in Figure 20.9 originates from station Hayfield NE and closes on station 21002, both in the North zone for Pennsylvania. The reduction of both the distances and angles to the SPCS83 grid were demonstrated in Section 20.8. Using these values, compute and adjust the traverse, and determine the state plane coordinates of all traverse stations.

Solution

1. The computed azimuth of the line station 21002 to Az Mk_2 is compared with its fixed control value. The difference (+9.0″) represents the traverse angular misclosure. This misclosure is divided by the number of angles (five) to get the correction per angle (−1.8″). (This calculation is shown at the bottom of Table 20.9.) It should be interesting to note that if the observed angles had been used in the computations, the angular misclosure would have been +10″, or 1″ greater than is appropriate. Also note that for lines this short the arc-to-chord corrections are minimal and could easily have been avoided without appreciably affecting the final solution.
2. Traverse computations are performed using the same steps as described in Chapter 10. The procedure, shown in Figure 20.12, includes (a) calculating departures and latitudes [columns (1) and (2)], (b) adjusting the departures and latitudes [columns (3) and (4)], and (c) determining the station coordinates [columns (5) and (6)]. Adjustment of departures and latitudes in this

TABLE 20.9 REDUCED HORIZONTAL DISTANCES, ANGLES TO THE RIGHT, ANGLE MISCLOSURE, AND ADJUSTED AZIMUTHS FOR EXAMPLE 20.9

Station	Reduced Horizontal Distance (m)	Corrected Angle to the Right	Preliminary Azimuth	Adjusted Azimuth
$AZ\ Mk_1$				
			161°58′03.0″ (fixed)	161°58′03.0″ (fixed)
HAYFIELD NE		84°58′57.9″		
	3732.227		66°57′00.9″	66°56′59.1″
B		249°22′16.8″		
	5175.146		136°19′17.7″	136°19′14.1″
C		176°50′03.8″		
	4116.122		133°09′21.5″	133°09′16.1″
D		133°31′45.7″		
	4993.608		86°41′07.2″	86°41′00.0″
21002		101°53′18.8″		
			8°34′26.0″	8°34′17.0″ (fixed)
$AZ\ Mk_2$				

Angular misclosure = 8°34′26.0″ − 8°34′17″ = +9″
Correction per angle = −9.0″/5 = −1.8″

```
                                             Unbalanced
Course       Length        Azimuth          Dep          Lat
~~~~~~~~~~~~~~~~~~~~~~~~~~~~~~~~~~~~~~~~~~~~~~~~~~~~~~~~~~~~~~
 A-B      3,732.227     66°56'59.1" 3434.2528   1461.3098
 B-C      5,175.146    136°19'14.1" 3574.0730  -3742.7447
 C-D      4,116.122    133°09'16.1" 3002.7618  -2815.2943
 D-E      4,993.608     86°41'00.0" 4985.2439    288.9023
          ----------                ---------   ---------
  Sum = 18,017.103                 14996.3515  -4807.8269


Misclosure in Departure = 14,996.3315 - 14,996.1680 = 0.1635
Misclosure in Latitude  = -4,807.8269 - -4,808.1110 = 0.2841


           Balanced                             Coordinates
       Dep          Lat       Point          X                 Y
~~~~~~~~~~~~~~~~~~~~~~~~~~~~~~~~~~~~~~~~~~~~~~~~~~~~~~~~~~~~~~~~~~~~~
    3434.2189  1461.2509      A      745,212.637      127,939.400
    3574.0261 -3742.8264      B      748,646.856      129,400.651
    3002.7245 -2815.3592      C      752,220.882      125,657.825
    4985.1986   288.8236      D      755,223.606      122,842.465
                              E      760,208.805      123,131.289


         Linear misclosure    = 0.3278
         Relative Precision = 1 in 55,000
```

Figure 20.12 Modified excerpt of compass rule adjustment done using WOLFPACK.

example has been done by compass rule, but any method could be used including least squares. In the adjustment, the differences in eastings (X) and northings (Y) between control points were computed and checked against their fixed values to obtain the misclosures in departure (+0.164 m) and latitude (+0.284 m). An adjustment was then made to correct these computed differences to the required totals. The relative precision of the traverse was 1:55,000. Had the original distance and angle observations been used in the computations instead of their reduced equivalents, the relative precision of the traverse is only 1:10,000. This demonstrates the importance of making proper observational reductions before attempting an adjustment.

In summary, the following steps are necessary for performing traverse computations in state plane coordinates:

1. Obtain a starting and closing azimuth, and, if necessary, reduce them to their grid-equivalent azimuths.
2. Analyze the scale factor for the project. A mean of the published scale factors may be adequate for the project. This can be done by analyzing the number of significant figures in the longest measured length versus the number of common digits in the scale factors. To avoid rounding errors there should be one more common digit in the scale factors than there is in the longest observed distance.
3. Analyze the elevation factor for the project. A mean factor may be adequate in terrain with small relief. Again the elevation factor for the highest and lowest station elevations in the project should have one more common digit than the number of significant figures in the longest observed length to avoid rounding errors.
4. If a project scale factor and elevation factor can be used, compute a combined factor for the project.
5. Reduce all horizontal distances to their grid equivalents.

For lines under 8 km and lower-order surveys, steps 6 thru 8 are typically skipped.

6. Using preliminary azimuths derived from unreduced angles and grid distances, compute approximate coordinates.
7. Analyze the magnitude of the arc-to-chord correction for each line using the approximate coordinates.
8. Apply the arc-to-chord corrections to the observed angles.
9. Compute and adjust the traverse.
10. Compute the final adjusted SPCS83 coordinates for the new stations. If adjusted ground distances are required, apply the inverse of the combined factor to each line.

Procedures for computing traverses in SPCS27 follow these same steps.

■ 20.10 SURVEYS EXTENDING FROM ONE ZONE TO ANOTHER

Surveys in border areas often cross into different zones or even abutting states. This does not represent a problem, however, because adjacent zones overlap by appreciable distances, as shown in Figures 20.3 and 20.4.

The general procedure for extending surveys from one zone to another requires that the survey proceed from the first zone into the overlap area with the second. Then the geodetic latitudes and longitudes are computed for two intervisible stations using their grid coordinates in the first zone. (Recall that this conversion is called the *inverse problem*.) Using the geodetic positions of the two points, their state plane coordinates in the new zone are then computed. (This is the *direct problem*.) Finally the grid azimuth for the line in the new zone can be obtained from the new coordinates of the two points.

Suppose, for example, that a survey being computed in SPCS83 originates in southern Wisconsin, which uses the Lambert conformal conic projection, and extends into northern Illinois, which uses the Transverse Mercator grid. With Wisconsin South Zone SPCS83 coordinates of two intervisible points in the overlap area of the two zones known, Equations (20.3) are solved for the geodetic latitudes and longitudes of the points.

With the geodetic latitudes and longitudes of the two points known, Equation (20.4) is solved using constants for the appropriate Illinois zone entered to obtain their E and N coordinates in that zone. From these coordinates, the grid azimuth of the line joining the intervisible points can be calculated by using Equation (11.5) and the survey can continue into Illinois.

If immediate coordinate values in the new zone are not required, the entire survey can be computed in one zone. The inverse problem can then be used to compute geodetic coordinate values of the points followed by the direct problem to compute grid coordinates in the second zone. For example, the survey in the previous paragraph could be computed entirely in the Wisconsin South Zone. Following this, the inverse problem can be used to compute the geodetic coordinates for the stations in northern Illinois. These geodetic coordinates can then be used to determine their grid coordinates equivalents in the appropriate Illinois zone.

It should be remembered that the zone limits do not mark the end of the map projection, but simply the extents of the zone were 1:10,000 precisions are maintained between grid and ellipsoidal lengths. For example, the zones in Pennsylvania can be used to perform traverse computations in neighboring New Jersey, Ohio, Maryland, New York, and so on. Once the Pennsylvania grid coordinates for the points are determined, their geodetic equivalents can be determined using inverse computations and converted to the appropriate state zone with direct computations.

Extending surveys from zone to zone in SPCS27 follows the same procedure. Solving the direct and inverse problems that are necessary in this procedure is most conveniently handled using the computer programs described previously.

■ 20.11 CONVERSIONS BETWEEN SPCS27 AND SPCS83

Many stations that have coordinates known in SPCS27 were not included as part of the adjustment of NAD83 and thus they do not have SPCS83 coordinates. Coordinates in SPCS83 can be determined for these stations in different ways, depending upon conditions and on the level of accuracy required in the conversion process. The most accurate procedure is to readjust the original survey data to control points whose positions were included in the NAD83 general adjustment and thus known in SPCS83. This process is only possible if the original survey was tied to one or more of these control points. If such points do not exist, they can be established using GPS or by traversing as described earlier in this chapter.

A second method involves knowing coordinate values for stations in both NAD27 and NAD83. These stations should encompass other stations that are to be converted. As discussed in Section 11.8, a two-dimensional conformal coordinate transformation can be used to transform all the coordinates known in NAD27 into NAD83.

Another method, which yields an intermediate level of accuracy, computes *expected* changes for all points in a large area based on known changes for a selected network of points in the region. Changes of the selected points, which should be uniformly distributed throughout the area, are obtained by computing differences between their SPCS27 and SPCS83 coordinates. Mathematical functions (polynomials) are then used to predict changes of all other points in the area based on the pattern of changes for the known points. This procedure gives reasonably good accuracy, suitable for many needs. The NGS has two software packages available in their website[2] that will perform this type of conversion: the software package NADCON, which was developed by the NGS, and CORPSCON, which is an enhanced version of NADCON from the U.S. Corps of Engineers.

A final lower-order conversion procedure uses either linear interpolation of changes in E and N coordinates between points of known change, or an average change for a given area. It produces results suitable for certain purposes—for example, correcting coordinate grids of small-scale maps and nautical charts.

■ 20.12 THE UNIVERSAL TRANSVERSE MERCATOR PROJECTION

The Universal Transverse Mercator (UTM) system is another important map projection that has worldwide use. Originally developed by the Department of Defense primarily for artillery use, it provides worldwide coverage from 80°S latitude to 80°N latitude. Each zone has a 6° longitudinal width; thus 60 zones are required to encircle the globe. The UTM system is a Transverse Mercator map projection and thus uses the equations presented in this chapter. It has recently taken on added importance for surveyors, since UTM coordinates in metric units are now being included along with state plane and geodetic coordinates for all published NAD83 station descriptions. UTM grids are also being included on all maps in the national mapping program and UTM coordinates are being used more frequently for referencing positions of data entered into Land and Geographic Information Systems.

UTM zones are numbered easterly from 1 through 60, beginning at longitude 180°W. The conterminous United States is covered from zone 10 (west coast) through zone 20 (east coast). The central meridian for each zone is assigned a false

---

[2]Both NADCON and CORPSCON are available for downloading at http://www.ngs.noaa.gov/PC_PROD/pc_prod.shtml.

easting E_0 of 500,000 m. A false northing N_b of zero is applied for the northern hemisphere of each zone, and 10 million m is assigned for the southern hemisphere to avoid negative Y coordinates. To specify the position of any point in the UTM system, the zone number must be given as well as its northing and easting.

In the UTM system, each zone overlaps adjacent ones by 0°30′. Because zone widths of 6° are considerably larger than those used in state plane systems, lower accuracies result and 1 part in 2500 ($k_0 = 0.9996$) applies at the center and edges of zones. Equations for calculating X and Y coordinates in the UTM system are the same as those for the Transverse Mercator projection. As with the state plane systems, tables giving formulas and constants for the system are available. Also, like the state plane coordinate system, the datum of reference must be specified. Because UTM coordinates are available for all points within the NAD83, calculations between very widely spaced points can readily be made. This is convenient and entirely consistent with current capabilities for conducting surveys of global extent with new devices such as GPS. The Universal Transverse Mercator map projection with a graphic showing zone boundaries is including in the Excel® spreadsheet contained on the CD that accompanies this book.

20.13 OTHER MAP PROJECTIONS

The Lambert conformal conic and Transverse Mercator map projections are designed to cover areas extensive in east-west and north-south directions, respectively. However, these systems do not conveniently cover circular areas or long strips of the Earth that are skewed to the meridians. Two other systems, the *oblique stereographic* and the *Oblique Mercator* projections, satisfy these problems.[3]

20.13.1 Oblique Stereographic Map Projection

The oblique stereographic projection can be divided into two classes: *tangent plane* and *secant plane*. In either case, as illustrated in Figure 20.13, the projection point P (the origin) is on the ellipsoid where a line perpendicular to the map plane and passing through center point O intersects the ellipsoid. In the tangent plane system, ellipsoid points a and b are projected outward to a' and b', respectively, on the map plane. For the secant plane system, ellipsoid points c and d are projected inward to c' and d' on the map plane. (If they were outside of the secant points, projection would be outward.) The oblique stereographic map projection is conformal and thus preserves the shapes of objects.

Oblique stereographic projections are not employed in the United States typically, but are used in Canada and other parts of the world. As discussed in Section 19.6.6, these projections are also used to convert geodetic coordinates determined by GPS into map projection coordinates for use in the localization process. If point P is the North or South Pole, the projection is called *polar stereographic;* if it is on the equator, *equatorial stereographic*.

---

[3]The oblique Mercator projection is also called the Hotine skew orthomorphic projection, named after the English geodesist Martin Hotine.

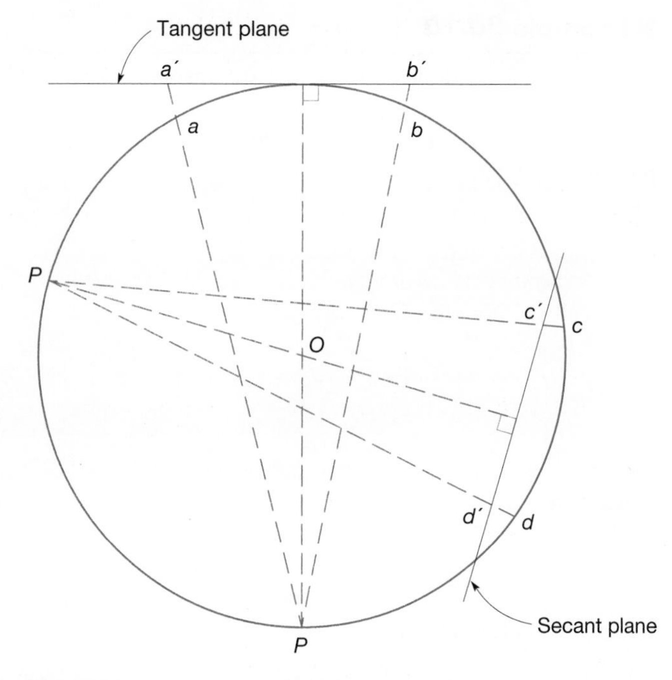

Figure 20.13 Tangent plane and secant plane horizon stereographic map projections.

The defining parameters of this projection are the latitude and longitude of the grid origin (φ_0, λ_0) and the scale factor at the grid origin k_0. It uses common functions

$$\chi(\varphi) = 2\tan^{-1}\left[\tan\left(\frac{\pi}{4} + \frac{\varphi}{2}\right)\left(\frac{1 - e\sin\varphi}{1 + e\sin\varphi}\right)\right]^{\frac{e}{2}} - \frac{\pi}{2} \tag{20.60}$$

$$M(\varphi) = \frac{\cos\varphi}{\sqrt{1-e^2\sin^2\varphi}} \tag{20.61}$$

where e is the eccentricity of the ellipsoid as defined by Equation (19.2a) or (19.2b). Using Equation (20.6), the zone constants for the projection are

$$\chi_0 = \chi(\varphi_0) \tag{20.62}$$

$$m_0 = m(\chi_0) \tag{20.63}$$

Using the latitude φ and longitude λ of the point, and the semimajor axis a for the ellipsoid, the equations for the direct problem are

$$\chi = \chi(\varphi) \tag{20.64}$$

$$m = m(\varphi) \tag{20.65}$$

$$A = \frac{2ak_0m_0}{\cos\chi_0[1 + \sin\chi_0\sin\chi + \cos\chi_0\cos\chi\cos(\lambda-\lambda_0)]} \tag{20.66}$$

$$E = A\cos\chi\sin(\lambda-\lambda_0) \tag{20.67}$$

$$N = A[\cos\chi_0\sin\chi - \sin\chi_0\cos\chi\cos(\lambda-\lambda_0)] \tag{20.68}$$

$$k = \frac{A\cos\chi}{am} \tag{20.69}$$

■ Example 20.10

The following geodetic coordinates are observed using RTK-GPS methods. What are the oblique stereographic map projection coordinates for Station A using a grid origin of (41°18′15″N,76°00′00″W) and $k_0 = 1$? (Use WGS84 ellipsoidal parameters.)

Station	Latitude	Longitude	Height (m)
A	41°18′09.88223″N	75°59′58.05637″W	282.476
B	41°18′21.11176″N	76°00′37.35445″W	296.571
C	41°18′19.33293″N	75°59′40.39279″W	313.814
D	41°18′09.67030″W	75°59′44.19645″W	304.205

Solution

The zone constants are

By Equation (20.62):

$$\chi_0 = 2\tan^{-1}\left[\tan\left(\frac{\pi}{4} + \frac{41°18'15''}{2}\right)\left(\frac{1 - e\sin 41°18'15''}{1 + e\sin 41°18'15''}\right)^{\frac{e}{2}}\right] = 41°06'48.66298''$$

By Equation (20.63):

$$m_0 = \frac{\cos 41°18'15''}{\sqrt{1 - e^2\sin^2 41°18'15''}} = 0.752314$$

By Equation (20.64):

$$\chi = 2\tan^{-1}\left[\tan\left(\frac{\pi}{4} + \frac{41°18'09.88223''}{2}\right)\left(\frac{1 - e\sin 41°18'09.88223''}{1 + e\sin 41°18'09.88223''}\right)^{\frac{e}{2}}\right]$$
$$= 41°06'43.54972''$$

By Equation (20.65):

$$m = \frac{\cos 41°18'09.88223''}{\sqrt{1 - e^2\sin^2 41°18'09.88223''}} = 0.752330$$

By Equation (20.66):

$$A = \frac{2(6378137)(1)(0.752314)}{\cos\chi_0[1 + \sin\chi_0\sin\chi + \cos\chi_0\cos\chi\cos(-75°59'58.05637'' + 76°)]}$$
$$= 6{,}368{,}873.344 \text{ m}$$

By Equation (20.67):

$$E = A\cos 41°06'43.54972''\sin(-75°59'58.05637'' + 76°) = 45.218 \text{ m}$$

By Equation (20.68):

$$N = A[\cos \chi_0 \sin \chi - \sin \chi_0 \cos \chi \cos(-75°59'58.05637'' + 76°)] = -157.869 \text{ m}$$

By Equation (20.69):

$$k = \frac{6368873.344 \cos 41°06'43.54972''}{6378137(0.752330)} = 1.00000$$

Note in the example that the direct problem was not defined with false easting or northing, and thus one of the resultant coordinates has a negative value. For localization, as discussed in Section 19.6.6, this is not a problem since the user never views the coordinates. However, a false easting and northing could be added to Equations (20.67) and (20.68) to provide positive coordinates values in a limited region.

The inverse problem converts the map projection coordinates of (N,E) to geodetic latitude and longitude. The equations used in the inverse problem for the oblique stereographic map projection are

$$\rho = \sqrt{E^2 + N^2} \tag{20.70}$$

$$c = 2\tan^{-1}\left(\frac{\rho \cos \chi_0}{2ak_0 m_0}\right) \tag{20.71}$$

$$\chi = \sin^{-1}\left(\cos c \sin \chi_0 + \frac{N \sin c \cos \chi_0}{\rho}\right) \tag{20.72}$$

$$\lambda = \lambda_0 + \tan^{-1}\left(\frac{E \sin c}{\rho \cos \chi_0 \cos c - N \sin \chi_0 \sin c}\right) \tag{20.73}$$

$$\phi = 2\tan^{-1}\left[\tan\left(\frac{\pi}{4} + \frac{\chi}{2}\right)\left(\frac{1 + e \sin \phi}{1 - e \sin \phi}\right)^{\frac{e}{2}}\right] - \frac{\pi}{2} \tag{20.74}$$

Using $\phi = \chi$ in the first iteration, Equation (20.73) is iterated until the change in ϕ becomes negligible. The scale factor is computed using Equation (20.69). Computations for the oblique stereographic map projection are demonstrated in a Mathcad® worksheet on the CD that accompanies this book.

20.13.2 Oblique Mercator Map Projection

The oblique Mercator projection is designed for areas whose major extent runs obliquely to a meridian, such as northwest to southeast. The projection is conformal and is developed by projecting points from the ellipsoid to an imaginary cylinder that is oriented with its axis skewed to the equator. It is used as the state plane coordinate projection for the southeast portion (panhandle) of Alaska. Computations for the aforementioned map projections are demonstrated in the Excel® spreadsheet and Mathcad® worksheet on the CD that accompanies this book.

PROBLEMS

Asterisks (*) indicate problems that have answers given in Appendix G.

20.1 Discuss the advantages of placing surveys on state plane coordinate systems.

20.2 Name the two basic projections used in state plane coordinate systems. What are their fundamental differences? Which one is preferred for states whose long dimensions are north-south? East-west?

20.3 Why do the states of Alaska, New York, and Florida use both the Lambert conformal conic and the Transverse Mercator projections for their state plane coordinate systems?

20.4 Describe how NAD27 state plane coordinate values can be converted to NAD83 state plane coordinates.

20.5 What corrections must be made to measured slope distances prior to computing state plane coordinates?

20.6 What corrections must be made to measured azimuths and angles prior to computing state plane coordinates?

20.7 Develop a table of SPCS83 elevation factors for geodetic heights ranging from 0 ft to 5500 ft. Use increments of 500 ft and an average radius for the Earth of 6,371,000 m.

20.8 Similar to Problem 20.7, except for geodetic heights from 0 m to 3200 m using 200 m increments.

20.9 Explain how surveys can be extended from one state plane coordinate zone to another, or from one state to another.

20.10 Develop a table similar to Table 20.1 for a range of latitudes from 40°30′ to 40°39′ in the Pennsylvania North Zone with standard parallels of 40°53′ N and 41°57′ N, and a grid origin at (40°10′ N, 77°45′ W).

20.11* The Pennsylvania North Zone SPCS83 state plane coordinates of points *A* and *B* are as follows:

Point	*E*(m)	*N*(m)
A	457,774.581	119,370.079
B	457,891.976	119,727.630

Calculate the grid length and grid azimuth of line *AB*.

20.12 Similar to Problem 20.11, except points *A* and *B* have the following New Jersey SPCS83 state plane coordinates:

Point	*E*(m)	*N*(m)
A	135,995.711	19,995.711
B	136,089.917	19,894.644

20.13* What are the SPCS83 coordinates and convergence angle for a station in the North zone of Pennsylvania with geodetic coordinates of 41°12′33.0745″ N and 76°23′48.9765″W?

20.14 Similar to Problem 20.13 except that the station's geodetic coordinates are 40°50′26.3484″ N and 76°12′40.6496″.

20.15 What is the scale factor for the station in Problem 20.13?

20.16 What is the scale factor for the station in Problem 20.14?

20.17* What are the SPCS83 coordinates for a station in New Jersey with geodetic coordinates of 39°02′26.0416″ N and 74°15′36.4908″ W?

20.18 Similar to Problem 20.17, except that the geodetic coordinates of the station are 40°51′24.0876″ N and 74°29′36.9641″ W.

20.19 What are the convergence angle and scale factor at the station in Problem 20.17?

20.20 What are the convergence angle and scale factor at the station in Problem 20.18?

20.21* What are the geodetic coordinates for a point *A* in Problem 20.11?

20.22 Similar to Problem 20.21, except for a point *B* in Problem 20.11?

20.23* What are the geodetic coordinates for a point *A* in Problem 20.12?

20.24 Similar to Problem 20.23, except for a point *B* in Problem 20.12.

20.25 In computing state plane coordinates for a project area whose mean orthometric height is 263 m, an average scale factor of 0.9999856 was used. The average geoidal separation for the area is −28 m. The given distances between points in this project area were computed from SPCS83 state plane coordinates. What horizontal length would have to be observed to lay off these lines on the ground? (Use 6,371,000 m for an average radius for the Earth.)

(a) *2834.79 ft

(b) 665.74 ft

20.26 Similar to Problem 20.25, except that the mean project area elevation was 888 m, the geoidal separation −22.5 m, the scale factor 0.99997053, and the computed lengths of lines from SPCS83 were:

(a) 645.997 m

(b) 265.074 m

20.27 The horizontal ground lengths of a three-sided closed polygon traverse were measured in feet as follows: AB = 1280.02, BC = 856.93, and CA = 810.61 ft. If the average orthometric height of the area is 156.080 ft and the average geoid separation is −31.05 m, calculate ellipsoid lengths of the lines suitable for use in computing SPCS83 coordinates. (Use 6,371,000 m for an average radius for the Earth.)

20.28 Assuming a scale factor for the traverse of Problem 20.27 to be 0.99995048, calculate grid lengths for the traverse lines.

20.29 For the traverse of Problem 20.27, the grid azimuth of a line from *A* to a nearby azimuth mark was 150°31′16″ and the clockwise angle measured at *A* from the azimuth mark to *B*, 80°52′36″. The measured interior angles were A = 41°12′26″, B = 38°32′50″, and C = 100°14′53″. Balance the angles and compute grid azimuths for the traverse lines. (*Note:* Line *BC* bears easterly.)

20.30 Using grid lengths of Problem 20.28 and grid azimuths from Problem 20.29, calculate departures and latitudes, linear misclosure, and relative precision for the traverse.

20.31 If station *A* has SPCS83 state plane coordinates E = 702,960.064 m and N = 171,676.036 m, balance the departures and latitudes computed in Problem 20.30 using the compass rule, and determine SPCS83 coordinates of stations *B* and *C*.

20.32* What is the combined factor for the traverse of Problems 20.27 and 20.28?

20.33 The horizontal ground lengths of a four-sided closed polygon traverse were measured as follows: AB = 479.549 m, BC = 830.616 m, CD = 685.983 m, and DA = 859.689 m. If the average orthometric height of the area is 1250 m, the geoidal separation is −30.0 m, and the scale factor for the traverse 0.9999574, calculate grid lengths of the lines for use in computing SPCS83 coordinates. (Use 6,371,000 m for an average radius of the Earth.)

20.34 For the traverse of Problem 20.33, the grid bearing of line *BC* is N57°39′48″ W. Interior angles were measured as follows: A = 120°26′28″, B = 73°48′56″, C = 101°27′00″, and D = 64°17′26″. Balance the angles and compute grid bearings for the traverse lines. (*Note:* Line *CD* bears southerly.)

20.35 Using grid lengths from Problem 20.33 and grid bearings from Problem 20.34, calculate departures and latitudes, linear misclosure, and relative precision for the traverse. Balance the departures and latitudes by the compass rule. If the SPCS83 state plane coordinates of point B are $E = 255{,}096.288$ m and $N = 280{,}654.342$ m, calculate SPCS83 coordinates for points C, D, and A.

20.36 The traverse in Problems 10.9 through 10.11 was performed in the Pennsylvania North Zone of SPCS83. The average elevation for the area was 314.56 m and the average geoidal separation was −30.25 m. Using the data in Table 20.1 and a mean radius for the earth, reduce the observations to grid and adjust the traverse. Compare this solution with that obtained in Chapter 10. (Use 6,371,000 m for an average radius of the Earth.)

20.37 The traverse in Problems 10.12 through 10.14 was performed in the New Jersey zone of SPCS83. The average elevation for the area was 124.9 m and the average geoidal separation was −29.7 m. Using the data in Tables 20.3 and 20.4, and a mean radius for the Earth, reduce the observations to grid and adjust the traverse. Compare this solution with that obtained in Chapter 10.

20.38 The traverse in Problem 10.23 was performed in the Pennsylvania North Zone of SPCS 1983. The average elevation of the area was 67.2 m and the average geoidal separation was −28.5 m. Using 6,371,000 m for the mean radius of the Earth, reduce the observations to grid and adjust the traverse using the compass rule. Compare this solution with that obtained in Problem 10.23.

20.39 The traverse in Problem 10.24 was performed in the Pennsylvania North Zone of SPCS83. The average elevation for the area was 67.89 m and the average geoidal separation was −30.23 m. Using the mean radius of the Earth, reduce the observations to grid and adjust the traverse using the compass rule. Compare this solution with that obtained in Problem 10.24.

20.40* If the geodetic azimuth of a line is 65°06′59.8″, the convergence angle is 1°24′32.2″, and the arc-to-chord correction is −1.2″, what is the equivalent grid azimuth for the line?

20.41 If the geodetic azimuth of a line is 308°06′16.97″, the convergence angle is −1°48′36.8″, and the arc-to-chord correction is −1.5″, what is the equivalent grid azimuth for the line?

20.42 Using the values given in Problems 20.40 and 20.41, what is the obtuse grid angle between the two azimuths?

20.43 The grid azimuth of a line is 102°37′08″. If the convergence angle at the endpoint of the azimuth is 2°05′52.9″ and the arc-to-chord correction is 0.7″, what is the geodetic azimuth of the line?

20.44 Similar to Problem 20.43, except the convergence angle is −1°02′20.7″ and the arc-to-chord correction is −0.6″.

20.45 What is the arc-to-chord correction for the line from A to B in Problems 20.21 and 20.22?

20.46 What is the geodetic azimuth of the line from A to B in Problem 20.45?

20.47 Using the defining parameters given in Example 20.10, compute oblique stereographic map projection coordinates for Station B.

20.48 Similar to Problem 20.47, except for Station C.

20.49 Similar to Problem 20.47, except for Station D.

20.50 Create a computational program that performs the direct and inverse problems for the Universal Transverse Mercator map projection.

20.51 Create a computational program that reduces distances from the ground to a mapping grid.

BIBLIOGRAPHY

Bunch, B. W. 2002. "A New Projection: Developing and Adopting a Single Zone State Plane Coordinate System for Kentucky, Part 1." *Professional Surveyor* 22 (No. 4): 26.

_____. 2002. "A New Projection: Developing and Adopting a Single Zone State Plane Coordinate System for Kentucky, Part 2." *Professional Surveyor* 22 (No. 5): 34.

GIA. 2006. "How Things Work: Scale, Elevation, Grid, and Combined Factors Used in Instrumentation." *Professional Surveyor* 26 (No. 2): 47.

Hartzell, P., L. Strunk, and C. Ghilani. 2002. "Pennsylvania State Plane Coordinate System: Converting to a Single Zone." *Surveying and Land Information Science* 62 (No. 2): 95.

Snay, R. A. 1999. "Using the HTDP Software to Transform Spatial Coordinates Across Time and Reference Frames." *Surveying and Land Information Systems* 59 (No. 1): 15.

Snyder, J. P. 1987. *Map Projections—A Working Manual.* U.S. Government Printing Office, Washington, D.C.

Stachurski, R. 2002. "History of American Projections: The American Projection, Part 1." *Professional Surveyor* 22 (No. 4): 16.

_____. 2002. "History of American Projections: The American Projection, Part 2." *Professional Surveyor* 22 (No. 5): 32.

Stem, J. E. 1989. "State Plane Coordinate System of 1983." *NOAA Manual NOS NGS 5* Rockville, Md.: National Geodetic Information Center.

21

Boundary Surveys

■ 21.1 INTRODUCTION

The oldest types of surveys in recorded history are boundary surveys, which date back to about 1400 B.C. when plots of ground were subdivided in Egypt for taxation purposes. Boundary surveys still are one of the main areas of surveying practice. From Biblical times[1] when the death penalty was assessed for destroying corners, to the colonial days of George Washington[2] who was licensed as a land surveyor by William and Mary College of Virginia, and through the years to the present, natural objects (i.e., trees, rivers, rock outcrops, etc.) and man-made objects (i.e., fences, wooden posts, iron, steel or concrete markers, etc.) have been used to identify land parcel boundaries.

As property increased in value and owners disputed rights to land, the importance of more accurate surveys, monumentation of the boundaries, and written records became obvious. When Texas became a state in 1845, its public domain amounted to about 172,700,000 acres, which the U.S. government could have acquired by payment of the approximately $13,000,000 in debts accumulated by the Republic of Texas. However, Congress allowed the Texans to retain their land and pay their own debts—a good bargain at roughly 7.6 cents/acre! Land at Waikiki in Honolulu has sold for more than $5,000,000/acre, and in Tokyo prices have reached over $30 million per acre.

The term *land tenure system* applies to the manner in which rights to land are held in any given country. Such a system, as a minimum, must provide (1) a means for transferring or changing the title and rights to the land, (2) permanently

---

[1]"Cursed be he that removeth his neighbor's landmark. And all the people shall say Amen." Deut. 27:17.

[2]"Mark well the land, it is our most valuable asset." George Washington.

monumented or marked boundaries that enable parcels to be found on the ground, (3) officially retained records defining who possesses what rights to the land, and (4) an official legal description of each parcel. In the United States a two-tier land tenure system exists. At the federal level, records of surveys and rights to federal land are maintained by the U.S. *Bureau of Land Management* (*BLM*). At the state and local levels, official records concerning land tenure are held in county courthouses.

Land titles in the United States are now transferred by written documents called *deeds* (*warranty, quitclaim,* or *agreement*), which contain a description of the property. Property descriptions are prepared as the result of a land survey. The various methods of description include (1) metes and bounds, (2) block-and-lot number, (3) coordinate values for each corner, and (4) township, section, and smaller subdivisions of the United States *Public Land Survey System* (*PLSS*) commonly referred to as the *aliquot part.* Often a property description will combine two or more of these methods. The first three methods are discussed briefly in this chapter; the PLSS is covered in Chapter 22.

■ 21.2 CATEGORIES OF LAND SURVEYS

Activities involved in the practice of land surveying can be classified into three categories: (1) *original surveys* to subdivide the remaining unsurveyed U.S. public lands, most of which are in Alaska, (2) *retracement surveys* to recover and monument or mark boundary lines that were previously surveyed, and (3) *subdivision surveys* to establish new smaller parcels of land within lands already surveyed. The last two categories are described in this chapter; the first is discussed in Chapter 22.

In establishing new property lines, and especially in retracing old ones, surveyors must exercise acute judgment based on education, practical experience, and knowledge of land laws. They must also be accurate and articulate in making observations. This background must be bolstered by tenacity in searching the records of all adjacent property as well as studying descriptions of the land in question. In fieldwork, surveyors must be untiring in their efforts to find points called for by the deed. Often it is necessary to obtain *parole evidence*—that is, testimony from people who have knowledge of accepted land lines and the location of corners, reference points, fences, and other information about the correct lines.

Modern-day land surveyors are confronted with a multitude of problems, created over the past two centuries under different technology and legal systems that now require professional solutions. These include defective compass and chain surveys; incompatible descriptions and plats of common lines for adjacent tracts; lost or obliterated corners and reference marks; discordant testimony by local residents; questions of riparian rights; and a tremendous number of legal decisions on cases involving property boundaries.

The responsibility of a professional surveyor is to weigh all evidence and try to establish the originally intended boundary between the parties involved in any property-line dispute, although without legal authority to force a compromise or settlement. Fixing title boundaries must be done by agreement of adjacent owners or court action. Surveyors are often called upon to serve as expert witnesses in proceedings to establish boundaries, but to do so they should be registered.

Because of the complicated technical judgment decisions that must be made, the increasing cost of land surveyors' professional liability insurance for "errors and omissions" has become a major part of operating expenses. Some states demand that a surveyor have it for the protection of all parties.

■ 21.3 HISTORICAL PERSPECTIVES

In the eastern part of the United States, individuals acquired the first land titles by gifts or purchase from the English Crown. Surveys and maps were completely lacking or inadequate, and descriptions could be given in only general terms. The remaining land in the 13 colonies was transferred to the states at the close of the Revolutionary War. Later this land was parceled out to individuals, generally in irregular tracts. Boundary lines were described by *metes and bounds* (see Section 21.4).

Many original transfers and subsequent ownerships and subdivisions were not recorded. Those that were usually had scanty or defective descriptions, since land was cheap and abundant. Trees, rocks, and natural landmarks defining the corners, as on the first example metes-and-bounds description (see Section 21.4) were soon disturbed. The intersection of two property lines might be described only as "the place where John killed a bear" or "the bend in a footpath from Jones's cabin to the river."

Numerous problems in land surveying stem from the confusion engendered by early property titles, descriptions, and compass surveys. The locations of thousands of corners have been established by compromise after resurveys or by court interpretation of all available evidence pertinent to their original or intended positions. *Squatters' rights, adverse possession,* and *riparian changes* have fixed other corners. Many boundaries are still in doubt, particularly in areas having marginal land where the cost of a good retracement survey exceeds the property's value.

The fact that four corners of a field can be found and the distances between them agree with the "calls" in a description does not necessarily mean that they are in the proper place. Title or ownership is complete only when the land covered by a deed is positively identified and located on the ground.

Land law from the time of the Constitution has been held as a state's right, subject to interpretation by state court systems. Many millions of land parcels have been created in the United States over the past four centuries under different technology and legal systems. Some of the countless problems passed on to today's professional surveyors, equipped with immensely improved equipment, are discussed in this chapter and in Chapter 22.

Land surveying measurements and analysis follow basic plane surveying principles. But a land surveyor needs years of experience in a given state to become familiar with local conditions, basic reference points, and legal interpretations of complicated boundary problems. Methods used in one state for prorating differences between recorded and measured distances may not be acceptable in another. Rules on when and how fences determine property lines are not the same in all states or even in adjacent ones.

The term *practical location* is used by the legal profession to describe an agreement, either explicit or implied, in which two adjoining property owners mark out an ambiguous boundary, or settle a boundary dispute. Fixed principles enter the process and the boundary established, called an *agreed-upon boundary,* can become permanent.

Different interpretations are given locally to (1) the superiority or definiteness of one distance over another associated with it, (2) the position of boundaries shown by occupancy, (3) the value of corners in place in a tract and its subdivisions, and (4) many other factors. Registration of land surveyors is therefore required in all states to protect the public interest.

■ 21.4 PROPERTY DESCRIPTION BY METES AND BOUNDS

As noted earlier, metes and bounds is one of the methods commonly used in preparing legal descriptions of property. Descriptions by metes (to measure, or assign by measure) and bounds (boundary lines or property limits) have a *point of commencement* (POC) such as a nearby existing corner of the PLSS. Commencing at this point, successive lengths and directions of lines are given that lead to the *point of beginning* (POB). The POB is usually a fence post, iron or steel rod, or some natural feature, which marks one corner of the property. Lengths and directions (bearings or azimuths) of successive lines from the point of beginning that enclose or bound the property are then given. Early distance units of chains, poles, and rods are now replaced by feet and decimals and sometimes by metric units. Bearings or azimuths may be geodetic, astronomic, magnetic, or grid. Care must be exercised to indicate clearly which of these is the basis of directions so that no confusion arises. In the past, *assumed* bearings or azimuths have sometimes been used, but many states no longer allow them because they are not readily reproducible. In some states, survey regulations call for exterior lines of new subdivisions to be based on the true meridian.

Surveyors write metes-and-bounds property descriptions and they are included in the legal documents that accompany the transfer of title to property. In preparing the descriptions, extreme care must be exercised. A single mistake in transcribing a numerical value, or one incorrect or misplaced word or punctuation mark, may result in litigation for more than a generation, since the intentions of the *grantor* (person selling property) and the *grantee* (person buying property) may then be unclear. If numbers are both spelled out and given as figures, words control in the case of conflicts unless other proof is available. There is a greater likelihood of transposing than misspelling—and lawyers prefer words!

The importance of permanent monuments to mark property is evident. In fact, some states require pipes, iron pins, and/or concrete markers set deep enough to reach below the frost line at all property corners before surveys will be accepted for recording. Actually, almost any suitable marker could be called for as a monument. A map attached to the description will contain a legend, which identifies all monuments. By scaling from the map, a rough check on the distances and directions in the description can be obtained.

To increase precision in property surveys and to simplify the process of preparing property descriptions and relocating the corners of established parcels, large cities and some states have established a network of control monuments to supplement stations of the National Spatial Reference System (NSRS). These points are then available as control for initiating topographic and construction surveys and can also serve as POCs for commencing property descriptions. Some states have embarked on statewide programs in which a relatively dense grid of control monuments is being set on a county-by-county basis—for example, the State of Florida established a 6-mi grid. State plane coordinates are determined for the monumented positions and they are tied to corners of the PLSS.

Description of land by metes and bounds in a deed should always contain the following information in addition to the recital:

1. *Point of commencement (POC)*. This is an established reference point such as a corner of the PLSS or NSRS monument to which the property description is tied or referenced. It serves as the starting point for the description.

2. *Point of beginning (POB)*. This point must be identifiable, permanent, well referenced, and one of the property corners. Coordinates, preferably state plane, should be given if known or computable. Note that a POB is no more important than others and a called-for monument in place at the next corner establishes its position, even though bearing and distance calls to it may not agree.

3. *Definite corners*. Such corners are clearly defined points with coordinates if possible.

4. *Lengths and directions of the property sides*. All lengths in feet and decimals (or metric units), and directions by angles, true bearings, or azimuths must be stated to permit computation of any misclosure error. Omitting the length or direction of a closing line to the POB and substituting a phrase "and thence to the point of beginning" is not acceptable. The survey date is required and particularly important if directions are referred to magnetic north.

5. *Names of adjoining property owners*. These are helpful to show the intent of a deed in case an error in the description leaves a gap or creates an overlap. However, called-for monuments in place will control title over calls for adjoiners unless precluded by *senior rights* (see Section 21.7).

6. *Areas*. The included area is normally given as an aid in the valuation and identification of a piece of property. Areas of rural land are given in acres or hectares and those of city lots in square feet or square meters. Because of differences in measurements and depending on the adjustment method used for a traverse (compass rule, least squares, etc.), one surveyor's calculated area, directions, and distances may differ slightly from another's. The expression "more or less," which may follow a computed area, allows for minor errors, and avoids nuisance suits for insignificant variations.

Note that items 4, 5, and 6 above establish the *shape, intent,* and *size,* respectively, of the parcel. Without item 5, the description would be strictly "metes" and would lack intent. Also if items 4 and 6 were excluded, the description would be purely "bounds" and would lack the important measurements needed to

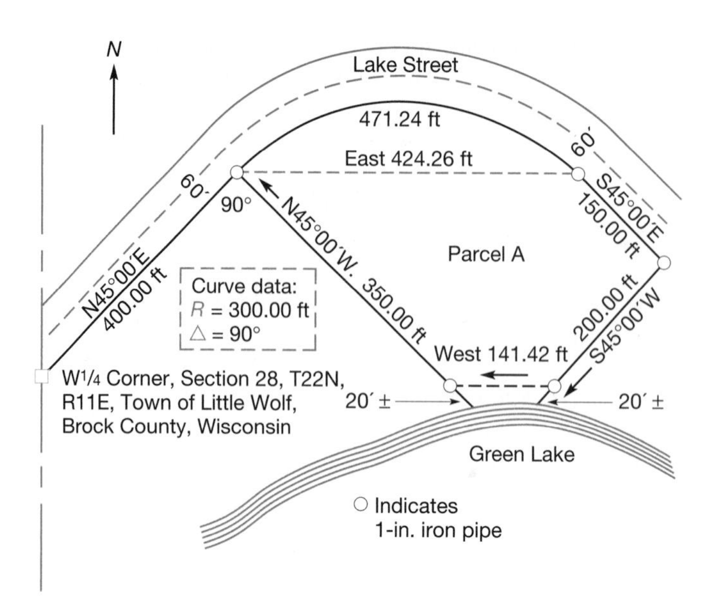

Figure 21.1 Metes-and-bounds tract.

establish the shape and size of the parcel. A partial metes-and-bounds description for the tract shown in Figure 21.1 is given as an example.

> That part of the SW 1/4 of the NW 1/4 of Section 28, T 22 N, R 11 E, 4th P.M., Town of Little Wolf, Brock County, Wisconsin, described as follows: Starting at the point of commencement, which is a stone monument at the W 1/4 corner of said Section 28; thence N45 degrees 00 minutes E, four hundred (400.00) feet along the Southeasterly R/W line of Lake Street to a 1 inch iron pipe at the point of beginning of this description, said point also being the point of curvature of a tangent curve to the right having a central angle of 90 degrees 00 minutes and radius of three hundred (300.00) feet; thence Easterly, four hundred seventy one and twenty four hundredths (471.24) feet along the arc of the curve, the long chord of which bears East, four hundred twenty- four and twenty six hundredths (424.26) feet, to a 1 inch iron pipe at the point of tangency thereof, said arc also being the aforesaid Southwesterly R/W line of Lake Street; thence continuing along the Southerly R/W line of Lake Street, S 45 degrees 00 minutes E, one hundred fifty (150.00) feet to a 1 inch iron pipe; thence S 45 degrees 00 minutes W, two hundred (200.00) feet to a 1 inch iron pipe located N 45 degrees 00 minutes E, twenty (20) feet, more or less, from the water's edge of Green Lake, and is the beginning of the meander line along the lake; thence West one hundred forty one and forty two hundredths (141.42) feet along the said meander line to a 1 inch iron pipe at the end of the meander line; said pipe being located N 45 degrees 00 minutes W, twenty (20) feet, more or less from the said water's edge; thence N 45 degrees 00 minutes W, three hundred fifty (350.00) feet to a 1 inch iron pipe at the point of beginning . . . including all lands lying between the meander line herein described and the Northerly shore of Green Lake, which lie between true extensions of the Southeasterly and Southwesterly boundary lines of the parcel herein described, said parcel containing 2.54 acres, more or less. Bearings are based on astronomic north.

Many metes-and-bounds descriptions have been prepared, which because of a variety of shortcomings, have created later problems. To illustrate, portions of two *old* metes-and-bounds descriptions from the eastern United States follow. The first, part of an early deed registered in Maine, is:

> Beginning at an apple tree at about 5 minutes walk from Trefethen's Landing, thence easterly to an apple tree, thence southerly to a rock, thence westerly to an apple tree, thence northerly to the point of beginning.

With numerous apple trees and an abundance of rocks in the area, the dilemma of a surveyor trying to retrace the boundaries many years later is obvious.

The second, a more typical old description of a city lot showing lack of comparable precision in angles and distances, follows:

> Beginning at a point on the west side of Beech Street marked by a brass plug set in a concrete monument located one hundred twelve and five tenths (112.5) feet southerly from a city monument No. 27 at the intersection of Beech Street and West Avenue; thence along the west line of Beech Street S 15 degrees 14 minutes 30 seconds E fifty (50) feet to a brass plug in a concrete monument; thence at right angles to Beech Street S 74 degrees 45 minutes 30 seconds W one hundred fifty (150) feet to an iron pin; thence at right angles N 15 degrees 14 minutes 30 seconds W parallel to Beech Street fifty (50) feet to an iron pin; thence at right angles N 74 degrees 45 minutes 30 seconds E one hundred fifty (150) feet to place of beginning; bounded on the north by Norton, on the east by Beech Street, on the south by Stearns, and on the west by Weston.

■ 21.5 PROPERTY DESCRIPTION BY BLOCK AND LOT SYSTEM

As urban areas grow, adjoining parcels of land are subdivided to create streets, blocks, and lots according to an orderly and specific plan. Each new subdivided parcel, called a *subdivision,* is assigned a name and annexed by the city. Block and lot number, tract and lot number, or subdivision name and lot number identifies the individual lots within the subdivided areas. Examples are:

> Lot 34 of Tract 12314 as per map recorded in book 232, pages 23 and 24 of maps, in the office of the county recorder of Los Angeles County.
>
> Lot 9 except the North twelve (12) feet thereof, and the East twenty-six (26) feet of Lot 10, Broderick's Addition to Minneapolis. [Parts of two lots are included in the parcel described.]
>
> That portion of Lot 306 of Tract 4178 in the City of Los Angeles, as per Map recorded in Book 75, pages 30 to 32 inclusive of maps in the office of the County Recorder of said County, lying Southeasterly of a line extending Southwesterly at right angles from the Northeasterly line of said lot, from a point in said Northeasterly line distant Southeasterly twenty-three and seventy-five hundredths (23.75) feet from the most Northerly corner of said Lot.

The block-and-lot system is a short and unique method of describing property for transfer of title. Standard practice calls for a map or plat of each subdivision to be filed with the proper office. The plat must show the types and locations of monuments, dimensions of all blocks and lots, and other pertinent information such as the locations and dimensions of streets and easements, if any. These subdivision plats are usually kept in map books in the city or county recorder's office.

Lot and block descriptions typically are created simultaneously and thus are not subject to *junior* or *senior rights* (described in Section 21.7). In performing resurveys to find or reestablish lot corners, therefore, any excess or deficiency found in the measurements is prorated equally among the lots.

Figure 21.2 is an example of a small hand-drafted block-and-lot subdivision. Figure 18.13 shows a portion of a subdivision map of blocks and lots produced using a CADD system.

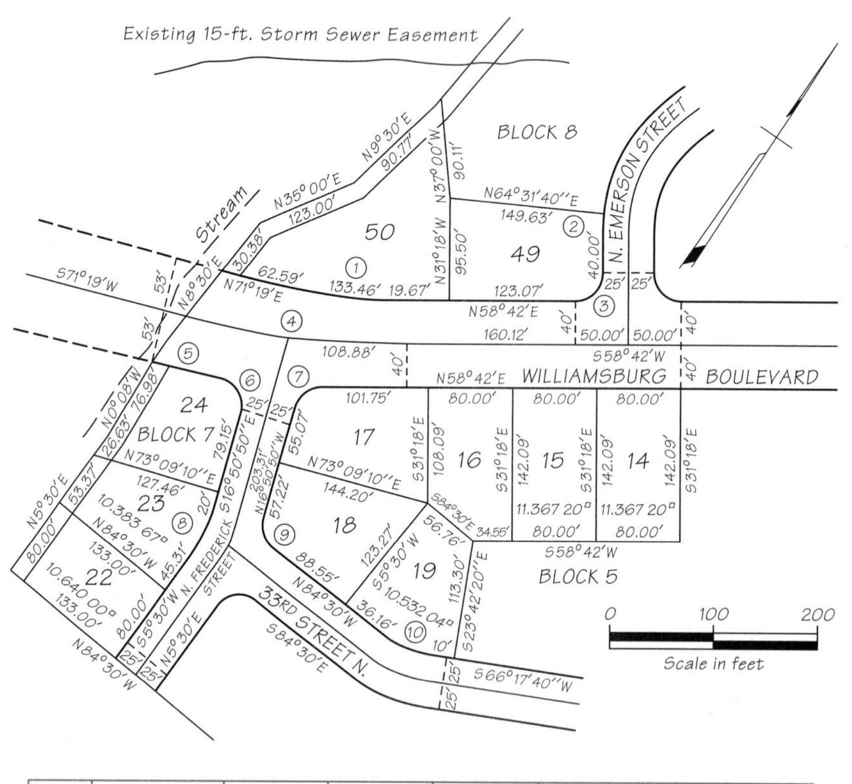

NO	DELTA (I)	TAN	ARC	RADIUS	CHORD	CHORD BEAR
1	12°37′	67.00′	133.46′	606.07′	133.19′	N65°00′30″E
2	5°49′40″	7.67′	15.33′	150.72′	15.32′	N28°23′10″W
3	90°00′	25.00′	39.27′	25.00′	35.36′	N13°42′E
4	12°37′	105.57′	210.28′	954.93′	209.85′	S65°00′30″W
5	5°50′15″	33.59′	67.11′	658.72′	67.08′	N68°23′30″E
6	97°40′27″	28.60′	42.62′	25.00′	37.64′	S65°41′05″E
7	75°32′50″	19.37′	32.96′	25.00′	30.63′	N20°55′35″E
8	22°20′50″	19.75′	39.00′	100.00′	38.76′	S5°40′25″E
9	67°39′10″	16.75′	29.52′	25.00′	27.83′	N50°40′25″W
10	29°12′20″	36.38′	71.16′	139.62′	70.41′	S80°53′50″W

Figure 21.2 Small subdivision plat.

■ 21.6 PROPERTY DESCRIPTION BY COORDINATES

State plane coordinate systems provide a common reference system for surveys in large regions, even entire states (see Chapter 20). Several advantages result from using them on property surveys. One of the most significant is that they greatly facilitate the relocation of lost and obliterated corners. Every monument, which has known state plane coordinates, becomes a "witness" to other corner markers whose positions are given in the same system. State plane coordinates also enable the evaluation of adjoiners with less fieldwork. As cities and counties develop *land information systems* (LISs) and *geographic information systems* (GISs) (see Chapter 28), descriptions by coordinates are becoming more common. Many local and state governments currently require state plane coordinates on the corners of recorded subdivision plat boundaries, and on the exteriors of large boundary surveys. It is only a matter of time before state plane coordinates are required on all descriptions of land.

A coordinate description of corners may be used alone, but is usually prepared in conjunction with an alternative method. An example of a description by coordinates of a parcel in California follows.

> A parcel of tide and submerged land, in the state-owned bed of Seven Mile Slough, Sacramento County, California, in projected Section 10, T 3 N, R 3 E, Mt. Diablo Meridian, more particularly described as follows:
>
> BEGINNING at a point on the southerly bank of said Seven Mile Slough which bears S62°37′ E, 860 feet from a California State Lands Commission brass cap set in concrete stamped "JACK 1969," said point having coordinates of X = 2,106,973.68 and Y = 164,301.93 as shown on Record of Survey of Owl Island, filed October 6, 1969, in Book 27 of Surveys, Page 9, Sacramento County Records, thence to a point having coordinates of X = 2,107,196.04 and Y = 164,285.08; thence to a point having coordinates of X = 2,107,205.56, Y = 164,410.72; thence to a point having coordinates of X = 2,106,983.20, Y = 164,427.57; thence to the point of beginning. Coordinates, bearings, and distances in the above description are based on the California Coordinate System, Zone II.

When the preceding description was prepared, the writer could not have anticipated that an ambiguity would later arise concerning the datum of reference. From the dates given in the description, of course it can be concluded that the coordinates refer to NAD27. In preparing coordinate descriptions nowadays, the datum upon which the coordinates are based should be identified as being either in NAD27 or NAD83, as appropriate, to avoid any future confusion.

Earthquakes in Alaska, California, and Hawaii, and the subsidence caused by the withdrawal of oil and groundwater in many states, have caused ground shifts that move corner monuments and thereby change their coordinates. If undisturbed relative to their surroundings, the monuments, rather than the coordinates, then have greater weight in ownership rights.

■ 21.7 RETRACEMENT SURVEYS

Retracement surveys are run for the purpose of relocating or reestablishing previously surveyed boundary lines. *They are perhaps the most challenging of all types of surveys.* The rules used in retracement surveys are guided by case law and as such can vary from state to state and with time. However, the fundamental precept governing retracement surveys is that the monuments as originally placed and agreed to by the grantee and grantor constitute the correct boundary location. *The objective of resurveys therefore is to restore boundary markers to their original locations, not to correct them,* and this should guide all of the surveyor's actions.

In making a retracement survey, written evidence of title for the parcel involved should first be obtained. This will normally be in the form of a deed, but could also be obtained from an abstract or title policy. Even if the deed is available, it is good practice to trace it back to its creation in order to ensure that no transcription errors have been made and to check for possible modifiers (such as the term "surface measure"). Deeds of all adjoining properties should also always be obtained and matched to (1) determine if any gaps or overlaps exist and (2) understand any junior-senior rights (defined later on) that might possibly apply. The possibility that the written title could be supplanted by an *unwritten conveyance* must also be investigated. In the absence of any alterations of the written title by unwritten means, an evaluation of all evidence related to the written conveyance should be made in order to properly establish the property boundaries.

Various types of evidence are considered and used when retracing boundary surveys. When conflicts exist between the different types, the order of importance, or weight, generally assigned in evaluating that evidence, is as follows:

1. *Senior rights.* When parcels of land are conveyed in sequence, the one created first (senior) receives all that was specified in the written documents, and in case of any overlap of descriptions, the second (junior) receives the remainder. In case of overlaps in other subsequent conveyances, the elder receives the benefits.
2. *Intent of the parties.* The intent of the grantee and grantor at the time of conveyancing must be considered in resurveys. Usually the best evidence of intent is contained in the written documents themselves.
3. *Call for a survey.* If the written documents describe a survey, an attempt should be made to locate the stakes or monuments placed as a result of that survey.
4. *Monuments.* If the written documents describe original monuments that were set to mark the boundaries, these must be searched out. When there are conflicts in monuments, natural ones such as trees or streams receive more weight than artificial ones such as stakes and iron pipes.
5. *Measurements.* Courts have consistently ruled that measurements called for in a description merely describe the positions of the corners. Consequently they generally receive the least weight in interpreting a conveyance. When measurements are evaluated as evidence, the order of importance that generally applies to them is (1) distance, (2) direction, (3) area, and (4) coordinates. However, in some states, the order of distances and directions is reversed.

It should be noted that the foregoing are "general rules" for evaluating conflicting evidence, but it is possible, in certain circumstances, for a supposedly inferior element to control a superior one.

Good judgment is especially important when old lines are being restored and the original corners are lost. Then all possible evidence that applies to the original location must be found and a decision made as to what evidence is good and what can or should be discarded. All found evidence should be recorded in the resurvey field notes and reasons for using or rejecting any of it noted. Then if the survey is questioned in the courts, all actions taken can be justified. Familiarity with state and local laws and past court decisions affecting surveys in an area is valuable when making evaluations of evidence.

A corner that has been preserved is the best evidence of the original location of a line in question. The original notes are used as guides in locating these monuments, but a marker's actual position is the governing factor. Thus if a monument is found that is and has been accepted for years as the location of a particular corner, the surveyor should also accept it.

A monument should not be assumed lost unless every possible source of information has been exhausted and no trace of it can be found. Even then the surveyor should hesitate before disturbing settled possessions. It may be possible, for example, that where one or more corners are lost, all concerned parties have acquiesced in lines or corners based upon some other corner or landmark. These acquiesced corners may not be the original ones, but it would be unreasonable to discredit them when the people concerned do not question them. In a legal controversy, the law as well as common sense will normally declare that a boundary line, long acquiesced in, is better evidence of where the real line should be than one established by a survey done long after the original monuments have disappeared.

In retracement surveys, a determination must be made as to whether directions cited in the documents are astronomic or magnetic. If they are magnetic, the declination at the time of the original survey must be determined so that astronomic bearings or azimuths can be determined and lines retraced following them. A good assumption to adopt in retracing lines is that the boundaries are where the description says they are. If they aren't, some indications of their actual locations will probably be found when the distances and directions in the description are followed.

Testimony of persons who remember boundary locations is always valuable, but not always reliable. Therefore when such testimony is taken, a careful search must be made to find some corroborative evidence. Old fences or decayed wood at the location where a stake was purportedly originally set are examples of extremely valuable corroborative evidence. When two possibilities exist for establishing a boundary, evidence for rejecting one is often as important as evidence for accepting the other.

In retracement surveys, measurements made with a total station instrument between found monuments may not agree with distances on record. This situation provides a real test for surveyors. Perhaps the original chain or tape had an uncorrected systematic error, or a mistake was made. Alternatively, the markers may have been disturbed, or are the wrong ones. If differences exist between distances of record and measured values for found monuments, it may be helpful to determine a "scale factor,"[3] which relates original recorded distances to actual

---

[3]A "scale factor" can be obtained by carefully measuring the distance between two found monuments whose positions appear to be undisturbed and reliable, and dividing the distance between these monuments as recorded on the original survey by the measured distance. As an example, if the recorded distance between two found monuments was 750.00 ft, and the measured distance was 748.62 ft, then the scale factor would be $748.62/750.00 = 0.99816$.

measurements. In searching for monuments and evidence, this scale factor can then be applied to lay off the actual distances noted by the original surveyor.

Measurements may indicate that bearings and angles between adjacent sides also do not fit those called for in the writings. These discrepancies could be due to a faulty original compass survey, incorrect corner marks, or other causes. If, while conducting a resurvey, discrepancies are found between adjacent descriptions that cannot be resolved, they should be called to the attention of the owners so that steps can be taken to harmonize the land actually owned with that contained in the titles. Once the location of the original boundary is decided upon, it should be carefully marked and witnessed so that it can be easily found in the future.

21.8 SUBDIVISION SURVEYS

Subdivision surveys consist in establishing new smaller parcels of land within larger previously surveyed tracts. In these types of surveys, one or only a few new parcels may be created, in which case they may be described using the metes-and-bounds system. Conversely, in areas where new housing is planned, a block-and-lot subdivision survey can be conducted, thus creating many small lots simultaneously. Laws governing subdivision surveys vary from state to state and surveyors must follow them carefully in performing these types of surveys.

Whether a single new lot is created or a block-and-lot survey performed, generally a closed traverse is first run around the larger tract, with all corners being occupied if possible. Fences, trees, shrubbery, hedges, common (party) walls, and other obstacles may necessitate running the traverse either inside or outside the property. If corners cannot be occupied, stub (side shot) measurements can be made to them and their coordinates computed, from which side lengths and bearings can be computed (see Section 10.12). All measurements should be made with a precision suited to the specifications and land values.

Before traversing around the larger tract, it is necessary to first establish a reference direction for one of its lines. Also, if state plane coordinates are to be used in the new parcel's description, starting coordinates must be determined for one of the parcel's corners. A reference direction can be established by making an astronomic observation. Alternatively, as illustrated in Figure 21.3, both a reference direction and coordinates can be transferred to the parcel by including one of its lines in a separate traverse initiated at existing nearby control monuments.

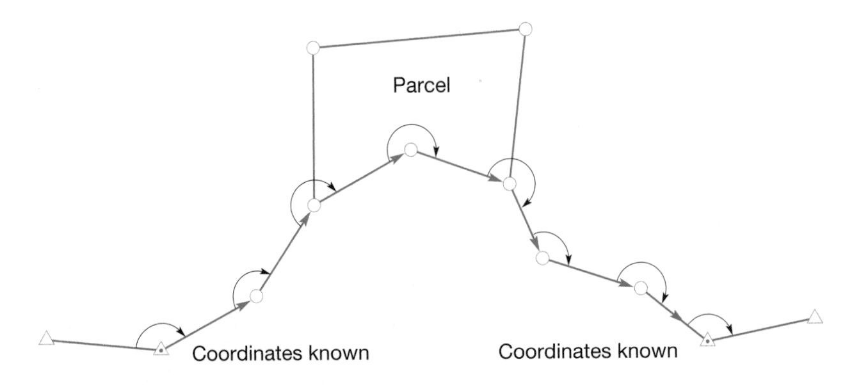

Figure 21.3 Transfer of direction and coordinates to a parcel by traverse.

A check is secured on the traverse by closing back on the starting station, or on a second nearby monument, as indicated in the figure.

After the traverse around the larger exterior parcel has been completed and adjusted, the new parcels can be surveyed. Where a single new parcel is being created, its corners are established according to the owner's specifications within requirements of the statutes. From the survey, the parcel description is prepared, certified, and recorded.

Block-and-lot subdivisions must not only conform to state statutes, but in addition many municipalities have laws covering these types of surveys. Regulations may specify minimum lot size, allowable misclosures for surveys, types of corner marks to be used, minimum width of streets and the procedure for dedicating them, rules for registry of plats, procedures for review, and other matters. Often several jurisdictions and agencies, each with its own laws and regulations, may have authority over the subdivision of land. In these cases, if standards conflict, the most stringent usually applies. The mismatched street and highway layouts of today could have largely been eliminated by suitable subdivision regulations and by a thorough review of them in past years.

A portion of a small block-and-lot subdivision is shown in Figure 21.2. (Some lot areas have been deleted so that their calculations can become end-of-chapter problems.) Computers with appropriate software such as coordinate geometry (see Chapter 11) and CAD greatly reduce the labor of computing subdivisions. They are especially valuable for large plats and designs with curved streets. Automatic plotters using stored computer files make drafting the final plat map accurate, simple, and fast.

Critical subdivision design and layout considerations include creating good building sites, an efficient street and utility layout, and assured drainage. Furthermore, subdivision rules and regulations must be followed and the developer's desires met as nearly as possible. A subdivision project involves a survey of the *exterior* boundaries of the tract to be divided, followed by a topographic survey, design of the subdivision, and layout of the interior of the tract. Following is a brief outline of the steps to be accomplished in these procedures.

1. Exterior survey
   (a) Obtain recorded deed descriptions of the parent tract of land to be subdivided, and of all adjoiners, from the Register of Deeds office. Note any discrepancies between the parent tract and its adjoiners.
   (b) Search for monuments marking corners of the parent tract, those of its adjoiner's where necessary, and, where appropriate, for U.S. Public Land Survey monuments to which the survey may be referred or tied. Resolve any discrepancies with adjoiners.
   (c) Make a closed survey of the parent tract and adequately reference it to existing monuments.
   (d) Compute departures, latitudes, and misclosures to see whether the survey meets requirements. Balance the survey if the misclosure is within allowable limits.
   (e) Resolve, if possible, any encroachments on the property or differences between occupation lines and title lines, so there will be no problems later with the final subdivision boundary.

2. Interior survey, design, and layout
   (a) Perform a topographic survey of the area within the tract. Include existing utility infrastructure, which may extend into the subdivision. From the survey data, prepare a topographic map for evaluating drainage, and for use in preliminary street layout and lot design.
   (b) Develop a preliminary plat showing the streets and the blocks and lots into which the tract is to be subdivided. Compute departures and latitudes on every block and lot to ensure their mathematical closures.
   (c) Obtain the necessary approvals of the preliminary plat.
   (d) Prepare a final plat map that will conform to state and city platting regulations, and include whatever certificates may be required.
   (e) Set block and lot corners of the tract according to the approved preliminary plat. Block corners should be located first, and lot corners set by measuring along lines between block ends. (In some cases final stakeout may be delayed until utilities are placed so corners are not destroyed in this process.)
   (f) Have the certificates executed and witnessed and the final plat approved and recorded.

21.9 PARTITIONING LAND

A common problem in property surveys is partitioning land into two or more parcels for sale or distribution to family members, heirs, etc. Prior to partitioning, a boundary survey of the parent tract is run, departures and latitudes computed, the traverse balanced, and the total enclosed area calculated. Computational procedures involved in partitioning land vary depending on conditions. Some parent tract shapes and partitioned parcel requirements permit formula solutions, often by using analytical geometry. Others require trial-and-error methods. These procedures were discussed in Chapter 11 and Section 12.8 where examples were computed to illustrate the procedures.

Typical partitioning problems consist of cutting a certain number of acres from a parent parcel, or dividing the parent parcel into halves, thirds, and so on. Required *cutoff lines* to separate a certain area from the parcel may have (1) a specific starting point (distance from one corner of the tract polygon and run to the midpoint, or any other location on the opposite side) or (2) a required direction (parallel with, perpendicular to, or on a designated bearing angle from a selected line). These cases can often be handled by trial-and-error solutions involving an initial assumption such as the cutoff line direction or the starting point. Certain problems are amenable to solution using coordinate formulas for the intersection of two lines (see Chapter 11).

Other partitioning problems may call for dividing a parcel into an easterly and westerly half, or a northerly and southerly half. But such descriptions could be ambiguous as illustrated in Figure 21.4. This figure shows that a statement "the southerly half" of tract *ABCD* can have a number of meanings—the most southerly half, half the frontage, or half the actual acreage. An important consideration in land partitioning is the final shape of each lot. Connecting midpoints *G* and *H* leaves the southerly "half" smaller than the northerly "half," but provides equal

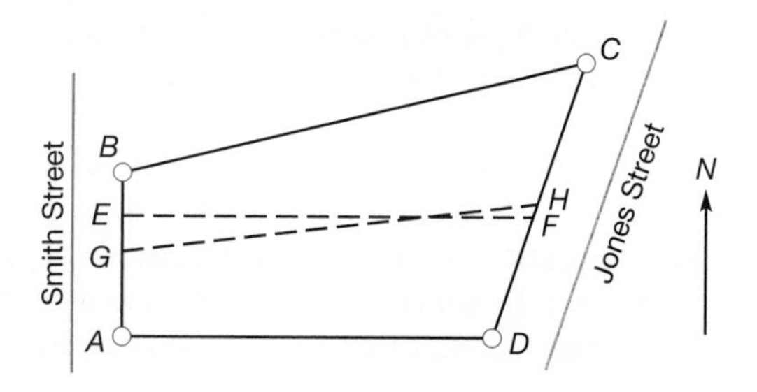

Figure 21.4
"The southerly half."

frontage for both parts on the two streets. Course *EF* parallel with *AD*, produces one trapezoidal lot, but a poorly shaped northerly parcel with meager frontage on Smith Street. In either case, the intent of the deed must be clearly stated. This is best accomplished by describing the dividing line by metes-and-bounds, which clearly depicts the intent.

21.10 REGISTRATION OF TITLE

To remedy difficulties arising from inaccurate descriptions and disputed boundary claims, some states provide for registration of property titles under rigid rules. The usual requirements include marking each corner with standardized monuments referenced to established points and recording a plat drawn to scale and containing specified items. The court under certain conditions then guarantees titles.

A number of states have followed Massachusetts' example and maintain separate land courts dealing exclusively with *land titles*. As the practice spreads, the accuracy of property surveys will be increased and transfer of property simplified.

Title insurance companies search, assemble, and interpret official records, laws and court decisions affecting ownership of land, and then insure purchasers against loss regarding title defects and recorded liens, encumbrances, restrictions, assessments, and easements. Defense in lawsuits is provided by the company against threats to a clear title from claims shown in public records and not exempted in the policy. The locations of corners and lines are not guaranteed; hence, it is necessary to establish, on the ground, the exact boundaries called for by the deed and title policy. Close cooperation between surveyors and title insurance companies is necessary to prevent later problems for their clients.

Many technical and legal problems are considered before title insurance is granted. In some states, title companies refuse to issue a policy covering a lot if fences in place are not on the property line and exclude from the contract "all items that would be disclosed by a property survey."

To guide surveyors in their conduct of land title surveys, the American Land Title Association (ALTA) and the National Society of Professional Surveyors (NSPS) have established a set of standards. Named the "ALTA/ACSM Land Title Surveys,"[4] they set forth concise guidelines on what must be included in property surveys for title insurance purposes and also state the following regarding point positional tolerance: *Relative Positional Accuracy* may be tested by: (1) comparing the

[4]A copy of the current ALTA/ACSM Land Title Surveys is available at http://www.acsm.net/alta.html.

relative location of points in a survey as measured by an independent survey of higher accuracy or (2) the results of a minimally constrained, correctly weighted least square adjustment of the survey. *Allowable Relative Positional Accuracy for Measurements Controlling Land Boundaries on ALTA/ACSM Land Title Surveys* is ±[0.07 ft (20 mm) + 50 ppm] at the 95% confidence level. Other accuracy criteria specified relate to the required precisions of instruments used and acceptable field procedures. Benefits derived from these guidelines are clarification of the exact requirements of land title surveys so that uniformly high-quality results are obtained.

■ 21.11 ADVERSE POSSESSION AND EASEMENTS

Adverse rights can generally be applied to gain ownership of property by occupying a parcel of land for a period of years specified by state law and performing certain acts.[5] To claim land or rights to it by *adverse possession,* its occupation or use must be (1) actual, (2) exclusive, (3) open and notorious, (4) hostile, and (5) continuous. It may also be necessary for the property to be held under *color of title* (a claim to a parcel of real property based on some written instrument, though a defective one). In some states all taxes must be paid. The time required to establish a claim of adverse possession varies from a minimum of 5 years in California to a maximum of 60 years for urban property in New York. The customary period is 20 years.

The occupation and use of land belonging to a neighbor, but outside his or her apparent boundary line as defined by a fence, may lead to a claim of adverse possession. Continuous use of a street, driveway, or footpath by an individual or the general public for a specified number of years results in the establishment of a right-of-way privilege, which cannot be withheld by the original owner.

An *easement* is a right, by grant or agreement, which allows a person or persons to use the land of another for a specific purpose. It always implies an interest in the land on which it is imposed. *Black's Law Dictionary* lists and defines 18 types of easements; hence, the exact purpose of an easement should be clearly stated. The discussion of property surveys has necessarily been condensed in this text, but it provides helpful information to readers while deterring inexperienced people from attempting to run boundary lines. For more extensive coverage, references are listed in the Bibliography.

■ 21.12 CONDOMINIUM SURVEYS

The word condominium is derived from the prefix *con* meaning "together," and from classical Roman law, *dominium* meaning "ownership." In the United States, the term *condominium* refers to a type of property ownership, where individual units within a multiple-unit building are owned separately. Every unit owner receives a deed describing their property, and is able to buy, mortgage, or sell their unit independent of the other owners. Thus, legal descriptions based on surveys are required. The condominium concept of ownership is relatively new in the United States, as compared with other countries. They have been present in Europe since the Middle Ages and appeared in this country in the later part of the 19th century.

---

[5]Except in a few special circumstances, adverse rights cannot be claimed against public lands.

The number of condominiums in the United States has been growing rapidly as more families discover the many benefits this type of living offers. Condominium ownership has tax advantages, investment benefits, and most of all, eliminates rent increases. This form of ownership can be an economical solution to rising land values, building costs, and maintenance expenses. It can also provide shared recreational facilities and other amenities that might otherwise be unaffordable.

A *condominium association* is the entity responsible for the operation of a condominium and the unit owners are members of the association. The document creating the association is called the *Articles of Incorporation,* which describes the purpose, powers, and duties of the association. The *By-laws* provide for the administration of the association, including meetings, quorums, voting, and other rules. The document that establishes a condominium is known as the *Condominium Declaration,* and once it is filed in the public records, the condominium is legally created. The Declaration contains important information including a legal description of the property, descriptions of the units, designation of *common elements* (those jointly owned and used by all units such as sidewalks, stairways, swimming pool, tennis courts, etc.), and identification of *limited common elements* (those reserved for the exclusive use of a particular unit such as a designated parking space). It also describes any *covenants* or restrictions on the use of the units, common elements, and limited common elements.

Although condominium ownership often applies to multi-story residential buildings, it is also used in commercial and industrial situations. The condominium concept has been applied to mobile home lots, travel trailer and camper sites, boat slips and docks, horse stables, shopping centers, and other types of properties. Special types of condominiums include: *timeshare,* in which the owner purchases an interest in a unit for a specified time period each year; *mixed-use,* which includes both residential and commercial units; and *multiple condominium community,* which is a development containing several separate condominiums that share a single common recreation area.

Condominium surveys differ from ordinary land surveys in several ways. They also bear many similarities with some property surveys, especially those for creating subdivision plats. Each state adopts statutory laws and promulgates rules that govern the procedures and requirements for creating condominiums. In many states, the statute is known as the *Condominium Act.* Preparation of the required materials for a condominium project is a joint effort, which typically includes an architect, engineer, attorney, and surveyor. The architect prepares the building plans and specifications; the engineer designs the construction plans; and the attorney creates the legal documentation for the condominium and the association. The surveyor assembles necessary information; prepares the required condominium plat, graphic plans, and descriptions; and performs the surveys needed for describing the parcel boundary and for locating the "as-built" improvements. It is important for the information shown on the graphic plans to agree with the provisions described in the Declaration.

Figures 21.5 through Figure 21.8 illustrate a four-sheet sample set of plans for a proposed multi-story residential condominium. The main purpose of these graphic plans is to clearly and accurately represent the locations of the units, common areas, and limited common areas of the condominium parcel. Figure 21.5 is the *Boundary*

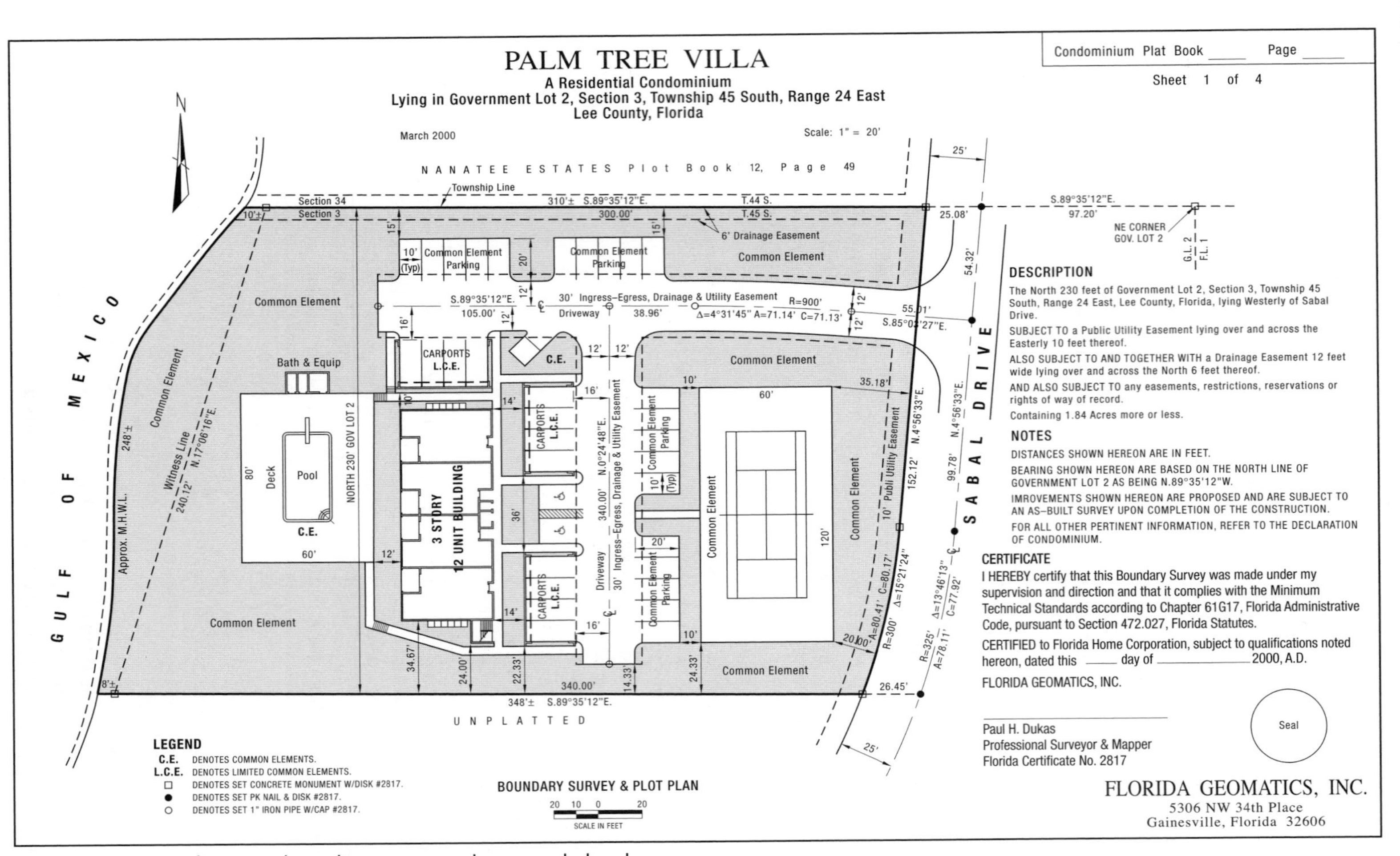

Figure 21.5 Condominium boundary survey and proposed plot plan.

Survey and Plot Plan. It shows the exterior boundary survey of the condominium parcel, gives a description for the parcel, and locates the proposed improvements with dimensions given from parcel boundary lines. Also included on this diagram are some general notes and the Surveyor's *Certificate* for the boundary survey. The Boundary Survey and Plot Plan is typically a small-scale drawing and is therefore generally unsuitable for showing sufficient detail and dimensions for all improvements. Thus, it is necessary to attach additional sheets at larger scales.

Figure 21.6 is the *Building and Carport Plot Plan*. It is a larger-scale graphic that not only depicts the residential building and carport areas, but also delineates the common and limited common elements and labels the individual units with identification numbers. Again, some general notes are included for clarification purposes. Although the scale of this drawing is larger than the Boundary Survey and Plot Plan, it is still too small to effectively show necessary details and dimensions for the individual unit areas. Thus, *Typical Unit Plans* are prepared at a still larger scale, as shown in Figure 21.7. These show the interior floor plans of the units, and their *perimetrical boundaries* (the perimeter or horizontal dimensions encompassing the vertical planes of the interior surfaces of the walls bounding the unit). In addition, this drawing includes a *Typical Wall Section* showing the elevation of the ground floor and heights of the units and building. Typically, the elevation of the ground floor of the building is referenced to a well-established vertical datum. Also included on the plan is a description of the boundaries for the units, and some general notes.

Figure 21.8 details a *Typical Carport and Storage Plan,* with dimensions showing the sizes of the storage and parking areas assigned to individual units. A tabular list of the *Undivided Share of Common Elements* for each unit is also shown on this sample plan, but alternatively it could be included only in the declaration. Computation of the undivided shares is sometimes prorated on the basis of a unit's area to the total area of all units. Another method uses the number of bedrooms in the unit relative to the total number of bedrooms for all units. Figure 21.8 also shows an *As-Built Dimension Table* used to record the actual field measurements of selected portions of each unit or limited common element. An ideal time to measure as-built dimensions is during construction just after the exterior unit walls are completed, but before the interior room partitions are added. Not all as-built dimensions are tabularized. Some measurements, such as the building ties from the boundary lines, may be revised directly on the appropriate plan sheets to reflect the as-built location of the building. If the difference between a measured distance and its corresponding proposed distance is within the construction tolerance, then the dimension need not be revised. For example, if the measured width of a driveway pavement was within ± 0.1 ft of the proposed width, then the proposed dimension generally would not be revised. Should the size or location of an improvement differ substantially from that proposed, then it would be changed on the graphic to reflect the as-built condition.

The last item on Figure 21.8 is the sample *Surveyor's Certificate*. This is executed only upon "substantial completion" of construction of the proposed improvements. Definitions for "substantial completion" vary. As a general rule, if a local building department issues a certificate of occupancy or other similar permit, then this signifies the construction of improvements as being substantially

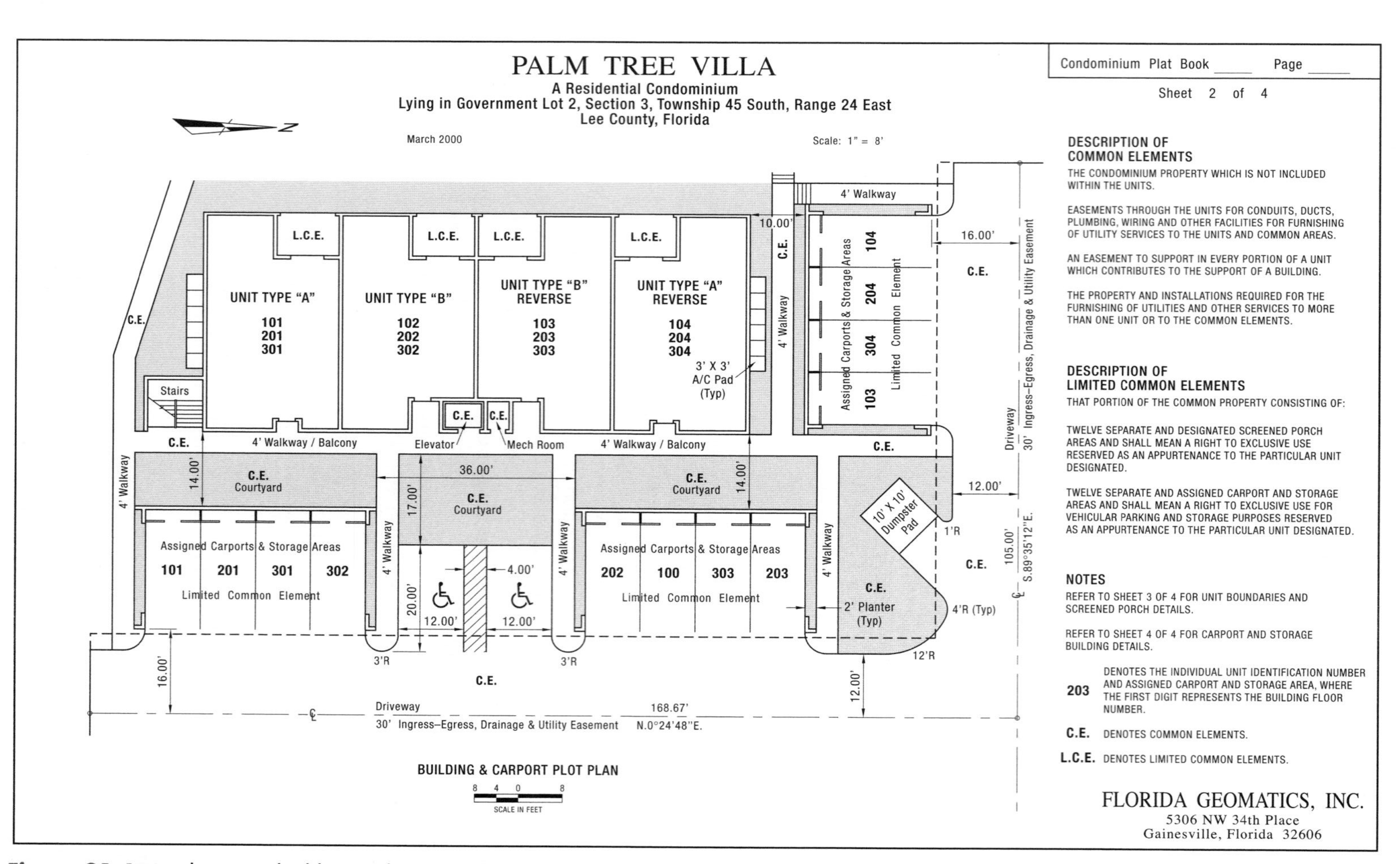

Figure 21.6 Condominium building and carport plan.

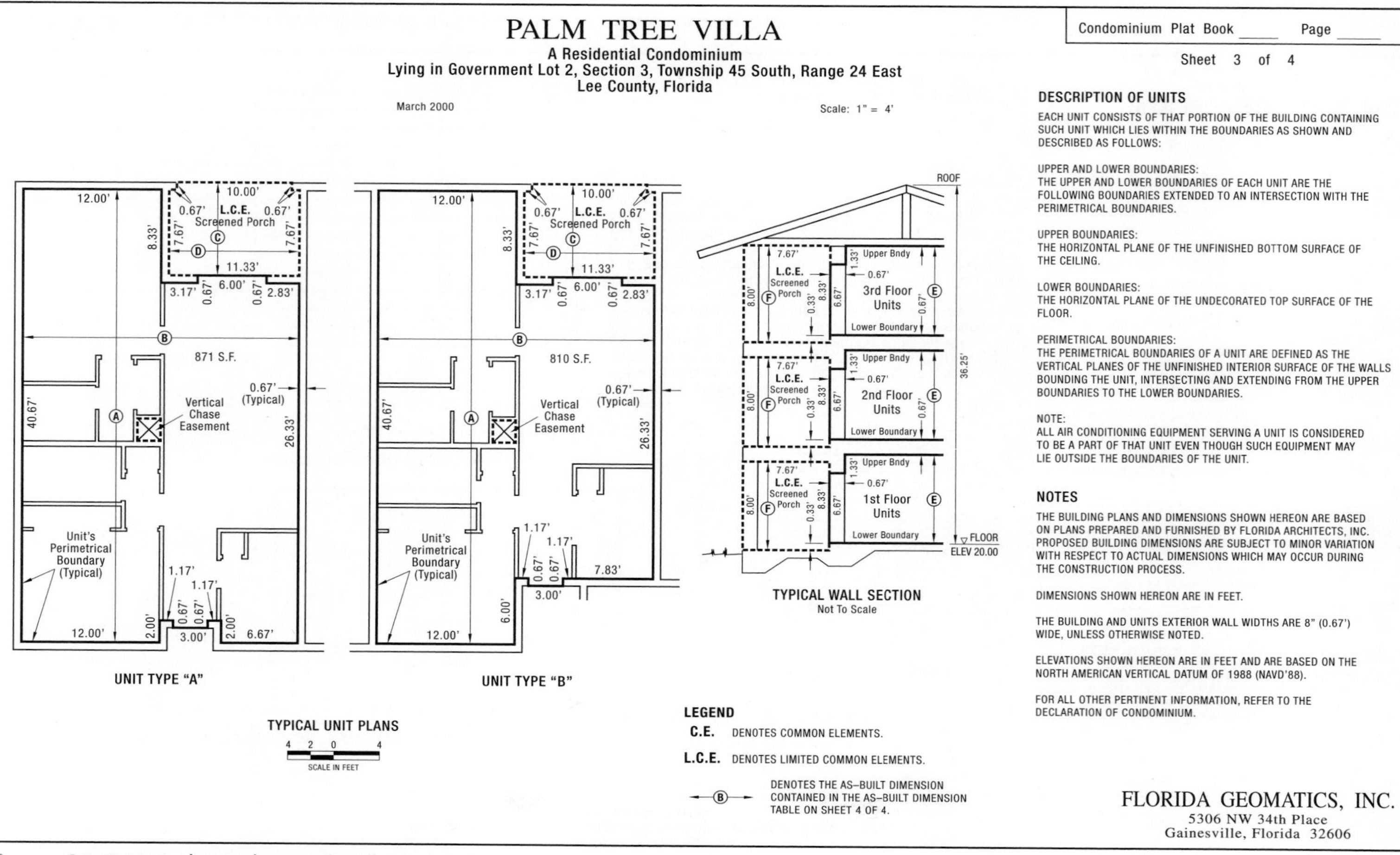

Figure 21.7 Typical unit plans and wall section.

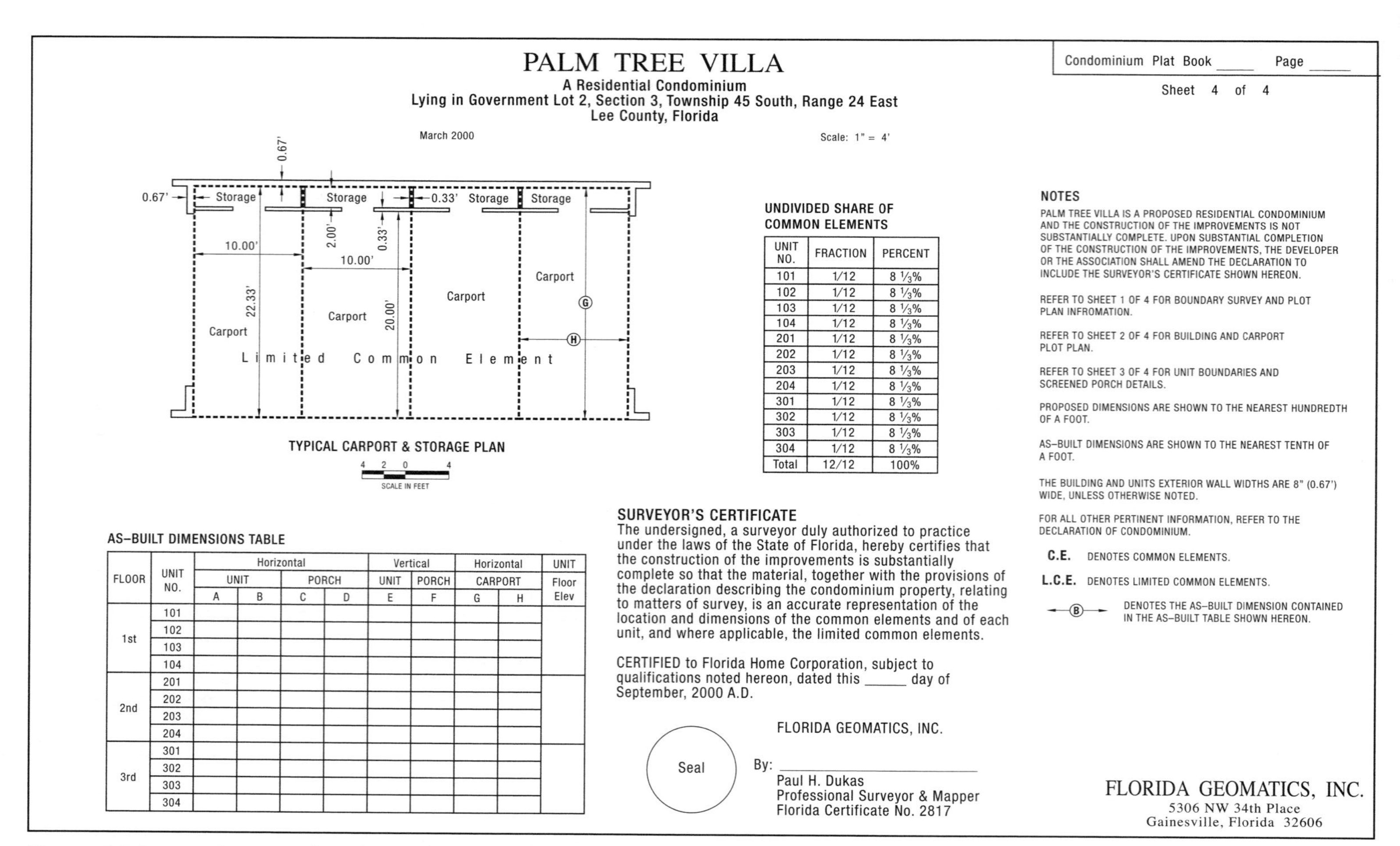

Condominium Plat Book _____ Page _____

Sheet 4 of 4

PALM TREE VILLA

A Residential Condominium
Lying in Government Lot 2, Section 3, Township 45 South, Range 24 East
Lee County, Florida

March 2000

Scale: 1" = 4'

TYPICAL CARPORT & STORAGE PLAN

UNDIVIDED SHARE OF COMMON ELEMENTS

UNIT NO.	FRACTION	PERCENT
101	1/12	8 1/3%
102	1/12	8 1/3%
103	1/12	8 1/3%
104	1/12	8 1/3%
201	1/12	8 1/3%
202	1/12	8 1/3%
203	1/12	8 1/3%
204	1/12	8 1/3%
301	1/12	8 1/3%
302	1/12	8 1/3%
303	1/12	8 1/3%
304	1/12	8 1/3%
Total	12/12	100%

NOTES

PALM TREE VILLA IS A PROPOSED RESIDENTIAL CONDOMINIUM AND THE CONSTRUCTION OF THE IMPROVEMENTS IS NOT SUBSTANTIALLY COMPLETE. UPON SUBSTANTIAL COMPLETION OF THE CONSTRUCTION OF THE IMPROVEMENTS, THE DEVELOPER OR THE ASSOCIATION SHALL AMEND THE DECLARATION TO INCLUDE THE SURVEYOR'S CERTIFICATE SHOWN HEREON.

REFER TO SHEET 1 OF 4 FOR BOUNDARY SURVEY AND PLOT PLAN INFROMATION.

REFER TO SHEET 2 OF 4 FOR BUILDING AND CARPORT PLOT PLAN.

REFER TO SHEET 3 OF 4 FOR UNIT BOUNDARIES AND SCREENED PORCH DETAILS.

PROPOSED DIMENSIONS ARE SHOWN TO THE NEAREST HUNDREDTH OF A FOOT.

AS-BUILT DIMENSIONS ARE SHOWN TO THE NEAREST TENTH OF A FOOT.

THE BUILDING AND UNITS EXTERIOR WALL WIDTHS ARE 8" (0.67') WIDE, UNLESS OTHERWISE NOTED.

FOR ALL OTHER PERTINENT INFORMATION, REFER TO THE DECLARATION OF CONDOMINIUM.

C.E. DENOTES COMMON ELEMENTS.

L.C.E. DENOTES LIMITED COMMON ELEMENTS.

—(B)— DENOTES THE AS-BUILT DIMENSION CONTAINED IN THE AS-BUILT TABLE SHOWN HEREON.

AS-BUILT DIMENSIONS TABLE

FLOOR	UNIT NO.	Horizontal UNIT		Horizontal PORCH		Vertical UNIT	Vertical PORCH	Horizontal CARPORT		UNIT Floor Elev
		A	B	C	D	E	F	G	H	
1st	101									
	102									
	103									
	104									
2nd	201									
	202									
	203									
	204									
3rd	301									
	302									
	303									
	304									

SURVEYOR'S CERTIFICATE

The undersigned, a surveyor duly authorized to practice under the laws of the State of Florida, hereby certifies that the construction of the improvements is substantially complete so that the material, together with the provisions of the declaration describing the condominium property, relating to matters of survey, is an accurate representation of the location and dimensions of the common elements and of each unit, and where applicable, the limited common elements.

CERTIFIED to Florida Home Corporation, subject to qualifications noted hereon, dated this _____ day of September, 2000 A.D.

FLORIDA GEOMATICS, INC.

Seal

By: ______________________

Paul H. Dukas
Professional Surveyor & Mapper
Florida Certificate No. 2817

FLORIDA GEOMATICS, INC.
5306 NW 34th Place
Gainesville, Florida 32606

Figure 21.8 Typical carport plan, shares of common element, as-built data, and certificate.

complete. However, if the certificate of occupancy is issued only for the building and a proposed improvement (for example, the pool and deck) is not substantially complete, then the surveyor's certificate should exclude those improvements from the certificate and label those amenities as "under construction" or "proposed."

Because this sample condominium is in the proposed stage, a statement of that fact is necessary. Note that this statement appears in the "Notes" of both Figures 21.5 and 21.8. For convenience many Declarations include a reduction of the full-sized condominium plat and graphic plans. In this case, careful consideration of the graphic scales, text sizes, and line weights should be exercised when preparing the original full-sized drawings.

■ 21.13 GEOGRAPHIC AND LAND INFORMATION SYSTEMS

As noted in Chapter 28, surveyors are playing a major role in the development and implementation of modern *Geographic Information Systems* (GISs) and *Land Information Systems* (LISs), and this activity will continue in the future. These systems include computerized data banks that contain descriptive information about the land such as its shape, size, location, topography, ownership, easements, zoning, flood plain extent if any, land use, soil types, existence of mineral and water resources, and much more. The information is available for rapid retrieval and is invaluable to surveyors, government officials, lawyers, developers, planners, environmentalists, and others.

Boundary surveys are fundamental to GISs and LISs since knowledge about the land is meaningless unless its position on Earth is specified. In most modern systems being developed, positional data are established by associating attributes about the land with individual lots or tracts of ownership. Perhaps the most fundamental information associated with each individual parcel is its legal description, which gives the unique location on the Earth of the parcel and thus provides the positional information needed to support the system. As described in previous sections, legal descriptions are written instruments, based on measurements and prepared to exacting standards and specifications. Thus the land surveyor's role in modern LISs and GISs is an important one.

■ 21.14 SOURCES OF ERROR IN BOUNDARY SURVEYS

Some sources of error in boundary surveys follow:

1. Errors in measured distances and directions.
2. Corners not defined by unique monuments.
3. Judgment errors in evaluating evidence.

■ 21.15 MISTAKES

Some typical mistakes in connection with boundary surveys are:

1. Failure to perform closed traverse surveys around parcels or not closing on a control station.
2. Not properly adjusting errors of closure.
3. Use of the wrong corner marks.

4. Failure to check deeds of adjacent property as well as the description of the parcel in question.
5. Failure to describe "intent" in deed descriptions, or preparing ambiguous deed descriptions.
6. Omission of the length or direction of the closing line in parcel descriptions.
7. Magnetic bearings not properly corrected to the date of the new survey.

PROBLEMS

Asterisks (*) indicate problems that have answers given in Appendix E.

21.1 Define the following terms:
(a) Retracement surveys
(b) Subdivision surveys
(c) Color of title
(d) Adverse possession

21.2 What are the essential elements of metes-and-bounds property descriptions?

21.3 Visit your county courthouse and obtain a copy of a metes-and-bounds property description. Write a critique of the description, with suggestions on how the description could have been improved.

21.4 From the publications listed in the Bibliography, find the meaning of prescription, appurtenance, eminent domain, and escrow.

21.5 Write a metes-and-bounds description for the exterior boundary of the condominium parcel shown in Figure 21.5.

21.6 Write a metes-and-bounds description for the house and lot where you live. Draw a map of the property.

21.7 What are the advantages of the block and lot system of describing property?

21.8 From the metes-and-bounds description of the lot in the town of Little Wolf, described in Section 21.4, compute the lot's misclosure.

21.9 What setbacks from property lines are required in the front, rear, and on the sides in the residential areas of your city?

21.10 What major advantage does the coordinate method of property description have over other methods?

21.11 What is the primary objective in performing retracement surveys?

21.12* List in their order of importance the following types of evidence when conducting retracement surveys: (a) measurements, (b) call for a survey, (c) intent of the parties, (d) monuments, and (e) senior rights.

21.13 In performing retracement surveys, list (in their order of importance), the four different types of measurements called for in a description for your state.

21.14 List in order the steps that must be performed in making subdivision surveys.

21.15 Identify all types of pertinent information or data that should appear on the plat of a completed property survey.

21.16 Utility poles occupy two corners of a rectangular lot. Explain how to survey the lot, locate boundaries, and find the area.

21.17 Two disputing neighbors employ a surveyor to check their boundary line. Discuss the surveyor's authority if (a) the line established is agreeable to both clients and (b) the line is not accepted by one or both of them.

21.18 What time period is required to claim adverse rights to property in your state?

21.19 Compute the misclosures of lots 16 and 17 in Figure 21.2. On the basis of your findings, would this plat be acceptable for recording? Explain.

21.20* Compute the areas of lots 18 and 19 of Figure 21.2.

21.21 Determine the misclosure of lot 50 of Figure 21.2, and compute its area.

21.22 For the accompanying figure; using a line perpendicular to AB through x, divide the parcel into two equal parts, and determine lengths xy and By.

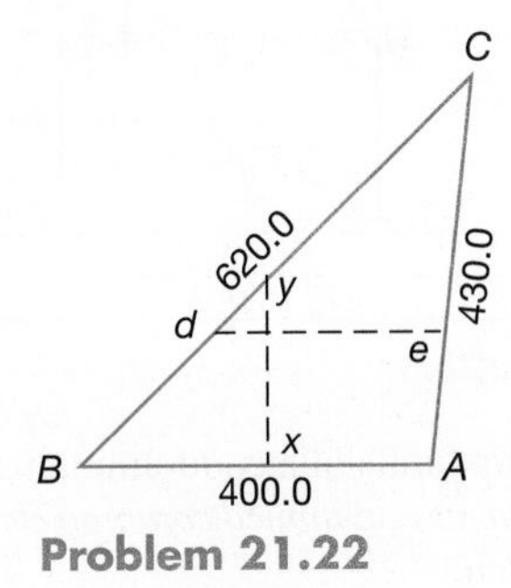

Problem 21.22

21.23 For the figure of Problem 21.22, calculate the length of line de, parallel to BA, which will divide the tract into two equal parts. Give lengths Bd, de, and eA.

21.24 Prepare a metes-and-bound description for the parcel shown in the accompanying figure. Assume all corners are marked with 1-in.-diameter steel rods and a 20-ft meander line setback from Indian Lake.

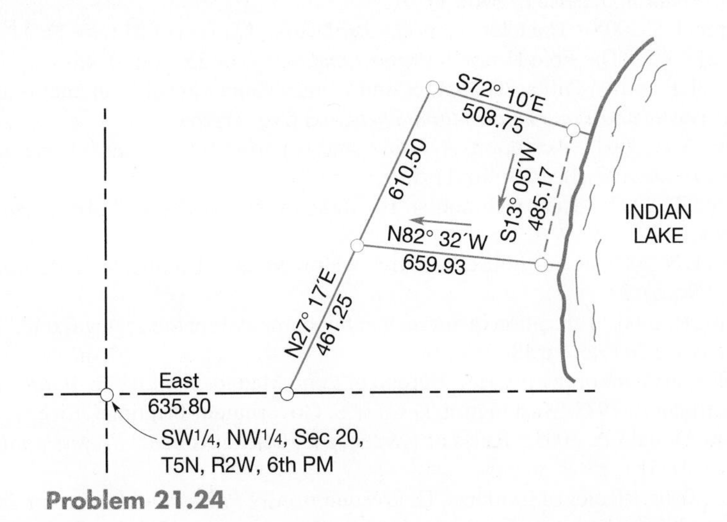

Problem 21.24

21.25 Draw a plat map of the parcel in Problem 21.24 at a convenient scale. Label all monuments and the lengths and directions of each boundary line on the drawing. Include a title, scale, north arrow, and legend.

21.26 Prepare a metes-and-bounds description for the property shown in the accompanying figure. Assume all corners are marked with 2-in. diameter iron pipes.

21.27 Create a 1-acre tract on the westerly side of the parcel in Problem 21.26 with a line parallel to the westerly property line. Give the lengths and bearings of all lines for both new parcels.

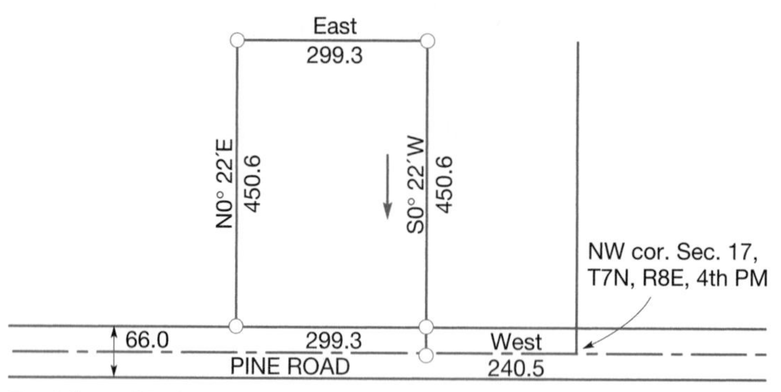

Problem 21.26

21.28 Discuss the ownership limits of a condominium unit.

21.29 Define *common elements* and *limited common elements* in relation to condominiums. Give examples of each.

21.30 What types of measurements are typically made by surveyors in performing work for condominium developments?

BIBLIOGRAPHY

Dronick, G. J., and W. W. Parks. 2007. "U.S. and Pennsylvania Law of Writings as Indispensable Evidence to Prove Ownership of Real Property." *Surveying and Land Information Science* 67 (No. 1): 51.

Estopinal, S. 2005. "The Class Action." *Professional Surveyor* 25 (No. 7): 44.

———. 2005. "The Free House." *Professional Surveyor* 25 (No. 6): 46.

Hartzell, P. and W. Parks. 2003. "U.S. and Pennsylvania Law of Monuments as Evidence." *Surveying and Land Information Science* 63 (No. 3): 169.

Kellie, A. C. 2004. "Accretion, Avulsion, and Riparian Boundaries." *Surveying and Land Information Science* 64 (No. 1): 5.

Liuzzo, T. 2007. "Encroachments: To State or Not to State." *American Surveyor* 4 (No. 1): 32.

Lucas, J. N. 2003. "Courthouse Research—How Much is Enough?" *Professional Surveyor* 23 (No. 5): 22.

Schultz, R. 2006. "Education in Surveying: Fundamentals of Surveying Exam." *Professional Surveyor* 26 (No. 3): 38.

U.S. Department of the Interior, Bureau of Land Management. 1973. *Manual of Surveying Instructions* 1973. Washington, D.C.: U.S. Government Printing Office.

Wilson, Donald A. 2005. "Rules of Evidence I: Judicial Notice." *Professional Surveyor* 25 (No. 3): 51.

———. 2005. "Rules of Evidence II: Presumptions." *Professional Surveyor* 25 (No. 5): 50.

———. 2005. "Rules of Evidence II: Exceptions to the Hearsay Rule." *Professional Surveyor* 25 (No. 9): 48.

———. 2005. "Rules of the Game: Rules for Investigation." *Professional Surveyor* 25 (No. 11): 46.

———. 2006. "Second Thoughts: Undoing a Survey." *Professional Surveyor* 26 (No. 1): 43.

22
Surveys of the Public Lands

22.1 INTRODUCTION

The term *public lands* is applied broadly to the areas that have been subject to administration, survey, and transfer of title to private owners under the public-lands laws of the United States since 1785. These lands include those turned over to the federal government by the colonial states and the larger areas acquired by purchase from (or treaty with) the Native Americans or foreign powers that had previously exercised sovereignty.

Thirty states, including Alaska, constitute the public-land survey states that have been, or are being, subdivided into rectangular tracts (see Figure 22.1). The area of these states represents approximately 72% of the United States. Title to the vacant lands, and therefore direction over the surveys within their own boundaries, was retained by the colonial states, the other New England and Atlantic coast states (except Florida), and later by the states of West Virginia, Kentucky, Tennessee, Texas, and Hawaii. In these areas the U.S. public-land laws have not been applicable.

The beds of navigable bodies of water are not public domain and are not subject to survey and disposal by the United States. Sovereignty is in the individual states.

The survey and disposition of the public lands were governed originally by two factors:

1. A recognition of the value of grid-system subdivision based on experience in the colonies and another large-scale systematic boundary survey—the 1656 Down Survey in Ireland.
2. The need of the colonies for revenue from the sale of public land. Monetary returns from their disposal were disappointing, but the planners' farsighted vision of a grid system of subdivision deserves commendation.

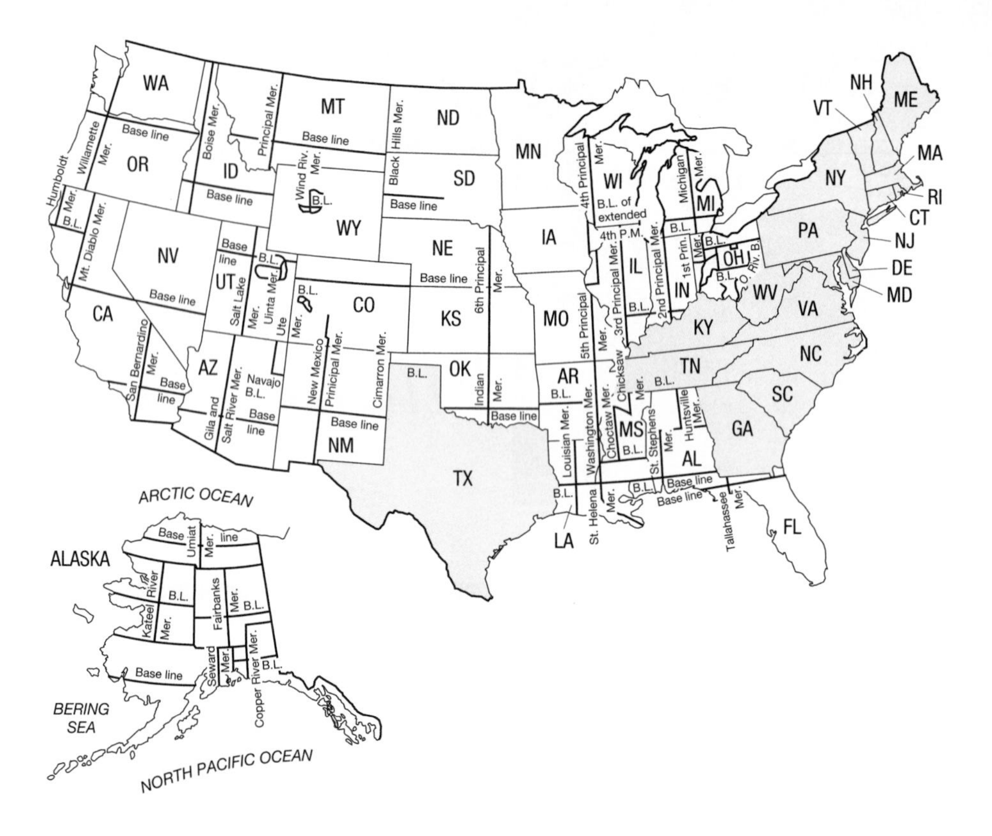

Figure 22.1 Areas covered by the public-lands surveys, with principal meridians shown. Areas excluded are shaded. (Hawaii, although not shown on this map, would also be shaded. Texas has a rectangular system similar to the U.S. public land system.)

Although nearly a billion acres of public land has either been sold or granted since 1785, approximately one-third of the area of the country is still federally owned. The U.S. *Bureau of Land Management* (BLM), within the Department of the Interior, was created in 1946 as a merger of the U.S. *Grazing Service* and the U.S. *General Land Office* and is responsible for surveying and managing a significant portion of these federal lands.

■ 22.2 INSTRUCTIONS FOR SURVEYS OF THE PUBLIC LANDS

The U.S. *Public Lands Survey System* (PLSS) was inaugurated in 1784 and the territory immediately northwest of the Ohio River in what is now eastern Ohio served as a test area. Sets of instructions for the surveys were issued in 1785 and 1796. Manuals of instructions were later issued in 1855, 1881, 1890, 1894, 1902, 1930, 1947, and 1973.[1]

In 1796 General Rufus Putnam was appointed as the first U.S. *Surveyor General* and at that time the numbering of sections changed to the system now in use (see the excerpt from the 1973 *Manual of Surveying Instructions* below, and

[1]The 1973 **Manual of Instructions** is available at http://pub4.ca.blm.gov:80/webmanual/id1_1.htm and begins at http://pub4.ca.blm.gov:80/webmanual/id2.htm.

Figure 22.8). Supplementary rules were promulgated by each succeeding surveyor general "according to the dictates of his own judgment" until 1836, when the General Land Office (GLO) was reorganized. Copies of changes and instructions for local use were not always preserved and sent to the GLO in Washington. As a result, no office in the United States has a complete set of instructions under which the original surveys were supposed to have been made.

Most of the later public-lands surveys have been run by the procedures to be described in this chapter, or variations of them. The task of present-day surveyors consists primarily of retracing the original lines set by earlier PLSS surveyors and/or further subdividing sections. To accomplish these tasks, they must be thoroughly familiar with the rules, laws, equipment, and conditions that governed their predecessors in a given area.

Basically, the rules of survey stated in the 1973 *Manual of Surveying Instructions* are as follows:[2]

> The public lands shall be divided by north and south lines run according to the true[3] meridian, and by others crossing them at right angles, so as to form townships six miles square. . . .
>
> The corners of the townships must be marked with progressive numbers from the beginning; each distance of a mile between such corners must be also distinctly marked with marks different from those of the corners.
>
> The township shall be subdivided into sections, containing as nearly as may be, six hundred and forty acres each, by running parallel lines through the same from east to west and from south to north at the distance of one mile from each other (originally at the end of every two miles but amended in 1800), and marking corners at the distance of each half mile. The sections shall be numbered, respectively, beginning with the number one in the northeast section, and proceeding west and east alternately through the township with progressive numbers until the thirty-six be completed.

Additional rules of survey covering field books, subdivision of sections, adjustment for excess and deficiency, and other matters are given in the manuals and special instructions. Private surveyors, on a contract basis, were paid $2/mi of line run until 1796 and $3/mi thereafter. Sometimes the amount was adjusted in accordance with the importance of a line, the terrain, location, and other factors. From this meager fee surveyors had to pay and feed a party of at least four while on the job and in transit to and from distant points. They had to brush out and *blaze* (mark trees by scarring the bark) the line, set corners and other marks, and provide satisfactory notes and one or more copies of completed plats. The contract system was completely discarded in 1910. Public-land surveyors are now appointed.

Since meridians converge, it is evident that the requirements that "lines shall conform to the true meridians and townships shall be 6 mi square" are mathematically impossible. An elaborate system of subdivision was therefore worked out as a practical solution.

---

[2]Since new surveys (particularly in Alaska) as well as retracements are still being carried out, the 1973 manual employs more than one tense, and this chapter does also.

[3]As previously noted, the true meridian may be taken as either astronomic or geodetic.

Two principles furnished the legal background for stabilizing lines on the land:

1. Boundaries of public lands established by duly appointed surveyors are unchangeable.
2. Original township and section corners established by surveyors must stand as the true corners they were intended to represent, whether or not in the place shown by the field notes.

Expressed differently, the original surveyors had an official plan with detailed instructions for its layout and presumably set corners to the best of their ability. After title passed from the United States, their established corners (monuments), regardless of errors, became the lawful ones. Therefore, if monuments have disappeared, *the purpose of resurveys is to determine where they were, not where they should have been*. Correcting mistakes or errors now would disrupt too many accepted property lines and result in an unmanageable number of lawsuits.

In general, the procedure in surveying the public lands provides for the following subdivisions:

1. Division into quadrangles (tracts) approximately 24 mi on a side (after about 1840)
2. Division of quadrangles into townships (16), approximately 6 mi on a side
3. Division of townships into sections (36), approximately 1 mi square
4. Subdivision of sections (usually by local surveyors)

It will be helpful to keep in mind that the purpose of the grid system was to obtain sections 1 mi on a side. To this end, surveys proceeded from south to north and east to west, and all discrepancies were thrown into the sections bordering the north and west township boundaries to get as many regular sections as possible.

Although the general method of subdivision outlined here was normally followed, detailed procedures were altered in surveys made at different times in various areas of the country. As examples, instructions for New Mexico said only township lines were to be run where the land was deemed unfit for cultivation and in Wisconsin the first four correction lines north of the baseline were 60 mi apart rather than 24.

Another current example relates to the surveys in Alaska, where the area's sheer vastness requires changes. When Alaska gained statehood in 1958, only 2 percent of its 375 million acres had been surveyed. Priorities were set for conducting the remaining surveys and plans developed that called for subdividing some 155 million acres to be transferred to that state and native Alaskans. To accelerate the project, the 18,651 townships in Alaska were first established on protraction diagrams and latitude and longitude determined for each corner. In executing the surveys, markers are being set at 2-mi intervals in most areas and GPS is being utilized extensively. But even with this modern technology and relaxed procedures, with so much area involved, it will take many years to complete the job.

Distances given in the instructions were in *chains* and miles. The particular chain referred to is the *Gunter's chain,* which was 66 ft long. It was selected for two reasons: (1) it was the best measuring device available to surveyors in the

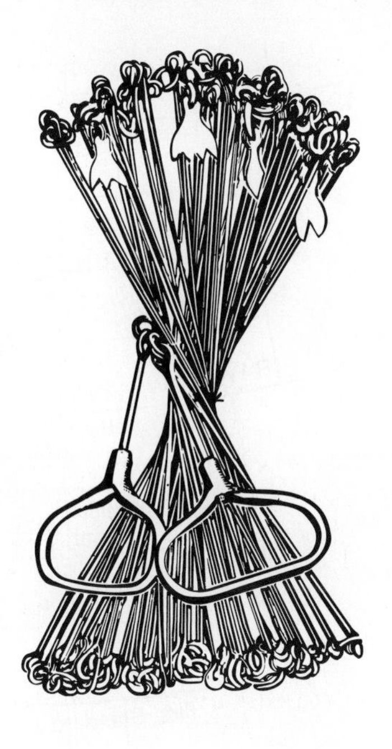

Figure 22.2
Gunter's chain.

United States at the inception of the PLSS and (2) it had a convenient relationship to the rod, mile, and acre, i.e., 1 chain (ch) = 4 rods, 80 ch = 1 mile, and 10 square chains (ch^2) = 1 acre. The Gunter's chain was introduced in this text as a unit of length in Section 2.2.

Figure 22.2 illustrates a Gunter's chain. It had 100 links, each link equal to 0.66 ft or 7.92 in. The links were made of heavy wire, had a loop at each end, and were joined together by three rings. The outside ends of the handles fastened to the end links were the 0 and 66 ft marks. Tags with one, two, three, and four teeth were fastened to the 10th, 20th, 30th, and 40th links, respectively, from both ends of the chain. The 50th link was marked with a round tag. These tags saved time when measuring partial chain lengths. With its many connecting link and ring surfaces subject to frictional wear, use elongated the chain and its length had to be adjusted by means of bolts in the handles. Distances measured with Gunter's chains were recorded either in chains and *links* or in chains and decimals of chains—for example, 7 ch 94.5 lk or 7.945 ch. Decimal parts of links were estimated.

Gunter's chains are no longer manufactured and are seldom, if ever, used today. Nevertheless, the many chain surveys on record oblige the modern practitioner to understand the limits of accuracy possible with this equipment, and the conversion of distances recorded in chains and links to feet or meters. Descriptions of field procedures for conducting PLSS surveys given in the following sections of this chapter are taken from the *1973 Manual of Surveying Instructions*. Again, because lengths are frequently given in chains, familiarity with this unit of measure is essential to understanding the material presented. Of course, surveyors involved in PLSS work today would likely use either total station instruments and measure distances electronically or employ GPS equipment, but the cited distances and the same basic principles described still apply.

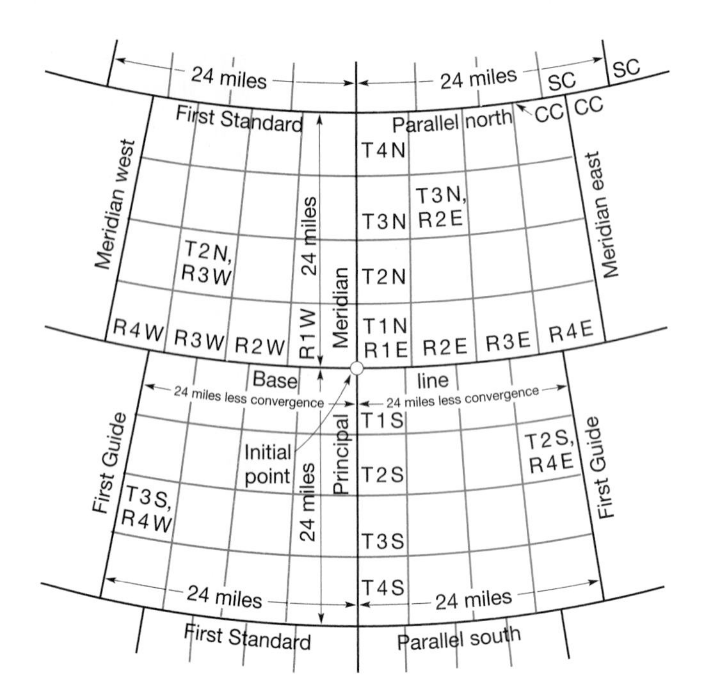

Figure 22.3 Survey of quadrangles. (Only a few of the standard corners and closing corners are identified.)

■ 22.3 INITIAL POINT

Thomas Jefferson recognized the importance of property surveys and served as chairman of a committee to develop a plan for locating and selling the western lands. His report to the Continental Congress in 1784, adopted as an ordinance on May 20, 1785, called for survey lines to be run and marked before land sales. Many of today's property disputes would have been eliminated if all property lines had been resurveyed and monuments checked and/or set before sales became final!

Subdivision of the public lands became necessary in many areas as settlers moved in and mining or other land claims were filed. The early hope that surveys would precede settlement was not fulfilled.

As settlers pressed westward, in each area where a substantial amount of surveying was needed, an *initial point* was established within the region to be surveyed. It was located by astronomical observations. The manual of 1902 was the first to specify an indestructible monument, preferably a copper bolt, firmly set in a rock ledge if possible and witnessed by rock bearings. In all, thirty-seven initial points have been set, five of them in Alaska. An initial point is illustrated near the center of Figure 22.3.

■ 22.4 PRINCIPAL MERIDIAN

From each initial point, a true north-south line called a *principal meridian* (Prin. Mer. or PM) was run north and/or south to the limits of the area to be covered. Generally, a solar attachment—a device for solving mechanically the mathematics of the astronomical triangle—was used. Monuments were set for section and quarter-section corners every 40 ch and at the intersections with all meanderable bodies of water (streams 3 ch or more in width, and lakes covering 25 acres or more).

The line was supposed to be within 3′ of the cardinal direction. Two independent sets of linear measurements were required to check within 20 lk (13.2 ft)/80 ch, which corresponds to a precision ratio of only 1/400. The allowable difference between sets of measurements is now limited to 7 lk/80 ch (precision ratio of 1/1140).

Areas within a principal meridian system vary greatly, as depicted in Figure 22.1.

■ 22.5 BASELINE

From the initial point, a baseline was extended east and/or west as a true parallel of latitude to the limits of the area to be covered. As required on the principal meridian, monuments were set for section and quarter-section corners every 40 ch and at the intersections with all meanderable bodies of water. Permissible closures were the same as those for the principal meridian.

Baselines are actually circular curves on the Earth's surface and were run with chords of 40 ch by the (1) solar method, (2) tangent method, or (3) secant method. These are briefly described as follows:

1. **Solar method.** An observation is made on the sun to determine the direction of astronomic north. A right angle is then turned off and a line extended 40 ch, where the process is repeated. The series of lines so established, with a slight change in direction every half mile, closely approaches a true parallel. Obviously, if the sun is obscured, the method cannot be used.

2. **Tangent method.** This method of laying out a true parallel is illustrated in Figure 22.4. A 90° angle is turned to the east or to the west, as may be required

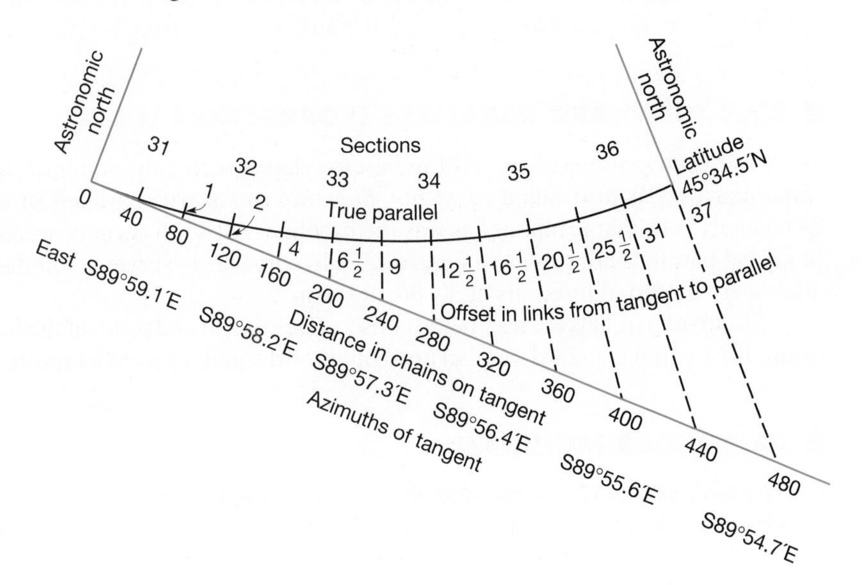

Figure 22.4 Layout of parallel by tangent method. (Adapted from 1973 *Manual of Surveying Instructions*)

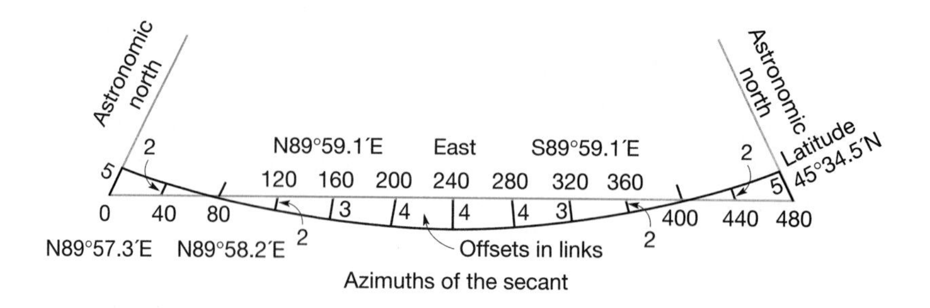

Figure 22.5 Layout of parallel by secant method. (Adapted from 1973 *Manual of Surveying Instructions*)

from an astronomic meridian, and corners are set every 40 ch. At the same time, proper offsets, which increase with increasing latitudes, are taken from *Standard Field Tables* issued by the BLM, and measured north from the tangent to the parallel. In the example shown, the offsets in links are 1, 2, 4, 6 1/2, 9, 12 1/2, 16 1/2, 20 1/2, 25 1/2, 31, and 37. The error resulting from taking right-angle offsets instead of offsets along the converging lines is negligible. The main objection to the tangent method is that the parallel departs considerably from the tangent, so both the tangent and the parallel must sometimes be brushed out to clear sight lines.

3. **Secant method.** This method of laying out an astronomic parallel is shown in Figure 22.5. It actually is a modification of the tangent method in which a line parallel to the tangent at the 3 mi (center) point is passed through the 1 and 5 mi points to produce minimum offsets.

Fieldwork includes establishing a point on the true meridian, south of the beginning corner, at a distance taken from the *Standard Field Tables* for the latitude of a desired parallel. The proper bearing angle from the same table is turned to the east or west from the astronomic meridian to define the secant, which is then projected 6 mi. Offsets, which also increase with increasing latitudes, are measured north or south from the secant to the parallel. Advantages of the secant method are that the offsets are small and can be measured perpendicular to the secant without appreciable error. Thus, the amount of clearing is reduced.

■ 22.6 STANDARD PARALLELS (CORRECTION LINES)

After the principal meridian and the baseline have been run, *standard parallels* (Stan. Par. or SP), also called *correction lines,* are run as true parallels of latitude 24 mi apart in the same manner as was the baseline. All 40 ch corners are marked. Standard parallels are shown in Figure 22.3. In some early surveys, standard parallels were placed at intervals of 30, 36, or 60 mi.

Standard parallels are numbered consecutively north and south of the baseline; examples are first standard parallel north and third standard parallel south.

■ 22.7 GUIDE MERIDIANS

Guide meridians (GM) are run due north (astronomic) from the baseline and the standard parallels at intervals of 24 mi east and west of the principal meridian, in the same manner as was the principal meridian and with the same limits of error.

Before work is started, the chain or tape must be checked by measuring 1 mi on the baseline or standard parallel. All 40-ch corners are marked.

Because meridians converge, a *closing corner* (CC) is set at the intersection of each guide meridian and standard parallel or baseline (see Figure 22.3). The distance from the closing corner to the *standard corner* (SC), which was set when the parallel was run, is measured and recorded in the notes as a check. Any error in the 24-mi-long guide meridian is put in the northernmost half mile.

Guide meridians are numbered consecutively east and west of the principal meridian; examples are first guide meridian west and fourth guide meridian east. Correction lines and guide meridians, established according to instructions, created *quadrangles* (or tracts) whose nominal dimensions are 24 mi on a side. These are shown in Figure 22.3.

■ 22.8 TOWNSHIP EXTERIORS, MERIDIONAL (RANGE) LINES, AND LATITUDINAL (TOWNSHIP) LINES

Division of a quadrangle, or tract, into townships is accomplished by running *range* (R) and *township* (T or Tp) lines.

Range lines are astronomic meridians through the standard township corners previously established at intervals of 6 mi on the baseline and standard parallels. They are extended north to intersect the next standard parallel or baseline and closing corners set (see Figures 22.3 and 22.6). Township lines are east-west lines that connect township corners previously established at intervals of 6 mi on the principal meridian, guide meridians, and range lines.

The angular amount by which two meridians converge is a function of latitude and the distance between the meridians. The linear amount of convergence

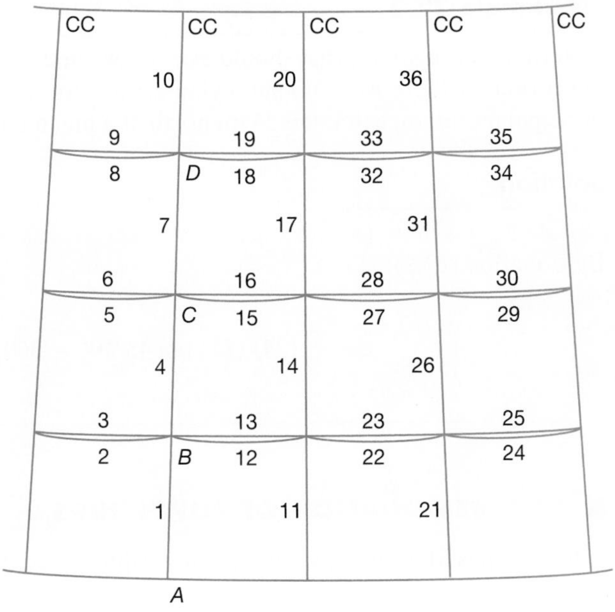

Figure 22.6 Order of running lines for the subdivision of a quadrangle into townships.

is a function of the same two variables, plus the length that the meridians are extended. Formulas for angular and linear convergence of meridians (derived in various texts on geodesy), are as follows:

$$\theta = 52.13\, d \tan \phi \tag{22.1}$$

and

$$c = \frac{4}{3} Ld \tan \phi \text{ (slight approximation)} \tag{22.2}$$

where θ is the angle of convergence, (in seconds); d the distance between meridians, (in miles), on a parallel; ϕ the mean latitude; c the linear convergence, (in feet); and L the length of meridians, (in miles).

Example 22.1

Compute the angular convergence at 40°25′ North latitude between two adjacent guide meridians.

Solution

By Equation(22.1) (guide meridians are 24 mi apart),

$$\theta = 51.13(24) \tan 40°25' = 1065'' = 17'45''$$

Example 22.2

Determine the distance that should exist between the standard corner and its closing corner (if there were no surveying errors) for a range line 12 mi east of the principal meridian, extended 24 mi north at a mean latitude of 43°10′.

Solution

By Equation (22.2),

$$c = \frac{4}{3} (24)\,(12) \tan 43°10' = 360.18 \text{ ft}$$

22.9 DESIGNATION OF TOWNSHIPS

A township is identified by a unique description based on the principal meridian governing it.

North and south rows of townships are called *ranges* and numbered in consecutive order east and west of the principal meridian as indicated in Figure 22.3.

East and west rows of townships are named *tiers* and numbered in order north and south of the baseline. By common practice, the term tier is usually replaced by the township in designating the rows.

An individual township is identified by its number north or south of the baseline, followed by the number east or west of the principal meridian. An example is Township 7 South, Range 19 East, of the Sixth Principal Meridian. Abbreviated, this becomes T 7 S, R 19 E, 6th PM.

■ 22.10 SUBDIVISION OF A QUADRANGLE INTO TOWNSHIPS

The method to be used in subdividing a quadrangle into townships is fixed by regulations in the *Manual of Surveying Instructions*. Under the old regulations, township boundaries were required to be within 21 min of the cardinal direction. Later this was reduced to 14 min to keep interior lines within 21 min of the cardinal direction.

The detailed procedure for subdividing a quadrangle into townships can best be described as a series of steps designed to ultimately produce the maximum number of regular sections with minimum unproductive travel by the field party. The order of running the lines is shown by consecutive numbers in Figure 22.6. Some details are described in the following steps:

1. Begin at the southeast corner of the southwest township, point A, after checking the chain or tape against a 1 mi measurement on the standard parallel.
2. Run north on the astronomic meridian for 6 mi (line 1 of Figure 22.6), setting alternate section and quarter-section corners every 40 ch. Set township corner B.
3. From B, run a random line (line 2 of Figure 22.6) due west to intersect the principal meridian. Set temporary corners every 40 ch.
4. If the random line has an excess or deficiency of 3 ch or less (allowing for convergence) and a falling north or south of 3 ch or less, the line is accepted. It is then corrected back (line 3), and all corners are set in their proper positions. Any excess or deficiency is thrown into the most westerly half mile. The method of correcting a random line having an excess of 1 ch and a north falling of 2 ch is shown in Figure 22.7.

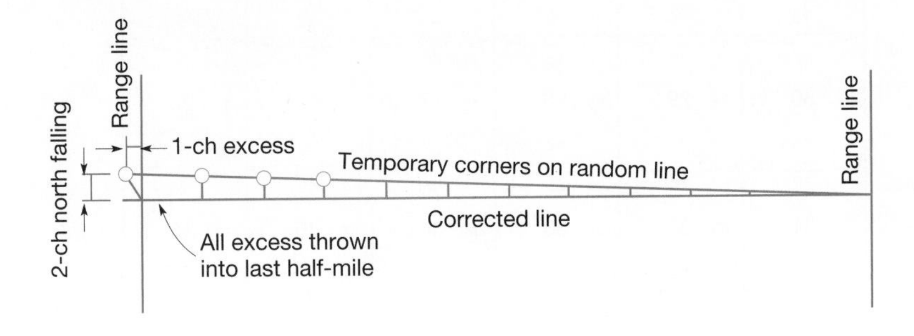

Figure 22.7 Correction of random line for excess and falling.

5. If the random line misses the corner by more than the permissible 3 ch, all four sides of the township must be retraced.
6. The same procedure is followed until the southeast corner *D* of the most northerly township is reached. From *D*, range line 10 is continued as an astronomic meridian to intersect the standard parallel or baseline, where a closing corner is set. All excess or deficiency in the 24 mi is thrown into the most northerly half mile.
7. The second and third ranges of townships are run the same way, beginning at the south line of the quadrangle.
8. While the third range is being run, random lines are also projected to the east and corrected back and any excess or deficiency is thrown into the most westerly half mile. (On this line, all points may have to be moved diagonally to the corrected line, instead of just the last point, as in Figure 22.7.)

■ 22.11 SUBDIVISION OF A TOWNSHIP INTO SECTIONS

Sections are now numbered from 1 to 36, beginning in the northeast corner of a township and ending in the southeast corner, as shown in Figure 22.8. The method used to subdivide a township can be described most readily as a series of steps to

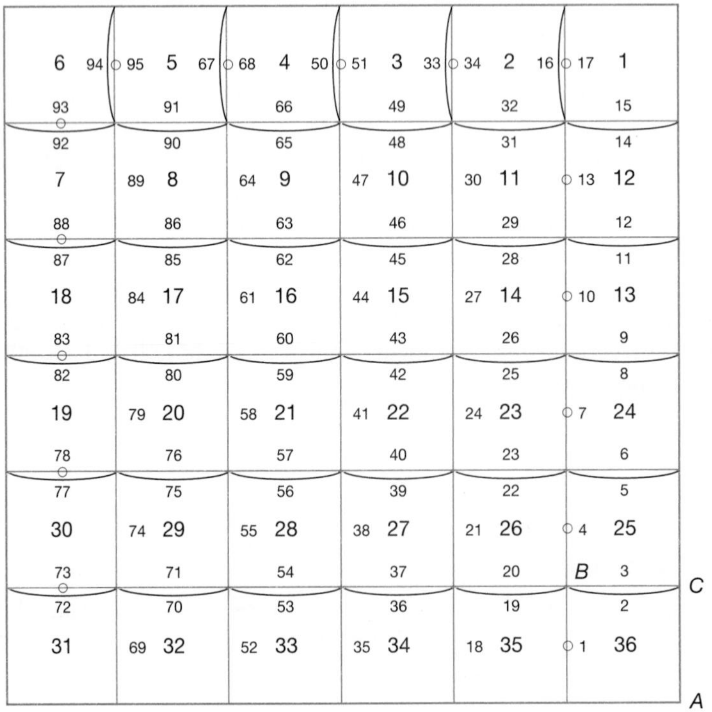

Figure 22.8 Order of running lines for the subdivision of a township into sections.

produce the maximum number of regular sections 1 mi on a side. Lines were run in the following order:

1. Set up at the southeast corner of the township, point *A*, and observe the astronomic meridian. Retrace the range line northward and the township line westward for 1 mi to compare the meridian, needle readings, and taped distances with those recorded.
2. From the southwest corner of section 36, run north *parallel with the east boundary of the township.* Set quarter-section and section corners on line 1 (Figure 22.8).
3. From the section corner just set, run a random line *parallel with the south boundary of the township* eastward to the range line. Set a temporary quarter corner at 40 ch.
4. If the 80 ch distance on the random line is within 50 lk, falling or distance, the line is accepted. The correct line is calculated and a quarter-corner located at the *midpoint* of line *BC* connecting the previously established corner *C* and the new section corner *B*.
5. If the random line misses the corner by more than the permissible 50 lk, the township lines must be rechecked and the source of the error determined.
6. The east range of sections is run in a similar manner until the southwest corner of section 1 is reached. From this point a continuation of the northward line is run to connect with the north township line section corner. A quarter corner is set 40 ch from the south section corner (on line 17, corrected back by later manuals). All discrepancies in the 6 mi are thrown into the last half-mile.
7. Successive ranges of sections across the township are run until the first four have been completed. All north-south lines are parallel with the township east side. All east-west lines are run randomly parallel with the south boundary line and then corrected back.
8. When the fifth range is being run, random lines are projected to the west as well as to the east. Quarter corners in the west range are set 40 ch from the east side of the section, with all excess or deficiency resulting from the errors and convergence being thrown into the most westerly half-mile.
9. If the north side of the township is a standard parallel, the northward lines which are run parallel with the east township boundary are projected to the correction line and closing corners set. The distance to the nearest corner is measured and recorded.
10. Bearings of interior north-south section lines for any latitude can be obtained by applying corrections from tables for the convergence at a given distance from the east boundary.

By placing the effect of meridian convergence into the westernmost half mile of the township and all errors to the north and west, 25 regular sections nominally 1 mi^2 are obtained. Also, the south half of sections 1, 2, 3, 4, and 5; the east half of sections 7, 18, 19, 30, and 31; and the southeast quarter of section 6 are all regular size.

22.12 SUBDIVISION OF SECTIONS

A section was the basic unit of the General Land Office system but land was often patented in parcels smaller than a section. Local surveyors and others performed subdivision of sections as the owners took up the land. The BLM provides guidelines on the proper and intended way a section should be subdivided. To divide a section into quarter sections (nominally 160 acres), straight lines are run between opposite quarter-section corners previously established or reestablished. This rule holds whether or not the quarter section corners are equidistant from the adjacent section corners. Due principally to underground ore deposits, which caused large local attraction errors in compass directions, one quarter section in a Wisconsin township contains 640 acres!

To divide a quarter section into quarter-quarter sections (nominally 40 acres), straight lines are run between opposite quarter-quarter-section corners established at the midpoints of the four sides of the quarter section. The same procedure is followed to obtain smaller subdivisions.

If the quarter sections are on the north or west side of a township, the quarter-quarter-section corners are placed 20 ch from the east or south quarter-section corners—or by single-proportional measurement (see Section 22.19) on line if the total length on the ground is not equal to that on record.

22.13 FRACTIONAL SECTIONS

In sections made fractional by rivers, lakes, or other bodies of water, lots are formed bordering on the body of water and numbered consecutively through the section (see Section 8 in Figure 22.9). Boundaries of lots usually follow the quarter section and quarter-quarter-section lines, but extreme lengths or narrow widths are avoided, as are areas of fewer than 5 acres or more than 45 acres.

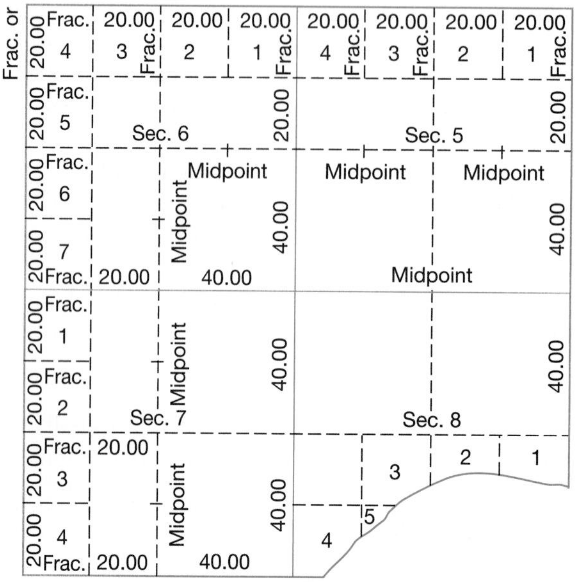

Figure 22.9 Subdivision of regular and fractional sections.

Quarter-sections along the north and west boundaries of a township, made irregular by discrepancies of measurements and convergence of the range lines, are usually numbered as lots (see Figure 22.9). Lot lines are not actually run in the field. Like quarter-section lines, they are merely indicated on the plats by *protraction* (subdivisions of parcels on paper only). Areas needed for selling the lots are computed from the plats.

■ 22.14 NOTES

Specimen field notes for each of the several kinds of lines to be run are shown in various instruction manuals. Actual recording had to closely follow the model sets. The original notes, or copies of them, are maintained in a land office in each state for the benefit of all interested persons.

■ 22.15 OUTLINE OF SUBDIVISION STEPS

Pertinent points in the subdivision of quadrangles into townships, and townships into sections, are summarized in Table 22.1.

■ 22.16 MARKING CORNERS

Various materials were approved and used for monuments in the original surveys. These included pits and mounds, stones, wooden posts, charcoal, and bottles. A zinc-coated, alloyed iron pipe with 2 1/2 in. outside diameter, 30 in. long, with brass cap is now standard. The bottom end of the pipe is split for several inches to form flanges that help hold it in place in the ground. Substitutions for this standard monument are permitted when authorized. Specially manufactured markers are now becoming commonplace. One type has a breakaway top so that if it is accidentally hit, for example by a plow or bulldozer, the upper part of the stake will break off but the lower part will remain in place. Another type uses a small, attached magnetic device to aid in recovery with a metal detector. In rock outcrops, a 3 1/4 in. diameter brass tablet with 3 1/2 in. stem is specified.

Stones and posts were marked with one to six notches on one or two faces. The arrangements identify a monument as a particular section or township corner. Each notch represents 1 mi of distance to a township line or corner. Quarter-section monuments were marked with the fraction "1/4" on a single face. In prairie country, where large stones and trees were scarce, a system of pits and mounds was used to mark corners. Different groupings of pits and mounds, 12 in. deep and 18 in. square, designated corners of several classes. However, unless some other type of mark perpetuated these markers, these corners were unfortunately lost in the first plowing.

■ 22.17 WITNESS CORNERS

Whenever possible, monuments were witnessed by two or three adjacent objects such as trees and rock outcrops. *Bearing trees* were blazed on the side facing the corner and marked with scribing tools.

When a regular corner fell in a creek, pond, swamp, or other place where it was impractical to place a mark, *witness corners* (WC) were set on a line(s) leading to the corner. Letters WC were added to all witness corner markers, and the witness corners were also witnessed.

TABLE 22.1 SUBDIVISION STEPS

Item	Subdivision of a Quadrangle	Subdivision of a Township
Starting point	SE corner of SW township	SW corner of SE section (36)
Meridional lines		
Name	Range line	Section line
Direction	Astronomic North	North, parallel with east range line
Length	6 mi = 480 ch	1 mi = 80 ch
Corners set	Quarter-section and section corners at 40 and 80 ch alternately	Quarter-section corner at 40 ch; section corner at 80 ch
Latitudinal lines		
Name	Township line	Section line
Direction of random	True east-west parallel	East, parallel with south side of section
Length	6 mi less convergence	1 mi
Permissible error	3 ch, length or falling	50 lk, length of falling
Distribution of error		
Falling	Corners moved proportionately from random to true line	Corner moved proportional from random to true line
Distance	All error thrown into west quarter-section	Error divided equally between quarter-sections

(*Work repeated until north side of area is reached. Subdivision of last area on the north of the range of townships and sections follows.*)

Case I. When Line on the North Is a Standard Parallel

Item	Subdivision of Quadrangle	Subdivision of Township
Direction of line	Astronomic north	North, parallel with east range line
Distribution of error in length	Placed in north quarter-section	Placed in north quarter-section
Corner placed at end	Closing corner	Closing error
Permissible errors	Specified in *Manual of Surveying Instructions*	Specified in *Manual of Surveying Instructions*

Case II. When Line on the North Is Not a Standard Parallel

Item	Subdivision of Quadrangle	Subdivision of Township
Direction of line	No case	Random north and correct back to section corner already established
Distribution of error in length		Same as Case I

TABLE 22.1 (CONTINUED)

[Other ranges of townships and sections continued until all but two are laid out.]

Location of last two ranges	On east side of tract	On west side of township
Next-to-last range subdivided	As before	As before
Last range		
Direction of random	True east	Westerly, parallel with south side of section
Nominal length	6 mi less convergence	1 mi less convergence
Correction of temporary corners	Corners moved proportionately from random to true line	Corners moved proportionately from random to true line
Distribution of closure error	Corners moved westerly (or easterly) to place error in west quarter-section	Corners are placed on the true line so total error falls in west quarter-section

■ 22.18 MEANDER CORNERS

A *meander corner* (MC) was established on survey lines intersecting the bank of a stream having a width greater than 3 ch, or a lake, bayou, or other body of water of 25 acres or more. The distance to the nearest section corner or quarter-section corner was measured and recorded in the notes. A monument was set and marked MC on the side facing the water, and the usual witness noted. If practical, the line was carried across the stream or other body of water by triangulation to another corner set in line on the farther bank.

A traverse joined successive meander corners along the banks of streams or lakes and followed as closely as practical the sinuosities of the bank. Calculating the position of the new meander corner and comparing it with its known position on a surveyed line checked the traverse.

Meander lines follow the mean high-water mark and are used only for plotting and protraction of the area. They are not boundaries defining the limits of property adjacent to the water.

■ 22.19 LOST AND OBLITERATED CORNERS

A common problem in resurveys of the public lands is the replacement of lost or obliterated corners. This difficult task requires a combination of experience, hard work, and ample time to reestablish the location of a wooden stake or post, set perhaps 150 years ago and with all witness trees long since cut or burned by apathetic owners.

An *obliterated corner* is one for which there are no remaining traces of the monument or its accessories, but whose location has been perpetuated or can be recovered beyond reasonable doubt. The corner may be restored from the acts or testimony of interested landowners, surveyors, qualified local authorities, witnesses, or from written evidence. Satisfactory evidence has value in the following order:

1. Evidence of the corner itself.
2. Bearing trees or other witness marks.

3. Fences, walls, or other evidence showing occupation of the property to the lines or corners.
4. Testimony of living persons.

A *lost corner* is one whose position cannot be determined beyond reasonable doubt, either from traces of the original marks or from acceptable evidence or testimony that bears on the original position. It can be restored only by rerunning lines from one or more independent corners (existing corners that were established at the same time and with the same care as the lost corner). *Proportionate measurements* distribute the excess or deficiency between a recently measured distance d separating the nearest found monuments that straddle the lost point and the record distance D given in the original survey notes between these monuments. Then the distance x from one of the found monuments required to set the lost point is calculated by proportion as $x = X(d/D)$, where X is the record distance from that monument.

Single-proportionate measurement follows the procedure just described and is used to relocate lost corners that have a specific alignment in one direction only. These include standard corners on baselines and standard parallels, intermediate section corners on township boundaries, all quarter-section corners and meander corners established originally on lines carried across a meanderable body of water. Corners on true lines of latitude such as baselines and standard parallels must be offset (south) from the proportion line to maintain the curvature of the latitudinal line.

Example 22.3

Figure 22.10 illustrates a lost quarter-section corner a on the line between sections 2 and 3. Section corners b and c have been found. The record distances for lines ba and ac are 40.00 and 39.57 ch, respectively. The observed distance between found corners b and c was 5246.25 ft. Describe the process for restoring lost corner a.

Solution

1. Since point a is a quarter-section corner, it is replaced on line bc by single-proportionate measurement.
2. Distance ba that must be laid off from section corner b to restore lost corner a is

$$(ba) = \frac{40}{79.57} \times 5246.25 = 2637.30 \text{ ft.}$$

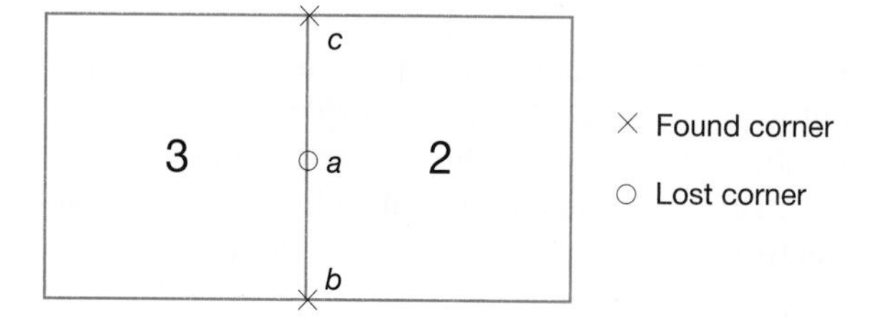

Figure 22.10 Example of single-proportionate measurement.

Double-proportionate measurements are used to establish lost corners located originally by specific alignment in two directions, such as interior section corners and corners common to four townships. The general procedure for single-proportionate measurement is used, but in two directions. It will establish two temporary points: one on the north-south line and another on the east-west line. The lost corner is then located where lines run from the two points, in the cardinal directions of north-south and east-west, intersect.

Example 22.4

Figure 22.11 illustrates lost corner *a* that is common to sections 22, 23, 26, and 27. Corners *b*, *c*, *d*, and *e* have been found. Record and measured distances are as follows:

Record		Measured	
Line	**Distance (CH)**	**Line**	**Distance (FT)**
ba	40.00	*bd*	7925.49
ca	40.00	*ec*	5293.24
da	79.20		
ea	39.72		

Describe the process of restoring lost corner *a*.

Solution

1. Corner *a* is an interior section corner, which is constrained in alignment in two directions. Thus it must be restored by double-proportionate measurement.
2. First establish temporary point *f* by laying off distance *bf* along line *bd*, where *bf* is computed as

$$bf = \frac{40.00}{119.36} \times 7925.49 = 2659.56 \text{ ft}$$

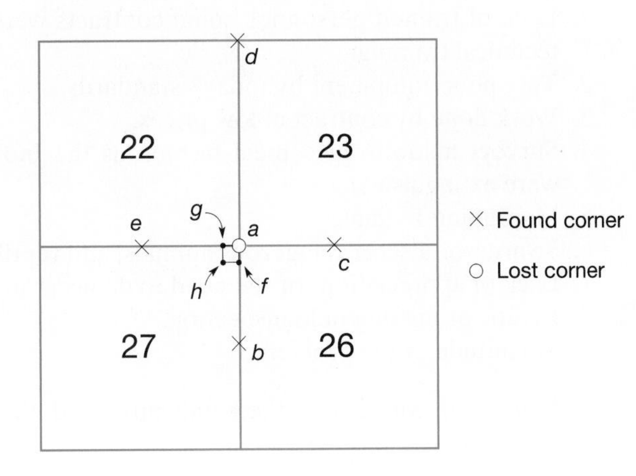

Figure 22.11 Example of double-proportionate measurement.

3. Then locate temporary point g by laying of distance eg along line ec, where eg is computed as

$$eg = \frac{39.72}{79.72} \times 5293.24 = 2637.32 \text{ ft}$$

4. Establish point h, the restored lost corner, where an east-west line through f intersects a north-south line through g.

When the original surveys were run, *topographic calls* (distances along each line from the starting corner to natural features such as streams, swamps, and ridges) were recorded. Using the recorded distances to any of these features found today and applying single- or double-proportionate measurements to them may help locate an obliterated corner or produce a more reliable reestablished lost corner.

■ 22.20 ACCURACY OF PUBLIC LANDS SURVEYS

The accuracy required in the early surveys was a very low order. Frequently it fell below what the notes indicated. A small percentage of the surveys were made by unscrupulous surveyors drawing on their imaginations in the comparative comfort of a tent; no monuments were set, and the notes serve only to confuse the situation for present-day surveyors and landowners. A few surveyors threw in an extra chain length at intervals to assure a full measure.

Many surveys in one California county were fraudulent. In another state, a meridian 108-mi long was run without including the chain handles in its 66-ft length. When discovered, it was rerun without additional payment.

The poor results obtained in various areas were primarily caused by the following:

1. Lack of trained personnel; some contracts were given to persons with no technical training.
2. Very poor equipment by today's standards.
3. Work done by contract at low prices.
4. Surveys made in piecemeal fashion as the Indian titles and other claims were extinguished.
5. Marauding Indians.
6. Swarms of insects, dangerous animals, and reptiles.
7. Lack of appreciation for the need to do accurate work.
8. Erratic or missing field inspection.
9. Magnitude of the problem.

In general, considering the handicaps listed, the work was reasonably well done in most cases.

■ 22.21 DESCRIPTIONS BY TOWNSHIP SECTION, AND SMALLER SUBDIVISION

Description by the U.S. Public Land Survey System offers a means of defining boundaries uniquely, clearly, and concisely. Several examples of acceptable descriptions are listed.

Sec. 6, T 7 S, R 19 E, 6th PM.
Frac. Sec. 34, T 2 N, R 5 W, Ute Prin. Mer.
The SE$\frac{1}{4}$, NE$\frac{1}{4}$, Sec. 14, Tp. 3 S, Range 22 W, SBM [San Bernardino Meridian].
E$\frac{1}{2}$ of N$\frac{1}{4}$ of Sec. 20, T 15 N, R 10 E, Indian Prin. Mer.
E 80 acres of NE$\frac{1}{4}$ of Sec. 20, T 15 N, R 10 E, Indian Prin. Mer.

Note that the last two descriptions do not necessarily describe the same land. A California case in point occurred when the owner of a southwest quarter section, nominally 160 acres but actually 162.3 acres, deeded the westerly portion as "the West 80 acres" and the easterly portion as "the East 1/2. "

Sectional land that is privately owned may be partitioned in any legal manner at the option of the owner. The metes-and-bounds form is preferable for irregular parcels. In fact, metes and bounds are required to establish the boundaries of mineral claims and various grants and reservations.

Differences between the physical and legal (or record) ground locations and areas may result because of departures from accepted procedures in description writing, loose and ambiguous statements, or dependence on the accuracy of early surveys.

■ 22.22 BLM LAND INFORMATION SYSTEM

As noted in Section 22.1, approximately one-third of the United States land area remains in the public domain. The Bureau of Land Management (BLM) is responsible for managing a significant portion of this vast acreage. Rapid access to accurate information related to this public land is in demand now more than ever before. As an example of its importance, consider that millions of dollars worth of oil and mineral royalties can he gained or lost by relatively small changes in property boundaries.

To assist in the monumental task of managing this enormous quantity of diverse land, the BLM and Forest Service provide business solutions for the management of cadastral records and land parcel information in a Land Information System (LIS—see Chapter 28) called the *National Integrated Land Survey* (NILS).[4] The LIS can be used to aid in making resource management decisions such as

[4] Information on this project can be obtained at http://www.blm.gov/nils/.

processing applications for mineral leases, designating utility corridors, issuing land use permits, locating wildlife habitat improvements, preparing timber sales, evaluating alternatives in environmental assessments and land use plans, and generating reports and maps.

The NILS concept provides users with tools to manage land records and cadastral data. The NILS project has four major components: (1) *survey management,* (2) *measurement management,* (3) *parcel management,* and (4) the *GeoCommunicator.* The survey management component consists of an integrated set of automation objects that will be embedded into compatible data collection software packages. This software will support the capture and input of feature measurements, and metadata (see Section 28.8) directly into a GIS database format. The measurement management system will allow users to further enhance the data set by performing a weighted least-squares adjustment (see Chapter 16) of new features. This will enable creation of a higher-quality network database in both PLSS and metes-and-bounds environments. The parcel management system will provide a process for managing land records and cadastral data stored in the database model. The GeoCommunicator is a website for sharing information about data and activities of interest to land managers. This system allows users to discover information that meets their needs with the goal of GeoCommunicator to facilitate data sharing and collaborative efforts among land managers.

As part of the NILS project, the BLM created a *Geographic Coordinate Data Base* (GCDB).[5] This database contains a digital layer of information on the U.S. Public Land Survey System (PLSS) and provides the positional components necessary for correlating all other information in the LIS. Included in the GCDB are geographic coordinates of all PLSS corners and estimates of their reliability, identifications of the surveyors who set the corners, the types of corners set and dates placed, any records of resurveys, a full record of ownership of each parcel, ownership of abutting parcels, and much more information. This effort was started in 1989 and continues today. The GCDB files for many townships are available for download via BLM websites. The national GCDB website can be found at http://ww.blm.gov/gcdb/.

■ 22.23 SOURCES OF ERROR

Some of the many sources of error in retracing the public-lands surveys follow.

1. Discrepancy between length measured with an early surveyor's chain and one obtained with present-day equipment.
2. Change in magnetic declination, local attraction, or both.
3. Lack of agreement between field notes and actual measurements.
4. Change in watercourses.
5. Nonpermanent objects used for corner marks.
6. Loss of witness corners.

[5]More information on the GCDB can be found at http://www.blm.gov/gcdb/.

■ 22.24 MISTAKES

Some typical mistakes in the retracement of boundaries in public-land surveys are:

1. Failing to follow the general rules and special instructions of procedure governing the original survey.
2. Neglecting to calibrate measured lengths against record distances for marks in place.
3. Treating corners as lost when they are actually obliterated.
4. Resetting corners without exhausting every means of relocating the original corners.
5. Failing to recognize that restored corners must be placed in their original locations regardless of deviations by the original surveyor from the general rules and special instructions.

PROBLEMS

Asterisks (*) indicate problems that have answers given in Appendix G.

22.1* Convert 79.89 chains to feet.

22.2 What steps in the subdivision of public lands are left to local surveyors?

22.3 In which states are the public lands surveys not applicable?

22.4 List the advantages and disadvantages of running baselines and standard parallels by each of the three methods described in Section 22.5.

22.5 Why are the boundaries of the public lands established by duly appointed surveyors unchangeable, even though incorrectly set in the original surveys?

22.6 What is the convergence in feet of meridians for the following conditions?
(a)* 12 mi apart, extended 18 mi, at mean latitude 45°20′ N.
(b) 24 mi apart, extended 24 mi, at mean latitude 38°45′ N.

22.7 What is the angular convergence, in seconds, for the two meridians defining a township exterior at a mean latitude of:
(a) 32°30′ N?
(b) 44°30′ N?

22.8 What is the nominal distance in miles between the following?
(a)* First Guide Meridian East, and the west Range Line of R10E.
(b) SE corner of Sec. 2, T 6 S, R 5 E, Indian PM, and the NW corner of Sec. 17, T 6 S, R 3 E, Indian PM.

22.9 Discuss when meander corners are to be set in a public land survey.

Sketch and label pertinent lines and legal distances, and compute nominal areas of the parcels described in Problems 22.10 through 22.12.

22.10 E 1/2, NE 1/4, Sec. 8, T 3 S, R 2 E, Salt River PM.

22.11 SW 1/4, NE 1/4, Sec. 21, T 2 N, R 1 W, Fairbanks PM.

22.12 NE 1/4, SW 1/4, NW 1/4, Sec. 28, T 1 S, R 3 E, 6th PM.

22.13 What are the nominal dimensions and acreages of the following parcels?
(a)* SE 1/4, SE 1/4, Sec. 36.
(b) S 1/2, Sec. 9.
(c) S 1/2, SE 1/4, NW 1/4, Sec. 26.

22.14 How many rods of fence are required to enclose the following?
(a) A parcel including the NE 1/4, NE 1/4, Sec. 29, and the NW 1/4, NW 1/4, Sec. 28, T2N, R3E?
(b) A parcel consisting of Secs. 14, 22, and 23 of T 2 N, R 1 W?

22.15 What lines of the U.S. public-land system were run as random lines?
22.16 In subdividing a township, which section line is run first? Which is ran last?
22.17 Corners of the SE 1/4 of the NW 1/4 of Sec. 22 are to be monumented. If all section and quarter-section corners originally set are in place, explain the procedure to follow, and sketch all lines to be run and corners set.
22.18 The quarter-section corner between Secs. 15 and 16 is found to be 39.86 ch from the corner common to Secs. 9, 10, 15, and 16. Where should the quarter-quarter-section corner be set along this line in subdividing Sec. 15?
22.19 As shown in the figure, in a normal township the exterior dimensions of Sec. 6 on the west, north, east, and south sides are 80, 78, 81, and 79 ch, respectively. Explain with a sketch how to divide the section into quarter sections. (See the following figure.)

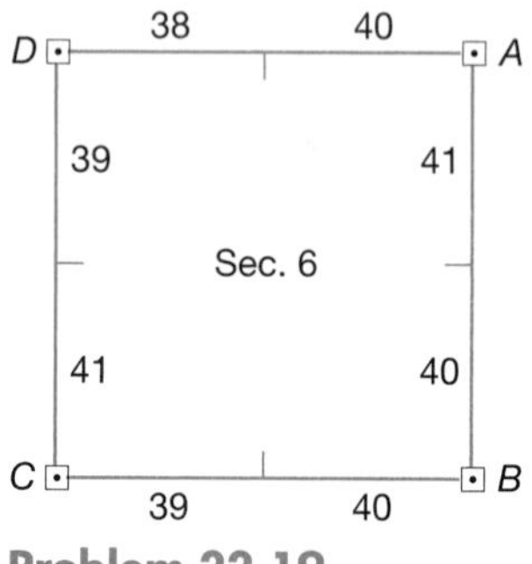

Problem 22.19

22.20 The problem figure shows original record distances. Corners A, B, C, and D are found, but corner E is lost. Measured distances are $AB = 10{,}603.27$ ft and $CD = 10{,}718.42$ ft. Explain how to establish corner E. (See the following figure.)

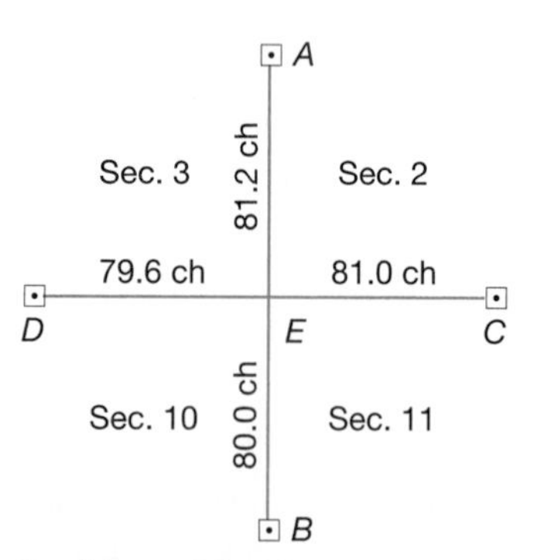

Problem 22.20

To restore the corners in Problems 22.21 through 22.24, which method is used: single proportion or double proportion?

22.21* Township corners on guide meridians; section corners on range lines.
22.22 Section corners on section lines; township corners on township lines.
22.23 Quarter-section corners on range lines.
22.24 Quarter-quarter-section corners on section lines.
22.25 Why are meander lines not accepted as the boundaries defining ownership of lands adjacent to a stream or lake?

22.26 What is a witness corner?
22.27 Explain the difference between "obliterated corner" and "lost corner."
22.28 The southern boundary of a township lies on a standard parallel at latitude 41°30′ N. What is the theoretical length of its northern boundary?
22.29 Why are the areas of many public-lands sections smaller than the nominal size?
22.30 Visit the NILS website and briefly describe the four components of NILS.
22.31 Visit the BLM website at http://www.blm.gov/nils/, and prepare a paper on the NILS project.

BIBLIOGRAPHY

Boyle, J. 2005. "The PLSS: Historical and Cultural Perspectives." *Professional Surveyor* 25 (No. 8): 20.

Brinker, D. 2002. "Understanding 'Protracted' Surveys." *Surveying and Land Information Science* 62 (No. 4): 255.

Broadus, J. 2001. "Examine Easements Carefully." *Point of Beginning* 26 (No. 5): 54.

———. 2001. "Reversionary Rights and Condemnations." *Point of Beginning* 26 (No. 8): 48.

Brown, C. M., W. G. Robillard, and D. A. Wilson. 1994. *Evidence and Procedures for Boundary Location*, 3rd Ed. New York: John Wiley & Sons, Inc.

Ciampa, D. and W. Parks. 2002. "Expected Positions of Standard and Closing Corners in Public Land Survey System Townships." *Surveying and Land Information Science* 62 (No. 4): 239.

Clinkscales, V. 2001. "How to Become a Surveyor." *Point of Beginning* 26 (No. 4): 36.

Craig, B. A. and J. Wahl. 2003. "Cadastral Survey Accuracy Standards." *Surveying and Land Information Science* 63 (No. 2): 87.

Duffy, M. 2003. "Three Points to One." *Point of Beginning* 28 (No. 7): 30.

Elam, E. 2002. "Searching for the Stone." *Point of Beginning* 27 (No. 5): 40.

Gay, P. 2002. "The Big and Small of Title Registration." *Point of Beginning* 27 (No. 9): 38.

Hartell, P. and W. Parks. 2003. "U.S. and Pennyslvania Law of Monuments as Evidence." *Surveying and Land Information Science* 63 (No. 3): 169.

Linklater, Andro. 2002. *Measuring America—How an Untamed Wilderness Shaped the United States and Fulfilled the Promise of Democracy*. New York: Walker and Company.

McHenry, T. 2002. "Fluid Challenges—How Water Boundaries Differ from their Upland Counterparts." *Point of Beginning* 27 (No. 10): 22.

———. 2002. "Fluid Challenges—A Lesson on Water Boundaries: Part II." *Point of Beginning* 27 (No. 11): 46.

———. 2002. "Fluid Challenges—A Lesson on Water Boundaries: Part III." *Point of Beginning* 27 (No. 12): 40.

Penry, J. 2003. "High Accuracy Surveying." *Professional Surveyor* 23 (No. 7): 28.

Robillard, W. et. al. 2002. *Evidence and Procedures for Boundary Location*. New York: John Wiley & Sons, Inc.

Thurow Glen and Steven Frank. 2001. "Coming to Terms with the Model Law: The Search for a New Definition." *Surveying and Land Information Systems* 61 (No. 1): 39.

U.S. Department of the Interior, Bureau of Land Management. 1973. *Manual of Surveying Instructions*. Washington, D.C.: U.S. Government Printing Office.

U.S. Department or Interior, Bureau of Land Management. 1974. *Restoration of Lost or Obliterated Corners and Subdivision of Sections*. Washington, D.C.: U.S. Government Printing Office.

23 Construction Surveys

■ 23.1 INTRODUCTION

Construction is one of the largest industries in the United States and thus surveying, as the basis for it, is extremely important. It is estimated that 60 percent of all hours spent in surveying are on location-type work, giving line and grade. Nevertheless, insufficient attention is frequently given to this type of survey.

An accurate topographic survey and site map are the first requirements in designing streets, sewer and water lines, and structures. Surveyors then lay out and position these facilities according to the design plan. A final "as-built" map, incorporating any modifications made to the design plans, is prepared during and after construction and filed. Such maps are extremely important, especially where underground utilities are involved, to assure that they can be located quickly if trouble develops and that they will not be disturbed by later improvements.

Construction surveying involves establishing both *line* and *grade* by means of stakes and reference lines that are placed on the construction site. These guide the contractor so that proposed facilities are constructed according to a plan. Placement of the stakes is most often done by making the fundamental observations of horizontal distances, horizontal and vertical angles, and differences in elevation using the basic equipment and methods described in earlier chapters of this text. However, the global positioning system (GPS) is also being used with increasing frequency for construction surveys (see Section 23.10). Other specialized equipment, including laser alignment devices and reflectorless electronic distance measuring equipment (see Section 23.2), have also been developed which greatly facilitates construction surveying.

All surveyors, engineers, and architects who may be involved with planning, designing, and building constructed facilities should be familiar with the fundamental procedures involved in construction surveying. This chapter describes procedures applicable for some of the more common types of construction projects. Chapters 24, 25, and 26 cover the subjects of horizontal curves, vertical curves, and volume computations, respectively. These topics are all pertinent to construction surveys, particularly those for transportation routes.

Construction surveying is perhaps best learned on the job and consists in adapting fundamental principles to the undertaking at hand. Since each project may involve unique conditions and present individual problems, coverage in this text is limited to a discussion of the fundamentals.

■ 23.2 SPECIALIZED EQUIPMENT FOR CONSTRUCTION SURVEYS

As noted previously, the placement of stakes for line and grade to guide construction operations is most often accomplished using the surveyor's standard equipment—levels, tapes, total station instruments, and GPS receivers. However, recent advances in modern technology have produced some additional new instruments that have improved, simplified, and greatly increased the speed with which certain types of construction surveying can be accomplished. *Visible laser-beam* alignment instruments and *reflectorless total stations* are among the new innovations. These are described briefly in the subsections that follow.

23.2.1 Visible Laser-Beam Instruments

The fundamental purpose of laser instruments is to create a visible line of known orientation, or a plane of known elevation, from which measurements for line and grade can be made. Two general types of lasers are described here:

Single-beam lasers, as shown in Figures 23.1 and 23.2, project visible reference lines ("string lines" or "plumb lines") that are utilized in linear and vertical alignment applications such as tunneling, sewer pipe placement, and building construction.

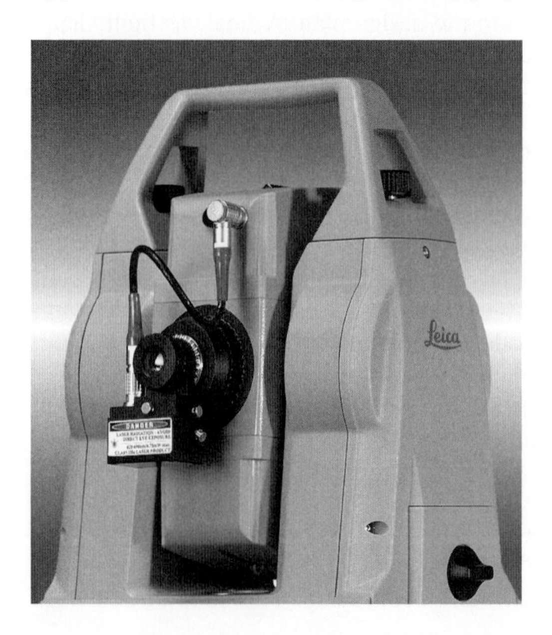

Figure 23.1 Laser-beam incorporated with a total station instrument. (Courtesy Leica Geosystems)

Figure 23.2 Rotating laser providing finished grades to machine control system in grader. (Courtesy Topcon Positioning Systems)

The instrument shown in Figure 23.1 is a single-beam type laser that has been combined with a total station instrument. This combination provides flexibilities that are convenient for a variety of construction layout applications. The laser beam is projected collinear with the instrument's line of sight, a feature that facilitates aligning it in prescribed horizontal alignments and/or along planned grade lines. The instrument can be used to project string lines for distances up to about 1000 m. With the zenith angle set to either 90° or 270°, if the total station instrument is rotated about its vertical axis, the laser will generate a horizontal plane. Also if it is turned about its horizontal axis, the laser will define a vertical plane.

The instrument shown in Figure 23.2 projects a visible laser beam a distance of 5 m below and 100 m above the instrument along the plumb line. These instruments are useful for alignment of objects in vertical structures, or in finished grading. A similar type of single-beam laser projects a visible laser-beam at a selected grade—a device that is especially useful in aligning pipelines.

Rotating-beam lasers are merely single-beam lasers with spinning optics that rotate the beam in azimuth, thereby creating planes of reference. They expedite the placement of grade stakes over large areas such as airports, parking lots, and subdivisions and are also useful for topographic mapping.

Figure 23.3 shows a rotating-beam type laser. It projects a beam up to 350 m while rotating at 600 rpm. One or more receivers attached to grade rods or staffs can pick up the laser signal. The instrument is self-leveling and quickly set up. If somehow bumped out of level, the laser beam shuts off and does not come back on

Figure 23.3 Sokkia LP30 laser plane level. (Courtesy Sokkia Corporation)

until it is releveled. It can be operated with the laser plane oriented horizontally for setting footings, floors, etc., or the beam can be turned 90° and used vertically for plumbing walls or columns.

Because laser beams are not readily visible to the naked eye in bright sunlight, special detectors attached to a hand-held rod are often used. To lay out horizontal planes with either of these devices, the height of the instrument above datum, *HI*, must be established. Then the height on a graduated rod that a reference mark or detector must be set is the difference between the *HI* and the plane's required elevation.

23.2.2 Reflectorless Total Stations

As described in Chapter 8, total station instruments simultaneously observe horizontal and vertical angles as well as slope distances. Their built-in microcomputers reduce the observed slope distances to their horizontal and vertical components and display the results in real-time. These features make total stations very convenient for construction stakeout.

Recent improvements in electronic distance measurement (EDM) technology include a device that does not require a reflector. Rather, these instruments employ a pulsed laser light, which is capable of measuring distances of up to 100 m or more to objects without the use of a reflector. Alternatively, they can be used with reflectors, or reflective sheets that can be applied to surfaces, a procedure which enables them to measure longer distances. Figure 23.4(a) shows a total station instrument equipped with this technology, and Figure 23.4(b) depicts a similar hand-held pulsed laser EDM instrument. Both devices are useful in observing distances to inaccessible locations, a feature that is particularly useful in assembling and checking the placement of structural members in bridges, buildings, and other large fabricated objects.

(a)

(b)

Figure 23.4 Pulsed laser instruments: (a) total station and (b) hand-held EDM instrument. (Courtesy Leica Geosystems)

■ 23.3 HORIZONTAL AND VERTICAL CONTROL

The importance of a good framework of horizontal and vertical control in a project area cannot be overemphasized. It provides the basis for positioning structures, utilities, roads, etc., in each of the stages of planning, design, and construction. Too often surveyors and engineers have skimped on establishing a suitable network of control points and they have also failed to preserve them through proper monumentation, references, and ties.

The surveyor in charge should receive copies of the plans well in advance of construction to become familiar with the job and have time to "tie out" or "transfer" any established control points that might be destroyed during building operations. The methods of Figures 9.4 and 9.5 are especially applicable and should be used with intersection angles as close to 90° as possible.

On most projects, additional horizontal and vertical control is required to supplement any already available in the job area. The control points must be:

1. Convenient for use, that is, located sufficiently close to the item being built so that work is minimized and accuracy enhanced in transferring alignment and grade.
2. Far enough from the actual construction to ensure working room for the contractor and to avoid possible destruction of stakes.
3. Clearly marked and understood by the contractor in the absence of a surveyor.
4. Supplemented by guard stakes to deter removal and referenced to facilitate restoring them. Contracts usually require the owner to pay the cost of setting initial control points and the contractor to replace damaged or removed ones.

5. Suitable for securing the accuracy agreed on for construction layout (which may be to only the nearest foot for a manhole, 0.01 ft for an anchor bolt, or 0.001 ft for a critical feature).

Construction stakes can be set in their required horizontal positions by making observations of horizontal angles or horizontal distances from established control points. Radial stakeout by angle and distance from one control point is often most expedient, but the choice will depend on the project's nature and extent. Frequent checks should be made on points set. This can be done with observations from other control stations or by checking distances from nearby points to verify their correctness in position.

Grade stakes and reference elevations are most often set using a leveling instrument whose *HI* has been established by differential leveling. For convenience, enough benchmarks are generally placed on construction sites so that at least one is readily accessible at any location in the area. Then the *HI* of the level can be established with just a single backsight to the benchmark. After grade stakes are set, a closing foresight is taken back to the benchmark for a check. However, this practice can be dangerous since the instrument operator will have a tendency to expect the closing foresight to agree with the initial backsight, and therefore could read it carelessly. As a result, a serious mistake in the initial backsight could go undetected, resulting in a faulty setting of grade stakes. Therefore, even though it requires more time, it is recommended that level circuits for setting grade stakes always begin on one benchmark and close on another.

■ 23.4 STAKING OUT A PIPELINE

Pipelines are used to carry water for human consumption, storm water, sewage, oil, natural gas, and other fluids. Pipes that carry storm runoff are called *storm sewers*; those that transport sewage are called *sanitary sewers*. Flow in these two types of sewers is usually by gravity, and therefore their alignments and grades must be carefully set. Flow in pipes carrying city water, oil, and natural gas is generally under pressure, so usually pipes need not be aligned to as high an order of accuracy.

In pipeline construction, trenches are usually opened along the required alignment to the prescribed depth (slightly undercut if pipe bedding is required), the pipe is installed according to plan, and the trench backfilled. Pipeline grades are fixed by a variety of existing conditions, topography being a critical one. A profile like that of Figure 5.12 is usually used to analyze the topography and assist in designing the grade line for each pipe segment. To minimize construction difficulties and costs, excavation depths are minimized, but at the same time a certain minimum cover over the pipeline must be maintained to protect it from damage by heavy loading from above and to prevent freezing in cold climates. Minimum grades also become an important design factor for pipes under gravity flow. Accordingly, a grade of at least 0.5 percent is recommended for storm sewers, but slightly higher grades are needed for sanitary sewers. In designing pipe grade lines, other existing underground elements often must be avoided, and due regard must also be given to the grades of connecting lines and the vertical clearances needed to construct manholes, catch basins, and outfalls.

Prior to staking a pipeline, the surveyor and contractor should discuss details of the project. An understanding must be reached concerning the planned trench width, where the installation equipment will be placed, and how and where the excavated material will be stockpiled. Then a reference offset line can be appropriately established that will (1) meet the contractor's needs, (2) be safe from destruction, and (3) not interfere with operations.

The alignment and grade for the pipeline are taken from the plans. An offset reference line parallel to the required centerline is established, usually at 25- or 50-ft stations when the ground is reasonably uniform. Marks should be closer together on horizontal and vertical curves than on straight segments. For pipes of large diameter, stakes may be placed for each pipe length—say, 6 or 8 ft. On hard surfaces where stakes cannot be driven, points are marked by paint, spikes, or scratch marks.

Either *batter boards* or laser beams guide precise alignment and grade for pipe placement. Figure 23.5 shows one arrangement of a batter board for a sewer line. It is constructed using 1 × 4 in., 1 × 6 in., or 2 × 4 in., or boards nailed to 2 × 4 in. posts, which have been pointed and driven into the ground on either side of the trench. Depending upon conditions, these may be placed at 50 ft, 25 ft, or any other convenient distance along the sewer line. The top of the batter board is generally placed a full number of feet above the *invert* (flow line or lower inside surface) of the pipe. Nails are driven into the board tops so a string stretched tightly between them will define the pipe centerline. A graduated pole or special rod, often called a *story pole*, is used to measure the required distance from the string to the pipe invert. Thus, the string gives both line and grade. It can be kept taut by hanging a weight on each end after wrapping it around the nails.

In Figure 23.5, instead of a fixed batter board, a two-by-four carrying a level vial can be placed on top of the offset-line stake, which has been set at some even number of feet above the pipe's invert elevation. Measurement is then made from the underside of the leveled two-by-four with a tape or graduated pole to establish the flow line.

If laser devices are used for laying pipes, the beam is oriented along the pipe's planned horizontal alignment and grade and the trench opened. Then with the

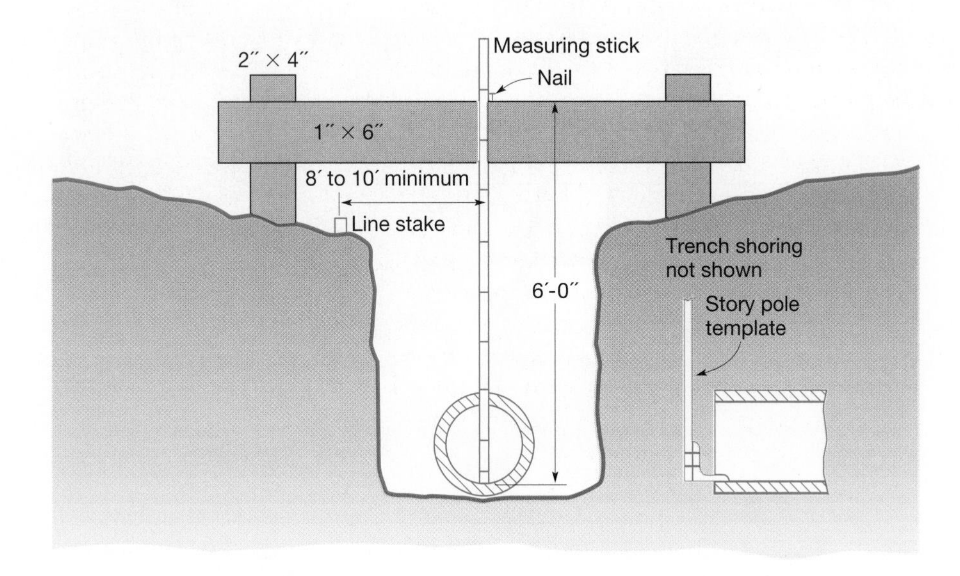

Figure 23.5 Batter board for sewer line.

beam set at some even number of feet above the pipe's planned invert, measurements can be made using a story pole to set the pipe segments. Thus, the laser beam is equivalent to a batter board string line. On some jobs that have a deep wide cut, the laser instrument is set up in the trench to give line and grade for laying pipes. If the pipe is large enough, the laser beam can be oriented inside it.

■ 23.5 STAKING PIPELINE GRADES

Staking pipeline grades is essentially the reverse of running profiles, although in both operations the centerline must first be marked and stationed in horizontal location. The actual profiling and staking are on an offset line.

Information conveyed to the contractor on stakes for laying pipelines usually consists of two parts: (1) giving the depth of cut (or fill), normally only to the nearest 0.1 ft, to enable a rough trench to be excavated, and (2) providing precise grade information, generally to the nearest 0.01 ft, to guide in the actual placement of the pipe invert at its planned elevation. Cut (or fill) values for the first part are vertical distances from ground elevation at the offset stakes to the pipe invert. After the pipe's grade line has been computed and the offset line run, cuts (or fills) can be determined by a leveling process, illustrated in Figure 23.6 and the corresponding field notes given in Plate B.6 in Appendix B. The process is summarized as follows:

1. List the stations staked on the pipeline (column 1 of the field notes).
2. Compute the flow line or invert elevation at each station (column 6).
3. Set up the level and get an *HI* by reading a plus sight on a BM; for example, $HI = 2.11 + 100.65 = 102.76$ (see Plate B.6 and Figure 23.6).

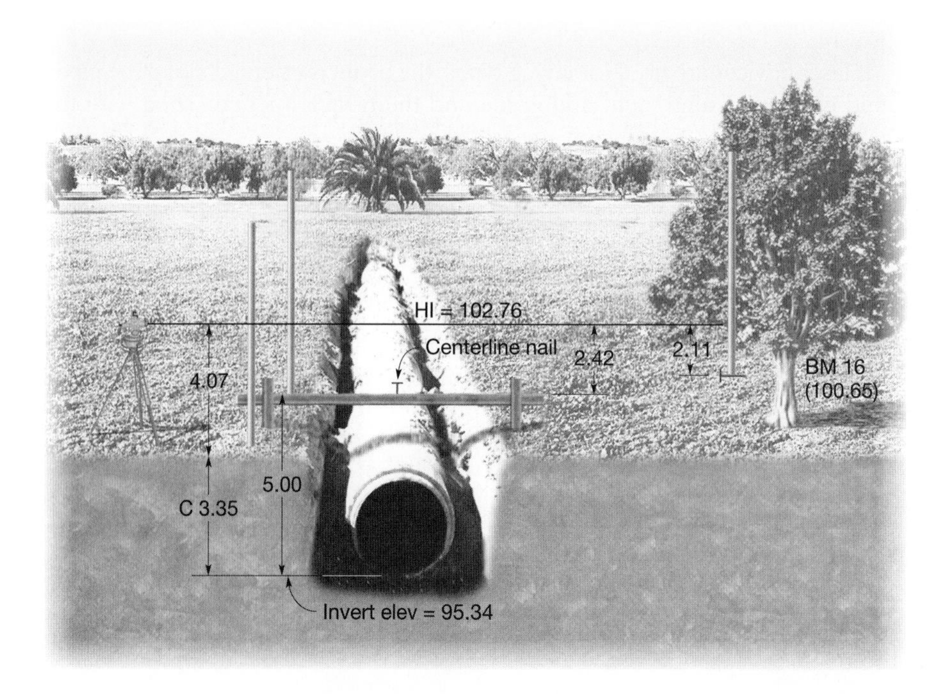

Figure 23.6 Leveling process to determine cut or fill and set batter boards for laying pipelines.

4. Obtain the elevation at each station from a rod reading on the ground at every stake (column 4)—for example, 4.07 at station 1 + 00 (see Plate B.6 and Figure 23.6)—and subtract it from the *HI* (column 5); for example, 102.76 − 4.07 = 98.69 at station 1 + 00.
5. Subtract the pipe elevation from the ground elevation to get cut (+) or fill (−) (column 7); for example, 98.69 − 95.34 = C 3.35 (see Plate B.6 and Figure 23.6).
6. Mark the cut or fill (using a permanent marking felt pen or keel) on an offset stake facing the centerline; the station number is written on the other side.

In another variation, which produces the same results, *grade rod* (difference between *HI* and pipe invert) is computed, and *ground rod* (reading with rod held at stake) is subtracted from it to get cut or fill. For station 1 + 00, grade rod = 102.76 − 95.34 = 7.42 and 7.42 − 4.07 = C 3.35.

After the trench has been excavated based on cuts and fills marked on the stakes, batter boards are set. Marks needed to place them can be made with a pencil or felt pen on the offset stakes during the same leveling operation used to obtain cut and fill information. Figure 23.6 also illustrates the process. Suppose that at station 1 + 00, the batter board will be set so its top is exactly 5.00 ft above the pipe invert. The rod reading necessary to set the batter board is obtained by subtracting the pipe invert elevation plus 5.00 ft from the *HI*; thus 102.76 − (95.34 + 5.00) = 2.42 ft (see Figure 23.6). The rod is held at the stake and adjusted in vertical position by commands from the level operator until a rod reading of 2.42 ft is obtained; then a mark is made at the rod's base on the stake. (To facilitate this process, a rod target or a colored rubber band can be placed on the rod at the required reading.) The board is then fastened to the stake with its top at the mark using nails or C clamps, and a carpenter's level is used to align it horizontally across the trench. A nail marking the pipe centerline is set by measuring the stake's offset distance along the board.

If a laser is to be employed, this same leveling procedure can be used to establish the elevation of the laser beam at some desired vertical offset distance above the pipe's invert. The procedure is used to establish the height of the laser instrument and also to set another identical offset elevation at a station forward on line. Then the laser beam is aimed at that target to establish the required grade line.

■ 23.6 STAKING OUT A BUILDING

The first task in staking out a building is to locate it properly on the correct lot by making measurements from the property lines. Most cities have an ordinance establishing setback lines from the street and between houses to improve appearance and provide fire protection.

Stakes may be set initially at the exact building corners as a visual check on the positioning of the structure, but obviously such points are lost immediately when excavation is begun on the footings. A set of batter boards and reference stakes, placed as shown in Figure 23.7, is therefore erected near each corner, but out of the way of construction. The boards are nailed a full number of feet above the footing base, or at first-floor elevation. (The procedure of setting boards at a

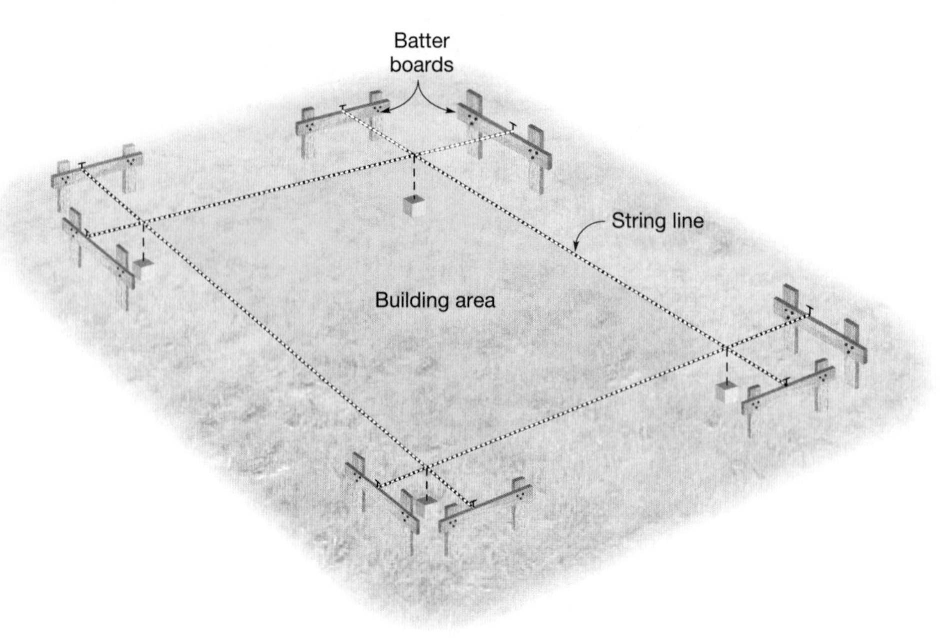

Figure 23.7 Batter boards for building layout.

desired elevation was described in the preceding section.) Nails are driven into the batter-board tops so that strings stretched tightly between them define the outside wall or form line of the building. The layout is checked by measuring diagonals and comparing them with each other (for symmetric layouts) or to their computed values. Figure 23.8 illustrates the placement on a lot and staking of a slightly more complicated building. The following are recommended steps in the procedure:

1. Set hubs *A* and *B* 5.00 ft inside the east lot line, with hub *A* 20.00 ft from the south lot line and hub *B* 70.00 ft from *A*. Mark the points precisely with nails.
2. Set a total station instrument over hub *A*, backsight on hub *B*, and turn a clockwise angle of 270° to set batter-board nails 1 and 2 and stakes *C* and *D*.
3. Set the instrument over hub *B*, backsight on hub *A*, and turn a 90° angle. Set batter-board nails 3 and 4 and stakes *E* and *F*.
4. Measure diagonals *CF* and *DE* and adjust if the error is small or restake if large.
5. Set the instrument over *C*, backsight on *E*, and set batter-board nail 5. Plunge the instrument and set nail 6.
6. Set the instrument over *D*, backsight on *F*, and set nail 7. Plunge and set nail 8.
7. Set batter-board nails 9, 10, 11, 12, 13, and 14 by measurements from established points.
8. Stretch the string lines to create the building's outline and check all diagonals.

(*Note:* When each batter board is set, it must be placed with its top at the proper elevation.)

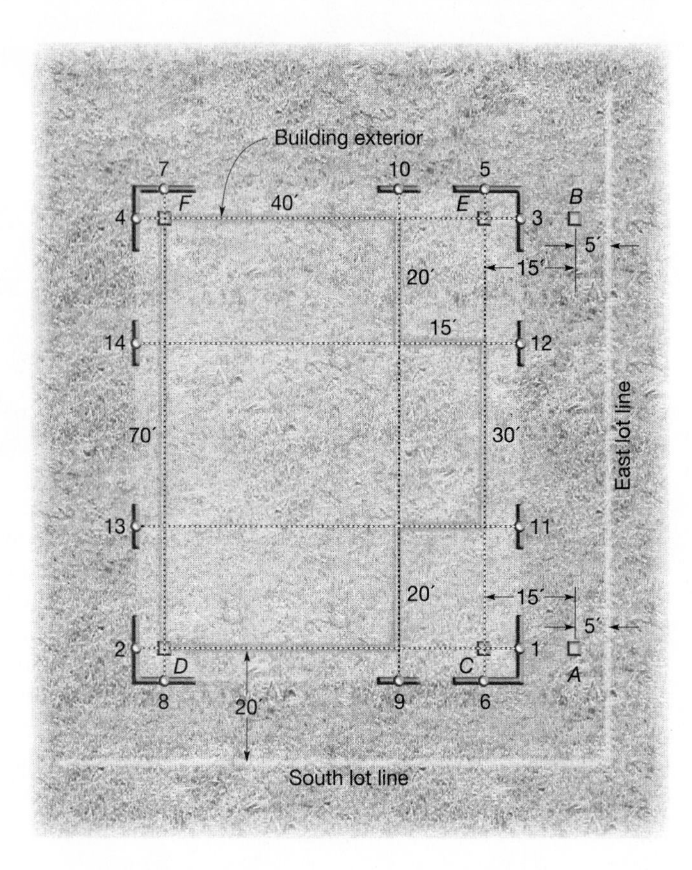

Figure 23.8 Location of building on lot and batter-board placement.

As an alternative to this building stakeout procedure, radial methods (described in Section 9.9) can be used. This can substantially reduce the number of instrument setups and stakeout time required. In the radial method, coordinates of all building corners are computed in the same coordinate system as the lot corners. Then the total station instrument is set on any convenient control point and oriented in azimuth by sighting another intervisible control point. Angles and distances, computed from coordinates, are then laid off to mark each building corner. Measuring the distances between adjacent points and also the diagonals checks the layout. (An example illustrating radial stakeout of a circular curve is given in Section 24.11.) After constructing the batter boards and setting the cross pieces at the desired elevations, the alignment nails on the batter boards can be set by pulling taut string lines across established corners. For example in Figure 23.8, with corners *D* and *F* marked, a line stretched across these two points enables placing nails 7 and 8 on the boards. With the strings in place after setting batter-board nails, diagonals between corners should again be checked.

Another method of laying out buildings is to stake two points on the building, occupy one of them with the total station instrument, take a backsight on the other, and stake all (or many) of the remaining points from that setup using precalculated angles and distances. In some cases, advantage can be taken of symmetrical layouts

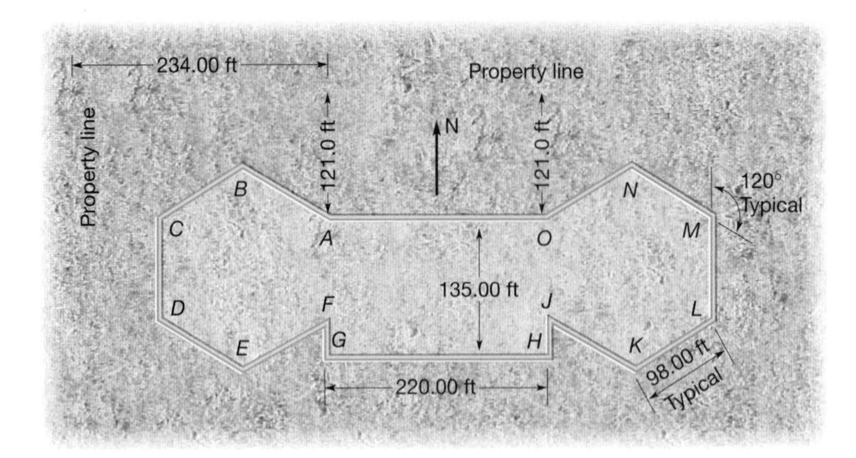

Figure 23.9
Building layout.

to save considerable time. Figure 23.9 shows an unusual symmetrical building shape, which was laid out rapidly using only two setups (at points *A* and *O*). With this choice of stations, half the corners could be set from each setup, and the same calculated angles and distances could be used (see the field notes of Figure 23.10). Again if this method is used, it is essential that enough dimensions be checked between marked corners to ensure that no large errors or mistakes were made.

To control elevations on a building construction site, a benchmark (two or more for large projects) should be set outside the construction limits but within easy sight distance. Rotating lasers (see Section 23.2.1) can be used to control elevations for the tops of footings, floors, and so on.

BUILDING LAYOUT

Sta.	Angle	To the	Distance
	⊼ @ Point *A*		
O	0°00′	Rt	220.00 ft
F	90°00′	Rt	98.00 ft
G	90°00′	Rt	135.00 ft
E	120°00′	Rt	169.74 ft
D	150°00′	Rt	196.00 ft
C	180°00′	Rt	169.74 ft
B	210°00′	Rt	98.00 ft
	⊼ @ Point *O*		
A	0°00′	Lt	220.00 ft
J	90°00′	Lt	98.00 ft
H	90°00′	Lt	135.00 ft
K	120°00′	Lt	169.74 ft
L	150°00′	Lt	196.00 ft
M	180°00′	Lt	169.74 ft
N	210°00′	Lt	98.00 ft

Figure 23.10
Precalculated angles and distances for building layout.

Permanent foresights are helpful in establishing the principal lines of the structure. Targets or marks on nearby existing buildings can be used if movement due to thermal effects or settlement is considered negligible. On formed concrete structures, such as retaining walls, offset lines are necessary because the outside wall face is obstructed. Two-by-two-in. hubs with tacks can first mark the positions of such things as interior footings, anchor bolts for columns, and special piping or equipment. Survey disks, scratches on bolts or concrete surfaces, and steel pins can also be used. Batter boards set inside the building dimensions for column footings have to be removed as later construction develops.

On multistory buildings, care is required to ensure vertical alignment in the construction of walls, columns, elevator shafts, structural steel, and so on. One method of checking the plumb of constructed members is to carefully aim a total station's line of sight on a reference mark at the base of the member. The line of sight is then raised to its top. For an instrument that has been carefully leveled and that is in proper adjustment, the line of sight will define a vertical plane as it is raised. It should not be assumed that the instrument is in good adjustment; therefore, the line should be raised in both the direct and reversed positions. It is necessary to check plumb in two perpendicular directions when using this procedure. To guide construction of vertical members in real-time, two instruments can be set up with their lines of sight oriented perpendicular to each other, and verticality monitored as construction progresses. Alternatively, lasers can be used to guide and monitor vertical construction.

If the surveyor does not give sufficient forethought to the basic control points required, the best method to establish them, and the most efficient approach to staking out a building, the job can be a time-consuming and difficult process. The number of instrument setups should be minimized to conserve time, and calculations made in the office if possible, rather than in the field while a survey party waits.

■ 23.7 STAKING OUT HIGHWAYS

Alignments for highways, railroads, and other transportation routes are designed after careful study of existing maps, air photos, and preliminary survey data of the area. From alternative routes, the one that best meets the overall objectives while minimizing costs and environmental impacts is selected. Before construction can begin, the surveyor must transfer that alignment (either the centerline or an offset reference line) to the ground.

Normally staking will commence at the initial point where the first straight segment (*tangent*) is run, placing stakes at *full stations* (100-ft intervals) if the English system of units is used, or at perhaps 30- or 40-m spacing if the metric system is employed. Stationing (this subject is described in Section 5.9.1) continues until the planned alignment changes direction at the first *point of intersection* (PI). The deflection angle is measured there and the second tangent stationed forward to the next PI, where the deflection angle there is measured. The process continues to the terminal point. Staking continuously from the initial point to the terminus may result in large amounts of accumulated error on long projects. Therefore work should be checked by making frequent ties to intermediate horizontal control points and adjustments should be made as necessary. Alternatively, on smaller projects the alignments can be run from both ends to a point near the middle.

After tangents are established, horizontal curves (usually circular arcs) are inserted at all PIs according to plan. The subject of horizontal alignments, including methods for computing and laying out horizontal curves, is discussed in detail in Chapter 24. Vertical alignments are described in Chapter 25.

After the centerline or reference line (including curves) has been established, the PIs, intermediate *points on tangent* (POTs) on long tangents, and points where horizontal curves begin (PCs), and end (PTs), are referenced using procedures described in Section 9.5. Points used in referencing must be located safely outside the construction limits. Referencing is important because the centerline points will be destroyed during various phases of construction and will need to be replaced several times. Benchmarks are also established at regular spacing (usually not more than about 1000 ft apart) along the route. These are placed on the right of way, far enough from the centerline to be safe from destruction, but convenient for access.

After the centerline or reference line has been established, stakes marking the right-of-way should be set. This is normally done by carefully measuring perpendicular offsets from the established reference line. The right-of-way is staked at every change in its width, at all changes in alignment, including each PC and PT, and at sufficient other intermediate points along the tangents so that it is clearly delineated.

When the reference line and right-of-way have been staked, the limits of actual construction are marked so that the contractor can clear to them. Following this, some contractors want points set on the right-of-way with subgrade elevations, showing cut or fill to a given elevation, for use in performing rough grading and preliminary excavation of excess material.

To guide a contractor in making final excavations and embankments, *slope stakes* are driven at the *slope intercepts* (intersections of the original ground and each side slope), or offset a short distance, perhaps 4 ft (see Figure 23.11). The cut or fill at each location is marked on the slope stake. Note that there is no cut or fill at a slope stake—the value given is the vertical distance from the ground elevation at the slope stake to grade.

Grade stakes are set at points that have the same ground and grade elevation. This happens when a grade line changes from cut to fill, or vice-versa. As shown in Figure 23.12, three *transition sections* normally occur in passing from cut to fill

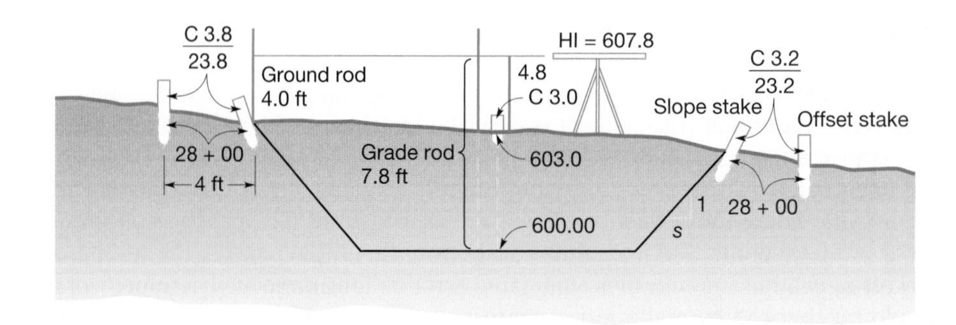

Figure 23.11 Slope stakes (shoulders and ditches not shown).

Sta.	L	CL	R
(a) 61 + 20	C 8.2 / 28.2	C 3.9	C 0.0 / 20.0
(b) 61 + 70	C 4.0 / 24.0	C 0.0	F 6.4 / 29.6
(c) 61 + 95	C 0.0 / 20.0	F 3.3	F 5.7 / 28.5

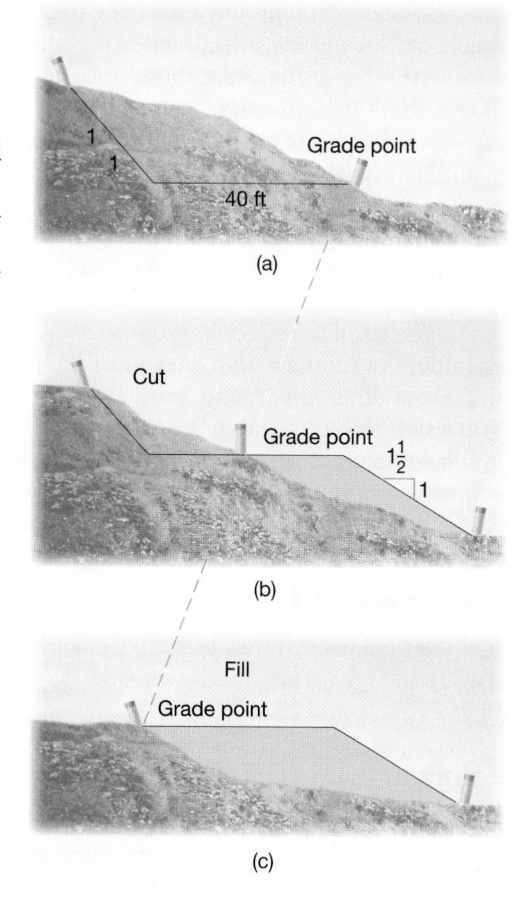

Figure 23.12 Grade points at transition sections.

(or vice versa) and a grade stake is set at each one. A line connecting grade stakes, perhaps scratched out on the ground, defines the change from cut to fill, as we will see in line *ABC* in Figure 26.1.

Slope stakes can be set at slope intercept locations predetermined in the office from cross-sectional data. (Methods for determining slope intercepts from cross-sections are described in Chapter 26.) Many of today's data collectors can compute slope intercepts in the field based on a load DTM of the project area, horizontal and vertical curves of the alignment, design templates of the cut and fill sections of the road, and grade shots by a total station from a control point. The data collectors provide the operator with the location of the slope stake and the amount of cut or fill to the centerline of the alignment.

If predetermined slope intercepts are used, the ground elevation at each stake must still be checked in the field to verify its agreement with the cross section. If a significant discrepancy in elevation exists, the stake's position must be adjusted by a trial-and-error method, as shown in Example 23.1. The amount of cut or fill marked on the stake is computed from the actual difference in elevation between the ground at the slope stake and grade elevation.

If slope intercepts have not been precalculated from cross-section data, slope stakes are located by a trial-and-error method based on mental calculations involving the *HI*, grade rod, ground rod, half roadway width, and side slopes. One or two trials are generally sufficient to fix the stake position within an allowable error of 0.3 to 0.5 ft for rough grading. The infinite number of ground variations prohibits use of a standard formula in slope staking. An experienced surveyor employs only mental arithmetic, without scratch paper or hand calculator. Whether using the method to be described or any other, systematic procedures must be followed to avoid confusion and mistakes.

Example 23.1 lists the sequential steps to be taken in slope staking, *assuming for simplicity, academic conditions of a level roadway*. In practice, travel lanes and shoulders of modern highways have lateral slopes for drainage, then a steeper slope to a ditch in cut, and another slope up the hillside to the slope intercept. Transition sections may have half-roadway widths in cuts different from those in fills to accommodate ditches, and flatter side slopes for fills that tend to be less stable than cuts. But the same basic steps still apply and can be extended by students after learning the fundamental approach.

Example 23.1

List the field procedures, including calculations, necessary to set slope stakes for a 40 ft wide level roadbed with side slopes of 1:1 in cut and 1-1/2:1 in fill (see Figures 23.11 and 23.12).

Solution

1. Compute the cut at the centerline stake from profile and grade elevations (603.0 − 600.0 = C 3.0 in Figure 23.11). Check in field by grade rod minus ground rod = 7.8 − 4.8 = C 3.0 ft. Mark the stake C 3.0/0.0. (On some jobs the center stake is omitted and stakes are set only at the slope intercepts.)
2. Estimate the difference in elevation between the left-side slope-stake point (20+ ft out) and the center stake. Apply the difference—say, +0.5 ft—to the center cut and get an estimated cut of 3.5 ft.
3. Mentally calculate the distance out to the slope stake, 20 + 1(3.5) = 23.5 ft, where 1 is the side slope.
4. Hold the zero end of a cloth tape at the center stake while the rodperson goes out at right angles with the other end and holds the rod at 23.5 ft. [The right angle can be established by prism (see Figure 16.10) or by using a total station instrument.]
5. *Forget all previous calculations to avoid confusion of too many numbers and remember only the grade-rod value.*
6. Read the rod with the level and get the cut from grade rod minus ground rod, perhaps 7.8 − 4.0 = C 3.8 ft.
7. Compute the required distance out for this cut, 20 + 1(3.8) = 23.8 ft.
8. Check the tape to see what is actually being held and find it is 23.5 ft.
9. The distance is within a few tenths of a foot and close enough. Move out to 23.8 ft if the ground is level and drive the stake. Move farther out if the

ground slopes up, since a greater cut would result, and thus the slope stake must be beyond the computed distance, or not so far if the ground has begun to slope down, which gives a smaller cut.

10. If the distance has been missed badly, make a better estimate of the cut, compute a new distance out, and take a reading to repeat the procedure.
11. In going out on the other side, the rodperson lines up the center and left-hand slope stake to get the right-angle direction.
12. To locate grade stakes at the road edge, one person carries the zero end of the tape along the centerline while the rodperson walks parallel, holding the 20-ft mark until the required ground-rod reading is found by trial. Note that the grade rod changes during the movement but can be computed at 5- or 10-ft intervals. The notekeeper should have the grade rod listed in the field book for quick reference at full stations and other points where slope stakes are to be set.
13. Grade points on the centerline are located using a starting estimate determined by comparing cut and fill at back and forward stations.

Practice varies for different organizations, but often the slope stake is set 4 ft beyond the slope intercept. It is marked with the required cut or fill, distance out from the centerline to the slope-stake point, side-slope ratio, and base half-width noted on the side facing the centerline. Stationing is given on the backside. A reference stake having the same information on it may also be placed 6 ft or farther out of the way of clearing and grading. On transition sections, grade-stake points are marked.

As previously stated, total station instruments, with their ability to automatically reduce observed slope distances to horizontal and vertical components, speed slope staking significantly, especially in rugged terrain where slope intercept elevations differ greatly from centerline grade. Many data collectors allow the user to input the "design template" (see Section 26.3), from which the data collector rapidly determines the positions of the slope stakes using field observed data. GPS receivers operating in the real-time kinematic mode (see Chapter 15) can also be advantageously used in these types of terrain if satellite visibility exists.

On many larger projects, machine control (see Section 15.9) is often employed in construction stakeout with GPS, laser levels, and total stations. In this environment, the machine operator sees a visual representation on a screen display original grades, planned grades and alignments, and position of this machine. The machine operator is guided by the machine control panel to excavate the alignment without the need for stakes. This process is covered in more detail in Section 23.11.

Slope staking should be done with utmost care, for once cut and fill embankments are started, it is difficult and expensive to reshape them if a mistake is discovered.

After rough grading has shaped cuts and embankments to near final elevation, finished grade is constructed more accurately from *blue tops* (stakes whose tops are driven to grade elevation and then marked with a blue keel or spray paint).

These are not normally offset, but rather driven directly on centerline or shoulder points. The procedure for setting blue tops at required grade elevation is described in Section 25.7.

Highway and railroad grades can often be rounded off to multiples of 0.05 or 0.10 percent without appreciably increasing earthwork costs or sacrificing good drainage. Streets need a minimum 0.50 percent grade for drainage from intersection to intersection, or from midblock both ways to the corners. They are also crowned to provide for lateral flow to gutters. Drainage profiles, prepared to verify or construct drainage cross sections, can be used to locate drainage structures and easements accurately. An experienced engineer, when asked a question regarding the three most important items in highway work, thoughtfully replied "drainage, drainage, and drainage." Good surveying and design must satisfy this requirement.

To ensure unobstructed drainage after construction, culverts must be placed in most fill sections so that water can continue to flow in its normal pattern from one side of the embankment to the other. In staking culverts, their locations, skew angles, if any, lengths, and invert elevations are taken from the plans. Required pipe alignments and grades are marked using stakes, offset from each end of the pipe's extended centerline. The invert elevation (or an even number of feet above or below it) is noted on the stake. This field procedure, like setting slope stakes, requires marking a point on the stake where a rod reading equals the difference between the required grade and the current *HI* of a leveling instrument.

After the subgrade has been completed, if the highway is being surfaced with rigid concrete pavement, paving pins will be necessary to guide this operation. They are usually about 1/2-in. diameter steel rods, driven to mark an offset line parallel to one edge of the required pavement. This line is usually staked at 50-ft increments, but closer spacing may be used on sharp curves. The finished grade (or one parallel to it but offset vertically above) is marked on the pins using tape, or affixing a special stringline holding device. Again in this operation, elevations are set by marking the stake where a rod reading equals the difference between the finished grade (or a vertically offset one) and a current *HI*. (The need for frequent project benchmarks at convenient locations is obvious.)

Utility relocation surveys may be necessary in connection with highway construction; for example, manhole or valve-box covers have to be set at correct grade before earthwork begins so they will conform to finished grade. Here differential elevation resulting from the transverse surface slope must be considered. Utilities are located by centerline station and offset distance.

Location staking for railroads, rapid-transit systems, and canals follows the same general methods outlined here for highways.

■ 23.8 OTHER CONSTRUCTION SURVEYS

For planning and constructing causeways, bridges, and offshore platforms, it is often necessary to perform hydrographic surveys (see Section 17.12). These types of projects require special procedures to solve the problem of establishing horizontal positions and depths where it is impossible to hold a rod or reflector. Modern surveying equipment and procedures, and sonar mapping devices, are used to plot dredging cross sections for underwater trenching and pipe layout. Today more pipelines are crossing wider rivers, lakes, and bays than ever before. Mammoth pipeline projects to transport crude oil, natural gas, and water have introduced numerous problems and solutions. Permafrost, extremely low temperatures, and the need to provide animal crossings are examples of special problems associated with Alaskan pipeline construction.

Large earthwork projects such as dams and levees require widespread permanent control for quick setups and frequent replacement of slope stakes, all of which may disappear under fill in one day. Fixed signals for elevation and alignment painted or mounted on canyon walls or hillsides can mark important reference lines. Failures of some large structures, such as at the Teton Dam, demonstrate the need for monitoring them periodically so that any necessary remedial work can be done.

Underground surveys in tunnels and mines necessitate transferring lines and elevations from the ground above, often down shafts. Directions of lines in mine tunnels can be most conveniently established using north-seeking gyros. In another, still-practiced method, two heavy plumb bobs hung on wires (and damped in oil or water) from opposite sides of the surface opening can be aligned by total station there and in the tunnel. (A vertical collimator will also provide two points on line below ground.) A total station or laser is "wiggled-in" (see Section 8.16) on the short line defined by the two plumb-bob wires, a station mark set in the tunnel ceiling above the instrument, and the line extended. Later setups are made beneath spads (surveying nails with hooks) anchored in the ceiling. Elevations are brought down by taping or other means. Benchmarks and instrument stations are set on the ceiling, out of the way of equipment.

Surveys are run at intervals on all large jobs to check progress for periodic payments to the contractor. And finally, an *as-built* survey is made to determine compliance with plans, note changes, make terminal contract payment, and document the project for future reference.

Airplane and ship construction requires special equipment and methods as part of a unique branch of surveying called *optical tooling*. The precise location and erection of offshore oil drilling platforms many miles from a coast utilizes new surveying technology, principally the global positioning system.

■ 23.9 CONSTRUCTION SURVEYS USING TOTAL STATION INSTRUMENTS

The procedures described here apply to most total station instruments, although some may require interfaced data collectors to perform the operations described.

Before using a total station for stakeout, it is necessary to orient the instrument. Depending on the type of project, *horizontal* or both *horizontal and vertical* orientation may be needed. For example, if just the lot corners of a subdivision are being staked, then only horizontal orientation (establishing the instrument's position and direction of pointing) is needed. If grade stakes are to be set, then the instrument must also be oriented vertically (its *HI* determined).

With total station instruments, three methods are commonly used for horizontal orientation: (1) *azimuth*, (2) *coordinates*, and (3) *resection*. The first two apply where an existing control point is occupied and the latter is used when the instrument is set up at a noncontrol point. In azimuth orientation, the coordinates of the occupied control station and the known azimuth to a backsight station are entered into the instrument. If the occupied station's coordinates have been downloaded into the instrument prior to going into the field, it is only necessary to input its point number. The backsight station is then sighted and when completed, the azimuth of the line is transferred to the total station by a keyboard stroke, whereupon it appears in the display.

The coordinate method of orientation uses the same approach, except that the coordinates of both the occupied and the backsight station are entered. Again these data could have been downloaded previously so that it would only be necessary to key in or select the identifiers of the two stations. The instrument computes the backsight line's azimuth from the coordinates, displays it, and prompts the operator to sight the backsight station. Upon completion of the backsight, the azimuth is transferred to the instrument with a keystroke and it appears on the display.

In the resection procedure, a station whose position is unknown is occupied and the instrument's position determined by sighting two or more control stations (see Sections 11.7 and 11.10). This is very convenient on projects where a certain point of high elevation in an open area gives good visibility to all (or most) points to be staked. As noted, two or more control points must be sighted. Observations of angles, or of angles and distances, are made to the control stations. The microprocessor then computes the instrument's position by the methods discussed in Sections 11.7 and 11.10. This method is often used in large open-pit mines.

Project conditions will normally dictate which orientation procedure to use. Regardless of the procedure selected, after orientation is completed, a check should be made by sighting another control point and comparing the observed azimuth and distance against their known values. If there is a discrepancy, the orientation procedure should be repeated. It is also a good idea to recheck orientation at regular intervals after stakeout has commenced, especially on large projects. In fact, if possible a reflector should be left on a control point just for that purpose.

Vertical orientation of a total station (i.e., determining its HI) can be achieved using one of two procedures. The simplest case occurs if the elevation of the occupied station is known, as then it is only necessary to *carefully* measure and add the hi (height of instrument above the point) to the elevation of the point. If the occupied station's elevation is unknown, then another station of known elevation must be sighted. The situation is illustrated in Figure 23.13, where the instrument is located at station A of unknown elevation, and station B whose elevation is known is sighted. From slope distance S and zenith angle z, the instrument computes V. Then its HI is

$$HI = elev_B + h_r - V \tag{23.1}$$

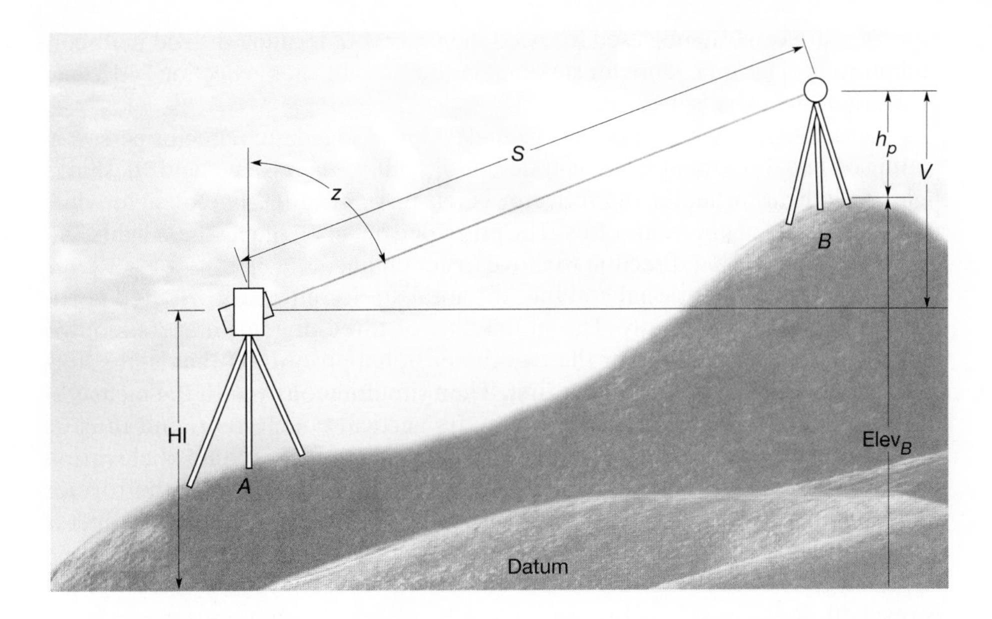

Figure 23.13 Vertical orientation of total station.

where h_r is the reflector height above station B. As with horizontal orientation, it is good practice to check the instrument's vertical orientation by sighting a second vertical control point.

Once orientation is completed, project stakeout can begin. In general, staking is either a two- or three-dimensional problem. Staking lots of a subdivision or layout of horizontal construction alignments is generally two-dimensional. Slope staking, blue-top setting, pipeline layout, and batter-board placement require both horizontal position and elevation and are therefore three-dimensional.

For two-dimensional stakeout, after the file of coordinates for all control stations and points to be staked is downloaded and the instrument is oriented horizontally, the identifier of a point to be staked is entered in the instrument through the keyboard. The microprocessor immediately calculates the horizontal distance and azimuth required to stake the point. The difference between the instrument's current direction of pointing and that required is displayed. The operator turns the telescope until the difference becomes zero to achieve the required direction. With total stations having robotic capabilities, the instrument will swing in direction to the proper azimuth without any further operator intervention.

Following azimuth alignment, the distance to the point must be laid out. To do this, the reflector is directed onto the azimuth alignment and a horizontal distance reading taken, whereupon the difference between it and that required is displayed. The reflector is then directed inward or outward, as necessary, until the distance difference is zero and the stake placed there. A two-way radio is invaluable for communicating with the reflector person in this operation. Additionally, a small tape measure can often be used to speed the process of locating the rod in its correct position. This procedure for stakeout is discussed further in Section 24.13, and an example problem is presented.

Special tracking systems have been developed to aid the reflector person in getting on line. For example, some total stations utilize "constraint" and "flashing" lights to indicate whether the reflector is left or right of the line of sight while others use lights of different colors. The prism person, upon seeing these lights, immediately knows what direction to move to get on line.

For three-dimensional staking, the total station must be oriented vertically as well as horizontally. The initial part of three-dimensional stakeout is exactly like that described for the two-dimensional procedure; that is, the horizontal position of the stake is set first. Then simultaneously with the measurement of the stake's horizontal position, its vertical component, and thus its elevation is determined. The difference ΔZ between the required elevation and the stake's elevation is displayed with a plus or minus sign, the former indicating fill, the latter cut. This information is communicated to the reflector person for marking the stake, or incrementally driving it further down until the required grade, or if desired, some even number of feet above or below grade is reached.

With the high order of accuracy possible using total station instruments, stakes quite distant from the instrument can be laid out and thus many points set from a single setup. Often, in fact, an entire project can be staked from one location. This is made possible in many cases because of the flexibility that resection orientation provides in instrument placement. It should be remembered that if long distances are involved in three-dimensional staking, Earth curvature and refraction should be considered (see Section 4.4). Total stations typically have some method, such as a toggle switch or keyboard menu, that allows the operator to compensate for curvature and refraction. Also with total stations, each point is set independently of the others and thus no inherent checks are available. Checks should therefore be made by either repeating the observations, checking placements from different control stations, or measuring between staked stations to ascertain their relative accuracies.

■ 23.10 CONSTRUCTION SURVEYS USING GPS EQUIPMENT

Any of the GPS surveying methods discussed in Chapters 14 and 15 could be used on construction projects. Specifically, static surveys can be used to establish project control and kinematic surveys can be used to produce maps for planning and design. Finally, real-time kinematic (RTK) surveys (see Chapter 15) can be used to locate construction stakes.

The base receiver does not have to occupy a station with known GPS coordinates. Instead the application software can be instructed to determine its position in *autonomous* mode. This process determines the base station receiver coordinates using point-positioning methods yielding low accuracy coordinates. However all points determined from this station will have GPS-type accuracies relative to the base station coordinates. The localization process, discussed next, then transforms this set of low accuracy GPS coordinates into the project control reference frame eliminating the inaccuracies of the autonomous base station coordinates.

As discussed in Sections 15.8 and 19.6.6, care must be taken to ensure that points located using GPS are placed in the same reference frame as the project coordinates. As discussed in Section 15.9, sufficient project control known in the local reference frame must be established at the perimeter of the construction project. Then prior to staking any points, the GPS receiver must occupy this control and determine their coordinates in the WGS84 reference frame; these are GPS coordinates. Using the project coordinates and the GPS coordinates, transformation parameters (see Section 19.6.6) are computed so that the GPS-derived coordinates can be transformed into the local project reference frame. Additional control points must also be established at critical locations in the construction project to provide convenient location of base station receivers, total stations (in canopy conditions), and laser levels for finish work. While the type of radio and antenna will determine the range of the base station radio, high-powered base station radios typically have a working range of 10 km. Thus sufficient control must be established to support the radio's range.

In construction staking using RTK surveying, a minimum of two receivers is needed. Each is equipped with a radio. One receiver occupies a nearby control station and the other, called the "rover," is moved from one point (to be set) to another. The points set must have their project coordinates known before setting any stakes. The base radio broadcasts the raw GPS signals from the base receiver to the rover. At the rover, the application software processes the GPS signals from both receivers in real-time using relative positioning techniques (see Section 13.9). This determines the location of the rover relative to the base station. If the observed coordinates do not agree with the required values for the point being staked, the GPS controller will indicate the direction and distance that the rover must be moved. The rover's position is adjusted until agreement is reached and the stake is set at this location.

Although excellent horizontal accuracies can be achieved using GPS typically, elevations are less reliable. GPS-determined ellipsoid heights can be accurate to within a few centimeters, but to get an orthometric height (elevation related to datum), the geoidal height must be applied as discussed in Section 19.5. Geoidal heights are not precisely known. Models are available that yield values which are generally accurate to within a few centimeters in most areas. However, they can be off by several decimeters and even meters in mountainous regions such as the Pacific Northwest. For this reason, if very precise elevations are required in construction staking, GPS is unsatisfactory. As shown in Figure 23.14, string lines have traditionally been used to guide finishing work. However, today laser levels are being combined with GPS to achieve millimeter accuracies in both horizontal and vertical alignments.

GPS is particularly useful in staking widely spaced points, especially in areas where terrain or vegetation makes it difficult to conduct traditional ground surveys. Staking subdivisions containing large parcels in rugged terrain and setting slope

Figure 23.14 A string line guiding a paver. (Courtesy Topcon Positioning Systems)

stakes in rugged areas where deep cuts and fills exist, are examples of situations where GPS can be very convenient for construction surveying. Of course, GPS requires an obstruction-free view of the satellites.

■ 23.11 MACHINE CONTROL

In recent years, research has led to *stakeless* construction where GPS receiver, robotic total stations, and laser levels are used to guide earth-moving equipment in real-time. Data necessary for this operation include a *digital elevation model* (DEM) or *digital terrain model* (DTM) (see Section 17.8), design plans of the alignments, grades, and design templates developed in the same three-dimensional coordinate system such as shown in Figure 18.10. With GPS, robotic total stations, and laser levels to guide the equipment operators and an on-board computer that continually updates cut and fill information, the grading can be accomplished without the need for construction stakes and with limited assistance of grade foremen. Machine control has been implemented on dozers, hoes, pans, graders, and trucks.

Using machine control, the surveyor's role in construction surveying shifts to tasks such as establishing the project reference coordinate systems, creating a DTM (see Section 17.14) of the existing surface for the design and grading work, managing the electronic design on the job site, calibrating the surveying equipment with respect to the construction site, and developing the necessary data for system operation. As stated in Section 15.9, GPSs radios have a working radius of 10 km from the base station typically; laser levels must be within 1500 ft of the machine; and total stations have a working radius of 1000 ft. The surveyor must ensure that sufficient control exists in the project to ensure that the construction process can continue seamlessly as work progresses.

As discussed in the previous section, the project design is normally performed in a project reference coordinate system. Thus the GPS receiver must be localized (see Sections 15.8 and 19.6.6) before any excavation is performed. It is important in this process to have control surrounding the job. Once localization is performed and accepted, the application software will convert the GPS coordinates to the project's horizontal and vertical reference frames. Additionally, the GPS receivers must be referenced to the cutting edges of the construction equipment. Since the GPS receiver is often placed on the blade of the construction vehicle, this often means measuring the height of the antenna reference point above the cutting edge of the blade on a regular basis to account for wear. Often the machine operator can be instructed on how to perform this function.

GPS can provide heights to a few centimeters. Thus it is sufficiently accurate for rough grading. However, in finished grading, a robotic total station or laser level is required. Some manufacturers have combined GPS with a robotic total station to provide a finished grade solution, while another has combined a laser level with the GPS antenna. In either case, as previously stated, the range of the vehicle from the base station is limited and thus additional control will be required to control the final grades. Additional control is also required where canopy conditions exist, such as under overpasses. In these situations, surveyors may need to switch machine guidance from GPS to total stations or may even need to set stakes to guide the construction process through the canopy.

Three-dimensional machine control is also possible with a robotic total station. In this system, a multi-faceted, 360° prism replaces the GPS receiver on the construction vehicle. Similar to using GPS, a DTM and grading plan of the site are loaded into the system. The robotic total station then tracks the prism mounted on the construction vehicle and provides the system with the prism's position and elevation. This in turn guides the operator during the excavation and finishing process. Similar to GPS guidance, the offset from the prism to the cutting edge of the equipment must be measured and entered into the system on a regular basis to account for wear.

Since robotic total stations typically work in a local coordinate system, there is no need for localization using a robotic total station. However the range of a robotic total station is generally limited to only 1000 ft (300 m). Additionally there must be continuous line of sight between the robotic total station and construction vehicle. Thus many more control stations must be added to the site when using a robotic total station. For example to cover the typical range (10 km) of a GPS base station radio using a robotic total station would require 33 control stations. Another drawback of this system is that an additional robotic total station must be located and oriented for each construction vehicle. While a GPS system requires an additional machine control system for each construction vehicle in the base station's range, it does not require continuous line of sight from the base station nor any additional base station receivers and radios.

Similar to machine control, site communication applications on trucks allow managers to monitor quantities of excavation, hauling distances, and any surplus or fill earthwork. The system provides managers with daily reports on earthwork quantities, machine maintenance, and timetable gains or losses. Because of these features, many companies are finding that they can complete projects in a timely fashion often receiving bonuses for completing the project on or ahead of schedule.

■ 23.12 AS-BUILT SURVEYS WITH LASER SCANNING

As mentioned in Section 23.8, as-built surveys are performed at the completion of a construction project to ensure that project specifications are met and note any changes to plans. In many cases these surveys are performed with traditional surveying equipment. However in projects that involve extensive detail, danger to instrument-operator, or interruption of daily-commerce, laser scanning can provide superior results in a fraction of the time. Figure 23.15 shows a rendered image of a bridge that was surveyed for renovations. In the bridge survey, enormous quantities of data were collected from an on-shore location. The digital image of the bridge is shown in the lower-right inset. The rendered, three-dimensional image of the bridge allows designers to obtain accurate measurements between points in the image. Figure 17.12 depicts the point-cloud image of

Figure 23.15 Rendered and rotated image of bridge shown in lower-right corner. (Courtesy Christopher Gibbons, Leica Geosystems, Inc.)

a refinery with the path of a new pipe shown in white. This three-dimensional image allows engineers to design the new pipe alignment so that existing obstructions are cleared. A traditional survey would have either lacked the detail provided by the three-dimensional laser-scanned image or cost considerably more to locate all the existing elements. Using laser-scanning technology in these projects can save thousands of dollars. Today, some total stations can also perform laser-scanning tasks. However these units work at a much slower rate typically than the dedicated laser scanner.

■ 23.13 SOURCES OF ERROR IN CONSTRUCTION SURVEYS

Important sources of error in construction surveys are:

1. Inadequate number and/or location of control points on the job site.
2. Errors in establishing control.
3. Observational errors in layout.
4. Failure to double-center in laying out angles or extending lines, and failure to check vertical members by plunging the instrument.
5. Careless referencing of key points.
6. Movement of stakes and marks.

■ 23.14 MISTAKES

Typical mistakes often made in construction surveys are:

1. Lack of foresight as to where construction will destroy points.
2. Notation for cut (or fill) and stationing on stake not checked.
3. Wrong datum for cuts, whether cut is to finished grade or subgrade.
4. Arithmetic mistakes, generally due to lack of checking.
5. Use of incorrect elevations, grades, and stations.
6. Failure to check the diagonals of a building.
7. Carrying out computed values to too many decimal places (one good hundredth is better than all the bad thousandths).
8. Reading the rod on top of stakes instead of on the ground beside them in profiling and in slope staking.

PROBLEMS

Asterisks (*) indicate problems that have answers given in Appendix G.

23.1 Describe the types of construction projects where reflectorless instruments are useful.

23.2 Discuss how line and grade can be set with a total station instrument.

23.3 Describe how a plumbing level can be used to ensure verticality in the construction of a tall building.

23.4 In what types of construction is a rotating beam laser level most advantageous?

23.5 Discuss the jobs not performed when using machine control in a construction project.

23.6 What is a story pole, and how is it used in pipeline layout?

23.7 Describe how RTK GPS can be used in sewer line layout.

23.8 State two conflicting requirements that enter the decision on how far offset stakes should be set beyond the construction line.

23.9 A sewer pipe is to be laid from station 10 + 00 to station 12 + 75 on a 0.50 percent grade, starting with invert elevation 553.64 ft at 10 + 00. Calculate invert elevations at each 50-ft station along the line.

23.10* A sewer pipe must be laid from a starting invert elevation of 748.96 ft at station 9 + 25 to an ending invert elevation 732.12 ft at station 12 + 75. Determine the uniform grade needed, and calculate invert elevations at each 50-ft station.

23.11 Grade stakes for a pipeline running between stations 0 + 00 and 5 + 64 are to be set at each full station. Elevations of the pipe invert must be 1168.25 ft at station 0 + 00 and 1192.05 ft at 6 + 37, with a uniform grade between. After staking an offset centerline, an instrument is set up nearby, and a plus sight of 4.06 taken on BM *A* (elevation 1173.63 ft). The following minus sights are taken with the rod held on ground at each stake: (0 + 00, 5.51); (1 + 00, 5.67); (2 + 00, 5.03); (3 + 00, 7.16); (4 + 00, 7.92); (5 + 00, 8.80); (5 + 64, 9.10); and (A, 4.06). Prepare a set of suitable field notes for this project (see Plate B.6 in Appendix B) and compute the cut required at each stake. Close the level circuit on the benchmark.

23.12 If batter boards are to be set exactly 8.00 ft above the pipe invert at each station on the project of Problem 23.11, calculate the necessary rod readings for placing the batter boards. Assume the instrument has the same HI as in Problem 23.11.

23.13 How are streets and street grades arranged for drainage in a city with flat terrain?

23.14 By means of a sketch, show how and where batter boards should be located: (a) for an I-shaped building and (b) for an L-shaped structure.

23.15 A building in the shape of an L must be staked. Corners *ABCDEF* all have right angles. Proceeding clockwise around the building, the required outside dimensions are $AB = 80.00$ ft, $BC = 30.00$ ft, $CD = 40.00$ ft, $DE = 40.00$ ft, $EF = 40.00$ ft, and $FA = 70.00$ ft. After staking the batter boards for this building and stretching string lines taut, check measurements of the diagonals should be made. What should be the values of *AC*, *AD*, *AE*, *FB*, *FC*, *FD*, and *BD*?

23.16* Compute the floor area of the building in Problem 23.15.

23.17* The design floor elevation for a building to be constructed is 1070.13. An instrument is set up nearby, leveled, and a plus sight of 6.37 taken on BM *A* whose elevation is 1069.98 ft. If batter-boards are placed exactly 1.00 ft above floor elevation, what rod readings are necessary on the batter-board tops to set them properly?

23.18 Compute the diagonals necessary to check the stakeout of the building in Figure 23.8.

23.19 Why is important to check the plumb of a building using a total station instrument in both faces of the instrument?

23.20 Should a street or highway be designed with a grade of 0.00 percent? Explain.

23.21 Discuss the importance of tying in and referencing critical centerline points on highway construction surveys.

23.22 What is a transition section in highway alignments?

23.23 Discuss the advantages of combining digital elevation models with design templates in staking out highway alignments with a data collector.

23.24 Describe a field procedure for setting slope stakes.

23.25 Discuss the procedure and advantages of using total station instruments with data collectors for slope staking.

23.26 How are alignments transferred from the surface into mines?

23.27 A highway centerline subgrade elevation is 660.67 ft at station 12 + 00 and 668.73 ft at 17 + 00 with a smooth grade in between. To set blue tops for this portion of the centerline, a level is set up in the area and a plus sight of 7.19 ft taken on a benchmark whose elevation is 665.96 ft above sea level. From that HI, what rod readings will be necessary to set the blue tops for the full stations from 12 + 00 through 17 + 00?

23.28 Similar to Problem 23.27, except the elevations at stations 12 + 00 and 17 + 00 are 1405.65 and 1411.55 ft, respectively, the BM elevation is 1406.36 ft, and the backsight is 6.66 ft.

23.29 Describe the procedure for aligning a laser in a pipe layout project.

23.30 What are the jobs of a surveyor in a project using machine control?

23.31 Discuss the checks that should be made when laying out a building using coordinates.

23.32 Why is localization important in a RTK-GPS stakeout survey?

23.33 How are finished grades established in machine control projects?

23.34 What is the minimum number of control points needed to establish finish-grades using a robotic total station on a machine-controlled project that is 12 mi in length?

23.35 What advantages does GPS have over robotic total stations when using machine control? What are the disadvantages?

23.36 Discuss the advantages of using laser-scanning technology when planning for a new pipeline in a refinery.

23.37 Find the websites of at least two regional companies that specialize in construction.

23.38 Do an article review on an application of three-dimensional laser scanning.

BIBLIOGRAPHY

Blackford, Rippin. 2005. "The Total Solution? Are Total Stations Here to Stay?" *Point of Beginning* 30 (No. 11): 26.

———. 2005. "Advanced Construction Surveying—Mountain Style." *Professional Surveyor* 25 (No. 5): 26.

Bryant, Matt. 2006. "3D Machine Control: Where Does the Surveyor Fit In?" *Professional Surveyor* 26 (No. 1): 18.

Carris, M. 2005. "Making the Most of Machine Calibration." *Point of Beginning* 30 (No. 12): 24.

Cosworth, C. 2006. "Conforming to Design." *Point of Beginning* 31 (No. 6): 18.

Garret, Justin. 2007. "Reservoir of Lessons Learned." *Professional Surveyor* 27 (No. 2): 18.

Gourley, B. and K. Stewart. 2001. "Merging Quality and Efficiency." *Point of Beginning* 26 (No. 6): 26.

Harris, Clay. 2007. "Whole New Ball Game." *Professional Surveyor* 27 (No. 2): 26.

Hoechst, J. 2006. "Surveying on the Fast Track." *Point of Beginning* 31 (No. 5): 16.

Hohner, L. 2006. "Three Men and a Total Station." *Point of Beginning* 31 (No. 5): 22.

Hohner, Leica N. 2007. "A Way to Grow." *Point of Beginning* 32 (No. 4): 26.

Jacobs, Geoff. 2004. "3D Laser Scanning: An Ultra-Fast, High-Definition, Reflectorless Topographic Survey." *Professional Surveyor* 24 (No. 5): 38.

Jacobs, Geoff. 2005. "High Definition Surveying—3D Laser Scanning: Uses in Transportation." *Professional Surveyor* 25 (No. 4): 26.

———. 2006. "Performing Classic As-Builts with Laser Scanning." *Professional Surveyor* 26 (No.3): 20.

Lawson, Ray. 2007. "Laser Scanning Hits the Road Running." *Professional Surveyor* 27 (No. 2): 22.

Nelson, J., et. al. 2003. "Horizontal Drilling in Mesa Verde." *Professional Surveyor* 23 (No. 10): 6.

Roth, A., and I. Gonzalez. 1989. "Rotating Beam Laser Leveling Systems—Applications for Surveyors." *Point of Beginning* 14 (No. 1): 78.

Schaal, R., A. Larocca, and M. Santos. 2005. "Using GPS to Monitor Movement of a Cable-Stayed Bridge." *Professional Surveyor* 25 (No. 7): 18.

Warne, T. R. 1990. "Contractor Staking: Saving Public Works Dollars." *ASCE, Journal of Surveying Engineering* 116 (No. 2): 82.

Wrock, D. 2001. "Big Dig's Precision Floatout." *Professional Surveyor* 21 (No. 8): 8.

24 Horizontal Curves

■ 24.1 INTRODUCTION

Straight (tangent) sections of most types of transportation routes, such as highways, railroads, and pipelines, are connected by curves in both the horizontal and vertical planes. An exception is a transmission line, in which a series of straight lines is used with abrupt angular changes at tower locations if needed.

Curves used in horizontal planes to connect two straight tangent sections are called *horizontal curves*. Two types are used: *circular arcs* and *spirals*. Both are readily laid out in the field with standard surveying equipment. A *simple curve* [Figure 24.1(a)] is a circular arc connecting two tangents. It is the type most often used. A *compound curve* [Figure 24.1(b)] is composed of two or more circular arcs of different radii tangent to each other, with their centers on the same side of the alignment. The combination of a short length of tangent (less than 100 ft) connecting two circular arcs that have centers on the same side [Figure 24.1(c)] is called a *broken-back* curve. A *reverse curve* [Figure 24.1(d)] consists of two circular arcs tangent to each other, with their centers on opposite sides of the alignment. Compound, broken-back, and reverse curves are unsuitable for modern high-speed highway, rapid transit, and railroad traffic and should be avoided if possible. However, they are sometimes necessary in mountainous terrain to avoid excessive grades or very deep cuts and fills. Compound curves are often used on exit and entrance ramps of interstate highways and expressways, although easement curves are generally a better choice for these situations.

Easement curves are desirable, especially for railroads and rapid transit systems, to lessen the sudden change in curvature at the junction of a tangent and a circular curve. A *spiral* makes an excellent easement curve because its radius decreases uniformly from infinity at the tangent to that of the curve it meets. Spirals are used to connect a tangent with a circular curve, a tangent with

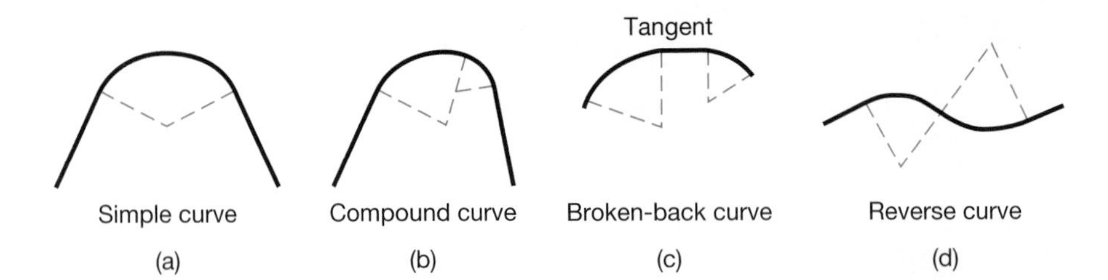

Figure 24.1 Circular curves.

a tangent (double spiral), and a circular curve with a circular curve. Figure 24.2 illustrates these arrangements.

The effect of centrifugal force on a vehicle passing around a curve can be balanced by *superelevation,* which raises the outer rail of a track or outer edge of a highway pavement. Correct transition into superelevation on a spiral increases uniformly with the distance from the beginning of the spiral and is in inverse proportion to the radius at any point. Properly superelevated spirals ensure smooth and safe riding with less wear on equipment. As noted, spirals are used for railroads and rapid-transit systems. This is because trains are constrained to follow the tracks, and thus a smooth, safe, and comfortable ride can only be assured with properly constructed alignments that include easement curves. On highways, spirals are less frequently used because drivers are able to overcome abrupt directional changes at circular curves by steering a spiraled path as they enter and exit the curves.

Although this chapter concentrates on circular curves, methods of computing and laying out spirals are introduced in Section 24.18.

■ 24.2 DEGREE OF CIRCULAR CURVE

The rate of curvature of circular curves can be designated either by their *radius* (for example a 1500-m curve or a 1000-ft curve), or by their *degree of curve.* There are two different designations for degree of curve, the *arc definition,* and *chord definition,* both of which are defined using the English system of units. By the arc definition, degree of curve is the central angle subtended by a circular *arc* of 100 ft [see Figure 24.3(a)]. This definition is preferred for highway work. By the chord definition, degree of curve is the angle at the center of a circular arc subtended by a chord of 100 ft [see Figure 24.3(b)]. This definition is convenient

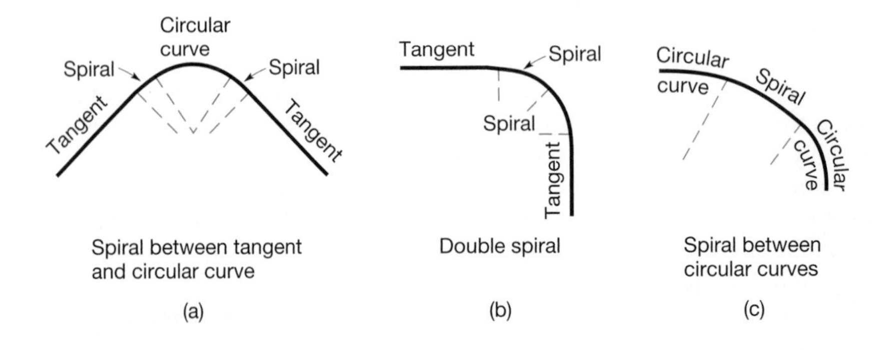

Figure 24.2 Use of spiral transition curves.

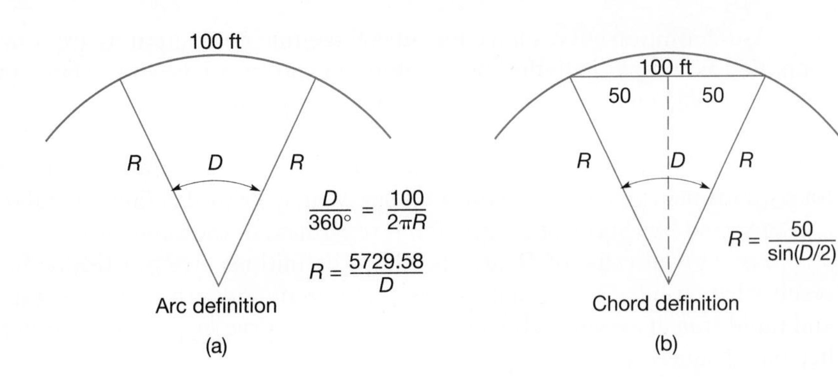

Figure 24.3 Degree of circular curve.

for very gentle curves and hence is preferred for railroads. The formulas relating radius R and degree D of curves for both definitions are shown next to the illustrations.

Using the equations given in Figure 24.3, radii of arc and chord definition curves for values of D from 1° to 10° have been computed and are given in columns (2) and (5) of Table 24.1. Although radius differences between the two definitions appear to be small in this range, they are significant.

When metric units are used, degree of curve can still be specified. As an example, a curve having a radius of exactly 700 m would have a degree of curve (arc definition) of

$$\frac{5729.58}{700 \times 3.28083} = 2°29'41''$$

TABLE 24.1 FUNCTIONS OF CIRCULAR CURVES (LENGTHS IN FEET)

	Arc Definition			Chord Definition		
Degree of Curve *D* (1)	**Radius *R* (2)**	**True Chord Full Sta (3)**	**True Chord Half-Sta (4)**	**Radius *R*(5)**	**Arc Length Full Sta (6)**	**True Chord Half-Sta (7)**
1	5729.58	100.00	50.00	5729.65	100.00	50.00
2	2864.79	99.99	50.00	2864.93	100.01	50.00
3	1909.86	99.99	50.00	1910.08	100.01	50.00
4	1432.39	99.98	50.00	1432.69	100.02	50.01
5	1145.92	99.97	50.00	1146.28	100.03	50.01
6	954.93	99.95	50.00	955.37	100.05	50.02
7	818.51	99.94	50.00	819.02	100.06	50.02
8	716.20	99.92	49.99	716.78	100.08	50.03
9	636.62	99.90	49.99	637.27	100.10	50.04
10	572.96	99.88	49.98	573.69	100.13	50.05

Arc-definition curves have the advantage that computations are somewhat simplified as compared to the chord definition and, as will be shown later, the formula for curve length is exact, which simplifies preparing right-of-way descriptions. A disadvantage with the arc definition is that most measurements between full stations are shorter than a full 100-ft tape length, but this is of little significance if a total station instrument is used for stakeout. With the chord definition, full stations are separated by chords of exactly 100 ft regardless of the value of D.

For a given value of D, arc and chord definitions give practically the same result when applied to the flat curves common on modern highways, railroads, and rapid-transit systems. However, as degree of curve increases, the differences become greater.

■ 24.3 DEFINITIONS AND DERIVATION OF CIRCULAR CURVE FORMULAS

Circular curve elements are shown in Figure 24.4. The *point of intersection* PI, of the two tangents is also called the *vertex V*. In stationing, the *back tangent* precedes the PI, the *forward tangent* follows it. The beginning of the curve, or *point of curvature* PC, and the end of the curve, or *point of tangency* PT, are also sometimes called BC and EC, respectively. Other expressions for these points are tangent to curve, TC, and curve to tangent, CT. The curve radius is R. Note that the radii at the PC and PT are perpendicular to the back tangent and forward tangent, respectively.

The distance from PC to PI and from PI to PT is called the *tangent distance, T*. The line connecting the PC and PT is the *long chord* LC. The *length of the curve, L*, is the distance from PC to PT, *measured along the curve for the arc definition, or by 100-ft chords for the chord definition*.

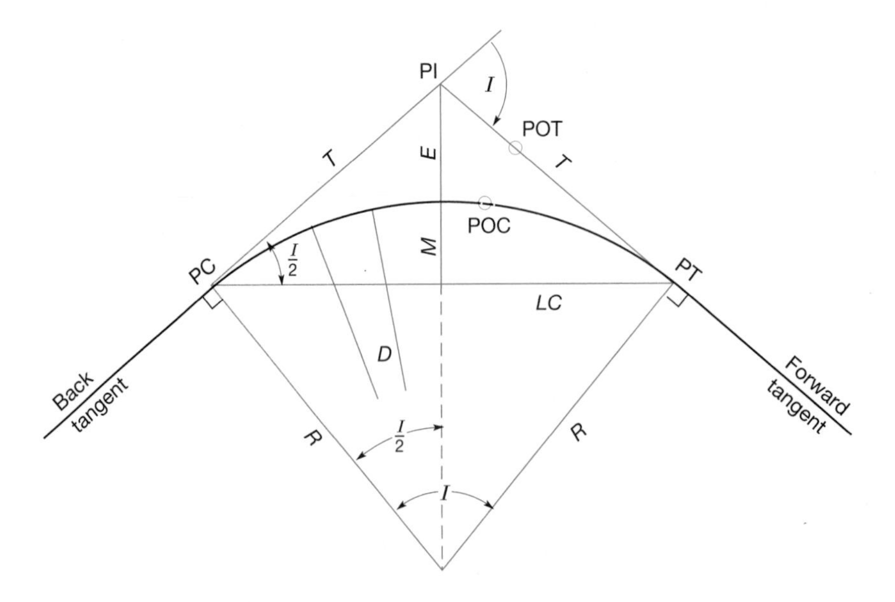

Figure 24.4 Circular curve elements.

The external distance E is the length from the PI to the curve midpoint on a radial line. The middle ordinate M is the (radial) distance from the midpoint of the long chord to the curve's midpoint. Any *point on curve* is POC; any *point on tangent,* POT. The degree of any curve is D_a (arc definition) or D_c (chord definition). The change in direction of two tangents is the intersection angle I, which is also equal to the central angle subtended by the curve.

By definition, and from inspection of Figure 24.4, relations for the arc definition follow:

$$L = RI\,(I \text{ in radians}) \tag{24.1}$$

$$L = 100\frac{I^\circ}{D^\circ}\,(\text{ft}) \tag{24.2a}$$

$$L = \frac{I^\circ}{D^\circ}\,(\text{sta}) \tag{24.2b}$$

$$R = \frac{5729.58}{D}\,(\text{ft}) \tag{24.3}$$

$$T = R\tan\frac{I}{2} \tag{24.4}$$

$$LC = 2R\sin\frac{I}{2} \tag{24.5}$$

$$\frac{R}{R+E} = \cos\frac{I}{2} \quad\text{and}\quad E = R\left[\frac{1}{\cos(I/2)} - 1\right] \tag{24.6}$$

$$\frac{R-M}{R} = \cos\frac{I}{2} \quad\text{and}\quad M = R\left(1 - \cos\frac{I}{2}\right) \tag{24.7}$$

Other convenient formulas that can be derived are:

$$E = T\tan\frac{I}{4} \tag{24.8}$$

$$M = E\cos\frac{I}{2} \tag{24.9}$$

Although curves are normally calculated by computers, if a hand-held calculator is used, it is expedient to compute R, T, E, and M in the sequence of Equations (24.3), (24.4), (24.8), and (24.9), because the previously computed value is in the calculator and available for each succeeding calculation.

The formulas for T, L, LC, E, and M also apply to a chord-definition curve. However, L calculated by Equation (24.2a) implies the total length as if observed along the 100-ft chords of an inscribed polygon. The following formula is used for relating R and D for a chord definition curve:

$$R = \frac{50}{\sin(D/2)}\,(\text{ft}) \tag{24.10}$$

Note that for the equations given above, Equations (24.2a), (24.2b), (24.3), and (24.10) involve degree of curvature and thus assume distances in feet, while either metric or English units can be used in all others.

24.4 CIRCULAR CURVE STATIONING

Normally, an initial route survey consists of establishing the PIs according to plan, laying out the tangents, and establishing continuous *stationing* along them from the start of the project, through each PI, to the end of the job. (Stationing was described in Section 5.9.1). The beginning point of any project is assigned a station value and all other points along the reference line are then related to it. If the beginning point is also the end point of a previous adjacent project, its station value may be retained and the new job referenced to that stationing. Otherwise, an arbitrary value such as 100 + 00 for English unit stationing, or 10 + 000 for metric stationing is assigned. Assigning a starting stationing of 0 + 00 is generally not done to avoid the possibility that future revisions to the project could extend it back beyond the starting point and hence result in negative stationing. In the English system, staking is usually in *full stations* (100 ft apart), although *half stations* (50 ft apart) or even *quarter stations,* (25 ft apart) can be set depending upon conditions. In metric stationing, full stations are generally 1 km apart, but stakes may be set at 40, 30, 20, or even 10 m apart, depending upon conditions. Staking at the closer spacing is usually done in urban situations, on sharp curves, or in rugged terrain, while the stakes are placed farther apart in relatively flat or gently rolling rural areas.

After the tangents have been staked and stationed, the intersection angle (I) is observed at each PI and curves computed and staked. The station locations of points on any curve are based upon the stationing of the curve's PI. To compute the PC station, tangent distance T is subtracted from the PI station, and to calculate the PT station, curve length L is added to the PC station.

Example 24.1

Assume that $I = 8°24'$, the station of the PI is 64 + 27.46, and terrain conditions require the minimum radius permitted by the specifications of, say, 2864.79 ft (arc definition). Calculate the PC and PT stationing and the external and middle ordinate distances for this curve.

Solution

By Equation (24.1), $L = 2864.79(8°24')\dfrac{\pi}{180} = 420.00 \text{ ft}$

By Equation (24.2a), $D° = 100\dfrac{8°24'}{420} = 2°00'$

By Equation (24.4), $T = 2864.79 \times \tan\left(\dfrac{8°24'}{2}\right) = 210.38 \text{ ft}$

Calculate stationing:

$$\begin{aligned} \text{PI station} &= 64 + 27.46 \\ -T &= \underline{-2 + 10.38} \\ \text{PC station} &= 62 + 17.08 \\ +L &= \underline{4 + 20.00} \\ \text{PT station} &= 66 + 37.08 \end{aligned}$$

Also by Equation (24.5), $LC = 2 \times 2864.79 \sin\left(\dfrac{8°24'}{2}\right) = 419.62$ ft

And by Equation (24.8), $E = 210.38 \times \tan\left(\dfrac{8°24'}{4}\right) = 7.71$ ft

Finally by Equation (24.9), $M = 7.71 \times \cos\left(\dfrac{8°24'}{2}\right) = 7.69$ ft

---

Calculation for the stations of the PC and PT should be arranged as shown. Note that the stationing of the PT cannot be obtained by adding the tangent distance to the station of the PI, although the location of the PT on the ground is determined by measuring the tangent distance from the PI. Points representing the PC and PT must be carefully marked and placed exactly on the tangent lines at the correct distance from the PI so other computed values will fit their fixed positions on the ground.

If route surveys are originally staked as a series of tangents having continuous stationing, as described previously, then an adjustment has to be made at each PT after curves are inserted. This is necessary because the length around the curve from PC to PT is shorter than the distance along the tangents from the PC to the PI to the PT. Thus for final stationing at the PT, there is a "station equation," which relates the stationing back along the curve to that forward along the tangent. For Example 24.1, it would be $66 + 37.08$ back $= 66 + 37.84$ ahead, where $66 + 37.08 = \text{PI} - T + L$, and $66 + 37.84 = \text{PI} + T$. The difference between the ahead and back stations represents the amount the route was shortened by inserting the curve. If the curves are run in and stationed at the time of staking the original alignment, continuous stationing along the route results and station equations at PTs are avoided.

The curve used in a particular situation is selected to fit ground conditions and specification limitations of minimum R or maximum D. Normally the value of the intersection angle, I, and the station of the PI are available from field observations on the preliminary line. Then a value of R or D, suitable for the highway or railroad, is chosen. Most of today's highways are designed using a minimum value for the radius R. Sometimes the distance E or M required to miss a stream or steep slope outside or inside the PI is observed and R or D computed holding that distance fixed. Tangent distance governs infrequently. (One exception is to make a railroad, bus, or subway station fall on a tangent rather than a superelevated curve.) The length of curve practically never governs.

■ 24.5 GENERAL PROCEDURE OF CIRCULAR CURVE LAYOUT BY DEFLECTION ANGLES

Except for unusual cases, the radii of curves on route surveys are too large to permit swinging an arc from the curve center. Circular curves are therefore laid out by more practical methods, including (1) deflection angles, (2) coordinates, (3) tangent offsets, (4) chord offsets, (5) middle ordinates, and (6) ordinates from the long chord. Layout by deflection angles has been the standard approach, although with the advent of total station instruments, the coordinate method is used typically.

Layout of a curve by deflection angles can be done by either the *incremental chord method* or the *total chord method*. In years past, the incremental chord method was almost always used as it could be readily accomplished with a theodolite and tape. The method can still be used when a total station instrument is employed, although then the distances are observed by taping rather than electronically. (Taping is still efficient in staking the stations along alignments because of the relatively short distance increments involved.) The total chord method was not practical until the advent of total stations, but with these instruments it is now conveniently employed even though longer distance measurements are involved.

The incremental chord method is illustrated in Figure 24.5. Assume that the instrument is set up over the PC (station 62 + 17.08 in Example 24.1). For this illustration, assume that each full station is to be marked along the curve, since cross sections are normally taken, construction stakes set, and computations of earthwork made at these points. (Half-stations or any other critical points can also be established, of course.) The first station to be set in this example is 63 + 00. To mark that point from the PC, a backsight is taken on the PI with zero set on the instrument's horizontal circle. Deflection angle δ_a to station 63 + 00 is then turned and two tapepersons measure chord c_a from the PC and set 63 + 00 at the end of

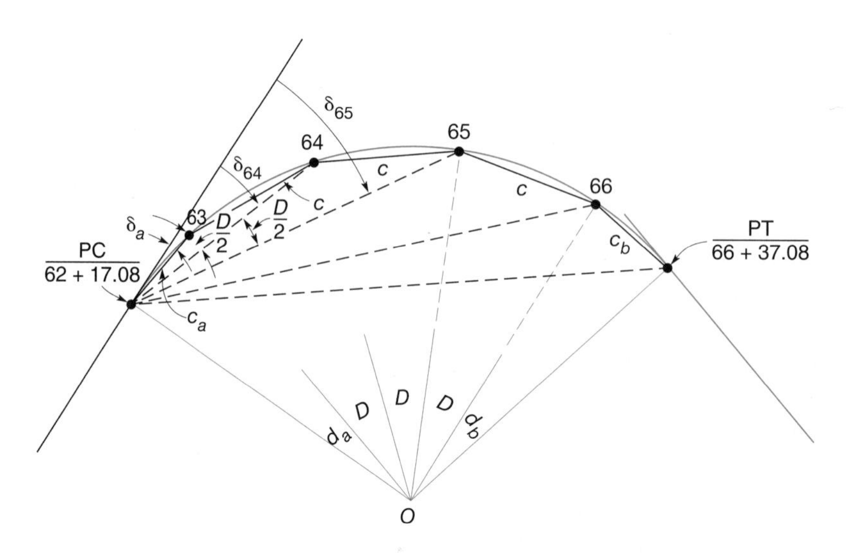

Figure 24.5 Circular curve layout by deflection angles.

the chord on the instrument's line of sight. With station 63 + 00 set, the tapepersons next measure the chord length c from it and stake station 64 + 00, where the line of sight of the instrument, now set to δ_{64}, intersects the end of that chord. This process is repeated until the entire curve is laid out. In this procedure it is seen that the accuracy in the placement of each succeeding station depends upon the accuracies of all those stations previously set.

The total chord method can also be described with reference to Figure 24.5. In this procedure, a total station instrument is set up at the PC, a backsight taken on the PI, and zero indexed on the horizontal circle. To set station 63 + 00, the deflection angle δ_a is turned with the instrument and the reflector is placed on line and adjusted until its distance from the instrument is c_a. Then the stake is set. To set station 64 + 00, deflection angle δ_{64} is turned, the reflector is placed on this line of sight and adjusted in position until the total chord from the PC to station 64 + 00 is obtained and the stake set.

This procedure is repeated, with each station being set independently of the others until the entire curve is staked. This method of staking a curve has some drawbacks. One is that in some areas vegetation or other obstructions can block sight lines along the chords. Another is that each station is set independently and thus there is no check at the end of the curve. For these reasons, the incremental chord approach is often preferred over the total chord method.

■ 24.6 COMPUTING DEFLECTION ANGLES AND CHORDS

From the preceding discussion it is clear that deflection angles and chords are important values that must be calculated if a curve is to be run by the deflection-angle method. To stake the first station, which is normally an odd distance from the PC (shorter than a full-station increment), *subdeflection angle* δ_a and *subchord* c_a are needed. These are shown in Figure 24.6. In this figure, central angle d_a

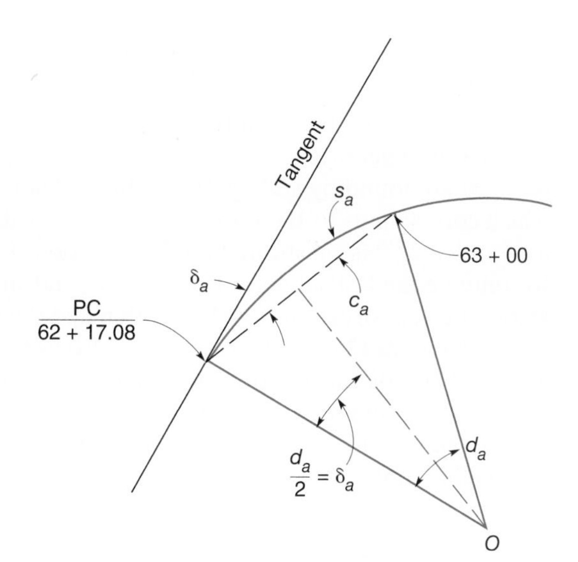

Figure 24.6 Subchords and subdeflections.

subtended by arc s_a from the PC to 63 + 00 is calculated by proportion according to the definition of D as

$$\frac{d_a}{s_a} = \frac{D}{100} \text{ from which } d_a = \frac{s_a D}{100} \text{(degrees)} \tag{24.11a}$$

where s_a is the difference in stationing between the two points. Equation (24.11a) is based on curves defined in English units using the arc definition for degree of curvature. For curves computed in English or metric units, the equivalent expression is

$$\frac{d_a}{s_a} = \frac{I}{L} \text{ from which } d_a = \frac{s_a I}{L} \text{(degrees)} \tag{24.11b}$$

A fundamental theorem of geometry helpful in circular curve computation and stakeout is that *the angle at a point between a tangent and any chord is equal to half the central angle subtended by the chord*. Thus, subdeflection angle δ_a needed to stake station 63 + 00 is $d_a/2$, or

$$\delta_a = \frac{s_a D}{200} \text{(degrees)} \tag{24.12a}$$

Also recognizing that I and L are constants for any particular curve, Equation (24.11b) can be rewritten as

$$\delta_a = s_a k \text{(degrees)} \tag{24.12b}$$

where k is $I/(2L)$.

The length of the subchord c_a can be represented in terms of δ_a and the curve radius as

$$\sin \delta_a = \frac{c_a}{2R} \text{ from which } c_a = 2R \sin \delta_a \tag{24.13}$$

Since the arc between full stations subtends the central angle D, from the earlier stated geometric theorem, deflection angles to each full station beyond 63 + 00 are found by adding $D/2$ to the previous deflection angle. Full chord c, which corresponds to 100 ft of curve length, is calculated using Equation (24.13), except that $D/2$ is substituted for δ_a. Equations (24.12) and (24.13) are also used to compute the last subdeflection angle δ_b and subchord c_b, but the difference in stationing s_b between the last full station and the PT replaces arc length s_a.

Equations (24.12) and (24.13) are also used for computing deflection angles and total chords for the total chord method of staking. Here s is simply the difference in stationing between the station being set and the PC.

For curves up to about 2°00′ (arc definition), the lengths of arcs and their corresponding chords are nearly the same. On sharper curves, chords are shorter than corresponding arc lengths. This is verified by the data in columns (3) and (4) of Table 24.1, which give true chord lengths for full- and half-station increments for varying values of D (arc definition).

Computations for deflection angles and chords on chord-definition curves use the same formulas, but R is calculated by Equation (24.10). Note that for a given degree of curve, R is longer for chord definition than for arc definition. Also arc lengths for full stations are longer than their nominal 100.00-ft value and true subchords are longer than their *nominal values* (differences in stationing). A check of columns (6) and (7) in Table 24.1 verifies these facts.

Example 24.2

Compute subdeflection angles and subchords δ_a, c_a, δ_b, and c_b, and calculate chord c of Example 24.1.

Solution

By Equation (24.12a),

$$\delta_a = 82.92 \times 0.0100° = 0.8292° = 0°49'45''$$

$$\delta_b = 37.08 \times 0.0100° = 0.3708° = 0°22'15''$$

$$\text{(Note that D/200} = 2°/200 = 0.01°)$$

By Equation (24.13),

$$c_a = 2(2864.79) \sin 0°49'45'' = 82.92 \text{ ft}$$

$$c_b = 2(2864.79) \sin 0°22'15'' = 37.08 \text{ ft}$$

$$c = 2(2864.79) \sin 1°00'00'' = 99.99 \text{ ft}$$

24.7 NOTES FOR CIRCULAR CURVE LAYOUT BY DEFLECTION ANGLES AND INCREMENTAL CHORDS

Based on principles discussed, the deflection angle and incremental chord data for stakeout of the complete curve of Examples 24.1 and 24.2 have been computed and listed in Table 24.2. Normally, as has been done in this case, the data are prepared

TABLE 24.2 DEFLECTION ANGLE AND INCREMENTAL CHORD DATA FOR EXAMPLE CURVE

Station	Incremental Chord	Deflection Increment	Deflection Angle
66 + 37.08 (PT)	37.08	0°22′15″	4°12′00″✓
66 + 00	99.99	1°00′00″	3°49′45″
65 + 00	99.99	1°00′00″	2°49′45″
64 + 00	99.99	1°00′00″	1°49′45″
63 + 00	82.92	0°49′45″	0°49′45″
62 + 17.08 (PC)			

for stakeout from the PC, although field conditions may not allow the curve to be completely run from there. This problem is discussed in Section 24.9.

Values of deflection angles are normally carried out to several decimal places for checking purposes and to avoid accumulating small errors when D is a non-integer number, such as, perhaps, 3°17′24″. Note in Table 24.2 that the deflection angle to the PT is 4°12′, exactly half the I angle of 8°24′. This comparison affords an important check on the calculations of all deflection angles.

Field notes for the curve of this example are recorded in Figure 24.7 as they would appear in a field book. Notes run up the page to simplify sketching while looking in a forward direction. Computers can conveniently perform all necessary calculations and list the notes for curve stakeout by deflection angles.

In many cases, it is desirable to "back in" a curve by setting up over the PT instead of the PC. One setup is thereby eliminated and the long sights are taken on the first measurements. In precise work it is better to run in the curve from both ends to the center, where small errors can be adjusted more readily. On long or very sharp circular curves, or if obstacles block sights from the PC or PT, setups on the curve are necessary (see Section 24.9).

ALIGNMENT OF

Station	Chord	Total Def.	Calc. Bearing	Mag. Bearing	Curve Data
P.O.T. 68	100.00				
67	62.92				
			N24°42′E	N24°45′E	
P.T. 66+37.08	37.08	4°12′00″			
					I=8°24′
66	99.99	3°49′~~48″~~ 45″			R=2864.79′
					D=2°00′
P.O.C. 65	99.99	2°49′45″			L=420.00′
					T=210.38′
64	99.99	1°49′45″			E=7.71′
					M=7.69′
63	82.92	0°49′45″			LC=419.62′
P.C. 62+17.08	17.08	0°00′00″			
			N16°18′E	N16°30′E	
62	100.00				
61	100.00				
60	100.00				
P.O.T. 59					

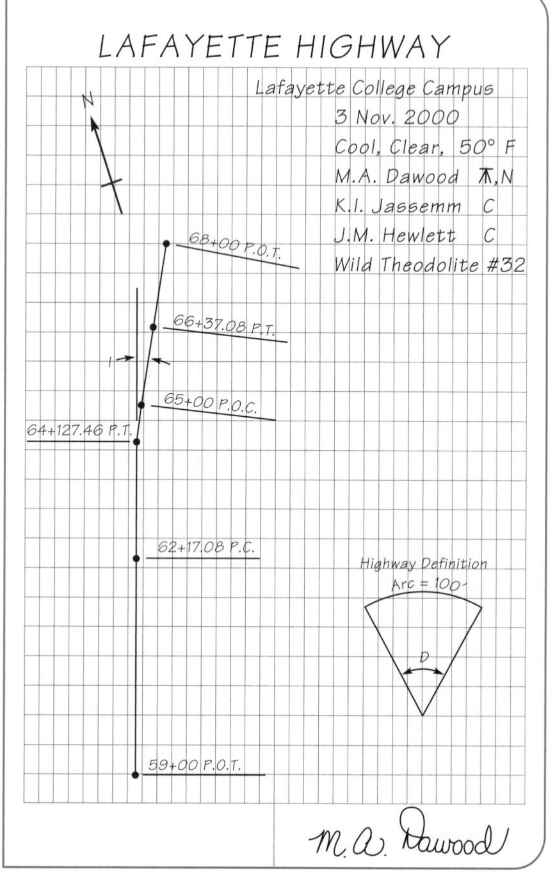

Figure 24.7 Field notes for horizontal curve in Examples 24.1 and 24.2.

■ 24.8 DETAILED PROCEDURES FOR CIRCULAR CURVE LAYOUT BY DEFLECTION ANGLES AND INCREMENTAL CHORDS

Regardless of the method used to stake intermediate curve points, the first steps in curve layout are: (1) establishing the PC and PT, normally by measuring tangent distance T from the PI along both the back and forward tangents and (2) measuring the total deflection angle at the PC from PI to PT. This latter step should be performed whenever possible, since the observed angle must check $I/2$; if it does not, an error exists in either observation or computation and time should not be wasted running an impossible curve.

It is also good practice to stake the midpoint of a curve before beginning to set intermediate points, especially on long curves. The midpoint can be set by bisecting the angle $180° - I$ at the PI and laying off the external distance from there. A check of the deflection angle from the PC to the midpoint should yield $I/4$. When the staking of intermediate points along the curve has reached the midpoint of the curve, a chord check measurement should be made to it.

The remaining steps in staking intermediate curve points by the deflection angle incremental chord method are presented with reference to the curve of Examples 24.1 and 24.2. With the instrument set up and leveled over the PC, it is oriented by backsighting on the PI, or on a point along the back tangent, with 0°00′ on the circle. The subdeflection angle of 0°49′45″ is then turned. Meanwhile, the 17-ft mark of the tape is held on the PC. The zero end of the (add) tape is swung until the line of sight hits a point 0.08 ft ahead from the zero mark. This is station 63 + 00. To stake station 64 + 00, the rear tapeperson next holds the 99-ft mark on station 63 and the forward tapeperson sets station 64 at distance 99.99 ft by direction from the instrument operator, who has placed an angle of 1°49′45″ on the circle. An experienced forward tapeperson will walk along the extended first full chord, know or estimate the chord offset, and from an outside-the-chord position will be holding the tape end and a stake within a foot or so of the correct location when the instrument operator has the deflection angle ready.

After placing the final full station (66 + 00 in this example), to determine any misclosure in staking a curve, the closing PT should be staked using the final deflection angle and subchord. This will rarely agree with the PT established by observing distance T along the forward tangent from the PI because of accumulated errors. Any misclosure ("falling") should be observed; then the field precision can be expressed as a numerical ratio like that used in traverse checks. The observed falling distance is the numerator, and $L + 2T$ the denominator. If the misclosure of this example was 0.25 ft, the precision would be $0.25/(420.00 + 2 \times 210.38) = 1/3300$.

■ 24.9 SETUPS ON CURVE

Obstacles that prevent visibility along curve chords and extremely long sight distances sometimes make it necessary to set up on the curve after it has been partially staked. The simplest procedure to follow is one that permits use of the same notes computed for running the curve from the PC. In this method the following rule applies: *The instrument is moved forward to the last staked point, backsighted on a station with the telescope inverted, and the circle set to the deflection angle from the PC for the station sighted. The telescope is then plunged to the normal position and deflection angles previously computed from the PC for the various stations are used.*

The example of the preceding sections is used to illustrate this rule. If a setup is required at station 65, place 0°00′ on the instrument and sight the PC with the telescope inverted. Plunge, set the circle to read the deflection angle 3°49′45″, and stake station 66.

To prove this geometric rule, assume the instrument were set on station 65 + 00 and the PC backsighted with 0°00′00″ on the circle. If the telescope were plunged and turned 2°49′45″ in azimuth, the line of sight would then be tangent to the curve. To stake the next station (66 + 00), an additional angle of $D/2$ or 1°00′00″ would need to be turned. The sum of 2°49′45″ and 1°00′00″ is of course the deflection angle to station 66 + 00 from the PC. The rule also applies for backsights to any previously set stations, not just the PC. Thus with the instrument at 65 + 00, station 63 + 00 could be sighted with 0°49′45″ on the circle and then turned to 3°49′45″ to set station 66 + 00. Further study of the geometry illustrated in Figure 24.5 should clarify this procedure.

■ 24.10 METRIC CIRCULAR CURVES BY DEFLECTION ANGLES AND INCREMENTAL CHORDS

Most foreign countries and many highway departments in the United States use metric units for observations and stationing on their projects. As noted earlier, in the metric system, circular curves are designated by their radius values rather than degree of curve. Otherwise, as illustrated by the following example, computations for laying out a curve using metric units by the deflection angle incremental chord method follow the same procedures as for the English system of units and stationing.

■ Example 24.3

Assume that a metric curve will be used at a PI where $I = 8°24'$. Assume also that the station of the PI is 6 + 427.464, and that terrain conditions require a minimum radius of 900 m. Calculate the PC and PT stationing, and other defining elements of the curve. Also compute notes for staking the curve using 20-m increments.

Solution

By Equation (24.1), $L = 900\left(8°24' \times \dfrac{\pi}{180°}\right) = 131.947 \text{ m}$

By Equation (24.4), $T = 900 \times \tan\left(\dfrac{8°24'}{2}\right) = 66.092 \text{ m}$

Calculate stationing:

$$
\begin{aligned}
\text{PI station} &= 6 + 427.464 \\
-T &= \underline{066.092} \\
\text{PC station} &= 6 + 361.372 \\
+L &= \underline{131.947} \\
\text{PT station} &= 6 + 493.319
\end{aligned}
$$

Also by Equation (24.5), $LC = 2 \times 900 \sin\left(\frac{8°24'}{2}\right) = 131.829$ m

And by Equation (24.8), $E = 66.092 \times \tan\left(\frac{8°24'}{4}\right) = 2.423$ m

Finally by Equation (24.9), $M = 2.423 \times \cos\left(\frac{8°24'}{2}\right) = 2.416$ m

The arc distance from the PC to the station 6 + 380 is (6380 − 6361.372) = 18.628 m. The arc distance for the final stationing is 6493.319 − 6480 = 13.319 m. All other stations have 20-m stationing intervals. Table 24.3 and Figure 24.8 depict the curve data and field notes necessary to stake the curve in this example.

By Equation (24.12b),

$$
\begin{aligned}
k &= 8°24'/(2 \times 131.947) = 0.03183096° \\
\delta_a &= 0.03183096 \times 18.628 = 0.59295° = 0°35'34.6'' \\
\delta &= 0.03183096 \times 20 = 0.63662° = 0°38'11.8'' \\
\delta_b &= 0.03183096 \times 13.319 = 0.42396° = 0°25'26.2''
\end{aligned}
$$

TABLE 24.3 DEFLECTION ANGLE AND INCREMENTAL CHORD DATA FOR EXAMPLE CURVE

Station	Incremental Chord	Deflection Increment	Deflection Angle
6 + 493.319 (PT)	13.318	0°25′26.2″	4°12′00″✓
6 + 480	19.999	0°38′11.8″	3°46′34″
6 + 460	19.999	0°38′11.8″	3°08′22″
6 + 440	19.999	0°38′11.8″	2°30′10″
6 + 420	19.999	0°38′11.8″	1°51′58″
6 + 400	19.999	0°38′11.8″	1°13′46″
6 + 380	18.268	0°35′34.6″	0°35′35″
6 + 361.372 (PC)			

ALIGNMENT OF LAFAYETTE HIGHWAY

Station	Chord	Total Def. ° ′ ″		Curve Data
(PT)				
6+493.319		4 12 00		I = 8°24′
	13.318			R = 900 m
6+480		3 46 34		k = 0.03183096
	19.999			L = 131.947 m
6+460		3 08 22		T = 66.092 m
	19.999			LC = 131.829 m
6+440		2 30 10		E = 2.423 m
	19.999			M = 2.416 m
6+420		1 51 58		
	19.999			
6+400		1 13 46		
	19.999			
6+380		0 35 35		
	18.628			
6+361.372				
(PC)				

Figure 24.8 Field notes for horizontal curve in Example 24.3.

By Equation (24.13),

$$c_a = 2(900)\sin 0°35'34.6'' = 18.628 \text{ m}$$

$$c = 2(900)\sin 0°38'11.8'' = 19.999 \text{ m}$$

$$c_b = 2(900)\sin 0°25'26.2'' = 13.318 \text{ m}$$

■ 24.11 CIRCULAR CURVE LAYOUT BY DEFLECTION ANGLES AND TOTAL CHORDS

If field conditions permit and a total station instrument is available, curves may be conveniently laid out by deflection angles and total chords. By using this method, the field party size is reduced from three to two, or possibly even a single person if a robotic total station instrument is available. Deflection angles are calculated and laid off as in the preceding example, but the chords are all measured electronically as radial distances (total chords) from the PC or other station where the instrument is placed. If stakeout is planned from the PC, total chords from there are the dashed lines of Figure 24.5. They are calculated by Equation (24.13), except that the deflection angle for each station is substituted for δ_α to obtain the corresponding chord. The total chords necessary to stake the curve of Example 24.2 using a total station instrument set up at PC are 82.92 ft for 63 + 00, 182.89 ft for 64 + 00, 282.80 ft for 65 + 00, 382.63 ft for 66 + 00, and 419.62 ft for the PT, which is the long chord LC given by Equation (24.5). The same deflection angles given in Table 24.2 apply.

To stake curves using a total station, the instrument is placed in its tracking mode. The deflection angle to each station is turned and the required chord to that station entered in the instrument. The instrument operator directs the person with the reflector to the proper alignment. The reflector is then moved forward or back as necessary, until the proper total chord distance is achieved, where the stake is set. It is often convenient to carry a short tape when staking out stations to quickly move to the final position from a nearby trial position. If intermediate setups are required on the curve using this method, the instrument is oriented as described in Section 24.9. New radial chords to be measured from the intermediate station would then have to be calculated.

Although curves can be staked rapidly with total stations using this method, as noted earlier, an associated danger is that each stake is set individually and therefore does not depend on previous stations. Thus a check at the end of the curve is not achieved like it is in the incremental chord method and mistakes in angles or distances at intermediate stations could go undetected. Large blunders can usually be discovered by visual inspection of the curve stakes, but quickly taping the chords between adjacent stations gives a better check.

■ 24.12 COMPUTATION OF COORDINATES ON A CIRCULAR CURVE

Today, because of the availability of total station instruments with data collectors, circular curves are often staked using the coordinate method. For this procedure, coordinates of the points on the curve to be staked must first be determined in some reference coordinate system. Although they are most often based upon an established map projection such as the State Plane Coordinate System or the Universal Transverse Mercator projection (see Chapter 20), often an arbitrary project coordinate system will suffice. This section describes the process of determining coordinates for stations on circular curves.

In Figure 24.9, assume that the azimuth of the back-tangent going from A to V is known, the coordinates of the PI (point V) are known, and that the defining parts of the curve have been computed using Equations (24.1) thru (24.10). Using the tangent distance and azimuth of the back tangent, the departure and latitude are computed by Equations (10.1) and (10.2), where Az_{VA} is the back azimuth of line AV. The coordinates of A (the PC) are then:

$$\begin{aligned} X_A &= X_V + T \sin Az_{VA} \\ Y_A &= Y_V + T \cos Az_{VA} \end{aligned} \qquad (24.14)$$

With the coordinates of the PC known, coordinates of points on the curve can be computed using the same deflection angles and subchords used to stake out the curve by the total chord method. Deflection angles are added to the azimuth of AV to get azimuths of the chords to all stations to be set. Using the total chord length and chord azimuth for each station, departures and latitudes are calculated, and added to the coordinates of A (the PC) to get the station coordinates. With

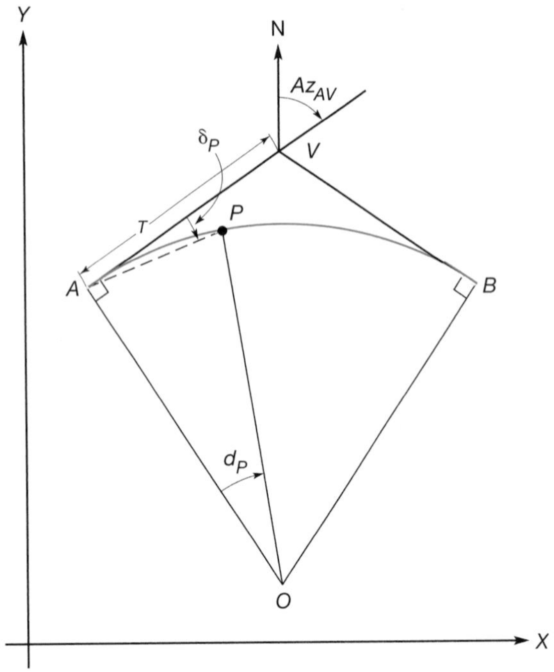

Figure 24.9 Geometry for computing coordinates of curve points.

coordinates known for all curve points, they can be staked with the total station occupying any convenient point whose coordinates are also known in the same system. Often the PC, PT, PI, or curve midpoint is used.

It is sometimes convenient to stake a circular curve by placing the instrument at the center of the curve, i.e., point O of Figure 24.9. In this case, the coordinates of the curve's center point are computed and then the coordinates of the stations to be set can be conveniently computed using radial lines from that point. From Figure 24.9, the azimuth of the radius going from A to the center of the curve is

$$Az_{AO} = Az_{AV} + 90° \tag{24.15a}$$

Equation (24.15a) is valid for a curve that lies right of the back tangent. For a curve that bends to the left, the proper expression is

$$Az_{AO} = Az_{AV} - 90° \tag{24.15b}$$

Using the appropriate azimuth from Equations (24.15), and the radius of the curve, R, the coordinates of center point O of Figure 24.9 are

$$\begin{aligned} X_O &= X_A + R \sin Az_{AO} \\ Y_O &= Y_A + R \cos Az_{AO} \end{aligned} \tag{24.16}$$

The azimuth of the radius line from O to any station P on the curve is

$$Az_{OP} = Az_{OA} + d_P \quad \text{(24.17)}$$

where d_P is determined in Equation (24.11). Then the coordinates of P are

$$\begin{aligned} X_P &= X_O + R \sin Az_{OP} \\ Y_P &= Y_O + R \cos Az_{OP} \end{aligned} \quad \text{(24.18)}$$

To stake the curve points, the total station instrument is set on the curve's center point, a backsight taken on point A, and the azimuth of line OA indexed on the horizontal circle. Then to stake any point such as P of Figure 24.9, the azimuth of OP is placed on the instrument's circle and the stake placed on the line of sight at a distance R from the instrument. Staking a curve from the center point has the advantage of providing an easy method of computing and laying out a line, which is offset from the reference line. This is done by simply using either a longer or shorter radius in Equations (24.18) to compute the coordinates of the offset line and then using that same radius value in staking. A disadvantage in staking a curve from its center is that radius values of the curves typically used on transportation routes are rather long which means that the instrument operator and prism person will usually be relatively far apart. Also with these longer values, obstructions that block lines of sight will often exist.

Example 24.4 in the following section demonstrates the method of computing coordinates of curve points using deflection angles and total chords.

■ 24.13 CIRCULAR CURVE LAYOUT BY COORDINATES

The coordinate method can be used to advantage in staking circular curves, especially if a total station instrument or GPS receivers are employed. In this procedure, the coordinates of each curve station to be staked are calculated as described in the preceding section. The total station instrument can then be placed at the PC, PT, curve midpoint, curve center point, or any other nearby control station, which gives a good vantage point of the entire area where the curve will be laid out. Azimuths and distances to each station are computed by inversing, using the coordinates of the occupied station and those of each curve station. The instrument is oriented by backsighting another visible control station. Then each curve point is staked by laying out its computed distance along its calculated azimuth.

Figure 24.10 illustrates a situation where a curve is being staked by the coordinate method. The total station instrument is placed at control station B because all curve points are visible from there. After a backsight on control station A, distances and directions are used to stake all curve points. The computations necessary for staking this curve by the coordinate method are illustrated with the following example.

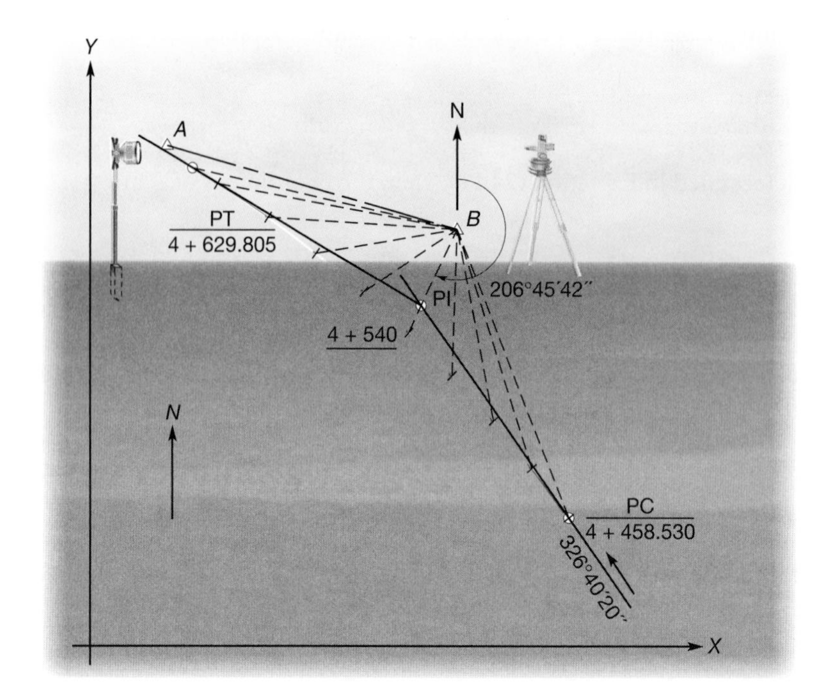

Figure 24.10 Layout of circular curve by coordinates with a total station instrument.

Example 24.4

Two tangents intersect at a PI station of 4 + 545.500 whose coordinates are $X = 5723.183$ m and $Y = 3728.947$ m. The intersection angle is 24°32′ left, and the azimuth of the back tangent is 326°40′20″. A curve with radius, R of 400 m will be used to join the tangents. Compute the data needed to stake the curve at 20 m increments by coordinates using a total station instrument. For staking, the instrument will be set at station B, whose coordinates are $X = 5735.270$ m and $Y = 3750.402$ m, and a backsight taken on station A, whose coordinates are $X = 5641.212$ m and $Y = 3778.748$ m.

Solution

By Equation (24.1), $L = 400 \times 24°32' \left(\frac{\pi}{180°}\right) = 171.275 \text{ m}$

By Equation (24.4), $T = 400.000 \tan(12°16') = 86.970 \text{ m}$

Curve stationing:

$$\begin{aligned} \text{PI} &= 4 + 545.500 \\ -\text{T} &= \underline{86.970} \\ \text{PC} &= 4 + 458.530 \\ +\text{L} &= \underline{171.275} \\ \text{PT} &= 4 + 629.805 \end{aligned}$$

TABLE 24.4 COMPUTATIONS FOR STAKING CURVE OF EXAMPLE 24.4 BY COORDINATES

Station (1)	Station Difference (2)	Total Deflection (3)	Total Chord (4)	Chord Azimuth (5)	ΔX (6)	ΔY (7)	X (8)	Y (9)
4 + 458.530							5770.967	3656.280
4 + 460	1.470	0°06′19″	1.470	326°34′01″	−0.810	1.227	5770.157	3657.507
4 + 480	21.470	1°32′16″	21.468	325°08′04″	−12.272	17.614	5758.695	3673.894
4 + 500	41.470	2°58′12″	41.452	323°42′08″	−24.539	33.408	5746.428	3689.688
4 + 520	61.470	4°24′09″	61.410	322°16′11″	−37.580	48.569	5733.387	3704.849
4 + 540	81.470	5°50′06″	81.330	320°50′14″	−51.362	63.060	5719.605	3719.340
4 + 560	101.470	7°16′02″	101.199	319°24′18″	−65.851	76.843	5705.116	3733.123
4 + 580	121.470	8°41′59″	121.004	317°58′21″	−81.011	89.885	5689.956	3746.165
4 + 600	141.470	10°07′55″	140.734	316°32′25″	−96.803	102.153	5674.164	3758.433
4 + 620	161.470	11°33′52″	160.376	315°06′28″	−113.189	113.616	5657.777	3769.896
4 + 629.805	171.275	12°16′00″	169.970	314°24′20″	−121.427	118.933	5649.540	3775.213

Note: All lengths and coordinate values are in meters.

A tabular solution for curve point coordinates is given in Table 24.4. The differences in stationing from one curve point to the next are listed in column (2). Total deflection angles are calculated using Equation (24.12b) and tabulated in column (3). Total chords, computed from Equation (24.13) using these total deflection angles, are tabulated in column (4). From the azimuth of the back tangent and the deflection angles, the azimuth of each total chord is calculated and tabulated in column (5). The coordinates of the PC are computed using Equations (24.14) as

$$X_{PC} = 5723.183 + 86.970 \sin(326°40'20'' - 180°) = 5770.967 \text{ m}$$

$$Y_{PC} = 3728.947 + 86.970 \cos(326°40'20'' - 180°) = 3656.280 \text{ m}$$

Using their lengths and azimuths, the departure ΔX and latitude ΔY of each total chord are calculated. These are added to the coordinates of the PC to obtain the coordinates of the curve points. Values of ΔX and ΔY are tabulated in columns (6) and (7), and the X and Y coordinates are listed in columns (8) and (9) of Table 24.4.

A check on the PT coordinates of Table 24.4 can be obtained by computing them independently using the azimuth and tangent length of the forward tangent. The azimuth of the forward tangent is calculated by subtracting the I angle from the back tangent's azimuth,

$$Az = 326°40'20'' - 24°32' = 302°08'20''$$

Then the X and Y coordinates of the PT are

$$X_{PT} = 5723.183 + 86.970 \sin(302°08'20'') = 5649.540 \text{ m (Check!)}$$
$$Y_{PT} = 3728.947 + 86.970 \cos(302°08'20'') = 3775.213 \text{ m (Check!)}$$

Computations for the lengths and azimuths of the radial lines needed to stake curve points from station B are listed in Table 24.5. Column (1) gives the curve stations and columns (2) and (3) list the differences ΔX and ΔY between each curve point's X and Y coordinates and those of station B. Radial lengths L computed from Equation (11.4) and azimuths computed by Equation (11.5) are tabulated in columns (4) and (5). While the calculations needed to stake a curve by coordinates may seem rather extensive, they are routinely handled by computers.

For orienting the instrument it is necessary to calculate the azimuth of line BA. This, also by Equation (11.5), is

$$Az_{BA} = \tan^{-1}\left(\frac{5641.212 - 5735.270}{3778.748 - 3750.402}\right) + 360° = 286°46'16''$$

After backsighting station A, 286°46′16″ is indexed on the total station's horizontal circle. Then each curve point is staked by measuring its radial distance and azimuth taken from Table 24.5. The radial lines are shown dashed in Figure 24.11. Note that to stake station 4 + 540, for example, a distance of 34.789 m is observed on an azimuth of 206°45′42″, as shown in the figure.

TABLE 24.5 COMPUTATIONS FOR RADIAL LENGTHS AND AZIMUTHS FOR STAKING CURVE OF EXAMPLE 24.4 BY COORDINATES

Station (1)	ΔX (2)	ΔY (3)	L (4)	Az (5)
4 + 458.530 (PC)	−85.730	24.811	89.248	286°08′27″
4 + 460	−77.493	19.494	79.907	284°07′13″
4 + 480	−61.106	8.031	61.631	277°29′14″
4 + 500	−45.314	−4.238	45.512	264°39′25″
4 + 520	−30.154	−17.280	34.754	240°11′05″
4 + 540	−15.665	−31.063	34.789	206°45′42″
4 + 560	−1.883	−45.553	45.592	182°22′01″
4 + 580	11.158	−60.714	61.731	169°35′11″
4 + 600	23.425	76.508	80.014	162°58′36″
4 + 620	34.887	92.895	99.230	159°24′58″
4 + 629.805 (PT)	35.697	94.122	100.664	159°13′48″

Note: The units of ΔX, ΔY, and L are meters.

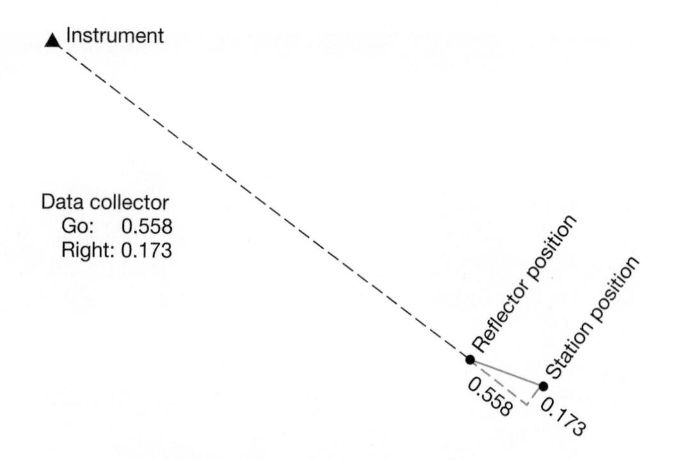

Figure 24.11 Setting a station with a tape from a "stake-out" position.

The process of staking a curve can be greatly simplified by using a data collector equipped with a "stakeout" option. When operating in this mode, before going into the field a file of point identifiers (IDs) and their corresponding coordinates for the job is downloaded into the data collector. In the field, the operator must input (1) the occupied station ID, (2) the backsight station ID, (or the azimuth of the backsight line), and (3) the ID of the point to be staked. The backsight is then taken, which automatically orients the instrument. Next the prism person walks to the estimated location of the station to be set, and the instrument operator sights the reflector. The stakeout software determines the coordinates of the reflector, and informs the operator of the distances and directions that the reflector must be moved, from the prism-person's perspective, to establish the station. For example, the software may say, "GO 0.558 RIGHT 0.173." This indicates that the rodperson should move the reflector 0.558 m from the instrument and 0.173 m to their right. Or the data collector may say "COME 0.558 LEFT 0.173." This means that the rod should be moved 0.558 m toward the instrument and 0.173 m to the left. A small tape is useful in quickly determining the location of the station to be set. Figure 24.11 shows the final measurements needed to set a station with a stakeout command of "GO 0.558 RIGHT 0.173." If a robotic total station and RPU (see Section 8.6) are available to set a curve, the instrument will automatically swing in azimuth to the desired line of sight. This equipment enables one person to lay out a curve.

As noted earlier, in staking circular curves by the coordinate method, any point can be selected for the instrument station as long as its coordinates are known. It can be a point on curve, another established control point, or a new point can be set by traversing (see Chapter 9). Alternatively, the instrument can be set at any point of unknown location that provides good vantage and its position can quickly be determined by the resection method (see Sections 11.7 and 23.9). In any case, after the instrument has been set up and oriented, a check should be made to assure its accuracy by measuring to another station of known coordinates. Any significant discrepancy should be reconciled before the curve is laid out.

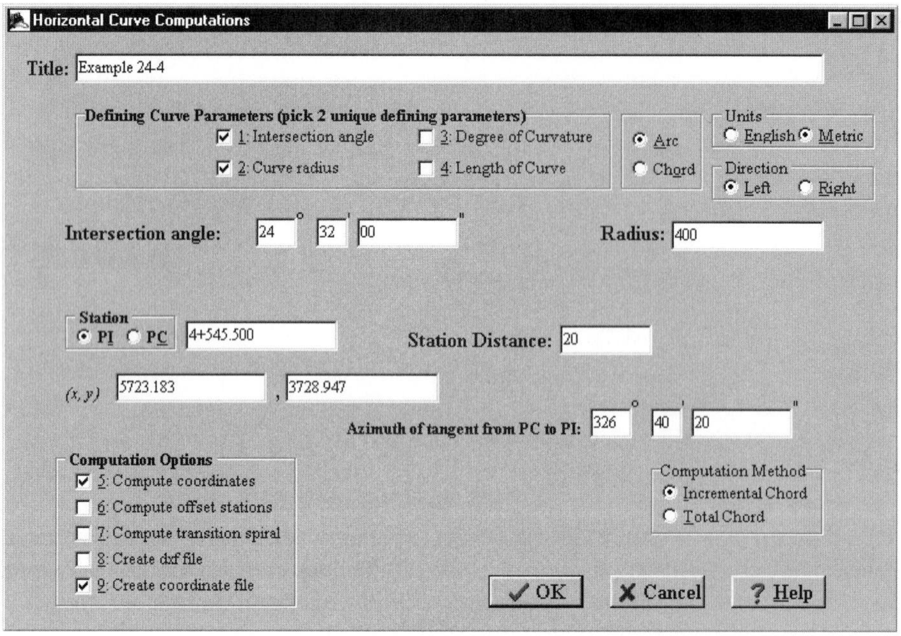

Figure 24.12 Data entry screen for horizontal curve computations in WOLFPACK.

It is important to note once again that each curve point is staked independently of the others and thus there is no misclosure at the end to verify the accuracy of the work. Thus the work must be done very carefully and the layout checked. To check, chord distances between successive stations can be quickly measured with a tape or all stations could be checked from an instrument setup at a second station of known coordinates.

The WOLFPACK software that accompanies this book can be used to establish curve-staking notes. The data entry screen for the horizontal curve computations option is shown in Figure 24.12. With this software the user can compute coordinates for the curve and have these coordinates saved to a coordinate file for uploading to a data collector. The figure shows a completed data entry screen for computing the curve in Example 24.4. Two *Computation Options* are selected: "Compute coordinates" and "Create coordinate file." As additional options are selected, additional data entry boxes will be displayed. The help file that comes with the software describes each option.

■ 24.14 CURVE STAKEOUT USING GPS AND ROBOTIC TOTAL STATIONS

As discussed in Sections 23.10 and 23.11, horizontal curves can also be staked out using RTK GPS. However, when using GPS, it is important to first perform a localization procedure as discussed in Sections 15.8 and 23.10 to place the GPS coordinates into the project coordinate system. This procedure requires control points known in the project coordinate system that encompass the project area.

As discussed in Section 23.10, the base station coordinates can be established initially using the autonomous mode or by a prior GPS survey (see Chapters 14 and 15). Since only relative positions of stations are required, the transformation process will remove any inaccuracies in the base station coordinates. However, failure to apply or perform this transformation when using GPS in stakeout, or extrapolation of stakeout points beyond the area encompassed by the project control, can result in serious errors.

Using GPS, the surveyor is guided by the survey controller (data collector) to each stake location where it is witnessed with a hub. Some survey controllers allow the operator to orient with respect to north or the position of the sun. The operator is then informed the amount to walk. The controller continuously updates the receiver's position, at a 1 sec epoch rate typically, and continues to guide the operator until the receiver is over the location of the station. A stake can then be set and the position of the station can be relocated on the hub and witnessed with a tack.

Horizontal layout is also possible using GPS and a machine control system. As discussed in Section 23.11, a GPS receiver is located on the construction vehicle and used in conjunction with a DTM and finished grading plan to guide the vehicle through the construction process. A base station must be located within radio range of the rover, within 10 km typically, to supply the rover with the base receiver observations.

Machine control systems are also available for robotic total stations. In this process, the total station is set up on a point with known coordinates and referenced to another. A DTM and grading plan are loaded into the machine control system and the robotic total station guides the construction equipment by tracking a multifaceted, 360° prism mounted on the construction vehicle.

The robotic total station requires sufficient control stations in the construction site to provide both location and orientation. Since the typical range of this system is about 1000 ft (300 m) and since there must be continuous line of sight between the total station and prism, this system requires more control stations in the project area than does a GPS machine control system. However, it typically works in the project coordinate system and thus does not require either localization or control that surrounds the project area. Furthermore, it is unaffected by canopy conditions and typically provides the only solution to machine control in these situations. The robotic total station provides both horizontal and vertical positioning to the machine control system. Additionally it offers the benefit of having sufficient accuracy to control finish grading without the need of a laser level. Machine control for this system is so accurate that it has provided guidance to curbing machines without the use of a stringline.

■ 24.15 CIRCULAR CURVE LAYOUT BY OFFSETS

For short curves, when a total station instrument is not available, and for checking purposes, one of four offset-type methods can be used for laying out circular curves: *tangent offsets* (TO), *chord offsets* (CO), *middle ordinates* (MO), and *ordinates from the long chord*. Figure 24.13 shows the relationship of CO, TO, and MO. Visually

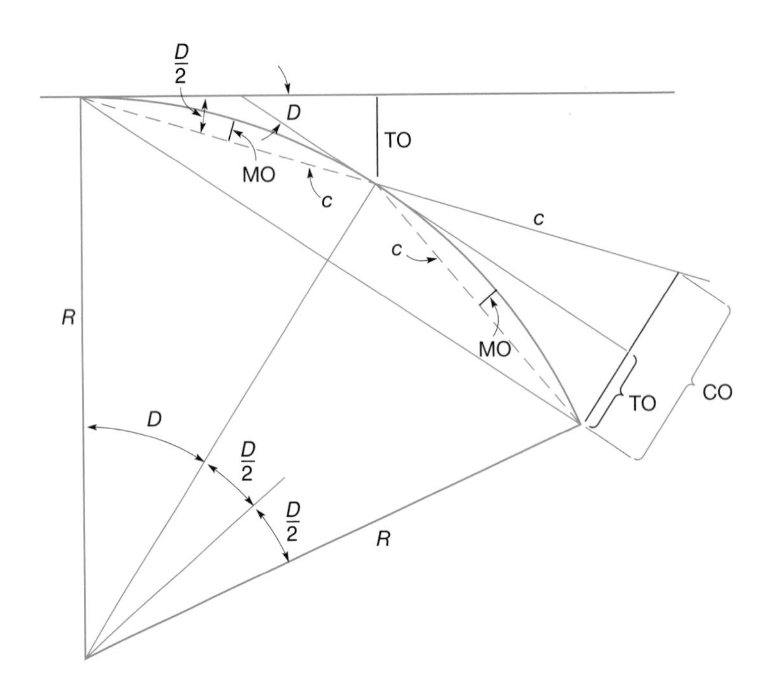

Figure 24.13 Circular curve offsets.

and by formula comparison, the chord offset for full station is

$$\text{CO} = 2c \sin\frac{D}{2} = \frac{c^2}{R} = TO \approx 8\,MO \tag{24.19}$$

Since $\sin 1° = 0.0175$ (approx.), $\text{CO} = c(0.0175)D$, where D is in degrees and decimals.

The middle ordinate m for any subchord is $R(1 - \cos\delta)$, with δ being the deflection angle for the chord. A useful equation in the layout or checking of curves in place is

$$D\,(\text{degrees}) = m\,(\text{inches}) \text{ for a 62-ft chord (approx.)} \tag{24.20}$$

The geometry of the tangent-offset method is shown in Figure 24.14. The figure illustrates that the curve is most conveniently laid out in both directions from PC and PT to a common point near the middle of the curve. This procedure avoids long observations and provides a check where small adjustments can be made more easily if necessary. To lay out a curve using this method, tangent distances are measured to established temporary points A, B, and C in Figure 24.14. From these points, right-angle observations (tangent offsets) are made to set the curve stakes. Tangent distances (TD) and tangent offsets (TO) are calculated using chords and angles in the following formulas:

$$\text{TD} = c\cos\delta \tag{24.21}$$

$$\text{TO} = c\sin\delta \tag{24.22}$$

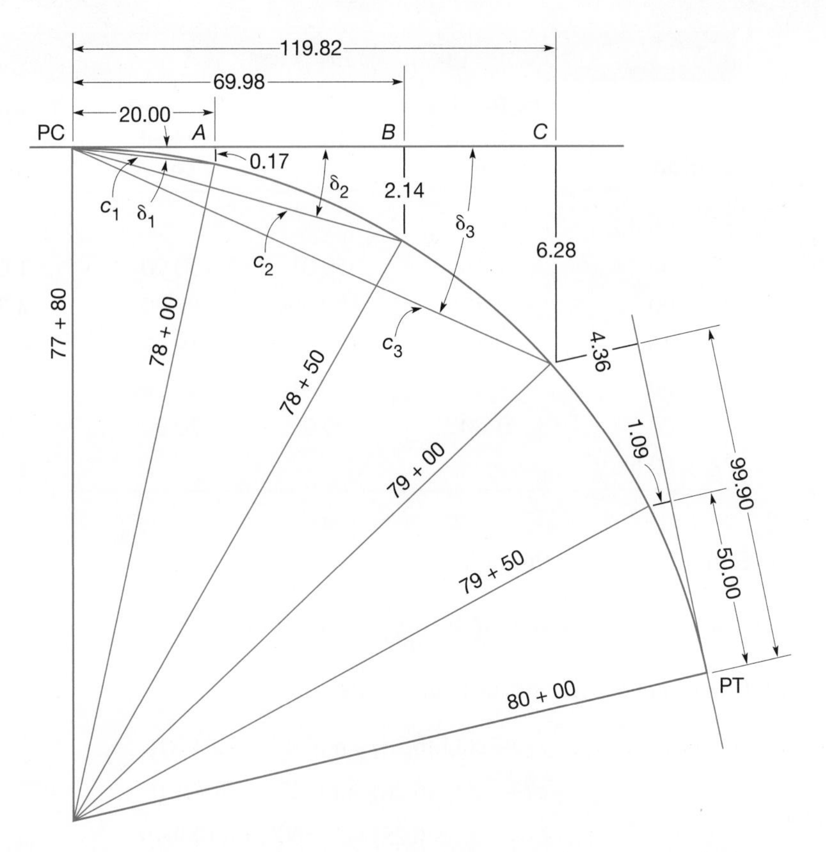

Figure 24.14 Circular curve layout by tangent offsets.

where δ angles are calculated using either Equation (24.12a) or (24.12b), and chords c are determined from Equation (24.13). These procedures are seldom used by today's surveyors.

■ Example 24.5

Compute and tabulate the data necessary to stake, by tangent offsets, the half stations of a circular curve having $I = 11°00'$, $D_c = 5°00'$ (chord definition), and PC = 77 + 80.00.

Solution

By Equation (24.2a), the curve length is $L = 100(11/5) = 220$ ft.

Therefore the PT station is (77 + 80) + (2 + 20) = 80 + 00. Intermediate stations to be staked are 78 + 00, 78 + 50, 79 + 00, and 79 + 50, as shown in Figure 24.14.

By Equation [(24.12(a)], δ angles from the PC are

$$\delta_1 = 0.025(20) = 0.50° = 0°30'$$
$$\delta_2 = 0.025(70) = 1.75° = 1°45'$$
$$\delta_3 = 0.025(120) = 3.00° = 3°00'$$

where $D/200 = 0.025$

TABLE 24.6 TANGENT OFFSET DATA FOR EXAMPLE 24.5

Station	Deflection Angle δ	Chord c	Tangent Offset $c \cos \delta$	Tangent Distance $c \sin \delta$
80 + 00 (PT)				
79 + 50	1°15′	50.01	50.00	1.09
79 + 00	2°30′	100.00	99.90	4.36
79 + 00	3°00	119.98	119.82	6.28
78 + 50	1°45′	70.01	69.98	2.14
78 + 00	0°30′	20.00	20.00	0.17
77 + 80 (PC)				

By Equation (24.10), the radius is

$$R = \frac{50}{\sin 2°30'} = 1146.28 \text{ ft}$$

By Equation (24.13), chords from the PC are

$$c_1 = 2(1146.28) \sin 0°30' = 20.00 \text{ ft}$$
$$c_2 = 2(1146.28) \sin 1°45' = 70.01 \text{ ft}$$
$$c_3 = 2(1146.28) \sin 3°00' = 119.98 \text{ ft}$$

Now, using Equations (24.21) and (24.22), tangent distances and tangent offsets are calculated. Chords, angles, tangent distances, and tangent offsets to stake points from the PT are computed in the same manner. All data for the problem are listed in Table 24.6. The tangent distances tabulated are lengths from the PC or PT that must be observed to establish points A, B, C, etc., and the tangent offsets are distances from these points needed to locate the curve stakes. Accurate curve layout by tangent offsets generally requires a total station to turn the right angles from the tangent. This involves more instrument setups and a greater amount of time than stakeout by deflection angles or coordinates. However, rough layouts can be done using a tape and right-angle prism.

■ 24.16 SPECIAL CIRCULAR CURVE PROBLEMS

Many special problems arise in the design and computation of circular curves. Three of the more common ones are discussed here, and each can be solved using the coordinate geometry formulas given Chapter 11.

24.16.1 Passing a Circular Curve Through a Fixed Point

One problem that often occurs in practice is to determine the radius of a curve connecting two established tangents and going through a fixed point such as an underpass,

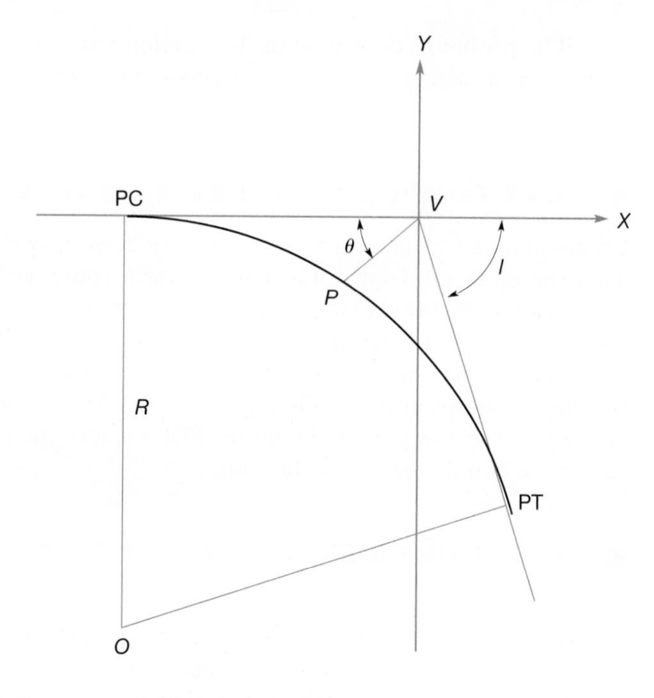

Figure 24.15 Passing a circular curve through a point.

overpass, or existing bridge. The problem can be solved by establishing an *XY* coordinate system, as shown in Figure 24.15, where the origin occurs at *V* (the PI) and *X* coincides with the back tangent. Coordinates of the radius point in this system are $X_0 = -R\tan(I/2)$ and $Y_0 = -R$. From observations of distance *PV* and angle θ the coordinates X_P and Y_P of point *P* can be determined. Then the following equation for a circle, obtained by substitution into Equation (11.9), can be written:

$$\left(X_P + R\tan\frac{I}{2}\right)^2 + (Y_P + R)^2 = R^2 \qquad (24.23)$$

With X_P, Y_P, and $I/2$ known, a solution for *R* can be found. The equation is quadratic, but can be solved using Equation (11.3).

24.16.2 Intersection of a Circular Curve and a Straight Line

Another frequently encountered problem involving curve computation is the determination of the intersection point of a circular curve and a straight line. An example is illustrated in Figure 11.6. In typical cases, coordinates X_A, Y_A, X_B, and Y_B are known, as well as *R*. Procedures for solving the problem are outlined in Section 11.5 and a worked example is presented.

24.16.3 Intersection of Two Circular Curves

Figure 11.7 illustrates another common problem: computing the intersection point of two circular curves. This can be handled by coordinate geometry, as discussed in Section 11.6. Coordinates X_A, Y_A, X_B, and Y_B are typically determined through survey and R_1 and R_2 are selected based on design or topographic constraints. A worked example is given in Section 11.6.

The problems described in this section and the preceding one arise most often in the design of subdivisions and interchanges and in calculating right-of-way points along highways and railroads.

■ 24.17 COMPOUND AND REVERSE CURVES

Compound and reverse curves are combinations of two or more circular curves. They should be used only for low-speed traffic routes and in terrain where simple curves cannot be fitted to the ground without excessive construction costs. Special formulas have been derived to facilitate computations for such curves and are demonstrated in texts on route surveying. A compound curve can be staked with instrument setups at the beginning PC and ending PT or perhaps with one setup at the *point of compound curvature* (PCC) where the two curves join. Reverse curves are handled in similar fashion.

■ 24.18 SIGHT DISTANCE ON HORIZONTAL CURVES

Highway safety requires certain minimum sight distances in zones where passing is permitted and in nonpassing areas to ensure a reasonable stopping distance if there is an object on the roadway. Specifications and tables list suitable values based on vehicular speeds, the perception and reaction times of an average individual, the braking distance for a given coefficient of friction during deceleration, and type and condition of the pavement.

A minimum stopping sight distance of 450 to 550 ft is desirable for a speed of 55 mi/hr. An approximate formula for determining the available horizontal sight distance on a circular curve can be derived from Figure 24.16, in which the clear sight distance past an obstruction is the length of the long chord AS, denoted by C; and the required clearance is the middle ordinate PM, denoted by m. Then in similar triangles SPG and SOH,

$$\frac{m}{SP} = \frac{SP/2}{R} \quad \text{and} \quad m = \frac{(SP)^2}{2R}$$

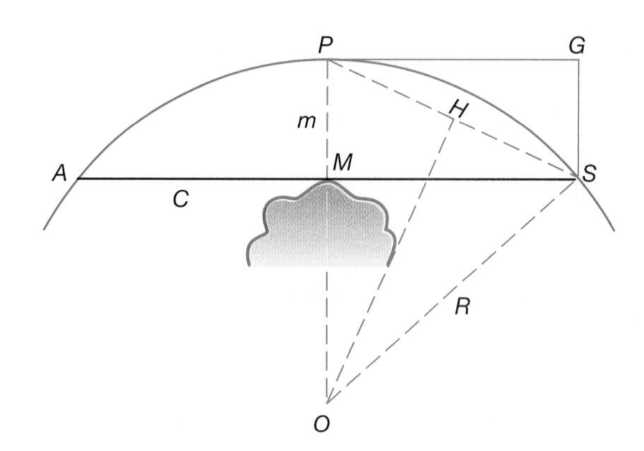

Figure 24.16 Horizontal sight distance on a circular curve.

Usually m is small compared with R, and SP may be assumed equal to $C/2$. Then

$$C = \sqrt{8mR} \tag{24.24}$$

If distance m from the centerline of a highway to the obstruction is known or can be measured, the available sight distance C is calculated from the formula. Actually cars travel on either the inside or the outside lane, so sight distance AS is not exactly the true stopping distance, but the computed length is on the safe side and satisfactory for practical use. If the calculation reveals a sight distance restriction, the obstruction might possibly be removed or a safe speed limit posted in the area.

24.19 SPIRALS

As noted in Section 24.1, spirals are used to provide gradual transitions in horizontal curvature. Their most common use is to connect straight sections of alignment with circular curves, thereby lessening the sudden change in direction that would otherwise occur at the point of tangency. Since spirals introduce curvature gradually, they afford the logical location for introducing superelevation to offset the centrifugal force experienced by vehicles entering curves.

24.19.1 Spiral Geometry

Figure 24.17 illustrates the geometry of spirals connecting tangents with a circular curve of radius R and degree of curvature D. The entrance spiral at the left begins on the back tangent at the TS (*tangent to spiral*) and ends at the SC (*spiral to curve*). The circular curve runs from the SC to the beginning of the exit spiral at the CS (*curve to spiral*), and the exit spiral terminates on the forward tangent at the ST (*spiral to tangent*).

The entrance and exit spirals are geometrically identical. Their length, L_S, is the arc distance from the TS to the SC or CS to ST. The designer selects this length to provide sufficient distance for introducing the curve's superelevation.

If a tangent to the entrance spiral (and curve) at the SC is projected to the back tangent, it locates the *spiral point of intersection* SPI. The angle at the SPI between the two tangents is the spiral angle Δ_S. From the basic property of a spiral, i.e., *its radius changes uniformly from infinity at the TS to the radius of the circular curve at the SC*, it follows that the spiral's degree of curve changes uniformly from 0° at the TS to D at the SC. Since the change is uniform, the average degree of curve over the spiral's length is $D/2$. Thus from the definition of degree of curve, spiral angle Δ_S is

$$\Delta_S = L_S\frac{D}{2} \tag{24.25}$$

where L_S is in stations and Δ_S and D are in degrees.

Assume in Figure 24.18 that M is at the midpoint of the spiral, so its distance from the TS is $L_S/2$. If this reasoning is continued, the degree of curvature at M is $D/2$, the average degree of curvature from the TS to M is $(D/2)/2 = D/4$, and the spiral angle Δ_M is

$$\Delta_M = \frac{L_S}{2}\frac{D}{4} = \frac{L_S D}{8} \tag{a}$$

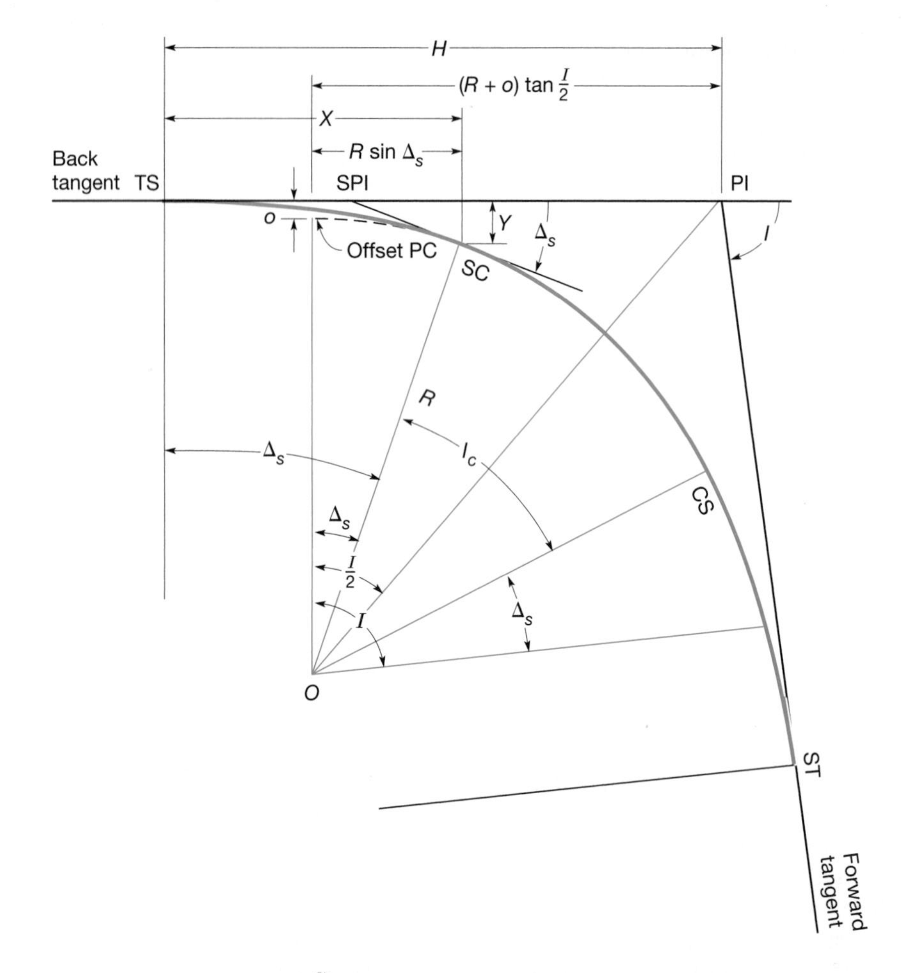

Figure 24.17 Spiral geometry.

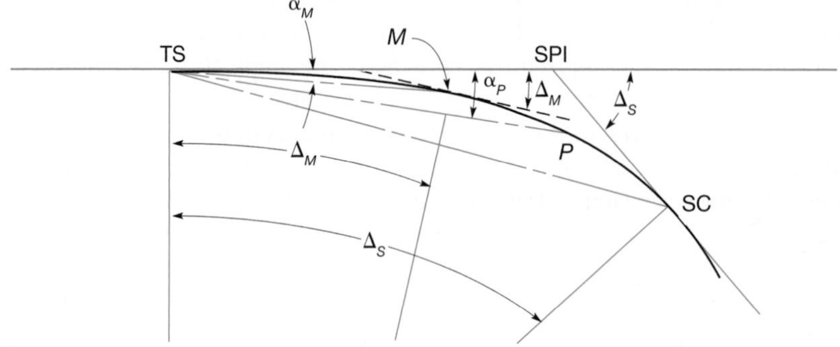

Figure 24.18 Spiral deflection angles.

Solving for D in Equation (24.25) and substituting into Equation (a) gives

$$\Delta_M = \frac{L_S}{8}\frac{2\Delta_S}{L_S} = \frac{\Delta_S}{4} \qquad \textbf{(b)}$$

According to Equation (b), at $L_S/2$, the spiral angle is $\Delta_S/4$. This demonstrates another basic property of a spiral: *spiral angles at any point are proportional to the square of the distance from TS to the point, or*

$$\Delta_P = \left(\frac{L_P}{L_S}\right)^2 \Delta_S \tag{24.26}$$

where Δ_P is the spiral angle at any point P whose distance from the TS is L_P.

24.19.2 Spiral Calculation and Layout

Deflection angles and chords in a manner similar to that used for circular curves can be used to lay out spirals. In Figure 24.18, α_P is the deflection angle from the *TS* to any point *P*. From calculus it can be shown that for the gradual spirals used on transportation routes, deflection angles are very nearly one-third their corresponding spiral angles. Thus

$$\alpha_P = \left(\frac{L_P}{L_S}\right)^2 \frac{\Delta_S}{3} \tag{24.27}$$

In Equation (24.27), L_P is the distance from the *TS* to *P*, which is simply the difference in stationing from the *TS* to point *P*.

In Figure 24.17 coordinates *X* and *Y* give the position of the *SC*. In this coordinate system, the origin is at the *TS* and the *X*-axis coincides with the back tangent. Approximate formulas for computing *X* and *Y* are:

$$X = L_S(100 - 0.0030462\ \Delta_S^2)\,(\text{ft}) \tag{24.28}$$

$$Y = L_S(0.58178\Delta_S - 0.000012659\ \Delta_S^3)\ (\text{ft}) \tag{24.29}$$

where L_S is in stations and Δ_S is in degrees. More accurate formulas for computing *X* and *Y* coordinates of any station, *P*, which is a distance L_P along the spiral, are

$$\begin{aligned} X_P &= L_P\left(1 - \frac{\Delta_P^2}{10} + \frac{\Delta_P^4}{216} - \frac{\Delta_P^6}{9360} + \frac{\Delta_P^8}{685{,}440}\right) \\ Y_P &= L_P\left(\frac{\Delta_P}{3} - \frac{\Delta_P^3}{42} + \frac{\Delta_P^5}{1320} - \frac{\Delta_P^7}{75{,}600} + \frac{\Delta_P^9}{6{,}894{,}720}\right) \end{aligned} \tag{24.30}$$

where Δ_P, defined in Equation (24.26), is expressed in radian units and L_P is the stationing distance, in either feet or meters, from the *TS* to point *P*. If L_S and Δ_S are substituted for L_P and Δ_P, respectively, in Equations 24.30, then coordinates *X* and *Y* of the *SC* will result. Equations (24.30) can be used to compute spiral coordinates in either metric or English units. The spiral could be staked by offsets using these coordinates, or they could be used with Equations (11.4) and (11.5) to calculate the deflection angles and chord distances necessary to stake out the spiral.

When a spiral is inserted ahead of a circular curve, as illustrated in Figure 24.17, the circular curve is shifted inward by an amount *o*, known as the *throw*. This can be visualized by constructing the circular curve back from the *SC* to the offset *PC* (the point where a tangent to the curve is parallel to the back tangent). The perpendicular distance from the offset *PC* to the back tangent is the throw, which from Figure 24.17 is

$$o = Y - R(1 - \cos\Delta_S) \tag{24.31}$$

To lay out a spiral in the field, distance H of Figure 24.17 from PI to TS is measured back along the tangent to locate the TS. Then the SC can be established by laying out X and Y. From the geometry of Figure 24.17 distance H is

$$H = X - R \sin \Delta_S + (R + o) \tan \frac{I}{2} \tag{24.32}$$

The following example will illustrate the computations required in laying out a spiral by the deflection-angle method.

Example 24.6

A spiral of 300-ft length is used for transition into a 3°00′ circular curve (arc definition). Angle I at the PI station of 20 + 00 is 60°00′. Compute and tabulate the deflection angles and chords necessary to stake out this spiral at half stations.

Solution

By Equation (24.3), $R = 5729.58/3.00 = 1909.86$ ft

By Equation (24.25), $\Delta_S = 3(3.00)/2 = 4.5° = 4°30'$

By Equations (24.28) and (24.29),

$$X = 3[100 - 0.0030462(4.5)^2] = 299.81 \text{ ft}$$
$$Y = 3[0.58178(4.5) - 0.000012659(4.5)^3] = 7.86 \text{ ft}$$

[*Note:* These same results can be obtained using Equations 24.30 with L_P = 300.00 ft and $\Delta_P = (4.5° \times \pi/180°)$.]

By Equation (24.31),

$$o = 7.86 - 1909.86\,(1 - \cos 4°30') = 1.97 \text{ ft}$$

By Equation (24.32),

$$H = 299.81 - 1909.86 \sin 4°30' + (1909.86 + 1.97)\tan 30° = 1253.75 \text{ ft}$$

Calculate stationing:

PI station	=	20 + 00.00
$-H$	=	−12 + 53.75
TS station	=	7 + 46.25
$+L_S$	=	3 + 00.00
SC station	=	10 + 46.25

The deflection angles calculated using Equation (24.27) are listed in column (3) of Table 24.7. The values for L_P used in the equation are given in column (2). If a total station is used for stakeout with a setup at the TS, the total chords of column (2) are used with the deflection angles of column (3). If a theodolite and tape were used, the deflection angles, and incremental chords listed in column (4) would be used. The chords in columns (2) and (4) are simply station differences between points on the spiral, and are nearly exact for the relatively flat curvature of highway and railroad spirals. After progressing through the spiral, the SC is finally staked. The falling between its position and

TABLE 24.7 DATA FOR STAKING SPIRAL OF EXAMPLE 24.6

Station (1)	Total Chord Distance from TS (ft) (2)	Deflection Angle (3)	Incremental Chord (ft) (4)
10 + 46.25 (SC)	300.00	1°30.0′	46.25
10 + 00	253.75	1°04.4′	50.00
9 + 50	203.75	0°41.5′	50.00
9 + 00	153.75	0°23.6′	50.00
8 + 50	103.75	0°10.8′	50.00
8 + 00	53.75	0°02.9′	50.00
7 + 50	3.75	0°00.0′	3.75
7 + 46.25 (TS)			

the *SC* set by coordinates should be measured, the precision calculated, and a decision made to accept or repeat the work.

To continue staking the alignment, the instrument is moved forward to the *SC*. With $2\Delta_S/3$ set on the horizontal circle, a backsight is taken on the TS and the line of sight plunged.[1] Turning the instrument to 0°00′00″ orients the line of sight tangent to the circular curve, ready for laying off deflection angles. The circular curve is computed and laid out by methods given in earlier sections of this chapter, except that its intersection angle I_c, as illustrated in Figure 24.17, is given by

$$I_c = I - 2\Delta_S \tag{24.33}$$

When staking reaches the circular curve's end, the *CS* is set. The exit spiral is calculated by the same methods described for the entrance spiral and laid out by staking back from the *ST*.

Spirals can be computed and staked by several different methods. In this brief treatment, only one commonly applied procedure has been discussed. Students interested in further study of spirals can consult a text on route surveying listed in the Bibliography at the end of this chapter.

24.20 COMPUTATION OF "AS-BUILT" CIRCULAR ALIGNMENTS

Most highways that exist today were carefully designed and then constructed according to plan. Therefore their centerlines "as-built" are precisely known and coordinates for critical points on their alignments are on file for future use.

[1]From Equation (24.27), the angle at the *TS* in triangle *TS-SPI-SC* of Figure 24.18 is $\Delta_S/3$. Also from Figure 24.18, the angle at the SPI in this triangle is $180° - \Delta_S$. It follows therefore that with the instrument at the *SC*, after backsighting the *TS*, the angle that must be turned to get the line of sight tangent to the spiral and circular curve is $2\Delta_S/3$.

However, some roads have their origins from "cartways" that through the years were periodically upgraded and improved in place. Thus it is possible that no formal plans or records of their alignments exist. Yet the boundary lines of adjoining properties may be referenced to the centerline and thus it becomes important for it to be precisely established. Also, it is sometimes desirable to determine the parameters of an as-built roadway to check adherence to contract specifications. In these cases, the coordinates of critical points on the approximate centerline of the facility, both on the curves and on the tangents, must be determined. One further important application of this type of problem relates to railroad abandonment programs. Here the rails have served for years as monuments for delineating right-of-way. Therefore, before their removal it is important to obtain the coordinates of important points along their alignments for use in future work related to establishing property lines adjoining the railroad right-of-way.

The procedure of establishing the coordinates of an existing or as-built alignment is illustrated in Figure 24.19. In the figure, assume that a traverse was used to establish coordinates for points A thru F along the existing alignment, as shown. From points B thru E, a least-squares fit of points to Equation (11.10) can be performed, which will establish the coordinates of the center point O and the radius R of the circle. In this example, the matrices of coefficients, unknowns, and observations are

$$\mathbf{A} = \begin{bmatrix} 2X_B & 2Y_B & -1 \\ 2X_C & 2Y_C & -1 \\ 2X_D & 2Y_D & -1 \\ 2X_E & 2Y_E & -1 \end{bmatrix} \quad \mathbf{X} = \begin{bmatrix} X_O \\ Y_O \\ f \end{bmatrix} \quad \mathbf{L} = \begin{bmatrix} X_B^2 + Y_B^2 \\ X_C^2 + Y_C^2 \\ X_D^2 + Y_D^2 \\ X_E^2 + Y_E^2 \end{bmatrix} \tag{24.34}$$

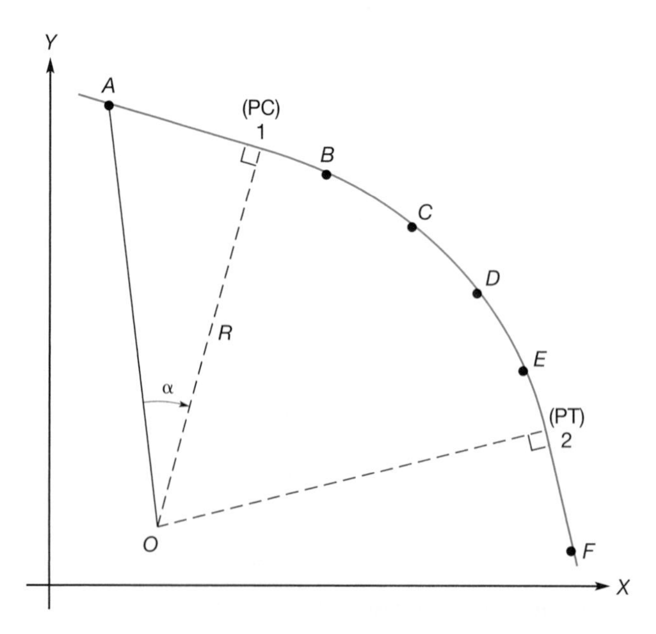

Figure 24.19 Geometry of an "as-built" survey.

Using Equations (15.6) or (15.7), the least-squares solution for X_O, Y_O, and f are determined. The radius R can then be computed from

$$R = \sqrt{X_O^2 + Y_O^2 - f} \tag{24.35}$$

Example 24.7

In reference to Figure 24.19, the following coordinates were determined during an "as-built" survey. What are the defining parameters for the curve and the coordinates of the *PC* and *PT*?

Point	*X* (ft)	*Y* (ft)	Point	*X* (ft)	*Y* (ft)
A	5354.86	7444.14	*D*	8084.03	6071.29
B	6975.82	6947.93	*E*	8431.38	5542.00
C	7577.11	6572.60	*F*	8877.96	4268.90

Solution

Substituting the coordinate values for points *B* thru *E* into Equation (24.34) yields the matrices *A* and *L* as

$$\mathbf{A} = \begin{bmatrix} 2 \times 6975.82 & 2 \times 6947.93 & -1 \\ 2 \times 7577.11 & 2 \times 6572.60 & -1 \\ 2 \times 8084.03 & 2 \times 6071.29 & -1 \\ 2 \times 8431.38 & 2 \times 5542.00 & -1 \end{bmatrix} = \begin{bmatrix} 13{,}951.64 & 13{,}895.86 & -1 \\ 15{,}154.22 & 13{,}145.20 & -1 \\ 16{,}168.06 & 12{,}142.58 & -1 \\ 16{,}862.76 & 11{,}084.00 & -1 \end{bmatrix}$$

$$\mathbf{L} = \begin{bmatrix} 6975.82^2 + 6947.93^2 \\ 7577.11^2 + 6572.60^2 \\ 8084.03^2 + 6071.29^2 \\ 8431.38^2 + 5542.00^2 \end{bmatrix} = \begin{bmatrix} 96{,}935{,}795.96 \\ 100{,}611{,}666.71 \\ 102{,}212{,}103.30 \\ 101{,}801{,}932.70 \end{bmatrix}$$

Substituting the *A* and *L* matrices into Equation (15.6) yields the solution

$$X_O = 5587.375$$
$$Y_O = 4053.992$$
$$f = 37{,}351{,}007.416$$

And finally by Equation (24.35), the radius of the circle is

$$R = \sqrt{(5587.375)^2 + (4053.992)^2 - 37{,}351{,}007.416} = 3209.766$$

Figure 24.20 shows the data file and the results of the least-squares fit of points to a circle from the software WOLFPACK, which accompanies this book.

From this solution, the coordinates of the PC can be determined by computing the length and azimuth of *AO* of Figure 24.19 using Equations (11.4) and (11.5)

```
Data file for Example 24.7 using WOLFPACK
B 6975.82 6947.93 **Point id, X, Y
C 7577.11 6572.60
D 8084.03 6071.29
E 8431.38 5542.00

Results of Least Squares Adjustment of Points
*****  Least Squares Fit Of Circle  *****
Center of Circle at: X = 5,587.375
                     Y = 4,053.992
With A Radius of:          3,209.766

Misfit of Points to Best Fit Curve
Station    Distance From Curve
====================
B             0.009
C            -0.026
D             0.028
E            -0.010
```

Figure 24.20 Data file and adjustment results for Example 24.7 using WOLFPACK.

and then solving right triangle $AO1$ for angle α using the cosine function. For this example, the values are

By Equations (11.4) and (11.5),

$$OA = \sqrt{(5354.86 - 5587.375)^2 + (7444.14 - 4053.992)^2} = 3398.11$$

$$Az_{OA} = \tan^{-1}\left(\frac{5354.86 - 5587.375}{7444.14 - 4053.992}\right) + 360° = 356°04'35''$$

From the cosine function,

$$\alpha = \cos^{-1}\left(\frac{3209.766}{3398.11}\right) = 19°09'56''$$

From Figure 24.19, the azimuth of line $O1$ is

$$Az_{O1} = 356°04'35'' + 19°09'56'' - 360° = 15°14'31''$$

Now the coordinates of the PC can be determined using the following equations:

$$X_{PC} = X_O + R \sin Az_{O1} = 5587.375 + 843.838 = 6431.21$$
$$Y_{PC} = Y_O + R \cos Az_{O1} = 4053.992 + 3096.859 = 7150.85$$

In similar fashion, the coordinates for the PT can be computed. From the azimuths of lines $O1$ and $O2$, the intersection angle can be determined using Equation (11.11). Additionally, the long chord can be determined using Equation (11.4). From these parameters, the tangent, external ordinate, and middle ordinate distances can be computed using Equations (24.4), (24.8), and (24.9), respectively. These steps are left as an exercise for the student.

24.21 SOURCES OF ERROR IN LAYING OUT CIRCULAR CURVES

Some sources of error in curve layout are:

1. Errors in setting up, leveling, and reading the instrument.
2. Total station (or theodolite) out of adjustment.

3. Bull's-eye bubble out of adjustment on prism pole used for stakeout with a total station.
4. Measurement errors in laying out angles and distances.

24.22 MISTAKES

Typical mistakes that occur in laying out a curve in the field are:

1. Failure to take equal numbers of direct and reversed measurements of the deflection angle at the PI before computing or laying out the curve.
2. Using 100.00-ft chords to lay out arc-definition curves having D greater than 2°.
3. Taping subchords of nominal length for chord-definition curves having D greater than 5° (a nominal 50-ft subchord for a 6° curve requires a measurement of 50.02 ft).
4. Failure to check curve points after staking them using either the total chord or the coordinate method.
5. A mistake in the backsight when staking by the coordinate method.
6. Failure to stake the PT independently by measuring the tangent distance forward from the PI.
7. Incorrect orientation of the instrument's horizontal circle.
8. Failure to properly establish the setup information in an automatic data collector.
9. Misidentification of a station, or stations.

PROBLEMS

Asterisks (*) indicate problems that have answers given in Appendix G.

24.1 What features make a spiral curve a particularly useful easement curve?

24.2 For the following circular curves having a radius R, what is their degree of curvature by (1) arc definition and (2) chord definition?

*(a) 300.00 ft
(b) 1500.00 ft
(c) 4000.00 ft

Compute L, T, E, M, LC, R, and stations of the PC and PT for the circular curves in Problems 24.3 through 24.6. Use the chord definition for the railroad curve and the arc definition for the highway curves.

24.3* Railroad curve with $D_c = 2°00'$, $I = 18°00'$, and PI station = 25 + 50.00 ft.
24.4 Highway curve with $D_a = 2°30'$, $I = 12°00'$, and PI station = 22 + 45.89 ft.
24.5 Highway curve with $R = 700.000$ m, $I = 13°10'$, and PI station = 5 + 784.850 m.
24.6 Highway curve with $R = 400.000$ m, $I = 8°15'$, and PI station = 3 + 508.900 m.

Tabulate R or D, T, L, E, M, PC, PT, deflection angles, and incremental chords to lay out the circular curves at full stations (100 ft or 30 m) in Problems 24.7 through 24.14.

24.7 Highway curve with $D_a = 4°00'$, $I = 24°30'$, and PI station = 42 + 26.80 (ft)
24.8 Railroad curve with $D_c = 3°30'$, $I = 13°00'$, and PI station = 56 + 98.40 (ft)
24.9 Highway curve with $R = 650$ m, $I = 12°00'$, and PI station = 4 + 178.595
24.10 Highway curve with $R = 725$ m, $I = 10°00'$, and PI station = 4 + 284.964.
24.11 Highway curve with $R = 1000$ ft, $I = 60°00'$, and PI station = 85 + 40.00.
24.12 Highway curve with $L = 400$ m, $R = 600$ m, and PI station = 5 + 318.682.

24.13 Highway curve with $T = 237.00$ ft, $R = 1520$ ft, and PI station = 196 + 82.96.
24.14 Railroad curve with $T = 165.00$ ft, $D_c = 2°15'$, and PI station = 60 + 12.95.

In Problems 24.15 through 24.18 tabulate the curve data, deflection angles, and total chords needed to lay out the following circular curves at full-station increments using a total station instrument set up at the PC.

24.15 The curve of Problem 24.7
24.16 The curve of Problem 24.8
24.17 The curve of Problem 24.9
24.18 The curve of Problem 24.10
24.19 A streetcar line on the center of an 80-ft street makes a 75°24′ turn into another street of equal width. The corner curb line has $R = 12$ ft. What is the largest R that can be given a circular curve for the track centerline if the law requires it to be at least 10 ft from the curb?

Tabulate all data required to lay out by deflection angles and incremental chords, at the indicated stationing, for the circular curves of Problems 24.20 and 24.21.

24.20 The R for a highway curve (arc definition) will be rounded off to the nearest larger multiple of 100 ft. Field conditions require M to be approximately 24 ft to avoid an embankment. The PI = 94 + 18.70 and I = 20°00′ with stationing at 100 ft.
24.21 For a highway curve, R will be rounded off to the nearest multiple of 10 m. Field measurements show that T should be approximately 85 m to avoid an overpass. The PI = 6 + 356.400 and $I = 13°00'$ with stationing at 30 m.
24.22 A highway survey PI falls in a pond, so a cutoff line $AB = 273.45$ ft is run between the tangents. In the triangle formed by points A, B, and PI, the angle at $A = 16°28'$ and at $B = 22°16'$. The station of A is 35 + 60.30. Calculate and tabulate curve notes to run, by deflection angles and incremental chords, a 4°30′ (arc definition) circular curve at half-station increments to connect the tangents.
24.23 In the figure, a single circular highway curve (arc definition) will join tangents XV and VY and also be tangent to BC. Calculate R, L, and the stations of the PC and PT.

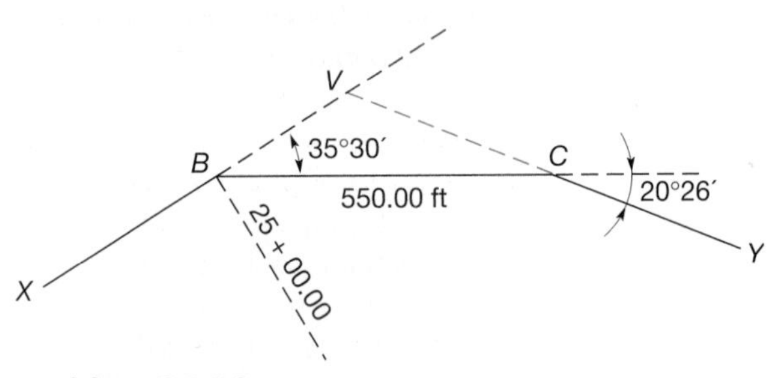

Problem 24.23

24.24* Compute R_x to fit requirements of the figure and make the tangent distances of the two curves equal.

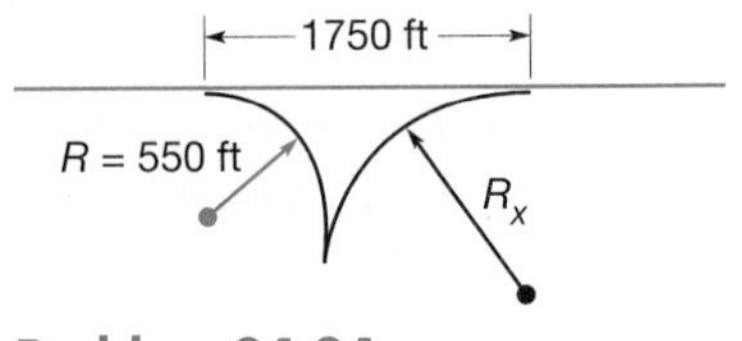

Problem 24.24

24.25 After a backsight on the PC with 0°00′ set on the instrument, what is the deflection angle to the following circular curve points?

*(a) Setup at curve midpoint, deflection to the PT.
(b) Instrument at curve midpoint, deflection to the 3/4 point.
(c) Setup at 1/4 point of curve, deflection to 3/4point.

24.26 In surveying a construction alignment, why should the I angle be measured by repetition?

24.27 A highway curve (arc definition) to the right, having $R = 600$ m and $I = 14°30'$, will be laid out by coordinates with a total station instrument setup at the PI. The PI station is 4 + 960.500, and its coordinates are $X = 65{,}304.654$ m and $Y = 44{,}127.368$ m. The azimuth (from north) of the back tangent proceeding toward the PI is 96°37′42″. To orient the total station, a backsight will be made on a POT on the back tangent. Compute lengths and azimuths necessary to stake the curve at 30-m increments.

24.28 In Problem 24.27, compute the XY coordinates at 30-m increments.

24.29 A running track must consist of two semicircles and two tangents, and be exactly 1320 ft along its centerline. The two curves are to constitute one-half its total length. Calculate L, R, and D_a for the curves.

24.30 Make the computations necessary to lay out the curve of Problem 24.8 by the tangent offset method. Approximately half the curve is to be laid out from the PC and the other half from the PT.

What sight distance is available if there is an obstruction on a radial line through the PI inside the curves in Problems 24.31 and 24.32?

24.31* For Problem 24.7, obstacle 15 ft from curve.

24.32 For Problem 24.12, obstacle 20 ft from curve.

24.33 If the misclosure for the curve of Problem 24.7, computed as described in Section 24.8, is 0.13 ft, what is the field layout precision?

24.34 Assume that a 200-ft entry spiral will be used with the curve of Problem 24.7. Compute and tabulate curve notes to stake out the alignment from the TS to ST at full stations using a total station and the deflection-angle, total chord method.

24.35 Same as Problem 24.34, except use a 100-ft spiral for the curve of Problem 24.8.

24.36 Same as Problem 24.34, except for the curve of Problem 24.9, with a 50-m entry spiral using a staking increment of 30 m, and a total station instrument.

24.37 Compute the area bounded by the two arcs and tangent in Problem 24.24.

24.38 In an as-built survey, the XY coordinates of three points on the centerline of a highway curve are determined to be A: (3770.52, 4913.84); B: (3580.80, 4876.37); C: (3399.27, 4809.35). What are the radius and coordinates for the center of the curve?

24.39 In Problem 24.38, if the (x, y) coordinates of two points on the centerline of the tangents are (3042.28, 4616.77) and (4435.66, 4911.19), what are the coordinates of the PC, PT, and the curve parameters L, T, and I?

24.40 Explain how GPS receivers can be used to stake out a curve.

24.41 Write a computational program to calculate circular curve notes for layout with a total station by deflection angles and total chords from a setup at the PC.

24.42 Develop a computational program to calculate the coordinates of the stations on a circular curve.

BIBLIOGRAPHY

Blackford, Rippin. 2006. "A Machine Control Primer for Surveyors." *Professional Surveyor* 26 (No. 1): 8.

25

Vertical Curves

25.1 INTRODUCTION

Curves are needed to provide smooth transitions between straight segments (tangents) of grade lines for highways and railroads. Because these curves exist in vertical planes, they are called vertical curves. An example is illustrated in Figure 25.1, which shows the profile view of a proposed section of highway to be constructed from A to B. A grade line consisting of three tangent sections has been designed to fit the ground profile. Two vertical curves are needed: curve *a* to join tangents 1 and 2, and curve *b* to connect tangents 2 and 3. The function of each curve is to provide a gradual change in grade from the initial (back) tangent to the grade of the second (forward) tangent. Because parabolas provide a constant rate of change of grade, they are ideal and almost always applied for vertical alignments used by vehicular traffic.

Two basic types of vertical curves exist: *crest* and *sag*. These are illustrated in Figure 25.1. Curve *a* is a crest type, which by definition undergoes a negative change in grade—that is, the curve turns downward. Curve *b* is a sag type, in which the change in grade is positive and the curve turns upward.

There are several factors that must be taken into account when designing a grade line of tangents and curves on any highway or railroad project. They include (1) providing a good fit with the existing ground profile, thereby minimizing the depths of cuts and fills, (2) balancing the volume of cut material against fill, (3) maintaining adequate drainage, (4) not exceeding maximum specified grades, and (5) meeting fixed elevations such as intersections with other roads. In addition, the curves must be designed to (a) fit the grade lines they connect, (b) have lengths sufficient to meet specifications covering a maximum rate of change of grade (which affects the comfort of vehicle occupants), and (c) provide sufficient sight distance for safe vehicle operation (see Section 25.11).

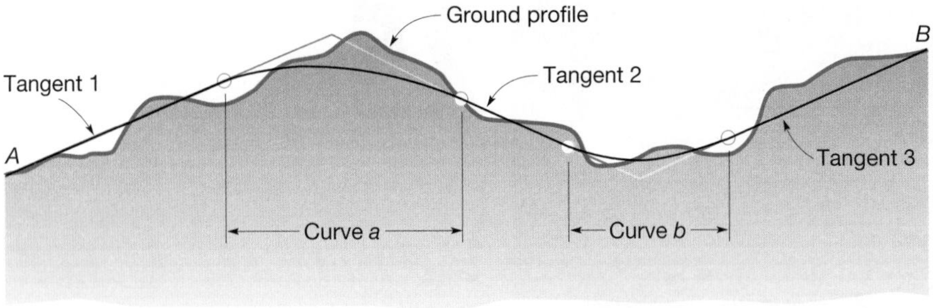

Figure 25.1 Grade line and ground profile of a proposed highway section.

Elevations at selected points (e.g., full or half-stations in the English system of stationing, or 20-, 30-, or 40-m in the metric system) along vertical parabolic curves are usually computed by the *tangent-offset* method. It is simple, straightforward, conveniently performed with calculators and computers, and self-checking. After the elevations of curve points have been computed, they are staked in the field to guide construction operations so the route can be built according to plan.

25.2 GENERAL EQUATION OF A VERTICAL PARABOLIC CURVE

The general mathematical expression of a parabola, with respect to an XY rectangular coordinate system, is given by

$$Y_P = a + bX_P + cX_P^2 \tag{25.1}$$

where Y_P is the ordinate at any point p on the parabola located a distance X_P from the origin of the curve, and a, b, and c are constants. Figure 25.2 shows a

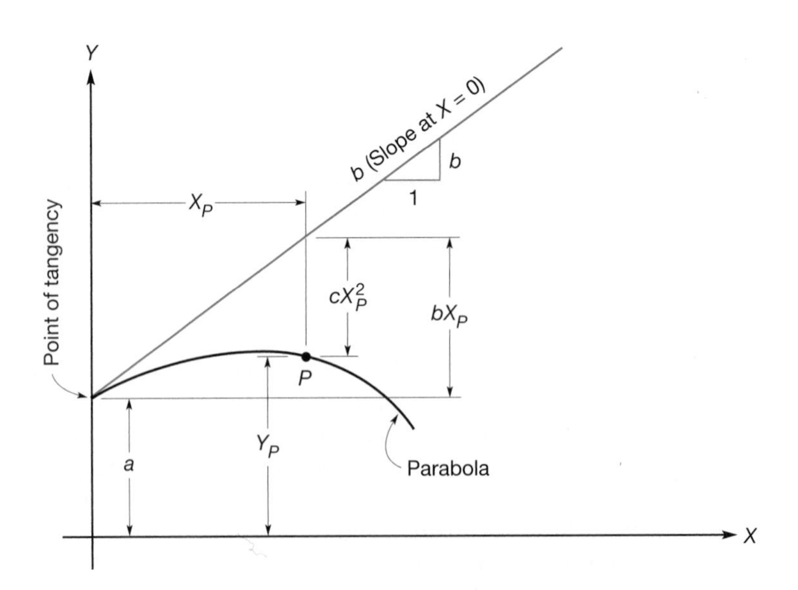

Figure 25.2 Terms for a parabola.

parabola in a XY rectangular coordinate system and illustrates the physical significance of the terms in Equation (25.1). Note from the figure that a is the ordinate at the beginning of the curve where $X = 0$, b is the slope of a tangent to the curve at $X = 0$, bX_P the change in ordinate along the tangent over distance X_P, and cX_P^2 the parabola's departure from the tangent (*tangent offset*) in distance X_P. When the terms a, bX_P, and cX_P^2 are combined as in Equation (25.1) and shown in Figure 25.2, they produce Y_P, the elevation on curve at X_P. For the crest curve of Figure 25.2, b has positive algebraic sign and c is negative.

■ 25.3 EQUATION OF AN EQUAL TANGENT VERTICAL PARABOLIC CURVE

Figure 25.3 shows a parabola that joins two intersecting tangents of a grade line. The parabola is essentially identical to that in Figure 25.2, except that the terms used are those commonly employed by surveyors and engineers. In the figure, BVC denotes the beginning of vertical curve, sometimes called the VPC (vertical point of curvature); V is the vertex, often called the VPI (vertical point of intersection); and EVC denotes the end of vertical curve, interchangeably called the VPT (vertical point of tangency). The percent grade of the back tangent (straight segment preceding V) is g_1, that of the forward tangent (straight segment following V) is g_2. The curve length L is the horizontal distance (in stations) from the BVC to the EVC. The curve of Figure 25.3 is called equal tangent because the horizontal distances from the BVC to V and from V to the EVC are equal, each being $L/2$. Proof of this is given in Section 25.5.

On the XY axis system in Figure 25.3, X values are horizontal distances measured from the BVC and Y values are elevations measured from the vertical datum

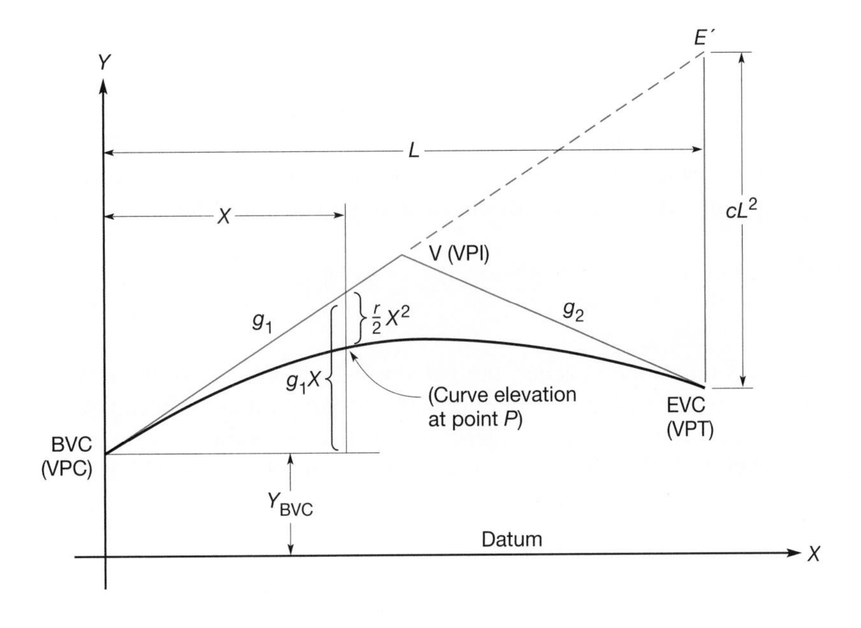

Figure 25.3 Vertical parabolic curve relationships.

of reference. By substituting this surveying terminology into Equation (25.1), the parabola can be expressed as

$$Y = Y_{BVC} + g_1X + cX^2 \tag{25.2}$$

In Equation (25.2), if the English system of units is used, Y is in ft, g_1 is percent grade, and for consistency of units, X must therefore be in 100-ft stations. If the metric system is used, Y is in meters, g_1 is again in percent grade and thus X must be in units of 100 meters, or 1/10th stations, where full stations are 1 km apart. The correspondence of terms in Equation (25.2) to those of Equation (25.1) is $a = Y_{BVC}$ (elevation of BVC) and $bX_P = g_1X$ (change in elevation along the back tangent with increasing X). To express the constant c of Equation (25.2) in surveying terminology, consider the tangent offset from E' on the extended back tangent to the EVC (dashed line in Figure 25.3). Its value (which is negative for the crest curve shown) is cL^2, where L (horizontal distance from BVC to E') is substituted for X. From the figure, cL^2 can be expressed in terms of horizontal lengths (in stations) and percent grades as follows:

$$cL^2 = g_1\left(\frac{L}{2}\right) + g_2\left(\frac{L}{2}\right) - g_1L \tag{a}$$

Solving Equation (a) for constant c gives

$$c = \frac{g_2 - g_1}{2L} \tag{b}$$

Substituting Equation (b) into Equation (25.2) results in the following equation for an equal-tangent vertical curve in surveying terminology:

$$Y = Y_{BVC} + g_1X + \left(\frac{g_2 - g_1}{2L}\right)X^2 \tag{25.3}$$

The *rate of change of grade, r*, for an equal-tangent parabolic curve equals the total grade change from BVC to EVC divided by length L (in stations for the English system, or 1/10th stations for metric units), over which the change occurs, or

$$r = \frac{g_2 - g_1}{L} \tag{25.4}$$

As mentioned earlier, the value r (which is negative for a crest curve and positive for a sag type) is an important design parameter because it controls the rate of curvature and hence rider comfort. To incorporate it in the equation for parabolic curves, Equation (25.4) is substituted into Equation (25.3),

$$Y = Y_{BVC} + g_1X + \left(\frac{r}{2}\right)X^2 \tag{25.5}$$

Figure 25.3 illustrates how the terms of Equation (25.5) combine to give the curve elevation at point P. Because the last term of the equation is the curve's offset from the back tangent, this formula is commonly called the *tangent offset equation*.

■ 25.4 HIGH OR LOW POINT ON A VERTICAL CURVE

To investigate drainage conditions, clearance beneath overhead structures, cover over pipes, or sight distance, it may be necessary to determine the elevation and location of the low (or high) point on a vertical curve. At the low or high point, a tangent to the curve will be horizontal and its slope equal to zero. Based on this fact, by taking the derivative of Equation (25.3) and setting it equal to zero, the following formula is readily derived

$$X = \frac{g_1 L}{g_1 - g_2} \tag{25.6}$$

where X is the distance from the BVC to the high or low point of the curve (in stations in the English system of units, and in 1/10th stations in the metric system), g_1 is the tangent grade through the BVC, g_2 the tangent grade through the EVC, and L the curve length (in stations or 1/10th stations).

If g_2 is substituted for g_1 in the numerator of Equation (25.6), distance X is measured back from the EVC. By substituting Equation (25.4) into Equation (25.6), the following alternate formula for locating the high or low point results

$$X = \frac{-g_1}{r} \tag{25.7}$$

■ 25.5 VERTICAL CURVE COMPUTATIONS USING THE TANGENT OFFSET EQUATION

In designing grade lines, the locations and individual grades of the tangents are normally selected first. This produces a series of intersection points V, each defined by its station and elevation. A curve is then chosen to join each pair of intersecting tangents. The parameter selected in vertical-curve design is length L. Having chosen it, the station of the BVC is obtained by subtracting $L/2$ from the vertex station. Adding L to the BVC station then determines the EVC station.

Stationing for the points on curve that are computed and staked are those that are *evenly divisible by the selected staking increment.* Thus if full stations is the staking increment selected for a curve in the English system of units, each full station, i.e., 10 + 00, 11 + 00, 12 + 00, etc., would be staked, but 10 + 50, 11 + 50, etc. would not be staked. For example, if the staking increment is half stations in the English system, then 15 + 00, 15 + 50, 16 + 00, 16 + 50, and so on, would be staked and not 15 + 25, 15 + 75, 16 + 25, 16 + 75, etc. In the metric system, if 40 m is the staking increment, then 2 + 400, 2 + 440, 2 + 480, 2 + 520, and so on would be staked, but not 2 + 420, 2 + 460, 2 + 500, etc.

Computations for vertical parabolic curves are normally done in tabular form.

25.5.1 Example Computations Using the English System of Units

Following are example computations for an equal-tangent vertical curve in the English system of units. The curve is a crest type.

Example 25.1

A grade g_1 of +3.00% intersects grade g_2 of −2.40% at a vertex whose station and elevation are 46 + 70 and 853.48 ft, respectively. An equal-tangent parabolic curve 600-ft long has been selected to join the two tangents. Compute and tabulate the curve for stakeout at full stations. (Figure 25.4 shows the curve.)

Solution

By Equation (25.4),

$$r = \frac{-2.40 - 3.00}{6} = -0.90\% \text{ station}$$

Stationing:

$$\begin{aligned} V &= 46 + 70 \\ -L/2 &= \underline{3 + 00} \\ BVC &= 43 + 70 \\ +L &= \underline{6 + 00} \\ EVC &= 49 + 70 \end{aligned}$$

$$\text{Elev}_{\text{BVC}} = 853.48 - 3.00(3) = 844.48 \text{ ft}$$

The remaining calculations utilize Equation (25.5) and are listed in Table 25.1.

A check on curve elevations is obtained by computing the first and second differences between the elevations of full stations, as shown in the right-hand columns of the table. Unless disturbed by rounding errors, all second differences (rate of change) should be equal. For curves in the English system of units at

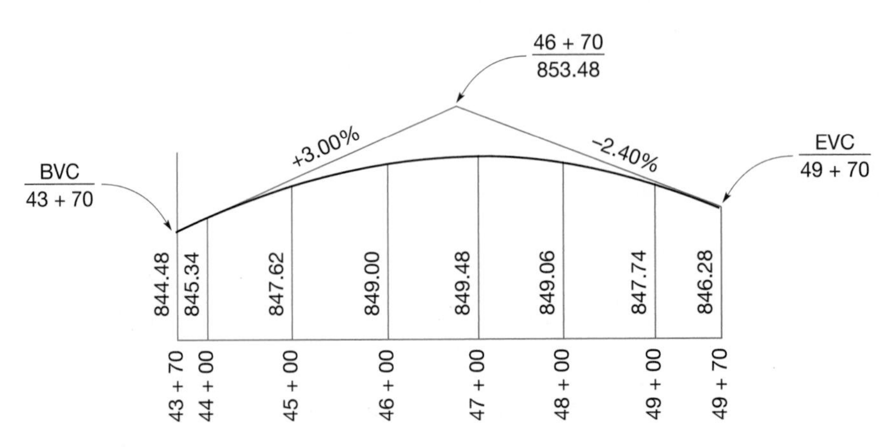

Figure 25.4 Crest curve of Example 25.1.

TABLE 25.1 NOTES FOR CURVE OF EXAMPLE 25.1

Station	X (sta)	g_1X	$\frac{rX^2}{2}$	Curve Elevation	First Difference	Second Difference
49 + 70 (EVC)	6.0	18.00	−6.20	846.28		
49 + 00	5.3	15.90	−2.64	847.74	−1.32	
48 + 00	4.3	12.90	−8.32	849.06	−0.42	−0.90
47 + 00	3.3	9.90	−4.90	849.48	0.48	−0.90
46 + 00	2.3	6.90	−2.38	849.00	1.38	−0.90
45 + 00	1.3	3.90	−0.76	847.62	2.28	−0.90
44 + 00	0.3	0.90	−0.04	845.34		
43 + 70 (BVC)	0.0	0.00	−0.00	844.48		

Check: $EVC = ELEV_V - g_2\left(\frac{L}{2}\right) = 853.48 - 2.40(3) = 846.28$ (Check!)

full-station increments, the second differences should equal r; for half-station increments they should be $r/4$.

It is sometimes desirable to calculate the elevation of the curve's center point. This can be done using $X = L/2$ in Equation (25.5). For Example 25.1, it is

$$Y_{\text{center}} = 844.48 + 3.00(3) - \left(\frac{0.90}{2}\right)(3)^2 + 849.43 \text{ ft}$$

This can be checked by employing the property of a parabolic curve, which is *the curve center falls halfway between the vertex and the midpoint of the long chord* (line from *BVC* to *EVC*). The elevation of the midpoint of the long chord (*LC*) is simply the average of the elevations of the *BVC* and *EVC*. For Example 25.1 it is

$$Y_{\text{midpoint } LC} = \frac{844.48 + 846.28}{2} = 845.38 \text{ ft}$$

By the property just stated, the elevation of the curve center for Example 25.1 is the average of the vertex elevation and that of the midpoint of the long chord, or

$$Y_{\text{center}} = \frac{845.38 + 853.48}{2} = 849.43 \text{ ft (check)}$$

25.5.2 Example Computations Using the Metric System

The following example illustrates the computations for an equal-tangent vertical curve when metric units are used. The curve is a sag type.

Example 25.2

A grade g_1 of −3.629% intersects grade g_2 of 0.151% at a vertex whose station and elevation are 5 + 265.000 and 350.520 m, respectively. An equal-tangent parabolic curve of 240-m length will be used to join the tangents. Compute and tabulate the curve for staking at 40-m increments.

Solution

By Equation (25.4),

$$r = \frac{0.151 + 3.629}{2.4} = 1.575$$

[*Note:* L used in Equation (25.4) is in units of m/100, or 1/10th stations.]
Stationing:

$$\begin{aligned} VPI \text{ Station} &= 5 + 265 \\ -L/2 &= \quad 120 \\ BVC \text{ Station} &= 5 + 145 \\ +L &= \quad 240 \\ EVC \text{ Station} &= 5 + 385 \end{aligned}$$

$$\text{Elev}_{BVC} = 350.520 + 3.629(120/100) = 354.875$$

The remaining calculations employ Equation (25.5) and are listed in Table 25.2.

TABLE 25.2 NOTES FOR CURVE OF EXAMPLE 25.2

Station	$X\left(\frac{m}{100}\right)$	g_1X	$\frac{rX^2}{2}$	Curve Elevation (m)	First Difference	Second Difference
5 + 385.000 (EVC)	2.400	−8.710	4.536	350.701		
5 + 360.000	2.150	−7.802	3.640	350.713	−0.223	
5 + 320.000	1.750	−6.351	2.412	350.936	−0.475	0.252
5 + 280.000	1.350	−4.899	1.435	351.411	−0.727	0.252
5 + 240.000	0.950	−3.448	0.711	352.138	−0.979	0.252
5 + 200.000	0.550	−1.996	0.238	353.117	−1.232	0.252
5 + 160.000	0.150	−0.544	0.018	354.348		
5 + 145.000 (BVC)	0.000	−0.000	0.000	354.875		

Check: $\text{EVC} = \text{ELEV}_V - g_2\left(\frac{L}{2}\right) = 350.520 + (0.151 \times 120/100) = 350.701$

Note that the second differences are all equal, which checks the computations. [The value of 0.252 is $r/6.25$, where 6.25 is $(100\ \text{m}/40\ \text{m})^2$.]

■ Example 25.3

Compute the station and elevation of the curve's high point in Example 25.1.

Solution

By Equation (25.7), $X = -3.00/-0.90 = 3.3333$ stations.
Then the station of the high point is

$$\text{station}_{\text{high}} = (43 + 70) + (3 + 33.33) = 47 + 03.33$$

By Equation (25.3), the elevation at this point is

$$844.48 + 3.00(3.333) + \frac{-2.40 - 3.00}{2(6)}(3.3333)^2 = 849.48$$

Note that in using Equation (25.7) and all other equations of this chapter, correct algebraic signs must be applied to grades g_1 and g_2.

By applying the same equations and procedures, the station and elevation of the low point of the curve of Example 25.2 are 5 + 375.413 and 350.694 m, respectively. The calculations are left as an exercise.

■ 25.6 EQUAL TANGENT PROPERTY OF A PARABOLA

The curve defined by Equations (25.3) and (25.5) has been called an equal-tangent parabolic curve, which means the vertex occurs a distance $X = L/2$ from the BVC. Proof of this property is readily made with reference to Figure 25.5, which illustrates a sag curve. In the figure, assume the horizontal distance from BVC to V is an unknown value X; thus the remaining distance from V to EVC is $L - X$. Two expressions can be written for the elevation of the EVC. The first, using Equation (25.3) with $X = L$, yields

$$Y_{EVC} = Y_{BVC} + g_1 L + \left(\frac{g_2 - g_1}{2L}\right)L^2 \tag{c}$$

The second, using changes in elevation that occur along the tangents, gives

$$Y_{EVC} = Y_{BVC} + g_1 X + g_2(L - X) \tag{d}$$

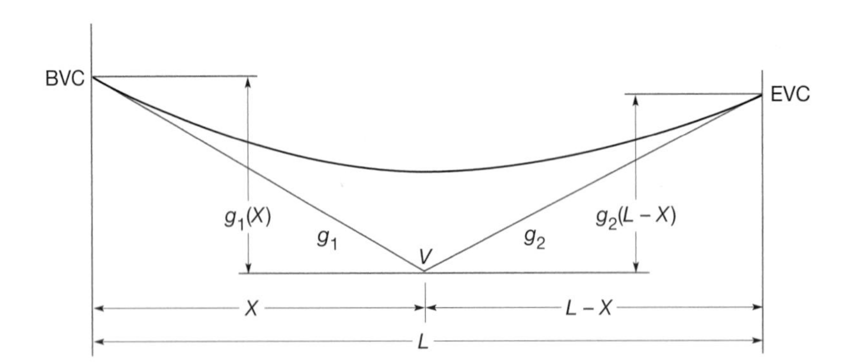

Figure 25.5 Proof of equal-tangent property of a parabola.

Equating Equations (c) and (d) and solving, $X = L/2$. Thus distances *BVC* to *V* and *V* to *EVC* are equal—hence the term *equal-tangent parabolic curve.*

■ 25.7 CURVE COMPUTATIONS BY PROPORTION

The following basic property of a parabola can be used to simplify vertical curve calculations: *Offsets from a tangent to a parabola are proportional to the squares of the distances from the point of tangency.* Calculating the offset E at the midpoint of the curve, and then computing offsets at any other distance X from the BVC by proportion according to the following formula conveniently utilize this property.

$$offset_X = E\left(\frac{X}{1/2L}\right)^2 \tag{25.8}$$

The value of E in Equation (25.8) is simply the difference in elevation from the curve's midpoint to the VPI. Computation of the curve's midpoint elevation was discussed in Section 25.4 and for Example 25.1 it was 849.43 ft. To illustrate the use of Equation (25.8), the offset from the tangent to the curve for station 47 + 00, where $X = 3.3$ stations, will be computed:

$$\text{offset}_{47+00} = (849.43 - 853.48)\left(\frac{3.3}{3}\right)^2 = -4.90 \text{ ft}$$

In the above calculation, 3.3 in the numerator is X in stations, and 3 in the denominator is $L/2$ also in stations. The value -4.90 checks the tangent offset listed in Table 25.1 for station 47 + 00. This simplified procedure is convenient for computing vertical curves in the field with a hand-held calculator.

■ 25.8 STAKING A VERTICAL PARABOLIC CURVE

Prior to initiating construction on a route project, the planned centerline, or an offset one, will normally be staked at full or half-stations, as well as other critical horizontal alignment points such as PCs and PTs. Then, as described in Section 23.7, slope stakes will be set out perpendicular to the centerline at or near the slope intercepts to guide rough grading. Excavation and embankment construction then proceed and continue until the grade is near plan elevation. The centerline stations are then staked again, this time using sharpened 2-in. square wooden hubs, usually about 10-in. long. These are known as "blue tops," so called because when their tops are driven to grade elevation, they are colored blue. Contractors request blue tops when excavated areas are still slightly high and embankments somewhat low. After blue tops are set to mark the precise grade, final grading is completed.

To set blue tops at grade, a circuit of differential levels is run from a nearby benchmark to establish the *HI* of a leveling instrument in the project area. The difference between the *HI* and any station's grade is the required rod reading on that stake. Assume, for example, a *HI* of 856.20 ft exists and station 45 + 00 of Example 25.1 is to be set. The required rod reading is 856.20 − 847.62 = 8.58 ft. With the stake initially driven firmly into the ground and the rod held on its top, suppose a reading of 8.25 is obtained. The stake then must be driven down an additional 8.58 − 8.25 = 0.33 ft further and the notetaker so indicates. After the stake is driven, a distance estimated to be somewhat less than 0.33 ft, the rod reading is checked. This is repeated until the required reading of 8.58 is achieved, whereupon the stake is colored blue using keel or spray paint.

This process is continued until all stakes are set. The required rod reading at station 46 + 00, for a *HI* of 856.20, is 7.20 ft. If a rock is encountered and the stake cannot be driven to grade, a vertical offset of, say, 1.00 ft above grade can be marked and noted on the stake.

When the level is too far away from the station being set, a turning point is established and the instrument brought forward to establish a new *HI*. Whenever possible, level circuit checks should be made by closing on nearby benchmarks as blue top work on the project progresses. Also, when quitting for the day or when the job is finished, the level circuit must always be closed to verify that no mistakes were made.

■ 25.9 MACHINE CONTROL IN GRADING OPERATIONS

As stated in Section 23.11, GPS provides sufficient accuracy for most rough grading operations. However it is not sufficient to provide final grades on most projects. Thus a GPS machine control using GPS must be augmented with laser levels to provide vertical accuracies under 3 mm. The laser level requires a sensor on the construction vehicle and a rotating laser (see Figure 23.3). The level must be positioned over a known calibrated point within 1 foot horizontally. The laser covers a radius of about 1500 ft. Continuous line of sight must be maintained between the rotating laser level and the sensor on the construction vehicle.

As stated in Section 23.11, a robotic total station with multifaceted, 360° prism can be used to establish horizontal and vertical positioning. This system is limited to a range of about 1000 ft from the total station to the construction vehicle. However it can provide both horizontal and vertical accuracies at a sufficient level for most construction applications.

■ 25.10 COMPUTATIONS FOR AN UNEQUAL TANGENT VERTICAL CURVE

An unequal-tangent vertical curve is simply a pair of equal-tangent curves, where the EVC of the first curve is the BVC of the second. This point is called CVC, point of compound vertical curvature. In Figure 25.6, a −2.00 percent grade

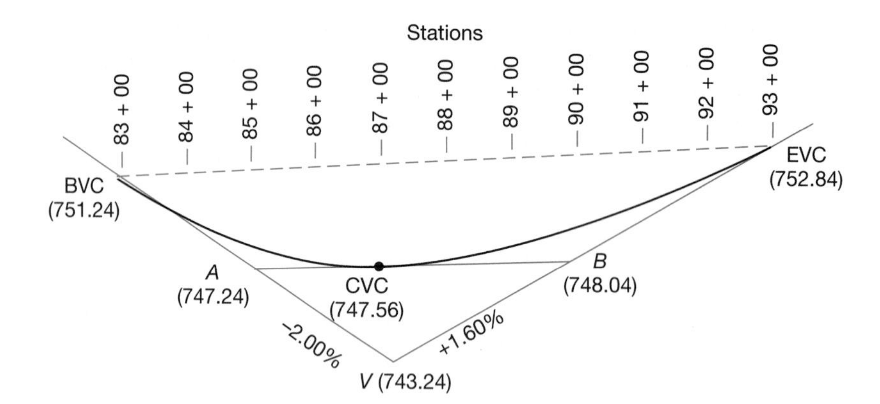

Figure 25.6 Unequal-tangent vertical curve.

intersects a +1.60 percent grade at station 87 + 00 and elevation 743.24 ft. A vertical curve of length $L_1 = 400$ ft is to be extended back from the vertex and a curve of length $L_2 = 600$ ft run in forward to closely fit ground conditions.

To perform calculations for this type of curve, connect the midpoints of the tangents for the two curves, stations 85 + 00 and 90 + 00, to obtain line *AB*. Point *A* is the vertex of the first curve and is located a distance of $L_1/2$ back from *V*. Point *B* is the vertex of the second curve which is $L_2/2$ forward from *V*. Compute elevations for *A* and *B* and, using them, calculate the grade of *AB* by dividing the difference in elevation between *B* and *A* by the distance in stations separating these two points. From grade *AB*, determine the *CVC* elevation.

Now compute two equal tangent vertical curves, one from *BVC* to *CVC* and another from *CVC* to *EVC*, by the methods of Section 25.4. Since both curves are tangent to the same line *AB* at point *CVC*, they will be tangent to each other and form a smooth curve.

Example 25.4

For the configuration of Figure 25.6, compute and tabulate the notes necessary to stake the unequal-tangent vertical curve at full stations.

Solution

1. Calculate elevations of *BVC*, *EVC*, *A*, *B*, and *CVC*, and grade *AB*.

$$\begin{aligned}
\text{Elev}_{BVC} &= 743.24 + 4(2.00) = 751.24 \text{ ft} \\
\text{Elev}_{A} &= 743.24 + 2(2.00) = 747.24 \text{ ft} \\
\text{Elev}_{EVC} &= 743.24 + 6(1.60) = 752.84 \text{ ft} \\
\text{Elev}_{B} &= 743.24 + 3(1.60) = 748.04 \text{ ft} \\
\text{Grade}_{AB} &= \left(\frac{748.04 - 747.24}{5}\right) = +0.16\% \\
\text{Elev}_{CVC} &= 747.24 + 2(0.16) = 747.56 \text{ ft}
\end{aligned}$$

These elevations are shown in Figure 25.6.

2. In computing the first curve, the grade of AB will be g_2 in the formulas and for the second curve it will be g_1. The rates of change of grade for the two curves are, by Equation (25.4),

$$r_1 = \frac{0.16 - (-2.00)}{4} = +0.54\%/\text{station}$$

$$r_2 = \frac{1.60 - 0.16}{6} = +0.24\%/\text{station}$$

3. Equation (25.5) is now solved in tabular form and the results are listed in Table 25.3.

TABLE 25.3 NOTES FOR CURVE OF EXAMPLE 25.3

Station	X (sta)	g_1X	$rX^2/2$	Curve Elevation	First Difference	Second Difference
93 + 00 (EVC)	6	0.96	4.32	752.84 ✓	1.48	
92 + 00	5	0.80	3.00	751.36	1.24	0.24
91 + 00	4	0.64	1.92	750.12	1.00	0.24
90 + 00	3	0.48	1.08	749.12	0.76	0.24
89 + 00	2	0.32	0.48	748.36	0.52	0.24
88 + 00	1	0.16	0.12	747.84	0.28	0.24
87 + 00 (CVC)	4	−8.00	4.32	747.56 ✓	−0.11	
86 + 00	3	−6.00	2.43	747.67	−0.65	0.54
85 + 00	2	−4.00	1.08	748.32	−1.19	0.54
84 + 00	1	−2.00	0.27	749.51	−1.73	0.54
83 + 00 (BVC)	0	0.00	0.00	751.24		

Vertical-curve computations by themselves are quite simple, hardly a challenge to a modern computer. But when vertical curves are combined with horizontal curves, spirals, and superelevation in complex highway designs, programming saves time. Figure 25.7 shows a completed data entry box for computing the first vertical curve of Example 25.3 using the WOLFPACK software that accompanies this book. The software employs the equations discussed in this chapter to compute staking notes for a vertical curve. The computed results are written to a file in table form like those given in Tables 25.1, 25.2, and 25.3, so they can be printed and taken into the field for staking purposes.

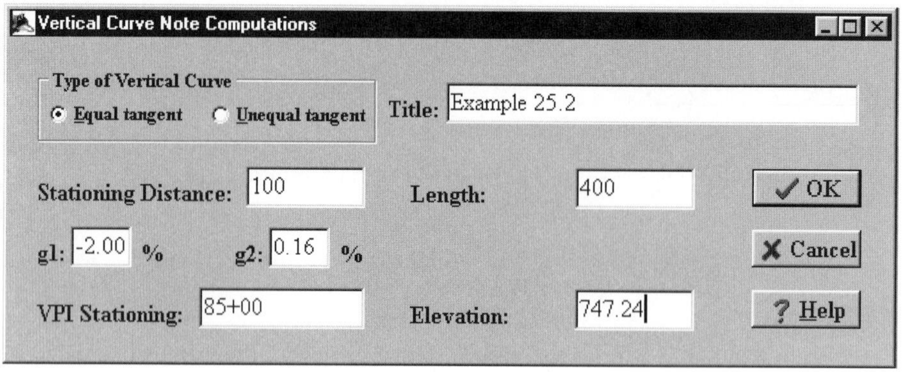

Figure 25.7 Data entry screen in WOLFPACK for computation of first vertical curve staking notes in Example 25.3.

25.11 DESIGNING A CURVE TO PASS THROUGH A FIXED POINT

The problem of designing a parabolic curve to pass through a point of fixed station and elevation is frequently encountered in practice. For example, it occurs where a new grade line must meet existing railroad or highway crossings or when a minimum vertical distance must be maintained between the grade line and underground utilities or drainage structures.

Given the station and elevation of the VPI and grades g_1 and g_2 of the back and forward tangents, respectively, the problem consists of calculating the curve length required to meet the fixed condition. It is solved by substituting known quantities into Equation (25.3) and reducing the equation to its quadratic form containing only L as an unknown. Two values will satisfy the quadratic equation, but the correct one will be obvious.

Example 25.5

In Figure 25.8, grades $g_1 = -4.00$ percent and $g_2 = +3.80$ percent meet at VPI station 52 + 00 and elevation 1261.50. Design a parabolic curve to meet a railroad crossing, which exists at station 53 + 50 and elevation 1271.20.

Solution

In referring to Figure 25.8 and substituting known quantities into Equation (25.3), the following equation is obtained:

$$1271.20 = \left[1261.50 + 4.00\left(\frac{L}{2}\right)\right] + \left[-4.00\left(\frac{L}{2} + 1.5\right)\right] + \left[\frac{3.80 + 4.00}{2L}\left(\frac{L}{2} + 1.5\right)^2\right]$$

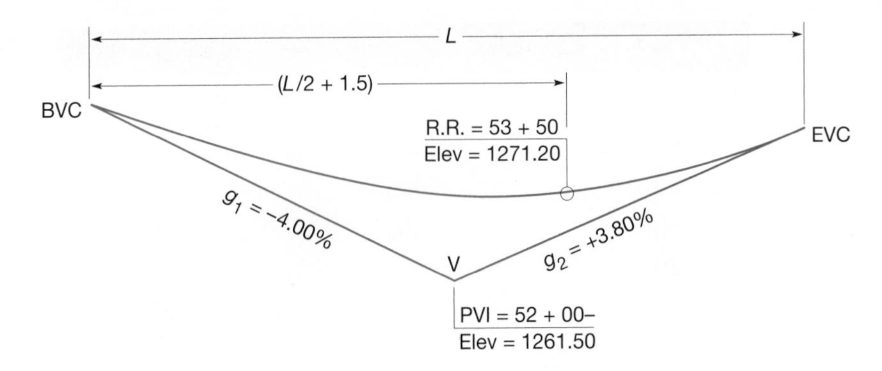

Figure 25.8 Designing a parabolic curve to pass through a fixed point.

In this expression, the value of X for the railroad crossing is $L/2 + 1.5$ and the terms within successive brackets are Y_{BVC}, g_1X, and $(r/2)X^2$, respectively. Reducing the equation to quadratic form gives

$$0.975L^2 - 9.85L + 8.775 = 0$$

Solving by use of Equation (11.3) for L gives 9.1152 stations. To check the solution, $L = 9.1152$ stations and $X = [(9.1152/2) + 1.5]$ stations are used in Equation (25.3) to calculate the elevation at station 53 + 50. A value of 1271.20 checks the computations.

■ 25.12 SIGHT DISTANCE

The vertical alignments of highways should provide ample sight distance for safe vehicular operation. Two types of sight distances are involved: (1) stopping sight distance (the distance required, for a given "design speed",[1] to safely stop a vehicle thus avoiding a collision with an unexpected stationary object in the roadway ahead) and (2) passing sight distance (the distance required for a given design speed, on two-lane two-way highways to safely overtake a slower moving vehicle, pass it, and return to the proper lane of travel leaving suitable clearance for an oncoming vehicle in the opposing lane). For either condition, as speed increases, required sight distance also increases. All highways should provide safe stopping sight distances for their entire extent at their given design speed, and if this cannot be achieved in certain sections, signs must be posted to reduce travel speeds to levels consistent with the available sight distances. Passing sight distance should be provided at frequent intervals along any section of highway to allow faster moving vehicles to pass slower moving ones. In sections of highway that do not provide ample passing sight distances, appropriate centerline markings and signs are used to inform drivers of this condition. The American Association of State Highway and Transportation Officials (AASHTO) has recommended minimum sight

---

[1]Design speed is defined as the maximum safe speed that can be maintained over a specified section of highway when conditions are so favorable that the design features of the highway govern. Once selected, all of the pertinent features of the highway, especially those involving safety, should be related to that speed.

TABLE 25.4 MINIMUM SIGHT DISTANCES FOR VARYING DESIGN SPEEDS

Design Speed (mph)	Stopping Sight Distance (ft)	Passing Sight Distance (ft)
40	275	1500
50	400	1800
60	525	2100
70	625	2500

distances for both stopping and passing for various design speeds. Table 25.4 lists these values for some commonly used design speeds.

Sight distances must be carefully considered in the design of vertical alignments of highway projects. Given the grades of two intersecting tangent sections, the length of vertical curve used to provide a transition from one to the other fixes the sight distance. A longer curve provides a greater sight distance.

The formula for length of curve L necessary to provide sight distance S on a *crest vertical curve, where S is less than L* is:

$$L = \frac{S^2(g_1 - g_2)}{2(\sqrt{h_1} + \sqrt{h_2})^2} \tag{25.9}$$

In Equation (25.9) the units of S and L are stations if the English system is used and one-tenth stations in the metric system. Also the units of h_1 (the height of the driver's eye) and h_2 (the height of an object sighted on the roadway ahead) are in feet for the English system and meters for the metric system. For design, AASHTO recommends 3.50 ft (1.070 m) for h_1. Recommended values for h_2 are 0.50 ft (0.150 m) for stopping and 4.25 ft (1.300 m) for passing. The lower value for h_2 represents the size of an object that would damage a vehicle and the higher value represents the height of an oncoming car.

Then for a crest curve having grades of $g_1 = +1.40$ percent and $g_2 = -1.00$ percent, by Equation (25.9) the length of curve needed to provide a 525-ft stopping sight distance is:

$$L = \frac{(5.25)^2(1.40 + 1.00)}{2(\sqrt{3.50} + \sqrt{0.50})^2} = 4.97 \text{ stations}$$

Since S is greater than L and thus not in agreement with the assumption used in deriving the formula, a different expression must be employed. If the vehicle is off the curve but on the tangent leading to it and *S is greater than L*, the applicable sight distance formula is:

$$L = 2S - \frac{2(\sqrt{h_1} + \sqrt{h_2})^2}{g_1 - g_2} \tag{25.10}$$

Then in the preceding example, the length of curve necessary to provide 525 ft of stopping sight distance is:

$$L = 2(5.25) - \frac{2\left(\sqrt{3.50} + \sqrt{(0.05)}\right)^2}{1.40 + 1.00} = 4.96 \text{ stations}$$

In this solution, the sight distance of 525 feet is greater than the computed curve length of 4.96 stations (496 ft) and thus the conditions are met.

Sag vertical curves also limit sight distances because they reduce lengths ahead that can be illuminated by headlights during night driving. Equations that apply in computing lengths of sag vertical curves based upon headlight criteria are:

(a) S less than L:

$$L = \frac{S^2(g_2 - g_1)}{4 + 3.5S} \tag{25.11}$$

(b) S greater than L:

$$L = 2S - \frac{4 + 3.5S}{g_1 - g_2} \tag{25.12}$$

As discussed in Section 24.17, horizontal curves may also limit visibility and sight distances for them can be computed as well. For a combined horizontal and vertical curve, the sight distance that governs is the smaller of the two values computed independently for each curve.

■ 25.13 SOURCES OF ERROR IN LAYING OUT VERTICAL CURVES

Some sources of error in staking out parabolic curves are:

1. Making errors in measuring distances and angles when staking the centerline.
2. Not holding the level rod plumb when setting blue tops.
3. Using a leveling instrument which is out of adjustment.

■ 25.14 MISTAKES

Some typical mistakes made in computations for vertical curves include the following:

1. Arithmetic mistakes.
2. Failure to properly account for the algebraic signs of g_1 and g_2.
3. Subtracting offsets from tangents for a sag curve or adding them for a crest curve.
4. Failure to make the second-difference check.
5. Not completing the level circuit back to a benchmark after setting blue tops.

PROBLEMS

Asterisks (*) indicate problems that have answers given in Appendix G.

25.1 Why are vertical curves needed on the grade lines for highways and railroads?

25.2 What is meant by the "rate of grade change" on vertical curves, and why is it important?

Tabulate station elevations for an equal-tangent parabolic curve for the data given in Problems 25.3 through 25.9. Check by second differences.

25.3 A +1.25% grade meets a −2.75% grade at station 44 + 25 and elevation 682.34 ft, 800-ft curve, stakeout at half stations.

25.4 A −2.50% grade meets a +1.50% grade at station 4 + 200 and elevation 293.585 m, 300-m curve, stakeout at 30-m increments.

25.5 A 375-ft curve, grades of $g_1 = -2.60\%$ and $g_2 = +0.50\%$, VPI at station 36 + 40, and elevation 905.35 ft, stakeout at full stations.

25.6 A 600-ft curve, grades of $g_1 = -4.00\%$ and $g_2 = -3.00\%$, VPI at station 66 + 50, and elevation 5560.00 ft, stakeout at full stations.

25.7 A 200-m curve, $g_1 = +4.00\%$, $g_2 = -2.50\%$, VPI station = 2 + 175, VPI elevation = 1157.830 m, stakeout at 40-m increments.

25.8 A 200-ft curve, $g_1 = -1.25\%$, $g_2 = +1.25\%$, VPI station = 46 + 00, VPI elevation = 895.00 ft, stakeout at quarter stations.

25.9 An 90-m curve, $g_1 = +1.00\%$, $g_2 = -0.50\%$, VPI station = 6 + 280, VPI elevation = 550.600 m, stakeout at 10-m increments.

Field conditions require a highway curve to pass through a fixed point. Compute a suitable equal-tangent vertical curve and full-station elevations for Problems 25.10 through 25.12.

25.10* Grades of $g_1 = -2.50\%$ and $g_2 = +1.00\%$, VPI elevation 750.00 ft at station 30 + 00. Fixed elevation 753.00 ft at station 30 + 00.

25.11 Grades of $g_1 = -2.60\%$ and $g_2 = +1.40\%$, VPI elevation 2430.00 ft at station 315 + 00. Fixed elevation 2436.50 ft at station 314 + 00.

25.12 Grades of $g_1 = +4.50\%$ and $g_2 = +2.00\%$, VPI station 6 + 300 and elevation 205.930 m. Fixed elevation 205.620 m at station 6 + 400. (Use 100-m stationing.)

25.13 A −1.10% grade meets a +0.60% grade at station 36 + 00 and elevation 800.00 ft. The +0.60% grade then joins a 2.40% grade at station 39 + 00. Compute and tabulate the notes for an equal-tangent vertical curve, at half-stations, that passes through the midpoint of the 0.60% grade.

25.14 When is it advantageous to use an unequal-tangent vertical curve instead of an equal-tangent one?

Compute and tabulate full-station elevations for an unequal-tangent vertical curve to fit the requirements in Problems 25.15 through 25.18.

25.15 A +3.00% grade meets a −2.50% grade at station 60 + 00 and elevation 1086.00 ft. Length of first curve 400 ft, second curve 600 ft.

25.16 Grade $g_1 = +2.00\%$, $g_2 = +4.50\%$, VPI at station 62 + 00 and elevation 650.00 ft, $L_1 = 600$ ft and $L_2 = 500$ ft.

25.17 Grades g_1 of +5.00% and g_2 of −1.50% meet at the VPI at station 4 + 300 and elevation 154.960 m. Lengths of curves are 100 m and 200 m. (Use 30-m stationing.)

25.18 A −2.40% grade meets a +1.75% grade at station 95 + 00 and elevation 320.64 ft. Length of first curve is 400 ft, of second curve, 200 ft.

25.19* A manhole is 10 ft from the centerline of a 20-ft wide street that has a 6-in. parabolic crown. The street center at the station of the manhole is at elevation 205.68 ft. What is the elevation of the manhole cover?

25.20 A 40-ft wide street has an average parabolic crown from the center to each edge of 1/4 in./ft. How much does the surface drop from the street center to a point 4 ft from the edge?

25.21 Determine the station and elevation at the high point of the curve in Problem 25.3.

25.22 Calculate the station and elevation at the low point of the curve in Problem 25.4.

25.23 Compute the station and elevation at the low point of the curve of Problem 25.5.

25.24 What are the station and elevation of the high point of the curve of Problem 25.7?

25.25 What additional factor must be considered in the design of crest vertical curves that is not of concern in sag curves?

25.26* Compute the sight distance available in Problem 25.3. (Assume $h_1 = 3.50$ ft and $h_2 = 4.25$ ft.)

25.27 Similar to Problem 25.26, except $h_2 = 2.50$ ft.

25.28 Similar to Problem 25.26, except for the data of Problem 25.7, where $h_1 = 1.2$ m and $h_2 = 0.5$ m.

25.29 In determining sight distances on vertical curves, how does the designer determine whether the cars or objects are on the curve or tangent?

What is the minimum length of a vertical curve to provide a required sight distance for the conditions given in Problems 25.30 through 25.32?

25.30* Grades of +3.40% and −2.30%, sight distance 600 ft, $h_1 = 3.50$ ft and $h_2 = 4.25$ ft.

25.31 A crest curve with grades of +2.50% and −2.00%, sight distance 500 ft, $h_1 = 4.50$ ft, and $h_2 = 1.50$ ft.

25.32 Sight distance of 200 m, grades of +1.40% and −1.50%, $h_1 = 1.1$ m and $h_2 = 1.3$ m.

25.33* A backsight of 6.85 ft is taken on a benchmark whose elevation is 567.50 ft. What rod reading is needed at that HI to set a blue top at grade elevation of 572.55 ft?

25.34 A backsight of 4.52 ft is taken on a benchmark whose elevation is 658.28. A foresight of 2.18 ft and a backsight of 5.04 ft are then taken in turn on TP_1 to establish a HI. What rod reading will be necessary to set a blue top at a grade elevation of 660.38?

25.35 Develop a computational program that performs the vertical-curve computations.

25.36 Report on an article in a trade magazine that discusses current methods used in construction surveying.

BIBLIOGRAPHY

American Association of State Highway and Transportation Officials. 1990. *A Policy on Geometric Design of Highways and Streets*. Washington, D.C.: AASHTO.

26

Volumes

■ 26.1 INTRODUCTION

Persons engaged in surveying (geomatics) are often called on to determine volumes of various types of material. For example, quantities of earthwork and concrete are needed on many types of construction projects. Volume computations are also required to determine the capacities of bins, tanks, reservoirs, and buildings, and to check stockpiles of coal, gravel, and other materials. The determination of quantities of water discharged by streams and rivers, per unit of time, is also important.

The most common unit of volume is a cube having edges of unit length. Cubic feet, cubic yards, and cubic meters are used in surveying calculations, with cubic yards and cubic meters being most common for earthwork. (*Note:* $1\ \text{yd}^3 = 27\ \text{ft}^3$; $1\ \text{m}^3 \approx 35.3144\ \text{ft}^3$). The *acre-foot* (the volume equivalent to an acre of area, 1 ft deep) is commonly used for large quantities of water, while cubic feet per second (ft^3/sec) and cubic meters per second (m^3/sec) are the usual units for water flow measurement.

■ 26.2 METHODS OF VOLUME MEASUREMENT

Direct measurement of volumes is rarely made in surveying, since it is difficult to actually apply a unit of measure to the material involved. Instead, indirect measurements are obtained by measuring lines and areas that have a relationship to the volume desired.

Three principal systems are used: (1) the cross-section method, (2) the unit-area (or borrow-pit) method, and (3) the contour-area method.

■ 26.3 THE CROSS-SECTION METHOD

The cross-section method is employed almost exclusively for computing volumes on linear construction projects such as highways, railroads, and canals. In this procedure, after the centerline has been staked, ground profiles called cross sections are taken (at right angles to the centerline), usually at intervals of full or half stations if the English system of units is being used, or at perhaps 10, 20, 30, or 40 m if the metric system is being employed. Cross sectioning consists of observing ground elevations and their corresponding distances left and right perpendicular to the centerline. Readings must be taken at the centerline, at high and low points, and at locations where slope changes occur to determine the ground profile accurately. This can be done in the field using a level, level rod, and tape. Plate B.5 in Appendix B illustrates a set of field notes for cross sectioning.

Much of the fieldwork formerly involved in running preliminary centerline, getting cross-section data, and making slope-stake and other measurements on long route surveys is now being done more efficiently by photogrammetry. Research has shown that earthwork quantities computed from photogrammetric cross sectioning agree to within about 1 or 2 percent of those obtained from good-quality field cross sections. It is not intended to discuss photogrammetric methods in this chapter; rather, basic field and office procedures for determining and calculating volumes will be presented briefly. Chapter 27 discusses the subject of photogrammetry.

In Section 17.8 the subject of terrain representation by means of digital elevation models (DEMs) was introduced and concepts for deriving triangulated irregular networks (TINs) from DEMs were presented. It was noted that once a TIN model is created for a region, profiles and cross sections anywhere within the area could be readily derived using a computer. This can be of significant advantage where the general location of a proposed road or railroad has been decided upon, but the final alignment is not yet fixed. In those situations, controlling points and breaklines can be surveyed in the region where the facility is expected to be located and a DEM for the area generated. Either ground or photogrammetric methods can be employed for deriving the terrain data. From the DEM information, a TIN model can be created and the computer can provide cross sections for the analysis of any number of alternate alignments automatically.

After cross sections have been taken and plotted, *design templates* (outlines of base widths and side slopes of the planned excavation or embankment) are superimposed on each plot to define the excavation or embankment to be constructed at each cross-section location. Areas of these sections, called *end areas*, are obtained by computation or by planimeter (see Section 12.9.4). Using computers, end areas are calculated directly from field cross-section data and design information. From the end areas, volumes are determined by the *average-end-area*, or *prismoidal* formula, discussed later in this chapter.

Figure 26.1 portrays a section of planned highway construction and illustrates some of the points just discussed. Centerline stakes are shown in place, with their stationing given in the English system of units. They mark locations where cross sections are taken, in this instance at full stations. End areas, based on the planned grade line, size of roadway, and selected embankment and excavation slopes, are superimposed at each station and are shown shaded. Areas of

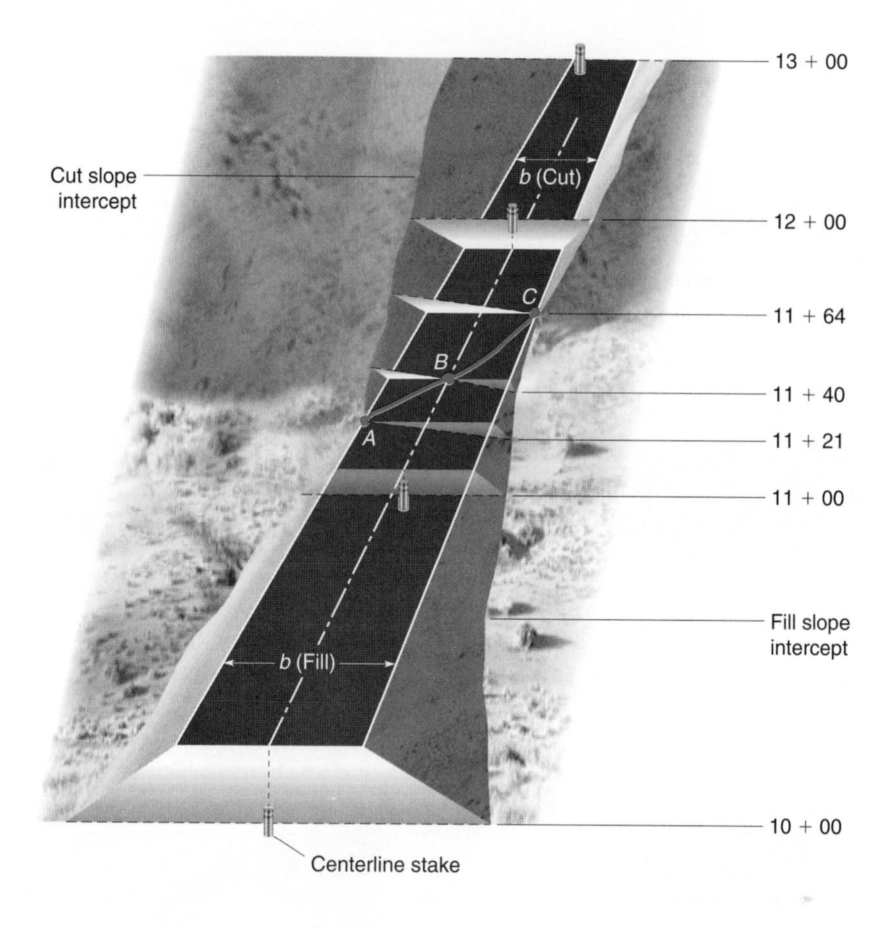

Figure 26.1 Section of roadway illustrating excavation (cut) and embankment (fill).

these shaded sections are determined, whereupon volumes are computed using formulas given in Section 26.5 or 26.8. Note in the figure that embankment, or *fill,* is planned from stations 10 + 00 through 11 + 21, a transition from fill to excavation, or *cut,* occurs from station 11 + 21 to 11 + 64, and cut is required from stations 11 + 64 through 13 + 00.

■ 26.4 TYPES OF CROSS SECTIONS

The types of cross sections commonly used on route surveys are shown in Figure 26.2. In flat terrain the *level section* (a) is suitable. The *three-level section* (b) is generally used where ordinary ground conditions prevail. Rough topography may require a *five-level section* (c), or more practically an *irregular section* (d). A *transition section* (e) and a *side-hill section* (f) occur when passing from cut to fill and on side-hill locations. In Figure 26.1, transition sections occur at stations 11 + 21 and 11 + 64, while a *side-hill section* exists at 11 + 40.

The width of base b, the finished roadway, is fixed by project requirements. As shown in Figure 26.1, it is usually wider in cuts than on fills to provide for drainage ditches. The side slope s [the horizontal dimension required for a unit vertical rise and illustrated in Figure 26.2(a)] depends on the type of soil

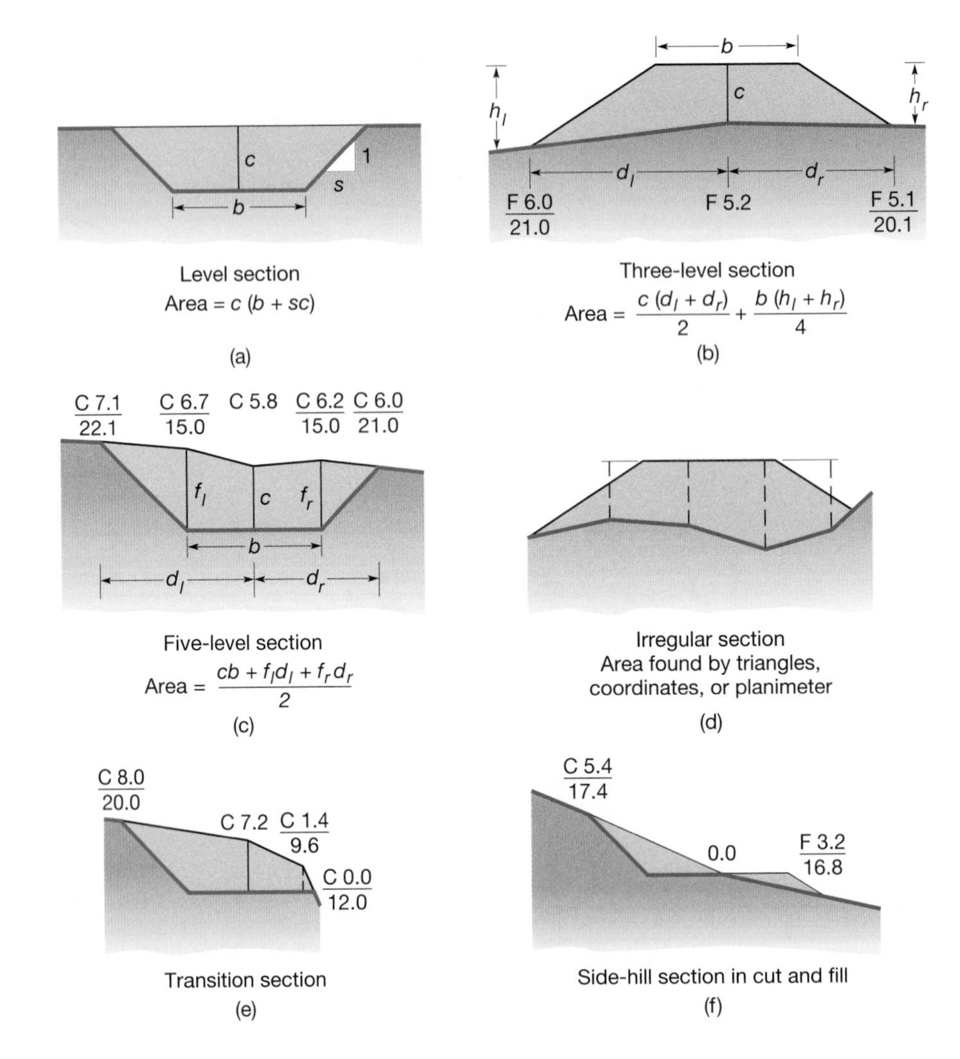

Figure 26.2
Earthwork sections.

encountered. Side slopes in fill usually are flatter than those in cuts where the soil remains in its natural state.

Cut slopes of 1:1 (1 horizontal to 1 vertical) and fill slopes of 1-1/2:1 might be satisfactory for ordinary loam soils, but 1-1/2:1 in excavation and 2:1 in embankment are common. Even flatter proportions may be required—one cut in the Panama Canal area was 13:1—depending on soil type, rainfall, and other factors. Formulas for areas of sections are readily derived and listed with some of the sketches in Figure 26.2.

■ 26.5 AVERAGE END AREA FORMULA

Figure 26.3 illustrates the concept of computing volumes by the average-end-area method. In the figure, A_1 and A_2 are end areas at two stations separated by a horizontal distance L. The volume between the two stations is equal to

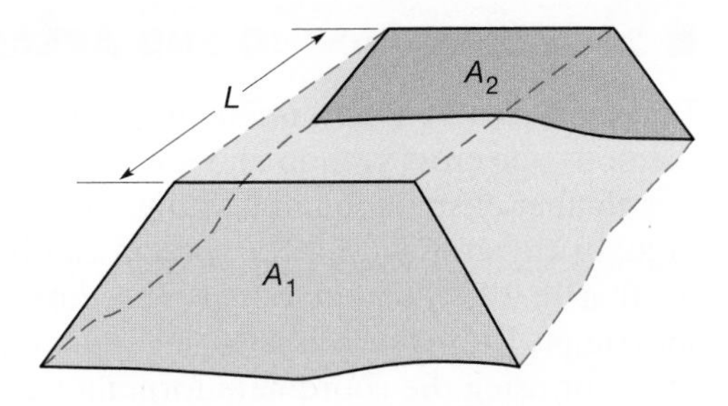

Figure 26.3
Volume by average-end-area-method.

the average of the end areas multiplied by the horizontal distance L between them. Thus,

$$V_e = \frac{A_1 + A_2}{2} \times \frac{L}{27}(\text{yd}^3) \quad \textbf{(26.1a)}$$

or

$$V_e = \frac{A_1 + A_2}{2} \times L(\text{m}^3) \quad \textbf{(26.1b)}$$

In Equation (26.1a), V_e is the average end area volume in cubic yards, A_1 and A_2 are in square feet, and L is in feet. In Equation (26.1b), A_1 and A_2 are in m^2, L is in m, and V_e is in m^3. Equation [26.1(b)] also applies to computing volumes in acre-ft, where A_1 and A_2 are in acres, and L is in ft.

If L is 100 ft, as for full stations in the English system of units, Equation (26.1a) becomes

$$V_e = 1.852(A_1 + A_2)\ \text{yd}^3 \quad \textbf{(26.2)}$$

Equations (26.1) and (26.2) are approximate and give answers that generally are slightly larger than the true prismoidal volumes (see Section 26.8). They are used in practice because of their simplicity and contractors are satisfied because pay quantities are generally slightly greater than true values. Increased accuracy is obtained by decreasing the distance L between sections. When the ground is irregular, cross sections must be taken closer together.

■ Example 26.1

Compute the volume of excavation between station 24 + 00, with an end area of 711 ft^2, and station 25 + 00, with an end area of 515 ft^2.

Solution

By Equation (26.2), $V = 1.852(A_1 + A_2) = 1.852(711 + 515) = 2270\ \text{yd}^3$.

■ 26.6 DETERMINING END AREAS

End areas can be determined either graphically or by computation. In graphic methods, the cross section and template are plotted to scale on grid paper; then the number of small squares within the section can be counted and converted to area, or the area within the section can be measured using a planimeter (see Section 12.9.4). Computational procedures consist of either dividing the section into simple figures such as triangles and trapezoids, computing and summing these areas, or using the coordinate formula (see Section 12.5). These computational methods are discussed in the sections that follow. Most such calculations are now done by computer—usually by the coordinate method, which is general and readily programmed.

26.6.1 End Areas by Simple Figures

To illustrate the procedures of calculating end areas by simple figures such as triangles or trapezoids, assume the following excerpt of field notes (in the English system of units) applies to the cross section and end area, shown in Figure 26.4. In the notes, Lt indicates that the readings were started on the left side of the reference line as viewed facing in the direction of increasing stationing.

HI = 879.29 ft

	867.3	870.9	874.7	876.9	869.0	872.8
24 + 00 Lt	$\frac{12.0}{50}$	$\frac{8.4}{36}$	$\frac{4.6}{20}$	$\frac{2.4}{CL}$	$\frac{10.3}{12}$	$\frac{6.5}{50}$

In this excerpt of field notes, the top numbers are elevations (in ft) obtained by subtracting rod readings (middle numbers) from the leveling instrument's HI. Bottom numbers are distances from centerline (in ft), beginning from the left. Assume the design calls for a level roadbed of 30-ft width, cut slopes of 1-1/2:1, and a subgrade elevation at station 24 + 00 of 858.9 ft. A corresponding *design template* is superimposed over the plotted cross section in Figure 26.4. Subtracting the subgrade elevation from cross-section elevations at *C*, *D*, and *E* yields the

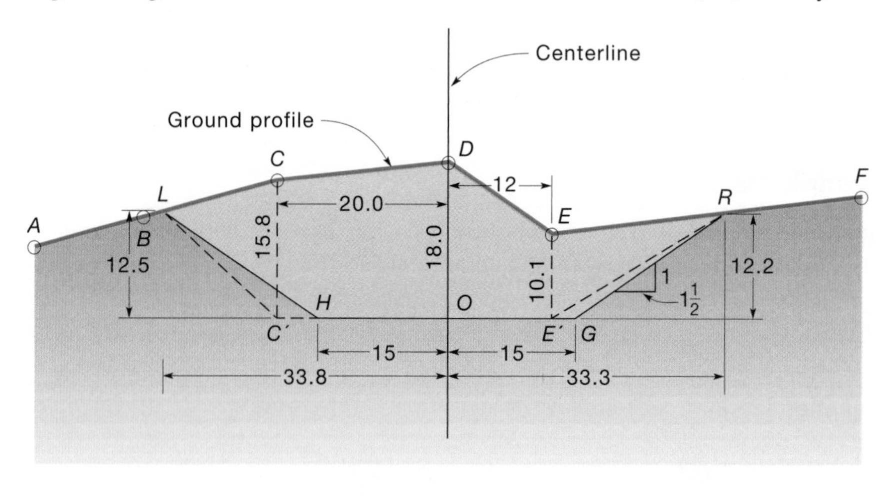

Figure 26.4 End-area computation.

ordinates of cut required at those locations. Elevations and distances out from centerline to the slope intercepts at L and R must be either scaled from the plot or computed. Assuming they have been scaled (methods for computing them are given in Section 26.7), the following tabulation of distances from centerline and required cut ordinates at each point to subgrade elevation was made:

Station	H	L	C	D	E	R	G
24 + 00	$\frac{0}{15}$	$\frac{C12.5}{33.8}$	$\frac{C15.8}{20}$	$\frac{C18.0}{0}$	$\frac{C10.1}{12}$	$\frac{C12.2}{33.3}$	$\frac{0}{15}$

Numbers above the line in the fractions (preceded by the letter C) are cut ordinates in feet; those below the line are corresponding distances out from the centerline. Fills are denoted by the letter F. Using C instead of plus for cut, and F instead of minus for fill, eliminates confusion. From the cut ordinates and distances from centerline shown, the area of the cross section in Figure 26.4 is computed by summing the individual areas of triangles and trapezoids. A list of the calculations is given in Table 26.1. (Refer to Figure 26.4 for triangle and trapezoid designations.)

26.6.2 End Areas by Coordinates

The coordinate method for computing end areas can be used for any type of section and has many engineering applications. The procedure was described in Section 12.5 as a way to determine the area contained within a closed polygon traverse.

To demonstrate the method in end-area calculation, the example of Figure 26.4 will be solved. Coordinates of each point of the section are calculated in an axis system having point O as its origin, using the earlier listed data on cuts and distances from centerline. In computing coordinates, distances to the right of centerline and cut values are considered plus; distances left and fill values are minus. Beginning with point O and proceeding clockwise around the figure,

TABLE 26.1 END AREA BY SIMPLE FIGURES

Figure	Computation	Area
$ODCC'$	[(18.0 + 15.8)20]/2	338.0
$C'CL$	[(15.8)13.8]/2	109.0
HLC'	[−(5)12.5]/2	−31.2
$ODEE'$	[(18.0 + 10.1)12]/2	168.6
$EE'R$	[(10.1)21.3]/2	107.6
$E'RG$	[(3)12.2]/2	18.3
		Area = 710 ft^2

TABLE 26.2 END AREAS BY COORDINATES

Point	X	Y	Plus +	Minus −
O	0	0	↙	↘
H	−15	0	0	0
L	−33.8	12.5	0	+188
C	−20	15.8	−250	+534
D	0	18.0	0	+360
E	12	10.1	216	0
R	33.3	12.2	336	−146
G	15	0	183	0
O	0	0	0	0
			+485	+936
			+936	
			Σ = 1421	
		Area = 1421 ÷ 2 = 710 ft²		(nearest ft²)

the coordinates of each point are listed in sequence. *Point O is repeated at the end* (see Table 26.2). Then Equation (12.7) is applied, with products of diagonals downward to the right (↘) considered minus, and diagonal products down to the left (↙) plus. Algebraic signs of the coordinates must be considered. Thus a positive product (↙) having a negative coordinate will actually be minus. The total area is obtained by dividing the absolute value of the algebraic summation of all products by 2. The calculations are illustrated in Table 26.2.

It is necessary to make separate computations for cut and fill end areas when they occur in the same section (as at station 11 + 40 of Figure 26.1), since they must always be tabulated independently for pay purposes. Payment is normally made only for excavation (its unit price includes making and shaping the fills) except on projects consisting primarily of embankment such as levees, earth dams, some military fortifications, and highways built up by continuous fill in flat areas.

■ 26.7 COMPUTING SLOPE INTERCEPTS

The elevations and distances out from the centerline to the slope intercepts can be calculated using cross-section data and the cut or fill slope values. For example in Figure 26.4, intercept *R* occurs between ground profile point *E* (distance 12 ft right and elevation 869.0) and point *F* (distance 50 ft right and elevation 872.8). The cut slope is 1-1/2: 1, or 0.67 ft/ft. A more detailed diagram, illustrating the geometry for calculating slope intercept *R*, is given in Figure 26.5.

The slope along ground line *EF* is $(872.8 - 869.0)/38 = 0.10$ ft/ft, where 38 ft is the horizontal distance between the points. The elevation of G' (point

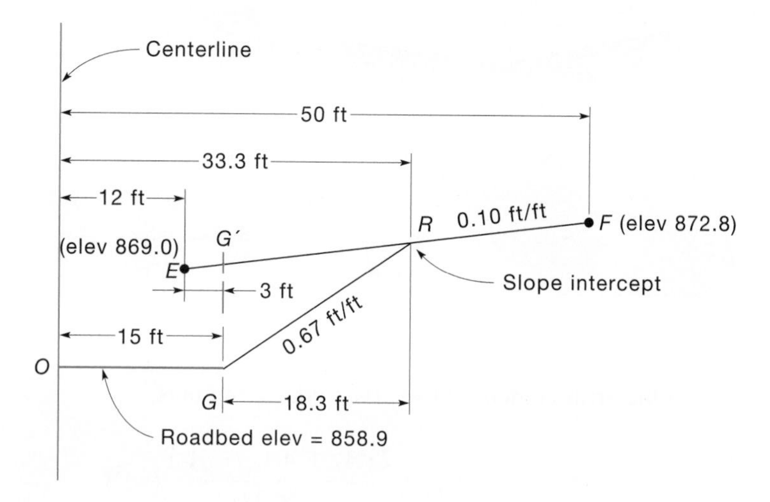

Figure 26.5 Computation of slope intercept *R* of Figure 26.4.

vertically above *G*) is 869.0 + 0.10(3) = 869.3; thus ordinate *GG′* is (869.3 − 858.9) = 10.4 ft. Lines *EF* and *GR* converge at a rate equal to the difference in their slopes (because they are both sloping upward), or 0.67 − 0.10 = 0.57 ft/ft. Dividing ordinate *GG′* by this convergence yields horizontal distance *GR*, or 10.4/0.57 = 18.3 ft. Adding 18.3 to distance *OG* yields 18.3 + 15 = 33.3 ft, which is the distance from centerline to slope intercept *R*. Finally to obtain the elevation of *R*, the increase in elevation from *E* to *R* is added to the elevation of *E*, or 0.10(21.3) + 869.0 = 871.1. Thus the cut ordinate at *R* equals 871.1 − 858.9 = 12.2 ft. Recall that 33.3 and 12.2 were the *X* and *Y* coordinates, respectively, used in the end-area calculations of Section 26.6.

The elevation and distance from centerline of the slope intercept *L* of Figure 26.4 are calculated in a similar manner, except the rate of convergence of lines *CB* and *HL* is the sum of their slopes because *CB* slopes downward and *HL* upward. Calculations of slope intercepts are somewhat laborious, but routine when programmed for solution by computer. If a computer is not used for computing end areas and volumes, an alternate procedure is to plot the cross sections and templates, determine the end area by planimeter, and scale the slope intercepts from the plot. Slope intercepts are essential since the placement of slope stakes that guide construction operations is based on them.

■ 26.8 PRISMOIDAL FORMULA

The prismoidal formula applies to volumes of all geometric solids that can be considered prismoids. A prismoid, illustrated in Figure 26.6, is a solid having ends that are parallel but not congruent and trapezoidal sides that are also not congruent. Most earthwork solids obtained from cross-section data fit this classification. However, from a practical standpoint, the differences in volumes computed by the average-end-area method and the prismoidal formula are usually so small as to be negligible. Where extreme accuracy is needed, such as in expensive rock cuts, the prismoidal method can be used.

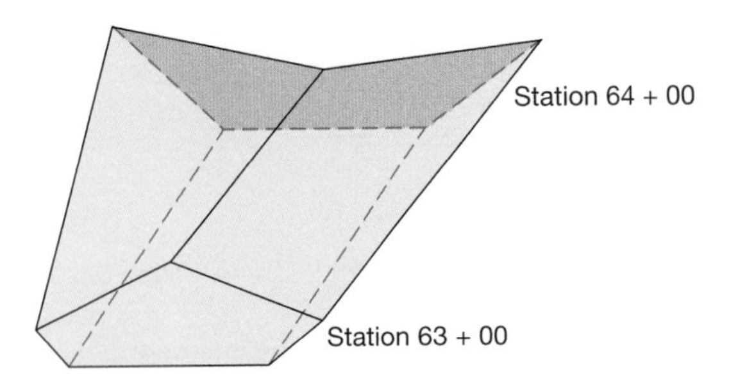

Figure 26.6 Sections for which the prismoidal correction is added to the end-area volume.

One arrangement of the prismoidal formula is

$$V_P = \frac{L(A_1 + 4A_m + A_2)}{6 \times 27}(\text{yd}^3) \quad \textbf{(26.3)}$$

where V_P is the prismoidal volume in cubic yards, A_1 and A_2 are areas of successive cross sections taken in the field, A_m is the area of a "computed" section midway between A_1 and A_2, and L is the horizontal distance between A_1 and A_2. Prismoidal volumes in m^3 can be obtained by using a slight modification of Equation (26.3), i.e., the conversion factor 27 in the denominator is dropped, and L is in meters, A_1, A_m and A_2 are in m^2.

To use the prismoidal formula, it is necessary to know area A_m of the section halfway between the stations of A_1 and A_2. This is found by the usual computation *after averaging the heights and widths of the two end sections*. Obviously, the middle area is not the average of the end areas, since there would then be no difference between the results of the end-area formula and the prismoidal formula.

The prismoidal formula generally gives a volume smaller than that found by the average-end-area formula. For example, the volume of a pyramid by the prismoidal formula is $Ah/3$ (the exact value), whereas by the average-end-area method it is $Ah/2$. An exception occurs when the center height is great but the width narrow at one station and the center height small but the width large at the adjacent station. Figure 26.6 illustrates this condition.

The difference between the volumes obtained by the average-end-area formula and the prismoidal formula is called the *prismoidal correction* C_p. Various books on route surveying give formulas and tables for computing prismoidal corrections, which can be applied to average-end-area volumes to get prismoidal volumes. A prismoidal correction formula, which provides accurate results for three-level sections is

$$C_P = \frac{L}{12 \times 27}(c_1 - c_2)(w_1 - w_2)(\text{yd}^3) \quad \textbf{(26.4)}$$

where C_P is the volume of the prismoidal correction in cubic yards, c_1 and c_2 are center heights in cut (or in fill), and w_1 and w_2 are widths of sections (from slope intercept to slope intercept) at adjacent sections. If the product of $(c_1 - c_2)$ $(w_1 - w_2)$ is minus, as in Figure 26.6, the prismoidal correction is added rather than subtracted from the end-area volume. For sections other than three-level, Equation (26.4) may not be accurate enough and therefore Equation (26.3) is recommended.

■ Example 26.2

Compute the volume using the prismoidal formula and by average end areas for the following three-level sections of a roadbed having a base of 24 ft and side slopes of $1\frac{1}{2}$:1.

Solution

Station	L	C	R	Area
12 + 00	$\frac{C7.8}{23.7}$	$\frac{C5.3}{0}$	$\frac{C7.4}{23.0}$	$\frac{5.3(23.7 + 23.0)}{2} + \frac{24(7.8 + 7.4)}{4} = 215.0 \text{ ft}^2$
12 + 50	$\frac{C6.5}{21.8}$	$\frac{C6.0}{0}$	$\frac{C7.5}{23.2}$	$\frac{6.0(21.8 + 23.2)}{2} + \frac{24(6.5 + 7.5)}{4} = 219.0 \text{ ft}^2$
13 + 00	$\frac{C5.8}{24.8}$	$\frac{C6.6}{0}$	$\frac{C7.0}{23.5}$	$\frac{6.6(24.8 + 23.5)}{2} + \frac{24(5.8 + 7.0)}{4} = 236.2 \text{ ft}^2$

Using Equation (26.3) yields a volume of $\frac{100(215.0 + 4(219.0) + 236.2)}{6(27)} = 819.2 \text{ yd}^3$

Using Equation (26.1a) yields $\frac{100(215.0 + 236.2)}{2(27)} = 835.6 \text{ yd}^3$

Using Equation (26.5) yields a prismoidal correction of $\frac{100}{12(27)}(5.3 - 6.6)(46.7 - 48.3) = 0.6 \text{ yd}^3$

Note that the difference between the volume computed by the prismoidal formula and the average end area is only 1.9 percent. The prismoidal correction applied to Equation (26.1a) yields a volume of 835 yd^3.

■ 26.9 VOLUME COMPUTATIONS

Volume calculations for route construction projects are usually done by computer and arranged in tabular form. To illustrate this procedure, assume that end areas listed in columns (2) and (3) of Table 26.3 apply to the section of roadway illustrated in Figure 26.1. By using Equation [26.1(a)], cut and fill volumes are computed and tabulated in columns (4) and (5).

The volume computations illustrated in Table 26.3 include the transition sections of Figure 26.1. This is normally not done when preliminary earthwork volumes are being estimated (during design and prior to construction) because

TABLE 26.3 TABULAR FORM OF VOLUME COMPUTATION

Station (1)	End Area (ft²) Cut (2)	End Area (ft²) Fill (3)	Volume (yd³) Cut (4)	Volume (yd³) Fill (5)	Fill Volume +25% (yd³) (6)	Cumulative Volume (yd³) (7)
10 + 00		992				0
				2614	3268	
11 + 00		421				−3268
				190	238	
11 + 21	0	68				−3506
			12	29	37	
11 + 40	34	31				−3531
			79	14	17	
11 + 64	144	0				−3469
			553			
12 + 00	686					−3916
			2967			
13 + 00	918					+51

the exact locations of the transition sections and their configurations are usually unknown until slope staking occurs. Thus for calculating preliminary earthwork quantities, an end area of zero would be used at the station of the centerline grade point (station 11 + 40 of Figure 26.1), and transition sections (stations 11 + 21 and 11 + 64 of Figure 26.1) would not appear in the computations. After slope staking (procedures for slope staking are described in Section 23.7) the locations and end areas of transition sections are known, and they should be included in final volume computations, especially if they significantly affect the quantities for which payment is made.

In highway and railroad construction, excavation or cut material is used to build embankments or fill sections. Unless there are other controlling factors, a well-designed grade line should nearly balance total cut volume against total fill volume. To accomplish a balance, either fill volumes must be expanded or cut volumes shrunk.[1] This is necessary because, except for rock cuts, embankments are compacted to a density greater than that of material excavated from its natural state and to balance earthwork this must be considered. (Rock cut expands to occupy a greater fill volume; thus either the cut must be expanded or the fill shrunk to obtain a balance.) The rate of expansion depends on the type of material and can

[1]Expansion of fill volumes is generally preferred, since payment is usually based on actual volumes of material excavated.

never be estimated exactly. However, samples and records of past projects in the immediate area are helpful in assigning reasonable factors. Column (6) of Table 26.3 lists expanded fills for the example of Figure 26.1, where a 25 percent factor was applied.

To investigate whether or not an earthwork balance is achieved, *cumulative volumes* are computed. This involves adding cut and expanded fill volumes algebraically from project beginning to end, with cuts considered positive and fills negative. Cumulative volumes are listed in column (7) of Table 26.3. In this example, there is a cut volume excess of 51 yd^3 between stations 10 + 00 and 13 + 00 or, in other words, there is a surplus of that much excavation.

To analyze the movement of earthwork quantities on large projects, *mass diagrams* are constructed. These are plots of cumulative volumes for each station as the ordinate, versus the stations on the abscissa. Horizontal (balance) lines on the mass diagram then determine the limit of economic haul and the direction of movement of material. Mass diagrams are described more thoroughly in books on route surveying. If there is insufficient material from cuts to make the required fills, the difference must be *borrowed* [obtained from borrow pits or other sources such as by "daylighting" curves (flattening cut slopes to improve visibility)]. If there is excess cut, it is wasted or perhaps used to extend and flatten the fills.

For projects with more than a few cross sections, computer programs are available and are generally used for earthwork computations, but surveyors and engineers must still understand the basic methods.

■ 26.10 UNIT-AREA, OR BORROW-PIT, METHOD

On many projects, except long linear route constructions, the quantity of earth, gravel, rock, or other material excavated or filled can often best be determined by the borrow-pit method. The quantities computed form the basis for payment to the contractor or supplier of materials. The volume of coal or other loose materials in stockpiles can be found in the same way.

As an example, assume the area shown in Figure 26.7 is to be graded to an elevation of 358.0 ft for a building site. Notes for the fieldwork are shown in

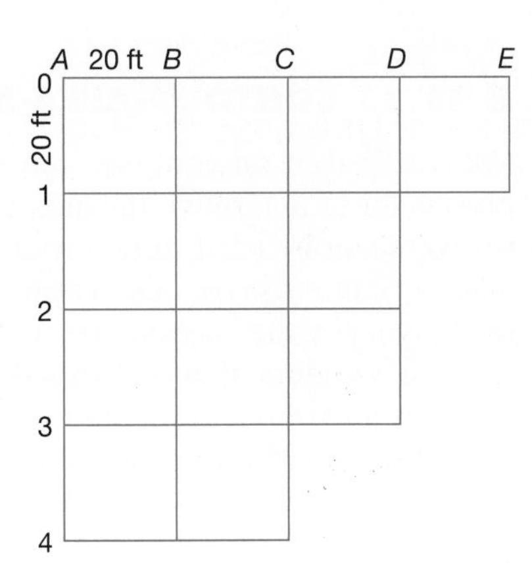

Figure 26.7
Borrow-pit leveling.

Plate B.2 of Appendix B. The area to be covered in this example is staked in squares of 20 ft, although 10, 50, 100, or more feet could be used, with the choice depending on project size and accuracy desired. A total station instrument and tape, or only a tape, may be used for the layout. A benchmark of known or assumed elevation is established outside the area in a place not likely to be disturbed.

After the area is laid out in squares, elevations are determined at all grid intersection points. For this, a level is set up at any convenient location, a plus sight taken on the benchmark, and minus sights read on each grid intersection. If the terrain is not too rough, it may be possible to select a point near the area center and take sights on all grid intersections from the same setup, as in the example of Plate B.2. For rough terrain, it may be most convenient to determine the elevations by radial surveying from one well-chosen setup using a total station instrument (see Section 16.9.1).

Letters and numbers designate grid intersection points, such as *A*-1, *C*-4, and *D*-2. For site grading to a specified elevation, say 358.0 ft, the amount of cut or fill at each grid square corner is obtained by subtracting 358.0 from its ground elevation. For each square, then, the average height of the four corners of each prism of cut or fill is determined and multiplied by the base area, $20 \times 20 \text{ ft} = 400 \text{ ft}^2$, to get the volume. The total volume is found by adding the individual values for each block and dividing by 27 to obtain the result in cubic yards.

To simplify calculations, the cut at each corner multiplied by the number of times it enters the volume computation can be shown in a separate column. The column sum is divided by 4 and multiplied by the base area of one block to get the volume. In equation form, this procedure is given as

$$V = \sum(h_{i,j}n)\left(\frac{A}{4 \times 27}\right)(\text{yd}^3) \tag{26.5}$$

where $h_{i,j}$ is the corner height in row i and column j, and n the number of squares to which that height is common. The corner at *C*-4, for example, is common to only one square, *D*-2 is common to two, *D*-1 is common to three, and *C*-1 is common to four. $\Sigma(h_{i,j}n)$ is the sum of the products of the height and the number of common squares, and A is the area of one square. An example illustrating the use of Equation (26.5) is given in the field notes of Plate B.2.

■ 26.11 CONTOUR-AREA METHOD

Volumes based on contours can be obtained from contour maps by using a planimeter to determine the area enclosed by each contour. Alternatively, CAD software can be used to determine these areas. Then the average area of the adjacent contours is obtained using Equation (26.1b) and the volume obtained by multiplying by the contour spacing (i.e., contour interval). Use of the prismoidal formula is seldom, if ever, justified in this type of computation. This procedure is the basis for volume computations in CAD software (see Section 17.14).

Instead of determining areas enclosed within contours by planimeter, they can be obtained using the coordinate formula [Equation (12.7) or (12.8)]. In this procedure a tablet digitizer like the one shown in Figure 28.8 is first used to measure

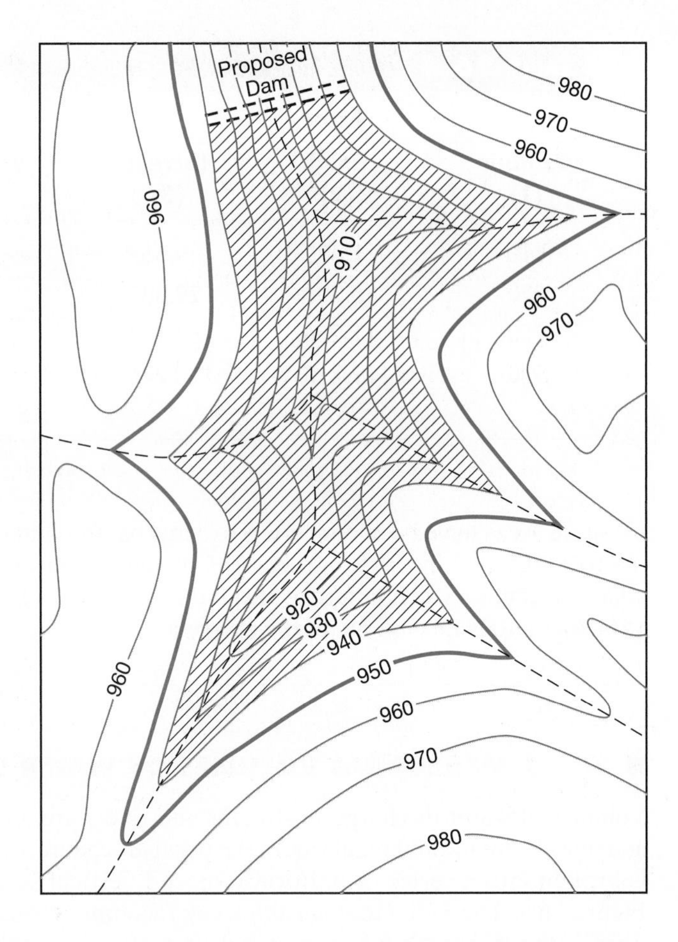

Figure 26.8 Determining the volume of water impounded in a reservoir by the contour-area method.

the coordinates along each contour at enough points to define its configuration satisfactorily.

The contour-area method is suitable for determining volumes over large areas, for example, computing the amounts and locations of cut and fill in the grading for a proposed airport runway to be constructed at a given elevation. Another useful application of the contour-area method is in determining the volume of water that will be impounded in the reservoir created by a proposed dam.

■ Example 26.3

Compute the volume of water impounded by the proposed dam illustrated in Figure 26.8. Map scale is 500 ft/in. and the proposed spillway elevation 940 ft.

Solution

The crosshatched portion of Figure 26.8 represents the area that will be inundated with water when the reservoir is full. The solution is presented in Table 26.4. Column (2) gives the area enclosed within each contour (determined by using a tablet digitizer and the coordinate method) in square inches, and in column

TABLE 26.4 VOLUME COMPUTATION BY CONTOUR-AREA METHOD

	Area		
Contour (1)	(in.^{2}) (2)	(acres) (3)	Volume (acre-ft) (4)
910	1.683	9.659	—
920	5.208	29.889	197.7
930	11.256	64.598	472.4
940	19.210	110.246	874.2
			Σ = 1544.3

(3) these areas have been converted to acres based on map scale, that is, 1 in.^{2} = $[(500)^2]/43{,}560$ (1 in.^{2} = 5.739 acres). Column (4) gives the volumes between adjacent contours computed by Equation [26.1(b)]. The sum of column (4), 1544.3 acre-ft, is the volume of the reservoir.

26.12 MEASURING VOLUMES OF WATER DISCHARGE

Volumes of water discharge in streams and rivers are a matter of vital concern and must be monitored regularly. In the usual procedure, the stream's cross-section is broken into a series of uniformly spaced vertical sections, as illustrated in Figure 26.9. The U.S. Geological Survey recommends using from 25 to 30 sections, with not more than 5 percent of the total flow occurring in any particular section. Depths and current velocities are measured at each ordinate using a *current meter*. (There are various types available.) The discharge volume for each section is the product of its area and average current velocity. The sum of all section discharges is the total volume of water passing through the stream at the cross-section location. Units of section areas and current velocities can be either ft^2 and ft/sec, respectively, with the discharge in ft^3/sec; or in m^2 and m/sec, respectively, giving the volume in m^3/sec.

Current velocities can be measured at every 0.1 of the depth at each ordinate and the average taken. Alternatively, a good average results from the mean of the 0.2 and 0.8 depth velocities, or a single measurement at the 0.6 depth point. For

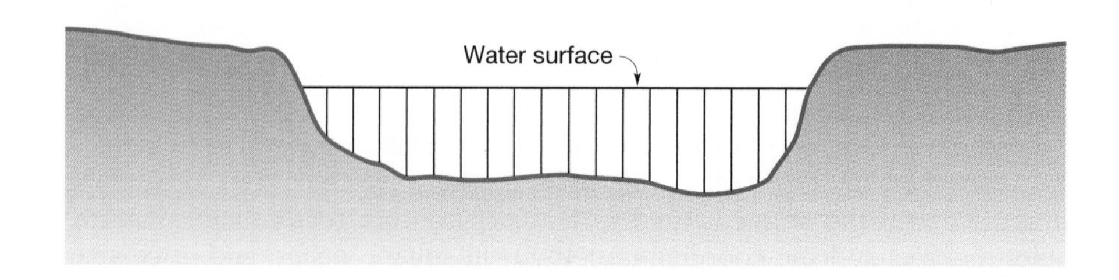

Figure 26.9 Vertical sections for making stream discharge measurements.

depths up to 2-1/2 ft, the U.S. Geological Survey uses the 0.6 method; for deeper sections the 0.2 and 0.8 procedure is employed.

The cross section should be taken at right angles to the stream and in a straight reach with solid bottom and uniform flow. In shallow streams, measurements can be made by wading, in which case, the current meter is held upstream free from eddies caused by the wader's legs. In deeper streams and rivers, measurements are taken from boats, bridges, or overhead cable cars. In these situations, the current meter, with a heavy weight attached to its bottom, is suspended by a cable and thus doubles as a lead line for measuring depths.

26.13 SOURCES OF ERROR IN DETERMINING VOLUMES

Some common errors in determining areas of sections and volumes of earthwork are:

1. Making errors in measuring field cross-sections, e.g., not being perpendicular to the centerline.
2. Making errors in measuring end areas.
3. Failing to use the prismoidal formula where it is justified.
4. Carrying out areas of cross sections beyond the limit justified by the field data.

26.14 MISTAKES

Some typical mistakes made in earthwork calculations are:

1. Confusing algebraic signs in end-area computations using the coordinate method.
2. Using Equation (26.2) for full-station volume computation when partial stations are involved.
3. Using average end-area volumes for pyramidal or wedge-shaped solids.
4. Mixing cut and fill quantities.

PROBLEMS

Asterisks (*) indicate problems that have answers given in Appendix G.

26.1 Why is the roadway in cut normally wider than the same roadway in fill?

26.2 Prepare a table of end areas versus depths of fill from 0 to 20 ft by increments of 2 ft for level sections, a 24-ft wide level roadbed, and side slopes of 1-1/2:1.

26.3 Similar to Problem 26.2, except use side slopes of 2:1.

Draw the cross sections and compute V_e for the data given in Problems 26.4 through 26.7.

26.4* Two level sections 100 ft apart with center heights 4.2 and 5.6 ft in fill, base width 30 ft, side slopes 1-1/2:1.

26.5 Two level sections of 30-m stations with center heights of 3.14 and 2.56 m. in cut, base width 10 m, side slopes 1-1/2:1.

26.6 The end area at station 36 + 00 is 305 ft^2. Notes giving distance from centerline and cut ordinates for station 36 + 60 are C 4.8/18.6; C 5.4; C 6.2/23.2 Base is 24 ft.

26.7 An irrigation ditch with $b = 12$ ft and side slopes of 2:1. Notes giving distances from centerline and cut ordinates for stations 52 + 00 and 53 + 00 are C 2.4/12.8; C 3.0; C 3.5/13.4; and C 3.3/16.2 C 3.8; C 4.2/15.2.

26.8 Why must cut and fill volumes be totaled separately?

26.9* For the data tabulated, calculate the volume of excavation in cubic yards between stations 10 + 00 and 15 + 00.

Station	Cut End Area (ft^2)
10 + 00	257
11 + 00	329
12 + 00	468
13 + 00	402
14 + 00	244
15 + 00	108

26.10 For the data listed, tabulate cut, fill, and cumulative volumes in cubic yards between stations 10 + 00 and 20 + 00. Use an expansion factor of 1.25 for fills.

	End Area (ft^2)	
Station	Cut	Fill
10 + 00	0	
11 + 00	175	
12 + 00	326	
13 + 00	375	
14 + 00	150	
14 + 60	0	0
15 + 00		148
16 + 00		266
17 + 00		302
18 + 00		259
19 + 00		179
20 + 00		143

26.11 Calculate the section areas in Problem 26.4 by the coordinate method.

26.12 Compute the section areas in Problem 26.5 by the coordinate method.

26.13 Determine the section areas in Problem 26.7 by the coordinate method.

26.14* Compute C_P and V_P for Problem 26.4. Is C_P significant?

26.15 Calculate C_P and V_P for Problem 26.7. Would C_P be significant in rock cut?

26.16 From the following excerpt of field notes, plot the cross section on graph paper and superimpose on it a design template for a 24-ft wide level roadbed with fill slopes of 1-1/2:1 and a subgrade elevation at centerline of 630.25 ft. Determine the end area graphically by counting squares.

HI = 864.96 ft						
20 + 00 Lt	$\frac{5.2}{50}$	$\frac{5.6}{32}$	$\frac{4.8}{0}$	$\frac{5.4}{15}$	$\frac{6.4}{32}$	$\frac{7.3}{50}$

26.17 For the data of Problem 26.16, determine the end area by plotting the points in a CAD package, and listing the area.

26.18 For the data of Problem 26.16, calculate slope intercepts, and determine the end area by the coordinate method. Check by computing areas of triangles and trapezoids.

26.19 From the following excerpt of field notes, plot the cross section on graph paper and superimpose on it a design template for a 40-ft wide level roadbed with cut slopes of 3:1 and a subgrade elevation of 1240.88 ft. Determine the end area graphically by counting squares.

HI = 102.04 ft						
46 + 00 Lt	$\frac{8.5}{60}$	$\frac{8.2}{42}$	$\frac{6.6}{0}$	$\frac{5.8}{22}$	$\frac{6.5}{39}$	$\frac{7.5}{60}$

26.20 For the data of Problem 26.19, calculate slope intercepts and determine the end area by the coordinate method. Check by computing areas of triangles and trapezoids.

26.21* Complete the following notes and compute V_e and V_P. The roadbed is level, the base is 30 ft.

Station			
Station 88 + 00	$\frac{C\,6.4}{27.8}$	C 3.6	$\frac{C\,5.7}{26.4}$
Station 89 + 00	$\frac{C\,3.1}{21.2}$	C 4.9	$\frac{C\,4.3}{23.6}$

26.22 Similar to Problem 26.21, except the base is 24 ft.

26.23 Calculate V_e and V_P for the following notes. Base in fill is 24 ft, base in cut is 30 ft, side slopes are 1-1/2:1.

Station				
12 + 90	$\frac{C\,3.4}{20.1}$	$\frac{C\,2.0}{0}$	$\frac{0.0}{6.0}$	$\frac{F\,2.0}{13.0}$
12 + 30	$\frac{C\,2.2}{18.3}$	$\frac{0.0}{0}$	$\frac{F\,3.0}{14.5}$	

26.24 Calculate V_e, C_P, and V_P for the following notes. Base in cut is 40 ft, side slopes are 2-1/2:1.

Station			
46 + 00	$\frac{C\,4.2}{32.6}$	$\frac{C\,3.0}{0}$	$\frac{C\,3.6}{30.8}$
45 + 00	$\frac{C\,2.4}{27.2}$	$\frac{C\,3.0}{0}$	$\frac{0.0}{20.0}$

For Problems 26.25 and 26.26, compute the reservoir capacity (in acre-ft) between highest and lowest contours for areas on a topographic map.

26.25*

Elevation (ft)	860	870	880	890	900	910
Area (ft^2)	1370	1660	2293	2950	3550	4605

26.26

Elevation (ft)	1015	1020	1025	1030	1035	1040
Area (ft^2)	1850	1957	2088	2155	2236	2672

26.27 A calibrated polar planimeter gives an average reading of 1.398 revolutions of the roller over a 4-in. diameter circle; 20-ft contours planimetered on a reservoir site map to a scale of 1 in. = 400 ft give the values tabulated. Calculate the reservoir capacity between highest and lowest contours in acre-ft.

Contour	720	740	760	780	800
Planimeter reading	0.200	1.240	1.817	2.608	4.596

26.28 State two situations where prismoidal corrections are most significant.

26.29 Write a computer program to calculate slope intercepts and end areas by the coordinate method, given cross-section notes and roadbed design information. Use the program to calculate the slope intercepts for the data of Problem 26.16.

26.30* Distances (ft) from the left bank, corresponding depths (ft), and velocities (ft/sec), respectively, are given for a river discharge measurement. What is the volume in ft^3/sec? 0, 1.0, 0; 10, 2.3, 1.30; 20, 3.0, 1.54; 30, 2.7, 1.90; 40, 2.4, 1.95; 50, 3.0, 1.60; 60, 3.1, 1.70; 74, 3.0, 1.70; 80, 2.8, 1.54; 90, 3.3, 1.24; 100, 2.0, 0.58; 108, 2.2, 0.28; 116, 1.5, 0.

26.31 Prepare a computational program that computes the volumes in Problem 26.9.

26.32 Prepare a computational program that computes the end-areas in Problem 26.20.

BIBLIOGRAPHY

Chen, C., and H. Lin. 1990. "Estimating Pit Excavation Volume Using Cubic Spline Volume Formula." *ASCE, Journal of Surveying Engineering* 117 (No. 2): 51.

Chen, C., and H. Lin. 1992. "Estimating Excavation Volumes Using New Formulas." *Surveying and Land Information Systems* 52 (No. 2): 104.

Vijay, R. et. al. 2005. "Computation of Reservoir Storage Capacity and Submergence Using GIS. *Surveying and Land Information Science* 65 (No. 4): 255.

27

Photogrammetry

■ 27.1 INTRODUCTION

Photogrammetry may be defined as the science, art, and technology of obtaining reliable information from photographs. It encompasses two major areas of specialization: *metrical* and *interpretative*. The first area is of principal interest to those involved in surveying (geomatics), since it is applied in determining spatial information including distances, elevations, areas, volumes, cross sections, and data for compiling topographic maps from measurements made on photographs. *Aerial* photographs (exposed from aircraft) are normally used, although in certain special applications, *terrestrial* photos (taken from Earth-based cameras) are employed.

Interpretative photogrammetry involves recognizing objects from their photographic images and judging their significance. Critical factors considered in identifying objects are the shapes, sizes, patterns, shadows, tones, and textures of their images. This area of photogrammetry was traditionally called *photographic interpretation* because initially it relied on aerial photos. More recently, other sensing and imaging devices such as multispectral scanners, thermal scanners, radiometers, and side-looking airborne radar have been developed which aid greatly in interpretation. These instruments sense energy in wavelengths beyond those which the human eye can see, or that standard photographic films can record. They are often carried in aircraft as remote as satellites; hence, the term *remote sensing* is now generally applied to the interpretative area of photogrammetry.

In this chapter, metrical photogrammetry using aerial photographs will be emphasized because it is the area of specialization most frequently applied in surveying work. Remote sensing, however, has also become very important in small-scale mapping and in monitoring our environment and managing our natural resources. This subject is discussed further in Section 27.20.

Metrical photogrammetry is accomplished in different ways depending upon project requirements and the type of equipment available. Simple analyses and computations can be made by making measurements on paper prints of aerial photos using engineer's scales, and assuming that the photos are "truly vertical," i.e., the camera axis coincided with a plumb line at the time of photography. These methods produce results of lower order, but they are suitable for a variety of applications. Other more advanced techniques, including *analog, analytical,* and *softcopy* methods, do not assume vertical photos and provide more accurate determinations of the spatial locations of objects. The analog procedure relies on precise optical and mechanical devices to create models of the terrain that can be measured and mapped. The analytical method is based upon precise measurements of the photographic positions of the images of objects of interest, followed by a mathematical solution for their locations. Softcopy instruments utilize digital images in computerized procedures that are highly automated. Each of these types of metrical analysis procedures are described in sections that follow.

■ 27.2 USES OF PHOTOGRAMMETRY

Photography dates back to 1839, and the first attempt to use photogrammetry in preparing a topographic map occurred a year later. Photogrammetry is now the principal method employed in topographic mapping and compiling other forms of spatial data. For example, the U.S. Geological Survey uses the procedure almost exclusively in compiling its maps. Cameras, films, and other photogrammetric instruments and techniques have been improved continually, so that spatial data collected by photogrammetry today meets very high accuracy standards. Other advantages of this method are the (1) speed of collecting spatial data in an area, (2) relatively low cost, (3) ease of obtaining topographic details, especially in inaccessible areas, and (4) reduced likelihood of omitting details in spatial data collection.

Photogrammetry presently has many applications in surveying and engineering. For example, it is used in land surveying to compute coordinates of section corners, boundary corners, or points of evidence that help locate these corners. Large-scale maps are made by photogrammetric procedures for many uses, one being subdivision design. Photogrammetry is used to map shorelines in hydrographic surveying, to determine precise ground coordinates of points in control surveying, and to develop maps and cross sections for route and engineering surveys. Photogrammetry is playing an increasingly important role in developing the necessary data for modern land and geographic information systems (LIS and GIS).

Photogrammetry is also being successfully applied in many nonengineering fields, such as geology, forestry, agriculture, conservation, planning, archeology, military intelligence, traffic management, and accident investigation. It is beyond the scope of this chapter to describe all the varied applications of photogrammetry. Use of the science has increased dramatically in recent years and its future growth for solving measurement and mapping problems is assured.

■ 27.3 AERIAL CAMERAS

Aerial mapping cameras are perhaps the most important photogrammetric instruments, since they expose the photographs on which the science depends. To understand photogrammetry, especially the geometrical foundation of its equations, it is essential to have a fundamental understanding of cameras and how they operate. Aerial cameras must be capable of exposing a large number of photographs in rapid succession while moving in an aircraft at high speed; so a short cycling time, fast lens, efficient shutter, and large-capacity magazine are required.

Single-lens frame cameras are the type most often used in metrical photogrammetry. These cameras expose the entire frame or format simultaneously through a lens held at a fixed distance from the focal plane. Generally they have a format size of 9 × 9 in. (23 × 23 cm) and lenses with focal lengths of 6 in. (152.4 mm), although 3-1/2, 8-1/4, and 12 in. (90, 210, and 305 mm) focal lengths are also used. A single-lens frame camera, together with its view finder and electronic controls, is shown in Figure 27.1.

The principal components of a single-lens frame camera are shown in the diagram of Figure 27.2. These include the *lens* (the most important part), which gathers incoming light rays and brings them to focus on the focal plane; the *shutter* to control the interval of time that light passes through the lens; a diaphragm to regulate the size of lens opening; a *filter* to reduce the effect of haze and distribute light uniformly over the format; a *camera cone* to support the lens-shutter-diaphragm assembly with respect to the focal plane and to prevent stray light from striking the film; a *focal plane,* the surface on which the film lies when exposed; *fiducial marks* (not shown in Figure 27.2 but illustrated later), four or eight in number, which are essential to define the geometry of the photographs; a *camera body* to house the drive mechanism that cocks and trips the shutter, flattens the film, and advances it between exposures; and a *magazine,* which holds the supply of exposed and unexposed film.

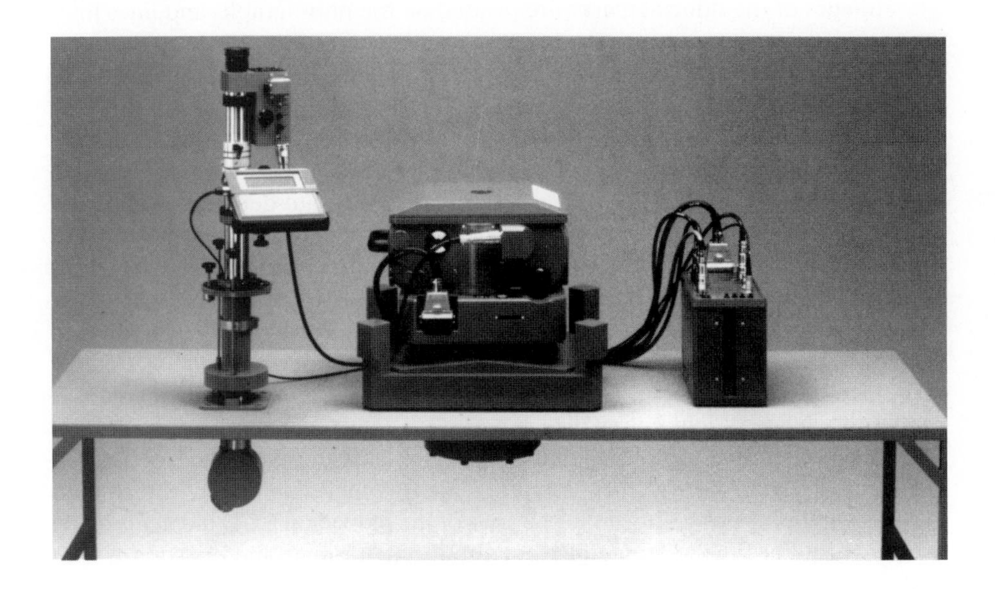

Figure 27.1 Aerial camera with view finder and electronic controls. (From *Elements of Photogrammetry: With Applications in GIS,* 3rd ed., by Wolf & Dewitt, 2000; Courtesy Carl Zeiss, Inc. and McGraw-Hill Book Co., Inc.)

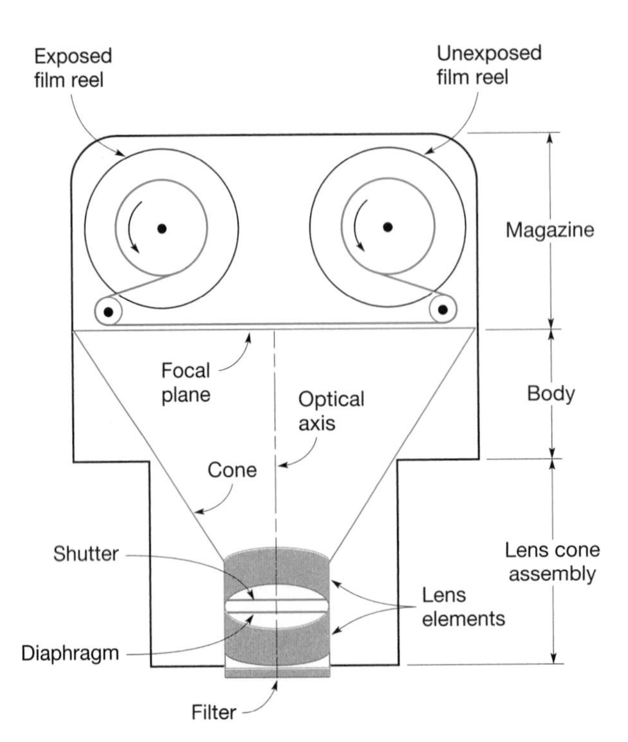

Figure 27.2 Principal components of a single lens frame aerial camera.

An aerial camera shutter can be operated manually by an operator, or automatically by the electronic control mechanism, so that photos are taken at specified intervals. A level vial attached to the camera enables an operator to keep the optical axis of the camera lens, which is perpendicular to the focal plane, nearly vertical in spite of any slight tip and tilt of the aircraft. Polyester roll film is used with magazine capacities of 200 ft or more.

Images of the fiducial marks are printed on the photographs and lines joining opposite pairs intersect at or very near the *principal point,* defined as the point where a perpendicular from the emergent nodal point of the camera lens strikes the focal plane. Fiducial marks may be located in the corners, on the sides, or preferably in both places, as shown in Figures 27.4 and 27.5.

A new type of camera is now used for obtaining images in digital form. Instead of film, these cameras employ an array of solid state detectors, which are placed in the focal plane. The most common type of detector is the *charge-coupled device* (CCD). The array is composed of tiny detectors arranged in contiguous rows and columns, as shown in Figure 27.3. Each detector senses the energy received from its corresponding ground scene and this constitutes one "picture element" (*pixel*) within the overall image. The principle of operation of CCDs is fundamentally quite simple. At any specific pixel location, the CCD element is exposed to incident light energy which builds up an electric charge proportional to the intensity of the incoming light. The electric charge is amplified, converted from analog to digital form, and stored in a file together with its row and column location within the array. Currently, the sizes of the individual CCD elements being manufactured are in the range of from about 5 to 15 micrometers square with

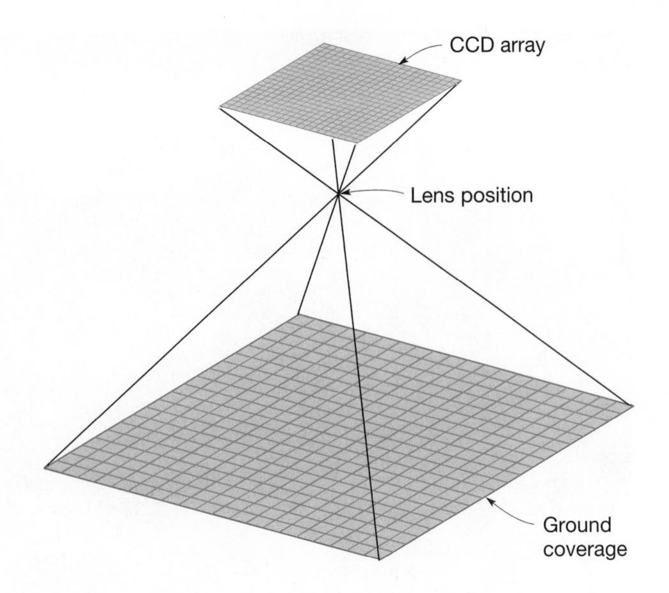

Figure 27.3 Geometry of a digital frame camera. (From *Elements of Photogrammetry: With Applications in GIS*, 3rd ed., Wolf & Dewitt, 2000; Courtesy McGraw-Hill Book Co., Inc.)

arrays consisting of from 500 rows and columns (250,000 pixels) for inexpensive cameras, to more than 4,000 rows and columns. Obviously, significant storage and data handling capabilities are necessary in acquiring and processing digital images.

Aerial mapping cameras, whether the film- or digital-type, are calibrated to get precise values for the focal length and lens distortions. Flatness of the focal plane, relative position of the principal point with respect to the fiducial marks, and fiducial mark locations are also specified. These calibration data are necessary for precise photogrammetric work.

■ 27.4 TYPES OF AERIAL PHOTOGRAPHS

Aerial photographs exposed with single-lens frame cameras are classified as *vertical* (taken with the camera axis aimed vertically downward, or as nearly vertical as possible) and *oblique* (made with the camera axis intentionally inclined at an angle between the horizontal and vertical). Oblique photographs are further classified as *high* if the horizon shows on the picture, and *low* if it does not. Figures 27.4 and 27.5 show examples of vertical and low oblique photographs, respectively. As illustrated by these examples, aerial photos clearly depict all natural and cultural features within the region covered such as roads, railroads, buildings, rivers, bridges, trees, and cultivated lands.

Vertical photographs are the principal mode of obtaining imagery for photogrammetric work. Oblique photographs are seldom used for metrical applications, but are advantageous in interpretative work and for reconnaissance.

■ 27.5 VERTICAL AERIAL PHOTOGRAPHS

A *truly vertical photograph* results if the axis of the camera is exactly vertical when exposure is made. Despite all precautions, small tilts, generally less than 1° and rarely greater than 3°, are invariably present and the resulting photos are called

Figure 27.4 Vertical aerial photograph. (Courtesy Pennsylvania Department of Transportation)

near-vertical or *tilted* photographs. Although vertical photographs look like maps to laypersons, they are not true orthographic projections of the Earth's surface. Rather, they are perspective views and the principles of perspective geometry must be applied to prepare maps from them. Figure 27.6 illustrates the geometry of a vertical photograph taken at exposure station L. The photograph, considered a *contact print positive,* is a 180° exact reversal of the negative. The positive shown in Figure 27.6 is used to develop photogrammetric equations in subsequent sections.

Distance oL (Figure 27.6) is the camera focal length. The x and y reference axis system for measuring photographic coordinates of images is defined by straight lines joining opposite-side fiducial marks shown on the positive of Figure 27.6. The x axis, arbitrarily designated as the line most nearly parallel with the direction of flight, is positive in the direction of flight. Positive y is 90° counterclockwise from positive x.

Vertical photographs for topographic mapping are taken in strips, which normally run lengthwise over the area to be covered. The strips or *flight lines* generally have a *sidelap* (overlap of adjacent flight lines) of about 30 percent. *Endlap*

Figure 27.5 Low-oblique aerial photograph showing state capital and downtown Madison, Wisconsin. (Courtesy State of Wisconsin, Department of Transportation)

(overlap of adjacent photographs in the same flight line) is usually about 60 ± 5 percent. Figures 27.19(a) and (b) illustrate endlap and sidelap. An endlap of 50% or greater is necessary to assure that all ground points will appear in at least two photographs and that some will show in three. Images common to three photographs permit *aerotriangulation* to extend or densify control through a strip or block of photographs using only minimal existing control.

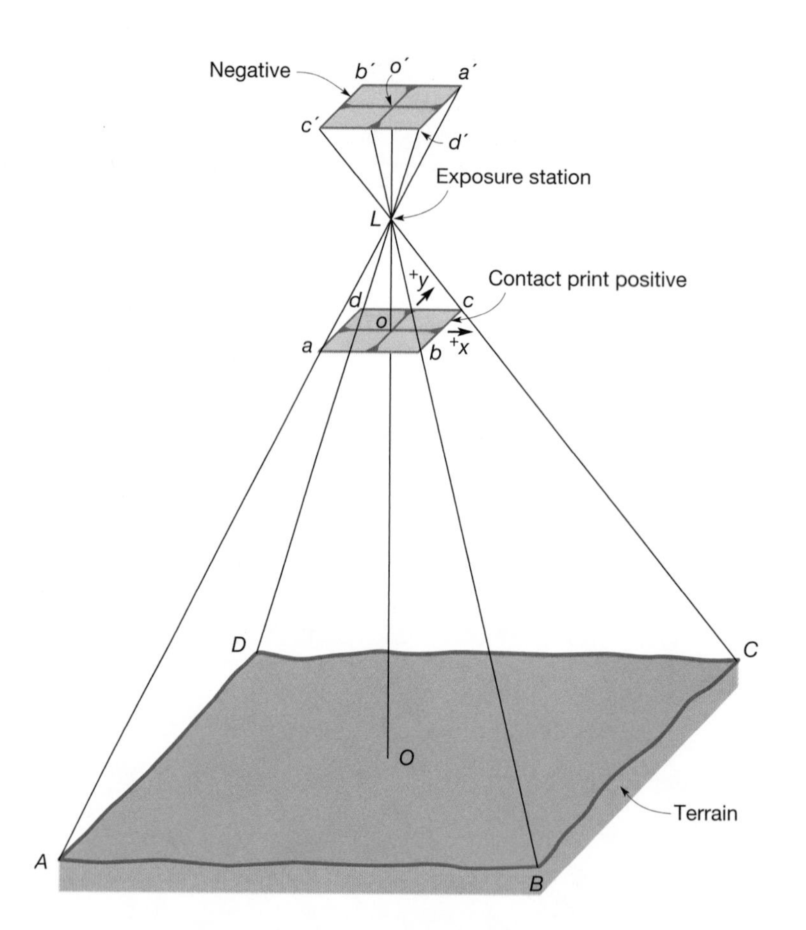

Figure 27.6 Geometry of a vertical aerial photograph.

■ 27.6 SCALE OF A VERTICAL PHOTOGRAPH

Scale is ordinarily interpreted as the ratio of a distance on a map to that same length on the ground. On a map it is uniform throughout because a map is an orthographic projection. The scale of a vertical photograph is the ratio of a photo distance to the corresponding ground distance. Since a photograph is a perspective view, scale varies from point to point with variations in terrain elevation.

In Figure 27.7, L is the exposure station of a vertical photograph taken at an altitude H above datum. The camera focal length is f and o is the photographic principal point. Points A, B, C, and D, which lie at elevations above datum of h_A, h_B, h_C, and h_D, respectively, are imaged on the photograph at a, b, c, and d. The scale at any point can be expressed in terms of its elevation, the camera focal length, and the flying height above datum. From Figure 27.7, from similar triangles Lab and LAB, the following expression can be written:

$$\frac{ab}{AB} = \frac{La}{LA} \tag{a}$$

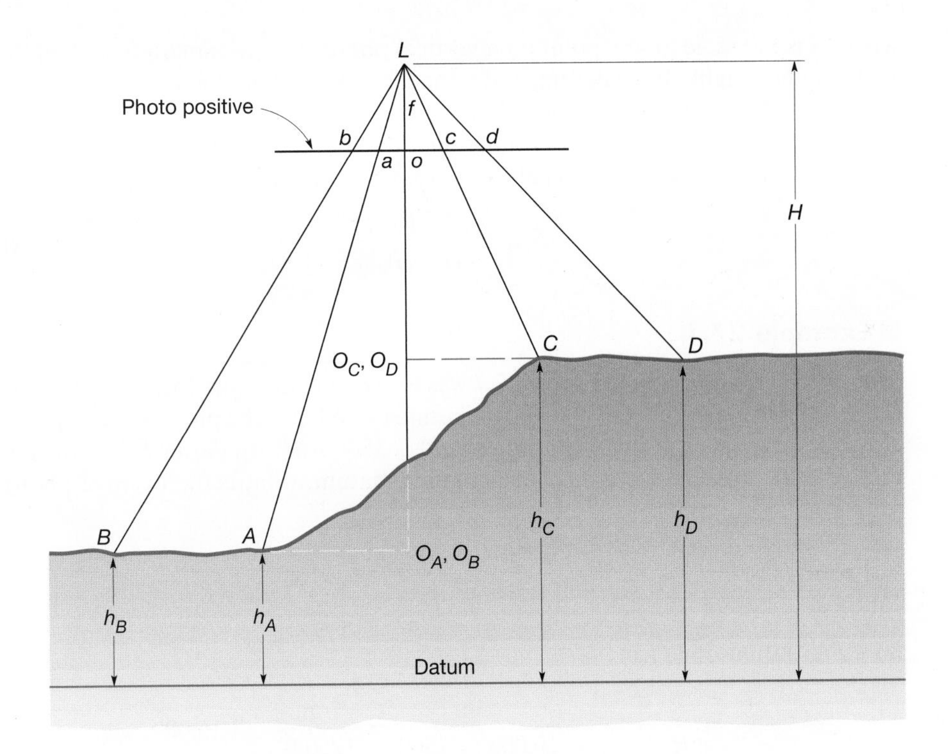

Figure 27.7
Scale of a vertical photograph.

Also from similar triangles *Loa* and *LOA*, a similar expression results:

$$\frac{La}{LA} = \frac{f}{H - h_A} \quad \textbf{(b)}$$

Equating (a) and (b), recognizing that *ab/AB* equals photo scale at *A* and *B*, and considering *AB* to be infinitesimally short, the equation for the scale at *A* is

$$S_A = \frac{f}{H - h_A} \quad \textbf{(c)}$$

Scales at *B*, *C*, and *D* may be expressed similarly as $S_B = f/(H - h_B)$, $S_C = f/(H - h_C)$, and $S_D = f/(H - h_D)$.

It is apparent from these relationships that photo scale increases at higher elevations and decreases at lower ones. This concept is seen graphically in Figure 27.7, where ground lengths *AB* and *CD* are equal, but photo distances *ab* and *cd* are not, *cd* being longer and at larger scale than *ab* because of the higher elevation of *CD*. In general, by dropping subscripts, the scale *S* at any point whose elevation above datum is *h* may be expressed as

$$S = \frac{f}{H - h} \quad \textbf{(27.1)}$$

where S is the scale at any point on a vertical photo, f is the camera focal length, H the flying height above datum, and h the elevation of the point.

Use of an average photographic scale is frequently desirable, but must be accepted with caution as an approximation. For any vertical photographs taken of terrain whose average elevation above datum is h_{avg}, the average scale S_{avg} is

$$S_{avg} = \frac{f}{H - h_{avg}} \tag{27.2}$$

Example 27.1

The vertical photograph of Figure 27.7 was exposed with a 6-in. focal length camera at a flying height of 10,000 ft above datum. (a) What is the photo scale at point a if the elevation of point A on the ground is 2500 ft above datum? (b) For this photo, if the average terrain is 4000 ft above datum, what is the average photo scale?

Solution

(a) From Equation (27.1),

$$S_A = \frac{f}{H - h_A} = \frac{6 \text{ in.}}{10{,}000 - 2500} = \frac{1 \text{ in.}}{1250 \text{ ft}} = 1{:}15{,}000$$

(b) From Equation (27.2),

$$S_{avg} = \frac{f}{H - h_{avg}} = \frac{6 \text{ in.}}{10{,}000 - 4000} = \frac{1 \text{ in.}}{1000 \text{ ft}} = 1{:}12{,}000$$

The scale of a photograph can be determined if a map is available of the same area. This method does not require the focal length and flying height to be known. Rather, it is necessary only to measure the photographic distance between two well-defined points also identifiable on the map. Photo scale is then calculated from the equation

$$\text{photo scale} = \frac{\text{photo distance}}{\text{map distance}} \times \text{map scale} \tag{27.3}$$

In using Equation (27.3), the distances must be in the same units, and the answer is the scale at the average elevation of the two points used.

Example 27.2

On a vertical photograph, the length of an airport runway measures 4.24 in. On a map plotted to a scale of 1:9600, it extends 7.92 in. What is the photo scale at the runway elevation?

Solution

From Equation (27.3),

$$S = \frac{4.24}{7.92} \times \frac{1}{9600} = \frac{1}{17{,}900} \text{ or } 1 \text{ in.} = 1490 \text{ ft}$$

The scale of a photograph can also be computed readily if lines whose lengths are common knowledge appear in the photograph. Section lines, a football or baseball field, and so on, can be measured on the photograph and an approximate scale at that elevation ascertained as the ratio of measured photo distance to known ground length. With an approximate photographic scale known, rough determinations of the lengths of lines appearing in the photo can be made.

■ Example 27.3

On a certain vertical aerial photo, a section line (assumed to be 5280 ft long) is imaged. Its photographic length is 3.32 in. On this same photo, a rectangular parcel of land measures 1.74 by 0.83 in. Calculate the approximate ground dimensions of the parcel and its acreage.

Solution

1. Approximate photo scale

$$\frac{3.32}{5280} = \frac{1 \text{ in.}}{1590 \text{ ft}} \quad \text{or} \quad 1 \text{ in.} = 1590 \text{ ft}$$

2. Parcel dimensions and area:

$$\text{length} = 1590 \times 1.74 = 2770 \text{ ft}$$
$$\text{width} = 1590 \times 0.83 = 1320 \text{ ft}$$
$$\text{area} = \frac{2770 \times 1320}{43{,}560} = 84 \text{ acres}$$

■ 27.7 GROUND COORDINATES FROM A SINGLE VERTICAL PHOTOGRAPH

Ground coordinates of points whose images appear in a vertical photograph can be determined with respect to an arbitrary ground-axis system. The arbitrary X and Y ground axes are in the same vertical planes as photographic x and y, respectively, and the system's origin is in the datum plane vertically beneath the exposure station. Ground coordinates of points determined in this manner are used to calculate horizontal distances, horizontal angles, and areas.

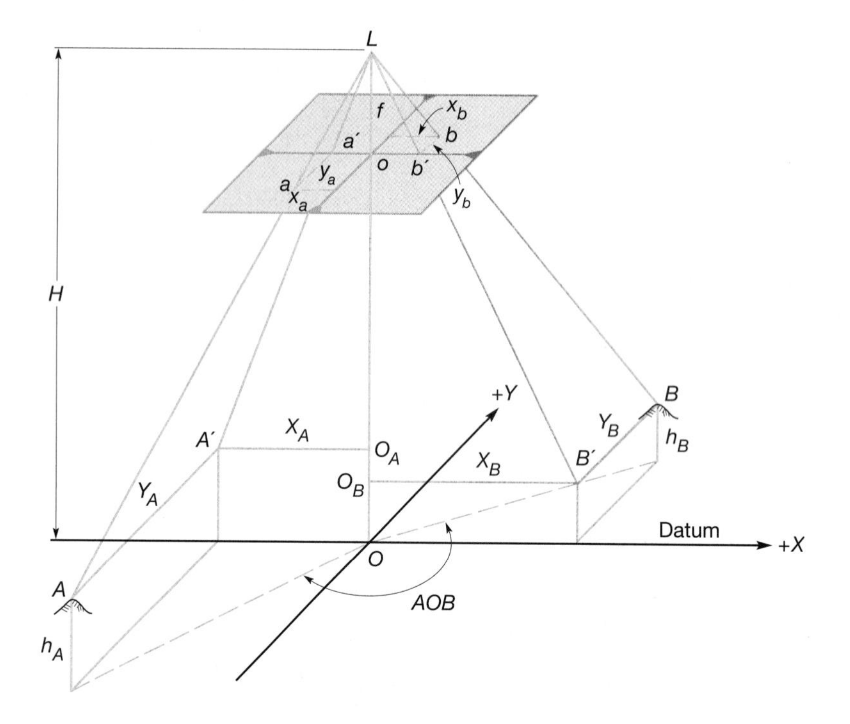

Figure 27.8
Ground coordinates from a vertical photograph.

Figure 27.8 illustrates a vertical photograph taken at flying height H above datum. Images a and b of ground points A and B appear on the photograph. The measured photographic coordinates are x_a, y_a, x_b, and y_b; the ground coordinates are X_A, Y_A, X_B, and Y_B. From similar triangles LO_AA' and Loa',

$$\frac{oa'}{O_AA'} = \frac{f}{H - h_A} = \frac{x_a}{X_A}$$

Then

$$X_A = \frac{(H - h_A)x_a}{f} \tag{27.4}$$

Also from similar triangles $LA'A$ and $La'a$,

$$\frac{a'a}{A'A} = \frac{f}{H - h_A} = \frac{y_a}{Y_A}$$

and

$$Y_A = \frac{(H - h_A)y_a}{f} \tag{27.5}$$

Similarly,

$$X_B = \frac{(H - h_B)x_b}{f} \tag{27.6}$$

$$Y_B = \frac{(H - h_B)y_b}{f} \tag{27.7}$$

Note that Equations (27.4) through (27.7) require point elevations h_A and h_B for their solution. These are normally either taken from existing contour maps or they can be obtained by differential or trigonometric leveling. From the X and Y coordinates of points A and B, the horizontal length of line AB can be calculated using Equation (14.4).

If X and Y coordinates of all corners of a parcel are computed in this way, the parcel area can be determined from those coordinates by the method discussed in Chapter 12. The advantage of calculating lengths and areas by the coordinate formulas, rather than by average scale as in Example 27.3, is that better accuracy results because differences in elevation, which cause the photo scale to vary, are more rigorously taken into account.

■ 27.8 RELIEF DISPLACEMENT ON A VERTICAL PHOTOGRAPH

Relief displacement on a vertical photograph is the shift or movement of an image from its theoretical datum location caused by the object's relief—that is, its elevation above or below datum. Relief displacement on a vertical photograph occurs along radial lines from the principal point and increases in magnitude with greater distance from the principal point to the image.

The concept of relief displacement in a vertical photograph taken from a flying height H above datum is illustrated in Figure 27.9, where the camera focal length is f and o is the principal point. Points B and C are the base and top, respectively, of a pole with images at b and c on the photograph. A is an imaginary point on the datum plane vertically beneath B with corresponding imaginary position a on the photograph. Distance ab on the photograph is the image displacement due to h_B, the elevation of B above datum, and bc is the image displacement because of the height of the pole.

From similar triangles of Figure 27.9, an expression for relief displacement is formulated. First, from triangles LO_AA and Loa,

$$\frac{r_a}{R} = \frac{f}{H}$$

and rearranging,

$$r_aH = fR \tag{d}$$

Also from similar triangles LO_BB and Lob,

$$\frac{r_b}{R} = \frac{f}{H - h_B} \quad \text{or} \quad r_b(H - h_B) = fR \tag{e}$$

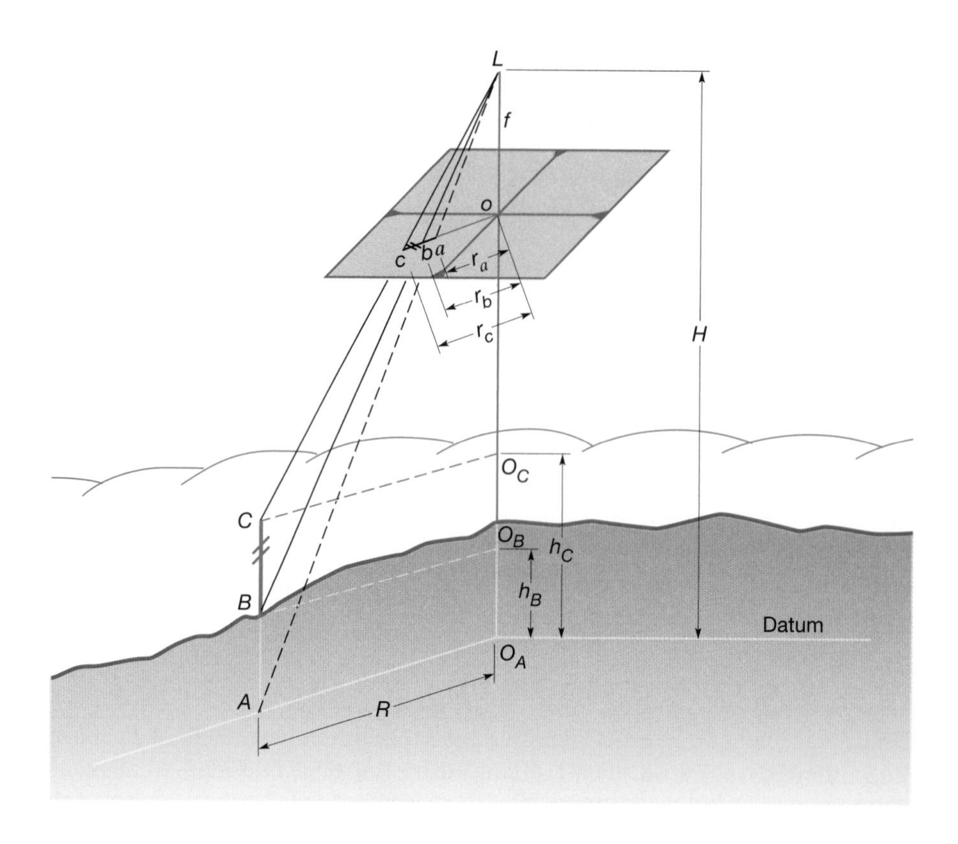

Figure 27.9 Relief displacement on a vertical photograph.

Equating (d) and (e),

$$r_a H = r_b(H - h_B)$$

and rearranging,

$$r_b - r_a = \frac{r_b h_B}{H}$$

If $d_b = r_b - r_a$ is the relief displacement of image b, then $d_b = r_b h_b / H$. Dropping subscripts, the equation can be written in general terms as

$$d = \frac{rh}{H} \tag{27.8}$$

where d is the relief displacement, r the photo radial distance from the principal point to the image of the displaced point, h the height above datum of the displaced point, and H the flying height above that same datum.

Equation (27.8) can be used to locate the datum photographic positions of images on a vertical photograph. True horizontal angles may then be scaled directly from the datum images and if the photo scale at datum is known, true horizontal lengths of the lines can be measured directly. The datum position is located by scaling the calculated relief displacement d of a point along a radial line to the principal point (inward for a point whose elevation is above datum).

Equation (27.8) can also be applied in computing heights of vertical objects such as buildings, church steeples, radio towers, trees, and power poles. To determine heights using the equation, images of both the top and bottom of an object must be visible.

Example 27.4

In Figure 27.9 radial distance r_b to the image of the base of the pole is 75.23 mm and radial distance r_c to the image of its top is 76.45 mm. The flying height H is 4000 ft above datum, and the elevation of B is 450 ft. What is the height of the pole?

Solution

The relief displacement is $r_c - r_b = 76.45 - 75.23 = 1.22$ mm. Selecting a datum at the pole's base and applying Equation (27.8),

$$d = \frac{rh}{H} \text{ so } 1.22 = \frac{76.45h}{4000 - 450}$$

Then

$$h = \frac{3550 \times 1.22}{76.45} = 56.6 \text{ ft}$$

The relief displacement equation is particularly valuable to photo interpreters, who are usually interested in relative heights rather than absolute elevations.

Figure 27.4 vividly illustrates relief displacements. This vertical photo taken over the national capital shows the relief displacement of the Washington Monument in the upper right-hand portion of the format. This displacement, as well as that of other buildings throughout the photograph, occurs radially outward from the principal point. Note, for example, the displacement of the tall buildings in the lower right portion of the photo.

27.9 FLYING HEIGHT OF A VERTICAL PHOTOGRAPH

From previous sections it is apparent that the flying height above datum is an important parameter in solving basic photogrammetry equations. For rough computations, flying heights can be taken from altimeter readings if available. An approximate H can also be obtained by using Equation (27.1) if a line of known length appears on a photograph.

Example 27.5

The length of a section line (known to be 5280 ft) is measured on a vertical photograph as 4.15 in. Find the approximate flying height above the terrain if $f = 6$ in.

Solution

Assuming the datum at the section line elevation, Equation (27.1) reduces to

$$\text{scale} = \frac{f}{H} \quad \text{and} \quad \frac{4.15}{5280} = \frac{6}{H}$$

from which

$$H = \frac{5280 \times 6}{4.15} = 7630 \text{ ft above the terrain}$$

If the images of two ground control points a and b appear on a vertical photograph, the flying height can be determined more precisely from the Pythagorean Theorem,

$$L^2 = (X_B - X_A)^2 + (Y_B - Y_A)^2$$

Substituting Equations (27.4) through (27.7) into this above expression,

$$L^2 = \left[\frac{(H - h_B)x_b - (H - h_A)x_a}{f}\right]^2 + \left[\frac{(H - h_B)y_b - (H - h_A)y_a}{f}\right]^2 \tag{27.9}$$

where L is the horizontal length of ground line AB, H the flying height above datum, $h_A$and h_B are the elevations of the control points above datum, and x and y the measured photo coordinates of the control points.

In Equation (27.9) all variables except H are known. Hence a direct solution can be found for the unknown flying height. The equation is quadratic, so there are two solutions, but the incorrect one will be obvious and can be discarded.

■ 27.10 STEREOSCOPIC PARALLAX

Parallax is defined as the apparent displacement of the position of an object with respect to a frame of reference due to a shift in the point of observation. For example, a person looking through the viewfinder of an aerial camera in an aircraft as it moves forward sees images of objects moving across the field of view. This apparent motion (parallax) is due to the changing location of the observer. By using the camera format as a frame of reference, it can be seen that parallax exists for all images appearing on successive photographs due to forward motion between exposures. Points closer to the camera (of higher elevation) will appear to move faster and have greater parallaxes than lower ones. For 60% endlap, the parallax of images on successive photographs should average approximately 40% of the focal plane width.

Parallax of a point is a function of its relief and consequently measuring it provides a means of calculating elevations. It is also possible to compute X and Y ground coordinates from parallax.

Movement of an image across the focal plane between successive exposures takes place in a line parallel with the direction of flight. Thus to measure parallax, that direction must first be established. For a pair of overlapping photos, this is done by locating positions of the principal points and *corresponding principal points* (that is, principal points transferred to their places in the overlap area of the other photo). A line on each print ruled through these points defines the direction of flight. It also serves as the photographic x-axis for parallax measurement. The y-axis for making parallax measurements is drawn perpendicular to the flight line passing through each photo's principal point. The x coordinate of a point is scaled on each photograph with respect to the axes so constructed and the parallax of the point is then calculated from the expression

$$p = x - x_1 \tag{27.10}$$

Photographic coordinates x and x_1 are measured on the left-hand and right-hand prints, respectively, with due regard given for algebraic signs.

Figure 27.10 illustrates an overlapping pair of vertical photographs exposed at equal flight heights H above datum. The distance between exposure stations L and L_1 is called B, the *air base*. The inset figure shows the two exposure stations

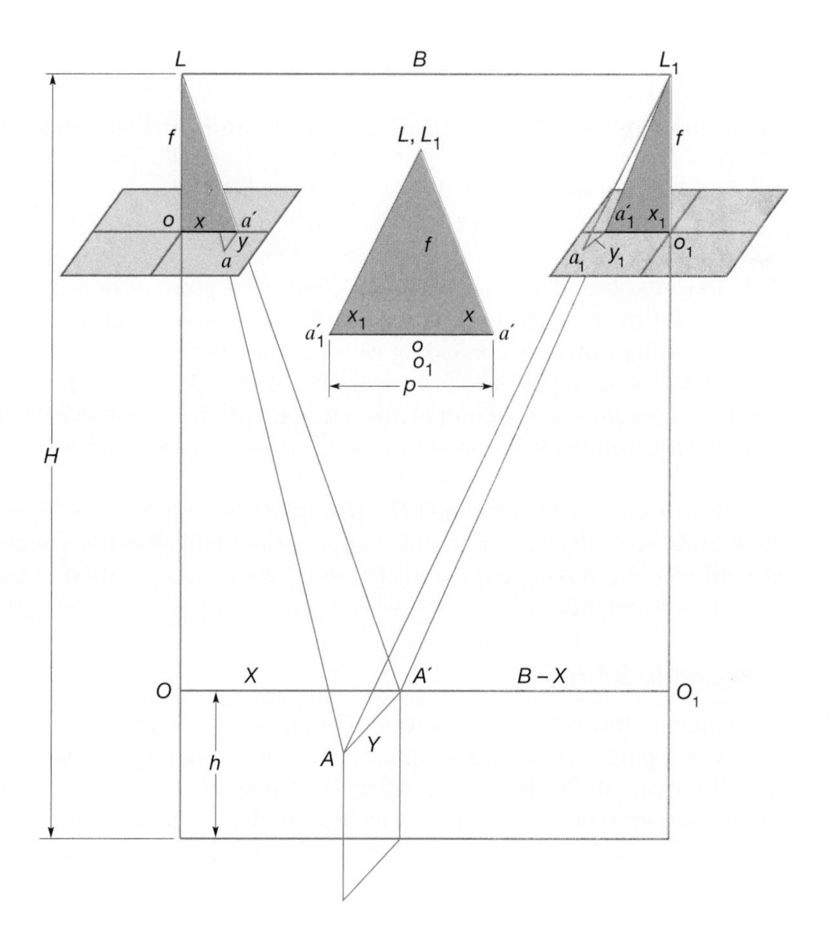

Figure 27.10 Stereoscopic parallax relationships.

L and L_1 in superposition to make the similarity of triangles $La_1'a'$ and $LA'L_1$ more easily recognized. When these two similar triangles are equated, there results

$$\frac{p}{f} = \frac{B}{H - h}$$

from which

$$H - h = \frac{Bf}{p} \tag{27.11}$$

Also from similar triangles LOA' and Loa',

$$X = \frac{x}{f}(H - h) \tag{f}$$

Substituting Equation (27.11) into (f) gives

$$X = \frac{B}{p}x \tag{27.12}$$

and from triangles LAA' and Laa', with substitution of Equation (27.11) yields

$$Y = \frac{B}{p}y \tag{27.13}$$

In Equations (27.12) and (27.13), X and Y are ground coordinates of a point with respect to an origin vertically beneath the exposure station of the left photograph, with positive X coinciding with the direction of flight. Positive Y is 90° counterclockwise to positive X. The parallax of the point is p; x and y are the photographic coordinates of a point on the left-hand print; H is the flying height above datum; h the point's elevation above the same datum; and f the camera lens focal length.

Equations (27.11) through (27.13), commonly called the *parallax equations,* are useful in calculating horizontal lengths of lines and elevations of points. They also provide the fundamental basis for the design and operation of stereoscopic plotting instruments.

Example 27.6

The length of line AB and elevations of points A and B, whose images appear on two overlapping vertical photographs, are needed. The flying height above datum was 4050 ft and the air base was 2410 ft. The camera had a 6-in. focal length. Measured photographic coordinates (in inches) on the left-hand image are $x_a = 2.10$, $x_b = 3.50$, $y_a = 2.00$, and $y_b = -1.05$; on the right-hand image, $x_{1a} = -2.25$ and $x_{1b} = -1.17$.

Solution

From Equation (27.10),

$$p_a = x_a - x_{1a} = 2.10 - (-2.25) = 4.35 \text{ in.}$$

$$p_b = x_b - x_{1b} = 3.50 - (-1.17) = 4.67 \text{ in.}$$

By Equations (27.12) and (27.13),

$$X_A = \frac{B}{p_a} x_a = \frac{2410 \times 2.10}{4.35} = 1160 \text{ ft}$$

$$X_B = \frac{2410 \times 3.50}{4.67} = 1810 \text{ ft}$$

$$Y_A = \frac{B}{p_a} y_a = \frac{2410 \times 2.00}{4.35} = 1110 \text{ ft}$$

$$Y_B = \frac{2410 \times (-1.05)}{4.67} = -542 \text{ ft}$$

By Equation (11.4), length AB is

$$AB = \sqrt{(1810 - 1160)^2 + (-542 - 1110)^2} = 1780 \text{ ft}$$

By Equation (27.11), the elevations of A and B are

$$h_A = H - \frac{Bf}{p_a} = 4050 - \frac{2410 \times 6}{4.35} = 726 \text{ ft}$$

$$h_B = H - \frac{Bf}{p_b} = 4050 - \frac{2410 \times 6}{4.67} = 954 \text{ ft}$$

■ 27.11 STEREOSCOPIC VIEWING

The term *stereoscopic viewing* means seeing an object in three dimensions. This is a process that requires normal *binocular* (two-eyed) vision. In Figure 27.11, two eyes L and R are separated by a distance b called the *eye base.* When the eyes are focused on point A, their optical axes converge to form angle ϕ_1, and when sighting on B, ϕ_2 is produced. Angles ϕ_1 and ϕ_2 are called *parallactic angles* and the brain associates distances d_A and d_B with them. The depth $d_B - d_A$ of the object is perceived from the brain's unconscious comparison of these parallactic angles.

If two photographs of the same subject are taken from two different perspectives or camera stations, the left print viewed with the left eye and

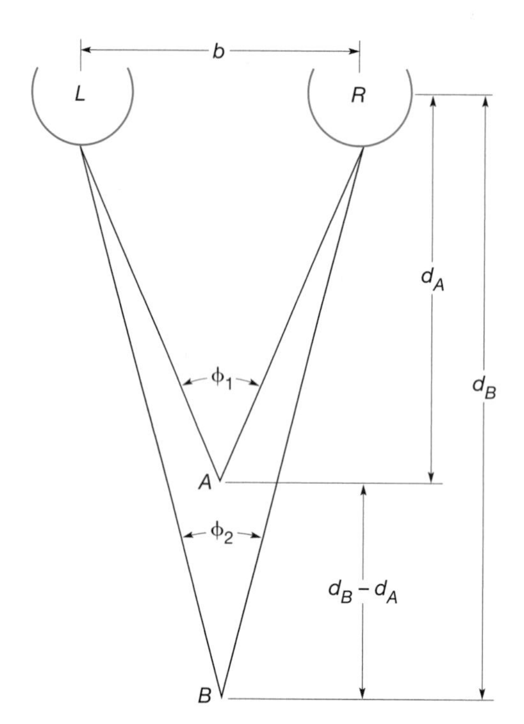

Figure 27.11 Parallactic angles in stereoscopic viewing.

simultaneously the right print seen with the right eye, a mental impression of a three-dimensional model results. In normal stereoscopic viewing (not using photos), the eye base gives a true impression of parallactic angles. While looking at aerial photographs stereoscopically, the exposure station spacing simulates an eye base so the viewer actually sees parallactic angles comparable with having one eye at each of the two exposure stations. This creates a condition called *vertical exaggeration,* which causes the vertical scale of the three-dimensional model to appear greater than its horizontal scale; that is, objects are perceived to be too tall. The condition is of concern to photo interpreters who often estimate heights of objects and slopes of surfaces when viewing air photos stereoscopically. The amount of vertical exaggeration varies with percent endlap and the camera's format dimensions and focal length. A factor of about 4 results if endlap is 60% and the camera has a 9-in. (23-cm) format with 6-in. (152-mm) focal length.

The *stereoscope* shown in Figure 27.12 permits viewing photographs stereoscopically by enabling the left and right eyes to focus comfortably on the left and right prints, respectively, assuming proper orientation has been made of the overlapping pair of photographs under the instrument. Correct orientation requires the two photographs to be laid out in the same order they were taken: with the stereoscope so set that the line joining its lens centers is parallel with the direction of flight. The print spacing is varied, carefully maintaining this parallelism, until a clear three-dimensional view (stereoscopic model) is obtained.

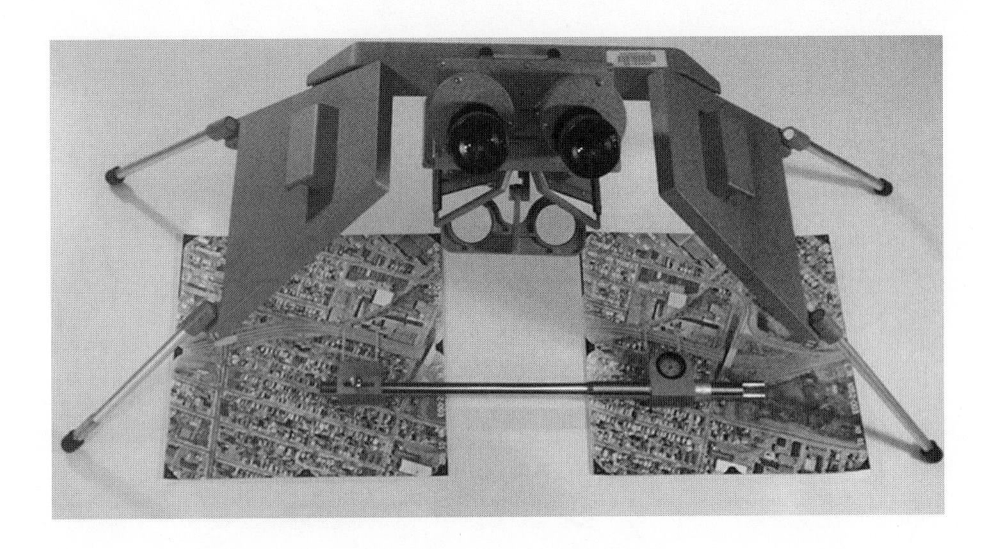

Figure 27.12 Folding mirror stereoscope with parallax bar.

■ 27.12 STEREOSCOPIC MEASUREMENT OF PARALLAX

The parallax of a point can be measured while viewing stereoscopically with the advantage of speed and, because binocular vision is used, greater accuracy. As the viewer looks through a stereoscope, imagine that two small identical marks etched on pieces of clear glass, called *half-marks,* are placed over each photograph. The viewer simultaneously sees one mark with the left eye and the other with the right eye. Then the positions of the marks are shifted until they seem to fuse together as one mark that appears to lie at a certain elevation. The height of the mark will vary or "float" as the spacing of the half-marks is varied; hence it is called the *floating mark.* Figure 27.13 demonstrates this principle and also illustrates that the floating mark can be set exactly on particular points such as A, B, and C by placing the half-marks (small black dots) at a and a', b and b', and c and c', respectively.

Based on the floating mark principle, the parallax of points is observed stereoscopically with a parallax bar, as shown beneath the stereoscope in Figure 27.12. It is simply a bar to which two half-marks are fastened. The right mark can be moved with respect to the left one by turning a micrometer screw to register the displacement on a dial. When the floating mark appears to rest on a point, a micrometer reading is taken and added to the parallax bar *setup constant* to obtain the parallax.

When a parallax bar is used, two overlapping photographs are oriented properly for viewing under a mirror stereoscope and fastened securely with respect to each other using drafting tape. The parallax bar constant for the setup is determined by measuring the photo coordinates for a discrete point and applying Equation (27.10) to obtain its parallax.

The floating mark is placed on the same point, the micrometer read, and the constant for the setup found by

$$C = p - r \qquad \textbf{(27.14)}$$

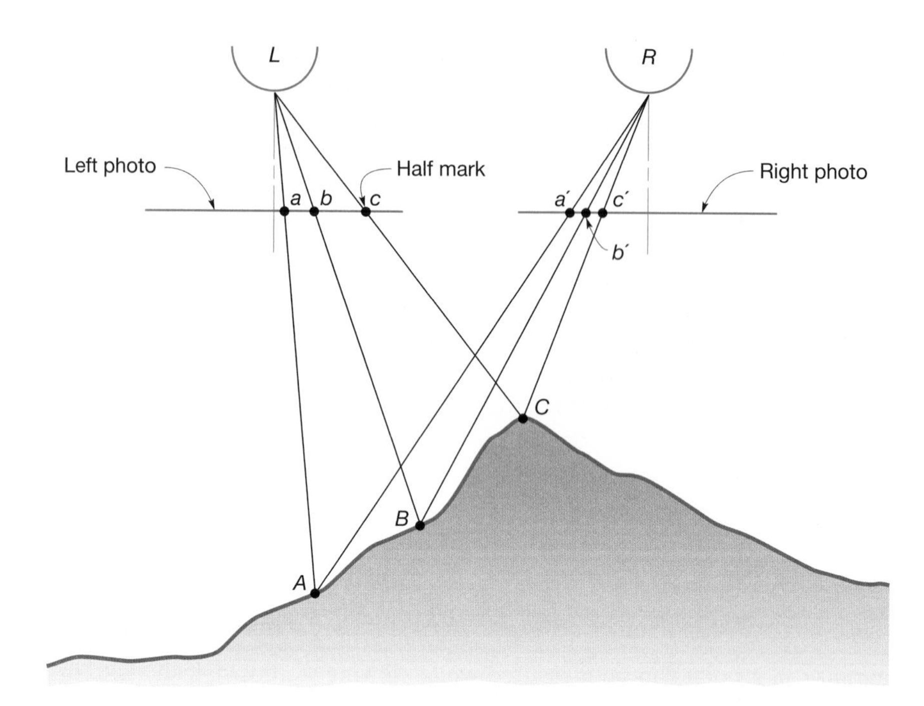

Figure 27.13 Principle of the floating mark.

where C is the parallax bar setup constant, p is the parallax of a point determined by Equation (27.10), and r the micrometer reading obtained with the floating mark set on that same point.

Once the constant has been determined, the parallax of any other point can be computed by adding its micrometer reading to the constant. Thus, a single measurement gives the parallax of a point. Each time another pair of photos is oriented for parallax measurements, a new parallax bar setup constant must be determined. A major advantage of the stereoscopic method is that parallaxes of nondiscrete points can be determined. Thus elevations of hilltops, depressions, and so on, in fields can be calculated using Equation (27.11), even though their x coordinates cannot be measured for use in Equation (27.10).

■ 27.13 ANALYTICAL PHOTOGRAMMETRY

Analytical photogrammetry involves the rigorous mathematical calculation of ground coordinates of points using computers. Input data consists of camera parameters (i.e., the lens focal length, its distortion characteristics, and the principal point location); observed photo coordinates of the images of all points whose ground coordinates are to be determined, as well as those of a limited number of well-distributed ground control points; and the ground coordinates of the control points. The photo coordinates are measured with respect to the coordinate system illustrated in Figure 27.6. Extremely precise instruments called *comparators* are used and values are recorded to the nearest micrometer. Unlike the elementary methods presented in earlier sections of this chapter that assume vertical photos and equal flying heights, analytical photogrammetry rigorously accounts for these variations.

Analytical photogrammetry generally involves the formation of large, rather complex systems of redundant equations, which are then solved using the method of least squares. The concepts have existed for many years, but it was not until the advent of computers that the procedures became practical. The formation of the equations used in analytical photogrammetry is beyond the scope of this text, but interested students can find their derivations, and illustrations of their use, in textbooks that specialize in photogrammetry.[1]

As noted previously, accuracies attainable using analytical photogrammetry are very high and are frequently expressed as a ratio of the flying height of the photography used. Accuracies within about 1/10,000th to 1/15,000th of the flying height above ground are routinely obtained in computed X, Y, and Z coordinates. Thus, for photos taken from 6,000 ft above ground, coordinates accurate to within about ± 0.4 to 0.6 ft can be expected.

Analytical photogrammetry forms the basis for two of the types of stereoplotters: *analytical* and *softcopy*, which are discussed in the following section.

■ 27.14 STEREOSCOPIC PLOTTING INSTRUMENTS

Stereoscopic plotting instruments, also simply called *stereoplotters*, are devices designed to provide accurate solutions for X, Y, and Z object space coordinates of points from their corresponding image locations on overlapping pairs of photos. The fact that the photos may contain varying amounts of tilt and have differing flying heights is of no consequence, because these instruments rigorously account for the position and orientation of the camera for each exposure. Stereoplotters are used to take cross sections, record digital elevation models, compile topographic maps, and generate other types of spatially related topographic information from overlapping aerial photographs.

Stereoplotters can be classified into four different categories: (1) *optical projection*, (2) *mechanical projection*, (3) *analytical*, and (4) *digital* or "softcopy" systems. Regardless of the type, all instruments contain optical and mechanical elements, and the newer versions either have built-in or interfaced computers. Optical and mechanical projection stereoplotters are now obsolete, but nevertheless still often preferred for introducing beginning students to the subject because of their relatively simple concept and design. Modern photogrammetric offices use either analytical plotters or "softcopy" systems. The four types of stereoplotters are described in the subsections that follow.

27.14.1 Direct Optical Projection Stereoplotters

Figure 27.14 illustrates the design concepts of a direct optical projection stereoplotter. In Figure 27.14(a) an overlapping pair of aerial photos is exposed. *Diapositives* (positives developed on film or glass plates) are prepared to exacting standards from the negatives and placed in the projectors of the stereoplotter, as shown in Figure 27.14(b). With the diapositives in place, light rays are projected

---

[1]See Wolf, P. R. and B. A. Dewitt. *Elements of Photogrammetry: With Applications in GIS*, 3rd Ed., 2000, McGraw-Hill Book Co., Inc., New York.

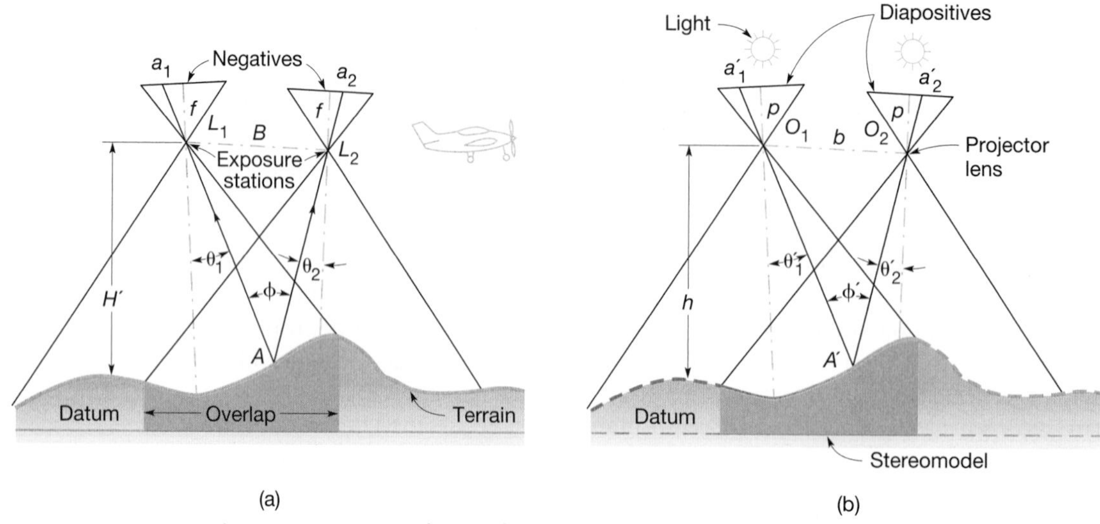

Figure 27.14 Fundamental concepts of stereoscopic plotting instrument design. (a) Aerial photography; (b) Stereoscopic plotter. (From *Elements of Photogrammetry: With Applications in GIS*, 3rd ed., 2000, by Wolf & Dewitt, Courtesy McGraw-Hill Book Co., Inc.)

through them and the positions and orientations of the projectors are adjusted until the rays from corresponding images intersect below to form a model of the overlap area of the aerial photos. The model of the terrain is called a *stereomodel* and once formed, it can be viewed, measured, and mapped.

In order to accomplish the steps outlined in the preceding paragraph, it is necessary that direct optical projection stereoplotters incorporate the following components: (1) a projection system (to project the light rays which create a stereomodel), (2) a viewing system (which enables an operator to see the stereomodel in three dimensions), and (3) a measuring/tracing system (for measuring or mapping the stereomodel). Projectors used in optical projection stereoplotters function like ordinary slide projectors. However, they are much more precise and can be adjusted in angular orientation and position to recreate the spatial locations and attitudes that the aerial cameras had when the overlapping photos were exposed. This produces a "true" model of the terrain in the overlap area. The scale of the stereomodel is, of course, greatly reduced, and is the ratio of the *model base* (distance b between projector lenses) to the *air base* (actual distance B between the two exposure stations).

Plotter viewing systems must provide a stereoscopic view and hence be designed so that the left and right eyes see only projected images of the corresponding left and right diapositives. One method of accomplishing this is to place a blue filter in one projector and a red filter in the other. The operator wears a pair of spectacles with corresponding blue and red lenses. This system of viewing stereoscopically is called the *anaglyphic* method. Another system, called the *stereo image alternator* (SIA), operates with rapidly rotating shutters, located on the projectors, and a viewing eyepiece. The shutters are synchronized so that the left and right eyes can see only the images from the corresponding left and right projectors. A third method, similar to the anaglyphic system and called the *polarized platen viewing* (PPV) system, uses filters of opposite polarity. With each of these viewing systems, the operator must work in a dark room—a disadvantage of the direct optical projection type of stereoplotter.

Various measuring and tracing systems have been devised for stereoplotters. In optical projection instruments, light rays carrying corresponding images are intercepted on the *tracing table platen* (a small white circular disk). A light emitted through a pinhole in the platen, which can be seen by both the left and right eyes and fuses, provides the floating mark. It can be made to appear to rest exactly on a point of the model by raising or lowering the tracing table. A tracing pencil located directly below the floating mark permits positioning the point planimetrically. A counter linked to the tracing table responds to up-and-down motion, so the elevation for any setting of the floating mark is read directly.

A stereoplotter operator, preparing to measure or map a stereomodel, must go through a three-stage orientation process consisting of *interior orientation*, *relative orientation*, and *absolute orientation*. Interior orientation ensures that the projected light rays are geometrically correct, i.e., angles θ_1' and θ_2' of Figure 27.14(b) (i.e., the angles between the projected light rays and the axis of the projector lens) must be identical to corresponding angles θ_1 and θ_2, respectively, in Figure 27.14(a) (i.e., the angles between the incoming light rays and the camera axis). Preparing the diapositives to exacting specifications, and centering them carefully in the projectors accomplish this.

After the diapositives have been placed in the projectors and the lights turned on, corresponding light rays will not intersect to form a clear model because of tilts in the photographs and unequal flying heights. To achieve intersections of corresponding light rays, the projectors are moved linearly along the X-, Y-, and Z-axes and also rotated about these axes until they duplicate the relative tilts and flying heights that existed when the photographs were taken. This process is called *relative orientation* and when accomplished, parallactic angle ϕ' of Figure 27.14(b) for each corresponding pair of light rays will be identical to its corresponding parallactic angle ϕ of Figure 27.14(a), and a perfect three-dimensional model will be formed.

The model is brought to required scale by making the rays of at least two, but preferably three, ground control points intersect at their positions plotted on a manuscript map prepared at the desired scale. It is leveled by adjusting the projectors so the counter reads the correct elevations for each of a minimum of three, but preferably four, corner ground control points when the floating mark is set on them. *Absolute orientation* is a term applied to the processes of scaling and leveling the model.

When orientation is completed, a map can be made from the model, or cross sections and other spatial information compiled. In mapping, planimetric details are located first by bringing the floating mark into contact with objects in the model and tracing them. A pencil directly beneath the reference mark records their locations on the manuscript map below. Contours are traced by setting the elevation counter successively at each contour elevation and moving the reference mark over the model while keeping it in contact with the terrain. Again, the pencil draws the contours. When a manuscript is completed, it is examined for omissions and mistakes and field-checked. The final map is then traced from the manuscript by inking or scribing.

27.14.2 Mechanical Projection Stereoplotters

Mechanical projection stereoplotters use two precisely made metal space rods to simulate projected light rays. The diapositives are viewed stereoscopically through binoculars via an optical train of lenses and prisms. A floating mark, composed of a pair of half-marks superimposed in the optical paths of the left and right eye, is available for making measurements of the stereomodel. The angle of intersection of the space rods [angle ϕ' in Figure 27.14(b)] can be changed by turning a hand screw or foot disk, and the space rods can be impelled in the X and Y directions, either manually or by means of hand wheels. Because the space rods drive the viewing optics, these motions enable an operator to scan the stereomodel and set the floating mark on any point of interest.

With mechanical projection stereoplotters, the same orientation processes (interior, relative and absolute) must be followed as were described for optical projection instruments. When this is completed, the stereomodel can be measured or mapped. When the operator places the floating mark on a point, its X, Y, and Z coordinates are determined and depending upon the instrument's measuring/tracing system, the point's location can either be plotted directly on a map or recorded in a digital file. Instruments equipped with a coordinatograph or pantograph allow planimetry to be traced directly on a manuscript. Figure 27.15 shows

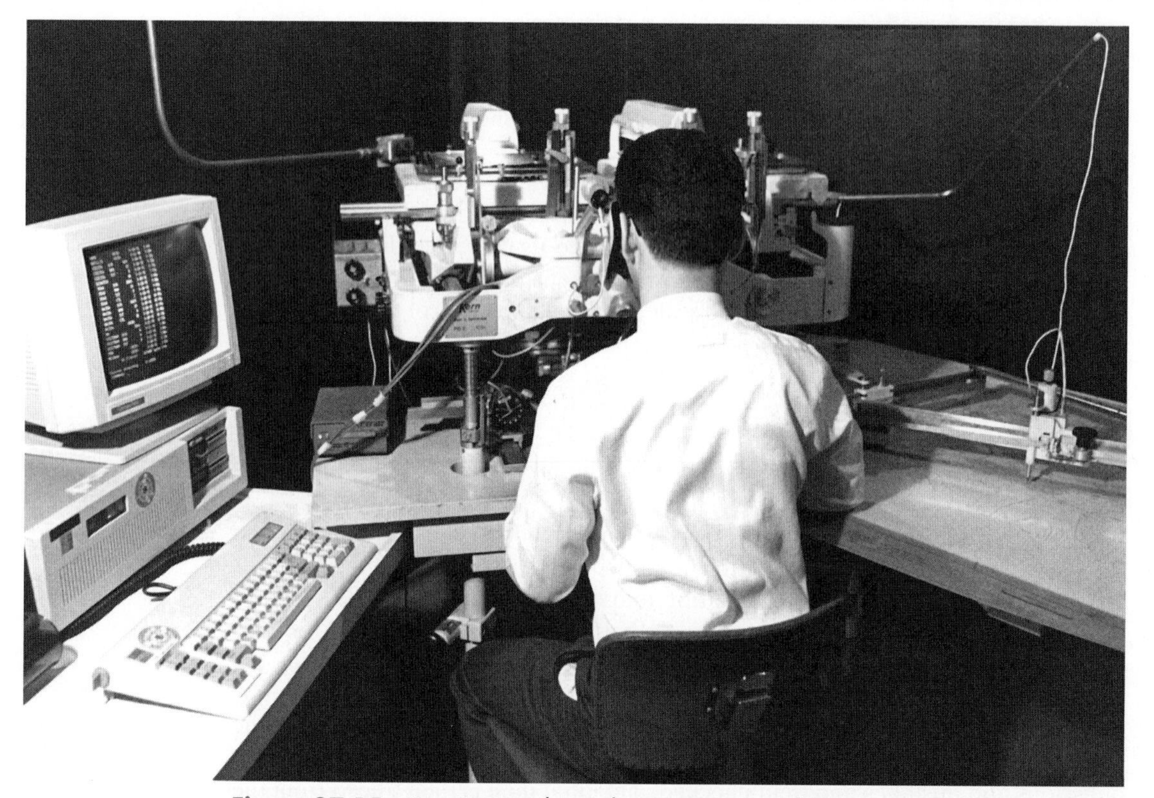

Figure 27.15 Kern PG2 mechanical projection stereoplotter equipped with pantograph for direct manuscript plotting and interfaced with computer for digital mapping. (Courtesy Department of Civil and Environmental Engineering, University of Wisconsin-Madison)

a Kern PG2 mechanical projection stereoplotter with an attached pantograph on the plotting table at the right side for direct map compilation.

Computer graphics systems have been incorporated with mechanical projection stereoplotters to greatly facilitate mapping and recording measurements. With these systems, after an operator sets the floating mark on a point in the stereomodel, a foot pedal is depressed and the object's *X*, *Y*, and *Z* coordinates are automatically recorded and stored within a computer file. Cross sections, profiles, and digital elevation models (DEMs) are rapidly and conveniently digitized in this way. Planimetric maps can also be compiled using these systems. In this mode the operator indicates the type of object being mapped by keying in a certain identifying code number. For example, a tree might be identified with the number 5, while 7 could indicate a building corner. To create lines such as fences, roads, and the like, connectivity between points is also keyed in. Then when the floating mark is set and the foot-pedal depressed, this information is recorded along with the object's ground coordinates. As the mapping progresses, a current view of work accomplished appears on the computer's monitor. When the map is completed, a hard copy can be obtained by transferring the digital data file to an automatic plotter. Maps compiled in digital form have many advantages. They can be plotted at any scale, are easily edited and updated, and can be integrated with other digital data for use in CADD or Geographic Information System (GIS) projects. The computer and monitor shown to the left of the PG2 stereoplotter of Figure 27.15 is being used to support an on-line computer graphics system.

Advantages of mechanical projection instruments over the optical types are that they (1) can be operated in a lighted room by an operator who sits rather than stands, (2) are more versatile, (3) have greater stability, and (4) produce more accurate results.

27.14.3 Analytical Stereoplotters

Analytical plotters combine a precise stereoscopic system for measuring photo coordinates, a digital computer, and sophisticated analytical photogrammetry software. In using an analytical plotter, an operator looks through a binocular viewing system and sees the stereomodel formed from a pair of overlapping photos. The floating mark, which again consists of half-marks superimposed within the optics of the viewing system, is placed on points whose ground positions are desired. When the mark is precisely positioned, the *x* and *y* photo coordinates of the point from both photos are measured by means of encoders and fed directly to the computer. The computer uses these photo coordinates, together with camera parameters and ground coordinates of control points that have been input into the system, to calculate the point's *X*, *Y*, and *Z* ground coordinates in real time. For this calculation, the analytical procedures described in Section 27.13 are employed. These ground coordinates are then stored in a file within the system's computer. Of course, before extracting information from the stereomodel, analytical plotters must be oriented by following the same basic processes as used for optical and mechanical projection instruments.

Because they have no optical or mechanical limitations, analytical plotters have great versatility. They can handle any type of photography, including vertical, tilted, oblique, and terrestrial photos. Also, in comparison to optical and

Figure 27.16 Zeiss P3 analytical plotter. (Courtesy Tom Pantages)

mechanical projection plotters, analytical plotters can provide superior accuracy for several reasons. First, they do not form model points by intersecting projected light rays or mechanical space rods, and thus they introduce no errors due to imperfections in these projection systems. Second, they can effectively correct for various systematic errors such as those caused by camera lens distortions, film shrinkage, and atmospheric refraction. Third, in almost every phase of their operation, they can take advantage of redundant observations and incorporate the method of least squares into the solution of the equations. An added advantage is that the data extracted are in digital form and thus readily compatible for direct use in CADD systems or for entry into the databases of land and geographic information systems. Figure 27.16 shows an analytical plotter with a monitor screen on the right to give the operator a visual record of work performed and to permit review and editing of the digitized data.

27.14.4 Softcopy Stereoplotters

Softcopy plotters are the latest development in the evolution of stereoscopic plotting instruments. These systems utilize digital or "softcopy" images. The images can be acquired by using a digital camera of the type described in Section 27.3, but more often they are obtained by scanning the negatives of aerial photos taken with film cameras. Scanners convert the contents of an aerial photo into an array of pixels arranged in rows and columns. Each pixel is identified by its row and column and is assigned a digital number, which corresponds to its *gray level* or degree of darkness at that particular element. Figure 27.17 shows the PhotoScan, developed jointly by Carl Zeiss and Intergraph Corporation for digitizing photographs.

A softcopy photogrammetric system requires a computer with a high-resolution graphics display. The computer must be capable of efficiently manipulating large files of digital images and must also be able to display the left and right photos of a stereopair simultaneously. Also the equipment must include a stereoscopic viewing mechanism, i.e., one that enables the operator to see only the left image with the left eye and only the right image with the right eye. Manual

Figure 27.17 Zeiss/Intergraph PS1 PhotoScan. (Courtesy Carl Zeiss)

operation of a softcopy system is similar to that of an analytical plotter. The operator moves the floating mark about the stereomodel and places it on any point whose position is desired. Once the mark is set, the row and column locations of the pixels at its location identify the photo coordinates of the point to the computer, which can then calculate, in real time, its ground X, Y, and Z coordinates by solving equations of analytical photogrammetry.

One type of softcopy photogrammetric workstation, the digital video plotter (DVP), is shown in Figure 27.18. As seen in the figure, it is simply a standard personal computer with a stereoscopic viewing system attached. With this device, the left and right photos are displayed simultaneously on the left and right portions, respectively, of the computer's monitor. An operator looking through the stereoscope can view the stereomodel in three dimensions. Using the cursor of an interfaced digitizer or the keyboard's arrow keys, the left and right images can be shifted on the screen with respect to each other. The floating marks, made of two reference pixels, one on each image, can also be moved about the screen. While viewing stereoscopically, an operator shifts the imagery and moves the floating mark until it appears to rest exactly on the point of interest. This identifies to the computer the image pixels, which correspond to the point in both the left and right photos. The computer converts the pixel row and column locations to x and y

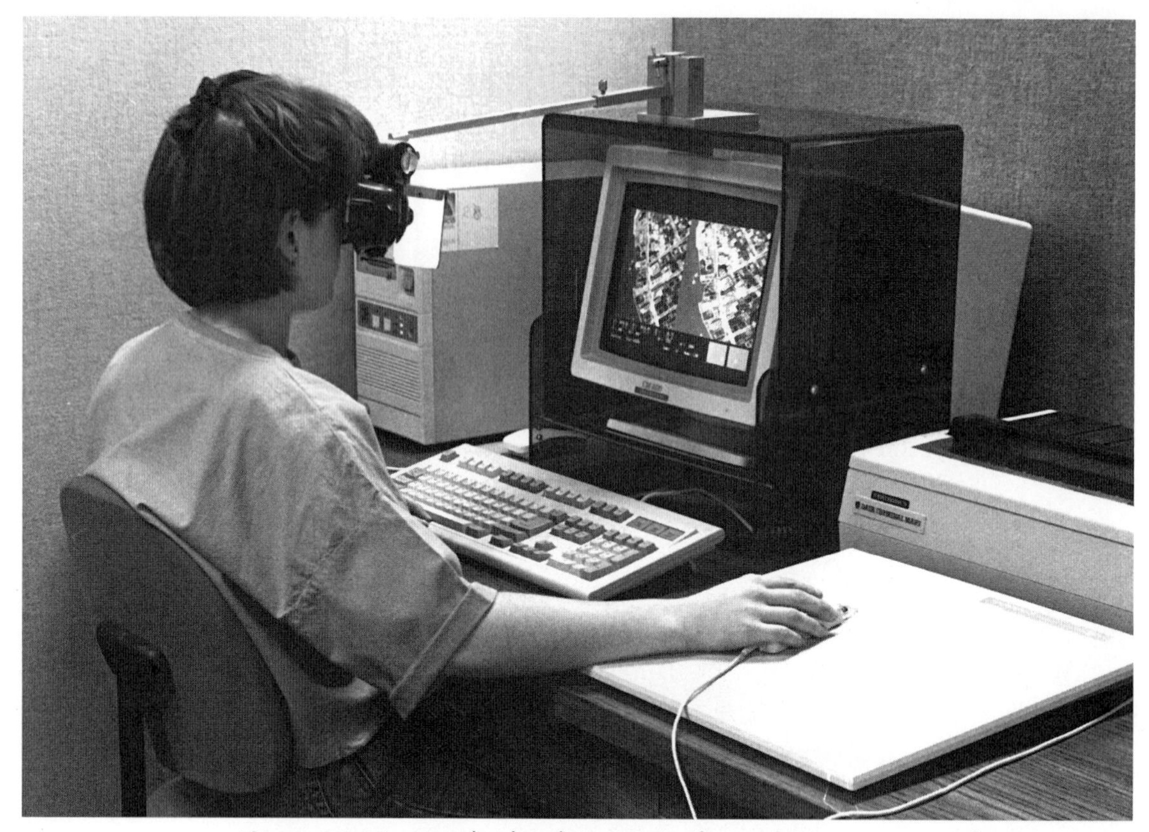

Figure 27.18 Digital Video Plotter (DVP) softcopy photogrammetric workstation. (Courtesy Dr. P. A. Gagnon, Laval University, Quebec, Canada)

photo coordinates and then calculates the point's ground coordinates by employing standard analytical photogrammetry equations.

Other softcopy photogrammetric systems are also available. Some rely on polarizing lenses for stereoviewing, while others, such as the Intergraph Image Station Z shown in Figure 1.6, use a system of electronically synchronized shutters. All are controlled by sophisticated software. Some advanced systems have become almost totally automated. They employ a process called *digital image correlation.* In this operation the computer automatically matches points in the left photo of the stereopair to their corresponding or *conjugate* points in the right photo. This is done by comparing the patterns of image densities in the point's immediate area on both photos. Thus, the operator's process of making measurements while stereoviewing is eliminated.

Softcopy photogrammetry systems are efficient, as well as versatile. Not only are they capable of producing maps, cross sections, digital elevation models, and other digital topographic files, but they can also be employed for a variety of image interpretation problems and they can support the production of mosaics and orthophotos (see Section 27.15). Also, digital maps produced by softcopy systems are created in a computer environment and are therefore in formats compatible for CADD applications and for in the databases of geographic information systems.

Softcopy systems have the added advantage that their major item of hardware is a computer rather than an expensive single-purpose stereoplotter, so it can be used for many other tasks in addition to stereoplotting.

A digital image viewing and measuring system has been incorporated on the CD that accompanies this book in the WOLFPACK software. This package allows you to digitize digital images in the bitmap (bmp) format. The resulting image coordinates can be placed into the photo coordinate system using the *interior orientation* option. This option requires the calibrated fiducial coordinates of the camera, in the file camera.fid, and transforms the digitized image coordinates into the photo coordinate system. Furthermore, by observing a minimum of three imaged points whose ground coordinates are known (in the file ground.crd), an *exterior orientation* can be performed. The exterior orientation determines the camera location and orientation parameters at the time of exposure. Finally if this procedure is performed on common points lying in the overlapping region of two photos, the ground coordinates of these points can be determined using the *space intersection* option. All the required points (fiducial points, ground control, and photo-identifiable points) for the analytical process should be observed in a single digitizing session so that repetition of the procedures is not necessary. This photogrammetric process is described in the help file that accompanies the WOLFPACK software.

While these functions demonstrate the rudimentary operations of softcopy systems, they allow the reader to experience digital photogrammetry. The CD also contains suitable aerial imagery, a file of calibrated fiducial coordinates (camera.fid), and another of ground control (ground.crd). The fiducial coordinates on the image are labeled F1 through F8 and the ground control is also circled and labeled in the imagery to aid the user in identify the points. The calibrated focal length of the aerial camera was 153.742 mm. Problems at the end of this chapter refer to this material.

■ 27.15 ORTHOPHOTOS

As implied by their name, orthophotos are orthographic representations of the terrain in picture form. They are derived from aerial photos in a process called *differential rectification,* which removes scale variations and image displacements due to relief and tilt. Thus the imaged features are shown in their true planimetric positions.

Instruments used for differential rectification have varied considerably in their designs. The first-generation instruments were basically modified stereoscopic plotters with either optical or mechanical projection. Optical projection instruments derived an orthophoto by systematically *scanning* a stereomodel and photographing it in a series of adjacent narrow strips. *Rectification* (removal of tilt) was accomplished by leveling the model to ground control prior to scanning, and scale variations due to terrain relief were removed by varying the projection distance during scanning. As the instrument automatically traversed back and forth across the model, exposure was made through a narrow slit onto an orthonegative. An operator, viewing the model in three dimensions, continually monitored the scans and adjusted the projection distance to keep the exposure slit in contact with the stereomodel. Because the model itself had uniform scale throughout, the resulting *orthonegative* (from which the orthophoto was made) was also of uniform scale. Orthophoto systems based on modified mechanical projection stereoplotters functioned in a similar fashion. These instruments are seldom used today.

Contemporary orthophoto production is done using softcopy photogrammetric systems in a procedure called *digital image processing.* These systems employ digital images, which, as described in Section 27.14.4, may be obtained either by using digital cameras or by scanning negatives obtained with film cameras. As noted earlier, a digital image consists of a *raster* (grid) of tiny pixels, each of which is assigned a digital value corresponding to its gray level, and each having its photo location given in terms of its row and column within the raster. The digital image is input to the system's computer, which uses analytical photogrammetry equations to modify each pixel location according to the tilt in the photograph and the scale at that point in the stereomodel. Through this process all pixels are modified to locations they would have on a truly vertical photo and all are brought to a common scale. The modified pixels are then printed electronically to produce an orthophoto.

Orthophotos combine the advantages of both aerial photos and line maps. Like photos, they show features by their actual images rather than as lines and symbols, thus making them more easily interpreted and understood. Like maps, orthophotos show the features in their true planimetric positions. Therefore true distances, angles, and areas can be scaled directly from them. *Orthophotomaps* (maps produced from orthophotos) are used for a variety of applications, including planning and engineering design. They have been particularly valuable in cadastral and tax mapping, because the identification of property boundaries is greatly aided through visual interpretation of fence lines, roads, and other evidence. Because they are in digital form, they are also ideal for use as base maps and for analyses in geographic information systems.

Orthophotos can generally be prepared more rapidly and economically than line or symbol planimetric maps. With their many significant advantages, orthophotos have superseded conventional maps for many uses.

■ 27.16 GROUND CONTROL FOR PHOTOGRAMMETRY

As pointed out in preceding sections, almost all phases of photogrammetry depend on ground control (points of known positions and elevations with identifiable images on the photograph). Ground control can be basic control—traverse, triangulation, trilateration, or GPS monuments already in existence and marked prior to photography to make them visible on the photos; or it can be *photo control*—natural points having images recognizable on the photographs and positions that are subsequently determined by ground surveys originating from basic control. Instruments and procedures used in ground surveys were described in earlier chapters. Ordinarily, photo-control points are selected after photography to ensure their satisfactory location and positive identification. Premarking points with artificial targets is sometimes necessary in areas that lack natural objects to provide definite images.

As discussed in Section 27.14, scaling and leveling stereomodels for mapping with stereoplotters requires a practical minimum of three horizontal control points and four vertical points in each model. For large mapping jobs, therefore, the cost of establishing the required ground control is substantial. In these situations, *analytical aerotriangulation* (see Section 27.13) is used to establish many of the needed control points from only a sparse network of ground-surveyed points. This reduces costs significantly.

Currently GPS is being used for real-time positioning of the camera at the instant each photograph is exposed. The kinematic GPS surveying procedure is being employed (see Chapter 15), which requires two GPS receivers. One unit is stationed at a ground control point; the other is placed within the aircraft carrying the camera. The integer ambiguity problem is resolved using *on-the-fly* techniques (see Section 15.2). During the flight, camera positions are continuously determined at time intervals of a few seconds using the GPS units and precise timing of each photo exposure is also recorded. From this information, the precise location of each exposure station, in the ground coordinate system, can be calculated. Many projects have been completed using these methods and they have produced highly accurate results, especially when supplemented with only a few ground control points. It is now possible to complete photogrammetric projects with only a few ground photo control points used for checking purposes.

■ 27.17 FLIGHT PLANNING

Certain factors, depending generally on the purpose of the photography, must be specified to guide a flight crew in executing its mission of taking aerial photographs. Some of them are (1) boundaries of the area to be covered, (2) required scale of the photography, (3) camera focal length and format size, (4) endlap, and (5) sidelap. Once these elements have been fixed, it is possible to compute the entire flight plan and prepare a flight map on which the required flight lines have been delineated. The pilot then flies the specified flight lines by choosing and correlating headings on existing natural features shown on the flight map. In the most modern systems, the flight planning is done using a computer and the coordinates of flight lines are calculated. Then the aircraft is automatically guided by an on-board GPS system along the planned flight lines.

The purpose of the photography is the paramount consideration in flight planning. For example, in taking aerial photos for topographic mapping using a stereoplotter, endlap should optimally be 60% and sidelap 30%. The required scale and contour interval of the final map must be evaluated to settle flying height. Enlargement capability from photo scale to map compilation scale is restricted for stereoplotting and generally should not exceed about 5 if satisfactory accuracy is to be achieved. By these criteria, if required map scale is 200 ft/in., photo scale becomes fixed at 1000 ft/in. If the camera focal length is 6 in., flying height is established by Equation (27.2) at $6 \times 1000 = 6000$ ft above average terrain. Some organizations may push this factor higher than 5, but it should be done with caution.

The *C factor* (the ratio of flight height above ground to contour interval which is practical for any specific stereoplotter) is a criterion often used to select the flying height in relation to the required contour interval. To ensure that their maps meet required accuracy standards, many organizations employ a C factor of about 1200 to 1500. Other organizations may push the value somewhat higher, but again this should be done with caution. By this criterion, if a plotter has a C factor of, say, 1200, and a map is to be compiled with a 5-ft contour interval, a flight height of not more than $1200 \times 5 = 6000$ ft above the terrain should be sustained.

Information ordinarily calculated in flight planning includes (1) flying height above mean sea level, (2) distance between exposures, (3) number of photographs per flight line, (4) distance between flight lines, (5) number of flight lines, and (6) total number of photographs. A flight plan is prepared based on these items.

Example 27.7

A flight plan for an area 10-mi wide and 15-mi long is required. The average terrain in the area is 1500 ft above datum. The camera has a 6 in. focal length with 9×9 in. format. Endlap is to be 60%, sidelap 25%. The required scale of the photography is 1:12,000 (1000 ft/in.).

Solution

1. Flying height above datum from Equation (27.2):

$$\text{scale} = \frac{f}{H - h_{\text{avg}}} \text{ so } \frac{1}{1000} = \frac{6}{H - 1500} \text{ and } H = 7500 \text{ ft}$$

2. Distance between exposures, d_e: Endlap is 60%, so the linear advance per photograph is 40% of the total coverage of 9 in. $\times$ 1000 ft/in. = 9000 ft. Thus the distance between exposures is $0.40 \times 9000 = 3600$ ft.
3. Total number of photographs per flight line:

$$\text{length of each flight line} = 15 \text{ mi} \times 5280 \text{ ft/mi} = 79{,}200 \text{ ft}$$

$$\text{number of photos per flight line, } N_{\text{Photos/Line}} = \frac{79{,}200 \text{ ft}}{3600 \text{ ft/photo}} + 1 = 23$$

   Adding two photos on each end to ensure complete coverage, the total is $23 + 2 + 2 = 27$ photos per flight line.
4. Distance between flight lines: Sidelap is 25%, so the lateral advance per flight line is 75% of the total photographic coverage,

$$\text{distance between flight lines, } d_s = 0.75 \times 9000 \text{ ft} = 6750 \text{ ft}$$

5. Number of flight lines:

$$\text{width of area} = 10 \text{ mi} \times 5280 \text{ ft/mi} = 52{,}800 \text{ ft}$$

$$\text{number of spaces between flight lines} = \frac{52{,}800 \text{ ft}}{6750 \text{ ft/line}} = 7.8 \text{ (say 8)}$$

$$\text{total flight lines, } N_{\text{Lines}} = 8 + 1 = 9$$

$$\text{planned spacing between flight lines} = \frac{52{,}800}{8} = 6600 \text{ ft}$$

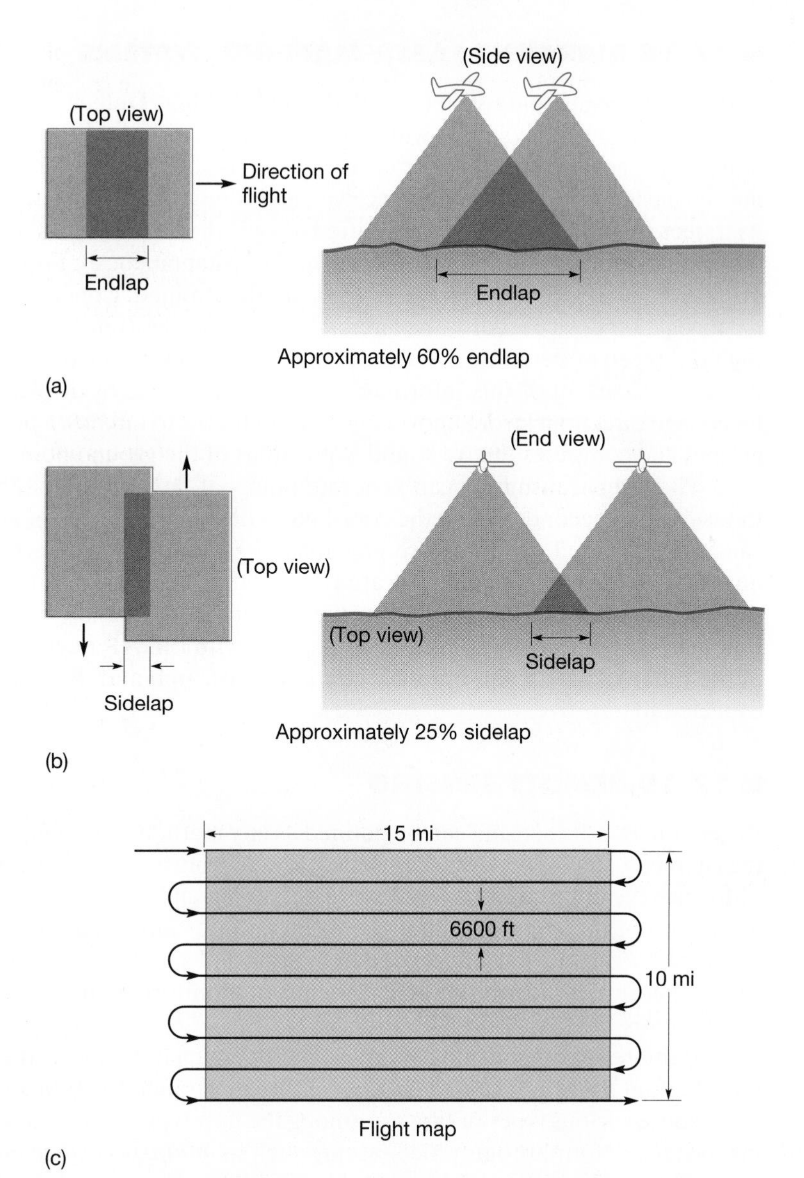

Figure 27.19
(a) Endlap,
(b) sidelap, and
(c) flight map

(*Note:* The first and last flight lines should either coincide with or be near the edges of the area, thus providing a safety factor to ensure complete coverage.)

6. Total number of photos required:

$$\text{total photos} = 27 \text{ per flight line} \times 9 \text{ flight lines} = 243 \text{ photos}$$

Figures 27.19(a) and 27.19(b) illustrate endlap and sidelap, respectively, and (c) shows the flight map.

■ 27.18 AIRBORNE LASER-MAPPING SYSTEMS

Airborne laser-mapping systems called *LiDAR* (**Li**ght **D**etection **A**nd **R**anging) have recently been developed which show great promise for the future. These systems, which are carried in airborne vehicles, consist of a laser scanning device, an inertial navigation system, a GPS receiver, and computer. As the aircraft flies along its trajectory, laser pulses are transmitted toward the terrain below, reflected from the ground or other objects and detected nearly instantaneously. From these pulses, distances and angles to reflective objects are determined. Concurrently, the inertial navigation device records the aircraft's attitude angles (pitch, yaw, and roll) and the GPS receiver determines the X, Y, and Z positions of the detector. The computer processes all of this information to determine vector displacements (distances and directions) from known positions in the air to unknown positions on the ground and computes the X, Y, and Z positions of the ground points.

The laser transmitter can generate pulses at an extremely rapid rate, i.e., thousands per second, so that the coordinates of a dense pattern of ground points can be determined. Not only are positions of ground points determined, but an image of the ground is also generated. The data can be used to produce digital elevation models (DEMs) and from them contour maps and other topographic products can be produced. Accuracies possible with LiDAR devices are currently in the range of 10–15 cm, but with continued research and development, this is expected to improve.

■ 27.19 REMOTE SENSING

In general, remote sensing can be defined as any methodology employed to study the characteristics of objects using data collected from a remote observation point. More specifically, and in the context of surveying and photogrammetry, it is the extraction of information about the earth and our environment from imagery obtained by various sensors carried in aircraft and satellites. Satellite imagery is unique because it affords a practical means of monitoring our entire planet on a regular basis.

Remote sensing imaging systems operate much the same as the human eye, but they can sense or "see" over a much broader range than humans. Cameras that expose various types of film are among the best types of remote sensing imaging systems. Nonphotographic systems such as *multispectral scanners* (MSS), *radiometers, side-looking airborne radar* (SLAR), and *passive microwave* are also employed. Their manner of operation, and some applications of the imagery, is briefly described here.

The sun and other sources emit a wide range of electromagnetic energy called the *electromagnetic spectrum*. X-rays, visible light rays, and radio waves are some familiar examples of energy variations within the electromagnetic spectrum. Energy is classified according to its wavelength (see Figure 27.20). Visible light (that energy to which our eyes are sensitive) has wavelengths from about 0.4 to 0.7 μm and thus, as illustrated in the figure, comprises only a very small portion of the spectrum.

Within the wavelengths of visible light, the human eye is able to distinguish different colors. The primary colors (blue, green, and red) consist of wavelengths

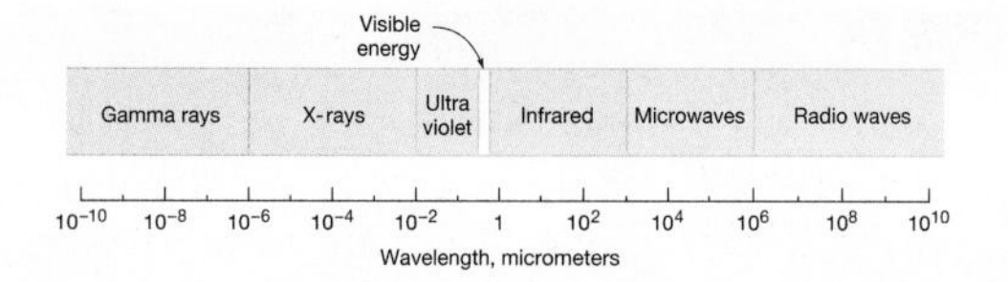

Figure 27.20 Classification of electromagnetic spectrum by wavelength.

in the ranges of 0.4–0.5, 0.5–0.6, and 0.6–0.7 μm, respectively. All other hues are combinations of the primary colors. To the human eye, an object appears a certain color because it reflects energy of wavelengths producing that color. If an object reflects all wavelengths of energy in the visible range, it will appear white, and if it absorbs all wavelengths, it will be black. If an object absorbs all green and red energy but reflects blue, that object will appear blue.

Just as the retina of the human eye can detect variations in wavelengths, photographic films or *emulsions* are also manufactured to have wavelength sensitivity variations. Normal color emulsions are sensitive to blue, green, and red energy; others respond to energy in the near-infrared range. These are called *infrared* (IR) emulsions. They make it possible to photograph energy that is invisible to the human eye. An early application of IR film was in camouflage detection, where it was found that dead foliation or green netting reflected infrared energy differently than normal vegetation, even though both appeared green to the human eye. Infrared film is now widely used for a variety of applications, such as detection of crop stress or identification and mapping of tree species.

Nonphotographic imaging systems used in remote sensing are able to detect energy variations over a broad range of the electromagnetic spectrum. MSS systems, for example, are carried in satellites and can operate within wavelengths from about 0.3–14 μm. In a manner similar to the way humans detect colors, MSS units isolate incoming energy into discrete spectral categories, or *bands,* then convert them into electric signals that can be represented by digits. These devices capture a digital image (see Section 27.3)—that is, the scanned scene below the satellite's path is recorded as a series of contiguous rows and columns of pixels. The digits associated with each pixel represent intensities of the various bands of energy within them. This digital format is ideal for computer processing and analysis and enables prints to be made by electronic processing. The bands of a scene can be analyzed separately, which is extremely useful in identifying and interpreting imaged objects. For certain applications, it is useful to combine two or more bands into a composite. The geometry of nonphotographic images differs from that of perspective photos, and thus methods for analyzing them also differ.[2]

Figure 27.21 shows an image obtained with the MSS system carried in an early Landsat satellite. It was taken at an altitude of 560 miles and shows a large portion of southeastern Wisconsin, including the city of Milwaukee. A portion of Lake Michigan is shown on the right side of the figure. Imagery of this type is useful for a variety of applications. As examples, geologic formations over large areas

[2]Information on the geometry of nonphotographic imaging systems can be found in Lillesand and Kiefer, *Remote Sensing and Image Interpretation.* (See the bibliography at the end of this chapter.)

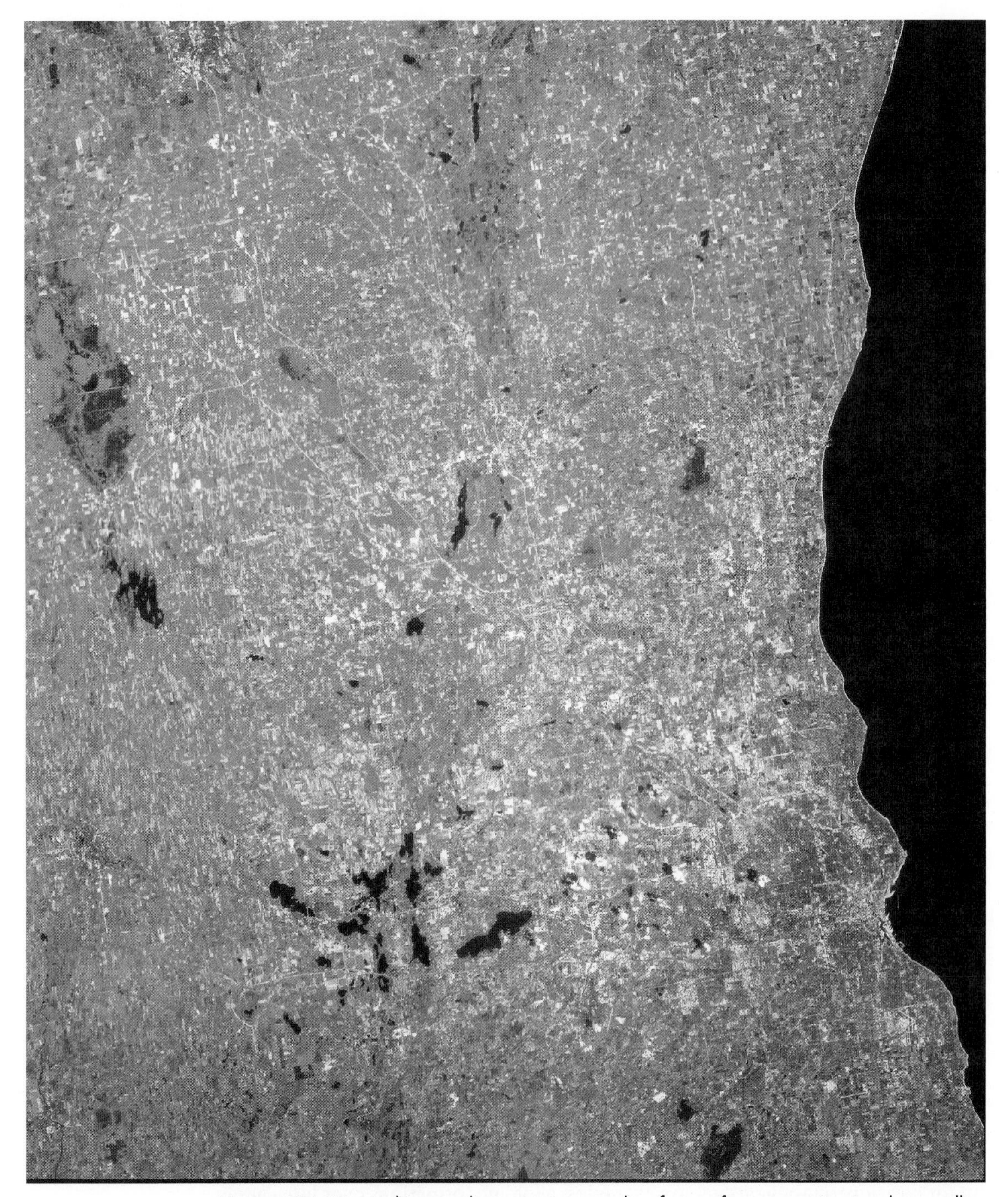

Figure 27.21 Multispectral scanner image taken from a first-generation Landsat satellite over Milwaukee, Wisconsin.

can be studied; the number of lakes in the area, their relative positions, shapes, and acreages can be determined readily; acreages of croplands and forests, with a breakdown of coniferous and deciduous tree types, can be obtained; and small-scale planimetric maps showing these different land-use classifications can be prepared. All of these tasks are amenable to computer processing.

The resolution of the earlier Landsat MSS imaging systems (like that used to acquire Figure 27.21) was 80 m; each pixel represents a square of 80 m on the ground. This was improved to 30 m for the *Thematic Mapper* (TM) imaging systems carried in the second generation of Landsat satellites. Figure 27.22 shows an image taken by Landsat TM near Madison, Wisconsin. It clearly illustrates the improved resolution as compared to Figure 27.21. Note, for example, the clarity of the roads and the detail with which urban areas and agricultural crops are shown. Images of this type have been applied for land-use mapping; measuring and monitoring various agricultural crops; mapping soils; detecting diseased crops and trees; locating forest fires; studying wildlife; mapping the effects of natural disasters such as tornadoes, floods, and earthquakes; analyzing population growth and distribution; determining the locations and extent of oil spills; monitoring water quality and detecting the presence of pollutants; and accomplishing numerous other tasks over large areas for the benefit of humankind.

In recent years a significant amount of research and development has been directed towards the design of satellite imaging systems with improved resolution and geometric properties. The goals have been to enable smaller objects to be identified and analyzed, and to improve the mapping capabilities of the systems, thus making them useful for many additional types of applications. The *Enhanced Thematic Mapper Plus* (ETM+) imaging system aboard the Landsat Satellite (Landsat 7), which was launched in April 1999, has a resolution capability of 15 m. This satellite operates at an altitude of 438 mi, and the imaging system covers a ground swath beneath its orbit that is 115 mi wide.[3] The French *Systeme Pour d'Observation de la Terre* (SPOT) satellite has an imaging system with 10 m resolution and a nadir ground swath width of 37 miles. Its imaging system can be aimed at angles up to 27° off-nadir. With this feature, the same areas that were imaged on earlier passes can be covered again on subsequent passes from different orbits, thereby achieving stereoscopic coverage. This imagery is therefore suitable not only for small-scale planimetric mapping, but also for determining elevations.[4]

In September 1999, Space Imaging launched the first commercial imaging satellite, IKONOS.[5] The most remarkable characteristic of its imaging system is its resolution of 1 m. This satellite is in orbit at an altitude of 423 miles above Earth, and the width of its image at the nadir is approximately 7 miles. The imaging system can be aimed off-nadir, either from side to side, or fore and aft. This not only enables pinpointing coverage on areas of interest, but it can also be used to obtain stereoscopic coverage. Thus the imagery is suitable for both planimetric

---

[3]Images from Landsat 7 and all previous Landsat satellites are available from the U.S. Geological Survey, Earth Science Information Center (ESIC), 12201 Sunrise Valley Drive, Reston, VA 20192. Contact can also be made by calling (888) ASK-USGS [(888) 275-8747] or at http://mapping.usgs.gov/esic/.

[4]Information on SPOT imagery can be obtained at http://www.spotimage.fr.

[5]Information about IKONOS satellite imagery can be obtained from Space Imaging by calling (800) 232-9037 or by visiting http://www.spaceimaging.com.

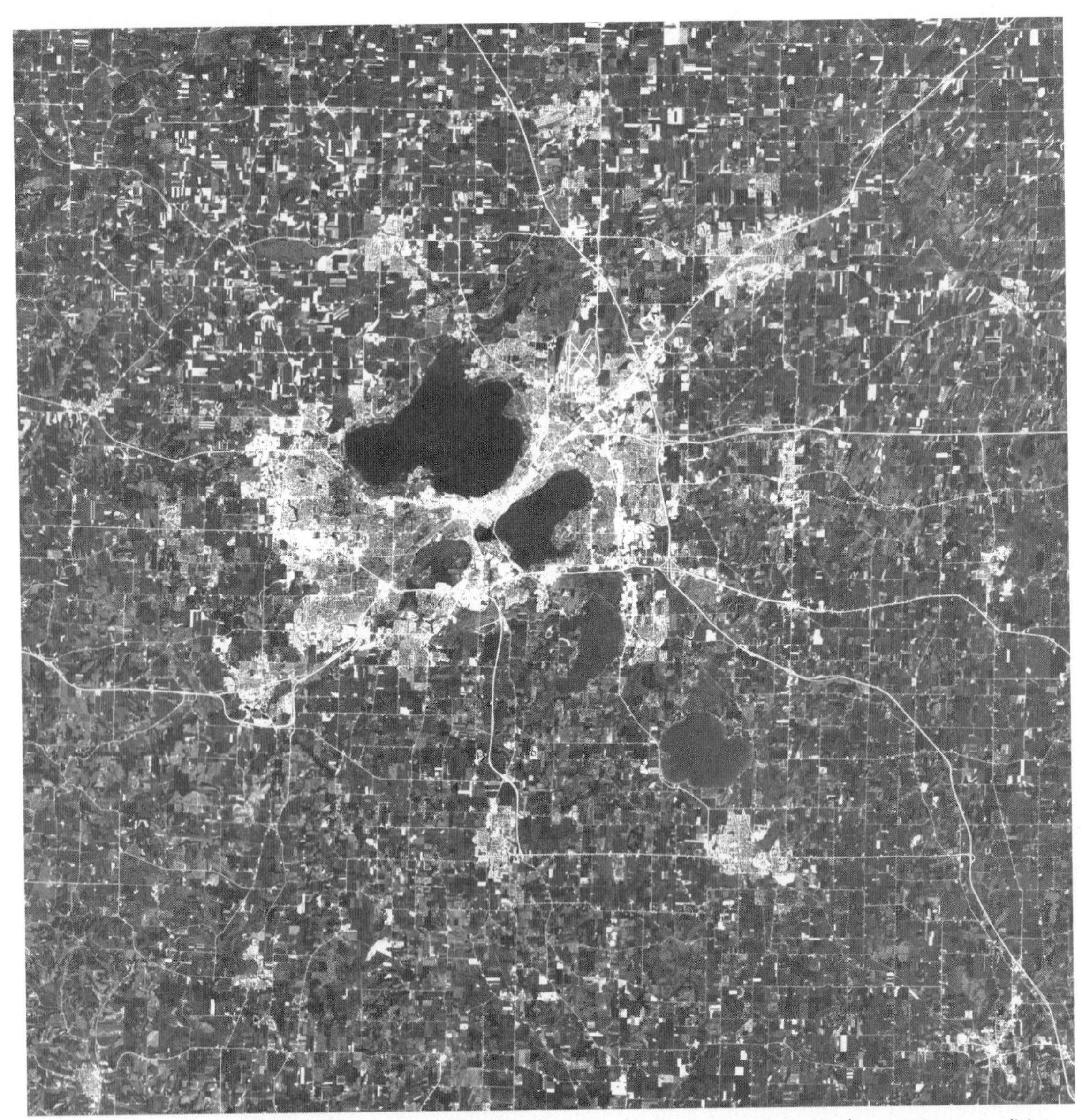

Figure 27.22 Landsat Thematic Mapper image taken near Madison, Wisconsin. (Note the improved resolution compared to the Landsat MSS image of Figure 27.21.) (Courtesy Environmental Remote Sensing Center, University of Wisconsin—Madison)

mapping and for determining elevations of ground points. Figure 27.23 is a 1-m image taken from IKONOS on October 11, 1999, over the city of San Francisco. It features Aquatic Park and Fisherman's Wharf. Note that the level of detail that can be resolved from this image is far superior to those of Figures 27.21 and 27.22. Individual houses, trees, small boats, and even automobiles can be identified. This imagery is useful for many types of applications. To mention just a few

Figure 27.23 1-m resolution image obtained from the IKONOS satellite showing Aquatic Park and Fisherman's Wharf in San Francisco. (Courtesy Spaceimaging.com)

possibilities of interest to surveyors and engineers, the imagery is suitable for (a) preparing site maps and preliminary plans for proposed construction projects; (b) planning for locations of GPS stations in large-area control surveys, or planning aerial photography; (c) generating detailed information layers such as land cover, hydrography, transportation networks, etc., for use in geographic information systems; and (d) using the images as reference frames for performing GIS analyses. There are also a variety of additional applications in many other fields such as forestry, geology, agriculture, etc.

In the future, those engaged in surveying (geomatics) will be called on to prepare maps and to extract a variety of other positional types of information from satellite images. Remote sensing will play a significant future role in providing data to assess the impacts of human activities on our air, water, and land resources. It can provide valuable information to assist in making sound decisions and formulating policies related to resource management, land-use, and land development activities.

27.20 SOURCES OF ERROR IN PHOTOGRAMMETRY

Some sources of error in photogrammetric work are:

1. Measuring instruments not standardized or calibrated.
2. Inaccurate location of principal and corresponding principal points.
3. Failure to use camera calibration data.
4. Assumption of vertical exposures when photographs are actually tilted.
5. Presumption of equal flying heights when they were unequal.
6. Disregard of differential shrinkage or expansion of photographic prints.
7. Incorrect orientation of photographs under a stereoscope or in a stereoscopic plotter.
8. Faulty setting of the floating mark on a point.

27.21 MISTAKES

Some mistakes that occur in photogrammetry are:

1. Incorrect reading of measuring scales.
2. Mistake of units—for example, inches instead of millimeters.
3. Confusion in identifying corresponding points on different photographs.
4. Failure to provide proper control or use of erroneous control coordinates.
5. Attachment of an incorrect sign (plus or minus) to a measured photographic coordinate.
6. Blunder in computations.
7. Misidentification of control-point images.

PROBLEMS

Asterisks (*) indicate problems that have answers given in Appendix G.

27.1 Describe the difference between vertical, low-oblique, and high-oblique aerial photos.

27.2 Discuss the advantages of softcopy stereoplotters over optical stereoplotters.

27.3 Define the terms (a) metric photogrammetry and (b) interpretative photogrammetry.

27.4 Describe briefly how a digital camera operates.

27.5 The distance between two points on a vertical photograph is ab and the corresponding ground distance is AB. For the following data, compute the average photographic scale along the line ab.

*__(a)__ ab = 2.41 in.; AB = 4820 ft

(b) ab = 5.29 in.; AB = 13,218 ft

(c) ab = 107.389 mm; AB = 536.943 m

27.6 On a vertical photograph of flat terrain, section corners appear a distance d apart. If the camera focal length is f compute flying height above average ground in feet for the following data:

(a) d = 1.85 in.; f = 3-1/2 in.

(b) d = 82.184 mm; f = 153.20 mm

27.7 On a vertical photograph of flat terrain, the scaled distance between two points is ab. Find the average photographic scale along ab if the measured length between the same line is AB on a map plotted at a scale of S_{map} for the following data.

(a) ab = 1.47 in.; AB = 3.52 in.; S_{map} = 1:6000

(b) ab = 41.53 mm; AB = 6.23 mm; S_{map} = 1:20,000

27.8 What are the average scales of vertical photographs for the following data, given flying height above sea level H, camera focal length f, and average ground elevation h?

*__(a)__ H = 7300 ft; f = 152.4 mm; h = 1250 ft

(b) H = 6980 ft; f = 6.000 in.; h = 1004 ft

(c) H = 2610 m; f = 152.4 mm; h = 324 m

27.9 The length of a football field from goalpost to goalpost scales 49.15 mm on a vertical photograph. Find the approximate dimensions (in meters) of a large rectangular building that also appears on this photo and whose sides measure 20.5 mm by 6.8 mm. (*Hint:* Football goalposts are 120 yards apart.)

27.10* Compute the area in acres of a triangular parcel of land whose sides measure 48.78 mm, 84.05 mm, and 69.36 mm on a vertical photograph taken from 6050 ft above average ground with a 152.4 mm focal length camera.

27.11 Calculate the flight height above average terrain that is required to obtain vertical photographs at an average scale of S if the camera focal length is f for the following data:

(a) S = 1: 8000; f = 152.4 mm

(b) S = 1: 6000; f = 88.9 mm

27.12 Determine the horizontal distance between two points A and B whose elevations above datum are h_A = 1560 ft and h_B = 1425 ft, and whose images a and b on a vertical photograph have photo coordinates x_a = 2.95 in., y_a = 2.32 in., x_b = −1.64 in., and y_b = −2.66 in. The camera focal length was 152.4 mm and the flying height above datum 7500 ft.

27.13* Similar to Problem 27.12, except that the camera focal length was 3-1/2 in., the flying height above datum 4075 ft, and elevations h_A and h_B, 983 ft and 1079 ft, respectively. Photo coordinates of images a and b were x_a = 108.81 mm, y_a = −73.73 mm, x_b = −87.05 mm, and y_b = 52.14 mm.

27.14 On the photograph of Problem 27.12, the image c of a third point C appears. Its elevation h_C = 1365 ft, and its photo coordinates are x_c = 2.96 in. and y_c = −3.02 in. Compute the horizontal angles in triangle ABC.

27.15 On the photograph of Problem 27.12, the image d of a third point D appears. Its elevation is h_D = 1195 ft, and its photo coordinates are x_d = 56.86 mm and y_d = 63.12 mm. Calculate the area, in acres, of triangle ABD.

27.16 Determine the height of a radio tower, which appears on a vertical photograph for the following conditions of flying height above the tower base H, distance on the photograph from principal point to tower base r_b, and distance from principal point to tower top r_t.

*__(a)__ H = 2425 ft; r_b = 3.18 in.; r_t = 3.34 in.

(b) H = 6600 ft; r_b = 96.83 mm; r_t = 98.07 mm

27.17 On a vertical photograph, images a and b of ground points A and B have photographic coordinates x_a = 3.27 in., y_a = 2.28 in., x_b = −1.95 in., and y_b = −2.50 in. The horizontal distance between A and B is 5283 ft, and the elevations of A and B above datum are 646 ft and 756 ft, respectively. Using Equation (27.9), calculate the flying height above datum for a camera having a focal length of 152.4 mm.

27.18 Similar to Problem 27.17, except x_a = −52.53 mm, y_a = 69.67 mm, x_b = 26.30 mm, y_b = −59.29 mm, line length AB = 4706 ft, and elevations of points A and B are 925 and 875 ft, respectively.

27.19* An air base of 3205 ft exists for a pair of overlapping vertical photographs taken at a flying height of 5500 ft above MSL with a camera having a focal length of 152.4 mm. Photo coordinates of points A and B on the left photograph are

$x_a = 40.50$ mm, $y_a = 42.80$ mm, $x_b = 23.59$ mm, and $y_b = -59.15$ mm. The x photo coordinates on the right photograph are $x_a = -60.68$ mm and $x_b = -70.29$ mm. Using the parallax equations, calculate horizontal length AB.

27.20 Similar to Problem 27.19, except the air base is 6940 ft, the flying height above mean sea level is 12,520 ft, the x and y photo coordinates on the left photo are $x_a = 37.98$ mm, $y_a = 50.45$ mm, $x_b = 24.60$ mm, and $y_b = -46.89$ mm, and the x photo coordinates on the right photo are $x_a = -52.17$ mm and $x_b = -63.88$ mm.

27.21 Calculate the elevations of points A and B in Problem 27.19.

27.22 Compute the elevations of points A and B in Problem 27.20.

27.23 List and briefly describe the four different categories of stereoscopic plotting instruments.

27.24 Name the three stages in stereoplotter orientation, and briefly explain the objectives of each.

27.25 What advantages does a softcopy plotter have over an analytical plotter?

27.26 What kind of images do softcopy stereoplotters require? Describe two different ways they can be obtained.

27.27 Compare an orthophoto with a conventional line and symbol map.

27.28 Discuss the advantages of orthophotos as compared to maps.

Aerial photography is to be taken of a tract of land that is X mi square. Flying height will be H ft above average terrain, and the camera has focal length f. If the focal plane opening is 9 × 9 in., and minimum sidelap is 30 percent, how many flight lines will be needed to cover the tract for the data given in Problems 27.29 and 27.30?

27.29* $X = 8$; $H = 4000$; $f = 152.4$ mm.

27.30 $X = 30$; $H = 10{,}000$; $f = 6$ in.

Aerial photography was taken at a flying height H ft above average terrain. If the camera focal plane dimensions are 9 × 9 in., the focal length is f and the spacing between adjacent flight lines is X ft, what is the percent sidelap for the data given in Problems 27.31 and 27.32?

27.31* $H = 4500$; $f = 152.4$ mm; $X = 4700$.

27.32 $H = 6800$; $f = 88.9$ mm; $X = 13{,}500$.

Photographs at a scale of S are required to cover an area X mi square. The camera has a focal length f and focal plane dimensions of 9 × 9 in. If endlap is 60 percent and sidelap 30 percent, how many photos will be required to cover the area for the data given in Problems 27.33 and 27.34?

27.33 $S = 1{:}\,6000$; $X = 6$; $f = 152.4$ mm

27.34 $S = 1{:}\,14{,}400$; $X = 40$; $f = 89.0$ mm.

27.35 Describe a system that employs GPS and that can reduce or eliminate ground control surveys in photogrammetry?

27.36 To what wavelengths of electromagnetic energy is the human eye sensitive? What wavelengths produce the colors blue, green, and red?

27.37 Discuss the uses and advantages of satellite imagery.

Problems 27.38 through 27.42 involve using WOLFPACK with images 5 and 6 on the CD that accompany this book. The ground coordinates of the paneled points are listed in the file "ground.crd." The coordinates of the fiducials are listed in the file "camera.fid." To do these problems, digitize the eight fiducials and paneled points 21002, 4, 41, GYM, WIL1A, WIL1B, and RD on both images. After digitizing the points, perform an interior orientation to compute photo coordinates for the points on images 5 and 6. The focal length of the camera is 153.742 mm.

27.38 Using photo coordinates for points 4 and GYM on image 5, determine the scale of the photo.

27.39 Using photo coordinates for points 4 and GYM on image 5, determine the flying height of the camera at the time of exposure.

27.40 Using photo coordinates for points 4 and GYM on image 5 and 6, determine the ground coordinates of points WIL1A and WIL1B using Eq. (27.12) and Eq. (27.13).

27.41 Using the exterior orientation option in WOLFPACK, determine the exterior orientation elements for image 5.

27.42 Using the exterior orientation option in WOLFPACK, determine the exterior orientation elements for image 6.

BIBLIOGRAPHY

Al-Tahir, Raid and Asim Ali. 2004. "Assessing Land Cover Changes in the Coastal Zone Using Aerial Photography." *Surveying and Land Information Science* 64 (No. 2): 107.

American Society for Photogrammetry and Remote Sensing. 2004. *Manual of Photogrammetry*, 5th ed. Bethesda, MD.

———. 1997. *Manual of Remote Sensing, A Series, Earth Observing Platforms and Sensors,* 3rd ed., (CD Rom), Bethesda, MD.

———. 1996. *Digital Photogrammetry: An Addendum to the Manual of Photogrammetry,* Bethesda, MD.

Bartorelli, J. 2002. "Photogrammetry 101." *Point of Beginning* 27 (No. 11): 36.

Butler, J. and J. Olson. 2005. "Northwest Team Refines LiDAR for Airport Obstruction Surveys." *Professional Surveyor* 25 (No. 3): 8.

Crabtree, J. 2006. "The Enduring Importance of Standards in Aerial Mapping." *Professional Surveyor Supplement* 26 (No. 3): 4.

Craun, K. 2006. "ASPRS: Serving the Geospatial Community for 72 Years." *Professional Surveyor* 26 (No. 5): 22.

Kunta-Mensah, P. and R. Hintz. 2001. "Airborne GPS—Photogrammetry Comes of Age." *Surveying and Land Information Systems* 61 (No. 2): 93.

Liberty, E. and J. Barnard. 2006. "History in the Making: Intelligent Lidar Imaging and Surveying for Historical Site Preservation." *Professional Surveyor* 26 (No. 10): 24.

Lillesand, T. M., et al. 2003. *Remote Sensing and Image Interpretation,* 5th ed. New York: John Wiley & Sons.

Meade, M. 2002. "Mapping at a Grand Scale." *Point of Beginning* 27 (No. 11): 32.

Mugnier, C. 2006. "Its Not Just Your Father's Photogrammetry Anymore." *Professional Surveyor* 26 (No. 5): 8.

Pappa, R. S. 2002. "Close-range Photogrammetry and Next Generation Spacecraft." *Professional Surveyor* 22 (No. 6): 6.

Schuckman, K. 2006. "Aerial Perspective: Licensure of Photogrammetrists." *Professional Surveyor* 26 (No. 4): 32.

Smith, D. and D. Roman. 2001. "A New High Resolution DEM for the Northwest United States." *Surveying and Land Information Systems* 61 (No. 2): 103.

Smith, Scot, et al. 2004. "Assessment of the Use of Remote Sensing Techniques for Locating and Mapping Ordinary High Water Lines." *Surveying and Land Information Science* 64 (No. 2): 113.

Tan, Willie. 2004. "The 3-point Resection Problem in Photogrammetry." *Surveying and Land Information Science* 64 (No. 3): 177.

Wolf, P. R., and B. A. Dewitt, 2000. *Elements of Photogrammetry: With Applications in GIS,* 3rd ed. New York: McGraw-Hill.

28

Introduction to Geographic Information Systems

■ 28.1 INTRODUCTION

The term *Geographic Information System* (GIS) first appeared in published literature in the mid-1960s. But although the term is relatively new, many of its concepts have long been in existence. For example, the *map overlay* concept, which is one of the important tools used in GIS spatial analysis (see Section 28.9), was used by French cartographer Louis-Alexandre Berthier more than 200 years ago. He prepared and overlaid a series of maps to analyze troop movements during the American Revolution. Dr. John Snow demonstrated another early example illustrating the use and value of the overlay concept in 1854. He overlaid a map of London showing where cholera deaths had occurred with another, giving locations of wells in that city to demonstrate the relationship between those two data sets. These early examples illustrate fundamentals that still comprise the basis of our modern GIS—that is, making decisions based on the simultaneous analysis of data of differing types, all located spatially in a common geographic reference system. However, the full capabilities and benefits of our modern GISs could not occur until the advent of the computer.

In general, a geographic information system can be defined as a system of hardware, software, data and organizational structure for collecting, storing, manipulating, and spatially analyzing "geo-referenced" data and displaying information resulting from those processes. A more detailed definition (Hanigan, 1988) describes a GIS as "any information management system that can:

1. Collect, store, and retrieve information based on its spatial location;
2. Identify locations within a targeted environment that meet specific criteria;

3. Explore relationships among data sets within that environment;
4. Analyze the related data spatially as an aid to making decisions about that environment;
5. Facilitate selecting and passing data to application-specific analytical models capable of assessing the impact of alternatives on the chosen environment; and
6. Display the selected environment both graphically and numerically either before or after analysis."

The thread that is common to both definitions just given is that in a GIS, decisions are made based on spatial analyses performed on data sets that are referenced in a common geographical system. The geographic referencing system used could be a state plane or UTM coordinate system, latitude and longitude, or other suitable local coordinate system. In any GIS, the accuracy of the spatial analyses and, hence, the validity of decisions reached as a result of those analyses are directly dependent on the quality of the spatially related data used. It is therefore important to realize at the outset of this chapter that the surveyor's role in developing accurately positioned data sets is a critical one in GIS activity.

A generalized concept of how data of different types or "layers" are collected and overlaid in a GIS is illustrated in Figure 28.1. In that figure, maps A through G represent some of the different layers of spatially related information that can be digitally recorded and incorporated into a GIS database and that include parcels of different landownership A, zoning B, floodplains C, wetlands D,

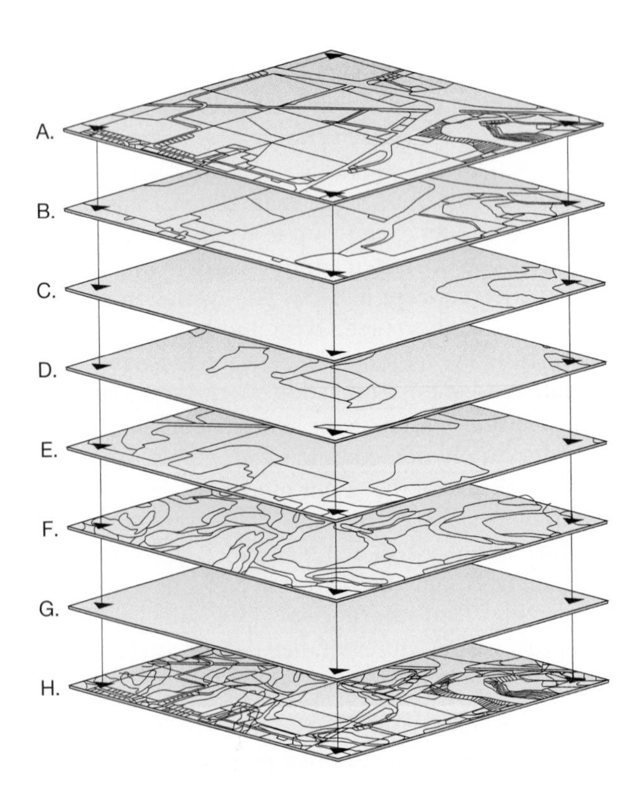

Figure 28.1 Concept of layers in a geographic information system. Map layers shown are (A) parcels; (B) zoning; (C) floodplains; (D) wetlands; (E) land cover; (F) soils; (G) reference framework; and (H) composite overlay. (Courtesy Land Information and Computer Graphics facility, College of Agricultural and Life Sciences, University of Wisconsin-Madison)

land cover E, and soil types F. Map G is the geodetic reference framework, consisting of the network of survey control points in the area. Note that these control points are found in each of the other layers thereby providing the means for spatially locating all data in a common reference system. Thus composite maps that merge two or more different data sets can be accurately created. For example in Figure 28.1, bottom map H is the composite of all layers.

A GIS merges conventional database management software with software for manipulating spatial data. This combination enables the simultaneous storage, retrieval, overlay, and display of many different spatially related data sets in the manner illustrated by Figure 29.1. These capabilities, coupled with sophisticated GIS software to analyze and query the data sets that result from these different overlay and display combinations, provide answers to questions that never before were possible to obtain. As a result, GISs have become extremely important in planning, design, impact assessment, predictive modeling, and many other applications.

GISs have been applied in virtually every imaginable field of activity, from engineering to agriculture, and from the medical science of epidemiology to wildlife management. Flood forecasting on a large regional basis, such as statewide, is one particular example that illustrates some of the benefits that can be derived from using GIS. Critical location-related data entered into the GIS to support statewide flood forecasting would include the state's topography; soil; land cover; number, sizes, and locations of drainage basins; existing stream network with streamgauging records; locations and sizes of existing bridges, culverts, and other drainage structures; data on existing dams and the water impoundment capacities of their associated reservoirs; and records of past rainfall intensity and duration. Given these and other data sets, together with a model to estimate runoff, a computer can be used to perform an analysis and predict locations of potential floods and their severity. In addition, experiments can be conducted in which certain input can be varied. Examples may include (1) input of an extremely intense rainfall for a lengthy duration in a given area to assess the magnitude of the resulting flood and (2) the addition of dams and other flood-detention structures of varying sizes at specific locations, to analyze their impact on mitigating the disaster. Many other similar examples could be given. Obviously, GISs are very powerful tools, and their use will increase significantly in the future.

The successful implementation of GISs relies on people with backgrounds and skills in many different disciplines, but none are more important than the contributions of those engaged in surveying (geomatics). Virtually every aspect of surveying, and thus all material presented in the preceding chapters of this book, bear upon GIS development, management, and use. However, of special importance are the global positioning system (Chapters 13 through 15), topographic surveying (Chapter 17), mapping (Chapter 18), control surveys and reference frames (Chapter 19), map projection coordinate systems (Chapter 20), boundary or cadastral surveys (Chapter 21), the U.S. Public Land Survey System (Chapter 22), and photogrammetry and remote sensing (Chapter 27). In addition to surveying (geomatics) specialists, personnel in the fields of computer science, geography, soil science, forestry, landscape architecture, and many others play important roles in GIS development.

28.2 LAND INFORMATION SYSTEMS

The terms *Geographic Information System* (GIS) and *Land Information System* (LIS) are sometimes used interchangeably. They do have many similarities, but the distinguishing characteristic between the two is that a LIS has its focus directed primarily toward land records data. Information stored within a LIS for a given locality would include a spatial database of land parcel information derived from property descriptions in the U.S. Public Land System; other types of legal descriptions such as metes and bounds or block and lot that apply to parcels in the area; and other cadastral data. It might include the actual deeds and other records linked to the spatial data. Information on improvements and parcel values would also be included.

Land Information Systems and Geographic Information Systems can share data sources such as control networks, parcel ownership information, and municipal boundaries. However, a GIS will usually incorporate data over a broader range and might include layers such as topography, soil types, land cover, hydrography, depth to groundwater, and so on. Because of its narrower focus, there is a tendency to consider an LIS as a subset of a GIS.

LISs are used to obtain answers to questions about who has ownership or interests in the land in a certain area, the particular nature of those interests, and the specific land affected by them. They can also provide information about what resources and improvements exist in a given area and give their values. Answers to these questions are essential in making property assessments for taxation, transferring title to property, mortgaging, making investment decisions, resolving boundary disputes, and developing roads, utilities, and other services on the land that require land appraisal and property acquisitions. The data are also critical in policy development and land-use planning.

28.3 GIS DATA SOURCES AND CLASSIFICATIONS

As noted earlier, the capabilities and benefits of any GIS are directly related to the content and integrity of its database. Data that are entered into a GIS come from many sources and may be of varying quality. To support a specific GIS, a substantial amount of new information will generally need to be gathered expressly for its database. More than likely, however, some of the data will be obtained from existing sources such as maps, engineering plans, aerial photos, satellite images, and other documents and files that were developed for other purposes. Building the database is one of the most expensive and challenging aspects of developing a GIS. In fact, it has been estimated that this activity may represent from about 60–80 percent of the total cost of implementing a GIS.

Two basic data classifications are used in GISs: (1) *spatial* and (2) *nonspatial*. These are described in the sections that follow.

28.4 SPATIAL DATA

Spatial data, sometimes interchangeably called *graphic data,* consists in general of natural and cultural features that can be shown with lines or symbols on maps, or seen as images on photographs. In a GIS these data must be represented and spatially located, in digital form, using a combination of fundamental elements called "simple spatial objects." The formats used in this representation are either *vector* or *raster*. The "relative spatial relationships" of the simple spatial objects are given by their *topology*.

These important GIS topics: (1) simple spatial objects, (2) data formats, and (3) topology are described in the following subsections.

28.4.1 Simple Spatial Objects

The simple spatial objects most commonly used in spatially locating data are illustrated in Figure 28.2 and described as follows:

1. *Points* define single geometric locations. They are used to locate features such as houses, wells, mines, or bridges [see Figure 28.2(a)]. Their coordinates give the spatial locations of points, commonly in state plane or UTM systems (see Chapter 20).
2. *Lines* and *strings* are obtained by connecting points. A line connects two points, and a string is a sequence of two or more connected lines. Lines and strings are used to represent and locate roads, streams, fences, property lines, etc. [see Figure 28.2(b)].
3. *Interior areas* consist of the continuous space within three or more connected lines or strings that form a closed loop [see Figure 28.2(c)]. For example, interior areas are used to represent and locate the limits of governmental jurisdictions, parcels of landownership, different types of land cover, or large buildings.
4. *Pixels* are usually tiny squares that represent the smallest elements into which a digital image is divided [see Figure 28.2(d)]. Continuous arrays of pixels, arranged in rows and columns, are used to enter data from aerial photos, orthophotos, satellite images, etc. Assigning a numerical value to each pixel specifies the distributions of colors or tones throughout the image. Pixel size can be varied and is usually specified by the number of *dots per inch* (dpi). As an example, 100 dpi would correspond to squares having dimensions of 1/100 in. on each side. Thus 100 dpi yields 10,000 pixels per square inch.

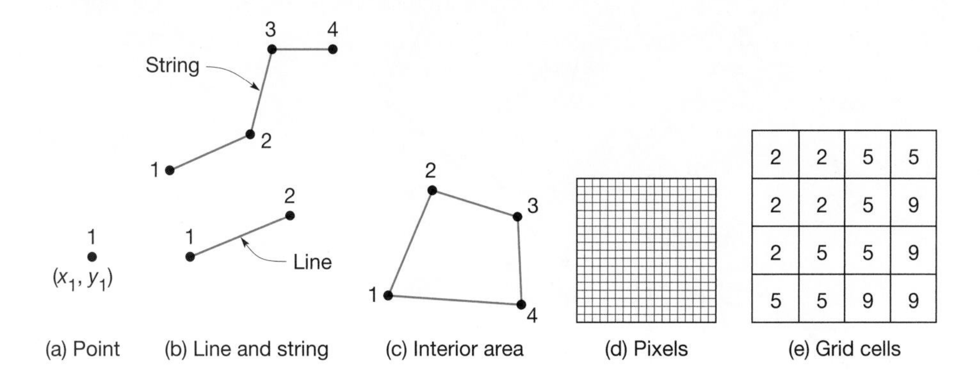

Figure 28.2 Simple spatial objects used to represent data digitally in a GIS.

5. *Grid cells* are single elements, usually square, within a continuous geographic variable. Similar to pixels, their sizes can be varied, with smaller cells yielding improved resolution. Grid cells may be used to represent slopes, soil types, land cover, water table depths, land values, population density, and so on. The distribution of a given data type within an area is indicated by assigning a numerical value to each cell—for example, showing soil types in an area using the number 2 to represent sand, 5 for loam, and 9 for clay, as illustrated in Figure 28.2(e).

28.4.2 Vector and Raster Formats

The simple spatial objects described in Section 28.4.1 give rise to two different formats for storing and manipulating spatial data in a GIS—*vector* and *raster*. When data are depicted in the vector format, a combination of points, lines, strings, and interior areas is used. The raster format uses pixels and grid cells.

In the vector format, points are used to specify locations of objects such as survey control monuments, utility poles, or manholes; lines and strings depict linear features such as roads, transmission lines, or boundaries; and interior areas show regions having common attributes; for example, governmental entities or areas of uniform land cover. An example illustrating the vector format is given with Figure 28.3 and Table 28.1. Figure 28.3 shows two adjacent land parcels, one designated parcel I, owned by Smith, and the other identified as parcel II, owned by Brown. As shown, the configuration consists of points, lines, and areas.

Vector representation of the data can be achieved by creating a set of tables that list these points, lines, and areas (Table 28.1). Data within the tables are linked using *identifiers* and related spatially through the coordinates of points. As illustrated in Table 28.1(a), all points in the area are identified by a reference number. Similarly, each line is described by its endpoints, as shown in Table 28.1(b), and

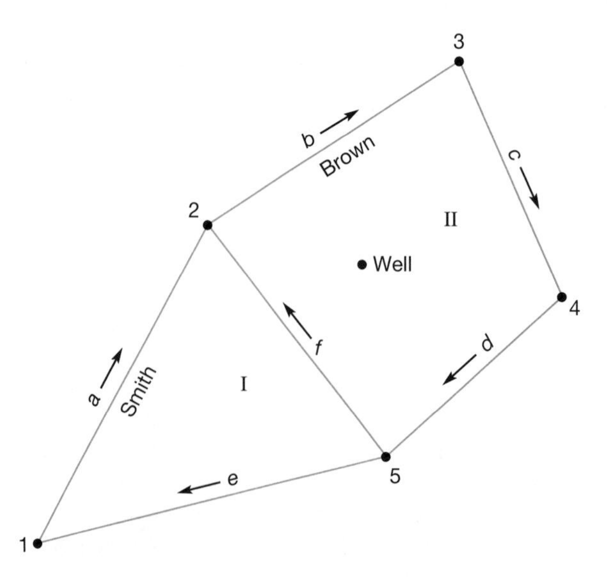

Figure 28.3 Vector representation of a simple graphic record.

TABLE 28.1 VECTOR REPRESENTATION OF FIGURE 28.3

(a)		(b)		(c)	
Point Identifier	**Coordinates**	**Line Identifier**	**Points**	**Area Identifier**	**Lines**
1	$(X, Y)_1$	*a*	1, 2	I	*a, f, e*
2	$(X, Y)_2$	*b*	2, 3	II	*b, c, d, f*
3	$(X, Y)_3$	*c*	3, 4		
4	$(X, Y)_4$	*d*	4, 5		
5	$(X, Y)_5$	*e*	5, 1		
Well	$(X, Y)_{well}$	*f*	5, 2		

the endpoint coordinates locate the various lines spatially. As shown in Table 28.1(c), areas in Figure 28.3 are defined by the lines that enclose them. As before, coordinates of line endpoints locate the areas and enable determining their locations and computing their magnitudes.

Another data type can also be represented in vector formats. For example, consider the simple case illustrated with the land cover map shown in Figure 28.4(a). In that figure, areas of different land cover (forest, marsh, etc.) are shown with standard topographic symbols (see Figure 18.5). A vector representation of this region is shown in Figure 28.4(b). Here lines and strings locate boundaries of regions having a common land cover. The stream consists of the string connecting points 1 through 10. By means of tables similar to Table 28.1 the data of this figure can also be entered into a GIS using the vector format. Having considered these simple vector representations, imagine the magnitude and complexity of entering data in vector format to cover a much larger area such as that shown in the map of Figure 17.2.

As an alternative to the vector approach, data can be depicted in the raster format using grid cells (or pixels if the data are derived from images). Each equal-sized cell (or pixel) is uniquely located by its row and column numbers and is coded with a numerical value or *code* that corresponds to the properties of the specific area it covers. In the raster format, a point would be indicated with a single grid cell, a line would be depicted as a sequence (linear array) of adjacent grid cells having the same code, and an area having common properties would be shown as a group of identically coded contiguous cells. Therefore, it should be appreciated that in general the raster method yields a coarser level of accuracy or definition of points, lines, and areas than the vector method.

In the raster format, the size of the individual cells defines the *resolution,* or precision, with which data are represented. Smaller the area covered by each cell, the higher the resolution for any given image. Examples illustrating raster representation and the degradation of resolution with increasing grid size are given in Figures 28.4(c) and 28.4(d). In these figures the land cover data from Figure 28.4(a) have been entered as two different raster data sets. Each cell has been assigned a

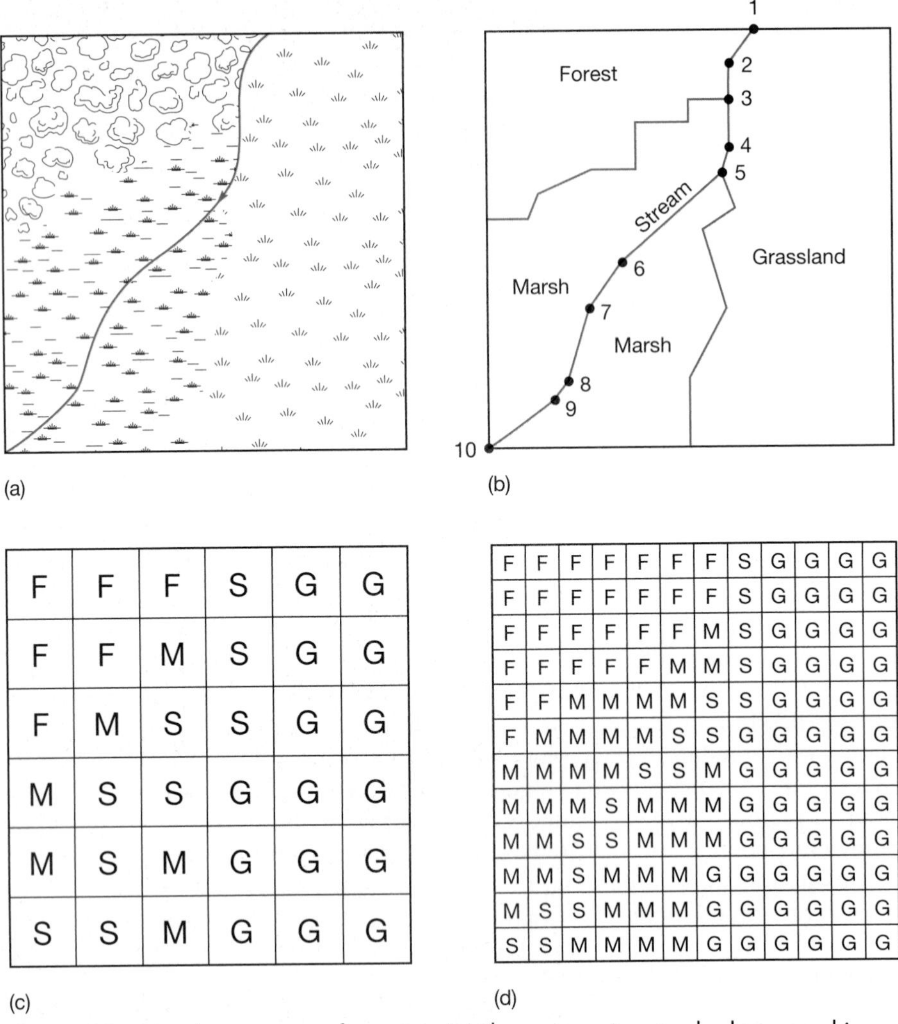

Figure 28.4 Land cover maps of a region. (a) The region using standard topographic symbols. (b) Vector representation of the same region. (c) Raster representation of the region using a coarse-resolution grid cell. (d) Raster representation using a finer-resolution grid cell.

value representing one of the land cover classes, that is, *F* for forest, *G* for grassland, *M* for marsh, and *S* for stream. Figure 28.4(c) depicts the area with a relatively large-resolution grid and, as shown, it yields a coarse representation of the original points, lines, and areas. With a finer-resolution grid, such as in Figure 28.4(d), the points, lines, and areas are rendered with more precision. However, it is important to note that as grid resolution increases, so does the volume of data (number of grid cells) required to enter the data.

Despite the coarser resolution present in a raster depiction of spatial features, this format is still often used in GISs. One reason is that many data are available in raster format. Examples include aerial photos, orthophotos, and satellite images. Another reason for the popularity of the raster format is the ease with which it enables collection, storage, and manipulation of data using computers. Furthermore, various refinements of raster images are readily made using available "image processing" software programs. Finally, for many data sets such as wetlands and soil types, boundary locations are rather vague and the use of the raster format does not adversely affect the data's inherent accuracy.

28.4.3 Topology

Topology is a branch of mathematics that describes how spatial objects are related to each other. The unique sizes, dimensions, and shapes of the individual objects are not addressed by topology. Rather, it is only their *relative relationships* that are specified.

In discussing topology, it is necessary to first define *nodes, chains,* and *polygons*. These are some additional simple spatial objects that are commonly used for specifying the topological relationships of information entered into GIS databases. Nodes define the beginnings and endings of chains, or identify the junctions of intersecting chains. Chains are similar to lines (or strings) and are used to define the limits of certain areas or delineate specific boundaries. Polygons are closed loops similar to areas and are defined by a series of connected chains. Sometimes in topology, single nodes exist within polygons for labeling purposes.

In GISs the most important topological relationships are:

1. *Connectivity*. Specifying which chains are connected at which nodes.
2. *Direction*. Defining a "from node" and a "to node" of a chain.
3. *Adjacency*. Indicating which polygons are adjacent on the left and which are adjacent on the right side of a chain.
4. *Nestedness*. Identifying what simple spatial objects are within a polygon. They could be nodes, chains, or other smaller polygons.

The topological relationships just described are illustrated and described by example with reference to Figure 28.3. For example in the figure, through connectivity, it is established that nodes 2 and 3 are connected to form the chain labeled *b*. Connectivity would also indicate that at node 2, chains *a*, *b*, and *f* are connected. Topological relationships are normally listed in tables and stored within the database of a GIS. Table 28.2(a) summarizes all of the connectivity relationships of Figure 28.3.

Directions of chains are also indicated topologically in Figure 28.3. For example, chain *b* proceeds from node 2 to node 3. Directions can be very important in a GIS for establishing such things as the flow of a river or the direction traffic moves on one-way streets. In a GIS, often a consistent direction convention is followed—that is, proceeding clockwise around polygons. Table 28.2(b) summarizes the directions of all chains within Figure 28.3.

The topology of Figure 28.3 would also describe, through adjacency, that Smith and Brown share a common boundary, which is chain *f* from node 5 to node 2, and that Smith is on the left side of the chain and Brown is on the right. Obviously the chain's direction must be stated before left or right positions can be declared. Table 28.2(c) lists the adjacency relationships of Figure 28.3. Note that

TABLE 28.2 TOPOLOGICAL RELATIONSHIPS OF ELEMENTS IN THE GRAPHIC RECORD OF FIGURE 28.3

(a) Connectivity		(b) Direction			(c) Adjacency			(d) Nestedness	
Nodes	Chain	Chain	From Node	To Node	Chain	Left Polygon	Right Polygon	Polygon	Nested Node
1-2	*a*	*a*	1	2	*a*	0	I	II	Well
2-3	*b*	*b*	2	3	*b*	0	II		
3-4	*c*	*c*	3	4	*c*	0	II		
4-5	*d*	*d*	4	5	*d*	0	II		
5-1	*e*	*e*	5	1	*e*	0	I		
5-2	*f*	*f*	5	2	*f*	I	II		

a zero has been used to designate regions outside of the polygons and beyond the area of interest.

Nestedness establishes that the well is contained within Brown's polygon. Table 28.2(d) lists that topological information.

The relationships expressed through the identifiers for points, lines, and areas of Table 28.1 and the topology in Table 28.2 conceptually yield a "map." With these types of information available to the computer, the analysis and query processes of a GIS are made possible.

■ 28.5 NONSPATIAL DATA

Nonspatial data, also often called *attribute* or *descriptive data,* describe geographic regions or define characteristics of spatial features within geographic regions. Nonspatial data are usually alphanumeric and provide information such as color, texture, quantity, quality, and value of features. Smith and Brown as the property owners of parcels I and II of Figure 28.3 and the land cover classifications of forest, marsh, grassland, and stream in Figure 28.4 are examples. Other examples could include the addresses of the owners of land parcels, their types of zoning, dates purchased, and assessed values; or data regarding a particular highway, including its route number, pavement type, number of lanes, lane widths, and year of last resurfacing. Nonspatial data are often derived from sources such as documents, files, and tables.

In general, spatial data will have related nonspatial attributes and thus some form of linkage must be established between these two different types of information. Usually this is achieved with a *common identifier* that is stored with both the graphic and the nongraphic data. Identifiers such as a unique parcel identification number, a grid cell label, or the specific mile point along a particular highway may be used.

Figure 28.5 Vector data overlaid on a raster image background. (Courtesy Tom Pantages)

28.6 DATA FORMAT CONVERSIONS

In manipulating information within a GIS database, it is often necessary to either integrate vector and raster data or convert from one form to the other. Integration of the two types of data, i.e., using both types simultaneously, is usually accomplished by displaying vector data overlaid on a raster image background, as illustrated in Figure 28.5. In that figure, vector data representing the dwellings (points) that exist within the different subdivisions (areas) are overlaid on a satellite image of the same area. This graphic was developed as part of a population growth and distribution study being conducted for a municipality. The combination of vector and raster data is useful to provide a frame of reference and to assist GIS operators in interpreting displayed data. Sometimes it is necessary or desirable to convert raster data to vector format or vice versa. Procedures for accomplishing these conversions are described in subsections that follow.

28.6.1 Vector-to-Raster Conversion

Vector-to-raster conversion is also known as *coding* and can be accomplished in several ways, three of which are illustrated in Figure 28.6. Figure 28.6(a) is an overlay of the vector representation of Figure 28.4(b) with a coarse raster of grid cells. In one conversion method, called *predominant type coding,* each grid cell is assigned the value corresponding to the predominant characteristic of the area it covers. For example, the cell located in row 3, column 1 of Figure 28.6(a) overlaps

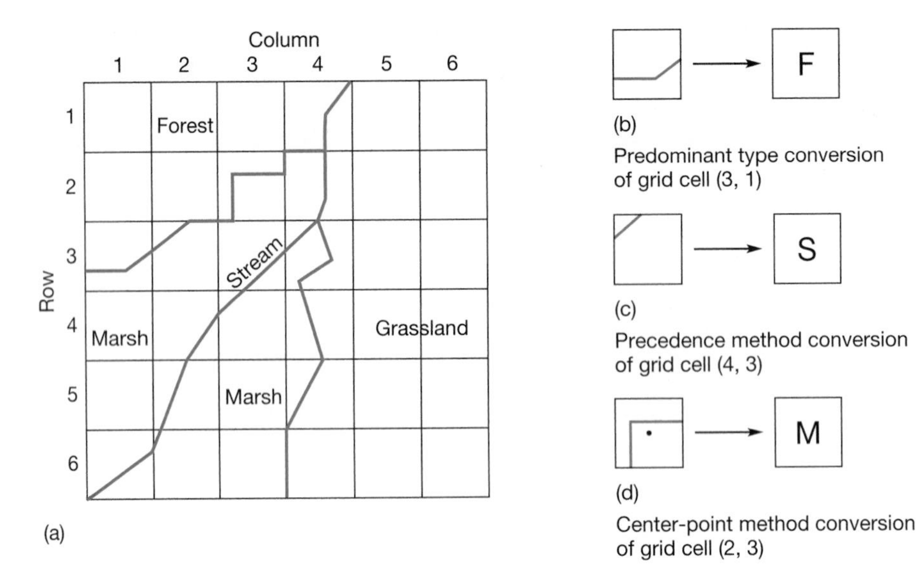

Figure 28.6 Methods for converting data from vector to raster format. (a) Vector representation of Figure 28.4(b) overlaid on a coarse raster format. (b), (c), and (d) Vector-to-raster conversion by predominant type, precedence, and center-point method, respectively.

two polygons, one of forest (type *F*), and one of marsh (type *M*). As shown in Figure 28.6(b), since the largest portion of this cell lies in forest, the cell is assigned the value *F*—the predominant type.

In another coding method, called *precedence coding,* each category in the vector data is ranked according to its importance or "precedence" with respect to the other categories. In other words, each cell is assigned the value of the highest-ranked category present in the corresponding area of the vector data. A common example involves water. While a stream channel may cover only a small portion of a cell area, it is arguably the most important feature in that area. Also, it is important to avoid breaking up the stream. Thus for the cell in row 4, column 3 of Figure 28.6(a), which is illustrated in Figure 28.6(c), water would be given the highest precedence and the cell coded *S* even though most of the cell is covered by marsh.

Center-point coding is a third technique for converting from vector-to-raster data. Here a cell is simply assigned the category value at the vector location corresponding to its center point. An example is shown in Figure 28.6(d), which represents the cell in row 2, column 3 of Figure 28.6(a). Here, since marsh exists at the cell's center, the entire cell is designated as category *M*. Note that the grid cell of row 3, column 4 would be classified by predominant type as grassland, by precedence as stream, and by center point as marsh. This illustrates how different conversion processes can yield different classifications for the same data.

The precisions of these vector-to-raster conversions depend on the size of the grid used. Obviously, using a raster of large cells would result in a relatively inaccurate representation of the original vector data. On the other hand, a fine-resolution grid can very closely represent the vector data, but would require a large amount of computer memory. Thus the choice of grid resolution becomes a tradeoff between computing efficiency and spatial precision.

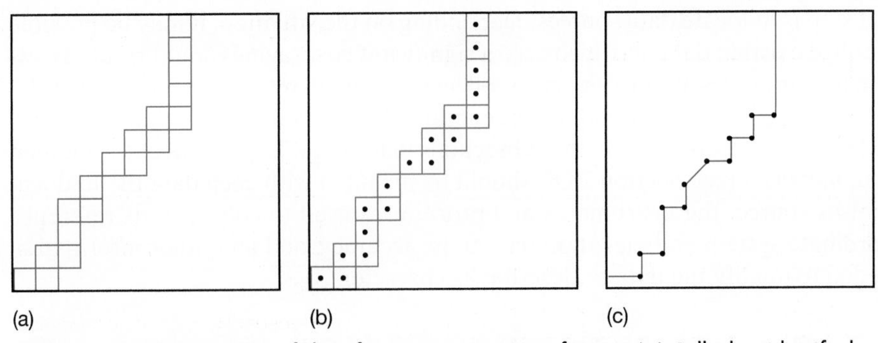

Figure 28.7 Conversion of data from raster-to-vector format. (a) Cells that identify the stream line of Figure 28.4(a). (b) Cell centers. (c) Vector representation of stream line recreated by connecting adjacent cell centers.

28.6.2 Raster-to-Vector Conversion

Raster-to-vector conversions are more vaguely defined than vector to raster. The procedure involves extracting lines from raster data, which represent linear features such as roads, streams, or boundaries of common data types. Whereas the approach is basically a simple one and consists in identifying the pixels or cells through which vector lines pass, the resulting jagged- or "staircase"-type outlines are not indicative of the true lines. One raster-to-vector conversion example is illustrated in Figure 28.7(a), which shows the cells identifying the stream line of Figure 28.4(b). Having selected these cells, a problem that must be resolved is how to fit a line to these jagged forms. One solution consists in simply connecting adjacent cell centers [see Figure 28.7(b)] with line segments. Note, however, that the resulting line [see Figure 28.7(c)] does not agree very well with the original stream of Figure 28.4(a). This example illustrates that some type of "line smoothing" is usually necessary to properly represent the gently curved boundaries that normally occur in nature. However, fitting smooth lines to the jagged cell boundaries is a complicated mathematical problem that does not necessarily have a unique solution. The decision ultimately becomes a choice between accuracy of representation and cost of computation.

No matter which conversion is performed, errors are introduced during the process and some information from the original data is lost. Use of smaller grid cells improves the results. Nevertheless, as illustrated by the example of Figure 28.7, if a data set is converted from vector-to-raster and then back to vector (or vice versa), the final data set will not likely match the original. Thus, it is important for GIS operators to be aware of how their data have been manipulated and what can be expected if conversion is performed.

■ 28.7 CREATING GIS DATABASES

Several important factors must be considered prior to developing the database for a GIS. These include the types of data that need to be obtained, optimum formats for these data, the reference coordinate system that will be used for spatially relating all data, and the necessary accuracy of each data type. Provisions for updating the database must also be considered. Having made these decisions, the next step is to locate data sources. Depending on the situation, it may be possible to utilize existing data, in which case a significant cost savings could result. However, in many cases, it is necessary to collect new data to meet the needs of the GIS. Different methods are available for generating the digital data needed to support a GIS. These are discussed in the subsections that follow. Regardless of the method used, *metadata* (see Section 28.8) should be included with each data file to document its source, the instruments and procedures used to collect it, its reference coordinate system and elevation datum, its accuracy, and any other information needed to qualify the data or describe its character.

28.7.1 Generating Digital Data from Field Surveys

Spatially related information needed to support a GIS is often generated by conducting new field surveys expressly for that purpose. Any of the equipment and procedures described in preceding chapters that are capable of locating objects in space can be used for this work. However, total station instruments interfaced with data collectors and GPS equipment are particularly convenient because they can rapidly and efficiently provide coordinates of points directly in a reference coordinate system that is suitable for the GIS, and because necessary identification codes can also be entered at the time the data is collected. Both planimetric positions and elevation data (preferably in the form of digital elevation models) can be obtained with these instruments. It must be remembered, however, that elevations obtained with GPS equipment are related to the ellipsoid and thus must be converted to orthometric heights (see Section 13.4.3). The field data can be downloaded into a computer, processed using COGO or other software if necessary, and entered directly into the GIS database. The cost of generating digital data in this manner is relatively high, but the data are generally very accurate.

Code-based GPS receivers offer several advantages over other instruments in collecting of mapping data for a GIS. These instruments are relatively inexpensive and when the data is reduced using differential GPS (DGPS) techniques, the positional accuracies obtained are often sufficient for many uses. Other advantages are speed, ease of use and the ability to enter attribute data about a point. For instance when collecting the position of a utility pole for inclusion into a GIS, additional information such as type (electric, cable, telephone, etc.), pole number, diameter, height, and condition can be entered while its position is being determined. These code-based units can also be used to collect the alignment of a feature such as a utility line, road, or sidewalk by simply entering the epoch rate and traversing the alignment either on foot or in a vehicle. Again, attribute information about the alignment can be entered while the user traverses it. If additional accuracy is required, the surveyor can use carrier phase-shift GPS receivers and kinematic reduction techniques to quickly obtain the data. However, as discussed in Chapter 15, in many situations kinematic surveys require some project planning for successful completion.

28.7.2 Digitizing from Aerial Photos with Stereoplotters

Information from aerial photographs can also be entered directly into a GIS database in digital form using an analytical or softcopy stereoplotter (see Section 27.14).

In this procedure, both planimetric features and elevations can be recorded, and high accuracy can be achieved. The data are registered to the selected reference coordinate system and vertical datum by *orienting* the stereoplotter to ground control points prior to beginning the digitizing. Then to record planimetric features, an operator views the stereomodel, points to objects of interest, enters any necessary feature identifiers or codes, and depresses a key or foot pedal to transfer the information to a file in an interfaced computer. To digitize elevations, a DEM is read directly from the three-dimensional stereomodel and stored in a computer file.

Accuracies of data obtained using this procedure will depend mainly on the scale and quality of the aerial photography, the accuracy of the ground control points used to orient the stereoplotter, and the experience and capabilities of the stereoplotter operator. Other factors that may affect accuracy to a lesser degree include camera lens distortions, atmospheric refraction, differential shrinkage and expansion of the materials upon which the photographic products are printed, and optical and/or mechanical imperfections in the stereoplotting or digitizing equipment.

Data sets generated by digitizing stereomodels will usually need to be checked carefully to ensure that all desired features have been included. Also the data must be corrected or "cleaned" before being used in a GIS. In this process, unwanted points and line portions must be removed and "unclosed" polygons, which result from imprecise pointing when returning to a polygon's starting node, must be closed (see Section 18.8.2). Finally, thin polygons or "slivers" created by lines being inadvertently digitized twice, but not precisely in the same location, must be eliminated. This editing process can be performed by the operator or with a program that can find and remove certain features that fall within a set of user-defined tolerances.

Digitizing from stereomodels and editing the data can usually be accomplished in less time and with lower cost than obtaining the data by field surveys, especially where relatively large areas must be covered. Of course, some field surveying is needed to establish the ground control needed to orient the stereomodels. If experienced people do the work carefully, the resulting accuracy of data obtained by digitizing stereomodels is usually very good.

28.7.3 Digitizing Existing Graphic Materials

If sources such as maps, orthophotos, plans, diagrams, or other graphic documents already exist that will meet the needs of the GIS database, these can be conveniently and economically converted to digital files using a tablet digitizer. Many GIS software packages provide programs to support the procedure directly. A tablet digitizer, as shown in Figure 28.8, contains an electronic grid and an attached cursor. Movement of the cursor across the grid creates an electronic signal unique to the cursor's position. This signal is relayed to the computer, which records the digitizer's coordinates for the point. Data identifiers or attribute codes can be associated with each point through the computer's keyboard or by pressing numerical buttons on the cursor.

The digitizing process begins by securing the source document to the tablet digitizer. If the document is a map, the next step is to register its reference coordinate system to that of the digitizer. This is accomplished by digitizing a series of

Figure 28.8 Tablet digitizer interfaced with personal computer. (Courtesy Tom Pantages)

reference (tic) marks on the map for which geographic coordinates such as state plane, UTM, or latitude and longitude are precisely known. With both reference and digitizer coordinates known for these tic marks, a coordinate transformation (see Section 11.8) can be computed. This determines the parameters of scale change, rotation, and translation necessary to convert digitized coordinates into the reference geographic coordinate system. After this, any map features can be digitized, whereupon their coordinates are automatically converted to the selected system of reference and they can then be entered directly into the database.

Both planimetric features and contours can be digitized from the map. Planimetric features are recorded by digitizing the individual points, lines, or areas that identify them. As described in Section 28.4.2, this process creates data in a vector format. Elevation data can be recorded as a digital elevation model (DEM) by digitizing critical points along contours. From these data, *triangulated irregular networks* (TIN models) can be derived using the computer (see Section 17.8). From the TIN models, point elevations, profiles, cross sections, slopes, aspects (slope directions), and contours having any specified contour interval can be derived automatically using the computer.

Data files generated in this manner can be obtained quickly and relatively inexpensively. Of course the accuracy of the resultant data can be no better than the accuracy of the document being digitized, and its accuracy is further diminished by differential shrinkages or expansions of the paper or materials upon which the document is printed and by inaccuracies in the digitizer and the digitizing process.

28.7.4 Keyboard Entry

Data can be entered into a GIS file directly using the keyboard on a computer. Often data input by this method are nonspatial, such as map annotations or numerical or tabular data. To facilitate keying in data, an intermediate file having a simple format is sometimes created. This file is then converted into a GIS-compatible format using special software.

Metes-and-bounds descriptions (see Section 21.4) can also be computed using coordinate geometry techniques (see Chapter 11). The resulting coordinates can be used to facilitate entry of the deed description into the GIS file.

28.7.5 Existing Digital Data Sets

Massive quantities of digital information are now being generated by a wide variety of offices and agencies involved in GIS activities. At the federal level, the U.S. Geological Survey, the National Oceanic and Atmospheric Administration, the Bureau of Land Management, the Environmental Protection Agency, and other organizations are developing digital information. The *digital line graphs* (DLGs) and *digital elevation models* (DEMs) produced by the U.S. Geological Survey (see Section 18.3) are examples of available digital files. In addition to federal agencies, offices of state governments, counties, and cities are involved in this work. As a result of this proliferation of information, an initiative known as the *National Spatial Data Infrastructure* (NSDI) has evolved at the federal level. The NSDI encompasses policies, standards, and procedures for organizations to cooperatively produce and share geographic data. The *National Geospatial Data Clearinghouse* is a component of the NSDI that provides a pathway to find information about available spatially referenced data.[1]

Of course, before using existing data, information about its content, source, date, accuracy, and other characteristics must be scrutinized to determine if it is suitable for the GIS at hand. This requirement underscores the need for maintaining good quality metadata (see Section 28.8) for all digital files. Also, existing digital data must often undergo conversion of file structures and formats to be usable with specific GIS software. Because of differences in the way data are represented by different software, it is possible that information can be lost or that spurious data can creep in during this process.

28.7.6 Scanning

Scanners are instruments that automatically convert graphic documents into a digital format. As discussed in Section 28.14.4, they are used to digitize the contents of aerial images to support softcopy photogrammetry. In GIS work, scanners are used not only to digitize aerial photos, but also to convert larger documents such as maps,

---

[1]Information about available geospatial data can be obtained by writing to the National Spatial Data Infrastructure, U. S. Geological Survey, 508 National Center, Reston, VA 20192, or by contacting them at http://nsdi.usgs.gov.

plans, and other graphics into digital form. The principal advantages of using scanners for this work is that the tedious work of manual digitizing is eliminated and the process of converting graphic documents to digital form is greatly accelerated.

Scanners accomplish their objective by measuring the amount of light reflected from a document and assigning this information to pixels. This is possible because different areas of a document will reflect light in proportion to their tones, from a maximum for white through the various shades of gray to a minimum for black. For example, the scanner of Figure 28.9 uses a linear array of light sensors to capture the varying intensities of reflected light, line by line, as the document is fed through the system. This creates a raster data set. Its pixel size can be varied and made as small as 1/500 in.[2] (500 dpi). Large complicated documents can

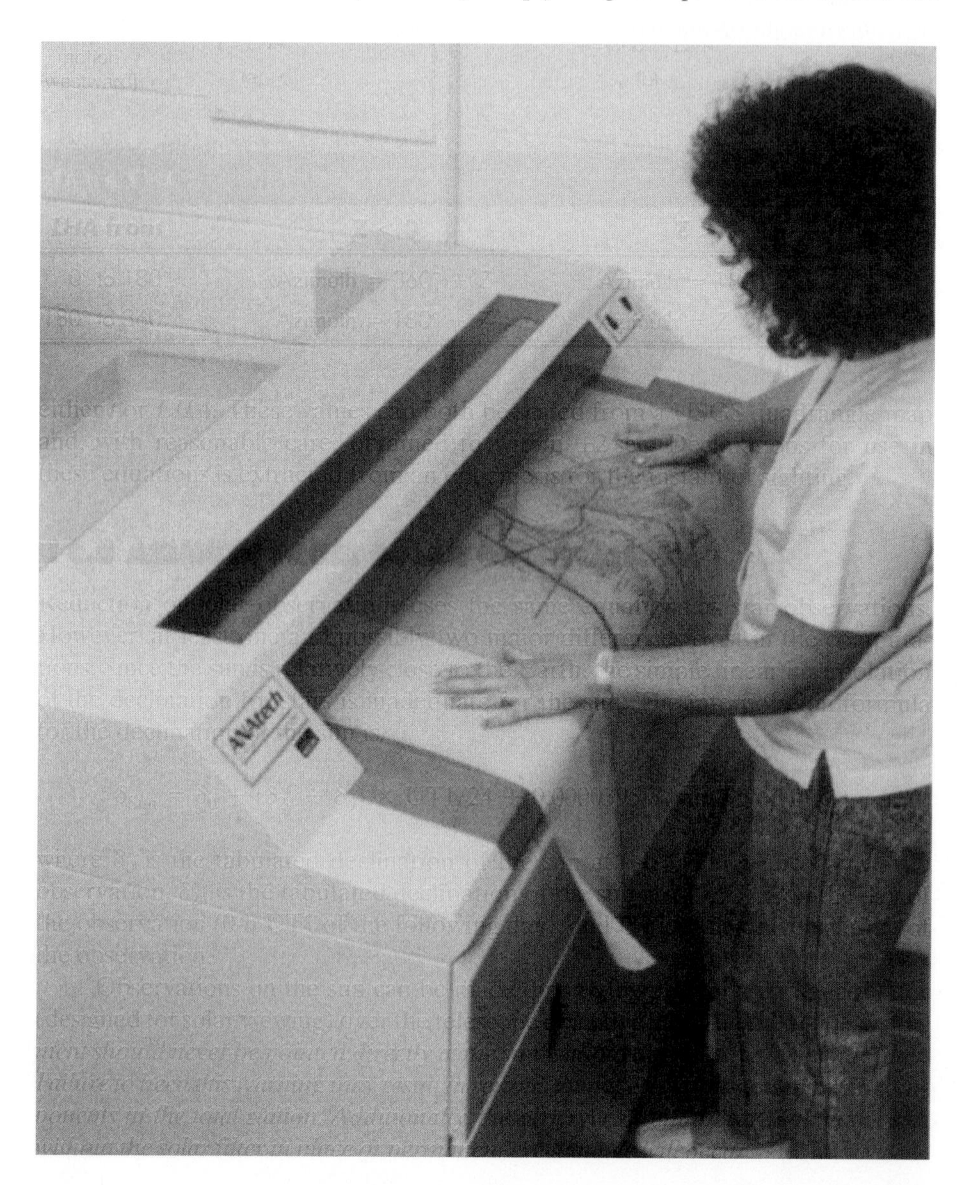

Figure 28.9 Intergraph ANAtech Eagle 4050 scanner. (Courtesy Wisconsin Department of Transportation)

be scanned in a matter of a few minutes. The data are stored directly on the hard drive of an interfaced computer and can be viewed on a screen, edited, and manipulated. Editing is an important and necessary step in the process, because the scanner will record everything, including blemishes, stains, and creases.

Documents such as subdivision plats, topographic maps, engineering drawings and plans, aerial photos, and orthophotos can be digitized using scanners. Then, if necessary, the raster data can be converted to vector form using techniques described in Section 28.6.2.

Accuracy of the raster file obtained from scanning depends somewhat on the instrument's precision, but pixel size or resolution is generally the major factor. A smaller pixel size will normally yield superior resolution. However, there are certain trade-offs that must be considered. Whereas a large pixel size will result in a coarse representation of the original, it will require less scanning time and computer storage. Conversely, a fine resolution, which generates a precise depiction of the original, requires more scanning time and computer storage. An additional problem is that at very fine resolution, the scanner will record too much "noise"—that is, impurities such as specks of dirt. For these reasons and others, this is the least-preferred method of capturing data in a GIS.

■ 28.8 METADATA

Metadata, often simply defined as "data about data," describes the content, quality, condition, and other characteristics about geospatial data and provides a record of changes or modifications that have been made to that data. It normally includes information such as who originally created the data, when was it generated, what equipment and procedures were used in collecting the data, and its original scale and accuracy. Once created, data can travel almost instantaneously through a network and be transformed, modified, and used for many different kinds of spatial analyses. It can then be retransmitted to another user, and then to another, etc.[2] It is important that each change made to any data set be documented by updating its associated metadata.

Although generating the original metadata and updating it as changes are made may be burdensome and add cost, in the long run it is worth the effort because it preserves the value of the data and extends its useful life. If it is not done, prospective users may not trust the data and as a result they may fail to take advantage of it and incur the cost of duplicate data collection.

The *Federal Geographic Data Committee* (FGDC) has developed metadata standards that provide a common set of terms and definitions for describing geospatial data, and outline a consistent and systematic approach to documenting data characteristics.[3] The primary benefit to be realized by following these standards is that all users, regardless of their backgrounds or specialty areas, will have a common understanding of the source, nature, and quality of any data set.

---

[2]These metadata standards may be obtained from the FGDC Secretariat, U.S. Geological Survey, 590 National Center, Reston, VA 22092, or information can be obtained at http://www.fgdc.gov.

[3]An example of a data exchange website, PASDA, with metadata can be found at http://www.pasda.psu.edu/.

■ 28.9 GIS ANALYTICAL FUNCTIONS

Most GISs are equipped with a set of basic analytical functions that enable data to be manipulated, analyzed, and queried. These functions, coupled with appropriate databases, provide GISs with their powerful capabilities for supplying information that so significantly aids in planning, management, and decision-making.

The specific functions available within the software of any particular GIS system will vary. They enable data to be stored, retrieved, viewed, analyzed spatially and computationally, and displayed. Some of the more common and useful spatial analysis and computational functions are (1) proximity analysis, (2) boundary operations, (3) spatial joins, and (4) logical operations. These are briefly described in the subsections that follow.

28.9.1 Proximity Analysis

This spatial analysis function creates new polygons that are geographically related to nodes, lines, or existing polygons, and usually involves processes called *buffering. Point buffering*, also known as *radius searching,* is illustrated in Figure 28.10(a). It involves the creation of a circular buffer zone of radius *R* around a specific node. Information about the new zone can then be gathered and analyses made of the new data. A simple example illustrates its value. Assume that well water that was

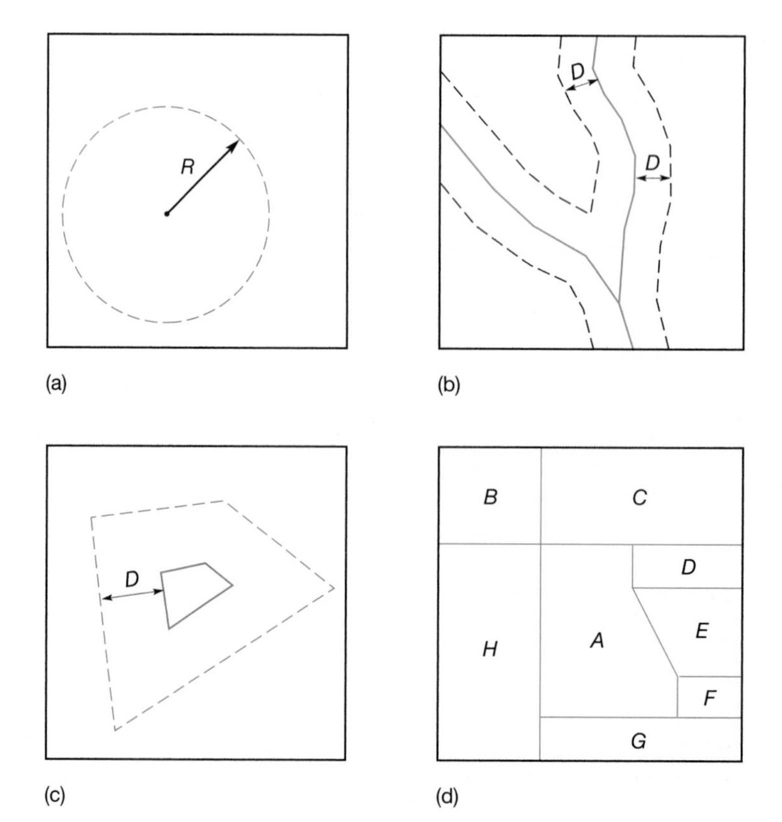

Figure 28.10 GIS spatial analysis functions. (a) Point buffering. (b) Line buffering. (c) Polygon buffering. (d) Adjacency analysis.

polluted by an accidental spill has just been discovered. With appropriate databases available, all dwellings within a specified radius of the well can be located, the names, addresses, and telephone numbers of all individuals living within the point buffer zone tabulated, and the people quickly alerted to the possibility of their water being contaminated.

Line buffering, illustrated in Figure 28.10(b), creates new polygons along established lines such as streams and roads. To illustrate the use of line buffering, assume that to preserve the natural stream bank and prevent erosion, a zoning commission has set the construction setback distance from a certain stream at D. Line buffering can quickly identify the areas within this zone. *Polygon buffering,* illustrated in Figure 28.10(c), creates new polygons around an existing polygon. An example of its use could occur in identifying those landowners whose property lies within a certain distance D of the proposed site of a new industrial facility. Many other examples could be given, which illustrate the value of buffering for rapidly extracting information to support management and decision-making.

28.9.2 Boundary Operations

If the topological relationships discussed in Section 28.4.3 have been entered into a database, certain analyses regarding relative positioning of features, usually called boundary operations, can be performed. *Adjacency* and *connectivity* are two important boundary operations that often assist significantly in management and decision-making. An example of adjacency is illustrated in Figure 28.10(d) and relates to a zoning change requested by the owner of parcel A. Before taking action on the request, the jurisdiction's zoning administrators are required to notify all owners of adjacent properties B through H. If the GIS database includes the parcel descriptions with topology and other appropriate attributes, an adjacency analysis will identify the abutting properties and provide the names and addresses of the owners.

Connectivity involves analyses of the intersections or connections of linear features. The need to repair a city water main serves as an example to illustrate its value. Suppose that the decision has been made that these repairs will take place between the hours of 1:00 P.M. and 4:00 P.M. on a certain date. If infrastructure data are stored within the city's GIS database, all customers connected to this line whose water service will be interrupted by the repairs can be identified and their names and addresses tabulated. The GIS can even print a letter and address labels to facilitate a mailing announcing details of the planned interruption to all affected customers. Many similar examples could be given to illustrate benefits that can result from adjacency and connectivity.

28.9.3 Spatial Joins

Spatial joins, also called *overlaying,* are one of the most widely used spatial analysis functions of a GIS. As indicated in Figure 28.1, GIS graphic data are usually divided into layers, with each containing data in a single category of closely related features. Nonspatial data or "attributes" are often associated with each category. The individual layers are spatially registered to each other through a common reference network or coordinate system. Any number of layers can be entered into a GIS database and could include parcels, municipal boundaries, public land survey system, zoning, soils, road networks, topography, land cover, hydrology, and many others.

Having these various data sets available in spatially related layers makes the overlay function possible. Its employment in a GIS can be compared to using a collection of Mylar overlays in traditional mapping. However, much greater efficiency and flexibility are possible when operating in the computer environment of a GIS, and not only can graphic data be overlaid, but attribute information can be combined as well.

Many examples could be given to illustrate the applications and benefits of the GIS spatial join or overlay process. Consider one case where the land in a particular area suitable for development must be identified. To perform an in-depth analysis of this situation, the evaluation would normally have to consider numerous variables within the area, including the topography (slope and aspect of the terrain), soil type, land cover, landownership, and others. Certain combinations of these variables could make land unsuitable for development. Figure 28.11 illustrates the simple case of land suitability analysis involving only two variables, slope and soil type. Figure 28.11(a) shows polygons within which the average slope is either 5 or 10 percent. Figure 28.11(b) classifies the soils in the area as E (erodible) or S (stable). The composite of the two data sets, which results from a *polygon-on polygon* overlay, is shown in Figure 28.11(c). It identifies polygon I, which combines 5 percent slopes and low-erodibility class S soils. Since this combination does not present potential erosion problems, considering those variables, the area within polygon I is suitable for development, while areas II, III, and IV are not.

Another GIS overlay function is that of point in polygon. Here the question involves which point features are located in certain polygons where layers are combined. For example, to predict possible well contamination, a GIS operator may want to know which wells are located in an area of highly permeable soil. A similar overlay process, *line in polygon*, identifies specific linear features within polygons of interest. An example of its application would be the identification of all bituminous roads, paved more than 15 years ago, in townships whose roadway maintenance budgets are less than $250,000. Such information would be valuable

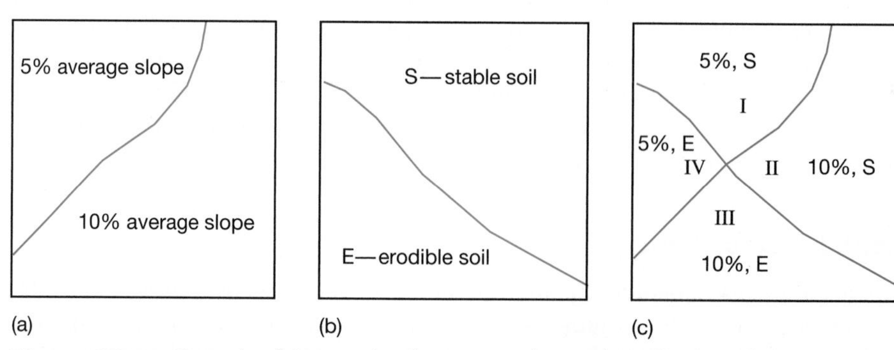

Figure 28.11 Example of GIS overlay function used to evaluate land suitability. (a) Polygons of differing slopes. (b) Varying soil types in the area. (c) Overlay of (a) and (b), identifying polygon I as an area combining lower slopes with stable soils that would be suitable for development.

to support decisions concerning the allocation of state resources for local roadway maintenance.

The GIS functions just described can be used individually, as has been illustrated with the examples, or they can be employed in combination. The following example illustrates the simultaneous application of line buffering, adjacency, and overlay. The situation involves giving timely notice to affected persons of an impending flood that is predicted to crest at a specific stage above a particular river's normal elevation. Here it is first necessary to identify the lands that lie at or below the expected flood stage. This can be done using line buffering, where the thread of the river is the line of reference. However, the width of the buffer zone is variable and is determined by combining elevation data in the TIN model with the buffering process. The adjacency and overlay functions are then used to determine which landowners are next to or within the flood zone. Then the names, addresses, and telephone numbers of property owners and dwellers within and adjacent to the affected area can be tabulated. These people can then be notified of the impending situation, and emergency preparations such as construction of temporary levees can be performed, or, if necessary, the area can be evacuated. Should evacuation be required, the GIS may even be used to identify the best and safest escape routes. Data sets necessary for such analyses would include the topography in the area, including the river's location, normal stage, and floodplain cross sections; census data; property ownership; and the transportation network.

28.9.4 Logical Operations

Typically, attribute data that is related to the features present in the GIS are stored in a database. Thus, the database can be used to perform logical operations on the data. For instance, a city can construct a GIS database that contains the time streetlights are installed and their rated life. The manager can then query the system to show all lights that have passed their rated life cycle and schedule maintenance personnel to replace these lights. Today GISs are being used in large buildings to help managers keep records of maintenance and schedule routine maintenance jobs. The number of useful logical queries that can be performed in a GIS is limited only by the data contained in the GIS database and the imagination of the user.

28.9.5 Other GIS Functions

In addition to the spatial analysis functions described in preceding subsections, many other functions are available with most GIS software. Some of these include the capability of computing (1) the number of times a particular type of point occurs in a certain polygon; (2) the distances between selected points or from a selected line to a point; (3) areas within polygons; (4) locations of polygon centroids; and (5) volumes within polygons where depth or other conditions are specified. A variety of different mapping functions may be performed using GISs. These may include (a) performing map scale changes; (b) changing the reference coordinate system from, for example, the state plane to the UTM system;

(c) rotating the reference grid; and (d) changing the contour interval used to represent elevations.

Most GISs are also capable of performing several different digital terrain analysis functions. Some of these include (1) creating TIN models or other DEMs from randomly spaced *XYZ* terrain data; (2) calculating profiles along designated reference lines and determining cross sections at specified points along the reference line; (3) generating perspective views where the viewpoint can be varied; (4) analyzing visibility to determine what can or cannot be seen from a given vantage point; (5) computing slopes and aspects; and (6) making sun intensity analyses.

Output from GISs can be provided in graphic form as charts, diagrams, and maps; in numerical form as statistical tabulations; or in other files that result from computations and manipulations of the geographic data. These materials can be supplied in either printed (hardcopy) form or on diskettes or tapes (softcopy).

■ 28.10 GIS APPLICATIONS

As stated earlier, and as indicated by the examples in preceding sections of this chapter, the areas of GIS applications are widespread. Further evidence of the diversity of GIS applications can be seen by reviewing the bibliography at the end of this chapter. The technology is being used worldwide, at all levels of government, in business and industry, by public utilities, and in private engineering and surveying offices. Some of the more common areas of application occur in (1) land-use planning; (2) natural resource mapping and management; (3) environmental impact assessment; (4) census, population distribution, and related demographic analyses; (5) route selection for highways, rapid-transit systems, pipelines, transmission lines, etc.; (6) displaying geographic distributions of events such as automobile accidents, fires, crimes, or facility failures; (7) routing buses or trucks in a fleet; (8) tax mapping and mapping for surveying and engineering purposes; (9) subdivision design; (10) infrastructure and utility mapping and management; (11) urban and regional planning; and many others.

As the use of GIS technology expands, there will be a growing need for trained individuals who understand the fundamentals of these systems. Users should be aware of the manner in which information is recorded, stored, managed, retrieved, analyzed, and displayed using a GIS. System users should also have a fundamental understanding of each of the GIS functions, including their bases for operation, their limits, and their capabilities. *Perhaps of most importance, users must realize that information obtained from a GIS can be no better than the quality of the data from which it was derived.*

From the perspective of those engaged in surveying (geomatics), it is important to underscore again the fact that the fundamental basis of GISs is a database of spatially related, digital data. Since accurate position determination and mapping are the surveyor's forte, in the future surveyors will continue to play key roles in designing, developing, implementing, and managing these systems. Their input will be particularly essential in establishing the necessary basic control frameworks, conducting ground and aerial surveys to locate geographic features and their attributes, compiling maps, and assembling the digital data files needed for these systems.

PROBLEMS

Asterisks (*) indicate problems that have answers given in Appendix G.

28.1 Describe the concept of layers in a geographic information system.

28.2 Discuss the role of a geographic reference framework in a GIS.

28.3 List the fundamental components of a GIS.

28.4 List the fields within surveying and mapping that are fundamental to the development and implementation of GISs.

28.5 Discuss the importance of metadata to a GIS.

28.6 Name and describe the different simple spatial objects used for representing graphic data in digital form. Which objects are used in raster format representations?

28.7 What are the primary differences between a GIS and a LIS?

28.8 How many pixels are required to convert the following documents to raster form for the conditions given:

(a)* A 384-in.-square map scanned at 200 dpi.

(b) A 9-in.-square aerial photo scanned at 1200 dpi.

(c) An orthophoto of 11 $\times$ 17 in. dimensions scanned at 300 dpi

28.9 Explain how data can be converted from:

(a) Vector to raster format

(b) Raster to vector format

28.10 For what types of data is the vector format best suited?

28.11 Discuss the compromising relationships between grid cell size and resolution in raster data representation.

28.12 Define the term topology and discuss its importance in a GIS.

28.13 Develop identifier and topology tables similar to those of Tables 28.1 and 28.2 in the text for the vector representation of (see the following figures):

(a) Problem 28.13 (a)

(b) Problem 28.13 (b)

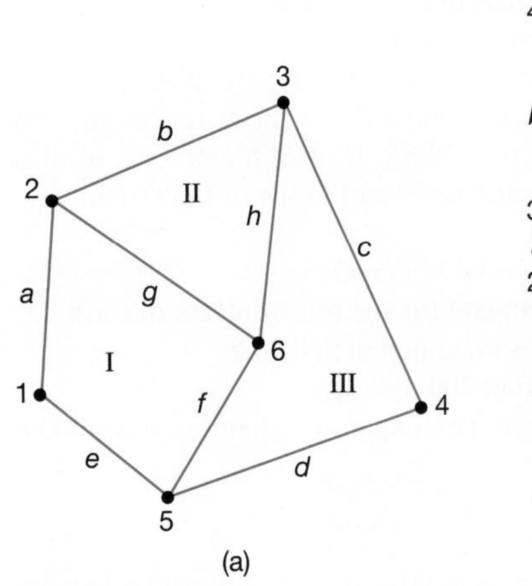

(a)

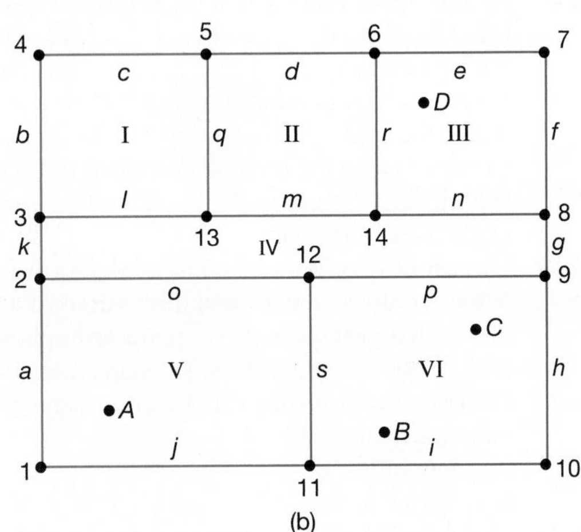

(b)

Problem 28.13

28.14 Compile a list of linear features for which the topological relationship of adjacency would be important.

28.15 Prepare a raster (grid cell) representation of the sample map of:
(a) Problem 28.15(a), using a cell size of 0.10-in. square (see accompanying figure).
(b) Problem 28.15(b), using a cell size of 0.20-in. square (see accompanying figure).

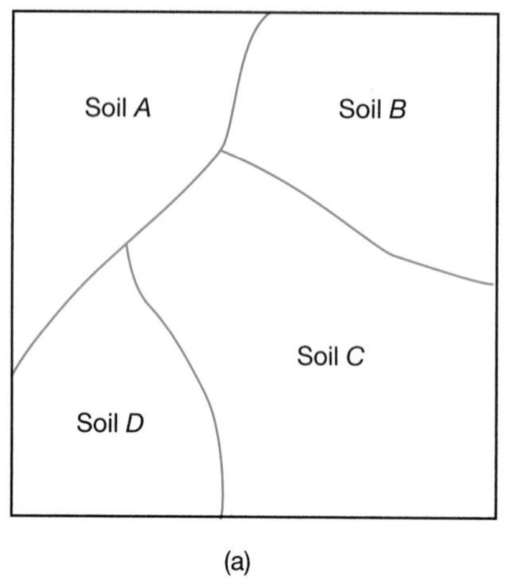

(a)

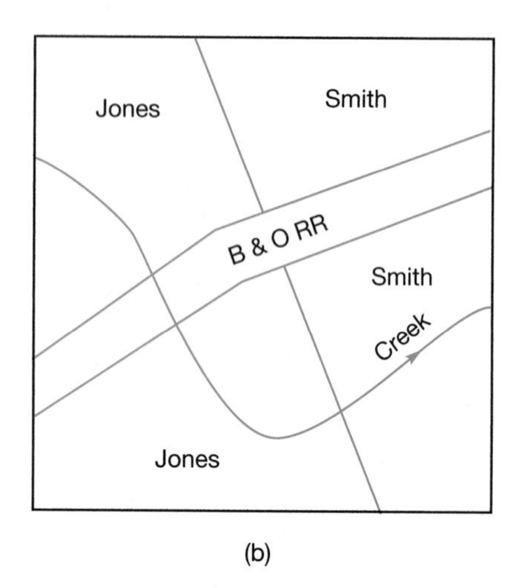

(b)

Problem 28.15

28.16 Discuss the advantages and disadvantages of using the following equipment for converting maps and other graphic data to digital form: (a) tablet digitizers and (b) scanners.

28.17 Explain the concepts of the following terms in GIS spatial analysis, and give an example illustrating the beneficial application of each: (a) adjacency and (b) connectivity.

28.18 If data were being represented in vector format, what simple spatial objects would be associated with each of the following topological properties?
(a) Connectivity **(b)** Direction
(c) Adjacency **(d)** Nestedness

28.19 Prepare a transparency having a 0.10-in grid, overlay it onto Figure 28.4(a), and indicate the grid cells that define the stream. Now convert this raster representation to vector using the method described in Section 28.6.2. Repeat the process using a 0.20-in grid. Compare the two resulting vector representations of the stream and explain any differences.

28.20 Discuss how spatial and nonspatial data are related in a GIS.

28.21 What are the actual ground dimensions of a pixel for the following conditions:
(a) A 1:10,000 scale, 9-in. square orthophoto scanned at 500 dpi?
(b)* A 748-in. square, 1:24,000 map scanned at 200 dpi?

28.22 Describe the following GIS functions, and give two examples where each would be valuable in analysis:
(a) line buffering
(b) spatial joins

28.23 Go to the PASDA website or a similar website in your state and download an example of:
(a) An orthophoto
(b) Zoning
(c) Floodplains and wetlands
(d) Soil types

28.24 Compile a list of data layers and attributes that would likely be included in an LIS.

28.25 Compile a list of data layers and attributes that would likely be included in a GIS for:

- **(a)** Selecting the optimum corridor for constructing a new rapid-transit system to connect two major cities
- **(b)** Choosing the best location for a new airport in a large metropolitan area
- **(c)** Routing a fleet of school buses
- **(d)** Selecting the fastest routes for reaching locations of fires from various fire stations in a large city

28.26 In Section 28.9.3, a flood-warning example is given to illustrate the value of simultaneously applying more than one GIS analytical function. Describe another example.

28.27 Consult the literature on GISs and, based on your research, describe an example that gives an application of a GIS in:

- **(a)** Natural resource management
- **(b)** Agriculture
- **(c)** Engineering
- **(d)** Forestry

BIBLIOGRAPHY

Binge, M. 2002. "A Practical Geographic Information Sciences Guide for Survey Professionals." *Point of Beginning* 26 (No. 4): 59.

———. 2002. "The Value of GIS." *Point of Beginning* 27 (No. 6): 20.

Cone, L. 2003. "The National Integrated Land System." *Surveying and Land Information Science* 63 (No. 4): 227.

Corbley, K. 2002. "Before the Levee Breaks." *Point of Beginning* 27 (No. 7): 24.

Cosworth, C. 2006. "Creating a New GIS Solution." *Point of Beginning* 31 (No. 11): 16.

Cowen, D. J. and W. J. Craig. 2004. "A Retrospective Look at the Need for a Multipurpose Cadastre." *Surveying and Land Information Science* 63 (No. 4): 205.

Flynn, R. 2002. "Find the Software Needle in the GIS Haystack." *Point of Beginning* 27 (No. 6): 18.

Greulich, Gunther. 2005. "The Global Land Information Explosion: GIS-ACSM-GLIS-GSDI-GITA-URISA-GE." *Surveying and Land Information Science* 64 (No. 4): 253.

Hanigan, F. 1988. "GIS by Any Other Name Is Still . . ." *The Forum* (No. 1): 6.

Jeffress, G. 2003. "The Value of Cadastral Surveying to Efficient Land Administration." *Surveying and Land Information Science* 63 (No. 4): 253.

Joffe, Bruce A. 2001. "Surveyors and GIS Professional Reach Accord." *Surveying and Land Information Systems* 61 (No. 1): 35.

Jones, R. 2006. "Redefining Spatial Boundaries." *Point of Beginning* 31 (No. 8): 20.

Lembo, A. R. O'Rourke, and B. Moses. 2003. "Coordinate Improvement of Vector GIS Data Using Rubbersheeting Methods." *Surveying and Land Information Science* 63 (No. 1): 47.

Luccio, M. 2005. "GIS Monitor Supplement: GISCorps' Response to Hurricane Katrina." *Professional Surveyor* 25 (No. 11): 44.

Lyle, S. 2003. "New Approach in Using eRTK-GPS for Direct Georeferencing of Aerial Images in a GIS Application." *Surveying and Land Information Science* 63 (No. 3): 161.

Monsef, H. A., M. El-Chawaby, and S. Smith. 2004. "GIS Analysis System for Investigating Sulphide Mineralization in South Sinai, Egypt." *Surveying and Land Information Science* 64 (No. 4): 243.

Pellegrino, M. 2003. "Mapping Missouri's Land of the Ozarks." *Point of Beginning* 28 (No. 6): 32.

Rambough, R. and J. Jackson. 2005. "Intersect: Are Surveyors the Right People to Collect GIS Field Data?" *Professional Surveyor* 25 (No. 11): 36.

——— 2005. "Intersect: How Do You Make a GIS Person Understand the Importance of Data Accuracy?" *Professional Surveyor* 25 (No. 10): 20.

Rameriz, R. J. 2005. "Updating Geospatial Data: A Theoretical Framework. *Surveying and Land Information Science* 65 (No. 4): 245.

Scheepmaker, S. and K. Stewart. 2006. "Prospering Township of Langley Takes GIS to a Higher Level." *Professional Surveyor* 26 (No. 8): 8.

Schutzberg, A. 2003. "Everything You Wanted to Know About Data Models (But Were Afraid to Ask)." *Professional Surveyor* 23 (No. 9): 26.

Strack, B. 2005."Does GIS Really Mean 'Get It Surveyed?' Indy Firm Says 'Why Not!'" *Professional Surveyor* 25 (No. 7): 8.

von Meyer, N. 2003. "Implementing the FGDC Cadastral Standard in ArcGIS." *Surveying and Land Information Science* 63 (No. 4): 215.

Weber, S. 2002. "Using GPS for the Utility Field Asset Management and the Creation of a Comprehensive GIS." *Point of Beginning* 27 (No. 12): 36.

Zimmer, R. 2003. "Interdependence of GIS Layers." *Professional Surveyor* 23 (No. 9): 41.

———. 2003. "The Surveying-GIS Revolution: Introducing the Survey Analyst Extension from ESRI, Part 1." *Professional Surveyor* 23 (No. 6): 42.

A

Dumpy Levels, Transits, and Theodolites

■ A.1 INTRODUCTION

Circa 1950, an important new development occurred in the evolution of leveling instruments—the automatic or "self-leveling" feature was introduced (see Section 4.10). This feature significantly reduced the time expended in setting up leveling instruments, and as a result leveling operations could be performed much more efficiently. Automatic levels therefore gradually replaced the "dumpy level," which up to that time had been the dominant leveling instrument. A few dumpy levels are still in use and therefore these instruments are discussed in Section A.2.

Another important development in surveying instrumentation also occurred at about the midpoint of the 20th century—the first electronic distance measurement (EDM) instruments appeared. The earliest versions were bulky and required large power sources for operation. However, through research and development their sizes were reduced and their capabilities enhanced. Soon small EDM instruments were being attached to theodolites, so that both distances and angles could be measured from a single setup—a feature which again greatly increased surveying efficiency. Later automatic angle-reading capabilities were incorporated into theodolites and the theodolite and EDM instrument were interfaced with a microcomputer to create what is now known as the *total station instrument* (see Chapter 8). With their many advantages, total station instruments rapidly replaced their predecessors, the transit and theodolite. Again, a few transits and theodolites are still in use, and thus these instruments are briefly described in Sections A.3 through A.5 in this Appendix.

A.2 THE DUMPY LEVEL

A dumpy level is shown in Figure A.1. As illustrated, the telescope (which as described in Section 4.7, consists of four main parts: the objective lens, the negative lens, the reticle, and the eyepiece) is rigidly fastened to the *dumpy bar,* also called the *level bar*. The bar is centered on an accurately machined vertical *spindle* that sits in a conical *socket* of the *leveling head*. This spindle and socket arrangement allows the dumpy bar to revolve in a horizontal plane when the instrument is properly adjusted. A *clamp screw* enables the telescope's line of sight to be locked in a given direction, but a *tangent screw* allows a fine adjustment to be made in its direction of pointing.

The dumpy level has its level vial (tube type) set in the dumpy bar (see Figure A.1) and it is thereby somewhat protected. The vial always remains in the same vertical plane as the telescope, but screws at each end permit vertical adjustment or replacement of the vial. (Level vials are described in detail in Section 4.8.)

Four large leveling screws carry the conical socket into which the vertical spindle of the dumpy bar fits. They rest on the *base plate,* which is screwed onto the top of the tripod. The four leveling screws are in two pairs at right angles to each other and are used to center the bubble when leveling the instrument. To level the instrument, after the tripod has been firmly pressed into the ground, the telescope is rotated until it is over two opposite screws, as in the direction AB of Figure A.2. Using the thumb and first finger of each hand to adjust the opposite screws simultaneously "approximately" centers the bubble. The telescope is then turned so that it is aligned over the other two leveling screws and the process repeated. These two steps are repeated until the bubble stays centered in each direction. Working with each pair of screws about three times should complete

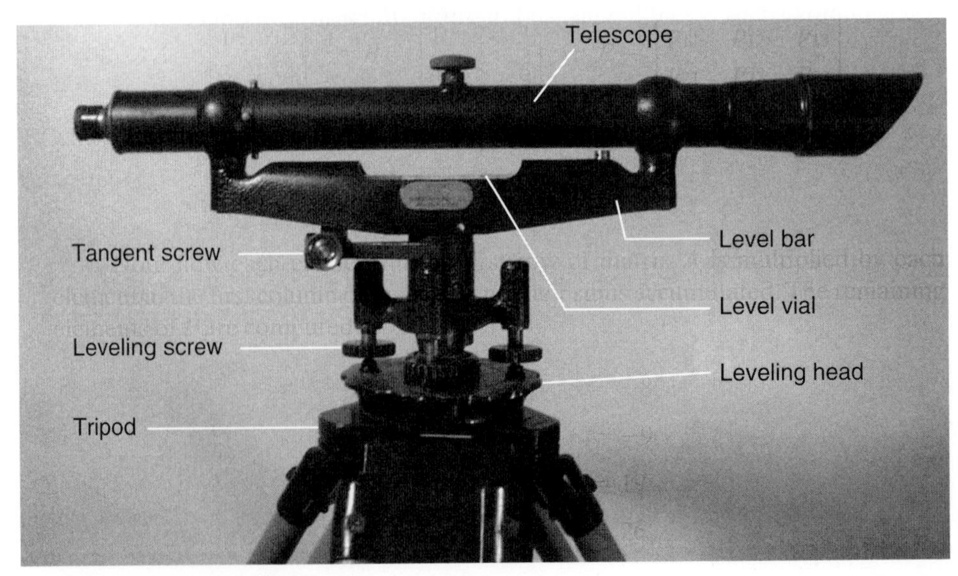

Figure A.1 Dumpy level.

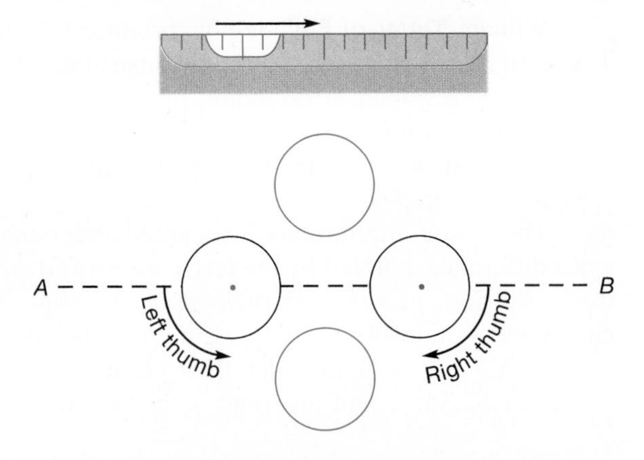

Figure A.2
Use of leveling screws with a four-screw leveling head.

the job. A simple useful rule in centering a bubble, illustrated in Figure A.2, is: *A bubble follows the direction of the left thumb when turning the leveling screws.* Final precise adjustment can be made with one screw only. Leveling screws should be snug, but not wrench-tight to avoid damage to threads and/or base plate.

If the bubble runs when the telescope is turned 180° in azimuth after the instrument has been leveled, the level vial is out of adjustment. It can be adjusted by following the procedures described in Section 4.15.3. The adjustment is made by turning the *level adjusting nuts* on one end of the level vial (see Figure A.1). When the bubble is in adjustment and the leveling process is completed as described here, the line of sight generates a horizontal plane as the telescope is revolved about its vertical axis. The dumpy level can be used for all types of leveling including differential, profile, and construction leveling. Care should be exercised to ensure that the bubble is centered whenever a rod reading is taken.

Another older leveling instrument called the *wye level* is similar in many respects to the dumpy level. Its telescope rests in supports on the level bar called *wyes*. Its advantage is that the telescope can be removed from the wyes, a procedure which facilitates adjusting the instrument.

■ A.3 INTRODUCTION TO THE TRANSIT AND THEODOLITE

The transit and theodolite are predecessors of the total station instrument. These instruments are fundamentally equivalent and can accomplish basically the same tasks. The most important application of transits and theodolites was observing horizontal and vertical (or zenith) angles, but they could also be used to obtain horizontal distances by stadia and subtense bars and determine elevations of points by stadia, accomplish low-order differential leveling, and establish alignments.

The main components of a transit and theodolite include a sighting telescope and two graduated circles mounted in mutually perpendicular planes. Prior to measuring angles, the *horizontal* circle is oriented into a horizontal plane by means of level vials, which automatically puts the other circle in a vertical plane. Horizontal and zenith (or vertical) angles can then be observed directly in their respective planes of reference.

William Young of Philadelphia produced the first American transit in 1831. The term *transit* was adopted for the instrument because its telescope could be *transited,* or reversed in direction, by rotating it about a "horizontal axis." In Europe, the name *transiting theodolite* was adopted for this type of angle-measuring instrument. Europeans eventually dropped the adjective and retained the name *theodolite.*

There is no internationally accepted understanding among surveyors on the exact difference denoted by the terms *transit* and *theodolite*. The most commonly used criterion, however, is their general design, particularly their graduated circles and systems for reading them. Transits have an "open-circle" design (see Figures A.3 and A.4), in which their graduated circles, made of metal, are visible to an operator and are read with the aid of verniers (see Section A.4.2).

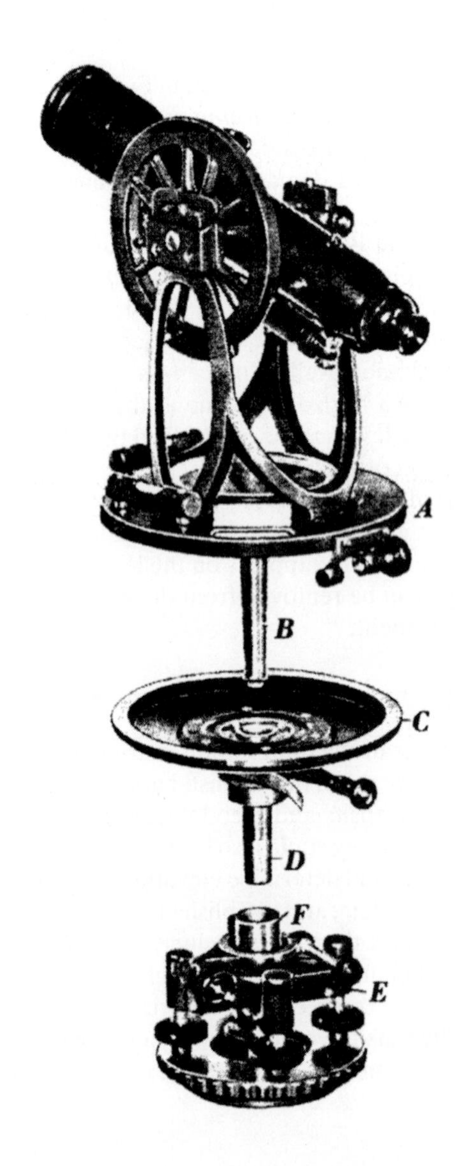

Figure A.3 Three main assemblies of the transit, from top to bottom: alidade, lower plate, and leveling head. Transit parts identified by letter are: (A) upper plate, (B) inner spindle, (C) lower plate, (D) outer spindle, (E) leveling head, and (F) socket. (Courtesy W. & L.E. Gurley)

Figure A.4
Assembled transit. (Courtesy Tom Pantages)

Theodolites, on the other hand, feature an enclosed design (see Figures A.6 and A.7). Theodolites have graduated circles made of glass. They are not directly visible to an operator and must be read by means of an internal microscopic optical system, hence these instruments are called *optical-reading theodolites*. Later versions, called *digital theodolites,* could automatically resolve the circle readings using electronic systems and display them in digital form. Other distinctions between transits and theodolites are described subsequently in this section.

In general, theodolites are capable of greater precision and accuracy in angle observations than transits. Because of this and other advantages, during the period from the 1960s through the mid-1980s, theodolites were purchased much more frequently than transits in the United States. Since the mid-1980s, total station instruments with their automatic angle and distance readout capabilities and built-in microprocessors for real-time data processing have rapidly replaced both transits and theodolites.

A.4 THE TRANSIT

In the subsections that follow, the parts of a transit and their functions are described. Procedures for handling, setting up, and using the instrument are also discussed.

A.4.1 Parts of a Transit

Transits have three main assemblies: the (1) *alidade,* (2) *lower plate,* and (3) *leveling head*. These three assemblies are shown in their relative positions from top to bottom in Figure A.3 and assembled in the cutaway diagram of Figure A.4. Specific parts within the assemblies are identified in the figures. Following is a detailed description of each of these assemblies and parts. Reference to the figures will lead to a better understanding of the descriptions.

Alidade Assembly. The alidade assembly contains the *telescope, vertical circle,* and *upper plate* (A in Figure A.3). A vertical *spindle B* is attached to the upper plate, which enables the assembly to revolve about a vertical axis. The tapered design of the spindle assures that despite wear, unless damaged by dirt or an accident, it will still seat and center properly. Two level vials are attached to the upper plate: the *altitude bubble,* which is parallel with the telescope, and the *azimuth bubble,* which is at right angles to it (see Figure A.4). Also two *verniers,* referred to as *A* and *B*, are mounted on the upper plate and set 180° apart. Provisions are made for adjusting the verniers and level vials. Two vertical *standards* are cast as an integral part of the alidade to support the horizontal *cross arms* of the telescope in bearings. The telescope revolves in a vertical plane about the centerline through the arms, called the *horizontal axis*.

The telescope, similar to that of a dumpy level, contains an eyepiece, a reticle, and an objective lens system. A sensitive telescope level vial is attached to the telescope tube so the transit can be used as a leveling instrument on work where lower magnification and lesser sensitivity of the telescope vial are satisfactory. The telescope is said to be in the *normal* or *direct* position when the telescope level vial is below it. Turning the telescope 180° about its horizontal axis puts the level vial above, and the instrument is said to be in a *plunged, inverted,* or *reversed* mode. To permit use of the telescope for leveling in either the normal or inverted position, a *reversion vial* (curved and graduated on both its top and bottom so it is usable in both positions) is desirable.

The vertical circle is supported by the cross arm and turns with the telescope as it is revolved. The circle normally is divided into 1/2° spaces with readings to the nearest minute obtained from a vernier having 30 divisions. The vernier is mounted on one standard with provisions for adjustment. If properly set, it should read zero when the telescope bubble is centered. If out of adjustment, a constant *index error* is read from the circle with the bubble centered and must be applied to all vertical angles, with appropriate sign, to get correct values.

The alidade assembly also contains a *compass box* and holds the *upper-tangent screw*. A *vertical-circle clamp* (for the horizontal axis) is tightened to hold the telescope horizontal or at any desired inclination. After the clamp is set, a limited range of vertical movement is possible by manipulating the *vertical-circle tangent screw* (also called the vertical slow-motion screw).

Lower Plate Assembly. The *lower plate* (*C* in Figure A.3) is a horizontal circular plate graduated on its upper face. Its underside is attached to a vertical, hollow, tapered spindle *D* into which the alidade's spindle fits precisely. The upper plate completely covers the lower plate, except for two openings where the verniers exactly meet the graduated circle.

The *upper clamp* on the lower plate assembly fastens the upper and lower plates together. A small range of movement is possible after clamping by using the *upper-tangent screw* (located on the alidade assembly).

Leveling Head Assembly. The leveling head assembly (*E* in Figure A.3) consists of a bottom horizontal plate, which has a threaded collar to fit on a tripod, and a "spider" with four *leveling screws*. The leveling screws, set in cups to prevent scoring the bottom plate, are partly or completely enclosed for protection against dirt and damage. A socket (*F* in Figure A.3) of the leveling head includes a *lower clamp* to fasten the lower plate. The *lower tangent screw* (also on the leveling head) is used to make precise settings after the lower clamp is tightened. The base of the socket is fitted into a ball-and-socket joint resting on the bottom plate of the leveling head, on which it slides horizontally. A *plummet chain* attached to the center of the spindle holds a plumb-bob string. An *optical plummet,* which is a telescope through the vertical center (spindle), is available on some transits. Like those used on total station instruments, it points vertically when the instrument is level and is viewed at right angles (horizontally) by means of a prism for ease of observation.

A recapitulation of the use of the various clamps and tangent screws may be helpful to the beginner. The "vertical-circle clamp and tangent screw" on one standard controls movement of the telescope in the vertical plane. The "upper clamp" fastens the upper and lower plates together, and an "upper-tangent screw" permits a small differential movement between them. A "lower clamp" fastens the lower plate to the socket, after which a "lower-tangent screw" turns the plate through a small angle. If the upper and lower plates are clamped together, they will, of course, move freely as a unit until the lower clamp is tightened.

Tripods used for transits may be either fixed- or adjustable-leg types.

A.4.2 Circle Scales and Verniers

The *horizontal circle* of the lower plate may be divided in various ways, but generally it is graduated into either 30′ or 20′ spaces. For convenience in measuring angles to the right or left, graduations are numbered from 0° to 360° both clockwise and counterclockwise. Figure A.5 shows these arrangements.

Vertical circles of most transits are graduated into 30′ spaces. They are usually numbered from zero at the bottom (for a horizontal sight) to 90° in both directions (for vertical sights), and then back to zero at the top. This facilitates reading either elevation or depression angles with the telescope in either a direct or an inverted position.

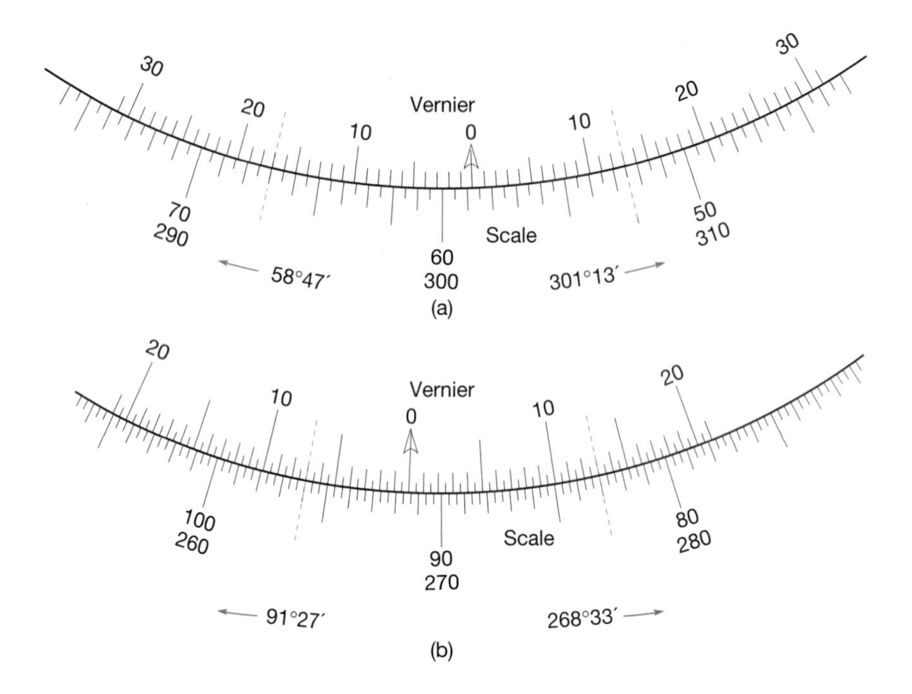

Figure A.5
Transit verniers. [Least count of (a) is 1′; least count of (b) is 1/2′.]

The horizontal and vertical circles of transits are read by means of *verniers*. A vernier is a short auxiliary scale set parallel to and beside a primary scale. It enables reading fractional parts of the smallest main-scale divisions without interpolation. Figure A.5(a) and (b) show two different vernier and angle scale combinations commonly used on transits. Both are double verniers; that is, they can be read clockwise or counterclockwise from the index (zero) mark. Considering the clockwise vernier scale of Figure A.5(a), it is constructed so that 30 of its divisions cover 29 half-degree (30′) divisions on the graduated circle. Thus each vernier division equals $29/30 \times 30' = 29'$. The difference between the length of one main-scale division and one vernier division is therefore 01′. This is the so-called *least count* of this vernier. In general, the least count of a vernier is given by

$$\text{least count} = d/n \tag{A.1}$$

where d is the value of the smallest main-scale division, and n the number of vernier divisions that span $(n - 1)$ main-scale units. By Equation (A.1), the least count of the vernier of Figure A.5(a) is $30'/30 = 01'$. This verifies the intuitive determination given above. *An observer cannot make readings using a vernier without first determining its least count.*

For the vernier of Figure A.5(a), if the index mark were perfectly aligned with a certain main-scale division, say the 58°30′ mark (reading clockwise), and then the vernier advanced 01′ clockwise, the first vernier division would then be aligned with the first main-scale division to the left of the index mark and the reading would be 58°31′. If the circle were moved another 01′, the second vernier division would then be aligned with the second main-scale division left of the index mark and the reading would be 58°02′, etc. In Figure A.5(a) the seventeenth vernier division aligns with a main-scale division (shown by a dashed line); thus, the circle has been advanced 17′ beyond 58°30′ and the clockwise reading is 58°47′.

With the double verniers shown in Figure A.5(a) and (b), an observer can measure angles both clockwise and counterclockwise. This is particularly convenient for measuring deflection angles, which can be right or left. With double verniers there will be two matching lines, one for the clockwise angle, and the other for the counterclockwise angle. The counterclockwise circle reading in Figure A.5(a) is 301°00′ + 13′ = 301°13′. Note that the sum of the clockwise and counterclockwise angles should be 360°00′.

In Figure A.5(b) the smallest circle graduation is 20′, and 40 vernier divisions span 39 on the circle. Thus by Equation (A.1), the least count is 20′/40 = 1/2 min = 30″. The reading of the clockwise angle is 91°20′ + 07′ = 91°27′; for the counterclockwise circle it is 268°20′ + 13′ = 268°33′.

A.4.3 Properties of the Transit

Transits are designed to have a proper balance between magnification and resolution of the telescope, least count of the vernier, and sensitivity of the plate and telescope bubbles. An average length of sight of about 300 ft is assumed in design. Thus a standard 1′ instrument has the following properties:

Magnification, 18−30 ×
Field of view, 1′−1°30′
Resolution, 3−5″
Minimum focus, about 3–8 ft
Sensitivity of plate levels per 2-mm division, 60−100″
Sensitivity of telescope vial per 2-mm division, 30−60″
Weight of instrument head without tripod, 11–18 lb

Transit reticles usually include vertical and horizontal center hairs and two additional stadia hairs, one above and the other below the horizontal center hair. Short stadia lines, used on glass reticles, avoid confusion between the center hair and the stadia hairs.

A.4.4 Handling, Setting Up, and Using a Transit

A transit is a precision instrument, and as such it should be handled carefully. Procedures for handling and caring for a total station instrument were discussed in Section 8.5, and these same precautions should be taken with a transit.

For setting a transit over a point, a plumb bob normally is used because most transits are not equipped with optical plummets. After the tripod has been placed so that the instrument is approximately over the point, the plumb-bob string is attached to a hook at the bottom of the spindle using a slip-knot. This permits raising or lowering the bob without retying, and it also avoids knots. A small slide attachment is also useful in accomplishing this purpose. If the instrument is being set up for the purpose of measuring an angle, the plumb bob must be brought directly over a definite point such as a tack in a wooden stake, and the plates leveled. The tripod legs can be moved in, out, or sideways to approximately level the plates before the leveling screws are used. Shifting the legs affects the position of the plumb bob, however, which makes it more difficult to set up a transit than a level.

Two methods are used to bring the plumb bob within about 1/4-in. of the proper point. In the first method, the transit is set over the mark and one or more legs are moved to bring the plumb bob into position. One leg may be moved circumferentially to level the plates without greatly disturbing the plummet. Beginners sometimes have difficulty with this method because at the start the transit center is too far off the point, or the plates are badly out of level. Several movements of the tripod legs may then fail to both level the plates and center the plumb bob while maintaining a convenient height of instrument. If an adjustable-leg tripod is used, one or two legs can be lengthened or shortened to bring the bob directly over the point.

In the second method, which is particularly suited to level on uniformly sloping ground, the transit is set up near the point and the plates are approximately leveled by moving the tripod legs as necessary. Then, with one tripod leg held in the left hand, another under the left armpit, and the third supported in the right hand, the transit is lifted and placed over the mark. A slight shifting of one leg should bring the plumb bob within perhaps 1/4-in. of the proper position and leave the plates practically level.

Loosening all four leveling screws and sliding them on the bottom plate using the ball-and-socket shifting-head device, which permits a limited movement, precisely centers the plummet. To assure mobility in any direction, the shifting head should be approximately centered on the bottom plate before setting up the instrument and when boxing it.

A transit is accurately leveled by means of the four leveling screws in somewhat the same manner described in Section A.2 for a dumpy level. However, each level vial on the upper plate is first set over a pair of opposite screws, and because there are two vials available, the telescope position need not be changed in the initial leveling process. After both bubbles are carefully centered, the telescope is rotated 180°. As explained in Section 8.20 and illustrated in Figure 8.25, if the bubbles run, they will indicate *double* the dislevelment. They should therefore be brought back halfway, and the telescope rotated back to its original direction. If the bubbles remain in their halfway positions, even though not centered, the instrument is level; if not, the process is repeated until the bubbles stay in the same location no matter what direction the telescope is pointed. If after leveling the instrument, the bubbles are very far off center, it may be desirable to adjust the vials, which can be done using the procedure described in Section 8.19.1.

If the plumb bob is still over the mark after leveling, the instrument is ready for use. But if the plates were initially badly out of level, or the leveling screws were not uniformly set to begin with, the plummet will move off the mark during leveling. The screws must then be loosened and shifted again, and the transit releveled. It is evident that time can be saved by exercising reasonable care when first setting up the tripod so that its head is approximately level to avoid excessive manipulation and possible binding of the screws.

As noted earlier, transits can be used for observing horizontal and vertical angles, and also for setting out alignments and determining distances by stadia. Transits are so-called *repeating instruments* because horizontal angles can be measured by repetition using them, i.e., an angle can be repeated any number of times and the total added on the plates. To eliminate instrumental errors, equal numbers of the individual angle are measured in both the direct and reversed mode. The final angle is then taken as the average. Advantages of the repeating procedure are (1) better accuracy obtained through averaging and (2) disclosure of mistakes and errors by comparing values of the single and multiple readings. The procedure of observing horizontal angles by repetition using total station instruments was described in Section 8.8. The same procedure applies with transits. Methods for measuring vertical angles with total station instruments were discussed in Section 8.13, but again these also apply to transits. Finally, the method of determining distances by stadia was presented in Section 16.9.2.

■ A.5 THE THEODOLITE

The theodolite is discussed in the subsections that follow. In particular, two types of instruments are described; *repeating theodolites* (see Figure A.6) and *directional theodolites* (see Figure A.7). Characteristics of these instruments, and procedures for handling and setting them up, are discussed.

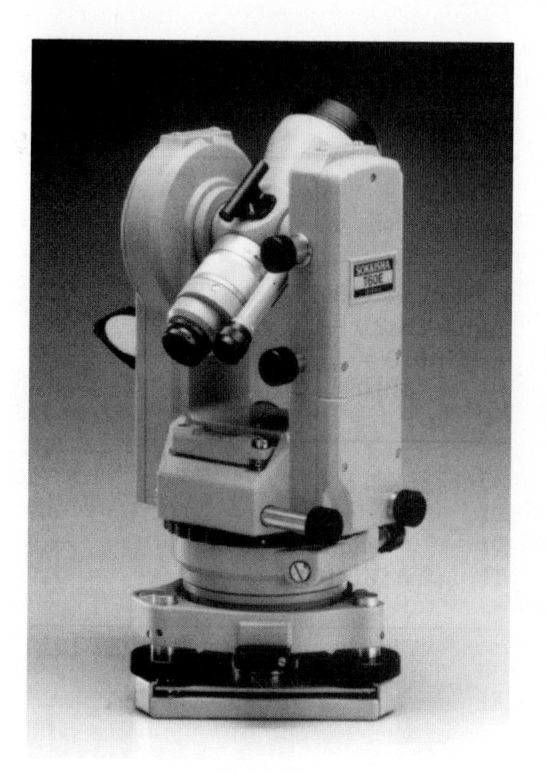

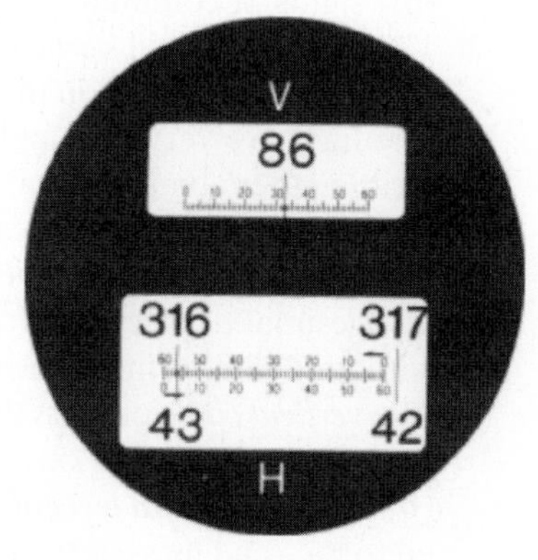

Figure A.6
(a) Lietz T60E optical-reading repeating theodolite and (b) reading system for optical-reading repeating theodolite. (Courtesy Sokkia Corporation)

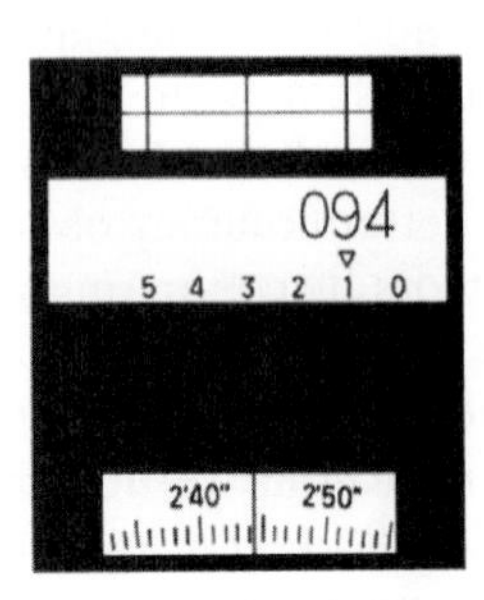

Figure A.7
(a) Wild T-2 optical-reading directional theodolite and (b) reading system for the Wild T-2. (Courtesy Leica Geosystems)

A.5.1 Characteristics of Theodolites

Theodolites differ from transits in appearance (they are generally more compact, lightweight, and "streamlined") and in design by a number of features, the more important of which are as follows:

1. The *telescopes* are short, have reticles etched on glass, and are equipped with rifle sights or collimators for rough pointing.
2. The *horizontal* and *vertical circles* are made of glass with graduation lines and numerals etched on the circles' surfaces (except for digital theodolites). The lines are very thin and more sharply defined than can be achieved by scribing them on the metal circles used on transits. Precisely graduated circles with small diameters can be obtained, and this is one reason the instruments are so compact. The circles are normally divided into conventional sexagesimal degrees and fractions (360°), but instruments can be obtained with graduations in centesimal *grads* or *gons* (full circle divided into 400^g).
3. The *vertical circle* of most theodolites is precisely indexed with respect to the direction of gravity in one of two ways: (a) by an *automatic compensator* or (b) by a *collimation level* or *index level,* usually the coincidence type connected to the reading system of the vertical circle. Both provide a more accurate plane of reference for measuring vertical angles than the plate levels used on transits. Vertical circle readings are *zenith angles,* i.e., 0° occurs with the telescope pointing vertically, and either 90° (in the direct mode) or 270° (in the reversed mode) is read when it is horizontal.
4. The circle *reading systems* consist of a microscope with the optics inside the instrument (except for digital theodolites which have automatic angle

reading capabilities). A reading eyepiece is generally adjacent to the telescope eyepiece or located on one of the standards. Some instruments have optical micrometers for fractional reading of circle intervals (the micrometer scale is visible through the microscope); others are direct-reading. With most theodolites a mirror located on one standard can be adjusted to reflect light into the instrument and brighten the circles for daytime use. They can be equipped with a battery-operated internal lighting system for night and underground operation. Some theodolites can also use the battery-operated system in lieu of mirrors for daytime work.

5. Rotation about the *vertical axis* occurs within a steel cylinder or on precision ball bearings, or a combination of both.
6. The *leveling head* consists of three screws or cams.
7. The *bases* or *tribrachs* of theodolites are often designed to permit interchange of the instrument with sighting targets, prisms, and EDM instruments without disturbing centering over the survey point.
8. An *optical plummet,* built into the base or alidade of most theodolites, replaces the plumb bob and permits centering with better accuracy.
9. A compass can be attached to a theodolite as an accessory, but it is not an integral part of the instrument, as it is with transits.
10. The tripods used with theodolites are the wide-frame type, and most have adjustable legs. Some are all metal and feature devices for preliminary leveling of the tripod head and mechanical centering ("plumbing") to eliminate the need for a plumb bob or optical plummet.

A.5.2 Repeating Theodolites

As noted earlier, theodolites are divided into two basic categories: the *repeating* type and the *directional* model. Repeating theodolites are equipped with a double-vertical axis, usually cylindrical in shape, or a repetition clamp. The double-vertical axis is similar to the double-spindle arrangement used on transits. This design enables horizontal angles to be repeated any number of times and added directly on the instrument's circle.

Figure A.6(a) shows the Lietz T60E, which is an example of a repeating-type theodolite. The optical reading system, typical of these types of repeating theodolites, is illustrated in Figure A.6(b). The repeating theodolite enables angles to be read directly to the nearest minute, with estimation possible to 0.1′. The instrument has a vertical-circle automatic compensator, a telescope with a standard eyepiece of 30° magnification, an optical plummet, and a plate bubble sensitivity of 30″/2-mm division. This instrument is representative of many others of this type.

The reading system of Lietz, and many other theodolites, consists of a graduated glass scale having a span of 1′ which appears superimposed on the degree divisions of the main circle. This scale is read directly by means of a microscope whose small eyepiece can be seen beside the main telescope in the figure. To take a reading, it is simply necessary to observe which degree number lies within the 1′ span of the glass scale and select the minute indicated by the index mark. The vertical- and clockwise horizontal-circle readings indicated in Figure A.6 are 86°32.5′ and 316°56.5′, respectively. (The counterclockwise horizontal circle reading is 43°03.5′.) Thus, on this instrument the horizontal and vertical circles can be viewed and read simultaneously through the reading microscope.

A.5.3 Directional Theodolites

A directional theodolite is used for reading *directions* rather than angles. With this type of instrument, after a sight has been taken on a point, the direction indicated on the circle is read. An observation on the next mark gives a new direction, so the angle between the lines can be found by subtracting the first direction from the second.

Directional theodolites have a single vertical axis and therefore cannot measure angles by the repetition method. They do, however, have a *circle-orienting drive* to make a rough setting of the horizontal circle at any desired position. On all directional theodolites each reading represents the *mean* of two diametrically opposed sides of the circle, made possible because the operator simultaneously views both sides of it through internal optics. This reading procedure, equivalent to averaging readings of the A and B verniers of a transit, automatically compensates for eccentricity errors (see Section 8.20.1).

A typical directional theodolite, the Wild T2, is shown in Figure A.7(a). It has a micrometer that permits reading the horizontal and vertical circles directly to 1″, with estimation possible to the nearest 0.1″, a vertical control bubble for orienting the vertical circle, an optical plummet, and a plate bubble with 20″/2-mm division sensitivity. This instrument is representative of many other similar ones in the directional theodolite category.

Figure A.7(b) illustrates the reading system for the vertical circle of the Wild T-2 instrument of Figure A.7(a). The horizontal circle has a similar system. By means of internal prisms, an operator looking through the eyepiece of the microscope simultaneously sees the two diametrically opposed portions of either the vertical (or horizontal) circle. Only one circle can be viewed at a time, and an optical switch enables the choice to be made. The main circles are graduated in 5-minute intervals. Within Figure A.7(b), there are three rectangles or "frames". The upper frame shows the 5′ graduations on the two diametrically opposed sides of the vertical circle, one side above the other and separated by a horizontal line. After completing the foresight in turning an angle, these opposite 5′ graduations in the upper frame will not coincide, but rather they will be offset. Coincidence is obtained by turning the micrometer knob until the top vertical lines match those on the bottom, as is shown in the upper frame of the figure. After setting the coincidence, the middle frame gives the degrees portion of the angle reading, plus the minutes part to the nearest 10′. The bottom frame enables the remaining minutes portion and seconds part of the reading to be made. In the example of Figure A.7(b), the middle frame provides a reading of 94°10′, and the bottom frame gives 2′44.3″. Thus the final value is 94°12′44.3″. The horizontal circle is read in the same manner. There are various other similar types of reading systems used in theodolites of different manufacture, and operators should become familiar with the system on their particular instrument.

A.5.4 Handling, Setting Up, and Using a Theodolite

Theodolites should be carefully lifted from their carrying cases by grasping the standards (some instruments are equipped with handles for this purpose) and the instrument securely fastened to the tripod by means of a tribrach. Beginners can use a plumb bob to approximate the required setup position, but precise centering over the point is done by means of an optical plummet which provides a line of sight directed downward collinear with the theodolite's vertical axis. The instrument must be level for the optical plummet to define a vertical line. Most theodolite tribrachs have a relatively insensitive bull's-eye bubble to facilitate rough preliminary leveling before beginning final leveling with the plate bubble. Some tribrachs also contain an optical plummet.

The setup process using a theodolite is the same as that described in Section 8.5 for total station instruments. It was noted earlier that theodolites have a three-screw leveling head and a single-plate bubble. The procedure to follow with this type of leveling head is also described in Section 8.5 and illustrated in Figure 8.4.

Theodolites can accomplish all of the tasks that can be done with transits. Procedures for measuring horizontal and vertical angles are described for total station instruments in Chapter 8, but the same methods apply with theodolites.

B

Example Noteforms

MEASURING DISTANCES

Sta.	Fwd. Dist.	Back Dist.	Mean	Error	Ratio
A					
	321.18	321.22	321.20	.04 / 321.20 =	1 / 8,000
B					
	276.54	276.60	276.57	.06 / 276.57 =	1 / 4,600
C					
	100.30	100.29	100.30	.01 / 100.30 =	1 / 10,000
D					
	306.77	306.81	306.79	.04 / 306.79 =	1 / 7,700
E					
	255.47	255.50	255.48	.03 / 255.48 =	1 / 8,500
A					

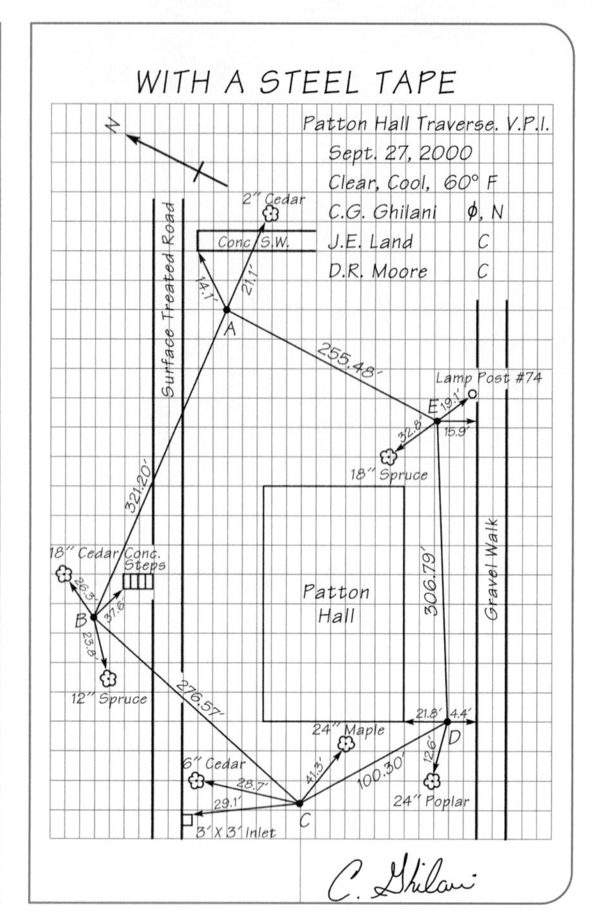

Plate B.1

BORROW-PIT LEVELING

Point	+ Sight	HI	− Sight	Elev.	Cut
BM Road	4.22	364.70		360.48	
A,0			5.2	359.5	1.5
B,0			5.4	359.3	1.3
C,0			5.7	359.0	1.0
D,0			5.9	358.8	0.8
E,0			6.2	358.5	0.5
A,1			4.7	360.0	2.0
B,1			4.8	359.9	1.9
C,1			5.2	359.5	1.5
D,1			5.5	359.2	1.2
E,1			5.8	358.9	0.9
A,2			4.2	360.5	2.5
B,2			4.7	360.0	2.0
C,2			4.8	359.9	1.9
D,2			5.0	359.7	1.7
A,3			3.8	360.9	2.9
B,3			4.0	360.7	2.7
C,3			4.6	360.1	2.1
D,3			4.6	360.1	2.1
A,4			3.4	361.3	3.3
B,4			3.7	361.0	3.0
C,4			4.2	360.5	2.5
BM Road			4.22	360.48	

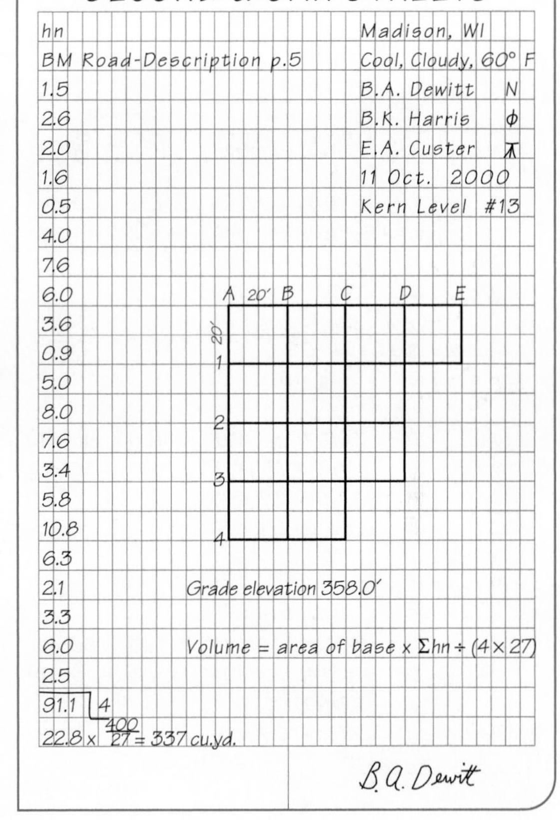

Plate B.2

TRAVERSING WITH A

Instrument at sta 101

$h_e = 5.8$ $h_r = 5.3$

Sta. Sighted	D/R	Horiz. Circle	Zenith Angle	Horiz. Dist.	Elev. Diff.
104	D	0°00′00″	86°30′01″	324.38	+19.84
102	D	82°18′19″	92°48′17″	216.02	−10.58
104	R	180°00′03″	273°30′00″		
102	R	262°18′18″	267°11′41″		

Instrument at sta 102

$h_e = 5.5$ $h_r = 5.5$

Sta. Sighted	D/R	Horiz. Circle	Zenith Angle	Horiz. Dist.	Elev. Diff.
101	D	0°00′00″	87°11′19″	261.05	+10.61
103	D	95°32′10″	85°19′08″	371.65	+30.43
101	R	180°00′02″	272°48′43″		
103	R	275°32′08″	274°40′50″		

Instrument at sta 103

$h_e = 5.4$ $h_r = 5.4$

Sta. Sighted	D/R	Horiz. Circle	Zenith Angle	Horiz. Dist.	Elev. Diff.
102	D	0°00′00″	94°40′48″	371.63	−30.42
104	D	49°33′46″	90°01′54″	145.03	− 0.08
102	R	180°00′00″	265°19′14″		
104	R	229°33′47″	269°58′00″		

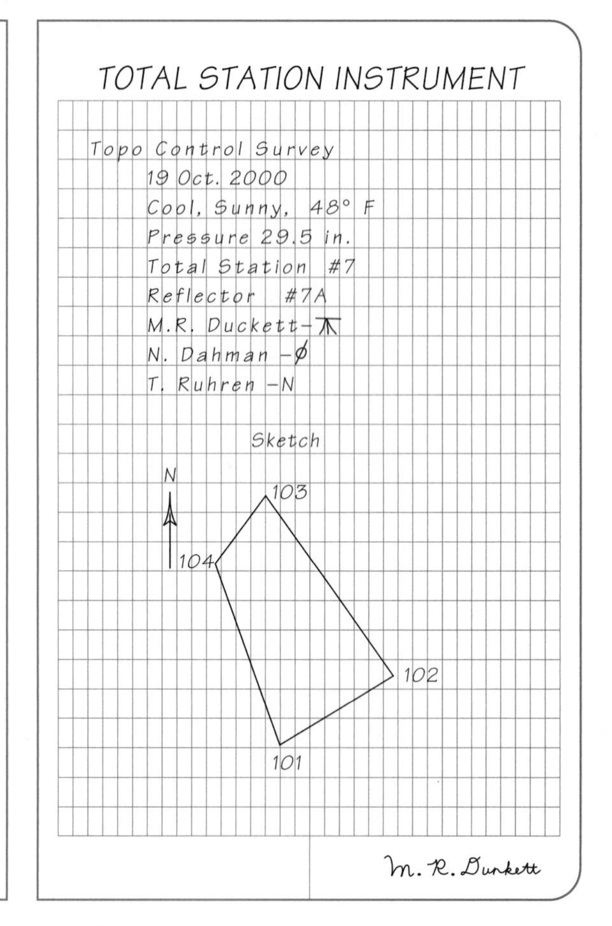

Plate B.3

DOUBLE DIRECT ANGLES

Hub	Dist.	Single ∡	Double ∡	Avg. ∡
A		38°58.0′	77°56.8′	38°58.4′
	321.31′			
B		148°53.6′	297°47.0′	148°53.5′
	276.57′			
C		84°28.1′	168°56.2′	84°28.1′
	100.30′			
D		114°40.3′	229°20.9′	114°40.4′
	306.83′			
E		152°59.4′	305°58.6′	152°59.3′
	255.48′			
A				
Σ	1260.49′			539°59.7′
			Misclosure	0°00.3′

Σ interior ∡s = (N-2) 180°
= (5-2) 180°
= 540°00′

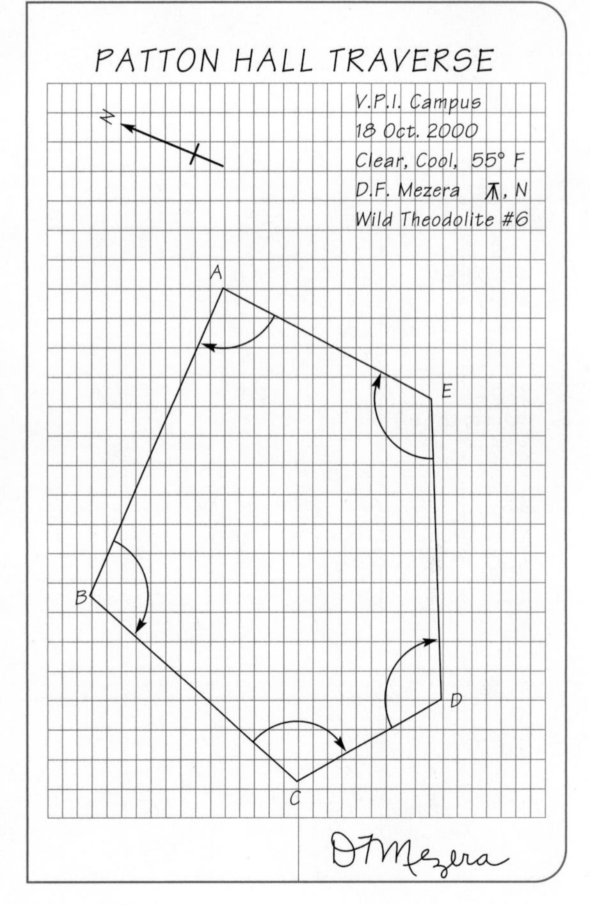

Plate B.4

CROSS-SECTION LEVELING

Sta.	+ Sight	H.I.	− Sight	Elev.	
5+00			9.5		
4+00			12.6		
TP 1	10.25	106.61	1.87	96.36	
3+00			2.1		
2+50			5.8		
2+00			7.4		
1+35			9.7		
1+00			5.6		
0+50			7.6		
0+00			8.5		
BM Pod	8.51	98.23		89.72	

HONOLULU-KAILUA HIGHWAY

Diamond Highway
25 Oct. 2000
Warm, Sunny 70°
A.C. Chun ⊼
R.E. Neilan N
W.E. Grube ϕ, C
M.L. Hagawa C
Lietz level #10

(Each entry: elevation / rod reading / distance)

99.2 / 7.4 / 52	101.5 / 5.1 / 30	97.4 / 9.2 / 10	97.1 / 9.5	95.8 / 10.8 / 12	97.0 / 9.6 / 28	103.8 / 2.8 / 45	
102.3 / 4.3 / 48	99.9 / 6.7 / 24	98.4 / 8.2 / 8	94.0 / 12.6	100.1 / 6.5 / 10	101.5 / 5.1 / 25	98.7 / 7.9 / 50	
	95.2 / 3.0 / 50	95.8 / 2.4 / 25	96.6 / 1.6 / 10	96.1 / 2.1	94.4 / 3.8 / 8	91.1 / 7.1 / 31	95.7 / 2.5 / 48
95.1 / 3.1 / 48	92.8 / 5.4 / 32	89.5 / 8.7 / 15	93.8 / 4.4 / 8	92.4 / 5.8	90.7 / 7.5 / 10	93.4 / 4.8 / 25	96.6 / 1.6 / 50
	92.3 / 5.9 / 54	90.0 / 8.2 / 30	90.8 / 7.4 / 10	90.8 / 7.4	91.3 / 6.9 / 9	93.2 / 5.0 / 25	95.9 / 2.3 / 40
	85.4 / 12.8 / 48	88.9 / 9.3 / 25	85.7 / 12.5 / 10	88.5 / 9.7	89.8 / 8.4 / 8	91.7 / 6.5 / 15	94.1 / 4.1 / 45
	88.6 / 9.6 / 52	97.2 / 1.0 / 28	92.2 / 6.0 / 12	92.6 / 5.6	95.8 / 2.4 / 10	93.6 / 4.6 / 28	95.4 / 2.8 / 50
	90.0 / 8.2 / 50	97.0 / 1.2 / 25	92.7 / 5.5 / 8	90.6 / 7.6	94.4 / 3.8 / 9	95.4 / 2.8 / 24	95.5 / 12.7 / 42
	88.6 / 9.6 / 50	96.1 / 2.1 / 25	92.0 / 6.2 / 10	89.7 / 8.5	93.5 / 4.7 / 8	97.0 / 1.2 / 25	91.5 / 6.7 / 50

BM Pod-Kalini Valley, Oahu, Ewa-makai corner Hibiscus and Kiawe Drives, Spike in 30" monkey pod tree, 2 ft. above ground.

Ruth E. Neilan

Plate B.5

8″ SEWER STAKEOUT

(1) Station	(2) +Sight	(3) HI	(4) −Sight	(5) Ground Elev.	(6) Pipe Flow Line
BM 16	2.11	102.76		100.56	
0+00			6.21	96.55	96.55
+00℄			3.20	99.56	96.55
+50℄			3.91	98.85	95.95
1+00℄			4.07	98.69	95.34
+31℄			8.22	94.54	94.97
+50℄			4.01	98.75	94.74
2+00℄			4.52	98.24	94.14
+33.7℄			5.03	97.73	93.73
+33.7			9.03	93.73	93.73
BM 16			2.11√	100.65	

Flowline Calculations

Line drops 50′(1.206%) = 0.60′ per 50′

Example

Sta. 0+50 = 96.55 − 0.60 = 95.95

1+00 = 96.55 − 1.21 = 95.34

1+31 = 96.55 − 131 (1.206%) = 94.97

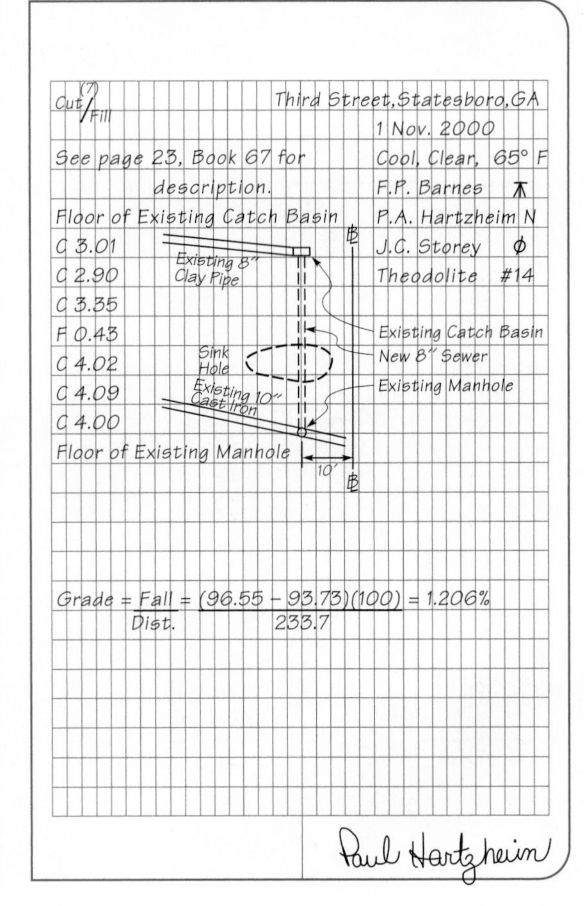

Plate B.6

C

Astronomical Observations

C.1 INTRODUCTION

Astronomical observations in surveying (geomatics) consist of observing positions of the sun or certain stars. The principal purpose of this procedure in plane surveying today is to determine the direction of the astronomic meridian (astronomic north). The resulting azimuth is needed to establish directions of new property lines so parcels can be adequately described; to retrace old property boundaries whose descriptions include bearings; to specify directions of tangents on route surveys; to orient map sheets; and for many other purposes. This procedure is rapidly being replaced with GPS where the coordinates of two points are established on the ground using either static or kinematic GPS procedures as discussed in Chapters 14 and 15.

The latitudes and longitudes of points can also be determined by making astronomical observations. However, this is seldom done today for two reasons: (1) the field procedures and computations involved, especially for longitude, are quite difficult and time consuming especially if accurate results are expected, and (2) the use of the global positioning system (GPS) has now made the determination of latitudes and longitudes a rather routine operation. Thus in this appendix, only astronomical methods for determining azimuth are discussed. For a more thorough presentation of this subject, readers are directed to previous editions of this book.

To expand on the definition of the astronomic meridian given in Section 7.4, at any point it is a line tangent to, and in the plane of, the great circle which passes through the point and the Earth's north and south geographic poles. This is illustrated in Figure C.1, where P and P' are the poles located on the Earth's axis of rotation. Arc PAP' is the great circle through A and line AN, the astronomic

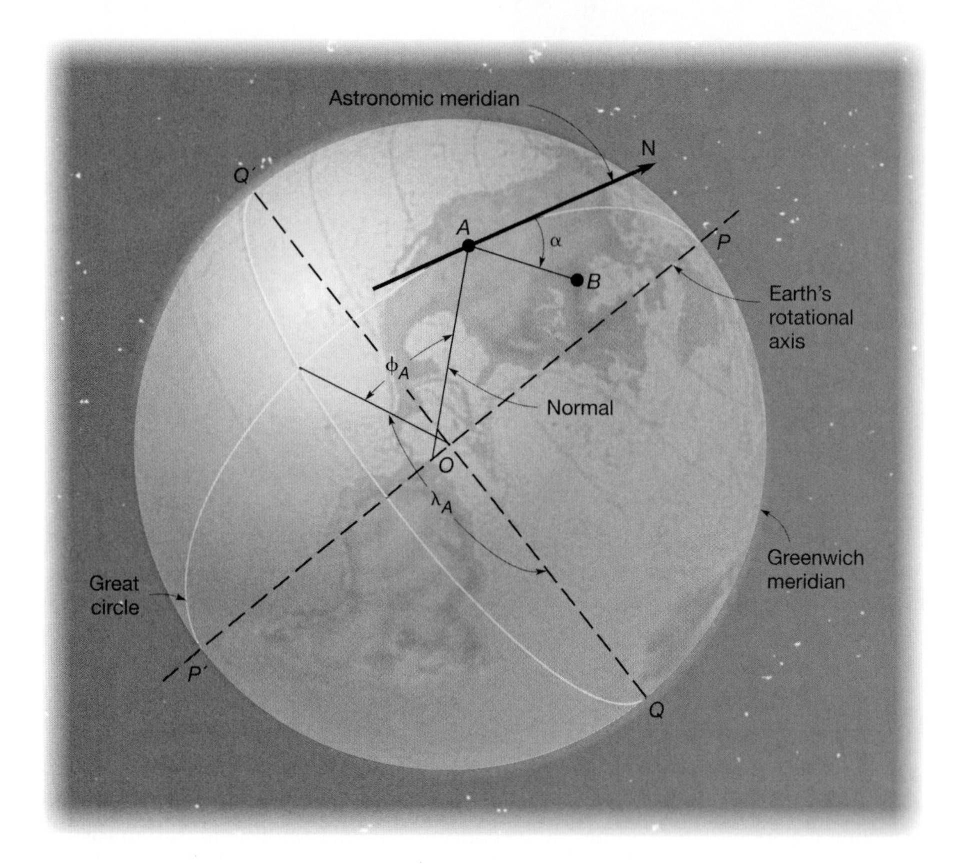

Figure C.1 Astronomic meridian, astronomic azimuth, latitude and longitude.

meridian (tangent to the great circle at A in plane $POP'A$). With an astronomic meridian established, the astronomic azimuth α of any line, such as AB of Figure C.1, can readily be obtained by determining horizontal angle NAB.

Astronomical observations are not necessarily required on every project where azimuths or bearings are needed. If a pair of intervisible control monuments from a previous survey exists in the area, and the azimuth or bearing is known for that line, new directions can be referenced to it. Also, as noted earlier, GPS (described in Chapters 13 through 15) can be used to establish the positions of the two points of a project line from which the azimuth is determined.

■ C.2 OVERVIEW OF USUAL PROCEDURES FOR ASTRONOMICAL AZIMUTH DETERMINATION

In Figure C.2, imagine that P is on the extension of the Earth's polar axis, and that S is a star, which appears to rotate about P due to the Earth's rotation on its axis. Point N is on the horizon and vertically beneath P, and therefore line AN represents true north. For the situation shown in this figure, the general field procedures employed by surveyors to define the direction of astronomic north consist of the following steps: (1) a total station is set up and leveled at one end of the line whose azimuth is to be determined, like point A of Figure C.2; (2) the

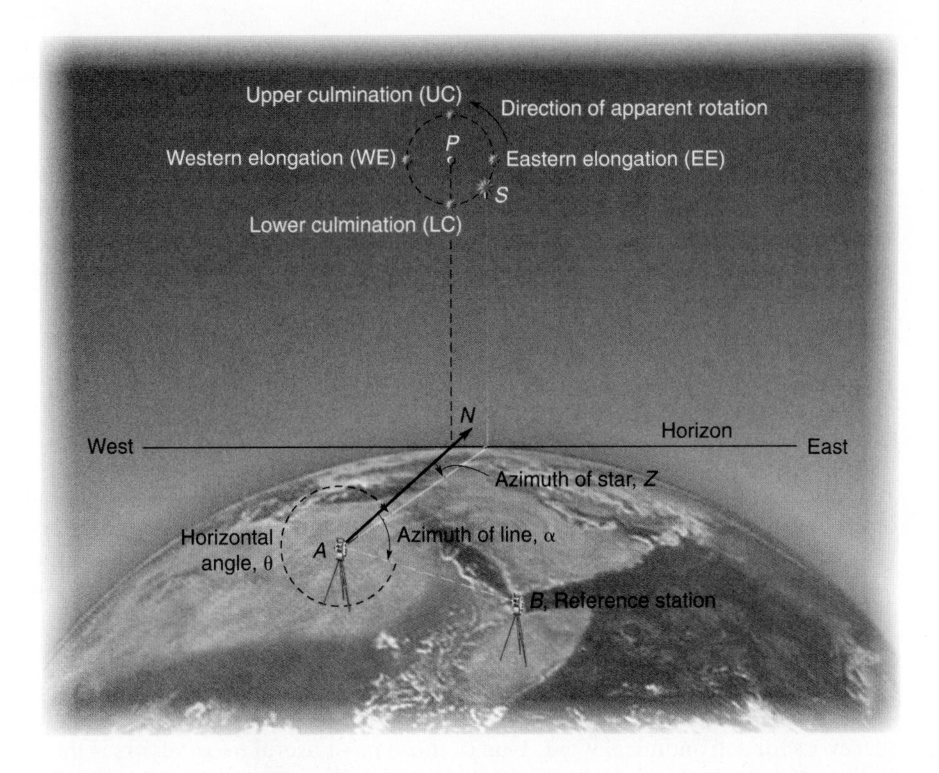

Figure C.2
Azimuth of a line on the ground from the azimuth of a star.

station at the line's other end, like *B* of Figure C.2, is carefully sighted and the instrument's horizontal circle indexed to 0°00′00″; (3) the telescope is turned clockwise and the star *S* carefully sighted; (4) the horizontal, and sometimes vertical, circles of the instrument are read at the instant of pointing on the star; (5) the precise time of pointing is recorded; and (6) the horizontal angle is recorded from the reference mark to the star, like angle θ of Figure C.2 from *B* to *S*. Office work involves (a) obtaining the precise location of the star in the heavens at the instant sighted from an ephemeris (almanac of celestial body positions); (b) computing the star's azimuth (angle *Z* in Figure C.2) based on the observed and ephemeris data; and (c) calculating the line's azimuth by applying the measured horizontal angle to the computed azimuth of the star as

$$\alpha = 360° + Z - \theta \tag{C.1}$$

Any visible celestial body for which ephemeris data are available can be employed in the procedures outlined. However, the sun and, in the northern hemisphere, *Polaris* (the North Star) are almost always selected.[1] The sun permits observations to be made in lighted conditions during normal daytime working hours; Polaris is preferred for higher-order accuracy.

[1]In the southern hemisphere, the star *Sigmus Octantis* and the stars in the constellation *Southern Cross* are commonly used for astronomical observations.

Accuracies attainable in determining astronomical azimuths depend on many variables, including (1) the precision of the instrument used, (2) ability and experience of the observer, (3) weather conditions, (4) quality of the clock or chronometer used to measure the time of sighting, (5) celestial body sighted and its position when observed, and (6) accuracy of ephemeris and other data available. In the northern hemisphere Polaris observations provide the most accurate results and, with several repetitions of measurements utilizing first-order instruments, accuracies to within $\pm 1''$ are possible. Sun observations yield a lower order of accuracy, but values accurate to within about $\pm 10''$ or better can be obtained if careful repeated measurements are made.

■ C.3 EPHEMERIDES

As noted previously, ephemerides are almanacs containing data on the positions of the sun and various stars versus time. Nowadays, ephemeris data are most conveniently obtained through the Internet. The U.S. Bureau of Land Management, for example, maintains an ephemeris of the sun and Polaris on their website.[2] Table C.1 on pages 888 and 889, which applies to December 2000, was taken from this website. The data in this table is used in connection with some of the example problems presented later in this chapter.

A variety of ephemerides are also published annually and are available to surveyors for astronomical work. One of those most useful to surveyors is the *The Sokkia Celestial Observation Handbook and Ephemeris,* published annually by Sokkia Corporation.[3] It not only contains tabulated data for the sun and Polaris, but also for several other of the brighter stars in the heavens. This booklet also includes a substantial amount of explanatory material, plus worked examples that demonstrate the use of the tabulated data and the computational procedures. Other published ephemerides are *The Apparent Place of Polaris and Apparent Sidereal Time,* published by the U.S. Department of Commerce; *The Nautical Almanac,* published by the U.S. Naval Observatory; and *Apparent Places of Fundamental Stars,* published by Astronomisches Rechen-Institut, Heidelberg, Germany.

In addition to published ephemerides, computer programs are also available which solve for the positions of celestial bodies. Their major advantages are that they provide accurate results without table lookup and can be used year after year. However, they must occasionally be updated.

Values tabulated in ephemerides are given for *universal coordinated time* (UTC), which is also Greenwich civil time, so before extracting data from them, standard or daylight times normally recorded for observations must be converted. This topic is discussed further in Section C.5.

---

[2]The Internet address for obtaining the ephemeris is http://www.cadastral.com/.

[3]The Sokkia ephemeris is authored by Drs. Richard Elgin, David Knowles, and Joseph Senne and is available from the Sokkia Corporation, 9111 Barton, Box 2934, Overland Park, Kansas 66021; telephone: (800) 255-3913.

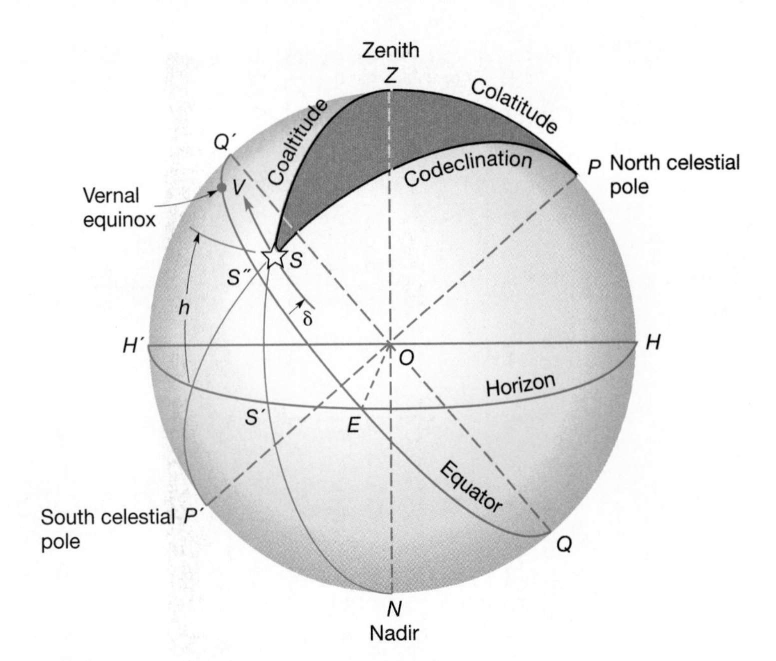

Figure C.3
Celestial sphere.

C.4 DEFINITIONS

In making and computing astronomical observations, the sun and stars are assumed to lie on the surface of a *celestial sphere* of infinite radius having the Earth as its center. Because of the Earth's rotation on its axis, all stars appear to move around centers that are on the extended Earth's rotational axis, which is also the axis of the celestial sphere. Figure C.3 is a sketch of the celestial sphere and illustrates some terms used in field astronomy. Here O represents the Earth and S a heavenly body, as the sun or a star whose apparent direction of movement is indicated by an arrow. Students may find it helpful to sketch the various features on a basketball or globe. Definitions of terms pertinent to the study of field astronomy follow.

The *zenith* is located where a plumb line projected upward meets the celestial sphere. In Figure C.3, Z designates it. Stated differently, it is the point on the celestial sphere vertically above the observer.

The *nadir* is the point on the celestial sphere vertically beneath the observer and exactly opposite the zenith. In Figure C.3, it is at N.

The *north celestial pole* is point P where the Earth's rotational axis, extended from the north geographic pole, intersects the celestial sphere.

The *south celestial pole* is point P' where the Earth's rotational axis, extended from the south geographic pole, intersects the celestial sphere.

A *great circle* is any circle on the celestial sphere whose plane passes through the center of the sphere.

A *vertical circle* is any great circle on the celestial sphere passing through the zenith and nadir, and represents the line of intersection of a vertical plane with the celestial sphere. In Figure C.3, $ZSS'N$ is a vertical circle.

The *celestial equator* is the great circle on the celestial sphere whose plane is perpendicular to the axis of rotation of the Earth. It corresponds to the Earth's

TABLE C.1 EPHEMERIS DATA AT HTTP://WWW.CADASTRAL.COM/.

2000	SUN Declination			. . . For 0 hrs Universal Time . . .							. . . Polaris . . . Declination			0 hrs UT . . .					
				. . . GHA . . .			Eq of Tm		Semi-Di					. . . GHA . . .			. . . TUC*		
Date	d	m	s	d	m	s	m	s	m	s	d	m	s	d	m	s	h	m	s
Dec 1 FR	−21	48	47.9	182	45	19.8	+11	01.32	16	13.2	89	16	08.98	31	36	42.7	21	49	58.0
Dec 2 SA	−21	57	53.4	182	39	38.3	+10	38.55	16	13.4	89	16	09.35	32	36	02.1	21	46	01.3
Dec 3 SU	−22	06	33.5	182	33	47.8	+10	15.18	16	13.5	89	16	09.70	33	35	22.8	21	42	04.6
Dec 4 MO	−22	14	47.9	182	27	48.5	+09	51.23	16	13.7	89	16	10.04	34	34	44.4	21	38	07.8
Dec 5 TU	−22	22	36.4	182	21	40.9	+09	26.72	16	13.8	89	16	10.36	35	34	06.4	21	34	11.0
Dec 6 WE	−22	29	58.7	182	15	25.3	+09	01.69	16	14.0	89	16	10.67	36	33	28.2	21	30	14.2
Dec 7 TH	−22	36	54.7	182	09	02.1	+08	36.14	16	14.1	89	16	10.96	37	32	49.4	21	26	17.4
Dec 8 FR	−22	43	24.0	182	02	31.7	+08	10.12	16	14.3	89	16	11.24	38	32	09.4	21	22	20.7
Dec 9 SA	−22	49	26.5	181	55	54.5	+07	43.63	16	14.4	89	16	11.52	39	31	28.1	21	18	24.1
Dec 10 SU	−22	55	02.0	181	49	10.9	+07	16.72	16	14.5	89	16	11.81	40	30	45.7	21	14	27.6
Dec 11 MO	−23	00	10.4	181	42	21.1	+06	49.41	16	14.6	89	16	12.12	41	30	02.9	21	10	31.1
Dec 12 TU	−23	04	51.5	181	35	25.7	+06	21.71	16	14.7	89	16	12.45	42	29	20.6	21	06	34.6
Dec 13 WE	−23	09	05.2	181	28	24.9	+05	53.66	16	14.8	89	16	12.80	43	28	40.0	21	02	37.9
Dec 14 TH	−23	12	51.3	181	21	19.2	+05	25.28	16	14.9	89	16	13.16	44	28	01.7	20	58	41.1
Dec 15 FR	−23	16	09.7	181	14	08.9	+04	56.59	16	15.0	89	16	13.51	45	27	26.0	20	54	44.1

Dec 16 SA	−23	19	00.2	181	06	54.6	+04	27.64	16	15.1	89	16	13.84	46	26	52.2	20	50	47.0
Dec 17 SU	−23	21	22.9	180	59	36.5	+03	58.44	16	15.2	89	16	14.13	47	26	19.5	20	46	49.9
Dec 18 MO	−23	23	17.6	180	52	15.4	+03	29.02	16	15.3	89	16	14.43	48	25	46.9	20	42	52.7
Dec 19 TU	−23	24	44.1	180	44	51.5	+02	59.43	16	15.3	89	16	14.68	49	25	13.7	20	38	55.6
Dec 20 WE	−23	25	42.5	180	37	25.5	+02	29.70	16	15.4	89	16	14.92	50	24	39.5	20	34	58.5
Dec 21 TH	−23	26	12.7	180	29	57.8	+01	59.85	16	15.5	89	16	15.16	51	24	04.3	20	31	01.5
Dec 22 FR	−23	26	14.7	180	22	29.1	+01	29.94	16	15.5	89	16	15.39	52	23	28.2	20	27	04.5
Dec 23 SA	−23	25	48.4	180	14	59.9	+00	59.99	16	15.6	89	16	15.64	53	22	51.6	20	23	07.6
Dec 24 SU	−23	24	53.8	180	07	30.7	+00	30.05	16	15.6	89	16	15.89	54	22	15.2	20	19	10.7
Dec 25 MO	−23	23	31.1	180	00	02.2	+00	00.15	16	15.7	89	16	16.15	55	21	39.4	20	15	13.7
Dec 26 TU	−23	21	40.1	179	52	34.8	−00	29.68	16	15.7	89	16	16.42	56	21	04.8	20	11	16.7
Dec 27 WE	−23	19	21.0	179	45	09.3	−00	59.38	16	15.8	89	16	16.69	57	20	31.5	20	07	19.6
Dec 28 TH	−23	16	33.8	179	37	46.0	−01	28.93	16	15.8	89	16	16.96	58	19	59.7	20	03	22.3
Dec 29 FR	−23	13	18.5	179	30	25.7	−01	58.29	16	15.8	89	16	17.23	59	19	29.5	19	59	25.0
Dec 30 SA	−23	09	35.4	179	23	08.8	−02	27.41	16	15.9	89	16	17.47	60	19	00.4	19	55	27.6
Dec 31 SU	−23	05	24.4	179	15	55.8	−02	56.28	16	15.9	89	16	17.70	61	18	32.3	19	51	30.1

(Courtesy Jerry Wahl—Cadastral Survey, Bureau of Land Management)
*Universal time of upper culmination at Greenwich.

equator enlarged in diameter. Half of the celestial equator is represented by $Q'EQ$ in Figure C.3.

An *hour circle* is any great circle on the celestial sphere that passes through the north and south celestial poles. Therefore, hour circles are perpendicular to the plane of the celestial equator. They correspond to meridians (longitudinal lines) and are used to observe hour angles. In Figure C.3, $PSS''P'$ is an hour circle.

The *horizon* is a great circle on the celestial sphere whose plane is perpendicular to the direction of the plumb line. In surveying, the plane of the horizon is determined by a level vial. Half of the horizon is represented by $H'EH$ in Figure C.3.

A *celestial meridian,* interchangeably called *local meridian,* is that unique hour circle containing the observer's zenith. It is both an hour circle and a vertical circle. The intersection of the celestial meridian plane with the horizon plane is line $H'OH$ in Figure C.3, which defines the direction of true north. Thus, it is the astronomic meridian line used in plane surveying. Since east is 90° clockwise from true north, line OE in the horizon plane is a true east line. The celestial meridian is composed of two branches: the *upper branch* contains the zenith and is the semicircle $PZQ'H'P'$ in Figure C.3, and the *lower branch* includes the nadir and is arc $PHQNP'$.

A *diurnal circle* is the complete path of travel of the sun or a star in its apparent daily orbit about the Earth. Four terms describe specific positions of heavenly bodies in their diurnal circles (see Figure C.2): (1) *lower culmination*—the body's position when it is exactly on the lower branch of the celestial meridian; (2) *eastern elongation*—where the body is farthest east of the celestial meridian with its hour circle and vertical circle perpendicular; (3) *upper culmination*—when it is on the upper branch of the celestial meridian; and (4) *western elongation*—when the body is farthest west of the celestial meridian with its hour circle and vertical circle perpendicular.

An *hour angle* exists between a meridian of reference and the hour circle passing through a celestial body. It is measured by the angle at the pole between the meridian and hour circle, or by the arc of the equator intercepted by those circles. Hour angles are measured westward (in the direction of apparent travel of the sun or star) from the upper branch of the meridian of reference.

The *Greenwich hour angle* of a heavenly body at any instant of time is the angle, measured westward, from the upper branch of the meridian of Greenwich to the meridian over which the body is located at that moment.[4] In the ephemeris of Table C.1, it is designated by GHA. Local hour angle (LHA) is similar to GHA, except it is observed from the upper branch of the observer's celestial meridian.

A *meridian angle* is like a local hour angle, except it is measured either eastward or westward from the observer's meridian, and thus its value is always between 0° and 180°.

The *declination* of a heavenly body is the angular distance (measured along the hour circle) between the body and the equator; it is plus when the body is north of the equator, and minus when south of it. Declination is usually denoted by δ in formulas, and represented by arc $S''S$ in Figure C.3.

---

[4]The meridian of Greenwich, England, is internationally accepted as the reference meridian for specifying longitudes of points on Earth and for giving positions of celestial bodies.

The *polar distance* or *codeclination* of a body is equal to 90° minus the declination. In Figure C.3, it is arc *PS.*

The *position* of a heavenly body with respect to the Earth at any moment may be given by its Greenwich hour angle and declination.

The *altitude* of a heavenly body is its angular distance measured along a vertical circle above the horizon, $S'S$ in Figure C.3. It is generally obtained by measuring a vertical angle with a total station, (or theodolite), and correcting for refraction, and parallax if the sun is observed. Altitude is usually denoted in formulas by h.

The *coaltitude* or *zenith distance* is arc ZS in Figure C.3 and equals 90° minus the altitude.

The *astronomical* or *PZS triangle* (darkened in the figure) is the spherical triangle whose vertices are the pole P, zenith Z, and astronomical body S. Because of the body's movement through its diurnal circle, the three angles in this triangle are constantly changing.

The *azimuth* of a heavenly body is the angle observed in the horizon plane, clockwise from either the north or south point, to the vertical circle through the body. An azimuth from north is represented by arc HS' in Figure C.3, and it equals the Z angle of the PZS triangle.

The *latitude* of an observer is the angular distance, measured along the meridian, from the equator to the observer's position. In Figure C.3, it is arc $Q'Z$. It is also the angular distance between the polar axis and horizon, or arc HP. Depending on the observer's position, latitude is measured north or south of the equator. Formulas in this book denote it as ϕ. *Colatitude* is $(90° - \phi)$.

The *vernal equinox* is the intersection point of the celestial equator and the hour circle through the sun at the instant it reaches zero declination (about March 21 each year). For any calendar year, it is a fixed point on the celestial sphere (the astronomer's zero-zero point of coordinates in the sky), and moves with the celestial sphere just as the stars do. In Figure C.3, V designates it.

The *right ascension* of a heavenly body is the angular distance measured eastward from the hour circle through the vernal equinox to the hour circle of a celestial body. It is arc VS'' in Figure C.3. Right ascension frequently replaces Greenwich hour angle as a means of specifying the position of a star with respect to the Earth. In this system, however, the Greenwich hour angle of the vernal equinox must also be given.

■ C.5 TIME

Four kinds of time may be used in making and computing an astronomical observation.

1. *Sidereal time.* A sidereal day is the interval of time between two successive upper culminations of the vernal equinox over the same meridian. Sidereal time is star time. At any location for any instant, it is equal to the local hour angle of the vernal equinox.

2. *Apparent solar time.* An apparent solar day is the interval of time between two successive lower culminations of the sun. Apparent solar time is sun time, and

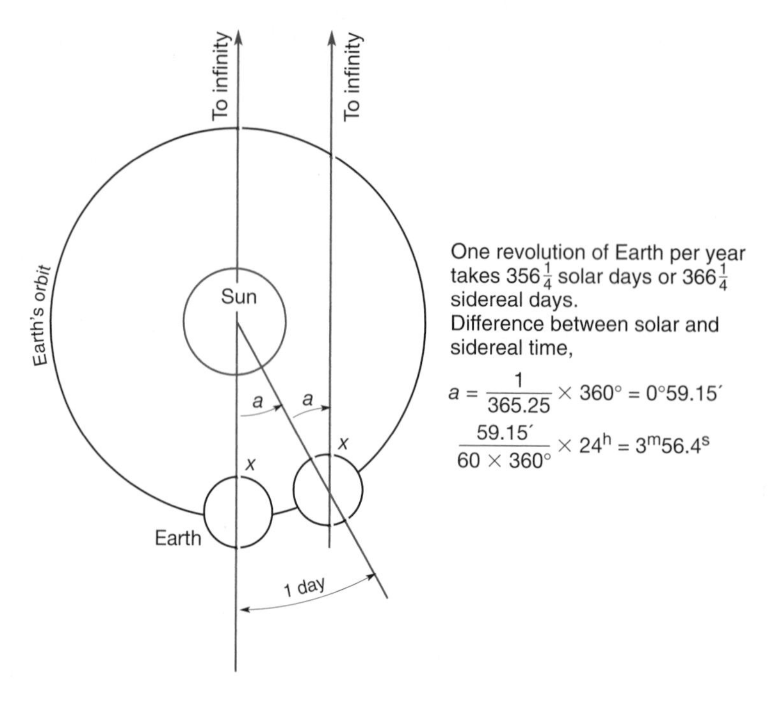

Figure C.4
The PZS triangle for Polaris at any hour angle.

the length of a day varies somewhat because the rate of travel of the sun is not constant. Since the Earth revolves about the sun once a year, there is one less day of solar time in a year than sidereal time. Thus, the length of a sidereal day is shorter than a solar day by approximately 3 min 56 sec. The relationship between sidereal and solar time is illustrated in Figure C.4. (*Note:* The Earth's orbit is actually elliptical, but for simplicity it is shown circular in the figure.)

3. *Mean solar,* or *civil, time.* This time is related to a fictitious sun, called the "mean" sun, which is assumed to move at a uniform rate. It is the basis for watch time and the 24-h day.

The *equation of time* is the difference between "apparent" solar and "mean" solar time. Its value changes continually as the true or apparent sun gets ahead of, and then falls behind, the mean sun. Values for each day of the year are given in ephemerides (see Table C.1). If the apparent sun is ahead of the mean sun, the equation is plus; if behind, it is minus. *Local apparent time is obtained by adding the equation of time to local civil time.*

4. *Standard time.* This is the mean time at meridians 15° or 1 h apart, measured eastward and westward from Greenwich. Eastern Standard Time (EST) at the 75th meridian differs from universal time (UT), or Greenwich Civil Time (GCT), by 5 h (earlier, since the sun has not yet traveled from the meridian of Greenwich to the United States). Standard time was adopted in the United States in 1883, replacing some 100 local times used previously. *Daylight Saving Time* (DST) in any zone is equal to the standard time in the adjacent zone to the *east;* thus, Central Daylight Time (CDT) is equivalent to Eastern Standard Time.[5]

---

[5]Daylight saving time officially begins at 2:00 A.M. on the first Sunday of April and ends at 2:00 A.M. on the last Sunday of October each year.

TABLE C.2 LONGITUDES OF STANDARD MERIDIANS IN THE UNITED STATES AND TIME DIFFERENCES FROM GREENWICH

Standard Time Zone (and Abbreviation)	Longitude of Standard Meridian	Corrections in Hours, to Add to Obtain UT	
		Standard Time	Daylight Time
Atlantic (AST)	60°	4	3
Eastern (EST)	75°	5	4
Central (CST)	90°	6	5
Mountain (MST)	105°	7	6
Pacific (PST)	120°	8	7
Yukon (YST)	135°	9	8
Alaska/Hawaii (AHST)	150°	10	9
Bering Sea (BST)	165°	11	10

As previously noted, sun and star positions tabulated in ephemerides are given in UT. Observation times, on the other hand, may be recorded in the standard or daylight times of an observer's location and must therefore be converted to UT. Conversion is based on the longitude of the standard meridian for the time zone. Table C.2 lists the different time zones in the United States, the longitudes of their standard meridians, and the number of hours to be added for converting standard and daylight time to UT.

In making civil time conversions based on longitude differences, the following relationships are helpful:

$$360^\circ \text{ of longitude} = 24 \text{ h}$$
$$15^\circ \text{ of longitude} = 1 \text{ h}$$
$$1^\circ \text{ of longitude} = 4 \text{ min (of time)}$$

C.6 TIMING OBSERVATIONS

In the United States, the *National Institute of Standards and Technology* (NIST), formerly the National Bureau of Standards, broadcasts time signals from station WWV in Fort Collins, Colorado, on frequencies of 2.5, 5, 10, 15, and 20 MHz. These signals can be received with short-wave radios, including inexpensive *time kubes* especially designed for this purpose. They can also be received over the telephone by dialing (303) 499-7111. To broaden coverage, signals are also transmitted from station WWVH in Hawaii on the same frequencies. These signals are broadcast as audible ticks with a computerized voice announcement of UT at each minute. In Canada, EST is broadcast from station CHU on frequencies of 3.33, 7.335, and 14.67 MHz. This can be converted to UT by adding 5 h.

The time that is broadcast by WWV is actually *coordinated universal time* (UTC), whereas the time used for tabulating sun and star positions in ephemerides is a corrected version known as UT1. UTC is a uniform time at Greenwich that, unlike UT1, does not vary with changing rates of rotation and other irregular motions of the Earth. Leap seconds are added to UTC as necessary to account for the gradual slowing of the Earth's rotation rate and thus keep UTC within ∓ 0.7 sec of UT1 at all times. For precise astronomical work, a difference correction (DUT) can be added to UTC to obtain UT1. The required DUT correction is given by means of double ticks broadcast by WWV and CHU during the first 15 sec following each minute tone. Each double tick represents a 0.1 sec correction. A plus correction is applied for double ticks that occur during the first 7 sec after the minute tone, while a negative correction is made for double ticks that are heard during sec 9 through 15. Thus if double ticks occurred for the first 5 sec after the minute tone, +0.5 sec would be added to broadcast UTC to obtain UT1. If double ticks were heard during the ninth and tenth seconds, −0.2 sec would be added to UTC. Because the DUT correction is quite small, it can be ignored for most observations on Polaris or other stars with very high declinations. The correction should be considered, however, for more accurate observations on the sun and stars of lower declination.

Digital watches, watches with sweep-second hands, and stopwatches are all suitable for timing most astronomical observations in surveying. Hand calculators and data collectors equipped with time modules are especially convenient since they can serve not only as timing devices, but also for recording data and computing. Regardless of the timepiece used, it should be checked against WWV before starting observations, and either set to agree exactly with UTC, or the number of seconds it is fast or slow recorded. The time a check is made should also be recorded. When all observations are completed, the clock check should be repeated and any change recorded. Then intermediate observation times can be corrected in proportion to the elapsed time since the original check. With a stopwatch, checks can be made before and after each individual observation.

■ C.7 COMPUTATIONS FOR AZIMUTH FROM POLARIS OBSERVATIONS BY THE HOUR ANGLE METHOD

In this method, only the horizontal circle reading and precise time need to be recorded when the star is sighted. A vertical circle reading for at least one pointing is recommended, however, to ensure that the correct star has been sighted. To make the observations, the instrument is set up and leveled on one end of a line whose azimuth is to be determined. In the usual field procedure, the line's other end is first sighted and then the horizontal angle measured to the star. To eliminate the effects of instrumental errors, equal numbers of direct and reversed observations are taken and the results averaged.

Computations after fieldwork require the solution for angle Z in the astronomical (PZS) triangle (see Figure C.3). Two formulas for Z that apply in the hour angle method, derived from laws of spherical trigonometry, are:

$$Z = \tan^{-1}\left(\frac{\sin t}{\cos \phi \tan \delta - \sin \phi \cos t}\right) \tag{C.2}$$

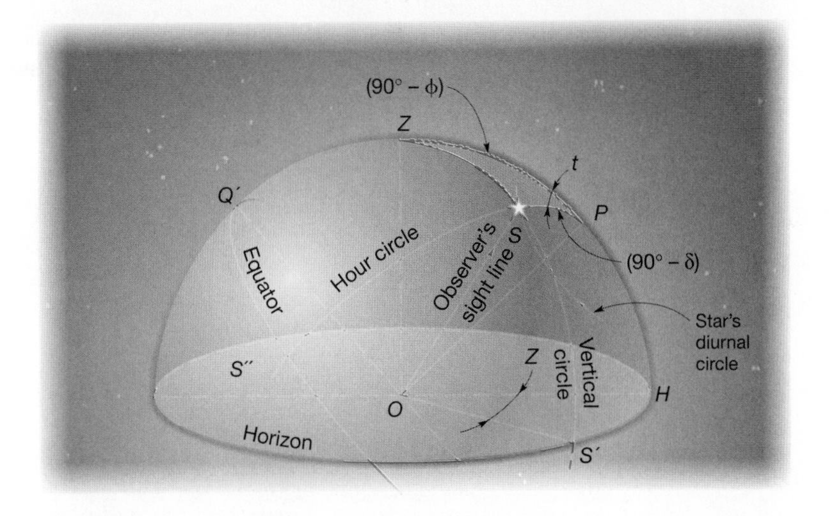

Figure C.5
Computation of the meridian angle *t*.

and

$$Z = \tan^{-1}\left(\frac{-\sin(LHA)}{\cos\phi\tan\delta - \sin\phi\cos(LHA)}\right) \quad \textbf{(C.3)}$$

The geometry upon which these equations are based is shown more clearly in Figure C.5, where the *PZS* triangle is again darkened. The latitude ϕ of the observer's position is arc *HP*; thus arc *PZ* is $(90° - \phi)$, or *colatitude.* Declination δ of the star is arc $S''S$, so *SP* is $(90° - \delta)$, or *polar distance.* Angle *ZPS* in Figure C.5 is *t*, the *meridian angle,* which is related to the (LHA) of the star. Diagrams such as those of Figure C.6 are helpful in understanding and determining *t* angles and *LHA*s. These diagrams show the north celestial pole *P* at the center of the star's diurnal circle *as viewed from the observer's position within the sphere.* On the diagrams, west is to the left of the pole, east is to the right, and the apparent rotation of the stars is counterclockwise. Angle λ between the meridian of Greenwich (G) and the local meridian (L) through the observer's position is the longitude of the station occupied. The star's Greenwich hour angle (GHA) for the observation time is taken from an ephemeris. Sketching λ and *GHA* approximately to scale on diagrams such as those of Figure C.6 immediately makes clear the star's position. From Figure C.6, it can be seen that the *LHA* is

$$LHA = GHA - \lambda \quad \textbf{(C.4)}$$

As shown in Figure C.6(a), the *LHA* is between 0° and 180° when the star is west of north, and as seen in Figure C.6(b), it is between 180° and 360° if the star is east of north. Also, $t = LHA$ if the star is west of north, and $t = (360° - LHA)$ if the star is east of north. The relationships between the *LHA* of a star, the sign of *Z* that is obtained using Equation (C.5), and the azimuth of the star are shown in Table C.3.

Note that the latitude of an observer's position is used directly in Equations (C.2) and (C.3) and that the station's longitude is also needed to compute

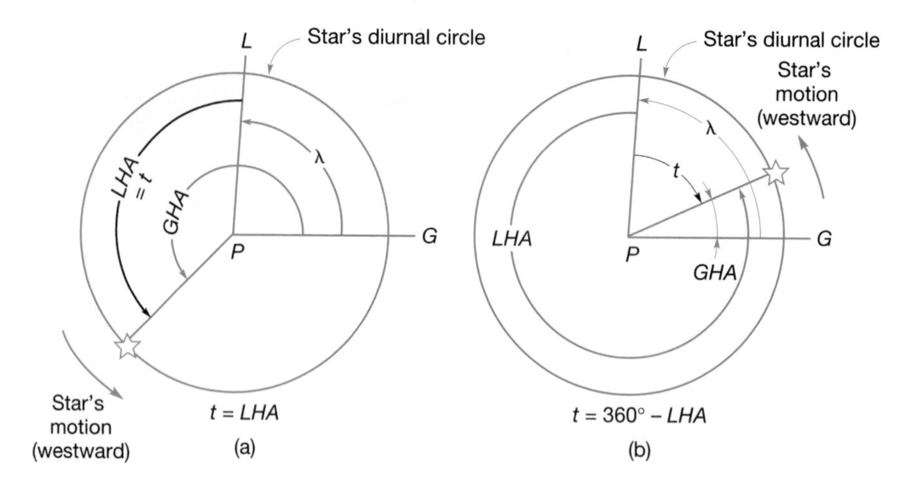

Figure C.6 View of the sun (a) just prior to coincidence of the vertical cross hair and (b) at coincidence.

TABLE C.3 RELATIONSHIP BETWEEN *LHA*, SIGN OF *Z*, AND AZIMUTH OF STAR

LHA from	$Z < 0$	$Z > 0$
0° to 180°	Azimuth $= 360° + Z$	Azimuth $= 180° + Z$
180° to 360°	Azimuth $= 180° + Z$	Azimuth $= Z$

either t or LHA. These values can both be scaled from a USGS quadrangle map and, with reasonable care, obtained to within ±2 sec. Declinations for use in these equations is extracted from an ephemeris for the instant of sighting.

■ C.8 AZIMUTH FROM SOLAR OBSERVATIONS

Reduction of solar observations uses the same equations as star observations. However, at least one and possibly two major differences exist in the computations. Since the sun is relatively close to the Earth, the simple linear interpolation of the declination for stars is inadequate for the sun. The interpolation formula for the declination of the sun is

$$\delta_{\text{Sun}} = \delta_0 + (\delta_{24} - \delta_0) \times \text{UT1}/24 + 0.0000395\, \delta_0 \sin(7.5 \times \text{UT1}) \quad \textbf{(C.5)}$$

where δ_0 is the tabulated declination of the sun at 0-h UT1 on the day of the observation, δ_{24} is the tabulated declination of the sun at 24-h UT1 on the day of the observation (0-h UT1 of the following day), and UT1 is the universal time of the observation.

Observations on the sun can be made directly by placing a dark glass filter (designed for solar viewing) over the telescope objective lens. *A total station instrument should never be pointed directly at the sun without an objective lens solar filter. Failure to heed this warning may result in serious damage to sensitive electronic components in the total station. Additionally, the observer should never look at the sun without the solar filter in place or permanent eye damage will occur.*

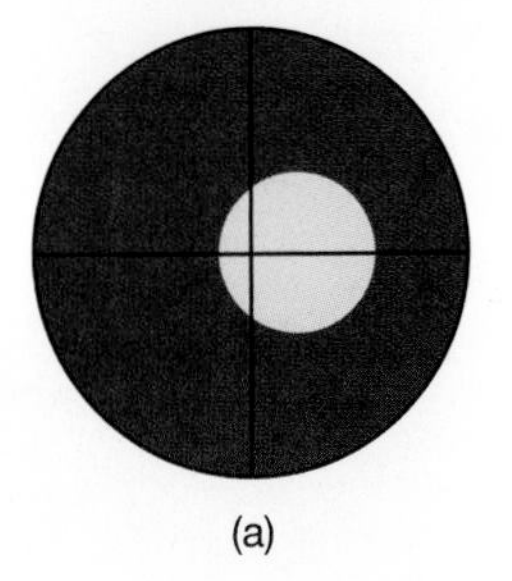
(a)

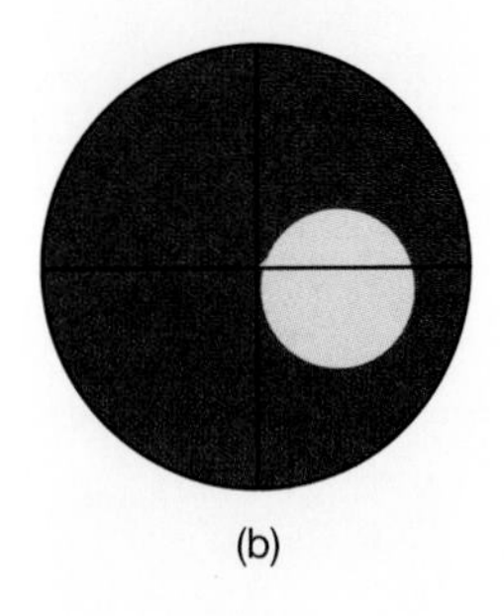
(b)

Figure C.7
View of the sun using a Roelof solar prism (a) just prior to and (b) at the time of the observation.

The second major difference depends on the method of pointing on the sun. Because the objective lens filter only allows the observer to see the instrument's crosshair in the sun's illuminating circle, the most precise pointing will occur at the trailing edge of the sun as shown in Figure C.7. Thus, the observer should place the trailing edge of the sun near the vertical crosshair as in Figure C.7(a), then wait as the sun moves and record the time when the trailing edge of the sun just touches the vertical crosshair as shown in Figure C.7(b). When using this field procedure, or any field procedure that involves the sun's edges, a correction must be made to the horizontal angle for the sun's semi-diameter. The sun's semi-diameter varies with the distance from the Earth to the sun, and values are tabulated for each day in the ephemeris (see Table C.1). The correction that must be applied for semi-diameter is computed as

$$C_{SD} = \frac{\text{Sun's semi-diameter}}{\cos h} \tag{C.6}$$

The correction for the sun's semi-diameter can be avoided by using a *Roelof solar prism*. The Roelof solar prism is a device designed specifically for sighting the sun. It is easily mounted on the objective end of the telescope, and by means of prisms, produces four overlapping images of the sun in the pattern shown in Figure C.8. While viewing the sun, an observer can accurately center the vertical crosshair in the small diamond-shaped area in the middle of the field of view. Because of symmetry, this is equivalent to sighting the sun's center.

To compute the azimuth of a line from observations on the sun, the hour angle equation [either Equation (C.2) or Equation (C.3)] can be employed. These are the same ones used for Polaris observations. Again, latitude and longitude can be taken from a USGS quadrangle map, and declination extracted from an ephemeris for the time of observation.

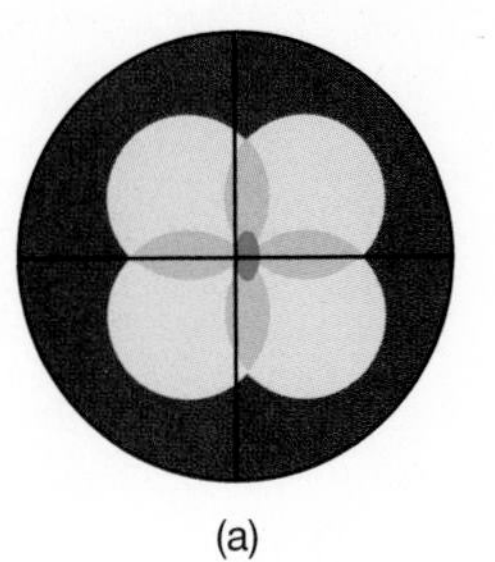
(a)

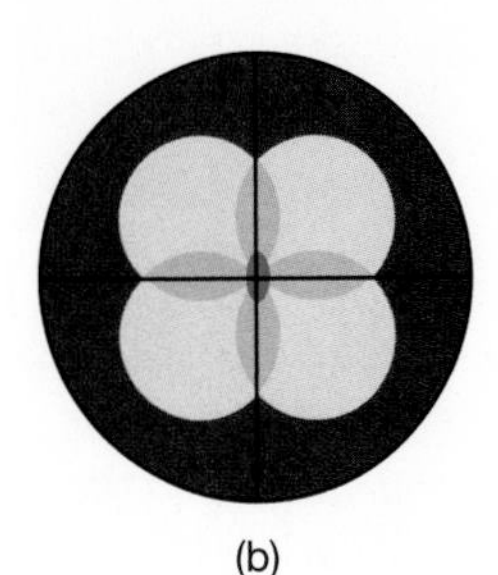
(b)

Figure C.8

D

Using the Worksheets on the Companion Disk

■ D.1 INTRODUCTION

The Mathcad® worksheets contained on the companion disk demonstrate many of the computational exercises presented in this book. The 42 worksheets allow you to modify the values of variables in the equations and see instantaneous changes in the results. Additionally, they further discuss topics presented in this book. These sheets require Mathcad® version 11.0 or higher. For readers who do not own Mathcad® 11.0 or higher, these worksheets have been converted to html files and can be viewed with a web browser. However, the html files are not computationally dynamic; that is, they can only display the equations at the time of creation and do not allow changes to the variables or equations.

To use either the worksheets or html files, you must install them on your computer with the installation program shown in Figure D.1. This software will automatically load when the CD is placed in your drive. However, if it does not, you should browse the CD and run the "Ghilani.exe" file. If you have a copy of Mathcad®, install the worksheets in the "Handbook" directory under the Mathcad® program file—that is, the default directory for the installation of Mathcad® handbooks. If you do not own Mathcad®, you should install the Mathcad® html files. After installation, a link to Mathcad® or the html files will be placed in the "Elementary Surveying" menu item of your Program menu. For the remainder of this appendix, only the worksheets will be discussed, since the html files do not allow interactivity.

As shown in Figure D.2, when using the Mathcad® worksheets, you will find the link to the electronic book (E-books) under the Help menu. Select the menu item entitled "Support files for Elementary Surveying: An Introduction to Geomatics" to open the electronic book. If the menu item is missing, use the "Open Book . . ." menu item to manually browse for the file "ElemSurv.hbk."

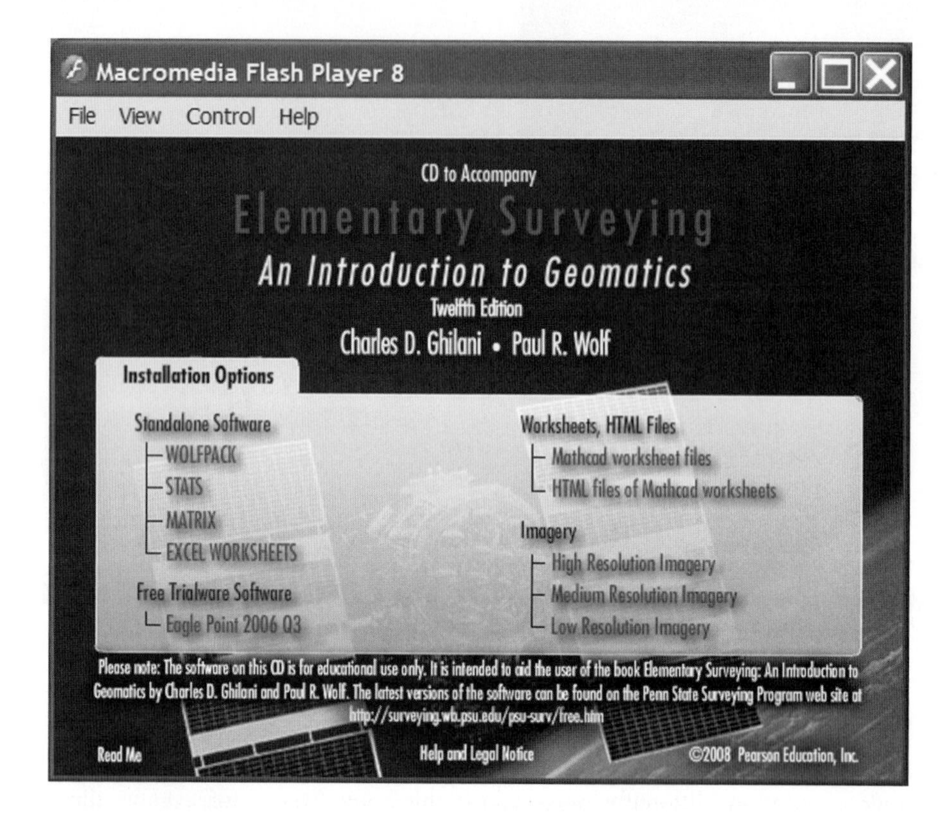

Figure D.1
Program screen to install software.

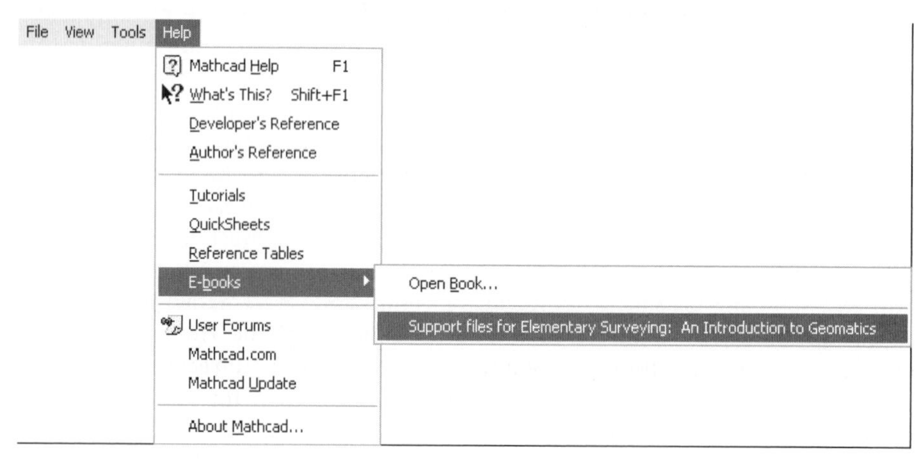

Figure D.2
Opening support files in Mathcad®.

■ D.2 USING THE FILES

Sixteen of the 28 chapters in this book have associated Mathcad® worksheets and html files. In some chapters, there are several associated worksheets, such as those for Chapter 20 shown in Figure D.3. These additional worksheets provide further information on the map projections that are briefly mentioned in Section 20.13 of this book. They allow you to explore map projections that are not commonly used in the United States.

20: State Plane Coordinates
- Properties of Map Projections
- Mercator Projection
- Mercator projection using files
- Transverse Mercator Projection
- Lambert Conformal Conic Projection
- Albers Equal-Area Map Projection
- Oblique Mercator Projection
- Computations using PA Tables
- Computations using NJ Tables
- Grid Reductions of Observations
- Computations of a Traverse

Figure D.3 Worksheets demonstrating map projection computations.

Other chapters with more than one worksheet include 11, 16, 19, and 27. In Chapter 11, besides demonstrating coordinate geometry problems presented in the chapter, the reader can view a least-squares solution of a two-dimensional conformal coordinate transformation, discussed in Section 11.8. In Chapter 16, several Mathcad® worksheets demonstrate the least-squares method. The first worksheet shows the least-squares method for fitting points to a line and circle using the equations discussed in Section 11.2 of the book. These adjustments, along with the two-dimensional conformal coordinate transformation, are not formally covered in Chapter 16, but instead the theory of the least-squares method is applied. Additionally both the horizontal plane survey and differential-leveling network adjustments are demonstrated in separate worksheets. In Chapter 19, there are worksheets discussing the basics of geodesy, geodetic reductions of traditional observations, direct and inverse geodetic problems, three-dimensional geodetic computations, and transformation of ITRF 2000 coordinates with velocity vectors to NAD 83 coordinates for a particular epoch in time. Finally, in Chapter 27, not only are the varying photogrammetric problems in the chapter demonstrated, but there are also two worksheets that cover the two-dimensional affine and projective coordinate transformations which are commonly used in photogrammetric computations.

While some worksheets obtain their data from values assigned to variables directly in the worksheet, others, such as the least-squares worksheets, obtain their data from text files generated using a text editor such as Notepad. For example, Figure D.4 shows the first screen in the support file for Chapter 3: Theory of Errors in Observations. Note the line that states "data := data3-1.txt" (there is an accompanying disk icon). This indicates that the contents of the file "data3-1.txt" are being read into the variable "data." The resultant variable "data" is partially listed on the right side of the window. (The values for this file come from Table 3.1 in this book.) The data file contains one observed value per line.

As previously stated, you can use a text editor such as Notepad to create your own data files for other statistical problems. Once a file is created and saved to disk, you can change the worksheet's selected data file by right-clicking the "data" variable and selecting "Properties" in the resulting pull-down menu (see Figure D.5). This will display the "Component Properties" dialog box shown

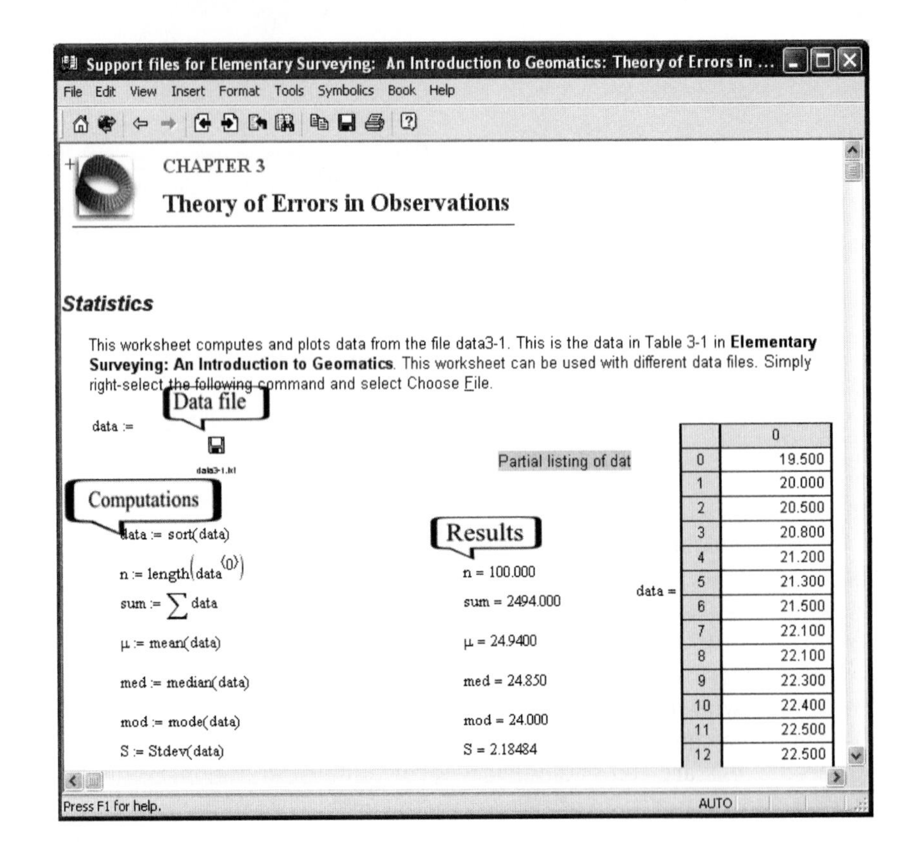

Figure D.4 Statistical computations for data from Table 3.1 of this book.

Figure D.5 Data entry pop-up menu.

in Figure D.6. Click the "Browse" button to locate the desired data file and then click "OK." The worksheet will then automatically update its computations and graphical plots to match the data in the newly specified file.

In Figure D.4, the left side of the worksheet shows the calls to the statistical functions contained in Mathcad® under the heading "Computations." Near the right side of the window, the column labeled "Results" displays the values that are assigned to each variable. For example, the mean of the data is 24.90, the median is 24.85, and the mode is 24.0. You can refer to the Mathcad® Help system to learn more about Mathcad® variables, functions, and their use in worksheets.

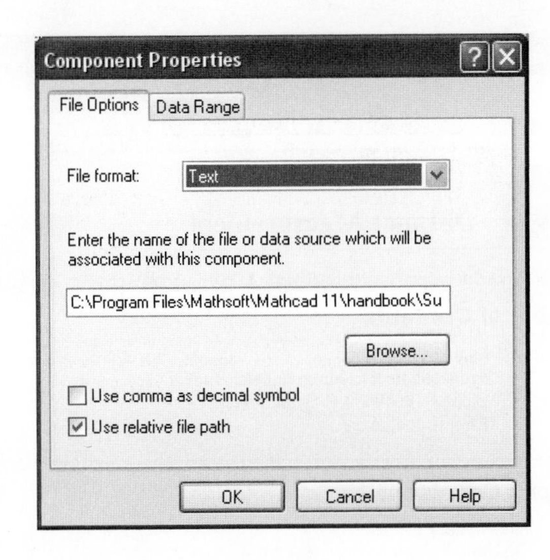

Figure D.6 Component Properties box displaying the file name in the middle of the box.

Figure D.7 shows the top of a worksheet that obtains its data from variables entered directly in the worksheet. This listing demonstrates the use of tape-correction formulas from Example 6.1. The variables near the top of the worksheet contain the calibration data for a 30-m tape as given in the example. Immediately following the calibration data is the field data for the length of 21.151 m. Once the calibrated and field data are entered into the appropriate variables, corrections are computed using Equations (6.3) through (6.6) in the book. Finally, the sum of the corrections is determined in the variable C_{total}. Similar problems can be solved using this worksheet by modifying the field and calibration data as appropriate.

■ D.3 USING THE WORKSHEETS AS AN AID IN LEARNING

You should not use these worksheets to solve assigned homework problems, since you will only truly learn by solving problems on your own. Instead, you should use these worksheets as a method for testing your understanding and checking your computations. As can be seen in Figure D.7, a major advantage of using the worksheets is that intermediate calculations can be viewed and compared with hand-computed results. These comparisons allow you to determine the location of computational errors.

Another advantage of the worksheets is that they demonstrate some of the common programming routines used in surveying. For example, the worksheets on least-squares adjustments demonstrate the parsing of values from data files, computation of coefficients, building of matrices, and matrix methods used to solve the problem and determine post-adjustment statistics. Many of these routines can be emulated in higher-level programming languages such as Basic, C, Fortran, or Pascal. Additionally, many of these programming sheets can be modified to solve other problems that may be encountered in future studies.

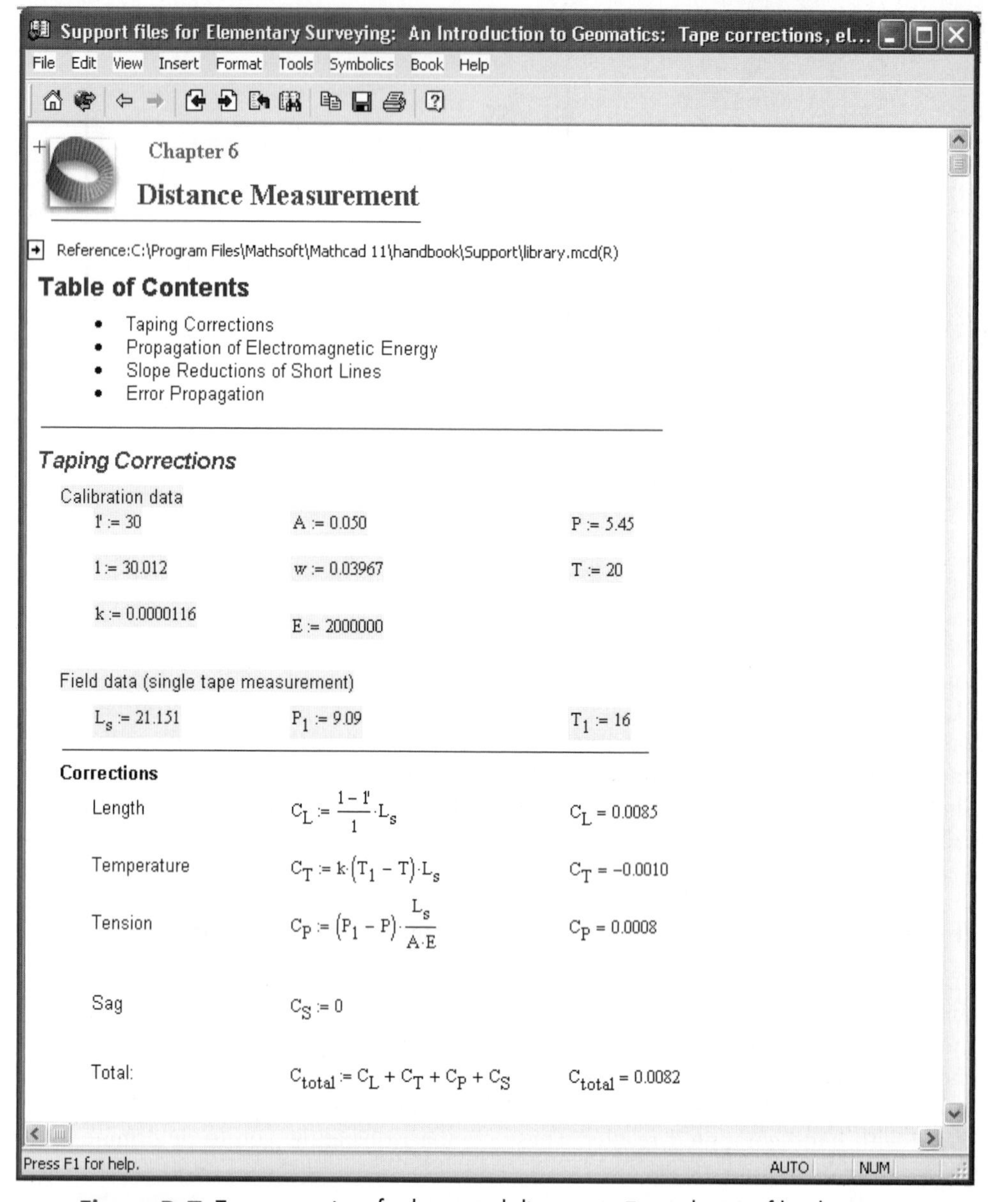

Figure D.7 Tape corrections for last taped distance in Example 6.1 of book.

E

Introduction to Matrices

■ E.1 INTRODUCTION

Matrix algebra enables users to express complicated systems of equations in a compact and easily manipulated form. It also provides a systematic mathematical method for solving systems of equations that can be easily programmed. Throughout this book, matrices have been used to solve systems of equations. Nowhere is this more evident than in Chapter 16 on least-squares adjustments. Matrices are frequently encountered in surveying, geodesy, and photogrammetry. This appendix provides readers with a basic understanding of matrices and their manipulations.

■ E.2 DEFINITION OF A MATRIX

A matrix is a set of number of symbols arranged in an array with m rows and n columns. This arrangement allows users to systematically express large systems of equations. For example, assume we have the three equations with three unknown parameters x, y, and z. The system of equations may appear as

$$\begin{aligned} 3x + 5y - 7z &= -24 \\ 2x - y + 6z &= 33 \\ 9x + 4y - 2z &= 12 \end{aligned} \tag{E.1}$$

This system of equations can be represented in matrix form as

$$\begin{bmatrix} 3 & 5 & -7 \\ 2 & -1 & 6 \\ 9 & 4 & -2 \end{bmatrix} \begin{bmatrix} x \\ y \\ z \end{bmatrix} = \begin{bmatrix} -24 \\ 33 \\ 12 \end{bmatrix} \tag{E.2}$$

Equation (E.2) can be represented in compact matrix notation as

$$AX = L \tag{E.3}$$

As can be seen in Equation (E.2), each coefficient from Equation (E.1) has been placed in order in the first matrix called A; each unknown parameter has been placed in a single row of the second matrix called X; and similarly, each constant has been placed in a single row of the last matrix called L. Thus the A matrix is often called the *coefficient matrix,* the X matrix the *unknown matrix,* and the L matrix the *constants matrix*. Once in this form, a system of equations can be manipulated and solved algebraically using matrix methods.

■ E.3 THE DIMENSIONS OF A MATRIX

The number of rows and columns in each matrix expresses the dimensions or size of a matrix. For example, the A matrix in Equation (E.2) has three rows and three columns. It is said to have dimensions of 3 by 3 and is known as a *square matrix*. The X and L matrices have three rows, but only one column. Their dimensions are 3 by 1 and are also known as *vectors*. In general, a matrix can have m rows and n columns. When m is not equal to n, the matrix is known as a *rectangular matrix*. As previously stated by example, a *square matrix* is formed when the number of rows m equals the number of columns n.

Individual elements of a matrix can be designated by their row and column locations in the matrix. The row-column identifiers are known as *indices*. For example, the A matrix in Equation (E.2) has a value of 3 in row 1 and column 1. The index of 3 is thus 1,1 indicating that 3 is in the first row and first column of A. The elements of matrices are generally written in lower-case letters with subscripts representing their index. The indices of the elements are generally written without the intervening comma. For example, a_{11} has a value of 3. With this in mind, the entire matrix A can be written as

$$A = \begin{bmatrix} a_{11} & a_{12} & a_{13} \\ a_{21} & a_{22} & a_{23} \\ a_{31} & a_{32} & a_{33} \end{bmatrix} \tag{E.4}$$

In Equation (E.4), each element of the A matrix from Equation (E.2) has been replaced by its elemental name. Thus, in reference to Equation (E.2), a_{11} is 3, a_{12} is 5, a_{13} is -7, and so on.

When a matrix is square, such as Equation (E.4), the elements that have equal row and column indices are known as *diagonal elements*. Thus a_{11}, a_{22}, and a_{33} are the diagonal elements of the A matrix in (E.4). In their entirety, they are known as the *diagonal* of matrix A. Their sum is known as the trace of matrix A. Only square matrices have diagonals. Matrices that are not square in dimensions, such as X and L in Equation (E.2), do not have diagonals.

E.4 THE TRANSPOSE OF A MATRIX

The *transpose* of a matrix is a process where each column of the transpose matrix is a row in the original matrix. That is, column 1 of the transpose matrix is row 1 of the original matrix, column 2 of the transpose matrix is row 2 of the original matrix, and so on. The transpose of the A matrix in Equation (E.2) is

$$A^{T} = \begin{bmatrix} 3 & 2 & 9 \\ 5 & -1 & 4 \\ -7 & 6 & -2 \end{bmatrix} \tag{E.5}$$

Notice in Equation (E.5) that placing a "T" as a superscript indicates the transpose of A. Also note that the first column of A^{T} is the first row of the A matrix in Equation (E.2). Similarly, the second column is the second row and the third column is the third row. As seen in Chapter 16, the transpose of the coefficient matrix is used to create the *normal equations*.

E.5 MATRIX ADDITION

Two matrices can be added or subtracted when they have the same dimensions. As an example, assume that we have two matrices, A and B, which have dimensions of 3 by 2. The addition or subtraction of the two matrices is performed element by element. The following example illustrates this procedure.

$$A + B = \begin{bmatrix} -1 & 4 \\ 2 & 3 \\ 4 & 8 \end{bmatrix} + \begin{bmatrix} 5 & -3 \\ 6 & 7 \\ -2 & -5 \end{bmatrix} = \begin{bmatrix} -1+5 & 4-3 \\ 2+6 & 3+7 \\ 4-2 & 8-5 \end{bmatrix} = \begin{bmatrix} 4 & 1 \\ 8 & 10 \\ 2 & 3 \end{bmatrix} = C \tag{E.6}$$

Notice in Equation (E.6) that the resulting matrix has the same dimensions as the original two matrices, A and B. Also note that addition is performed element by element. The difference between A and B is written as

$$A - B = \begin{bmatrix} -1 & 4 \\ 2 & 3 \\ 4 & 8 \end{bmatrix} - \begin{bmatrix} 5 & -3 \\ 6 & 7 \\ -2 & -5 \end{bmatrix} = \begin{bmatrix} -1-5 & 4-(-3) \\ 2-6 & 3-7 \\ 4-(-2) & 8-(-5) \end{bmatrix} = \begin{bmatrix} -6 & 7 \\ -4 & -4 \\ 6 & 13 \end{bmatrix} = C \tag{E.7}$$

In Equation (E.7), each element of the C matrix is found by subtracting the individual elements of the B matrix from the A matrix.

E.6 MATRIX MULTIPLICATION

Matrix multiplication requires that the two matrices being multiplied have the same *inner dimensions*. That is, if A has dimensions of m rows by i columns and is to be multiplied by B, then B must have dimensions of i by n. Notice that A has i

columns and B has i rows. These are the inner dimensions of the product AB. Their resulting product, AB, will have dimensions of m rows and n columns. The *outer dimensions* of the product AB are m and n. This can be expressed as

$$ {}_mA_i^iB^n = {}_mP^n \tag{E.8} $$

where P is the product of AB having dimensions of m by n.

When the number of rows of A does not equal the number of columns in B, then the product BA cannot be performed. Thus matrix multiplication is not commutative. That is, the product of AB does not necessarily equal BA. In fact, when m does not equal n in Equation (E.8), it can't even be performed. The following matrix multiplications are possible.

$$ {}_3A_2^2B^4 = {}_3P^4 $$
$$ {}_1A_3^3B^2 = {}_1P^2 $$
$$ {}_6A_2^2B^6 = {}_6P^6 $$

The following matrix multiplications are not possible.

$$ {}_2B_3^4A^2 $$
$$ {}_3B_1^2A^3 $$

The reason why these multiplications are not possible is best explained by understanding how matrix multiplication is performed. To obtain the first element of the product matrix P, we must multiply the first row of A by the first column B. This is best demonstrated with an example. Suppose we wanted to find the product AB using the following two matrices

$$ A = \begin{bmatrix} 1 & -2 \\ 3 & 4 \end{bmatrix} \quad B = \begin{bmatrix} 1 & 2 & 3 \\ 4 & 5 & 6 \end{bmatrix} \quad P = \begin{bmatrix} p_{11} & p_{12} & p_{13} \\ p_{21} & p_{22} & p_{23} \end{bmatrix} $$

where P is the product of AB. Then p_{11} is computed as

$$ p_{11} = 1 \times 1 - 2 \times 4 = -7 $$

Note how each element in the first row of matrix A is multiplied by each element in the first column of matrix B and their sums accumulated. The remaining elements of P are computed as

$$ p_{12} = 1 \times 2 - 2 \times 5 = -8 $$
$$ p_{13} = 1 \times 3 - 2 \times 6 = -9 $$
$$ p_{21} = 3 \times 1 + 4 \times 4 = 19 $$
$$ p_{22} = 3 \times 2 + 4 \times 5 = 26 $$
$$ p_{23} = 3 \times 3 + 4 \times 6 = 33 $$

Thus the product of AB is

$$\begin{bmatrix} -7 & -8 & -9 \\ 19 & 26 & 33 \end{bmatrix}$$

It should be understandable now that the product of BA can't be performed, since there are not enough elements in the first column of A to pair with the elements in the first row of B.

Using matrix multiplication, the representation of Equation (E.1) as Equation (E.2) can now be verified. That is, the product of the first row of the A matrix in Equation (E.2) with the elements of the X matrix results in the first equation of Equation (E.1). Similarly, the product of the second row of the A matrix in Equation (E.2) with the X matrix results in the second equation and the use of the third row of the A matrix results in the third equation.

■ E.7 MATRIX INVERSE

A *matrix inverse* is similar to division when working with numbers. When A is multiplied by its inverse, the resulting matrix is known as the *identity matrix I*. The identity matrix has values of 1 for the diagonal elements and zeros for all other elements. Thus an identity matrix with dimensions of 3 by 3 is

$$I = \begin{bmatrix} 1 & 0 & 0 \\ 0 & 1 & 0 \\ 0 & 0 & 1 \end{bmatrix}$$

When the identity matrix of appropriate dimension is multiplied by another matrix, say B, the resulting product is equivalent to matrix B. Thus the identity matrix shares the properties of 1 in simple arithmetic multiplication. To solve Equation (E.2), we need to determine the inverse of A and multiply it by L to obtain X or

$$X = A^{-1}L \tag{E.9}$$

There are several methods of inversing a matrix. While these methods are beyond the scope of this book, these methods often employ the same *elementary row transformations* that are used in mathematics to solve a system of equations. Software is readily available that can perform this operation. Once the inverse of the matrix is known, it can be used to solve a system of equations. For example, the inverse of the A matrix in Equation (E.2) expressed to five decimal places is

$$A^{-1} = \begin{bmatrix} -0.20952 & -0.17143 & 0.21905 \\ 0.55238 & 0.54286 & -0.30476 \\ 0.16190 & 0.31429 & -0.12381 \end{bmatrix}$$

When the inverse of A is multiplied times L, the resulting matrix X is

$$X = A^{-1}L = \begin{bmatrix} x \\ y \\ z \end{bmatrix} = \begin{bmatrix} 2 \\ 1 \\ 5 \end{bmatrix}$$

Thus solution of Equation (E.2) is where x is 2, y is 1, and z is 5. The reader should check this by inserting these values into Equation (E.1) and confirming that the constants on the right side of the equation are determined.

The software program MATRIX is on the CD that accompanies this book. The reader can use software to solve other matrix problems presented in this book. This software reads files of matrices as explained in the Help system of the program. Once the files are read into the software, you can use the menu options to manipulate the files using options under the Numerical operations menu item. The reader should refer to the Help system to learn more on use of this software.

F

U.S. State Plane Coordinate System Defining Parameters

F.1 INTRODUCTION

As discussed in several chapters, the U.S. State Plane Coordinate System (SPCS) is used in many applications in surveying (geomatics). In Chapter 20, the equations necessary to compute SPCS map projection coordinates are presented. This appendix provides the defining SPCS parameters for use in the United States. Section F.2 provides the defining parameters for states using the Lambert conformal conic map projection. Section F.3 provides the defining parameters for states using the Transverse Mercator map projection.

F.2 DEFINING PARAMETERS FOR STATES USING THE LAMBERT CONFORMAL CONIC MAP PROJECTION

State Zone	AR North	AR South	CA 401	CA 402	CA 403	CA 404	CA 405	CA 406
φ_S	34°56′	33°18′	40°00′	38°20′	37°04′	36°00′	34°02′	32°47′
φ_N	36°14′	34°46′	41°40′	39°50′	38°26′	37°15′	35°28′	33°53′
φ_0	34°20′	32°40′	39°20′	37°40′	36°30′	35°20′	33°30′	32°10′
λ_0	92°00′W	92°00′W	122°00′W	122°00′W	120°30′W	119°00′W	118°00′W	116°15′W
N_b	0.000	400,000.000	500,000.0	500,000.0	500,000.0	500,000.0	500,000.0	500,000.0
E_0	400,000.000	400,000.000	2,000,000.0	2,000,000.0	2,000,000.0	2,000,000.0	2,000,000.0	2,000,000.0

State Zone	CO North	CO Central	CO South	CT 600	FL North	IA North	IA South	KS North
φ_S	39°43′	38°27′	37°14′	41°12′	29°35′	42°04′	40°37′	38°43′
φ_N	40°47′	39°45′	38°26′	41°52′	30°45′	43°16′	41°47′	39°47′
φ_0	39°20′	37°50′	36°40′	40°50′	29°00′	41°30′	40°00′	38°20′
λ_0	105°30′W	105°30′W	105°30′W	72°45′W	84°30′W	93°30′W	93°30′W	98°00′W
N_b	304,800.6096	304,800.6096	304,800.6096	152,400.3048	0.0	1,000,000.0	0.0	0.0
E_0	914,401.8289	914,401.8289	914,401.8289	304,800.6096	600,000.0	1,500,000.0	500,000.0	400,000.0

State Zone	KS South	KY North	KY South	LA North	LA South	LA Offshore	MD 1900	MA Mainland
φ_S	37°16′	37°58′	36°44′	31°10′	29°18′	26°10′	38°18′	41°43′
φ_N	38°34′	38°58′	37°56′	32°40′	30°42′	27°50′	39°27′	42°41′
φ_0	36°40′	37°30′	36°20′	30°30′	28°30′	25°30	37°40′	41°00′
λ_0	98°30′W	84°15′W	85°45′W	92°30′W	91°20′W	91°20′W	77°00′W	71°30W
N_b	400,000.0	0.0	500,000.0	0.0	0.0	0.0	0.0	750,000.0
E_0	400,000.0	500,000.0	500,000.0	1,000,000.0	1,000,000.0	1,000,000.0	400,000.0	200,000.0

State Zone	MA Island	MI North	MI Central	MI South	MN North	MN Central	MN South	MT 2500
φ_S	41°17′	45°29′	44°11′	42°06′	47°02′	45°37′	43°47′	45°00′
φ_N	41°29′	47°05′	45°42′	43°40′	48°38′	47°03′	45°13′	49°00′
φ_0	41°00′	44°47′	43°19′	41°30′	46°30′	45°00′	43°00′	44°15′
λ_0	70°30′W	87°00′W	84°22′W	84°22′W	93°06′W	94°15′W	94°00′W	109°30′W
N_b	0.0	0.0	0.0	0.0	100,000.0	100,000.0	100,000.0	0.0
E_0	500,000.0	8,000,000.0	6,000,000.0	4,000,000.0	800,000.0	800,000.0	800,000.0	600,000.0

State Zone	NE 2600	NY Long Island	NC 3200	ND North	ND South	OH North	OH South	OK North
φ_S	40°00′	40°40′	34°20′	47°26′	46°11′	40°26′	38°44′	35°34′
φ_N	43°00′	41°02′	36°10′	48°44′	47°29′	41°42′	40°02′	36°46′
φ_0	39°50′	40°10′	33°45′	47°00′	45°40′	39°40′	38°00′	35°00′
λ_0	100°00′W	74°00′W	79°00′W	100°30′W	100°30′W	82°30′W	82°30′W	98°00′W
N_b	0.0	0.0	0.0	0.0	0.0	0.0	0.0	0.0
E_0	500,000.0	300,000.0	609,601.2199	600,000.0	600,000.0	600,000.0	600,000.0	600,000.0

State Zone	OK South	OR North	OR South	PA North	PA South	SC 3900	SD North	SD South
φ_S	33°56′	44°20′	42°20′	40°53′	39°56′	32°30′	44°25′	42°50′
φ_N	35°14′	46°00	44°00′	41°57′	40°58′	34°50′	45°41′	44°24′
φ_0	33°20′	43°40′	41°40′	40°10′	39°20′	31°50′	43°50′	42°20′
λ_0	98°00′W	120°30′W	120°30′W	77°45′W	77°45′W	81°00′W	100°00W	100°20W
N_b	0.0	0.0	0.0	0.0	0.0	0.0	0.0	0.0
E_0	600,000.0	2,500,000.0	1,500,000.0	600,000.0	600,000.0	609,600.0	600,000.0	600,000.0

State Zone	TN 4100	TX North	TX North Central	TX Central	TX South Central	TX South	UT North	UT Central
φ_S	35°15′	34°39′	32°08′	30°07′	28°23′	26°10′	40°43′	39°01′
φ_N	36°25′	36°11′	33°58′	31°53′	30°17′	27°50′	41°47′	40°39′
φ_0	34°20′	34°00′	31°40′	29°40′	27°50′	25°40′	40°20′	38°20′
λ_0	86°00′W	101°30′W	98°30′W	100°20′W	99°00′W	98°30′W	111°30′W	111°30′W
N_b	0.0	1,000,000.0	2,000,000.0	3,000,000.0	4,000,000.0	5,000,000.0	1,000,000.0	2,000,000.0
E_0	600,000.0	200,000.0	600,000.0	700,000.0	600,000.0	300,000.0	500,000.0	500,000.0

State Zone	UT South	VA North	VA South	WA North	WA South	WV North	WV South	WI North
φ_S	37°13′	38°02′	36°46′	47°30′	45°50′	39°00′	37°29′	45°34′
φ_N	38°21′	39°12′	37°58′	48°44′	47°20′	40°15′	38°53′	46°46′
φ_0	36°40′	37°40′	36°20′	47°00′	45°20′	38°30′	37°00′	45°10′
λ_0	111°30′W	78°30′W	78°30′W	120°50′W	120°30′W	79°30′W	81°00′W	90°00′W
N_b	3,000,000.0	2,000,000.0	1,000,000.0	0.0	0.0	0.0	0.0	0.0
E_0	500,000.0	3,500,000.0	3,500,000.0	500,000.0	500,000.0	600,000.0	600,000.0	600,000.0

State Zone	WI Central	WI South	PR VI 5200
φ_S	44°15′	42°44′	18°02′
φ_N	45°30′	44°04′	18°26′
φ_0	43°50′	42°00′	17°50′
λ_0	90°00′W	90°00′W	66°26′W
N_b	0.0	0.0	200,000.0
E_0	600,000.0	600,000.0	200,000.0

F.3 DEFINING PARAMETERS FOR STATES USING THE TRANSVERSE MERCATOR MAP PROJECTION

State Zone	AL East	AL West	AK 5001/O.M.	AK 5002	AK 5003	AK 5004	AK 5005	AK 5006
$1{:}k_0$	25,000	15,000	10,000	10,000	10,000	10,000	10,000	10,000
φ_b	30°30′	30°00′	57°00′	54°00′	54°00′	54°00′	54°00′	54°00′
λ_b	85°50′W	85°50′W	133°40′W	142°00′W	146°00′W	150°00′W	154°00′W	158°00′W
E_0	200,000.0	600,000.0	5,000,000.0	500,000.0	500,000.0	500,000.0	500,000.0	500,000.0
N_b	0.0	0.0	5,000,000.0	0.0	0.0	0.0	0.0	0.0

State Zone	AK 5007	AK 5008	AK 5009	AZ East	AZ Central	AZ West	DE 700	FL East
$1{:}k_0$	10,000	10,000	10,000	10,000	10,000	15,000	200,000	17,000
φ_b	54°00′	54°00′	54°00′	31°00′	31°00′	31°00′	38°00′	24°20′
λ_b	162°00′W	166°00′W	170°00′W	110°10′W	111°55′W	113°45′W	75°25′W	81°00′W
E_0	500,000.0	500,000.0	500,000.0	213,360.0	213,360.0	213,360.0	200,000.0	200,000.0
N_b	0.0	0.0	0.0	0.0	0.0	0.0	0.0	0.0

State Zone	FL West	GA East	GA West	HI 5101	HI 5102	HI 5103	HI 5104	HI 5105
1:k_0	17,000	10,000	10,000	30,000	30,000	100,000	100,000	0
φ_b	24°20′	30°00′	30°00′	18°50′	20°20′	21°10′	21°50′	21°40′
λ_b	82°00′W	82°10′W	82°10′W	155°30′W	156°40′W	158°00′W	155°30′W	155°30′W
E_0	200,000.0	200,000.0	700,000.0	500,000.0	500,000.0	500,000.0	500,000.0	500,000.0
N_b	0.0	0.0	0.0	0.0	0.0	0.0	0.0	0.0

State Zone	ID East	ID Central	ID West	IL East	IL West	IN East	IN West	ME East
1:k_0	19,000	19,000	15,000	40,000	17,000	30,000	30,000	10,000
φ_b	41°40′	41°40′	41°40′	36°40′	36°40′	37°30′	37°30′	43°40′
λ_b	112°10′W	114°00′W	115°45′W	88°20′W	90°10′W	85°40′W	87°05′W	68°30′W
E_0	200,000.0	500,000.0	800,000.0	300,000.0	700,000.0	100,000.0	900,000.0	300,000.0
N_b	0.0	0.0	0.0	0.0	0.0	250,000.0	250,000.0	0.0

State Zone	ME West	MS East	MS West	MO East	MO Central	MO West	NV East	NV Central
1:k_0	10,000	20,000	20,000	15,000	15,000	17,000	10,000	10,000
φ_b	42°50′	29°30′	29°30′	35°50′	35°50′	36°10′	34°45′	34°45′
λ_b	70°10′W	88°50′W	90°20′W	90°30′W	92°30′W	94°30′W	115°35′W	116°40′W
E_0	900,000.0	700,000.0	700,000.0	250,000.0	250,000.0	250,000.0	200,000.0	500,000.0
N_b	0.0	0.0	0.0	0.0	0.0	0.0	8,000,000.0	6,000,000.0

State Zone	NV West	NH 2800	NJ/NY East 2900	NM East	NM Central	NM West	NY East	NY Central
1:k_0	10,000	30,000	10,000	11,000	10,000	12,000	10,000	16,000
φ_b	34°45′	42°30′	38°50′	31°00′	31°00′	31°00′	38°50′	40°00
λ_b	118°35′W	71°40′W	74°30′W	104°20′W	106°15′W	107°50′W	74°30′W	76°35′W
E_0	800,000.0	300,000.0	150,000.0	165,000.0	500,000.0	830,000.0	150,000.0	250,000.0
N_b	4,000,000.0	0.0	0.0	0.0	0.0	0.0	0.0	0.0

State Zone	NY West	RI 3800	VT 4400	WY East	WY East Central	WY West Central	WY West
1:k_0	16,000	160,000	28,000	16,000	16,000	16,000	16,000
φ_b	40°00	41°05′	42°30′	40°30′	40°30′	40°30′	40°30′
λ_b	78°35′W	71°30′W	72°30′W	105°10′W	107°20′W	108°45′W	110°05′W
E_0	350,000.0	100,000.0	500,000.0	200,000.0	400,000.0	600,000.0	800,000.0
N_b	0.0	0.0	0.0	0.0	100,000.0	0.0	100,000.0

G

Answers to Selected Problems

CHAPTER 2

2.4(a) 10,738.26 ft
2.6(a) 1476 ft
2.10(a) 29.20 ac.
2.13(a) 69°18′00″, 1.20951 radians
2.16(a) 39°41′54″ = 44.1092 grad,
91°30′16″ = 101.6716 grad,
48°47′50″ = 54.2191 grad
Σ = 200 grad

2.5(a) 1666.94 m
2.7(b) 35.208 ha
2.12(a) 200,365.68 ft
2.14(a) 261
2.19 From Section 2.7, to prevent loss of data

CHAPTER 3

3.6(a) 125.475 (b) ±0.003 (c) ±0.001
3.11(a) 125.4704 ≤ m ≤ 125.4804, 100%
3.11(b) 125.4695 ≤ m ≤ 125.4813, 100%
3.15(a) 130°32′40″ (b) ±3.6″ (c) ±1.8″
3.18 ±0.026
3.22(a) 126.13 ± 0.017 ft
3.24(a) 193.14 ft
3.27(a) 189,510 ± 30 ft^2 or 4.3506 ± 0.0006 ac
3.28(a) A = 49°27′33″

CHAPTER 4

4.2 0.270 m; 6.750 m
4.12 −0.002 ft
4.21 0.029 ft

4.8 31,200 ft or 5.9 mi
4.15 308.100 ft
4.29 313.644 m; 312.812 m

CHAPTER 5

5.6 0.000 mm
5.10 BM $8_{\text{Adj}} = 304.881$; Second order, Class I
5.16 −1.62 ft; 6.80 ft
5.18 25 m
5.20 5272.064 ft
5.26 3.00%
5.31 Yes
5.32 ±6.6 mm

CHAPTER 6

6.2(a) 2.2 ft/pace; (b) 250 ft
6.5 See Section 6.14
6.7 86.07 ft
6.12 235.78 ft
6.22 236.80 ft
6.26 0.00000133 sec
6.28(a) 10.0021 m
6.32 2302.35 ft
6.35 703.784 m
6.39 ±6.0 mm

CHAPTER 7

7.7 N82°46′51″E, S13°23′32″E, S13°15′06″W, N12°48′27″W; 83°49′37″, 26°38′38″, 153°56′27″, 95°35′18″
7.10 272°24′00″
7.13 N81°54′54″E
7.16 Az_{CD}: 284°17′09″; Brg_{CD}: N75°42′51″W
7.26(a) 11.8° W
7.30 2°35′E
7.33 N22°03′E

CHAPTER 8

8.14 6″
8.20 265°36′54″
8.24 19.2″
8.26(a) 55″
8.29 17″

CHAPTER 9

9.9(a) 720°
9.14 14″
9.15 −10″
9.18 ±6.3″
9.21 −7″
9.23 −5″; Second Order, Class I

CHAPTER 10

10.2 −21″, +3″
10.5 +1″ per angle; $\text{Az}_{DE} = 329°56'59''$; $D = 128°05'09''$
10.16 Angle at station *D*
10.25 *CD*

CHAPTER 11

11.3 $m = 1.0424$; $b = -538.170$ m
11.5 57°35′20″
11.7 0.705 m

11.9 (6932.18, 4868.39)
11.13 (8748.77, 7903.09)
11.15 (4322.85, 3050.76)
11.19 (5701.44, 3384.06)
11.21(a) 0.30483

CHAPTER 12

12.1 280,800 sq. units
12.5 26,500 ft^2 or 0.608 ac.
12.11 419,700 ft^2 or 9.634 ac.
12.16 83,230 m^2 or 83.23 ha

CHAPTER 13

13.4 See Section 13.3, paragraph 3
13.21 (320559.446, −4921314.168, 4031328.395)
13.24 (40°26′32.11896″N, 88°23′42.09876″W, 247.842 m)
13.28 314.149 m
13.30 95.888 m

CHAPTER 14

14.2(a) 25 min.
14.9(b) 5 seconds
14.17 3, 15
14.20 c
4.33(d) 0.0386

CHAPTER 15

15.3 See Section 15.4, paragraph 7
15.10 See Section 15.4, paragraph 4
15.15 See Section 15.6, paragraph 2
15.26 See Section 15.9, paragraph 4

CHAPTER 16

16.4 532.686
16.7 184.311 and 164.651
16.10 ±0.007
16.28 0.77, 0.64, 0.19
16.27 Ray1: (41°13′58.16049″N; 75°58′16.63403″W; 387.622 m)
16.31 $-0.164401\, dx_{\text{Steve}} - 0.986394\, dy_{\text{Steve}} + 0.164401\, dx_{\text{Frank}} + 0.986394\, dy_{\text{Frank}} = 0.016721 + v$
16.34 31°57′42″, ±0.024, ±0.016

CHAPTER 17

17.6 6 ft
17.9 1.7 in.
17.15 Overhead obstructions
17.27 100.5
17.33 (9991.088, 4565.431, 282.862)
17.39 0.47 in.

CHAPTER 18

18.2 320 in. or 26.7 ft
18.10 ±10 ft
18.15 ±40 ft
18.18 30 ft/in.
18.24 6.3 mm
18.25 side AB: (5431.49, 4472.79)

CHAPTER 19

19.4 436 m
19.9 6,364,728.497 m; 6,387,950.747 m
19.11 272.624 m
19.15 219°23′25.4″
19.21 2456.101 m; 2458.659 m
19.25 268°18′26.1″
19.30 204°32′47.0″; 85°56′09.6″

CHAPTER 20

20.11 376.330 m, 18°10′35.8″
20.13 (713470.678, 116,661.975) 0°53′42.4″
20.17 (170764.593, 23030.985)
20.21 (41°13′44.4392″N, 79°26′47.3885″W)
20.23 (39°00′48.0890″N, 74°39′42.1545″W)
20.25(a) 2834.94 ft
20.32 0.99994789
20.40 63°42′26.4″

CHAPTER 21

21.12 e, c, b, d, a
21.20 11,700 ft^2; 10,220 ft^2

CHAPTER 22

22.1 5273 ft
22.6(a) 291.4 ft
22.8(a) 30 mi
22.13(a) 20 ch sq, 40 acres
22.21 Single proportion; single proportion

CHAPTER 23

23.10 −4.811%
23.16 4000 ft^2
23.17 5.22 ft

CHAPTER 24

24.2(a) 19°05′55″; 19°11′17″
24.3 L = 900.00 ft; T = 453.76 ft; E = 35.71 ft; M = 35.27 ft; LC = 896.35 ft; R = 2864.93 ft; PC = 20 + 96.24; PT_{back} = 29 + 96.26; PT_{For} = 25 + 50.02
24.24 1392.04 ft
24.25(a) I/2
24.31 414.59 ft
24.33 1:14,500

CHAPTER 25

25.10 685.71 ft
25.19 205.18 ft
25.26 787 ft
25.30 663 ft
25.33 1.80 ft

CHAPTER 26

26.4 680.6 yd^3
26.9 6154 yd^3
26.14 0.9 yd^3; 678.8 yd^3
26.21 655 yd^3; 653 yd^3
26.25 3.086 ac-ft
26.30 400 ft^3/sec

CHAPTER 27

27.5(a) 1/2000 in./ft
27.8(a) 1 in./1010 ft
27.10 24.76 ha
27.12 6774 ft
27.16(a) 116 ft
27.19 3409 ft
27.29 12 flight lines
27.31 30%

CHAPTER 28

28.8(a) 5,898,240,000 pixels, or 5.76 MB

INDEX

SINGLE PC LICENSE AGREEMENT AND LIMITED WARRANTY

READ THIS LICENSE CAREFULLY BEFORE OPENING THIS PACKAGE. BY OPENING THIS PACKAGE, YOU ARE AGREEING TO THE TERMS AND CONDITIONS OF THIS LICENSE. IF YOU DO NOT AGREE, DO NOT OPEN THE PACKAGE. PROMPTLY RETURN THE UNOPENED PACKAGE AND ALL ACCOMPANYING ITEMS TO THE PLACE YOU OBTAINED THEM [[FOR A FULL REFUND OF ANY SUMS YOU HAVE PAID FOR THE SOFTWARE]]. ***THESE TERMS APPLY TO ALL LICENSED SOFTWARE ON THE DISK EXCEPT THAT THE TERMS FOR USE OF ANY SHAREWARE OR FREEWARE ON THE DISKETTES ARE AS SET FORTH IN THE ELECTRONIC LICENSE LOCATED ON THE DISK:***

1. **GRANT OF LICENSE and OWNERSHIP:** The enclosed computer programs and data ("Software") are licensed, not sold, to you by Prentice-Hall, Inc. ("We" or the "Company") and in consideration [[of your payment of the license fee, which is part of the price you paid]] [[of your purchase or adoption of the accompanying Company textbooks and/or other materials,]] and your agreement to these terms. We reserve any rights not granted to you. You own only the disk(s) but we and/or our licensors own the Software itself. This license allows you to use and display your copy of the Software on a single computer (i.e., with a single CPU) at a single location for *academic* use only, so long as you comply with the terms of this Agreement. You may make one copy for back up, or transfer your copy to another CPU, provided that the Software is usable on only one computer.

2. **RESTRICTIONS ON USE AND TRANSFER:** You may *not* transfer, distribute or make available the Software or the Documentation, except to instructors and students in your school in connection with the Course. You may *not* reverse engineer, disassemble, decompile, modify, adapt, translate or create derivative works based on the Software or the Documentation. You may be held legally responsible for any copying or copyright infringement which is caused by your failure to abide by the terms of these restrictions.

3. **TERMINATION:** This license is effective until terminated. This license will terminate automatically without notice from the Company if you fail to comply with any provisions or limitations of this license. Upon termination, you shall destroy the Documentation and all copies of the Software. All provisions of this Agreement as to limitation and disclaimer of warranties, limitation of liability, remedies or damages, and our ownership rights shall survive termination.

4. **LIMITED WARRANTY AND DISCLAIMER OF WARRANTY:** Company warrants that for a period of 60 days from the date you purchase this SOFTWARE (or purchase or adopt the accompanying textbook), the Software, when properly installed and used in accordance with the Documentation, will operate in substantial conformity with the description of the Software set forth in the Documentation, and that for a period of 30 days the disc(s) on which the Software is delivered shall be free from defects in materials and workmanship under normal use. The Company does *not* warrant that the Software will meet your requirements or that the operation of the Software will be uninterrupted or error-free. Your only remedy and the Company's only obligation under these limited warranties is, at the Company's option, return of the disc for a replacement of the disk. THIS LIMITED WARRANTY IS THE ONLY WARRANTY PROVIDED BY THE COMPANY AND ITS LICENSORS, AND THE COMPANY AND ITS LICENSORS DISCLAIM ALL OTHER WARRANTIES, EXPRESS OR IMPLIED, INCLUDING WITHOUT LIMITATION, THE IMPLIED WARRANTIES OF MERCHANTABILITY AND FITNESS FOR A PARTICULAR PURPOSE. THE COMPANY DOES NOT WARRANT, GUARANTEE OR MAKE ANY REPRESENTATION REGARDING THE ACCURACY, RELIABILITY, CURRENTNESS, USE, OR RESULTS OF USE, OF THE SOFTWARE.

5. **LIMITATION OF REMEDIES AND DAMAGES:** IN NO EVENT, SHALL THE COMPANY OR ITS EMPLOYEES, AGENTS, LICENSORS, OR CONTRACTORS BE LIABLE FOR ANY INCIDENTAL, INDIRECT, SPECIAL, OR CONSEQUENTIAL DAMAGES ARISING OUT OF OR IN CONNECTION WITH THIS LICENSE OR THE SOFTWARE, INCLUDING FOR LOSS OF USE, LOSS OF DATA, LOSS OF INCOME OR PROFIT, OR OTHER LOSSES, SUSTAINED AS A RESULT OF INJURY TO ANY PERSON, OR LOSS OF OR DAMAGE TO PROPERTY, OR CLAIMS OF THIRD PARTIES, EVEN IF THE COMPANY OR AN AUTHORIZED REPRESENTATIVE OF THE COMPANY HAS BEEN ADVISED OF THE POSSIBILITY OF SUCH DAMAGES. IN NO EVENT SHALL THE LIABILITY OF THE COMPANY FOR DAMAGES WITH RESPECT TO THE SOFTWARE EXCEED THE AMOUNTS ACTUALLY PAID BY YOU, IF ANY, FOR THE SOFTWARE OR THE ACCOMPANYING TEXTBOOK. BECAUSE SOME JURISDICTIONS DO NOT ALLOW THE LIMITATION OF LIABILITY IN CERTAIN CIRCUMSTANCES, THE ABOVE LIMITATIONS MAY NOT ALWAYS APPLY TO YOU.

6. **GENERAL:** THIS AGREEMENT SHALL BE CONSTRUED IN ACCORDANCE WITH THE LAWS OF THE UNITED STATES OF AMERICA AND THE STATE OF NEW YORK, APPLICABLE TO CONTRACTS MADE IN NEW YORK, AND SHALL BENEFIT THE COMPANY, ITS AFFILIATES AND ASSIGNEES. HIS AGREEMENT IS THE COMPLETE AND EXCLUSIVE STATEMENT OF THE AGREEMENT BETWEEN YOU AND THE COMPANY AND SUPERSEDES ALL PROPOSALS OR PRIOR AGREEMENTS, ORAL, OR WRITTEN, AND ANY OTHER COMMUNICATIONS BETWEEN YOU AND THE COMPANY OR ANY REPRESENTATIVE OF THE COMPANY RELATING TO THE SUBJECT MATTER OF THIS AGREEMENT. If you are a U.S. Government user, this Software is licensed with "restricted rights" as set forth in subparagraphs (a)–(d) of the Commercial Computer-Restricted Rights clause at FAR 52.227-19 or in subparagraphs (c)(1)(ii) of the Rights in Technical Data and Computer Software clause at DFARS 252.227-7013, and similar clauses, as applicable.

Should you have any questions concerning this agreement or if you wish to contact the Company for any reason, please contact in writing:
Robin Short
Pearson Prentice Hall
One Lake Street
Upper Saddle River, New Jersey 07458

Abbreviations

Construction Surveys

Bb	batter boards
BL	building line
CB	catch basin
CG	centerline of grade
CL	centerline
C	cut
CS	curve to spiral
esmt	easement
F	fill
FG	finish grade
FH	fire hydrant
FL	fence line
FS	finished surface
GC	grade change
GP	grade point
GR	grade rod (ss notes)
L, Lt	left (X-sect notes)
MH	manhole
PC	point of curvature
PI	point of intersection
PL	property line
PP	power pole
PT	point of tangency
pvmt	pavement
R, Rt	right (X-sect notes)
R/W	right-of-way
SC	spiral to curve
SD	storm drain
SG	subgrade
spec	specifications
Sq	square
ss	slope stake; side slope
Std	standard
Str Gr	straight grade
X sect	cross section

Control Surveys; GPS Surveys

BM	benchmark
BS	backsight
CORS	continuously operating reference station
DGPS	differential GPS
EDM	electronic distance measurement
FS	foresight
GPS	global positioning system
HARN	high accuracy reference network
HDOP	horizontal dilution of precision
NGRS	National Geodetic Reference System
OTF	on-the-fly initialization (kinematic GPS)
PDOP	positional dilution of precision
RTDGPS	real-time differential GPS
RTK	real-time kinematic (GPS)
SNR	signal-to-noise ratio
VDOP	vertical dilution of precision

Property Surveys

A	area
CF	curb face
ch "X"	chiseled cross
CI	cast iron
diam	diameter
Dr	drive
ER	end of return
Ex	existing
H & T	hub and tack
HC	house connection sewer
IB	iron bolt; iron bar
IP	iron pipe; iron pin
L & T	lead and tack
MHW	mean high water
MLLW	mean low low water
MLW	mean low water
Mon	monument
P	pipe; pin
PLS	professional land surveyor
Rec	record
St	street
Std Surv Mon	standard survey monument
2″ × 2″	two-inch square stake
"X"	crosscut in stone
yd	yard

Public Lands Surveys

AMC	auxiliary meander corner
bdy, bdys	boundary; boundaries
BT	bearing tree
CC	closing corner
ch, chs	chain; chains
cor, cors	corner; corners
corr	correction
decl	declination
dist	distance
frac	fractional (sec, etc.)
GM	guide meridian
lk, lks	link; links (Gunter's chain)
mer	meridian
mkd	marked
Mi Cor	mile corner
MC	meander corner
MS	mineral survey
Prin Mer, PM	principal meridian
R, Rs	range; ranges
R 1 W	range 1 west
SC	standard corner
Sec, Secs	section; sections
SMC	special meander corner
Stan Par, SP	standard parallel
T, Tp, Tps	township; townships
T 2 N	township 2 north
USMM	U.S. mineral monument
WC	witness corner